THE
BUILDING ESTIMATOR'S
REFERENCE BOOK

A Reference Book Setting Forth Detailed
Procedures And Cost Guidelines For Those
Engaged In Estimating Building Trades

EDITED BY

WILLIAM H. SPRADLIN, JR.
GENERAL CONTRACTOR
AND
CERTIFIED CONSTRUCTION ESTIMATOR

FRANK R. WALKER COMPANY

PUBLISHERS
5030 N. Harlem Ave.
Chicago, Ill. 60656
Phone: 312-867-7070

We will be grateful to readers of this volume who will kindly call our attention to any errors, typographical or otherwise, discovered therein. We also invite constructive criticism and suggestions that will assist in making future editions of this book more complete and useful.

FRANK R. WALKER COMPANY

IMPORTANT
How to Obtain the Best Results From the Estimating and Cost Data in This Book

In compiling a book of estimating and cost data that is used in all parts of America and many foreign countries, it is impossible to quote material prices and labor costs that will apply in every locality without some computation on the part of the user.

For this reason all labor costs state the quantity of work a man should perform per hour or per day, together with the number of hours required to complete a certain unit of work. In each instance a certain basic price has been used, but the user is urged to insert local wage scales and material prices, as by so doing accurate estimates are sure to result.

This method of tabulation has been used throughout the book and an example of the method used is given below.

When framing and placing wood floor joists up to 2"x8" in buildings of regular construction, a carpenter should frame and place 550-600 B.F. per 8-hr. day, at the following labor cost per 1,000 B.F.:

	Hours	Rate	Total	Rate	Total
Carpenter	13.9	$....	$....	$16.47	$228.93
Laborer	4.5	....	....	12.54	56.43
Cost per 1,000 B.F.			$....		$285.36

In order that this data will prove of the utmost value to YOU in your business, the blank spaces have been left for you to insert your local wage scale as shown below:

	Hours	Rate	Total	Rate	Total
Carpenter	13.9	$12.00	$166.80	$16.47	$228.93
Laborer	4.5	9.00	40.50	12.54	56.43
Cost per 1,000 B.F.			$207.30		$285.36

No attempt has been made to include a factor for overhead and profit on any of the costs included in this book as this is a variable item and must be established by each contractor on an individual project basis.

Because of the wide variance in State sales tax laws, this cost has not been included and the user must apply the prevailing rate to obtain the total material cost.

To accomodate the maximum number of contractors, the insurance and taxes applicable to labor costs *have not* been included in any of the labor unit costs contained in this book as some contractors prefer to incorporate these costs as a part of overhead, while others maintain them as direct job costs.

Neither the Frank R. Walker Co. nor the editor guarantees or warrants the correctness or sufficiency of the information contained in this book; nor do either assume any responsibility or liability in connection with the use of this book or of any products, process or pricing described in this book.

TABLE OF CONTENTS

13. DOORS AND WINDOWS

14. FINISHES

15. SPECIALTIES

16. CONVEYING SYSTEMS

PREFACE

The Frank R. Walker Company has been responding to the needs of the construction industry since 1915 by publishing The Building Estimator's Reference Book. Through the years, publishing activities have increased by selection of additional authoritative books to assist the Contractor and Estimator in developing an expanded knowledge in producing a standardized format for estimating and bidding construction projects. These systems apply equally for General Contractors and the specialty type subcontractors.

The role of the Estimator has been elevated to incorporate many management functions and the publications have been reviewed and updated to provide specific information on these subjects to enhance the career of the Contractor/ Estimator.

Recognizing technical books requires constant review by specially trained people, the Frank R. Walker Company, in 1981, contracted for Senior Editing services with William H. (Harry) Spradlin, Jr..

Harry Spradlin has been a General Contractor in the Metropolitan Area of Washington, DC since 1946 specializing in commercial, federal, municipal, and design/build projects. He is a member of The Associated General Contractors of America, National President of The American Society of Professional Estimators, Cost Consultant to the Smithsonian Institution, and provides estimating services for many leading architectural firms throughout the United States.

Harry Spradlin has served as a Faculty Member in the Department of Civil Engineering at The Catholic University of America since 1977 conducting classes in Estimating and Construction Operational Management. Harry serves as a Consultant to attorneys, bonding companies, Owners, and contractors involved in construction disputes. He also is an Arbitrator and Mediator for the American Arbitration Association.

Frank R. Walker Company maintains an Educational Division under the direction of Harry Spradlin to provide seminar services on a variety of construction subjects.

The dedication to the industry and the excellence of quality are reflected in each publication.

Eugene R. Callahan
Publisher

Chicago, IL

CHAPTER 1

THE ROLE OF THE CONTRACTOR

THE GENERAL CONTRACTOR
CONSTRUCTION MANAGEMENT
BIDDING FOR A CONTRACT
NEGOTIATING A CONTRACT
CONTRACT DOCUMENTS
SETTING UP THE JOB
CARRYING OUT THE WORK
GETTING PAID
DEFINITIONS
VALUE ENGINEERING

The best of buildings are the combined efforts of several parties. Traditionally these have been centered on the owner, who is, or represents, the user of the building; the architect, who designs the building; and the contractor, who constructs the building. The owner may have as his consultants a realtor, both for land purchase and building management, and the financiers, who may provide short term cash for building construction and later the final long term mortgage when the project is completed.

The architect consults with the structural and mechanical engineers, the site planners and the interior space planners, any or all of which may or may not be part of his immediate office. He also consults with the governmental agencies who control zoning, building codes and city planning, although he usually does not take out the building permit. He also relies on the cooperation of the producers, who manufacture the products which he designs into his structure.

The contractor has as his team the various subcontractors and the producers and suppliers of the equipment, materials, services, appliances, tools and machinery called for by the architect and required to complete the job. The contractor must also cooperate with the governing officials as he must take out all building permits and conduct his operation without violating local laws governing everything from blocking traffic to waking up the neighbors. And then there are also the trade unions, the local utilities, and the ever-necessary bankers who, we hope, are always friendly but often are not as sympathetic as we'd like.

As land becomes scarce, as money rates soar, as ecological concerns are enacted into governmental restrictions the exclusive club of owner-architect-contractor has been forced to let in others, some welcome, some not, so that the triangle has now become more of a circle, Thus, in large projects today the owner-user may be the large landowner who wishes to develop his holdings; government wishing to rebuild the inner city; or a large insurance company that needs to keep its investments active. The architect's concern may no longer be the individual building but an entire complex, a neighborhood, a whole new town. And the general contractor may find himself in demand less for his craftsmanship abilities than for his managerial capabilities.

But no matter how small or how complex the project is, someone must have a need (owner), someone else must develop the needs into satisfactory plans (architect), and someone must interpret the plans into an actual building (the contractor).

THE GENERAL CONTRACTOR

Erecting a building is a very complex undertaking today, and seldom is one firm capable of doing all phases of the work. Yet the owner or developer usually prefers to let one contract and make one firm responsible for the completion of the project. That firm is then known as the General Contractor. Usually he assumes this role when the owner asks for firm price bids. He is the one who will compile all material, labor and service and other costs called for on the architect's drawings and specifications. Where his own firm is not able to do the work called for, he will turn to other contractors who have the expertise in those fields. These firms are then known as subcontractors, and if the project is awarded to the general they will enter into contracts with him and be responsible to him rather than directly to the owner. The general will ask for bids from many subcontractors in order to put together not only the best price but also the best team.

The amount of work to be performed by specialty type subcontractors varies to suit the needs of each project as determined by the General Contractor.

This is a major decision that is not predicated solely upon economic reasons as each time some portion of the work is assigned to a subcontractor, there is introduced into the construction team, another member whose strengths and weaknesses are directly translated into the total performance plan. It must be recognized that the General Contractor who attempts to "broker" a project by maximizing the utilization of subcontractors with the intent of limiting his financial responsibility, will find that other hazards have been created as there will result a considerable loss in coordination and production as the General Contractor has relinquished his right and ability of direct control in the assignment of personnel, materials and equipment, and has placed himself in the position of being subject to the responsiveness of a subcontractor who may not be motivated with the same dedication to the project as the General Contractor.

Initially, it may appear that the large dollar value assigned for subcontractor performance insulates the General from financial loss on that portion of the project but, this is usually found to be false economy because immediately upon signing the subcontract document, the General Contractor has inherited all of the faults and deficiencies that may exist in the structure of the company of the subcontractor. Unfortunately these can be extensive and serious as they become apparent during the life of the contract, to the extent that they adversely impact the project schedule, subjecting the General Contractor to delay claims from other subcontractors as well as the Owner, in addition to his own direct costs which increase in direct ratio to delayed completion.

In the most severe instances, the General may be forced to terminate the subcontract and here again, excess costs will be incurred due to reprocurement costs and the associated delays which will result. It is an accepted practice to employ certain subcontractors who possess expertise in a particular discipline however, they must be carefully selected on financial capacity, technical ability, and the proven ability to perform consistent with the requirements of the contract documents. When this policy prevails, there will usually result a successful project reflecting credit to the Owner, Architect/Engineer, and the contracting team.

It is not possible to establish, with precision, the average percentage of work which is performed by subcontractors. However, surveys conducted by The Associated General Contractors of America indicates a variance of from 40% to 70%.

The General Contractor must retain the financial ability to perform the work volume he has under contract, including allowances for unanticipated delays in receiving payments for work performed, which seem to be increasing at an alarming rate, thus placing an additional burden on the financial resources of the General.

There must be in-house personnel, experienced in the particular type of projects which the firm is engaged and, of equal importance, there must exist managerial skills to coordinate the field construction activities and the administrative functions into a total effort. The General Contractor is expected to provide many services to the Owner and the Architect prior to award of the contract as well as during construction, and his knowledge and contribution may be the deciding factor in receiving an award or expanding a project under construction.

How much of the contract will be sublet will vary with each general contractor. The American Subcontractor's Association claims that 90% of the work force in the building construction industry is employed by subcontractors. It is certainly quite possible for a subcontractor to erect a greater portion of the project than the general.

In some instances the general is forced to sublet work. Some manufacturers insist that their products be installed by contractors licensed to them. Often, such as in curtain wall construction, the specifications will call for the supplier to install his product in order to put the responsibility for any corrective work on one source.

Obviously if anything like 90% of the contract is to be sublet to others, the general contractor must have managerial abilities, for he must schedule and coordinate all the firms involved so that the work proceeds without delay.

In addition to building and managerial skills the general contractor must be financially responsible. He must not only meet his own payroll and overhead costs, but must pay his subcontractors and suppliers as well in advance of being reimbursed, hopefully with some profit, by the owner. Usually the general will put in a request for payment each month for that work completed the month before. These requests may have to be accompanied with waivers of lien for the amount requested from each sub and supplier. In submitting such a waiver the sub gives up his right to file a claim against the property, and to obtain a waiver the general will have had to pay the sub or supplier at least the major portion of his request. Usually there is a retainage of at least 5% to 10% by the general to his subs and in turn by the owner to the general. It will not be released until the project is substantially completed.

A contractor whose financial affairs are in good order and who has a contract with a responsible client who in turn has secured adequate construction and mortgage loans to insure the completion of the project should have little trouble in securing a loan from his bank to finance these month to month payments, for the contract will serve as the loan security. The amount needed for this interim financing is often stated as *one* dollar for every *ten* dollars in the contract. But before quoting a final price for a project the contractor should ascertain the length of period he must finance, estimate the amount involved and add the cost of the money to the job cost. It should not have to come out of the profit.

The successful general contractor should have a broad knowledge of building construction. His initial contact with a prospective client may be about site selection, or sources of construction finance or the advantages of one type of construction over another. He must develop contacts with architects, bankers, realtors, mortgage brokers, soil engineers and the local enforcers of building and zoning codes. His knowledge in these allied fields may be the deciding factor in his being considered for the job.

In most municipalities and many states, a license is required to act as either a Subcontractor or General Contractor. There is no general rule governing licensing. A state license may not fully qualify a contractor to practice in a municipality within that state. Many suburban areas will require additional fees of out of town contractors to work in their area. One must always inquire of the laws governing the site of construction.

Often a license or permit bond is required guaranteeing a public body that the contractor will comply with all applicable statutes and ordinances.

If land development is part of the contract, subdivison bonds may be required guaranteeing proper installation of roads, sewers and other utilities, even today extending to provision of schools and park systems.

And to practice, one must also enter into proper relationship with the unions who may also require a wage bond guaranteeing union scale and welfare fund payments.

Finally, one must also consider the type of business relationship within his firm, whether to practice as an individual, within a partnership, or as a corporation. Your lawyer and banker should be consulted on this.

CONSTRUCTION MANAGEMENT

In Construction Management, usually referred to as CM, the general contractor enters into a contract with the owner prior to the bidding period and acts in an advisory role. Bids are taken under his supervision, but separate contracts are let for each phase directly between the contractor for that phase and the owner.

The construction manager is paid a flat fee by the owner and this fee is completely separate from other job expenses. CM fees are based on several choices including:

a)   a stated number of dollars plus expenses. The expense items would include the hourly wages paid the principals billed at an agreed upon fixed rate, plus employees' time billed at an agreed upon multiple to cover overhead and profit, plus a multiple of expense for any outside consultants.

b)   same as above but without a fixed fee, all billing being on a multiple of direct expense.

c)   an agreed upon fixed fee.

d)   A percentage of the total construction cost.

The services rendered by the CM are varied, but generally he is hired early in the planning and will work with the architect in setting up job budgets, bidding and construction schedules, site development, and the selection of building systems. He may make detailed value engineering studies; that is, he will not only determine what material or procedure is least costly, but which ones will give optimum quality within the budget. This is discussed later in this chapter. He will cooperate with the architect in arranging his specifications to include the proper extent of work for each phase to be bid, and he will prepare prequalification criteria for the bidders and will conduct prebid conferences to acquaint the bidders with the extent of work covered under other contracts, what facilities at the job site are available to them and what they must furnish themselves, the time scheduling and special procedures to be followed.

When the bids are received the CM will evaluate them and either recommend that one be accepted or that that phase should be modified and rebid.

When the construction is started the CM will coordinate all work on the site to coincide with the agreed to schedules, budgets and quality standards. He will recommend corrective procedures and job changes that he feels are necessary, but his decisions are subject to the approval of the architect and owner. He will maintain an office at the job site with sufficient personnel to inspect the work, record job progress, keep up-to-date records of drawing and specification changes, shop drawings, samples, operating manuals and other records which will be delivered to the owner upon completion of the project. He will set up and maintain the project accounting system which on larger jobs may be computer based and which will give instant reports on the project cost status, payment status, an analysis of each contract and the project cash flow.

The construction management approach has gained wide acceptance and may be encountered in federally sponsored construction. It gives the owner access to complete data of all phases of the contract. It frees the contractor from many of the risks of a lump sum contract. But if it limits his losses it also limits his profits as any job savings that his expertise may bring about accrue to the owner not the CM.

The CM format is not found attractive to all owners especially, the ones that desire to minimize their direct activity with the project during the construction period and in these cases it is more desirable to employ the services of a General Contractor on a lump sum contract basis.

BIDDING FOR A CONTRACT

Construction contracts are usually awarded in one of two ways—competitive bidding or negotiation.

Competitive bidding is the method used in the majority of instances, but it is only truly effective when complete working drawings and specifications are available and when contractors are screened so all who bid are of the same caliber. To be asked to bid on sketchy plans and vague specifications is generally a waste of time for the more capable contractors for they will most always be underbid by those who will sacrifice quality for price. To get on the bidding list of an architect's office that is known to produce detailed drawings and tight specifications, and insist upon their being carried out, is one of the most valuable contacts a contractor aspiring to fine work can make. The architect has already spent many hours working with the owner making most of the critical decisions, and by the time bids are taken, the contractor's responsibilities are clearly set forth. He can proceed, if awarded the contract, knowing that the finished project will most probably satisfy the owner, and satisfied owners are probably the most valuable asset a contractor can have.

In private work the architect, often with the owner's help, will limit the list of bidders so you can judge your competition and know that if your price is low and reasonable you will get the contract. Public work must usually be open to any "qualified" bidder, and "qualifications" are generally concerned more with financial status than reputation for quality. In order to propose a low bid against unknown competition a good contractor must be certain his organization can bring to the project experience and efficiencies uniquely his own. Otherwise he most certainly will be underbid by those desperate enough for work to bid without profit just to get a job.

Private work is sometimes announced in the newspaper or listed in trade publications, but a real estate agent or a banker can often steer a contractor to jobs when he is not contacted directly by an architectural firm. Public work is almost always publicly advertised.

Often to get on a bidding list a contractor must be prequalified. There are various ways this is done, by organizations, governmental bodies and banks but, in general, the information that must be submitted will follow that contained in AIA Document A-305 "Contractor's Qualifications Statement." This requires the firm, whether it is a corporation, a partnership or an individual, provide details on: length of time in business, names of principals, percent of work performed by own staff, record of and reasons for any past failures to complete a job, major projects now under construction, major projects completed in past five years, a record of experience of firm employees, bank references, trade references, name of bonding company, and a statement of financial conditions including current assets, liabilities and the auditor's name. However, often just being a member in good standing of a local contractor's association plus a bank reference will satisfy an inquiry.

Once on a bidding list, a contractor will receive an Invitation to Bidders for each prospective job. This form will include such information as job location, type of proposal, the number of sets of drawings and specifications to be furnished to bidders, what to do should discrepancies or errors be found in drawings and specifications, bulletins to be issued later, when and how bids will be opened, and when and where bids are due. A proposal sheet is usually attached so only the blanks need be filled out, but sometimes the contractor must furnish his own.

Often the Invitation to Bid is accompanied by an Instruction to Bidders further defining the job restrictions such as completion dates, visiting job site, special conditions, etc.

Bid deposits may be required, the amount being either a Lump Sum or a percentage of the proposal. The Bid Bond guarantees that the contractor bidding, if awarded the contract, will enter into the contract and furnish a Performance Bond, if required. If he does not honor his bid he forfeits the amount of the bond, or the difference between his bid and the amount the owner can contract the work for with another contractor, whichever sum is less. Bid Bonds and the later Performance, Material, Labor, Maintenance, Completion, Supply and subcontractor bonds often are encountered in public work, often waived in private work where the contractor's reputation is deemed sufficient and the cost of bonds unwarranted. The cost of the Bid Bond is a cost borne by the contractor. Performance Bonds, unless made a part of the contract, are usually billed to the owner. Bonds are further discussed under "Definitions" in this chapter.

NEGOTIATING A CONTRACT

The other type of contract, one that is negotiated, allows the owner control over the selection of his general contractor. It is often used on complex work, i.e., remodeling jobs where the space is occupied and must be done piecemeal; interior work where strict control of subcontractors is desired; contracts in which mechanical equipment is a major factor and owner may prefer one manufacturer over another; work which must proceed in great haste where the architect may not be able to produce complete plans in the time allotted and work will be let in sections: foundation work, then steel framing, then masonry, etc. Often the contractor will proceed as with bid jobs in submitting a Lump Sum price, usually on the safe side and known as an UPSET or GUARANTEED FIGURE. This will sometimes include the contractor's profit, but often he will be reimbursed for that which he has spent in completing the work and his profit and overhead will be a stated fixed dollar amount or a percentage of the job total.

A variation of the negotiated contract is "phased construction". In this arrangement the project may be started before all the plans are fully developed. Each phase of the job, such as foundations, masonry, carpentry, etc., is bid separately and at the time just prior to when that phase is required to be installed. The advantages are that the project can be started earlier than if it must wait until all the design details are settled; the bidders need not add contingencies to cover unknowns such as wage and material cost rises or adverse job conditions for the work will proceed immediately and these factors are known; the bidding time is increased from three or four weeks to several months, and more "shopping around" is possible; and there is better control of the budget in that if any phase comes in over the budget it can be redesigned and rebid and if an earlier phase is contracted over the budget later stages can be modified so that the total job can still conform to the budget.

As the phases are bid the successful contractors may be assigned to a general contractor in the same manner as a lump-sum contract; or the general may act in the role of a project manager in which each contractor for each phase will have a direct contract with the owner. This variation is referred to as "multiple bidding".

CONTRACT DOCUMENTS

Once it is determined which contracting firm is to do the job, a formal contract will be drawn up. The Contract Documents usually should include the Owner-Contractor Agreement; the General Conditions of the contract; Supplementary Conditions of the contract (if any); the Working Drawings, giving all sheet numbers with revisions; Specifications, giving page numbers; and Addenda or Bulletins issued prior to contract. After the contract is signed, any changes are added as Modifications to Contract and would take the form of written amendments, Change Orders or architectural interpretations.

The Owner-Contractor Agreement is a legal affair, but for many jobs standard forms are used. If the architect furnishes the Owner-Contractor Agreements he will most probably use one printed by the American Institute of Architects. These are:

a) The Stipulated Sum Agreement A-101, four pages long or A-107, a shortened form. A special form A-101 CM is available for construction management.

b) The Cost of the Work Plus a Fee Agreement A-111, eight pages long.

Agreements A-101 and A-111 assume that the AIA Document A-201, "General Conditions of the Contract for Construction," 18 pages long, is made a part of the contract. These 14 articles are discussed later in this chapter; a shortened version of the General Conditions is included in Document A-107.

The Stipulated Sum Agreement will include the date of the agreement, the owner's name, contractor's name, the description of work, the architect's name, the contract sum, the time of commencement and completion, final payment and a list of the contract documents.

The Cost Plus a Fee Agreement will include date of agreement, owner's name, contractor's name, description of work, architect's name, the contractor's duties and status, time of commencement and completion, cost of work and guaranteed maximum cost (this may be deleted and if included may be supplemented by a provision for distribution of savings), the contractor's fee, provisions for changes in the work, reimbursable costs, non reimbursable costs, discounts, rebates and refunds, subcontractors, accounting records, applications for payment, payments to contractor, termination of contract and a list of the contract documents.

For less complicated Stipulated Sum contracts, where the contractor furnishes the form, one of the forms published by the Frank R. Walker Company may be useful. These include a one page proposal form #147 which includes an acceptance paragraph for the owner to sign to make a short contract; a one page contract agreement form #158; and a longer four page agreement form #160 which covers many of the provisions found in the usual general conditions. If the contract is for a construction manager, AIA form B-801 may be used.

Large corporations and governmental bodies will usually have contracts drawn up by their legal staffs. The contractor may want his own lawyer to review these contracts.

A third type of contract, but one only for both experienced owners and contractors, is the contract for a COST PLUS a variable sum. Here, a fixed cost for the building is agreed upon. If the contractor is able to complete a project at less than the agreed upon figure, the savings is split between owner and contractor; if the cost runs higher, the overage is also split. Often, not only cost is included with penalties and rewards, but also completion dates.

PROPOSAL

_____ 19___

_____

Dear Sir:
    The undersigned proposes to furnish all materials and perform all labor necessary to complete the following:

_____

_____

_____

_____

_____

_____

_____

_____

_____

_____

_____

_____

_____

_____

_____

_____

_____

    All of the above work to be completed in a substantial and workmanlike manner for the sum of_____

_____($_____) Dollars

Payments to be made each_____as the work progresses to the value of_____(_____%)
per cent of all work completed. The entire amount of contract to be paid within_____days after completion.

    Any alteration or deviation from the above specifications involving extra cost of material or labor will only be executed upon written orders for same, and will become an extra charge over the sum mentioned in this contract. All agreements must be made in writing.

    The Contractor agrees to carry Workmen's Compensation and Public Liability Insurance, also to pay all Sales Taxes, Old Age Benefit and Unemployment Compensation Taxes upon the material and labor furnished under this contract, as required by the United States Government and the State in which this work is performed.

Respectfully submitted,

Contractor

By

ACCEPTANCE

    You are hereby authorized to furnish all materials and labor required to complete the work mentioned in the above proposal, for which the undersigned agrees to pay the amount mentioned in said proposal, and according to the terms thereof.

_____

Date_____19___

_____

PRACTICAL
MFD. IN U.S A. Form 147

FRANK R. WALKER CO., PUBLISHERS, CHICAGO

Short Form Proposal

Frank R. Walker Co.   Form 147

Contract Agreement

THIS AGREEMENT, made this_____ day of_____A.D. 19_____,
by and between_____hereinafter called the
Owner, and_____hereinafter called
the Contractor.

For the consideration hereinafter named, the said Contractor covenants and agrees with said Owner, as follows:

FIRST. The Contractor agrees to furnish all material and perform all work necessary to complete the_____

for the above named structure, according to the plans and specifications (details thereof to be furnished as needed) of
_____Architect, and to the full satisfaction of said Architect or Owner.

SECOND. The Contractor agrees to promptly begin said work as soon as notified by said Architect or Owner, and to complete the work as follows:_____

THIRD. The Contractor shall take out and pay for Workmen's Compensation and Public Liability Insurance, also Property Damage and all other necessary insurance, as required by the Owner, Architect or by the State in which this work is performed.

FOURTH. The Contractor shall pay all Sales Taxes, Old Age Benefit and Unemployment Compensation Taxes upon the material and labor furnished under this contract, as required by the United States Government and the State in which this work is performed.

FIFTH. No extra work or changes under this contract will be recognized or paid for, unless agreed to in writing before the work is done or the changes made.

SIXTH. This contract shall not be assigned by the Contractor without first obtaining permission in writing from the Architect or Owner. All Sub-contracts shall be subject to the approval of the Architect or Owner.

IN CONSIDERATION WHEREOF, the said Owner agrees that he will pay to the said Contractor, in_____
_____payments, the sum of_____
_____Dollars
for said materials and work, said amount to be paid as follows: _____per cent (_____%) of all labor
and material which has been placed in position by said Contractor, to be paid on or about the_____
of the following month, except the final payment, which the said Owner shall pay to the said Contractor within_____
_____days after the Contractor shall have completed his work to the full satisfaction of the said
Architect or Owner.

The Contractor and the Owner for themselves, their successors, executors, administrators and assigns, hereby agree to the full performance of the covenants of this agreement.

IN WITNESS WHEREOF, they have executed this agreement the day and date written above.

Witness:

_____

_____   Owner.

_____

MFD. IN U.S.A. FORM 158   _____
                                                            Contractor.

FRANK R. WALKER CO. PUBLISHERS, CHICAGO

Short Form Contract Agreement

Frank R. Walker Co.   Form 158

A bonus is given the contractor for each day of delivery earlier than the specified date, but a penalty is deducted for each day of failure to deliver on the specified date. Not only must a contractor be knowledgeable about his ability to perform under these conditions, he should also be well acquainted with the owner's reputation since what constitutes "completion" by a certain date can be a matter for the courts.

The General Conditions of the Contract usually refer to the American Institute of Architects Document A-201. These are often accompanied by Supplementary Conditions which modify and extend the General Conditions. Sometimes these two are combined and rewritten in a form more exactly fitting the special job conditions. But in any case, they set forth the rights, responsibilities and relations of all parties to the contract and every contractor should be thoroughly acquainted with them.

The latest edition of the AIA form contains fourteen articles. Briefly they cover the following:

Article 1. Defines the contract documents, the work, the project, the execution, correlation, intent and interpretations of the contract documents, and sets forth the copies and ownership of these documents.

Article 2. Defines the architect and his administration of the contract. (A further discussion of administration is included later in this chapter.)

Article 3. Defines the owner and the information and services required of him, his right to stop work and to carry out work if contractor defaults.

Article 4. Defines the contractor; his responsibility to review the contract documents; his supervision and construction procedures; his furnishing and control of labor and materials; his responsibility for a warranty, for taxes, for permits, fees and notices; cash allowances, if any; the furnishing of a superintendent; his responsibility for those performing the work; the preparation of a progress schedule; the maintenance on the job site of drawings and specifications; the preparation of shop drawings and furnishing of samples; the use of the site; cutting and patching of work; cleaning up; proper channel for communications, and the indemnification of both owner and architect against all claims, damages, losses and expenses within the limits set forth.

Article 5. Defines the subcontractors, the award of subcontracts, subcontractual relations and payments to subcontractors.

Article 6. Sets forth the owner's right to award separate contracts, the mutual responsibilities of contractors, the problems of cutting and patching under separate contracts and the owner's right to clean up.

Article 7. Includes miscellaneous provisions such as the law of the place, successors and assigns, what constitutes written notice, how claims for damages are to be made, right of owner to require performance and labor and material bonds and how they will be paid for, rights and remedies, payments of royalties and patents, tests that may be required, interest on money not paid when due, and arbitration of claims and disputes.

Article 8. Defines contract time, progress and completion and delays and extensions.

Article 9. Discusses payments and completion sum, the filing of a schedule of values to be used as a basis for contractor's applications, progress payments, certificates for payment, payment withheld, failure of payment, and substantial completion and final payment.

Article 10. Covers protection of persons and property including safety precautions and programs, safety of persons and property, and emergency precautions.

Article 11. Lists the insurance requirements including contractor's liability insurance, owner's liability insurance, property damage insurance and loss of use insurance.

Article 12. Covers changes in the work, including change orders, how to file claims for additional cost, minor changes in the work and field orders.

Article 13. Discusses uncovering and correction of work and acceptance of defective or non-conforming work by the owner.

Article 14. Discusses termination of the contract by the contractor or by the owner.

The Supplementary Conditions may include two lists; those that amend the General Conditions and those that made additions to it.

A Special Edition, AIA Document A-201/CM is available for jobs using construction Management. For U.S. Department Of Health, Education And Welfare federally assisted construction projects the standard A-201 form is supplemented by form A-201/SC which adds three articles which modify the 14 conditions, and conditions such as equal opportunity employment and special insurance requirements.

SETTING UP THE JOB

Having been awarded the contract, the general contractor must then proceed to sign up his subcontractors. The standard AIA Owner-Contractor Agreement calls for the successful bidder to submit a list of subcontractors for approval prior to award of contract, and the owner and architect must object at that time to any firm on the list. If there is an objection, the contractor has two alternatives: to submit an acceptable substitute with the change in cost, if any, in which case the owner may at his discretion accept or reject the bid; or, the contractor may withdraw his bid without penalties.

Additional expenses occasioned by owner's or architect's rejection of subcontractors after award of contract, will be added by change order to the contract sum and paid for by the owner. There is conflict on this issue and extreme caution is to be observed.

The contractor must also submit a progress schedule and a schedule of values based on his subcontracts.

The AIA agreement further sets forth the contents of agreements between subcontractors and the general contractor and the method of paying the subcontractors. AIA form A-401 may be used, or the shorter forms #133 or #144 published by Frank R. Walker Company.

The general contractor with the cooperation of his subs, the architect and owner, will file for and usually pay for all permits. Often it is necessary to have selected certain subcontractors before filing for permit as the permit may require the license number of certain trades.

The Supplementary General Conditions will usually set forth the initial site requirements—office, telephone, toilets, water, light, heat, barricades, temporary partitions, etc., which the general contractor must provide. He must also make arrangements with local utilities.

CARRYING OUT THE WORK

The administration of the contract is clearly set forth in the AIA standard contractor agreements. The general contractor is responsible for executing the work in strict accordance with the contract documents and for turning the completed project over to the owner in full conformance with them. He must provide the continual supervision necessary to check all labor, materials and procedures to accomplish this.

Sub-Contract Agreement

THIS AGREEMENT, made this _____ day of _____ A.D. 19____,
by and between _____ hereinafter called the
Contractor, and _____ hereinafter called
the Sub-contractor.

For the consideration hereinafter named, the said Sub-contractor covenants and agrees with said Contractor, as follows:

FIRST. The Sub-contractor agrees to furnish all material and perform all work necessary to complete the _____

_____

_____

_____

_____

_____

for the above named structure, according to the plans and specifications (details thereof to be furnished as needed) of
_____ Architect, and to the full satisfaction of said Architect.

SECOND. The Sub-contractor agrees to promptly begin said work as soon as notified by said Contractor, and to complete
the work as follows: _____

_____

_____

_____

THIRD. The Sub-contractor shall take out and pay for Workmen's Compensation and Public Liability Insurance, also
Property Damage and all other necessary insurance, as required by the Owner, Contractor or by the State in which this work
is performed.

FOURTH. The Sub-contractor shall pay all Sales Taxes, Old Age Benefit and Unemployment Compensation Taxes upon the
material and labor furnished under this contract, as required by the United States Government and the State in which this
work is performed.

FIFTH. No extra·work or changes under this contract will be recognized or paid for, unless agreed to in writing before
the work is done or the changes made.

SIXTH. This contract shall not be assigned by the Sub-contractor without first obtaining permission in writing from
the Contractor.

IN CONSIDERATION WHEREOF, the said Contractor agrees that he will pay to the said Sub-contractor, in _____
_____ payments, the sum of_____

_____ Dollars

for said materials and work, said amount to be paid as follows: _____ per cent (_____%) of all labor
and material which has been placed in position by said Sub-contractor, to be paid on or about the _____
of the following month, except the final payment, which the said Contractor shall pay to the said Sub-contractor within
_____ days after the Sub-contractor shall have completed his work to the full satisfaction of the said
Architect or Owner.

The Contractor and the Sub-contractor for themselves, their successors, executors, administrators and assigns, hereby agree
to the full performance of the covenants of this agreement.

IN WITNESS WHEREOF, they have executed this agreement the day and date written above.

Witness:

_____

_____

_____

Sub-Contractor	
By	
Contractor	
By	

PRACTICAL   FORM 133   FRANK R. WALKER CO. PUBLISHERS, CHICAGO
MFD. IN U.S.A.

Short Form Sub-Contractor Agreement

Frank R. Walker Co.   Form 133

The architect serves in an advisory capacity to both contractor and owner, interpreting drawings, and specifications, approving shop drawings and samples, checking conformity of the job with contract documents, making periodic inspections to determine progress and approving payments to the contractors, and approving final acceptance of the contract. The architect will also have his consultants visit the site at certain intervals.

Often the owner will want more supervision than the architect includes in his standard contract and a "clerk of the works" is hired. He is usually paid by the owner, but works under the architect. He usually keeps complete job records and reports daily progress. But he does not supplant the architect and the architect's decisions are the final ones, nor does he relieve the contractor of his responsibilities of supervision.

In addition, public building inspectors will probably visit the site periodically. Any objections they raise to apparent contradictions between codes and the contract documents should be reported in writing to the architect. At certain periods the contractor will be required to make tests and he must notify the proper authorities and the architect when these are to occur.

GETTING PAID

Applications for payment are usually made monthly. The Frank R. Walker Company form #591 is an excellent one for this. It provides a column for the list of subcontractor names, what work the contract is for, the total amount of each contract, the amount previously requested and finally the balance to complete. The bottom of the form provides a resume of total amounts less percentages retained, and a space for notarizing. Waivers of lien showing that the contractor has paid for all materials, labor and subcontractor billings are usually required to accompany the request for payment. The Frank R. Walker Company has printed forms for a Contractor's Affidavit (#592) and Waiver of Lien (#593).

The architect then checks the application and if it agrees with what he feels to be proportionate to the amount of the contract, he will issue a certificate of payment. Walker publishes a certificate for payment (form #496) in duplicate. His decision in this may be reviewed by either the company writing the performance bond or the company supplying the interim financing. As all this takes time for the contractor to prepare, the architect to check and approve and the owner to process, the contractor must keep all his billings and payments current. With the high cost of money today, short term loans to tide a contractor over for a month because he didn't get around to filing a proper or complete request for payment can eat away at profits as surely as having to correct mistakes on the job.

When Standard AIA Contract forms are used then the AIA application for payment will also be utilized to retain the same document format. Upon final completion and acceptance, the contractor should be prepared to file final waivers along with any written guarantees, certificates of inspections, operating instructions, etc., that may have been called for.

DEFINITIONS

In reviewing standard contract forms, certain terms may be used which have special meaning within the profession. Some of these are explained below:

Addenda—Modifications to the contract documents issued by the architect's office during the bidding period. Often identified as "Bulletins."

Alternates: Additions or subtractions to a contract sum for substitutions asked for by the architect, which the contractor must submit with his proposal.

FORM 591
MFD IN U S A

SWORN STATEMENT FOR CONTRACTOR AND SUBCONTRACTOR TO OWNER

State of .... }
County of .... } ss.

The affiant, ... being first duly sworn, on oath deposes
and says that he is (1)...
contract with (2).. owner .... for
(3) ....
on the following described premises in said County, to-wit:...

That, for the purpose of said contract, the following persons have been contracted with, and have furnished, or are furnishing and preparing
materials for, and have done or are doing labor on said improvement. That there is due and to become due them, respectively, the amounts set
opposite their names for materials or labor as stated. That this statement is made to said owner........ for the purpose of procuring from said
owner........ (4) Partial—Final Payment on said contract, and is a full, true and complete statement of all such persons, and of the amounts paid,
due and to become due them.

(1) A member of the firm of, or officer of the corporation of, naming same. If a subcontractor so state and name the contractor. (2) Name of the owner or
owners. (3) What the contract or subcontract is for. (4) Partial or Final Payment.

NAME AND ADDRESS	CONTRACT FOR	AMOUNT OF CONTRACT	TOTAL PREVIOUS REQUESTS	AMOUNT OF THIS REQUEST	BALANCE TO COMPLETE

AMOUNT OF ORIGINAL CONTRACT	$		TOTAL AMOUNT REQUESTED	$
EXTRAS TO CONTRACT	$		LESS % RETAINED	$
TOTAL CONTRACT AND EXTRAS	$		NET AMOUNT EARNED	$
CREDITS TO CONTRACT	$		AMOUNT OF PREVIOUS PAYMENTS	$
NET AMOUNT OF CONTRACT	$		AMOUNT DUE THIS PAYMENT	$
			BALANCE TO COMPLETE	

It is understood that the total amount paid to date plus the amount requested in this application shall not exceed   % of the cost of
work completed to date.
I agree to furnish Waivers of Lien for all materials under my contract when demanded.

Signed...

....

Subscribed and sworn to before me this .... day of .... 19.....

PRACTICAL
© FRANK R. WALKER CO., PUBLISHERS.
CHICAGO   The above sworn statement should be obtained by the owner before each and every payment.   Notary Public

Application For Payment
Sworn Statement For Contractor and Subcontractor To Owner

Frank R. Walker Co.   Form 591

Approved Equal: A rather loose term indicating that the contractor may substitute another product for that which the architect specifies, providing it is "equal" in all respects. It is well to accompany a proposal with a list of such substitutions upon which a contractor may have based his figure, as it is the architect not the contractor who has the final say as to what is "equal."

Bid: A proposal stating the sum for which the contractor will complete a project.

Bid Bond: A bond furnished by the contractor that guarantees he will sign a contract on the basis of his bid.

Cash Allowance: Sums set forth by the architect for specific items which the contractor must include in his proposal. Upon completion of a job the contractor must make an accounting of these sums returning any unspent amounts to the owner. Often used for buying hardware, face brick, light fixtures, and other special items.

Certificate of Occupancy: Issued by a government authority authorizing occupation of a building.

Change Order: An authorization by the architect to change the basic contract because of additions or deductions from it.

Cost Breakdown (or schedule of values): A schedule prepared by the contractor showing how the contract sum is divided among the various divisions of the work, which is filed before work is started. The architect will use this as a basis for approving applications for payment and waivers. It should be accurate and kept up to date, reflecting any changes in the contract.

Extras: Construction costs outside of contract sum. A contractor should not proceed with any extras until a Change Order is issued by the architect.

Final Acceptance: When the owner accepts a project as complete.

Payment and Performance Bond: A bond assuring that the contractor will complete a project in accordance with contract documents including discharging all financial obligations.

Letter of Intent: Written prior to drawing up complete contract documents to tell the contractor a formal agreement is forthcoming. Often this authorizes certain preliminary work such as filing for building permits and site clearance.

Liens: Legal claims against an owner for labor and materials for construction of a project.

Liquidated Damages: A sum agreed to by owner and contractor, chargeable to the contractor, for damages suffered by the owner because of contractor's failure to fulfill his contract obligations.

Maintenance Bond: Guarantees owner that contractor will rectify defects in workmanship or materials reported to him within a specified period following final acceptance of the work under contract. Often this is included as part of the Performance Bond and usually runs for one year. May also be issued for longer periods covering specific phases of the work, such as roofing, curtain wall, paving, etc.

Punch List: The architect's list of work yet to be done or to be corrected by the contractor.

Retainage: A percentage set forth in the Owner-Contractor Agreement to be withheld from each payment to the contractor. Usually 10%, reduced to a total of 5% upon substantial completion.

Separate Contract: A contract let by the owner directly to a contractor other than the general contractor. Usually for special equipment, landscaping, interior furnishings, etc.

Shop Drawings: Drawings prepared at the contractor's expense showing how specific items shall be fabricated and/or installed. These usually are prepared by the subcontractors and are standard for such items as steel, reinforcing, miscellaneous and ornamental iron, cut stone, millwork, partitions, door frames, etc. They must be submitted to the architect for his approval.

Subcontractor Bonds: A performance bond given by a subcontractor to the general contractor guaranteeing performance of contract and payment of labor and material bills.

Substantial Completion: A state reached when the project is sufficiently complete to allow the owner to occupy it, but not necessarily finally accept it.

Superintendence: The work of the contractor's representative, not to be confused with that of the architect (administration) or clerk of the works.

Supply Bonds: A guarantee to the owner by a manufacturer that the materials delivered comply with the contract documents.

Unit Prices: Amounts asked for on the proposal for furnishing materials per unit of measurement. That is, cost of concrete per cu. yd., floor tile per sq. ft., or steel per ton.

Upset Price: An amount agreed to by the contractor as a maximum cost to perform a specific project. Used on Cost Plus a Fee jobs.

Warranty: A guarantee furnished by the contractor or a manufacturer covering quality, workmanship and performance for a specified period.

VALUE ENGINEERING

As we noted above, value engineering may be part of the services assigned to a general contractor under a construction management contract. Value engineering, or value analysis as it is often called, is a systematic approach to determine the total value of a product or procedure to a project. It constantly investigates the possibilities of equivilent performance at a lower cost. It is thus primarily a design consideration and to be most effective it should be applied early in the development of a project, although the same considerations can be applied to construction methods in the field during actual building.

To achieve maximum value one must first pinpoint those values essential to the project or procedure and determine those items which are secondary or which contribute nothing to the function but do add to the costs. Costs must be extended to include original cost, operating costs, maintenance costs and replacement costs. Items that do not contribute directly to the primary function but increase costs lower overall value.

Every job has areas where costs seem "out of line" or "come in too high" and it is to these areas that value engineering might first be applied. However it is often in the most commonly repeated items that low value may be determined and large savings found possible.

Once the item or procedure is pinpointed for analysis the system can be applied. First the primary function must be defined. This is best done with two words - a verb and a noun. As an example one might say the function of a door is to "provide (verb) access (noun)". But a cased opening would do the same. The function must be more exactly defined. If it is an exterior door its primary function may be to "provide security"; the entrance to a beauty salon may be to "express elegance" and security is of secondary importance; the entrance to a fire stair may be to "exclude fire" and security has no function. The door to a private secretary's office may only function to "contain noise".

Having defined the primary function it is time to investigate what else might do the job as well. Perhaps the secretary needs no door if an acoustic ceiling and carpet on the floor is provided or the general noise level is overcome by piped-in music.

It is then time to assign exact costs, and to determine the cost/worth ratio. This ratio is the total cost of the item divided by the cost necessary to perform only the primary function. A similar cost/worth ratio would be established for each alternative and a final decision can be made. The results obviously will be only as reliable as are the cost figures supplied. This is where the estimator can make a valuable contribution.

Value engineering is most often a team effort involving the designers, the suppliers, the builders and the users. It must have broad representation as it has as its objective the total job. The door supplier will not suggest substituting carpet, and all the subsystems of the building trades must be breached for the good of the whole. The team member must then have an open mind and a creative approach. He must be knowledgeable about past procedures, codes and union demands, but must also have enough technical training to be able to investigate and evaluate new products and procedures and sufficient interest to know what is available. Further he must be cooperative and a good team player with the patience to listen and the ability to communicate. And finally he must remain objective to the end which is to identify and eliminate unnecessary costs without compromising essential quality.

Value engineering was first applied to industrial production about thirty years ago and only more recently used by the building trades. It would seem that it is an idea whose time has come as the public, faced with ever increasing prices and even greater increases in maintenance and operating costs, will demand of the construction industry, as it already has of the automobile industry, a more efficient, higher, true value product.

Value engineering is an important development and all interested in the initial stages of construction, and this certainly includes the estimator, should have some knowledge of it. There are several books available both from private and governmental sources. There are also seminars and workshops given periodically, some addressed directly to the construction industry. Local contracting and architectural societies should be contacted about these.

PRACTICAL

Job No. ....

CERTIFICATE FOR PAYMENT TO CONTRACTOR

Certificate

No. ....

Date   TO $.....................

STATEMENT		
Contract		
Extras		
TOTAL		
Deductions		
TOTAL		
Total Comp. Wk.		
Less Retention		
Prev. Payments		
AMOUNT DUE		

THIS IS TO CERTIFY, that

Contractor for ....

is entitled to a .... payment of

.... Dollars

under terms of this contract.

Received Amount of above Certificate.

....

Contractor

FORM 596   FRANK R. WALKER CO., PUBLISHERS, CHICAGO   MFD. IN U.S.A.

CONTRACTOR'S AFFIDAVIT

State of_____⎫
                ⎬ ss.   Office of _____
County of_____⎭   _____

_____

being duly sworn on his oath deposes and says that he is_____

the contractor for the_____

for the building erected for_____ , owner,

on the premises described as follows, to-wit: _____

_____

_____

_____

All the bills for labor and material are fully paid and discharged_____

_____

_____

_____

That____he____makes this affidavit for the purpose of procuring from_____

_____a_____payment

of_____Dollars and_____Cents

upon _____contract for said labor or material or both.

        SUBSCRIBED AND SWORN to before me

this_____day of_____A. D., 19_____   _____

_____

_____

PRACTICAL FORM 592   NOTARY PUBLIC

FRANK R. WALKER CO., PUBLISHERS, CHICAGO   MFD. IN U. S. A.

WAIVER OF LIEN
MATERIAL OR LABOR

State of_____⎫
                ⎬ ss.   _____19_____
County of_____⎭

TO ALL WHOM IT MAY CONCERN:

    Whereas _____ the undersigned _____

ha ____ been employed by _____

to furnish _____

for the Building known as _____

City of _____

Lot No. _____ Section _____ Township _____ Range_____

County of_____State of_____

NOW, THEREFORE, KNOW YE, That _____ the undersigned

for and in consideration of the sum of _____ Dollars

and other good and valuable considerations, the receipt whereof is hereby acknowledged, do hereby waive and release any and all lien, or

claim or right to lien on said above described building and premises under the Statutes of the State of _____

relating to Mechanics' Liens, on account of labor or materials, or both, furnished or which may be furnished, by the undersigned to or

on account of the said _____

_____ for said building or premises.

Given under____ hand ____ and seal ____ this _____ Day of _____ A. D., 19_____

_____   _____ (SEAL)

_____ (SEAL)

PRACTICAL FORM 593

FRANK R. WALKER CO., PUBLISHERS, CHICAGO   MFD. IN U. S. A.

CHAPTER 2

CONSTRUCTION FINANCE

COST OF MONEY
SOURCES OF MONEY
MORTGAGE LOANS
MORTGAGE BANKER
SELLING THE LENDER
THE COMMITMENT
SHORT TERM LOANS
INTERIM FINANCING

Before proceeding with the details of estimating the various building materials and labor that go into a project, let us take a quick look at the "material" that makes it all possible—money.

THE COST OF MONEY

A recent news article was headlined "Typical Home Total Cost— $100,000". One might quibble about the definition of "typical", but what the article pointed out was that a $25,000 house of ten years ago now costs $45,000; but adding finance charges, based on a 20% down payment and a mortgage for twenty five years with interest charges of 6% ten years ago and 9% today, the houses really cost $43,661 and $99,636 respectively. The increase in the initial cost of the house in ten years was 80%; but the increase in the cost of financing was 120%. In the recent tight money market some mortgages were not at 9% but at 16% and more! It becomes obvious then that money must be used as efficiently as any other commodity. And by efficiently we mean not only spent wisely but carefully planned for, shopped for, drawn upon only at the exact time it is needed and paid for on time just as surely as if it were brick and mortar.

Our concern here is not with the money that the contractor needs to run his own business affairs, but the money to pay for the project itself. This breaks down into two types of financing—immediate cash to pay for the labor and materials as they are placed on the job and, once the project is completed, a loan against the project with repayment scheduled over many years. These are known as the short term or the construction loan and the long term or mortgage loan.

Project financing is not usually the direct concern of the contractor. It is the owner's or project developer's responsibility to secure both the short and long term loans. But it is essential that a contractor understand all phases of the project financing and his relationship to it.

First, few projects today are paid for from funds the owner already has accrued. He must find someone willing to loan the money to make the project possible and that someone must have enough faith in the project to be willing to wait for repayment of his loan from money generated by the success of the project. In order to "sell" his project to a loan source the owner must prepare a complete analysis of all aspects of it including reliable cost projections, and he will often turn to a contractor for this information. It is important that the contractor understands what such an estimate could commit him to, for what the lender "sees" is what the contractor later "gets" if the project loan is approved.

Second, the construction loan is the source from which the contractor will be paid, and he will have to work closely with the lender and his job inspectors to get his money on time. If the contractor is inefficient in compiling his requests for payment his men and suppliers will go unpaid unless the contractor borrows the money until the next period the lender makes his payouts. The interest the contractor will have to pay will cut into his profit as surely as if he had underestimated the materials for the job.

Third, the owner may be very naive in construction finance. He may build only one project in his entire life. The local bank may be his only financial contact. Now the local banks are an excellent starting place to seek construction funds; they know the owner and the community and the owner's best interests are presumably the bank's best interests. But the bank may have other business loans with the owner, and they may not be willing, or even legally able, to make an additional loan of the size, at the rate and for the length of time the owner requires. Thus the project may die in the bank's board room. The owner may lack the time and the sophistication to search further, and so the contractor looses the job as surely as if he had been underbid by a competitor. But there are other money sources besides banks and an aggressive contractor will develope these other contacts so that he can steer the owner to them and thereby perhaps salvage a good job that otherwise might never get off the drawing boards.

Fourth, as we have discussed in Chapter 1, the triangle of the owner—architect—contractor has expanded into more of a circle with each often over-lapping into the other's territory. The contractor may become the owner either as the project developer or as the official landlord who then leases the project with perhaps an option to buy. Such arrangements can be very profitable and can be used in slack times to keep a contractor's force busy. But a good contractor does not always make a good landlord or even a good client for his company, and it is not the purpose of this chapter to encourage contractors to become entrepreneurs.

SOURCES OF MONEY

The main sources of money are:
a) Commercial Banks—short and intermediate term financing
b) Savings and Loans—small residential loans
c) Insurance Companies—restricted long term loans
d) Pension and Trust Funds—long term loans. These funds are often administered by commercial banks.
e) Real Estate Investment Trusts (REITS)—long and short term loans, land loans, second mortgages, sell and leaseback deals.
f) Governmental Agencies—FHA and VA loans for residential work (guarantee loans)
g) Government Sponsored Bonds—Industrial Revenue Bonds for projects that increase employment. Local governments issue tax-exempt bonds secured by the project income.
h) Mortgage Banking Firms—Long and short term loans, land loans, second mortgages, sell and leaseback deals.

While each source has an area it is most interested in they all compete to a degree. Many projects will draw on two or more sources. Short term loans are for three years or less. Long term loans are seldom for less than ten years and may run up to thirty years.

One thing all loan sources have in common is that they are "selling" their money, not "buying" your project. When the project is presented to them they not only want to know the physical details but how the project will generate the money to pay back the loan. The essence of a good presentation is the proof that the loan can be repaid.

MORTGAGE LOANS

While it is the construction, or short term loan, that is drawn on first, and land and "front money" loans may preceed it, it is the mortgage, or long term, loan that is shopped for first. Once the owner has the mortgage loan lined up the construction loan is more easily secured as it will eventually be paid off by the mortgage loan when the project is completed. Further the interest charged for the mortgage loan may decide whether the project can generate the income to make the necessary payments. If one fails to get a commitment for the mortgage interest and the interest rates soar as they have in recent years the debt service may take all the projected profits.

Small projects can usually be financed both long and short term at local banks and savings and loan associations. These institutions make no direct charge for reviewing proposals and can usually give a yes or no answer in a few days so one can shop several sources at the same time with no obligations. Once a loan is approved and accepted there will be certain charges made, called closing costs, and a part of these costs known as "points" or "origination fees" or "the discount" cover the cost of setting up the loan. These points are a one-time charge made as a percentage of the loan and are deducted from the amount of the loan. If the points are 2% then two dollars is deducted from every hundred dollars of the loan. The usual range of points is 1%–3%, but in states where there are legal limits set on the interest that may be charged, points have been quoted as high as 7% as this is a way to get around the usury laws. Such points are common to all loan sources and are one of the important items to shop for.

MORTGAGE BANKER

For larger projects and projects that generate income one should check out all the money sources. It isn't likely that even the most gregarious contractor would know representatives from all the institutional sources, or that he could keep abreast of their current interests and capabilities if he did. For this one must consult a specialist and that specialist is the mortgage banker as we shall call him here to distinguish him from a commercial banker.

The mortgage broker is an intermediator between borrowers and lenders although his firm may have some funds of their own to invest. He is most often compared to a "match-maker" who brings two parties together into a "marriage". To do this he must carefully evaluate both sides, shopping for the best terms for the owner and the projects most compatible with the portfolio of the lender. For consummating the transaction he charges a flat fee of around 2% of all the loans placed by him. This fee is in addition to the usual closing costs which are charged by the lender. Both expenses are paid by the borrower. Many lenders require a 1% fee at the time loan application is filled. This fee is refundable if the lender rejects the loan but becomes earned once the lender approves the application, and is credited to the fee due the lender at the time of settlement.

But there are other benefits in working with a mortgage broker for he can advise on the techniques of financing as well as the sources. There are numerous options available and the owner with his accountant and lawyer will want to review them all and select what may suit his particular situation best. Decisions to be made

Constant Annual Percent

Monthly Payment In Arrears

Description: This table shows the percent of the principal amount of a loan needed each year to pay off the loan when the actual payments are monthly and are paid in arrears. Divide the percent by 12 to get the level monthly payment per $100 that includes both interest and principal.

Example: The Constant Annual Percent needed to pay off a 15%, 30 year loan if payments are made monthly and in arrears is 15.18%. Divide by 12 to get the actual monthly payment. The constant annual payment for a $50,000 loan is $7,590. The monthly payment is $632.50.

Interest Rate	5 yr	6 yr	7 yr	8 yr	9 yr	10 yr	11 yr	12 yr	13 yr	14 yr	15 yr	16 yr	17 yr	18 yr	19 yr	20 yr
10.00	25.50	22.24	19.93	18.21	16.90	15.86	15.03	14.35	13.78	13.30	12.90	12.56	12.26	12.00	11.78	11.59
10.25	25.65	22.39	20.08	18.37	17.06	16.03	15.20	14.52	13.96	13.48	13.08	12.74	12.45	12.20	11.98	11.78
10.50	25.80	22.54	20.24	18.53	17.23	16.20	15.37	14.69	14.14	13.67	13.27	12.93	12.64	12.39	12.17	11.99
10.75	25.95	22.69	20.39	18.69	17.39	16.37	15.54	14.87	14.31	13.85	13.46	13.12	12.84	12.59	12.37	12.19
11.00	26.10	22.85	20.55	18.86	17.56	16.54	15.72	15.05	14.50	14.03	13.64	13.31	13.03	12.79	12.57	12.39
11.25	26.25	23.00	20.71	19.02	17.72	16.71	15.89	15.23	14.68	14.22	13.83	13.51	13.23	12.98	12.77	12.60
11.50	26.40	23.15	20.87	19.18	17.89	16.88	16.07	15.40	14.86	14.41	14.02	13.70	13.42	13.18	12.98	12.80
11.75	26.55	23.31	21.03	19.34	18.06	17.05	16.24	15.58	15.04	14.59	14.21	13.89	13.62	13.39	13.18	13.01
12.00	26.70	23.47	21.19	19.51	18.23	17.22	16.42	15.77	15.23	14.78	14.41	14.09	13.82	13.59	13.39	13.22
12.25	26.85	23.62	21.35	19.67	18.40	17.40	16.60	15.95	15.42	14.97	14.60	14.29	14.02	13.79	13.60	13.43
12.50	27.00	23.78	21.51	19.84	18.57	17.57	16.78	16.13	15.60	15.16	14.80	14.49	14.22	14.00	13.80	13.64
12.75	27.16	23.94	21.67	20.01	18.74	17.75	16.96	16.32	15.79	15.36	14.99	14.68	14.42	14.20	14.01	13.85
13.00	27.31	24.09	21.84	20.17	18.91	17.92	17.14	16.50	15.98	15.55	15.19	14.88	14.63	14.41	14.22	14.06
13.25	27.46	24.25	22.00	20.34	19.08	18.10	17.32	16.69	16.16	15.74	15.39	15.09	14.83	14.62	14.44	14.28
13.50	27.62	24.41	22.16	20.51	19.26	18.28	17.50	16.87	16.36	15.94	15.58	15.29	15.04	14.83	14.65	14.49
13.75	27.77	24.57	22.33	20.68	19.43	18.46	17.68	17.06	16.55	16.13	15.78	15.49	15.25	15.04	14.86	14.71
14.00	27.93	24.73	22.49	20.85	19.61	18.64	17.87	17.25	16.75	16.33	15.99	15.70	15.45	15.25	15.08	14.93
14.25	28.08	24.89	22.66	21.02	19.78	18.82	18.05	17.44	16.94	16.53	16.19	15.90	15.66	15.46	15.29	15.15
14.50	28.24	25.05	22.83	21.19	19.96	19.00	18.24	17.63	17.14	16.73	16.39	16.11	15.87	15.68	15.51	15.36
14.75	28.40	25.22	22.99	21.37	20.14	19.18	18.43	17.82	17.33	16.93	16.60	16.32	16.09	15.89	15.72	15.59
15.00	28.55	25.38	23.16	21.54	20.31	19.37	18.62	18.02	17.53	17.13	16.80	16.53	16.30	16.11	15.94	15.81
15.25	28.71	25.54	23.33	21.71	20.49	19.55	18.80	18.21	17.73	17.33	17.01	16.74	16.51	16.32	16.16	16.03
15.50	28.87	25.71	23.50	21.89	20.67	19.73	18.99	18.40	17.93	17.53	17.21	16.95	16.72	16.54	16.38	16.25
15.75	29.03	25.87	23.67	22.06	20.85	19.92	19.19	18.60	18.12	17.74	17.42	17.16	16.94	16.76	16.60	16.48
16.00	29.19	26.04	23.84	22.24	21.04	20.11	19.38	18.79	18.33	17.94	17.63	17.37	17.16	16.98	16.83	16.70
16.25	29.35	26.20	24.01	22.42	21.22	20.29	19.57	18.99	18.53	18.15	17.84	17.58	17.37	17.20	17.05	16.93
16.50	29.51	26.37	24.18	22.59	21.40	20.48	19.76	19.18	18.73	18.36	18.05	17.80	17.58	17.42	17.27	17.15
16.75	29.67	26.53	24.35	22.77	21.58	20.67	19.96	19.39	18.93	18.56	18.26	18.01	17.81	17.64	17.50	17.38
17.00	29.83	26.70	24.53	22.95	21.77	20.86	20.15	19.59	19.14	18.77	18.47	18.23	18.03	17.86	17.72	17.61
17.25	29.99	26.87	24.70	23.13	21.95	21.05	20.35	19.79	19.34	18.98	18.69	18.45	18.25	18.08	17.95	17.84
17.50	30.15	27.04	24.88	23.31	22.14	21.24	20.54	19.99	19.55	19.19	18.90	18.66	18.47	18.31	18.17	18.06
17.75	30.31	27.21	25.05	23.49	22.33	21.43	20.74	20.19	19.75	19.40	19.11	18.88	18.69	18.53	18.40	18.29
18.00	30.48	27.37	25.23	23.67	22.51	21.63	20.94	20.39	19.96	19.61	19.33	19.10	18.91	18.76	18.63	18.52
18.25	30.64	27.54	25.40	23.86	22.70	21.82	21.14	20.60	20.17	19.82	19.55	19.32	19.13	18.98	18.86	18.76
18.50	30.80	27.71	25.58	24.04	22.89	22.01	21.34	20.80	20.38	20.04	19.76	19.54	19.36	19.21	19.09	18.99
18.75	30.97	27.89	25.76	24.22	23.08	22.21	21.54	21.01	20.59	20.25	19.98	19.76	19.58	19.44	19.32	19.22
19.00	31.13	28.06	25.93	24.41	23.27	22.41	21.74	21.21	20.80	20.47	20.20	19.98	19.81	19.67	19.55	19.45
19.25	31.30	28.23	26.11	24.59	23.46	22.60	21.94	21.42	21.01	20.68	20.42	20.21	20.03	19.89	19.78	19.69
19.50	31.46	28.40	26.29	24.78	23.65	22.80	22.14	21.63	21.22	20.90	20.64	20.43	20.26	20.12	20.01	19.92
19.75	31.63	28.57	26.47	24.96	23.84	23.00	22.34	21.84	21.43	21.11	20.86	20.65	20.49	20.35	20.24	20.16
20.00	31.81	28.75	26.65	25.15	24.04	23.20	22.55	22.04	21.65	21.33	21.08	20.88	20.72	20.58	20.48	20.39
20.25	31.96	28.92	26.83	25.34	24.23	23.39	22.75	22.25	21.86	21.55	21.30	21.10	20.94	20.82	20.71	20.63
20.50	32.13	29.10	27.01	25.52	24.42	23.59	22.96	22.46	22.08	21.77	21.53	21.33	21.17	21.05	20.95	20.88
20.75	32.30	29.27	27.20	25.71	24.62	23.80	23.16	22.68	22.29	21.99	21.75	21.56	21.40	21.28	21.18	21.10

Interest Rate	21 yr	22 yr	23 yr	24 yr	25 yr	26 yr	27 yr	28 yr	29 yr	30 yr	31 yr	32 yr	33 yr	34 yr	35 yr	40 yr
10.00	11.41	11.26	11.13	11.01	10.91	10.82	10.73	10.66	10.59	10.54	10.48	10.44	10.39	10.36	10.32	10.19
10.25	11.62	11.47	11.34	11.22	11.12	11.03	10.95	10.88	10.82	10.76	10.71	10.66	10.62	10.58	10.55	10.43
10.50	11.82	11.68	11.55	11.43	11.34	11.25	11.17	11.10	11.04	10.98	10.93	10.89	10.85	10.81	10.78	10.67
10.75	12.03	11.88	11.76	11.65	11.55	11.46	11.39	11.32	11.26	11.21	11.16	11.12	11.08	11.05	11.02	10.91
11.00	12.23	12.09	11.97	11.86	11.77	11.68	11.61	11.54	11.48	11.43	11.39	11.35	11.31	11.28	11.25	11.14
11.25	12.44	12.30	12.18	12.08	11.98	11.90	11.83	11.77	11.71	11.66	11.62	11.58	11.54	11.51	11.48	11.38
11.50	12.65	12.51	12.40	12.29	12.20	12.12	12.05	11.99	11.94	11.89	11.85	11.81	11.77	11.74	11.72	11.62
11.75	12.86	12.73	12.61	12.51	12.42	12.35	12.28	12.22	12.16	12.12	12.08	12.04	12.01	11.98	11.95	11.87
12.00	13.07	12.94	12.83	12.73	12.64	12.57	12.50	12.44	12.39	12.35	12.31	12.27	12.24	12.22	12.19	12.11
12.25	13.28	13.16	13.05	12.95	12.87	12.79	12.73	12.67	12.62	12.58	12.54	12.51	12.48	12.45	12.43	12.35
12.50	13.50	13.37	13.26	13.17	13.09	13.02	12.96	12.90	12.85	12.81	12.78	12.74	12.71	12.69	12.67	12.59
12.75	13.71	13.59	13.48	13.39	13.31	13.24	13.18	13.13	13.09	13.05	13.01	12.98	12.95	12.93	12.91	12.84
13.00	13.93	13.81	13.71	13.62	13.54	13.47	13.41	13.36	13.32	13.28	13.25	13.22	13.19	13.17	13.15	13.08
13.25	14.14	14.03	13.93	13.84	13.77	13.70	13.64	13.59	13.55	13.51	13.48	13.45	13.43	13.41	13.39	13.32
13.50	14.36	14.25	14.15	14.07	13.99	13.93	13.87	13.83	13.79	13.75	13.72	13.69	13.67	13.65	13.63	13.57
13.75	14.58	14.47	14.37	14.29	14.22	14.16	14.11	14.06	14.02	13.99	13.96	13.93	13.91	13.89	13.87	13.81
14.00	14.80	14.69	14.60	14.52	14.45	14.39	14.34	14.30	14.26	14.22	14.19	14.17	14.15	14.13	14.11	14.06
14.25	15.02	14.92	14.82	14.75	14.68	14.62	14.57	14.53	14.49	14.46	14.43	14.41	14.39	14.37	14.36	14.30
14.50	15.24	15.14	15.05	14.98	14.91	14.86	14.81	14.77	14.73	14.70	14.67	14.65	14.63	14.61	14.60	14.55
14.75	15.47	15.37	15.28	15.21	15.14	15.09	15.04	15.00	14.97	14.94	14.91	14.89	14.87	14.86	14.84	14.80
15.00	15.69	15.59	15.51	15.44	15.37	15.32	15.28	15.24	15.21	15.18	15.15	15.13	15.12	15.10	15.09	15.04
15.25	15.92	15.82	15.74	15.67	15.61	15.56	15.51	15.48	15.45	15.42	15.40	15.38	15.36	15.34	15.33	15.29
15.50	16.14	16.05	15.97	15.90	15.84	15.79	15.75	15.72	15.69	15.66	15.64	15.62	15.60	15.59	15.58	15.54
15.75	16.37	16.28	16.20	16.13	16.08	16.03	15.99	15.95	15.93	15.90	15.88	15.86	15.85	15.83	15.82	15.79
16.00	16.59	16.50	16.43	16.37	16.31	16.27	16.23	16.19	16.17	16.14	16.12	16.10	16.09	16.08	16.07	16.03
16.25	16.82	16.74	16.66	16.60	16.55	16.50	16.47	16.43	16.41	16.38	16.36	16.35	16.33	16.32	16.31	16.28
16.50	17.05	16.97	16.89	16.83	16.78	16.74	16.71	16.67	16.65	16.63	16.61	16.59	16.58	16.57	16.56	16.53
16.75	17.28	17.20	17.13	17.07	17.02	16.98	16.94	16.92	16.89	16.87	16.85	16.84	16.82	16.81	16.80	16.78
17.00	17.51	17.43	17.36	17.31	17.26	17.22	17.19	17.16	17.13	17.11	17.10	17.08	17.07	17.06	17.05	17.02
17.25	17.74	17.66	17.60	17.54	17.50	17.46	17.43	17.40	17.38	17.36	17.34	17.33	17.32	17.30	17.30	17.27
17.50	17.97	17.90	17.83	17.78	17.74	17.70	17.67	17.64	17.62	17.60	17.59	17.57	17.56	17.55	17.54	17.52
17.75	18.20	18.13	18.07	18.02	17.97	17.94	17.91	17.88	17.86	17.85	17.83	17.82	17.81	17.80	17.79	17.77
18.00	18.44	18.37	18.31	18.26	18.21	18.18	18.15	18.13	18.11	18.09	18.08	18.06	18.05	18.05	18.04	18.02
18.25	18.67	18.60	18.54	18.49	18.45	18.42	18.39	18.37	18.35	18.34	18.32	18.31	18.30	18.29	18.29	18.27
18.50	18.91	18.84	18.78	18.73	18.69	18.66	18.64	18.61	18.60	18.58	18.57	18.56	18.55	18.54	18.54	18.52
18.75	19.14	19.07	19.02	18.97	18.94	18.90	18.88	18.86	18.84	18.83	18.81	18.80	18.80	18.79	18.78	18.77
19.00	19.37	19.31	19.26	19.21	19.18	19.15	19.12	19.10	19.09	19.07	19.06	19.05	19.04	19.04	19.03	19.02
19.25	19.61	19.55	19.50	19.45	19.42	19.39	19.37	19.35	19.33	19.32	19.31	19.30	19.29	19.28	19.28	19.26
19.50	19.85	19.79	19.74	19.69	19.66	19.63	19.61	19.59	19.58	19.56	19.55	19.55	19.54	19.53	19.53	19.51
19.75	20.08	20.02	19.98	19.94	19.90	19.88	19.86	19.84	19.82	19.81	19.80	19.79	19.79	19.78	19.78	19.76
20.00	20.32	20.26	20.22	20.18	20.15	20.12	20.10	20.08	20.07	20.06	20.05	20.04	20.03	20.03	20.02	20.01
20.25	20.56	20.50	20.46	20.42	20.39	20.36	20.34	20.33	20.31	20.30	20.30	20.29	20.28	20.28	20.27	20.26
20.50	20.80	20.74	20.70	20.66	20.63	20.61	20.59	20.57	20.56	20.55	20.54	20.54	20.53	20.53	20.52	20.51
20.75	21.03	20.98	20.94	20.90	20.88	20.85	20.84	20.82	20.81	20.80	20.79	20.78	20.78	20.77	20.77	20.76

would include the desired length of the mortgage; whether to select a loan at lower rates but "locked in" for many years or one at higher rates but with prepayment provisions; what personality—that is, items not part of the building itself such as equipment, carpet, drapes and the like—should be included in the mortgage; what owner assets should be pledged; what existing property should be made part of the new mortgage—a "wrap-around" mortgage;—what "kickers" might be of-fered—offers to share income or equity with the lender if certain income or worth is exceeded; what advantages second and third mortgages might have; what method of repayment of the mortgage loan should be adopted—a constant covering interest and principal, the usual way; payments of interest only with the principal paid when the note is due; a constant payment of principal with interest figured on the declining balance; no payments on either interest or principal until the note is due but with the interest compounding; or a combination of the preceeding with refi-nancing at certain periods to fit the owner's needs. There are no fixed answers that cover all situations. These decisions can contribute as much to the success of the project as decisions about room arrangements or types of building materials.

Having determined the best package for the owner the mortgage banker will then shop his various money sources for the best deal. They will each have certain features that they will be looking for and the final deal most probably will be a compromise.

The mortgage banker's opinions are useful in other ways. Having reviewed nu-merous similar projects at the preliminary stage he can often make suggestions as to refinements in the physical arrangements as they relate to potential income that he has seen work out well in other situations. And finally in accepting or refusing a project he gives his opinion as to its probable success. If two or three mortgage bankers turn a project down when they are placing others the owner should review his figures as the risk elements must be greater than those in the normal market. The mortgage broker then need not be just a necessary evil interested in his finding fee, but he can be a creative addition to the building team.

SELLING THE LENDER

Whether one works directly with a loan organization or through a broker, one must be prepared to make an effective presentation of his project. While each type of project will emphasize different elements, all should include the following:

a) Physical analysis of the project including a surveyor's plot plan; an archi-tect's sketch plans and specifications; an artist's rendering of the developed site; and a contractor's estimate of the cost.

b) A market feasibility study.

c) A financial analysis of the ownership including personal statements.

d) A financial analysis of the project.

A great deal of thought must go into the presentation of the physical plant. Once the project is approved it is difficult to make major changes. A too glamourous presentation may backfire. It is utility not beauty that sells the lender. An impres-sive entrance framed by a tree lined drive circling a large fountain may be just an artist's window-dressing, and be the first thing abandoned when the final budgets are reviewed. But that may be the one feature a lender may remember and he may insist it be built. Another lender, with his eye on the cash flow, will see such embellishments as so much maintenance expense, and thus as a detriment rather than an asset.

In addition to the contractor's estimate the lender will make one of his own and may ask that the borrower have an independent source make up a third. This is usually done by a member of the American Institute of Real Estate Appraisers who carry the initials MAI (Member Appraisal Institute) after their name, or senior member of an appraisal society such as the American Society of Appraisers.

The owner's financial statement may be prepared by his certified public accountant, but often the lender will have his own standard form to fill out. The proposed contractor may also have to file a financial statement. Some loans will only be made if performance bonds are posted. Before presenting his proposal an owner will want to check with his lawyer and accountant as to the best form of ownership for the project—individual corporate, or partnership. Some buildings are owned "at arms length"; that is, a third party is brought in that actually owns the building and the promoter of the project becomes the tenant.

The market feasibility study show the compatibility of the project with the surroundings; means of public access; the effect on the environment and compliance with zoning regulations. Similar successful projects in the area should be noted, and the continuing need for additional such facilities proven. And finally the adaptability to another occupancy and resale value should be evaluated.

The most carefully scrutinized section of the presentation will probably be the projected value of the project based on income, or whether the project will generate enough income to repay the loan on time with interest. What is included here will vary with the type of project. Where rents are involved rent rolls should be projected. The lender in such cases may stipulate that a certain rent level be reached, usually around 80% of the projection, before the mortgage becomes final. If not the mortgage will be for a lower amount. This is called "setting a floor" for the mortgage. The spread may be as much as 20%, and if the owner does not reach the agreed upon rent level he must find another source to cover this amount. This will be discussed later.

THE COMMITMENT

Once a project has been accepted by a lending organization a commitment paper will be issued stating that they will lend a certain amount of money for a certain length of time at a certain rate of interest providing the borrower fulfills certain stated conditions and closes the loan by a certain date. This commitment, if the owner accepts it, will become the basis for the first mortgage on the project. There may be other mortgages either on the project as a whole or on some phase of it such as furnishings, equipment or the land itself. But the holder of the first mortgage takes precedence over all other claims except taxes, and he has the authority to take possession of the project upon proven default and to sell it and deduct the amount due him from the selling price which is usually determined by a public auction. If any excess cash remains it then goes to the holder of the second mortgage, then to the third, if any, and so on with the original developer the last to receive any distribution. Holders of other mortgages than the first cannot instigate the sale of the project. They are subordinate to the first mortgage holder and therefor carry a higher risk and charge a higher rate of interest.

The commitment will be quite specific and should not be thought of as a preliminary proposal but the basis of a contract one will live with for many years. This is why the presentation of the project must be carefully thought out and based on realistic goals.

SHORT TERM LOANS

Having received a commitment the owner must then work fast to finalize his plans. The commitment will carry a deadline date for acceptance which may be as little as a couple of weeks. It will also carry an expiration date by which time all closing documents shall be submitted to the loan company so that the final mortgage loan may go into effect. If this date is not met the owner will forfeit an agreed upon deposit which is usually around 1% of the mortgage amount, and is deposited with the lender's agent upon accepting the commitment. One must then proceed immediately to secure his construction loan, gap financing, if it is needed, and "front money".

The construction loan is a short term loan to cover the building costs during the erection of the project. With the mortgage loan commitment in hand the construction loan repayment is guaranteed providing that the project is completed in a satisfactory manner, so it is not necessary to sell the feasibility of the project all over again. But the construction loan lender does take the risk that the owner or the contractor may get in financial difficulties during the construction period, or that the projected budget proves inadequate. He may then have to complete the project himself. Because the mortgage amount may be less than the stated amount if certain income levels are not reached, as we discussed earlier, the construction loan will be based on the mortgage "floor" rather than the "ceiling" figure which may deduct as much as 20% from the amount needed to complete the construction. Further, the construction loan is seldom for more than 80% of the floor figure, and, like all loans, is subject to "points" and other closing costs, so perhaps another 2% will be deducted from the loan.

While a little "shopping around" may improve one's position, the construction loan will still fall short of the amount needed to complete the project. The missing funds must be supplied by the owner and represent his "equity" in the project. If the borrower can pledge sufficient liquid assets, that is cash, stocks or bonds or other unencumbered property, he may be able to raise the additional funds needed from the same source as the construction loan. If not he will have to take in a partner or borrow from those who take high risks. One such source are those who loan the difference between the floor and the ceiling of the mortgage. This is known as gap financing. They will issue a commitment to pay this difference if the income generated by the project is less than that specified to qualify for the full or ceiling amount. To obtain such a commitment one must pay in advance a flat fee, usually around 5% of the amount to be loaned. If the project reaches the income level to qualify for the full mortgage and this gap loan is not needed, this fee is not refundable. If it is needed the loan will become a second mortgage against the project, and must be repaid with interest.

In setting up the project budget there are other costs to be met besides the construction cost. First the land must be obtained. While this is sometimes leased or paid for under a mortgage arrangement, the first mortgage commitment may ask that the land title be "unencumbered". In addition to land costs there may be land development expenses such as surveys, soil investigation borings, utility extensions and connections and access roads or sidings, and purchasing expenses including attorney's and real estate agent's fees, taxes and permit and zoning fees. If the zoning for the land is to be changed it can be both time-consuming and expensive.

Then the architect must be paid. His fee will run from around 4% for a large industrial plant up to 15% for a fine residence or interior remodeling job. Some of this fee will not become due until the job is completed.

When the loan commitments are accepted they may require a good faith deposit of around 1% of the loan. Consulting fees must be paid to accountants and lawyers. At loan closings legal and recording fees and transaction taxes may be charged, and, if asked for, a performance bond obtained.

When the work is started there is always a lapse of time between paying for permits, materials and salaries and being reimbursed for these items from the construction loan. Usually each request for payment from the construction loan must be accompanied by waivers of lien. In actual practice the contractor may finance this period rather than the owner, but the lender may still demand that the owner show he has the money to cover these start-up costs.

How much front money, or equity, an owner will need to launch his project will vary with the type of project, the money market and the owner's reputation. It is often said that an owner with a proven need, a piece of property free of debt and an architect's set of plans can obtain all the financing he will need. We have used the term "owner" in our preceeding discussions as though the one who builds the project is the one who will occupy it. Today even large corporations may choose to be tenants rather than owners and will rely on developers to put the project together. The contractor may be asked to be the developer or at least to become a partner in it. Often the owner of the real estate and the architect will be included as "limited" partners, their limits of both liability and equity being proportional to the cost of the services or land contributed by them. When the construction loan is paid off by the mortgage loan they are also often bought out and the ownership reverts to the general partner. However some mortgage commitments guard against such arrangements by adding a penalty fee if there is any change of ownership within a stipulated time.

Rates of interest on loans for construction purposes will vary from week to week and community to community, but are usually tied to the prime rate, that is the interest charged by commercial banks to their preferred customers for short term loans. Currently construction loans are quoted at 2½ % to 4% over the prime. High risk loans will have considerably higher rates.

As stated before it is not the purpose of this chapter to encourage the contractor to become the project entrepreneur. But with the cost of money tied so closely to the success of the project from an income-producing standpoint, a contractor must know all the procedures in order to follow the time and budget schedules, to process his monthly draws on time, and to turn the completed project over to the owner in such condition as to be completely acceptable to the mortgage lender.

INTERIM FINANCING

This particular type of financing is not related to the Owner's activity in securing money to provide sufficient funds to pay for the construction costs of a project and the later long term mortgage financing which is repaid over an extended period from the proceeds of the revenue generated from the project.

The definition of the word interim is provisional - intervening - interlude or relating to a time period.

This is the precise type of financial assistance required by virtually all contractors, regardless of size, to provide additional short term temporary working capital to compensate for unanticipated delays in receiving payments from Owners for work performed and properly invoiced.

The source is usually a commercial bank that is responsive to the needs of the contractor. All banks are not receptive to this type of business and the contractor must establish his/her banking relations with a bank that has personnel possessing the knowledge of the intimate financial structure found in the construction industry.

Once the selection of the acceptable financial institution has been accomplished, then the contractor should identify a particular person within the bank with whom a personal relationship can be established so that there exists an on-going rapport between the contractor and the bank.

Next, it is the obligation of the contractor to prove, to the complete satisfaction of the bank, the credibility of the technical and financial assets of the company.

This can be accomplished by furnishing Certified (CPA) Financial Statements of the company along with a written detailed resume of the qualifications of each person directing the destiny of the company.

Progress, at this point, should place the contractor in the position to obtain from the bank a general commitment as to the limit and terms under which they would participate in granting short term loans (usually 30-90 days). The ideal situation is to obtain a letter of credit from the bank outlining their position so that at any time during the next twelve (12) months, a loan could be arranged without further review from a committee which eliminates several days or possible weeks of delay from the time the loan application is presented until the proceeds are available.

At the time a loan request is made, the position of the contractor is significantly enhanced if there is presented a cash flow analysis of all current projects projecting anticipated expenditures and income on a monthly basis so that it can be clearly defined as to the amount of money required and the availability of sufficient income to repay the loan.

It can be expected that interest on this type of loan will be from 1% to 6% above the current prime rate as banks consider this high risk business. As an example, if the current prime rate is 16% and the loan is assigned 3% over prime, this means the contractor is paying 19% for the use of the money. It is obvious that a contractor is forced to impose restrictions on the use of such costly money as it cannot be used as continuous working capital, which at best, has an earning capacity of about 10%, therefore there has to be found justification in the form of large discounts or the prompt payment of equipment invoices would reflect an early delivery date which would equate to a time and cost saving that would justify the interest rate.

The availability of interim financing provides a measure of security to the contractor in being assured that external financial assistance is available to augment the working capital of the company during periods of depressed cash flow.

CHAPTER 3

DRAWINGS AND SPECIFICATIONS

WORKING DRAWINGS
SYMBOLS
SPECIFICATIONS
ADDENDA
CHANGE ORDERS

The owner's dreams and the architect's visions must eventually be put down in a form the contractor can use as a basis for bidding and later constructing in enduring materials. This form has developed into the dual one of working drawings and specifications. Basically, the drawings present the design, the location and the dimensions of a project. The specifications give the quality required.

The drawings are done in pencil or ink on tracing paper or cloth which is easily reproducible, and prints are delivered to the contractor. The architect retains the original drawings in his office.

The specifications, except in very small jobs when they may be incorporated on the drawings, are printed and bound separately.

Both are made available to each contractor bidding the job. As bid periods are usually quite limited, it is the contractor's ability to interpret them that enables him to present a realistic and complete bid. This chapter presents an outline of what information a contractor should expect to find in a bid set and where it may be found.

THE WORKING DRAWINGS

Arrangement.—A typical complete project will find several drawing sheets, usually all of uniform size, bound into one job set. The first sheet should include a site plan, sometimes incorporating the roof plan, plus the complete job title, and sometimes an index of the drawings. This will be followed by the sheets showing architectural details, usually in the following order: basement plan, first floor plan, upper floor plans, exterior elevations, sections, interior elevations and details.

The architectural sheets are followed by the structural set including footing and foundation plans, basement framing plans, floor framing plans, roof framing plans, and then the structural details. These are followed by the heating and ventilating plans; then the plumbing plans and, finally, the electrical plans, each starting with the lowest floor and working upward.

Each sheet should have a title block in the lower righthand corner with the sheet number (usually "A-" for architectural, "S-" for structural, "M-" for heating and ventilating, "P-" for plumbing and "E-" for electrical); the number of sheets in each set (i.e., A-1 of 7); the date made plus each date it has been revised, and the initials of the person or persons who drew and approved the sheet.

Revisions are sometimes listed outside the title block. Some architects, rather than revising a drawing once it is printed, will issue supplementary drawings which must be attached to the appropriate set as they are issued.

Type of Drawings.—Most working drawings for building construction are based on Orthographic Projection, that is, by a parallel projection to a plane by lines perpendicular to the plane. In this way, all dimensions will be true. If the plane is horizontal, the projection is a plan; if vertical, it is an elevation for outside the building, or a sectional elevation if through the building.

Of course, the only descriptive drawing which presents a building as the eye sees it is the perspective. But perspectives are used mainly to study the building and present it to the client in a form he can easily understand. A perspective is seldom useful for presenting information on working drawings. However, some other pictorial presentations are helpful to the builder.

Two of these are the Isometric and the Cabinet Projection. Isometrics are drawings in which all horizontal and vertical lines have a true length, and those lines parallel on the object are also parallel on the drawing. Vertical lines are vertical, but horizontal lines are set at 30° or 60°. Isometrics are often used for piping diagrams and to show complicated intersections such as on sloped roofs.

Cabinet Drawings are those with the front face shown in true shape and size as if it were an Orthographic Projection, but they simulate a perspective. The sides are shown receding at 45° and at 1/2 scale. Variations of this are Oblique Drawings, where the angle and side scale may be anything that best shows the object, and Cavalier Drawings, where the side scale is at the same scale as the front. Cabinet Drawings get their name from the fact that they are often used for cabinet work.

Scale.—The architect's scale, that is, the inch divided into 1/8S, 1/16S, 1/32S, is standard for building construction in the United States. The engineer's scale, with the inch divided into tenths, is sometimes used in structural work or on site plans. However, it is advantageous to have architectural, structural and mechanical drawings all at the same scale for the same job.

The metric scale is divided into centimeters and millimeters, 2.54 centimeters equalling one inch. It is the common scale used in Europe, and will soon be adopted here.

Graphic scales are those devised by the draftsman and shown on the drawings with subdivisions comparable to the other scale. This is convenient if the drawing is to be reduced in size in reproduction and is meant to be scaled rather than dimensioned. It is seldom used on working drawings.

Plans and elevations are usually drawn at no less than 1/8 =1'-0"; plot plans at 1/16"=1'-0". Complicated plan areas, such as toilet rooms, are usually repeated at 1/4" to 1/2" scale. Wall sections and cabinet work must show more detail and are from 3/4" to 11/2" scale. Certain areas of sections will be further detailed at 3" scale, while moldings and decorative items may even be drawn at half to full scale.

Dimensioning.—While scales should be clearly identified for each drawing either in the title block or beneath each individual layout, contractors should never scale drawings but should insist that the architect furnish them with complete stated dimensions. This is the duty of the architect and the contractor is foolish to take responsibility for it. Blueprints are subject to stretching and shrinking with humidity and some reproductive methods reduce the drawings in size.

Notes on Drawings.—The drawings should not be cluttered with notes which should be in the specifications, but architects' offices will vary in this. The draftsman will often put notes on drawings to alert his own specification writer rather than the contractor. It is the specification that should govern, and all notes on drawings should be checked against the specifications and cross-referenced. If a contractor finds discrepancies during bidding, he should report this to the architect and ask for a written statement, known as an Addendum, to clarify it if it would affect his cost.

Schedules.—Schedules may appear either as part of the drawings or the specifications. Room finish schedules are often on the drawings next to the floor plan to which they refer. Schedules for lintels, columns, footings, doors and frames are usually on the drawings, while hardware schedules are usually in the specifications. Fixture schedules for mechanical and electrical work may be found either place. A room finish schedule blank, Frank R. Walker Co. #401, is shown on page 34.

A GARDEN SHED
6' x 6' x 8' HIGH
WITH PEAKED ROOF
SCALE: 1/16"=1'0"

SHOWN AS AN ISOMETRIC,

LEFT SIDE FRONT RIGHT SIDE REAR

FLOOR PLAN   ROOF PLAN

IN ORTHOGRAPHIC PROJECTION,

45°   1/2 SCALE

AND AS A CABINET DRAWING

COPPER FINIAL
SLATE ROOF
1'-0"
ASPHALT SHINGLES
FACE BRICK
METAL LINTLE
1'-0"
D.H. WOOD SASH
WOOD CASEM'T
4'-0"
8'-0"
WOOD SHUTTERS & TRIM
BRICK SILL
3'-0"
WOOD SIDING
GRADE
GRADE
CONC. FND'N
2'-6"
4'-0"
CONC. FND'N

WEST ELEVATION
SCALE: 1/8"=1'-0"

W.I. VANE
WOOD SHAKES
STONE LINTLE & SILL
METAL WIND.
RAND. ASHLAR
GRADE
CONC. FND'N

THE GARDEN SHED SHOWING THE SAME WEST (LEFT SIDE) ELEVATION BUT WITH THREE DIFFERENT MATERIAL INDICATIONS AS MIGHT BE FOUND ON A WORKING DRAWING.

(BROKEN DIMENSION LINE)
(OVERALL DIMENSION LINE)
(PROPERTY LINE)

6'-0"   10'-0"

1'-9"
2'-6"
1'-9"

(MATERIAL INDICATION)
(WINDOW INDICATION)

6'-0"

(FOOTING LINE)
1'-6" 1'-6" 1'-6" 1'-6"
(BUILDING LINE)
(CENTER LINE)

N

FLOOR PLAN
SCALE: 1/4"=1'-0"
(DIRECTIONAL SIGN)
(TITLE BLOCK INFORMATION)

TITLE : FLOOR PLAN OF GARDEN SHED		
REVISION-DATE	PROJECT NAME ADDRESS OWNER'S NAME ADDRESS	ARCHITECT· ADDRESS
JOB:#____	DATE:____	DRAWN BY:____	SHT#__OF__

ROOM FINISH SCHEDULE

JOB NO.:
SHEET NO.:
DATE:

ROOM	#	FLOORS													BASE AND/OR WAINSCOT									WALLS						CEILING				TRIM											DECORATING							NOTES

Floors columns: CEMENT, ASPHALT TILE, RUBBER TILE, VINYL ASBESTOS T., VINYL, LINOLEUM, CERAMIC TILE, WOOD, SLATE, MARBLE, TERRAZZO, SEAMLESS, CARPET

Base and/or wainscot columns: CEMENT HIGH, RUBBER HIGH, VINYL HIGH, WOOD HIGH, CER. TILE HIGH, SLATE HIGH, MARBLE HIGH, TERRAZZO HIGH, SEAMLESS HIGH

Walls columns: PLASTIC, WALLBOARD, CER. TILE, CONC. BLK., WOOD

Ceiling columns: PLASTER, WALLBOARD, ACOUSTICAL, WOOD

Trim columns: WOOD CASING, METAL CASING, WOOD SILLS, METAL SILLS, MARBLE SILLS, PLASTIC SILLS, CHAIR RAIL, CORNICE, BUILT-IN CABINET

Decorating columns: PAINT FOR FLOORS, PAINT FOR WALLS, PAINT FOR CEILINGS, PAINT FOR TRIM, PAINT FOR DOORS, PAPER FOR WALLS, PAPER FOR CEILINGS

NOTES

PRACTICAL MFG. IN U.S.A. FORM 60

FRANK R. WALKER CO. PUBLISHERS CHICAGO

Reproduction.—"Blueprints," that is, parts with pencil lines showing in white on a blue background, may be required by many building departments for permit in that they are difficult to alter. Notes on them can be added only in colored pencil, either red or yellow. The contractor should limit the notes he makes on his construction set and authorize only one person to add them so that somebody's doodles don't end up in brick and mortar.

Today, many architectural offices are adding their own printing machines. These are usually of the ammonia process which produce black, blue, or maroon lines on a white background. They are much more easily marked on, although the added note will not be as obvious as on a blueprint, unless made in color. Small prints may be run off on one of the many types of office duplicating machines.

If it is desirable to reduce or enlarge a drawing, a photostat, a photographic process that produces negatives as positives, can be used.

There are processes that transfer blueprints to cloth drawings or to sepia prints, which can be altered, added to and printed just like an original tracing.

Interpreting Lines.—The architect will use various lines in making his drawings. Thick solid lines are used to outline the edges of the plans and elevations. Lines consisting of long and short dashes in sequence denote center lines. Short dashes usually indicate hidden lines, but are also used on windows to show where they are hinged if they are operable. Thin lines are used for dimensions. These terminate in arrows to show the limit of the given dimensions. Repeat marks or the notation "do" indicates the repetition of a given dimension. Dimension lines are often broken by a zigzag which shows the object in the drawing has been shortened.

SYMBOLS

Material indications, and items relative to plumbing, heating and ventilation, and electrical work have been standardized. Some architects may have a symbol schedule on their drawings, but most no longer do this. The accompanying schedule shows many of the usual symbols a contractor may encounter, but always check with the specifications for building materials as symbols may vary from draftsman to draftsman.

EARTH	ROCK	CINDER FILL
STONE CONCRETE	CINDER CONCRETE	CEMENT
CONCRETE BLOCK	TILE	GYPSUM BLOCK

COMMON BRICK   FACE BRICK   FIRE BRICK

CUT STONE   RUBBLE STONE   MARBLE

ROUGH WOOD   FINISH WOOD   METAL

REINFORCING BARS   LOOSE INSULATION   SOLID INSULATION

GENERAL OUTLETS:

CEILING	WALL	
○	—○	Fixture Outlet
Ⓑ	—Ⓑ	Blanked Outlet
Ⓒ	—Ⓒ	Clock Outlet
Ⓓ		Drop Cord Outlet
Ⓕ	—Ⓕ	Fan Outlet
Ⓙ	—Ⓙ	Junction Box
Ⓡ	—Ⓡ	Recessed Outlet
Ⓢ	—Ⓢ	Pull Switch
Ⓧ	—Ⓧ	Exit

▭	Surface Fluorescent
▭ R	Recessed Fluorescent
⊢——⊣	Bare Lamp Fluorescent

CONVENIENCE OUTLETS:

⊖	Duplex
⊖	Waterproof
⊖R	Radio
⊖	Range
⊙	Floor
▲	Special Purpose

SWITCH OUTLETS:

S	Single Pole Switch
S_2	Double Pole Switch
S_3	Three Way Switch
S_D	Automatic Door Switch
S_E	Electrolier Switch
S_K	Key Operated Switch
S_{CB}	Circuit Breaker
S_{MC}	Momentary Contact
S_{RC}	Remote Control

S_{WP}   *Weatherproof Switch*

S_F   *Fused Switch*

PANELS & CIRCUITS:

Lighting Panel

Power Panel

Two Wire Branch Circuit

Three Wire Branch Circuit

Under Floor Duct & Box

Generator

Motor

Controller

SIGNALLING SYSTEMS:

Push Button

Buzzer

Bell

Chime

Annunciator

Electric Door Opener

Telephone Switchboard

Interconnecting Telephone

◄ Outside Telephone

F⊃ Fire Alarm Bell

F Fire Alarm Station

✕ City Fire Alarm

FS Automatic Fire Alarm

W Watchman's Station

H Horn

N Nurse's Signal Plug

M Maid's Signal Plug

R Radio Outlet

TV Television Outlet

▭ Interconnection Box

PIPE FITTING SYMBOLS:

—————— Low Pressure Steam

—/—/—/— Medium Pressure Steam

—//—//—//— High Pressure Steam

—·— — — — Return Line

—o—o—o— Condensate Line

—·— —·— — Make Up Water

— — — — — Air Relief Line

———FOF——— Fuel Oil Flow

———*FOR*———   *Fuel Oil Return*

———*FOV*———   *Fuel Oil Vent*

———*A*———   *Compressed Air*

———*RL*———   *Refrigerant Liquid*

———*RD*———   *Refrigerant Discharge*

———*C*———   *Condenser Water Flow*

———*CR*———   *Condenser Water Return*

———*CH*———   *Chilled Water*

———*CHR*———   *Chilled Water Return*

———*H*———   *Humidification Line*

———*D*———   *Drain Line*

———*B*———   *Brine Supply*

HEATING SYMBOLS:

Exposed Radiator

Flush Enclosed Radiator

Projecting Enclosed Radiator

Unit Heater (Prop. Fan)

Unit Heater (Cent. Fan)

Unit Ventilator

Thermostatic Trap

,Float Trap

- Boiler Return
- Pressure Reducing Valve
- Air Line Valve
- Lock & Shield Valve
- Diaphragm
- Air Eliminator
- Strainer
- Thermometer
- Thermostat
- Gage
- ,Pressure Switch
- Expansion Valve

DUCTWORK SYMBOLS:

- Duct and Direction of Flow
- Inclined Drop
- Inclined Rise
- Access Door
- Adjustable Blank Off
- Adjustable Plaque

Automatic Damper

Canvas Connections

Deflecting Damper

Volume Damper

Supply Duct in Section

Exhaust Duct in Section

Supply & Exhaust Wall Grilles

Vanes

PIPING SYMBOLS:

Flanged Joint

Screwed Joint

Bell & Spigot Joint

Welded Joint

Soldered Joint

Flanged Elbow-Up

Flanged Elbow-Down

Flanged Tee-Up

Flanged Tee-Down

Flanged Reducer-Concentric

Flanged Reducer-Eccentric

Flanged Gate Valve

Flanged Globe Valve

Flanged Check Valve

Flanged Cock Valve

Flanged Safety Valve

Flanged Expansion Joint

Reducing Flange

Flanged Union

Flanged Sleeve

Abbreviations.—A recent task force of the American Institute of Architects made a study of words often abbreviated on working drawings and shop drawings and issued the following list as a guide so that such abbreviations can be standardized on all drawings used within the construction industry and be commonly recognized. In the list below the abbreviations are given in alphabetical order, not the words they represent, so that contractors who come across those abbreviations with which they are unfamiliar may more easily look them up.

AB	anchor bolt	ADJT	adjustable
ABV	above	AFF	above finished floor
AC	acoustical	AGG	aggregate
A/C	air conditioning	AL	aluminum
ACC	access	ALT	alternate
ACFL	access floor	ANC	anchor
ACPL	acoustical plaster	ANOD	anodized
ACR	acrylic plastic	AP	access panel
ACT	acoustical tile	APX	approximate
AD	area drain	ARCH	architect
ADD	addendum	ASB	asbestos
ADH	adhesive	ASC	above suspended ceiling
ADJ	adjacent	ASPH	asphalt

AT	asphalt tile		COMPO	composition
AUTO	automatic		COMPT	compartment
			CONC	concrete
			CONST	construction
BBD	bulletin board		CONT	continuous
BD	board		CONTR	contract
BEL	below		CORR	corrugated
BET	between		CPR	copper
BIT	bituminous		CPT	carpet
BJT	bed joint		CR	chromium
BLDG	building		CRG	cross grain
BLK	block		CRS	course
BLKG	blocking		CS	countersink
BM	bench mark		CSMT	casement
BOT	bottom		CST	cast stone
BP	back plaster		CT	ceramic tile
BPL	bearing plate		CTR	counter
BRG	bearing		CTSK	countersunk screw
BRK	brick		CX	connection
BRZ	bronze		CYD	cubic yard
BS	both sides			
BSMT	basement			
BUR	built up roofing		D	drain
BVL	beveled		DA	doubleacting
BW	both ways		DEM	demolish
			DEP	depressed
			DF	drinking fountain
CAB	cabinet		DH	double hung
CAD	cadmium		DIAG	diagonal
CB	catch basin		DIAM	diameter
CEM	cement		DIM	dimension
CER	ceramic		DIV	division
CFL	counterflashing		DL	dead load
CFT	cubic foot		DMT	demountable
CG	corner guard		DP	dampproofing
CHAM	chamfer		DPR	damper or dispenser
CHBD	chalkboard		DR	door
CHT	ceiling height		DRB	drainboard
CI	cast iron		DS	downspout
CIPC	cast-in-place concrete		DT	drain tile
CIR	circle		DTA	dovetail anchor
CIRC	circumference		DTL	detail
CJT	control joint		DTS	dovetail anchor slot
CK	caulk		DW	dumbwaiter
CLG	ceiling		DWG	drawing
CLL	contract limit line		DWR	drawer
CLR	clear			
CLS	closure			
CM	centimeter		E	east
CMT	ceramic mosaic tile		EB	expansion bolt
CMU	concrete masonry unit		EF	each face
COL	column		EL	elevation
COMB	combination		ELEC	electric
COMP	compression		ELEV	elevator

EMER	emergency
ENC	enclose
EP	electrical panelboard
EQ	equal
EQP	equipment
ESC	escalator
EST	estimate
EWC	electric water cooler
EXCA	excavate
EXH	exhaust
EXG	existing
EXMP	expanded metal plate
EXP	exposed
EXS	extra strong
EXT	exterior

FA	fire alarm
FAS	fasten
FB	face brick
FBD	fiberboard
FBO	furnished by others
FBRK	fire brick
FD	floor drain
FE	fire extinguisher
FEC	fire extinguisher cabinet
FF	factory finish
FFE	finished floor elevation
FFL	finished floor line
FGL	fiberglass
FHMS	flathead machine screw
FHS	fire hose station
FHWS	flathead wood screw
FIN	finish
FJT	flush joint
FLCO	floor cleanout
FLG	flashing
FLR	floor
FLUR	fluorescent
FLX	flexible
FN	fence
FND	foundation
FOC	face of concrete
FOF	face of finish
FOM	face of masonry
FOS	face of studs
FP	fireproof
FPL	fireplace or floor plate
FR	frame
FRA	fresh air
FRC	fire-resistant coating
FRG	forged
FRT	fire-retardant
FS	full size

FTG	footing
FUR	furred
FUT	future

GA	gauge
GB	grab bar
GC	general contractor
GCMU	glazed concrete masonry units
GD	grade
GF	ground face
GI	galvanized iron
GKT	gasket
GL	glass or glazing
GLB	glass block
GLF	glass fiber
GP	galvanized pipe
GPDW	gypsum dry wall
GPL	gypsum lath
GPPL	gypsum plaster
GPT	gypsum tile
GRN	granite
GSS	galvanized steel sheet
GST	glazed structural tile
GT	grout
GV	galvanized
GVL	gravel

HB	hose bibb
HBD	hardboard
HC	hollow core
HD	heavy duty
HDR	header
HDW	hardware
HES	high early-strength cement
HH	handhole
HJT	head joint
HK	hook
HM	hollow metal
HOR	horizontal
HT	height
HTG	heating
HVAC	heating/ventilating/ air conditioning
HWD	hardwood
HWH	hot water heater
HX	hexagonal

ID	inside diameter
ILK	interlock
INCIN	incinerator
INCL	included

INS	insulation		MI	malleable iron
INSC	insulating concrete		MIN	minimum
INSF	insulating fill		MIR	mirror
INT	interior		MISC	miscellaneous
INTM	intermediate		MLD	moulding
INV	invert		MM	millimeter
IPS	iron pipe size		MMB	membrane
			MO	masonry opening
			MOD	modular
J	joist		MOV	movable
JC	janitor's closet		MR	mop receptor
JF	joint filler		MRB	marble
JT	joint		MRD	metal roof deck
			MT	mounted
			MTFR	metal furring
KCPL	Keene's cement plaster		MTHR	metal threshold
KIT	kitchen		MTL	material
KO	knockout		MULL	mullion
KPL	kickplate		MWK	millwork
L	length		N	north
LAB	laboratory		NAT	natural
LAD	ladder		NI	nickel
LAM	laminated		NIC	not in contract
LAV	lavatory		NL	nailable
LB	lag bolt		NMT	nonmetallic
LBL	label		NOM	nominal
LC	light control		NR	noise reduction
LH	left hand		NRC	noise reduction coefficient
LL	live load		NTS	not to scale
LMS	limestone			
LP	lightproof			
LPT	low point		OA	overall
LT	light		OBS	obscure
LTL	lintel		OC	on center
LVR	louver		OD	outside diameter
LW	lightweight		OH	overhead
LWC	lightweight concrete		OHMS	ovalhead machine screw
			OHWS	ovalhead wood screw
			OJ	open-web joist
M	meter		OP	opaque
MAS	masonry		OPG	opening
MAX	maximum		OPH	opposite hand
MB	machine bolt		OPP	opposite
MBR	member		OPS	opposite surface
MC	medicine cabinet			
MECH	mechanical			
MED	medium		PAR	parallel
MET	metal		PB	panic bar
MFR	manufacturer		PBD	particle board
MFD	metal floor deck		PCC	precast concrete
MH	manhole		PCF	pounds per cubic foot

PE	porcelain enamel	RO	rough opening
PED	pedestal	ROW	right of way
PERF	perforated	RVS	reverse
PERI	perimeter	RVT	rivet
PFB	prefabricated	RWC	rainwater conductor
PFL	pounds per lineal foot		
PFN	prefinished		
PG	plate glass	S	south
PK	parking	SC	solid core
PL	plate or property line	SCH	schedule
PLAM	plastic laminate	SCN	screen
PLAS	plaster	SCT	structural clay tile
PNL	panel	SD	storm drain
PNT	painted	SEC	section
PRF	preformed	SFGL	safety glass
PSC	prestressed concrete	SG	sheet glass
PSF	pounds per square foot	SH	shelf
PSI	pounds per square inch	SHO	shoring
PT	point	SHT	sheet
PTC	post-tensioned concrete	SHTH	sheathing
PTD	paper towel dispenser	SIM	similar
PTN	partition	SKL	skylight
PTR	paper towel receptor	SL	sleeve
PV	paving	SNT	sealant
PVC	polyvinyl chloride	SP	soundproof
PVMT	pavement	SPC	spacer
PWD	plywood	SPEC	specification
		SPK	speaker
		SPL	special
QT	quarry tile	SQ	square
		SSK	service sink
		SST	stainless steel
R	riser	ST	steel
RA	return air	STA	station
RAD	radius	STD	standard
RB	rubber base	STG	seating
RBL	rubble stone	STO	storage
RBT	rabbet, rebate or rubber tile	STR	structural
RCP	reinforced concrete pipe	SUS	suspended
RD	roof drain	SYM	symmetrical
RE	reinforced	SYN	synthetic
REF	reference	SYS	system
REG	register		
REM	remove		
RES	resilient	T	tread
RET	return	TB	towel bar
REV	revision or revised	TC	terra cotta
RFG	roofing	TEL	telephone
RFL	reflected	T&G	tongue and groove
RFH	roof hatch	THK	thickness
RH	right hand	THR	threshold
RL	railing	TKBD	tackboard
RM	room	TKS	tackstrip

TOL	tolerance	VJ	v-joint
TPD	toilet paper dispenser	VNR	veneer
TPTN	toilet partition	VRM	vermiculite
TR	transom	VT	vinyl tile
TSL	top of slab		
TST	top of steel		
TV	television	W	west, width or wide
TW	top of wall	WB	wood base
TYP	typical	WC	water closet
TZ	terrazzo	WD	wood
		WG	wired glass
		WH	wall hung
UC	undercut	WHB	wheel bumper
UNF	unfinished	WI	wrought iron
UR	urinal	WIN	window
		WM	wire mesh
		WO	without
VAR	varnish	WP	waterproofing
VAT	vinyl asbestos tile	WPT	working point
VB	vapor barrier or vinyl base	WR	water repellent
VERT	vertical	WS	waterstop
VF	vinyl fabric	WSCT	wainscot
VG	vertical grain	WTW	wall to wall
VIN	vinyl	WWF	welded wire fabric

SPECIFICATIONS

Arrangement.—There has been much restudying of the proper way to set up specifications recently, but little standardization is seen yet among architects.

The most impressive effort thus far is that of the Construction Specification Institute, known as the Uniform System. This system not only sets forth an arrangement for specifications, but is equally applicable to data filing (most of the manufacturers' new catalogs today carry the CSI symbol, a rectangle enclosing three elipses, and the Uniform System number) and to a cost accounting system. The CSI form divides the specifications into four sections: bidding requirements, contract forms, general conditions, and specifications. The specifications section is broken down into 16 divisions; a typical specification prior to the CSI format might run to 60 sections. The present suggested divisions, under the fourth section "Specifications," are as follows:

Division 1. General Requirements, including summary of work, schedules and reports, samples and shop drawings, temporary facilities, cleaning up, project closeout, allowances and alternates.

Division 2. Site Work, including clearing of site, earthwork, piling, caissons, shoring and bracing, site drainage, site utilities, roads and walks, site improvements, lawns and planting, railroad work, and marine facilities.

Division 3. Concrete, including formwork, reinforcement, cast-in-place concrete, precast concrete, and cementitious decks.

Division 4. Masonry, including unit masonry, stone and masonry restoration.

Division 5. Metals, including structural, joists, decking, lightgage framing, miscellaneous metals, ornamental metals and special formed metals.

Division 6. Carpentry, including rough, finish, glue laminated and custom woodwork.

Division 7. Moisture Control, including waterproofing, dampproofing, insulation, shingles and roof tiles, preformed roofing and siding, membrane roofing, sheet metal work, wall flashing, roof accessories, and caulking and sealants.

Division 8. Doors, Windows and Glass, including metal doors and frames, wood doors, special doors, metal windows, wood windows, finish hardware, operators, weatherstripping, glass and glazing, curtain walls, and store fronts.

Division 9. Finishes, including lath and plaster, gypsum, tile, terrazzo, veneer stone, acoustics, wood flooring, resilient flooring, special flooring, special coatings, painting and wall coverings.

Division 10. Specialties, including chalk and tack boards, chutes, compartments and cubicles, demountable partitions, disappearing stairs, firefighting devices, fireplace equipment, flagpoles, folding gates, identifying devices, lockers, mesh partitions, postal specialties, retractable partitions, scales, storage shelving, sun control devices, telephone booths, toilet and bath accessories, vending machines, wardrobe specialties, and waste disposal units.

Division 11. Equipment, including that for banks, commerce, darkrooms, churches, education, food service, gymnasiums, industry, laboratories, laundries, libraries, medicine, mortuaries, parking, prisons, residences, and stage.

Division 12. Furnishings, including artwork, light control, cabinets, carpets, drapes, furniture, and seating.

Division 13. Special Construction, including audio, bowling, broadcasting, clean rooms, conservatory, incinerator, integrated ceiling, observatory, pedestal floor, prefabricated, chimney, vault, swimming pool and zoo.

Division 14. Conveying Systems, including elevators, dumbwaiters, hoists and cranes, lifts, material handling, escalators and tubes.

Division 15. Mechanical. Divided into general provisions, basic materials and methods, and plumbing, heating, ventilating and cooling covering water supply, soil and waste, roof drainage, plumbing fixtures, gas piping, special piping, fire extinguishing, fuel handling, steam, hot water, chilled water, dual temperature air tempering, refrigeration, and controls and instruments.

Division 16. Electrical, including general provisions, basic materials and methods, electrical service, distribution, fixtures, communication, TV, power, heating, controls and lighting protection.

This system has met with industry and professional acceptance and is the generally accepted form for specifications today. This book has been organized to reflect this in its chapter arrangement.

The CSI Uniform System listing of 16 divisions and its sub-divisions is as follows:

01	**GENERAL REQUIREMENTS**	02012	CORE DRILLING
01020	ALLOWANCES	02013	STANDARD PENETRATION TESTS
01100	ALTERNATIVES	02014	SEISMIC EXPLORATION
01200	PROJECT MEETINGS	02100	CLEARING
01300	SUBMITTALS	02101	STRUCTURE MOVING
01400	QUALITY CONTROL	02102	CLEARING AND GRUBBING
01500	TEMPORARY FACILITIES & CONTROLS	02103	TREE PRUNING
		02104	SHRUB AND TREE RELOCATION
01600	MATERIAL AND EQUIPMENT	02110	DEMOLITION
01700	PROJECT CLOSEOUT	02200	EARTHWORK
01999	MISCELLANEOUS	02210	SITE GRADING
		02211	ROCK REMOVAL
02	**SITE WORK**	02212	EMBANKMENT
02000	ALTERNATIVES	02220	EXCAVATING AND BACKFILLING
02010	SUBSURFACE EXPLORATION	02221	TRENCHING
02011	BORINGS	02222	STRUCTURE EXCAVATION

02223	ROADWAY EXCAVATION
02224	PIPE BORING AND JACKING
02225	TRENCH BACKFILL AND COMPACTION
02226	STRUCTURE BACKFILL & COMPACTION
02227	WASTE MATERIAL DISPOSAL
02230	SOIL COMPACTION CONTROL
02240	SOIL STABILIZATION
02250	TERMITE CONTROL
02252	VEGETATION CONTROL
02300	PILE FOUNDATIONS
02350	CAISSONS
02351	DRILLED CAISSONS
02352	EXCAVATED CAISSONS
02400	SHORING
02420	UNDERPINNING
02500	SITE DRAINAGE
02550	SITE UTILITIES
02600	PAVING & SURFACING
02610	PAVING
02620	CURBS & GUTTERS
02630	WALKS
02640	SYNTHETIC SURFACING
02700	SITE IMPROVEMENTS
02710	FENCES AND GATES
02720	ROAD AND PARKING APPURTENANCES
02730	PLAYING FIELDS
02740	FOUNTAINS
02750	IRRIGATION SYSTEM
02760	SITE FURNISHINGS
02800	LANDSCAPING
02810	SOIL PREPARATION
02820	LAWNS
02830	TREES, SHRUBS, AND GROUND COVER
02850	RAILROAD WORK
02851	TRACKWORK
02852	BALLASTING
02900	MARINE WORK
02910	DOCKS
02920	BOAT FACILITIES
02930	PROTECTIVE MARINE STRUCTURES
02931	FENDERS
02932	SEAWALLS
02933	GROINS
02934	JETTYS
02940	DREDGING
02950	TUNNELING
02960	TUNNEL EXCAVATION
02970	TUNNEL GROUTING
02980	SUPPORT SYSTEMS
02999	MISCELLANEOUS

03	**CONCRETE**
03000	ALTERNATIVES
03100	CONCRETE FORMWORK
03150	EXPANSION & CONTRACTION JOINTS
03200	CONCRETE REINFORCEMENT
03210	STEEL BAR & WELDED WIRE REINF.
03230	STRESSING TENDONS
03300	CAST-IN-PLACE CONCRETE
03305	CONCRETE CURING
03310	CONCRETE
03320	LIGHTWEIGHT CONCRETE
03321	INSULATING CONCRETE
03322	LIGHTWEIGHT STRUCTURAL CONCRETE
03330	HEAVYWEIGHT CONCRETE
03340	PRESTRESSED CONCRETE
03350	SPECIALLY FINISHED CONCRETE
03351	EXPOSED AGGREGATE CONCRETE
03352	BUSHHAMMERED CONCRETE
03353	BLASTED CONCRETE
03354	HEAVY-DUTY CONCRETE FLOOR FINISH
03355	GROOVED-SURFACE CONCRETE
03360	SPECIALLY PLACED CONCRETE
03370	GROUT
03400	PRECAST CONCRETE
03410	PRECAST CONCRETE PANELS
03411	TILT-UP WALL PANELS
03420	PRECAST STRUCTURAL CONCRETE
03430	PRECAST PRESTRESSED CONCRETE
03500	CEMENTITIOUS DECKS
03510	GYPSUM CONCRETE
03520	CEMENTITIOUS WOOD FIBER DECK
03999	MISCELLANEOUS

04	**MASONRY**
04000	ALTERNATIVES
04100	MORTAR
04150	MASONRY ACCESSORIES
04160	JOINT REINFORCEMENT
04170	ANCHORS AND TIE SYSTEMS
04180	CONTROL JOINTS
04200	UNIT MASONRY
04210	BRICK MASONRY
04220	CONCRETE UNIT MASONRY
04230	REINFORCED UNIT MASONRY
04240	CLAY BACKING TILE
04245	CLAY FACING TILE
04250	CERAMIC VENEER
04270	GLASS UNIT MASONRY
04280	GYPSUM UNIT MASONRY
04400	STONE
04410	ROUGH STONE
04420	CUT STONE
04422	MARBLE
04430	SIMULATED MASONRY
04435	CAST STONE

04440	FLAGSTONE
04450	NATURAL STONE VENEER
04500	MASONRY RESTORATION & CLEANING
04510	MASONRY CLEANING
04550	REFRACTORIES
04999	MISCELLANEOUS

05 METALS

05000	ALTERNATIVES
05100	STRUCTURAL METAL FRAMING
05120	STRUCTURAL STEEL
05130	STRUCTURAL ALUMINUM
05200	METAL JOISTS
05300	METAL DECKING
05400	LIGHTGAGE METAL FRAMING
05500	METAL FABRICATIONS
05510	METAL STAIRS
05520	HANDRAILS AND RAILINGS
05521	PIPE AND TUBE RAILINGS
05530	GRATINGS
05540	CASTINGS
05700	ORNAMENTAL METAL
05710	ORNAMENTAL STAIRS
05720	ORNAMENTAL HANDRAILS & RAILINGS
05730	ORNAMENTAL SHEET METAL
05800	EXPANSION CONTROL
05999	MISCELLANEOUS

06 WOOD & PLASTICS

06000	ALTERNATIVES
06100	ROUGH CARPENTRY
06110	FRAMING AND SHEATHING
06111	LIGHT WOODEN STRUCTURES FRAMING
06112	PREASSEMBLED COMPONENTS
06113	SHEATHING
06114	DIAPHRAGMS
06130	HEAVY TIMBER CONSTRUCTION
06131	TIMBER TRUSSED
06132	MILL-FRAMED STRUCTURES
06133	POLE CONSTRUCTION
06150	TRESTLES
06170	PREFABRICATED STRUCTURAL WOOD
06180	GLUED-LAMINATED CONSTRUCTION
06181	GLUE-LAMINATED STRUCTURAL UNITS
06182	GLUE-LAMINATED DECKING
06190	WOOD TRUSSES
06200	FINISH CARPENTRY
06220	MILLWORK
06240	LAMINATED PLASTIC
06300	WOOD TREATMENT
06400	ARCHITECTURAL WOODWORK
06410	CABINETWORK
06411	WOOD CABINETS: UNFINISHED

06420	PANELING
06421	ARCH. HARDWOOD PLYWOOD PANELING
06422	SOFTWOOD PLYWOOD PANELING
06430	STAIRWORK
06431	WOOD STAIRS AND RAILINGS
06500	PREFAB. STRUCTURAL PLASTICS
06600	PLASTIC FABRICATIONS
06999	MISCELLANEOUS

07 THERMAL & MOISTURE PROTECTION

07000	ALTERNATIVES
07100	WATERPROOFING
07110	MEMBRANE WATERPROOFING
07120	FLUID APPLIED WATERPROOFING
07121	LIQUID WATERPROOFING
07130	BENTONITE CLAY WATERPROOFING
07140	METAL OXIDE WATERPROOFING
07150	DAMPPROOFING
07160	BITUMINOUS DAMPPROOFING
07170	SILICONE DAMPPROOFING
07175	WATER REPELLENT COATING
07180	CEMENTITIOUS DAMPPROOFING
07190	VAPOR BARRIERS/RETARDANTS
07200	INSULATION
07210	BUILDING INSULATION
07211	LOOSE FILL INSULATION
07212	RIGID INSULATION
07213	FIBROUS & REFLECTIVE INSULATION
07214	FOAMED-IN-PLACE INSULATION
07215	SPRAYED-ON INSULATION
07230	HIGH AND LOW TEMPERATURE INSUL.
07240	ROOF AND DECK INSULATION
07250	PERIMETER AND UNDER-SLAB INSUL.
07300	SHINGLES & ROOFING TILES
07310	SHINGLES
07320	ROOFING TILES
07400	PREFORMED ROOFING & SIDING
07410	PREFORMED WALL & ROOF PANELS
07411	PREFORMED METAL SIDING
07420	COMPOSITE BUILDING PANELS
07440	PREFORMED PLASTIC PANELS
07460	CLADDING/SIDING
07461	WOOD SIDING
07462	COMPOSITION SIDING
07463	ASBESTOS-CEMENT SIDING
07464	PLASTIC SIDING
07500	MEMBRANE ROOFING
07510	BUILT-UP BITUMINOUS ROOFING
07520	PREPARED ROLL ROOFING
07530	ELASTIC SHEET ROOFING
07540	FLUID APPLIED ROOFING
07570	TRAFFIC TOPPING

07600	FLASHING & SHEET METAL
07610	SHEET METAL ROOFING
07620	FLASHING AND TRIM
07630	ROOFING SPECIALTIES
07631	GUTTERS AND DOWNSPOUTS
07660	GRAVEL STOPS
07700	FLASHING
07800	ROOF ACCESSORIES
07810	SKYLIGHTS
07811	PLASTIC SKYLIGHTS
07812	METAL-FRAMED SKYLIGHTS
07830	HATCHES
07840	GRAVITY VENTILATORS
07850	PREFABRICATED CURBS
07860	PREFABRICATED EXPANSION JOINTS
07900	SEALANTS
07950	GASKETS
07999	MISCELLANEOUS

08   DOORS & WINDOWS

08000	ALTERNATIVES
08100	METAL DOORS & FRAMES
08110	HOLLOW METAL WORK
08111	STOCK HOLLOW METAL WORK
08112	CUSTOM HOLLOW METAL WORK
08120	ALUMINUM DOORS AND FRAMES
08130	STAINLESS STEEL DOORS AND FRAMES
08140	BRONZE DOORS AND FRAMES
08200	WOOD & PLASTIC DOORS
08300	SPECIAL DOORS
08310	SLIDING METAL FIRE DOORS
08320	METAL-CLAD DOORS
08330	COILING DOORS
08350	FOLDING DOORS
08355	FLEXIBLE DOORS
08360	OVERHEAD DOORS
08370	SLIDING GLASS DOORS
08375	SAFETY GLASS DOORS
08380	SOUND RETARDANT DOORS
08390	SCREEN AND STORM DOORS
08400	ENTRANCES & STOREFRONTS
08450	REVOLVING DOORS
08500	METAL WINDOWS
08510	STEEL WINDOWS
08520	ALUMINUM WINDOWS
08530	STAINLESS STEEL WINDOWS
08540	BRONZE WINDOWS
08600	WOOD & PLASTIC WINDOWS
08610	WOOD WINDOWS
08620	PLASTIC WINDOWS
08550	SPECIAL WINDOWS
08700	HARDWARE & SPECIALTIES
08710	FINISH HARDWARE
08720	OPERATORS
08721	AUTOMATIC DOOR EQUIPMENT
08725	WINDOW OPERATORS
08730	WEATHERSTRIPPING & SEALS

08740	THRESHOLDS
08800	GLAZING
08810	GLASS
08811	PLATE GLASS
08812	SHEET GLASS
08813	TEMPERED GLASS
08814	WIRED GLASS
08815	ROUGH AND FIGURED GLASS
08820	PROCESSED GLASS
08821	COATED GLASS
08822	LAMINATED GLASS
08823	INSULATING GLASS
08830	MIRROR GLASS
08840	GLAZING PLASTICS
08850	GLAZING ACCESSORIES
08900	WINDOW WALLS-CURTAIN WALLS
08999	MISCELLANEOUS

09   FINISHES

09000	ALTERNATIVES
09100	LATH & PLASTER
09110	FURRING AND LATHING
09160	PLASTER
09167	GYPSUM PLASTER
09180	CEMENT PLASTER
09190	ACOUSTICAL PLASTER
09250	GYPSUM WALLBOARD
09260	GYPSUM WALLBOARD SYSTEMS
09280	ACCESSORIES
09300	TILE
09310	CERAMIC TILE
09320	CERAMIC MOSAICS
09330	QUARRY TILE
09340	MARBLE TILE
09350	GLASS MOSAICS
09360	PLASTIC TILE
09370	METAL TILE
09400	TERRAZZO
09410	PORTLAND CEMENT TERRAZZO
09420	PRECAST TERRAZZO
09430	CONDUCTIVE TERRAZZO
09440	PLASTIC MATRIX TERRAZZO
09500	ACOUSTICAL TREATMENT
09510	ACOUSTICAL CEILINGS
09511	ACOUSTICAL PANELS
09512	ACOUSTICAL TILES
09520	ACOUSTICAL WALL TREATMENT
09530	ACOUSTICAL INSULATION & BARRIERS
09540	SUSPENSION SYSTEMS
09550	WOOD FLOORING
09560	WOOD STRIP FLOORING
09580	PLYWOOD BLOCK FLOORING
09590	RESILIENT WOOD FLOOR SYSTEM
09600	WOOD BLOCK INDUSTRIAL FLOORING
09650	RESILIENT FLOORING
09651	CEMENTITIOUS UNDERLAYMENT
09660	RESILIENT TILE FLOORING

11600	LABORATORY EQUIPMENT
11630	LAUNDRY EQUIPMENT
11650	LIBRARY EQUIPMENT
11700	MEDICAL EQUIPMENT
11800	MORTUARY EQUIPMENT
11830	MUSICAL EQUIPMENT
11850	PARKING EQUIPMENT
11860	WASTE HANDLING EQUIPMENT
11861	PACKAGED INCINERATORS
11862	WASTE COMPACTORS
11863	BINS
11864	PULPING MACHINES & SYSTEMS
11870	LOADING DOCK EQUIPMENT
11871	DOCK LEVELERS
11872	LEVELING PLATFORMS
11873	PORTABLE RAMPS & BRIDGES
11874	SEALS & SHELTERS
11875	DOCK BUMPERS
11880	DETENTION EQUIPMENT
11900	RESIDENTIAL EQUIPMENT
11970	THEATER & STAGE EQUIPMENT
11990	REGISTRATION EQUIPMENT
11999	MISCELLANEOUS

12	**FURNISHINGS**
12000	ALTERNATIVES
12100	ARTWORK
12110	MURALS
12120	PHOTO MURALS
12300	CABINETS AND STORAGE
12500	WINDOW TREATMENT
12550	FABRICS
12600	FURNITURE
12670	RUGS & MATS
12700	SEATING
12710	AUDITORIUM SEATING
12730	STADIUM SEATING
12735	TELESCOPING BLEACHERS
12800	FURNISHING ACCESSORIES
12999	MISCELLANEOUS

13	**SPECIAL CONSTRUCTION**
13000	ALTERNATIVES
13010	AIR-SUPPORTED STRUCTURES
13050	INTEGRATED ASSEMBLIES
13100	AUDIOMETRIC ROOM
13250	CLEAN ROOM
13350	HYPERBARIC ROOM
13400	INCINERATORS
13440	INSTRUMENTATION
13450	INSULATED ROOM
13500	INTEGRATED CEILINGS
13540	NUCLEAR REACTORS
13550	OBSERVATORY
13600	PREFABRICATED BUILDINGS
13700	SPECIAL PURPOSE ROOMS & BLDGS
13750	RADIATION PROTECTION
13770	SOUND & VIBRATION CONTROL

13800	VAULTS
13850	SWIMMING POOLS
13999	MISCELLANEOUS

14	**CONVEYING SYSTEMS**
14000	ALTERNATIVES
14100	DUMBWAITERS
14200	ELEVATORS
14201	ELEVATOR HOISTING EQUIPMENT
14202	ELEVATOR OPERATION
14203	ELEVATOR CARS AND ENTRANCES
14300	HOISTS & CRANES
14400	LIFTS
14430	PLATFORM AND STAGE LIFTS
14500	MATERIAL HANDLING SYSTEMS
14550	CONVEYORS & CHUTES
14570	TURNTABLES
14600	MOVING STAIRS & WALKS
14610	ESCALATORS
14700	PNEUMATIC TUBE SYSTEMS
14800	POWERED SCAFFOLDING
14999	MISCELLANEOUS

15	**MECHANICAL**
15000	ALTERNATIVES
15010	GENERAL PROVISIONS
15050	BASIC MATERIALS AND METHODS
15060	PIPE AND PIPE FITTINGS
15075	HOSE
15080	PIPING SPECIALTIES
15100	VALVES AND COCKS (MANUAL)
15120	CONTROL VALVES
15140	PUMPS
15160	VIBRATION ISOLAT. EXPANSION COMP.
15170	METERS AND GAGES
15175	TANKS
15180	INSULATION
15200	WATER SUPPLY & TREATMENT
15220	PUMPS AND PIPING
15230	BOOSTER PUMPING EQUIPMENT
15240	WATER RESERVOIRS AND TANKS
15250	WATER TREATMENT
15270	DISTRIBUTION AND METERING SYSTEMS
15300	WASTE WATER DISPOSAL & TREATMENT
15310	SEWAGE EJECTORS
15320	GREASE INTERCEPTORS
15330	BASINS AND MANHOLES
15340	SEWERAGE
15350	LIFT STATIONS
15360	SEPTIC TANK SYSTEMS
15380	SEWAGE TREATMENT
15400	PLUMBING
15420	EQUIPMENT
15440	SYSTEM ACCESSORIES (SPECIAL)
15450	SPECIAL FIXTURES AND TRIM
15451	WATER COOLERS

SPECIFICATIONS

15452	WASHFOUNTAINS
15460	PLUMBING FIXTURES
15480	POOL EQUIPMENT
15500	FIRE PROTECTION
15510	SPRINKLER EQUIPMENT
15520	CO2 EXTINGUISHING EQUIPMENT
15530	STANDPIPE & FIRE HOSE EQUIPMENT
15540	PRESS. EXTING. & FIRE BLANKETS
15550	FIRE EXTINGUISHER CABINETS
15560	HOOD AND DUCT FIRE PROTECTION
15600	POWER OR HEAT GENERATION
15610	FUEL HANDLING EQUIPMENT
15611	OIL STORAGE TANKS AND PIPING
15612	BOTTLED GAS TANKS AND PIPING
15613	OIL PIPING
15614	GAS PIPING
15615	STOKERS AND CONVEYORS
15616	ASH REMOVAL SYSTEM
15617	BREECHINGS
15618	EXHAUST EQUIPMENT
15619	DRAFT CONTROL EQUIPMENT
15630	BOILERS
15640	BOILER FEEDWATER EQUIPMENT
15650	REFRIGERATION
15658	REFRIGERANT PIPING SYSTEM
15660	COMPRESSORS
15670	CONDENSING UNITS
15680	CHILLERS
15690	EVAPORATORS
15698	COMMERCIAL ICE MAKING EQUIPMENT
15699	REFRIGERATION ACCESSORIES
15700	LIQUID HEAT TRANSFER
15710	HOT WATER SPECIALTIES
15720	STEAM SPECIALTIES
15730	HEAT EXCHANGERS
15740	TERMINAL UNITS
15760	PACKAGED HEAT PUMP
15770	PACKAGED HEATING AND COOLING
15780	HUMIDITY CONTROL
15800	AIR DISTRIBUTION
15810	FURNACES
15811	INFRA-RED HEATERS
15820	FANS
15835	AIR CURTAINS
15840	DUCTWORK
15850	SPECIAL DUCTWORK EQUIPMENT
15860	DUCT ACCESSORIES
15870	OUTLETS
15880	AIR TREATMENT EQUIPMENT
15890	SOUND ATTENUATORS
15900	CONTROLS & INSTRUMENTATION
15910	CONTROL PIPING, TUBING, & WIRING
15920	CONTROL PANELS
15930	PRIMARY CONTROL DEVICES
15950	SEQUENTIAL CONTROLS
15960	RECORDING DEVICES
15970	ALARM DEVICES
15980	SPECIAL PROCESS CONTROLS
15999	MISCELLANEOUS
16	**ELECTRICAL**
16000	ALTERNATIVES
16010	GENERAL PROVISIONS
16100	BASIC MATERIALS AND METHODS
16110	RACEWAYS
16120	CONDUCTORS
16130	OUTLET BOXES
16133	CABINETS
16134	PANELBOARDS
16140	SWITCHES AND RECEPTACLES
16150	MOTORS
16160	STARTERS
16170	DISCONNECTS (MOTOR AND CIRCUIT)
16180	OVERCURRENT PROTECTIVE DEVICES
16190	SUPPORTING DEVICES
16199	ELECTRONIC DEVICES
16200	POWER GENERATION
16210	GENERATOR
16220	ENGINE
16230	COOLING EQUIPMENT
16240	EXHAUST EQUIPMENT
16250	STARTING EQUIPMENT
16260	AUTOMATIC TRANSFER EQUIPMENT
16300	POWER TRANSMISSION
16310	SUBSTATION
16320	SWITCHGEAR
16330	TRANSFORMER
16340	VAULTS
16350	MANHOLES
16360	RECTIFIERS
16370	CONVERTERS
16380	CAPACITORS
16400	SERVICE & DISTRIBUTION
16410	ELECTRIC SERVICE
16411	UNDERGROUND SERVICE
16420	SERVICE ENTRANCE
16421	EMERGENCY SERVICE
16430	SERVICE DISCONNECT
16440	METERING
16450	GROUNDING
16460	TRANSFORMERS
16470	DISTRIBUTION SWITCHBOARDS
16480	FEEDER CIRCUIT
16490	CONVERTERS
16491	RECTIFIERS
16500	LIGHTING
16510	INTERIOR LIGHTING FIXTURES
16515	SIGNAL LIGHTING
16530	EXTERIOR LIGHTING FIXTURES

16531	STADIUM LIGHTING	16780	TELEVISION SYSTEMS
16532	ROADWAY LIGHTING	16850	HEATING & COOLING
16550	ACCESSORIES	16858	SNOW MELTING CABLE AND MAT
16551	LAMPS	16859	HEATING CABLE
16552	BALLASTS AND ACCESSORIES	16860	ELECTRIC HEATING COIL
16570	POLES AND STANDARDS	16865	ELECTRIC BASEBOARD
16600	SPECIAL SYSTEMS	16870	PACKAGED ROOM AIR
16610	LIGHTING PROTECTION		CONDITIONERS
16620	EMERGENCY LIGHT AND POWER	16880	RADIANT HEATERS
16640	CATHODIC PROTECTION	16890	ELECTRIC HEATERS (PROP FAN
16700	COMMUNICATIONS		TYPE)
16710	RADIO TRANSMISSION	16900	CONTROLS & INSTRUMENTATION
16720	ALARM AND DETECTION	16910	RECORDING AND INDICATING
	EQUIPMENT		DEVICES
16740	CLOCK AND PROGRAM	16920	MOTOR CONTROL CENTERS
	EQUIPMENT	16930	LIGHTING CONTROL EQUIPMENT
16750	TELEPHONE & TELEGRAPH	16940	ELECTRICAL INTERLOCK
16760	INTERCOMMUNICATION	16950	CONTROL OF ELECTRIC HEATING
	EQUIPMENT	16960	LIMIT SWITCHES
16770	PUBLIC ADDRESS EQUIPMENT	16999	MISCELLANEOUS

Subdivision.—Each section of the specification division will be subdivided into the "Scope," or "Work Included;" "Materials," listing the quality of all materials to be furnished; and finally, "Fabrication and Erection," defining the quality of workmanship desired.

"Scope" will include those items to be furnished and installed, those items to be furnished but installed by others, those items furnished by others and installed under this division, and finally, related work in other divisions of the specifications. "Scope" will also list some general requirements such as those for shop drawings, samples, tests, methods of delivery and storage and the like that may be required. However detailed, the "Scope" section will not include each and every item. The drawings must always be checked against "Scope" which often carries the phrase "including, but not limited to, the following."

"Materials" will list the materials to be used in one of several ways, often found in combination. The "closed specification" will list a single trade name and this must be furnished. The "contractor's option specification" (or "bidder's choice") will list more than one trade name and the contractor may choose which of those listed he prefers to use. The "substitute bid specification" may list a choice of trade names and the bid must be based on one of the products included, but the contractor is allowed to suggest a substitute at the time of submitting his bid, naming the amount he would subtract for using the alternate product.

A variation of this is the "product approval specification" which asks the contractor to submit any substitutions prior to submitting his bid. If the architect approves the substitute, it will be put in an addenda sent to all contractors. The "or approved equal" type specification which names a product but qualifies it by "or approved equal" is the most common type in general practice and provides the widest competition and rewards to the contractor who knows his products, prices and sources.

However, it is the architect who rules whether the product is "equal." If the contractor questions its approval, he should bring his substitute to the architect's attention upon being selected as the low bidder so any adjustments can be made before award of contract should the architect rule his substitution is not equal.

The "product description specification" is used to describe bulk materials and some manufacturers' articles, the latter particularly in government work. The "performance specification" describes not the material, but what job it is required to do, either in matters of strength, mechanical ability, or similar measurable result. Specifications based on "reference standards" refer to specifications published by national organizations related to the building industry. These are often encountered. The more commonly referred standards are:

ACI	American Concrete Institute
AISC	American Institute of Steel Construction
ASA	American Standards Association
ASCE	American Society of Civil Engineers
ASHVE	American Society of Heating and Ventilating Engineers
ASTM	American Society for Testing Materials
AWI	Architectural Woodwork Institute
AWS	American Welding Society
BSI	Building Stone Institute
CS	Commercial Standards (U.S. Dept. of Commerce)
CSI	Construction Specifications Institute, Inc.
IES	Illuminating Engineering Society
ILI	Indiana Limestone Institute
MIA	Marble Institute of America
MLMA	Metal Lath Manufacturers Association
NBFU	National Board of Fire Underwriters
NFPA	National Fire Protection Association
NLMA	National Lumber Manufacturers Association
NTMA	National Terrazzo and Mosaic Association
PC	Producers Council Inc.
PCA	Portland Cement Association
SPA	Southern Pine Association
SJI	Steel Joist Institute
SCPI	Structural Clay Products Institute
TMA	Tile Manufacturers Association
UL	Underwriters' Laboratories, Inc.

These specifications are general in nature and usually apply only in part to the specific job. A contractor should build up a library of these publications to check specification references to them. They are informative and worthwhile.

In addition, many architects will refer to the published specifications of a named manufacturer. These are easily available and generally are included in regards to installation procedures.

"Fabrication and Erection" will set performance standards for both shop and field work and will include, if required, sections on protection, cleaning, guarantees, warranties, maintenance and operating instructions. The specifications of trade organizations and manufacturers' installation instructions are often made a part of this section.

Alternates.—The architect often may wish to investigate at the time of bidding the cost either upward or downward, of a substitute product, material or procedure, or adding or subtracting work from the contract. Such requests for alternate prices are usually noted at the end of the section of the specifications to which they refer, and are repeated at the end of Division 1 of the specifications and on the proposal form. On the typical Lump Sum proposal form, the Alternate follows the Statement of the Lump Sum price in such form as:

If the following Alternates are accepted,	Add:	Deduct:
Alternate No. 1	$_____	$_____
Alternate No. 2	$_____	$_____

Obviously a request for a lot of Alternates complicates bidding and sometimes the extent of the Alternate is not clear. Since the architect will accept the price change for an Alternate with the same responsibility as attached to the Lump Sum figure, the contractor must be certain of all the possible contingencies involved should the change be accepted.

Often an Alternate is put only under the one section that makes the basic substitution, but other subcontractors' work is also affected by the change. It is the contractor's responsibility to foresee this. As an example, under Carpentry, an architect might include:

"Alternate No. 1—For substituting one 3'-0"x6"-8"x1¾" door with frame for the two 3'-0"x6'-8"x1¾" doors and frame shown between the living and dining rooms, deduct $_____."

Obviously, the door cost will be less but the wall construction will be more, and no mention may occur under that division.

Cash Allowances.—Sometimes, the architect may not have a final decision from an owner on certain items and, rather than leave them out of the Lump Sum proposal, he gives a definite budget amount in the specification that is to be included in the bid. This is often encountered with the selection of brick and hardware and other finish items. As an example:

"Facing brick shall be included under an allowance of $75 per M, fob, job site."

Thus, every bidder will include the same material cost, rather than guessing just what the "face brick" the owner may choose will cost. Most always, cash allowances are for the cost of material and applicable taxes delivered to the site only. Labor to set, overhead and profit on the material should be included in the Lump Sum price unless specifically noted otherwise.

Upon completion of the job, the actual cost of the items furnished under the Allowance shall be tabulated with any savings credited to the owner, any overruns added to the contract cost.

Unit Prices.—Where quantity of materials is in doubt, but quality is known, the specification may ask for Unit Prices. As opposed to Allowances, Unit Prices are set by the bidder, not the architect; they are firm and binding on the bidder, the Lump Sum price being adjusted up or down due only to variation in the quantity actually used on the job.

Unit Prices are included on the proposal form for each material per unit of measure. Unit Prices are often asked for concrete per sq. yd., piling per lineal ft., partition block per sq. ft., etc. These prices should be complete with all costs, profit and overhead included.

Unit Prices must be quoted with care, even though the amount of work covered by them is small, as they will be compared directly with those submitted by other bidders and the owner may expect the same Unit Prices to hold on additional work.

ADDENDA

Few plans and specifications are perfect when they leave the architect's office and the contractors bidding the job should call to the architect's attention all discrepancies they note. These, plus changes the architect and owner may wish to make after the plans and specifications have been issued, but before bids are turned in, are incorporated in the Addenda. This usually takes the form of a printted sheet or sheets the same size as the specification to which it should be attached. Addenda eventually becomes a part of the Contract Documents.

Often, even changes to the drawings can be described and included this way. But sometimes additional drawings or sketches must be issued, or a drawing is reissued with a revision date. Reissued drawings, however, usually should be avoided during bidding.

Each Addendum is given a number and the changes it covers will be listed in the order the divisions occur in the specification, with drawing changes last.

Sometimes, Addenda must be acknowledged by signing a receipt; sometimes they are sent registered mail, but usually they are acknowledged on the proposal form in a space provided for it. The single Addendum or several Addenda are also included in the Owner-Contractor Agreement.

Upon receipt of an Addendum, the contractor should mark his specifications and drawings with appropriate notes, or with a stamp on each affected page saying "See Addendum #_____." The Addendum will usually identify the change by division and page number and a reference to location on the page. As an example:

3:01—**Scope:** Add item "h) Remove existing wire fence along north edge of property."

6:02—**Materials:** Delete paragraph referring to concrete masonry lintels and add " . . . lintels in block work shall be units of sizes and shapes shown on the lintel schedule."

Dwg. Sheet A-2: Move rolling shutter door from center of Bay C-D to just north of Column B.

Dwg. Sheet A-4; delete ladder to roof.

CHANGE ORDERS

Change orders are modifications issued after the contract is signed. Errors on drawings and specifications are usually uncovered during bidding. A contractor who sees where a drawing is not complete and therefore does not include what he knows must be furnished in his price, is not only playing dangerously with his reputation, but may very well end up having to pay for the item anyway to complete the job or face a lengthy law suit.

However, some Change Orders are necessary due to job conditions, changes required when plans are submitted for permits, substitutions made to satisfy the owner or his mortgage holder, products not being available, or because, as the job progresses, certain changes for the benefit of the job become obvious.

The contractor must cooperate in making these changes. This is set forth in Article 12 of the General Conditions. Sometimes no cost is involved. Most often, an extra is called for. This may be handled by a firm price change being submitted, or an order to proceed on a time and material basis. The Change Order is issued by the architect but signed by the owner and contractor. The format of the change order will follow closely that of the Addendum, except that costs of changes and tabulation of the new contract amount are added. The Frank R. Walker Company publishes a standard form, L-101, in triplicate, for contract change orders, a sample of which is shown on the following page.

SUMMARY

As can be imagined, practices as to what appears on drawings and what is stated in specifications will vary from job to job. There is no legal precedent that one is more binding than the other, although some specifications will state which is to be followed in case of duplication.

In general, sizes, quantities, design and location will be best taken from the drawings. Materials, quality, procedures and general requirements are most reliably defined in the specifications. Never rely solely on one or the other. A draftsman's delineation of a stone wall in elevation may not show at all the coursing called for in the specifications, while the specifier may make no special mention of a stone sill condition carefully detailed on the drawings. One must always complement the other.

CONTRACT CHANGE ORDER

JOB
CONTRACT JOB NO.

FOR

CHANGE
ORDER   **No.**

DATE

TO

REVISED CONTRACT AMOUNT

PREVIOUS CONTRACT AMOUNT $

AMOUNT OF THIS ORDER $

TOTAL CONTRACT AND EXTRAS $

The work covered by this order shall be performed under the same Terms and Conditions as that included in the Original Contract.

CHANGES APPROVED

AMOUNT OF ORIGINAL CONTRACT $

CHANGE NO.

By

By   By

PRACTICAL Form L-101
MFD IN U S A   FRANK R. WALKER CO . PUBLISHERS, CHICAGO

CHAPTER 4

SETTING UP THE ESTIMATE

THE DETAILED ESTIMATE
THE ESTIMATOR
OFFICE OVERHEAD
JOB OVERHEAD
BONDS
INSURANCE
PROFIT
ESTIMATE CHECK LIST
DETERMINING COSTS

Estimating is one of the most important aspects of a building contractor's business. In nearly every instance it is necessary for a contractor either to estimate the cost of the work, or to bid in competition with others before a contract for the work is awarded. If a contractor's bid is too low and a contract is awarded on the basis of it, the contractor often must complete the work without profit and sometimes at an actual loss in time and money.

A contractor must know his limitations and those of his organization as to the size, complexity, and quality of work that he can efficiently handle. To exceed these limitations often leads to a poor contract which can be costly not only monetarily, but also in making dissatisfied customers and loss of future opportunities for bidding.

Most contractors fall into categories as to the type of work which their organizations can handle efficiently, expeditiously, and profitably. A heavy construction contractor specializing in structures such as airport runways, highways, bridge abutments, etc., usually should forego bidding on such work as churches, banks, hospitals or modern office buildings. A contractor whose organization and personnel are geared to a particular category of construction can find himself in difficulty if he strays from his specialty.

Many contractors in business today are former employees of contracting firms who have broken away from them to go into business for themselves. Many of these people doubtless were exceptional mechanics, foremen or superintendents, and had a thorough understanding of how the work should be carried out in the field, but lack business ability and training to estimate costs accurately. Some of these contractors make a reasonable profit on their work and remain in business, but many more of them fail and eventually go back to working for someone else.

There are various reasons for these failures, but probably the most important one is the inability of the person estimating costs to come up with realistic and profitable estimates. The procedures involved in preparing an accurate, detailed cost estimate necessary for a realistic bid are relatively simple and most contractors know them well. Yet almost all instances of failure of contracting firms are the result of poor estimating practices, and the consequent submission of unsound and unprofitable bids. There is no substitute for complete and detailed material take-offs and pricing in preparing a bid, and in cross-checking estimated costs with actual costs so that adjustments in pricing can be made for future bidding.

Another common reason for failures in the contracting business is overexpansion. Again, the contractor must know the capabilities of his office and supervisory staff and how much they can adequately handle at any one time. The successful contractor does not take on more work than his staff can take care of efficiently.

Bonding and insurance companies have certain minimum requirements which must be met by contractors applying for surety bonds or contractual liability insurance. Some of the most important considerations which such companies investigate before they will bond an applicant are as follows:

1. The technical ability of the contractor applying for a bond, whether or not he understands his business thoroughly, and his capabilities in the preparation of detailed cost estimates.

2. The reputation of the contractor for honesty and his standing among those with whom he does business.

3. The financial ability of the contractor, and whether or not he has sufficient capital at his disposal to carry on his business, purchase materials and supplies, meet his payrolls and current expenses, etc.

Bonding companies seldom get the opportunity of bonding the best contracting firms. On public work projects the bidding is usually open to all bidders and in some instances as many as fifty or more bids may be received on one job. In most cases, the lowest bidder is awarded the contract, and he often may be the poorest risk because of substandard estimating procedures, lack of technical ability to perform the work as specified, and lack of financial responsibility.

On privately owned projects, the risk to the bonding company may not be as great as it is on public work as the bidding list can be controlled and only contractors with the proper credentials may be asked to submit proposals. Architectural and engineering firms which handle work for private owners are often asked to select contractors to bid on a project. They will attempt to choose bidders whom they know can perform the work equally well and will carry out the intent of the drawings and specifications in the best interest of the owner.

The above illustrates the fact that somebody besides the contractor attaches importance to a correct and detailed estimate. A well prepared estimate has just as much to do with the success of a contractor as the manner in which he purchases his materials, lets his subcontracts or carries out his work on the job.

THE DETAILED ESTIMATE

If twenty estimators or contractors were furnished the same set of plans and specifications and told to prepare an estimate of cost from same so that the different methods might be compared, it is safe to assume there would be not more than two estimates in the entire twenty that had been prepared on the same basis or from the same units.

Why should there be such a variation in methods with only one result to be obtained? It is due to the absence of a uniform method of estimating and the lack of detailed information regarding the cost of the work to be performed. Each contractor has developed a system of his own, making it almost impossible for one contractor to check another's figures.

This is the result of each contractor developing unit price costing records reflecting the particular experience and expertise of his organization which may vary considerably between contractors. This is not to say that one estimate is correct or another one which is at variance is incorrect. In the final analysis of estimates and bids they should be accepted as representing the projected final cost for performing certain prescribed work within a stipulated time schedule, taking into consideration all of the variables which are anticipated to be encountered during the construction of the project.

To prove the validity of the bid, one must consider the extreme risks that the contractor is willing to assume by furnishing Payment and Performance Bonds and to enter into a legal contract dedicating the technical and financial resources of the firm to a successful completion of the contract in hopes of realizing a modest profit.

Estimating is a laborious and costly operation, and under no circumstances

PRACTICAL
FORM 514

GENERAL ESTIMATE

ESTIMATE NO._____

BUILDING_____

SHEET NO._____

LOCATION_____

ESTIMATOR_____

ARCHITECTS_____

CHECKER_____

SUBJECT_____

DATE_____

DESCRIPTION OF WORK	NO. PIECES	DIMENSIONS	EXTENSIONS	EXTENSIONS	TOTAL ESTIMATED QUANTITY	UNIT PRICE M'T'L	TOTAL ESTIMATED MATERIAL COST	UNIT PRICE LABOR	TOTAL ESTIMATED LABOR COST

MFD. IN U.S.A.

FRANK R. WALKER CO., PUBLISHERS, CHICAGO

General Estimate Form

Frank R. Walker Co.

Form 514

should an estimator or contractor produce a less than totally professional estimate which will stand the test of proving each element contained in the estimate.

The detailed estimate means precisely just that. It should be prepared in the very same sequence which the project will be constructed, thereby providing an added insurance against the error of omission which could easily occur in an undisciplined estimate where quantities were measured and priced on a random selective approach rather than through gaining the continuity of the logical sequence of construction events which will reflect the interface of all trades employed on the project.

The estimate should serve the purpose for which it was intended.

The conceptual type estimates are considerably less accurate and can be prepared faster, at a reduced cost, but they are only required to produce accuracy within a wide range to be useful in feasibility studies or general budget allowances.

The detailed estimate contains a quantity analysis and cost allowance for every item of work or expense required by the plans and specifications. It is to be identified so that reference could be made to a specific item long after the preparation of the estimate. This identification also makes the estimate more useful to others who may have a need to review the original estimated costs. Any attempt to "short cut" or utilize the hazardous method of square foot or cubic foot estimates for use as a detailed bid type estimate will ultimately result in disaster for a contractor.

The use of the detailed estimate extends far beyond being the basis of a bid. It serves as the guide for the purchase of materials, and the award of subcontracts. It also provides the information for the establishment of a cost control system so that the project progress can be monitored to determine if the estimated material and the labor costs are consistent with the actual costs incurred during performance of the contract.

As an example, assume that the detailed estimate reflected the following labor cost allowances for each of the listed items of work:

	Labor Allowance	Quantity
Footing Excavation	$3,000.00	300 cu. yds.
Form Footings	$1,200.00	1,200 sq. ft.
Pour Footing Concrete	$1,500.00	300 cu. yds.

As each of these phases of work started the labor chargeable to each would be allocated on the project daily time sheet, Walker Form No. 505 so that as work progressed, there would be an accurate cost record which could be reviewed daily, weekly or monthly depending upon the magnitude of the operation. In this way, periodic projections could be made. As an example, the 50% completion review of each of the items reflected the following conditions:

Item	Labor Allowance	Expended To Date	Remaining Cost	+ − (loss)
Footing Excavation	$3,000.00	$1,800.00	$1,800.00	($600.00)
Form Footings	1,200.00	400.00	400.00	400.00
Pour Footings	1,500.00	600.00	600.00	300.00

The contractor is not to view the combination of these three work items as reflecting a net $100.00 savings above the estimated costs but, should be abruptly alerted that there is a serious overrun of 20% in the labor budget for footing excavation. This item should be reviewed in detail to determine if a remedy is available to improve labor production for the remaining 50% of work. Should it be found that there is no solution to the problem then it is to be assumed that the project anticipated profits will be reduced by the $600.00 additional cost. However,

SUMMARY OF ESTIMATE

FORM 115

BUILDING	LOCATION		ESTIMATE NO.
ARCHITECT	OWNER		DATE
CUBICAL CONTENTS	NO. OF STORIES	COST PER CUBIC FOOT	ESTIMATOR
FLOOR, AREA, SQUARE FEET	COST PER SQUARE FOOT		CHECKER

CLASSIFICATION	TOTAL ESTIMATED MATERIAL COST	TOTAL ESTIMATED LABOR COST	TOTAL SUB-BIDS	TOTAL	ADJUSTMENTS
1. GENERAL CONDITIONS AND OFFICE OVERHEAD					
2. JOB CONDITIONS AND JOB OVERHEAD					
3. CONSTRUCTION PLANT, TOOLS AND EQUIPMENT					
4. DEMOLITION AND SITE CLEARANCE					
5. EXCAVATION, GRADING AND DEWATERING					
6. SHEETING, SHORING AND BRACING					
7. PILING AND CAISSON WORK					
8. SITE DEVELOPMENT					
9. CONCRETE FORM WORK					
10. CAST IN PLACE CONCRETE					
11. PRECAST CONCRETE AND CEMENTITIOUS DECK					
12. BRICK, TILE, CONCRETE & GLASS UNIT MASONRY					
13. UNIT MASONRY PARTITIONS & FIREPROOFING					
14. CUT, ROUGH, NATURAL & SIMULATED STONE					
15. STRUCTURAL METALS					
16. OPEN WEB JOISTS AND METAL DECKING					
17. MISCELLANEOUS METALS					
18. ORNAMENTAL METALS					
19. ROUGH CARPENTRY AND ROUGH HARDWARE					
20. FINISH CARPENTRY					
21. CUSTOM MILLWORK					
22. WATERPROOFING AND DAMP-PROOFING					
23. BUILDING INSULATION					
24. SHINGLE AND ROOFING TILE					
25. MEMBRANE ROOFING					
26. PREFORMED ROOFING AND SIDING					
27. ROOFING ACCESSORIES					
28. SHEET METAL FLASHINGS					
29. SEALANTS AND CAULKING					
30. METAL DOORS AND FRAMES					
31. WOOD AND PLASTIC DOORS					
32. SPECIAL DOORS					
33. METAL WINDOWS					
34. WOOD AND PLASTIC WINDOWS					
35. FINISHED HARDWARE & WEATHER STRIPPING					
36. GLASS, GLAZING AND MIRRORS					
37. CURTAIN WALL AND STORE FRONT SYSTEMS					
38. LATHING AND FURRING					
39. PLASTERING AND STUCCO WORK					
40. GYPSUM DRYWALL					
41. CERAMIC, MOSAIC, QUARRY, MARBLE & SLATE TILE					
42. TERRAZZO AND SEAMLESS FLOORING					
43. WOOD FLOORING					
44. RESILIENT FLOORING					
45. ACOUSTICAL TILES AND PANELS					
46. SPRAY ON FIRE PROTECTION					
47. PAINTING AND SPECIAL FINISHES					
48. SPECIAL BUILDING PARTITIONING					
49. BUILDING SPECIALTIES, EQUIP. & FURNISHINGS					
50. ELEVATORS & MECHANICAL TRANSPORT					
51. PLUMBING					
52. FIRE PROTECTION & SPRINKLER SYSTEMS					
53. HEATING, VENTILATION & AIR CONDITIONING					
54. ELECTRICAL WORK					
55. COMMUNICATION SYSTEMS					
56. TOTALS					
57.		TOTAL COST			
58.		PROFIT			
59.		SURETY BOND			
60.		AMOUNT OF BID			

MFD. IN U.S.A.   FRANK R. WALKER CO., PUBLISHERS, CHICAGO

valuable cost information has been developed as these actual cost records become the basis of unit costs for estimating future projects as the above example would indicate that excavation should be increased from $10.00 to $12.00 per cu. yd. and each of the other items unit prices could be reduced thus, producing a more accurate and competitive bid.

It is strongly suggested that a positive estimating format be established utilizing the several pre-printed forms as published by The Frank R. Walker Company as a method of producing the most accurate and useful estimates possible.

THE ESTIMATOR

The estimator is the most important man in any contractor's organization because it is up to him to prepare estimates that will either make or break the contractor. The most efficient job organization or the best purchasing department in existence cannot make money on contracts taken below cost.

On the other hand, it does not take a good salesman or contract man to obtain all the work you can handle if your estimates are way below your competitor's. Remember, the low man is wrong more often than he is right,—so the efficient organization is one that can get the work at the right price. It is not an easy task, but a business built upon the principles of service, dependability, good materials and workmanship, and fair price, will be in business many years after the get-the-work-at-any-price contractor has made his last assignment in favor of his creditors.

The estimator should be a practical man. He should possess a thorough understanding of job conditions; how the work will be carried on in the field, and the operations necessary to assemble the materials and put them in place in the structure. He must be able to visualize. He must be able to take a set of blueprints and draw a mental picture of the building during construction. He must know every branch of the work to be handled, the materials required, and the labor operations necessary to convert a pile of gravel, sand, brick, cement, lumber, steel, glass, etc. into a completed building.

On the other hand not every visualizer can be a successful estimator. Your job superintendent or foreman may be able to take a set of blueprints and lay the entire job out in his mind and picture the building progressing from the foundation to the finished structure, but unless he possesses the knowledge necessary to compute the quantity of materials and the labor cost of putting them in place, he will never be able to prepare an accurate estimate of cost.

The following seven requisites are essential for the making of a good estimator:

First: He should be able to read plans and measure them accurately.

Second: He should possess a fair knowledge of arithmetic, including addition, subtraction, multiplication, division, together with a knowledge of the decimal system. Nearly all measurements and computations in the average building estimate are made in lineal feet, square feet or yards, and cubic feet or yards. These material quantities are then usually multiplied at a certain unit price, so a knowledge of decimals is necessary to obtain the various costs. More and more drawings will carry dimensions based on the metric system and one must be able to convert from one to the other.

Third: He should be able to visualize,— that is, he should be able to take a set of blueprints and draw a mental picture of the building; how it will look during construction and at completion. He must be able to place his equipment so that the work can be handled economically. In short, he must first construct the building on paper.

Fourth: He should possess an intimate knowledge of job conditions, the most practical methods of handling materials and labor on the job and the most economical methods of construction. While the estimator is generally an office man, he must possess much the same knowledge of construction methods as your superintendent or foreman.

Fifth: The knowledge and ability to assemble materials into workable units. A thousand brick in the wall means more than just the brick. It means a certain amount of sand, lime and cement, and possibly mortar color, to make the mortar. A cubic yard of concrete means the exact quantities of cement, sand and gravel must be computed. The same applies to nearly every complete unit in the building. The competent estimator must have this knowledge at his finger tips.

Sixth: He should possess an intimate knowledge of labor performances and operations and be able to convert them into costs in dollars and cents. This requires a working knowledge of what a man will do under given conditions and is obtained only by a careful study of different jobs completed under varying conditions.

Seventh: He should be endowed with at least an average amount of common sense.

Estimators may now be interested in joining The American Society of Professional Estimators which has been established to promote the profession. There are now some twenty-six active chapters in seventeen states and the District of Columbia. The Society headquarters can be reached at 5201 North 7th St. Phoenix AZ 85014. There are several categories of membership including student and affiliate for those who do not meet the requirement that to join as an estimator one must have five years active employment in the construction industry as an estimator. Education programs are promoted, conventions are held on both national and local levels, and newsletters are printed. The Society has recently established a Certification which is given after successfully passing an eight hour test which will be given periodically.

Office Overhead Expense

The item of adequate overhead expense is sometimes neglected by many contractors, but it is of such importance that every contractor should give it careful consideration.

Every contractor has a certain fixed expense that must be paid regardless of the amount of work done or contracts received, and these items should be charged to Office Overhead.

This includes such items as office rent, fuel, lights, telephone and telegraph, stationery, office supplies, advertising, trade journals and magazines, donations, legal and accounting expenses not directly chargeable to any one job, fire and liability insurance for the office, club and association dues, office employees, such as bookkeeper, stenographer, clerks, estimator, janitor services and salaries of executives, along with travel and entertainment expenses of a good will nature and depreciation on office furnishings and equipment.

These items should be estimated for a year and then reduced to a certain percentage of the total annual business, as follows:

Non-reimbursable Salaries*	$	$ 50,000
Employee Benefits*		10,000
Legal & Accounting Fees		3,000
Insurance*		6,000
Office Rent		5,000
Office Utilities		3,000
Office Supplies		2,000
Job Procurement & Promotion		7,000
Depreciation*		2,000
Maintenance*		1,500
Job Management* (incl. transportation)		5,000
Miscellaneous		5,500
		$100,000

*Indirect expenses only, that is, those not reimbursed by fees from individual jobs.

The above list is much larger than smaller contractors will have and considerably less than that carried by larger contractors but it will serve as an example of the items that should be included as overhead or general expense.

For example, suppose you have an overhead expense of $100,000 per year, and do an annual business approximating $1,250,000. On this basis your overhead or general expense would average 8 per cent of the total annual business.

With some concerns the overhead may run 6 per cent and with others 10 to 15 per cent of the total annual business but every contractor can approximate his overhead based on his average yearly business. Even an approximate is far better than not making any attempt in this direction.

Many smaller contractors do not maintain an office and probably attend to the greater part of their office duties themselves, so only such items of expense as they actually incur should be included.

Proposals for jobs taken on a time and material basis and provisions for extra work on jobs taken on a stipulated sum basis may require the contractor to state the percentage to be added for overhead, plus that for profit. In a recent job bid from five medium sized contractors in Chicago, these varied as follows: 8+4, 8+7, two at 10+5 and one at 10+10, so while overhead percentage varied only from 8 to 10%, the profit varied from 4 to 10%.

Whether these same percentages are used in figuring competitive bids are, of course, each contractor's secret. But certainly, some addition to cover office overhead should be made unless the organization is so successful that the year's overhead has already been absorbed by other jobs; or that the work is of such nature that it is primarily subcontracted and office expenses are minor.

All manufacturers, jobbers and retailers add their overhead expense to the cost of their products when computing the selling price and there is no reason why contractors should not do the same—always keeping their eye on the competition.

Office Furniture and Equipment.—All expenditures for office furniture, typewriters, calculating machines, filing cabinets, etc., should be kept as a separate account and a certain percentage of this amount charged off each year as depreciation. The depreciation, however, should be charged as a part of the overhead expense of conducting the business, as the cash value of the equipment decreases each year.

In a completely efficient office, construction equipment would be constantly utilized on the job and would be charged as a job cost, not office overhead. However, there are always slack periods when equipment is idle, or was bought for a specific job but still has a useful life and can be depreciated and charged to overhead until it can be put to use again. The Internal Revenue Department publishes fixed schedules of allowed depreciation.

Some contractors set up a "rental" cost for equipment owned by the firm but chargeable to individual jobs. Today the trend is to rent tools and equipment from outside firms specializing in this which simplifies bookkeeping—all costs are directly charged to the job and no depreciation need be carried—and frees the contractor's capital for more lucrative investment. A discussion of plant and equipment is included later in this chapter.

Insurance.—Insurance costs directly chargeable to office overhead would include fire and theft for office property, liability for office premises, workmen's compensation for office employees, and automobile insurance for trucks and cars owned by the firm.

As liability insurance is a must in today's construction, this is also often included as an office overhead, a necessary cost of doing business. However, if the policy has rather low limits, the contractor will find most specifications will require that he take out additional coverage, and this would be a job rather than overhead expense.

Surety bonds, performance bonds, property insurance for work under construction and workmen's compensation, for on-job employees can be charged directly to the job costs and are discussed under job expenses. The cost of bid bonds for work awarded to another contractor would obviously have to be absorbed by office overhead.

JOB OVERHEAD

Job overhead includes three divisions. First, those items the general contractor must furnish for the conduct of the job as a whole, generally termed "General Conditions" often supplemented with modifications and additions under "Special Conditions."

Secondly, there are items the general contractor may furnish to expedite the job which may benefit all or just a single trade. These are carried under "Plant and Equipment."

Finally, the contractor may be asked to include in his work certain auxiliary functions which require outside advisors. This may range from a simple plot survey to complete design services. These expenses are carried under "Professional Services."

Job Overhead: General Conditions

Each job will entail certain expenses which are chargeable to the job as a whole. These are set forth in the General Conditions, the 14 articles of the AIA usually made a part of the Contract, and the Supplementary Conditions and/or General Requirement Section of the Specifications. These will include such items as superintendence, temporary buildings, enclosures, barricades, offices, toilets, utilities, protection, clean-up, permits, surveys, photographs, tools and equipment, insurance and benefits, sales taxes, surety bonds, and warranties.

Sometimes, these are covered only in a general way and it is up to the contractor to visit the site to determine what is needed. Often the unions and local laws may be more demanding than the architect's specifications, and the contractor must know for what items he is liable.

As discussed above, job overhead and office overhead are best kept separately, although occasionally employees' insurance, tools and equipment and other items may be divided between office expense and job expense.

Many contractors figure a certain percentage of the cost of the job to take care of all overhead expense and plant charges, but this is not the correct method to use and does not give the contractor or estimator a reliable basis for computing the cost of this work.

Each item entering into the cost of job overhead expense should be estimated the same as any other branch of work.

Superintendence.—This should include the salary and traveling expenses of the job superintendent, assistant superintendent, engineers, timekeepers, material clerks, watchmen, waterboys, labor foremen and all other foremen required to be carried on the general contractor's payroll.

The proper method of computing these costs is to estimate the salaries of the various men required at a certain price per week or per month for the entire estimated time they will be required on the job.

Temporary Buildings and Enclosures.—This should include temporary fences, bridges, or platforms, temporary sheds and storerooms, temporary enclosures over hoisting engines, concrete mixers, etc., temporary runways, ladders and stairs in the building before the permanent stairs are installed, temporary dustproof partitions to isolate work areas from building space still occupied, and temporary closures for doors and windows before the permanent doors and windows are in place.

Temporary Office.—This should include the temporary job office for use by the contractor and architect during construction, together with all necessary expense in maintaining same, such as telephone, heat, light, stationery, and other items of expense. These are usually trailers, which may be rented. An 8' x 30' trailer, furnished, will cost around $5,000 but can be rented for around $200.00 per month.

Barricades and Signal Lights.—This should cover the cost of building and maintaining barricades during construction, cost of maintaining signal lights, etc.

Temporary Toilet.—Practically every job must have some provision for sanitary facilities. Local codes usually govern what must be supplied. Chemical, portable toilets will usually suffice. On large jobs, however, a more elaborate setup is usually specified requiring a complete water supply, drain, vent and fixture installation. This work is sometimes included in the plumbing specifications. Chemical toilets can be rented for about $65.00 per month.

In the case of a remodeling job where permission is granted to use the existing facilities, an item should be included in the estimate for periodic and final cleaning of the toilet room and its appurtenances.

Temporary Water.—On most jobs water for construction purposes must be provided by the general contractor for all trades. This item can become a high-cost operation especially in barren, undeveloped localities where either a well must be drilled or water must be hauled into the job. Where water mains are in existence, a plumber must be engaged to tap same, install water meter and run a water line to strategic locations on the job. This work is sometimes included in the plumbing specifications. Also included under this heading are the charges made by the local water department for the volume of water used. Often hydrants can be used at a set fee from the municipality.

Temporary Light and Power.—This item can vary from a small charge for tapping into existing adjacent power lines and running entrance cable into a meter setup from which extension cords may be run as desired to complex installations requiring long run pole lines, transformers, switch gear, etc., as in the case where electric mixers, hoists and other heavy duty electric power equipment is to be used. This work is sometimes included in the electrical specifications. Also included in this item are the power company charges for energy consumed and rental of transformers, etc.

Temporary Heat.—If a job is to run into or through a winter it is very likely that some temporary heat will be required to maintain proper working conditions for the continuance of the construction work. On small jobs a few salamanders may be enough to fulfill the requirements. Large jobs, however, usually require temporary hook-up of radiators to the new and as yet uncompleted heating system. When this occasion arises the boiler usually must be tended by steamfitters working 3 shifts and the labor cost is quite high. Also to be included in this item is the cost of all fuel consumed.

Repairing Streets and Pavements.—This item should cover the repair and replacement of all sidewalks and pavements damaged or destroyed during construction.

Damage to Adjoining Buildings.—There is always a chance of damage to adjacent structures, such as breaking windows, damaging foundations and walls improperly shored, damaging skylights, etc., and these items should be considered when preparing the estimate.

Protecting New Work from Damage During Construction.—During construction there are certain classes of work that must be protected from damage, such as cut stone, marble, terra cotta, granite, etc., also covering newly finished cement or wood floors to prevent damage pending the acceptance of the work by the owner.

Repairing New Work Damaged During Construction.—This item should cover patching of damaged plaster, replacing broken lights of glass, etc. Quite often architects specify the cost of such repair work to be prorated among the various contractors working on the job unless specific responsibility for the damages can be charged to an individual contractor.

Cutting and Patching for Other Trades.—Many specifications require the general or mason contractor to do all cutting and patching of walls and floors for the plumbing, heating and electrical contractors, where it is necessary for the mechanical trades to place sleeves or pipes through the floors or walls after the permanent work is in place.

Removing and Replacing Public Utilities.—In the larger cities it may be necessary to remove or relocate water hydrants, poles and wires of telephone, telegraph, or electric light companies, sewers and drains, or other underground pipes pending the installation of certain portions of the new construction work and in such instances it will be necessary for the contractor to pay for any damage incurred.

Removing Rubbish, Cleaning Floors, Windows, Etc.—During the construction of every building there is always a certain amount of rubbish, such as old concrete, broken bricks, tile, etc., that must be removed at intervals. The most satisfactory method is to estimate so many loads of rubbish at a certain price per load.

If finish floors are to be swept or scrubbed before acceptance of the building by the owner, it is advisable to make a certain allowance to provide for this work. $2.50 per 100 sq. ft. should cover this.

Some architects require the contractor to clean all windows before the building will be accepted. To arrive at this cost, count the windows and allow a certain price for cleaning each window, generally around 5¢ per sq. ft., 2 sides.

Building Permits.—The cost of building permits varies with the city in which the work is to be performed. Some cities base the cost of the permit on the total cost of the work, or the cubical content of the building, while others charge for water consumed, street frontage occupied, and the number of plumbing fixtures, for which an inspection charge is made.

As a general rule a building permit will cost from 1/10 to 1/4 of 1 per cent of the total cost of the building.

In Chicago permit fees are charged as follows:

(a) Single car garage, two car garage, and shelter sheds: For each 1,000 cubic feet or fractional part, $4.00; minimum charge, $10.00.

(b) New buildings or other structures, except those included in (a): For each 1,000 cubic feet of volume or fractional part, $1.25; minimum charge, $20.00. Volume computations under (a) and (b) shall include every part of the building from the basement to the highest point of the roof and include bay windows and other projections.

(c) Alterations and repairs to any structures: For the first $1,000.00 of estimated cost, $7.75; for each additional $1,000.00 of estimated cost, $3.00; minimum charge, $20.00.

(d) Shoring, raising, underpinning or moving any building: For each 1,000 cubic feet of volume or fractional part, 40 cts.; minimum charge, $20.00.

(e) Wrecking any building or other structure more than one story in height: For each 25 lineal feet of frontage or fractional part, $13.00; minimum charge, $20.00.

(f) Fire Escape, erection or alteration: Four stories or less in height, $20.00; each story above four, $2.75.

(g) Canopy or marquee, erection or alteration, $20.00.

(h) Chimneys, isolated or over 50 feet above any roof, $20.00.

(i) Tanks, above roof or tower: 400 gallon capacity or less, $13.00; over 400 gallon capacity, $25.00; structural supports for tank over 400 gallon capacity, $25.00.

(j) Elevators and escalators—Power operated, five floors or less in height, installed or altered, $20.00; each additional floor above five, $4.00; hand operated, $13.50; levelators, $13.50; theater curtains, $50.00; stage or orchestra platforms, $20.00.

Dumbwaiters—Power operated, five floors or less in height, installed or altered, $20.00; each additional floor above five, $4.00. Hand operated, five floors or less in height, $13.00; each additional floor above five, $2.75. Temporary construction towers over 50 feet in height $20.00.

(k) Amusement Devices, mechanical riding, sliding, sailing or swinging: Portable—for each assembly or installation, $12.50. Permanent—installed or altered, $105.00. Temporary seating stands, $70.00.

(l) Ventilating Systems—mechanical supply or exhaust: Capacity, 3,000 cubic feet of air per minute, $10.50; for each additional 1,000 cubic feet per minute, $3.00; increase in capacity, each 1,000 cubic feet per minute, $3.00. Capacity shall be the sum of supply and exhuast. Warm air furnaces, $12.50 each.

(m) Fences over 5 feet high, 100 lineal feet, $10.00; each additional 100 lineal feet or part thereof, $3.00.

(n) Tanks for flammable liquids: Classes I and II: Capacity 121 to 500 gallons, $21.00; each additional 1,000 gallons or fraction thereof, 65 cts.; Classes III and IV: Capacity 551 to 1,000 gallons, $21.00; each additional 1,000 gallons or fraction thereof, 40 cts.

(o) Temporary Platforms for Public Assembly Units, $21.00.

(p) Roof of any building, recoating or recovering, $6.50.

(q) Billboards and signboards—erection, construction or alteration: up to 150 sq. ft. $5.00; 151 to 225 sq. ft. $7.75; 226 to 375 sq. ft. $13.00; for each additional 375 sq. ft. or part thereof $13.00.

A fee of $20.00 shall be paid to the city collector for the approval of plans, and inspection, and test, of any plumbing within any building containing not more than five plumbing fixtures. An additional fee of $3.00 shall be paid for every plumbing fixture in excess of five within such building.

Surveys.—Where the contractor must furnish a survey of the property on which the building is to be erected, a charge of $300 to $400 is usually made for a lot up to 100' x 100'. In the downtown sections of large cities this charge will be substantially more. A topographical survey with 2' contours will run around $200 for an acre; $1500 for 10 acres; and around $50 per acre from 10 to 100.

Photographs.—Where it is necessary to have photographs taken at intervals to show the progress of the work, the cost will vary with the number of views made at one time. In the larger cities it will cost $8.00 to $10.00 for the first photograph and $4.00 to $5.00 for each additional negative. Additional 8" x 10" prints mounted on muslin about $1.00 each. A minimum charge of $50 to $75 per visit is usual.

In the smaller towns and cities this cost may be reduced one-third.

Social Security Tax.—The Federal Insurance Contributions Act, F.I.C.A., requires contractors to pay a tax on all payrolls, both job and office. At the present time the employer's share of this tax amounts to 6.70 per cent of the first $32,400 of taxable wages paid each employee for the year 1982.

The employer is also obliged to deduct 6.70 per cent from each employee's salary or wages but of course this portion of the tax does not increase the contractor's cost —except for the cost of additional clerical help maintaining these records. In many organizations extra employees have been required to keep these records.

Federal Unemployment Tax.—This tax applies to the first $6,000 of wages paid each employee during a calendar year. The rate is 3.2 but you may take a credit not in excess of 2.5 per cent of taxable wages for contributions you paid into State Unemployment Funds. Federal Unemployment Tax is imposed on employers and must not be deducted from wages of employees.

State Unemployment Tax.—This will vary from state to state. Note above that a credit is allowed against the federal rate.

Sales Taxes.—Many states have sales or use taxes applicable to building construction as well as to other commodities. This tax varies with the different states and there are different methods of levying same. In some states, the tax is levied on all building materials and the tax is paid by the contractor when paying for his materials, in other states the tax is applicable to the entire cost of materials entering into the construction of the building, and is paid by the contractor direct to the state. In either event, whether added to the cost of materials purchased by the contractor or whether paid by the contractor to the state, based on all or any portion of the cost of the building, the contractor should add this cost to his estimate, otherwise it will make a big dent in estimated profits. Note that some jobs will be tax exempt which could save as much as 5% on materials. However it is up to the client to establish his tax exemption and the contractor should qualify his bid accordingly.

Job Overhead: Plant and Equipment

Plant overhead includes equipment costs, the cost of transporting equipment to and from the job site, and costs of operating them. These costs may be in the form of depreciation allowances if contractor-owned, or as actual monthly or daily charges, if rented. Generally, equipment used by only one trade will be furnished as part of that subcontract and will not appear as general contractor's overhead.

At the other extreme, some items such as outside hoists and elevators on a hi-rise building will be used by all trades and would be charged similarly to a general conditions expense. But there are many items that will be used by several, although not all trades.

As an example, a job may require a rolling scaffold which would be used by the mechanical trades and electricians installing above ceiling work; by metal lathers and plasterers installing a fancy coved lobby ceiling; by acoustical tilers installing office ceilings; by drywall men installing storage room ceilings, and by painters decorating ceilings. If each trade brought its own scaffold, there would be 7 transportation charges and individual trades would probably be renting on a daily rate. If the general contractor furnishes the scaffold, one transport charge is involved and a less costly monthly rate can be had. This savings can lower the bid and help get the job.

If the contractor owns the scaffold, he can rent to his subs at the going rate and make a profit for himself as well as lowering the job cost. Another advantage in the general contractor's furnishing plant equipment is he can control the quality and safety of what is on the job. With the new Occupational Safety and Health Administration rules on job safety in effect, it is crucial that the general contractor control the quality of equipment on his job or face stiff fines.

If the contractor owns the equipment it may be depreciated for tax purposes. However with the current inflation rate the depreciation allowed by the Internal Revenue Service is not adequate to cover replacement costs and the tax savings may not be so attractive as they once were. On the other hand a high inflation rate may make it possible to buy equipment, use it and resell it for what one paid for it.

The advantages in renting, a short term agreement, or leasing, a long term agreement, include:

a) it eliminates tying up capital or borrowing against bank credit for equipment that most probably will not return so much money as if those assets were applied toward financing another job or taking advantage of cash discounts.

b) it offers flexibility in choosing equipment. One may try one product on one job and another on another job until one finds just the right piece of equipment for each situation. If the item doesn't perform to expectations it can be returned.

c) it eliminates personal property taxes on the equipment.

d) it eliminates maintenance costs and the expense of training personnel to handle those tasks as well as minimizes possible losses due to obsolescence and vandalism.

e) it eliminates costs of shipping, handling and storage and the tying up of trucks and drivers to move this equipment, as this will be done by the lessor.

The question then becomes whether to rent or buy and depreciate plant equipment. In a large community where several rental sources are available, renting is probably preferable, leaving the contractor's money for investment in other forms such as building up capital in order to take on larger jobs, or for investment in land on which a contractor cannot only hope to make a profit on resale, but also can insist his firm be used as contractor for any development on the land.

If the contractor does elect to purchase equipment, he should set up separate records for each item, taking into consideration the Internal Revenue Department's depreciation schedule allowed, the probable life, probable maintenance and storage cost, actual cost of transporting the item to and from the job, and the costs of operating the equipment on the job, entering an hourly rate chargeable for the use of the item on each job on his estimate.

Charges should be sufficient to replace the item when it is worn out, as well as to cover interest charges if a loan is taken out to purchase it and a separate fund (often known as a "sinking" fund) set aside to cover future replacement. These charges should not be carried as profit. If the equipment is idle for long periods the costs could be transferred to general overhead, but this is rather dodging the issue for, if the equipment costs are accurately charged against each job, a contractor can easily tell whether his costs of ownership exceed current rental costs, keeping in mind that the funds tied up in plant ownership could always be making money elsewhere.

Chapter 6 discusses plant and equipment costs in more detail.

Job Overhead: Prepurchase and Prepayment of Materials

With the present inflation rate constantly pushing the cost of building materials ever upward, and with spot shortages of some raw materials a constant threat, a contractor may more and more be confronted with owners who want to protect themselves and order out materials months in advance of their being installed on the job. If this is a stated part of the contract the estimator must make allowances for it.

In most cases the owner will reimburse the contractor upon his properly presenting his request for payment; but if he is not to be paid until the material is installed, the estimator must add the costs of interim financing. Then the material must be suitably stored. Few job sites today are vandalproof even if the material in question is weatherproof. Sometimes the material may arrive before the construction site is available. If the contractor must provide a shed, a watchman, or an off-the-site warehouse, or if the manufacturer is to store the material until it is needed at the site these costs must also be figured and added to the bid or specifically excluded and made part of an extra work order. There are insurance and handling costs that must also be assigned either to the bid or to the owner. And it must be kept in mind that these costs will grow if the job schedule is not adhered to.

Job Overhead: Professional Services

Sometimes an owner may approach a contractor to act as his advisor on matters other than field construction. This can include finding a suitable site; lining up financing; handling rezoning or other legal and governmental problems such as sewer and road extensions, utilities and tax assessments; hiring of architects and engineers to design the project; landscapers to beautify the grounds, and decorators to furnish the interiors.

Some of the initial time spent on such consultations will be charged as good will under general office expense; then, if the job progresses, the contractor should be reimbursed not only for fees for outside services, but also for his and his staff's time spent in administering these services.

A contractor should move slowly in assuming these auxiliary functions. All too often they require much more time, responsibility and cash investment than anticipated; then his major responsibility, the contracting business, is left without adequate supervision and in a cash bind, since reimbursement for preliminary phases often is held up until ground is broken and the interim loan is in effect.

Further, a contractor relies on the good will of the community for his future business. Choosing one real estate office, one architect, one lawyer, one banker, one decorator, over another may limit his future prospects more than the immediate job can justify.

However, a good contractor does tend to be a good citizen and as his experience grows, he is a valuable source of information on building in the broad sense in the community. This knowledge is an asset and it is entirely proper to charge for it on an hourly basis as for any other skill.

BONDS

Bonds are usually issued to cover a single job and are a direct job cost. Some covering only a single trade may be including in a subcontract sum, such as warranties or roofing bonds. Bonds taken out by the general contractor are usually carried under "Overhead: General Conditions." Some of the types of bonds encountered are discussed below.

Rates on the several types and classes of coverage described below vary greatly, depending upon location, type of operations, type of structure, etc. Rates applying to each form of necessary coverage for specific projects are readily obtainable when details of the job exposure are known.

Bid Bonds.—Bids are invited by advertisements and the bidder may have to submit with his bid a certified check, usually for 5 per cent of the bid, or a bid bond, usually for 10 per cent of the bid. This requirement is made to insure that bids will be submitted only by qualified contractors who will not only enter into the contract but qualify for performance and payment bonds.

The bid bond guarantees that as the contract is awarded, the principal will, within a specified time, sign the contract and furnish the required performance and payment bonds. If the principal fails to furnish the required performance bond, there will be a default under the bid bond, and the damage will be the difference between the principal's bid and that of the next highest bidder, not exceeding the penalty of the bid bond.

Surety companies generally follow the practice of authorizing a bid bond only after the performance bond on the contract has been underwritten and approved, so that there have been few defaults under bid bonds.

Perfomance and Payment Bond.—Some owners and most public work usually require that each contractor to whom a contract is awarded shall furnish bonds which contain two obligations: performance to indemnify the owner against loss resulting from the failure of the contractor to complete the work in accordance with the plans and specifications; and payment to guarantee payment of all bills incurred by the contractor for labor or materials for the work.

The federal government, under the Miller act, requires that a contractor furnish two separate bonds, one for the performance and one for the payment of labor and materials. Some states follow the same procedure.

The risk of the surety under the construction contract bond is that should the contractor fail in the performance of his obligation to complete the job and pay the bills for labor or materials. The underwriting procedure is designed to weigh the hazards involved against the contractor's qualifications. If the contractor's ability is deemed sufficient to warrant the assumption that he will be able to complete the work and pay the bills, the application is approved. If not, it is declined.

Some of the most common causes of failure on the part of construction contractors which are unforeseen price rises are abnormal labor costs, fire or other loss not adequately covered by insurance, unforeseen restriction or withdrawal of credit, weather or sub-surface conditions which increase the cost but not the contract price, over extension beyond the contractor's financial capacity, unavailability of materials, failure of subcontractors, death or disability of key men in the contractor's organization, a poor accident record, diversion of funds from one job to pay off overdue accounts on other jobs, inexperience in the type of work involved or dishonesty on the part of the contractor's employees.

Costs of performance bonds depend on amount of contract and length of coverage, as well as the contractor's reputation and whether the coverage is for a portion or complete cost of work.

Bonds are usually issued for a 24 month period and will run on an average job at $12.00 per $1,000.00 of contract cost for the first $500,000; $7.25 per $1,000.00 for the next $2,000,000.00; $5.75 per $1,000.00 for the next $2,500,000.00; $5.25 per $1,000.00 for the next $2,500,000.00 and $4.85 per $1,000.00 over $7,500,000.00. Should the bond need to be extended beyond two years, figure 1% of the regular premium per month.

Maintenance Bonds.—These may be required to guarantee an owner that the general contractor agrees to correct all defects of workmanship and materials for a specified time following completion. Often the above performance bond can be arranged to cover this period which usually lasts one to three years.

License or Permit Bonds.—These are bonds given by a contractor to a public body guaranteeing compliance with local codes and ordinances. If the contractor regularly operates within the area requiring such bonds, this cost should be carried under Office Overhead as it is a normal cost of doing business item. But if it is required for a specific job, it can be charged directly as a job expense.

Subdivision Bonds.—With so many restrictions being applied to uses of sewers and water supplies, shortages of natural gas and restrictions on fuel emissions, more and more public bodies are requiring bonds be filed that guarantee all site improvements and utilities as well as individual buildings themselves will be constructed in accordance with governing codes and ordinances. Unless the contractor is the developer, this is usually filed directly by the owner and he, in turn, is already protected by the contractor's performance bond to the extent that the drawings and specifications comply with the ordinances. As specifications may also include such catchall phrases as "and in accordance with all local codes and ordinances . . . " following detailed instructions of procedures to be followed which may be contrary to such codes, and as there is a growing legal history on developments that have been refused licenses after completion because they do not comply, the contractor should inform himself of all local requirements before entering into any contract so as not to end up in the middle of an expense lawsuit.

Supply and Subcontractor Bonds.— These are guarantees from manufacturers and installers for specific items and workmanship furnished, the cost of which is usually borne by them and included in the sub-price for that trade rather than in Overhead.

Insurance.—A reliable insurance agent is one of the most valuable "subs" a contractor can add to his team. Almost everything can be insured, and has been—at a price. Some types of insurance are required by law; some are made part of the contract; others may be indicated because of some job peculiarity. The following list discusses the more usual types encountered. Rates will obviously vary all over the country and from one part of town to the other, from one suburb to the next. There will be some overlap between insurance costs chargeable to Office Overhead and to Job Overhead, but it is best to keep them as separate as possible.

INSURANCE

Workmen's Compensation.—This insurance provides the contractor's employees with the benefits specified by the various state and federal compensation laws in the event of injury or death arising out of the employment.

Most compensation acts contain at least some provisions specifying benefits for the most common occupational diseases.

Rates vary widely between the several states, the average being from less than 3% in Indiana to over 22% in the District Of Columbia. Sixteen states, primarily in the Southeast average 4–6%; eight, primarily in the West, average over 10 and the remaining twenty five average 6–10%. In those states where medical benefits are limited, it is advisable to carry full or extra-legal medical coverage.

Certain compensation laws are restricted and provide benefits for only specified occupations. The workmen's compensation policy may be broadened to voluntarily pay compensation to employees injured in other occupations.

In addition to providing the statutory benefits, the standard workmen's compensation policy affords employer's liability insurance against claims for bodily injury or disease arising out of employment in the states named in the policy. This coverage against liability not within the scope of the compensation law is subject to $25,000 standard limit which can be increased for a moderate premium charge.

General Workmen's Compensation and Public Liability Insurance Base Rates		
	Approximate Percent Rates	
Jobs Classification	Range	Average
Excavation, Grading, Backfilling, etc.	2 –40	8
Piles, Caissons, Foundations	5 – 36	15
Concrete	1.5 – 20	6
Masonry	2 – 20	6
Carpentry and Millwork	1 – 20	6
Lathing and Plastering	1 – 17	5
Painting and Decorating	2 – 12	6
Sash and Glazing	2 – 18	6
Roofing	4 – 30	13
Acoustical Ceilings	2 – 20	6

Jobs Classification	Approximate Percent Rates	
	Range	Average
Siding - Metal or Composition	2 – 50	12
Marble, Slate, Tile	1 – 28	4
Doors and Hardware	2 – 18	6
Structural Steel	3 – 50	20
Miscellaneous Metals	2 – 18	6
Electric and Hydraulic Elevators	....	6
Heating, Ventilating and Air Conditioning	1 – 12	4
Plumbing and Sewerage	1 – 10	4
Sprinklers	1 – 10	4
Electrical	1 – 9	3

Notes for this table: The rates given in this table represent approximate nationwide average rates. Each state has its own rate for each job classification which should be checked out prior to submitting a proposal. For preliminary estimating purposes, a composite average would run from 7 to 8 percent of the cost of the labor.

An injured employee or his dependents, in the event of death, generally have the right of choosing the benefits of the state where the accident occurred or the state where the employee was hired. In addition, a number of states impose penalties upon an employer who, although subject to the compensation act, has failed to provide insurance coverage. Accordingly, it is important that you discuss these points with your insurance representative before entering a state not covered by your policy and before hiring anyone in such other state.

The general insurance rates in the preceding table are representative rates to be used as a guide only. Rates and experience modifications are subject to frequent and sometimes radical changes and should be verified with an insurance representative for each specific project.

Major job classifications are listed in the table above. Consult your insurance representative for specific classifications.

Reference has been made herein to base or manual rates and to experience modification of the manual rates or premium. A program of Retrospective Rating is available on an optional basis. This rating plan, superimposed on experience rated premiums is usually attractive and appropriate to the larger contractors. Subject to certain state limitations as to combinations permissible, it is available on an intrastate or interstate basis for Workmen's Compensation and the Liability lines either separately or in combination.

Your insurance representative will be glad to explain the details of this rating plan as well as any other features of coverage or rating on these various lines of insurance.

Contractor's Public Liability Insurance.—This policy protects against liability for bodily injury to a member of the public arising out of the contractor's operations and against property damage liability. It can be had under either a schedule general liability form policy or a comprehensive general liability policy form.

Under the schedule form, the several hazards to be insured are specifically named and the coverage may be applied either to specified locations and operations of the insured contractor. The comprehensive form is broader in nature and covers all the usual hazards of a general liability nature other than those specifically excluded.

Under either form, consideration should be given to at least the lines of coverage briefly outlined below:

Owners, Landlords and Tenants, Premises Exposure. This hazard includes such exposures as offices, warehouses and fixed locations in contrast to specific contracting operations as such. The usual rating basis is a charge per 100 sq. ft. of area and in the case of the usual contractor, the premium for this hazard is relatively minor compared to that for the operations hazard.

Products Liability. Completed operations coverage applying to accidents which occur after the contractor's operations have been completed. Rates per $1,000 of receipts or contract price vary over a wide range according to the operations being conducted. These rates are frequently the subject of individual account rate determination rather than the application of any base Manual rates.

Owners Protective. Covering the liability of the owner for injuries to the public arising out of operations of independent contractors. Characteristic rates for general construction operations in most states are .018 bodily injury and .011 property damage per $100 of contract cost. These rates apply to the first $500,000 of cost on each specific project. Corresponding rates for the next $500,000 of contract price are .009 and .006; for contract cost over $1,000,000–.002 and .002. Rates in Illinois and New York are somewhat higher.

Contractor's Protective. Covering liability of the contractor for injuries to the public arising out of operations of subcontractors. Rating basis and rate levels stated above for owners' protective are applicable under this line to sublet contract costs.

Additional features can be incorporated into the general liability policy either for application on an individual exposure basis or to be applied on an automatic or blanket basis.

Contractual Coverage. Applying to the liability of others assumed by the contractor. Rates depend on the nature of exposure and degree of liability assumed.

Accident Defined Coverage. Broadening the coverage to apply to continuous or repeated exposure giving rise to injury or damage in contrast to the "caused by accident" basis on which most General Liability Policies are written. Cost is usually expressed as a percentage of policy premium.

Personal Injury Coverage. Broadening the usual "bodily injury" nature to include such personal injury as libel, slander, false arrest, malicious prosecution, etc. Pricing varies with the degree of exposure and extent of coverage afforded.

Malpractice Coverage. Covering contractor for liability arising out of error or mistake in professional treatment in on-the-job hospital or medical facilities under his control. This line is individually rated.

Increased Limits of Liability. The general liability rates referred to herein and in subsequent pages are at standard limits of $5,000 per person, $10,000 per accident for bodily injury and at $5,000 per accident, $25,000 aggregate for property damage. The percentage increase factors for various higher limits may be secured from your insurance representative.

Automobile Liability Coverage. Bodily injury and property damage coverage for owned vehicles, non-owned vehicles, hired vehicles or independent contractor's may be secured under a schedule form or a comprehensive form of automobile liability policy. Coverage can be selected either on the basis of a listing of owned automobiles or on an automatic basis whereby coverage applies to all owned vehicles. The rating basis for owned vehicles is based on size and weight, usage of the vehicle and geographical territory involved.

Physical damage coverage (fire, theft and collision) on owned vehicles may be afforded either under the same policy which affords automobile liability coverage or as a separate policy.

Property Damage Insurance. The contractor should make a careful study of what his responsibilities are in regard to damage to other property and to the work completed so his policies will mesh with those carried by the owner.

If provision is not made in the specifications that fire insurance will be carried by the owner during construction, it is important that the contractor protect his interests by carrying some type of builders' risk insurance. This is normally written on a standard fire policy together with an extended coverage endorsement which extends the fire and lightning provisions of the standard fire policy to include windstorm, hail, explosion, riot, smoke damage and damage from aircraft and vehicles.

An additional endorsement, vandalism and malicious mischief, may be obtained. Some companies afford a broader coverage of an all risk nature although there are a number of exclusions.

There are two common builders' risk forms that may be written.

Builders' risk reporting form: This form affords coverage for the contractor from the time material is placed on the job site until the building is completed. Under this form the contractor reports to the insurance company the actual value of the building at the end of each month as construction progresses and pays a premium each month on the additional values that go into the building.

Completed value builders' risk form: This form of coverage today is normally preferred by a contractor as it gives him 100 per cent coverage without the necessity of keeping records as to the value of the building as it progresses or the necessity to report at the end of each month.

This form is written for the completed value of the building. The normal rate charged is about 55 per cent of the rate on the basis that at the beginning of the job there is little value and at the end you have the full value. Normally the premium under the two types of policies is remarkably close.

The interests of a general contractor can be covered and he may exclude or include the interests of subcontractors who also may be employed in the construction of the building.

Temporary structures, tools and equipment belonging to the contractor and on the premises of the building site may also be included in the coverage for the hazards insured.

Costs of such insurance vary by location, type of construction and available public fire protection and may be obtained from your insurance man locally.

Contractors equipment floater: Many contractors possess many thousands of dollars in contracting machinery and equipment. To properly protect these values such equipment and machinery is usually insured under a contractors' equipment floater policy.

Insurance under this form is quite flexible so that many situations may be insured to fit the contractor's particular needs. The insurance is normally written on a named peril policy which includes loss or damage by fire or lightning, explosion, windstorm, flood, earthquake, collapse of bridges and culverts, theft, landslide, collision and upset. Or if desired, a policy may be issued protecting from practically all risks of physical loss or damage. Such contracts usually write a deductible clause.

Installation Floater: the contractor who specializes in plumbing, heating, air conditioning or electrical equipment may find an installation floater more suitable to his needs. This type of policy may be written for an individual job or may be written to cover all the jobs he may secure during a year's time.

Coverage may be on a so-called "named peril" policy usually including the hazards of transportation of his material to the job site and covering fire and lightning, windstorm and hail, explosion, riot and civil commotion, smoke damage and aircraft and vehicle damage at the job site. He may also include coverage for vandalism and theft if desired.

PROFIT

The amount of profit to be added to the estimated cost of the work is a question which a contractor must answer individually for each bid. There is no set amount that can be added. It all depends on local conditions, competition, and how badly the job is wanted.

On small jobs, alterations, remodeling and similar work, a contractor is justified in adding 20 to 25 per cent to the actual cost for profit, but whether he is able to obtain this amount each contractor must decide for himself.

On new work, where it is possible to estimate cost with a fair degree of accuracy, a contractor is entitled to 10 per cent on the actual cost of the work (job overhead included in the actual cost of the job) but when submitting competitive figures it is safe to say there are many jobs let on a 5 per cent instead of a 10 per cent basis. A contractor is entitled to 10 per cent profit but getting it is another matter.

Another item that has caused considerable discussion is, should a contractor add a certain percentage to his estimate to take care of contingencies? Yes and No.

Yes—because it is not necessary to be quite so careful in the preparation of the estimate and if any items have been overlooked, there is a fund to take care of omissions. But a contractor who is not careful in the preparation of his estimates never knows how much should be added to provide for forgotten items or contingencies.

No—because if you add a percentage for contingencies, it is quite possible that one or more of your competitors has not, and another job is lost.

The only safe way is to be as careful as possible in listing all items from the plans and specifications. Include them all and it would not be necessary to add for contingencies. This includes everything to be furnished and all labor costs including job overhead expense, and a percentage for profit.

An item often overlooked but which can become a major expense is the finishing up of a job and making corrections after the job is completed.

Often, there is confusion on the job at the end and it is difficult to say who chipped the paint, scratched the door, broke the window, stole the light bulb, broke the door handle, backed into the fencing, etc., as most of the trades are on the job doing touch-up work, the owner is moving in, his decorator is hanging drapes and his children running wildly through the corridors. But the general contractor gets the complaint first and, in the interest of having a satisfied client, will probably assume the responsibility for correcting this.

In a warehouse, such problems may be minimal; in a fine residence, you might be called back daily for weeks.

Other factors an estimator must evaluate in setting up his bid are the weather conditions that are apt to prevail at the time of construction; possible rises in labor rates and the possibility of strikes if labor contracts expire during the construction period; material shortages and strike actions at the manufacturers level; the money supply and its cost; and public relations with the community in which the project is located and possible job stoppage or slowdowns because of environmental or safety concerns which the drawings and specifications fail to cover.

ESTIMATE CHECK LIST

Too many estimates are done under pressure against a deadline. If at all possible, a second party should check all arithmetic on the original estimate sheets and the original figures against those copies on the final bid sheet. It is not unusual to receive a bid from a subcontractor with an obvious error in the figures. This is an indication that the bid was not checked prior to tendering and it could result in a severe impact on the company offering the bid and the general contractor utilizing the bid as part of the general estimate.

Storage, transportation and handling costs are often overlooked. It is often well to qualify a bid as to the time of year a job is to be started. A bid given in March anticipating an April start and October finish should so state, for if the contract is delayed for any reason so that a job starts in October and runs through the winter, allowances must be added for temporary enclosures and heat, idle days or weeks, and less efficient work conditions.

Furthermore, with inflation such a major factor today, labor and material costs can escalate 4 to 6 per cent in six months.

The estimate should be carefully checked for plant equipment costs with each subcontractor being carefully informed as to what will be furnished to him. Architects specifications are usually very general on this matter, often stating in Work Included " ... all labor, materials, and related items ... " to cover plant equipment. It is left up to the general contractor to have a clear understanding with his subs on this.

Check your profit. How much of this is certain; how much contingent on optimism? Set forth a definite percentage for contingencies and try to hold to it. Keep the profit apart. If there is a questionable item, alert the owner and/or architect; then qualify your bid accordingly. Most proposals provide a space for this. Give a firm price on what is clear. Set up an allowance to cover what is in doubt with a proposal, stating what is to be charged against the allowance and why.

Finally, make certain all items are covered. Check not only the drawings and specifications, but also any bulletins addenda. An addendum changing wood paneling from full height to wainscot in an office means a savings in millwork. But now, of course, the drywall above needs to be taped and finished for painting. So both the drywall and painting contractors must be alerted, although the addendum only refers to a change in the millwork.

A check list of items is given below, based on the Uniform System for Construction.

It is intended as a reminder for the estimator, in order that the estimate may be compared with this list to see that all items have been included, and to check against his subcontractor bids.

Please bear in mind that not all of the items on the following pages will be found in every building because the list is intended to cover every type of structure from the smallest residence to the largest skyscraper, commercial or industrial plant.

Compare it with your estimate before submitting your bid as it may be the means of including some items that would otherwise be omitted from your estimate.

This may require a few minutes extra time but always remember it is much better to be safe than sorry.

Forgetting has cost many a contractor his profit.

DIVISION 1
GENERAL REQUIREMENTS

Summary of Work:
  Work under this Contract
  Work under Separate
    Contracts
  Work by Others
  Work by Owner
  Permits
  Taxes
  Warranties
  Fees and Notices

Indemnification
Insurance
  Liability
  Loss of Use
  Property Damage
Bonding
Superintendance
Timekeepers
Material Clerks
Foremen
Watchmen
Coordination of Trades
Cutting and Patching

Allowances
Alternates
Required
Proposed
Project Meetings
Preconstruction Conferences
Progress Meetings
Job Site Administration
Submittals
Construction Schedules
Progress Reports
Survey Data
Shop Drawings
Samples
Photographs
Operation and Maintenance
Data
Schedule of Values
Quality Control
Tests and Lab Services
Inspection Services
Temporary Facilities
Temporary Utilities
Temporary Electricity
Temporary Lighting
Temporary Heat &
Ventilation
Temporary Telephone Service
Temporary Water
Temporary Sanitary Facilities
Temporary First Aid Facilities
Construction Aids
Construction Elevators and Hoists
Temporary Enclosure
Swing Staging
Barriers
Fences
Tree and Plant Protection
Guardrails & Barricades
Security
Access Roads
Special Controls
Noise Control
Dust Control
Water Control
Pest Control
Rodent Control
Pollution Control
Traffic Regulation
Traffic Signals
Flagmen
Flares and Lights
Parking

Project Identification
Temporary Field Offices
Products
Transportation
Handling
Storage at Job
Storage at Warehouse
Protection
Project Closeout
Cleaning Up
Record Documents
Start-up
Punch List
Touch-up and Repair
Transportation Allowance

**DIVISION 2
SITE WORK**

Subsurface Exploration
Borings
Core Drilling
Penetration Tests
Clearing
Structure Moving
Clearing and Grubbing
Tree Pruning
Shrub and Tree Relocation
Demolition
Buildings
Appertanences
Interior
Earthwork
Site Grading
Rock Removal
Embankment
Excavating and Backfilling
Trenching, Backfilling, and
Compacting
Structural Excavation, Backfill,
and Compaction
Roadway Excavation, Backfill, and
Compaction
Pipe Boring and Jacking
Waste Material Disposal
Soil Compaction Control
Soil Stabilization
Soil Treatment
Termite Control
Vegetation Control
Pile Foundations
Pile Driving

Pile Load Tests
Wood Piles
Precast Concrete Piles
Prestressed Concrete Piles
Compacted Concrete Piles
Steel H-Section Piles
Steel Pipe Piles
Concrete-Filled Steel Shell Piles
Caissons
Drilled
Excavated
Shoring
Steel Sheeting
Walers and Shores
Cribbing
Piling and Lagging
Underpinning
Site Drainage
Subdrainage Systems
Foundation Drainage
Underslab Drainage
Drainage Structures
Drainage Pipe
Dewatering
Wellpoints
Relief Wells
Erosion Control
Site Utilities
Gas Distribution System
Gas Transmission Lines
Oil Distribution System
Oil Transmission Lines
Water Distribution System
Water Transmission Lines
Steam Distribution
Hot Water Distribution
Chilled Water Distribution
Water Wells
Sewage Lagoons
Paving and Surfacing
Mudjacking
Paving
Crushed Stone Paving
Asphalt Concrete Paving
Brick Paving
Portland Cement Concrete Paving
Bituminous Block Paving
Repair and Resurfacing
Pavement Sealing
Pavement Marking
Curbs and Gutters
Walks
Synthetic Surfacing

Site Improvements
Chain Link Fences
Wire Fences
Wood Fences
Guardrails
Signs
Traffic Signals
Culvert Pipe Underpasses
Playing Fields
Recreational Facilities
Fountains & Equipment
Irrigation System
Sprinkler Systems
Site Furnishings
Rubble Site Structures
Railroad Tie Structures
Lighting
Landscaping
Soil Preparation
Lawn Seeding & Sodding
Trees & Shrubs
Plants
Beds
Railroad Work
Trackwork
Ballasting
Service Facilities
Traffic Control
Marine Work
Docks
Boat Facilities
Protective Structures
Seawalls
Groins
Jetties
Dredging
Tunneling
Excavation
Grouting
Support Systems
Rockbolting

DIVISION 3
CONCRETE

Accessories
Inserts
Anchors
Ties
Hangers
Hardeners
Abrasives
Coloring

Formwork
  Liners and Coatings
  Wood Forms
  Prefabricated Forms
  Panel Forms
  Pan Forms
  Steel Forms
  Fiberglass Forms
  Stair Forms
Expansion and Contraction Joints
  Metal
  Plastic
Reinforcement
  Steel Bar
  Welded Fabric
  Stressing Tendons
Cast-In-Place Concrete
  Curing
  Concrete Types
    Premixes
    Lime
  Shrinkage Compensating
  Lightweight
  Insulating
  Lightweight Structrual
  Heavyweight
  Prestressed
  Post-tensioned
Specially Finished Concrete
  Exposed Aggregate
  Tooled
  Blasted
  Heavy Duty Floor Finishes
  Grooved Surface
Specially Placed Concrete
  Shotcrete
  Grout
  Catalyzed Metalic Grout
  Non-shrink Grout
  Non-corrosive Grout
  Epoxy Grout
Precast Concrete
  Wall Panels
  Tilt-up Panels
  Structural
  Deck Panels
  Prestressed
    Standard
    Custom
Cementitious Decks
  Gypsum
  Wood Fiber
Installation Equipment

Scaffolding
Machinery
Hoists
Temporary Heat
Tarpaulins
Straw
Special Handling and Storage

DIVISION 4
MASONRY

Mortar
  Cement and Lime
  Acid Resisting
  Premixed
Accessories
  Joint Reinforcement
  Anchors and Ties
  Control Joints
  Lintels
Unit Masonry
  Brick
    Common
    Face
    Enamelled
    Fire
    Standard Sized
    Special Sized
  Adobe
  Concrete Units
  Reinforced Units
  High-Lift Grouted
  Preassembled Panels
  Clay Backing Tile
  Clay Facing Tile
  Ceramic Veneer
  Terra Cotta Veneer
  Mechanically Supported Veneer
  Glass Unit
  Gypsum Unit
  Sound Absorbing
Stone
  Rough
  Cut and/or Carved
  Marble
  Limestone
  Granite
  Sandstone
  Simulated
  Cast
  Flagstone
  Natural Veneer
Restoration and Cleaning

Cleaning
Restoration
Pointing
Coating
Refactories
Flue Liners
Corrosion Resisting
Combustion Chamber
Installation Equipment
Scaffolding
Machinery
Special Handling and Storage

DIVISION 5
METALS

Structural Metal Framing
Structural Steel
Beams and Girders
Roof Trusses
Crane Tracks
Grillage
Bolts and Rivets
Welding
Structural Aluminum
Metal Joists
Standard Joists
Custom Joists
Aluminum
Framing Systems
Space Frames
Metal Decking
Roof
Floor
Lightgage Framing
Metal Studs
Metal Joists
Metal Fabrications
Metal Stairs
Handrails and Railings
Pipe and Tube Railings
Gratings
Castings
Manhole Covers
Fireplace Accessories
Bumpers
Joist Hangers and Straps
Wall Plates
Package Receivers
Window Guards
Sidewalk Doors
Nosings
Rough Hardware For Above

Ornamental Metals
Stairs
Handrails and Railings
Sheet Metal Fabrications
Cast Iron Work
Bronze
Aluminum
Stainless
Expansion Controls
Shop Painting
Field Painting
Material Handling
Trucking
Unloading
Erecting Machinery

DIVISION 6
CARPENTRY

Rough Carpentry
Framing
Trusses
Columns
Beams and Girders
Rafters
Floor and Ceiling Joists
Porch Framing
Rough and Exterior Stairs
Stud Walls
Bridging
Lintels
Sheathing
Roof
Wall
Insulating
Preassembled Components
Diaphragms
Accessories and Hardware
Heavy Timber Construction
Timber Trusses
Mill-Framed Structures
Pole Construction
Trestles
Prefabricated Structural Wood
Glue Laminated Structural Units
Glue Laminated Decking
Wood Trusses
Wood-Metal Joists
Finish Carpentry
Millwork
Exterior
Facias, Cornices and Trim
Columns

Shutters and Blinds
Railings
Stoops and Porch Work
Cupolas
Interior
Base and Quarterround
Chair and Picture Moldings
Cornices
Strip Paneling
Ceiling Beams
Closet Trim
Laminated Plastic
Wood Treatment
Pressure Treated Lumber
Preservative Treated Lumber
Fire Retardant Lumber
Architectural Woodwork
Custom Cabinet Work
Paneling
Solid Wood
Architectural Plywood
Ornamental Stair Work
Custom Trim
Decorative Wallboard

DIVISION 7
THERMAL AND MOISTURE PROTECTION

Waterproofing
Membrane
Elastomeric
Fluid Applied
Liquid
Bentonite Clay
Metal Oxide
Damproofing
Bituminous
Silicone
Water Repellant Coatings
Cementitious
Vapor Barriers
Bituminous
Laminated
Plastic
Insulation
Loose Fill
Rigid
Fibrous
Reflective
Foamed-in-place
Sprayed-On
High and Low Temperature

Roof and Deck
Perimeter
Under Slab
Shingles and Roofing Tiles
Asphalt
Asbestos-cement
Wood
Slate
Porcelain Enamel
Metal
Clay Tiles
Cement Tile
Preformed Roofing and Siding
Metal Siding
Composite Panels
Preformed Plastic
Cladding/Siding
Wood
Composition
Asbestos-cement
Plastic
Membrane Roofing
Built-Up Bituminous
Prepared Roll
Elastic Sheet
Fluid Applied
Traffic Topping
Flashing and Sheet Metal
Roofing
Flashing and Trim
Specialties
Gutters and Downspouts
Gravel Stops
Roof Accessories
Skylights
Plastic
Metal Framed
Hatches
Gravity Vents
Prefabricated Curbs
Expansion Joints
Sealants
Joint Fillers and Gaskets
Sealants and Caulking
Material Handling
Hoists
Scaffolds
Special Storage

DIVISION 8
DOORS AND WINDOWS

Metal Doors and Frames

Hollow Metal
Aluminum
Stainless Steel
Bronze
Wood and Plastic
Flush
Panel
Plastic Faced
Plastic Formed
Special Doors
Sliding Metal Fire Doors
Metal Clad
Coiling Doors
Coiling Grilles
Folding
Accordion
Flexible
Overhead
Sliding Glass
Safety Glass
Sound Retardant
Screen and Storm
Entrances and Storefronts
Aluminum
Stainless
Bronze
Revolving Doors
Metal Windows
Steel
Aluminum
Stainless
Bronze
Operating Hardware
Wood and Plastic Windows
Milled Wood
Preglazed
Plastic Faced
Plastic Formed
Hardware and Specialties
Finish Hardware
Operating Hardware
Automatic Door Equipment
Weatherstripping and Seals
Thresholds
Glazing
Float Glass
Sheet Glass
Tempered
Wired
Rough and Figured
Processed
Coated
Laminated
Insulating

Mirror
Plastics
Accessories
Window Wall/Curtain Wall
Steel
Aluminum
Stainless
Bronze
Wood
Installation Machinery
Scaffolding

**DIVISION 9
FINISHES**

Lath and Plaster
Furring and Lathing
Gypsum Plaster
Cement Plaster
Acoustical Plaster
Gypsum Wallboard
System Types
Required Accessories
Metal Furring
Metal Stud Partitions
Tile
Ceramic
Mosaic
Quarry
Marble
Glass
Plastic
Metal
Terrazzo
Portland Cement
Precast
Conductive
Plastic Matrix
Acoustical Treatments
Ceilings
Panels
Tiles
Wall Treatments
Insulation and Barrier
Ceiling Suspension Systems
Wood Flooring
Strip
Parquet
Plywood Block
Resilient Systems
Industrial Block
Resilient Flooring
Cementitious Underlayment
Asphalt Tile

Asbestos
Vinyl
Rubber
Sheet
Fluid Applied
Carpeting
Cushioning
Wool
Synthetic
Blend
Custom
Tile Cut
Special Flooring
Magnesium Oxychloride
Epoxy-marble Chip
Elastomeric Liquid
Heavy-duty Concrete Toppings
Armored Floor
Brick
Laminated Plastic
Special Coatings
Abrasion Resistant
Cementitious
Elastomeric
Fire-resistant
Sprayed Fireproofing
Aggregate Wall Coating
Painting
Exterior
Flat
Enamel
Wall
Trim
Porches
Stairs
Windows
Doors
Railings
Sheet Metal Work
Steel and Cast Iron Work
Interior
Preliminary Preparation
Flat
Semi-gloss
Enamel
Varnish
Plastic
Graining
Sanding Between Coats
Washing
Touch-up
Wall Coverings
Vinyl-Coated Fabric
Cork

Paper
Fabric
Asbestos
Flexible Wood
Prefinished Panels
Adhesives
Temporary Equipment
Ladders
Scaffolds
Rolling Platforms

DIVISION 10
SPECIALTIES

Chalkboards and Tackboards
Compartments and Cubicles
Hospital
Toilet
Metal
Plastic
Stone
Shower and Dressing
Louvers and Vents
Grilles and Screens
Wall and Corner Guards
Access Flooring
Fireplaces, Prefabricated
Flagpoles
Identifying Devices
Lockers
Protective Prefab Covers
Walkway
Car
Postal Specialties
Prefab Partitions
Special Shelving
Scales
Sun Control Devices
Telephone Enclosures
Toilet and Bath Accessories
Wardrobe Specialties

DIVISION 11
EQUIPMENT

Built-In Maintenance Equipment
Vacuum
Window Washing
Bank and Vault
Commercial
Checkroom
Darkroom
Ecclesiastical
Educational

Food Service
Athletic
Industrial
Laboratory
Laundry
Library
Medical
Mortuary
Musical
Parking
Waste Handling
Loading Dock
Detention
Residential
Theater and Stage
Registration

DIVISION 12
FURNISHINGS

Artwork
Cabinet Furniture
Window Treatments
Fabrics
Rugs and Mats
Seating

DIVISION 13
SPECIAL CONSTRUCTION

Air Supported Structures
Integrated Assemblies
Audiometric Rooms
Clean Rooms
Hyperbaric Rooms
Incinerators
Instrumentation
Insulated Rooms
Integrated Ceilings
Nuclear Reactors
Observatory
Prefabricated Structures
Radiation Protection
Sound and Vibration Controls
Vaults
Swimming Pools
Saunas

DIVISION 14
CONVEYING SYSTEMS

Dumbwaiters
Elevators

Hoists and Cranes
Lifts
Material Handling Systems
Turntables
Moving Stairs and Walks
Pneumatic Tube Systems
Powered Scaffolding

DIVISION 15
MECHANICAL

General Provisions
  Pemits
  Notices
  Fees
  Excavation and Backfill
  Tests and Balances
  Identification
  Maintenance Contracts
  Demonstration
Basic Materials and Methods
  Steel Pipe
  Cast Iron Pipe
  Copper Pipe
  Plastic Pipe
  Glass Pipe
  Stainless Steel Pipe
  Aluminum Pipe
  Hose
  Piping Specialties
  Gaskets
  Swivel Joints
  Strainers, Filters, and Driers
  Vent Caps
  Traps
  Vacuum Breakers
  Shock Absorbers
  Supports, Anchors, and Seals
  Anchors
  Wall Seal
  Flashing and Safing
  Hangers and Supports
  Gate Valves
  Blowdown Valves
  Butterfly Valves
  Ball Valves
  Globe Valves
  Refrigerant Valves
  Stop Cocks
  Curb Stops
  Hydrants
  Check Valves
  Swing Check Valves
  Backwater Valves

Vertical Check Valves
Stop and Check Valves
Faucets
Washer Outlets
Self Contained Control Valves
Pressure Regulating Valves
Pressure Relief Valves
Automatic Temperature and
  Pressure Relief Valves
Solenoid Valves
Tempering Controllers
Photo Lab Tempering Controller
Mixing Station
Refrigerant Control Valves and
  Specialties
Feed Water Regulator
Pumps
  Centrifugal
  Rotary
  Turbine
  Reciprocating
  Sump Pump
  Submersible Pump
  Pneumatic Ejector
  Compressors
  Vacuum Pumps
  Air Compressors
Vibration Isolation and Expansion
  Compensation
  Vibration Isolation
  Expansion Joints
  Flexible Connections
Meters and Gages
  Temperature Gages
  Pressure Gages
  Flow Measuring Devices
  Liquid Level Gages
Tanks
  Steel Tanks
  Plastic Tanks
  Cast Iron Tanks

INSULATION
General
Cold Water Piping
Chilled Water Piping
Refrigerant Piping
Hot Water Piping
Steam and Condensate Return
  Piping
Underground Piping
Outside Piping
Duct
Breeching

Equipment
Water Supply and Treatment
Pump and Piping System
Booster Pumping Equipment
Water Reservoirs and Tanks
Water Treatment
Filtration Equipment
Aeration Equipment
Water Softening Equipment
Chemical Feeding Equipment
Chlorinating Equipment
Metering and Related Piping
**Waste Water Disposal
and Treatment**
Sewage Ejectors
Grease Interceptors
Lift Stations
Septic Tanks
Drainage Fields
Sewage Treatment Equipment
Screens and Skimming Tanks
Sedimentation Tanks
Filtration Equipment
Aeration Equipment
Sludge Digestion Equipment
Plumbing
Water Supply System
Chilled Water Piping Systems
Distilled Water Piping System
Compressed Air Piping System
Oxygen Piping System
Helium Piping System
Nitrous Oxide Piping System
Vacuum Piping System
Laboratory Gas Piping System
Compressed Industrial Gas Piping
  Systems
Central Soap Piping System
Soil Piping System
Waste Piping System
Roof Drainage System
Chemical Waste Drainage System
Industrial Waste Drainage System
Process Piping Systems
Floor and Shower Drains
Roof Drains
Cleanouts and Cleanout Action
  Covers
Domestic Water Heaters
Aftercoolers & Separators
Stills
Anti-syphon Equipment
Sediment Interceptors
Laundry/Utility Units

Packaged Waste, Vent, or Water
  Piping Units
Domestic Water Conditioners
Special System Accessories
Gas Accessories
Plumbing Fixtures and Trim
Urinal Flush Valves
Special Fixtures & Trim
Fixture Carriers
Domestic Watercoolers
Wash Fountains
Showers
Receptors
Pool Equipment
Circulation and Filtration
  Equipment
Pool Drains, Inlets, and Outlets
Pool Cleaning Equipment
Chemical Treatment Equipment
Fountain Piping and Nozzles
Special Equipment
Fire Protection
Sprinkler Equipment
Foam Equipment
Halon Systems
CO_2 Equipment
Standpipe and Fire Hose
  Equipment
Fire Hose Connections
Fire Hose Cabinets and Accessories
Fire Hose Reels
Fire Hose
Portable Extinguishers
Fire Blankets
Fire Extinguisher Cabinets and
  Accessories
Hood and Duct Fire Protection
Non-electrical Alarm Equipment
Air Distribution
Furnaces
Direct Fired Furnaces
Cast Iron Furnaces
Steel Furnaces
Rooftop Furnaces
Direct Fired Unit Heaters
Direct Fired Duct Heaters/
  Reheaters
Fans
Centrifugal Fans
Propeller Fans
Attic Exhaust Fans
Fly Fans
Axial Flow Fans

Induced Draft Fans
Exhaust Fans
Power Roof Ventilators
Power Wall Ventilators
Roof Ventilators (connected to
  ductwork)
Air Handling Units (without coils)
Air Curtains
Ductwork
Low Pressure Steel Ductwork
High Pressure Steel Ductwork
Nonmetallic Ductwork
Special Ductwork
Prefabricated Insulated Ductwork
Flexible Ductwork
Duct Lining
Duct Hangers and Supports
Special Ductwork Systems
Tailpipe Exhaust Equipment
Dust Collection Equipment
Paint Spray Booth System
  Equipment
Fume Collection System
  Equipment
Breeching ; and Smokepipe
Duct Accessories
Manual Dampers
Gravity Backdraft Dampers
Barometric Dampers
Fire Dampers
Smoke Dampers
Turning Vanes
Distribution Devices
Duct Access Panels and Test Holes
Outlets
Wall and floor Diffusers
Ceiling Diffusers
Ceiling Air Distribution System
Light Troffer-Diffusers
Warm Air Baseboard
Cabinet Diffusers
Air Floors
Roof Mounted Air Inlets & Outlets
Air Inlet and Outlet Louvers
  (connect to ductwork)
Air Treatment Equipment
Disposable Filters
Permanent Filters
High Efficiency Filters
Roll Filters
Oil Bath Air Filters
Electronic Air Filters
Air Washers

Dust Collectors
Fume Collectors or Dispensers
Sound Attenuators
Special Devices
Power or Heat Generation
Fuel Handling Equipment
Oil Storage Tanks, Controls, and
Piping
L-P Gas Tanks, Controls, and
Piping
Ash Removal System
Lined Breechings
Lined Prefabricated Chimneys and
Stacks
Exhaust Equipment
Draft Control Equipment
Boilers
Cast Iron Boiler
Firebox Boiler
Scotch Marine Boiler
Water Tube Boilers
Absorption Boiler
Burners and Controls
Stokers
Fuel Preheaters
Boiler Accessories
Boiler Feedwater Equipment
Packaged Boiler Feed Pump
System
Deaerators
Refrigeration
Refrigerant Compressors
Centrifugal Compressor
Rotary Compressor
Reciprocating Compressor
Condensing Units
Air Cooled Condensing Units
Water Cooled Condensing Units
Evaporative Condensing Units
Chillers
Reciprocating Chillers
Air Cooled Chillers
Ethylene Glycol Chillers
Centrifugal Chillers
Absorption Chillers
Rotary Chillers
Cooling Tower (Propeller type)
Cooling Tower (Centrifugal type)
Ice Bank
Special Ice Making Equipment
Commercial Ice Making
Equipment
Evaporators

Unit Coolers
Condensers
Refrigeration Accessories
Liquid Heat Transfer
Hot Water Piping System
Chilled Water Piping System
Steam Supply and Return Piping
System
Radiant Heat System
Snow Melting System
Hot Water Specialties
Steam Specialties
Condensate Pump and Receiver
Set
Heat Exchangers
Storage Water Heater
Converter
Clean Steam Heat Exchanger
Water Heat Reclaim Equipment
Terminal Units
Induction Units
Radiant Panels
Coils
Baseboard Units
Finned Tube
Convectors
Radiators
Unit Heaters
Fan Coil Units
Unit Ventilators
Air Handling Units (with coils)
Packaged Heating and Cooling
Packaged Heat Pump
Humidity Control
Humidifiers
Centrifugal Type Humidifier
Dehumidifiers
Desicant Dehumidifiers
Process Heating
Storage Cells
Special Devices
Controls and Instrumentation
Electrical and Interlocks
Identification
Inspection, Testing, and Balancing
Control Piping, Tubing, and
Wiring
Control Air Compressor and Dryer
Control Panels
Instrument Panelboard
Primary Control Devices
Thermostats
Humidistats

Aquastats
Relays and Switches
Timers
Control Dampers
Control Valves
Control Motors
Sequence of Operation
Recording Devices
Alarm Devices
Special Process Controls

DIVISION 16
ELECTRICAL

General
Tests
Demonstration
Identification
Basic Materials and Methods
Raceways
Conduits
Bus Ducts
Underfloor Ducts
Cable Trays
Wires and Cables
Wire Connections and Devices
Pulling Cables
Outlet Boxes
Pull and Junction Boxes
Floor Boxes
Cabinets
Panelboards
Switches and Receptacles
Motors
Motor Starters
Disconnects (motor and circuit)
Overcurrent Protective Devices
Supporting Devices
Electronic Devices
Power Generation
Generator
Engine
Reciprocating Engine
Turbine
Cooling Equipment
Exhaust Equipment
Starting Equipment
Automatic Transfer Equipment
Power Transmission
Substation
Switchgear
Transformer
Vaults

Manholes
Rectifier
Converter
Capacitor
Service and Distribution
Electric Service
Underground Service
Service Entrance
Emergency Service
Service Disconnect
Primary Load Interrupter
Metering
Grounding
Transformers
Distribution Switchboards
Branch Circuit Panelboard
Feeder Circuit
Converters
Rectifiers
Lighting
Interior Lighting Fixtures
Luminous Ceiling
Signal Lighting
Exterior Lighting Fixtures
Stadium Lighting
Roadway Lighting
Accessories
Lamps
Ballasts and Accessories
Poles and Standards
Special Systems
Lightning Protection
Emergency Light and Power
Storage Batteries
Battery Charging Equipment
Cathodic Protection
Communications
Radio Transmission
Shortwave Transmission
Microwave Transmission
Alarm and Detection
Fire Alarm and Detection
Smoke Detector
Burglar Alarm
Clock and Program Equipment
Telephone
Telegraph
Intercommunication Equipment
Public Address Equipment
Television Systems
Master TV Antenna Equipment
Learning Laboratories
Heating and Cooling

Snow Melting Cable and Mat
Heating Cable
Electric Heating Coil
Electric Baseboard
Packaged Room Air Conditioners
Radiant Heaters
Duct Heaters
Electric Heaters (Prop Fan Type)

Controls and Instrumentation
Recording and Indicating Devices
Motor Control Centers
Lighting Control Equipment
Electrical Interlock
Control of Electric Heating
Limit Switches

DETERMINING COSTS

The detailed estimate is a projection into the future of the anticipated costs of a project reflecting all of the known conditions at the time of preparation.

It is most important to maintain a cost system which utilizes, as a base, each of the components contained in the estimated cost to make a comparison with the actual costs incurred during construction to determine if the estimated costs are realistic and accurate so that the project will produce the profit which was anticipated.

Equally important, the periodic review of actual cost versus estimated cost will serve as a data base to adjust, either by increasing or decreasing the unit costs, so that future estimates will be produced more competitively and, at the same time, preserving the profit margin.

Cost control systems can take many forms. They can be relatively simple and still produce accurate information or they can be designed in an elaborate and detailed format to provide information relating to cost for every item of material and labor productivity for the entire project.

The selection of the system is one of personal choice however, it must be recognized that any cost system requires considerable expenditure of time in accumulating the material and labor records and distribution to the proper categories of work, and to then calculate the costs of each and to record this information on progressive cost record forms so that monitoring of the actual costs can occur while the work is being performed. When this information is available, it is then possible, should an overrun appear, to review the situation to determine if a remedy is possible such as adjustment of crew balance, pursue another method of construction, or allocate additional supervision to a specific operation. The selection of the system should be one which serves the needs of the company and anything in excess of this is to be considered a burden and wasteful in both time and cost.

Walker Practical Forms Nos. 505, 506 and 550 will serve as a suitable base to establish an effective cost system.

ESTIMATE AND COST SUMMARY

PROJECT

LOCATION

ARCHITECT
ENGINEER

ESTIMATOR

EXTENSIONS

CHECKED

ESTIMATE NO.

SHEET NO.

DATE

CLASSIFICATION

DESCRIPTION

CODE

ESTIMATED COST

MATERIAL

UNIT QUANTITY UNIT COST AMOUNT

LABOR

HOURS AMOUNT

SUB-CONTRACTS
OTHER COSTS

TOTAL
ESTIMATE

% COM-
PLETED

ACTUAL COST

MATERIAL

QUANTITY USED UNIT COST AMOUNT

LABOR

HOURS AMOUNT

SUB-CONTRACTS
OTHER COSTS

TOTAL
COST

PROFIT
OR
(LOSS)

TOTALS

COMMENTS

FRANK R. WALKER CO., PUBLISHERS, CHICAGO

MFD. IN U.S.A. FORM 550

Form 550

DAILY TIME SHEET AND LABOR DISTRIBUTION

JOB NO.

NAME

LOCATION

REMARKS:

REPORT NO.

SHEET NO.

DATE

LABOR CLASSIFICATIONS

WEATHER

TEMPERATURE
8 A.M.

1 P.M.

OCCUP-ATION	EMPLOYEE'S NAME	EMPLOYEE'S NUMBER									HOURS	RATE	AMOUNT

DAILY CONSTRUCTION REPORT

JOB

LOCATION

WEATHER

TEMPERATURE 8 A.M.

1 P.M.

DAY

DATE

DAILY TIME SHEET AND LABOR DISTRIBUTION

REPORT NO.

SHEET NO.

JOB NO.

JOB

LOCATION

DATE

REMARKS:

LABOR   CLASSIFICATIONS

WEATHER

TEMPERATURE
8 A.M.

1 P.M.

OCCUP-ATION	EMPLOYEE'S NAME	EMPLOYEE'S NUMBER	Hrs.	AMOUNT	Hrs.	AMOUNT	Hrs.	AMOUNT	Hrs.	AMOUNT	Hrs.	AMOUNT	Hrs.	AMOUNT	Hrs.	AMOUNT	Hrs.	AMOUNT	Hrs.	AMOUNT	Hrs.	AMOUNT	HOURS	RATE	AMOUNT

TOTALS

PRACTICAL Form 505

MFD. IN U.S.A.

FRANK R. WALKER CO., PUBLISHERS, CHICAGO

Forms 505 and 506

CHAPTER 5

CRITICAL PATH METHOD

PLANNING
SCHEDULING
PROGRESS REPORTS

The Critical Path Method (CPM) is a management tool which can provide contractors with more precise planning and scheduling techniques by graphically showing each activity in a project and its relationship to other activities. Critical Path provides an accurate picture of where the project stands, what remains to be done and which jobs are critical to finishing on time.

Critical Path Method encourages decisions but can not make them. The preparation of the diagram helps the planner to understand the project through the various activities involved in the project.

CPM is a logical and organized planning system. It is important that the physical layout of the network reflects this same logical organization. The drawing size should be reasonable, using multiple sheets if necessary. Significant chains of activities should be centered to form a network backbone.

CRITICAL PATH METHOD

To produce an accurate and effective CPM Schedule it is necessary to include the major subcontractors and material suppliers as each of them can provide more detailed information concerning their procurement of supplies and materials and the installation duration so that this factual information can be integrated into the total schedule.

It is often the case that subcontractors and suppliers are lacking the managerial expertise of the General Contractor and will resist an attempt to confine their activities within a detailed time schedule, however, the Contractor, by skillful guidance, will be able to gain their confidence and cooperation in developing a schedule that causes little, if any, impact on any of the construction team.

There are basically three stages of the Critical Path Method: CPM controls planning, scheduling, and costing.

PLANNING

—The first step is to separate activities of the project and place them in logical sequence. Some activities must precede others, some must follow others and others may be performed at the same time. After the sequence is determined an arrow diagram is made.

The backbone of CPM is a graphic model of a project. The basic component is the arrow. Each arrow represents one activity in the project. The tail of the arrow represents the starting point of the activity and the head, the completion. The arrow is not drawn to scale and does not necessarily have to be a straight line.

Arrows are arranged to show the plan or logical sequence in which the activities of the project are to be accomplished. The logical flow chart or diagram is the result of a network of arrows indicating the decisions the planner makes to program the project.

If a project will consist of three consecutive activities, the network will appear as follows:

If B and C are concurrent activities, the diagram would be:

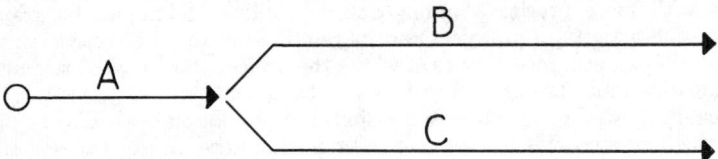

An activity D which follows activities B and C would be diagrammed:

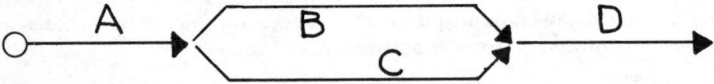

A description of the activity should be written above each line represented above by letters A, B, C and D. At this point, the contractor is provided with a clear picture of sequence of activities to successfully complete the project.

SCHEDULING

—Time estimates, usually referred to as activity duration, based on experience and knowledge, are determined for each activity and are shown below the arrow of the particular activity. Any convenient unit of time may be used, i.e., shifts, days or weeks, but it must be consistent throughout the network. The estimator assumes a normal work crew in estimating the duration of an activity. The Critical Path determines the length of the project. The longest path of activities through the network is the critical path as is shown below:

The project time to the two paths are summarized as follows:

Path I	*Path II*
A — 3 days	A — 3 days
B — 2 days	E — 3 days
C — 5 days	F — 6 days
D — 3 days	G — 3 days
H — <u>4 days</u>	H — <u>4 days</u>
Total 17 days	19 days

Here it is noticed that Path II will take 19 days and therefore is the Critical Path.

It is indicated by the heavy line in the diagram above and therefore is the latest completion date for the project. The activities B, C, D are not on the Critical Path and so are considered slack or float time.

It must be remembered that any change in the time for one of the activities on the Critical Path will result in an equal change in the over-all project time. Also, activities not on the Critical Path considered float time will have an Earliest Starting Time, Latest Starting Time, Earliest Finishing Time and Latest Finishing Time. Further, if an activity on the Critical Path is accelerated, an activity not on the original Critical Path may necessarily be moved to it.

With the time estimates completed, the contractor now has a clear, concise picture of the overall project. The network will show which activities are critical for the completion of the project on time.

Critical Path is a tool that may be used to evaluate costs and provides a systematic way to determine the most profitable duration for a project. The optimum time schedule lies somewhere between the "normal" time for the project and the "crash time."

A CPM diagram computed on a normal time basis—that is, without extra direct costs such as overtime labor, is the starting point for preparation of a series of diagrams, showing alternate methods of speeding the total project. The final, shortest schedule is the "Crash" schedule.

As activities on the project are shortened due to the crash program, the Critical Path may thereby shift, as the Critical Path is the longest path.

Referring to the diagram used previously, Path I and Path II were computed and Path II (A, E, F, G, H) was shown as the Critical Path. After revision to a crash program, the following network was developed:

The project time for the revised paths is as follows:

Path III	Path IV
A — 3 days	A — 3 days
E — 3 days	B — 2 days
F — 4 days	C — 5 days
G — 2 days	D — 4 days
H — 4 days	H — 4 days
Total 16 days	18 days

As a result of this crash program the Critical Path is now Path IV.

Ordinarily, as time is cut from the normal schedule, direct costs increase; thus, as the series of schedules are prepared, direct costs will be at a minimum for the normal schedule and at a maximum for the crash schedule.

Indirect costs ordinarily will reduce as the project time is shortened.

CPM can show contractors just how much they can shorten the project time before reaching the point of diminishing returns as extra speed-up measures are sometimes more costly than they are worth.

C.P.M. Network

The first step is to determine direct costs for the normal time Critical Path program. Next, the direct cost for the crash program is computed including overtime, extra equipment and other costs that may be required to effect the crash program. Between the crash program and the normal schedule lie a number of possible programs. A series of CPM diagrams for the various combinations of time and cost should be prepared.

If we chart the findings of the various diagrams broken down into direct costs, indirect costs and total costs, the optimum CPM program can be determined.

DAILY MATERIAL REPORT

JOB NO.	103		DATE	June 6, 19--
NAME OF WORK	Smith Store Building		WEATHER	Clear
LOCATION	656 N Main St, Arlington, Ill		TEMPERATURE	62 7 A.M. 83 2 P.M.

RAILROAD AND CAR NO.	CLASS OF WORK	MATERIALS RECEIVED AND FROM WHOM	QUANTITY
	Cem Floors	Ready-mixed concrete - Arlington Ready-Mix Co.	9 cy
	Masonry	Common brick - Arlington Builders Supply Co.	8000
	Rough Carpentry	62 Pcs - 2	
		91 " - 2	
		25 " - 1	

KIND OF MATERIAL	Concrete Sand	Gravel	Crushed Stone		SUMM
PREVIOUS					
TO-DAY					
TOTAL					

KIND OF MATERIAL	Fire Brick	Special Brick	Reinforcing Steel
PREVIOUS			
TO-DAY			
TOTAL			

KIND OF MATERIAL	Conc. Form Lumber	Framing Lumber	Ready-Mi Concrete
PREVIOUS	2500 fbm	—	56 cy
TO-DAY	—	3010 fbm	9
TOTAL	2500	3010	65

DAILY CONSTRUCTION REPORT

BUILDING	Smith Store Building	DATE	June 6, 19--
LOCATION	656 N Main St, Arlington, Ill.	WEATHER	Clear
ARCHITECTS	Jonnson and Anderson	TEMPERATURE	62 7 A.M. 83 2 P.M.

CLASS OF WORK	Fore-man	Mech. back	Labor-ers	Misc.	REMARKS
1. WRECKING—SHORING					
2. EXCAVATING—PUMPING		1	2		2. Bulldozer operator and 2 laborers backfilling around foundation walls
3. PILING OR CAISSONS					
4. FOUNDATIONS—RETAINING WALLS					6. 2 cement masons and 2 laborers placing cement floor in basement.
5. WATER—DAMPPROOFING					
6. CEMENT FLOORS—WALKS—PAVEMENTS		2	2		
7. BRICK, TILE, CONCRETE MASONRY	1	4	6		7. Laying face brick and common brick back-up on east elevation. Laying solid common brick south wall
8. CAST STONE—CUT STONE—GRANITE					
9. TERRA COTTA					
10. FIRE-PROOFING, TILE—GYPSUM					
11. ARCHITECTURAL CONCRETE					13. Framing and placing joists for first floor
12. REINFORCED CONCRETE					
(a) Forms					
(b) Reinforcing Steel					30. 2 ironworkers finished erection of lally column and girders
(c) Mix and Deposit Concrete					
13. ROUGH CARPENTRY	1	3	2		
14. FINISH CARPENTRY					31. 2 plumbers working on connection to water main 2 laborers backfilling sewer trenches.
15. WOOD FLOORS					
16. INSULATION, SOUND DEADENING					
17. WEATHER STRIPS—CAULKING					Owner visited job today. Seems well satisfied with progress.
18. LATHING					
19. PLASTERING					
20. FIRE DOORS—WINDOWS					Must have 1x6 O&M for rough flooring immediately. Lumber yard says they are tracing a car of this material which was due in two days ago. Better check with them.
21. STEEL DOORS—WINDOWS					
22. SHEET METAL WORK					
23. ROOFING					
24. TILE AND MOSAIC					
25. ASPHALT, CORK, LINOLEUM FLOORS					Also check with heating contractor. Should have started today.
26. ART MARBLE—SCAGLIOLA					
27. MARBLE—SLATE Interior					
28. GLASS AND GLAZING					
29. PAINTING AND DECORATING					
30. STRUCTURAL IRON—STEEL Erection		2			
31. MISC. IRON AND STEEL					
32. ORNAMENTAL IRON—STEEL					
33. PLUMBING—SEWERAGE—GAS FITTING		2	2		
34. HEATING AND VENTILATING					
35. AIR CONDITIONING					
36. ELECTRIC WIRING					
37. LIGHTING FIXTURES					
38. ELEVATORS—ESCALATORS					
39. SPRINKLER SYSTEM					
40. MAIL CHUTE					

FRANK R. WALKER CO., PUBLISHERS, CHICAGO

Glossary of Critical Path Method terms:

Activity—a work item which is a component of the project.

Activity times—Identity of start and finish times for each activity in the network.

Arrow—Graphic representation of a single activity in the CPM network.

Critical Path—Specific sequence of activities in network which sets the maximum time for completion of a project.

Event—Point in time that marks the start of finish of an activity.

Network—The connected sequence of arrows representing the entire project.

Project—The overall work being planned.

Bibliography and suggested additional reading:

Network Analysis for Planning and Scheduling Albert Batters—Saint Martin's Press

Operating Planning and Control Arch R. Dooley

An introduction to Critical Path Analysis

K. G. Lockyer—Pitman Publishing Corporation

CPM in Construction Management

James Jerome O'Brien—McGraw-Hill

The Critical Path Method

L. R. Shaffer, J. B. Ritter and W. L. Meyer—McGraw-Hill

Modern Management Methods

David M. Stires—Boston Materials Management Institute

PROGRESS REPORTS

The CPM schedule is an effective tool in monitoring the progress of a project and it can also be adapted to reflect manhours, dollars, equipment requirements, activities to be performed by the Owner or external contractor, and, of extreme importance, it provides the graphic base for a defense against claims for delay or interference by the Owner or other Contractors.

On projects that are smaller in size or less complicated, it is often found that there does not exist the need for the detailed information provided by the CPM and that the less complex Gannt or bar graph will provide sufficient information.

The Walker Form No. 112 can be used to record the activities of the various disciplines of work on the project.

CONSTRUCTION REPORT

JOB				LOCATION	
NUMBER OF RAINY DAYS	TOTAL PERIOD	TEMPERATURE RANGE	DAYS LOST	WEEK ENDING	

FORM 112

CONSTRUCTION PROGRESS REPORT

JOB NO. _____   REPORT NO. _____

PROJECT   LOCATION   SHEET ___ OF ___

CONTRACTOR   ADDRESS   WEEK ENDING

ITEMS OF CONSTRUCTION				PLANNED PROGRESS			BAR GRAPHS AND DATES OF SCHEDULED AND ACTUAL PROGRESS											% MT'L ON JOB
NO	DESCRIPTION	COST ALLOWANCE	PER CT TOTAL	START	COM-PLETE	DATE STARTED	10%	20%	30%	40%	50%	60%	70%	80%	90%	100%	
TOTAL OR CARRY FORWARD							PERCENT COMPLETE	THIS WEEK			PREVIOUS			TOTAL TO DATE			
WORK ORDER ISSUED		19		PLANNED													
CONTRACT COMPLETION DATE		19		ACTUAL													

USE OTHER SIDE FOR REMARKS

Form 112—8½x11, 112L—17x22

TOOLS AND EQUIPMENT

DETERMINING HOURLY RATE
MATERIAL HANDLING
HOISTING TOWERS
SCAFFOLDING
SIDEWALK BRIDGES
AIR COMPRESSORS

There may be no segment of contracting that puts to test an estimator's ability to visualize future job conditions than that of planning for equipment and plant at the construction site. There may have been a time when the estimator would only include that equipment specified in the general and special conditions such as barracades, scaffolds, dust chutes and the like. It was assumed that each sub-contractor would furnish the additional equipment he might need. This is still true in certain trades. The excavator will base his bid on furnishing and using specific equipment which will allow him to complete the work most expeditiously; the concrete firm will furnish their own mixers and vibrators; the mason will furnish his own saws and grinders.

But all trades will have certain needs in common and the estimator should forsee these needs and lump all shared equipment under the responsibility of the general. This will eliminate duplicate machinery and operators, unnecessary transport to and from the site and the set-up and dismantling costs. Further savings will be realized by taking advantage of reduced rental rates that apply to extended periods for the weekly rate or the monthly rate. Further the more equipment the general sets up, the more control he has over the safety conditions at the job site. Many a contractor has been fined, or worse had his job closed down, because a government inspector ruled job conditions unsafe.

CAUTION

The owning, rental or use of tools and equipment subjects the contractor to compliance with specific safety requirements which may be imposed by cities, states and the Federal Government. It is not unusual to detect considerable conflict or variance in the degree of protection required under the several sources of regulations. In such instances, it has been held that the contractor is required to comply with the maximum requirement.

A national safety guide is the Occupational Safety and Health Act (OSHA) established by the Department of Labor in 1970 and all tools and equipment utilized on a construction project should meet these standards. Purchase Orders or Rental Agreements should contain a specific clause relative to this compliance.

Failure to meet the provisions of the safety laws can subject the contractor to the penalty of fines and costly litigation.

DETERMINING HOURLY RATE

The general should consult with his subs to determine what their needs will be and then inform them what he will be furnishing. He may then have the subs deduct these costs from their bids or make an agreed upon back charge for rental rates.

Whether the equipment is rented or owned, the estimator should charge for it at a fixed hourly rate. This rate would include delivery, set-up and dismantling charges plus the basic rental. This would then be divided by the working hours that the equipment stayed at the site. In some cases one might elect to charge only for the hours that the equipment was in use at the job site. The remainder of the time would then be charged to office overhead. Operating costs would be added as a separate item only for the actual in-use hours.

Equipment rental rates vary greatly throughout the country, depending upon local practices and conditions. They may also vary depending upon the use to which the equipment is placed; such as hard service in rocky soil or for use in sand.

It is general practice in the industry to base rates upon one shift of 8 hours per day, 40 hours per week, or 176 hours per month of a 30 consecutive day period.

The lessor usually bears the cost of repairs due to normal wear and tear on non-tractor equipment. On rubber-tired hauling and tractor equipment, the lessee may be required to bear all costs for maintenance, and repair.

The equipment rented should be delivered to the lessee in good operating condition, and is expected to be returned to the lessor in the same condition as delivered, less normal wear.

National average rates quoted herein are f.o.b. the lessor's warehouse or shipping point. The lessee pays all freight or delivery charges from shipping point to destination and return. Lessee also pays all charges for unloading, assembling, dismantling and loading where required.

Rentals are usually paid in advance and subject to the terms and conditions of the lessor's rental contract. Rates do not include contractor's insurance, license fees, sales taxes or use taxes.

Where the contractor owns the equipment the estimator must arrive at an hourly cost to substitute for the basic rental. To do this he must consider the following:

a) original capital investment
b) anticipated life
c) depreciation
d) estimate of maintenance
e) estimate of taxes, insurance, interest and storage

Items c), d) and e) are expressed in percentages of the original capital investment, item a).

The estimator will need the cooperation of the firm accountant in arriving at this hourly cost figure as well as some knowledge of the possible future use of the equipment. The capital investment cost should include all original delivery, test and start-up charges. The expected life may be based on Internal Revenue tables. It is usually entered as 2000 hours per year. (40 hour week, 50 week year)

Depreciation cost is the original capital investment less any resale or scrap value. This is often so small that depreciation on all but highway trucks may be figured at 100%. However in these highly inflationary times, for equipment held but a short time, the inflation rate may approximate the depreciation rate so that the equipment would appear as having zero depreciation.

Maintenance allowances will vary from 40% to 100% of the investment cost, but most equipment will fall in the 65% to 75% bracket. Those items falling below 50% would include air tools, gas engines, small motors, mortar mixers, vibrators, drills, screw jacks, small power tools, towers, small tractors, trailers and trucks. Those items for which over 75% of investment cost should be used would include bituminous equipment, hoppers, crushers, derricks, electric generators, hydraulic jacks, pumps and rollers.

Taxes insurance and interest percentages should be worked out with the accountant. 10% used to be an acceptable figure for these items, but with todays rates 15% would probably be more realistic. 3% is generally added for storage. A formula to arrive at this entry would be:

$$\frac{p^* \text{ (useful life)}}{2} + p^{**} \text{ (useful life)} = \text{total percentage}$$

where p* is percentage for taxes, insurance and interest
p** is percentage for storage.

If the useful life is five years, p* is 15% and p** is 3% the total percentage to be entered under item e) would be:

$$\frac{15 (5 + 1)}{2} + 3 \times 5 = 45 + 15 = 60\%$$

Once items a) through e) are determined the hourly rate can be figured as follows:

$$\frac{\text{capital investment}}{\text{anticipated life in hours}} \quad \text{x sum of \% for items c), d) and e)}$$

As an example assume the contractor has purchased a $50,000 piece of equipment and has determined the following to be reasonable projections:

a) capital investment	$50,000.00
b) anticipated life 5 yrs at 2000 hrs	10,000 hrs.
c) depreciation $2500 scrap value	95%
d) maintenance	70%
e) taxes, insur., int. & stor.	60%

$$\text{Hourly Use Charge} = \frac{\$50,000}{10,000 \text{ hrs.}} \text{ x } 95\% + 70\% + 60\% = \$11.25$$

MATERIAL HANDLING

Manpower. Labor is too expensive today to be used where machines can be utilized. But in certain conditions it may be necessary.

A laborer is generally rated as being able to handle a maximum of one hundred pounds. For loads over that, or for oversized material, additional help must. be scheduled.

One man in one hour should be able to unload and pile up on the site 300 to 500 standard bricks per hour using his hands, or up to 1000 using special tongs. A fork lift truck handling a palletted load can handle well over twenty times as many units in the same time, and provides a wider coverage than a man. It can be seen that an estimator can allot considerable expense for temporary access for mechanical equipment if much material is involved, and still make substantial savings over using manual labor.

Man labor for unloading and piling other materials can be estimated as 80–120 per hour for sacks of cement; 100–200 per hour for block or tile; and 1000–2500 fbm per hour for lumber.

Hand shovelling of bulk materials will vary as to whether the load is transferred at the same or lower elevation or raised to a higher one. The following table is based on the amount one man can handle in one hour under dry conditions.

	Even Transfer		Upward Transfer	
Sand	3	tons/hr	2.6	tons/hr
Stone	2.12 " "		1.8 " "	
Gravel	2.3 " "		2 " "	
Top soil	2.6 " "		2.35 " "	

Wheelbarrows. Distribution over the job site will still often require hand powered wheelbarrows, trucks and buggies. Rental costs can be figured at $7.00 per day or $30.00 per month.

Power buggies are widely used on larger construction projects and they have proved to be a more economical method of transporting materials. The walk-behind models rent for $55.00/day or $444.00/month, not including gas.

Sizes, Weights and Capacities of Contractor's Wheelbarrows

Struck Capacity Cu. Ft.	Heaped Capacity Cu. Ft.	Approx. Weight Pounds	Type of Materials Used for
2½	3½	74	Dry Materials
3	4	78	Dry Materials
3½	4¼	75	Concrete-Mortar
4	5	79	Concrete-Mortar
4½	6	83	Concrete-Mortar

It requires about 150 shovels (No. 2 shovel) for one cu. yd.

Courtesy J. Emil Anderson & Son, Inc.

Helicopters are playing an increasingly important role in modern construction particularly in materials handling.

Chutes and Conveyors. The cheapest energy the estimator can plan for is gravity, and he should see the job is laid out so that material distribution will be by gravity in so far as possible. Straight line chutes can be rented in ten foot sections for $10.00 to $12.00 per week or may be built up out of plywood at the site. Prefabricated metal vertical chutes will cost $20.00 to $25.00 per foot erected. Gravity steel roller conveyors can be purchased in 10' lengths for around $175.00 for 16" widths. Powered conveyor belts will run two and a half times as much plus one must add for operating costs.

Portable Belt Conveyors.— Some applications in the construction field require the use of portable belt conveyors of various sizes and capacities. Their flexibility does not limit the user to a fixed receiving or discharging point, and when used in series, this flexibility is still further increased to carrying of materials over long horizontal distances, over obstructions, around corners, etc.

Portable conveyors are available in boom lengths to 120 ft. and belt widths of 18", 24", 30", and 36". Swivel wheels permit building of high capacity radial stock piles or discharge into any one of several bins or hoppers. Both single and multi-deck vibrating screens may be used at the discharge end to separate the material into two or three sizes. Feeder-trap and feeder-hopper units mounted at the tail end, permit charging by front end loaders, dozers, cranes, draglines, shovels or other conveyors.

Portable conveyors are used for conveying earth, clay, sand, gravel, stone, bulk cement, concrete, etc., from stock piles or to convey and elevate materials to trucks, etc. They can be used for conveying wet concrete from mixer to elevated structures where the incline is not too steep.

Conveyors are furnished with either plain or cleated belts but the cleated belts should not be used with wet materials, such as wet concrete, wet clay, etc.

The tables below list capacities of belt conveyors, wheelbarrows, mixers, and concrete carts.

Capacities of Standard Conveyors in Tons Per hour (Based on
the Assumption of Uniform and Continuous Loading)

Belt Width	Belt Speed Ft. per Min.	Earth Dry	Clay Dry	Sand	Stone Gravel	Bulk Cement
18"	300	112	92	150	160	150
24"	300	215	185	290	315	290
30"	300	350	300	470	510	470
36"	300	520	450	690	750	690

HOISTING TOWERS

Tubular steel hoisting towers are extensively used by contractors in all parts of the country for hoisting all types of building materials and for concreting, because of the ease with which they are erected and dismantled, their great strength and stability, as well as their economy.

Tubular hoisting towers are designed for handling material and concrete, with or without chuting. They are built in any height and with one or more compartments. Their many advantages both in construction and in performance, their durability and low maintenance, and their remarkable adaptability to various jobs, make them ideal as low cost equipment for contractors.

A tubular hoisting tower consists of a bottom section, the necessary number of intermediate tubular sections, each 6'-6" in height, and a top section containing the cathead with sheaves.

All intermediate sections of a given type of tower are identical and the top and bottom sections can be attached to any one of them.

The single and double (two compartment) towers are classified as follows:

Size of legs	Heavy Type 3-in.	Light Type 2-in.
Maximum height	350'-0"*	201'-0"
Live load capacity	5,000 lbs.	3,200 lbs.
Maximum size concrete bucket	35 cu. ft.	....

*Higher towers require special material.

The light type towers are known as class 3200 and the heavy type towers are known as class 5000.

The size of the cage platform in the single tower is 5'-10½" x 7'-6½". For double towers the size of one cage platform will be 5'-10½" x 7'-6½" and the other platform will be 5'-10" x 7'-2".

Erecting and Dismantling Steel Tubular Towers.—When estimating the cost of erecting and dismantling a steel tubular tower, the number of man-hours involved depends upon the height of the tower and the general conditions involved. For instance, it will take considerably longer to erect a tower 208'-0" high than one 65'-0" high, not only in time consumed but also in man-hours per section, taking into consideration the amount of material involved as well as the height to which the equipment must be raised.

Time given includes erection, dismantling and handling in man-hours per lin. ft. of height.

Light Type Single Tubular Towers .... 1-1.5 hr.
Light Type Double Tubular Towers .... 1.5-2 hr.
Heavy Type Single Tubular Towers .... 1.5-2 hr.
Heavy Type Double Tubular Towers... 2-2.5 hr.

Prices of light or heavy duty hoisting towers should be obtained from the manufacturer of such equipment, at the time that they are needed. Quotations for rental charges on a monthly basis can also be obtained for a specific project. A 50' tower will rent for about $1,500 per month, with each additional 10' about $40 extra per month.

SCAFFOLDING

Tubular Steel Scaffolding.—The most versatile type scaffolding is galvanized steel tubing with a male fitting on one end and a female fitting on the other to form interlocking units. These units are then coupled together with a patented coupling device. It can be adapted to a variety of uses and its simplicity has found general acceptance among the building trades. Basically, it consists of four parts; interlocking steel tubes of various sizes, a base plate or caster, and an adjustable or right angle coupler. All parts are galvanized as a permanent protection against rust.

This galvanized carbon steel tubing or pipe is approximately of 2" o.d. (1½" Standard Pipe) and weighs 2¾ lbs. per ft. Other sizes are 2½" o.d. (2" Standard Pipe Size) which weighs about 3¾ lbs. per ft.

Standard lengths for tube are 6'-0", 8'-0", 10'-0", 13'-0", 16'-0" and 20'-0", and for pipe are 6'-0", 8'-0", 10'-0" and 13'-0". Other sizes may be obtained but those mentioned are the sizes generally used by the building trades. Each section is fitted with a bayonet locking device.

This locking device between members is permanently affixed to the tubes, which method insures a perfectly rigid attachment which has been tested to a load of 20,000 lbs. in tension. The assembled tube and fittings make up a complete unit so designed as to have flush joints.

Standard couplers are made of drop-forged parts, except for the steel hinge pins on which the catch bolts hinge. The most widely used couplers are 2" x 2", both adjustable and standard, and 2" x 2½", both adjustable and standard, and 2½" x 2½", adjustable and standard. This type of scaffold is the most versatile of all built-up scaffolds.

Estimating costs of erecting scaffolding depend upon many conditions; the type of job to be done, whether interior or exterior; ground conditions; height and width, as well as load to be carried; and length of time it will be in use.

Costs of purchasing or renting exterior tubular steel scaffolding are quoted on the square footage of wall area actually covered, and may be obtained from manufacturers of the equipment. Low buildings will cost about $40.00 per hundred sq. ft. installed (rental). Costs for buildings over 5 stories will cost about $60.00 per hundred sq. ft. of surface installed (rental).

Costs for purchasing or renting interior tubular steel scaffolding are quoted on the number of cubic feet actually occupied by the scaffolding and may be obtained from the manufacturer or supplier of the equipment.

Typical materials and man-power lift used in present day constuction.

Sectional Steel Scaffolding.—Sectional steel scaffolding or prefabricated scaffolding is readily assembled and installed and is well adapted to various kinds of building and maintenance work. Various sizes of 5'-0" wide frame units, trusses and ladder scaffold components, make it simple to erect a suitable and a rigid, safe scaffold under any conditions.

The scaffold components are manufactured to resist corrosion and rusting. Because of their rugged construction they are not damaged during assembling or dismantling. This is a lighter type scaffold than the tube and coupler type and has an approved working height limit of approximately 125'-0".

The tubing used in the frames and trusses is carbon steel, the higher physical properties of which affords the opportunity to reduce weight without sacrificing strength.

The sectional steel scaffolding has found wide acceptance among bricklayers, plasterers and others of the building trades because of its adaptability to rolling towers and to exterior and interior construction and maintenance.

The spacing between frames may be either 3'-0", 5'-0", 6'-0", 7'-0" or 10'-0", and the sizes of the frames may vary from 3'-0" to 10'-0" in height, thereby allowing for a wide variety of uses. Braces are attached to the frames by means of wing nuts or automatic locks.

Costs of the component parts of sectional steel scaffolding may be obtained from the manufacturer or supplier.

Tubular Steel Scaffolding for Exterior Work

Suspended Scaffolding

The most economical method of scaffolding for masonry work on buildings with a structural steel or reinforced concrete skeleton framework is suspended scaffolding. The working platform is supported by regularly spaced scaffolding machines, each consisting of a pair of cable drum hoist mechanisms with connecting steel putlog, suspended by steel wire rope from steel or aluminum outrigger beams anchored to roof or floor framing above.

Standard width machines accommodate a 5'-0" width of solid planking. Machines are available to accommodate 8'-0" wide decking to permit wheeling of material on scaffold platform. Machines are also available for mechanized material handling with the inboard cable drums set back from the wall to provide a 5'-0" wide platform for material handling traffic and stockpiles on one side and a separate 20" wide working platform for masons on the other. All machines are available with or without steel frames to support overhead protective planking. For maximum safety, overhead planking should be used, together with toe boards and guard rails along outer edge of platform.

Courtesy of Cupples Products Division, H. H. Robertson Company

Tubular steel scaffolding used as an outrigger in curtain wall construction

Scaffolding machines are usually spaced 7'-0" apart, permitting the use of 16'-0" scaffold plank. A ratchet device on the drum hoist mechanisms provides positive locking while still permitting easy raising or lowering of scaffold to maintain bench-high location of work for maximum efficiency and best workmanship.

Outrigger beams are usually anchored to roof construction 7'-0" on centers, if building is not over 150'-0" high, the maximum lift capacity of scaffolding ma-

Sectional steel scaffolding

chines. For structures exceeding 150'-0" high, one or more additional installations will be required on lower floors. There are various methods of anchoring beams to building framework, such as securing with steel straps to anchor bolts set in concrete slabs or beams; bolting to inserts set in concrete: and anchoring to beams or joists with U-bolts and bars through holes in concrete slabs.

Conditions affecting the installation of suspended scaffolding vary so widely it is impossible to give unit costs that will be of any value in estimating. The manufacturer and erector will gladly quote a firm price on any specific job where this scaffolding method is applicable. This service is available nationwide.

Suspended Scaffolding for Masons

Ladder Type Scaffolding.—For interior scaffolding, particularly for low to medium height work, ladder type scaffolds are considered very efficient. The type with horizontal bracing is most desirable for interior work because they span over furniture, machinery or equipment. Scaffolding that can be assembled without bolts or nuts can be erected or dismantled more quickly and easily. Platforms that can be adjusted for height permit men to stand at the most efficient distance from

Courtesy Symons Corporation

Scaffolding used in the erection of forms for poured concrete decks their work. This latter feature is important when scaffolding for electricians, plumbers, heating and ventilating workmen, acoustical erectors, painters, maintenance men and others doing off-the-floor jobs.

Ladder type scaffolds are sold as complete scaffold units which include two ladders, two platform support trusses and one plywood platform (exterior type plywood, metal bound). The platforms are adjustable for height every three inches. Each unit is 6'-0" high by 6'-0" long (or longer if desired) by 2'-5" wide. They can be used separately or in combinations to scaffold larger areas. Casters can be furnished either with or without locks.

One of the outstanding advantages of scaffolds of this type is their maneuverability over permanent or semi-permanent objects, such as church pews, machinery or furniture. It isn't necessary to scaffold an entire room or auditorium because the scaffolds can be readily moved from place to place.

Scaffolds of this type sell at approximately $160.00 to $215.00 per unit, including the platform, plus the casters.

SIDEWALK BRIDGES

Where construction takes place adjacent to heavy traffic areas and it is impracticable to re-route sidewalk traffic into the street, a sidewalk bridge must be erected to protect the public. Rigid standards must be met in the design and erection of these structures.

Construction is as follows: 4½" o.d. steel columns set on heavy timber mudsills support 8" steel I-beams bridging the sidewalk. Beams are securely attached to columns by means of special steel column head castings. Columns are tied together longitudinally with two rows of pipe girts with an occasional diagonal pipe brace as required. Wood timbers, cut to a taper and set to slope towards the street, are fastened to the cross beams with U-bolts to support 6" steel beams, spaced approximately 16" on center, running the length of the sidewalk bridge. Two layers of 2" or 3" planking are then laid over the beams, the bottom layer running laterally and the top layer longitudinally. A parapet for the street side may be added if desired.

Most contractors lease sidewalk bridges on an erected and dismantled basis and the supplier usually quotes a lump sum figure for a specific job. Before submitting a bid, a firm quotation covering this work should be obtained, where required.

For preliminary estimating purposes only, and average conditions, sidewalk bridges may be figured as follows:

Erection and dismantling, per lin. ft.	$20.00-$36.00
Sidewalk bridge rental, per lin. ft., per month	4.75
Erect and dismantle parapet, per lin. ft.	9.00
Parapet rental, per lin. ft., per month	2.00

AIR COMPRESSORS

One sales and rental source for air compressors advertises over 1800 models are available. Add to this all the various attachments that are made to work with the compressor and one can see one must give some thought so as to match the unit to the tools for the most efficient operation.

Two and more tools can be operated from the same compressor. The rating of the compressor should be the delivery, but leaking hoses, poor tools and adverse job conditions could lower the cfm output. In some gang operations not all units will be in operation at the same time so that a unit sized for two may be adequate for three, providing the compressor is set up for more than 2 hose connections.

The following list will give some idea of air requirements for the more commonly used tools:

Breakers
35 # – 30 cfm
60 # – 40
80 # – 50

Diggers
20 #
– 20
cfm
25 # – 25
35 # – 35

Drills
Air
1" – 35 cfm
2" – 50
4" – 60
Wagon
3" – 160 cfm
3½" – 200
4" – 250

Hammers
Jack
25 # – 40 cfm
45 # – 90
75 # – 160
Chipping
light – 20 cfm
heavy– 30 cfm

Rivet
⅝" – 25 cfm
⅞" – 35

Saws
Circ.
12" – 50 cfm
Chain
30" – 90
36" – 140
48" – 160

Tampers
35 # – 30 cfm
60 # – 40
80 # – 50

Vibrators
2½" – 25 cfm
3" – 45
5" – 80

Wrenches (Impact)
⅝" – 15
¾" – 30
1½" – 75

Compressor rentals will vary as to the condition and rated size of the equipment, but should be available for following approximations:

fuel	cfm	per day	per week	per month
gas	85	$ 34	$110	$ 260
"	100	40	110	270
"	125	40	110	270
"	150	60	190	415
"	175	75	200	425
diesel	250	100	245	570
"	375	130	300	700
"	600	150	400	1,100
"	750	225	700	1,800
"	1000	280	925	2,800

Tools, hoses and other attachments must be added separately.

SITE WORK

02010   SUBSURFACE EXPLORATION

When preparing estimates on any structures involving foundations or earthwork, there is a need to obtain all information available concerning the conditions and characteristics of the soils to be encountered in the subsurface. Frequently some of this information is made available on the working drawings as an aid to the Contractors/Estimators but complete reliance on such limited information places the Contractor in a risk position especially when the drawings provide for a disclaimer of responsibility by the Architect/Engineer for the accuracy or completeness of the subsurface information. A site investigation is mandatory as many conditions can be determined through adequate visual observations of the particular site involved and adjacent property. There can be discovery of previous stream beds, evidence of prior dumping of debris, existence of abandoned mining operations, existence of flood plains, appearance of rock outcroppings and many other situations which would strongly influence the determination of construction methods and the costing involved.

In urban areas records are usually available from utility companies and highway authorities concerning subsurface conditions affecting their installations.

It must be recognized that any penetration of the ground surface has attendant risk and the accumulation of knowledge of the conditions to be encountered during excavation will tend to control the risk to a reasonable level thus avoiding exposure to a catastrophic situation.

Excavating test pits or caissons produces accurate information in cohesive soils inasmuch as this method permits visual inspection of the soil in place and samples for laboratory tests may be carved out of the pit walls in a relatively undisturbed condition. Costs of this work are high, however, and will be about the same as excavating costs for comparable pits and caissons.

However, most subsoil investigations require studies of the soil at greater depths than can be economically reached by test pits. For these projects, various methods of soil boring and sampling are used by which a relatively small diameter hole is sunk into the ground from which soil samples are obtained.

Hand Auger Boring Method.—The hand auger boring method, satisfactory for highway explorations at shallow depths, can be used only for preliminary investigations for foundations. Auger borings are used in cohesive soils and cohesionless soils above ground water. Depths up to 20 feet may be reached. Soil samples obtained through this method are disturbed to such an extent that little or no information is furnished about the character of the soil in its natural state. Where the auger hole is filled with water, sample disturbance is particularly critical. Auger holes are most useful for ground water determination.

Hand auger borings generally cost from $4.00 to $6.00 per lin. ft. depending on the soil, the amount of boring necessary, and the diameter of the hole which will vary from 4" to 8".

Subsoil Investigations

Split Spoon Sampling.—A split barrel sampler known as a split spoon is frequently used for obtaining representative samples. This method is described in detail in ASTM Method D-1586.

In securing split spoon samples, the drill hole is opened and cleaned to the desired sampling level by drilling bits and wash water, or alternatively the hole may be drilled with a regular power auger. Drill casing or drilling mud is used when it is necessary to prevent the soil above the sampling level from closing the drill hole. A drive pipe is attached to the upper end of the drill rod and a slip jar weight placed on the drive pipe.

The sampling spoon is then driven into the soil by repeated blows of a drop hammer. The number of blows required to drive the spoon are recorded on the logs. The spoon is then brought to the surface and the sample removed, classified and placed in a glass jar.

The ASTM method specifies a 2-inch OD x 1-⅜ inch ID split spoon driven 18" to 24" by a 140 lb. weight falling 30 inches. Various other size split spoons, drop weights and length of drops are employed by some companies. The split spoon method is most useful in granular or dense soils.

Thin-Wall Tube Sampling.—The method of sampling which has been found to give the least disturbed samples is the thin-wall tube method. These samples are most suitable for testing. This method is described in detail in ASTM Method D-1587.

In this method the hole is opened and cleaned as in the split spoon sampling method, but a thin-wall tube, usually with an outside diameter of 2", 3" or 5" is then attached to the end of the drill rod and lowered to the bottom of the boring. The tube is then pushed into the soil by a hydraulic piston arrangement with a rapid and continuous motion.

The thin-wall tube is then raised to the surface, cleaned, labeled, and sealed to prevent loss of moisture. It is then taken to the laboratory for testing. Here it is usually cut into short lengths and the sample in each length ejected and tested.

Samples are normally taken at 5' intervals or every time a soil change is expected as determined by watching the overflowing wash water and feeling the resistance to penetration of the drilling bit, irrespective of the distance between changes.

Cost of drilling and sampling is usually between $6.00 and $16.00 per lin. ft. depending on the depth, diameter, and difficulty of drilling, plus on and off charges for the equipment and men. Laboratory testing is performed at an additional cost, depending on what tests are needed—for density, moisture content, compression or mechanical analysis.

Core Borings.—In addition to soil samples, information is frequently wanted on the nature or thickness of underlying rock. Core borings are made for this purpose with rotating coring tools such as diamond drills, shot drills, or carbide bits. Continuous samples are generally recovered for examination and further testing. ASTM Standard D-2113, Diamond Core Drilling for Site Investigation, should be called for in performing core borings.

The cost of making core borings will depend upon the location of the work, quantity, and the difficulty encountered in getting to the holes.

The drilling costs per lin. ft. should be estimated at from $18.00 to $22.00 per foot of drilling with the lower costs on jobs of larger size. To the above prices must be added the on and off charges for the equipment and men, and testing charges.

Other Methods.—There are many methods used to investigate soil conditions for foundation purposes. Some are reliable and sufficient to use for design purposes while others produce inadequate or dangerously deceptive data. In problem soils the gathering of the basic information under controlled conditions is important but, to obtain definitive and factual information there is a need for professional engineering interpretation. In such instances the best approach is to obtain the services of a soils engineer with laboratory facilities to completely perform the subsurface investigation and to provide a written report of the findings and recommendations for remedy of the problems to be encountered.

Engineering and laboratory services are not inexpensive. However, when balanced against the magnitude of potential loss which could be encountered without this advance knowledge, the cost factor can usually be justified.

As engineering costs vary widely by geographic location and services required, it is not practical to attempt to provide cost range figures. Sufficient competition exists to allow the Contractor/Estimator to solicit proposals from several engineering firms providing such services which should be tailored to the specific needs of the contractor.

Seismic resistivity, vane shear tests, cone penetration test and nuclear methods are more frequently being used in subsurface investigations with this work being performed by specially qualified engineers having the necessary equipment and regularly engaged in this type of work. Cost estimates will be tendered to the contractor predicated upon the services required.

02100 CLEARING

Clearing of a construction site by the removal of trees and vegetation has become more costly because of the Federal and State restrictions imposed in response to the environmental protection regulations. Most areas prohibit the burning of the trees. Therefore, if the wood does not have a market value, the use of chipping machines is employed. This method requires additional personnel to cut the trees into manageable lengths, clean of branches, and transport to the chipping machine. Some recovery of cost can be achieved by the sale of the by-product chips to nurseries and landscape companies for use as a mulch. Stump disposal is very expensive, particularly where they cannot be buried at some location on the site and have to be transported to a distant disposal area, which must be left in a graded condition so as not to impact the environment.

Clearing, Cut & Chip	Unit	Avg. Price Range
Light -6" Diam.	Acre	$1,500.00 - $1,900.00
Medium - 12" Diam.	Acre	2,500.00 - 3,000.00
Heavy - 16" Diam	Acre	3,500.00 - 4,500.00
Grub Stump Plus Haul Charge	EA	$20.00 - $35.00

NOTE: If chipping is not required, average prices can be reduced by 25%.

Selective Cutting Down and Removing Trees

The cost of cutting down and removing trees will vary with their size and the method used in cutting and disposing of them.

The following production times are based on a crew of 3 or 4 men working together, cutting down trees by hand, removing branches and cutting the trunk into 4'-0" lengths. Labor costs of digging down around roots and removing stump are given extra. It is assumed the limited quantity would not require chipping.

Approx. Diam. of Tree	Approx. Height of Tree	Labor Hours	Extra Labor Removing Stump
8" to 12"	20'-0" to 25'-0"	9 to 12	8
14" to 18"	30'-0" to 40'-0"	12 to 16	10
20" to 24"	45'-0" to 50'-0"	20 to 24	12

If necessary to cut trunks and stumps into small pieces and load into trucks, add extra for this work. Add also for removal from the site.

Where many trees are to be removed, a power or chain saw will speed the work and reduce costs considerably. A chain saw will cut within 2 to 4 inches of the ground but if the stump must be removed, it is usually done by hand.

Chain saws are obtainable operated by gasoline or electric power or compressed air, and with 24", 36", 48" and larger saws, and rent for around $30.00 per day.

Using a chain saw with 3 men in the crew cutting trees, removing branches and cutting the trees into 2'-0" lengths, should average as follows:

Approx. Diam. of Tree	Approx. Height of Tree	Labor Hours	Extra Labor Removing Stump
8" to 12"	20'-0" to 25'-0"	2¼ to 3	8
14" to 18"	30'-0" to 40'-0"	3 to 4	10
20" to 24"	45'-0" to 50'-0"	5 to 6	12

A faster and more economical method of removing trees is by using a bulldozer or loader provided it would not cause damage to adjacent trees.

Trees up to 15 inches diameter or trees such as pine, which have small root systems, can be rapidly up-rooted by raising the dozer blade or shovel bucket to maximum height and pushing against the tree. It may be necessary to strike the tree a few times to loosen it.

Trees with diameters of 24 inches or more, or trees like elms, maples and oaks, which have heavy, wide-flaring root systems usually require root cutting before they can be pushed over. First make a trench-cut with the dozer blade or shovel bucket across the roots on the side opposite the direction in which the tree must be felled. Backfill the trench and form a ramp up to the trunk so the machine can strike as high on the tree as possible. Unless the tree is exceptionally large or well-rooted, it can now be pushed over. If not, try cutting the roots on each side of the tree perpendicular to the first cut and, if necessary, the roots on the side toward which it will be felled.

Once the tree is felled it can be pushed or dragged to a disposal area where it can be sawed into lengths for hauling or burned.

A WORD OF CAUTION—Always watch for falling limbs. In the case of pines, if the whipping action is too great, the top may snap off and kick back on to the machine or operator. Also as the tree begins to fall, the machine must be backed up quickly so that the upturning root system will not come up under the machine with the possibility of damaging the unit or hanging it up.

An experienced operator, using a medium to large size tractor, 80-Hp and up if crawler type, 160-Hp and up if rubber-tired, with ground conditions affording good traction and no problems regarding steep slopes, etc., should fell trees at the following rates:

Approx. Diam. of Tree	Approx. Height of Tree	Minutes Req'd. for Felling
8" to 12"	20'-0" to 25'-0"	5 to 10
15" to 24"	30'-0" to 50'-0"	15 to 20
27" to 36"	55'-0" to 75'-0"	25 to 30

Add for pushing felled trees to disposal area and labor disposing of same.

Stripping & Storing Topsoil

Following the site clearing operation the stripping and storing of topsoil, if available, will be the next sequence of construction. In the interest of economy the recovery of the *maximum* quantity of existing topsoil is important. However, any area cleared of trees and stumps will have a loss factor of from approximately 15% to 50% depending on the number and size of stumps removed. The remaining topsoil should be stripped, cleaned of foreign materials and stored in a convenient location and protected from erosion or contamination.

The depth of the topsoil influences the production rate as less quantity would be available in a given area if it were 4" in depth as opposed to 6" and yet, the same area would have to be stripped by the dozer. The selection of the storage area is important as it should not only provide protection for the topsoil material, but it also should minimize the haul time involved in the storing and redistribution of the material to the final location.

Strip & Store Topsoil

Description	Unit	Average Cost
With dozer, 4"-6" depth 100' haul - average conditions	CU. YD.	$0.90 - $1.10 Erosion protection additional
With dozer, 4"-6" depth 300' haul - average conditions	CU. YD.	$3.50 - $4.25 Erosion protection additional
If adverse conditions are present increase unit price 20%-33%		Price range depends on terrain and capacity of equipment

The redistribution of topsoil usually requires some preparation of the ground prior to placing the topsoil.

Construction activities cause rutting, depressions and general unevenness of the site area, and to avoid excessive quantities of topsoil being used, the site should be graded to the designed contours less the depth of the topsoil. Any preparatory work such as scarifying, raking or rolling should be accomplished before the introduction of the topsoil. Requirements for soil conditioning, fertilizer, etc., become cost items in addition to the basic spreading cost. When specifications have rigid conditions concerning the quality of topsoil, it may be necessary to submit samples to a laboratory for analysis to establish compliance.

Off-site purchase of topsoil is very expensive and can vary from $4.00 - $10.00 per cu. yd. depending on the quality and availability.

Spread and Prepare Topsoil

Description	Unit	Average Cost
Scarify subsoil - average conditions	SQ. FT.	$0.02
Spread, 300' with dozer, average conditions	CU. YD.	$1.50 - $2.50
Hand dress topsoil ($10.00 hourly rate)	CU. YD.	$1.50 - $2.00
Rolling, push roller	SQ. FT.	$0.03 - $0.05

Caution is to be exercised in determining if the specified depth of topsoil is loose or compacted depth. Depending on the characteristics of the material involved, topsoil will compact from 33% to 50%. This could mean to obtain 4" compacted depth would require 6" of loose material. An Estimator not taking this into consideration would discover a 50% shortage from the estimated quantities.

If topsoil is to be placed in a series of small restricted areas or at the perimeter of buildings, walks, drives or other obstructions, additional hand labor will be required to place, grade and hand compact the material.

02110   DEMOLITION

Demolition can be segregated into several main classification types i.e. demolition of entire buildings or structures, controlled removals for historic preservation work, and general building demolition of items such as doors and frames, windows, floors, ceilings, partitions, etc.

In most areas of the country there are specialty contractors whose sole business is the demolition of entire buildings and structures. These contractors will use a wide range of heavy equipment, even explosives for the controlled demolition work. It is always best to solicit a quotation from one of these firms for this type of demolition, however, the following per cubic foot unit prices will provide some budget guidelines in the absence of quotations:

Building Type	Unit Price
Low Rise - Steel Frame	$ .18 per/cu. ft.
Low Rise - Concrete Frame	$ .25 per/cu. ft.
Low Rise - Masonry Wall Bearing	$ .20 per/cu. ft.

Demolition associated with historical preservation work is highly specialized and, since each project has it's own particular set of intricacies, it is extremely difficult to produce accurate historical cost data that can then be saved for use in future estimates.

Of less intricate nature than the controlled demolition of historical preservation work, but closely related thereto, is the general building demolition work referred to above. It is this type of work that in the following text we will address for the various types and items of demolition and their appropriate labor production tables and costs.

The estimator must always take into consideration factors such as, working space conditions, is the building occupied or not, method and route in removing debris to the trash receptacle or truck, special equipment requirements, temporary bracing and shoring, dust and security control partitions, length of the haul to the dump site, and special use permits and dump fees.

Demolition Using Compressed Air Tools

Compressed air is used extensively in all classes of construction work including such operations as concrete breaking, masonry chipping, excavating, compacting earth, driving sheet pilings, etc. To reduce the overall operating cost (ownership or rental rates), it is always wise to use an air compressor that is capable of handling at least 2 pneumatic tools. The use of 2 tools lowers the rental rate per hour for the machine down to a level half of what it would be if only one operator was using the machine to do an operation. For example: If you had 1,000 sq. ft. of concrete sidewalk to demolish, using one compressor with two hammers and two men would require the same total number of labor hours as one man would take. However, your equipment rental hours would be half of that required if you used only one man on the air compressor.

The following chart has been constructed to show the classification of work to be demolished by an air hammer, the daily production of work per crew of two (2) laborers 8 hours each, and the total number of man hours required per 100 sq. ft. or lin. ft. whichever is applicable. Keep in mind that the production rates shown in the following table only include the material break-up, not transport/removal to the trash receptacle.

Item Description	Crew Daily Production	Man Hours Per 100 Sq. Ft.	Man Hours Per 100 Lin. Ft.
Wall Footing - 2' wide x 1' thick	125 lin. ft.	—	12.80
Wall Footing - 3' wide x 1'-6" thick	65 lin. ft.	—	24.62
Slab on Grade - 4" with mesh	450 sq. ft.	3.56	—
Slab on Grade - 8" with mesh	250 sq. ft.	6.40	—
Concrete Foundations Walls			
Lightly Reinforced:			
6" Thick	200 sq. ft.	8.00	—
8" Thick	160 sq. ft.	10.00	—
10" Thick	135 sq. ft.	11.85	—
12" Thick	110 sq. ft.	14.55	—
Header Curb	225 lin. ft.		7.11
Curb & Gutter	175 lin. ft.		9.14

Partition Removal

The following labor production table gives the daily crew (2 laborers 8 hours each) output and the man hours required per 100 sq. ft. for the demolition of various types of partitions. The labor production hours does not include the transport/removal of demolished materials to the trash receptacle.

Partitions 10' High

Type Of Partitions	Daily Production Sq. Ft.	Man Hours Per 100 Sq. Ft.
Metal Stud & Drywall	850	1.88
Plaster & Lath on Wood Stud	650	2.46
4" Concrete Masonry Units	650	2.46
8" Concrete Masonry Units	500	3.20
12" Concrete Masonry Units	400	4.00
4" Brick	500	3.20

Partitions 15' High

Type Of Partitions	Daily Production Sq. Ft.	Man Hours Per 100 Sq. Ft.
Metal Stud & Drywall	800	2.00
Plaster & Lath on Wood Stud	600	2.67
4" Concrete Masonry Units	600	2.67
8" Concrete Masonry Units	450	3.56
12" Concrete Masonry Units	375	4.27
4" Brick	450	3.56

Door & Frame Removal

The following labor production rates are for the removal of doors and frames from walls to be demolished. Keep in mind that if a frame is to be removed from a wall that is to remain in place, more care must be used to prevent needless damage, and thus costly repairs, to the wall. As a guide the hours shown in the following table could be doubled for the removal of metal door frames from masonry walls that will remain in place. The production times shown in the table does not include any transport/removal of the demolished materials to the trash receptacle.

Item Description	Man Hours Per Unit
Single wood door & frame in stud partition	1.00
Pair - wood doors & frame in stud partition	1.25
Single door and frame in masonry partition	1.25

Pair doors and frame in masonry partition	1.50
Single hollow metal door and frame in stud partition	1.50
Pair hollow metal doors & frame in masonry partition	2.50
Single hollow metal door & frame in masonry partition	2.25

Window Removal

The following table shows various types of window materials and the required man hours per unit for their removal. The production times shown in the table does not include any transport/removal of the demolished materials to the trash receptacle.

Type	Size	Man Hours Per Unit
Wood	up to 3' x 5'	.75
Wood	up to 5' x 6'	1.25
Aluminum	up to 3' x 5'	1.25
Aluminum	up to 5' x 6'	1.75
Steel	up to 3' x 5'	2.00
Steel	up to 5' x 6'	2.50

Floor Finishes Removal

The following table shows various types of finish flooring materials and the required man hours per 100 sq.ft. for their removal. The production times shown in the table does not include any transport/removal of the demolished materials to the trash receptacle.

Type of Flooring	Man Hours Per 100 Sq. Ft.
Vinyl composition tile	2.00
Wood parquet	2.50
Wood strip	2.75
Carpet & pad - tack strip	0.40
Carpet - glued down	1.00
Ceramic tile	2.50

Ceiling Finishes Removal

The following table shows various types of ceiling finish materials and the required man hours per 100 sq. ft. for their removal. The production times shown in the table does not include any transport/removal of the demolished materials to the trash receptacle.

Type of Material	Man Hours Per 100 Sq. Ft.
Drywall	1.00
Suspended plaster	3.00
Suspended acoustical tile and grid	1.25
Plywood	1.00

*If the ceiling finishes are more than 10' above the finished floor, add a 10% labor factor multiplier for every 2' increment over 10'.

Wall Finish Removal

The following table shows various types of wall finish materials and the required man hours per 100 sq. ft. for their removal. The production times shown in the table does not include any transport/removal of the demolished materials to the trash receptacle.

Type Of Material	Man Hours Per 100 Sq. Ft.
Drywall	.60
Plaster	1.00
Wood paneling	.85
Ceramic tile	2.25

Roofing Material Removal

The following table shows various types of roofing materials and the required man hours per 100 sq. ft. of their removal. The production times shown in the table does not include any transport/removal of the demolished materials to the trash receptacle.

Item Description	Man Hours Per 100 Sq. Ft.
Built up roofing - 4 ply w/gravel	3.50
Rigid insulation board	.75
Shingles - composition strip	.90
Shingles - wood	1.25
Shingles - slate	1.30
Sheathing - plywood	.95
Sheathing - 1" plank	1.10
Sheathing - 2" plank	1.50

Siding Removal

The following table shows various types of siding materials and the required man hours per 100 sq. ft. for their removal. The production times shown in the table does not include any transport/removal of the demolished materials to the trash receptacle.

Item Description	Man Hours Per 100 Sq. Ft.
Metal - horizontal	2.75
Metal - vertical	3.00
Wood - horizontal	3.00
Wood - vertical	3.25
Shingle - wood or asbestos	3.50
Plywood	1.50

Core Drilling

One of the quickest and neatest ways of cutting a hole in a concrete slab or wall is by core drilling. This particular rig uses a circular shaped diamond drill bit that is operated by a large drilling machine mounted on a stand. The drill needs water injected into the core drilling bit to provide lubrication and cooling. However, the water is not a problem as the amount of water that is used is small and certainly controllable to prevent any damage to surrounding finishes.

Core drill bits come in diameters from 1" to 18" in size and the depth of cut allowed by the average coring bit is about 12", however, special bits are made to allow boring through columns or walls up to 24" in thickness or more.

Core drilling is usually performed by specialty subcontractors whose only business is to provide this service to the general contracting field. It is best to obtain a firm quotation for any core drilling that is required for your project. However, in the event that a quick budget estimate is needed, core drilling is usually priced by the unit inch, which is the diameter in inches x the depth in inches. For example a hole 4" in diameter and 10" deep would be equal to 40 unit inches. Currently the going rate per unit inch is approximately $1.00 to $1.25 depending on the quantity of holes to be drilled, location and set-up time required. Horizontal holes will cost about 10% to 15% more than a vertical hole.

02200   EARTHWORK

It is not a difficult matter to compute the number of cu. yds. of earth to be excavated or removed from the premises on a specific project, but it is quite another matter to estimate the actual cost of performing this work, because of the many unknown items entering into the cost. This includes such items as type of soil, whether or not water will be encountered, pumping, whether the banks of the excavated portion will be self-supporting or whether it will be necessary to brace and sheet them, disposal of excavated earth, length of haul to dump, etc.

Estimating Quantities of Excavation.—Excavating is measured by the cu. yd. containing 27 cu. ft.

General or Mass Excavation.—When computing the cubical contents of any basement or portion to be excavated take all measurements from outside face of wall footings and add 6" to 1'-0" on each side to allow for placing and removing forms and then allow for sloping the banks so they will stand reasonably stable without being supported by bracing, sheet piling, etc. The amount of slope necessary to provide a "safe" hole will vary with the kind of soil to be excavated, depth of excavation, etc. Digging in previously undisturbed material and assuming that no water or unstable conditions exist most estimators, when taking off quantities for excavation, use a 1 : 1 slope, that is one horizontal to one vertical, for sand and gravel; a 1 : 2 slope for ordinary clay and a 1 : 3 or 1 : 4 slope for stiff clay.

In some cases job conditions may not permit the sloping of banks and then it is usually necessary to sheet and brace the banks to prevent accidents from occurring resulting in damage to adjoining property and injury to workmen on the job.

On jobs having column footings which project beyond the wall footings, additional cuts must be made into the banks at the column locations to accommodate the additional width of the footings. The volume of this cut is computed by taking the column footing length plus 1'-0" or 2'-0" for work space plus "layback" of banks on both ends and multiplying by the amount of projection beyond the wall footing to obtain the area of the additional cut, which is then multiplied by the depth of the cut to obtain the volume of additional excavation required.

The depth of the general cut is usually taken from the underside of the topsoil, previously removed, to the underside of the floor slab. Where sand, gravel or other filling material is to be placed under the concrete floor the depth should be taken to the underside of the filling material.

Example: Assume you are figuring the general or mass excavation for a building

100'-0" x 100'-0" at grade. The wall footings extend 6" beyond the face of the foundation walls and on each side there are 6 column footings which are 6'-0" long and project 2'-0" beyond the face of the wall footings. Also assume the depth of the excavation to be 8'-0" and that the soil will stand at a slope of 1 : 2 (one horizontal to two vertical). The volume of the excavation should be figured as follows:

100'-0" building length plus 2 x 0'-6" for wall footings plus 2 x 0'-6" for work space plus an average of 4'-0" for slope equals a length of 106'-0". Since the building is square the main volume would be 106'-0" x 106'-0" x 8'-0" or 89,888 cu. ft.

For the 24 column footings which project beyond the wall footings, the length of each additional cut would be the 6'-0" footing length plus 2 x 0'-6" for work space plus an average of 4'-0" for slope or 11'-0" and the width of each cut would be the 2'-0" projection. Therefore the additional excavation required to accommodate the column footing projections would be 24 x 11'-0" x 2'-0" x 8'-0" or 4,224 cu. ft.

The total volume required to be excavated would then be 89,888 cu. ft. plus 4,224 cu. ft. equals 94,112 cu. ft. or 3,486 cu. yds.

In some cases a portion of or the entire building site may be high and require a grading cut to produce the required contours. In these instances the grading cut should be figured first and the general or mass excavation computed afterwards taking the depth from the underside of the grading cut to the bottom of the floor slab or fill under the floor.

Trench and Pit Excavation.—To complete the building excavation quantity take-off, there are usually additional items of excavation to figure for footings, foundations, pits, trenches, etc., which extend below the level of the general or mass excavation, or which occur in unexcavated areas. This kind of excavation is always estimated separately from the general work as it is usually a more expensive operation.

In many instances a good portion of this work may be done with power equipment, with some handwork required to clean out corners and grade the bottoms. On the other hand there may be portions of the work which must be done entirely by hand and which would be much more costly than machine work. In view of the wide variance in unit cost between machine work and handwork, the estimator should decide, when he is taking off the quantities, into which category each portion of the work will fall and list the machine work separately from the handwork so that each may be properly priced.

Quantities for this work are measured and listed in the same manner as for general excavation, except the slope factor may usually be ignored, unless the additional depth exceeds 4'-0" or if the ground will not stand safely for this depth with vertical banks. Under these conditions, allowances for sloping the banks should be made in the same manner as for general excavation.

Backfill.—After all excavating items have been figured the amount of backfill should be computed.

An easy method of doing this is by computing the displacement volume of the construction which is to be built within the limits of the excavation, that is, the volume of footings, piers, the basement volume figured from the underside of the fill under the floor to the elevation of the top of the general cut, etc., and deducting this volume from the sum of the general or mass excavation and the trench and pit excavation. The remainder is the volume of backfill required.

Another method of figuring backfill, usually more difficult than the one previously described, is that of computing the volume from the actual dimensions of the spaces requiring backfill.

Some specifications require interior backfills to be made with sand, gravel, bank-run gravel, etc., while the exterior backfill may be excavated material. In this case it is necessary to keep the two items separate as the interior backfill material will probably have to be purchased and brought in from outside sources.

Backfill very often is specified to be compacted by mechanical tampers and a degree of compaction is also specified. In these instances the backfill operations must be considered very carefully as this is expensive work.

Excavation and Fill for New Site Grades.—Excavation and/or filling for new site grades, including cuts and fills for drives, walks, parking areas, etc., vary from simple to complex according to the site ground conditions. Estimating quantities for this work may be rather difficult and some surveying or engineering knowledge will greatly help at this point.

A number of methods are available for determining the volumes of cut and fill required and the method which is most adaptable to the particular problem will have to be chosen by the estimator after considering the problem with the information given and the accuracy desired.

Cross Sectional Method

One of the easiest and most frequently used methods of computing grading cuts and fills when the plot plan shows both original and proposed contours is that of gridding or dividing up the area to be graded into squares, rectangles, triangles or combinations of these figures, of regular and convenient dimensions, thus forming a series of truncated prisms. The depths of cut or fill at the corners of these prisms are computed and the volume of each prism may be determined by averaging these depths and multiplying the average by the cross sectional area of the prism. The results of these computations are totaled, keeping the cuts separate from the fills, and the total excavation and/or fill required is obtained.

Example: Assume the contour sketch shown on the next page to be a 400.0' x 200.0' portion of a site grading plan showing both present and proposed contour lines. Since both the present and proposed surfaces have fairly regular slopes, as indicated by the even spacing of contour lines, a grid spacing of 50 feet in both directions has been chosen and grid lines drawn as shown, labeled alphabetically for the vertical lines and numerically for the horizontal lines.

The next operation is to figure the present ground elevation for each grid line intersection by interpolating between the present contour lines. The results are marked on the drawing in the upper right quadrant of each respective intersection.

Following this, the proposed ground elevations are similarly figured for each intersection, interpolating between the proposed contour lines and entering the results in the upper left quadrant of each respective intersection.

When this is completed the depth of cut or fill at each intersection can be readily determined by subtracting one elevation from the other and noting whether a cut or fill is required. This determination should be made for each intersection and the results entered in the proper quadrant, using the lower left for cuts and the lower right for fills.

When all intersections have a depth value entered for either cut or fill the plan should be examined closely to see where the grading operation will change from cut to fill or vice versa. In the sketch it will be seen that advancing from left to right on horizontal grid line "1" a change from fill to cut occurs between vertical grid lines "D" and "E" and by interpolating between the fill and cut depths the zero point "a" can be established approximately 36 feet to the right of intersection "D-1."

Similarly, zero points "b, d, e and g" are established on the horizontal grid lines and zero points "c and f" on the vertical grid lines. These points are then connected by a line "a, d, c, d, e, f, g" and for all practical purposes it may be assumed that all points on this line are at the proposed elevation and require neither cut nor fill. In this case the line forms the boundary between cuts and fills. The change from cut to fill and vice versa could, of course, occur many times in actual problems depending on the irregularity of the present and/or proposed contour lines.

After the previously described operations have been completed, the actual volume of cut and/or fill required can be computed.

The volume of cut or fill in each prism in the grid system is obtained by averaging the corner depth values of each prism and multiplying by its cross sectional area. Thus, for the square prism "A-1, B-1, B-2, A-2" the volume of fill would be ¼ x (2.4'+1.6'+1.6'+2.5') x 50.0' x 50.0' or 5,063 cu. ft. All grid prisms are similarly treated, keeping cut and fill results separated, except those prisms through which the zero line passes. Therefore for square prism "D-1, E-1, E-2, D-2" it will be seen that this volume has been divided into two trapezoidal prisms "D-1, a, b, D-2" and "a, E-1, E-2, b." To compute the respective fill and cut volumes of these prisms they must be divided into triangular prisms first and then the corner depths may be averaged and multiplied by their respective cross sectional areas. Thus for the triangular prism "D-1, a, b" the volume of fill would be ⅓ x (0.5'+0.0'+0.0') x ½ x 50.0' x 36.0' or 150 cu. ft. Similarly, the volume of fill for the triangular prism "D-1, b, D-2" would be ⅓ x (0.5'+0.0'+0.3') x ½ x 50.0' x 21.0' or 140 cu. ft. The volume of cut required for the trapezoidal prism "a, E-1, E-2, b" is figured in the same manner. In the case of square cross section "C-2, D-2, D-3, C-3" it will be seen that the zero line crosses one corner of the square and divides it into a triangle "c, D-3, d" and a five-sided figure "C-2, D-2, c, d, C-3." The five-sided figure should be further divided into three triangles "C-2, D-2, c," "C-2, c, C-3" and "C-3, c, d" and the volumes of fill for each triangular prism would be as follows:

⅓ x (0.7'+0.3'+0.0') x ½ x 50.0' x 30.0' or 250 cu. ft.
⅓ x (0.7'+0.0'+0.5') x ½ x 50.0' x 50.0' or 500 cu. ft.
⅓ x (0.5'+0.0'+0.0') x ½ x 36.0' x 20.0' or 60 cu. ft.

When there are several prisms with the same cross section to be either cut or filled the volume of such can be computed as one solid by assembling them as follows: Multiply each corner depth by the number of typical prisms in which it occurs and then add these results and divide by 4. This is then multiplied by the cross sectional area of one prism. For example the volume bounded by A-1, D-1, D-2, C-2, C-4, B-4, B-5, A-5, A-1 can be obtained by one computation because it is composed of a series of prisms having the same cross section. In the summation of depths, those at A-1, D-1, D-2, C-4, B-5 and A-5 are taken but once, those at B-1, C-1, C-3, A-4, A-3 and A-2 are multiplied by 2, at C-2 and B-4 the depths are multiplied by 3 and at B-2 and B-3 they are multiplied by 4. All of these values are then totaled and divided by 4 and then multiplied by the cross sectional area of one prism giving the volume of fill required.

When a volume has been computed for each grid division of the site, either individually or collectively, it is only a matter of summing up the cuts and/or fills to obtain the total excavation and/or fill required to bring the site to the proposed grades.

End Area Method

It often happens that a cut and fill operation is to be performed over a long and narrow area such as for roads, levees, ditches, etc. In this case the usual method of computing the earthwork quantities is by means of end areas.

In jobs of this sort a profile plan and a topographic plan is generally furnished to the contractor as part of the bidding documents. Using this information, sections are taken perpendicular to the longitudinal center line of the work, usually at each 100 foot station, at each radical change in either present or proposed grades and where the cut changes to fill or vice versa. These sections are plotted to scale

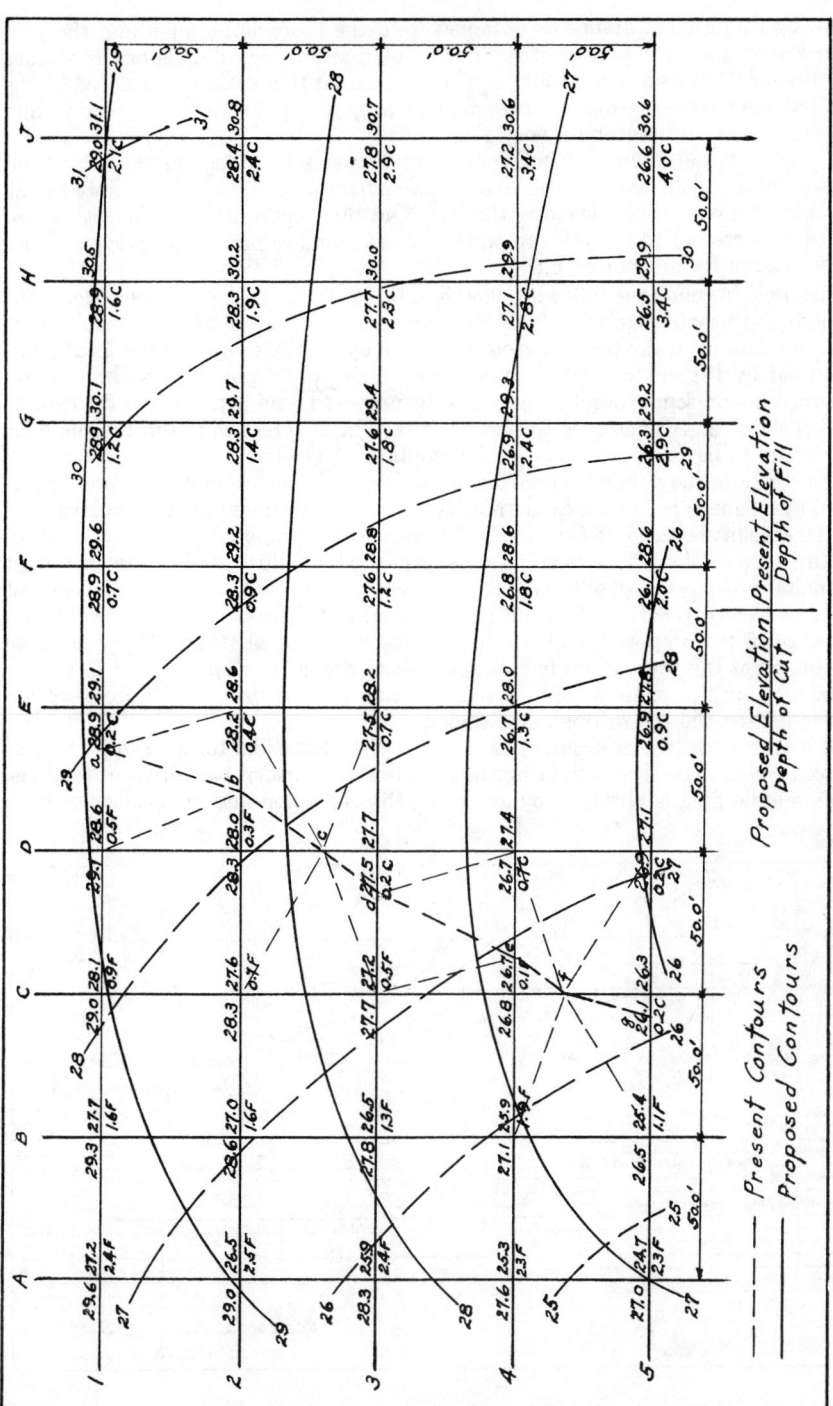

Method of Cross Sectioning a Contour Plan.

showing the original grades as obtained from the topographic plan and the proposed grades as obtained from the profile plan and design details of the surface required. After this has been done it will be apparent that areas are enclosed by the plotted lines representing the present and proposed grades at each section and these areas should be computed.

To obtain the volume of cut and/or fill between each section the usual and easiest method is to average the areas of two adjacent sections and multiply by the distance between them, keeping the cuts and fills separate. This method is not quite accurate but the results are on the safe side and in practically every case are close enough for estimating purposes.

Example: Assume the following sketch to be a 300 foot portion of the topographic plan and profile plan of a proposed road location. The topographic plan shows the undulations of the present ground surface by the contours and the location of the road by the center and shoulder lines. The profile plan shows the relation between the present ground surface and the proposed road bed grade at the centerline of the road. Also assume the road bed is to be 20.0 feet wide and the shoulder slopes are to be 1½ : 1, that is 1½ horizontal to 1 vertical.

To compute the volumes of earthwork involved in this proposed grading operation, by means of the end area method, sections must be taken perpendicular to the center-line of the proposed road and the present and proposed grade lines plotted. In this case sections have been taken and plotted, as illustrated in the sketch, at even stations 15+00, 16+00, 17+00, and 18+00. Sections have also been taken at stations 15+07, 15+37, 16+50 and 16+65 because at these points the cut or fill at a shoulder of the proposed road is zero. A section is taken at station 17+65 because at this point the present grade makes a radical change in slope.

In plotting the sections any convenient scale may be used and if cross-section paper is used the job will be much easier.

If a greater degree of accuracy is required the prismoidal formula must be employed which necessitates figuring the middle area between two adjacent sections. This middle area is obtained by averaging the corresponding dimensions of both

Topographic and Profile Plans.

Method of Plotting End Areas.

end areas and computing an area from the average dimensions; it should not be taken as the average of the two end areas. To obtain the volume between adjacent sections using the prismoidal formula, multiply the length between sections by one-sixth of the sum of the two end areas and 4 times the middle area combined.

Most specifications call for the fills to be placed in layers and then specify some degree of compaction to be attained. This should be considered by the estimator as additional equipment such as sheepsfoot rollers, pneumatic rollers, tandem rollers, etc., will be required plus the labor of operation.

As each section is plotted, the area enclosed by the present and proposed grade lines is computed and the result tabulated. Final computations are made as follows:

Station	Distance Between Stations	Area of Sections in Sq. Ft.				Volume in Cut	Cu. Ft. Fill
		Actual Cut	Average Cut	Actual Fill	Average Fill		
15+00		24.38		—			
	7.0 ft.		20.41		—	142.87	—
15+07		16.44		—			
	30.0 ft.		8.22		7.46	246.60	223.80
15+37		—		14.92			
	63.0 ft.		—		51.23	—	3,227.49
16+00		—		87.54			
	50.0 ft.		—		53.45	—	2,672.50
16+50		—		19.36			
	15.0 ft.		7.56		9.68	113.40	145.20

Station	Distance Between Stations	Actual Cut	Area of Sections in Sq. Ft. Average Cut	Actual Fill	Average Fill	Volume in Cut	Cu. Ft. Fill
16+65		15.11		—			
	35.0 ft.		26.72		—	935.20	—
17+00		38.33		—			
	65.0 ft.		59.77		—	3,885.05	—
17+65		81.20		—			
	35.0 ft.		62.34		—	2,181.90	—
18+00		43.48		—			
						7,505.02	6,268.99

Rough Grading of Site.—An item for rough grading consisting of the total area cut and/or filled, less buildings, roads, walks, etc., should be included in the estimate to cover the expense of dressing up the surface after the required grades have been approximately attained and preparatory to the spreading of top soil, if required.

Spreading Top Soil.—In most instances the specifications call for spreading top soil over site areas not covered by other construction, usually to a depth of 6 inches.

Finish Grading of Site.—Fine grading of top soil, usually including a hand raking operation, is generally called for and should be listed in the estimate as an item, the quantity being the area used in computing the volume of top soil to be spread.

Other Grading.—Other grading items to be considered by the estimator are the following: grading the subgrade for floor slabs, walks, paving, etc. Also some contractors include an item for grading bottoms of footings, when they are to be dug by machine, and also include an amount for labor squaring and cleaning the holes.

Disposal of Excess or Borrow.—After all excavation, backfill, grading cuts and fills, etc., have been computed and listed in the estimate, the total cuts and the total fills should be compared to determine whether there will be an excess of materials to dispose of or a deficiency of materials to be made up by borrowing or purchasing from outside sources and the results of this comparison should be listed in the estimate. The top soil comparison should be kept separate from the other materials because the cost of any top soil which must be purchased usually is much more expensive than ordinary fill material. Also any excess of top soil may find a ready market and be sold thereby reducing the overall cost of the earthwork. The difference between the cuts and fills will give the bank or compacted volume required. In the case of a deficiency of materials this compacted volume quantity must be adjusted for shrinkage due to compaction as the material will probably be bought by loose measure. In the same manner all excess bank measurement volumes must be adjusted for swell due to bulking to obtain the true volume which must be handled in the disposal operation. The amount of increase for gravels and sand is from 5 to 12 percent and for clays and loams from 10 to 30 percent.

Pumping or De-watering.—Finally, careful consideration should be given to the probability of pumping or de-watering operations. In some parts of the country this is no problem whatsoever, quite the reverse being the case. However, in most localities ground or climatic conditions make it necessary to include pumping or de-watering as an item in the estimate.

This item is quite variable, being affected by the season of the year as well as the locale. If only rainwater run-off is expected, an allowance for a few small pumps may be sufficient. If, however, the job is in close proximity to a body of water or springs are present, etc., it may be necessary to de-water part or all of the operational area by means of a well point system in which case the contractor should consult a company which specializes in this work. See section on wellpoint system.

Estimating the Labor Cost of Excavation.—Before an excavating estimate is made up or a bid submitted, the estimator should visit the site and ascertain the exact conditions surrounding the work to be performed. The natural grade of the lot should be noted to compute the depth of the excavation from grade to underside of floor slab; the nature of the soil should be determined, whether sand, loam, clay or rock, and whether or not water is likely to be encountered.

Water usually means an additional cost—pumping out the hole day and night until the foundations are completed. Rock usually means blasting, with a greatly increased cost for drilling, dynamiting, excavating and loading.

The safest method of determining the kind of soil is to dig a test hole the approximate depth of the basement, and from this you should obtain a fairly accurate idea of the conditions to be encountered.

In most areas the soil strata, as well as the ground-water table, is fairly consistent, and its character may be ascertained by talking to contractors or engineers who have worked in the local area for some time. In most larger jobs, the plans will show the results of test borings ordered by the Owner or Architect/Engineer which indicate the character of the sub-surface.

If the lot is low and the basement does not cover the entire lot, part of the excavated soil may be used for filling. On the other hand, if the natural grade of the lot is above the sidewalk or established grade, it may be necessary to remove a certain amount of soil from the entire lot.

In large structures having basements and sub-basements, the excavation may extend 40 to 50 feet below grade. This requires the construction of steep runways so that trucks may get in and out of the hole, also an extra truck or bulldozer may be required to help the loaded trucks out of the hole. These items must all be taken into consideration when pricing any excavating job.

If it is necessary to remove the excavated soil from the premises, locate the nearest available dump, and ascertain just how the soil may be handled at the dump, whether it is necessary to spread it, etc.

These are all items that affect the cost and should be given careful consideration by the estimator.

In most localities there are contractors who make a specialty of excavating and are familiar with the kind of soil encountered in different parts of the town. These contractors usually quote a price per cu. yd. for all of the general excavation or they may quote a lump sum for the entire job.

Trench, pier and all hand excavating is usually performed by the general contractor.

METHODS OF EXCAVATING

Excavating costs will vary with the kind of soil and the method used in removing and hauling to a final destination.

Except on very small jobs, practically all mass excavation today is performed by power operated equipment, i.e., power shovels, cranes with dragline or clamshell buckets, bulldozers, tractor excavators, scrapers and trenching machines, while ordinary grading and backfilling is performed by bulldozers and high shovel tractor excavators.

In shallow excavations, where the excavated earth may be spread over the site not more than 50 feet from the excavation, bulldozers or loaders (a tractor equipped with a shovel that may be elevated to load directly into trucks or over the sides of the excavation) are frequently used. This type of excavator is also used for excavating basements, loading directly into trucks.

There is also a "back-digger" type of power shovel commonly called a "hoe" that is extensively used for basement excavation as it is equipped with a shovel that "pulls" toward the machine which is always located on the bank above the excavation, working from the sides and ends of the excavation.

In large excavations, the use of a regular power shovel is the most satisfactory and economical method to use. The power shovel will excavate and load the trucks as fast as they can be handled in and out of the hole and will enable the job to proceed much faster.

If it is desirable to keep the digging machine on the bank above the excavation and the depth is greater than a hoe can handle, a crane equipped with a dragline bucket or clam bucket is usually used.

For individual pier footings, small pits and hard-to-reach spots, a crane equipped with a clamshell bucket or a small back-hoe may be used.

On medium to large size grading jobs, self-propelled or tractor hauled scrapers are used quite often and perform cutting and filling operations very creditably.

The tables of quantities and production times given in the following table are average. Experienced excavators will perform more work than given in the tables but an inexperienced man will perform less, so average conditions have been used.

HAND EXCAVATION

Most construction projects require extensive use of hand labor to accomplish the selective types of excavation which cannot be performed by machine excavators.

In the excavation of footings it is frequently found to be cost effective to combine the use of machines for the bulk excavation and then to "hand dress" the bottoms, sides, and corners to produce the precise horizontal and vertical planes required. Similarly this is true in finalizing the subgrade elevation under slabs on grade, walks and patio slabs.

The quantities and production rates given in the following table are for average conditions.

Material & Operation	Avg. No. Cu. Yds. Per 8-Hour Day	Labor Hours Per Cu. Yd.
Sandy loam		
Small footings, hand	6	1.33
Trenches 3' to 4' deep, hand	6	1.33
Piers 6' deep, hand 1 lift	5	1.60
Piers 12' deep, hand 2 lifts	3	2.66
Heavy soil and clay		
Small footings, hand	5	1.60
Trenches 3' to 4' deep, hand	5	1.60
Piers 6' deep, hand, 1 lift	4	2.00
Piers 12' deep, hand, 2 lifts	2	4.00
Excavation with ½ cu.yd. backhoe, casting to side		
Trench footing	300	—
Pier footing 6' deep	200	—
Pier footing 12' deep	150	—

Hand Trimming Trench, Piers And Slabs After Machine Excavation

Area	Sq. Ft. Per 8 Hour Day	Sq. Ft. Per Hour
Bottom and side areas	240	30
Fine grade under slabs	600	75

Backfilling By Hand With Compaction Material From Stockpile

Ordinary Light Soils	Cu. Yd. Per 8 Hour Day	Labor Hours Per Cu. Yd.
Soil distribution laborer	12	.67
Compactor operator	12	.67

*(No allowance for compactor equipment cost)

MACHINE EXPLORATION

Excavating Using a 1 Cu. Yd. Tractor Shovel

Class of Work	Ave. No. Cu. Yds. per 8-Hr. Day	Tractor and Operator Hours Per 100 Cu. Yds.
Excavating ordinary soil and placing in piles on premises...........	325 to 375	2.30
Excavating heavy soil and clay and placing in piles on premises...........	250 to 300	2.90
Loading loose sand or gravel into trucks...........	400 to 450	1.90
Excavating ordinary soil and loading into trucks...........	400 to 450	1.90
Excavating heavy soil and clay and loading into trucks	300 to 350	2.50
Backfilling loose earth—Bulldozing...........	500 to 600	1.45

The above table is based upon a 50-foot haul from excavation to dump. For each additional 50 feet in length of haul, add 1½ to 2 hrs. tractor time per 100 cu. yds. or consider the use of trucks.

Excavating Using a 2¼ Cu. Yd. Tractor Shovel

Class of Work	Ave. No. Cu. Yds. per 8-Hr. Day	Tractor and Operator Hours Per 100 Cu. Yds.
Excavating ordinary soil and placing in piles on premises...........	600 to 650	1.30
Excavating heavy soil and clay and placing in piles on premises...........	450 to 500	1.70
Loading loose sand or gravel into trucks...........	775 to 825	1.00
Excavating ordinary soil and loading into trucks...........	775 to 825	1.00
Excavating heavy soil and clay and loading into trucks	575 to 625	1.35
Backfilling loose earth—bulldozing...........	925 to 975	0.85

The above table is based upon a 50-foot haul from excavation to dump. For each additional 50 feet in length of haul, add ¾ to 1 hrs. tractor time per 100 cu. yds. or consider the use of trucks.

ESTIMATING THE COST OF OWNING AND OPERATING CONSTRUCTION EQUIPMENT

The following information on the costs of owning and operating power shovels, hoes, cranes, motor trucks, tractors and scrapers, is included to give the estimator a general idea of the costs involved in making up an estimate for this type of work.

The equipment purchase prices given in the following text are only a relative guideline to be used in the tables for "Obtaining the Approximate Cost of Ownership and Operation" of various pieces of equipment. Of major importance is the format for determining these costs, not the price of the equipment listed herein, which will vary greatly depending on locality, method of purchase, number of pieces bought at one time, etc.

The price f.o.b. factory should cover the complete machine with all variable equipment and accessories such as various attachments, light plant, magnet generators, clamshell bucket, etc.

Total Investment or Cost of Equipment.—Depreciation, interest, taxes and insurance are directly related to the initial investment in the various types of construction equipment. In addition, certain other costs may be estimated by their normal relationship to this figure. Therefore, the proper determination of the total cost is a basic requirement.

Approximate Prices of Power Shovels, Draglines, Clamshells and Lift Cranes

The following are approximate prices, f.o.b. factory, and are given so the estimator will have some idea of the investment required in shovels and cranes of various sizes:

Power Shovels

Size of Shovel

¾ cu. yd	$ 80,000.00
1 cu. yd	96,000.00
1¼ cu. yd	120,000.00
1¾ cu. yd	170,000.00
3¼ cu. yd	250,000.00
4¼ cu. yd	300,000.00
5½ cu. yd	450,000.00

Dragline Excavators Including a Medium Duty Bucket
Size of Dragline

½ cu. yd	$ 72,000.00
¾ cu. yd	87,000.00
1 cu. yd	109,000.00
1¼ cu. yd	145,000.00
2½ cu. yd	215,000.00
3 cu. yd	260,000.00
3½ cu. yd	350,000.00

Clamshell Excavators Including a Medium Duty Bucket
Size of Clamshell

½ cu. yd	$ 70,000.00
¾ cu. yd	87,000.00
1 cu. yd	108,000.00
1¼ cu. yd	144,000.00
2½ cu. yd	216,000.00
3 cu. yd	260,000.00
3½ cu. yd	350,000.00

Lift Cranes
Capacity

12 ton	$ 69,000.00
20 ton	86,000.00
22 ton	108,000.00
40 ton	144,000.00
44 ton	210,000.00
45 ton	258,000.00
75 ton	342,000.00

Approximate Rental Costs For Excavating Equipment With Operator

Type	Size	Day	Week	Month
Backhoe	½ cu. yd.	$ 370	$1,000	$ 4,200
	1 " "	600	1,700	6,900
	2 " "	900	4,000	16,000
Compactor*	1000# Ram.	47	225	800
	1000# Vib.	39	160	750
	5000# Vib.	180	750	2,800
	2000# Drum	80	350	1,300
Crane-Cable Crawler				
	30 T	560	2,500	9,500
	40 T	640	2,900	11,800
	60 T	720	3,200	13,100
Crane-Cable Truck				
	40 T	760	3,300	14,000
	60 T	900	3,900	16,400
	80 T	1,100	4,800	19,600
Crane Hydro Truck				
	5 T	470	2,100	8,400
	15 T	525	2,300	9,500
	30 T	570	2,500	10,300
Drill*	40# Rock	25	75	200
	65# Rock	30	85	240
Scraper	11 cu. yd.	500	2,250	9,000
	22 " "	850	3,900	15,800
Tractor - Dozer	105 HP	400	1,900	7,600
	180 "	450	2,100	8,600
Loader				
	1½ cu. yd.	400	1,900	7,600
	2¼ " "	480	2,300	9,100
Trencher	16 "	330	1,400	4,500
	24 "	400	1,600	5,800

*No Operator
The above costs do not include move charges or permits

Basic Power Shovel Unit and Front End Attachments.—Naturally, the ideal situation would be to have the most suitable power shovel, crane or dragline, as the case may be, for each job that is undertaken. Although very sound in principle, this is seldom economically practical or profitable. More frequently, the contractor must take into consideration the handling of a wide range of work entailing varying conditions, requiring more or less frequent and easy conversion from, for example, shovel to dragline or crane to clamshell, etc., by simply changing attachments on the basic unit to meet the varying job requirements. Power cranes and shovels, therefore, are designed to accommodate a wide range of front-end attachments and working tools to meet the various job requirements.

The attachments are classified into three basic groups: The shovel boom, the crane type boom and the hoe boom. These attachments generally are interchangeable in the field.

Since the basic purpose of the shovel boom is for shovel work and the hoe boom for hoe work, the crane type boom serves for all other materials handling tools some of which are as follows: dragline buckets, clamshell buckets, orange peel buckets, hooks, hook blocks, tongs, grabs, clamps, grapples, concrete buckets, skull crackers and pile driving leads.

Crane type booms can be extended for extremely high lifts by inserting standard sections in the center of the boom, and by boom tip extensions (jibs). These jibs are used primarily to extend horizontal reach with boom raised vertically close to maximum elevation. They also provide an increase in vertical lifting range. Since the jib adds weight at the outer lifting radius and extends the working range, it has a limiting effect on lifting capacity.

Average Useful Life of Power Shovels, Hoes and Cranes.—A reasonable allowance for the exhaustion, wear and tear of property used in business, including a reasonable allowance for obsolescence is permitted in figuring depreciation on your equipment.

The figures tabulated below are taken in part from Government Bulletin No. 456, "Depreciation Guide Lines and Rules," of the U. S. Treasury Dept. Internal Revenue Service.

Average Useful Life of Power Shovels and Hoes

⅜-¾ cu. yds.	5 years	or	10,000 hours
1-1½ cu. yds.	6 years	or	12,000 hours
2 cu. yds. and over	8 years	or	16,000 hours

Average Useful Life of Draglines, Clamshells and Cranes

Cu. Yds	Tons	Years	Hours
⅜-¾	2½-5	5	10,000
1-1½	10-15	9	18,000
2 and over	20 and over	12	24,000

The grouping shown for lifting cranes is a separate grouping and does not necessarily relate to the machines by size in cu. yds. For example, many ¾ cu. yd. excavators (clamshell or dragline) would have a greater rating than 5 tons and might be rated in tons to fall in a different group than they do as excavators.

Depreciation.—The straight line method of figuring depreciation is used. This method is in general adequate for this purpose, however if it is desired to take changing prices for equipment into account in estimating, an annual appraisal to determine a current value should be considered.

Therefore, for the years and hours used, the following percentage of depreciation for the various groups are used:

Years	Hours	% per Year	% per Hour
5	10,000	20 %	.01 %
6	12,000	16.67%	.00833%
8	16,000	12.5 %	.00625%
9	18,000	11.11%	.00555%
10	20,000	10 %	.005 %
12	24,000	8.33%	.00416%

Interest, Taxes and Insurance.—Interest, taxes, insurance, storage are usually charged at 20% of the average investment, of which 15% is interest (to be adjusted for current rates) and 5% covers taxes, insurance, storage and incidentals.

Average investment must be established, based on the number of years used for depreciation. Since the first year is considered as 100% and since the investment is considered at the beginning of each year, the method of calculating average investment for a 5 year depreciation period is as follows:

1st year	100% of Total Investment
2nd year	80% of Total Investment
3rd year	60% of Total Investment
4th year	40% of Total Investment
5th year	20% of Total Investment
	300% of Total Investment

Average investment equals 300% divided by 5 years equals 60% of Total Investment.

The percent of Total Investment to use for Average Investment for 5, 6, 8, 9, 10 and 12 year periods are as follows:

No. Years Depreciation	Percent Average Investment per Year
5	60 %
6	58.33%
8	56.25%
9	55.56%
10	55 %
12	54.17%

Therefore, the percentage of Total Investment per year and per hour to use for interest, insurance, taxes and storage, based on 20 percent of average investment for the groups considered here is as follows:

Interest, Insurance, Taxes, Storage
Based on 20 percent of Average investment per year.

Percent Average Investment per Year	Years	Hours	Percent of Total Investment Per Year	Per Hour
60 %	5	10,000	12 %	.006 %
58.33%	6	12,000	11.666%	.005833%
56.25%	8	16,000	11.25 %	.005625%
55.56%	9	18,000	11.112%	.005556%
55 %	10	20,000	11 %	.0055 %
54.17%	12	24,000	10.834%	.005417%

Fixed Costs, Repairs, Maintenance and Supplies.—These are difficult to estimate. The figures used are not based on actual records but are believed to be a fair representative average. In the case of shovels and hoes, this cost is based on 100 percent of the total investment spread over the life of the machine, and includes repairs, replacement parts and repair labor for normal operation, and includes such items as ropes, dipper and bucket teeth, oil and grease for lubricating machinery, as well as seasonal overhaul. They do not include engine fuel or lubricating oil.

In the case of draglines and clamshells, 80% of the total investment spread over the economic life of the machine is a fair average cost of these items.

In the case of lifting cranes 60% of the total investment spread over the economic life of the machinery may be used.

The owner of an excavator should anticipate three things in regard to repairs and maintenance. First, the average figures used presume proper maintenance. Without this, repair bills will greatly exceed the estimates given here. Second, it is assumed the equipment will be used within its specified rating. Third, these expenses are not uniform each year. Provision should be made for a periodic general overhaul, and it should be expected that both the amount of repairs required and the losses caused by shutdowns will increase as the equipment becomes older.

Repairs, Maintenance, and Supplies, Including Labor
Associated With Them
Shovels and Hoes
Based on 100 percent of Total Investment Spread over Life of Machine.

Cu. Yds.	Years	Hours	Percent Of Total Investment Per Year	Per Hour
⅜-¾	5	10,000	20 %	.01 %
1-1½	6	12,000	16.67%	.00833%
2-3½	8	16,000	12.5 %	.00625%
4¼-up	10	20,000	10 %	.005 %

For example: Assume a ¾-cu. yd. shovel cost $80,000.00.
Repairs, maintenance and supplies equal $80,000.00 x .0001 = $8.00 per hr.

Draglines and Clamshells
Based on 80 percent of Total Investment Spread over Life of Machine.

Cu. Yds.	Years	Hours	Percent Of Total Investment Per Year	Percent Of Total Investment Per Hour
⅜-¾	5	10,000	16 %	.008 %
1-1½	9	18,000	8.89%	.00445%
2-3½	12	24,000	6.66%	.00333%

Lifting Cranes
Based on 60 percent of Total Investment Spread over Life of Machine.

Tons	Years	Hours	Percent Of Total Investment Per Year	Percent Of Total Investment Per Hour
2½-5	5	10,000	12 %	.006 %
10-15	9	18,000	6.67%	.00333%
20 and over	12	24,000	5 %	.0025 %

Operating Costs, Engine Fuel and Lubricating Oil.—A formula for estimating the approximate amount of diesel fuel consumption for equipment of this type is as follows:

$$\frac{BHP \times Factor \times lbs.\ fuel\ per\ hp\ hr.}{Weight\ of\ fuel\ per\ gallon} = gal.\ per\ hr.$$

Where
BHP=Brake hp. of engine—or rated hp.
Factor=Factor for this use 50 to 60%
Diesel=0.5 lbs. per brake horsepower hour.
Diesel=7.3 lbs. per gallon (U.S.)

From this formula we obtain an approximate diesel fuel consumption of 0.034 to 0.041 gallons per horsepower hour (based on 50% to 60% factor respectively), and suggest using an average of about 0.040 gal. per horsepower hour.

Shovels normally will consume a greater amount of fuel than the other types of machines considered here. Therefore the larger consumption rate indicated should be used for estimating fuel for shovels and the smaller rate for draglines and clamshells. Machines used for lifting crane service only, usually operate intermittently and fuel consumption is difficult to estimate.

Example: A 100 horsepower engine in shovel service is estimated to consume per hour:

100 x .040=4.0 gals. diesel fuel per hr.

Lubricating oil is considered with fuel as it varies with the size and type of engine. It usually includes a complete change every 100 hours plus make-up oil between changes. Allow 15% of fuel costs.

Labor Operating Costs.—The labor rates as well as the number in the operating crew varies in different parts of the country and on different jobs, to such an extent that each contractor must estimate this cost for himself. Some costs relating to crew costs besides rates of pay are Federal Insurance Contribution Act, Workmen's Compensation Insurance, Unemployment Compensation, Overtime, Paid Holidays, Contractor's Contributions to Union Welfare and Pension Funds, etc.

Power Shovel Yardages.—To estimate yardage production on a job is a real problem. If all conditions were the same on all jobs, it would be a simple matter, but jobs are as different as night and day. There are many factors which must be considered, many of which can be learned only through experience.

The type of material, the depth of cut for maximum effect, no delays in operation, a 90° swing for the shovel, all affect the output, let alone the conditions existing on the job.

The following table gives an approximate idea of the difference in maximum output possible, subject to the conditions as listed above.

Hourly Shovel Output in Cubic Yards

Class of Material	3/8	1/2	3/4	1	1¼	1½	1¾	2	2½
Moist Loam or
Light Sandy Clay	85	115	165	205	250	285	320	355	405
Sand and Gravel	80	110	155	200	230	270	300	330	390
Common Earth	70	95	135	175	210	240	270	300	350
Clay,
Hard and Tough	50	75	110	145	180	210	235	265	310
Rock, Well Blasted	40	60	95	125	155	180	205	230	275
Common Earth, with
Rocks and Roots	30	50	80	105	130	155	180	200	245
Clay,
Wet and Sticky	25	40	70	95	120	145	165	185	230
Rock,
Poorly Blasted	15	25	50	75	95	115	140	160	195

Shovel Capacity spans the numeric columns.

The quantities given in the above table are based on bank measure which means cubic yards removed from the bank rather than cubic yards in the hauling unit. There is a big difference between a yard of dirt in the bank and a yard of dirt in the truck. This is because of the swell of the material or its increase in volume due to voids when it is dug or loosened. For instance, common excavation will swell from 10 to 30 percent. Another condition is that the machine is working in a depth of cut suitable for maximum digging efficiency. The optimum depth of cut for various sizes of shovels, may be defined as that depth which produces the greatest output and at which the dipper comes up with a full load.

The output figures are based on continuous operations, 60 minutes a working hour, without any delays for adjustments, lubrication, operator stopping for any reason, etc.

The figures are based on each dipper-full of material being swung through an arc of 90° before dumping. This is important as a swing of either a lesser number or greater number of degrees than 90° will either save time or consume more time and affect output capacity accordingly.

Tables are based on all materials being loaded into hauling units, with the further understanding that these hauling units are properly sized to shovel capacity and are provided in sufficient quantities to take away all the material the shovel can dig.

While these figures represent a very comprehensive job and as realistic a picture as can be presented on this involved subject, a word of caution is in order. These figures are general only and should not be considered as guaranteed outputs, or used by anyone as a basis for figuring and bidding jobs.

How Degrees of Swing and Depth of Cut Affect Shovel Yardage Output.— The quantites in the above table were based on the optimum depth of cut and a swing of 90° was specified.

Since variations in either of these conditions will have considerable effect on the yardage output, the following chart shows what happens when swing and depth of cut vary from the conditions on which the previous basic hourly yardages were established.

Table Giving Effect of Depth of Cut and Angle of Swing
On Power Shovel Output

Depth of Cut In % Of Optimum	Angle of Swing in Degrees						
	45°	60°	75°	90°	120°	150°	180°
40	.93	.89	.85	.80	.72	.65	.59
60	1.10	1.03	.96	.91	.81	.73	.66
80	1.22	1.12	1.04	.98	.86	.77	.69
100	1.26	1.16	1.07	1.00	.88	.79	.71
120	1.20	1.11	1.03	.97	.86	.77	.70
140	1.12	1.04	.97	.91	.81	.73	.66
160	1.03	.96	.90	.85	.75	.67	.62

Dragline Yardages.—As in the case of power shovels and hoes, dragline yardage production is affected by the type of material to be excavated, the depth of cut, the swing before unloading, the type of unloading (unloaded into hauling units, cast onto spoil banks, etc.), the degree of continuity in the operation, etc.

The following table gives an approximate idea of the difference in maximum production possible subject to conditions listed thereafter:

Hourly Short Boom Dragline Output in Cubic Yards

Bucket Capacity Class of Material	⅜	½	¾	1	1¼	1½	1¾	2	2½
Moist Loam or Light Sandy Clay.......	70	95	130	160	195	220	245	265	305
Sand and Gravel............	65	90	125	155	185	210	235	255	295
Common Earth	55	75	105	135	165	190	210	230	265
Clay, Hard and Tough........	35	55	90	110	135	160	180	195	230
Clay, Wet and Sticky..........	20	30	55	75	95	110	130	145	175

The dragline is working in the optimum depth of cut for maximum efficiency.
The dragline is working a full 60 minutes each hour—no delays
The dragline is making a 90° swing before unloading.
The bucket loads are being dumped into "properly sized" hauling units.
The proper type bucket is being used for the job.
The dragline is being used within the working radius recommended by the manufacturer for machine stability.

How Degrees of Swing and Depth of Cut Affect Dragline Yardage Output.—The following table gives the effect of depth of cut and angle of swing on dragline yardage production. Variations in degrees of swing, in particular, have a marked effect on production and this is important to keep in mind when laying out a job as the shorter the swing the more the yardage.

Table Giving Effect of Depth of Cut and Angle of Swing
on Dragline Output

Depth of Cut In % of Optimum	Angle of Swing in Degrees							
	30	45	60	75	90	120	150	180
20	1.06	.99	.94	.90	.87	.81	.75	.70
40	1.17	1.08	1.02	.97	.93	.85	.78	.72
60	1.24	1.13	1.06	1.01	.97	.88	.80	.74
80	1.29	1.17	1.09	1.04	.99	.90	.82	.76
100	1.32	1.19	1.11	1.05	1.00	.91	.83	.77

Depth of Cut In % of Optimum	|	|	|	Angle of Swing in Degrees	|	|	|	|
	30	45	60	75	90	120	150	180
120	1.29	1.17	1.09	1.03	.985	.90	.82	.76
140	1.25	1.14	1.06	1.00	.96	.88	.81	.75
160	1.20	1.10	1.02	.97	.93	.85	.79	.73
180	1.15	1.05	.98	.94	.90	.82	.76	.71
200	1.10	1.00	.94	.90	.87	.79	.73	.69

Clamshell Production.—The clamshell excavator is not to be considered a high production machine but rather a machine to be used where the work is beyond the scope of other types of equipment. For example, a condition which usually requires a clamshell is where digging is vertical or practically straight down as in digging pier holes, shafts, etc. Digging in trenches which are sheathed and cross-braced generally calls for a clamshell because the vertical action of the bucket enables it to be worked through the cross-bracing. Jobs requiring accurate dumping or disposal of materials are usually calmshell jobs. Also, for high dumping jobs, whether it be charging a bin, building a stockpile or wherever the material must be dumped well above the machine level, the machine is well adapted. In general, the clamshell can operate vertically and dig or spot dump below, at or above the level of the machine.

The clamshell is only effective where the materials to be handled are relatively soft or loose.

Conditions are so variable on clamshell operation that a table showing typical production has little value.

However, in an effort to establish some point to work from, the following table has been prepared showing maximum production to be expected from clamshell excavators operating under the following conditions:

The clamshell is engaged in open digging such as a basement, large footing, etc., with the permissible cut at least a full bucket depth.

The depth of the excavation is not over 10'-0".

The quantities are in terms of bank measure.

There is no wasted time, 60 minutes of digging each hour.

The clamshell is making a 90° swing before unloading.

The bucket loads are being dumped into "properly sized" hauling units.

The clamshell is being used within the working radius recommended by the manufacturer for machine stability.

Hourly Short Boom Clamshell Handling Capacity in Cubic Yards

Bucket Capacity	3/8	1/2	3/4	1	1¼	1½	1¾	2	2½
Class of Material									
Moist Loam or Sandy Clay	50	65	95	120	140	155	170	190	225
Sand and Gravel	45	60	85	110	130	140	160	175	205
Common Earth	40	55	70	95	115	125	145	160	185

It must be thoroughly understood that the above production is based on the most ideal conditions with 100% job and management factors.

How Degrees of Swing Affect Clamshell Yardage.—The output of a clamshell operating at a steady pace is affected by the swing before unloading in the same manner as a dragline. The following table gives the effect of angle of swing on clamshell yardage production.

			Angle of Swing in Degrees				
30°	45°	60°	75°	90°	120°	150°	180°
1.32	1.19	1.11	1.05	1.00	.91	.83	.77

Dragline Excavator At Work

Job and Management Factors.—Perfect conditions seldom exist on any construction job, therefore it is necessary to rationalize these figures by some other factor to compensate for the fact that actual job conditions differ widely from the perfect conditions assumed so far.

There are two sets of factors on every job which have a great deal to do with the output of equipment on the job. They are the job factor and the management factor.

Job factors are the physical conditions pertaining to a specific job which affect the production rate, other than the class of material to be handled. These may be divided into three general headings.

1.   Topography and the dimensions of the work, which include the depth of cut and whether the work will require much moving within the cut or from one cut to another.

2.   Surface and weather conditions, which include in some cases the difference between summer and winter work, and also the question of drainage of either surface or underground water.

3.   Specification requirements which control the manner in which the work must be handled; or indicate the sequence of various operations; and which also show the amount of bank sloping or cutting close to finished line and grade that is required.

The above conditions are all inherent in the job itself and can be taken into account by the contractor in making up the bid.

Another group of factors which affect output are the management factors. These cover conditions which pertain to the efficiency of the operation and thus affect output. These are grouped under the head of the management factor because, in general, they are made up of items which management can determine and control. These include—

1.   Selection, training and direction of men. The quantities given in the tables indicate what can be reasonably expected by an experienced operator who is willing to work. The availability of trained men and the incentive to produce must be considered.

2.   Selection, care and repair of equipment. The quality of the inspection and preventive maintenance program can have much to do with the lost time on the job.

3.   Planning, laying out the job, supervising and coordinating the operations is a very important factor in increasing efficiency. Providing the right number and size of hauling units is one of the first things to show its effect.

Foremanship and supervision are important. When everything is working smoothly, there is added incentive for high production rates.

Because both job and management factors must be considered jointly, that is, the output yardages must be modified by both a job and a management factor, the following table consolidates both.

To estimate the yardage, select the proper combined job and management factor from the following table. First, select a job factor classification of Excellent, Good, Fair or Poor, whichever applies to your job conditions and follow straight across the Management Factor Classification of Excellent, Good, Fair or Poor, whichever applies to the management conditions. The resulting figure is the combination job-management factor by which the yardage figures from the basic output tables (as modified by the conversion factors for depth of cut and swing) must be multiplied to give the yardage factor that can reasonably be expected under the job-management conditions that will exist on the job.

Usually pit and quarry and heavy construction jobs will fall into groups of 1, 2 or 3, whereas highway grading will usually fall into groups 2, 3 or 4 because of the variations in cuts, required machine travel, close cuts, etc.

Combination Job and Management Factors

	Management Factors			
	1	2	3	4
Job Factors	Excellent	Good	Fair	Poor
1. Excellent............	.84	.81	.76	.70

2. Good...................	.78	.75	.71	.65
3. Fair.....................	.72	.69	.65	.60
4. Poor....................	.63	.61	.57	.52

Power Shovel Excavating

Modern shovels, powered by gasoline or diesel motors, have tremendous capacity when conditions permit working without interruptions.

The cost of power shovel excavating will vary with the size of the job, size of the shovel, and the speed with which the trucks are handled in and out of the hole. Seldom can the shovel work to capacity on building work, due to the congestion of trucks waiting to get in or out of the hole or due to trucks being delayed at the dump.

The size of the job has considerable bearing on the costs, as it costs just as much to get the shovel to the job and remove it at completion for 1,000 cu. yds. as for a 10,000 cu. yd. job.

Where it is possible to use a ¾-cu. yd. shovel to capacity (figuring a 100% dipper load each cycle), it should excavate 135 cu. yds. an hr. but the dipper is not always filled to capacity and there are delays in loading, moving shovel from place to place, cleaning up, bad weather, etc., so the average on building work will probably run 50 to 60 per cent of the rated capacity of the shovel.

Basis for Computing Daily Output of Diesel Powered Shovel.— The following method is used in figuring the output of a power shovel in basement excavation, based on a 1 cu. yd. shovel:

Dipper capacity.. 1 Cu. Yd.

Dipper efficiency, average material..90 percent

Load carried ....0.9 Cu. Yds.

Average operating cycle..20 Seconds

Cycles per hour..180 Cycles

Cu. Yds. per hour, operating 100% time..162 Cu. Yds.

Operating efficiency in basement excavation due to haulage service,
    machine delays and cleanup work, approximately...................... 50 to 60 percent

Average output per 8-hr. day... 640 to 760 Cu. Yds

Operating efficiency on heavy construction, roads, dams, open
    cut excavation, etc., including machine delays and cleanup
    work, approximately 66⅔ percent.

Average output per 8-hr. day..864 Cu. Yds.

Dragline Excavating

The cost of dragline excavating will vary, in the same manner as shovel excavating, according to the size of the job, size of the machine and the capacity and speed of the hauling units although the hauling hazards are reduced substantially by the trucks not having to be loaded in the hole thereby eliminating the necessity of trucking up a ramp.

As may be determined by comparing the basic output tables for shovels and draglines, the dragline has about 75 to 80 percent of the basic output capacity of shovels in materials it can handle. However, the dragline definitely has a place in excavating for construction work by virtue of its generous digging reach, its ability to dig under extremely wet conditions since it may stand on the top on dry, firm footing, easier to haul from with no ramps to climb, etc., all conditions which, under certain circumstances, would make shovel digging very expensive if not totally impossible.

In excavating for building construction the dragline is affected by job and man-

Clamshell Excavator

agement factors the same as shovels and the same basis for computing daily output may be used as follows:

Basis for Computing Daily Output of Dragline Excavator.—The following is based on a dragline having a 1 cu. yd. bucket.

Bucket capacity .... 1 Cu. Yd.
Bucket efficiency, average material ....90 percent
Load carried ....0.9 Cu. Yd.
Average operating cycle..24 Seconds
Cycles per hour.. 150 Cycles
Cu. Yds. Per hour, operating 100% time .... 135 Cu. Yds.
Operating efficiency in basement excavation due to machine delays, traffic delays to hauling units
 and clean up work, approximately.. 50 to 60 percent
Average output per 8-hr. day .... 540 to 650 Cu. Yds.
Operating efficiency on heavy construction, roads, dams, open cut excavation, etc., including
 machine delays and clean up work, approximately .... 662/3 percent
Average output per 8-hr. day .... 720 Cu. Yds.

Power Hoe Excavating

The hoe may be said to be a cross between a dragline and a shovel as it incorporates some of the characteristics of each and overcomes some of the limitations of each. The hoe, of course, is primarily a unit to dig below machine level. It will, however, dig harder material than the clamshell or dragline because the weight of the boom itself may be used to force the dipper into the material. It is, however, limited in digging by the length of the boom and stick. The hoe dipper may be controlled more accurately than the dragline bucket and is therefore better suited to close limit work.

In building construction the hoe is used to dig trenches, footings and basements. On small residence basements the hoe offers many advantages. It digs straight, vertical side walls (in soil which will stand); it cuts a level floor; it trims corners neatly and squarely; it can dig sewer and waterline trenches; it always works from the top on dry, safe ground; it reduces hand trim to a minimum.

In general, the process of digging a basement with a hoe is to dig a trench around the four sides of the basement and scoop out the center as you go.

On small work it is necessary to have low-cost, simple and easy means of moving the shovel from job to job. This is best done by using a single-purpose trailer which can be loaded or unloaded in 15 to 20 minutes. The cost of moving the hoe from job to job will vary with the distance between jobs.

On time studies made on several small basement jobs, where the basements contained from 275 to 350 cu. yds. of excavation each, a hoe power excavator equipped with a ¾-cu. yd. dipper averaged from 66 to 88 cu. yds. per hour.

An example of the cost of excavating a small basement containing approximately 350 cu. yds. of excavation, with the soil placed around the excavation, is as follows:

	Hours	Rate	Total	Rate	Total
Move-In Charge		$....	$....	$100.00	$100.00
Excavating					
Hoe operator	5	....	....	17.95	89.75
Labor, clean-up...........................	5	....	....	12.54	62.70
Hoe charge..................................	5	....	....	38.00	190.00
Total Direct Cost........................			$....		$442.45
Cost per cu. yd............................			....		1.27

The above costs do not include removal of excavated earth from site or backfilling.

Courtesy U. S. Homes

Power hoe being used to install sewer pipe

Data on Diesel Power Shovel Costs and Operation

Description				Size of Shovel			
	¾ Cu. Yd.	1 Cu. Yd.	1¼ Cu. Yd.	1¾ Cu. Yd.	3¼ Cu. Yd.	4¼ Cu. Yd.	5½ Cu. Yd.
Cu. Yds. per Hour Based on 100% Time, Cu. Yds.	135	180	225	315	585	765	950
Cu. Yds. per Hour Based on 100% Time and 90% Dipper Efficiency	121	162	202	284	525	690	855
Output per Hour Based on 66⅔% Efficiency, Cu. Yds.	90	120	150	210	390	510	635
Output per Hour Based on 50% Efficiency, Cu. Yds.	68	90	112	157	292	382	475
Output per 8-Hour Day, Based on 66⅔% Efficiency, Cu. Yds.	720	960	1200	1680	3120	4080	5075
Output per 8-Hour Day, Based on 50% Efficiency, Cu. Yds.	544	720	1000	1260	2340	3060	3800
Average Cost—Diesel Shovel, Including Freight and Unloading	$80,000	$96,000	$120,000	$170,000	$250,000	$300,000	$450,000
Depreciation, Percent per Year	20%	16⅔%	16⅔%	16⅔%	12½%	10%	10%
Depreciation per Hour Based on 2,000 Hours per Year	$8.00	$8.00	$10.00	$14.17	$15.63	$15.00	$22.50
Interest, Taxes, Insurance, 20% per Year of 2,000 Hours	$4.80	$5.60	$7.00	$9.92	$14.06	$16.50	$24.75
Hourly Consumption of Fuel, Approximate Gals. per Hour	4.5	4.5	5.2	6.4	12.0	15.5	18.0
Fuel Cost per Hour*	$ 6.75	$ 6.75	$ 7.80	$ 9.60	$18.00	$23.25	$ 27.00
Engine Lub. Oil Cost, 15% of Fuel	$ 1.01	$ 1.01	$ 1.17	$ 1.44	$ 2.70	$ 3.49	$ 4.05
Cost per Hour, Repairs, Maintenance and Supplies, Including Rope, Grease, etc.	$ 8.00	$ 8.00	$10.00	$14.17	$15.63	$15.00	$ 22.50
Total Hourly Cost, Not Including Labor, Supervision, Compensation, Unemployment, Soc. Sec. etc.	$28.56	$29.36	$35.97	$49.30	$66.02	$73.24	$100.80

*Fuel costs based on diesel fuel at $1.50 per gal.

Digging plan for excavating small basements using a hoe. The order of the letters indicates the order in which the sections are removed. The arrows indicate the direction the hoe travels in taking out each section.

Truck and trailer time to deliver the hoe and removing same at completion of job will vary with length of haul.

In medium to large work and at moderate depths, hoe production can approach that of shovels. Output falls off considerably, however, at greater depths.

For the smaller machines, the size hoe is usually chosen for the width trench it will cut, but for the larger sizes, the capacity of the machine is chosen on a production basis.

Data on typical hoe bucket dimensions and weight is given in the following table.

Dipper Capacity in Cu. Yds. Struck	Bucket Outside Width in Inches	Total Width Added for Two Side Cutters in Inches	Weight In Lbs.
⅜	20 to 24	4 or 6	850
½	24 to 28	4 to 9	950
¾	28 to 39	4 to 9	1,400
1	33 to 45	4 to 9	1,700
1¼	39 to 45	4 to 9	2,000
1½	39 to 45	4 to 9	2,300
2	47 to 55	4 to 7	3,200

Production for Power Shovel Excavating Using a ⅜-Cu. Yd. Shovel Based on an Average Daily Output of 300 Cu. Yds.							
	Number of Cubic Yards of Excavation in Job						
	1,000	1,500	2,000	2,500	3,000	4,000	5,000
Actual Operating Time of Shovel							
Foreman*	27	40	54	67	80	107	134
Shovel Operator	27	40	54	67	80	107	134
*Foreman in charge of Job, handling and expediting trucks, etc.							

Production for Power Shovel Excavating Using a ½-Cu. Yd. Shovel Based on an Average Daily Output of 400 Cu. Yds.							
	Number of Cubic Yards of Excavation in Job						
	1,000	1,500	2,000	2,500	3,000	4,000	5,000
Actual Operating Time of Shovel							
Foreman*	20	30	40	50	60	80	100
Shovel Operator	20	30	40	50	60	80	100
*Foreman in Charge of Job, handling and expediting trucks, etc.							

**Production for Power Shovel Excavating Using a ¾-Cu. Yd. Shovel
Based on an Average Daily Output of 600 Cu. Yds.**

	Number of Cubic Yards of Excavation in Job						
	1,000	1,500	2,000	2,500	3,000	4,000	5,000
Actual Operating Time of Shovel							
Foreman*	14	20	27	34	40	54	68
Shovel Operator	14	20	27	34	40	54	68

*Foreman in Charge of Job, handling and expediting trucks, etc.

**Production for Power Shovel Excavating Using a 1-Cu. Yd. Shovel
Based on an Average Daily Output of 720 Cu. Yds.**

	Number of Cubic Yards of Excavation in Job						
	1,000	1,500	2,000	2,500	3,000	4,000	5,000
Actual Operating Time of Shovel							
Foreman*	11	17	22	28	34	45	56
Shovel Operator	11	17	22	28	34	45	56

*Foreman in charge of Job, handling and expediting trucks, etc.

**Production for Dragline Excavating Using a ½-Cu. Yd. Dragline
Based on an Average Daily Output of 320 Cu. Yds.**

	Number of Cubic Yards of Excavation in Job						
	1,000	1,500	2,000	2,500	3,000	4,000	5,000
Actual Operating Time							
Foreman*	25	38	50	63	75	100	125
Dragline Operator	25	38	50	63	75	100	125

*Foreman in Charge of Job, handling and expediting trucks, etc.

**Production for Dragline Excavating Using a 1-Cu. Yd. Dragline
Based on an Average Daily Output of 560 Cu. Yds.**

	Number of Cubic Yards of Excavation in Job						
	1,000	1,500	2,000	2,500	3,000	4,000	5,000
Actual Operating Time							
Foreman*	18	27	36	45	54	72	90
Dragline Operator	18	27	36	45	54	72	90

*Foreman in Charge of Job, handling and expediting trucks, etc.

Example Excavating Estimate

The following example is based on a basement containing 5,000 cu. yds. excavation, average conditions, using a 1¼-cu. yd. power shovel, loading into trucks, based on an average of 900 cu. yds. per 8-hr. day.

	Hours	Rate	Total	Rate	Total
Bringing Shovel to Job		$. ...	$. ...	$100.00	$ 100.00
Move-In Charge					

Excavating	Hours	Rate	Total	Rate	Total
Foreman	48	....	....	18.45	885.60
Shovel operator	48	....	....	17.95	861.60
Shovel charge	48	....	....	36.00	1,728.00
Total Direct Cost			$....		$3,575.20
Cost per cu. yd				....	.72

Add cost of trucks for removing excavated soil, overhead expense and profit.

The above method for estimating shovel excavation can also be used for hoe, dragline and clamshell excavation by merely substituting the proper production values and equipment charges for those pertaining to shovels.

Hauling Excavated Materials in Trucks

Where an excess of excavated materials, over and above the amount required for backfill, exists, it will have to be removed from the job by trucks.

The usual procedure followed in these cases is to determine, beforehand, the approximate quantity of material required for backfill, etc., and this amount subtracted from the estimated excavation quantity will give the volume of excavated material to be hauled away. The excess material should be loaded and hauled as the excavation progresses so that no rehandling is necessary.

Types of Trucks.—Truck hauling units are mainly of two types, namely: conventional highway hauling units and off-the-highway heavy duty hauling equipment.

The conventional highway hauling unit is the most frequently used. For the average contractor it is the most economical type of truck to use as it is adaptable to practically every job and may be used for other hauling purposes as well.

For the contractor specializing in large earth moving projects, the high production, heavy duty, off-the-highway hauling unit is the most economical equipment to use because it is specifically designed for this purpose and not being limited by state highway regulations, as to weight and size, has a high ratio of net weight to payload, the additional weight being found in its more rugged construction, such as heavier frame, axles, body, hoist, engine, etc.

Hourly Cost of Ownership and Operation of Trucks.—To intelligently estimate hauling costs the contractor must know what the equipment costs to own and operate and this may be determined by the same method previously described for excavating equipment.

Attention is called to the manner in which tire costs, original, replacement and repair are handled. Recognizing the fact that the economic life of tires will not be as long as the life of the vehicle, the original value of the tires is deducted from the total price before computing the hourly charge for depreciation and is then treated as an operating cost.

The hourly cost of tire replacement and tire repairs varies greatly according to operating conditions, road surfaces and tire loadings. While tire costs in ordinary trucking operations normally are figured on a mileage basis, in construction work, with its rough roads, short hauls and relatively low speeds (below 30 mph.), they are figured on an hourly basis and are computed by dividing the replacement cost by the estimated tire life in hours. The average life expectancy of truck tires in construction work is from 3,000 to 3,500 hours. Tire repairs should run about 15 percent of the hourly tire cost.

The following tables give data on various sizes of rear dump trucks for both conventional highway hauling units and heavy duty off-highway equipment.

Table Showing How Hourly Cost of Ownership and Operation May Be Derived for Conventional Highway Rear Dump Trucks Used in Construction Work Based on an Economic Life of 3 Yrs. of 2,000 Hours Each

Capacity in Tons	5	10	15	20
Capacity in Cu. Yds., Struck	3	6	9	12
Approximate Delivered Price	$20,000	$30,000	$55,000	$70,000
Original Tire Value	1,800	2,400	3,400	4,000
Depreciation, .0167% of Total Investment Less Tires	$ 3.04	$ 4.61	$ 8.62	$ 11.02
Interest, Taxes and Insurance, .0067% of Total Investment	1.34	2.01	3.69	4.69
Hourly Tire Replacement Cost, Based on 3,000-Hr. Life	.60	.80	1.13	1.33
Hourly Cost of Tire Repairs, 15% of Hourly Tire Replacement Cost	.09	.12	.17	.20
Maintenance and Repairs, .004% of Total Investment	.80	1.20	2.20	2.80
Fuel Cost per Hour, .04 Gal. per Hp	6.60	8.10	9.00	9.60
Oil and Grease, 15% of Fuel	.99	1.22	1.35	1.44
Total Hourly Cost, Not Including Labor, Supervision, Unemployment, Soc. Sec., Overhead Expense and Profit	$ 13.46	$ 18.06	$ 26.16	$ 31.08

Table Showing How Hourly Cost of Ownership and Operation May be Derived for Off-Highway Rear Dump Trucks Based on an Economic Life of 3 Yrs. of 2,000 Hours Each

Capacity in Tons	10	18	27	45	62
Capacity in Cu. Yds., Struck	6.5	10.5	18	40	40
Horsepower	130	225	335	500	670
Approximate Delivered Price	$67,000	$116,000	$174,000	$290,000	$400,000
Original Tire Value	3,200	7,200	9,600	16,600	28,000
Depreciation, .01% of Total Investment Less Tires	6.38	10.88	16.44	27.34	37.20
Interest, Taxes and Insurance, .0067% of Total Investment	4.49	7.77	11.66	19.43	26.80
Hourly Tire Replacement Cost, Based on 3,000-Hr. Life	1.07	2.40	3.20	5.53	9.33
Hourly Cost of Tire Repairs, 15% of Hourly Tire Replacement cost	.16	.36	.48	.83	1.40
Maintenance and Repairs, .004% of Total Investment	2.68	4.64	6.96	11.60	16.00
Fuel Cost per Hour, .025 Gal. per Hp	4.88	8.44	12.56	18.75	25.13
Oil and Grease, 15% of Fuel	.73	1.27	1.88	2.81	3.77
Total Hourly Cost*	$ 20.39	$ 35.76	$ 53.18	$ 86.29	119.63

*The foregoing tables are for the purpose of showing the items which must be taken into consideration in figuring hourly costs but do not necessarily reflect current prices of equipment, tires or fuel.

Cost of Hauling Excavated Material in Trucks.—Hauling costs vary with the capacities of the trucks and time of the hauling cycle.

Truck capacity, for proper sizing, should be at least 4 times the dipper or bucket capacity of the excavator. This is important as the efficiency of the excavator is seriously affected by undersized hauling units due to the increased hazards of truck delays.

The following table gives theoretical spotting time cycles for various sizes of power shovel excavators working at 100 percent efficiency and loading trucks of 4 times dipper size capacity.

Haul Units Needed to Spot Under Shovel per
Hour in Medium Digging

Size Excavator Dipper	Minimum Haul Unit Capacity at 4 Times Dipper Size	Approximate Shovel Cycle in Seconds 90° Swing No Delays	Loading Time for 4-Dipper Truck in Seconds	Time of Spotting Cycle for Steady Operation in Minutes	Number of Spots Required per Hour for Steady Operation
⅜	1½ Yd.	19	76	1.26	48
½	2 Yd.	19	76	1.26	48
¾	3 Yd.	20	80	1.33	45
1	4 Yd.	21	84	1.40	43
1¼	5 Yd.	21	84	1.40	43
1½	6 Yd.	23	92	1.53	39
2	8 Yd.	25	100	1.66	36
2½	10 Yd.	26	104	1.73	35

For dragline operation the above time values may be increased from 20 to 30 percent.

The hauling time cycle consists of several operations, namely; spotting of truck under excavator; loading truck by excavator; traveling to dumping area; dumping; returning to excavator.

Some of these operations, such as spotting, loading and dumping, can be minimized by good management and supervision. Traveling time, however, will vary with the haul distance, traffic conditions, road conditions, etc., and must be carefully analyzed for each job.

When the hauling time cycle has been estimated, the number of loads each truck is able to haul per day can be determined and multiplying this by the capacity of the truck will give the theoretical daily haul of each truck in cu. yds., assuming 100 percent job efficiency. This amount should be multiplied by the expected job efficiency factor to obtain the net yardage hauled. Divide the daily truck cost by the net yardage hauled and the result will be the hauling cost per cu. yd.

The number of trucks required to keep the excavator going can also be determined by dividing the hauling time cycle by the loading time cycle using 100 percent excavator efficiency as a basis. This is done to be sure that a hauling unit is under the excavator at all times. While an excavating job may actually work out to a 50 or 60 percent efficiency, there usually are times on the job when the excavator is working at full capacity and to take advantage of this, sufficient, properly sized hauling units must be on the job. In large operations many contractors maintain standby units to step in when one of the hauling fleet trucks breaks down or is otherwise unusually delayed.

In sizing trucks to conform to dipper or bucket capacities, the struck capacity of each is used, assuming that the heaping of loads on both trucks and dippers or buckets will offset each other and produce bank measure quantities.

However, all estimating should use actual job data rather than theoretical figures. Some materials do not follow the usual swell ratios and job efficiency ratings vary widely. Following book figures to the letter in excavating, or any other branch of construction work, can be disastrous. Book values, at best, can only give what can be done under a definite set of conditions. The contractor must exercise his judgment in estimating this work and base his figure on his own particular set of conditions, i.e., equipment, supervision, operators, local materials and local weather.

Courtesy U. S. Homes

Tandem Powered Wheel Tractor-Scraper

Tractors and Scrapers

The tractor is probably the most used piece of earth moving equipment on the construction job. It can perform a wide variety of operations and equipped with the proper accessories is a high production and efficient tool for such jobs as site clearing, grading, shallow excavating, etc. In addition the tractor is often called upon to act as the "strong man" in freeing stuck trucks, hauling or dragging equipment, sheds, etc., into place, miscellaneous hoisting using a snatch block and cable, etc. In general, the tractor is almost indispensable on the job.

Equipped with bulldozer apparatus, the tractor, either rubber-tired or crawler type, has demonstrated its ability to do a wider variety of jobs than any other earth moving tool.

The bulldozer usually is employed to open up the job, which may consist of anything from clearing brush and vegetation to the removal of average size trees (up to 30 inches in diameter), stumps, rocks, paving, old floor slabs, old foundations, etc.

Stripping and stockpiling top soil on small quantity jobs with a haul distance from 200 to 400 feet or less is most efficiently done by bulldozer. On larger jobs this is done more economically by scrapers.

On small quantity, shallow excavation jobs with a short haul, tractor dozers are moderately efficient.

Crawler type bulldozers are reasonably economical on pushing distances up to 200 feet. Under the same conditions the rubber-tired dozer's economical push distance is about 400 feet.

Most backfilling jobs fall to the lot of the bulldozer where they out-perform any other piece of equipment.

Tractors also perform a valuable service when they are called upon to provide the additional power for push loading scrapers. On a production scraper job, dozers helping to load these units more than pay for themselves.

On residence and other small jobs, a loader usually can dig the basement and do all exterior grading cuts and fills with maximum efficiency.

Another application of the bulldozer is spreading fill hauled in and dumped by trucks or cast by other excavating equipment.

Tractors are also the ideal power source for hauling sheepsfoot rollers and pneumatic rollers used in compacting fills.

Scraper Application.—Scrapers are high production earth moving units which are capable of digging their own load, hauling the load and then spreading the load in controlled layers. Furthermore, the scraper is a precision tool, as a skilled operator can cut a grade to within 0.1 ft. and can spread fill to the same degree of accuracy.

Scrapers may be powered by three types of prime movers: two wheeled or four wheeled rubber-tired units or crawler type tractors.

Scrapers, as a rule, are not used to any great extent in residence building or on any work in congested city areas but for jobs such as suburban and rural schools, hospitals, factories, housing projects, in fact any construction work with a medium to large site grading problem, scrapers are probably the most efficient pieces of excavating equipment for the job.

The scrapers can be used to strip and stockpile top soil, after the site has been cleared, and reclaim the top soil from the stockpile and spread it where required at the end of the job. Many times basement excavation may be carried on in conjunction with site grading cuts and fills by scrapers.

The basic efficiency of scrapers is relatively unaffected by depth of cut, length of haul or type of soil. Quantities being sufficient to warrant the use of scraper equipment, crawler tractor drawn scrapers are economical at haul distances up to 500 feet and with rubber-tired, high speed prime movers can compete with trucks on longer hauls.

Capacities of scrapers range from about 7 cu. yds. struck measure to approximately 40 cu. yds. heaped measure.

Scrapers powered by crawler tractors are capable of self-loading in most soils but generally it is more economical to use pusher assistance from a tractor to obtain heaped loads in the shortest time. With few exceptions, pusher assistance is essential for self-propelled, rubber-tired scrapers. Pusher loading is generally accomplished in an average of about 1 minute in 100 ft. of travel. To determine the number of scrapers one pusher can handle, divide the scraper cycle by the pusher cycle.

Estimating the Cost of Owning and Operating Tractors and Scrapers.—In order to properly prepare an estimate on work to be done by tractors and/or scrapers, the contractor must know what to charge for the use of the equipment. He should work out an hourly charge for the equipment, to be used in pricing his estimates, so that the cost of ownership and operation is carried proportionately by each job and does not have to come out of profit.

Hourly Cost of Ownership.—The cost of ownership has as its basis the total delivered cost of the piece of equipment and is composed of several items, namely, depreciation, interest, insurance and taxes.

In figuring the hourly depreciation cost most contractors usually take the full delivered price of the machine, less cost of tires, and divide by the expected economic life of the machine in hours. The life expectancy figure commonly accepted is 5 years of 2,000 hours each or 10,000 hours.

Interest, insurance, taxes, and storage are computed on the basis of average investment. The average investment in a piece of equipment which is to be fully depreciated in 5 years is 60 percent of the full delivered cost, as is shown on previous pages under "Power Shovels and Cranes." The assumed rate for interest, insurance, taxes, and storage is a total of 20 percent which, when applied to the average investment and then divided by 2,000 hours, amounts to .006 percent of the full delivered cost to give the hourly cost of interest, insurance, taxes, and storage.

Total hourly ownership cost is then obtained by totaling the hourly cost of depreciation and the hourly cost of interest, insurance, taxes, and storage.

Hourly Cost of Operation.—This cost includes tire replacement, tire repairs, fuel, lubrication, mechanical repairs, maintenance of blades and cable, etc.

The hourly costs of tire replacements and tire repairs are computed in the same manner as previously described, under "Hourly Cost of Ownership and Operation of Trucks", and are based on a tire life of 3,000 hours of operation with repair costs assumed to be 15 percent of the hourly tire cost.

Fuel costs may be figured on the following basis: Assuming an average machine in good repair, operating under average conditions, diesel fuel consumption of 0.04 gallons per horsepower per hour is used.

Experience has shown that lubricants and lubricating labor amount to approximately 50 percent of the fuel costs per hour.

Cost of repairs and maintenance, including major overhauls at the end of the working season, should be determined by the contractor from his own experience record.

Many contractors find that total repairs, maintenance and overhauls for the economic life of the machine runs from 70 percent of the full delivered price for the larger machines to 90 percent for the smaller machines. This percentage divided by the economic life of the machine in hours, 10,000, will give the hourly cost of mechanical repairs, maintenance, etc., and amounts from .007 percent to .009 percent of the total delivered cost.

The hourly cost of maintenance of blades and cables is figured separately from the cost of general maintenance and repair as this item will vary depending on the soil conditions encountered. This cost may be estimated by dividing the replacement cost of blades and cables by their expected life in hours. Under average conditions this life may be expected to be about 450 hours; under favorable conditions, such as digging soft loam, etc., this life may be increased to about 800 hours; under unfavorable conditions, where sharp rocks and shale are being loaded, the expected life should be reduced to about 250 hours.

Approximate Prices of Crawler Type Tractors
with Bulldozer Equipment

Horsepower	Weight in Lbs.	Approximate Price
50	9,800	$ 36,000
65	14,600	54,000
75	19,500	72,000
105	25,800	90,000
200	45,900	143,000

Approximate Prices of Crawler Type Loaders

Capacity in Cu. Yds.	Horsepower	Approximate Price
1	65	$ 54,000
1½	80	71,000
2¼	130	92,000
3	190	140,000

Approximate Prices of Self-Propelled Scrapers

Capacity in Cu. Yds. Struck	Heaped	Horsepower	Approximate Price	Original Tire Value
7	9	150	$148,000.00	$ 8,600.00
14	20	330	234,000.00	25,000.00
24	34	450	350,000.00	36,000.00

Approximate Prices of Tractor-Elevating Scrapers

Capacity in cu. yds.	Horsepower	Approximate Price F.O.B. Factory	Original Tire Value
11	150	$123,000.00	$11,000.00
22	330	245,000.00	27,000.00
34	450	370,000.00	39,000.00

Data on Cost of Ownership and Operation of Bulldozers, Loaders and Scrapers.—The following data gives the hourly cost of ownership and operation for various sizes of tractor and scraper equipment and is based on the preceding price and horsepower information. It does not include operator's wages, supervision, compensation and liability insurance or Social Security taxes.

Hourly Cost of Crawler Type Bulldozers

Horsepower	50	65	75	105	200
Approx. Delv'd. Cost.	$36,000	$54,000	$72,000	$90,000	$143,000
Depreciation, .01%	3.60	5.40	7.20	9.00	14.30
Interest, Insurance and Taxes. .006%	2.16	3.24	4.32	5.40	8.58
Fuel, .04 Gal. per HP*	3.00	3.90	4.50	6.30	12.00
Lubrication, Including Labor, 50% of Fuel	1.50	1.95	2.25	3.15	6.00
Mech. Repairs and Maintenance, .007%	2.52	3.78	5.04	6.30	10.01
Blade and Cable Replacement	.45	.47	.49	.51	.55
Total Hourly Cost †	$ 13.23	$ 18.74	$ 23.80	$ 30.66	$ 51.44

Hourly Cost of Crawler Type Loaders

Bucket Capacity, Cu. yds.	1	1 ½	2 ¼	3
Horsepower	65	80	130	190
Approx. Delv'd. Cost	$54,000	$71,000	$92,000	$140,000
Depreciation, .01%	5.40	7.10	9.20	14.00
Interest, Insurance and Taxes, .006%	3.24	4.26	5.52	8.40
Fuel, .04 Gal. per HP.*	3.90	4.80	7.80	11.40
Lubrication, Including Labor, 50% of Fuel	1.95	2.40	3.90	5.70
Mech. Repairs and Maintenance, .007%	3.78	4.97	6.44	9.80

Blade
Replacement.. .40   .42   .44   .48
Total Hourly Cost †.................................... $ 18.67 $ 23.95 $ 33.30 $ 49.78

Hourly Cost of Self-propelled Scrapers

Capacity, Struck in Cu. Yds..................	7	14	24
Capacity, Heaped in Cu. Yds.................	9	20	34
Horsepower...	150	330	450
Approx. Delv'd. Cost..............................	$148,000	$234,000	$350,000
Original Tire Value	8,600	25,000	36,000
Depreciation, .01% of Total Investment Less Tires	13.94	20.90	31.40
Interest, Insurance and Taxes, .006% of Total Investment..................	8.88	14.04	21.00
Tire Replacement, 3,000-Hr. Life......................................	2.87	8.34	12.00
Tire Repairs, 15%...................................	.43	1.25	1.80
Fuel, .04 Gal. per Hp.*	9.00	19.80	27.00
Lubrication, Incl. Labor, 50% of fuel	4.50	9.90	13.50
Mech. Repairs and Maintenance, .007% of Total Investment.................	10.36	16.38	24.50
Blade and Cable Replacement.......................................	.65	.70	.80
Total Hourly Cost †...............................	$ 50.63	$ 91.31	$ 132.00

* Based on $1.50 per gallon for diesel fuel.
†Not including Labor, Supervision, Unemployment, Soc. Sec., Overhead Expense

Bulldozer Production

In an average day's work, the bulldozer may be called upon to do several of the many operations it is capable of and therefore time will be lost in moving from one task to another. This time loss must be considered in estimating bulldozer work or the job may not be profitable. Inasmuch as each job has its own peculiar problems, no table can be set up which will reflect the time losses to be experienced and the estimator must analyze each job carefully and adjust the production rates to conform to the conditions present.

The following table gives the approximate hourly production to be expected from various sized bulldozers and is based on having good weather conditions, efficient, well-maintained machinery and steady operation. It has been established that a well organized and supervised job will result in approximately 50 work-productive minutes each hour or a job efficiency of 83 percent.

Approximate Hourly Production of Bulldozers
Based on a Job Efficiency of 83 Percent

	Horsepower				
Description of Work	50	65	75	105	200
Simple clearing, vegetation,	900	1,000	1,150	1,250	1,450
light brush, tree saplings, in	to	to	to	to	to
sq. ft. per hour.	1,000	1,100	1,250	1,350	1,550
Moderately difficult clearing,	300	350	400	600	700
thick brush and tree sapp-	to	to	to	to	to
lings, in sq. ft. per hour.	350	400	450	650	750
Add extra for trees.					

Description of Work	Horsepower				
	50	65	75	105	200
Strip top soil or shallow excavation of ordinary soil and placing in piles of premises, in cu. yds. per hour.	15 to 20	17 to 22	20 to 25	22 to 27	27 to 32
Backfill, ordinary soil placed slow enough to permit compacting, in cu. yds., per hour.	33 to 38	37 to 42	43 to 48	50 to 55	60 to 65
Spread material dumped in piles by trucks or cast by other excavating equipment, in cu. yds. per hour.	42 to 47	45 to 50	55 to 60	60 to 65	70 to 75

Scraper Production

Relative scraper production is mainly governed by the cycle time of operation, that is, the length of time a scraper will take to load, haul, spread-dump and return to start over again. In properly organized jobs with good, adequately powered equipment, a portion of this cycle is usually termed the fixed time because it is relatively unaffected by the length and grades of the haul and consists of the following: loading, approximately 1.0 to 1.5 minutes, spread-dumping, about 0.5 minutes, two turns of 0.25 minutes each, and shifting and acceleration of 0.5 to 1.0 minute or a total fixed time of 2.5 to 3.5 minutes. The other part of the cycle is the actual time of hauling the load to the fill area plus the time of returning empty to the beginning of a new cut. This traveling time is the big variable in the operation as it is affected by length of haul, road conditions and grades to be negotiated and must be estimated accurately. Scrapers powered by rubber-tired tractors can reach a speed of 30 m.p.h. when empty and on a good level road but this will often be reduced to 10 to 20 m.p.h. when loaded. It must also be remembered that adverse grades will greatly reduce the speed and on the other hand favorable grades will increase the speed. The traveling time plus the fixed time will be the cycle time of operation.

Yardage production should be figured in terms of bank measure. Therefore, when estimating scraper production, the struck capacity of the machine is usually used as the pay yardage volume, anticipating the scraper to be heap loaded. The difference between struck and heaped capacity is generally assumed to be the amount the material will swell in being disturbed.

The following table gives the daily scraper production for various sized machines operating at several different time cycles and is based on ordinary soil conditions, good weather, good equipment and an 83 percent job efficiency.

Daily Scraper Production for Various Operation Cycles
Based on a 50-Minute Hour and an 8-Hour Day

Operation Cycle in Minutes	Struck Capacity of Scraper in Cu. Yds.		
	7	14	24
3	930	1860	3200
4	700	1400	2400
5	560	1120	1920
6	465	930	1600
7	400	800	1370
8	350	700	1200

| 9 | 310 | 620 | 1070 |
| 10 | 280 | 560 | 960 |

Example Estimate for Scraper Excavation

Based on a site grading cut containing 8,000 cu. yds., excavated material to be spread-dumped in low areas requiring fill, good conditions permitting a 50-minute hour, an operation cycle of 6 minutes and using two 14 cu. yd., struck measure, self-propelled scrapers.

Bringing Scrapers to Job	Hours	Rate	Total	Rate	Total
Move-in charge 2 ea		$. . . .	$. . . .	$125.00	$ 250.00
Excavating					
Foreman	35	. . . .	. . . .	18.45	645.75
Scraper operators	70	. . . .	. . . .	17.95	1,256.50
Scraper Hours	70	. . . .	. . . .	91.31	6,391.70
Total Direct Cost			$. . . .		$8,543.95
Cost per cu. yd.				. . . .	1.06

If a 105-Hp. bulldozer would be required to give pusher assistance in loading, to maintain the operation cycle time of 6 minutes, the additional cost per cu. yd. would be as follows:

	Hours	Rate	Total	Rate	Total
Move-in charge		$. . . .	$. . . .	$100.00	$ 100.00
Excavating					
Bulldozer operator	35	. . . .	. . . .	17.95	628.25
Bulldozer charge	35	. . . .	. . . .	30.66	1,073.10
Additional Direct Cost			$. . . .		$1,801.35
Additional cost per cu. yd.				. . . .	.23

Trenching and Ditching Machine Excavation

Designed strictly for the one purpose of cutting ditches, the trenching machine, when properly applied, is the fastest earthmover for its weight and horsepower in the construction field. Before a contractor can reap the benefits of this great potential output, he must first select a trenching machine of a type and size which fits the work to be done and, even more important, must put an operator on the machine who possesses both skill and mechanical knowledge.

There are two basic types of machines, the wheel type and the ladder type. The wheel type is generally considered to be the fastest and is by far the most common type in use today. Each type has advantages and disadvantages for performing certain classes of work.

Inasmuch as most trenching or ditching operations in connection with building construction will be foundations, service trenches, etc., the following data covers the cost and operation of a general purpose, crawler mounted, wheel machine of the type and size generally used for this work.

This machine is equipped with shifting-tilting boom that permits it to dig close to obstructions, such as foundations or curbs, and to dig a vertical trench even when the machine is on a slope.

The cost of excavating using a trenching or ditching machine will vary with the size of the job, width and depth of trenches, kind of soil, accessibility, etc. Most machines are designed to dig certain standard width trenches, so that trenches a trifle narrower or wider will not make any appreciable difference in either output or costs.

While the following table gives the rated capacity of the machine, it will be necessary to allow a factor for time lost moving machine, bad weather, etc., so it is not advisable to figure rated capacity as the average daily output of any machine.

Data on Crawler Mounted Ditching Machine
Operation and Cost

Wheel Size	Possible Cutting Widths	Maximum Depth Trench	Weight in Pounds	Approx. Cost F.O.B. Fcty.
16"x11'-6"	16"-24"	7'-0"	27,000	$130,000
24"x12'-4"	24"-30"	7'-6"	43,000	$170,000

Maximum digging speed is 34 f.p.m. Forward and reverse travel speeds are 0.46, 0.95, 1.7 and 2.9 m.p.h.

Digging speeds are controlled by a hydraulic transmission, permitting speed to be varied to match soil conditions. Wheel speeds are variable from 0 to 9.2 r.p.m.

Crawler Mounted Ditching Machine Operation and Cost

When figuring operating cost on a ditching machine, include fuel and oil consumption, depreciation, interest on investment, repairs, replacements, taxes, etc.

Description	Digging Width 12"	24"
Approximate Delivered Price	$130,000	$170,000
Depreciation per year, 18%	23,400	30,600
Interest 15% per year	19,500	25,500
Insurance and Taxes per year, 5%	6,500	8,500
Repairs and Replacement per year, 10%	13,000	17,000
Hourly Consumption of Fuel, Gals.	5	6
Daily Cost of Fuel, Oil, Grease, $1.50 gal. + 15%	69.00	82.80
Operating Cost per Mile (Based on 75 Mi. per Year, 200 Working Days per Year, Avg. Trench Depth, 5'-0")		
Depreciation	312.00	408.00
Interest	260.00	340.00
Taxes and Insurance	86.67	113.34
Repairs and Replacements	173.34	226.67
Fuel, Oil and Grease	184.00	220.80
Total machine cost per mile	$1,016.01	$1,308.81
Machine cost per lin. ft. trench	.193	.248

Cost of Digging 100 Lin. Ft. of Trench 5'-0" Deep,
Using a Crawler Mounted Ditching Machine,
Based on 350 Lin. Ft. per Hour

	Hours	Rate	Total	Rate	Total
Machine operator	0.29	$. ...	$. ...	$17.95	$ 5.21
Machine charge, 12" wide		....	....	.193	19.30
Machine charge, 24" wide		....	....	.248	24.80
Cost per 100 lin. ft., 12" wide					24.51
Cost per 100 lin. ft., 24" wide					30.01
Cost per lin. ft., 12" wide					.25
Cost per lin. ft., 24" wide					.30

To the above, add cost of trucking machine to job and removing same at completion, supervision, compensation and liability insurance, Social Security and Unemployment taxes, overhead expense and profit. Add for backfilling trenches, if required.

Dimensions of Crawler Mounted Wheel Ditcher

Volume of Trench Excavation
Cu. Yd. per 100 Lin. Ft.

Trench depth, in.	Trench width, in.						
	12	18	24	30	36	42	48
12	3.7	5.6	7.4	9.3	11.1	13.0	14.8
18	5.6	8.3	11.1	13.9	16.7	19.4	22.3
24	7.4	11.1	14.8	18.5	22.2	26.0	29.6
30	9.3	13.8	18.5	23.2	27.8	32.4	37.0
36	11.1	16.6	22.2	27.8	33.3	38.9	44.5
42	13.0	19.4	30.0	32.4	38.9	45.4	52.0
48	14.8	22.2	29.6	37.0	44.5	52.0	59.2
54	16.7	25.0	33.3	41.6	50.0	58.4	66.7
60	18.6	27.8	37.0	46.3	55.5	64.9	74.1
66	...	30.5	40.7	51.0	61.0	71.3	81.6
72	...	33.3	44.5	55.5	66.7	77.9	89.0
78	...	36.0	48.1	60.2	72.2	84.2	96.5
84	...	38.9	51.9	64.8	77.8	90.8	104.0
90	...	41.6	55.6	69.5	83.4	97.3	111.4
96	...	44.4	59.2	74.0	88.9	102.0	118.6

Drilling Holes Using an Earth Boring Machine.—Where a large number of holes are required for line construction, foundation work, pre-boring, guard rails, piling, fencing, draining, bridges, etc., an earth boring machine will perform the work much faster and at lower cost than most other methods.

These machines will drill holes through loam, clay, hard pan and shale and are furnished in various sizes by different manufacturers. Some machines will drill up to 8'-0" diameter holes to a depth of 120 ft. or more.

When drilling holes up to 18" in diameter in ordinary soil, a machine will drill at the rate of 1" depth in 1½ to 2 seconds, or drill a 3'-0" deep hole in 1 to 1¼ minutes. Moving the machine and spotting the holes takes some time, so this will govern the number of holes that may be drilled per day.

A machine that will drill holes to 24" diameter and 9'-0" deep costs approximately $14,000.00 f.o.b. factory. This does not include cost of truck, which will have to be added. The larger machines which are used for drilling caisson foundations may cost as much as $175,000.00

Based on 20 holes, 18" diam. and 3'-0" deep, an hour or 160 holes per 8-hr. day, the daily cost should average as follows:

	Hours	Rate	Total	Rate	Total
Operator	8	$. . . .	$. . . .	$17.95	$143.60
Machine charge	8	. . . .	. . . .	20.00	160.00
Truck charge	8	. . . .	. . . .	18.00	144.00
Total cost 160 holes			$. . . .		$447.60
Cost per 3'-0" hole					2.80
Cost per lin. ft.					.93

There is also a small portable post hole digger, mounted on two rubber tired wheels and equipped with a 5 to 6 h.p. gasoline engine.

This machine can be operated by one man and in ordinary ground it will bore a 9" hole 3'-0" deep in 1 to 1½ minutes. Such a machine costs $1100.00 to $1500.00.

Placing And Compacting Fills

Architects and engineers have become more rigid in their requirements as to the manner of placing and compacting fill materials. The usual specification for this work is to place the fill in from 6 to 12 inch layers and to compact each layer with some sort of roller equipment.

Placing and Compacting Site Grading Fills.—Fill material in site grading work is placed either by scrapers spread-dumping their loads, trucks dumping their loads more or less in piles or some sort of excavating equipment casting dipper or bucket loads as they are dug. In the case of the scraper spread-dumping fill material no further spreading is necessary as this can be controlled with precision by the scraper operator. In the other cases, however, a bulldozer will be required to spread out the fill material into layers of specified depth and hourly production for this work may be found in the table under "Bulldozer Production."

When the fill has been placed and spread in the specified depth of layer, the compacting operation follows and this is usually accomplished by rolling with a tractor drawn sheepsfoot roller. The degree of compaction may be specified by the number of passes to be made by the roller or by a percentage, such as 95%, of maximum density of the material obtained at optimum moisture content in a soil mechanics laboratory. The latter specification is usually called for in government work and for fills under paving. To obtain 95 percent of maximum density compaction may take as many as 12 passes of a sheepsfoot roller.

If rolling is not to interfere with production schedules, a rate must be established in cubic yards compacted per hour and applied to the rate of placing and spreading the fill material to determine the size and number of rollers required to keep the job moving. Where frequent turns are made or restricted areas require maneuvering these factors must be considered and the figures adjusted accordingly.

The following table gives rate of compaction, in cu. yds. per hr., for rolling fill with a 5'-0" wide sheepsfoot roller at 2.5 m.p.h. with the fill material placed in 12" layers for soil of various compaction factors and for 1 pass of the roller to 12 passes.

Courtesy U. S. Homes

After fill has been placed self-powered compactor is used to tamp earth

Rate of Sheepsfoot Roller Compaction in Cu. Yds. per Hr.
Based on 5'-0" Wide Roller, 2½ M.P.H. Speed,
12" Fill Layers and 100% Job Efficiency

Number of Passes of Roller	Percentage Factor of Pay Yd. to Loose Yd.		
	Hard Tough Clay	Medium Clay	Loam
	70%	80%	90%
1	1,711	1,956	2,200
2	856	978	1,100
3	570	652	733
4	428	489	550
5	342	391	440
6	285	326	367
7	244	279	314
8	214	245	275
9	190	217	244
10	171	196	220
11	156	178	200
12	143	163	183

For rollers of different widths adjust the above rates proportionately, that is, for 10'-0" wide roller double the rates, for 15'-0" wide roller triple the rates, etc.

For different depths of layers adjust the above rates proportionately, that is, for 6" layers reduce the rates to ½, for 9" layers reduce the rates to ¾, etc.

The above rates must also be adjusted to normal job efficiency and further adjusted to take care of lost time in maneuvering and turning. In small jobs as much as 10 percent of the time could be lost in turns, etc.

Example: Assume a fill, placed in 9" lifts, to be rolled 8 passes with a double drum sheepsfoot roller 10'-0" wide at 2.5 m.p.h. in soil with a compaction factor of 80 percent.

From the table, 8 passes in 80% material will compact 245 cu. yds. per hr. Adjusting this rate for a 10'-0" wide roller gives a rate of 490 cu. yds. per hr. and further adjusting for 9" fill layers results in a compaction rate of 368 cu. yds. per hr. Assuming that a 50-minute hour or a job efficiency of 83% is expected, the rate is reduced to 305 cu. yds. per hr. and with a time loss of 10% anticipated for turns and maneuvering the final result is a compaction rate of 275 cu. yds. per hr.

The approximate cost for compacting fill at rate of 275 cu. yds. per hr. is as follows :

	Hours	Rate	Total	Rate	Total
Hourly charge for 75-Hp. bulldozer with double drum sheepsfoot roller	1	$....	$....	$23.80	$23.80*
Bulldozer operator	1	....	....	17.95	17.95
Cost of Compacting 275 Cu. Yds.			$....		$41.75
Cost per cu. yd.			....		.15

*Does not include hauling equipment to and from job, compensation and liability insurance, Soc. Sec. taxes or overhead and profit.

The sheepsfoot roller method of compaction is applicable to all soils with cohesive qualities, such as loams and clays, and in very dry soils it may be necessary to add moisture to the material to obtain the correct degree of plasticity.

With non-cohesive materials, such as sands and gravels, the sheepsfoot roller usually is ineffective and rubber-tired pneumatic rollers are used instead.

Placing and Compacting Fills for Floors on Ground Inside Buildings.—Fill material, necessary to raise the elevation of the subgrade for floors inside buildings, may be placed in a number of ways.

For small jobs this is usually done after the building is enclosed, the fill being brought in through door and window openings where it is then spread and tamped by hand. Three laborers working together should spread and tamp 35 to 40 cu. yds. of fill per 8-hr. day at the following cost per cu. yd. :

	Hours	Rate	Total	Rate	Total
Labor	0.6	$....	$....	$12.54	$7.53

On larger jobs it may be possible to truck the fill material directly into the space and spread-dump it, using a small bulldozer to further spread into even layers. Compaction of material in this case can usually be accomplished by using a single drum sheepsfoot roller or small pneumatic roller for the open areas and by hand tamping at walls and columns. Trucking traffic over the various layers of fill is a big help in compacting the material. The roller can be drawn by the same bulldozer that spreads the material and with 2 laborers spreading and tamping at "hard-to-get" places a daily production of 95 to 100 cu. yds. placed and compacted should be realized at the following cost per cu. yd. :

	Hours	Rate	Total	Rate	Total
50-Hp. bulldozer and single drum sheepsfoot roller	0.08	$....	$....	$13.23	$1.06
Bulldozer operator	0.08	....	....	17.95	1.44
Labor	0.16	....	....	12.54	2.01
Cost per Cu. Yd.			$....		$4.51

*Does not include hauling equipment to and from job, compensation and liability insurance, Soc. Sec. taxes or overhead and profit.

Grading

Grading consists of dressing up ground surfaces, either by hand or machine, to conform to specified contours or elevations and is usually preparatory to a subsequent operation such as placing a floor slab, walk or drive, spreading top soil for planting areas, etc.

Grading costs vary according to the degree of accuracy demanded and the method by which it is done. Following are approximate costs on some of the various types of grading usually encountered in construction work.

Rough Grading.—After all backfill is in place and the site has been cut and/or filled to the approximate specified contours, rough grading of the site is done, by hand for small jobs and by machine for large areas, usually preparatory to the spreading of top soil. Tolerance for this type grading is usually plus or minus 0.1 ft.

On small jobs a laborer should rough grade 800 sq. ft. of ground surface per 8-hr. day at the following cost per 100 sq. ft. :

	Hours	Rate	Total	Rate	Total
Labor	1.0	$....	$....	$12.54	$12.54
Cost Per Sq. ft.			....		.13

On large jobs a bulldozer is usually employed to do rough grading and a laborer generally accompanies the machine to check the surface and direct the operation. A 75-hp. bulldozer should rough grade 6,500 to 7,500 sq. ft. per 8-hr. day at the following cost per 1,000 sq. ft. :

	Hours	Rate	Total	Rate	Total
75-Hp. bulldozer charge	1.15	$....	$....	$23.80	$27.37
Bulldozer operator	1.15	....	....	17.95	20.64

	Hours	Rate	Total	Rate	Total
Labor	1.15	. . . .	. . . .	12.54	14.42
Cost per 1,000 sq. ft.			$. . . .		$62.43
Cost per sq. ft.				. . . .	.06

Does not include hauling equipment to and from job, compensation and liability insurance, Soc. Sec. taxes or overhead and profit.

Grading for Slabs on Ground.—Grading for slabs on ground such as floors, walks, driveways, etc., usually is done by hand unless the job is quite large and can be organized so that some of the work is done by machine. This work generally must be done accurately with tolerances of no more than ½" allowed. A laborer should grade 500 to 600 sq. ft. per 8-hr. day at the following cost per 100 sq. ft. :

	Hours	Rate	Total	Rate	Total
Labor	1.5	$. . . .	$. . . .	$12.54	$18.81
Cost per sq. ft.				. . . .	.19

For sloping surfaces add about 50 percent to the above cost depending on steepness of pitch.

Finish Grading of Top Soil.—After top soil has been spread over areas specified, a finish grading operation, usually including hand-raking, must be performed prior to seeding or sodding. The tolerance on this work usually is plus or minus 1 inch. A laborer should finish grade and hand-rake 600 to 700 sq. ft. per 8-hr. day at the following cost per 100 sq. ft. :

	Hours	Rate	Total	Rate	Total
Labor	1.25	$. . . .	$. . . .	$12.54	$15.68
Cost per sq. ft.				. . . .	.16

For sloping surfaces, such as berms, ditch sides, etc., add 50 percent to the above cost.

Grading for Footing Bottoms.—When footing pits and trenches are dug by machine there is always cleanup, squaring and grading work to be done by hand and many contractors price this work by the sq. ft. of footing bottom. A laborer should clean up, square and grade 200 sq. ft. of footing bottom per 8-hr. day at the following cost per 100 sq. ft. :

	Hours	Rate	Total	Rate	Total
Labor	4.0	$. . . .	$. . . .	$12.54	$50.16
Cost per sq. ft.				. . . .	.50

Costs Of Excavating

Digging Fence Post Holes.—When digging 250 post holes about 3'-0" deep in black soil and sand, using an ordinary augur post-hole digger, a man will dig 4 holes an hour or 32 per 8-hr. day, at the following labor cost per hole :

	Hours	Rate	Total	Rate	Total
Labor	0.25	$. . . .	$. . . .	$12.54	$3.14

Loader Excavating under Favorable Conditions.—The following costs are for a theoretical job performed in early spring, where conditions were favorable, with no time lost because of bad weather or breakdowns during the period required to complete the work.

This is a basement excavation 80'-0"x80'-0" and 13'-0" deep. The top 5'-0" consists of loose clay and loam, while the lower 8'-0" is blue grey dolomite limestone in beds varying from 6" to 24".

A 1½-cu. yd. loader will be used, rented at $50.00 per hr. including operator and fuel.

Five 6-cu. yd. trucks will transport the excavated earth and rock to a dump about ¼-mile from the job. Trucks are estimated to cost $35.00 per hr. including gasoline and driver.

A 5½"x5" portable air compressor, capacity 110 cu. ft. per minute and a Type 12 Standard I-R Jackhammer is used for rock drilling. The rental of the compressor and jackhammer, including gasoline, operator and repairs to drills is estimated at $25.00 an hour.

Excavating Loose Clay and Loam.—The area of loose clay and loam is 80'-0"x80'-0"x5'-0", a total of 1,185 cu. yds. of excavation.

The average on this job is 135 cu. yds. an hr. at the following cost per 100 cu. yds.:

	Hours	Rate	Total	Rate	Total
Loader and Operator	.75	$....	$....	$50.00	$37.50
Cost per cu. yd.			....		.38

Five 6-cu. yd. trucks will haul 1,080 cu. yds. of excavated earth ¼-mile to a dump per day, or an average of 27 cu. yds. an hr. per truck, at the following cost per 100 cu. yds. :

	Hours	Rate	Total	Rate	Total
Truck and Driver	3.7	$....	$....	$35.00	$129.63
Cost per cu. yd.			....		1.30

Four men at the dump spread the excavated earth after it had been dumped by the trucks. Very little grading was necessary as the earth dump was a high-terraced fill around a building which required very little handling.

A man will spread about 10 cu. yds. an hr. at the following labor cost per cu. yd. :

	Hours	Rate	Total	Rate	Total
Labor	0.10	$....	$....	$12.54	$1.26

The total cost of excavating, hauling and spreading one cu. yd. of loose earth is as follows :

	Rate	Total
Loader excavation	$....	$ .38
Trucks hauling excavated earth	....	1.30
Labor spreading loose earth at dump	....	1.26
Cost per cu. yd.		$2.94

Note: Does not include equipment move-in charge.

Rock Excavation

After the top 5'-0" of soil has been removed, the excavation consists of blue grey dolomite limestone, hard, and stratified in beds varying from 6" to 24". The depth of the cut was 8'-0", which was taken out in two lifts of 4'-0" each. The last lift was drilled 5'-0" deep or 1'-0" below the desired level.

Total rock excavation is 80'-0"x80'-0"x8'-0", or 1,900 cu. yds.

Drilling Rock for Blasting.—The area drilled is 80'-0" x 80'-0", less a 5'-0" border on 3 sides and a 10'-0" border on one side. The rows were 5'-0" apart and the holes were 5'-0" on centers, staggered, and all of the holes will be 2" in diameter.

There will be 14 rows with 15 holes in a row or a total of 210 holes, as follows : 105 holes 4'-0" deep, containing 420 lin. ft. and 105 holes 5'-0" deep, containing 525 lin. ft., a total of 945 lin. ft. of 2" holes.

One man and a helper will drill 15 lin. ft. of 2" hole an hr. or 120 lin. ft. per 8-hr. day, at the following cost per 100 lin. ft. :

	Hours	Rate	Total	Rate	Total
Compressor, jackhammer					
and operator	7	$. . . .	$. . . .	$25.00	$175.00
Labor helping	7	. . . .	. . . .	12.54	87.78
Cost 100 lin. ft.			. . . .		$262.78
Cost per lin. ft.			. . . .		2.63
Cost per cu. yd. (1,900 cu. yds.)					1.30

Explosives.—The job will require 2.5 cases (50 lbs. each) of 40% dynamite, a total of 125 lbs. which is equivalent to ⅛-lb. per lin. ft. of hole or per cu. yd. of excavation, and the cost of explosives was as follows :

	Hours	Rate	Total	Rate	Total
125 lbs. dynamite		$. . . .	$. . . .	$.85	$106.25
220 electric detonators					
with 6' wire lead		. . . .	. . . .	.45	99.00
75 lin. ft. No. 14,					
R.C.S.B. lead wire		. . . .	. . . .	.12	9.00
Total cost			$. . . .		$214.25
Cost per hole			. . . .		1.02
Cost per cu. yd. (1,900 cu. yds.)			. . . .		.11

Rock Excavation.—There will be 1,900 cu. yds. of rock excavation, taken out in two lifts of 4'-0" each and the loader will handle 45 cu. yds. an hr. at the following cost per 100 cu. yds. :

	Hours	Rate	Total	Rate	Total
Loader					
and operator	2.3	$. . . .	$. . . .	$50.00	$115.00
Labor*	2.3	. . . .	. . . .	12.54	28.84
Powderman	2.0	. . . .	. . . .	17.95	35.90
Labor assisting	5.0	. . . .	. . . .	12.54	62.70
Total cost 100 cu. yds.			$. . . .		$242.44
Cost per cu. yd.			. . . .		2.43

*One laborer on bull chain placing chain on rocks too large for shovel to handle.

Two 6-cu. yd. trucks hauled the excavated rock to a dump ¼-mile from the job, averaging about 23 cu. yds. an hr. per truck, at the following cost per 100 cu. yds. :

	Hours	Rate	Total	Rate	Total
Truck and driver	4.35	$. . . .	$. . . .	$35.00	$152.25
Cost per cu. yd.			. . . .		1.52

Four men at the dump leveled 20 cu. yds. of excavated rock an hr. at the following labor cost per 100 cu. yds. :

	Hours	Rate	Total	Rate	Total
Labor	20	$. . . .	$. . . .	$12.54	$250.80
Cost per cu. yd.			. . . .		2.51

The total cost of rock excavation, including drilling, blasting, excavating, hauling and spreading, averaged as follows per cu. yd. :

	Total	Total
Drilling	$. . . .	$1.30
Explosives	. . . .	.11
Excavating	. . . .	2.43
Hauling	. . . .	1.52
Spreading and leveling	. . . .	2.51
Cost per cu. yd.	$. . . .	$7.87

Rock excavation conditions and costs vary widely. Always obtain information from sources at or near job site before bidding on this work. Check with one or more of the local powdermen—they will know what the conditions are and what production may be expected.

Drilling Rock Using Pneumatic Jackhammer Drills.—The hand held jackhammer has become a very popular tool for rock drilling on all types of construction jobs. Because they are furnished in sizes weighing from 30 lbs. to 80 lbs., many rock drilling conditions can be met. Holes up to 20 ft. in depth are easily drilled with the heavier machines. Most sizes of jackhammers can be furnished in three styles depending upon the conditions of drilling and the depth of holes being drilled, these being the dry style for shallow holes, blower style for deep holes, and wet for jobs where dust must be kept down to a minimum.

Vehicle mounted pneumatic hammer used to break up pavement

Hollow drill steel of various sizes and shapes are used, the most common being ⅞" and 1" hexagon. Bits are sometimes forged on the steel but the detachable bit is by far the most widely used.

When drilling holes in rock for trenches, blasting etc., two men operating drills should drill 125 to 175 lin. ft. of holes per 8-hr. day, at the following cost per 100 lin. ft. :

	Hours	Rate	Total	Rate	Total
Jackhammerman	11.0	$....	$....	$12.54	$137.94
Compressor operator	5.5	....	....	17.05	93.78
Compressor expense	5.5	....	....	13.00	71.50
Cost per 100 lin. ft.			$....		$303.22
Cost per lin. ft.			....		3.03

*Add only if required.   Note: Does not include bit charges.

Rock Drilling Using Wagon Drills.—Wagon drills consist of a one piece tubular frame on pneumatic or steel wheels and tilting tower upon which is mounted a heavy pneumatic drill. They are recommended for drilling holes downward at any angle from the horizontal to the vertical to depths of 25 to 30 feet. This type of drill is suitable for all kinds of excavation and trenching work and will usually do the job of at least 2 to 3 heavy hand held jackhammers and requires only one operator.

Rock Drilling Using Crawler Mounted Drills.—An innovation in rock drilling for blast holes, etc. is the completely mechanized, self-propelled drill unit. Many of the construction machinery manufacturers are now making such units. These units will do the work of two or three wagon drills and are taking over the field. The unit can move easily over rugged terrain and can tow its own air supply with it. The set-up time for each hole is greatly reduced because of the self-propelled features. It is not uncommon for a crawler mounted drill to drill more than 500 lineal feet per 8-hour shift in hard sedimentary rock. These can be rented with operator for $325 to $450.00 per day.

Excavating Using Pneumatic Diggers.—These tools can be used to advantage on all classes of excavation that requires picking of dirt, clay, etc., such as trench work, tunneling, caisson sinking and all kinds of building excavation; in fact, on all kinds of work in stiff clay or hard ground where power shovels, trenching machines, etc., cannot be used.

On work of this kind, two men using air picks and six men shoveling will loosen and shovel 35 to 40 cu. yds. per 8-hr. day, at the following labor cost per cu. yd. :

	Hours	Rate	Total	Rate	Total
Labor operating picks	0.4	$....	$....	$12.54	$ 5.02
Labor shoveling	1.2	....	....	12.54	15.05
Compressor operator	0.2	....	....	17.05	3.41
Compressor expense	0.2	....	....	13.00	2.60
Cost per cu. yd.			$....		$26.08

*Add only if required.

On larger excavations providing plenty of room for the men to work, a man with pneumatic pick will loosen 25 to 30 cu. yds. per 8-hr. day. After the clay or earth is loosened, a man will shovel 6 to 7 cu. yds. per 8-hr. day.

Comparative Cost By Hand

Excavating in stiff clay or tough ground, a man using a hand pick and shovel will loosen 2½ to 3 cu. yds. of dirt per 8-hr. day. The same man, picking only, will loosen 3½ to 4 cu. yds. of dirt per 8-hr. day, and when shoveling only, will remove 6 to 7 cu. yds. per 8-hr. day.

Using a hand pick, a man will loosen and shovel one cu. yd. of dirt at the following cost per cu. yd. :

	Hours	Rate	Total	Rate	Total
Labor	3	$....	$....	$12.54	$37.62

Backfill Tamping Using a Pneumatic Tamper.—A pneumatic tamper enables a workman to do nearly 10 times the amount of work he can do by hand. A backfill tamper strikes over 600 hard, snappy blows a minute, rams the fill hard, and works in and around pipe with far greater thoroughness than hand devices are able to do.

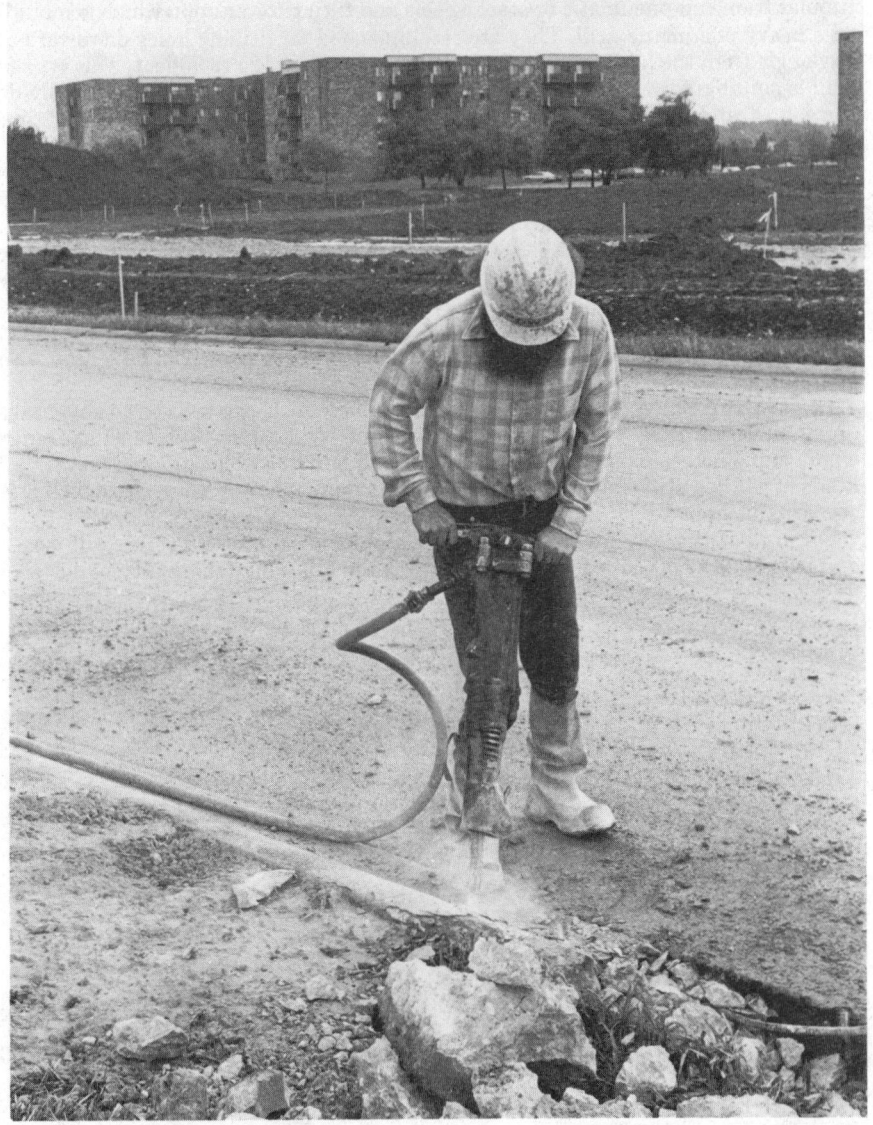

Breaking up old curbing using pneumatic jack hammer

In piers, trenches, etc., where the excavated earth can be shoveled directly into the trench, a man should backfill 16 to 20 cu. yds. per 8-hr. day.

Where the trenches have to be tamped, one man with a pneumatic tamper will tamp as much as 2 men can backfill, or approximately 35 to 40 cu. yds. per 8-hr. day.

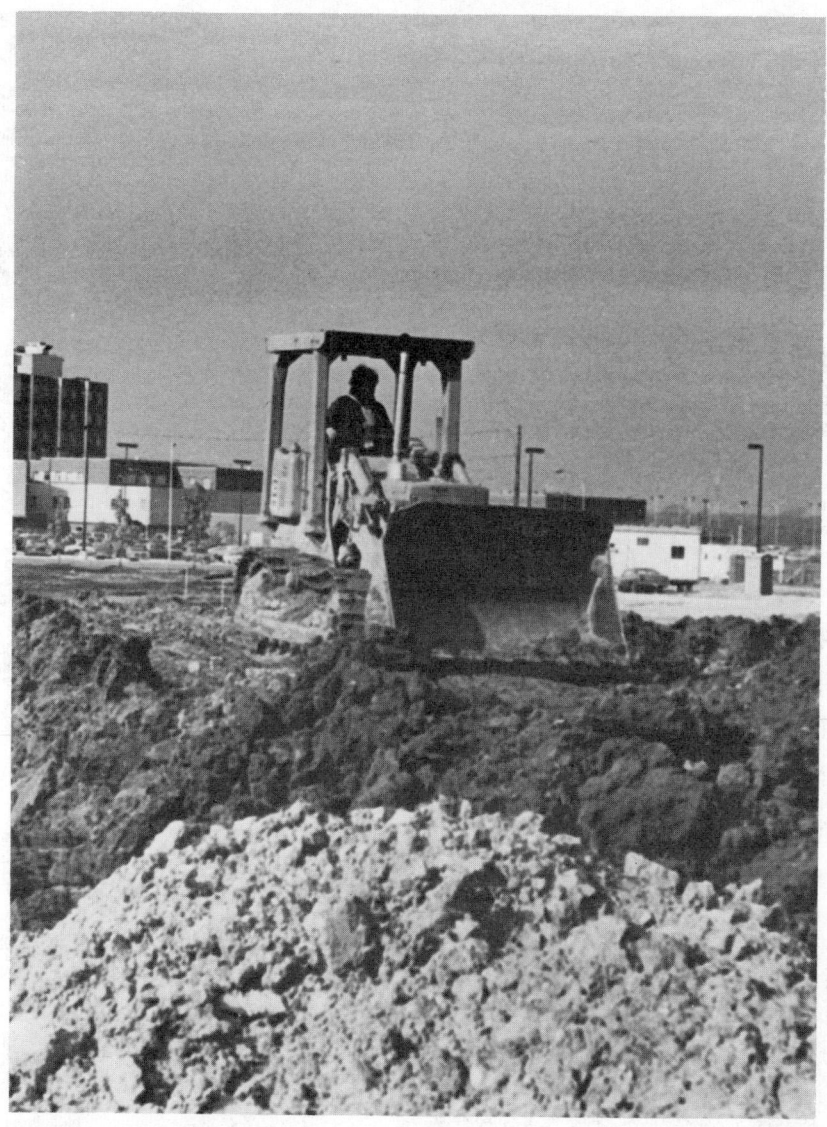

Courtesy U. S. Homes

Back filling is frequently accomplished using different size Caterpillar tractors

The cu. yd. cost should average as follows :

	Hours	Rate	Total	Rate	Total
Labor backfilling	0.4	$. . . .	$. . . .	$12.54	$ 5.02
Labor tamping	0.2	. . . .	. . . .	12.54	2.51
Compressor operator	0.2	. . . .	. . . .	17.05	3.41
Compressor expense & tamper	0.2	. . . .	. . . .	15.00	3.00
Cost per cu. yd.			$. . . .		$13.94

*Add only if required.

Breaking pavement to install underground electrical cable

Oversize hydraulic tamper finishing off sewer trench

Courtesy U. S. Homes

Excavating with Caterpillar tractor loader

Comparative Cost By Hand

A man will tamp by hand 8 to 10 cu. yds. of backfill per 8-hr. day, at the following cost per cu. yd. :

	Hours	Rate	Total	Rate	Total
Labor backfilling	0.4	$. . . .	$. . . .	$12.54	$ 5.02
Labor tamping	0.8	. . . .	. . . .	12.54	10.04
Cost per cu. yd.			$. . . .		$15.06

02250   SOIL TREATMENT

Often a soil poisoning will be made a part of the contract for termite control. Termites are found in most all states but are more prevalent in the south, south-west, southern Pacific, and Ohio Valley regions.

The contractor should know local laws as so many applicators have appeared on the pest control scene that strict licensing controls are in effect, and the work must be subcontracted.

Soil treatments are an added protection to, not a substitute for, preventive measures that can be incorporated in the building design. These include avoiding all contact between wood and soil; removing all wood chips and scraps from crawl spaces, backfill and under porches; ventilating crawl spaces and draining them so that they will stay dry; providing tight foundations, preferably poured concrete; building in shields at tops of foundations and where piping penetrates building construction; and providing access to those vulnerable areas which must have year-ly inspections and future retreatment of soil.

The usual chemicals encountered are chlordane, 1% emulsion, mixed 1 gal. of concentrate to 48 gals. of water; dieldrin, 0.5% emulsion mixed 1 to 36 gals. water; and benzene hexachloride, 0.8% emulsion mixed 1 to 15 gals. of water. When treating soils within 100 linear feet of a ground fed water supply (well), a special solution called Dursban TC is required to prevent poisoning the water supply.

The rate of application is usually 2 gals. of diluted emulsion per 5 lin. ft. of wall per each 12 in. of depth treated. The inside of a crawl space or on grade slab is usually trenched only 6" to 8" deep and the same width. The emulsion is poured in, then backfilled, and another application made for a total of 4 gals. per each 5 lin. ft. of wall. Remember to include piers also. Applications on exterior of walls is made at the same rate, a new application being made after each 12 in. of depth of backfill. A 3 foot deep foundation wall would thus require a 6 gal. emulsion per each 5 ft., while a 6 ft. deep basement wall would require twice that. Treatment of slab on grade areas should be at the rate of one (1) gallon emulsion for every ten (10) sq. ft. of slab area.

02300   PILE FOUNDATIONS

When estimating the cost of wood piles the item of overhead expense must be considered before the actual cost of driving the piles can be computed. This includes cost of moving the equipment to the job, setting it up ready to operate and dismantling and removing at completion.

This item will remain practically the same regardless of the size of the job, as the equipment move charge will be as much on a job having 100 piles as on a job containing 1,000 piles.

The actual cost of driving the piles will vary with soil conditions, length of piles and the amount of moving necessary to drive piles at the proper location on the site.

Setting up and Removing Equipment.—The cost of moving the pile driving equipment to the job must be taken into consideration, and then after it arrives at the job, there is the task of setting up the pile driver and getting ready to operate.

After the pile driver has been delivered at the job it should take about 3 days time for the crew to set it up ready to drive and this labor should cost as follows :

Setting up Equipment	Hours	Rate	Total	Rate	Total
Foreman	24	$....	$....	$18.62	$ 446.88
Engineer on pile driver	24	....	....	17.95	430.80
Oiler	24	....	....	12.54	300.96
2 pile driver men	48	....	....	16.47	790.56
4 men, general labor	96	....	....	12.54	1,203.84
Total labor cost			$....		$3,173.04

The above item will remain practically the same whether there are 100 or 1,000 piles to be driven.

After the job has been completed, the pile driver must be dismantled and removed from the job and the dismantling should cost as follows :

Dismantling Equipment	Hours	Rate	Total	Rate	Total
Foreman	16	$....	$....	$18.62	$ 297.92
Engineer on pile driver	16	....	....	17.95	287.20
Oiler	16	....	....	12.54	200.64
2 pile driver men	32	....	....	16.47	527.04
4 men, general labor	64	....	....	12.54	802.56
Total Labor Cost			$....		$2,115.36

WOOD PILES

Driving Wood Piles.—After the pile driver is set up and ready to operate, the labor cost of driving the piles will vary with the length of the piles and job conditions.

The following table gives the approximate number of piles of various lengths that should be driven per hour and per day :

Length of Piles	1 Hour	8 Hours
24 feet	4.5	36
26 feet	4.2	34
28 feet	4.0	32
30 feet	3.7	29
32 feet	3.5	28
34 feet	3.3	26
36 feet	3.1	25
38 feet	3.0	24
40 feet	2.8	22
45 feet	2.5	20
50 feet	2.3	18
55 feet	2.1	17
60 feet	2.0	16

The labor operating the pile driver per 8 hr. day should cost as follows :

	Hours	Rate	Total	Rate	Total
Foreman	8	$. ...	$. ...	$18.62	$ 148.96
Engineer on pile driver	8	....	....	17.95	143.60
Oiler	8	....	....	12.54	100.32
2 pile driver men	16	....	....	16.47	263.52
4 men, general labor	32	....	....	12.54	401.28
Labor cost per day			$. ...		$1,057.68

The above does not include time laying out the work or spotting piles, pumping, shoring, excavating or cutting off wood piles after they are driven, but includes only the actual time driving the piles. Neither does it include the cost of equipment, fuel, oil, and other supplies for the pile driver, which should be added.

Cutting off Wood Piles.—After the wood piles have been driven, the tops projecting above the ground will have to be cut off to receive the foundation which is to be placed upon them.

This cost will vary according to the size of the quarters in which the men are obliged to work, but on the average job it should require 1/6 to 1/4 hrs. to cut off each wood pile 12" or 14" in diameter with a chain saw, at the following labor cost per pile :

	Hours	Rate	Total	Rate	Total
Labor	0.22	$. ...	$. ...	$16.47	$3.62

To the above costs, add the hourly cost of chain saw, and the time for removing sawed-off ends of the piles from the premises.

Computing the Cost of Wood Piles.—As described above, the cost of driving wood piles will vary with the number of piles, length, kind of soil, and other job conditions. Costs of the piles fluctuate so rapidly that current costs must be verified for each project.

An example of an estimate showing a method for arriving at costs for driving 680 treated wood piles 40'-0" long is as follows :

	Local Rate	Total	Total
Trucking pile driver to job	$. ...	$. ...	$ 1,500.00
Setting up pile driver, 3 days for crew	....	....	3,173.04

	Local Rate	Total	Total
680 wood piles, 40'-0" long @ $160.00	....	....	108,800.00
Driving piles (22 per 8-hr. day) 31 days @ $1,057.68	....	....	32,788.08
Fuel, oil, misc. supplies, etc. 31 days @ $150.00	....	....	4,650.00
Dismantling pile driver at completion, 2 days	....	....	2,115.36
Trucking pile driver, job to yard	....	....	1,500.00
Equipment rental or depreciation allowance (36 days)	....	....	18,000.00
Cost 680 piles		$....	$172,526.48
Cost per pile		....	253.72
Cost per lin. ft.		....	6.34

Add cost of cutting off piles after driving, and for overhead & profit.

Driving Wood Piles on a Large Chicago Job

CONCRETE PILES

Due to the wide variations and conditions under which concrete piles are installed and used, it is difficult to give dependable cost figures without possessing the details of each installation. The different types of piles also vary considerably in cost. When estimating work containing concrete piles, it is always advisable to consult contractors specializing in this class of work. Most of them maintain offices in the principal cities and are usually willing to investigate and quote budget prices on any anticipated work. Figures can be obtained with more accuracy than can be given here.

There are two principal types of concrete piles, cast-in-place and precast. The cast-in-place pile is formed in the ground in the position in which it is to be used in the foundation. The precast pile is cast above ground and, after it has been properly cured, it is driven or jetted just like a wood pile.

In ordinary building foundation work the cast-in-place pile is more commonly used than the precast pile for the following reasons. For cast-in-place piles, the required length can readily be adjusted in the field as the job progresses. Thus there is no need to predetermine pile lengths and the required length is installed at each pile location. In contrast, it is necessary to predetermine the length of precast piles, and to provide for contingencies, it is generally required that piles be ordered longer than the actual anticipated length. Furthermore it is not possible to determine the pile length at each location as subsoil conditions at any construction site will show considerable variation. Therefore, the cost of waste piling for precast piles could be quite high. In addition, the cost of cutting off precast piles to the proper grade could be high. Precast piles are prestressed and are manufactured at established plants. Job site casting is relatively rare. It is difficult and costly to handle and transport precast piles especially in long lengths. Therefore a production plant must be relatively close to the job site.

In marine installations either in salt or in fresh water, the precast pile is used almost exclusively, because of the difficulty involved in placing cast-in-place piles in open water. For docks and bulkheads, the cast-in-place pile is sometimes used in the anchorage system. On trestle type structures such as highway viaducts, the precast pile is more commonly used. A portion of the pile often extends above the ground and serves as a column for the superstructure. Precast concrete piles are always reinforced internally so they can resist stresses produced by handling and driving. Cast-in-place piles are rarely reinforced since lateral support of even the poorest of soils is sufficient to overcome any bending moments induced in the pile by column action.

Raymond Concrete Piles*—The basic Raymond piles are Step-Taper and Pipe Step-Taper. They are steel-encased cast-in-place concrete piles installed by :
1) placing a steel shell closed at the tip over a steel mandrel,
2) driving the mandrel and shell to the required resistance and/or penetration,
3) withdrawing the mandrel, leaving the steel shell in place as driven,
4) inspecting the steel shell internally its full length,
5) removing excess shell, and
6) filling the steel shell with concrete to pile cut-off grade.

The steel shell not only maintains the driving resistance but also protects the fresh concrete from being mixed with the surrounding soil or from being distorted by soil pressures set up by the driving of adjacent piles.

Raymond Step-Taper Pile* shells are made of sheet steel in gages of 12 to 20 and in basic section lengths of 4, 8, 12 and 16 feet. Longer lengths can be furnished for special conditions. Step-Taper shells are helically corrugated to provide greater

*Raymond International, Inc.

strength against collapsing pressures. A driving ring is welded to the bottom of each shell section and joints between sections are screw-connected. The bottom section is closed with a flat steel plate welded to the driving ring (See details of Step-Taper Piles). Step-Taper Pile shells are manufactured in standard nominal diameters ranging from 8 5/8 inches to 18 3/8 inches but larger diameters can be made to meet special conditions.

STEP ON MANDREL
ENGAGES DRIVE RING

DRIVE RING

COLLAR FOR
SCREW JOINT

MANDREL

SHELL

FLAT CLOSURE PLATE
WELDED TO BOTTOM
DRIVE RING

Raymond Step Taper Piles

When sufficient shell sections are joined together to make up the required pile length, the pile diameter increases from tip to butt at the rate of one inch per section length. Thus the rate of taper or pile shape will vary with the section lengths used. Within practical limits, different section lengths can be combined in a single pile thus providing a wide range of available pile shapes. Nominal tip diameters usually range from 8 to 11 inches but larger tip diameters can be used. With the variety of diameters and section lengths available, the Step-Taper pile provides maximum flexibility in the choice of pile size and configuration to best meet subsoil conditions and loading requirements.

The length of a Step-Taper Pile can be readily adjusted as the work proceeds, to meet variable soil conditions. Thus waste costs are held to a minimum. It is unnecessary to install drive-test piles to predetermine pile lengths.

Internal reinforcement is not required for Step-Taper Piles except to resist uplift loads or high lateral loads (where batter piles are not used) or for unsupported pile lengths extending through air, water or very fluid soil.

Step-Taper Piles are suitable for all types of soils and can function as friction or point bearing piles. In most cases, piles are supported through a combination of friction and direct bearing. In many soils, tapered piles usually develop higher capacity than piles of no taper.

The methods and equipment used to install Raymond Piles permit driving of piles in the shortest possible time with corresponding savings in construction costs.

Step-Taper Piles are driven with a heavy rigid steel mandrel which provides effective hard driving and the development of high capacity piles. The maximum practical length for an all-shell Step-Taper Pile is about 140 feet. However, longer all-shell piles could be installed providing a pile driving rig of adequate capacity were available. Where exceptionally long, high capacity piles are required, the Raymond Pipe Step-Taper Pile is used.

Raymond Pipe Step-Taper Piles* are fundamentally the same as the Step-

*Raymond International Inc.

Taper Piles except that a closed-end steel pipe is used for the lower portion of the pile. In some cases, the mandrel used to drive the shells can extend through the pipe to the pile tip. The joint between pipe and shell is either a water-tight slip joint, a drive sleeve joint or a welded joint and is designed to transfer the driving force from the mandrel to the top of the pipe. If the mandrel extends for the full length of the pipe, the driving force is also carried down to the pile tip (See Detail Pipe Step-Taper Pile).

Raymond Pipe Step-Taper Piles

The diameter of the Step-Taper shell section just above the pipe is determined by the diameter of the pipe used and is either approximately the same size as the pipe or about one inch larger in diameter. Pipe diameters normally range from 10 to 14 inches although larger pipe sizes have been used.

The required pipe wall thickness would depend upon driving conditions, the length of the pipe portion and whether or not it were driven with an internal mandrel. For hard driving, high penetration resistance, long pipe portions or non-mandrel driven pipe, a heavier wall pipe is required for effective driving.

Pipe Step-Taper Piles are suitable for all types of soils and function as friction or point bearing piles. Through variations in lengths of pipe and shell portions, sizes of pipe and sizes and section lengths for the Step-Taper shells, the Pipe Step-Taper Pile also provides a wide range of pile size and configuration to best meet subsoil conditions and loading requirements. Pile lengths can readily be adjusted in the field to meet variable subsoil conditions.

The length of the Pipe Step-Taper Pile is not limited by the capacity of the pile rig. The full length pile can be driven in one piece—pipe and shell together—or the pile can be installed in two or more stages—the pipe first, followed by the Step-Taper shells. For a very long pipe portion, the pipe can be driven in sections which are joined by sleeves or welding.

Budget prices are obtainable for Raymond Concrete Piles from Raymond International Inc. The budget price will include all materials, labor, equipment, etc. for the complete installation of the piles.

Steel Pipe Piles.—Another kind of cast-in-place pile with a permanent shell is the steel pipe pile, both open and closed end.

The closed end pipe pile is simply a piece of steel pipe, closed at the bottom with a heavy boot, then driven into the ground and filled with concrete. The uses and allowable loads for such piles are about the same as other types of cast-in-place piles with driven shells. On some jobs the pipe shell is driven all in one piece, but

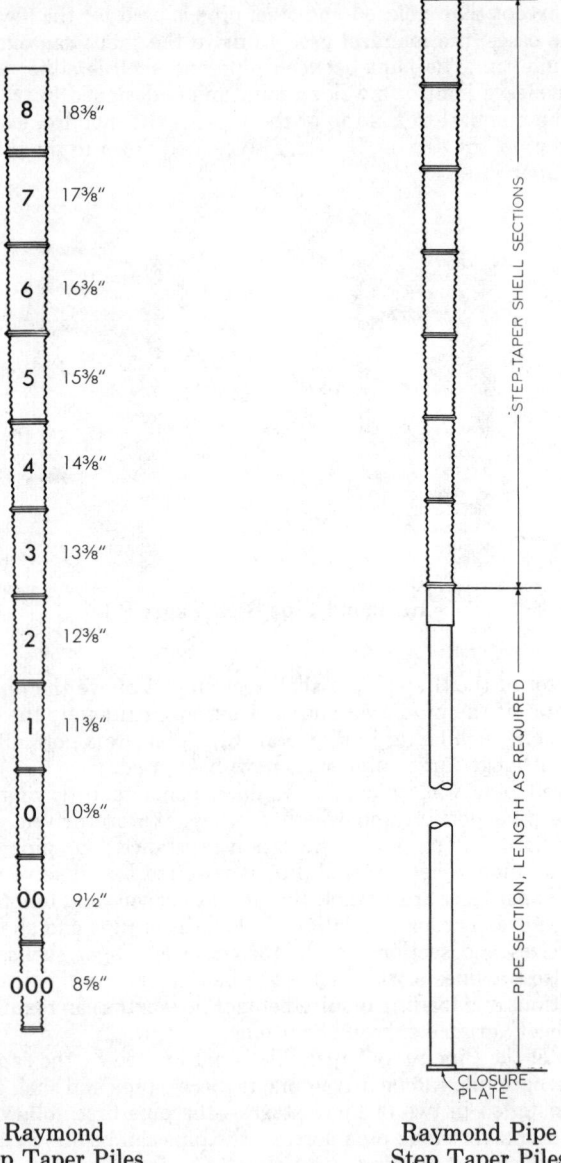

8	18⅜"
7	17⅜"
6	16⅜"
5	15⅜"
4	14⅜"
3	13⅜"
2	12⅜"
1	11⅜"
0	10⅜"
00	9½"
000	8⅝"

Raymond
Step Taper Piles

Raymond Pipe
Step Taper Piles

very often it is driven in sections which are welded together or fitted together with internal sleeves as the driving progresses. Adequate wall thickness of the pipe is necessary to develop the required stiffness and thus driveability of the pipe.

Open end steel pipe piles are usually driven to bearing on rock. Since the pipe is open at the bottom during driving, the interior fills with soil which must be removed before concreting. Cleaning is usually done with air and water jets, after which the pipe is re-driven to insure proper seating in the rock, remaining water is pumped out and the pipe pile is filled with concrete.

Open end pipe piles installed in this manner are usually allowed to carry relatively high working loads. They have been installed up to 24 inches in diameter and in lengths up to 200 feet.

Precast Concrete Piles.—Precast concrete piles are commonly used in marine structures, such as docks, piers, etc., where they are driven in open water.

Ordinarily they are designed for the particular job and made in a casting yard at or near the job site. The use of precast concrete piles usually delays the starting of actual pile driving because of the time involved in the casting and curing of this type of pile.

Precast concrete piles have the same advantages as other concrete piles over wood and steel piles. They are always reinforced internally to resist stresses produced by handling and driving.

Precast concrete piles have been driven in lengths up to 120 feet. They can be driven in sections with adequately designed joining systems. However, this is a slow and costly process and is only used where there is no alternative.

Prestressed precast concrete piles may be more economical than conventionally reinforced precast piles since less steel is necessary to adequately resist the handling and driving stresses.

Raymond *Prestressed Concrete Cylinder Piles—are centrifugally cast in 16-foot sections which are assembled end-to-end and post-tensioned to make up the pile thus providing a wide range of available pile shapes. These piles are cast with zero-slump concrete and strengths from 7,000 to 10,000 psi are attained. They are made with the Cen-Vi-Ro process which is a combination of centrifugal casting, vibrating and rolling which produces an extremely dense concrete having an absorption factor about 40 percent less than bed-cast concrete piles. The use of zero-slump concrete prevents migration of the coarse aggregate during the spinning process and thus the aggregates are uniformly distributed across the pile wall.

These piles can be made available in diameters from 36 to 90 inches but the standard sizes are 36, 54 and 66 inch O.D. Wall thicknesses are 5 inches for the 36 and 54-inch piles and 6 inches for the 66-inch piles.

Raymond Cylinder Piles are driven either open or closed-end. They are capable of carrying extremely high compression loads and resisting large lateral forces or bending moments. They are used for elevated highways, bridges, piers, offshore platforms, breakwaters, etc. For bridges and other trestle-type structures, the Cylinder Pile serves as both foundation pile and bridge pier or column.

Steel H-Beam Piles.—Rolled structural steel shapes are also used as bearing piles. The shape commonly used for this purpose is the H-beam. This type of pile has proved especially useful for trestle structures in which the pile extends above the ground and serves not only as a pile but also as a column.

Because of their small cross-sectional area, piles of this type can often be driven through dense soils to point bearing where it would be difficult to drive a pile of solid cross-section, such as a wood, cast-in-place, or precast concrete pile. This ability to penetrate dense soils easily works to a disadvantage in other soil conditions. Where piles are used to support loads by friction or where they are used primarily for compaction, a considerably longer H-beam is required to carry the

---

*Raymond International, Inc.

STANDARD SIZES
Raymond Cylinder Piles

O.D. in.	I.D. in.	W in.	Weight/Foot lbs	Number of Cables
36	26	5	524	8 to 16
54	44	5	829	12 to 24
66	54	6	1217	16 to 32

Raymond Prestressed Cylinder Piles

same load that can be supported by a pile of less length but greater cross-sectional area.

The upper ends of H-beam piles may be encased in concrete to prevent corrosion.

Steel H-beam piles, weighing 200 pounds per lin. ft., have been driven in lengths up to 130 feet and lighter sections to even greater lengths.

Estimating the Cost of Concrete Piles.—The cost of concrete piles will vary greatly, depending upon the number of piles in the job and the length, as it is the total footage of piles in the job rather than the number of piles that controls the price. For example, the price per foot for 100 piles 35 feet long would be much greater than the price per foot for 1,000 piles 25 feet long.

Contractors specializing in this work are usually willing to submit budget prices on any type of pile, which would include all labor, materials, equipment, etc. for the complete job. A detailed estimate would only be required when such budget prices were not available or for some reason, the estimator did not desire to seek out a budget price.

Regardless of the size of the job there is a fixed charge for moving equipment on and off the job that will amount to $5000.00 to $25,000.00, provided the equipment is located in the immediate vicinity of the contemplated work.

If it is necessary to ship plant and equipment from one city to another, it will be necessary to add additional freight and trucking amounting to $2000.00 to $4,000.00, making a total fixed charge of $7000.00 to $29,000.00, before figuring the cost of the piles.

Labor Driving Concrete Piles.—After the pile driver has been set up ready to operate, the number of piles driven per day and the labor cost driving same will vary with the length of the piles and job conditions. For instance, on a job where the piles were closely spaced and driven from a flat surface, a pile driver averaged 100 piles a day (which is exceptional), while on another job where the piles were more widely scattered and had to be driven from different elevations, it was with difficulty that 20 piles were driven per day.

A good average for heavy precast piles is about 10 piles per 8-hr. day.

Raymond tapered piles in which the shell is left in the ground can usually be placed at the rate of 20 to 30 piles per day, but other types of cast-in-place piles are usually placed at a slower rate, averaging 10 to 20 piles per 8-hr. day. Under favorable job conditions all of these averages may be considerably increased.

Cu. Ft. of Concrete in Concrete Piles of Various Sizes

	Diameter of Concrete Pile in Inches		
Length in Ft.	13" Cu. Ft.	15" Cu. Ft.	16" Cu. Ft.
1	.92	1.23	1.4
12	11	14.8	16.8
15	13.8	18.4	21
18	16.6	22.1	25.2
20	18.4	24.6	28
22	20.2	27	30.8
25	23	30.7	35
28	25.8	34.4	39.2
30	27.6	36.9	42
33	30.4	40.6	46.3
35	32.2	43	49
38	35	46.7	53.2
40	36.8	49.2	56

Driving 20' -0" Concrete Piles.—On a 12 story reinforced concrete building requiring 280 concrete piles 20' -0" long, the following costs were obtained.

There was considerable broken time on this job, due to bad weather, and while the job required 17½ days, only 112 hrs. actual driving time was required. This included moving the pile driver from one place to another, as some of the piles were driven from the street level and others from the basement floor level.

The pile driving crew averaged two 20' -0" concrete piles an hr. or 16 piles (320 lin. ft.) per 8-hr. day, at the following labor costs:

Setting up Pile Driver	Hours	Rate	Total	Rate	Total
Foreman	32	$....	$....	$18.62	$ 595.84
Engineer	32	....	....	17.95	574.40
Oiler	32	....	....	12.54	401.28
Linesman	32	....	....	16.47	527.04
Four pile driver men	128	....	....	16.47	2,108.16
Driving Piles					
Foreman	140	....	....	18.62	2,606.80
Engineer	140	....	....	17.95	2,513.00
Oiler	140	....	....	12.54	1,755.60
Linesman	140	....	....	16.47	2,305.80
Four pile driver men	560	....	....	16.47	9,223.20
Dismantling Pile Driver					
Foreman	20	....	....	18.62	372.40
Engineer	20	....	....	17.95	359.00
Oiler	20	....	....	12.54	250.80
Linesman	20	....	....	16.47	329.40

Four pile driver men	80	. . .	. . . .	16.47	1,317.60
Labor cost 280 piles	1,536		$. . . .		$25,240.32
Cost per 20'-0" pile				. . . .	90.15
Cost per lin. ft				. . . .	4.51

Does not include move charges for equipment, equipment costs, or pile material costs.

Labor Placing Concrete For Piles.—Practically 90 percent of all concrete used in concrete piles is ready-mix concrete, which is usually discharged directly into the piles. Ordinarily it requires ½ to 1 hour time for one man per cu. yd. of concrete.

Prices of Steel H-Beam Piles.—Prices of steel H-beam piles vary considerably, mainly due to the material cost of the rolled steel sections. Steel sections commonly used for this purpose may range from an 8-inch H-section weighing 36 lbs. per lin. ft. to a 14-inch H-section weighing 117 lbs. per lin. ft.

Assuming a unit cost of .30 cents per lb., delivered to job site from warehouse, the H-beam material will cost from $10.80 to $35.10 per lin. ft. If job is in an isolated location, additional transportation charges must be figured.

Various Types of Precast Concrete Piles

Under average conditions, the cost of driving steel H-beam piles will vary from $3.00 to $4.00 per lin. ft., not including moving the equipment on and off the job.

02350   CAISSONS

Caisson foundations are ordinarily used to support heavy structures where soil conditions near grade are unsatisfactory for the support of them. Where such a condition exists the foundations should be carried down through the unsatisfactory soil to material capable of carrying the imposed load without causing any detrimental settlement. The size and depth of caissons will vary with the loads to be carried and the distance to material of the required bearing capacity. Caissons bearing on rock in the Chicago Loop area vary from 90 to 190 feet in depth.

Caissons may be dug either with straight shafts to the required depth, or they may be belled at their bottoms to provide additional bearing area. The majority of caissons are now being installed using various types of drill rigs but under certain conditions hand dug caissons may be used. The drilling machines are capable of not only drilling the shaft, but mechanically forming a bell as well. The cost of excavating varies with the diameter and depth of the caisson, the site conditions, and with the type of soil being excavated.

There are contractors who specialize in installing the different types of caissons. On jobs of any size, it is advisable to obtain figures from contractors specializing in this work, as they are equipped to perform it much more economically than the contractor who has only an occasional job.

Hand Dug Caissons.—The hand method of excavation most commonly used is known as the Chicago method. This consists of excavating with hand tools and lining the hole with wood lagging held in place by steel rings. The wood lagging has beveled edges and is tongue grooved. The lagging is milled from either 2"x6" or 3"x6" mixed hard wood in 16' lengths. The 16' lengths are commonly cut into thirds so that a set of lagging is 5'-4" long. Three inch steel channels are formed in a half circle with a plate on each end to permit bolting the half sections together. Two complete rings are required for each set of lagging.

To determine the excavation quantity, multiply the cross-sectional area by the length allowing for the increased diameter due to the lagging. To determine the quantity of concrete required, multiply the cross-sectional area by lengths making proper corrections if the finished concrete surface is to be below the ground surface. If the caisson has a bell, the volume of it must be computed using the formula for a truncated cone — $V = \frac{1}{3}(A_1 + A_2 + A_1 + A_2)h$ where $A_1 + A_2$ are the area of the bases and h is the height of the bell.

Excavation quantity for 6'-0" caisson 106' deep
Cross-sectional area for 6'-4" = 31.5 ft.^{2}
$\quad$ 31.5 x 106 x 1/27 = 124 cu. yds.
Concrete quantity:
Cross-sectional area for 6'-0" = 28.3 ft.^{2}
$\quad$ 28.3 x 106 x 1/27 = 111 cu. yds.
Allow 5% for waste or a total of 117 cu. yds.

After determining the excavation quantities and the concrete quantities, it is then necessary to determine the amount of lagging and rings. By cutting the bevel and tongue and grooves in the lagging, the width of the lag-

Section Through Caisson
Showing Lagging in Place.

ging is reduced, so that it requires 7 pieces of 2 x 6 per foot of diameter of the caisson. For example, a 4' diameter caisson requires 28 pieces of lagging around the circumference. Therefore, using 2"x6" lagging, which is one board foot, it would require 28 board feet per foot of depth for a 4' caisson or 42 board feet using 3" thick lagging. To determine the amount of lagging required in a 6' diameter caisson, 106' deep using 2" lagging, multiply 6x7 or 42 board feet per foot of depth x the depth, which is 106' or 4452 board feet. If the lagging had been 3", it would be 6678. A caisson 106' deep would require 20 sections of 5'-4" lagging. Using 2 rings for each section of lagging, the caisson would require 40 rings.

The rings are usually made from 3" channels, weighing 4.2 lbs. per foot. Each ring will weigh approximately 80 lbs. plus 10 lbs. for the plates at the ends of the half circles, for a total of 90 lbs. This type of material costs approximately $55.00 per ring.

Labor Required for Caissons.—Labor costs of caissons excavating varies with the diameter and depth of caissons, whether hand or pneumatic tools are used, the type of soil, and whether or not water or sand is encountered. It is generally assumed that one caisson digger can hand dig approximately 4 to 5 yards per 8 hour shift and place the lagging and rings.

Only one man can work in a 4' diameter shaft and two men can work in a 5' diameter or larger shaft. In a 6' caisson, two men each taking out 4 yards should be able to excavate approximately 7' per shift. At the surface, a pneumatic tugger hoist and tripod would be required with two laborers to operate the hoist and dump the buckets. Generally, several caissons are constructed simultaneously so the cost involved in furnishing compressed air for the hoist is spread over three or more caissons.

In addition to the above, it will be necessary to allow for time of foreman, timekeepers, material clerks, engineers, cost of setting up and removing equipment, etc. which will cost from $4.00 to $6.00 per cu. yd. or it may be estimated separately under overhead expense and an amount allowed to cover the entire job.

To summarize the costs involved for constructing a caisson 6' in diameter and 106 ft. deep, while two others are being constructed simultaneously:

Time to excavate and lag:

    124 cu. yds. @ 4 cu. yds./shift/digger=31 shifts.

    Using two diggers, this would be 15½ shifts.

    Allow 16 shifts.

Labor	Hours	Rate	Total	Rate	Total
Caisson diggers	256	$....	$....	$13.54	$ 3,466.24
Hoist operator	128	....	....	17.05	2,182.40
Top laborers	256	....	....	12.54	3,210.24
Compressor operator	43	....	....	17.05	733.15
					$ 9,592.03
Material					
Lagging	4.452MBF	....	....	$760.00	$ 3,383.52

Material	Hours	Rate	Total	Rate	Total
Rings	40	....	....	55.00	2,200.00
					$ 5,583.52

Equipment					
Hoist	16 shifts	....	....	$ 25.00	$ 400.00
Air tools	32 "	....	....	40.00	1,280.00
Compressor	5-⅓ "	....	....	50.00	266.65
					$ 1,946.65

Removal of					
Excavated soil	124 cu. yds	....	....	$ 10.00	1,240.00
Grand Total			$....		$18,362.20
Cost per cu. yd.				....	$ 148.08

To the above cost it will be necessary to add for setting up and removing necessary plant and equipment, electric current and wiring, concrete furnishing and placing, overhead expenses, contingencies, miscellaneous expenses, and profit.

Work of this type in the Chicago area generally costs from $200.00 to $275.00 per cu. yd. of completed caisson.

Excavating for Caissons with Pneumatic Diggers.—When excavating for caissons 5'-0" to 8'-0" in diameter and up to 100'-0" deep, a man with a pneumatic digger will loosen and load into buckets 9 to 11 cu. yds. per 8-hr. shift, depending upon the toughness of the clay and the working space, at the following cost per cu. yd.:

	Hours	Rate	Total	Rate	Total
Labor operating digger and loading buckets	0.90	$....	$....	$13.54	$12.19
Compressor operator	0.45	....	....	17.05	7.67
Compressor expense w/tools	0.45	....	....	11.25	5.06
Cost per cu. yd.			$....		$24.92

*Add only if required.
Note: Compressor and operator time at half time (assuming 2 men digging).

When excavating caissons by hand, two "diggers" will loosen and load into buckets, 6 to 7 cu. yds. per 8-hr. shift, at the following labor cost per cu. yd.:

	Hours	Rate	Total	Rate	Total
Labor excavating	2.5	$....	$....	$13.54	$33.85

Drilled Caissons.—The use of drilling machines for the installation of caisson foundations has eliminated the hand digging method except in situations where adequate clearances to surrounding structures or utilities cannot be maintained. Drilled caissons have proved more economical than hand-dug caissons for several reasons, the most important of which is the fact that smaller shaft diameters may be used. The minimum size of a hand-dug caisson is 4' in diameter, which oftentimes provides a much greater cross-sectional area of concrete than is required for the load imposed upon it.

Another important reason is that drilled caissons may be installed in considerably less time than is required for hand-dug caissons, thereby reducing overall construction time and allowing earlier occupancy of the completed building. There are many firms specializing in this type of work throughout the country.

A typical job consisting of 100 caissons with 2'-0" shaft diameters, 5'-0" bells and 25' deep, might be estimated as follows, assuming normal clay digging. First, compute the total volume of excavation which in this case would be 418 cu. yds. To this figure add about 10% overbreak, making a total of 460 cu. yds. Assuming each drill

Courtesy Caisson Corporation

Belling tool for hard-pan caissons

rig can complete 5 caissons per day, the job would take 20 days. Allow approximately 10% extra time for weather and total of 23 days to complete the job.

Labor	Hours	Rate	Total	Rate	Total
Drill Rig Operator	184	$....	$....	$ 17.95	$ 3,302.80
Oiler	184	....	....	12.54	2,307.36
Caisson Laborer	184	....	....	13.54	2,491.36
Laborer	184	....	....	12.54	2,307.36
					$10,408.88

Equipment	Hours	Rate	Total	Rate	Total
Drill rig...	23 days	$. . . .	$. . . .	$400.00	9,200.00
Material					
Concrete...	460 cu.				
	yds.	. . . .	. . . .	$ 55.00	25,300.00
					44,908.88

Cost per cu. yd. based
on 418 cu. yds........................... $107.44

These costs are for the complete caisson installation. Allowances must be added for reinforcing steel, additional labor based on union requirements, temporary or permanent casings if necessary, soil and concrete testing, surveying, disposal of excavated materials, and equipment move and set-up charges, if these items are included in the caisson contractor's contract, and overhead and profit.

02400 SHORING

Wood sheet piling and bracing is usually estimated by the square foot, taking the number of sq. ft. of bank or trench walls to be braced or sheet pile, plus an additional amount for penetration at the bottom of the excavation, and an allowance for extension above the top, and estimating the cost of the work at a certain price per sq. ft. for labor and material required.

Sheet piling or bracing is required where the soil is not self supporting or where the excavation banksides cannot be sloped back. There is usually considerable vibration adjacent to railroad tracks and roadways which may cause the banks to cave in. Unless there is ample space on all sides of the excavations, the banks should be sheet piled to avoid damage to streets, alleys or adjacent buildings.

Trenches or foundation piers 5'-0" to 8'-0" deep and 3'-0" or 4'-0" wide may require bracing or sheet piling but this is much simpler than is ordinarily required for basement excavations.

The cost of this work will vary with the kind of soil, depth of excavation, amount of bracing or sheet piling required and the method used in placing same.

In ordinary excavations where no water is encountered, 2" or 3" square edge planks may be used but if running sand or water is encountered, tongue and grooved planks should be used, as it is easier to keep them in line and they are more water tight than square edge planks. It is customary to cut the bottom edge of each plank on a slight angle, so that in driving it is wedged against the preceding plank. The upper corners of the plank should also be cut off so that the effect of the driving will be concentrated along the vertical axis of the plank.

It requires 3 to 3½ ft. of lumber, b.m. to sheet pile one sq. ft. of trench bank 5'-0" to 8'-0" deep, but on large deep excavations it will require 6 to 8 ft. of lumber, b.m. per sq. ft. of wall.

Sheet piling is usually driven with a pneumatic hammer designed especially for this purpose.

On very large jobs an ordinary pile driver is used for driving the sheet piling but this is done only on jobs of sufficient size to warrant the cost of a pile driver and crew.

Bracing and Sheet Piling Trenches and Piers.—When excavating trenches and piers 5'-0" to 8'-0" deep, it is not always necessary to sheathe the banks solid but two or three lines of braces placed along the sides of the trench, as illustrated on the following page will often be sufficient.

Where necessary to sheathe the banks solid, it will require about 3 to 3½ ft. of lumber, b.m. per sq. ft. of trench bank.

An example of the method used in determining the quantity of lumber required for sheet piling a trench similar to the one illustrated, is as follows: Compute the quantity of lumber necessary to sheet pile a trench 50'-0" long and 7'-0" deep, sheathed on both sides. To the depth of 7'-0" must be added an amount for penetration. If 1'-0" will be sufficient, then 8'-0" lumber may be figured. If, however, 2'-0" penetration is required, then 10'-0" lumber must be figured leaving 1'-0" to extend above grade. Assuming 2'-0" penetration is required and 10'-0" lumber must be used this is equivalent to 100 lin. ft. of sheathing 10'-0" high, or 1,000 sq. ft., and the following lumber will be required:

	Ft. B.M.
150 pcs. 2"x8"-10'-0" sheathing	2,000
12 pcs. 4"x6"-16'-0" stringers or "wales"	384
26 pcs. 4"x6"-2'-0" braces spaced about 4'-0" apart	104
Total lumber required	2,488
Feet of lumber, b.m. per sq. ft. sheathing, (divide by 1,000)	2.49
Feet of lumber, b.m. per sq. ft. trench bank, (divide by 700)	3.55

This is based on using 2" plank for sheathing, 2 rows of 4"x6" stringers or "wales" placed near the top and bottom as illustrated, and 4"x6" or 6"x6" braces. Trenches over 10'-0" deep will require an additional line of stringers for each additional 4'-0" to 5'-0" depth.

Method of Bracing
Trench Excavation

Method of Sheet Piling
Trench Excavation

Most contractors, who do a volume of this work, use trench jacks in lieu of the 4"x6" or 6"x6" braces. Inasmuch as these jacks cost from $200.00 to $300.00 each, it would not be economical for the contractor who does only an occasional job of this sort to invest in this equipment as the savings in both material and labor are so slight they are practically negligible.

Sheet Piling for Basements and Deep Foundations.—On general basement excavation or large piers 8'-0" to 12'-0" deep, where it is necessary to brace any one or all of the outside banks, the following is an example of the lumber required for a bank 50'-0" long, 8'-0" deep, and containing 400 sq. ft. of bank to be sheathed:

	Ft. B.M.
80 pcs. 2"x8"-10'-0" sheathing	1,067

Ft. B.M.

6 pcs. 6"x8"-16'-0" (2 lines of stringers or "wales").................................. 384
10 pcs. 8"x8"-12'-0" top braces spaced about 5'-0" apart........................... 640
10 pcs. 8"x8"-6'-0" bottom braces spaced about 5'-0" apart..................... 320
10 pcs. 8"x8"-6'-0" bottom braces spaced about 5'-0" apart..................... 320
  Total lumber required 2,731
Feet, b.m. per ft. sq. ft. of sheet piling, (divide by 500)............................. 5.46
Feet, b.m. per sq. ft. of bank, (divide by 400).. 6.83

On deep excavations and piers 14'-0" to 20'-0" below grade 3" plank should be used for sheathing and either 6"x8" or 8"x8" lumber for stringers or "wales" and 8"x8" or 10"x10" timbers for bracing. Lines of stringers or "wales" should be spaced 4'-0" to 5'-0" apart, depending upon the earth pressure.

Methods of Bracing Basements or Deep Excavations.

An example of the lumber required to sheet pile and brace a bank 50'-0" long and 16'-0" deep, containing 800 sq. ft. of bank, is as follows:

Ft. B.M.

80 pcs. 3"x8"-18'-0" sheathing... 2,880
 9 pcs. 8"x8"-16'-0" (3 lines of stringers or "wales").................................... 768
10 pcs. 8"x8"-16'-0" top braces spaced 5'-0" apart... 853
10 pcs. 8"x8"-12'-0" Intermediate braces spaced 5'-0" apart...................... 640
10 pcs. 8"x8"-8'-0" bottom braces spaced 5'-0" apart.................................. 427
10 pcs. 8"x8"-6'-0" stakes spaced 5'-0" apart.. 320
  Total lumber required 5,888
Ft. b.m. per sq. ft. of sheet piling, (divide by 900)...................................... 6.54
Ft. b.m. per sq. ft. of bank (divide by 800) 7.36
Cost of lumber should be computed according to number of times it can be used on the job.

Where labor hours are given, either carpenters or laborers may be used, as required.

Lumber Required for Sheet Piling

Size of Timbers	Width Sq. Edge	100 Lin. Ft. Covers Sq. Ft.	Ft. B. M. per 100 Sq. Ft.	Width T & G	100 Lin. Ft. Covers Sq. Ft.	Ft. B. M. per 100 Sq. Ft.
2" x 8"	7½"	62½	215	7¼"	60	220
3" x 8"	7½"	62½	320	7¼"	60	330
2" x 10"	9½"	79	210	9¼"	77	215
3" x 10"	9½"	79	315	9¼"	77	320

Wood Bracing and Sheet Piling Banks By Hand

Quantities and Production are given per 100 sq. ft. of bank braced

Class of Work	Depth in Feet	Ft. B. M. Lbr. Req'd 100 Sq. Ft. Bank	No. Sq. Ft. Placed per 8-Hr. Day	Labor Hrs. Placing 100 Sq. Ft.	Labor Hrs. Removing 100 Sq. Ft.
Bracing Trenches	5- 8	100-150	150-200	4- 5	1 -1½
Trench Sheet Piling	5- 8	300-350	80- 90	9-10	2½-3
Trench Sheet Piling	10-15	325-375	65- 75	11-12	2½-3
Basement Sheet Piling	8-12	675-725	40- 50	17-18	5 -5½
Basement Sheet Piling	14-20	750-800	35- 45	20-21	5½-6

Cost of Bracing 100 Sq. Ft. of Trench Bank With Plank Bracing

	Hours	Rate	Total	Rate	Total
125 ft. lumber b.m.		$....	$....	$ 0.76	$ 95.00
Labor placing	4.50	....	....	12.54	56.43
Labor removing**	1.25	....	....	12.54	15.68
Cost per 100 sq. ft.			$....		$167.11
Cost per sq. ft.			....		1.67

Cost of Sheet Piling 100 Sq. Ft. of Trench Walls 5'-0" to 8'-0" Deep

	Hours	Rate	Total	Rate	Total
325 ft. lumber b.m.		$....	$....	$ 0.76	$247.00*
Labor placing	9.50	....	....	12.54	119.13
Labor removing**	2.75	....	....	12.54	34.49
Cost per 100 sq. ft.			....		$400.62
Cost per sq. ft.			....		4.01

Cost of Sheet Piling 100 Sq. Ft. of Trench Walls 10'-0" to 15'-0" Deep

	Hours	Rate	Total	Rate	Total
350 ft. lumber, b.m.		$....	$....	$ 0.76	$266.00*
Labor placing	11.50	....	....	12.54	144.21
Labor removing**	2.75	....	....	12.54	34.49
Cost per 100 sq. ft.			$....		$444.70
Cost per sq. ft.			....		4.45

Cost of 100 Sq. Ft. of Sheet Piling for Basements or Foundations 8'-0" to 12'-0" Deep

	Hours	Rate	Total	Rate	Total
700 ft. lumber, b.m.		$....	$....	$ 0.76	$532.00*
Labor placing	17.50	....	....	12.54	219.45
Labor removing**	5.25	....	....	12.54	65.84
Cost per 100 sq. ft.			$....		$817.29
Cost per sq. ft.			....		8.17

Cost of 100 Sq. Ft. of Sheet Piling for Basements or Foundations 14'-0" to 20'-0" Deep

	Hours	Rate	Total	Rate	Total
775 ft. lumber, b.m.		$....	$....	$ 0.76	$589.00*
Labor placing	20.50	....	....	12.54	257.07

Method of Sheet Piling and Bracing Piers and Pits.

	Hours	Rate	Total	Rate	Total
Labor removing**	5.75	. . . .	. . . .	12.54	72.11
Cost per 100 sq. ft.			$. . . .		$918.18
Cost per sq. ft.				. . . .	9.18

*Lumber costs may be reduced if it is removed & used again.
**Omit labor removing bracing or piling if left in place.

Driving Sheet Piling by Compressed Air.—Pneumatic pile drivers are very effective for driving sheet piling. They are essentially a heavy paving breaker equipped with a special fronthead which is adjustable for driving 2" to 3" piling. These machines weigh about 125 lbs. and will drive wooden sheet piling in any ground which the piling will penetrate, such as sand, gravel or shale and any gradation between soils.

Its driving speed varies from 2'-0" per minute in hard clay or shale to 9'-0" per minute in sand or gravel.

Under average conditions, two men working together on one machine should drive 100 lin. ft. (60 to 79 sq. ft.) per hr. at the following labor cost per 100 lin. ft.:

	Hours	Rate	Total	Rate	Total
Jackhammerman	1	$. . . .	$. . . .	$12.54	$12.54
Helper	1	. . . .	. . . .	12.54	12.54
Compressor expense & Tools	1	. . . .	. . . .	13.50	13.50
Cost per 100 lin. ft			$. . . .		$38.58
Cost per lin. ft				. . . .	3.86

Add for compressor engineer if required.

Cost of Sheet Piling 100 Sq. Ft. of Trench Banks 5'-0" to 8'-0" Deep, Using Pneumatic Hammers

	Hours	Rate	Total	Rate	Total
325 ft. lumber, b.m.		$....	$....	$ 0.76	$247.00*
Jackhammerman	2.00	....	....	12.54	25.08
Labor	4.00	....	....	12.54	50.16
Labor removing	2.75	....	....	12.54	34.49
Compressor & Tools	2.0	....	....	13.50	27.00
Compressor Engineer	2.0	....	....	17.05	34.10**
Cost per 100 sq. ft			$....		$417.83
Cost per sq. ft				....	4.18

Cost of Sheet Piling 100 Sq. Ft. of Trench Banks 10'-0" to 15'-0" Deep, Using Pneumatic Hammers

	Hours	Rate	Total	Rate	Total
350 ft. lumber. b.m.		$....	$....	$ .76	$266.00*
Jackhammerman	2.00	....	....	12.54	25.08
Labor	4.50	....	....	12.54	56.43
Labor removing	2.75	....	....	12.54	34.49
Compressor & Tools	2.0	....	....	13.50	27.00
Compressor Engineer	2.0	....	....	17.05	34.10**
Cost per 100 sq. ft			$....		$443.10
Cost per sq. ft				....	4.43

*Lumber cost will vary with number of times it may be used.
**Add only if required.

STEEL SHEET PILING

Steel sheet piling is used for supporting the soil in large excavations, cofferdams, caissons, and deep piers or trenches where wood sheet piling is impractical. It is driven with a pile driver the same as wood or concrete piles.

There are a number of different types of steel sheet piling on the market but the following will furnish an idea of the sizes, weights, etc., commonly used.

Sizes and Weights of United States Steel Sheet Piling

Section Number	Width Inches	Web Thick. Inches	Wt. Lbs. per Lin. Ft.	Wt. Lbs. per Sq. Ft. Wall
MZ38	18"	3/8"	57.0	38.0
MZ32	21"	3/8"	56.0	32.0
MZ27	18"	3/8"	40.5	27.0
MP102	15"	1/2"	40.0	32.0
MP101	15"	3/8"	35.0	28.0
MP113	16"	1/2"	37.3	28.0
MP112	16"	3/8"	30.7	23.0
MP110	16"	31/64"	42.7	32.0
MP116	16"	3/8"	36.0	27.0
MP115	19"5/8	3/8"	36.0	22.0

Estimating the Quantity of Steel Sheet Piling.—When estimating the quantity of steel sheet piling required for any job, take the entire girth of the basement, cofferdam or piers to be sheet piled, which gives the number of lineal feet required. Example: suppose you have a basement 100'x125'-0" to be sheet piled. Adding the four sides of the excavation, 100+100+125+125=450 lin. ft. Using MZ-27 Carnegie Illinois sheet piling 18" wide, 450'-0"÷1.50'=300, thus, it requires 300 pcs. to sheet pile the excavation. Using sheet piling 30'-0" long, 300x30=9,000 lin. ft. weighing 40.5 lbs. per lin. ft. equals 364,500 lbs.

A shorter method is as follows: 450 lin. ft. around excavation multiplied by 30'-0" long, equals 13,500 sq. ft. of piling at 27 lbs. per sq. ft. equals 364,500 lbs.

Either method is satisfactory, except it is necessary to know the number of pieces to be driven to compute the labor costs accurately.

Labor Driving Steel Sheet Piling.—The labor cost handling and driving steel sheet piling will vary with the size of the job, length of piling, kind of soil, etc., together with conditions encountered on the job, as described under "Wood Piles."

Example: Find the cost of driving 300 pcs. of MZ-27 sheet piling 30'-0" long.

	Total	Total
Trucking pile driver, yard to job	$....	$ 1,500.00
Setting up pile driver, 3 days for crew	....	3,173.04
300 pcs. MZ-27 piling, 364,500 lbs. at $ 0.40	....	145,800.00
Driving piling (30 per day) 10 days @ $1,057.68	....	10,576.80
Fuel, oil, misc. supplies, 10 days @ $150.00	....	1,500.00
Dismantling pile driver at completion, 2 days	....	2,115.36
Trucking pile driver, job to yard	....	1,500.00
Equipment Rental or depreciation allowance 16 days @ $500.00	....	8,000.00
Cost 300 pcs. sheet piling	$....	$174,165.20
Cost per pc.	....	580.55
Cost per sq. ft. (13,500 sq. ft.)	....	12.90

Sq. ft. costs will vary with weight of piling used, size of job, etc.
Add for overhead and profit.

02500   SITE DRAINAGE

Site drainage includes such items as subdrainage, foundation and underslab drainage, drainage structures, sanitary and storm drainage piping, and dewatering and wellpoints. Some of these will be subcontracted directly with specialists in the field. Most will probably be included in the plumbing subcontract, while others may involve local utilities. Some items the contractor may elect to handle directly.

Drainage trenches should be pitched. To find the cubic yards of earth excavation, first determine the average depth by multiplying the pitch by the required length of run, divide by two and add this to the depth at the start of the run. The total length x required width x the average depth = cubic feet. Divide by 27 to find cubic yards.

Subsoil drains are usually vitrified clay pipe, set with lowest point at the same elevation as the bottom of the footing, pitched a minimum of 6 in. in 100 lin. ft., laid ¼" apart and joint covered with 15# asphalt felt or #12 copper or aluminum mesh. 4" pipe is minimum, 6" preferred. One laborer can lay approximately 10 ft. of 4" pipe per hour, 8 ft. of 6". 4" pipe costs around $1.00 per ft; 6", $1.40. Excavation is, of course, extra and it is assumed the trench is properly sloped with a firm bottom before pipe is installed.

All sharp turns shall be formed with fittings which cost $2.50 for 4" elbows, $4.75 for 6". Wide angles can be made by bevelling tile ends into easy radius bends.

The clay drain tile is laid on a thin bed of washed gravel and after the joints have been covered with felt or a filter cloth more gravel is placed around and on top of the drain tile to allow water filtration into the drain tile itself. Approximately .75 to 1.0 cu.ft. of gravel per linear foot of drain is required for proper filtration.

In recent years, vitrified clay drain tile is being displaced by the use of 4" plastic tubing that is already perforated by the manufacturer. This 4" tubing comes in 25', 50', and 100' rolls, is very flexible, and since you are not handling small individual pieces as with the vitrified clay, this particular material is installed very rapidly. The joints are made by a coupling sleeve very much like that used with plumbing PVC piping. It is wise to lay a strip of filter cloth above the plastic tubing prior to installing the washed gravel to keep all sandy fines from entering the plastic tubing and creating a stoppage. All other grading preparation and gravels are installed around this particular material as with the vitrified clay tile.

The subsoil drains, whether perimeter or underslab, can be connected to an existing storm sewer, in some areas to a sanitary sewer, depending upon the existing terrain could be trenched to an outfall at grade level away from the structure, or connected to a sump pump area within the lower level of the building. In the event that the subsoil drainage piping terminates at a sump pump, a submergible pump with an automatic float control switch and discharge piping will be needed.

DRAINAGE STRUCTURES

There are a number of different types of drainage structures required in storm drain piping systems. These structures can be constructed of masonry, cast-in-place concrete, or precast concrete.

Curb inlets and yard drains are one type of structure that is used to collect surface water run-off and discharge it into the storm water piping system. The size of the inlet is determined by the area of run-off that will be collected by that particular curb inlet or yard drain, as well as the size of the discharge pipe leading away from the inlet structure itself. An average price for a curb inlet drainage structure is $250 to $300 per vertical linear foot and the average price of a yard drainage structure is $175 to $210 per vertical linear foot.

Storm drainage manhole structures are required at certain intervals along a storm pipe line depending upon the local codes and whenever a storm drainage pipe system changes direction. The majority of manholes are precast concrete since it is much quicker to set an entire manhole in the course of the day in lieu of keeping the open pit shored while constructing a cast-in-place concrete manhole and bottom chamber. Once the manhole structure is in place and the piping connected, the bottom of the manhole structure is channeled using brick and mortar to aid the water flow from the inlet pipe to the outlet pipe.

Headwalls and endwalls are usually cast-in-place reinforced concrete walls at the ends of pipe, where either water enters the pipe (headwall) or where the water discharges the pipe (endwall). Quite often at headwalls there is a concrete flume entrance slab and may include the installation of a trash rack (a grillage screen) at the entrance way to collect debris, thus preventing a pipe stoppage.

At the endwalls an outfall slab or rip-rap area is installed that prevents the discharge water from eroding the soil away. The price of headwalls and endwalls varies tremendously because each one is sized individually depending upon the size of the inlet or discharge pipe and the amount of earth and height of the earth bank above the pipe that is being retained by the headwall.

STORM PIPING

For other than sewers, concrete, asbestos cement and corrugated metal pipe are often used. The relative material costs per lineal foot are approximately as follows:

Size of Pipe	Concrete	Asbestos Cement	Corrugated Metal
6"	$ 3.00	$ 2.50	$. . . .
8"	3.25	3.00	3.70
10"	3.75	4.25	4.30

12"	4.00	6.00	5.25
18"	7.00	11.50	7.35
24"	11.00	19.00	12.50
30"*	15.00	27.00	15.00
36"*	22.00	36.00	23.00

*Reinforced

One cannot choose the type of pipe by price alone. While the specifications and job conditions usually dictate the selection, wage costs can also bear heavily, as the lighter weight pipe will lay much faster than the heavy weight and can be handled on the site much more easily.

For 12" pipe a 4 man crew in a day will lay around 100 lin. ft. of concrete pipe, 150 ft. of asbestos cement and 200 ft. of corrugated metal. For 36" pipe, a crew will lay around 20 ft. of concrete pipe, 30 ft. of asbestos and 60 ft. of corrugated metal.

WELLPOINT SYSTEM OF DE-WATERING

The following information on wellpoints for dewatering excavations is general information only. For specific applications consult a local de-watering contractor.

Wellpoint systems are well known for their ability to de-water previous water bearing soils, such as sand and gravel, so that excavation and construction of foundations can proceed in the dry.

In addition, wellpoint systems are used extensively in such work as soil stabilization, pressure relief for dams and levees and for water supply for municipalities and industrial plants.

A wellpoint system as used in construction work, usually consists of a series of properly sized wellpoints, surrounding or paralleling the area to be de-watered, connected to a header pipe by means of risers and swing piping. Header piping, in turn, is connected to one or more centrifugal pumps depending on the volume of water which must be handled.

Wellpoints are usually self-jetted into place to the correct depth and at the proper spacing to meet the requirements. Some soil conditions, however, require pre-drilling or "hole-punching" before the wellpoints can be installed.

The discharge from the pumps should be piped to an area where it will not interfere with the construction work. The entire installation should be located so that it will interfere as little as possible with other divisions of the work.

This system is usually installed in advance of the excavation but where the water level is low it is sometimes possible that a portion of the digging may be done at a cheaper rate if it is done prior to the installation of the system.

Any de-watering problem which is beyond the scope of ordinary pumping, or which the contractor or engineer thinks can be done more economically with a wellpoint system, should be referred to the engineering department of a wellpoint company. In this way the problem will receive expert analysis, at no cost to the contractor or engineer. It is essential for most contractors to obtain expert advice on the layout and installation of wellpoint systems because frequently a relatively small excavation with unusual soil, presents greater de-watering difficulties than a larger, deeper job in a well-graded medium sand.

After carefully considering the various factors involved, including soil characteristics, site accessibility, hydrology of area, size and depth of area, total dynamic head involved, excavation and construction schedules, labor and working conditions, power facilities, etc., the wellpoint company can make a layout recommendation for the job.

Layout recommendations are usually accompanied by a quotation giving rental rates and charges for all equipment required, together with an estimate of the labor

Method of De-Watering Using Wellpoint System

hours needed to install and remove the system, fuel and lubrication required per operating day, transportation charges for the equipment to and from the job and daily rates for salary and expenses of company demonstrator. Companies specializing in this type of work will furnish rental quotations on a "sufficient equipment" basis, i.e., a fixed rental rate regardless of the amount of equipment actually required to do the job. Some of these companies will submit lump sum contract proposals for de-watering projects covering furnishing, installation operation, maintenance and removal of all de-watering equipment with a guaranteed result.

It is sound practice to have a demonstrator from the wellpoint company on the job site to supervise the initial installation of the system. The cost of his time and expenses is more than compensated for by his working knowledge of wellpoint systems and their installation.

Operation of a wellpoint system is usually performed on a 24-hr. day, 7 day per week basis and must continue until all work, dependent on dry conditions, is completed.

The labor cost of operating the system must be determined by the contractor and is based on the length of time operation is estimated to be required, figuring around-the-clock operation.

From the above information the contractor or estimator should be able to arrive at a lump sum figure for the required wellpoint system to be used in his estimate.

As an example, take the case of a job in which an underground tank was to be constructed. The test borings indicated the soil to be sand with a ground water level at 6'-0" below the surface. The excavation for the circular tank was an average of 84'-0" in diameter and 18'-6" deep. To maintain dry conditions, the water level had to be lowered at least 12'-6". It was also estimated that a dry condition would have to be maintained for a period of 3 months.

The following example estimate is given to show how a detailed de-watering estimate might be put together.

Equipment First Month	Hours	Rate	Total	Rate	Total
52 wellpoints, 2"		$....	$....	$ 16.00	$ 832.00
260 lin. ft. header pipe, 6"		....	....	1.80	468.00
50 lin. ft. discharge pipe, 6"		....	....	1.00	50.00
2 valves, 6"		....	....	40.00	80.00
1 wellpoint pump		....	....	950.00	950.00
1 jetwell pump		....	....	1,025.00	1,025.00
Equipment Rental, Second Month					
52 wellpoints, 2"		....	....	10.00	520.00
260 lin. ft. header pipe, 6"		....	....	.90	234.00
50 lin. ft. discharge pipe, 6"		....	....	.50	25.00
2 valves, 6"		....	....	25.00	50.00
1 wellpoint pump		....	....	750.00	750.00
1 jetwell pump		....	....	850.00	850.00
Equipment Rental, Third Month					
52 wellpoints, 2"		....	....	10.00	520.00
260 lin. ft. header pipe, 6"		....	....	.90	234.00
50 lin. ft. discharge pipe, 6"		....	....	.50	25.00
2 valves, 6"		....	....	25.00	50.00
1 wellpoint pump		....	....	750.00	750.00
1 jetwell pump		....	....	850.00	850.00
2 weeks rental, 200 lin. ft. jetting hose		$....	$....	$ 200.00	$ 400.00
Labor	160	....	....	11.85	1,895.00
Supervisor, 5 days		....	....	200.00	1,000.00
Supervisor's expenses		....	....	500.00	500.00
Operating Expense					
Diesel fuel 91 days at 40 gals. per day		....	....	1.30	4,732.00
Lubricating oil, 91 days at 1 gal. per day		....	....	5.00	455.00
Pump operator, 3 shifts, regular time	1,560	....	....	17.05	26,598.00
Pump operator, 3 shifts, overtime	624	....	....	25.58	15,961.92
Removal of Equipment					
Labor	60	....	....	12.54	752.40
Transportation to and from job, 16,000 lbs. at $4.00 per cwt		....	....		640.00
Total cost of wellpoint system			$....		$61,197.32

Add for Insurance, Social Security, Unemployment Compensation, Overhead and Profit.

The jetwell pump is a double-duty unit and is connected into the system as a stand-by wellpoint pump after installation of wellpoints has been completed.

DEEP WELL SYSTEM OF DE-WATERING

A cheaper alternative to the wellpoint system of de-watering would be the use of deep wells, which are 18" to 36" diameter drilled shafts into the ground and lined with a perforated, corrugated metal pipe in which an electric submergible pump is installed. This pump can be discharged directly onto the ground or into another collector pipe discharge system.

The size (diameter and depth) of the drilled shafts, the distance spaced apart, and the size of the electric submergible pump is determined by the depth of the planned excavation, the elevation of the water table, and the subsurface water flow.

SUBMERSIBLE PUMP INSTALLATION

TWO WIRE SUBMERSIBLE INSTALLATION WITH CONVENTIONAL TANK

Fused Main Disconnect Switch

Pressure Gauge

Air Volume Control

Pressure Tank

Service Line

Conduit

Pressure Switch

Snifter Valve

Check Valve

Well Vent

Well Seal

Bleeder Valve

Tee

THREE WIRE SUBMERSIBLE INSTALLATION

For Two-Wire installation, conduit by-passes control box, goes directly into pressure switch.

For Three-Wire installation, conduit goes into control box, from there to pressure switch.

TWO WIRE SUBMERSIBLE INSTALLATION WITH AIR SEAL TANK (No air volume control required)

Fused Main Disconnect Switch

Pressure Tank

Pressure Gauge

Service Line

Pressure Switch

Snifter Valve

Check Valve

Well Vent

Well Seal

When installing the submergible pump, the use of rigid PVC piping is preferred over flexible canvas discharge hoses. Depending upon the depth of the shaft, a pump will be required to have sufficient head pressure to pump the water up to ground elevation. Therefore, any restrictions in collapsed discharge piping would create an additional burden for the pump, thus, losing pump efficiency.

Pump costs vary widely depending upon the size of the pump, the head pressure rating, and the GPM capacity. Pumps can be purchased or rented and this determination should be based upon an evaluation of both methods for the amount of time that the pumps will be in operation.

Upon completion of the project and termination of the need for the de-watering deep well system, the pump and piping are extracted from the hole and sand or gravel material is used to fill the shaft to prevent potential hazards of future cave-in.

The following two charts give the cost per vertical linear foot for drilled shafts and perforated corrugated metal pipe liner.

Drilled Shafts		Installation Of Perforated Corrugated Pipe	
Diameter of Shaft	Price Per Lin. Ft.	Diameter of Pipe	Price per Lin. Ft.
18"	$12.00	18"	$14.50
24"	15.00	24"	22.00
30"	20.00	30"	27.75
36"	26.00	36"	42.00

02550   SITE UTILITIES

Gas and water are usually brought to the property line or building by the local utility company. It is then picked up by the piping contractor doing the building work. Costs on this work are discussed under the Mechanical Chapter. The distribution of utilities on new subdivision work is a specialized field subject to many code, health and sanitation requirements.

Installing Service Pipes.—Many public utility companies, instead of breaking up sidewalks and digging trenches, are installing service pipes by means of portable compressors and pneumatic tools. Where no rock is encountered it has been found possible to drive pipe directly through the earth and into cellars, basements, etc.

This method consists of placing a pointed steel cap over the end of the pipe, which can be removed after the pipe holes through.

DRILLING WATER WELLS

In recent years pollution of streams, rivers and lakes by industrial wastes, has caused many municipalities to look elsewhere for sources of water. Many of these municipalities have had to install wells as an alternate source for their water supply because of this pollution. Industries sometimes also have to go to wells for their water supply where other sources are not available or are inadequate for their demands.

Large quantities of water are usually found in glacial deposits of sand and gravel. The quality of water from a glacial deposit is usually harder and of lower temperature than water derived from sand rock.

When it is necessary to locate large quantities of water for a town or city, it is customary to explore first by drilling a test hole to locate a water bearing strata, and to determine if there is an ample supply of water at that level. If a strata is found that is determined to be inadequate, blank pipe is then driven to the water bearing strata and a well screen is set in it.

Water travelling through a well screen has a tendency to precipitate the solids from the water on the screening which eventually will restrict the water flow through it. Voids in the sand and gravel surrounding the pipe and screen may also become filled with the precipitates from the water. Water with a high mineral content may close the screen openings in a relatively short period of time, while water with low mineral content may not restrict the flow of water appreciably for many years. When screens do become clogged up with precipitates they sometimes may be opened by treating them with acid. If this process is not effective, it may be necessary to install a new well.

Rock wells are installed by driving a pipe through the overburden above the rock to the rock surface, and continuing by drilling through the rock to the desired depth. It is sometimes possible to increase the quantity of water from a rock well in dense rock by shooting and fracturing the rock in the vicinity of the well. If the water bearing rock is already loose, porous, and creviced, shooting will probably not increase the water quantity.

Wells which extend into glacial sand and gravel deposits, will cost approximately $4.00 to $5.00 per inch in diameter, per lineal foot of vertical depth. For example a 12" diameter well would cost from $48.00 to $60.00 per lineal foot of depth.

Wells drilled into rock formations generally cost approximately $1.50 to $2.00 per inch of diameter, per lineal foot of depth, or a 6" diameter rock well would cost about $9.00 to $12.00 per lineal foot of depth. In the rock itself a pipe is not used, but if there is 100 ft. of earth or sand and gravel above the rock surface, it is necessary to include the cost of the pipe for this distance.

For wells with a capacity of 25 to 250 gallons per minute, submersible, multistage type pumps are generally used. Turbine type pumps are usually used for wells of more than 250 gallons per minute capacity.

Submersible pumps cost approximately $900.00 for a 10 gallon per minute pump to $10,000 for a 250 gallon per minute pump.

Turbine type pumps cost approximately $2,000 to $8,000 depending upon the capacity and the depth of the well.

It is recommended that quotations for drilling a well at a specific site be asked for from local well drilling contractors who have knowledge of the area, and know the approximate depth to the water bearing strata.

02600 PAVING AND SURFACING

Bituminous Roadways and Parking Lots.—This section will be confined to on-site roadways and parking lot construction for commercial and residential projects. Street and highway pavements are covered by the chapter on Road and Highway Construction.

Roadways and parking lots may be constructed in any number of thicknesses depending upon the type of use, amount of traffic, gross vehicle weight anticipated, as well as the suitability of the subsoil conditions for the intended usage.

After careful attention is given to the designed grade contours and drainage patterns and the rough grading operation is completed, the paving contractor performs the final grading with a motor grader. This machine is a rubber-tired tractor with a long cutting blade that is variable controlled for depths and angles of cut.

Once the motor grader has completed the final grading, the roadway or parking lot bed is rolled and compacted using a large rubber-tired compactor. While compacting the paving subgrade, the roller operator watches for any soft spots in the subgrade and, if any are found, they are excavated and refilled with a select grade of backfill material. If these soft spots are missed and not repaired, the asphalt paving will flex and eventually break up over these soft areas causing costly repairs at a later date.

Now that the subgrade has been suitably prepared (graded and compacted) the base course of paving materials can be laid to the proper thickness and compacted in preparation for the bituminous base course or bituminous wearing course. This gravel base course can vary in thickness from 4" to over 20" depending on the paving composite design and is usually compacted to 98%. After the base course is installed, it may be sprayed with an asphalt tack coat just prior to the first bituminous paving course being laid.

The next step in the paving process is the laying of the bituminous paving materials to the proper thickness and rolling the surface to compact and smooth out the hot asphalt mix. The bituminous course can vary in thickness from 3" to 8" depending upon the requirements and can be laid in one (1) course or can be split into two (2) courses, e.g. binder course and wearing course. When the bituminous is laid in two (2) courses, the top wearing course is usually a finer aggregate mix allowing for a much smoother compacted top surface.

Laying the bituminous paving in two (2) courses will always cost a premium. However, there are some distinct advantages such as:

1. Allows early paving of a work site without construction abuse of a finished surface.
2. Should soft spots occur, they can be fixed prior to the finish wearing course installation.
3. Allows sufficient time to completely check all areas for low spots (birdbaths) that collect water - these can then be adjusted in the top wearing course thickness.

When estimating quantities of bituminous paving and base courses for a project, it is customary to figure all quantities in square yards (sq.yd.). Example: Length in feet x width in feet = Area in square feet divided by 9 = square yards.

Concrete Curb, Gutter & Paving

This section is not intended to cover road building but only small curb and gutter jobs, and small concrete paving projects.

Large jobs of this kind require a vast amount of equipment and the costs vary widely according to the amount of cut and fill required and the method of receiving and dispersing materials.

The item of plant and equipment should include the cost of transporting equipment to the job and removing same at completion, and includes the following items: bulldozers, loaders, power rollers, trucks, wood or metal forms for curb and gutter, etc. These items should be estimated on the basis of total transportation costs both ways and a rental charge for plant and equipment. The total should be divided by the number of sq. yds. of pavement in the job, as it is customary to submit bids on paving at a certain price per sq. yd.

Rolling Streets and Roads.—After the excavating and grading has been completed, the subgrade should be rolled to provide a solid foundation to receive the concrete base. A power roller, should roll 130 to 150 sq. yds. of subgrade an hr. at the following cost per 100 sq. yds.:

	Hours	Rate	Total	Rate	Total
Operator	0.7	$. . . .	$. . . .	$17.95	$12.57
Power roller	0.7	. . . .	. . . .	22.00	15.40
Cost per 100 sq. yds.			$. . . .		$27.97
Cost per sq. yd.			. . . .		.28

Hauling Sand, Gravel, Crushed Stone, Etc.—The cost of hauling paving materials will vary with the length of haul from pit or cars to point of delivery. Tables of hauling time are given under "Excavating."

Concrete Curb and Gutter

When estimating the cost of forms for concrete curb or curb and gutter, the size and shape must be considered and the amount of materials estimated.

Combined curb and gutter is usually 12" high at the outside edge (including pavement thickness), 24" to 30" wide and 6" to 7½" thick.

Separate concrete curb is usually 6" wide at the top, 8" to 9" wide at the bottom, and 24", 30" or 36" high.

Forms for Concrete Curb and Gutter.—When placing wood forms for separate concrete curb, two mechanics and a helper should set forms for 100 to 110 lin. ft. of 24" curb per 8-hr. day, at the following labor cost per 100 lin. ft.:

	Hours	Rate	Total	Rate	Total
Mechanics	15.0	$....	$....	$16.47	$247.05
Helper	7.5	....	....	12.54	94.05
Cost per 100 lin. ft.			$....		$341.10
Cost per lin. ft.			....		3.41

Two mechanics and a helper should set forms for 90 to 100 lin. ft. of 30" concrete curb per 8-hr. day, at the following labor cost per 100 lin. ft.:

	Hours	Rate	Total	Rate	Total
Mechanics	17.0	$....	$....	$16.47	$279.99
Helper	8.5	....	....	12.54	106.59
Cost per 100 lin. ft.			$....		$386.58
Cost per lin. ft.			....		3.87

When setting wood forms for combined concrete curb and gutter, two mechanics and a helper should set forms for 80 to 90 lin. ft. of curb and gutter per 8-hr. day, at the following labor cost per 100 lin. ft.:

	Hours	Rate	Total	Rate	Total
Mechanics	18.8	$....	$....	$16.47	$309.64
Helper	9.4	....	....	12.54	117.88
Cost per 100 lin. ft.			$....		$427.52
Cost per lin. ft.			....		4.28

Combined
Curb and
Gutter

Separate
Concrete
Curb

Wood forms for concrete curb and gutter are ordinarily built of 2" lumber. Including all bracing, it requires 2¼ to 2½ B.F. of lumber per sq. ft. of forms. This cost will run from 15 to 20 cents per sq. ft. and should be added to the above setting costs, taking into consideration the number of uses which are expected to be obtained from the forms. These forms may be used many times if properly taken care of.

Placing Ready-mix Concrete for Curb and Gutter.—A crew of 2 men should place about 1 cu. yd. of concrete an hr. at the following labor cost per cu. yd.:

	Hours	Rate	Total	Rate	Total
Labor	2	$....	$....	$12.54	$25.08

Concrete Roads and Pavements

Concrete roads and pavements could be thicker in the center than at the sides, so it will be necessary to obtain the average thickness of the pavement when estimating quantities of concrete materials.

Some pavements have uniform thickness, others have uniform center section and thickened edges. In each case the area of the cross section must be determined and this is multiplied by the length to find the volume of concrete required.

Labor Placing Edge Forms for Concrete Paving.—Wood edge forms for concrete paving usually consist of 2" plank, of a width equal to the slab thickness, set to line and grade, secured to wood stakes and braced.

Two mechanics and a helper should set to line and grade from 250 to 300 lin. ft. of edge forms per 8-hr. day at the following labor cost per 100 lin. ft.:

	Hours	Rate	Total	Rate	Total
Mechanics	3.0	$....	$....	$16.47	$49.41
Helper	1.5	....	....	12.54	18.81
Cost per 100 lin. ft.			$....		$68.22
Cost per lin. ft.			....		.68

The above costs are based on the forms being straight or set to long radius curves. If sharp curves are required the form must be either made from 1 inch boards or the back of a 2" plank must be kerfed and in both cases the stakes must be set much closer together.

Concrete paving edge forms set to short radius curves will cost 50 to 100 percent more than straight forms.

Approximate Sq. Yds. of Concrete Pavement Obtainable from One Cu. Yd. of Concrete

Thickness Inches	Number Sq. Yds.	Thickness Inches	Number Sq. Yds.
4	9	8	4.5
4½	8	8½	4.25
5	7	9	4
5½	6.5	9½	3.75
6	6	10	3.5
6½	5.5	10½	3.5
7	5	11	3.25
7½	4.8	11½	3
		12	3

Placing Ready-mix Concrete Pavements.—Using ready-mix concrete and wheeling the concrete into place with wheelbarrows the labor per cu. yd. of concrete should cost as follows:

	Hours	Rate	Total	Rate	Total
Labor...	1.2	$. . . .	$. . . .	$12.54	$15.05

If ready-mix concrete can be discharged from truck directly into place without wheeling, deduct 0.7-hr. labor time.

On larger jobs, containing 100 cu. yds. or more of concrete, better organization may be obtained and lower costs are the result. Assuming a pavement containing from 100 to 200 cu. yds. of ready-mixed concrete wheeled into place with buggies the labor per cu. yd. of concrete should cost as follows:

	Hours	Rate	Total	Rate	Total
Labor...	.9	$. . . .	$. . . .	$12.54	$11.29

If ready-mix concrete can be discharged from truck directly into place without wheeling, deduct 0.4-hr. labor time.

Finishing Concrete Pavements.—Finishing concrete pavements usually consists of a series of operations performed in the following sequence: strike-off and consolidation, straight-edging, floating, brooming and edging.

Labor Finishing Concrete Pavements.— Assuming an 8 inch concrete pavement being placed at the rate of 25 cu. yds. per hr. or 1,013 sq. ft. per hr. a crew of 10 cement masons should be able to keep up with the pour and perform all the above described finishing operations at the following labor cost per 100 sq. ft. :

	Hours	Rate	Total	Rate	Total
Cement Mason...................................	1.0	$. . . .	$. . . .	$16.62	$16.62
Helper..	.4	. . . .	. . . .	12.54	5.02
Cement mason foreman....................	.1	. . . .	. . . .	17.12	1.71
Cost per 100 sq. ft.........................			$. . . .		$23.35
Cost per sq. ft.			. . . .		.23

The above costs are based on a job which will run from 7,500 to 8,500 sq. ft. per 8-hr. day. On smaller jobs the costs will increase from 20 to 25 percent.

02700   SITE IMPROVEMENTS

02710   FENCES & GATES

Industrial chain link fences usually consist of 3" OD tubular end posts and 2" to 2½" OD intermediate posts depending upon the height of the fencing to be installed. Top and bottom rails are not always used because having a top rail on a chain link fence gives much more stability for intruders to gain a handhold. Therefore, the spring wire is used for top and bottom stability in preventing chain link fabric sag. This spring wire is at least a 6 gauge material. The wire fabric is normally a 9 gauge wire and can be galvanized, aluminized steel, or vinyl coated in a number of colors.

For chain link fencing over 6' in vertical height, 6 gauge wire fabric will be used in lieu of the 9 gauge fabric.

Wire arms are available either in single or double sided for the installation of 2 to 4 strands of barbed wire per side.

The maximum line post or intermediate post spacing is 10' on center and these posts as well as in the corner posts are usually set in a 9" diameter by 3' deep concrete footing.

Industrial Chain Link Fencing - Cost Per Linear Foot

Fabric	4' High 9 ga.	5' High 9 ga.	6' High 9 ga.	8' High 6 ga.
Galvanized	$7.00	$7.75	$8.25	$ 9.50
Aluminized Steel...............	7.25	8.50	9.00	10.50
Vinyl Coated .	8.40	9.25	9.90	11.50

Add for barbed wire:

Single side - 3 strand...$	.45/lin.ft.
Double side - 3 strand	.80/lin. ft.

Add for gates:

3' x 4'...$ 45.00 Each	
3' x 5'... 60.00 Each	
3' x 6'... 80.00 Each	
3' x 8'... 105.00 Each	

RESIDENTIAL CHAIN LINK FENCING

The standard corner posts are 2" to 2½" diameter, the standard line posts are 1⅝" diameter set at 10' on center. Normally with the residential fencing a top rail is added for stability.

Residential Chain Link Fencing - Cost Per Linear Foot

Fabric	3' High 11 ga.	4' High 11 ga.
Galvanized	$4.15	$4.75
Aluminized..	4.50	5.25
Vinyl Coated	4.70	5.50

Add for gates:

3' x 3'...$ 35.00 Each	
3' x 4'... 42.00 Each	

Wood Residential Fences - Cost Per Linear Foot

Type of Fence	Cost Per Lin. Ft.	Cost Per Each
Cedar Stockade Screen Fence 4' High..................................	$ 8.50	—
Gate 3' Wide x 4' High..	—	$55.00
Cedar Stockade Screen Fence 6' High..................................	12.00	—
Gate 3' Wide x 6' High..	—	80.00
Cedar Split Rail - 2 Rail 3' High...	6.00	—
Cedar Split Rail - 3 Rail 3'6" High.......................................	6.75	—
Redwood or Cedar Board on Board 4' High.........................	11.25	—
Redwood or Cedar Board on Board 6' High.........................	14.00	—
Gates 3' Wide x 4' High..	—	60.00
Gates 3' Wide x 6' High..	—	85.00

02720   ROAD AND PARKING APPURTENANCES.

Guard rails of formed steel are set on steel, wood or concrete posts. Rail sections are set 18" to the centerline above grade. Posts are set every 12'-6" with rails lapped 11½" and attached with 7-⅝" bolts. 12 ga. sections weigh less than 100 lbs. so 2 men can handle installation. Special terminal wings are furnished for exposed ends.

A budget price for this type of rail set on steel posts is $12.50 per lin. ft., around $11.00, if set on wood posts.

4'x8' timber rails set on wood posts will cost around $6.00 a foot. They do not offer the protection of the metal rail but may better fit into the landscape.

Wheel stops for parking lots may be either wood or concrete. The concrete ones are shaped on the bottom to allow drainage; wood ones, unless set on standards, tend to collect water. Costs per lin. ft. installed are $3.25 for 6x6 wood dowelled into paving, $3.25 for concrete dowelled into paving and $2.40 for 4x4 wood set on metal brackets.

4" striping in parking lots will cost about $0.15 per foot or $3.00 per stall.

02760   SITE FURNISHINGS

Playground Equipment

Type of Equipment	Cost Per Each
Bike Rack, 10' long	$ 325.00
Horizontal Monkey Ladders, 12' long	575.00
Parallel Bars, 10' long	375.00
See-saw 2 Units	450.00
See-saw 4 Units	750.00
Slides - Stainless Steel, 12' long, 6' high	875.00
Swings - 4 seats, 8' high	800.00
Swings - 8 seats, 8' high	1,250.00

Benches

Type	Cost Per Each
Precast concrete with wood backs and seats, 8' long	$ 600.00
Fiberglass, 8' long	275.00
Fiberglass supports with wood backs and seats, 8' long	240.00
Wood, 8' long	200.00

Planters

Type of Planter	Cost Per Each
Precast concrete, 24" dia. x 18" high	$ 225.00
Precast concrete, 48" dia. x 30" high	375.00
Fiberglass, 36" dia. x 30" high	300.00
Fiberglass, 60" dia. x 36" high	525.00

02800   LANDSCAPING

Landscaping requirements are influenced to a considerable degree by the aesthetic appreciation of existing vegetation and the economics of retaining and protection costs during the construction period. Some Owners and Architects are dedicated to the preservation of the maximum number of trees which do not directly interfere with construction in an effort to blend the new construction into the existing environment. This theory produces exceptional immediate beauty to any new construction project, but also involves additional costs to construction by requiring extensive protective barriers for trees and vegetation and restricts work areas causing a lowering of productivity for automotive and heavy construction equipment.

In instances where existing trees and vegetation are of low quality, or the Owner is unwilling to absorb the cost of retention, the site is cleared of all natural growth and provisions are made to provide new landscaping for the site at the conclusion of construction. The beauty of landscaping is subjective and may be limited to a combination of simple sodding, seeding, and minimal plantings. In other instances, it can progress to an elaborate extent requiring the special talents of a Landscape Architect.

Most Contractor/Estimators find they lack the source of materials, experienced personnel, and special equipment required in the landscaping operation. Therefore, they rely upon landscape contractors to perform these services. Frequently, specifications require landscaping be performed by a certified contractor who is primarily engaged in this type of work and has the resources and capabilities to provide guarantees and certifications in compliance with specification requirements.

Estimators attempt to preserve the integrity of their estimates by developing a general knowledge of the costs involved in landscaping in order to evaluate the proposals received from the subcontractor.

Generally, it has been found that nurseries providing service to homeowners and transit trade are not competitive in prices for commercial and large scale landscape projects.

Soliciting proposals from qualified landscaping subcontractors presents a unique problem for the Contractor/Estimator. It has been found that these specialized subcontractors recognize that their performance will not be required until near the completion of a project, which may be many months in the future, and are reluctant to tender firm price proposals prior to the scheduled bid date.

Estimators must take a special effort to obtain these proposals in a timely manner to allow sufficient time to make an analysis of the bid before it is incorporated into the total project estimate.

The establishment of a work performance schedule to coincide with the accepted local planting season will have a tendency to lower the subcontractors price and will also preserve their guarantee period. It is not unusual to provide for landscaping work slightly out of sequence in order to accommodate the planting season.

For the Estimator to develop the proper unit of measure for pricing landscape work, or determining the completeness of a proposal from a subcontractor, the following units are required.

1. Area of seed and fertilizing in square yards, or acres on large projects.
2. Topsoil in cubic yards with allowance for topsoil available from the site and the amount which must be purchased from external sources.
3. Area of sodding in square yards.
4. Number, size and species of shrubs.
5. Number, size and species of trees.
6. Detailed specifics of any special feature such as rock gardens, etc.
7. Allowance for maintenance for specified periods.
8. Allowance for replacement guarantee, if specified.

After the establishment of an accurate determination of quantities, quality and features required, the Estimator is then prepared to develop a budget type estimate for the work which can be used as a cost comparison with the proposals received from landscaping subcontractors.

For general guidance, the following table will provide average cost ranges, but due to quantity, cost fluctuation, and geographical conditions, estimators should verify with local landscape subcontractors.

Item	Unit	In-Place Cost
Edging, 1 x 4 Redwood	Lin. Ft.	$1.15
Erosion Control - Plastic Netting	Sq. Yd.	1.00
Marble Stone Chips - Hand Placed	100#/Bag	9.20

Move Existing Plants - replanting		
3'-0" high	Each	50.00
Seed, Fertilizer - Small Areas	Sq. Yd.	1.50
Mulch, Peat Moss 2" Deep	Sq. Yd.	1.40
Sodding	Sq. Yd.	2.50 - 4.00
TREES: Spruce 8'	Each	175.00
Birch 8'	Each	150.00
Pine-White 8'	Each	100.00
Dogwood 8'	Each	200.00
SHRUBS: Honeysuckle 4'	Each	35.00
Forsythia 4'	Each	35.00
Plant Bed By Hand	Sq. Ft.	3.75

Plant maintenance is necessary until rooting begins with the duration being determined by local weather conditions and the time of planting. This operation can extend for many months beyond the conclusion of the actual construction period, thus, allowances have to be made in the landscape estimate to accommodate this requirement. If the maintenance period extends beyond the conclusion of construction this will necessitate the dispatch of personnel, including transportation of supplies and equipment necessary to perform the maintenance work. It will be found the per trip cost will be considerably higher as opposed to the anticipated cost if the work had been performed simultaneously with the construction phase. These excess costs occur from the additional travel time involved for each trip which would include the driver's wage cost plus a fair and reasonable allowance for truck rental. It must also be recognized the actual work performance would be in an unsupervised mode which equates to low productivity.

PLANT GUARANTEE

Should the Prime Contractor assume the obligation of the guarantee, then the estimate should include an allowance for replanting shrubs and trees which may fail to root and grow after transplanting. To arrive at a reasonable allowance to cover this cost the sum of 10%-20% of the original planting allowance should be sufficient.

CHAPTER 8

CONCRETE

CSI DIVISION 3

03100 CONCRETE FORMWORK
03200 CONCRETE REINFORCEMENT
03300 CAST-IN-PLACE CONCRETE
03400 PRECAST CONCRETE
03500 CEMENTITIOUS DECKS

When estimating the total costs for placement of concrete in footings, foundations, walls, columns, slabs, etc., there are five (5) categories of labor operations to be considered, in addition to the material and equipment costs.

1. *FORM PREPARATION*: Since concrete is placed in a fluid consistency, it must be contained or formed to the desired shape prior to its hardening process. This forming operation would include such items as: a) earth excavation to the desired shape for footings and foundations together with supplemental wood forms, if necessary, for proper elevation, steps (changes in elevation) and in the event of unstable soil conditions, b) proper finish grading of earth surfaces for the placement of concrete slabs on grade to the correct thickness and elevation, c) the cutting, fabricating, placing, alignment and bracing of wood or metal forms for walls, columns, slabs, etc.

2. *EMBEDDED ITEMS PLACEMENT*: Labor will be required for the proper placement of all items to be cast into the concrete mix. Usually steel reinforcing (mesh or bars) will be necessary to insure the structural integrity of the concrete member. Non-structural items that frequently are installed into cast-in-place concrete would include: anchor bolts, plates or inserts for the future connection of other items of work, corner protection angles, and sleeves for the passage of pipes, conduit, ductwork, etc.

3. *PLACEMENT OF CONCRETE*: The ready-mixed concrete that is delivered to the job site by truck must be physically moved from the truck to the location of placement by the use of a chute, wheelbarrow, buggy, or by some mechanical means such as a conveyor, pump or crane. All of the previously mentioned means of moving the concrete mix will involve a labor operation, whether it be a group of laborers, an equipment operator, or a combination thereof. Once the concrete mix is in place, it is then necessary to tamp or vibrate the concrete for proper consolidation and "strike off" or screed the concrete to its designed level or elevation. In the event that only a small amount of concrete is required, or in remote areas where ready-mixed concrete is unavailable, it will be necessary to estimate the additional labor required to job mix the components (gravel, sand, cement and water) for the quantity of concrete needed.

4. *FINISH, CURE & PROTECTION*: Once the concrete mix is placed into the forms, vibrated, and screeded to the proper elevation, the next operation necessary is the proper finishing of the exposed surfaces, e.g. a steel troweled finish to a slab surface to provide a smooth, hard and consolidated surface, free of voids and unevenness. Concrete requires a chemical action to change it from a fluid consistency to the set or hardened condition, which generates a certain amount of heat. On hot days the heat generated by the setting process and the weather combine to produce a concrete mix that sets too fast and hampers workability for the proper finish; and will produce surface cracking, and, in some cases, for the full depth in thin slabs. To help slow the setting process, slabs for example, will be covered with wet burlap, polyethelene, or a chemical curing compound to

help retard the setting process and seal the moisture into the concrete, thereby allowing a slower curing time which produces a stronger end result. Beyond proper curing and protection in hot weather, concrete, because of its water content prior to its hardening process, must be kept from freezing by either covering with straw or some insulating material to keep its generated heat in; or by a supplemental heating process such as covering the area and installing a temporary heat source. Concrete work, and in particular slabs, must be protected from any form of precipitation prior to the setting process, because the water droplets landing on the slab cause: a) small craters in the finish surface, thereby ruining the quality of finish and b) a diluting of the cement ratio of the concrete mix which weakens at least the top wearing surface of the slab.

5. *FORM REMOVAL*: After the concrete has reached its hardened condition, the labor process of removing the forms can commence. This will entail stripping the nails from the forms, cleaning off the hardened concrete skim particles, preparing the forms for the next forming location by means of applying a form release agent (oiling), and physically moving the forms to the next area of the building for starting the forming process on another scheduled work item. The proper cleaning and oiling (coating the concrete contact surface) of form work is important because it will effect many re-uses of form materials, thereby reducing the costly expense of new materials and fabrication labor costs.

03100   CONCRETE FORMWORK

Concrete formwork should be estimated by computing the actual surface area of the form that comes into contact with the concrete; this is referred to as the "contact surface". The formwork quantities are then categorized into specific groups (wall form, column form, etc.) and in the proper unit of measure (square foot or linear foot) as follows:

Ribbon footing forms	-	Linear Foot
Slab edge forms (by height)	-	Linear Foot
Foundation edge forms	-	Square Foot
Wall Forms	-	Square Foot
Pier Forms	-	Square Foot
Column Forms	-	Square Foot
Beam bottom forms	-	Square Foot
Beam side forms	-	Square Foot
Supported slab forms	-	Square Foot

Example: Compute the quantity of formwork required for a free standing wall 50'-0" long, 4'-0" high, and 1'-0" thick. Forms will be required for both sides of the wall. The formwork and quantity is obtained as follows:
Face Form: 50'-0" (length) x 4'-0" (height) x 2 (sides) = 400 s.f.
End Form: 1'-0" (width) x 4'-0" (height) x 2 (ends) = 8 s.f.

WOOD FORMS

Forms for Concrete Footings.—There are two main types of footings which are used in most construction work, namely: column footings and wall footings. Column footings are usually square or rectangular masses of concrete designed to spread the column loading over an area to accommodate the soil bearing value and are located at the column centers. Wall footings are usually ribbons of concrete somewhat wider than the walls they carry and are similarly designed to spread the loading over a larger area than the actual wall bottom thickness.

Forms for footings which are formed completely, are usually simple and rough, using 2" planking for sides and 2"x4" stakes to hold the sides in place and 2"x4" struts for braces to the banks. This material may be reused numerous times in the course of the work.

Ribbon forms for partly formed footings usually consist of 2" thick lumber in as long lengths as practicable, secured at the correct elevation to 2"x4" stakes and tied across the tops at intervals of 4'-0" to 6'-0" with 1"x2" spreader-ties, except in the case of column footings which are usually diagonally tied at the corners with 1-inch boards and tied transversely, if required, with strap iron.

The following are itemized costs of the column and wall footing forms found in the average simple job. Complicated or "cut-up" jobs require up to 25 percent more material and up to 50 percent more labor.

Material Cost of 100 Sq. Ft. of Column Footing Forms Based on Footings 5'-0"x5'-0"x1'-6" Completely Formed

Material	B.F.	Rate	Total	Rate	Total
Sides, 2" planking	200	$....	$....	$0.06 *	$12.00*
Stakes and braces, 2"x4"	100	....	....	.055*	5.50*
Nails, etc.		....	....	2.00	2.00
Cost per 100 sq. ft.			$....		$19.50
Material cost per sq. ft.			....		.20

Labor Cost of 100 Sq. Ft. of Column Footing Forms Based on Footings 5'-0"x5'-0"x1'-6" Completely Formed

Placing Forms—18-22 sq. ft. per hr.	Hours	Rate	Total	Rate	Total
Carpenter	5.0	$....	$....	$16.47	$ 82.35
Labor helping	2.5	....	....	12.54	31.35
Stripping Forms—90-110 sq. ft. per hr.					
Labor	1.0	....	....	12.54	12.54
Cost per 100 sq. ft.			$....		$126.24
Labor cost per sq. ft.			....		1.26

Material Cost of 100 Lin. Ft. of Column Footing Ribbon Forms Based on Footings 5'-0"x5'-0"x1'-6" Partly Formed

Material	B.F.	Rate	Total	Rate	Total
Ribbons, 2"x4", 100 Lin. Ft.	67	$....	$....	$0.06 *	$ 4.02*
Stakes, 2"x4", 3'-0" long	80	....	....	0.055*	4.40*
Corner ties, 1"x6", 2'-0" long	20	....	....	0.06 *	1.20*
Nails, etc.		....	....	1.50	1.50
Cost per 100 Lin. Ft.			$....		$11.12
Material Cost per Lin. Ft.			....		.11

Labor Cost of 100 Lin. Ft. of Column Footing Ribbon Forms Based on Footings 5'-0"x5'-0"x1'-6" Partly Formed

Placing Forms—35 Lin. Ft. per hr.	Hours	Rate	Total	Rate	Total
Carpenter	3.0	$....	$....	$16.47	$49.41
Labor helping	1.5	....	....	12.54	18.81
Stripping Forms—450-500 Lin. Ft. per hr.					
Labor	0.21	....	....	12.54	2.64
Cost per 100 Lin. Ft.			$....		$70.86
Labor Cost per Lin. Ft.			....		.71

Material Cost of 100 Sq. Ft. of Wall Footing Forms Based on Footings 1'-0" Deep Completely Formed

Material	B.F.	Rate	Total	Rate	Total
Sides, 2"x12" planks	200	$....	$....	$0.06 *	$12.00*
Stakes and braces	75	....	....	0.055*	4.13*
Nails, etc.		....	....	1.20	1.20
Cost per 100 sq. ft.			$....		$17.33
Material cost per sq. ft.			....		.17

*Based on 6 uses of lumber.

Labor Cost of 100 Sq. Ft. of Wall Footing Forms Based on Footings 1'-0" Deep Completely Formed

Placing Forms—25-30 sq. ft. per hr.	Hours	Rate	Total	Rate	Total
Carpenter	3.50	$....	$....	$16.47	$57.65
Labor	1.75	....	....	12.54	21.95
Stripping Forms—70-75 ft. per hr.					
Labor	1.50	....	....	12.54	18.81
Cost per 100 sq. ft.			$....		$98.41
Labor cost per sq. ft.			....		.98

Material Cost of 100 Lin. Ft. of Wall Footing Ribbon Forms Based on Footings 1'-0" Deep Partly Formed

Material	B.F.	Rate	Total	Rate	Total
Ribbons, 2"x4", 100 lin. ft.	67	$....	$....	$0.06 *	$4.02*
Stakes, 2"x4", 2'-0" long	33	....	....	.055*	1.82*
Ties, 1"x2"x2'-0" long	4	....	....	.055*	.22*
Nails, etc.		....	....	1.20	1.20
Cost per 100 Lin. Ft.			$....		$7.26
Material Cost per Lin. Ft.			....		.07

Labor Cost of 100 Lin. Ft. of Wall Footing Ribbon Forms Based on Footings 1'-0" Deep Partly Formed

Placing Forms—40 Lin. Ft. per hr.	Hours	Rate	Total	Rate	Total
Carpenter	2.5	$....	$....	$16.47	$41.18
Labor helping	1.25	....	....	12.54	15.68
Stripping Forms—450-500 Lin. Ft. per Hr.					
Labor	0.21	....	....	12.54	2.64
Cost per 100 Lin. Ft.			$....		$59.50
Labor Cost per Lin. Ft.			....		.60

*Based on 6 uses of lumber.

On jobs where the wall footings are stepped down to a lower elevation, a riser form is required at each step. The cost of this riser form is about twice the cost of the regular footing side form.

Grooves are usually formed in the tops of footings and are called "Footing Keys." They are formed by forcing a beveled 2"x4" into the surface of the wet concrete. After the concrete has hardened the "key" forms are pried loose, cleaned and reused.

Labor Cost of 100 Lin. Ft. 2"x4" Footing Keys

	Hours	Rate	Total	Rate	Total
Carpenter placing	1.25	$. . . .	$. . . .	$16.47	$20.59
Labor stripping	.50	. . . .	. . . .	12.54	6.27
Cost per 100 lin. ft.			$. . . .		$26.86
Labor cost per lin. ft.				. . . .	.27

Forms for Pile Caps.—After piles are driven and cut off at the required elevation, they are capped with masses of concrete which are designed to transmit the total loading of the structure to the piles. These pile caps vary in size and shape from simple square caps covering one pile to complex many sided caps covering clusters of twenty or more piles.

Forms for pile caps are usually prefabricated in panels at the bench and are then set up and put together in the pile location. Side forms are usually made up from ⅝" plyform with 2" x 4" stiffners nailed to the back face. Five or 6 uses can usually be obtained from these forms.

After panels are set in place they are secured with 2"x4" braces and tied with strap iron.

The following is an itemized cost of making average pile cap forms, based on 5'-6"x5'-6"x3'-0", containing 66 sq. ft.:

Material Costs

Making Pile Cap Side Forms	B.F.	Rate	Total	Rate	Total
⅝" Plyform sq. ft.	66	$. . . .	$. . . .	$0.50	$33.00
20 pcs. 2x4x3'-0"	41	. . . .	. . . .	0.36	14.76
8 pcs. 2x4x5'-0"	27	. . . .	. . . .	0.36	9.72
Bracing					
14 pcs. 2x4x8'-0"	75	. . . .	. . . .	0.36	27.00
Nails		. . . .	. . . .	1.20	1.20
Cost per cap, 66 sq. ft.			$. . . .		$85.68
Material cost per sq. ft.				. . . .	1.30

Labor Costs

Making Pile Cap Side Forms	Hours	Rate	Total	Rate	Total
Carpenter	4	$. . . .	$. . . .	$16.47	$65.88
Labor	2	. . . .	. . . .	12.54	25.08
Cost per cap, 66 sq. ft.			$. . . .		$90.96
Labor cost per sq. ft.				. . . .	1.38

Divide above sq. ft. costs by number of times forms may be used on the job.

Labor Cost Setting and Removing Pile Cap Forms

	Hours	Rate	Total	Rate	Total
Carpenter	3.0	$. . . .	$. . . .	$16.47	$49.41
Labor	1.5	. . . .	. . . .	12.54	18.81
Labor stripping, cleaning, moving	1.0	. . . .	. . . .	12.54	12.54
Cost per cap, 66 sq. ft.			$. . . .		$80.76
Labor cost per sq. ft.				. . . .	1.23

Forms for Concrete Piers.—When column footing tops are lower than the bottom of the floor slab on the ground, piers are usually designed to carry the column loading to the footings.

Forms for this work are usually made up in side panels at the job bench, hauled to the footings, erected, clamped and braced.

Side panels are usually made of ⅝" plyform with 2"x4" stiffners. Clamping may be accomplished by using either 2"x4" lumber or metal column clamps properly spaced to withstand the pressure developed when placing the concrete.

Bracing is required to hold the form in position and usually consists of 2"x4" or 2"x6" lumber, secured to the form at grade level, spanning the pit and spiked to stakes driven into the ground. Bracing below this point is done by wedging 2"x4" struts between the form and the earth banks.

The following are itemized costs of making average pier forms; based on a pier 2'-0"x2'-0"x4'-0" containing 32 sq. ft.:

Material Costs

Making Pier Forms	B.F.	Rate	Total	Rate	Total
⅝" Plyform Sq. Ft.	32	$....	$....	$0.50	$16.00
12 pcs. 2x4x4'-0"..............................	32	....	....	0.36	11.52
Clamps 8 pcs. 2x4x3'-0"....................	16	....	....	.36	5.76
Nails, etc.	16	....	....	.35	.35
Cost 32 sq. ft.			$....		$33.63
Material cost per sq. ft.				....	1.05

Labor Costs

Making Pier Forms	Hours	Rate	Total	Rate	Total
Carpenter..	1.0	$....	$....	$16.47	$16.47
Labor..	0.5	....	....	12.54	6.27
Cost 32 sq. ft.			$....		$22.74
Labor cost per sq. ft.				....	.71

Divide sq. ft. costs by number of times forms may be used.

Labor Cost Erecting Concrete Pier Forms

	Hours	Rate	Total	Rate	Total
Carpenter..	1.0	$....	$....	$16.47	$16.47
Labor..	0.5	....	....	12.54	6.27
Stripping, including cleaning and moving Labor..	0.5	....	....	12.54	6.27
Cost per pier, 32 sq. ft.			$....		$29.01
Labor cost per sq. ft.				....	.91

Forms for Concrete Walls

Forms for concrete walls must be designed and constructed to withstand the horizontal pressures exerted by the fluid concrete against them. This pressure against any given point of the form will vary and be influenced by any one or all of the following factors:

1. Rate of vertical rise in filling the forms.
2. Temperature of the concrete and the weather.
3. Proportions of the mix and its consistency.
4. Method of placement and degree of vibration.

When wall forms are filled so rapidly that the concrete in the bottom of the form has not had sufficient time to start its initial setting process, then regardless of any other factor, the full height of the fluid concrete must be considered as a column of liquid with a weight of approximately 150 lbs. per cubic foot having a head pressure equal to its fluid height. Example: At 10' below the top of the concrete, the pressure will amount of 10 (feet) x 150 lbs. (weight) = 1,500 lbs per sq. ft. exerted against the form sides at the bottom.

If the wall form has not been adequately tied and braced to withstand this pressure, then the form will break open allowing the concrete to spill out of the form.

To negate the necessity of overdesigning form work, which is a costly measure, or the possibility of form breakage, a more prudent method for filling a wall form would be to place the concrete in layers or "lifts" at the rate of about 3 vertical feet per hour. This allows the concrete a chance to start the transformation from the fluid state to a hardened mass that supports itself and relieves pressure from the form surface.

The temperature at which concrete is placed has an influence on the amount of time required for the setting process. Concrete placed at 40° F takes almost twice the time to set than concrete placed at 80° F therefore, concrete placed in wall forms during cold weather will require more control of the amount of concrete placed within a given time so that the forms will not be stressed beyond their design capability.

The proportions and consistency of the mix (water/cement ratio per cubic yard of concrete) has an effect on the rate of setting time as well as the strength of the concrete mix, i.e., the more the cement content and/or the less water needed to provide a workable consistency (slump) per cubic yard will mean a shorter duration for setting and yield a stronger concrete.

When placing concrete into a form, care should be exercised to reduce the shock loading of the concrete hitting against the form material. This can be achieved by using a "tremie" (a flexible enclosed chute in the shape of an elephant's trunk) that confines the concrete until it reaches the bottom of the form or the level of the preceding lift. The tremie will also keep the concrete mix from segregating and should be used whenever placing concrete that could "free fall" a distance more than 4 feet. This segregation is caused by the large aggregate within the concrete mix hitting the reinforcing steel and shaking the cement/sand matrix coating off; the result being an improperly mixed concrete with inferior strength and possible voids (honeycomb).

Mechanical vibration of the concrete mix is accomplished after each lift is placed and is necessary for the proper consolidation (elimination of voids) and to ensure that the fine matrix of the mix reaches the form surface for a smooth wall finish. Although vibration of the concrete mix is desirable as well as necessary, it can have some adverse effects on formwork if accomplished improperly. If the vibrator is allowed to pass through the just placed concrete lift into the preceding lift, it will disturb the preceding lifts "setting process" causing it to liquefy, thereby, increasing the head pressure against the formwork. Over vibrating the concrete will also imbalance the large aggregate to fine aggregate proportions within those areas of the mix.

Concrete Wall Forms Made in Panels or Sections.—Many contractors, especially those performing smaller work, use foundation wall forms made up into panels 2'x6', 2'x8', 4'x6' or 4'x8' etc., in size. The frame for these sections is usually 2"x4" lumber, with intermediate braces spaced 12" to 16" on centers. The corners are reinforced with galvanized straps or strap iron, and the entire frame is sheathed with ⅝"or ¾" plyform plywood.

Forms of this kind can be used a large number of times, especially where the walls are fairly straight.

The labor cost of making panel forms will vary with the method used and the number made up at one time. A carpenter should complete one panel (28 to 32 sq. ft.) in ¾ to 1-hr. or 8 to 10 panels per 8-hr. day.

The cost of a panel 4'-0"x8'-0" in size, using ⅝" "Plyform" plywood should cost as follows:

Material Costs

	B.F.	Rate	Total	Rate	Total
2 pcs. 2"x4"-8'-0"	10.67	$. . . .	$. . . .	$ .36	$ 3.84
7 pcs. 2"x4"-4'-0"	18.67	. . . .	. . . .	.36	6.72
Plyform plywood, 4'-0"x8'-0"-⅝" sq. ft.	32.00	. . . .	. . . .	.50	16.00
Nails, etc.		. . . .	. . . .	.80	.80
Cost per panel 32 sq. ft.			$. . . .		$27.36
Material cost per sq. ft.				. . . .	.86

Labor Costs

	Hours	Rate	Total	Rate	Total
Carpenter	0.75	$. . . .	$. . . .	$16.47	$12.35
Labor	0.25	. . . .	. . . .	12.54	3.14
Cost per panel			$. . . .		$15.49
Labor cost per sq. ft.				. . . .	.49

Many contractors specializing in concrete work and especially foundations for residences, use panel forms exclusively and get up to 25 uses from the forms with the usual amount of repairs. These conditions should be given consideration when pricing any wall work using panel forms.

Symons Steel-Ply Forms*.—Symons Steel-Ply Forms are lightweight panels with a rolled steel frame completely enclosing a plastic coated plywood face. The plywood has a high strength face, which is 12% stronger than standard plywood. And due to the plastic coating, effects of moisture are reduced 40%. The welded steel frames are painted to prevent rusting and with reasonable care should last for a number of years.

Panels are 2'-0" wide and come in lengths ranging from 2'-0" to 8'-0" high in one foot increments. Fillers are made in widths 4" to 22", from 2'-0" to 8'-0" high. All steel fillers are available in 1", 1½", and 2" widths. The average weight of the Steel-Ply Forms is 6 pounds per square foot. Outside corners are steel angles that lock adjoining forms together. Inside corners are 6" x 6".

The side rails of the forms are rolled exclusively for Symons, and have a yield strength in excess of 55,000 p.s.i. "L" shaped cross members on all forms are on one foot centers and have a yield strength in excess of 60,000 p.s.i. The frame is 2½" in width.

In the Symons system, only two basic pieces are necessary for erection—the tie, and one piece connecting hardware called the Wedge-Bolt. The same Wedge-Bolt is also used for connection of any Symons accessories to the form. In erecting forms, the panel tie is placed in the dado slot between the forms.

A wedge-bolt is inserted through the adjoining forms and tie loop. A second wedge-bolt is inserted through the slot of the first, automatically tying and joining the forms in one operation. The Symons System also uses an easy, efficient method for the erection of corners, pilasters, walers, bracing, and scaffolding. Standard panel ties have 1" breakbacks; 1½" and 2" are also available with wood cones. Safe load capacity of panel ties is 3,000 lbs. and 4,000 lbs.; flat ties 4,000 lbs. and 4,500 lbs. at 1.5 to 1.0 factor of safety.

In stripping, panels may be quickly released from the finished wall. Wedge-bolts are easily loosened with a hammer tap, and panel ties then removed by giving tie loops a half twist, or in the case of flat ties, a vertical hammer blow.

---

*Symons Corporation, Des Plaines, Ill.

Used panels may be rented for about 80¢ per sq. ft. for the first month and 70¢ per sq. ft. for each subsequent month. New panels rent for about 90¢ per sq. ft. for the first month and 80¢ per sq. ft. for each subsequent month. Rental costs should be checked with local suppliers when needed. Minimum rental period is one month. Should the contractor wish to purchase the forms, a portion of the paid rentals shall be applicable to the purchase price. Rentals are charged from the date shipped from distribution center and stop on date of return. The contractor pays transportation charges both ways.

A typical foundation wall 8" thick and 8' high containing 2500 sq. ft. of plywood faced steel wallforms and using a crew of three carpenters to four laborers would average as follows:

Labor Costs

Erecting, Aligning and Bracing Panels	Hours	Rate	Total	Rate	Total
Carpenters	16	$....	$....	$16.47	$263.52
Laborers helping	21	....	....	12.54	263.34
Stripping., Cleaning and Oiling for Reuse					
Carpenters	8	....	....	16.47	131.76
Laborers helping	16	....	....	12.54	200.64
Total Cost					
2,500 Sq. Ft			$....		$859.26
Cost Per Sq Ft					.34

To the above one must add form rental and foreman.

The Symons Company has made several recent, on the job time studies of the placement of their forms, aligning them and stripping them. They report that a wall similar to the above example and using a crew of three lead and four supporting men should work out to a productivity of 1.355 manhours per 100 sq. ft. In five hours this team should complete some 2500 sq. ft. of contact area.

A study of a curved wall forming a sewage treatment structure and using a crew of one carpenter to one laborer showed a productivity of 2.91 manhours per 100 sq. ft. of contact area. Setting of reinforcement and pouring of concrete are not included in the above figures.

Symons' Versiform.*—Concrete forming system is a steel framed, plywood faced gang forming system for forming massive concrete structures and architectural concrete. This sytem is adaptable to use in walls, columns, piers, culverts, etc., and can be had with four different standard facing materials.

Versiform panels are available for rental in 8' x 4', 8' x 2', 4' x 4', 4' x 2', 2' x 4', and 2' x 2' sized panels, with special sized panels available for purchase only. A complete line of components and accessory hardware is available, for forming virtually any shape or size concrete structure.

Versiform panels may be used in conjunction with other forming systems manufactured by the Symons Corporation.

An on the job time study of Versiform panels and walers on continuous straight walls showed a productivity of 2.64 man hours per 100 sq. ft. of contact area.

Uni-Form† Steel Panel Forms.—Uni-Forms are prefabricated panels consisting of a steel frame 2'x1', 2'x2', 2'x3', 2'x4', 2'x5', 2'x6', 2'x7', 2'x8', with a plywood facing. The panels are also furnished in fractional widths. The steel frame consists

---

*Symons Corporation
Des Plaines, Illinois

†Universal Form Clamp Co., Chicago

of a special rolled tee angle, with an equal legged tee strut to provide additional structural rigidity to the portion of the frame which acts as a plywood protector.

The plywood face is 5-ply Plyform for use in concrete forms. The plywood facing provides a convenient nailing surface. It is advisable to oil the panels after each use and with proper care Uni-Forms have been used as many as 50 times before turning the plywood.

Uni-Form panels require alignment on one side only and employ a simplified method of forming inside corners and fillers.

Typical Method of Waling and Bracing Symons Steel-ply Forms

On a foundation wall job for 47 houses requiring 2,350 sq. ft. of forms per house, using the 7' Uni-Form panels, the following labor costs were obtained:

Erection	Hours	Rate	Total	Rate	Total
Carpenter foreman	6	$....	$....	$16.97	$ 101.82
Carpenters	30	....	....	16.47	494.10
Labor	18	....	....	12.54	225.72
Stripping, Cleaning and Moving					
Carpenters	10	....	....	16.47	164.70
Labor	8	....	....	12.54	100.32
Total cost 2,350 sq. ft.			$....		$1,086.66
Labor Cost per sq. ft.				....	.46

Ganged Symons Forms Showing Waling, Bracing and Scaffold Attachment

Courtesy Symons Corporation

Symons forms being set to form room walls in circular high-rise hotel

A carpenter erected and braced 78.3 sq. ft. of forms per hr. and required 0.6 hr. labor time to each hr. carpenter time. A carpenter stripped 235 sq. ft. of forms an hr. A laborer cleaned, oiled, and moved 294 sq. ft. of forms an hr.

On another foundation wall job for 50 houses, requiring 1,550 sq. ft. of forms per house (forms 2'-0"x7'-0" and 1'-0"x7'-0" sizes), the following labor costs were obtained per 100 sq. ft. of forms, contact area:

Uni-Form Concrete Wall Forms

Erecting and Bracing Panels	Hours	Rate	Total	Rate	Total
Carpenter	2.6	$. . . .	$. . . .	$16.47	$42.82
Stripping Forms and Moving					
Carpenter	0.5	. . . .	. . . .	16.47	8.24
Labor	0.8	. . . .	. . . .	12.54	10.03
Oiling Panels					
Labor	0.5	. . . .	. . . .	12.54	6.27
Cost per 100 sq. ft			$. . . .		$67.36
Labor Cost per sq. ft.			. . . .		.67

Uni-Form Concrete Wall Forms

A carpenter erected and braced 38.75 sq. ft. of forms per hr. A carpenter stripped 194 sq. ft. of forms per hr. A laborer oiled 194 sq. ft. of forms per hr. A laborer carried from one job to the other or loaded into trucks, 129 sq. ft. of forms per hr.

Based on the two preceding examples using Uni-Form prefabricated form panels, you can ascertain that even though the panels, type of construction, and classification of labor was similar, the production time on the first example project was considerably faster. There are many factors that will affect the production time at the project, e.g., weather, adaptability of materials used, accessibility to the work area, layout of the work, efficiency and experience of the workmen, etc., and the estimator must evaluate all of these variables when preparing the estimate.

Built-In-Place Wood Forms For Concrete Walls

For those situations where the wall forms are to be built-in-place, the estimator should be familiar with form design parameters so that he can ascertain the correct quantity of form materials necessary for the work.

The following data and tables on form design have been furnished by The American Plywood Association of Tacoma, Washington:

CONCRETE FORMWORK

Grade-Use Guide for Concrete Forms*

Use these terms when you specify plywood	DESCRIPTION	VENEER GRADE	
		Faces	Inner Plies
APA B-B PLYFORM Class I & II**	Specifically manufactured for concrete forms. Many reuses. Smooth, solid surfaces. Mill-oiled unless otherwise specified.	B	C
APA High Density Overlaid PLYFORM Class I & II**	Hard, semiopaque resin-fiber overlay, heat-fused to panel faces. Smooth surface resists abrasion. Up to 200 reuses. Light oiling recommended between pours.	B	C Plugged
APA STRUCTURAL I PLYFORM**	Especially designed for engineered applications. All Group 1 species. Stronger and stiffer than PLYFORM Class I and II. Recommended for high pressures where face grain is parallel to supports. Also available with High Density Overlay faces.	B	C or C Plugged
Special Overlays, proprietary panels and Medium Density Overlaid plywood specifically designed for concrete forming**	Produce a smooth uniform concrete surface. Generally mill treated with form release agent. Check with manufacturer for design specifications, proper use, and surface treatment recommendations for greatest number of reuses.		

*Commonly available in ⅝" and ¾" panel thicknesses (4'x8' size).
**Check dealer for availability in your area.

Concrete Pressures For Column And Wall Forms

Pour Rate (ft./hr.)	Pressures of Vibrated Concrete (psf) (a), (b)			
	50° F		70° F	
	Columns	Walls	Columns	Walls
1	330	330	280	280
2	510	510	410	410
3	690	690	540	540
4	870	870	660	660
5	1050	1050	790	790
6	1230	1230	920	920
7	1410	1410	1050	1050
8	1590	1470	1180	1090
9	1770	1520	1310	1130
10	1950	1580	1440	1170

Notes: (a) Maximum pressure need not exceed 150h, where h is maximum height of pour.
(b) Based on concrete with density of 150 pcf and 4 in. slump.

Concrete Pressures For Slab Forms

Depth of Slab (in.)	Concrete Pressure (psf)	
	Non-Motorized Buggies (a)	Motorized Buggies (b)
4	100	125
5	113	138
6	125	150
7	138	163
8	150	175
9	163	188
10	175	200

Notes: (a) Includes 50 psf load for workmen, equipment, impact, etc.
(b) Includes 75 psf load for workmen, equipment, impact, etc.

Allowable Pressures On Plyform Class I For Architectural Applications
(deflection limited to 1/360th of the span)

Face Grain Across Supports Allowable Pressures (psf) (a)

Support Spacing (inches)	Plywood Thickness (inches)					
	½	⅝	¾	⅞	1	1-⅛
4	2935	3675	4495	4690	5075	5645
8	970	1300	1650	1805	1950	2170
12	410	575	735	890	1185	1345
16	175	270	370	475	645	750
20	100	160	225	295	410	490
24			120	160	230	280
32					105	130
36						115

(a) Plywood continuous across two or more spans.

Face Grain Parallel To Supports Allowable Pressures (psf) (a)

Support Spacing (inches)	Plywood Thickness (inches)					
	½	⅝	¾	⅞	1	1-⅛
4	1670	2110	2610	3100	4140	4900
8	605	810	1005	1190	1595	1885
12	215	360	620	740	985	1165
16		150	300	480	715	845
20		105	210	290	400	495
24			110	180	225	320

(a) Plywood continuous across two or more spans.

Allowable Pressures On Plyform Class I
(deflection limited to 1/270th of the span)

Face Grain Across Supports Allowable Pressures (psf) (a)

Support Spacing (inches)	Plywood Thickness (inches)					
	½	⅝	¾	⅞	1	1-⅛
4	2935	3675	4495	4690	5075	5645
8	970	1300	1650	1805	1950	2170
12	430	575	735	890	1185	1345
16	235	325	415	500	670	770
20	135	210	285	350	465	535
24		110	160	215	295	340
32					140	170
36					105	120

(a) Plywood continuous across two or more spans.

Face Grain Parallel To Supports Allowable Pressures (psf) (a)

Support Spacing (inches)	Plywood Thickness (inches)					
	½	⅝	¾	⅞	1	1-⅛
4	1670	2110	2610	3100	4140	4900
8	605	810	1005	1190	1595	1885
12	270	405	620	740	985	1165
16	115	200	375	525	715	845
20		125	210	290	400	495
24			135	185	255	320

(a) Plywood continuous across two or more spans.

Maximum Spans for Lumber Framing, Inches*
Douglas Fir-Larch No. 2 Or Southern Pine No. 2 (KD)

Equivalent Uniform Load (lb/ft)	Continuous Over 2 or 3 Supports (1 or 2 Spans) Nominal Size								Continuous Over 4 or More Supports (3 or More Spans) Nominal Size							
	2x4	2x6	2x8	2x10	2x12	4x4	4x6	4x8	2x4	2x6	2x8	2x10	2x12	4x4	4x6	4x8
400	36	53	70	90	109	50	79	101	41	60	78	100	122	62	91	118
600	30	43	57	73	89	44	66	88	31	49	64	82	99	51	74	98
800	24	38	50	63	77	39	58	76	25	39	52	66	81	44	64	85
1000	21	33	43	55	67	35	51	68	21	34	44	57	69	39	58	76
1200	19	29	38	49	60	32	47	62	19	30	39	50	61	35	53	69
1400	17	27	35	45	54	30	43	57	17	27	36	46	56	31	49	64
1600	16	25	32	41	50	27	41	54	16	25	33	42	52	28	44	58
1800	15	23	30	39	47	25	38	51	15	24	31	40	48	26	40	53
2000	14	22	29	37	45	23	36	48	14	22	29	38	46	24	37	49
2200	13	21	28	35	43	22	34	45	14	21	28	36	44	22	35	46
2400	13	20	26	34	41	20	32	42	13	20	27	34	42	21	33	44
2600	12	19	26	33	40	19	31	40	13	20	26	33	40	20	31	41
2800	12	19	25	32	38	19	29	38	12	19	25	32	39	19	30	39
3000	12	18	24	31	37	18	28	37	12	19	24	31	38	18	29	38
3200	11	18	23	30	36	17	27	35	12	18	24	30	37	18	28	36
3400	11	17	23	29	36	17	26	34	11	18	23	30	36	17	27	35
3600	11	17	22	29	35	16	25	33	11	17	23	29	35	16	26	34
3800	11	17	22	28	34	16	24	32	11	17	22	29	35	16	25	33
4000	10	16	22	28	34	15	24	31	11	17	22	28	34	15	24	32
4500	10	16	21	27	32	14	22	29	10	16	21	27	33	14	23	30
5000	10	15	20	25	31	13	21	28	10	16	20	26	32	14	22	28

*Spans are based on PS-20 lumber sizes, Single member stresses were multiplied by a 1.25 duration-of-load factor for 7-day loads. Deflection limited to 1/360th of the span with ¼" maximum. Spans are center-to-center of the supports.

Maximum Spans for Lumber Framing, Inches*
Hem-Fir No. 2

Equivalent Uniform Load (lb/ft)	Continuous Over 2 or 3 Supports (1 or 2 Spans) Nominal Size								Continuous Over 4 or More Supports (3 or More Spans) Nominal Size							
	2x4	2x6	2x8	2x10	2x12	4x4	4x6	4x8	2x4	2x6	2x8	2x10	2x12	4x4	4x6	4x8
400	33	48	63	80	97	48	73	96	36	53	70	90	109	56	81	107
600	27	39	51	65	80	41	59	78	27	43	57	72	88	45	66	88
800	22	34	44	57	69	35	51	68	22	35	46	59	72	39	58	76
1000	19	29	39	50	60	31	46	61	19	30	40	51	62	35	51	68
1200	17	26	35	44	54	29	42	55	17	27	36	45	55	31	47	62
1400	15	24	32	41	49	27	39	51	16	25	33	42	51	27	43	57
1600	14	23	30	38	46	24	36	48	15	23	30	39	47	25	39	52
1800	14	21	28	36	43	22	34	45	14	22	29	36	44	23	36	47
2000	13	20	27	34	41	21	33	43	13	21	27	35	42	21	33	44
2200	12	19	26	33	40	19	31	40	13	20	26	33	40	20	31	41
2400	12	19	25	31	38	18	29	38	12	19	25	32	39	19	30	39
2600	12	18	24	30	37	18	28	36	12	18	24	31	38	18	28	37
2800	11	18	23	30	36	17	26	35	11	18	24	30	37	17	27	36
3000	11	17	23	29	35	16	25	33	11	17	23	29	36	17	26	34
3200	11	17	22	28	34	16	24	32	11	17	22	29	35	16	25	33
3400	10	16	22	27	33	15	24	31	11	17	22	28	34	15	24	32
3600	10	16	21	27	32	15	23	30	10	16	22	27	33	15	23	31
3800	10	15	20	26	32	14	22	29	10	16	21	27	33	15	23	30
4000	10	15	20	25	31	14	22	29	10	16	21	27	32	14	22	29
4500	10	14	19	24	29	13	21	27	10	15	20	26	31	13	21	28
5000	9	13	18	23	28	12	20	26	9	15	20	25	30	13	20	26

*Spans are based on PS-20 lumber sizes, Single member stresses were multiplied by a 1.25 duration-of-load factor for 7-day loads. Deflection limited to 1/360th of the psan with ¼" maximum. Spans are center-to-center of the supports.

LATERAL CONCRETE PRESSURES FOR VARIOUS TEMPERATURES

Estimating Material Requirements for Wood Forms Built-In-Place for Concrete Walls

The following examples of wall form material estimates assumes: 1. Concrete at 50° F with 4" slump, 2. Vertical pour rate of 3'-0" to 4'-0" per hour, and 3. A maximum form pressure of 750 PSF.

Lumber Required for Concrete Wall Forms 6'-0" High

Based on a wall 40'-0" long and 6'-0" high, containing 480 sq. ft. of forms.

	B.F.	Rate	Total	Rate	Total
Studs, 82 pcs. 2"x4"-6'-0"					
(studs 12" on centers)	328	$. . . .	$. . . .	$.36	$118.08
Sheathing ⅝" Plyform sq. ft	480	. . . .	. . . .	.50	240.00
Plates, 80 lin. ft. 2"x4"	53	. . . .	. . . .	.36	19.08
Wales, 3 sets per side					
12 pcs. 2"x4"-40'-0"	320	. . . .	. . . .	.36	115.20
Bracing and stakes, 2"x4"	70	. . . .	. . . .	.36	25.20
Cost 480 sq. ft		. . . .	$. . . .		$517.56
Cost per sq. ft			. . . .		1.08
Material not subject to reuse	Quan.	Rate	Total	Rate	Total
Nails	20 lbs.	$. . . .	$. . . .	$.60	$12.00
Form ties, (3,000 lb safe load)					
24" o.c. along wales	63	. . . .	. . . .	.40	25.20
Total Cost 480 sq. ft			$. . . .		$37.20
Cost per sq. ft			. . . .		.08

Assuming 3 uses of form lumber 1.08 ÷ 3=36¢. per use for form lumber, plus 8¢ per sq. ft. for nails and form ties, or a total of 44¢ cts. per sq. ft. for material.

Lumber Required for Concrete Wall Forms 8'-0" High

Based on a wall 40'-0" long and 8'-0" high, containing 640 sq. ft. of forms.

	B.F.	Rate	Total	Rate	Total
Studs, 82 pcs. 2"x4"-8'-0"					
(12" on centers)	438	$. . . .	$. . . .	$.36	$157.68
Sheathing, ⅝" plyform sq. ft	640	. . . .	. . . .	.50	320.00
Plates, 80 lin. ft. 2"x4"	53	. . . .	. . . .	.36	19.08
Wales, 4 sets per side, 16 pcs.					
2"x4"-40'-0"	427	. . . .	. . . .	.36	153.72
Bracing and stakes, 2"x4"	80	. . . .	. . . .	.36	28.80
Cost 640 sq. ft			$. . . .		$679.28
Cost per sq. ft			. . . .		1.06

Material not subject to reuse:	Quan.	Rate	Total	Rate	Total
Nails	30 lbs.	$....	$....	$.60	$18.00
Form ties (3,000 lb safe load)					
24" o.c. wales	84	....	....	.40	33.60
Total cost 640 sq. ft			$....		$51.60
Cost per sq. ft			....		.08

Assuming 3 uses of form lumber, 1.06 ÷ 3=35.4¢ per use of form lumber, plus 8¢ per sq. ft. for nails and form ties, or a total of 43.4¢ per sq. ft. for material.

Lumber Required for Concrete Wall Forms 12'-0" High

Based on a wall 40'-0" long and 12'-0" high, containing 960 sq. ft. of forms.

	B.F.	Rate	Total	Rate	Total
Plates 4 pcs. 2"x4"-20'-0"	53	$....	$....	$.36	$ 19.08
Studs, 82 pcs. 2"x4"-12'-0"					
(12" on centers)	656	....	....	.36	236.16
Sheathing, ⅝" plyform sq. ft	960	....	....	.50	480.00
Wales, 6 sets per side wales @					
24" on centers					
24 pcs. 2"x4"-40'-0"	640	....	....	.36	230.40
Bracing and stakes, 2"x4"	98	....	....	.36	35.28
Cost 960 sq. ft			$....		$1,000.92
Cost per sq. ft			....		1.04

Material not subject to reuse:	Quan.	Rate	Total	Rate	Total
Nails	40 lbs.	$....	$....	$.60	$24.00
Form ties (3,000 lb. safe					
load) 24" o.c. along					
wales	126	....	....	.40	50.40
Total cost 960 sq. ft			$....		$74.40
Cost per sq. ft			....		.08

Assuming 3 uses of form lumber, 1.04 ÷ 3=34.7¢ per use for form lumber, plus 8¢ per sq. ft. for nail and form tiles, or a total of 42.7¢ per sq. ft. for material.

In comparing the three preceeding example material estimates for wall forming on 6, 8 and 12 foot high walls, you will note that all three are in the same cost range. This will hold true for lower or higher walls as long as the design parameters for temperature, slump, pour rate, and allowable form pressure remain the same.

If the temperature of the concrete is lower, the mix is wetter, or it is advantageous to place the concrete into the forms at a faster pour rate, then the forms will have to be designed accordingly to withstand the increased pressure per square foot on the form surface.

The design changes would involve one or all of the following:

1. Stronger form tie (increased tensile strength).
2. Closer horizontal spacing for form ties.
3. Increased stud size to counteract bending under increased loads.
4. Closer on-center spacing of stud members to counteract deflection of the plyform, or, use thicker plyform material.
5. Increased size of wale members.

FORM TIES TYPICAL

5/8" PLYFORM

1'-0" 2'-0" 2'-0" 1'-0"

1'-0"

2'-0"

2x4 PLATE

2 - 2x4 WALES @ 24" O.C.

PRE-PLACED FOOTING

2x4 STUDS @ 12"–16" O.C.

Basis for Computing Labor Costs.—After checking the labor costs on a large number of jobs over a period of years, it has been found that a carpenter will frame and erect about 325 to 400 ft. of lumber, b.m. per 8-hr. day, and will require ⅜ to ½ hr. laborer time carrying lumber, etc., to each hour carpenter time.

If there are no laborers employed on the job and all lumber is handled and carried by carpenters, a carpenter should handle, frame and erect about 225 to 275 ft. of lumber b.m. per 8-hr. day.

When removing or "stripping" concrete forms, a carpenter or laborer should remove 750 to 1,000 ft. of lumber, b.m. per 8-hr. day.

High walls necessitating work from a scaffold requires 10 to 50 per cent more time per square foot of forms because of the additional handling and hoisting of lumber, additional bracing, and the fact that the men are working on a scaffold.

The sq. ft. costs on the following pages have been computed on this basis.

Production Times for 100 Sq. Ft. of Wood Forms for Concrete Walls

Class of Work	Height Of Wall in Feet	BF. Lbr. per Sq. Ft. of Forms	No. Sq. Ft. of Forms 8-Hr. Day	Carpenter Hours	Labor* Hours	Removing Forms	
						No. Sq. Ft. 8-Hr. Day	Labor Hours
Wall Forms Built-in-Place............	4 to 6	2.6	190–210	4.0	2.0	360	2.25
Wall Forms Built-in-Place............	7 to 8	2.6	165–190	4.5	2.3	350	2.20
Wall Forms Built-in-Place............	9 to 10	2.6	150–160	5.3	2.5	325	2.50
Wall Forms Built-in-Place............	11 to 12	2.6	125–135	6.0	3.0	300	2.70

*If Union rules do not permit laborer working with carpenter, figure labor time as additional carpenter time.

When foundation walls rest directly on the earth without a concrete footing, mud sills of 2" or 3" plank are usually employed to serve as layout lines and supports for the wall panel forms required. This will increase the cost of 4'-0" high panel wall forms about 10 percent for both material and labor. Similarly, 8'-0" high wall forms should be increased about 5 percent.

Labor Cost of 100 Sq. Ft. Built in Place Concrete Wall Forms 4'-0" to 6'-0" High, Requiring 2.6 BF of Lumber Per Sq. Ft. of Forms

	Hours	Rate	Total	Rate	Total
Carpenter	4.0	$....	$....	$16.47	$ 65.88
Labor	2.0	....	....	12.54	25.08
Labor stripping-cleaning	2.25	....	....	12.54	28.22
Cost 100 sq. ft			$....		$119.18
Labor cost per sq. ft			....		1.19

Labor Cost of 100 Sq. Ft. Built-in-Place Concrete Wall Forms 8'-0" High, Requiring 2.6 BF of Lumber Per Sq. Ft. of Forms

	Hours	Rate	Total	Rate	Total
Carpenter	4.50	$....	$....	$16.47	$ 74.12
Labor helping	2.30	....	....	12.54	28.84
Labor stripping-cleaning	2.20	....	....	12.54	27.59
Cost 100 sq. ft			$....		$130.55
Labor cost per sq. ft			....		1.31

Labor Cost of 100 Sq. Ft. Built-in-Place Concrete Wall Forms 10'-0" High, Requiring 2.6 BF of Lumber Per Sq. Ft. of Forms

	Hours	Rate	Total	Rate	Total
Carpenter	5.3	$....	$....	$16.47	$ 87.29
Labor helping	2.5	....	....	12.54	31.35
Labor stripping-cleaning	2.5	....	....	12.54	31.35
Cost 100 sq. ft			$....		$149.99
Labor cost per sq. ft			....		1.50

Labor Cost of 100 Sq. Ft. Built-in-Place Concrete Wall Forms 12'-0" High, Requiring 2.6 BF of Lumber Per Sq. Ft. of Forms

	Hours	Rate	Total	Rate	Total
Carpenter	6.0	$....	$....	$16.47	$ 98.82
Labor helping	3.0	....	....	12.54	37.62
Labor stripping-cleaning	2.7	....	....	12.54	33.86
Cost 100 sq. ft			$....		$170.30
Labor cost per sq. ft			....		1.70

Pilasters.—Walls which are thickened for short intervals of 12" to 36" in length and then return to their former thickness are known as pilasters at that thickened area.

The cost of forming the face and sides of pilasters runs from 25 to 50 percent more than plain wall forms for both material and labor.

Radial Wall Forms.—Wall forms which are built to radii or other curves require closer stud spacing, more bracing and more labor laying out than straight wall forms.

Formwork for radial walls will cost from 50 to 100 percent more than forms for straight walls.

Method of Forming for Radial Concrete Walls.

Forming Openings in Concrete Walls.—Openings in concrete walls for doors, windows, louvers, etc., are formed by placing bulkheads in the wall forms to form the sills, jambs and heads of the openings required. These forms usually require from 2¼ to 3 BF of lumber per sq. ft. of contact area, depending upon the size of the opening.

The following is an itemized labor cost of boxed opening forms, based on 4'-0"x4'-0" openings in a 1'-0" wall requiring 2½ BF of lumber per sq. of contact area.

Labor Cost of 100 Sq. Ft. of Boxed Opening Forms

	Hours	Rate	Total	Rate	Total
Carpenter, setting forms	7.0	$....	$....	$16.47	$115.29
Labor Helping	3.5	....	....	12.54	43.89
Labor stripping forms	3.0	....	....	12.54	37.62
Cost 100 sq. ft			$....		$196.80
Labor cost per sq. ft			....		1.97

Forming Setbacks in Concrete Walls.—When concrete foundation walls extend some height above ground level, it is often required that the portion above grade be faced with masonry. This necessitates the portion above grade being reduced in thickness sufficient to accommodate the facing material.

This reduction in thickness is accomplished by placing a "setback' form in the regular wall forms.

Setback forms are usually made up of 2" framing lumber faced with ⅝" or ¾" plyform, and usually require 1¾ to 2¼ BF of lumber per sq. ft. of contact area, depending mainly upon the thickness of the setback.

The following is an itemized labor cost of 100 sq. ft. of setback forms requiring 2 BF of lumber per sq. ft. of contact area:

	Hours	Rate	Total	Rate	Total
Carpenter	6.5	$....	$....	$16.47	$107.06
Labor helping	3.3	....	....	12.54	41.38

Labor stripping	2.7	....	....	12.54	33.86
Cost 100 sq. ft			$....		$182.30
Labor cost per sq. ft				....	1.82

Forming A Continuous Haunch on Concrete Walls—When an intermediate support on a wall face is required to provide bearing for masonry, slabs or future construction and reducing the wall thickness is prohibited by the design, a continuous haunch is commonly used.

A Continuous haunch is usually formed of 2" framing lumber, faced with ⅝" or ¾" plyform and requires from 3½ to 4 BF of lumber per sq. ft. of contact area. The following is an itemized labor cost of 100 sq. ft. of continuous haunch forms:

	Hours	Rate	Total	Rate	Total
Carpenter	12.0	$....	$....	$16.47	$197.64
Labor helping	6.0	....	....	12.54	75.24
Labor stripping forms	2.5	....	....	12.54	31.35
Cost 100 sq. ft			$....		$304.23
Labor cost per sq. ft				....	3.04

Form Coating

For clean, easy stripping of forms and to prolong the life of form facings, the contact surface of all forms should receive a coating which will prevent concrete from bonding to the form face.

The old method of brushing or spraying form faces with paraffin oil is still used by some contractors and gives adequate results for the average job, such as foundation walls, unexposed columns, beams and slabs, etc.

Forms should always be coated before erection, but where this is impracticable, such as for built-in-place wall forms, pan and joist forms for combination slabs, etc., oiling should be done in advance of setting reinforcing steel and care should be taken not to get any oil on concrete, masonry or steel bearing surfaces.

On large areas, such as flat slab decks, a laborer can coat 400 to 500 sq. ft. of surface per hour, but production is considerably less when coating forms for columns, beams, combination slabs, etc. A fair overall average for coating forms is about 300 to 350 sq. ft. per hour.

For better class work, when using plywood, plastic-coated plywood or lined forms, where smooth concrete surfaces are required, there are numerous patented coatings and lacquers available for this purpose. While the initial cost of these coatings is much higher than paraffin oil, better results are obtained and the forms may be used a couple of times without recoating, depending on the care used in applying the coating and in handling the forms when erecting or stripping.

Prices for these coating materials vary from $5.00 to $13.00 per gallon, but coverage and durability also vary. Some coatings require special thinners before application while others can be thinned with comparatively inexpensive mineral spirits.

Application may be either by brushing, spraying or dipping and a man can coat about 400 to 500 sq. ft. of form surface per hour.

Formfilm*—Formfilm is a quick-drying plastic type plywood form coating with a high nitro-cellulose content. It is completely water-repellent, resistant to abrasion, and impervious to the alkaline action of concrete. It is recommended for use on untreated plywood forms to produce smooth concrete surfaces free from grain markings. It is not intended for use on metal or plastic forms.

Formfilm must be applied to concrete forms which are free from oil, grease, dirt or foreign matter. It may be applied by brush, spray, roller, squeegee or dipping.

---

*W. R. Grace & Co.

Formfilm is not ready mixed because job temperatures vary widely and affect the viscosity. It is usually thinned with Formfilm Thinner to the proper working consistency.

Concrete Column Forms.

Reinforced concrete columns may be either square, rectangular, or round in shape. The forms for square or rectangular columns are usually constructed of wood while metal or fiber molds are used for round columns.

12" MAX.

3/4" plywood with vertical 2 x 4 cleats.

Height of column in feet	STORY HEIGHTS 10	12	14	16	18	20
	33	33	33	33	33	33
	27	27	27	27	27	27
	24	24	24	24	24	24
	21	21	21	21	21	21
10	9 / 6	20	20	20	20	20
12		13 / 6	19	19	19	19
14			18 / 6	18	18	18
				15	17	17
16				9 / 6	16	16
					15	15
18					6	12
						12
20						6

Each horizontal line represents a column clamp and numbers between lines indicate suggested spacing in inches.

Column Clamp Spacing, in Inches, Based on Strength of Materials and Limited by ⅛" Deflection.

Forms for reinforced concrete columns should be estimated by the sq. ft. obtained by multiplying the girth of the column by the height. Example: obtain the number of sq. ft. of forms in a column 18" square and 12'-0" high. Proceed as follows: 18+18+18+18=72" or 6'-0", the girth. Multiply the girth by the height, 6'-0"x12'-0"=72 sq. ft. of forms.

The forms for all square or rectangular columns should be estimated in this manner.

To prevent the column forms from spreading while the concrete is placed and until it hardens, wood or metal clamps should be run around the column at intervals, depending upon the height of the column.

In recent years the use of wood clamps has been almost entirely eliminated by the use of metal clamps.

The pressure of the fresh concrete has a greater effect on the design of column forms than on wall forms on account of the smaller amount of concrete to be placed. The contractor usually finds it more convenient and economical to fill a group of medium size columns in as short time as possible, perhaps 30 minutes, when the full liquid pressure of the fresh concrete will be acting: namely, 150 lbs. per sq. ft. for every foot in depth of the concrete.

Column Form Design

There are two principal methods of erecting column forms; one is to unite all the four sides on two or three horses near the site where the form is to be erected and have 4 or 6 laborers hoist the complete form over the column dowels and stand it up temporarily; or in case of heavier columns, a simple 4"x6" pole derrick may be used for erection. Sometimes 3 sides of the column form are united and shoved into place without lifting the heavy box over the dowels, while the other and decidedly more expensive method is to erect each column form side separately.

As an example of the lumber required for a column form, take a concrete column 20" square and 11'-0" high underneath the lowest beam, and assuming a full liquid pressure of 150 lbs. per cu. ft. acting for the entire height of the column. Assume the clear story height as 12'-6". The area of the column form as usually taken off by the estimator will be 4 sides x 1.667 x 12'-6" high equals 83.33 sq. ft.

Using steel column clamps the following lumber would be required for a 20"x20" concrete column form 12'-6" high.

	Feet, B.M.
2 pcs. ¾" plywood, 1'-9½"x8'-0"	32
2 pcs. ¾" plywood, 1'-9½"x4'-6"	18
2 pcs. ¾" plywood, 1'-8"x8'-0"	32
2 pcs. ¾" plywood, 1'-8"x4'-6"	18
Total plywood required for 83.33 sq. ft. of forms sq. ft.	100
20 pcs. 1x6x1'-9" cleats	18
Lumber required for 83.33 sq. ft. forms	118
Lumber required per sq. ft.	1.42
Bracing lumber per sq. ft. forms	.58
Total lumber required per sq. ft. forms	2.00

In order to brace the column form and hold it plumb it will require 4 pcs. 2"x4"-16'-0" to 20'-0" long, which will add another 48 BF to the lumber required. This bracing lumber may be used at least 3 or 4 times and eventually, for other purposes than column forms. Bracing lumber is included in the above table.

The number of steel column clamps required may be somewhat larger than usually recommended by the manufacturers depending upon the speed with which the column forms are filled. The spacing in the lower part of the column should be about 9"; about 1'-0" near the middle and the first three spacings from the top should be 1'-3" to 1'-8" apart.

An average of one clamp per foot of column height will be safe in this particular instance.

Inasmuch as plywood is furnished in panels 4'-0" wide and 8'-0" long, there is considerable waste unless columns are of sizes that will cut from the plywood sheets without waste, which is unusual.

Another method of stiffening plywood column form panels is by using 2"x4" vertical cleats. This method permits greater spacing of clamps as shown in the table and the total lumber requirements for a 20" column using ¾" plywood sheathing and 2"x4" vertical cleats would be as follows:

	BF
Plywood sheathing	100
Cleats	112
Bracing	48
Lumber required for 83⅓ sq. ft. forms	260
Lumber per sq. ft. of forms	3.1

When round concrete columns are used, metal or fibre molds are used for supporting the wet concrete until it has "set." Some columns consist of a straight shaft extending from floor to ceiling, while others have a cone head at the top of the column. The forms for round concrete columns should be estimated at a certain price per column, stating whether columns have plain shafts or cone heads, as it costs considerably more to erect the latter. There are several firms in the country who make a specialty of renting these column molds and will quote a lump sum price on the job.

Labor Cost of Wood Forms for Square Concrete Columns.—The labor cost of wood forms for square or rectangular concrete columns will average about the same regardless of the type of building, but the variations are due to the method used in assembling and building up the column section, their size, method of clamping, etc.

On jobs of any size it is customary to have a mill on the job, where the column forms are built to exact size, including cut-outs for beams or girders, and then sent to the floor on which they are to be used, ready for erection.

Here is an important item when figuring column form costs. How many times can they be used on the job without cutting down or extending their length or must they be reworked for each floor? This is a matter each estimator or contractor must decide for himself from a study of the plans and for this reason costs are given on making up the column forms and separate labor costs covering the erection.

Labor Cost of 100 Sq. Ft. Concrete Column Forms Using Plywood
Sheathing with Adjustable Steel Clamps and 2"x4" Vertical Cleats

Making Forms	Hours	Rate	Total	Rate	Total
Carpenter	4.0	$....	$....	$16.47	$ 65.88
Labor	2.0	....	....	12.54	25.08
Assembling, Erecting, Plumbing and Bracing					
Carpenter	3.0	....	....	16.47	49.41
Labor helping	3.0	....	....	12.54	37.62
Removing Clamps and Forms					
Labor	2.0	....	....	12.54	25.08
Cost per 100 sq. ft			$....		$203.07
Labor cost per sq. ft				....	2.03

Roos Adjustable Metal Column Clamps
Hinged Bar Type—2 Hinged Units=1 Complete Clamp

Clamp Size	Bar Size	Weight Per Clamp	Net Concrete Sizes			
			With 1" Column Lumber		With 2" Column Lumber	
			Minimum	Maximum	Minimum	Maximum
36"	5/16" x 2 1/2" x 36"	40 lbs.	8 1/2" x 8 1/2"	26" x 26"	6 1/2" x 6 1/2"	24" x 24"
48"	3/8" x 2 1/2" x 48"	58 lbs.	12" x 12"	38" x 38"	10" x 10"	36" x 36"
60"	3/8" x 3" x 60"	85 lbs.	22 1/2" x 22 1/2"	48 1/2" x 48 1/2"	20 1/2" x 20 1/2"	46 1/2" x 46 1/2"
36" x 60"	1 Leg, 5/16" x 2 1/2" x 36" / 1 Leg, 3/8" x 3" x 60"	65 lbs.	7 1/2" (36" Leg) / 23 1/2" (Leg)	25" (36" Leg) / 49 1/2" (60" Leg)	5 1/2" (36" Leg) / 21 1/2" (60" Leg)	23" (36" Leg) / 47 1/2" (60" Leg)

Metal Forms for Round Columns

The cost of metal forms for round concrete columns will vary according to the diameter of the column, height, whether a plain shaft or one having a conical head, also whether it is necessary to cut out for beams and girders framing into the metal column molds.

Round Metal Column Molds

Where it is necessary to cut the metal molds or "heads" to frame into wood forms for concrete beams and girders, an extra labor allowance should be made.

When plain round metal forms up to 24" in diameter are used without caps, which do not require cutting and fitting at head for concrete beams and girders, the labor erecting, removing, and oiling one column mold should cost as follows:

	Hours	Rate	Total	Rate	Total
Carpenter	2	$....	$....	$16.47	$32.94
Laborer	2	....	....	12.54	25.08
Total per column			$....		$58.02

If the column has a conical head, necessitating placing a metal cap in addition to the column shaft, the labor erecting, removing, and oiling one column should cost as follows:

	Hours	Rate	Total	Rate	Total
Carpenter	2.5	$....	$....	$16.47	$41.18
Laborer	2.5	....	....	12.54	31.35
Total per column			$....		$72.53

If necessary to cut and fit at column heads for concrete beams and girders, add 1 to 2 hrs. time per column.

The above labor costs are for columns up to 16'-0" in length. For columns 16'-0" to 20'-0" long, add 1 to 2 hrs. time.

Leasing or Renting Metal Column Molds from the Manufacturers.—Where metal column molds are leased or rented by the manufacturer to the contractor to

Cu. Ft. of Concrete and Sq. Ft. of Forms Required Per Lin. Ft. of Height For Round and Square Concrete Columns

Diam. in Inches	Round Columns			Square Columns			Square in Inches
	Area Sq. In.	Cu. Ft. Conc.	Sq. Ft. Forms	Area Sq. In.	Cu. Ft. Conc.	Sq. Ft. Forms	
12	113	0.78	3.16	144	1.00	4.00	12
13	133	0.92	3.40	169	1.18	4.33	13
14	154	1.07	3.67	196	1.36	4.67	14
15	177	1.22	3.93	225	1.57	5.00	15
16	201	1.40	4.20	256	1.78	5.33	16
17	227	1.57	4.45	289	2.05	5.67	17
18	255	1.77	4.71	324	2.25	6.00	18
19	284	1.97	4.97	361	2.52	6.33	19
20	314	2.18	5.24	400	2.78	6.67	20
21	346	2.40	5.50	441	3.06	7.00	21
22	380	2.64	5.76	484	3.36	7.33	22
23	416	2.88	6.10	529	3.67	7.67	23
24	452	3.14	6.30	576	4.00	8.00	24
25	491	3.41	6.54	625	4.34	8.33	25
26	531	3.69	6.81	676	4.70	8.67	26
27	573	3.97	7.08	729	5.05	9.00	27
28	616	4.27	7.33	784	5.43	9.33	28
29	661	4.59	7.60	841	5.85	9.67	29
30	707	4.91	7.86	900	6.25	10.00	30
31	755	5.24	8.08	961	6.68	10.33	31
32	804	5.58	8.38	1024	7.10	10.67	32
33	855	5.94	8.67	1089	7.56	11.00	33
34	908	6.30	8.90	1156	8.02	11.33	34
35	962	6.68	9.17	1225	8.50	11.67	35
36	1018	7.06	9.44	1296	9.00	12.00	36

be used on a specified job and returned in good condition upon completion of the work, the following prices will prove a fair average for columns up to 24" in diameter.

For a plain metal mold for round concrete columns without cone head, the monthly rental should figure about $7.50 per lin. ft. of column form to be rented.

For plain round metal forms for concrete columns having cone head, add $45.00 per column per month rental.

To the above prices add delivery charges to the project site.

Cu. Ft. of Concrete in Flared Caps For Round Concrete Columns

Diam. of Col. Cap at Top	Diam. of Col. Cap at Bottom	Cu. Ft. Conc. Including Column Diam.	Cu. Ft. Conc. Excluding Column Diam.
4'-0"	12"	9.75	8.5
4'-0"	14"	9.70	8.0
4'-0"	16"	9.60	7.6
4'-0"	18"	9.40	7.0
4'-0"	20"	9.30	6.5
4'-0"	22"	9.00	5.8
4'-0"	24"	8.80	5.3
4'-0"	26"	8.70	4.9
4'-6"	16"	13.6	11.2
4'-6"	18"	13.5	10.6
4'-6"	20"	13.3	10.0
4'-6"	22"	13.1	9.3
4'-6"	24"	12.9	8.6
4'-6"	26"	12.6	7.8
4'-6"	28"	12.3	7.1
4'-6"	30"	11.9	6.4
5'-0"	20"	18.2	14.3
5'-0"	22"	18.0	13.4
5'-0"	24"	17.8	12.6
5'-0"	26"	17.4	11.7
5'-0"	28"	17.1	10.9
5'-0"	30"	16.8	10.0
5'-0"	32"	16.4	9.1
5'-0"	34"	15.8	8.2
5'-0"	36"	15.3	7.3

Sonotube* Fibre Forms.—Sonotube fibre forms are spirally wound, laminated fibre tubes which were developed to provide a fast and economical method of forming for round concrete columns, piers, etc., and particularly adaptable as forms for encasing wood or steel piling. They provide a fast, economical method of forming all types of round concrete columns. They also permit new design possibilities for architects, provide new structural features for engineers, and save time, labor and money.

Sonotube fibre forms are available in standard sizes from 8" to 48" inside diameter, with wall thickness ranging from .200" to .500"

They may be obtained in any length up to 48 ft. and may be easily cut to length on the job with an ordinary hand or power saw.

---

*Sonoco Products Company, Hartsville, S.C.

There are three types of Sonotube fibre forms to meet virtually any combination of job requirements. The seamless is the premium form with specially finished inner ply which produces a smoother, continuous concrete surface. Regular A-coated is the standard form for exposed columns. For exposed columns not exceeding 12 ft. in height, the light wall A-coated is recommended.

Sonotube fibre forms are lightweight, easily handled and require few men for erection manually or with simple block and tackle. Minimum bracing is required, only fixing in place at bottom and light lumber bracing to keep them erect and plumb. No clamping is necessary.

Under normal conditions, for average size columns, 12" to 18" diameter, 10-ft. to 14-ft. in length, a carpenter and a helper should handle, cut to length, erect and brace about 12 Sonotube column forms per 8-hr. day at the following labor cost per column:

	Hours	Rate	Total	Rate	Total
Carpenter	0.7	$....	$....	$16.47	$11.53
Helper	0.7	....	....	12.54	8.78
Cost per column			$....		$20.31

Sonotube fibre forms strip easiest and most economically from two to ten days after concrete is placed. Stripping may be accomplished with the least effort by using an electric hand saw, with blade set to cut slightly less than the wall thickness of Sonotube, and make two or three evenly spaced vertical cuts from bottom to top, after which form may be pulled free from column. Sonotube forms are for one time use only.

Under normal conditions, a carpenter and a helper should strip 25 to 30 average size Sonotube column forms per 8-hr. day at the following labor cost per column:

	Hours	Rate	Total	Rate	Total
Carpenter	0.3	$....	$....	$16.47	$4.94
Helper	0.3	....	....	12.54	3.76
Cost per column			$....		$8.70

Approximate Lineal Foot Prices of Sonotube Fibre Forms with "A" Coating

Inside Dia. and Wall Thickness	Price Per Lin. Ft.	Approx Wt. Per Foot
8" x .200"	$ 1.45	1.7
10" x .225"	2.00	2.3
12" x .225"	2.50	2.8
14" x .250"	3.10	3.6
16" x .300"	4.00	4.8
18" x .300"	5.10	5.4
20" x .375"	6.25	7.5
22" x .375"	7.00	8.2
24" x .375"	8.00	8.9
26" x .375"	9.25	9.6
28" x .375"	10.25	10.4
30" x .400"	11.50	11.8
36" x .400"	15.00	14.5
42" x .500"	28.00	18.6
48" x .500"	36.00	23.5

Sonotube forms are manufactured in twenty-two plants in the United States, and distributors maintain stocks in most principal cities. The above prices are f.o.b. the nearest shipping point.

Courtesy Symons Corporation

Symons adjustable column forms for poured-concrete high-rise construction

Anchoring Brick, Stone and Terra Cotta to a Concrete Backing

There are a number of methods of anchoring brick, stone and terra cotta to concrete, some of which consist of metal slots nailed to the column or beam forms, which form a slot in the concrete for placing anchors.

The dovetail anchor slot method of anchoring masonry to concrete consists of a metal slot dovetail in cross section and anchors with dovetail ends for cut stone, brickwork and terra cotta.

The dovetail anchor slot is nailed to the form with the open side against the wood, the ends are then closed with a wood plug or they may be obtained with foam filled slots. After the concrete is poured and the forms removed, the slot remains in the concrete available at any course height to receive the anchors.

The anchors are installed by merely inserting the dovetail end of the anchor edgewise in the slot, then turning anchor crosswise of the slot with the hand, the other end of anchor being imbedded in the facing material or mortar joint. The dovetail slot holds the end of anchor securely, making any other means of fastening, drilling or boring unnecessary.

FORMS FOR CONCRETE BEAMS, GIRDERS, SPANDREL BEAMS, AND LINTELS

Forms for reinforced concrete beams, girders, lintels, etc., should be estimated by the sq. ft. obtained by adding the dimensions of the three sides of the beam and multiplying by the length. Example: a beam or girder may be marked 12"x24" on the plans but perhaps 6" of this depth is included in the slab thickness, so the form area should be obtained by taking the dimensions of the beam or girder from the underside of the slab, i.e., 18"+12"+18"=48" or 4'-0" the girth, which multiplied by the length gives the number of sq. ft. of forms in the beam, viz., 4'-0"x18'-0"=72 sq.ft.

Spandrel beams or those projecting above or below the slab, should be measured in the same manner. A good rule to remember when measuring beam forms is to take the area of all forms that come in contact with the concrete.

When preparing an estimate on formwork, beams and girders should always be estimated separately, that is, the columns should be taken off, then the beams, girders and lintels, followed by the floor slabs, stairs, etc.

As an example of the lumber required for beam and girder forms, assume an inside beam 12" wide by 1'-6" deep and 19'-0" long.

The contact area of this beam would be 2 sides 1'-6" deep and a bottom 1'-0" wide by 19'-0" long, 1'-6"+1'-6"+1'-0"=4 sq. ft. girth multiplied by 19'-0" long, equals 76 sq. ft. of forms requiring the following lumber:

	B.F.
Beam soffit, 1 pc. ⅝" plyform 12"x19'-0" S.F.	19
Beam sides, 2 pcs. ⅝" plyform 22"x19'-0" S.F.	70
Beam Bottom Joists 4 pcs. 3"x4"x10'-0"	40
Beam Side Template 4 pcs. 2"x4"x10'-0"	27
Beam Side Ledger 4 pcs. 2"x4"x10'-0"	27
Beam Side Studs 40 pcs. 2"x4"x1'-4"t	36
Shores 6 pcs. 4"x4"x10'-0"	80
Shores Cross Head (T) 6 pcs. 4"x4"x3'-0"	24
Braces 12 pcs. 1"x4"x2'-6"	10
Lumber required for 76 sq. ft. of forms	333
Lumber required per sq. ft. of "contact area"	4.4

When beams or girders are more than 2'-0" deep, the designs given under Concrete Wall Forms should be consulted, which gives spacing and size of cleats or studs required, together with bracing and form ties necessary.

Labor Cost of Wood Forms for Concrete Beams and Girders.

The labor cost on beam and girder forms will vary considerably on different jobs, due to the amount of duplication, size of beams and girders and the methods of framing and placing them.

On typical buildings where the same size beam and girders are used on several floors, so that the forms may be used several times with only a small amount of labor with each reuse, minimum labor costs should prevail. Also, larger beams are more economically framed than very small ones but one of the most important cost factors is the method of fabricating the units and whether they are built up in a job "shop" using power saws. These are factors that cut costs and the contractor attempting to compete without using the most up-to-date methods and equipment available is bound to suffer financially.

The following labor costs are based on efficient management, economical design

PLACED MONOLITHIC CONCRETE

2x4 LEDGER

2x4 @ 12" O.C.

2x4 TEMPLATE

3x4 OR 4x4 RUNNER JOISTS

4x4 OR 4x6

1x4 BRACES

⅝" PLYFORM DECK

3x4 OR 4x4 DECK JOISTS

⅝" PLYFORM BEAM SIDE

⅝" PLYFORM BEAM BOTTOM

4x4 WOOD PROP

Method of Framing Wood Forms for Concrete Beams.

and labor-saving tools and equipment. If the old hand-saw methods are to be used, increase the costs given 10 to 20 per cent.

It requires 4 to 4½ B.F. of lumber to complete 1 sq. ft. of beam and girder forms, which includes uprights or "shores", beam soffits, bracing, etc.

A carpenter should frame and erect 250 to 275 B.F. of lumber per 8-hr. day, at the following labor cost per 1,000 BF:

	Hours	Rate	Total	Rate	Total
Carpenter	29.0	$....	$....	$16.47	$477.63
Helper	14.5	....	....	12.54	181.83
Cost per 1,000 ft. b.m.			$....		$659.46

If laborers are not permitted to carry the lumber and forms for the carpenters, the "helper" time given above should be figured as carpenter time.

A laborer should remove or "strip" 900 to 1,000 BF of lumber per 8-hr. day, at the following labor cost per 1,000 BF:

	Hours	Rate	Total	Rate	Total
Labor	9	$....	$....	$12.54	$112.86

Labor Cost of 100 Sq. Ft. of Inside Beam and Girder Forms Requiring 4½ BF of Lumber Per Sq. Ft. of Forms, Using 4"x4" Wood Shores

Making Beam Soffits and Sides	Hours	Rate	Total	Rate	Total
Carpenter	3.50	$....	$....	$16.47	$57.65
Labor helping	1.75	....	....	12.54	21.95
Assembling, Erecting, Shoring and Bracing					
Carpenter	5.50	....	....	16.47	90.59
Labor helping	2.75	....	....	12.54	34.49
Removing Forms					
Labor	2.75	....	....	12.54	34.49
Cost per 100 sq. ft			$....		$239.17
Labor cost per sq. ft				....	2.39

Labor Cost of 100 Sq. Ft. of Spandrel Beam or Lintel Forms Requiring 4 BF of Lumber Per Sq. Ft. of Forms, Using 4"x4" Wood Shores

Making Beam Soffits and Sides	Hours	Rate	Total	Rate	Total
Carpenter	3.50	$....	$....	$16.47	$ 57.65
Labor helping	1.75	....	....	12.54	21.95
Assembling, Erecting, Shoring and Bracing					
Carpenter	8.00	....	....	16.47	131.76
Labor helping	4.00	....	....	12.54	50.16
Removing Forms					
Labor	3.50	....	....	12.54	43.89
Cost per 100 sq. ft			$....		$305.41
Labor cost per sq. ft				....	3.05

If same forms may be reused without working over and cutting down, "Labor Making Forms" after first use may be reduced or omitted.

Forms for Upturned Concrete Beams.—When the top of a beam is higher than the floor slab level, the beam side forms for that portion above the floor must be supported on temporary legs which are removed after the concrete is poured and is still wet. It is also necessary to use spreaders to maintain the proper beam width.

While it is true that the additional form material required is negligible, the labor cost on this kind of beam side form will run 50 to 100 percent more than ordinary beam side forms.

FORMS FOR REINFORCED CONCRETE FLOORS

Forms for all types of reinforced concrete floors should be estimated by the sq. ft. taking the actual floor area or the area of the wood or metal forms that come in contact with the concrete.

Estimating the Cost of Form Lumber.— When computing the lumber cost per sq. ft. of forms, the estimator should study the plans carefully to determine just how many times it is possible to use the same form lumber in the construction of the building.

On 8 to 12 story buildings it may be possible to use the same lumber 4 to 5 times while on a 3 story building it might be necessary to buy sufficient lumber to form almost the entire building. These are items that affect the material costs and can only be determined by the estimator or contractor as they should be in position to know just how soon the forms can be stripped and how fast they will be required to carry on the work.

On many jobs it will be more economical to use high early strength cement instead of ordinary portland cement, which will enable you to strip forms in 3 days instead of 7 to 10 days, thus saving in the number of sets of forms required.

Form lumber that has been used 3 or 4 times will have little salvage value but each contractor should determine this for himself.

Types of Reinforced Concrete Floors.—There are a number of distinct types of reinforced concrete floor construction which require wood forms, centering and shores. Because of the variations in costs, some of the more commonly used types will be described in detail.

Flat Slab Construction.—Mill, factory, warehouse and industrial buildings designed for heavy floor loads are usually of flat slab construction, which consists of floors without beams or girders, except spandrel beams between columns on outside walls, and around stair wells, elevator shafts, etc. Buildings of this type usually have square or rectangular columns.

Beam and Girder Type Construction. —This is the oldest type of concrete structure, having square or round columns with beams and girders running between each row of columns and the floor slabs, carried by the beams and girders and then to the columns.

Pan Construction with Concrete Joists.—Apartment buildings, hotels, schools, hospitals, department stores, office buildings and other structures requiring light floor loads ordinarily use this type of construction, which consists of metal pans in conjunction with concrete joists. This type of floor is used with either round or square columns and beams that run at right angles to the joists.

Designing Forms for Reinforced Concrete Floors and Estimating Quantity of Lumber Required

The process of designing forms for reinforced concrete floors may be divided into several parts, such as determining the span of the sheathing; size and spacing of joists and girders or stringers, and determining the size and spacing of the upright supports or shores.

The spacing of joists for concrete floor forms is governed by the strength of the plywood used.

Forming Reinforced Concrete Floors Using Adjustable Shores.

When plywood panels are used for sheathing, the spacing of the joists must coincide with the sizes of the plywood panels, 4'-0"x8'-0", which means that the joists should be spaced 12", 16" or 19" on centers, depending upon the load to be carried.

Weight of Concrete Floors of Various Thicknesses

In determining the weight per sq. ft. of floor, wet concrete is figured at 150 lbs. per cu. ft. Dead load of form lumber and live load on forms while concrete is being placed is figured at 38 lbs. per sq. ft. which is sufficient for temporary construction.

Floor Thickness	2"	3"	4"	5"	6"	7"	8"	9"	10"	11"	12"
Conc. Wt. per Sq. ft., Lbs	25	37.5	50	62.5	75	87.5	100	112.5	125	137.5	150
Temp. Live Load per Sq. ft., Lbs.	40	37.5	38	37.5	38	37.5	38	37.5	38	37.5	38
Total Weight per Sq. Ft., Lbs	65	75.0	88	100.0	113	125.0	138	150.0	163	175.0	188

Table Giving Spacing of Stringers and Shores When Supporting Floor Forms for Metal Pan and Concrete Joist Construction

Using 2" plank to support the concrete joists 2'-1" or 2'-11" on centers, with stringers spaced 4'-0" apart to support the 2" plank.

TABLE FOR DESIGNING SLAB FORMS FOR REINFORCED CONCRETE FLOORS
BASED ON USING ¾" SHEATHING, WITH JOISTS SPACED 2'-0" ON CENTERS

Thickness of Concrete Floor Slab and Total Dead and Live Load per Sq. Ft.

Thickness Concrete Slab.............. Total Load per Sq. Ft., Lbs	2" 65	3" 75	4" 88	5" 100	6" 113	7" 125	8" 138	9" 150	12" 188
Distance Between Stringers When Joists are Spaced 2'-0" on Centers									
2"x4" Joists, S4S.............................	6'-0"	5'-6"	5'-6"	5'-0"	5'-0"	4'-6"	4'-6"	4'-0"	4'-0"
2"x6" Joists, S4S.............................	7'-6"	7'-0"	7'-0"	6'-6"	6'-6"	6'-0"	6'-0"	6'-0"	5'-6"
2"x8" Joists, S4S.............................	9'-6"	9'-0"	8'-6"	8'-6"	8'-0"	8'-0"	7'-6"	7'-6"	7'-0"
2"x10" Joists, S4S...........................	11'-0"	10'-6"	10'-0"	10'-0"	9'-6"	9'-6"	9'-0"	9'-0"	8'-6"
Distance Between Shores or Supports Using Stringers of Size and Spacing as Given Below									
4"x4" S4S Stringers, 5-ft. c.c.	5'-0"	4'-6"	4'-0"	4'-0"	4'-0"	3'-6"	3'-6"	....	....
4"x6" S4S Stringers, 6-ft. c.c.	6'-6"	6'-0"	6'-0"	6'-0"	6'-0"	5'-6"	5'-0"	5'-0"	4'-6"
3"x8" S4S Stringers, 6-ft. c.c.	8'-6"	7'-6"	7'-0"	7'-0"	7'-0"	6'-6"	6'-0"	6'-0"	5'-6"
2"x10" S4S Stringers, 7-ft. c.c.	7'-6"	6'-0"	5'-6"	4'-6"	4'-6"	4'-0"	....	....	....

The distance between shores or span of stringers is generally an even fraction of their length. Stringers 14'-0" long will permit a spacing of 4'-8" or 7'-0", while 18'-0" stringers will permit spans of 3'-7", 4'-6", 6'-0" and 9'-0".

Based on using yellow pine, Douglas fir or woods of equal strength.

The table gives the sizes of stringers and the distance between supports or shores when supporting various floor loads.

Size of Stringer	Thickness of Floor in Inches................................ Weight of Floor per Sq. Ft.*	6+2½ 86 lbs.	8+2½ 92 lbs.	10+2½ 98 lbs.	12+2½ 105 lbs.	14+2½ 112 lbs.
		Distance Between Shores or Supports				
4"x 4"S4S	....	5'-0"	4'-9"	4'-8"	4'-8"	4'-6"
3"x 6"S4S	....	6'-6"	6'-5"	6'-3"	6'-1"	6'-0"
4"x 6"S4S	....	7'-0"	7'-0"	6'-8"	6'-7"	6'-6"
3"x 8"S4S	....	8'-0"	8'-0"	7'-6"	7'-6"	7'-6"
2"x10"S4S	....	8'-0"	8'-0"	7'-6"	7'-6"	7'-6"

*Includes dead weight of concrete and 38 lbs. per sq. ft. for live load during construction. Joists 5" wide, 20" wide pans.

The above tables are based on using yellow pine, Douglas fir or woods of equal strength.

Data on Plywood Sheathing

When plywood is used for sheathing instead of ¾" boards, the following loads may be safely carried.

ALLOWABLE PRESSURES ON PLYFORM
CLASS I FOR ARCHITECTURAL APPLICATIONS
(deflection limited to 1/360th of the span)

FACE GRAIN ACROSS SUPPORTS
ALLOWABLE PRESSURES (psf) (a)

Support Spacing (inches)	PLYWOOD THICKNESS (inches)					
	½	⅝	¾	⅞	1	1-⅛
4	2935	3675	4495	4690	5075	5645
8	970	1300	1650	1805	1950	2170
12	410	575	735	890	1185	1345
16	175	270	370	475	645	750
20	100	160	225	295	410	490
24			120	160	230	280
32					105	130
36						115

(a) Plywood continuous across two or more spans.

FACE GRAIN PARALLEL TO SUPPORTS
ALLOWABLE PRESSURES (psf) (a)

Support Spacing (inches)	PLYWOOD THICKNESS (inches)					
	½	⅝	¾	⅞	1	1-⅛
4	1670	2110	2610	3100	4140	4900
8	605	810	1005	1190	1595	1885
12	215	360	620	740	985	1165
16		150	300	480	715	845
20		105	210	290	400	495
24			110	180	255	320

(a) Plywood continuous across two or more spans.

ALLOWABLE PRESSURES ON PLYFORM CLASS I
(deflection limited to 1/270th of the span)

FACE GRAIN ACROSS SUPPORTS ALLOWABLE PRESSURES (psf) (a)

Support Spacing (inches)	PLYWOOD THICKNESS (inches)					
	½	⅝	¾	⅞	1	1-⅛
4	2935	3675	4495	4690	5075	5645
8	970	1300	1650	1805	1950	2170
12	430	575	735	890	1185	1345
16	235	325	415	500	670	770
20	135	210	285	350	465	535
24		110	160	215	295	340
32					140	170
36					105	120

(a) Plywood continuous across two or more spans.

FACE GRAIN PARALLEL TO SUPPORTS ALLOWABLE PRESSURES (psf) (a)

Support Spacing (inches)	PLYWOOD THICKNESS (inches)					
	½	⅝	¾	⅞	1	1-⅛
4	1670	2110	2610	3100	4140	4900
8	605	810	1005	1190	1595	1885
12	270	405	620	740	985	1165
16	115	200	375	525	715	845
20		125	210	290	400	495
24			135	185	255	320

(a) Plywood continuous across two or more spans.

Carrying Capacity of 4"x4" Uprights or Shores.—The carrying capacity of a given size upright or shore is generally limited by two principal factors. The first is compression at right angles to the fibers which the shore exerts on the cross beam or stringer. The stress per sq. in. on the bearing area of the stringer on the shore should not greatly exceed 500 lbs. per sq. in. for Yellow Pine, Western Fir or similar lumber, otherwise there is a noticeable impression of the shore into the fibers of the stringer, especially when the lumber is green or water soaked after a long rain. This often amounts to ⅛". Hence a (4"x4", S4S), 3½"x3½" upright should never be loaded to more than 6,000 lbs., no matter how short it is, and a 4"x4" rough shore should not be loaded to more than 8,000 lbs.

The other principal limitation of strength is due to the length of the shore and the degree of crookedness in its length. Every experienced contractor knows that shores often show bows of 1" to 2" and if it is further considered that the shores are not cut off exactly square to their length, the carrying capacity of most shores are greatly limited by the eccentric loading impressed on them by the uneven bearing of the stringer and the bow in the shores.

The following table gives the permissible load on shores for various lengths and eccentricities:

Safe Load in Pounds on Uprights or Shores of Various Lengths

Length of Upright or Shore	4"x4" S4S Yellow Pine or Fir		4"x4" Rough Yellow Pine or Fir		4"x4" Rough Wisconsin Hemlock	
Eccentricity	8'-0"	10'-0"	8'-0"	10'-0"	8'-0"	10'-0"
(Bow+Top) 0"..................	6,000	6,000	8,000	8,000	8,000	8,000
(Bow+Top) 1"..................	5,700	4,800	8,000	7,100	7,300	5,600
(Bow+Top) 1½"..............	4,700	3,900	7,300	5,800	6,000	5,100
(Bow+Top) 2"..................	4,200	3,400	6,700	5,500	5,200	4,400
(Bow+Top) 2½"..............	3,600	3,000	5,300	4,500	4,800	4,100
(Bow+Top) 3"..................	3,400	2,800	5,100	4,300	4,500	3,800

Lintels as a rule produce quite large eccentricities on the shores due to the load from the adjoining slab, which comes through on the inside of the lintel forms onto the shore, therefore contractors should never allow a load larger than those given for 3" eccentricity for shores supporting lintels. One frequently sees lintels supported by shores spaced less than 2'-0" on centers, which is a clear waste of money.

Where the height of the uprights or shores are longer than given in the above table, it is advisable to cross brace the shores, in which case the vertical distance between braces plus 40 percent, may be taken as the post height. Example: Using a 16'-0" shore, with 2 rows of cross bracing at 6'-0" and 12'-0" above the floor. Distance between braces 6'-0" plus 40 percent, or 2.40 ft. making the capacity the same as a shore 8.40 or practically 8'-6" long. Thus an 8'-0" shore good for 8,000 lbs. would still be good for 8,000 lbs. when 16'-0" long, if cross braced by two rows of bracing 6'-0" and 12'-0" above the floor.

Method of Constructing Forms for Beam and Girder Type Solid Concrete Floor Slabs

Estimating the Quantity of Lumber Required for Forms for Beam and Girder Type Solid Concrete Floor Slabs.—The accompanying illustrations give detailed designs for forms from which the lumber required per sq. ft. of forms may be computed.

The accompanying illustration shows a 4" concrete floor slab having a 7'-6" clear span between beams. Using a joist spacing of 2'-0" as given in the table for joist spacing, 2"x8" joists will carry the required load.

For a floor area 2'-0"x7'-6" containing 15 sq. ft. the following lumber will be required:

	Feet, B.M.
Sheathing, 2'-0"x7'-6"=15 sq. ft. plus 20% waste	18
Joist, 1 pc. 2"x8"-8'-0", plus 10% for joints	12
Ledger boards, 2 pcs. 1"x4"-2'-0"	1
Total lumber required for 15 sq. ft. forms	31
Feet of lumber, b.m. required per sq. ft. of forms	2.1

For a 6" solid concrete slab having a clear span of 11'-0" between girders, ordinarily it is not economical to support joists having a 11'-0" span directly on the girders and it is better practice to provide a stringer or girt half-way between the concrete beams so that the joists have a span of only 5'-6". Using a joist spacing of 2'-0", a 2"x6" joist is sufficiently strong, as shown by the table of joist spacings. The same table shows that a 4"x6" supported by shores 6'-0" on centers will suffice.

A floor area 2'-0"x11'-0" containing 22 sq. ft. will require the following lumber:

	Feet, B.M.
Sheathing, 2'-0"x11'-0"=22 sq. ft. plus 20% waste	27
Joists, 1 pc. 2"x6"-12'-0", plus 10% for joints	13
Stringer, 1 pc. 4"x6"-2'-0"	4
Shore, 1/3, of a 4"x4"-10'-0", S4S, plus sill, wedges, 18÷3=	6
Total lumber required for 22 sq. ft. forms	50
Feet of lumber, b.m. required per sq. ft. of forms	2.25

If shores require bracing, add 0.1 ft. b.m. per sq. ft. of slab.

Assuming a solid concrete floor slab 16'-0" wide, 19'-0" long and 12" thick, containing 304 sq. ft. of slab forms.

Joists will be 2"x6" spaced 24" apart, supported by 3 lines of 3"x8"-16'-0" stringers, which in turn are supported by 4"x4" shores 5'-4" apart.

The following lumber will be required for 304 sq. ft. of forms:

Method of Constructing Forms for Beam and Girder Type Solid Concrete Floors

	Feet, B.M.
Sheathing, 16'-0"x19'-0" 304 sq. ft. plus 20% waste	365
Joists, 9 pcs. 2"x6"-14'-0"	126
Joists, 9 pcs. 2"x6"-7'-0"	63
Stringers, 3 pcs. 3"x8"-16'-0"	96
Shores, 9 pcs. 4"x4"-10'-0", including sills and wedges	153
Braces, 6 pcs. 1"x6"-16'-0"	48
Joist support at girders, 2 pcs. 1"x6"-16'-0"	16
Total lumber required for 304 sq. ft. forms	867
Lumber required per sq. ft. of forms, b.m.	2.85

The above quantities are based on the assumption that the ends of joists and stringers are supported by the beam and girder forms. If a wall bearing job without beams and girders, the following additional lumber will be required:

	Feet, B.M.
Stringers, 2 pcs. 2"x8"-16'-0" at wall	43
Shores, 6 pcs. 4"x4"-10'-0", including sills and wedges	102
Braces, 2 pcs. 1"x6"-16'-0"	16
Additional lumber required for 304 sq. ft. forms	161
Additional lumber, b.m. required per sq. ft.	0.53

Method of Constructing Forms for Flat Slab Concrete Floors

Estimating the Quantity of Lumber Required for Flat Slab Concrete Floor Forms.—Assume a flat slab having a 20'-0" span, with 7'-0" square drop heads, 8" thick as shown in illustration.

The maximum span of the joists is 5'-9" and the table of Joist Spacing shows that 2"x6" joists are sufficiently strong; also that 3"x8" stringers are sufficient. The load on the shores of the center stringer will be 5'-9"x6'-0"=34.5 sq. ft. at 138 lbs. per sq. ft. equals 4,761 lbs. Good practice requires a 4"x4" shore rough, although a 4"x4", S4S, with bracing will do.

The lumber required for a 20'-0"x20'-0" panel is as follows:

	Feet, B.M.
Sheathing, 400 sq. ft. plus 20% waste	480
Using 16'-0" long floor sheathing boards will require 9 joists (a double joist being placed at the joint) or for 20 ft. it will require 11¼ pcs 2"x6"-12'-0" joists (1¼x9)	135
Column Strips, 9 joists 2"x6"-10'-0"	90
Stringers, 3 pcs. 3"x8"-20'-0"	120
Stringers, 1 pc. 3"x8"-12'-0"	24
Shores, 13 pcs. 4"x4"-10'-0" plus sills, wedges, etc. @ 17 ft.	221
Horizontal bracing, 7 pcs. 1"x6"-20'-0"	70
X Bracing, not less than 1/3 horizontal bracing	23
Total lumber required for 400 sq. ft. forms	1,163

For Drop Heads

Joists, 5 pcs. 2"x6"-14'-0"	70
Stringers, 2 pcs. 2"x8"-8'-0"	21
Shores, 4 pcs. 4"x4" at 17 ft. b.m.	68
Total lumber required including drop heads	1,322
Feet of lumber, b.m. required per sq. ft. of forms	3.31

A 10'-0" shore is sufficient for a 12'-0" story height. For a 14'-0" story height it will be necessary to add 2'-0" to each of 17 shores or 34x1.33 ft.=45 ft. b.m. which adds 0.11 ft. b.m. per sq. ft. of floor. A 16'-0" story height will require 0.23 ft. b.m. per sq. ft. of floor.

For an 18'-0" story height, it will be necessary to add for another set of cross bracing in addition to the 0.33 ft. of lumber, b.m. required for shores. Another set of cross bracing will add 93 ft. of lumber b.m. for 400 sq. ft. of floor or 0.23 ft. b.m. additional, which with the 0.33 ft. required for shores, will increase the lumber required per sq. ft. to 3.31+.33+.23=3.87 ft. b.m. per sq. ft. of forms.

A 20'-0" story height will add another 0.11 ft. for extra length of shores, or a total of 3.98 ft. b.m. per sq. ft. of forms.

Where plywood is used for floor sheathing, there is at least a 10 percent waste in flooring, and while its use saves 480 ft. b.m. for sheathing, it is necessary to use more joists, because the joists should not be spaced over 19" on center, and the joists must be doubled every 8'-0" instead of 16'-0" where 16'-0" long sheathing boards are used.

In 400 sq. ft. of forms, the following additional lumber would be required:

	Feet, B.M.
Extra joists, 4 pcs. 2"x6"-12'-0"	48
Extra joists, 2 pcs. 2"x6"-10'-0"	20
Total extra lumber required for 400 sq. ft. forms	68
Extra lumber required per sq. ft. of forms	0.17

If adjustable shores are used instead of 4"x4" shores, stringers must be supported every 4'-0" in order to reduce the load to not much more than 3,000 lbs. per shore although some adjustable shores will carry up to 6,000 lbs. This will require 17 to 23 adjustable shores instead of 17 pcs. of 4"x4" shores (247 ft. b.m.) and 9 pcs. of horizontal bracing, 1"x6"-20'-0" (90 ft. b.m.) instead of 7 pcs. of 1"x6"-20'-0" (70 ft. b.m.) and the X bracing will also be increased by 7 ft. b.m.

Using adjustable shores, 4"x4" stringers can be used instead of 3"x8" stringers, which will result in the following saving:

		Feet, B.M.
Shores, 17 pcs. 4"x4"-10'-0"		247
Stringers, 4"x4" instead of 3"x8", saving		44
Total saving using adjustable shores		291
Extra 1"x6" required for bracing	27	
Extra sills, 6 pcs. 2"x10"-2'-0"	20	47
Total saving on 400 sq. ft. forms		244
Total saving per sq. ft. of forms b.m.		0.61

On the other hand it will be necessary to handle 17 to 23 shores weighing 1,000 to 1,400 lbs. while the 224 ft. saving weighs only 725 lbs.

For a 12" thick flat slab having a 20'-0" span, the lumber required can be approximated as follows:

	Feet, B.M.
Sheathing, per sq. ft forms	1.20
Joists, same spacing but using 2"x8" joists	0.75
Stringers, same as before	0.33
Shores, spacing will be 4'-6", requiring 21 shores	0.90

Bracing, 9 pcs. 1"x6"-20'-0" 0.23
X bracing, ⅓ of horizontal bracing... 0.08
Drop heads, same as before.. 0.25
    Total lumber, b.m. per sq. ft. for 12'-6" story height.................................. 3.74
    For each 2 ft. in story height, more or less, add or deduct per sq. ft......... .14

Weight of Combination Metal Pan and Concrete Joist Floors

Based on concrete joists 4", 5" and 6" wide, with a 2½" concrete slab on top, using metal pans of various depths and widths.

Concrete Thickness, Inches	2½"	2½"	2½"
Width of Concrete Joists, Inches	4"	5"	6"
Using	Weight per Sq. Ft. of Slab. Pounds		
Metal Pans 6" deep, 20" wide	45	48	50
Metal Pans 6" deep, 30" wide	41	43	45
Metal Pans 8" deep, 20" wide	51	54	57
Metal Pans 8" deep, 30" wide	45	48	50
Metal Pans 10" deep, 20" wide	56	60	64
Metal Pans 10" deep, 30" wide	50	52	55
Metal Pans 12" deep, 20" wide	63	67	72
Metal Pans 12" deep, 30" wide	54	57	61
Metal Pans 14" deep, 20" wide	..	75	80
Metal Pans 14" deep, 30" wide	..	62	66

Method of Constructing Forms for Floor of Combination Metal Pans and Concrete Joist Construction

For each ½" variation in slab thickness, add or subtract 6.25 lbs. per sq. ft. of floor.

Add 38 lbs. per sq. ft. to the above weights to take care of temporary dead and live load on floors while concrete is being poured.

Lumber Required for Floor Forms Using Metal Pans and Concrete Joist Construction.—Assume a slab of 19'-0" clear span and 16'-0" long, using 8" metal pans with 5" wide ribs spaced 2'-1" on centers and a 2½" concrete top over the entire slab.

According to the above table this construction weighs 52 lbs. per. sq. ft. and including a 38 lb. live load during construction, makes a total load of 90 lbs. per sq. ft.

It is customary to use 2"x8", S4S, planks to support the concrete ribs or joists. If the planks are continuous without patching at the end spans, a span of 4'-0" will support a load of 265 lbs. per lin. ft. with a deflection of .235 in. In this instance the load is 2.09x90 lbs. per sq. ft.=188 lbs. per lin. ft. which reduces the deflection to 0.166" or about 1/6", which is permissible.

It will require a stringer or girt every 4'-0" to support the planks and the load per lin. ft. on the stringers will be 360 lbs.

A 2"x10"-16'-0", S4S, will easily carry this uniform load per lin. ft. when supported by 4"x4" shores spaced 8'-0" apart.

The following lumber will be required for 304 sq. ft. of forms:

	Feet, B.M.
Planks, 9 pcs. 2"x8"-20'-0"	240
Stringers, 4 pcs. 2"x10"-16'-0"	107
Shores, 8 pcs. 4"x4"-10'-0", including sills and wedges	136
Joist supports at girders, 2 pcs. 1"x6"-16'-0"	16
Bracing, 4 pcs. 1"x6"-16'-0"	32
Bracing, 2 pcs. 1"x6"-20'-0"	20
Total lumber required for 304 sq. ft. forms	551
Lumber required per sq. ft. of forms, b.m.	1.8

The above quantities are based on the assumption that the ends of the joists and stringers are supported by the beam and girder forms. If a wall bearing job without beams or girders, requiring additional stringers and shores at walls, add the following:

	Feet, B.M.
Stringers, 2 pcs. 2"x10"-16'-0"	54
Shores, 4 pcs. 4"x4"-10'-0", including sills and wedges	68
Bracing, 2 pcs. 1"x6"-16'-0"	16
Additional lumber required for 304 sq. ft. forms	138
Additional lumber required per sq. ft. of forms	0.45

Lumber Required for Slab Forms.—The following tables give the approximate quantity of lumber required per sq. ft. of forms for floors of different thickness and with varying ceiling heights. These tables are only approximate as every building is different and the quantities will vary with the span, ceiling height, slab thickness and weight of the wet concrete to be supported.

Also, the following tables are based on using 4"x4" shores or uprights. If adjustable shores are used it will be necessary to deduct the lumber required for 4"x4"'s and add cost or rental of adjustable shores. Add cost of additional bracing as more shores and bracing are frequently required.

The difference in the quantity of lumber required per sq. ft. for floors having 10'-0" ceiling heights and those having 20'-0" ceiling heights is principally in the shoring and bracing. If adjustable shores are used, deductions can be made on the following basis: Floors 3" to 6" thick having 10'-0" ceiling heights, require ⅓, to ⅜ ft. of lumber per sq. ft. for shores; 7" to 9" slabs require ½ to ⅝ ft.; and 10" to 12" slabs require ⅝, to ¾ ft. of shoring lumber per sq. ft. of floor.

The above quantities can be used for all 10'-0" story heights, plus the additional lumber given in the tables for higher ceiling heights.

The forms are designed in yellow pine, Douglas fir, or woods of equal strength, with the loads computed as follows: wet concrete 150 lbs. per cu. ft., dead load of form lumber and live load on forms while concrete is being placed at 38 lbs. per sq. ft.

Number of Feet of Lumber Required for One Sq. Ft. of Flat
Slab Concrete Floor Forms

Ceiling Height in Feet	Thickness of Slab in Inches			
	6"	8"	10"	12"
	113 lbs.	138 lbs.	163 lbs.	188 lbs.
10	3.0	3.1	3.4	3.6
12	3.1	3.2	3.5	3.7
14	3.3	3.4	3.6	3.9
16	3.4	3.5	3.9	4.0
18	3.4	3.8	4.2	4.4
20	3.5	3.9	4.3	4.6

The above quantities include lumber required for forming drop heads.

If adjustable shores are used, deduct lumber quantities given in descriptive matter on previous pages.

Using Steel Truss Joists Instead of Shoring for Supporting Formwork for Reinforced Concrete Floors.—In steel frame buildings where clear floor space is essential, steel truss joists may be used for supporting the wood formwork, as shown in the accompanying illustration.

Steel Joists Used for Supporting Concrete Floor Forms

This is a simple method that is very effective as the salvage value of the truss joists is very high.

Sectional Steel Shoring.—Many contractors now use their sectional steel scaffolding units for shoring formwork for reinforced concrete slabs and beams. Prefabricated sectional steel scaffolding sections are as simple to erect for shoring as they are for scaffolding. The same scaffolding components—base plates, adjustable extension legs, welded 5-ft. wide frames varying in height from 3'-0" to 6'-6" and diagonal braces to provide spacing between frames from 2'-6" to 7'-0"—are assembled quickly to provide free standing shoring sections suited to the easy placing of stringers, joists and decking. In most instances, sectional steel shoring can be used wherever conventional shoring methods of adjustable shores or 4"x4" lumber are used, but is especially adaptable for shoring forms for large expanses of flat slab construction.

Height adjustments are quickly made by the use of 20" adjustable legs, which eliminates the need for cutting, fitting and wedging. These legs also facilitate stripping operations by the ease with which the shoring units may be lowered.

Sectional steel shoring also provides scaffolding within the shoring for forming and stripping thereby increasing job safety. This method also minimizes the use of wood reducing the fire hazard and increasing the number of reuses obtainable from form lumber.

Safe working loads for sectional steel shoring are dependent on the following: (1) spacing between frames in the same row; (2) distance between rows of frames and (3) position of the stringers on the frames. Sectional steel shoring will support the largest loads when the stringers are placed directly over the frame legs by means of inverted base plates or U- heads as may be seen in the illustration. If this is not practicable, stringers should be placed as close as possible to frame legs. If required, additional lateral rigidity may be attained by interlacing frames with standard bracing.

It is available in three types: Standard, 20K Heavy Duty for loads up to 10,000 pounds on each frame leg, and Extra Heavy Duty for loads up to 40,000 pounds on each leg.

The contractor, who does not own this equipment, may rent sectional steel shoring, engineered for a specific job, from the manufacturer on the same basis as scaffolding. This service is available nation-wide.

Since job conditions and requirements vary considerably for different projects, it is practically impossible to give unit labor production values on this method of shoring, but various contractors who have used this method, report savings up to 30 percent in the labor and materials-handling costs over conventional methods. With shoring labor, including bracing, ranging from 15 to 25 percent of the total form cost, this results in overall savings of 4½ to 7½ percent on the entire form cost.

Adjustable Shores for Reinforced Concrete Floors

There are a number of adjustable shores on the market to support reinforced concrete beam and slab forms. These shores are a combination of wood and metal, or all metal, which can be raised or lowered within certain limits.

Adjustable Shores:—Adjustable shores in a complete unit, as illustrated are made in 3 sizes: 8 ft. adjustable to 14 ft.; 7 ft. adjustable to 13 ft.; 5 ft. adjustable to 9 ft.

They may be purchased for about $35.00 for the smaller size and $45.00 for the larger sizes. Extension type shores are also available in two sizes: 8 ft. adjustable to 14 ft., at about $40.00 each; and 6 ft. adjustable to 10 ft. 6 in., at about $38.00 each. They can be easily extended by inserting a 4"x4" of any length in the top of the shore.

Rental stocks are maintained in most principal cities and may be rented at about $2.00 per month, per section, f.o.b. warehouse and return.

Each shore will support about 3,000 lbs. with an adequate factor of safety. In some cases, however, shores are spaced to carry a load of not over 2,200 lbs., as some forming authorities consider it cheaper to use more shores than to use heavier lumber to prevent excessive deflection between supports.

Adjustable Shores Versus 4"x4"s.—It is much easier to shore up a floor using adjustable shores instead of 4"x4"s, which require wedging and cutting or adding to for each change in story height. This additional labor all costs money at present wage scales. However, to offset this, the original cost of the adjustable shores is much more than the 4"x4"s but if they can be used often enough they will eventually show a saving. This will have to be considered when contemplating a purchase or rental basis.

Sectional Shoring Supporting Forms for Concrete Cap of Reservoir

Method of Shoring for Reinforced Concrete Construction Using Sectional Steel
Scaffolding Components

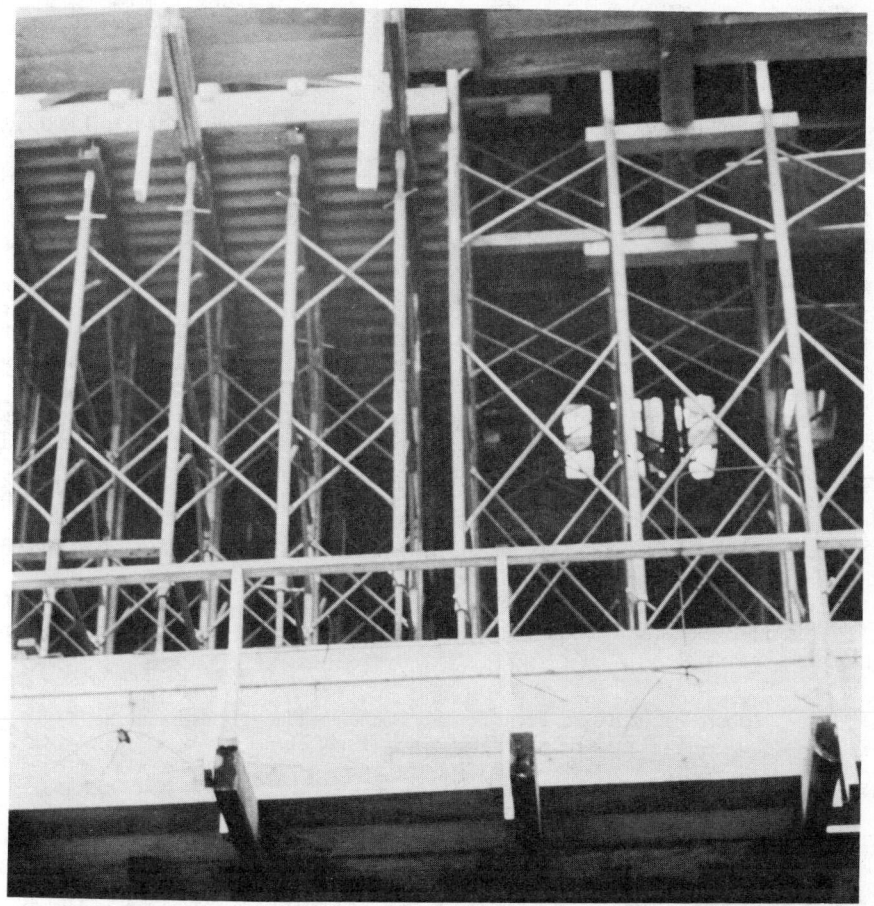

Courtesy Symons Corporation

Adjustable shores used to support reinforced concrete forms

Another thing that should be considered is the fact that a 4"x4" shore will usually carry a larger load than an adjustable shore, so in many cases it will require more adjustable shores than 4"x4"s to carry the same load. On the other hand, especially on light constructed floors, the adjustable shores may carry all the load required. This is an item that should be considered by the estimator before pricing the job.

The form costs on the following pages are computed on a rental basis, based on 10 days' time for each shore or 3 uses a month.

Labor cost of Flat Slab Forms.—Forms for flat slab floors are continuous—without breaks or offsets—except at column heads, stair wells, elevator shafts, etc., where it is necessary to frame for beams and girders.

The sq. ft. labor cost will vary with the amount of lumber required per sq. ft. of forms, as the heavier the slab and the higher the ceiling, the more lumber required per sq. ft.

Forms for exceedingly light floor slabs cost proportionately more per sq. ft. than heavy forms as the preliminary work is the same with both types. However, on work of this class a carpenter should frame and erect 375 to 425 BF of lumber per 8-hr. day, at the following labor cost per 1,000 BF:

	Hours	Rate	Total	Rate	Total
Carpenter	20	$....	$....	$16.47	$329.40
Labor	10	....	....	12.54	125.40
Cost per 1,000 BF		....	$....		$454.80

When removing or "stripping" forms, a laborer should remove 1,000 to 1,200 BF of lumber per 8-hr. day, at the following labor cost per 1,000 BF:

	Hours	Rate	Total	Rate	Total
Labor	7	$....	$....	$12.54	$87.78

Labor Framing for Slab Depressions at Column Heads.—On flat slab jobs there may be a depression at each column head 5'-0" to 7'-0" square and 4" to 8" deep. The labor cost framing for these depressions will cost about 80 percent more than for straight slab forms.

A satisfactory method of estimating this additional labor is to figure the entire floor area at the regular price and then compute the area of the depressions and figure them at 80 percent additional labor.

Example: Floor contains 10,000 sq. ft. of slab forms, including 40 column head depressions 5'-0"x5'-0", containing 25 sq. ft. each.
10,000 sq. ft. slab forms $2.00 per sq. ft. $20,000.00
Add for 40 depressions (5'-0"x5'-0") 1,000 sq. ft. at 80%
     more than plain slab forms or $1.60 per sq. ft. 1,600.00

This makes a total of $3.60 per sq. ft. for labor framing depressions at column heads.

Labor Cost of 100 Sq. Ft. of Flat Slab Concrete Floor Forms Requiring 3 BF of Lumber per Sq. Ft. of Forms

4"x4" Shores

	Hours	Rate	Total	Rate	Total
Carpenter	7.00	$....	$....	$16.47	$115.29
Labor	3.50	....	....	12.54	43.89
Labor removing forms	2.25	....	....	12.54	28.22
Cost per 100 sq. ft			$....		$187.40
Labor cost per sq. ft			....		1.87

Adjustable Shores

	Hours	Rate	Total	Rate	Total
Carpenter	6.00	$....	$....	$16.47	$ 98.82
Labor	3.00	....	....	12.54	37.62
Labor removing forms	2.25	....	....	12.54	28.66
Cost per 100 sq. ft			....		$164.66
Labor cost per sq. ft			....		1.65

For story heights over 16'-0", add 25% labor costs.
†Add for additional labor framing drop heads.

Labor Cost of 100 Sq. Ft. of Flat Slab Forms Requiring 4 Ft. of Lumber Per Sq. Ft.

4"x4" Shores

	Hours	Rate	Total	Rate	Total
Carpenter	9.0	$....	$....	$16.47	$148.23
Labor	4.5	....	....	12.54	56.43
Labor removing forms.	3.0	....	....	12.54	37.62
Cost per 100 sq. ft			$....		$242.28
Labor cost per sq. ft			....		2.42

	Adjustable Shores				
	Hours	Rate	Total	Rate	Total
Carpenter	8.0	$....	$....	$16.47	$131.76
Labor	4.0	....	....	12.54	50.16
Labor removing forms.	3.0	....	....	12.54	37.62
Cost per 100 sq. ft			$....		$219.54
Labor cost per sq. ft				....	2.20

For story heights over 16'0", add 25% to labor costs.
†Add for additional labor framing drop heads.

Forms for Beam and Girder Type Solid Concrete Floors

The labor cost of wood forms for beam and girder type floors will run somewhat higher than flat slab floors on account of the shorter spans, additional framing around beams, girders, etc.

On forms of this class, a carpenter should frame and erect 325 to 375 BF of lumber per 8-hr. day, at the following labor cost per 1,000 BF:

	Hours	Rate	Total	Rate	Total
Carpenter	23.0	$....	$....	$16.47	$378.81
Labor	11.5	....	....	12.54	144.21
Cost per 1,000 ft. b.m			$....		$523.02

An experienced laborer should remove or "strip" 1,000 to 1,200 BF. of lumber, per 8-hr. day, at the following labor cost per 1,000 BF.:

	Hours	Rate	Total	Rate	Total
Labor	7	$....	$....	$12.54	$87.78

Number of Feet of Lumber Required for One Sq. Ft. of Beam and Girder Type Solid Concrete Floor Forms

Ceiling Height in Feet	Thickness of Slab in Inches							
	3" 75 lbs.	4" 88 lbs.	5" 100 lbs.	6" 113 lbs.	7" 125 lbs.	8" 138 lbs.	10" 163 lbs.	12" 188 lbs.
10	2.0	2.1	2.4	2.5	2.6	2.6	2.7	2.7
12	2.1	2.2	2.4	2.5	2.6	2.6	2.7	2.8
14	2.1	2.2	2.4	2.5	2.6	2.6	2.7	2.8
16	2.2	2.3	2.5	2.6	2.7	2.7	2.8	2.9
18	2.4	2.5	2.6	2.7	2.8	2.8	2.9	3.0
20	2.5	2.6	2.7	2.8	2.9	2.9	3.0	3.1

Labor Cost of 100 Sq. Ft. of Slab Forms Between Beam and Girder Forms Requiring 2 Ft. of Lumber Per Sq. Ft. of Floor

	4"x4" Shores				
	Hours	Rate	Total	Rate	Total
Carpenter	5.0	$....	$....	$16.47	$ 82.35
Labor	2.5	....	....	12.54	31.35
Labor removing forms.	1.5	....	....	12.54	18.81
Cost per 100 sq. ft			$....		$132.51
Labor cost per sq. ft				....	1.33

	Adjustable Shores				
	Hours	Rate	Total	Rate	Total
Carpenter	4.6	$....	$....	$16.47	$ 75.76
Labor	2.3	....	....	12.54	28.84
Labor removing forms	1.5	....	....	12.54	18.81
Cost per 100 sq. ft			$....		$123.41
Labor cost per sq. ft				....	1.23

Labor Cost of 100 Sq. Ft. of Slab Forms Between Beam and Girder
Forms Requiring 2½ Ft. of Lumber Per Sq. Ft. of Floor

4"x4" Shores

	Hours	Rate	Total	Rate	Total
Carpenter	6.30	$....	$....	$16.47	$103.76
Labor	3.15	....	....	12.54	39.50
Labor removing forms	1.90	....	....	12.54	23.83
Cost per 100 sq. ft			$....		$167.09
Labor cost per sq. ft			....		1.67

Adjustable Shores

	Hours	Rate	Total	Rate	Total
Carpenter	5.70	$....	$....	$16.47	$ 93.88
Labor	2.85	....	....	12.54	35.74
Labor removing forms	1.90	....	....	12.54	23.83
Cost per 100 sq. ft			$....		$153.45
Labor cost per sq. ft			....		1.54

Labor Cost of 100 Sq. Ft. of Slab Forms Between Beam and Girder
Forms Requiring 3 Ft. of Lumber Per Sq. Ft. of Floor

4"x4" Shores

	Hours	Rate	Total	Rate	Total
Carpenter	7.6	$....	$....	$16.47	$125.17
Labor	3.8	....	....	12.54	47.65
Labor removing forms	2.2	....	....	12.54	27.59
Cost per 100 sq. ft			$....		$200.41
Labor cost per sq. ft			....		2.00

Adjustable Shores

	Hours	Rate	Total	Rate	Total
Carpenter	6.8	$....	$....	$16.47	$112.00
Labor	3.4	....	....	12.54	42.64
Labor removing forms	2.2	....	....	12.54	27.59
Cost per 100 sq. ft			$....		$182.23
Labor cost per sq. ft			....		1.82

For story heights over 16'-0", add 25% to labor costs.

Forms for Floors of Metal Pan and Concrete Joist Construction

Wood forms for floors of metal pan and concrete joist construction may be of much lighter construction than solid concrete floors, as the dead load is less.

Metal pans are usually furnished 20" to 30" wide, with 2"x6" or 2"x8" plank spaced 24" to 37" on centers, depending upon the width of the joists, which vary from 4" to 7".

The open deck construction is ordinarily used owing to the wide spacing of the floor boards.

In buildings having one-way beams and long floor spans, a carpenter should frame and erect 350 to 400 BF of lumber per 8-hr. day, at the following labor cost per 1,000 BF:

	Hours	Rate	Total	Rate	Total
Carpenter	21.4	$....	$....	$16.47	$352.46
Labor	10.7	....	....	12.54	134.18
Cost per 1,000 BF			$....		$486.64

Method of Constructing Wood Slab Forms for Floors of metal Pans and Concrete Joist Construction

Floors having beams and girders running in both directions with the floor panels averaging 16'-0"x16'-0" and smaller, a carpenter should frame and erect 325 to 375 BF of lumber per 8-hr. day, at the following labor cost per 1,000 BF:

	Hours	Rate	Total	Rate	Total
Carpenter	22.8	$....	$....	$16.47	$375.52
Labor	11.4	....	....	12.54	142.96
Cost per 1,000 ft. b.m			$....		$518.48

An experienced laborer should remove or strip 1,000 to 1,200 BF. of lumber per 8-hr. day, at the following labor cost per 1,000 BF:

	Hours	Rate	Total	Rate	Total
Labor	7	$....	$....	$16.47	$87.78

Number of Feet of Lumber Required for One Sq. Ft. of Floor Forms for Metal Pan and Concrete Joist Construction

The following table is based on the assumption the ends of joists and stringers are supported by the beam and girder forms.

Ceiling Height in Feet	Thickness of Floor Slab in Inches				
	8½" 85 lbs.	10½" 90 lbs.	12½" 100 lbs.	14½" 105 lbs.	16½" 115 lbs.
10	1.7	1.7	1.7	1.8	1.8
12	1.7	1.7	1.8	1.9	1.9
14	1.7	1.7	1.8	1.9	2.0
16	1.8	1.8	1.9	2.0	2.1
18	2.1	2.1	2.2	2.3	2.4
20	2.2	2.2	2.3	2.4	2.5

Number of Feet of Lumber Required for One Sq. Ft. of Floor Forms For Metal Pan and Concrete Joist Construction

The following table is based on the assumption it is a wall bearing job and will require wood girts or stringers together with supporting shores at both ends of joists and stringers.

Ceiling Height in Feet	Thickness of Floor Slab in Inches				
	8½" 85 lbs.	10½" 90 lbs.	12½" 100 lbs.	14½" 105 lbs.	16½" 115 lbs.
10	2.1	2.1	2.1	2.2	2.2
12	2.1	2.1	2.1	2.2	2.2
14	2.2	2.2	2.2	2.3	2.4
16	2.3	2.3	2.3	2.4	2.4
18	2.5	2.5	2.6	2.7	2.8
20	2.6	2.6	2.7	2.8	2.9

Using yellow pine, Douglas fir, or woods of equal strength.

Labor Cost of 100 Sq. Ft. of Metal Pan and Concrete Joist Slab Forms Requiring 1¾ Ft. of Lumber Per Sq. Ft.

Long spans and 1-way beams. For short spans and 2-way beams, add 10% to labor costs.

	4"x4" Shores				
	Hours	Rate	Total	Rate	Total
Carpenter	4.2	$....	$....	$16.47	$ 69.17
Labor	2.1	....	....	12.54	26.33
Labor removing forms	1.3	....	....	12.54	16.30
Cost per 100 sq. ft			$....		$111.80
Labor cost per sq. ft			....		1.12

	Adjustable Shores				
	Hours	Rate	Total	Rate	Total
Carpenter	3.8	$....	$....	$16.47	$ 62.59
Labor	1.9	....	....	12.54	23.83
Labor removing forms	1.3	....	....	12.54	16.30
Cost per 100 sq. ft			$....		$102.72
Labor cost per sq. ft			....		1.03

For story heights over 16'-0", add 25% to labor costs

Labor Cost of 100 Sq. Ft. of Metal Pan and Concrete Joist Slab Forms Requiring 2 Ft. of Lumber Per Sq. Ft.

Long spans and 1-way beams. For short spans and 2-way beams, add 10% to labor costs.

	4"x4" Shores				
	Hours	Rate	Total	Rate	Total
Carpenter	4.90	$....	$....	$16.47	$ 80.70
Labor	2.45	....	....	12.54	30.72
Labor removing forms	1.50	....	....	12.54	18.81
Cost per 100 sq. ft			$....		$130.23
Labor cost per sq. ft			....		1.30

Adjustable Shores

	Hours	Rate	Total	Rate	Total
Carpenter	4.30	$....	$....	$16.47	$ 70.82
Labor	2.15	....	....	12.54	26.96
Labor removing forms	1.50	....	....	12.54	18.81
Cost per 100 sq. ft			$....		$116.59
Labor cost per sq. ft				....	1.17

For story heights over 16'-0", add 25% to labor costs.

Labor Cost of 100 Sq. Ft. of Metal Pan and Concrete Joist Slab Forms Requiring 2½ Ft. of Lumber Per Sq. Ft.

4"x4" Shores

	Hours	Rate	Total	Rate	Total
Carpenter	6.0	$....	$....	$16.47	$ 98.82
Labor	3.0	....	....	12.54	37.62
Labor removing forms	1.8	....	....	12.54	22.57
Cost per 100 sq. ft			$....		$159.01
Labor cost per sq. ft				....	1.59

Adjustable Shores

	Hours	Rate	Total	Rate	Total
Carpenter	5.4	$....	$....	$16.47	$ 88.94
Labor	2.7	....	....	12.54	33.86
Labor removing forms	1.8	....	....	12.54	22.57
Cost per 100 sq. ft			$....		$145.37
Labor cost per sq. ft				....	1.45

For story heights over 16'-0", add 25% to labor costs.

Labor Cost of 100 Sq. Ft. of Metal Pan and Concrete Joist Slab Forms Requiring 3 Ft. of Lumber Per Sq. Ft.

4"x4" Shores

	Hours	Rate	Total	Rate	Total
Carpenter	7.0	$....	$....	$16.47	$115.29
Labor	3.5	....	....	12.54	43.89
Labor removing forms	2.2	....	....	12.54	27.59
Cost per 100 sq. ft			$....		$186.77
Labor cost per Sq. ft				....	1.87

Adjustable Shores

	Hours	Rate	Total	Rate	Total
Carpenter	6.4	$....	$....	$16.47	$105.41
Labor	3.2	....	....	12.54	40.13
Labor removing forms	2.2	....	....	12.54	27.59
Cost per 100 sq. ft			$....		$173.13
Labor cost per sq. ft				....	1.73

For story heights over 16'-0", add 25% to labor costs.

FORMS FOR REINFORCED CONCRETE STAIRS

The labor cost of framing and erecting concrete stair forms will vary with the type of stair, whether open or box string, straight runs from floor to floor or having intermediate landing platforms. Also, whether the stair is straight or winding and having square or bull nose treads and risers. The story height will also affect the labor costs on account of the shoring and bracing necessary to support the wet concrete.

It is rather difficult to estimate the cost of stair forms on a sq. ft. basis as it costs as much to frame a 3'-0" stair as one 4'-0" wide, while the latter contains one-third more forms per lin. ft. The better method is to allow a certain price per flight of stairs of each type as this will result in more accurate estimates than the sq. ft. method. However, where it is desirable to estimate the stair forms by the sq. ft. take the area of the soffit of the stairs and platforms. Example: a stair 4'-0" wide and 18'-0" long, contains 72 sq. ft. of forms.

Some contractors and estimators estimate the cost of stairs at a certain price per lin. ft. of riser, which includes forms, reinforcing steel, concrete and finishing.

Wood Forms for Straight Concrete Stairs.—On straight stairs extending from floor to floor without intermediate landing platforms and having an average story height of 10'-0" to 12'-0", two carpenters working together should lay out the stair, place rough stringers, mark off treads and risers on the rough string boards and set them in place in about 8 to 9 hours.

After the rough strings are in place it will require another 8 to 10 hrs. for 2 carpenters to sheath the stairs, cut, bevel and place risers, and place all necessary shoring and bracing ready for concrete.

The forms for an average flight of concrete stairs, 4'-0" wide and 18'-0" long, containing 72 sq. ft. of forms (soffit measurement) or 72 lin. ft. of risers, should cost as follows:

	Hours	Rate	Total	Rate	Total
Carpenter	36	$....	$....	$16.47	$592.92
Labor helping and removing forms	9	....	....	12.54	112.86
Cost per flight			$....		$705.78
Cost per sq. ft				....	9.80
Cost per lin. ft. riser				....	9.80
Cost per riser (18)				....	39.21

The lumber cost for this stair should run about as follows:

	BF	Rate	Total
Stringers, 2 pcs. 2"x12"-20'-0"	80	$0.36	$ 28.80
Risers, 18 pcs. 2"x8"-4'-0"	96	.36	34.56
Soffit sheathing, 4'-0"x18'-0"	72	.50	36.00
Joists supporting sheathing, 4 pcs. 2"x8"-18'-0"	96	.36	34.56
Shores or uprights, 6 pcs. 4"x4"-10'-0"	80	.36	28.80
Purlins or stringers, 3 pcs. 4"x6"-4'-0"	24	.36	8.64
Sills, wedges and bracing	30	.33	9.90
Cost per flight			$181.26
DCost per sq. ft. (72)			2.52

Wood Forms for Concrete Stairs Having Intermediate Landing Platforms. —Where the concrete stairs consist of two short flights with an intermediate landing platform between floors, the labor cost per story will run somewhat higher than for straight run stairs, on account of laying out and framing for two short flights instead of one long one.

Stairs of this type up to 4'-0" wide and 8'-0" to 10'-0" long from floor to platform, with each short flight containing 36 to 40 sq. ft. of forms (soffit measurement), and 8 to 10 risers, should cost about as follows for labor:

	Hours	Rate	Total	Rate	Total
Carpenter	23	$....	$....	$16.47	$378.81
Labor helping and removing forms	7	....	....	12.54	87.78
Cost per flight			$....		$466.59
Cost per sq. ft. (36)				....	12.96
Cost per lin. ft. riser (36)				....	12.96
Cost per riser (9)				....	51.84

Method of Constructing Wood Forms for Concrete Stairs

Another Method of Constructing Wood Forms for Concrete Stairs.

If the stairs are built between masonry walls and have their bearing in slots or chases left in the wall, it will require about 2 hrs. additional carpenter time cutting out stringers, if using a hand saw or ½ to ¾-hr. if using a power saw.

Winding stairs, stairs having bull nose treads and risers and other difficult or complicated construction may cost 2 or 3 times as much as given above. The estimator will have to use his judgment, depending upon detail of the stair.

When erecting formwork for concrete stairs over 4'-0" wide, it will be necessary to place additional "cut-out" stringers, to prevent the risers from bulging or giving way until the wet concrete has set. These "cut-out" stringers should be spaced 3'-0" to 4'-0" apart.

When figuring labor costs of landing platforms, double the labor cost of plain slab forms.

STEELFORMS FOR JOIST CONSTRUCTED FLOORS AND ROOFS

Where the same size steelforms are used on a number of floors, they are removed after the concrete has attained the desired strength and reused again on the same job until the work is complete. The removable steelforms are available and can be furnished by companies specializing in this type of work. These companies will erect and remove the steelforms with or without the supporting centering at a stipulated lump sum for the entire job.

Removable steelforms in flange and adjustable type are made from 14, 15 and 16 gauge metal that will stand hard usage. The flange type are nailed through the flanges while the adjustables are nailed through the sides. The flange type steelforms are furnished in 20" and 30" widths as standard widths in standard depths of 6", 8", 10", 12" and 14." The intermediate forms are in lengths of 1', 2' and 3' with either end caps, 6" ends or 1' straight ends as the closures. In addition to the straight ends, there are also single tapered end forms in 3' lengths for the various depths. Also available are adjustable straight side type forms that can be set to a depth of 8", 10", 12", 14" and 15". Likewise, these are furnished in standard 20" and 30" widths and have the same combination of 1', 2' and 3' intermediates, end forms and the 3' single tapered ends. For filling out spaces, filler forms are furnished in 10" and 15" widths matching the steelform depths as described. Tapered end forms are not furnished in the filler widths. Removable steelforms made from a lighter gauge metal do not withstand job abuse, particularly with men working over them before the concrete is placed and as a result, often dent badly so that it requires considerable extra concrete and gives a poor surface finish. In addition, there is added labor expense in reconditioning the forms prior to the next use.

For two-way concrete joist construction, one-piece metal domes are also available. The depths furnished in the 30" x 30" void are 8", 10", 12" and 14", and in the 19" x 19" void are 4", 6", 8", 10" and 12". Flanges for the 30/30 domes are 3" wide to make a 3 ft. module and for the 19/19 domes 2½" wide to make a 2 ft. module. The metal gauge ranges from 14 to 16 gauge.

The steelforms and steeldomes must be nailed to the wood supporting centering to hold them rigid and in line while the concrete is placed.

It is common practice to oil the forms prior to the installation of the reinforcing steel so that they can be removed easier from the ceiling and without concrete adhering to their surfaces.

Estimating Quantities of Steelforms—In estimating the area of floor and roof construction requiring removable or permanent forms, no deductions are to be made for beams or for tees of beams or for wide joists. Opening 50 square feet or

Method of Construcing Wood Forms for Concrete Stairs Having Intermediate
Platforms

over are to be deducted except when the company furnishing the steelforms pro-
poses to furnish the wood centering, in which case only openings 100 square feet or
over are to be deducted. Concrete joist construction may be supported by any one
of three types of structural systems and in addition to the aforementioned general
estimating rules the job area is subject to the following special rules: For a reinforced concrete frame, the areas are to be figured out-to-out of concrete frame; for a
structural steel frame, the areas are to be figured center to center spandrel beams;

and for bearing wall construction, the areas are to be figured clear inside brick walls plus a bearing on all walls of 6".

Labor Handling and Placing Steelforms—The labor cost of placing steelforms will vary with the type of form, type of building, quantity of equipment furnished, amount of horizontal and vertical movement of equipment, depressions, openings in the slabs, headers, etc., and the locality in which the work is being executed.

Steelforms are placed usually by carpenters and laborers. In a few localities, iron workers handle the steelforms from the point of removal from the concrete ceiling to the next point of steelform installation on the job. In most localities, however, laborers assist the carpenters and handle the steelforms to the next point of erection on the job ready for the carpenter to install the steelform to the wood centering. The local area practice of work assignment affects greatly the labor costs and should be the basis upon which the estimated costs are determined.

The following costs are based on the steelforms being placed by experienced men such as employed by companies specializing in the erection and removal of steelforms.

On straight run work using flange type steelforms such as factory buildings, garages, warehouses, etc. having typical spans without much cutting for headers, openings, etc., an experienced crew consisting of two (2) carpenters and four (4) laborers should handle and place steelforms over approximately 3,500 square feet of floor area in an eight hour day at the following direct labor cost per 100 square feet.

	Hours	Rate	Total	Rate	Total
Mechanic	.46	$. . . .	$. . . .	$16.47	$ 7.58
Labor	.91	. . . .	. . . .	12.54	11.41
Total Direct Labor Cost per 100 sq. ft.	1.37		$. . . .		$18.99
Direct Unit Labor Cost per sq. ft.			. . . .		.19

For school buildings, hospitals, apartments, office buildings, etc., where the concrete joist require fitting and adjusting around pipes, pipe sleeves, chases, conduits, the handling and placing costs of the flange type equipment could increase as much as 40–50 percent, depending on the amount of special work required.

Labor Handling and Removing or Stripping Steelforms

In removing or stripping flange type steelforms after the concrete has set, an experienced crew usually consisting of two (2) carpenters and three (3) laborers can remove the steelforms from the ceiling from about 4,000 sq. ft. of floor per 8 hr. day, at the following direct labor cost per 100 sq. ft.

Labor Handling and Removing or Stripping Steelforms

	Hours	Rate	Total	Rate	Total
Mechanic	.40	$. . . .	$. . . .	$16.47	$ 6.59
Labor	.60	. . . .	. . . .	12.54	7.52
Total Direct Labor Cost per 100 sq. ft.			$. . . .		$14.11
Direct Unit Labor Cost per sq. ft.			. . . .		.14

It should be noted that the above costs are based on flange type steelforms. The cost for handling and placing the one-piece dome is slightly higher. The cost for handling and placing the adjustable type steelform is about 33 1/3 percent higher.

Ceco Steelform Construction.*—Ceco Steelform Construction is a combination of concrete joist construction and thin top slabs. The steelforms are formed of 14, 15, or 16 gauge steel, depending on the width and depth of the form. These forms are placed on supporting wood centering and removed from the concrete after the concrete has reached sufficient strength. The type of forms made available to the industry by The Ceco Corporation are flange forms, the most commonly used type of form, followed by the one-piece metal dome, adjustable forms and the one-piece longforms.

The cost of the steelforms installed in a building will depend on a number of factors such as the type of steelform used, the size of the job, the number of reuses of the equipment, location, availability of the steelforms with respect to warehouse stocks, etc. Local working conditions also play a very important part in job costs.

The Ceco steelform service includes supplying the necessary steelforms and the labor for their erection and removal. Also offered in a complete service of furnishing and erecting all open centering for the support of the steelforms. In steeldome construction centering is usually included for the concrete solid slab areas that are of the same depth as the dome construction and within the floor areas containing the steeldomes. The quotations and contracts are generally on a lump sum basis for specific projects. It is always recommended that the steelform prices be obtained from a steelform supplier before finalizing and submitting a bid on this portion of the job costs.

FORMS OTHER THAN WOOD

Lightweight Steel Forming Material for Concrete.——High-strength corrugated steel forming material is often used for forming reinforced concrete floor and roof slabs. It comes either galvanized or uncoated and is manufactured from 100,000 p.s.i. tough-temper steel. It has a definite reliable structural strength nearly twice that of ordinary steel having equal weight. Used primarily in floor and roof systems having steel joists, junior beams, or purlins it is also used over pipe tunnels or similar installations where economies of using permanent forms can be realized. Either structural grade concrete or lightweight, insulating concrete may be used with corrugated steel forms, but in either case reinforcing bars or wire mesh should be added to satisfy flexure and temperature steel requirements.

flange type   adjustable type

longform type   steeldome type

*The Ceco Corporation

Method of Constructing Open Wood Centering and Placing Steelforms for Concrete Joist Floor Construction.

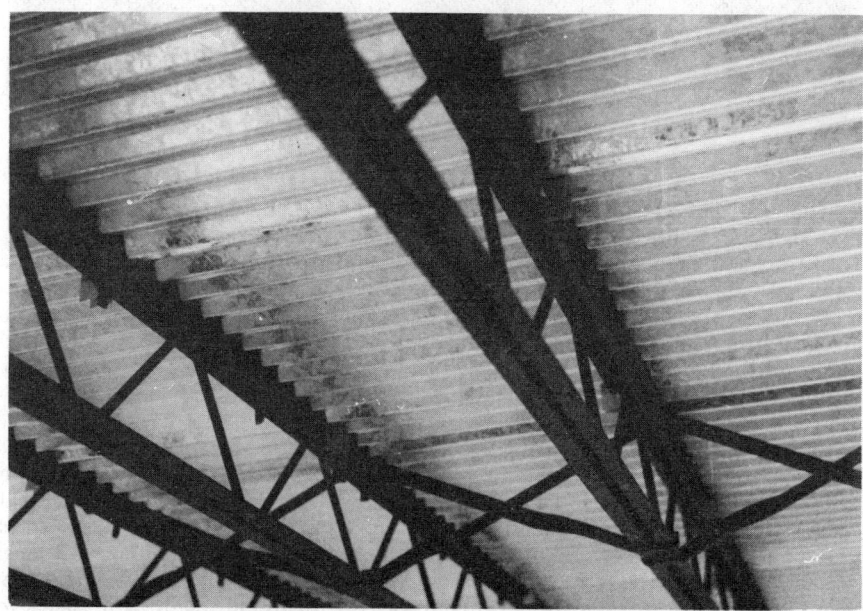

Placing Corrugated Steel Floor Panels.

Exposed Ceiling of Corrugated Steel Forms

Installing Corrugated Steel Forms

This material offers many advantages over other methods of flexible centering. The light, rigid sheets are quickly placed. No side pull is exerted on joists, instead, joist top chords are given lateral support by the stiffness of the material. Sheets in place provide a safe working platform. Mesh is easily and effectively placed. Elimination of sag between joists reduces concrete quantity. Uniform thickness of slab over joists and mid span permits monolithic finish. Little cleanup is required underneath.

For exposed construction it is available galvanized. For unexposed joist construction it is available uncoated.

Sheets are placed with corrugations normal to the supporting joists or beams with the end lap of sheets occurring over the joists or beams. The sheets are placed with edge lips up and are lapped a minimum of one corrugation with adjacent sheets. The ends of sheets should lap a minimum of 2". The sheets are fastened to the supporting members by means of clips, arc-welding or by nailing. For steel joists construction the most satisfactory method of attachment is arc-welding. The manufacturer can provide special curved washers for use in welding the sheets to the steel framework that provide fast, high-strength welds.

Estimating.—Corrugated steel forming material is sold by the square with the area determination based on sheet width times actual sheet length.

Erection Costs.—Corrugated steel forming material can be rapidly erected in all kinds of weather thereby providing cover to floors below. The corrugated sheets are large in area but extremely light in weight and can be easily handled by one man.

A well organized crew, working under favorable conditions, can place and fasten up to 10,000 square feet of corrugated steel forming material per eight hour day, but most jobs will average 3,000 to 5,000 square feet per day.

Using an average figure of 4,000 square feet per eight hour day for a four man crew, the labor costs for placing 1,000 square feet of corrugated steel forming material would be as follows:

	Hours	Rate	Total	Rate	Total
Ironworker	8	$. . . .	$. . . .	$17.70	$141.60
Cost per sq. ft			. . . .		.14

The above figures are based upon one man welding or fastening the deck in place and three men handling and placing the material. Costs do not include overhead and profit.

Computing Concrete Quantities.——When computing volume of concrete placed over corrugated steel decking, subtract ¼-in. from slab thickness to allow for corrugations of standard and ⅜-in. for heavy duty corrugated steel forming. Size of corrugations; standard, ½-in. depth x 2 3/16-in. pitch; heavy duty, ¾-in. depth x 3-in. pitch.

ARCHITECTURAL CONCRETE FORMWORK

Appearance is of paramount importance to the success of architectural concrete, therefore, every operation in the construction must be planned and executed with this in mind. The architect may design and specify but regardless of how detailed the specifications, it is the contractor and his superintendent who must have the know-how to detail and construct the forms and be on constant watch during all operations to see that the correct procedures are followed.

Forms for Architectural Concrete

Form design and construction must be approached with a different viewpoint from that for ordinary structural concrete. In structural concrete some leakage and irregularities in formwork may be tolerated, so long as the forms are strong enough to carry the weight of the concrete and the imposed loads and are built fairly true to line and grade. In architectural concrete, there must be no leakage through the forms and bulging and other irregularities must be so inconsequential as not to detract from the appearance of the finished surfaces.

Every conceivable precaution must be taken to prevent leakage through the forms. Chamfer strips at corners and edges are helpful for this purpose. Wherever possible, the forms should be designed so that pressure of the fresh concrete will tighten joints rather than tend to loosen them.

A common source of trouble is bulging of forms at external corners, resulting in both poor alignment and leakage with subsequent ragged corners, sometimes to the extent of exposing the aggregate at the corners.

The accompanying illustration gives one method of constructing the forms to provide locking of the corner. The wales are extended beyond the intersection far enough to permit two vertical strips to be nailed in the corners. Wedges are driven between these strips and the wales to tighten the corner.

Block slightly loose

Piece of 1"material nailed to
wale to support sheathing on
side of pilaster

Form Designed so that Pressure of Fresh Concrete Tightens It To Prevent
Leakage.

Double wales

Double wales

4"x4" Studs at
plywood joints

Plywood joint

Double wales

Double 2"x6" liners
at 8 to 10 ft. centers

Wedge

Form Construction To Provide Positive Corner Locking

Another method of locking the form corners is to use a tie instead of the wood
strips and wedges. Since there are no wedges to work loose, a set of wales can be set
and tightened on one level around the building without much chance of subsequent
loosening when tightening the wales above or below.

Various materials are used for forming architectural concrete to eliminate imperfections and irregularities in the finished surface. Committee 622 of the American
Concrete Institute has prepared a report entitled "Formwork for Concrete" which
discusses these forming materials and their application. Plastic coated plywood,
hardboard form lining, fiber or laminated pressed paper tubes, plaster waste molds,
metal forms, fiber glass forms, etc. are all used as forming material for architectural
concrete. Fiberglass has been used on large building projects with the forms being
reused as many as 60 or 70 times.

Plywood Form Sheathing.—Smooth surfaces required on much present day
work is obtained with plywood form sheathing. This should be of structural grade
made especially for form use and preferably should be oiled at the factory. If oiled
on the job, sufficient time should be allowed for the oil to penetrate and any surplus
oil should be wiped off before the plywood is used. Plywood ⅝" or ¾" thick can be
used directly against studs without backing when the outer plies are at right angles
to the studding. Structural plywood usually has five plies with the fibers in the two

ELEVATION

PLAN · SECTION

Form Layout for Use of Structural Plywood.

outer and middle plies in one direction and the fibers of the other two plies in the opposite direction. Thus the plywood is stronger and stiffer in one direction than the other. The result may not be serious on the first use of the plywood and under average placing speed, but with faster placing the deflection may be pronounced. With the second use of the plywood there will be noticeable deflection and on the third and additional uses, a deflection of ⅛" to ¼" can be expected with ¾" plywood on studs spaced 16" on centers.

Sometimes concrete form Presdwood is used as a form liner, which is usually nailed to decking or sheathing lumber as a backing. Board 3/16" thick should be used over solid backing while ¼" board may be used semi-structurally. Ordinary wood working tools are used in cutting and applying Concrete Form Presdwood.

Some manufacturers advertise that plywood may be used 10 times or more in the construction of formwork but where appearance is important, it is seldom possible to use it more than 3 or 4 times. However, it may be used for other purposes where appearance is not so important.

When using plywood for the facade or facing forms it must be thoroughly cleaned after each use. Where a special form coating is used to treat the plywood instead of oil, it should be applied with care and allowed to dry thoroughly.

Architectural concrete walls must be straight and plumb. This is essential. It means that the 2"x4" or preferably 2"x6" studs and wales must be of uniform depth and perfectly straight, otherwise the walls will be crooked—with the possibility of the work being rejected.

The top of the foundation walls on which the architectural concrete wall rests must be perfectly level and straight, using two 2"x6" wales or similar planks, braced at intervals of not less than 4'-0", in addition to all other bracing ordinarily required for foundation walls. It requires about 4½ BF of lumber per lin. ft. of wall for these wales and bracing. In order to hold the studs of the form for the architectural concrete wall in line, it is necessary to insert ⅝" greased stud bolts, spaced about 3'-0" on centers, into the top of the foundation wall, as illustrated. It requires about ⅛-hr. carpenter and 1/16-hr. labor time per lin. ft. of wall to brace the top of the foundation wall and to insert the bolts.

As a rule, it is necessary to obtain a perfect horizontal line of the concrete at the top of the foundation wall, which will be assured by nailing a 1"x2" strip on the outside form of the foundation wall and placing the concrete about 1" higher than the bottom of this strip. About two hours after placing the concrete, the surplus concrete (which generally contains a good deal of laitance), is scraped off to the level of the under side of the strip.

After the forms for the foundation walls have been stripped, the wales should be installed and blocked about 3'-0" apart against the foundation wall. The 2"x4" or 2"x6" studs should now be installed between wales and the wall. The studs should have 8d nails (protruding about 1") driven into the outside edge exactly 8" from the lower end and 6d nails (protruding less than ¾") exactly 8" from the lower edge. When inserting the studs between the wales and the foundation wall, the 8d nails form their support on the wales and the 6d nails furnish the support for the ¾" plywood. The studs are usually spaced 12" or 16" on centers but at every vertical joint in the plywood (usually 8'-0") a double stud or a stud 3½" wide should be used.

The studs are temporarily braced horizontally by nailing a 1"x6" board higher up and are braced vertically by wedging at the bottom between the studs and foundation wall. The first sheet of plywood is then inserted between the studs and the foundation wall and bolts, holding the wale tightened until the lowest part of the form is straight and plumb. The wales are spaced 24" vertically and the oiled tie bolts are spaced 2'-0", 2'-8" and 3'-4" apart, so that the holes do not interfere with the 16" stud spacing and are in proper relation to the ends of the 8'-0" long plywood sheets.

The arrangement of the holes for the tie bolts and the arrangement of the nails (6d box nails) by which the sheets of plywood are held to the studs are also shown in the accompanying illustration.

After 2 or 3 of the 4'-0" wide sheets of ¾" plywood have been fastened to the studs and the proper amount of horizontal walers put in place, the outside form is further strengthened by vertical wales consisting of 2 pcs. of 2"x6" spaced not more than 8'-0" on centers with about 2'-0" or more projecting over the top of the outside form. These wales are held in place and plumbed on top by two or more No. 9 or No. 10 wires, secured to column dowels, special bolts or other means provided in the foundations or in the floor. These vertical wales are bolted to every pair of horizontal wales they cross using ½" bolts. The form is now ready to support the vertical and horizontal reinforcing.

After the reinforcing steel has been placed, the inside form may be erected. The studs are nailed to a 1"x6" sill or plate, which has been nailed to the floor one day after concreting, and if the inside face of the wall is to have a finished appearance, requiring the use of plywood, the plywood should be in sheets 24" wide and 8'-0" long in order to facilitate the nailing of the plywood to the studs and the insertion of the tie bolts and spreaders.

The joints of the plywood should be filled with a mastic consisting of a mixture of 50 percent tallow and 50 percent cement for summer use, while for winter work, the tallow should be replaced by pump grease. The surplus filler should then be scraped off using No. 0 sandpaper.

Estimating Quantities for Architectural Concrete Forms.—When estimating the quantities for plain architectural concrete walls, do not make any deductions for ordinary door and window openings, as these must be formed with the wall and then boxes built in to allow for the door or window frame to be inserted at a later date.

The only exception to this rule is where there are large display windows in the first story of a commercial building or for skeleton type buildings where the piers and lintels are estimated at a higher rate than plain wall surfaces.

All ornamentation should be estimated separately, either by the lin. ft. or sq. ft. depending upon the class of work.

Ornamental work will require the use of wood or plaster waste molds for forming, in addition to the structural backing, so that the entire wall area should first be figured as plain wall surfaces and then all ornamentation figured extra.

Water tables, belt courses, cornices, door and window trim, copings, etc., should be estimated by the lin. ft. when less than 12" wide and by the sq. ft. when more than 12" wide. Always mention size and detail for each type of ornamentation, as this is necessary for pricing the work.

Columns, pilasters, etc., should be estimated separately, giving width, height and area in sq. ft. together with a detail of the forming, whether fluted, etc.

Window and Door Openings in Architectural Concrete Walls.—It is necessary to form for all door and window openings in concrete walls, i.e., the outside wall form is built up solid and then boxes or "bucks" are constructed the exact size of the door and window openings. They should be made of 2" lumber and held in place by 1"x4" strips nailed to both the outside and inside wall sheathing and braced where necessary using 2"x4"'s placed horizontally and diagonally. Where the opening is over 3'-0" wide it will be necessary to leave an opening in the bottom of the frame so that concrete can be filled to the proper level. After filling to the correct level, the opening is closed, nailing a piece of plank to the bottom of the frame. A few hours after the concrete has set, the plank is removed and the surplus concrete scraped off. To do this, it is necessary to leave an opening in the inside form at every opening.

Reveals for door and window openings should be designed so that standard size 2" lumber may be used.

Rustication Strips.—Rustication strips should be designed as narrow as possible and only about ¾" deep. When the strips are less than 1" wide, they should have one saw cut and when 2" wide, they should have 2 saw cuts in the back as illustrated. These strips should be nailed to a chalk line, using casing nails long enough to go through strip and sheathing. It is recommended that the nails be

Method of Forming Waste Molds and Other Ornamentation in
Architectural Concrete Forms.

withdrawn by pulling them through the sheathing before the form is removed and
allow the wooden strip to remain in place a few days after the form is removed or
long enough so that they can be withdrawn without injuring the edges of the
concrete.

BRACE

CONST. JOINT "C"

PLASTER
WASTE
MOLDS

CONST. JOINT "B"

STUDS 2"x4" S4S
SPACED 16" O.C.
WALES 2"x4" S1E
SPACED 28" O.C.
TIES ½" ROUND
SPACED 34" O.C.

PLASTER
WASTE
MOLD

CONST. JOINT "A"

SECTION 1-1

CONST JOINT "B"

PLASTER
WASTE
MOLD

4"x 4" SHORE
3 IN ALL

2"x 4"
STUDS

BLOCKING

LEAVE OPEN

1"x 4"

CONST. JOINT "A"

SECTION 2-2

Method of Forming for Architectural Concrete Waste Molds for Belt
Courses and Cornices

Ornamentation

Belt courses or other ornamentation where continuous members occur may be formed with wood moldings cut to the proper shape. While some contractors may prefer a waste mold for this particular detail, the wood mold has the advantage of being more easily held in alignment because the various members are broken at different points, while in a plaster waste mold the entire form would be cut in two at one point. Soft grained wood that does not warp or split easily should be used. Soft white pine is best with soft-grained Douglas fir a second choice. In the cornice illustrated, the various members are narrow enough so that common stock sizes can be used throughout. Note the generous use of saw-kerfs in the back side of members to prevent swelling and warping of the lumber. The lumber should be thoroughly oiled on all sides before building the forms as further precaution against swelling and warping.

Large, heavy wood members should be avoided in detailing for ornamental work as they will swell and in stripping, the corners of the projecting concrete will be fractured. Using several smaller pieces requires less lumber and the members may be arranged so that swelling is away from the concrete, making it easier to strip the form without damage to the concrete corners. Note the illustration where the left-hand detail shows a large member milled from a single wide board in contrast to the smaller pieces shown in the right-hand detail.

Wood Mold Used in Cornice Form.

In assembling forms one must keep constantly in mind the steps that can be taken to aid in removing them without injury to the concrete. Boxes, waste molds, wood molds, rustication strips or anything applied to the face of the forms should be nailed lightly so that when the forms are removed, these members will pull loose from the wall form and remain in the concrete. After these materials have dried out thoroughly and thus have shrunk, they can be removed without much difficulty and without injury to the concrete corners or edges.

As previously indicated, solid strips of wood, even though well oiled, may swell and result in considerable breakage of concrete corners when the strips are removed. Saw-kerfs on the back of the member will prevent this trouble by relieving any pressure against the concrete. They should be approximately two-thirds the depth of the member, not more than 1½" apart and, in general, there should be one kerf within ¾" of each edge.

Waste molds are the only solution for some highly ornamental work. They should be made only by experienced ornamental plaster workers who have been given instructions as to how they are to be fitted into the forms. Ordinarily, the waste molds are about 2" thick, reinforced on the back with 2"x4"s which are attached to the mold with burlap dipped in plaster and then wound around the wood. The 2"x4"s permit easy handling of the mold and are also used to attach the mold to the form.

When waste molds are large and have deep undercuts, it is usually best to wire them back to the studs or wales with enough wires to be sure that all points will be pressed firmly against the form. Any openings between the form and waste mold should be pointed with plaster of Paris or a patching plaster and nailheads driven through the face and around the edge of the waste molds should be countersunk and similarly pointed.

Ornamental Concrete Cornices.—Where it is necessary to use molds for belt courses, cornices, door and window reveals, columns, cornices, entrances, etc., it is more economical to build them of wood, if practicable. However, if plaster waste molds are required, the following will give some idea of their cost, although on this work, it is always advisable to obtain definite prices from a concern specializing in this work.

Plain molds consisting of a few straight lines, $8.50 per sq. ft.

Ordinary molds of stock design will cost from $8.50 to $12.00 per sq. ft.

Simple models will cost from $8.50 to $12.00 per sq. ft. plus the cost of waste molds.

Cornice molds of stock design about 1'-6" deep and 1'-3" wide cost about $15.00 to $20.00 per lin. ft.

Capitals for 1'-6" pilasters, stock designs, cost $125 to $200 each.

Joints in plaster waste molds should be patched with non-shrinking patching plaster and painted with shellac.

Individual pieces should not weigh more than 150 lbs. to be set without difficulty by 2 men.

Plaster waste molds are usually given 2 coats of shellac in the shop to make them waterproof and non-absorbent. Before concrete is placed all molds should be greased with a light yellow cup-grease, which may be cut with kerosene if too thick. The grease should be wiped into all angles of the mold and every bit of surplus grease carefully wiped off. Care must be taken not to drop oil, grease or shellac onto hardened concrete or reinforcing.

Construction Joints.—Much progress has been made on all types of concrete work in the production of satisfactory construction joints, both from the standpoint of good appearance and good bond between the successive lifts. On the other hand, some very unsightly joints for which there is no excuse occasionally are made on architectural concrete work. It is just as easy to produce a good joint as a poor one. The designer, of course, should indicate on his plans exactly where such joints should be located. Rustications or offsets will mask the joints and are advisable where they can be made to fit into the design.

One requirement that is sometimes neglected is to provide means of holding the form tightly against the hardened concrete. As a result, leakage occurs, discoloring the concrete below and the upper lift will project slightly beyond the lower lift. The illustration shows the use of ⅝" bolts cast in the concrete for fastening the form for the next lift. Some types of form ties can be used instead of bolts. The bolt or form tie assembly can be used over and over, as only the nut is left in the concrete if the bolt is used, or the insert if a form tie is used. In re-erecting the forms for the next lift, the contact surface of the form sheathing should not overlap the hardened concrete more than 1". More overlapping than this only presents more opportunity for leakage due to irregularities in the wall surface against which the sheathing is to be held.

Fig.3b- Second Stage
TYPICAL HORIZONTAL CONSTRUCTION JOINT

Fig.4b- Second Stage
CONSTRUCTION JOINT AT FLOOR LEVEL

Fig.3a- First Stage
TYPICAL HORIZONTAL CONSTRUCTION JOINT

Fig.4a- First Stage
CONSTRUCTION JOINT AT FLOOR LEVEL

HORIZONTAL CONSTRUCTION JOINTS

Method of Embedding Bolts and Placing Forms for Successive Lifts.

Since sufficient time must be allowed for the concrete between windows or other openings to settle and shrink before the concrete above is placed, the tops of such openings are "must" locations for construction joints. Similarly, it is desirable to locate a joint at the sill line. By so doing the joints are broken up into short lengths by the openings, making them relatively inconspicuous. Moreover, cracking at corners of openings is thereby minimized.

The construction joint should be only a straight thin line at the surface. If a rustication strip is not used at the joint, a ¾" wood strip to a point ½" above the bottom of this strip and just before the concrete becomes hard, the top should be lightly tamped to be sure that it is tight up against the strip after it has settled, or shrunk. All surplus concrete or laitance should then be removed.

In resuming the placing of concrete in the new lift, steps should be taken to get good bond and to avoid honeycomb. The hardened concrete should be clean and thoroughly saturated. A 6" layer of concrete in which the coarse aggregate is reduced about 50 percent, or a 2" layer of cement-sand grout should then be placed on the hardened concrete followed by the regular concrete.

Control Joints.—It is important, both for appearance and performance, that control joints be carefully installed so they will be perfectly straight and vertical. Care is required during placing of concrete to avoid knocking the joint out of alignment or otherwise injuring it. When control joints are to be caulked with a plastic compound, this should be done before the final clean down so that any compound smeared onto the surface is removed. Of course, excessive smearing should be avoided.

Form Ties.—Various makes of form ties are available for architectural concrete construction as shown in the accompanying illustration. Important features to look for in selecting form ties are: removal of all metal to a depth of at least 1½" from the face of the wall, minimum strength of 3,000 lbs. when fully assembled, adjustable length to permit tightening of forms around all openings and inserts and a design that will leave a hole at the surface not more than ⅞" in diameter and not less than ¾" deep so that it will hold a patch. Ties should not be fitted with cones or washers acting as spreaders, as tapered and shallow holes cannot be patched properly.

Labor Cost of Architectural Forms.—The labor fabricating, erecting and stripping forms is usually estimated on the basis of sq. ft. of contact area. It is customary on a job having a normal amount of ornamentation to take off the total contact area as though the wall were plain. Window and other openings, unless very large, are figured solid. The area thus obtained is priced as though the wall were unornamented. Separate allowance is made for window frames or "bucks", ornamentation, etc.

The average amount of lumber required for plain wall forms is 2¼ to 2½ ft. b.m. of lumber per sq. ft. of contact area, plus the sheathing, plywood or Concrete Form Presdwood, used for the facing materials. Figure one tie rod for approximately each 10 sq. ft. of form.

On plain walls up to 10 ft. story height, it requires about 11 to 13 hrs. carpenter time and 6 to 8 hrs. labor time per 100 sq. ft. of contact area. For walls 12 to 14 ft. high, add about 10 percent to these costs. This includes all time necessary for erecting and stripping the forms and hoisting it up for the next use. In addition, figure about 1 to 1¼ hrs. labor time for removing hardened concrete from plywood and coating same between each use.

Door or window frame boxes or "bucks" require 5 to 8 hrs. carpenter time for each opening, depending upon size and detail of same.

Rustication strips should be placed at the rate of 25 to 30 lin. ft. an hr. for one carpenter.

Types of Form Ties Available for Use in Architectural Concrete Construction.

Details of Architectural Concrete Facade

Ornamental curtain walls, fluted pilasters and piers, reveals, etc., will require about 10 to 12 hrs. carpenter time and 6 hrs. labor time per 100 sq. ft. in addition to the cost of plain wall forms.

Ornamental wood cornices built to special detail will require 24 to 28 hrs. carpenter time and 12 to 14 hrs. labor time per 100 sq. ft. in addition to cost of plain wall forms.

Ornaments projecting from the face of plain walls, such as special sills, balconies, etc., will require about 36 to 40 hrs. carpenter time and 12 to 15 hrs. labor time per 100 sq. ft. in addition to the cost of plain wall forms.

Where plaster waste molds are used, they may be set either by carpenters or plasterers, depending upon local labor union regulations.

When setting simple belt courses of plaster waste mold, allow 12 to 14 hrs. mechanic time and 6 to 7 hrs. helper time per 100 sq. ft.

When erecting and blocking cornices and other elaborate moldings using plaster waste molds, figure 20 to 25 hrs. mechanic time and 10 to 12 hrs. helper time per 100 sq. ft.

When placing letters, a carpenter should set 2 to 3 letters an hr. depending upon size.

Labor Cost of 100 Sq. Ft. of Plain Architectural Concrete Walls Using
Plywood for Facing Material

	Hours	Rate	Total	Rate	Total
Carpenter erecting forms	12.0	$....	$....	$16.47	$197.64
Labor helping......................................	6.0	....	....	12.54	75.24
Labor stripping and					
cleaning forms..............................	2.0	....	....	12.54	25.08
Labor coating plywood	0.33	....	....	12.54	4.14
Cost 100 sq. ft. forms....................			$....		$302.10
Labor cost per sq. ft				....	3.02

Details for Architectural Concrete Forms.—For the rapid and economical construction of architectural concrete forms, it is necessary to have complete working drawings, which should show the elevations in large scale, together with every piece of plywood, every stud, wale, bolt, lumber for reveals, window boxes or "bucks," etc., similar to the methods used for detailing cut stone, terra cotta or structural steel.

03200   CONCRETE REINFORCEMENT

Reinforcing steel is estimated by the pound or the ton, obtained by listing all bars of different sizes and lengths and extending the total to pounds. Reinforcing bars may be purchased from warehouse in stock lengths and all cutting and bending done on the job or they may be purchased cut to length and bent, ready to place in the building.

Example: Assume a retaining wall is reinforced with #5 bars spaced 6" on centers both vertically and horizontally. The wall is 40'-0" long and 15'-0" high and with bars spaced 6" on centers it will require the following:

  80   #5   bars   15'-0"   long = 1,200   lin.   ft. @ 1.043   lbs.=1,252   lbs.
  30   #5   bars   40'-0"   long = 1,200   lin.   ft. @ 1.043   lbs.=1,252   lbs.
      Total weight of steel required   =2,504   lbs.

If reinforcing steel is worth $40.00 per 100 lbs. in place, or $0.40 per lb., the cost of the steel in the wall ready for pouring would be 2,504 lbs. x 0.40 = $1,001.60

When purchased in stock lengths, the labor cost of unloading, sorting, handling and storing is less than where sent to the job cut to length and bent. The former method requires less storage space as the different size stock length bars are placed in compact piles, while steel cut to length and bent requires considerable storage space as all bars of each size and length must be placed in separate piles.

While it costs more to handle and store steel sent to the job cut to length and bent, the shop costs of cutting and bending are much less than the job costs, especially where the work is done by iron workers at present day wages. Practically all reinforcing steel used today is shop fabricated, i.e., cut to length and bent prior to delivery to the job site.

There has been considerable change in reinforcing bar design in recent years, with more steel going into the deformations and a trifle less into the bar proper.

The total weight of the bars remains the same as previously but the actual diameter of the bar is slightly less.

For this reason, instead of calling the bars ½", ¾", etc., they now are known as No. 3, 4, 5, 6, etc.

The following table gives the bar classifications:

Standard Sizes and Weights of Concrete Reinforcing Bars

Bar Designation (Numbers)	Unit Weight Pounds per Foot	Nominal Dimensions,—Round Sections		Perimeter Inches
		Inches Diameter	Cross Sectional area Square Inches	
2	0.167	.250	0.05	0.786
3	0.376	.375	0.11	1.178
4	0.668	.500	0.20	1.571
5	1.043	.625	0.31	1.963
6	1.502	.750	0.44	2.356
7	2.044	.875	0.60	2.749
8	2.670	1.000	0.79	3.142
9	3.400	1.128	1.00	3.544
10	4.303	1.270	1.27	3.990
11	5.313	1.410	1.56	4.430
14S	7.650	1.693	2.25	5.320
18S	13.600	2.257	4.00	7.090

Bar No. 2 in plain rounds only.

Basis for Estimating Price of Reinforcing Steel.—When estimating quantities and weights of reinforcing steel, all bars of each size should be listed separately to obtain the correct weight. No. 6 bars and over take the lowest price per lb. while bars smaller than No. 6 increase in price on a graduated scale depending upon the size of the bars. The smaller the bars the higher the price per lb.

Example: If base price is $22.00 per 100 lbs. what is price of No. 3 bars? The table shows that No. 3 bars take an increase of $5.75 per 100 lbs. over base or $27.75 per 100 lbs.

Prices are usually quoted f.o.b. warehouse, with freight allowed to destination.

Size Extras on Reinforcing Bars

Number	Extra per ton
14S and 18S	$ 40.00*
7 to 11	20.00
6	25.00
5	30.00
4	40.00
3	115.00

*Includes cutting extra

Prices on rail steel are theoretically the same as for billet steel but will vary with the scrap market on rails.

There are three specifications covering the use of reinforcing steel. A.S.T.M. designation A305, the Standard Specification for Deformed Reinforcing Bars, defines the type of bar; A.S.T.M. 15 and A.S.T.M. 16 define the type of steel from which they may be rolled.

Two widely used standard grades of reinforcement are available in American and Canadian markets. Billet Steel (A.S.T.M. Specifications A15) and Rail Steel (A.S.T.M. Specifications A16) the latter being rolled from rails of high carbon steel which have been slit into 3 longitudinal bars by a special machine.

Both grades are standard in building codes and in the design recommendations of the American Concrete Institute.

Quantity Extras.—In addition to the base price of steel there is a quantity extra, based on the number of tons of steel in the job. This extra is approximately as follows:

	Add per ton
300 tons and over	Base—no extra
100 to 300 tons	$ 5.00
50 to 99 tons	10.00
20 to 49 tons	15.00
Under 20 tons	25.00

Bending Reinforcing Steel.—The labor cost of bending reinforcing bars will vary with the size of the bars and the amount of bending necessary; the lighter the bars the higher the bending cost per ton.

Type L Bending.—Bending all No. 2 and No. 3 bars; all No. 4 stirrups and column ties; truss bars and all sizes of bars bent at more than 6 points. It also includes all bars bent in more than one plane and all radius bending with more than one radius in any bar, approximately $80.00 per ton.

Type H Bending.—Bending truss bars for beams and slabs, other than No. 2 and No. 3, radius bending and types not otherwise described, approximately $35.00 per ton.

When cut is made by means other than sawing or milling, $0.05 per end.

When cut is made by sawing or milling, $0.15 per end.

Spirals for Reinforced Concrete Columns.—Spirals should always be coiled and fabricated in the shop as it is all machine work.

Prices of spirals vary throughout the country in the same manner and for the same reasons as reinforcing steel. Quantity extras are figured independently of reinforcing steel tonnages involved.

Cost of Engineering Service.—The cost of making working drawings, bending details, setting plans, etc., will vary with the amount of work involved.

Where the engineers prepare the complete design for a reinforced concrete structure, including working drawings, placing plans, bar lists and bending details, add $30.00 per ton to the detailing charges given below.

For preparing order lists with or without bending details from plans on which all sizes and lengths are shown by others, $0.25 per 100 lbs. or $5.00 per ton.

Detailing and placing plans from designs made by others, cost approximately as follows:

	Price per ton
Less than 20 tons	$37.00
20 to 74.99	34.00
75 to 199.99	30.00
200 to 499.99	25.00
500 and over	20.00

Hauling Reinforcing Bars From Shops to Job.—The cost of hauling reinforcing steel from warehouse to job will vary with the distance of the haul and the size of the city.

For example, in the Chicago area, the trucking charge for deliveries within city limits is about 80 cts. per cwt. ; for jobs located in the Chicago suburbs, the trucking charge is about $1.00 per cwt. Minimum charge for both zones is $10.00.

PERCENTAGES AND WEIGHTS OF STANDARD STEEL REINFORCING SPIRALS[1]

Weights are in pounds per foot of height exclusive of spacers and ends[2][3]

Spirals above the stepped line are not collapsible for shipment

¼" ⌀ SPIRALS — PITCH IN INCHES

Core Diam. Inches	1¾ %	1¾ Wt.	2 %	2 Wt.	2¼ %	2¼ Wt.	F
12	0.95	3.65	0.83	3.19			1.59
13	0.88	3.96	0.77	3.47	0.64	3.34	1.74
14	0.82	4.28	0.71	3.71	0.64	3.34	1.86
15	0.76	4.56	0.67	4.02	0.59	3.54	2.01
16	0.71	4.85	0.62	4.23	0.56	3.82	2.12
17	0.67	5.16	0.59	4.85	0.52	4.00	2.27
18	0.64	5.53	0.56	4.84			2.42
19	0.60	5.77	0.53	5.10			2.55
20	0.57	6.09	0.50	5.34			2.67
21	0.54	6.35					2.78
22	0.52	6.71					2.94
23	0.50	7.05					3.09

⅜" ⌀ SPIRALS — PITCH IN INCHES

Core Diam. Inches	1¾ %	1¾ Wt.	2 %	2 Wt.	2¼ %	2¼ Wt.	2½ %	2½ Wt.	2¾ %	2¾ Wt.	F
12	2.09	8.02	1.83	7.03							3.55
13	1.93	8.71	1.69	7.63							3.84
14	1.79	9.35	1.57	8.21	1.40	7.32	1.17				4.14
15	1.67	10.00	1.47	8.82	1.30	7.80	1.17	7.02			4.44
16	1.57	10.71	1.37	9.35	1.22	8.33	1.10	7.51			4.73
17	1.48	11.48	1.29	9.93	1.15	8.85	1.03	7.93	0.94	7.24	5.03
18	1.40	12.09	1.22	10.52	1.09	9.42	0.98	8.47	0.89	7.68	5.32
19	1.32	12.70	1.16	11.03	1.03	9.91	0.93	8.95	0.84	8.09	5.61
20	1.25	13.34	1.10	11.73	0.98	10.47	0.88	9.40	0.80	8.85	5.91
21	1.20	14.11	1.05	12.33	0.93	10.93	0.84	9.88	0.76	8.94	6.21
22	1.14	14.70	1.00	12.90	0.89	11.48	0.80	10.32	0.73	9.42	6.50
23	1.09	15.38	0.96	13.53	0.85	11.98	0.77	10.86	0.70	9.88	6.80
24	1.05	16.12	0.92	14.13	0.82	12.43	0.73	11.20	0.67	10.29	7.10
25	1.00	16.65	0.88	14.65	0.78	13.00	0.71	11.82	0.64	10.67	7.39
26	0.97	17.46	0.85	15.30	0.75	13.50	0.68	12.23	0.62	11.17	7.68
27	0.93	18.08	0.82	15.93	0.72	14.10	0.65	12.63	0.59	11.44	7.97
28	0.90	18.81	0.79	16.52	0.70	14.63	0.63	13.18	0.57	11.92	8.27
29	0.87	19.51	0.76	17.05	0.67	15.00	0.61	13.68	0.55	12.72	8.57
30	0.84	20.15	0.73	17.52	0.65	15.60	0.59	14.16	0.53	12.72	8.86
31	0.81	20.78	0.71	18.20	0.63	16.15	0.57	14.62	0.52	13.33	9.16
32	0.79	21.60	0.69	18.86	0.61	16.67	0.55	15.03	0.50	13.67	9.46
33	0.76	22.02	0.67	19.42	0.59	17.11	0.53	15.38			9.76

½" ⌀ SPIRALS — PITCH IN INCHES

Core Diam. Inches	2 %	2 Wt.	2¼ %	2¼ Wt.	2½ %	2½ Wt.	2¾ %	2¾ Wt.	3 %	3 Wt.	3¼ %	3¼ Wt.	3½ %	3½ Wt.	F
12	3.33	12.79													6.30
13	3.08	13.89													6.82
14	2.86	14.92	2.54	13.26											7.35
15	2.67	16.00	2.37	14.20	2.13	12.78									7.87
16	2.50	17.05	2.22	15.14	2.00	13.64									8.40
17	2.35	18.10	2.09	16.10	1.88	14.48	1.71	13.18							8.93
18	2.22	19.18	1.97	17.02	1.78	15.38	1.62	14.00	1.48	12.79					9.45
19	2.10	20.21	1.87	18.00	1.68	16.16	1.53	14.72	1.40	13.47					9.97
20	2.00	21.36	1.78	19.00	1.60	17.09	1.45	15.48	1.33	14.18	1.23	13.12			10.50
21	1.90	22.32	1.69	19.85	1.52	17.85	1.39	16.34	1.27	14.92	1.17	13.75	1.09	12.81	11.02
22	1.82	23.48	1.61	20.79	1.45	18.71	1.32	17.02	1.21	15.61	1.12	14.45	1.04	13.42	11.54
23	1.74	24.55	1.54	21.70	1.39	19.60	1.26	17.76	1.16	16.35	1.07	15.09	1.00	14.10	12.08
24	1.67	25.64	1.48	22.72	1.33	20.42	1.21	18.59	1.11	17.04	1.02	15.67	0.95	14.59	12.60
25	1.60	26.62	1.42	23.62	1.28	21.30	1.16	19.32	1.07	17.82	0.98	16.31	0.91	15.15	13.12
26	1.54	27.72	1.37	24.65	1.23	22.14	1.12	20.04	1.03	18.55	0.95	17.00	0.88	15.84	13.64
27	1.48	28.32	1.31	25.42	1.18	22.90	1.08	20.83	0.99	19.11	0.91	17.69	0.85	16.52	14.17
28	1.43	29.48	1.27	26.15	1.14	23.82	1.04	21.17	0.95	19.66	0.88	18.41	0.82	17.14	14.70
29	1.38	30.62	1.22	27.16	1.10	24.65	1.00	22.41	0.92	20.63	0.85	19.06	0.79	17.72	15.22
30	1.33	31.38	1.18	28.32	1.06	25.70	0.97	23.27	0.89	21.38	0.82	19.99	0.76	18.23	15.73
31	1.29	33.08	1.15	29.48	1.03	26.39	0.94	24.10	0.86	22.04	0.79	20.25	0.74	18.98	16.28
32	1.25	34.19	1.11	30.30	1.00	27.33	0.91	24.89	0.83	22.69	0.77	21.05	0.71	19.40	16.80
33	1.21	35.34	1.08	31.34	0.97	28.08	0.89	25.51	0.81	23.49	0.75	21.75	0.69	20.00	

⅝" ⌀ SPIRALS — PITCH IN INCHES

Core Diam. Inches	2 %	2 Wt.	2¼ %	2¼ Wt.	2½ %	2½ Wt.	2¾ %	2¾ Wt.	3 %	3 Wt.	3¼ %	3¼ Wt.	3½ %	3½ Wt.	F
12	5.17	19.88													9.84
13	4.77	21.50													10.66
14	4.43	23.10	3.94	20.59											11.47
15	4.14	24.88	3.68	22.09	3.31	19.85									12.29
16	3.88	26.44	3.45	23.53	3.10	21.14									13.11
17	3.65	24.98	3.24	24.98	2.92	22.14	2.65	20.40							13.93
18	3.45	26.43	3.06	26.43	2.76	23.83	2.51	21.68	2.30	19.88					14.74
19	3.26	25.10	2.90	27.90	2.61	25.10	2.37	22.80	2.17	20.89					15.58
20	3.10	32.10	2.76	29.45	2.48	26.48	2.25	24.00	2.07	22.10	1.91	20.40			16.39
21	2.95	34.66	2.62	30.80	2.36	27.75	2.15	25.28	1.97	23.17	1.82	21.40	1.69	19.85	17.21
22	2.82	36.40	2.51	32.42	2.26	29.19	2.05	26.46	1.88	24.25	1.73	22.30	1.61	20.79	18.03
23	2.69	37.98	2.40	33.86	2.16	30.45	1.96	27.62	1.80	25.39	1.66	23.40	1.54	21.70	18.84
24	2.58	39.60	2.30	35.31	2.07	31.80	1.88	28.87	1.73	26.41	1.59	24.41	1.48	22.72	19.68
25	2.48	41.30	2.21	36.80	1.98	33.00	1.80	30.00	1.65	27.39	1.53	25.48	1.42	23.62	20.48
26	2.38	42.80	2.12	38.20	1.91	34.40	1.74	31.34	1.59	28.62	1.47	26.46	1.36	24.48	21.30
27	2.29	44.50	2.04	39.62	1.84	35.79	1.67	32.34	1.53	29.64	1.41	27.41	1.31	25.42	22.12
28	2.20	46.20	1.97	41.20	1.77	36.68	1.61	33.68	1.47	30.78	1.37	28.63	1.27	26.58	22.95
29	2.14	47.20	1.90	42.60	1.71	38.36	1.55	34.66	1.43	32.09	1.32	29.80	1.22	27.39	23.79
30	2.07	49.70	1.84	44.20	1.65	39.60	1.50	36.00	1.38	33.10	1.27	30.49	1.18	28.32	24.60
31	2.00	51.24	1.78	45.60	1.60	41.00	1.46	37.41	1.33	34.09	1.24	31.54	1.14	29.21	25.40
32	1.94	53.05	1.72	47.00	1.55	42.35	1.41	38.52	1.30	35.24	1.19	32.50	1.11	30.30	26.22
33	1.88	54.55	1.67	48.49	1.50	43.50	1.37	39.78	1.25	36.25	1.18	33.65	1.08	31.54	27.08

1 Spirals are sold on the basis of theoretical weights.

2 F = Weight to Add for Finishing. (This includes 1½ turns at top and 1¼ turns at bottom of spiral—ACI 318-56.) Total Wt. = (Wt. per Ft. x Height) + (F) + Wt. of Spacers.

3 Two spacers are required for Spirals 20" or less in diameter, 3 spacers for Spirals 20" to 30', and 4 spacers for Spirals over 30" in diameter and for ALL ¾" or larger spiral rods, regardless of spiral diameter. Allow ¾ pound per lin. ft. per spacer.

4 The four spiral rod sizes shown in the table have been approved by the U.S. Department of Commerce Simplified Practice Recommendation R53 and are carried in stock ready for fabrication.

Unloading Reinforcing Steel From Cars or Trucks.

—If steel can be unloaded direct from cars or trucks, without sorting or carrying to stock piles, a man will unload 4½ to 5½ tons per 8-hr. day, at the following labor cost per ton:

	Hours	Rate	Total	Rate	Total
Iron Worker	1.75	$....	$....	$17.58	$30.77

If necessary to carry the bars up to 50 ft. sort and place in stock piles, figure 2½ to 3 tons per 8-hr. day per man, at the following labor cost per ton:

	Hours	Rate	Total	Rate	Total
Iron Worker	3	$....	$....	$17.58	$52.74

Labor Placing Reinforcing Bars.—The labor cost of placing reinforcing steel will vary with the average weight of the bars, the manner in which it is placed; i.e., whether bars may be laid loose or whether it is necessary to tie them in place, and upon the class of labor employed.

In most of the larger cities there are concerns that make a specialty of setting reinforcing steel. These men become very adept at this work and under favorable conditions can place steel for 25% less than the average iron worker. For this reason costs are given both ways and the estimator should use his own judgment as to the class of workmen obtainable.

Setting Reinforcing Bars.—On jobs using bars ⅝" and smaller, where it is not necessary to tie them in place, a man should handle and place 700 to 900 lbs. per 8-hr. day, at the following labor cost per ton:

	Hours	Rate	Total	Rate	Total
Iron Worker	20.7	$....	$....	$17.58	$363.91

Experienced reinforcing steel setters should handle and place 900 to 1,100 lbs. of steel per 8-hr. day, at the following labor cost per ton:

	Hours	Rate	Total	Rate	Total
Iron worker	16	$....	$....	$17.58	$281.28

On jobs using bars ¾" and heavier, where it is not necessary to tie the bars in place, a man should handle and place 900 to 1,100 lbs. of steel per 8-hr. day, at the following labor cost per ton:

	Hours	Rate	Total	Rate	Total
Iron worker	16	$....	$....	$17.58	$281.28

An experienced reinforcing steel setter should handle and place 1,400 to 1,600 lbs. of steel per 8-hr. day, at the following labor cost per ton:

	Hrs	Rate	Total	Rate	Total
Iron worker	10.6	$....	$....	$17.58	$186.35

Setting Reinforcing Bars Tied in Place.—On jobs having concrete floors where it is necessary to tie the bars in place with tie wire, and where lightweight bars, (⅝" and under) are used, a man should handle, place and tie 600 to 800 lbs. per 8-hr. day, at the following labor cost per ton:

	Hours	Rate	Total	Rate	Total
Iron worker	23	$....	$....	$17.58	$404.34

An experienced reinforcing steel setter should handle, place and tie 800 to 1,000 lbs. of steel per 8-hr. day, at the following labor cost per ton:

	Hours	Rate	Total	Rate	Total
Iron worker	17.8	$....	$....	$17.58	$312.93

On jobs using heavy bars, (¾" and over) a man should handle, set and tie 800 to 1,000 lbs. of steel per 8-hr. day, at the following labor cost per ton:

	Hours	Rate	Total	Rate	Total
Iron worker	17.8	$....	$....	$17.58	$312.93

An experienced reinforcing steel setter should handle, set and tie 1,300 to 1,500 lbs. per 8-hr. day, at the following labor cost per ton:

	Hours	Rate	Total	Rate	Total
Iron worker	11.5	$....	$....	$17.58	$202.17

Welded Steel Fabric or Mesh Reinforcing

Welded steel fabric is a popular and economical reinforcing for concrete work of all kinds, especially driveways and floors. It may also be used for temperature reinforcing, beam and column wrapping, road and pavement reinforcing, etc.

It is usually furnished in square or rectangular mesh.

It is usually sold at a certain price per sq. ft. or sq. yd. depending upon the weight.

Prices of mesh vary with quantity ordered and freight from mill to destination. Typical prices for some styles usually stocked, with truck job site delivery are as follows: Approximate prices are per 100 sq. ft.

Style	Wt. Per 100 Sq. Ft.	Under 5 Tons
6x6-6x6 (W2.9)	42 lbs.	$15.00
6x6-8x8 (W2.1)	30 lbs.	10.50
4x4-8x8 (W2.1)	44 lbs.	15.75

Labor Placing Mesh Reinforcing In Walls.—When mesh reinforcing is used, furnished in sheets or rolls, a man should place 700 to 800 sq. ft. in walls per 8-hr. day, at the following labor cost per 100 sq. ft. :

	Hours	Rate	Total	Rate	Total
Steel setter	1.0	$....	$....	$17.58	$17.58
Cost per sq. ft			....		.176

Sizes and Weights of Welded Steel Fabric

Style	Weight per 100 Sq. Ft. Based on Net Width of 60"	Spacing of Wires in Inches		A.S. & W. Co. Steel Wire Gauge Number		Sect. Areas Sq. In. per Ft.	
		Longit.	Trans.	Longit.	Trans.	Longit.	Trans.
22-1616	13	2	2	16	16	.018	.018
22-1414	21	2	2	14	14	.030	.030
22-1313	28	2	2	13	13	.039	.039
22-1212	37	2	2	12	12	.052	.052
22-1111	48	2	2	11	11	.068	.068
22-1010	60	2	2	10	10	.086	.086
24-1414	16	2	4	14	14	.030	.015
24-1314	19	2	4	13	14	.039	.015
24-1212	28	2	4	12	12	.052	.026
212-38	105	2	12	3	8	.280	.021
212-06	166	2	12	0	6	.443	.029
216-812	46	2	16	8	12	.124	.007
216-711	55	2	16	7	11	.148	.008
216-610	65	2	16	6	10	.174	.011
216-510	75	2	16	5	10	.202	.011
216-49	89	2	16	4	9	.239	.013
216-38	104	2	16	3	8	.280	.015
216-28	119	2	16	2	8	.325	.015
216-17	139	2	16	1	7	.377	.018
33-1414	14	3	3	14	14	.020	.020
33-1212	25	3	3	12	12	.035	.035
33-1111	32	3	3	11	11	.046	.046
33-1010	41	3	3	10	10	.057	.057
33-99	49	3	3	9	9	.069	.069
33-88	58	3	3	8	8	.082	.082

Sizes and Weights of Welded Steel Fabric

Style	Weight per 100 Sq. Ft. Based on Net Width of 60"	Spacing of Wires in Inches		A.S. & W. Co. Steel Wire Gauge Number		Sect. Areas Sq. In. per Ft.	
		Longit.	Trans.	Longit.	Trans.	Longit.	Trans.
316-812	32	3	16	8	12	.082	.007
316-711	38	3	16	7	11	.098	.009
316-610	45	3	16	6	10	.116	.011
316-510	52	3	16	5	10	.135	.011
316-49	61	3	16	4	9	.159	.013
316-38	72	3	16	3	8	.187	.015
316-28	83	3	16	2	8	.216	.015
316-17	96	3	16	1	7	.252	.018
316-06	113	3	16	0	6	.295	.022
44-1414	11	4	4	14	14	.015	.015
44-1313	14	4	4	13	13	.020	.020
44-1212	19	4	4	12	12	.026	.026
44-1010	31	4	4	10	10	.043	.043
44-88	44	4	4	8	8	.062	.062
44-77	53	4	4	7	7	.074	.074
44-66	62	4	4	6	6	.087	.087
44-44	85	4	4	4	4	.120	.120
48-1313	11	4	8	13	13	.020	.010
48-1214	12	4	8	12	14	.026	.008
48-1212	14	4	8	12	12	.026	.013
48-1112	17	4	8	11	12	.034	.013
48-1012	20	4	8	10	12	.043	.013
48-912	23	4	8	9	12	.052	.013
48-812	27	4	8	8	12	.062	.013
48-711	33	4	8	7	11	.074	.017
412-1212	13	4	12	12	12	.026	.009
412-1112	16	4	12	11	12	.034	.009
412-1012	19	4	12	10	12	.043	.009
412-912	22	4	12	9	12	.052	.009
412-812	25	4	12	8	12	.062	.009
412-711	31	4	12	7	11	.074	.011
412-610	36	4	12	6	10	.087	.014
412-510	42	4	12	5	10	.101	.014
412-57	45	4	12	5	7	.101	.025
412-49	49	4	12	4	9	.120	.017
416-1012	18	4	16	10	12	.043	.007
416-912	21	4	16	9	12	.052	.007
416-812	25	4	16	8	12	.062	.007
416-711	30	4	16	7	11	.074	.009
416-610	35	4	16	6	10	.087	.011
416-510	40	4	16	5	10	.101	.011
416-49	48	4	16	4	9	.120	.013
416-38	56	4	16	3	8	.140	.015
416-28	64	4	16	2	8	.162	.015
66-1212	13	6	6	12	12	.017	.017
66-1010	21	6	6	10	10	.029	.029
66-99	25	6	6	9	9	.035	.035
66-88	30	6	6	8	8	.041	.041
66-77	36	6	6	7	7	.049	.049
66-66	42	6	6	6	6	.058	.058
66-55	49	6	6	5	5	.067	.067
66-46	50	6	6	4	6	.080	.058

Sizes and Weights of Welded Steel Fabric

Style	Weight per 100 Sq. Ft. Based on Net Width of 60"	Spacing of Wires in Inches		A.S. & W. Co. Steel Wire Gauge Number		Sect. Areas Sq. In. per Ft.	
		Longit.	Trans.	Longit.	Trans.	Longit.	Trans.
66-44	58	6	6	4	4	.080	.080
66-33	68	6	6	3	3	.093	.093
66-22	78	6	6	2	2	.108	.108
66-11	91	6	6	1	1	.126	.126
66-00	107	6	6	0	0	.148	.148
612-77	27	6	12	7	7	.049	.025
612-66	32	6	12	6	6	.058	.029
612-55	37	6	12	5	5	.067	.034
612-44	44	6	12	4	4	.080	.040
612-33	51	6	12	3	3	.093	.047
612-25	52	6	12	2	5	.108	.034
612-22	59	6	12	2	2	.108	.054
612-17	56	6	12	1	7	.126	.025
612-14	61	6	12	1	4	.126	.040
612-11	69	6	12	1	1	.126	.063
612-06	65	6	12	0	6	.148	.029
612-03	72	6	12	0	3	.148	.047
612-00	81	6	12	0	0	.148	.074

Labor Placing Steel Fabric on Floors.—When used for reinforcing floor slabs or for temperature reinforcing, a man should place 1,400 to 1,600 sq. ft. of fabric per 8-hr. day, at the following labor cost per 100 sq. ft. :

	Hours	Rate	Total	Rate	Total
Iron worker	0.53	$....	$....	$17.58	$9.32
Cost per sq. ft			....		.093

Reinforcement For Architectural Concrete Walls.—Although it is practically impossible to provide sufficient reinforcement to prevent cracking in a long wall and for that reason control joints are used, it is necessary and practical to provide adequate reinforcement between joints when properly spaced to resist volume change stresses. To be most effective small bars with a type of deformation to give maximum bond should be placed relatively close together rather than using larger bars at wider spacings. Generally #3 bars are preferable to larger sizes and the horizontal reinforcement should be at least equal to 0.25 percent of wall area. In the vertical direction 0.15 percent reinforcement is sufficient.

The size and spacing of bars recommended for walls of various thickness are given below:

Wall thickness in.	Horizontal reinforcement	Vertical reinforcement
6	#3 @ 8" in outside face of wall	#3 @ 8" in outside face of wall
8	#3 @ 6" in outside face of wall	#3 @ 8" in outside face of wall
10	#3 @ 10" in both faces of wall	#3 @ 12" in both faces of wall
12	#3 @ 8" in both faces of wall	#3 @ 12" in both faces of wall

Recommended Size and Spacing of Reinforcing Bars in Concrete Walls.

The bars should be tied at each intersection with a non-slip tie and where two curtains of steel are used it is advisable to include a detail drawing to show how the curtains are to be fastened together with tie spreaders at frequent intervals to make a rigid web of reinforcement.

Difficulty is sometimes experienced in placing concrete in walls because of a congestion of reinforcing bars and this condition should be avoided insofar as practicable.

Bars provided as beam reinforcement (not shown here) over openings should be continuous through control joints. Alternate wall bars and corner bars should be cut or stopped at control joints.

Cut or stop every other bar at control joints

Min.2-#5 bars over all openings

Corner bars to lap 2'-0" at corners

For openings less than 4'-6" wide, hook extra bars at control joints

For 6" or 8" walls use 1 #6 bar
for 10" or 12" walls use 2- #5 bars

A　　　　　　B　　　　　　C

Alternate Methods of Placing Reinforcing around Openings.

Labor Placing Reinforcing Bars.—The labor cost of placing reinforcing steel in architectural concrete walls is expensive on account of the small size of the rods, the fact they run both horizontally and vertically, must be wired together and the cramped quarters in which the men must work.

INNER CURTAIN

SPREADER

OUTER CURTAIN

Tie Spreader.

On walls 8 to 12 inches thick, a reinforcing steel setter will handle, place and wire 750 to 850 lbs. of reinforcing steel per 8-hr. day, at the following labor cost per ton:

	Hours	Rate	Total	Rate	Total
Iron worker......................................	20	$....	$....	$17.58	$351.60

Setting rebars prior to placing forms for cast-in-place concrete wall

03300   CAST-IN-PLACE CONCRETE

Reinforced and architectural concrete are among our most common building materials and are used on virtually all types of construction work.

They are used in all types of structures, from residential, apartment, hotel and hospital buildings, requiring comparatively light floor loads, to the heaviest constructed warehouses, factory buildings, wharves, and industrial buildings.

In fact, entire buildings are constructed of concrete, including foundations, foundation walls, columns, beams and girders, floor slabs, ornamental exterior walls, cornices, etc.

Regardless of the fact that it is one of our most common building materials, costs on all types of reinforced concrete construction are subject to wide variation. This is due to a number of causes, such as the function and size of building, whether flat slab or beam and girder type, whether solid concrete construction or used in conjunction with steel, building codes, insurance requirements, etc.

The methods used in erecting and moving the formwork, handling, mixing and placing the concrete, and the ability of the superintendent or foreman, all affect the final cost of the work.

Regardless of how carefully or accurately the job has been estimated, if the superintendent or foreman in charge of the work are not proficient, the actual costs are almost certain to run higher than the estimate—not because the work was estimated too low but because it was poorly handled on the job.

All labor costs given on the following pages are based on efficient management; economical layout and design of all formwork; use of modern labor-saving methods, such as bench fabrication of column, beam and girder forms where possible, and the use of labor saving tools such as power saws, band saws, electric hand saws, etc.

In these days of high wage scales, the efficient contractor must use methods and tools that will reduce costs to a minimum, or else he may be out of the running when it comes to obtaining work in competition and of performing it economically after he has obtained the job. Remember, good management, economical design, labor-saving tools and methods are absolutely necessary to a profitable contract.

Items to be Included in the Estimate.—Reinforced concrete should be estimated in the same manner and in the same units as the work is constructed on the job. This allows comparisons between estimated and actual costs and enables a contractor to tell where variations occur between estimated and actual costs.

The first operation on any reinforced concrete job is the erection of the temporary wood or metal forms to support the weight of the wet concrete until it has set sufficiently to be self-supporting. The members usually requiring forms are the columns, girders, beams, spandrel beams, lintels, floor slabs, stairs and platforms. Each class of formwork should be estimated separately and in detail.

All types of form work, including columns, beams, girders and the various type of floor slabs have been designed in the best engineering practice, giving lumber requirements, clamp spacing for columns, beam and girder lumber requirements, including shoring, together with a complete set of design tables for all types of concrete floor slabs, giving joist spacing, spans of stringers and shores required for any floor load.

After the forms are erected, it is necessary to place the reinforcing steel or wire mesh ready to receive the concrete. This may include unloading steel from trucks at the job, cutting steel to length and bending to special shapes for columns, beams, girders and floor slabs, and placing and tieing it in position ready to receive the concrete.

After the reinforcing steel is in place, the job is then ready for concrete. This consists of preparing runways, unloading cement and other materials, mixing the cement, sand and gravel into the concrete mass, hoisting, wheeling and placing it in position, removing runways, removing old concrete drippings and spillage, patching, finishing, cleaning and washing wheelbarrows or buggies, concrete mixer, etc.

Each of these items should be estimated in detail to include all material and labor operations necessary to make up the completed unit.

Weights of Miscellaneous Building Materials
Weight of Cement, Sand, Gravel, Crushed Stone, Concrete, Etc.

The following table gives the approximate weights of cement, concrete and concrete aggregates, and will assist the estimator in figuring freight or when purchasing materials by weight:

	Weight Lbs. Per Cu. Yd.
Bank sand	2,500
Torpedo sand	2,700
Crushed stone	2,500
Crushed stone screenings	2,500
Gravel	2,700
Roofing gravel	2,700

Cement weighs 376 lbs. per bbl.
(1-sack weighs 94 lbs. and bulks 1-cu. ft.)

Cinder concrete	Wt. per cu. ft.	112
Concrete of gravel, limestone, sandstone, trap rock, etc.	Wt. per cu. ft.	150

Quantity of Pit Run Sand and Gravel Required for One Cubic Yard of Concrete.—Pit run sand and gravel is not recommended for concrete work where strength is a factor because it is impossible to grade the aggregate. There is either too much gravel and too little sand or too little gravel and too much sand, usually the latter, to make a dependable concrete.

An aggregate containing approximately 50 percent sand and 50 per cent gravel (providing it does not contain any humic acid), may be considered acceptable aggregate and the strength will probably be only ten percent less than when using aggregates furnished by washing and grading plants.

It ordinarily requires about 1.25 cu. yds. of aggregate per cu. yd. of concrete, depending upon the coarseness of the sand.

CONCRETE

Concrete should be estimated by the cu. ft. or cu. yd. containing 27 cu. ft. The cu. yd. is the most convenient unit to use because practically all published tables giving quantities of cement, sand, gravel or crushed stone are based on the cu. yd. of concrete. The cu. yd. is also a convenient unit to use when estimating labor costs.

It was formerly the practice to specify concrete mixtures as so many volumes of fine and coarse aggregate to each volume of cement, as for example 1:1:2, 1:1.5:3, 1:2:4, 1:3:5, etc. The condition of the aggregate when measured was not specified, nor were the consistency of the concrete or the water content.

As used on the job, sand nearly always contains some surface moisture, usually about 4 or 5 percent by weight of the sand, although it may be considerably more just after rain or if the sand is used soon after washing it. The moisture causes the sand to bulk up in volume, the amount of bulking depending on the amount of moisture and the grading. This bulking which may be 20 to 30 percent or more, reduces the true volume of surface dry sand in a given volume of damp sand. For example, if a 1:2:4 mix is specified and 2 volumes of damp sand are measured, the volume of sand on a surface dry basis would be less. Suppose the sand is bulked 25 percent due to the moisture. Then the volume of dry sand in each cubic foot of damp sand will be 100 divided by 125, or 0.8 cu. ft. of damp sand would give us 2x0.8 or 1.6 cu. ft. dry sand.

The coarse aggregate carries very little surface moisture and the moisture has little effect on the volume. The result was that undersanded mixes were often used. In the above example the 1:2:4 mix if based on damp materials would actually be a 1:1.6:4 mixture of surface dry materials. The amount of sand in this case would be less than 29 percent of the total volume of aggregate. For average mixtures 35 to 45 percent is usually desirable for good workability when concrete is placed by hand.

To correct for the bulking in the above example the amount of damp sand should be increased to 1.25x2 or 2.5 cu. ft. and the mix would then be 1:2.5:4 based on damp materials as measured in the field.

As a result of the increased knowledge that has been made available, other methods of specifying concrete have been adopted, such as a mix by weight, a minimum cement factor, a maximum water content or the strength at a given age, or a combination of these.

The Importance of Mixing Water.—The hardening of concrete mixtures is brought about by chemical reactions between cement and water, the aggregates (sand, gravel, stone, etc.) being inactive ingredients used as fillers. So long as workable mixtures are used, the less water there is in the mix, the stronger, more watertight and more durable will be the concrete. Excess mixing water dilutes the paste made by the cement and water and makes weaker, more porous concrete. For any set of conditions of mixing, placing and curing and for given materials there is a definite relation between the strength of the concrete and the amount of water used in mixing. This relation can be determined by test and a water-cement ratio-strength curve can be developed representing the actual results on a specific job.

American Concrete Institute Standard for Selecting Proportions for Concrete Mixes

Committee 211 of the American Concrete Institute recommends certain practices for selecting proportions for concrete mixes. Following are extracts from this Standard which should be of value to all users of concrete.

BASIC RELATIONSHIP

Concrete proportions must be selected to provide necessary placeability, strength, and durability for the particular application. Well established relationships governing these properties are discussed briefly below.

Placeability (including satisfactory finishing properties) encompasses traits loosely accumulated in the terms "workability" and "consistency." For the purpose of this discussion, **workability** is considered to be that property of concrete which determines its capacity to be placed and consolidated properly and to be finished without harmful segregation. It embodies such concepts as moldability, cohesiveness, and compactability. It is affected by the grading, particle shape and proportions of aggregate, the amount of cement, the presence of entrained air, admixtures, and the consistency of the mixture. Procedures in this recommended practice permit these factors to be taken into account to achieve satisfactory placeability economically.

Consistency, loosely defined, is the wetness of the concrete mixture. It is measured in terms of slump—the higher the slump the wetter the mixture—and it affects the ease with which the concrete will flow during placement. It is related to, but not synonymous with, workability. In properly proportioned concrete, the unit water content required to produce a given slump will depend on several factors. Water requirement increases as aggregates become more angular and rough textured (but this disadvantage may be offset by improvements in other characteristics such as bond to cement paste). Required mixing water decreases as the maximum size of well graded aggregate is increased. It also decreases with the entrainment of air. Mixing water requirement may often be significantly reduced by certain admixtures.

Strength.—Strength is an important characteristic of concrete, but other characteristics such as durability, permeability, and wear resistance are often equally or more important. These may be related to strength in a general way but are also affected by factors not significantly associated with strength. For a given set of materials and conditions, concrete strength is determined by the net quantity of water used per unit quantity of cement. The net water content excludes water absorbed by the aggregates. Differences in strength, for a given water-cement ratio may result from changes in: maximum size of aggregate; grading, surface texture, shape, strength, and stiffness of aggregate particles; differences in cement types and sources; air content; and the use of admixtures which affect the cement hydration process or develop cementitious properties themselves. To the extent that these effects are predictable in the general sense, they are taken into account in this recommended practice. However, in view of their number and complexity, it should be obvious that accurate predictions of strength must be based on trial batches or experience with the materials to be used.

Durability.—Concrete must be able to endure those exposures which may deprive it of its serviceability—freezing and thawing, wetting and drying, heating and cooling, chemicals, de-icing agents, and the like. Resistance to some of these may be enhanced by use of special ingredients: low-alkali cement, pozzolans, or selected aggregate to prevent harmful expansion due to the alkali-aggregate reaction which occurs in some areas when concrete is exposed in a moist environment; sulfate resisting cement or pozzolans for concrete exposed to sea water or sulfate-bearing soils; or aggregate free of excessive soft particles where resistance to surface abrasion is required. Use of a low water-cement ratio will prolong the life of concrete by reducing the penetration of aggressive liquids. Resistance to severe weathering, particularly freezing and thawing, and to salts used for ice removal is greatly improved by incorporation of a proper distribution of entrained air. Entrained air should be used in all exposed concrete in climates where freezing occurs.

BACKGROUND DATA

To the extent possible, selection of concrete proportions should be based on test data or experience with the materials actually to be used. Where such background is limited or not available, estimates given in this recommended practice may be employed.

The following information for available materials will be useful:

Sieve analyses of fine and coarse aggregates
Unit weight of coarse aggregate
Bulk specific gravities and absorptions of aggregates
Mixing water requirements of concrete developed from experience with available aggregates

Relationships between strength and water-cement ratio for available combinations of cement and aggregate.

Estimates from Tables 3 and 4, respectively, may be used when the last two items of information are not available. As will be shown, proportions can be estimated without the knowledge of aggregate specific gravity and absorption (third item above).

PROCEDURE

Estimating the required batch weights for the concrete involves a sequence of logical, straightforward steps which, in effect, fit the characteristics of the available materials into a mixture suitable for the work. The question of suitability is frequently not left to the individual selecting the proportions. The job specifications may dictate some or all of the following:

Maximum water-cement ratio   Slump
Minimum cement content   Maximum size of aggregate
Air content   Strength
Other requirements relating to such things as strength over-design, admixtures, and special types of cement or aggregate

Regardless of whether the concrete characteristics are prescribed by the specifications or are left to the individual selecting the proportions, establishment of batch weights per cubic yard of concrete can best be accomplished in the following sequence:

Step 1. Choice of slump. If slump is not specified, a value appropriate for the work can be selected from Table 1. The slump ranges shown apply when vibration is used to consolidate the concrete. Mixes of the stiffest consistency that can be placed efficiently should be used.

Table 1—Recommended Slumps
For Various Types of Construction

Types of construction	Slump, in.	
	Maximum*	Minimum
Reinforced foundation walls and footings	3	1
Plain footings, caissons, and substructure walls	3	1
Beams and reinforced walls	4	1
Building columns	4	1
Pavements and slabs	3	1
Heavy mass concrete	3	1

*May be increased 1 in. for methods of consolidation other than vibration.

Step 2. Choice of maximum size of aggregate. Large maximum sizes of well graded aggregates have less voids than smaller sizes. Hence, concretes with the larger-sized aggregates require less mortar per unit volume of concrete. Generally, the maximum size of aggregate should be the largest that is economically available and consistent with dimensions of the structure. Ordinarily, the ratio of the nominal maximum aggregate size to the minimum dimension within which concrete must be placed should not exceed the value shown in Table 2.

Table 2—Recommended Maximum Ratio
of Nominal Aggregate Size to Minimum
Criterion Dimension

Criterion dimension		Recommended maximum ratio
Narrowest dimension between sides of forms		1/5
Depth of slab		1/3
Clear spacing between	In columns	2/3
individual reinforcing bars,	In all	
bundles of bars, or pre-	other	
tensioning strands	members	3/4

These limitations are sometimes waived if workability and methods of consolidation are such that the concrete can be placed without honeycomb or void. When high strength concrete is desired, best results may be obtained with reduced maximum sizes of aggregate since these produce higher strengths at a given water-cement ratio.

Step 3. Estimation of mixing water and air content. The quantity of water per unit volume of concrete required to produce a given slump is dependent on the maximum size, particle shape and grading of the aggregates, and on the amount of entrained air. It is not greatly affected by the quantity of cement. Table 3 provides estimates of required mixing water for concretes made with various maximum sizes of aggregate, with and without air entrainment. Depending on aggregate texture and shape, mixing water requirements may be somewhat above or below the tabulated values, but they are sufficiently accurate for the first estimate. Such differences in water demand are not necessarily reflected in strength since other compensating factors may be involved. For example, a rounded and an angular coarse aggregate, both well and similarly graded and of good quality, can be expected to produce concrete of about the same compressive strength for the same cement factor in spite of differences in water-cement ratio resulting from the different mixing water requirements. Particle shape per se is not an indicator that an aggregate will be either above or below average in its strength-producing capacity.

Table 3 indicates the approximate amount of entrapped air to be expected in non-air-entrained concrete, and shows the recommended levels of average air content for concrete in which air is to be purposely entrained for durability. Air-entrained concrete should always be used for structures which will be exposed to freezing and thawing, and generally for structures exposed to sea water or sulfates. When severe exposure is not anticipated, beneficial effects of air entrainment on concrete workability and cohesiveness can be achieved at air content levels approximately half those shown for air-entrained concrete.

When trial batches are used to establish strength relationships or verify strength-producing capability of a mixture, the least favorable combination of mixing water and air content should be used. This is, the air content should be the maximum permitted or likely to occur, and the concrete should be gauged to the highest permissible slump. This will avoid developing an over-optimistic estimate of strength on the assumption that average rather than extreme conditions will prevail in the field. For information on air content recommendations, see ACI 201, 301, and 302.

Table 3—Approximate Mixing Water and
Air Content Requirements for Different
Slumps and Maximum Sizes of Aggregates*

Slump, in.	Water, lb. per cu. yd. of concrete for indicated maximum sizes of aggregate						
	⅜ in.	½ in.	¾ in.	1 in.	1½ in.	2 in.	3 in.
Non-air-entrained concrete							
1 to 2	350	335	315	300	275	260	240
3 to 4	385	365	340	325	300	285	265
6 to 7	410	385	360	340	315	300	285
Air content, percent	3	2.5	2	1.5	1	0.5	0.3
Air-entrained concrete							
1 to 2	305	295	280	270	250	240	225
3 to 4	340	325	305	295	275	265	250
6 to 7	365	345	325	310	290	280	270
Air content, percent	8	7	6	5	4.5	4	3.5

*These quantities of mixing water are for use in computing cement factors for trial batches. They are maxima for reasonably well-shaped angular coarse aggregates graded within limits of accepted specifications.

Step 4. Selection of water-cement ratio. The required water-cement ratio is determined not only by strength requirements but also by factors such as durability and finishing properties. Since different aggregates and cements generally produce different strengths at the same water-cement ratio, it is highly desirable to have or develop the relationship between strength and water-cement ratio for the materials actually to be used. In the absence of such data, approximate and relatively conservative values for concrete containing Type I portland cement can be taken from Table 4(a). With typical materials, the tabulated water-cement ratios should produce the strengths shown, based on 28-day tests of specimens cured under standard laboratory conditions. The average strength selected must, of course, exceed the specified strength by a sufficient margin to keep the number of low tests within specified limits.

For severe conditions of exposure, the water-cement ratio should be kept low even though strength requirements may be met with a higher value. Table 4(b) gives limiting values.

Table 4(a)—Relationships Between
Water-Cement Ratio and Compressive Strength of Concrete

Compressive strength at 28 days, psi*	Water-cement ratio, by weight	
	Non-air-entrained concrete	Air-entrained concrete
6000	0.41	—
5000	0.48	0.40
4000	0.57	0.48
3000	0.68	0.59
2000	0.82	0.74

*Values are estimated average strengths for concrete containing not more than the percentage of air shown in Table 3. For a constant water-cement ratio, the strength of concrete is reduced as the air content is increased.

Strength is based on 6 x 12 in. cylinders moist-cured 28 days at 73.4 ± 3 F(23 ± 1.7 C) in accordance with Section 9(b) of ASTM C 31 for Making and Curing Concrete Compression and Flexure Test Specimens in the Field.

Relationship assumes maximum size of aggregate about ¾ to 1 in.; for a given source, strength produced for a given water-cement ratio will increase as maximum size of aggregate decreases.

Table 4(b)—Maximum Permissible
Water-Cement Ratios for
Concrete in Severe Exposures*

Type of structure	Structure wet continuously or frequently and exposed to freezing and thawing†	Structure exposed to sea water or sulfates
Thin sections (railings, curbs, sills, ledges, ornamental work) and sections with less than 1 in. cover over steel	0.45	0.40†
All other structures	0.50	0.45††

*Based on report of ACI Committee 201, "Durability of Concrete in Service," previously cited.
†Concrete should also be air-entrained.
††If sulfate resisting cement (Type II or Type V of ASTM C 150) is used, permissible water-cement ratio may be increased by 0.05.

Step 5. Calculation of cement content. The amount of cement per unit volume of concrete is fixed by the determinations made in Steps 3 and 4 above. The required cement is equal to the estimated mixing water content (Step 3) divided by the water-cement ratio (Step 4). If, however, the specification includes a separate minimum limit on cement in addition to requirements for strength and durability, the mixture must be based on whichever criterion leads to the larger amount of cement.

Step 6. Estimation of coarse aggregate content. Aggregates of essentially the same maximum size and grading will produce concrete of satisfactory workability when a given volume of coarse aggregate, on a dry-rodded basis, is used per unit volume of concrete. Appropriate values for this aggregate volume are given in Table 6. It can be seen that, for equal workability, the volume of coarse aggregate in a unit volume of concrete is dependent only on its maximum size and the fineness modulus of the fine aggregate. Differences in the amount of mortar required for workability with different aggregates, due to differences in particle shape and grading, are compensated for automatically by differences in dry-rodded void content.

The volume of aggregate, in cubic feet, on a dry-rodded basis, for a cubic yard of concrete is equal to the value from Table 6 multiplied by 27. This volume is converted to dry weight of coarse aggregate required in a cubic yard of concrete by multiplying it by the dry-rodded weight per cubic foot of the coarse aggregate.

Table 6—Volume of Coarse Aggregate
Per Unit of Volume of Concrete

Maximum size of aggregate, in.	Volume of dry-rodded coarse aggregate* per unit volume of concrete for different fineness moduli of sand			
	2.40	2.60	2.80	3.00
3/8	0.50	0.48	0.46	0.44
1/2	0.59	0.57	0.55	0.53
3/4	0.66	0.64	0.62	0.60
1	0.71	0.69	0.67	0.65
1½	0.75	0.73	0.71	0.69

Maximum size of aggregate, in.	Volume of dry-rodded coarse aggregate* per unit volume of concrete for different fineness moduli of sand			
	2.40	2.60	2.80	3.00
2	0.78	0.76	0.74	0.72
3	0.82	0.80	0.78	0.76

*Volumes are based on aggregates in dry-rodded condition as described in ASTM C 29 for Unit Weight of Aggregate.

These volumes are selected from empirical relationships to produce concrete with a degree of workability suitable for usual reinforced construction. For less workable concrete such as required for concrete pavement construction they may be increased about 10 percent. When placement is to be by pump, they should be reduced about 10 percent.

Step 7. Estimation of fine aggregate content. At completion of Step 6, all ingredients of the concrete have been estimated except the fine aggregate. Its quantity is determined by difference. Either of two procedures may be employed: the "weight" method or the "absolute volume" method.

If the weight of the concrete per unit volume is assumed or can be estimated from experience, the required weight of fine aggregate is simply the difference between the weight of fresh concrete and the total weight of the other ingredients. Often the unit weight of concrete is known with reasonable accuracy from previous experience with the materials. In the absence of such information, Table 7 can be used to make a first estimate. Even if the estimate of concrete weight per cubic yard is rough, mixture proportions will be sufficiently accurate to permit easy adjustment on the basis of trial batches as will be shown in the examples.

If a theoretically exact calculation of fresh concrete weight per cubic yard is desired, the following formula can be used:

$$U = 16.85 \, G_a \, (100 - A)$$
$$+ \, C(1 - G_a/G_c) - W \, (G_c - 1)$$

where

U = weight of fresh concrete per cubic yard, lb.
G_a = weighted average specific gravity of combined fine and coarse aggregate, bulk SSD*
G_c = specific gravity of cement (generally 3.15)
A = air content, percent
W = mixing water requirement, lb. per cu. yd.
C = cement requirement, lb. per cu. yd.

A more exact procedure for calculating the required amount of fine aggregate involves the use of volumes displaced by the ingredients. In this case, the total volume displaced by the known ingredients—water, air, cement, and coarse aggregate—is subtracted from the unit volume of concrete to obtain the required volume of fine aggregate. The volume occupied in concrete by any ingredient is equal to its weight divided by the density of that material (the latter being the product of the unit weight of water and the specific gravity of the material).

Step 8. Adjustments for aggregate moisture. The aggregate quantities actually to be weighed out for the concrete must allow for moisture in the aggregates. Generally, the aggregates will be moist and their dry weights should be increased

*SSD indicates saturated-surface-dry basis used in considering aggregate displacement. The aggregate specific gravity used in calculations must be consistant with the moisture condition assumed in the basic aggregate batch weights—i.e., bulk dry if aggregate weights are stated on a dry basis, and bulk SSD if weights are stated on a saturated-surface-dry basis.

by the percentage of water they contain, both absorbed and surface. The mixing water added to the batch must be reduced by an amount equal to the free moisture contributed by the aggregate—i.e., total moisture minus absorption.

Table 7—First Estimate of Weight of Fresh Concrete

Maximum size, of aggregate, in.	First estimate of concrete weight, lb. per cu. yd.*	
	Non-air-entrained concrete	Air-entrained concrete
⅜	3840	3690
½	3890	3760
¾	3960	3840
1	4010	3900
1½	4070	3960
2	4120	4000
3	4160	4040

*Values calculated by Eq. 7 for concrete of medium richness (550 lb. of cement per cu. yd.) and medium slump with aggregate specific gravity of 2.7. Water requirements based on values for 3 to 4 in. slump in Table 3. If desired, the estimated weight may be refined as follows if necessary information is available: for each 10 lb. difference in mixing water from the Table 3 values for 3 to 4 in. slump, correct the weight per cu. yd. 15 lb. in the opposite direction; for each 100 lb. difference in cement content from 550 lb., correct the weight per cu. yd. 15 lb. in the same direction; for each 0.1 by which aggregate specific gravity deviates from 2.7, correct the concrete weight 100 lb. in the same direction.

Step 9. Trial batch adjustments. The calculated mixture proportions should be checked by means of trial batches prepared and tested in accordance with ASTM C 192, "Making and Curing Concrete Compression and Flexure Test Specimens in the Laboratory," or full-sized field batches. Only sufficient water should be used to produce the required slump regardless of the amount assumed in selecting the trial proportions. The concrete should be checked for unit weight and yield (ASTM C 138) and for air content (ASTM C 138, C 173, or C 231). It should also be carefully observed for proper workability, freedom from segregation, and finishing properties. Appropriate adjustments should be made in the proportions for subsequent batches in accordance with the following procedure.

Re-estimate the required mixing water per cubic yard of concrete by multiplying the net mixing water content of the trial batch by 27 and dividing the product by the yield of the trial batch in cubic feet. If the slump of the trial batch was not correct, increase or decrease the re-estimated amount of water by 10 lb. for each required increase or decrease of 1 in. in slump.

If the desired air content (for air-entrained concrete) was not achieved, re-estimate the admixture content required for proper air content and reduce or increase the mixing water content of the above paragraph by 5 lb. for each 1 percent by which the air content is to be increased or decreased from that of the previous trial batch.

If estimated weight per cubic yard of fresh concrete is the basis for proportioning, re-estimate that weight by multiplying the unit weight in pounds per cubic foot of the trial batch by 27 and reducing or increasing the result by the anticipated percentage increase or decrease in air content of the adjusted batch from the first trial batch.

Calculate new batch weights starting with Step 4, modifying the volume of coarse aggregate from Table 6 if necessary to provide proper workability.

SAMPLE COMPUTATIONS

Two example problems will be used to illustrate application of the proportioning procedures. The following conditions are assumed:

Type I non-air-entraining cement will be used and its specific gravity is assumed to be 3.15.*

Coarse and fine aggregates in each case are of satisfactory quality and are graded within limits of generally accepted specifications.†

The coarse aggregate has a bulk specific gravity of 2.68* and an absorption of 0.5 percent.

The fine aggregate has a bulk specific gravity of 2.64,* an absorption of 0.7 percent, and fineness modulus of 2.8.

Example 1. Concrete is required for a portion of a structure which will be below ground level in a location where it will not be exposed to severe weathering or sulfate attack. Structural considerations require it to have an average 28-day compressive strength of 3500 psi.** On the basis of information in Table 1 as well as previous experience, it is determined that under the conditions of placement to be employed, a slump of 3 to 4 in. be used and that the available No. 4 to 1½-in. coarse aggregate will be suitable. The dry-rodded weight of coarse aggregate is found to be 100 lb. per cu. ft. Employing the sequence outlined in the Procedure section, the quantities of ingredients per cubic yard of concrete are calculated as follows:

Step 1. As indicated above, the desired slump is 3 to 4 in.

Step 2. The locally available aggregate graded from No. 4 to 1½ in., has been indicated as suitable.

Step 3. Since the structure will not be exposed to severe weathering, non-air-entrained concrete will be used. The approximate amount of mixing water to produce a 3-to 4-in. slump in non-air-entrained concrete with 1½-in. aggregate is found from Table 3 to be 300 lb. per cu. yd. Estimated entrapped air is shown as 1 percent.

Step 4. From Table 4(a), the water-cement ratio needed to produce a strength of 3500 psi in non-air-entrained concrete is found to be about 0.62.

Step 5. From the information derived in Steps 3 and 4, the required cement content is found to be 300/0.62 = 484 lb. per cu. yd.

Step 6. The quantity of coarse aggregate is estimated from Table 6. For a fine aggregate having a fineness modulus of 2.8 and a 1½ in. maximum size of coarse aggregate, the table indicates that 0.71 cu. ft. of coarse aggregate, on a dry-rodded basis, may be used in each cubic foot of concrete. For a cubic yard, therefore, the coarse aggregate will be 27 x 0.71 = 19.17 cu. ft. Since it weighs 100 lb. per cu. ft. the dry weight of coarse aggregate is 1917 lb.

Step 7. With the quantities of water, cement, and coarse aggregate established, the remaining material comprising the cubic yard of concrete must consist of sand and whatever air will be entrapped. The required sand may be determined on the basis of either weight or absolute volume as shown below:

Weight basis. From Table 7, the weight of a cubic yard of non-air-entrained concrete made with aggregate having a maximum size of 1½ in. is estimated to be

---

*The specific gravity values are not used if proportions are selected to provide a weight of concrete assumed to occupy 1 cu. yd.

†Such as the "Specifications for Concrete Aggregates," (ASTM C 33).

**This is not the specified strength used for structural design but a higher figure expected to be produced on the average. For the method of determining the amount by which average strength should exceed design strength, see "Recommended Practice for Evaluation of Compression Test Results of Field Concrete (ACI 214-65.)"

4070 lb. (For a first trial batch, exact adjustments of this value for usual differences in slump, cement factor, and aggregate specific gravity are not critical.) Weights already known are:

Water (net mixing)	300 lb.
Cement	484 lb.
Coarse aggregate	1917 lb. (dry)*
Total	2701 lb.

The weight of sand, therefore, is estimated to be
$$4070 - 2701 = 1369 \text{ lb. (dry)*}$$

Absolute volume basis. With the quantities of cement, water, and coarse aggregate established, and the approximate entrapped air content (as opposed to purposely entrained air) taken from Table 3, the sand content can be calculated as follows:

Volume of water	=	$\dfrac{300}{62.4}$	=	4.81 cu. ft.
Solid volume of cement	=	$\dfrac{484}{3.15 \times 62.4}$	=	2.46 cu. ft.
Solid volume of coarse aggregate	=	$\dfrac{1917}{2.68 \times 62.4}$	=	11.46 cu. ft.
Volume of entrapped air	=	0.01×27	=	0.27 cu. ft.
Total solid volume of ingredients except sand			=	19.00 cu. ft.
Solid volume of sand required	=	$27 - 19.00$	=	8.00 cu. ft.
Required weight of dry sand	=	$8.00 \times 2.64 \times 62.4$	=	1318 lb.

Batch weights per cubic yard of concrete calculated on the two bases are compared below:

	Based on estimated concrete weight, lb.	Based on absolute volume of ingredients, lb.
Water (net mixing)	300	300
Cement	484	484
Coarse aggregate (dry)	1917	1917
Sand (dry)	1369	1318

---

*Aggregate absorption is disregarded since its magnitude is inconsequential in relation to other approximations.

Note: These batch weights will require adjustment in the field to take into account moisture on aggregates. Also, some adjustment in proportions may be found desirable on the basis of actual field experience.

Example 2. Concrete is required for a heavy bridge pier which will be exposed to fresh water in a severe climate. An average 28-day compressive strength of 3000 psi will be required. Placement conditions permit a slump of 1 to 2 in. and the use of large aggregate, but the only economically available coarse aggregate of satisfactory quality is graded from No. 4 to 1 in. and this will be used. Its dry-rodded weight is found to be 95 lb. per cu. ft. Other characteristics are as indicated in the first part of this section.

The calculations will be shown in skeleton form only. Note that confusion is avoided if all steps of the Procedure section are followed even when they appear repetitive of specified requirements.

Step 1. The desired slump is 1 to 2 in.

Step 2. The locally available aggregate, graded from No. 4 to 1 in., will be used.

Step 3. Since the structure will be exposed to severe weathering, air-entrained concrete will be used. The approximate amount of mixing water to produce a 1 to 2-in. slump in air-entrained concrete with 1-in. aggregate is found from Table 3 to be 270 lb. per cu. yd. The recommended air content is 5 percent.

Step 4. From Table 4(a), the water-cement ratio needed to produce a strength of 3000 psi in air-entrained concrete is estimated to be about 0.59. However, reference to Table 4(b) reveals that, for the severe weathering exposure anticipated, the water-cement ratio should not exceed 0.50. This lower figure must govern and will be used in the calculations.

Step 5. From the information derived in Steps 3 and 4, the required cement content is found to be 270/0.50 = 540 lb. per cu. yd.

Step 6. The quantity of coarse aggregate is estimated from Table 6. With a fine aggregate having a fineness modulus of 2.8 and a 1 in. maximum size of coarse aggregate, the table indicates that 0.67 cu. ft. of coarse aggregate, on a dry-rodded basis, may be used in each cubic foot of concrete. For a cubic yard, therefore, the coarse aggregate will be 27 x 0.67 = 18.09 cu. ft. Since it weighs 95 lb. per cu. ft., the dry weight of coarse aggregate is 18.09 x 95 = 1719 lb.

Step 7. With the quantities of water, cement and coarse aggregate established, the remaining material comprising the cubic yard of concrete must consist of sand and air. The required sand may be determined on the basis of either weight or absolute volume as shown below.

Weight basis. From Table 7, the weight of a cubic yard of air-entrained concrete made with aggregate of 1 in. maximum size is estimated to be 3900 lb. (For a first trial batch, exact adjustments of this value for differences in slump, cement factor, and aggregate specific gravity are not critical.) Weights already known are:

Water (net mixing)	270 lb.
Cement	540 lb.
Coarse aggregate (dry)	1719 lb.
Total	2529 lb.

The weight of sand, therefore, is estimated to be
$$3900 - 2529 = 1371 \text{ lb. (dry)}$$

Absolute volume basis. With the quantities of cement, water, air, and coarse aggregate established, the sand content can be calculated as follows:

$$\text{Volume of water} = \frac{270}{62.4} = 4.33 \text{ cu. ft}$$

$$\text{Solid volume of cement} = \frac{540}{3.15 \times 62.4} = 2.75 \text{ cu. ft.}$$

$$\text{Solid volume of coarse aggregate} = \frac{1719}{2.68 \times 62.4} = 10.28 \text{ cu. ft.}$$

$$\text{Volume of air} = 0.05 \times 27 = \underline{1.35 \text{ cu. ft.}}$$

$$\text{Total volume of ingredients except sand} = 18.71 \text{ cu. ft.}$$

$$\text{Solid volume of sand required} = 27 - 18.71 = 8.29 \text{ cu. ft.}$$

$$\text{Required weight of dry sand} = 8.29 \times 2.64 \times 62.4 = 1366 \text{ lb.}$$

Batch weight per cubic yard of concrete calculated on the two bases are compared below:

	Based on estimated concrete weight, lb.	Based on absolute volume of ingredients, lb.
Water (net mixing)	270	270
Cement	540	540
Coarse aggregate (dry)	1719	1719
Sand (dry)	1371	1366

See Note at end of Example 1

READY-MIXED CONCRETE

In the United States more than 90 percent of all concrete being used in building projects is ready-mixed. Ready-mixed concrete may be defined as portland cement concrete manufactured for delivery to a purchaser in a plastic and unhardened state.

In some instances the concrete is mixed completely in a stationary mixer and then is transported to the job in trucks. This is known as central-mixed concrete. In other ready-mixed operations, the materials are dry batched and mixed en route to the job in truck mixers. This is known as transit mixed concrete. In another procedure, the concrete is mixed in a stationary mixer at the central plant only enough to intermingle the ingredients, usually about ½ minute, and the mixing is completed en route to the job. This is known as shrink-mixed concrete.

Truck mixers consist essentially of a mixer with a separate water tank and water measuring device mounted on a truck chassis. There are other truck conveyances which are similar but are without provisions for water.

ASTM C-94 requires that when a truck mixer is used either for complete mixing or to finish the partial mixing, each batch of concrete is to be mixed not more than 100 revolutions of the drum or blades at the speed of rotation designated by the manufacturer as the mixing speed. Any additional mixing is to be done at the agitating speed. The specification also requires that the concrete must be delivered and discharged from the truck mixer or agitator truck within 1½ hours after introduction of the water to the cement and aggregate or the cement to the aggregate.

Quality of Ready-Mixed Concrete. ASTM C-94 states that in the absence of applicable general specifications, the purchaser shall select one of the alternate bases for specifying the quality of concrete.

Alternate No. 1:

When the purchaser assumes responsibility for the proportioning of the concrete mixture, he shall also specify the following:

(1) Cement content in bags or pounds per cubic yard of concrete, or equivalent units.

(2) Maximum allowable water content in gallons per cubic yard of concrete, or equivalent units, including surface moisture on the aggregates, but excluding water of absorption.

(3) If admixtures are required, the type, name, and dosage to be used. The cement content shall not be reduced when admixtures are used under Alternate No. 1 without the written approval of the purchaser.

Alternate No. 2:

When the purchaser requires the manufacturer to assume full responsibility for the selection of the proportions for the concrete mixture, the purchaser shall also specify the following:

Requirements for compressive strength as determined on samples taken from the transportation unit at the point of discharge. The purchaser shall specify the requirements in terms of the compressive strength of standard specimens cured under standard laboratory conditions for moist curing. Unless otherwise specified the age at test shall be 28 days.

Alternate No. 3:

When the purchaser requires the manufacturer to assume responsibility for the selection of the proportions for the concrete mixture with the minimum allowable cement content specified, the purchaser shall also specify the following:

(1) Required compressive strength as determined on samples taken from the transportation unit at the point of discharge. The purchaser shall specify the requirements for strength in terms of tests of standard specimens cured under standard laboratory conditions for moist curing. Unless otherwise specified the age at test shall be 28 days.

(2) Minimum cement content in bags or pounds per cubic yard of concrete.

(3) If admixtures are required, the type, name, and dosage to be used. The cement content shall not be reduced when admixtures are used.

Note:—Alternate No. 3 can be distinctive and useful only if the designated minimum cement content is at about the same level that would ordinarily be required for the strength, aggregate size, and slump specified. At the same time, it must be an amount that will be sufficient to assure durability under expected service conditions, as well as satisfactory surface texture and density, in the event specified strength is attained with it.

The proportions arrived at by Alternates 1, 2, or 3 for each class of concrete and approved for use in a project shall be assigned to a designation to facilitate identification of each concrete mixture delivered to the project. This is the designation required and supplies information on concrete proportions when they are not given separately on each delivery ticket. A certified copy of all proportions as established in Alternates 1, 2, and 3 shall be on file at the batch plant.

Curing.—The protection of concrete during the early period to prevent loss of moisture at low temperatures is an important factor in the development of both strength and durability in concrete. Specifications require that the concrete be protected to prevent loss of moisture from the surface and to prevent temperatures at the surface from going below 50 deg. F. for periods of 5 days where normal portland cement is used, and 3 days where high early strength portland cement is used.

In fixing these limits it is recognized that under average conditions, curing of concrete does not cease immediately upon removal of the protection against loss of moisture. Where the conditions are extremely severe, such as for thin sections in hot dry air or very low temperatures, it may be desirable to increase somewhat the protection periods specified.

High Early Strength Portland Cement.—Most of the cement companies are now manufacturing what is designated as high early strength portland cement, which develops practically the same strength in 72 hours as is obtained with normal portland cement in 7 to 10 days.

High early strength cement is used in the same proportions and in exactly the same manner as ordinary portland cement and it is general knowledge with contractors that this cement is advantageous not only in special rush work but in every day reinforced concrete construction. High early strength cement at present costs $3.50 cents per barrel more than standard portland cement, or, if we assume that approximately 1½ barrels of cement per cu. yd. will be used, the cost per cu. yd. of concrete is increased by $5.25. If the average amount of concrete per sq. ft. of floor construction (including columns, beams, lintels, etc.) is 8", it takes 40 sq. ft. of floor for a yard of concrete, or the extra expense on account of the use of high early strength cement is 13.13 cents per sq. ft.

Under favorable conditions using high early strength portland cement, it is possible to strip part or an entire floor in three days so it may be possible to save half of the lumber required for a floor. Under "Concrete Formwork" it is shown that it takes not less than 3 to 4 BF of lumber for each sq. ft. of floor construction (including columns, girders, and lintels) in addition to the labor costs per sq. ft. of floor to make up the various sides, panels, shores, wedges, etc. required. If it is possible to save one-half of the lumber plus half the labor of the make up, the resulting saving per sq. ft. is greater than the additional 13.13 cts per sq. ft. for high-early cement.

During the winter there is the additional saving of two to three days for cold weather protection.

Air Entrained Concrete.—Most of the highway departments are specifying air-entrained concrete for pavements and many require it for bridges and other structures. In this concrete 3 to 5 or 6 percent of air is incorporated in the concrete in the form of minute separated air bubbles. Such concrete is more resistant to freezing and thawing and to salt action than normal concrete. It is produced by using air-entraining portland cement or by the addition of an air-entraining agent at the mixer. Originally developed to prevent scaling of pavement where salts are used for ice removal, air-entrained concrete is being widely adopted for all types of work and in all locations because of its better workability as well as its better resistance to weathering even where salts are not used.

In air-entrained concrete the proportion of sand and the amount of water is generally reduced from the normal mix in sufficient quantities to make up for the increased volume of concrete produced by the air. Thus the same cement factor is maintained. In the tables given on the previous pages, the sand can be reduced about 5 lbs. for each 1 percent of air. Thus, if 4 percent air is incorporated, the quantities of sand shown in these tables would be reduced by 20 lbs. per cubic yard of concrete. For each 1 percent of air the amount of water would be reduced by about 6 to 7 lbs. per cubic yard.

Important Suggestions Regarding the Purchasing of Materials for Job Mixed Concrete.—When purchasing sand, gravel or crushed stone for concrete, it is important to see that the aggregates and particularly the sand do not contain an abnormal amount of moisture, otherwise you will be short on materials, which will reduce profits or may even result in a loss, particularly if there are considerable quantities of aggregates to be purchased.

The following suggestions are important when placing orders:

1. Determine whether the sand has a high or low moisture content.
2. Determine whether the sand has a high or low specific gravity.
3. When sand is purchased at a certain price per cu. yd. make certain that you are not paying for water instead of sand, as it is not uncommon to see settlement of 4" in trucks or railroad cars from the time it is shipped to the time it arrives at the job. This is especially true where washed sand is loaded immediately after washing or after hard rains.
4. When sand is purchased by weight, also make certain you are not paying for water instead of sand. Ordinarily sand should not contain over 4 to 5 percent moisture, while wet sand may contain as high as 18 percent moisture. This makes a difference in the cost of sand.
5. Watch the grading of sand and percentages from fine to coarse.
6. Determine the grading of the gravel and its specific gravity.
7. When gravel is purchased by the cu. yd. stipulate measurement at job or destination, if possible.
8. When gravel is purchased by weight, weight at destination should govern instead of weight at loading point, when gravel may contain much moisture.
9. When purchasing crushed stone, obtain the grading and specific gravity.
10. When purchasing crushed stone, determine the weight per cu. yd. also the moisture content.

An easy method of determining the moisture content of sand, gravel or crushed stone, is to measure out one cu. ft. of each and weigh them. Then dry this same material and weigh it. The difference in weight is the moisture content. Example: If one cu. ft. of loose damp sand weighs 105 lbs. and after drying it weighs only 90 lbs. then the moisture content is 15 lbs. or 16 2/3 percent.

Quantities of Materials Required for Job-Mixed Concrete.—For estimating purposes, material quantities for job mixed concrete given in the following table may be used as a guide. The quantities shown in the table will produce concrete of medium consistency or with a slump of about 3" to 4".

Suggested Mixes for   Non-Air-Entrained   Concrete of Medium Consistency
(3 to 4 in. slump*)

Water-cement ratio, lb. per lb.	Maximum size of aggregate in.	Air Content (entrapped air) per cent	Water, lb. per cu. yd. of concrete	Cement, lb. per cu. yd. of concrete	With fine sand—fineness modulus = 2.50**			With coarse sand—fineness modulus = 2.90**		
					Fine aggregate, per cent of total aggregate	Fine aggregate, lb. per cu. yd. of concrete	Coarse aggregate, lb. per cu. yd. of concrete	Fine aggregate, per cent of total aggregate	Fine aggregate, lb. per cu. yd. of concrete	Coarse aggregate, lb. per cu. yd. of concrete
0.40	⅜	3	385	965	50	1240	1260	54	1350	1150
	½	2.5	365	915	42	1100	1520	47	1220	1400
	¾	2	340	850	35	960	1800	39	1080	1680
	1	1.5	325	815	32	910	1940	36	1020	1830
	1½	1	300	750	29	880	2110	33	1000	1990
0.45	⅜	3	385	855	51	1330	1260	56	1440	1150
	½	2.5	365	810	44	1180	1520	48	1300	1400
	¾	2	340	755	37	1040	1800	41	1160	1680
	1	1.5	325	720	34	990	1940	38	1100	1830
	1½	1	300	665	31	960	2110	35	1080	1990

W/C	Max size		Water	Cement	Fine %	Fine lb	Coarse lb	Fine %	Fine lb	Coarse lb
0.50	⅜	3	385	770	53	1400	1260	57	1510	1150
	½	2.5	365	730	45	1250	1520	49	1370	1400
	¾	2	340	680	38	1100	1800	42	1220	1680
	1	1.5	325	650	35	1050	1940	39	1160	1830
	1½	1	300	600	32	1010	2110	36	1130	1990
0.55	⅜	3	385	700	54	1460	1260	58	1570	1150
	½	2.5	365	665	46	1310	1520	51	1430	1400
	¾	2	340	620	39	1150	1800	43	1270	1680
	1	1.5	325	590	36	1100	1940	40	1210	1830
	1½	1	300	545	33	1060	2110	37	1180	1990
0.60	⅜	3	385	640	55	1510	1260	58	1620	1150
	½	2.5	365	610	47	1350	1520	51	1470	1400
	¾	2	340	565	40	1200	1800	44	1320	1680
	1	1.5	325	540	37	1140	1940	41	1250	1830
	1½	1	300	500	34	1090	2110	38	1210	1990
0.65	⅜	3	385	590	55	1550	1260	59	1660	1150
	½	2.5	365	560	48	1390	1520	52	1510	1400
	¾	2	340	525	41	1230	1800	45	1350	1680
	1	1.5	325	500	38	1180	1940	41	1290	1830
	1½	1	300	460	35	1130	2110	39	1250	1990
0.70	⅜	3	385	550	56	1590	1260	60	1700	1150
	½	2.5	365	520	48	1430	1520	53	1550	1400
	¾	2	340	485	41	1270	1800	45	1390	1680
	1	1.5	325	465	38	1210	1940	42	1320	1830
	1½	1	300	430	35	1150	2110	39	1270	1990

*Increase or decrease water per cubic yard by 3 per cent for each increase or decrease of 1 in. in slump, then calculate quantities by absolute volume method. For manufactured fine aggregate, increase percentage of fine aggregate, by 3 and water by 15 lb. per cubic yard of concrete. For less workable concrete, as in pavements, decrease percentage of fine aggregate by 3 and add water by 8 lb. per cubic yard of concrete.
**For definition of fineness modulus, see *Aggregates for Concrete*, available from the Portland Cement Association.

Suggested Mixes for Air-Entrained Concrete Of Medium Consistency
(3 to 4 in. slump*)

Water-cement ratio, lb. per lb.	Maximum size of aggregate in.	Air content, per cent	Water, lb. per cu. yd. of concrete	Cement, lb. per cu. yd. of concrete	With fine sand— fineness modulus = 2.50**			With coarse sand— fineness modulus = 2.90**		
					Fine aggregate, per cent of total aggregate	Fine aggregate, lb. per cu. yd. of concrete	Coarse aggregate lb. per cu. yd. of concrete	Fine aggregate, per cent of total aggregate	Fine aggregate, lb. per cu. yd. of concrete	Coarse aggregate, lb. per cu. yd. of concrete
0.40	⅜	7.5	340	850	50	1250	1260	54	1360	1150
	½	7.5	325	815	41	1060	1520	46	1180	1400
	¾	6	300	750	35	970	1800	39	1090	1680
	1	6	285	715	32	900	1940	36	1010	1830
	1½	5	265	665	29	870	2110	33	990	1990
0.45	⅜	7.5	340	755	51	1330	1260	56	1440	1150
	½	7.5	325	720	43	1140	1520	47	1260	1400
	¾	6	300	665	37	1040	1800	41	1160	1680
	1	6	285	635	33	970	1940	37	1080	1830
	1½	5	265	590	31	930	2110	35	1050	1990
0.50	⅜	7.5	340	680	53	1400	1260	57	1510	1150
	½	7.5	325	650	44	1200	1520	49	1320	1400
	¾	6	300	600	38	1100	1800	42	1220	1680
	1	6	285	570	34	1020	1940	38	1130	1830
	1½	5	265	530	32	980	2110	36	1100	1990

Water-cement ratio, lb. per lb.	Maximum size of aggregate in.	Air content, per cent	Water, lb. per cu. yd. of concrete	Cement, lb. per cu. yd. of concrete	With fine sand—fineness modulus = 2.50**			With coarse sand—fineness modulus = 2.90**		
					Fine aggregate, per cent of total aggregate	Fine aggregate, lb. per cu. yd. of concrete	Coarse aggregate, lb. per cu. yd. of concrete	Fine aggregate, per cent of total aggregate	Fine aggregate, lb. per cu. yd. of concrete	Coarse aggregate, lb. per cu. yd. of concrete
0.55	⅜	7.5	340	620	54	1450	1260	58	1560	1150
	½	7.5	325	590	45	1250	1520	49	1370	1400
	¾	6	300	545	39	1140	1800	43	1260	1680
	1	6	285	520	35	1060	1940	39	1170	1830
	1½	5	265	480	33	1030	2110	37	1150	1990
0.60	⅜	7.5	340	565	54	1490	1260	58	1600	1150
	½	7.5	325	540	46	1290	1520	50	1410	1400
	¾	6	300	500	40	1180	1800	44	1300	1680
	1	6	285	475	36	1100	1940	40	1210	1830
	1½	5	265	440	33	1060	2110	37	1180	1990
0.65	⅜	7.5	340	525	55	1530	1260	59	1640	1150
	½	7.5	325	500	47	1330	1520	51	1450	1400
	¾	6	300	460	40	1210	1800	44	1330	1680
	1	6	285	440	37	1130	1940	40	1240	1830
	1½	5	265	410	34	1090	2110	38	1210	1990
0.70	⅜	7.5	340	485	55	1560	1260	59	1670	1150
	½	7.5	325	465	47	1360	1520	51	1480	1400
	¾	6	300	430	41	1240	1800	45	1360	1680
	1	6	285	405	37	1160	1940	41	1270	1830
	1½	5	265	380	34	1110	2110	38	1230	1990

* and **. See references bottom of previous table on p. 321.

Labor Mixing and Placing Concrete

The labor cost of mixing and placing concrete will vary considerably according to the size of the job, the method used in mixing and placing the concrete, the proximity of the stock piles to the mixer, and the distance the concrete has to be transported after mixing.

On large jobs using conveyors and bins for feeding the aggregate to the mixer, the purchase or rental of this equipment, together with the labor cost of placing same, should be estimated separately and not included in the cu. yd. price of placing the concrete. This plant and equipment cost may be the same whether the job contains 1,000 or 10,000 cu. yds. of concrete, and unless it is kept as a separate item, the resulting labor costs will be misleading and valueless insofar as estimating the cost of future jobs are concerned.

Ready-mix concrete is used in all parts of the country and where the concrete is delivered to the job mixed and ready to place in the forms, it is often possible for the truck mixers to deposit the concrete directly into the trenches or wall forms without any handling on the job, other than spading, vibrating, spreading and leveling the concrete.

However, on many jobs this is not possible and it is necessary to have a receiving hopper where the truck mixers dump the concrete and it is then hauled from receiving hopper to place of deposit either by wheelbarrows, hand or power buggies. On large operations the concrete may be pumped from receiving hopper to place of deposit.

Another method of placing the concrete that is often used on large operations is to use a crane which picks up a batch of concrete in a bucket and places it directly over the footings, piers or forms.

Where any of the above methods are used, all mixer labor is eliminated but it usually requires one or two men at the receiving hopper or at the truck dump to assist the truck mixer, prevent spillage, cleanup, etc.

Most specifications prohibit the use of long spouts for chuting concrete from the mixer to the forms on account of the great danger of segregation. Buggies or buckets are generally desirable. Self-propelled buggies are available and are used on many of the larger jobs.

Weather conditions also influence the labor costs considerably because placing concrete in freezing weather usually necessitates heating the sand, gravel or stone and water, and protecting the concrete from freezing by the use of tarpaulins, straw, oil burning salamanders, or other artificial means, all of which add to the overall expense.

With so many small mixers on the market very little concrete is hand mixed today, but occasionally a job presents itself where it is more economical to mix by hand.

Mixes for Small Jobs

For small jobs where time and personnel are not available to determine proportions in accordance with the recommended procedure, mixes in Table 1 will usually provide concrete that is amply strong and durable if the amount of water added at the mixer is never large enough to make the concrete overwet. These mixes have been predetermined in conformity with the recommended procedure by assuming conditions applicable to the average small job, and for aggregate of medium specific gravity. Three mixes are given for each maximum size of coarse aggregate. For the selected size of coarse aggregate, Mix B is intended for initial use. If this mix proves to be oversanded, change to Mix C; if it is undersanded, change to Mix A. It should be noted that the mixes listed in the table are based on dry or surface-dry sand. If the sand is moist or wet, make the corrections in batch weight prescribed in the footnote.

The approximate cement content per cubic foot of concrete listed in the table will be helpful in estimating cement requirements for the job. These requirements are based on concrete that has just enough water in it to permit ready working into forms without objectionable segregation. Concrete should slide, not run, off a shovel.

Table 1—Concrete Mixes for Small Jobs

Maximum size of aggregate, in.	Mix designation	Cement	Approximate weights of solid ingredients per cu. ft. of concrete, lb.			
			Sand.*		Coarse aggregate	
			Air-entrained concrete†	Concrete without air	Gravel or crushed stone	Iron blast furnace slag
½	A	25	48	51	54	47
	B	25	46	49	56	49
	C	25	44	47	58	51
¾	A	23	45	49	62	54
	B	23	43	47	64	56
	C	23	41	45	66	58

Maximum size of aggregate, in.	Mix designation	Approximate weights of solid ingredients per cu. ft. of concrete, lb.				
			Sand.*		Coarse aggregate	
		Cement	Air-entrained concrete†	Concrete without air	Gravel or crushed stone	Iron blast furnace slag
1	A	22	41	45	70	61
	B	22	39	43	72	63
	C	22	37	41	74	65
1½	A	20	41	45	74	65
	B	20	39	43	77	67
	C	20	37	41	79	69
2	A	19	40	45	79	69
	B	19	38	43	81	71
	C	19	36	41	83	72

*Weights are for dry sand, If damp sand is used, increase tabulated weight of sand 2 lb. and, if very wet sand is used, 4 lb.

†Air-entrained concrete should be used in all structures which will be exposed to alternate cycles of freezing and thawing. Air-entrainment can be obtained by the use of an air-entraining cement or by adding an air-entraining admixture. If an admixture is used, the amount recommended by the manufacturer will, in most cases, produce the desired air content.

Power Buggy.—A method of transporting construction materials, such as wet concrete, earth, sand, gravel, cement, etc., is through the use of a power operated buggy, which requires only steering on the part of the operator.

They are furnished in various sizes by different manufacturers and the cost varies according to size and capacity.

The power buggy is mounted on a power driven three-wheel chassis and is steered by means of a rear caster wheel. The maximum speed over the ground conforms to a fast walking pace either on the level or up to a 20 percent incline.

This machine has a capacity of 1,500 lbs.; a bucket volume of 10 cu. ft. and will handle 8 to 9 cu. ft. of wet concrete but a larger volume of light, dry materials. This buggy weighs 560 lbs., is powered by a 7 hp. gasoline engine having a fuel capacity of 2¾ gals. which will operate the machine for eight hours. The machine is 5'-5½" in length overall, 2'-7½" wide and is furnished with either single or dual wheels and has a turning radius of 34¾".

Comparison of Hand and Power Buggy Operation.—The smaller power buggy having a capacity of 8 to 9 cu. ft. of wet concrete will probably do the work of three men pushing hand powered concrete buggies.

A large machine having a capacity of 10 to 13 cu. ft. of wet concrete will probably handle as much concrete as 4 to 5 men using hand operated buggies, but it may be that heavier runways will be required to handle the larger loads of concrete.

Sizes, Weights and Capacities of Concrete Carts or Buggies

Capacity Cu. ft.	Turning Radius	Width Overall	Length Overall	Maximum Height	Approx. Weight
12	72"	34"	87 "	54"	950
15	82"	41"	94½"	54"	1020
21	91"	54"	106½"	56"	1400

The following is an approximate comparison of cost between the two methods; based on wheeling 100 cu. yds. of concrete from a 16-S mixer:

Hand Method	Quan.	Rate	Total	Rate	Total
5 laborers ....Hours	40	$. . . .	$. . . .	$12.54	$501.60
Cost per cu. yd...................................			. . . .		5.02
Power buggy method					
Power buggy, fixed charge/(per)					
Day	1	$. . . .	$. . . .	$75.00	$ 75.00
Fuel, gals, and oil	5	. . . .	. . . .	1.25	6.25
Operator...................................Hours	8	. . . .	. . . .	12.54	100.32
Total cost per day			$. . . .		$181.57
Cost per cu. yd................................			. . . .		1.82

On some jobs, it may be that 2 power buggies would be required to keep the mixer operating at capacity, and in such instances would be double that given above—but in either instance the saving is considerable.

Capacities of Concrete Mixers Based on A. G. C. Standards
(in Sacks of Cement per Batch)

Size of mixer denotes number of cubic feet of mixed concrete, plus 10 percent excess, when mixer is level.

Concrete	Sizes of Standard Mixers				
Proportions	3½ - S	6 - S	11 - S	16 - S	28 - S
1:1½:3	½	1	3	4	8
1:1½:3½	½	1	2	3	7
1:2 :3	½	1	2	3	7
1:2 :3½	½	1	2	3	7
1:2 :4	½	1	2	3	6
1:2½:4	½	1	2	3	6
1:3 :5	½	1	1	2	5
1:3 :6	½	1	1	2	4

Mixing Concrete in a One Sack (6-S) Mixer.—Foundation concrete is usually mixed in the proportions of 1 :2 :4, 1 :3 :5 or 1 :3 :6, which is a one sack mix using a 6-S mixer.

Often a 1 :3 :5 mix is used for foundation walls where watertight concrete is not necessary. This requires 1 sack of cement, 3 cu. ft. of dry sand and 5 cu. ft. of gravel or crushed stone, per batch. Wheelbarrows ordinarily used for dry materials hold 3 to 3½ cu. ft. loaded level full, so each batch will require 1 barrow of sand and 2 barrows of gravel or 9 cu. ft. of dry material.

Where possible to start mixing first thing in the morning and continue all day without delays, a mixer should average 30 batches (6 to 7 cu. ft. of concrete each) per hr. or 240 batches (54 to 62 cu. yds.), per 8-hr day. Runs like this on foundation work are exceptional, because in addition to the actual mixing time, it will be necessary to allow for delays, shutdowns, etc., as maximum output cannot be figured for the entire job, as the men must be paid for time getting ready to start work in the morning, cleaning and washing wheelbarrows or buggies and concrete mixer at noon and at night; cutting and removing old concrete drippings and spillage; building and removing runways; unloading cement; delays caused by non-delivery of materials, etc.

Based on 6 to 7 cu. yds. of concrete an hr. the following men will be required in the gang: 2 laborers loading and wheeling gravel or crushed stone; 2 laborers wheeling sand and cement; 1 laborer attending mixer; 3 or 4 laborers wheeling concrete from mixer to place of deposit; 2 or 3 laborers dumping barrows, spreading, leveling and tamping concrete; 1 mixer engineer (if required), and 1 mason foreman in charge of work (if required).

Allowing for average conditions, including runways, cleaning and repairing mixer, engine, etc., the labor per cu. yd. of concrete should cost as follows:

	Hours	Rate	Total	Rate	Total
Labor	2.5	$....	$....	$12.54	$31.35

Under extremely favorable conditions, this cost may be reduced ½ to ¾ -hr. per cu. yd. Add for mason foreman and mixer engineer, if required by union rules.

If a power buggy is used for wheeling concrete from mixer to place of deposit, one power buggy will take away all the concrete a 6-S mixer can turn out. This will eliminate 2 to 3 laborers wheeling concrete and result in a saving of about ¼ -hr. labor time per cu. yd. of concrete.

Mixing Concrete Using a 11-S Mixer.—When using a 11-S mixer of 1 to 2 sack mix capacity, the costs will run about the same as given for 6-S or 16-S mixers, under the same conditions.

Mixing Concrete in a 16-S (3 Sack Batch) Mixer.—A mixer of this size will turn out about 16 cu. ft. of mixed concrete at each batch, depending upon the proportions.

Opinions vary as to amount of mixing required. The time of mixing specified by architects and engineers usually varies between 1½ and 2 minutes per batch. The time specified affects the labor costs to a considerable extent as the longer mixing time restricts the day's output.

If the mixer has a side-loader and is fed by wheelbarrows with the concrete wheeled direct from the mixer to place of deposit, in wheelbarrows or concrete buggies, the mixer should average one batch every 2 minutes or 30 batches (13 to 15 cu. yds.) an hr.

Using an average of 2 minutes per batch, the mixer should turn out 104 to 120 cu. yds. per 8-hr. day, depending upon the concrete proportions and the size of each batch. In addition to the actual mixing time, it will be necessary to allow for delays, shutdowns, etc., as maximum output cannot be figured for the entire job, as the men must be paid for time getting ready to start work in the morning, cleaning and washing wheelbarrows and concrete mixer at noon and at night; cutting and removing old concrete drippings and spillage; building and removing runways; delays caused by non-delivery of materials, etc.

Based on an average of 14 cu. yds. of concrete an hr. it will require 25 to 28 men in the gang, distributed as follows: 6 laborers loading and wheeling gravel or crushed stone; 3 laborers wheeling sand; 1 or 2 laborers on cement; 1 laborer feeding mixer; 1 laborer dumping concrete from mixer into wheelbarrows or buggies; 5 to 7 laborers wheeling concrete from mixer to place of deposit, depending upon length of haul; 3 to 4 laborers dumping concrete carts or barrows, spreading, spading and tamping concrete in place; 3 to 4 laborers general labor, such as building and removing runways, cleaning concrete mixer, wheelbarrows or buggies, picking up or baling cement sacks and other miscellaneous labor; 1 engineer on mixer and 1 mason or concrete foreman in charge of gang.

On the above basis, the labor per cu. yd. of concrete should cost as follows:

	Hours	Rate	Total	Rate	Total
Labor	2.0	$....	$....	$12.54	$25.08
Foreman	0.13	....	....	13.04	1.70
Mixer engineer	0.13	....	....	17.95	2.33
Cost per cu. yd.			$....		$29.11

In many instances concrete will be placed for 65 to 75 percent of the above costs, but this will merely represent a day's run and not the average for the entire job.

Using a power buggy to transport the concrete from the mixer or hopper to place of deposit, one or two power buggies should handle all the concrete a 16-S mixer can turn out. This will eliminate 3 to 4 laborers wheeling concrete and result in a saving of 1/4 to 1/3 hr. per cu. yd. of concrete.

Mixing Concrete in a 28-S (1 Cu. Yd.) Mixer.—If the concrete is mixed in a 28-S mixer, fed by wheelbarrows and the concrete wheeled from the mixer to place of deposit the labor costs per cu. yd. will run approximately the same as given for 16-S mixers, under similar conditions.

While it is true the daily output will be greatly increased, the labor cost feeding the mixer and distributing the concrete will be increased in proportion.

Labor Mixing and Placing One Cu. Yd. of Concrete
for Foundations

Method of Mixing and Placing	Common Labor	Hours per Cubic Yard Mixer Engr.	Hoist. Engr.	Foreman
Hand mixing using one mixing board	3.33*	..	..	0.33
Hand mixing using two mixing boards	3**	..	..	0.25
Using one sack (6-S) mixer and wheeling to place	2.5†	..	..	0.17
Using two sack (11-S) mixer and wheeling to place	2	0.125	..	0.125
Using 16-S mixer and wheeling to place	2	0.125	..	0.125
Using 28-S mixer and wheeling to place	2	0.083	..	0.08
Using conveyors, batcherplant, hoist and wheeling to place	1.25	0.06	0.06	0.06

*In extreme warm weather or southern states, add 1/2 to 2/3 hr.
**If necessary to load concrete into barrows and wheel from mixing board to place of deposit, add 1/2 to 3/4 hr. per. cu. yd.
†Under extremely favorable conditions, this may be reduced 1/4 to 1/2-hr.

Above time includes all lost time in connection with concrete foundations.

Add for labor erecting and dismantling hoisting towers, spouts, hoisting engines, mixers, etc., as this may vary from $500.00 to $5,000.00, depending upon size of job.

If power buggies are used for wheeling concrete from mixer or receiving hopper to place of deposit, it will usually result in a saving of 1/4 to 1/3-hr. per cu. yd. of concrete.

Loading Mixer Using a Front-End Loader.

Since labor costs are high, many new methods are being used to reduce costs.

One method employs a loader to load the sand and gravel from the stock piles and dump it into the loading skip at the mixer. This eliminates the men loading wheelbarrows with sand and gravel by hand, but there is usually one man necessary on the stock pile, plus the small loader and operator. When supplying a 6-S mixer, it would eliminate about 5 men loading and wheeling sand and gravel but would add one man for the stock piles, plus the loader and operator.

When supplying a 16-S mixer with a loader, it would eliminate about 9 men loading and wheeling sand and gravel but would add 1 man on the sand stock pile; 1 man on the gravel stock pile, plus the loader and operator.

When supplying a 11-S mixer in this manner, it would eliminate 6 to 7 men loading and wheeling sand and gravel but would add 1 man on the stock piles, plus the loader and operator.

Mixing and Placing Concrete Where Materials Are Delivered to Job Batch Proportioned.—Another method of handling concrete work is to have the concrete aggregates delivered to the job in trucks, correctly proportioned for one batch of concrete, including cement, sand and gravel or crushed stone.

When this method is used, the concrete aggregates for each batch are proportioned in the material yard and delivered to the job ready to mix. The truck backs up to the mixer and discharges the cement, sand and gravel or crushed stone into the loading hopper. A paving mixer with a large, flat loading hopper is commonly used for this purpose as it facilitates dumping the aggregates from the trucks into the loading hopper.

This method eliminates the necessity of all men loading and wheeling cement, sand and coarse aggregates, and the time required per cu. yd. of concrete should average as given in the following table.

Labor Mixing and Placing One Cu. Yd. of Concrete for Foundations When Aggregates Are Yard Proportioned and Delivered to the Job a Batch to a Load

		Hours per Cubic Yard		
Method of Mixing and Placing	Common Labor	Mixer Engr.	Hoist. Engr.	Fore-man
16-S mixer and wheeling to place.	1.25	0.125	..	0.125
28-S mixer and wheeling to place.	1.25	0.083	..	0.083

If power buggies are used for wheeling concrete from mixer to place of deposit, it will usually result in a saving of 1/4 to 1/3]hr. labor time per cu. yd. of concrete.

Above includes all lost time in connection with mixing and placing concrete.

Add for labor erecting and dismantling hoisting towers, hoisting engines, mixers, etc., as this may vary from $500.00 to $5,000.00, depending upon size of job.

Placing Concrete with a Crane and Bucket.—The crane and bucket has become ever increasingly popular on jobs as a means for transporting mixed concrete from the mixer or ready-mix truck to the forms, thereby eliminating the necessity of building and removing runways, labor wheeling concrete, etc. This is especially true when the crane and bucket is used in conjunction with ready-mixed concrete as the amount of labor is reduced to a minimum.

Labor Placing One Cu. Yd. of Foundation Concrete Using Crane and Bucket

	Quantity Placed per 8-Hr. Day cu. yds.	Common Labor	Hours per cu. Yd.		
Method Mixed			Crane Operator	Mixer Engineer	Fore-man*
16-S mixer, batched material	70 to 90	.70	.10	.10	.10
28-S mixer, batched material	150 to 170	.50	.05	.05	.05
Ready-mix	150 to 170	.40	.05	...	.05

Add for equipment cost or rental of crane and bucket.
*Add only if required.

Cold Weather Concreting

During the late fall and early spring and during all of the winter months it is necessary to use special precautions to assure proper hardening of the concrete. Careful contractors will use protection whenever the temperature falls below 50° F. At such time the most important protection required is to heat the sand and mixing water.

During freezing weather an extra boiler (old hoisting engine boiler, for example) having a rating of 25 to 50 hp. according to the size of the job (10,000 to 25,000 sq. ft. floor area) with a steam gauge set for 2 or more lbs. pressure per sq. in. will be needed with at least 2 perforated pipes for heating the coarse aggregate in addition to that required for heating the sand. On larger jobs or for fast concreting where aggregate bins are used, a grillage of steam pipes 4'-0" apart at the bottom of the bins will be sufficient. These pipes should have 1/4" perforations about 4" on centers. A 1 1/2" steam line should be run from the boiler to the water tank and another line run up the side of the building with valves and connections for steam hose to be used for cleaning out ice and snow.

Sufficient steam should be available so the concrete has a temperature of 50 to 70 degrees when leaving the mixer. This will promote the setting of the cement and permit earlier removal of the forms.

On small jobs the flame of an oil heater is allowed to project into the mixer to heat the materials.

Footings are generally protected by covering the fresh concrete with 8" to 12" of hay or straw and with tarpaulins or polyethelene thrown over the hay.

Foundation walls should be protected in mild weather by covering the top of the wall with hay and the sides of the forms with tarpaulins. During more severe weather place oil burning salamanders every 30 to 40 ft. apart and cover with tarpaulins; care being taken to keep the tarpaulins sufficient distance from the salamanders.

In skeleton construction it will be necessary to curtain the walls with tarpaulins for the story being concreted and in very severe weather it may be necessary to curtain two stories with tarpaulins, and to cover the top floor with hay or straw, held down by planks or timbers, which in turn support a layer of boards and planks with tarpaulins thrown over them. Stair and elevator openings, and openings for steam and plumbing pipes will allow heat to penetrate beneath the protecting layers and to heat the top of the floor.

It will require one salamander for each 200 to 250 sq. ft. of floor surface according to the severity of the weather and sufficient fuel should be placed on the floor in the early afternoon to last until next morning. Heat should be supplied for 2 to 5 days or more depending upon whether high-early-strength or normal portland cement is used and on the temperature maintained. Barrels of water should be placed on each floor for fire protection.

Ready Mixed Concrete

When using ready-mixed concrete on all classes of reinforced construction, such as floors, beams, girders, columns, etc, the cost of handling and placing the concrete will run approximately the same as for job mixed concrete, except for the labor handling materials, supplying and operating mixer, etc., as the trucks dump the ready-mixed concrete into a receiving hopper or bucket, where it is distributed by wheelbarrows, concrete buggies or hoisted into place.

In addition to saving about 3/4 -hr. labor time per cu. yd. of concrete for charging mixer, the use of ready-mixed concrete saves the cost of cement sheds and provides more room for storing other materials. It permits higher output per hour because ready-mixed concrete can usually be furnished in any quantity desired and it also saves the cost of installing and dismantling expensive equipment, which may easily cost $5,000.00 or more on a job of any magnitude.

It is necessary to have a fairly large size hopper on the job to receive the concrete from the trucks, as this hopper must be of sufficient capacity to supply the hoisting bucket as fast as it can handle the concrete, otherwise the work will be slowed down and with the crew idle part of the time the labor costs go up.

Courtesy Portland Cement Association

Placing ready mixed concrete using bucket

The labor cost of handling and placing ready-mixed concrete will vary with the amount handled and placed per day, the method of handling, hoisting and placing the concrete, and the class of work.

The following tables gives the labor hours necessary to handle and place one cu. yd. of ready-mixed concrete, using various methods:

Labor Placing One Cu. Yd. of Ready-Mixed Concrete for Concrete Foundations

	Hours per Cubic Yard		
Method and Quantity Placed per 8-Hr. Day	Common Labor	Hoisting Engineer	Fore-man
Placing concrete directly in forms, 1 truck servicing job, 50 to 70 cu. yds	0.35-0.50	. . . .	0.13
Placing concrete directly in forms, 2 trucks servicing job, 100 to 150 cu. yds	0.27-0.40	. . . .	0.08
Using wheelbarrows, 50 to 70 cu. yds	1.25-1.50	0.17*	0.17**
Using wheelbarrows, 100 cu. yds. and over	1.00-1.25	0.13*	0.13**
Using concrete buggies, 50 to 70 cu. yds	1.00-1.25	0.17*	0.17**
Using concrete buggies, 100 cu. yds and over	0.90-1.10	0.13*	0.13**
Using 13 cu. ft. power concrete buggies	0.65-0.75	0.13*	0.13**

*Where the concrete is hoisted and it is necessary to use a hoisting engineer, add as given in table.
**If union regulations require a mason or cement finisher foreman in charge of crew, add as given in table.

Labor Placing One Cu. Yd. of Ready-Mixed Concrete for Reinforced Floors, Beams, Girders, Columns, Etc.

	Hours per Cubic Yard		
Method of Handling and Placing the Concrete	Common Labor	Hoisting Engineer	Fore-man
Hoisting concrete in wheelbarrows on a material hoist and wheeling into place.	2.3	.11*	.11*
Hoisting concrete in concrete buggies on a material hoist and wheeling into place.	1.1	.08*	.08*
Thin floors 2 to 2½-in. thick. Hoisting concrete in wheelbarrows on a material hoist and wheeling into place	2.8	.11*	.11*
Thin floors 2 to 2½-in. thick. Hoisting concrete in concrete buggies on a material hoist and wheeling into place	1.6	.08*	.08*
Hoisting concrete in a concrete bucket, discharging into a floor hopper and wheeling into place, using wheelbarrows	1.2	.05*	.05*
Hoisting concrete in a concrete bucket, discharging into a floor hopper and wheeling into place, using concrete buggies	0.8	.04*	.04*
If power buggies are used for transporting concrete from hopper to place of deposit, deduct per cu. yd	0.25-0.33	. . .	. . .
Mobile crane with bucket on ground, hoisting concrete to floor or forms and depositing directly into forms	0.5	. . .	.04*

Method of Handling and Placing the Concrete	Hours per Cubic Yard		
	Common Labor	Hoisting Engineer	Fore-man
Add for rental of mobile crane and operator	...	.04*	...
For thin concrete floors, 2 to 2½-in. thick, add to the above costs	0.5	...	...

*Time of hoisting engineer and foreman will vary according to the number of cubic yards of concrete placed per day.

The above costs do not include cost of erecting and dismantling material hoists, concrete hoisting towers, etc., which may run from $500.00 to $5,000.00, depending on size of job and type of equipment used.

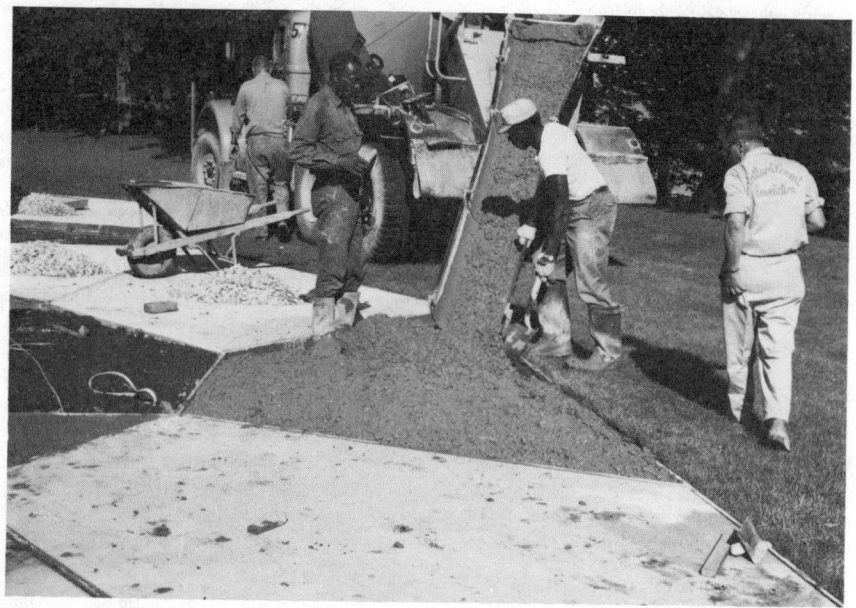

Courtesy Portland Cement Association
Ready mixed concrete placed directly from mixing truck

PLACING CONCRETE BY PIPELINE

The development and perfection of pumps for transporting freshly mixed concrete through pipelines has revolutionized the placing of substantial yardages of concrete in dams, tunnels, retaining walls, bridge decks, mill, factory, warehouse and apartment buildings where concrete is required at rates varying from 15 to 80 cu. yds. an hr. These machines for pumping concrete force it through a pipeline in much the same manner that a piston pump pumps water. It will transport freshly mixed workable concrete to distances of 1,000 ft. and to heights in excess of 120 ft. Many instances are known where a concrete pump has delivered concrete to the forms 1,300 ft. away, or 200 ft. high where a well designed mix is used.

Concrete pumps can handle efficiently concrete of any slump from ½" to 7" but the most dependable slump for general conditions is around 3" to 5". However, plastic concrete of about 6" slump can generally be pumped the maximum distance and height. The rate of pumping depends considerably upon the consistency of the concrete.

The concrete is delivered to the forms in better condition than when discharged from the mixer due to the added mixing received in the remixing hopper which

Placing ready mixed concrete using truck mounted pumper

Modern pumping rigs frequently replace bucket and cranes

receives the concrete from the mixer and keeps it in a uniform condition as it is fed through the cylinder of the concrete pump to the pipeline. During the pumping operation, the pipeline is full at all times which keeps the concrete from segregating, thereby eliminating one of the major objections found in some methods used to transport concrete to the forms. The troweling action of the walls of the pipeline in contact with the concrete as it passes through improves the quality of the concrete.

The smooth, even flow of concrete into the forms reduces the shock loading on the bracing and ties. Danger of "honeycomb" is reduced as there is no "slug" of pebbles dumped in, as in the case of a buggy, at the end of a batch which might not be properly spaded. Walls 120 ft. long and 15 ft. high have been made at one pour in 3 ft. lifts without signs of longitudinal jointing or uneven surfaces due to the dumping of wheelbarrows or buggies in such a way as to splash the form above the point of placement.

Runways for wheelbarrows or concrete buggies are not required when placing concrete by pipeline, although it is necessary to provide supports for the pipelines, which usually consists of "horses" or "bridges".

Mechanical maintenance will vary with the efficiency of the operator, the type of work, abrasiveness of materials, etc., as with all construction machinery.

In addition to replacements for the machine there will be pipeline maintenance which will be dependent upon the abrasiveness of materials. The couplings, which represent approximately three-fourths of the cost of pipe sections, will last indefinitely. The straight pipe will last from 50,000 to 150,000 cu. yds. and the elbows will last from 30,000 to 70,000 cu. yds., depending upon the abrasiveness of the concrete. The piping can be replaced as it becomes worn out and the old couplings reused.

Cost of Placing Concrete by Pipeline.—The cost of placing concrete by pipeline varies widely as do other methods of distribution, and is governed by the size of the job, area covered, the yardage of concrete that can be placed at one setup of the pipelines, distance from the pump, etc.

When placing concrete by ordinary methods, the cost of rental, erecting and dismantling the hoisting tower and spouts must be considered, also the cost of raising the concrete hopper and moving spouts for placing the concrete.

The same conditions must be considered with concrete pump distribution, which includes original cost of equipment, depreciation or salvage value, cost of setting up and removing pumping equipment, pipelines, etc.

Always bear in mind, the cost of mixing the concrete will be the same regardless of the method of distribution because where the concrete is mixed and dumped into a hoisting bucket using ordinary methods, it is deposited in the hopper. It is only the method of handling, hoisting and placing the mixed concrete that should be considered. This also applies where ready-mixed concrete is used.

The crew operating the pumping unit will vary with the size, area and yardage in the job. An operator is required on the concrete pump. A foreman and 6 to 12 laborers should place and remove pipelines, handle, spread and level the concrete, and the quantity will vary from 15 to 80 cu. yds. an hr. depending upon the size of the pumping equipment and the class of work.

Placing Concrete on a Commercial Building.—Assume a 5-story steel frame commercial building, 140 by 250 ft. having floors of metal pan construction, with pans from 6" to 14" deep and 2½" of concrete over the pans, the entire job containing approximately 5,200 cu. yds. of concrete.

Assume the concrete is ready-mixed, using ¾" gravel aggregate and having a 5" slump.

The concrete is delivered to the job and deposited directly into the remixing hopper on the pumper and pumped to the various floors.

Triple boom pumper placing ready mixed concrete on fifth floor

The labor per cu. yd. cost as follows:

	Hour	Rate	Total	Rate	Total
Pump operator	0.07	$....	$....	$17.95	$1.26
Hopper man	.07	....	....	12.54	.88
Labor vibrating,					
pipeline, etc	.49	....	....	12.54	6.15
Foreman	.07	....	....	13.04	.91
Cost per cu. yd			$....		$9.20

Average rate of pour 15 cu. yds. per hr. Maximum rate of pour 19 cu. yds. per hr.

The above costs do not include the cost of the pumper, depreciation, parts and replacements, gas and oil, labor installing and removing equipment, etc.

Placing Concrete by Pipeline on a Sewage Disposal Plant.—Assume a sewage disposal plant, size 370'-0"x380'-0", containing 15,000 cu. yds. of concrete consisting of heavily reinforced concrete in slabs and walls, all concrete to be placed by pipeline.

The mixer will discharge into the remixing hopper of a single pumper which is placed below grade in the excavation made for the plant addition. From this point the concrete is to be pumped through pipelines to the point of placement.

Workmen simultaneously vibrate concrete at two locations in double pumping operation

The floors of the aerating basins, 30 ft. wide and 120 ft. long are poured by radial spouting from the end of the pipeline, which is supported on horses. The piping is taken up as the concreting progresses and is laid in the next basin.

Staging supported on the forms will permit erection of the pipeline, pouring 112 cu. yds. of concrete in the wall and removal of the pipe line in one day. The walls of the aerating and settling basins are 15 ft. high and are poured in 3 ft. lifts. Valves at 20 ft. intervals are inserted in the 120 ft. of distributing line which permits the withdrawal of concrete at these points. Elephant trunks (tremie) direct the concrete into specially constructed chutes leading the concrete down through the maze of steel to the level being poured where it is discharged without spattering the forms above that point. Pouring was started from the valve nearest the pumper. When enough concrete had been withdrawn at a valve, it was closed and the next valve opened for concreting at this point. This procedure is continued until the lift is completed at the end of the pipeline.

With the aid of internal vibration, these high long walls are poured relatively free of blemishes or jointing.

The average labor cost of placing one cu. yd. of concrete is as follows:

Placing Concrete by Pipeline	Hours	Rate	Total	Rate	Total
1 pumper operator	.071	$....	$....	$17.95	$ 1.28
1 foreman	.071	....	....	13.04	.93
8 laborers*	.86	....	....	12.54	10.79
Cost per cu. yd.			$....		$13.00

*Labor on pumper includes time placing and removing pipeline.

During the pouring of the concrete in the walls, 2 additional men are required to operate vibrators. Vibrators would be required regardless of the distribution method used.

CAST IN PLACE CONCRETE

Concrete floors, sidewalks, and lightweight concrete floor fill are estimated by the sq. ft., taking the number of sq. ft. of any given thickness, as 2", 3",4", etc. to obtain the total cubic yards for placement.

The quantities are easily computed by obtaining the area of the various spaces to have concrete floors or walks and stating the number of sq. ft. of each thickness.

When preparing an estimate for concrete floors or sidewalks placed on the ground, there are distinct material and labor items to be considered as follows:

1. Labor grading and removing surplus earth under floors or walks or filling where present grade is too low.
2. Cost of sand, gravel or slag to be placed under concrete floor; labor spreading and tamping same in place.
3. Labor placing wood or metal screeds and forms.
4. Cost of vapor barrier and reinforcing material and placement.
5. Labor mixing and placing concrete.
6. Labor troweling finish surface and blocking off into squares, if required.
7. Labor and materials for curing and hardners, if required.

Grading for Concrete Floors and Walks.—On practically all cement floors and walks placed directly on the ground, it is necessary to do a certain amount of grading before the bed of sand, gravel or slag can be placed.

If it is not necessary to remove the excavated soil from the premises, a bulldozer or loader is probably the most economical method to use.

A medium size loader and operator should excavate and place in piles, 25 to 30 cu. yds. per hour at the following cost per cu. yd.:

	Hours	Rate	Total	Rate	Total
Loader w/Operator	0.045	$. . . .	$. . . .	$50.00	$2.25

This cost will vary with the kind of soil encountered, i.e., sand or loam, ordinary black soil, heavy soil or clay. Detailed excavating costs are given in chapter on "Site Work."

Number of Sq. Ft. of Soil of Various Thicknesses Obtainable
From One Cu. Yd.

	Depth of Excavation In Inches									
One cu. yd.	3"	4"	5"	6"	7"	8"	9"	10"	11"	12"
27 cu. ft.=	108	81	65	54	46	40	36	32	29	27

To obtain the cost of grading one sq. ft. of any thickness, divide the cu. yd. cost by the number of sq. ft. obtainable from one cu. yd. and the result is the sq. ft. cost for grading. Example: With excavating at $2.25 per cu. yd. find the sq. ft. cost of removing 5" of top soil. The table shows that one cu. yd. contains 65 sq. ft. of 5" fill. $2.25÷65=$0.0346 per sq. ft. for grading.

Hand Grading.—If the top soil is loaded by hand into wheelbarrows or power buggies, a man should loosen, shovel and load 5 to 6 cu. yds. per 8-hr. day, at the following labor cost per cu. yd.:

	Hours	Rate	Total	Rate	Total
Labor	1.45	$. . . .	$. . . .	$12.54	$18.18

Grading and Tamping Sand, Gravel or Slag Fill Under Concrete Floors or Walks.—Where sand, gravel or slag fill is placed under concrete floors or walks, the cost will vary with the thickness of the fill and job conditions.

It costs as much to grade a bed of slag or gravel 3" thick as one 9" thick, because it is only the surface that is graded and the only additional labor on the thicker fill is for handling, tamping, and spreading a larger quantity of material.

Where the fill varies from 3" to 4" thick, a man will spread, grade and tamp about 6 cu. yds. per 8-hr. day, at the following labor cost per cu. yd.:

	Hours	Rate	Total	Rate	Total
Labor	1.33	$....	$....	$12.54	$16.68

If the fill varies from 5" to 6" thick, a man will spread, grade and tamp about 8½ cu. yds. per 8-hr. day, at the following labor cost per cu. yd.:

	Hours	Rate	Total	Rate	Total
Labor	0.94	$....	$....	$12.54	$11.79

If the fill varies from 7" to 9" thick, a man will spread, grade and tamp about 10½ cu. yds. per 8-hr. day, at the following labor cost per cu. yd.:

	Hours	Rate	Total	Rate	Total
Labor	0.76	$....	$....	$12.54	$9.53

A man will spread, grade and tamp about 12½ cu. yds. of 10" or 12" fill per 8-hr. day, at the following labor cost per cu. yd.:

	Hours	Rate	Total	Rate	Total
Labor	0.64	$....	$....	$12.54	$8.03

All fill materials will shrink under compaction—some more than others—and the additional material required to make up the deficiency due to shrinkage, together with the additional labor for handling same, must be provided for in the estimate.

The best and easiest way to do this is to increase the quantity before pricing by applying the proper shrinkage factor for the material to the net computed volume.

For example: If an area 100'-0" x 50'-0" is to receive a gravel fill 6-inches thick after tamping, the gross volume of gravel would be 100'-0" x 50'-0" x 0'-6" x 1.12=2,800 cu. ft. or approximately 104 cu. yds. In the computation, 1.12 is the shrinkage factor for gravel.

The following table gives approximate values for shrinkage percentages and shrinkage factors for various fill materials:

Material	Shrinkage Percentage	Shrinkage Factor
Cinder	30	1.30
Crushed Limestone	20	1.20
Granulated Slag	15	1.15
Gravel	12	1.12
Sand	8	1.08

To obtain the cost of one sq. ft. of fill of any thickness, divide the cost per cu. yd. by the number of sq. ft. obtainable from one cu. yd., multiply by the shrinkage factor and the result is the sq. ft. cost.

Forms or Screeds for Concrete Sidewalks.—It is customary to use 2"x4" or 2"x6" lumber for forms at each side of a concrete sidewalk, the depth of the forms depending upon the thickness of the concrete. On ordinary sidewalk work, obtain the number of lin. ft. of walk and multiply by 2, which will be the number of lin. ft. of forms required. A sidewalk 4" thick will require 2"x4" lumber; 6" thick, 2"x6" lumber, etc.

Where concrete floors are placed on the ground it is customary to set wood screeds 6'-0" to 8'-0" on centers, so the floor will be level or have a uniform pitch. These screeds are removed as the work progresses and may be used many times.

They are usually of 2"x4" lumber, with stakes placed 3 to 4 ft. apart to hold the screeds in place.

Labor Placing Forms and Screeds for Sidewalks and Floors.—On all classes of sidewalk work where the forms are either 4", 6", or 8" wide, a carpenter should place and level 250 to 300 lin. ft. per 8-hr. day, at the following labor cost per 100 lin. ft.:

	Hours	Rate	Total	Rate	Total
Carpenter	3.0	$....	$....	$16.47	$49.41
Helper	1.5	....	....	12.54	18.81
Cost per 100 lin. ft			....		$68.22
Cost per lin. ft			....		.68
Cost per lin. ft. walk			....		1.36

The above includes time removing forms.

When placing screeds for concrete floors on the ground, 2 carpenters working together should drive stakes, place and level 650 to 700 lin. ft. of screeds per 8-hr. day, at the following labor cost per 100 lin. ft.:

	Hours	Rate	Total	Rate	Total
Carpenter	2.4	$....	$....	$16.47	$39.53
Helper	1.2	....	....	12.54	15.05
Cost per 100 lin. ft			$....		$54.58
Cost per lin. ft			....		.55

Estimating the Quantity of Concrete Required for Floors and Walks.—To compute the number of sq. ft. of concrete floor or sidewalk that may be obtained from one cu. yd. of concrete (27 cu. ft.), laid out into a cube 3'-0" wide, 3'-0" thick and 3'-0" high. The surface of this cube contains 9 sq. ft. and it is 36" high.

If the rough concrete is 4" thick, divide 36" (the height of the cube) by 4" (the thickness of the slab) and the result is 9; showing that it is possible to obtain 9 slabs, each containing 9 sq. ft. or 81 sq. ft. of 4" floor or walk per cu. yd. of concrete.

Number of Sq. Ft. of Concrete Floor of Any Thickness
Obtainable From One Cu. Yd. of Concrete

Thickness Inches	No. Sq. Ft.	Thickness Inches	No. Sq. Ft.	Thickness Inches	No. Sq. Ft.	Thickness Inches	No. Sq. Ft.
1	324	4	81	7	46	10	32
1¼	259	4¼	76	7¼	44	10¼	31
1½	216	4½	72	7½	43	10½	31
1¾	185	4¾	68	7¾	42	10¾	30
2	162	5	65	8	40	11	29½
2¼	144	5¼	62	8¼	39	11¼	29
2½	130	5½	59	8½	38	11½	28
2¾	118	5¾	56	8¾	37	11¾	27½
3	108	6	54	9	36	12	27
3¼	100	6¼	52	9¼	35	12¼	26½
3½	93	6½	50	9½	34	12½	26
3¾	86	6¾	48	9¾	33	12¾	25½

Mixing and Placing Concrete for Floors and Walks

Mixing Concrete By Hand.—In these days of high labor costs, concrete is seldom, if ever, mixed by hand but is usually mixed in a small mixer or ready-mix concrete is used and discharged direct into the forms or onto the floor.

Materials Required for 1 Cu. Yd. of Concrete

This table is based on use of average wet sand and on 3- to 4-in. slump. Quantities will vary according to grading of aggregate and workability desired. No allowance has been made for waste.   1 sack cement = 1 cu. ft.   4 sacks cement = 1 bbl.

| Size of Aggregate | Mixes | | | | | | Materials per Cu. Yd. of Concrete | | | |
| | Mixing Water* per sack of Cement U.S. Gal. | Portland Cement Sacks | Aggregates | | Yield of concrete per Sack of Cement Cu. Ft. | Water U.S. Gal. | Portland Cement Sacks | Aggregates | |
			Sand Cu. Ft.	Gravel or Crushed Stone Cu. Ft.				Sand Cu. Yd.	Gravel or Crushed Stone Cu. Yd.
Maximum size ¾ in.	5	1	2.50	2.75	4.3	31	6.25	0.58	0.64
Maximum size 1 in.	5	1	2.25	3.00	4.5	30	6.00	0.50	0.67
Maximum size 1½ in.	5	1	2.25	3.50	4.7	29	5.75	0.48	0.75

*Based on 6 gal. of water per sack of cement including water contained in average wet sand.

Materials Required for 100 Sq. Ft. of Surface Area for Various Thicknesses of Concrete*

Thickness	Quantity of Concrete	Mixes								
		1:2½:2¾			1:2¼:3			1:2¼:3½		
		Cement Sacks	Aggregate		Cement Sacks	Aggregate		Cement Sack	Aggregate	
In.	Cu. Yd.		Sand Cu. Yd.	Gravel or Crushed Stone Cu. Yd.		Sand Cu. Yd.	Gravel or Crushed Stone Cu. Yd.		Sand Cu. Yd.	Gravel or Crushed Stone Cu. Yd.
3	0.93	5.8	0.54	0.60	5.6	0.47	0.62	5.4	0.45	0.70
4	1.23	7.7	0.71	0.79	7.4	0.61	0.82	7.1	0.59	0.92
5	1.54	9.6	0.89	0.99	9.2	0.77	1.03	8.9	0.74	1.16
6	1.85	11.6	1.07	1.18	11.1	0.93	1.24	10.6	0.89	1.39
8	2.46	15.4	1.43	1.57	14.8	1.23	1.65	14.1	1.18	1.85
10	3.08	19.3	1.79	1.97	18.5	1.54	2.06	17.7	1.48	2.31
12	3.70	23.1	2.14	2.37	22.2	1.85	2.48	21.3	1.78	2.78

No allowance has been made for waste.
*100 sq. ft. of portland cement plaster ½ in. thick requires 1.6. sacks of cement and 0.15 cu. yd. of sand.

However, for that one in a million job, here is some data on hand mixing, if you can find the men who will work that hard.

As with other classes of concrete work, the cost of mixing and placing the concrete will depend upon the number of times it is turned on the board and the method used in mixing and placing; but on the smaller jobs it will require 3½ to 4 hrs. labor time to mix and place one cu. yd. of concrete, as this class of work does not proceed as fast as foundation work.

On work of this kind, the gang usually consists of 5 or 6 men, so it is not possible for one man to work at the same class of work all day, but must alternate between wheeling sand, gravel, cement, and mixing concrete, as the job may require.

Five men working together should handle all materials, mix, place and tamp 20 to 24 batches (½-cu. yd. each) or 10 to 12 cu. yds. per 8-hr. day, at the following labor cost per cu. yd.:

	Hours	Rate	Total	Rate	Total
Labor	3.64	$....	$....	$12.54	$45.65

Labor Mixing and Placing Concrete for Floors and Walks Using a Small Mixer.—A small concrete mixer of ½ or 1 sack (3½-S or 6-S) capacity is ordinarily used for floor and sidewalk work.

Where possible to use a full gang of men and keep the mixer in operation steadily, a mixer should discharge a batch every 2 minutes, but as in all other kinds of construction work there are always delays, such as engine adjustments, repairing mixer, delays caused by non-delivery of sand, gravel or cement, moving and placing runways, cleaning concrete barrows, etc., so the average for a job will be about 20 to 25 batches an hr. or 160 to 200 batches per 8-hr. day.

On the above basis, either a 3½-S or 6-S concrete mixer, the labor cost per cu. yd. should run as follows:

	Hours	Rate	Total	Rate	Total
Labor	2.5	$....	$....	$12.54	$31.35

Using a larger mixer will increase the output but the labor cost will increase in direct proportion. Many contractors have found that a one sack mixer is the most economical unit for floor and sidewalk work.

Labor Placing Ready-Mixed Concrete for Floors.—In order to place ready-mixed concrete for floors with the most economy it is necessary to have a crew large enough to handle the concrete quickly to eliminate waiting time charges by the material company and the supply of mixed concrete must be steady and of sufficient quantity to keep the crew busy.

A crew of 14 or 15 laborers, using buggies to wheel concrete, should place 22 to 27 cu. yds. of ready-mixed concrete per hr. at the following labor cost per cu. yd.:

	Hours	Rate	Total	Rate	Total
Labor	0.6	$....	$....	$12.54	$7.53

If the floor is small, such as for a residence, small addition, etc., the work must be organized in a different manner. It is obvious that a job of this sort is too small to keep a number of ready-mix trucks busy and therefore no more than 1 or 2 trucks will serve the placing operation.

Assuming a floor pour serviced by one 7 cu. yd. capacity ready-mix truck, delivering one load of concrete per hr., a crew of 6 or 7 laborers should be kept reasonably busy and still handle the ready-mixed load of concrete fast enough to prevent waiting time penalties.

Based on the above conditions the labor cost of placing one cu. yd. of floor concrete will be as follows:

	Hours	Rate	Total	Rate	Total
Labor	1.0	$. . . .	$. . . .	$12.54	$12.54

Estimating the Labor Cost Per Sq. Ft. of Floor or Sidewalk.—The labor costs given for mixing and placing concrete are based on a cu. yd. while floors and sidewalks are usually estimated by the sq. ft.

After the labor cost per cu. yd. has been obtained, divide this by the number of sq. ft. of floor obtainable from a cu. yd. of concrete and the result will be the labor cost per sq. ft. of floor. Example: Find the labor cost per sq. ft. of 4" floor. The table on previous pages shows that 81 sq. ft. of 4" floor can be obtained from one cu. yd. of concrete. If the labor cost is $12.54 per cu. yd. $12.54÷81=$0.155 per sq. ft.

Cost of any other thickness is obtained in the same manner.

Labor Applying A Finish on Reinforced Concrete Floors.—The cost of placing a finish on reinforced concrete floors is subject to wide fluctuations. One of the reasons is the custom of placing the concrete over as large an area as possible and then allowing the concrete to stand until the water disappears from the surface before starting finishing. However, the time required for cement to take its initial set depends upon the brand of cement used, as some cements are noted for being either "quick setting" or "slow setting."

The amount of water used in the concrete mix also has considerable bearing on the finishing cost because if "sloppy" concrete is used it requires much more time for the floor to set sufficiently to receive the finish than if concrete of a rather "stiff" consistency is used.

Architects and engineers specify the floor must be finished at the same time the concrete is placed, and in some instances it necessitates a large amount of overtime for finishers.

Where the gang places concrete right up until quitting time, it is a foregone conclusion that the finishers will have to work far into the night.

The item of overtime should be carefully considered by the estimator for as a usual thing the cement mason starts work when all other trades have quit for the day.

While under favorable conditions, some finishers may finish 175 to 200 sq. ft. of floor an hour on certain classes of work, the general average for a reinforced concrete job will vary from 70 to 80 sq. ft. an hr. or 560 to 640 sq. ft. per 8-hr. day.

The labor costs given are for straight time but the estimator should bear in mind that overtime costs time and a half or double time.

The labor per 100 sq. ft. of floor should cost as follows:

	Hours	Rate	Total	Rate	Total
Cement mason	1.3	$. . . .	$. . . .	$16.62	$21.61
Cost per sq. ft			. . . .		.22

Finishing Concrete Floors by Machine.—Over the years rapid strides have been made in the finishing of concrete floors by power operated machines. These machines are used for screeding, floating and troweling the finish floor.

When placing concrete floors, walks and pavements on the ground, the wood or metal floor screeds are placed in the usual manner and the concrete is struck off by using a power operated screeding machine, which makes 2½" transverse strokes on the header boards, leveling and compacting the mix. A steady pull forward by the operator, advances the machine at the rate of 10 ft. a minute or 600 lin. ft. per hour and will strike off a slab 6 to 20 feet in width.

Concrete Screeding Machine

A four man crew is usually required with the machine; two laborers spreading the concrete and two cement masons operating the screeding machine.

After the concrete slab has been screeded ready for troweling, allow sufficient time for the concrete to set hard enough to walk on before starting finishing.

For floating, the machine is equipped with 12 gauge steel trowels, which revolve on a 46" diameter. It usually requires 3 cement masons; 1 finisher to operate the machine and 2 finishers touching up corners, columns, edging, jointing, etc. On a well prepared slab, one finishing crew can float 4,000 sq. ft. of floor an hr. but this is not usually maintained due to job delays, etc.

A smaller finishing machine having a 34" diameter will float up to 3,000 sq. ft. of floor an hour under the same conditions as stated above.

After the floor has been floated, the heavy gauge steel floating trowels are removed and replaced with steel "finishing" trowels. The cement mason then guides the rotating, adjustable pitch trowels over the slab until a smooth, level surface is obtained.

Using the larger machines, and assuming a "four time over" was necessary to complete the job, it would require 4 hrs. of machine operation and assuming a ¾ to 1-hr. waiting period was necessary between each operation, it would require a total of 6¼ to 7 hrs. time to finish 4,000 sq. ft. of floor.

The labor screeding 4,000 sq. ft. of floor should cost as follows:

	Hours	Rate	Total	Rate	Total
Cement mason	3	$. . . .	$. . . .	$16.62	$49.86
Cost per sq. ft.			. . . .		.013

*Add Machine Rental Charges

After the floor has been screeded, the labor floating and troweling (4 times over), 4,000 sq. ft. of cement finish floor should cost as follows:

	Hours	Rate	Total	Rate	Total
Cement mason	32	$....	$....	$16.62	$531.84
Cost per sq. ft			....		0.133

*Add Machine Rental Charges

Color for Concrete Floors

Color for concrete is used for sidewalks, driveways, porches, tennis courts, floors, etc., the desired color being mixed into the cement topping which should be either ½, ¾ or 1-in. thick, depending upon service requirements.

The amount required varies with the different manufacturers but the following proportions generally will produce good results.

Pigments for COLORED CONCRETE Floors

Color desired	Commercial names of colors for use with cement	Approximate quantities required—lb. per sack of cement	
		Light shade	Medium shade
Greys,	Germantown lampblack*	½	1
blue-black	or carbon black* or	½	1
and black	black oxide of manganese*	1	2
	or mineral black	1	2
Blue	Ultramarine blue	5	9
Brownish red to dull			
brick red	Red oxide of iron	5	9
Bright red to vermillion	Mineral turkey red	5	9
Red sandstone to			
purplish red	Indian red	5	9
Brown to reddish-brown	Metallic brown (oxide)	5	9
Buff, colonial tint and	Yellow ochre or	5	9
yellow	yellow oxide	2	4
Green	Chromium oxide or	5	9
	greenish blue ultramarine	6	

* Only first-quality lampblack should be used. Carbon black is light in weight and requires very thorough mixing. Black oxide or mineral black is probably most advantageous for general use. For black use 11lb. of oxide for each sack of cement.

Approximate quantities required for 100 sq. ft. of floor:

	½" Top	¾" Top	1" Top
Gray portland cement	1.5 sacks	2.25 sacks	3.00 sacks
Clean sand	150 lbs.	225 lbs.	300 lbs.
Coarse aggregate, ⅜" max. size	263 lbs.	394 lbs.	525 lbs.
Conc. cement color*	13.5 lbs.	20.3 lbs.	27 lbs.

For 1" topping, use not more than 5 gal. mixing water, including water in aggregates. Smaller mixes in proportion.

Sand and coarse aggregate based on 100 lbs. per cu. ft.

Colors are usually packed in 9-lb. bags, 100-lb. drums and barrels.

When mixed in the proportions of 1 :2, use 9 lbs. of cement color to each bag of portland cement used in the topping.

It is advisable to establish the mixture from samples made under job conditions and allowed to dry. Careful measurement of every ingredient, including water, in each batch is essential to insure uniformity of color.

Final troweling should be done after the concrete has stiffened, as this assures a smooth, polished finish. Integral water-proofing or hardener can be added to mix materials, also polishing or a surface treatment can be applied, if desired.

Colorundum* Surface Colorant and Hardener for Concrete.—Colorundum is a ready-to-use powder containing non-crushable hardeners, dispersing agents and cementitious binders. It provides a durable, integral concrete color that does not wear off as paint and ordinary surface coatings do. When Colorundum is dusted on and trowelled into freshly poured concrete it produces bright, decorative surfaces with dense coloration and hard, abrasion resistant surfaces with long life.

Colorundum is for interior and exterior concrete floors and sidewalks. The hard wearing, highly decorative finish of Colorundum floors makes them ideal for use in factories, schools, hospitals, office buildings, showrooms, supermarkets, stores or any public building. Colorundum may be used inside or outside on traffic areas such as sidewalks, ramps, patios, terraces, sun decks and playgrounds. For best results, the slump of concrete monolithic slabs should not exceed 2½" and for toppings 3". Excess water causes laitance to rise to the surface, producing discoloration, and also causes separation which delays the finishing operation.

Do not use air entraining agents, or other admixtures in the mix.

When excess water disappears and the topping will hold up knee boards, uniformly dust on 2/3 of specified amount of Colorundum per 100 sq. ft. of surface.

After spreading, allow the dry material to wet up. It should then be wood floated and worked into the slab. The first floating should be discontinued as soon as the surface becomes wet. Floating should be resumed when surface moisture has disappeared. Do not steel trowel. Immediately apply the remaining 1/3 of the specified amount per 100 sq. ft. and thoroughly wood float. (For heavy duty finish, the quantity of Colorundum may be increased for the second dust coat.) Apply Colorundum exactly as it comes from the container; do not mix with cement or concrete.

Steel trowel to an even plane free from surface marks and voids. When the floor has obtained its initial set, give it a final steel trowel burnishing. Leave exterior areas, such as sidewalks, under wood float finish after the trowelling.

To cure, use an approved curing and sealing compound according to manufacturer's instructions or cover the surface after finish trowelling with polyethelene. This seals against evaporation of the water so vitally necessary to hydrate the cement, thereby developing optimum strength and density throughout the mass. It also protects the colored surface against spillages.

Use approximately 40 lbs. per 100 sq. ft. for normal traffic areas. For areas subject to extra heavy traffic, the quantity of Colorundum may be increased proportionately.

Gilco* Non-Shrink Grouting Compound.—Gilco Grout is a premixed compound containing iron particles, hardening and dispersing agents, binders and oxidizing agents. When mixed with water, it produces a mass which has controlled expansion to overcome the inherent shrinkage in cementitious mixtures. The me-

---

*W. R. Grace & Co.

---

*Gifford-Hill & Co., Inc.

tallic aggregate also provides a ductile surface with the "give and take" properties concrete must have to absorb the vibration of machinery, equipment and columns.

Gilco grout can be used wherever non-shrink concrete is mandatory: setting machinery on concrete foundations; full bearing surface grouting for building columns; and grouting anchor bolts, floor grids, steel sash and jambs.

Grouting should be continuous once it has started. Do not mix more grout at one time than can be rodded and placed in a period of 20 minutes. Add only enough water to make mix placeable—avoid an excess of water.

Anchor forms securely to prevent movement during placing and curing. Intersection of form and base slab can be filled with a mix of equal parts of cement and sand before placing grout. Be sure to allow adequate clearance between forms and base plate.

Remove waste material and water from anchor bolt holes. All oil, grease and paint in contact with grout should be thoroughly removed from base plates before grouting.

It is important to place grout quickly and continuously to avoid all effects of overworking, resulting in segregation, bleeding and break-down of initial set. Under no circumstances should the grout be retempered by introducing additional water after it has taken its initial set.

Lightweight Concrete Floor Fill

Concrete made of lightweight aggregate, such as cinders, slag, Haydite, pumice, vermiculite and other lightweight aggregates are often used for floor fill under marble and tile floors, between wood floor sleepers, screeds, etc.

The labor cost of placing concrete floor fill runs considerably higher than for ordinary concrete because the fill is usually only 2" or 3" thick, which necessitates covering a large floor area to place a comparatively small amount of concrete. In many instances it is necessary to pump or hoist the concrete to the upper floors in wheelbarrows or concrete buggies and then wheel it to the different parts of the building where it is to be used.

When placing concrete fill between wood floor sleepers or screeds, it is important that every precaution be taken to prevent the sleepers from being knocked out of level when wheeling over them.

When placing concrete floor fill, it is not advisable to use a large mixer because of the necessity of covering a large floor area to place a small amount of concrete. It usually requires considerable leveling and tamping, making it impractical to handle the concrete in large quantities. A concrete mixer of 1 or 2 sacks capacity is best adapted for this class of work or partial loads of ready-mix can be ordered.

Mixing and Placing Concrete Floor Fill.—Because of the additional labor required for spreading, grading and tamping concrete floor fill, the cu. yd. labor costs run higher than for foundation or sidewalk work.

There is also more labor required wheeling concrete from the mixer to the hoist and from the hoist to place of deposit, as floor fill is usually placed after the structural portions of the building have been completed, and it is not possible to carry on the work with much speed.

A crew of 13 men with a 6-S mixer should handle materials, mix and wheel into place 15 to 20 batches (containing 6 to 7 cu. ft. each) or 3½ to 5 cu. yds. of concrete an hour with the men distributed as follows : 2 men wheeling coarse aggregate from stock pile to mixer; 2 men wheeling sand and cement to mixer; 1 man attending mixer; 3 men wheeling concrete from mixer to hoist; 3 men wheeling concrete from hoist to place of deposit; 2 men dumping barrows and spreading concrete.

The labor per cu. yd. should cost as follows :

	Hours	Rate	Total	Rate	Total
Labor	3.00	$....	$....	$12.54	$37.62
Hoisting engineer	0.35	....	....	17.95	6.28
Cost per cu. yd			$....		$43.90

If union rules require a mixer engineer, deduct 0.35-hr. labor time and add 0.35-hr. engineer time per cu. yd.

If concrete is discharged from mixer directly into a hoist concrete bucket and wheeled from a floor hopper to place of deposit, deduct 0.50-hr. labor time.

If no hoisting is required, deduct 0.70-hr. labor time and omit hoisting engineer.

For concrete roof fill on flat roofs, deduct about 10 percent from the above costs. On pitch or gable roofs, add 25 to 50 percent, depending on steepness of slope, accessibility, etc.

After the cubic yard cost has been computed, to obtain the sq. ft. cost, divide the cu. yd. cost by the number of sq. ft. of fill obtainable from one cu. yd. and the result will be the labor cost per sq. ft. of fill. Example : Find the sq. ft. labor cost of 3" floor fill with the cu. yd. cost $43.90. The table on the previous pages gives 108 sq. ft. of 3" fill obtainable from one cu. yd. $43.90÷108=$0.406 per sq. ft.

To the above costs must be added the cost of setting screeds and all finishing operations.

Finishing Lightweight Concrete Floor and Roof Fill.

All concrete floor and roof fills require some finishing labor, if only setting of screeds and striking off fill at screed height.

Where fill is placed between wood sleepers, screeds are not required but the fill must be struck off and darbied so that no part of it is higher than the top of the sleepers. In addition, the fill should be spaded along both sides of each sleeper so that good anchorage is obtained.

For 2" thick concrete fill between wood floor sleepers, 2 cement masons working with a helper should strike off and darby 5,000 to 6,000 sq. ft. per 8-hr. day or about all the fill a small-mixer crew can mix and place in an 8- hr. day. The labor per 100 sq. ft. should cost as follows :

	Hours	Rate	Total	Rate	Total
Cement mason	0.30	$....	$....	$16.62	$4.99
Helper	0.15	....	....	12.54	1.88
Cost per 100 sq. ft			$....		6.87
Cost per sq. ft			....		.069

Finishing Concrete Floor Fill For Resilient Flooring.—Where concrete fill is used as a base for cork, asphalt, vinyl, rubber or vinyl asbestos floor covering, it is necessary to obtain a smooth level finish on the concrete to receive the finish floor material or all irregularities will show through on the finished surface.

To obtain the required degree of smoothness, screeds must be placed 4'- 0" to 6'-0" on centers and concrete fill must be struck off, darbied, floated and troweled.

If the fill is not over 2" thick, 2 cement masons working together should strike off, darby, float and trowel about 1,600 sq. ft. of floor per 8-hr. day at the following labor cost per 100 sq. ft. :

	Hours	Rate	Total	Rate	Total
Cement mason	1.0	$....	$....	$16.62	$16.62
Cost per sq. ft			....		.167

Finishing Concrete Roof Fill.—Concrete fill for flat roofs are usually struck off, darbied and floated. If fill is not over 2" to 3" thick, 2 cement masons with a helper should place screeds, strike off, darby and float 1,500 to 1,700 sq. ft. of roof fill per 8-hr. day at the following labor cost per 100 sq. ft.:

	Hours	Rate	Total	Rate	Total
Cement mason	1.0	$....	$....	$16.62	$16.62
Helper	0.5	....	....	12.54	6.27
Cost per 100 sq. ft			$....		$22.89
Cost per sq. ft				....	.23

Where concrete fill is placed on pitch or gable roofs, finishing is more difficult, screeds must be set closer together to help hold the fill from sliding down the slope and an additional helper is usually required. On work of this type, 2 cement masons and 2 helpers should place screeds, strike off, darby and float 900 to 1,000 sq. ft. of fill per 8-hr. day at the following labor cost per 100 sq. ft.:

	Hours	Rate	Total	Rate	Total
Cement mason	1.60	$....	$....	$16.62	$26.59
Helper	1.60	....	....	12.54	20.06
Cost per 100 sq. ft			$....		$46.65
Cost per sq. ft				....	.47

CONCRETE SLAB ON GRADE CONSTRUCTION

Previously this type of floor was built without adequate test data or established construction standards relating to conditions involving comfort—dampness and floor temperature. Many such floors have proved to be cold, especially at the outer edge. Some have also been damp enough to damage floor coverings and wall construction.

To determine the most satisfactory floor from the standpoint of comfort, the University of Illinois* studied the construction of concrete floor slabs which are laid on the ground and which are designed for use in climates where central heating is necessary.

Nine different types of concrete floors were tested in an effort to determine the proper design for slab on grade houses with respect to:

1. Heat losses of the different types of floors.
2. Temperatures at various points throughout the floors. This information was sought for proper placement of floor insulation.
3. The amount of moisture passing from the ground to the top of the concrete slab.

Insulation Requirements for Different Parts of the Country

*Reprinted by permission Small Homes Council from its circular F4.3, "Concrete Floors for Basementless Houses."

The floors selected for tests represented standard construction practices and, at the same time, permitted the use of varying amounts and types of insulation along the edges. The floor slabs were tested simultaneously under similar conditions in a specially constructed laboratory.

Methods of Constructing Concrete Floors for Basementless Houses.

When panel heating systems are used with concrete floors, the heat loss to the ground and at the edge is increased due to the higher temperatures maintained in the floors. To prevent excessive heat losses in such floors, a minimum of 2 inches of rigid waterproof insulation should be provided at the edge. In addition, the use of insulation under the entire heated floor area is recommended. Gravel and rock fills are desirable for drainage controls, but they have no insulating value.

The use of concrete floor slabs for basementless homes requires an insulating material around the perimeter of the slab (at the exterior walls) which will not be compressed by imposed loads or soil backfill and which will not be affected by soil acids, moisture, insects or vermin.

Fiberglass perimeter insulation and styrofoam are used successfully for this purpose and come in panels of various widths and thicknesses.

Cleaning Concrete Floors Using a Sanding Machine.—To clean droppings of paint, mortar, joint finishing cement, etc., from concrete floors before laying the finish floor surface, a power sanding machine does a satisfactory job at a considerable saving in labor.

This is accomplished by using a floor sander equipped with a sanding disk and open grit silicon carbon grain paper, which cuts through and removes the droppings satisfactorily.

A man operating a sanding machine will clean about 500 sq. ft. of floor an hour at the following cost per 100 sq. ft. :

	Hours	Rate	Total	Rate	Total
Machinery cost	0.2	$....	$....	$ 5.00	$1.00
Sanding disk		....	....	.75	.75
Labor	0.2	....	....	12.54	2.51
Cost per 100 sq. ft			$....		$4.26
Cost per sq. ft				....	0.043

When this work is performed by hand using a putty knife or steel scraper, a man will clean and remove droppings from 150 to 175 sq. ft. an hour, at the following cost per 100 sq. ft.:

	Hours	Rate	Total	Rate	Total
Labor	0.6	$....	$....	$12.54	$7.53
Cost per sq. ft				....	0.075

CONCRETE

Lightweight Concrete

There are two types of lightweight concrete. One type, in use for a considerable time, is produced by simply combining portland cement and a lightweight aggregate. The second is known as aerated concrete.

Aerated concrete is produced by a process depending upon the introduction of certain chemicals to generate gases which cause the mass to expand. The weight of aerated concrete may be controlled with a variation of from 20 lbs. per cu. ft. to approximately the weight of ordinary lightweight concrete. For average conditions a weight of 40 to 50 lbs. per cu. ft. with a compressive strength of about 500 lbs. per sq. in. is usually selected.

Aerated concrete is fireproof, has a low moisture absorption rate and provides good insulation against the transmission of heat and sound. It may be sawed or nailed as readily as other lightweight concrete.

All lightweight concretes are of decided advantage for partition walls and fireproofing, whether in monolithic or precast unit form and for floor fill and roof slabs. The estimator should consider the combined advantages of each type of lightweight concrete before selection.

Haydite Concrete.—Haydite concrete uses burnt clay or shale for coarse aggregate instead of gravel or crushed stone. This inert, cellular material weighs less than 50 lbs. per cu. ft. and when mixed with ordinary torpedo sand and portland cement, produces concrete weighing about 98 lbs. per cu. ft. instead of 150 lbs. for ordinary concrete.

Full structural strengths can be obtained with Haydite concrete and in many instances a redesign based on lighter weight concrete and reduced dead load more than pays for the slight additional cost of the aggregates.

Perlite Lightweight Concrete.—For lightweight concrete using Perlite as the aggregate, refer to "Concrete Floors and Walks."

Pottsco Lightweight Aggregate.—Pottsco lightweight aggregate is an expanded blast furnace slag. By special processes, molten slag is converted into hard, cellular clinker, which is crushed and screened to commercial sizes, ready for use with portland cement and water, without the addition of sand or other materials. It is supposed to be chemically inert and has no corrosive effect on structural or reinforcing steel, plaster or paint.

Pumice Lightweight Concrete.—For lightweight concrete using Pumice as the aggregate, refer to "Concrete Floors and Walks."

Vermiculite Lightweight Concrete.—Refer to "Vermiculite Concrete" section.

Waylite Lightweight Concrete.—For lightweight concrete using Waylite as the aggregate, refer to "Concrete Floors and Walks."

Reinforced Concrete

After the forms and reinforcing steel are in place, the job is ready for concrete. This should include the cost of gravel or crushed stone, sand and cement used in the concrete mixture as well as the labor cost of handling materials, mixing, hoisting, placing and curing the concrete.

For many years concrete used in reinforced concrete construction was specified to be mixed in the proportions of 1 :1 :2, 1 :1½ :3, 1 :2 :4, etc., depending upon the class of work for which it was used. More modern practice specifies that concrete used for specific purposes shall develop a compressive strength of 2,500, 3,000, 3,750 or 4,000 lbs. or more per sq. in. at 28 days.

On nearly all types of reinforced concrete building work, such as columns, beams, girders, floor slabs, etc., the coarse aggregate is graded from ¼" to 1" in size. Where conditions permit, however, coarse aggregate graded from ¼" to 1½" is used.

Estimating the Quantity of Concrete Required for Various Types of Concrete Floors

If all concrete floors were merely solid slabs of concrete it would be an easy matter to compute the quantity of concrete required per sq. ft. of floor but with so many combination floors consisting of metal pans, etc, having joists 4" to 7" wide, 4" to 15" deep and from 16" to 35" on centers, it requires considerable figuring to obtain accurate concrete quantities.

It is practically impossible to estimate accurately the amount of concrete required per sq. ft. of floor for the various combination joist floors, as they invariably have T beam construction at all beams and girders, solid concrete slabs usually extend in 6" to 1'-6" from the outside walls, the joists vary in width and are often doubled under partitions, around stair wells, elevator shafts, etc. The only accurate method is to figure the floors as though they were solid concrete and then deduct for the displacement of the tile or pans.

Quantity of Concrete Required for Steelform and Concrete Joist Floor Construction.—Floors of this type should be estimated by obtaining the entire slab volume as though of solid concrete and deduct for the displacement of the steelforms.

The deductions for steelform area in floors of this type is not simple since steelforms are furnished in a number of different sizes, with tapering sides and ends.

Steelforms furnished by different manufacturers vary slightly in shape but not sufficient to materially affect the concrete quantities. Ceco* steelforms have been used in the tables as being typical for general use.

Sizes and Cu. Ft. Displacement of Flange Type Steelforms

Ceco* flange type steelforms slope 1" in each 12" in height. Example: a 12" pan is 20" wide at the bottom and 18" wide at the top, because of the 1" slope on each side. See illustrations on following pages.

---

*The Ceco Corporation.

The sides of single tapered endforms slope 1" for each 12" in height and one end slopes ¼" for each 1" in height. For estimating purposes their displacement is the same as for standard steelforms in the following tables.

Depth of Steel-form in Inches	Width at Wide End	Width at Narrow End	Average Width	Length in Feet	Cubic Foot Displacement per 3-Ft. Section	Difference Between Tapered end forms and 3'-0" length of straight forms in Cu. Ft.
			Sizes and Cu. Ft. Displacement of Tapered End Forms			
6"	20"	16"	18"	3'-0"	2.192	0.244
8"	20"	16"	18"	3'-0"	2.900	0.322
10"	20"	16"	18"	3'-0"	3.594	0.399
12"	20"	16"	18"	3'-0"	4.274	0.475
14"	20"	16"	18"	3'-0"	4.944	0.549
16"	20"	16"	18"	3'-0"	5.553	0.666
6"	30"	25"	27½"	3'-0"	3.381	0.306
8"	30"	25"	27½"	3'-0"	4.481	0.406
10"	30"	25"	27½"	3'-0"	5.571	0.504
12"	30"	25"	27½"	3'-0"	6.649	0.602
14"	30"	25"	27½"	3'-0"	7.711	0.698
16"	30"	25"	27½"	3'-0"	8.706	1.188

Sizes and Cu. Ft. Displacement of Ceco* Flange Type Steelforms

Depth of Steelform in Inches	Width at Bottom	Width at Top	Average Width	Cubic Foot Displacement per Lin. Ft.
6"	10"	9"	9½"	0.396
8"	10"	8⅔"	9⅓"	0.518
10"	10"	8⅓"	9⅙"	0.636
12"	10"	8"	9"	0.750
14"	10"	7⅔"	8⅚"	0.859
16"	10"	7⅓"	8⅔"	0.964
6"	15"	14"	14½"	0.604
8"	15"	13⅔"	14⅓"	0.796
10"	15"	13⅓"	14⅙"	0.984
12"	15"	13"	14"	1.167
14"	15"	12⅔"	13⅚'"	1.345
16"	15"	12⅓"	13⅔"	1.482
6"	20"	19"	19½"	0.812
8"	20"	18⅔"	19⅓"	1.074
10"	20"	18⅓"	19⅙"	1.331
12"	20"	18"	19"	1.583
14"	20"	17⅔"	18⅚"	1.831
16"	20"	17⅓"	18⅔"	2.073
6"	30"	29"	29½"	1.229
8"	30"	28⅔"	29⅓"	1.629
10"	30"	28⅓"	29⅙"	2.025
12"	30"	28"	29"	2.417
14"	30"	27⅔"	28⅚"	2.803
16"	30"	27⅓"	28⅔"	3.298

*The Ceco Corporation.

Sizes and Cu. Ft. Displacement of Adjustable Steelforms

Depth of Steelform in Inches	Width at Bottom	Width at Top	Cubic Foot Displacement per Lin. Ft.
8"	10"	8"	0.514
10"	10"	8"	0.653
12"	10"	8"	0.792
14"	10"	8"	0.931
15"	10"	8"	1.000
8"	15"	13"	0.792
10"	15"	13"	1.000
12"	15"	13"	1.208
14"	15"	13"	1.417
15"	15"	13"	1.521
8"	20"	18"	1.069
10"	20"	18"	1.347
12"	20"	18"	1.625
14"	20"	18"	1.903
15"	20"	18"	2.042
8"	30"	28"	1.625
10"	30"	28"	2.042
12"	20"	28"	2.458
14"	30"	28"	2.875
15"	30"	28"	3.083

Cu. Ft. of Concrete Displaced by Adjustable Steelforms and Tapered Endforms

Depth of Steelform	Width of Steelform	Length of Each Row of Steelforms in Feet									
		10'	11'	12'	13'	14'	15'	16'	17'	18'	19'
8"	30"	15.33	16.96	18.58	20.21	21.83	23.46	25.08	26.71	28.33	29.96
10"	30"	19.33	21.38	23.42	25.46	27.50	29.54	31.59	33.63	35.67	37.71
12"	30"	23.33	25.79	28.25	30.71	33.16	35.62	38.08	40.54	43.00	45.45
14"	30"	27.33	30.21	33.08	35.96	38.83	41.71	44.58	47.46	50.33	53.21
15"	30"	29.33	32.42	35.50	38.58	41.66	44.75	47.83	50.91	54.00	57.08
		20'	21'	22'	23'	24'	25'	26'	27'	28'	29'
8"	30"	31.58	33.21	34.83	36.46	38.08	39.71	41.33	42.96	44.58	46.21
10"	30"	39.75	41.80	43.84	45.88	47.92	49.96	52.01	54.05	56.09	58.13
12"	30"	47.91	50.37	52.83	55.29	57.74	60.20	62.66	65.12	67.58	70.03
14"	30"	56.08	58.96	61.83	64.71	67.58	70.46	73.33	76.21	79.08	81.96
15"	30"	60.16	63.25	66.33	69.41	72.49	75.58	78.66	81.74	84.83	87.91

Cu. Ft. of Concrete Displaced by Flange Type Steelforms and Straight Endforms

Depth of Steelform	Width of Steelform	Length of Each Row of Steelforms in Feet									
		1'	2'	3'	4'	5'	6'	7'	8'	9'	10'
6"	10"	0.40	0.79	1.19	1.58	1.98	2.38	2.77	3.17	3.56	3.96
8"	10"	0.52	1.04	1.55	2.07	2.59	3.11	3.63	4.14	4.66	5.18
10"	10"	0.64	1.27	1.91	2.54	3.18	3.82	4.45	5.09	5.72	6.36

Cu. Ft. of Concrete Displaced by Flange Type Steelforms
and Straight Endforms (Cont.)

Depth of Steelform	Width of Steelform	Length of Each Row of Steelforms in Feet									
		1'	2'	3'	4'	5'	6'	7'	8'	9'	10'
12"	10"	0.75	1.50	2.25	3.00	3.75	4.50	5.25	6.00	6.75	7.50
14"	10"	0.86	1.72	2.58	3.44	4.30	5.15	6.01	6.87	7.73	8.59
16"	10"	0.96	1.92	2.88	3.84	4.80	5.75	6.71	7.67	8.63	9.59
		11'	12'	13'	14'	15'	16'	17'	18'	19'	20'
6"	10"	4.36	4.75	5.15	5.54	5.94	6.34	6.73	7.13	7.52	7.92
8"	10"	5.70	6.22	6.73	7.25	7.77	8.29	8.81	9.32	9.84	10.36
10"	10"	7.00	7.62	8.27	8.90	9.54	10.18	10.81	11.45	12.08	12.72
12"	10"	8.25	9.00	9.75	10.50	11.25	12.00	12.75	13.50	14.25	15.00
14"	10"	9.45	10.31	11.17	12.03	12.89	13.74	14.60	15.46	16.32	17.18
16"	10"	10.55	11.51	12.47	13.43	14.39	15.34	16.30	17.26	18.22	19.18

Depth of Steelform	Width of Steelform	Length of Each Row of Steelforms in Feet									
		1'	2'	3'	4'	5'	6'	7'	8'	9'	10'
6"	15"	0.60	1.21	1.81	2.42	3.02	3.62	4.23	4.83	5.44	6.04
8"	15"	0.80	1.59	2.39	3.18	3.98	4.78	5.57	6.37	7.16	7.96
10"	15"	0.98	1.97	2.95	3.94	4.92	5.90	6.89	7.87	8.86	9.84
12"	15"	1.17	2.33	3.50	4.67	5.84	7.00	8.17	9.34	10.50	11.67
14"	15"	1.35	2.69	4.04	5.38	6.73	8.07	9.42	10.76	12.11	13.45
16"	15"	1.48	2.96	4.44	5.92	7.40	8.88	10.36	11.84	13.32	14.80
		11'	12'	13'	14'	15'	16'	17'	18'	19'	20'
6"	15"	6.64	7.25	7.85	8.46	9.06	9.66	10.27	10.87	11.48	12.08
8"	15"	8.76	9.55	10.35	11.14	11.94	12.74	13.53	14.33	15.12	15.92
10"	15"	10.82	11.81	12.79	13.78	14.76	15.74	16.73	17.71	18.70	19.68
12"	15"	12.84	14.00	15.17	16.34	17.51	18.67	19.84	21.01	22.17	23.34
14"	15"	14.80	16.14	17.49	18.83	20.18	21.52	22.87	24.21	25.56	26.90
16"	15"	16.28	17.76	19.24	20.72	22.20	23.68	25.16	26.64	28.12	29.60

Depth of Steelform	Width of Steelform	Length of Each Row of Steelforms in Feet									
		1'	2'	3'	4'	5'	6'	7'	8'	9'	10'
6"	20"	0.81	1.62	2.44	3.25	4.06	4.87	5.68	6.50	7.31	8.12
8"	20"	1.07	2.15	3.22	4.30	5.37	6.44	7.52	8.59	9.67	10.74
10"	20"	1.33	2.66	3.99	5.32	6.66	7.99	9.32	10.65	11.98	13.31
12"	20"	1.58	3.17	4.75	6.33	7.92	9.50	11.08	12.66	14.25	15.83
14"	20"	1.83	3.66	5.49	7.32	9.16	10.99	12.82	14.65	16.48	18.31
16"	20"	2.07	4.14	6.21	8.28	10.35	12.42	14.49	16.56	18.64	20.70
		11'	12'	13'	14'	15'	16'	17'	18'	19'	20'
6"	20"	8.93	9.74	10.56	11.37	12.18	12.99	13.80	14.62	15.43	16.24
8"	20"	11.81	12.89	13.96	15.04	16.11	17.18	18.26	19.33	20.41	21.48
10"	20"	14.64	15.97	17.30	18.63	19.97	21.30	22.63	23.96	25.29	26.62
12"	20"	17.41	19.00	20.58	22.16	23.75	25.33	26.91	28.49	30.08	31.66
14"	20"	20.14	21.97	23.80	25.63	27.47	29.30	31.13	32.96	34.79	36.62
16"	20"	22.77	24.84	26.91	28.98	31.05	33.12	35.19	37.26	39.33	41.40

Cu. Ft. of Concrete Displaced by Flange Type Steelforms and Straight Endforms (Cont.)

Depth of Steelform	Width of Steelform	Length of Each Row of Steelforms in Feet									
		1'	2'	3'	4'	5'	6'	7'	8'	9'	10'
6"	30"	1.23	2.46	3.69	4.92	6.15	7.37	8.60	9.83	11.06	12.29
8"	30"	1.63	3.26	4.89	6.52	8.15	9.77	11.40	13.03	14.66	16.29
10"	30"	2.03	4.05	6.08	8.10	10.13	12.15	14.18	16.20	18.23	20.25
12"	30"	2.42	4.83	7.25	9.67	12.09	14.50	16.92	19.34	21.75	24.17
14"	30"	2.80	5.61	8.41	11.21	14.02	16.82	19.62	22.42	25.23	28.03
16"	30"	3.30	6.60	9.90	13.20	16.50	19.80	23.10	26.40	29.70	33.00
		11'	12'	13'	14'	15'	16'	17'	18'	19'	20'
6"	30"	13.52	14.75	15.98	17.21	18.44	19.66	20.89	22.12	23.35	24.58
8"	30"	17.92	19.55	21.18	22.81	24.44	26.06	27.69	29.32	30.95	32.58
10"	30"	22.28	24.30	26.33	28.35	30.38	32.40	34.43	36.45	38.48	40.50
12"	30"	26.59	29.00	31.42	33.84	36.26	38.67	41.09	43.51	45.92	48.34
14"	30"	30.83	33.64	36.44	39.24	42.05	44.85	47.65	50.45	53.26	56.06
16"	30"	36.30	39.60	42.90	46.20	49.50	52.80	56.10	59.40	62.70	66.00

Cu. Ft. of Concrete Displaced by Adjustable Steelforms and Straight Endforms

Depth of Steelform	Width of Steelform	Length of Each Row of Steelforms in Feet									
		1'	2'	3'	4'	5'	6'	7'	8'	9'	10'
8"	10"	0.51	1.03	1.54	2.06	2.57	3.08	3.60	4.11	4.63	5.14
10"	10"	0.65	1.31	1.96	2.61	3.27	3.92	4.57	5.22	5.88	6.53
12"	10"	0.79	1.58	2.38	3.17	3.96	4.75	5.54	6.34	7.13	7.92
14"	10"	0.93	1.86	2.79	3.72	4.66	5.59	6.52	7.45	8.38	9.31
15"	10"	1.00	2.00	3.00	4.00	5.00	6.00	7.00	8.00	9.00	10.00
		11'	12'	13'	14'	15'	16'	17'	18'	19'	20'
8"	10"	5.65	6.17	6.68	7.20	7.71	8.22	8.74	9.25	9.77	10.28
10"	10"	7.18	7.84	8.49	9.14	9.80	10.45	11.10	11.75	12.41	13.06
12"	10"	8.71	9.50	10.30	11.09	11.88	12.67	13.46	14.26	15.05	15.84
14"	10"	10.24	11.17	12.10	13.03	13.97	14.90	15.83	16.76	17.69	18.62
15"	10"	11.00	12.00	13.00	14.00	15.00	16.00	17.00	18.00	19.00	20.00

Depth of Steelform	Width of Steelform	Length of Each Row of Steelforms in Feet									
		1'	2'	3'	4'	5'	6'	7'	8'	9'	10'
8"	15"	0.79	1.58	2.38	3.17	3.96	4.75	5.54	6.34	7.13	7.92
10"	15"	1.00	2.00	3.00	4.00	5.00	6.00	7.00	8.00	9.00	10.00
12"	15"	1.21	2.42	3.62	4.83	6.04	7.25	8.46	9.66	10.87	12.08
14"	15"	1.42	2.83	4.25	5.67	7.09	8.50	9.92	11.34	12.75	14.17
15"	15"	1.52	3.04	4.56	6.08	7.61	9.13	10.65	12.17	13.69	15.21
		11'	12'	13'	14'	15'	16'	17'	18'	19'	20'
8"	15"	8.71	9.50	10.30	11.09	11.88	12.67	13.46	14.26	15.05	15.84
10"	15"	11.00	12.00	13.00	14.00	15.00	16.00	17.00	18.00	19.00	20.00
12"	15"	13.29	14.50	15.70	16.91	18.12	19.33	20.54	21.74	22.95	24.16
14"	15"	15.59	17.00	18.42	19.84	21.26	22.67	24.09	25.51	26.92	28.34
15"	15"	16.73	18.25	19.77	21.29	22.82	24.34	25.86	27.38	28.90	30.42

Cu. Ft. of Concrete Displaced by Adjustable Steelforms and
Straight Endforms (Cont.)

Depth of Steelform	Width of Steelform	Length of Each Row of Steelforms in Feet									
		1'	2'	3'	4'	5'	6'	7'	8'	9'	10'
8"	20"	1.07	2.14	3.21	4.28	5.35	6.41	7.48	8.55	9.62	10.69
10"	20"	1.35	2.69	4.04	5.39	6.74	8.08	9.43	10.78	12.12	13.47
12"	20"	1.63	3.25	4.88	6.50	8.13	9.75	11.38	13.00	14.63	16.25
14"	20"	1.90	3.81	5.71	7.61	9.52	11.42	13.32	15.22	17.13	19.03
15"	20"	2.04	4.08	6.13	8.17	10.21	12.25	14.29	16.34	18.38	20.42
		11'	12'	13'	14'	15'	16'	17'	18'	19'	20'
8"	20"	11.76	12.83	13.90	14.97	16.04	17.10	18.17	19.24	20.31	21.38
10"	20"	14.82	16.16	17.51	18.86	20.21	21.55	22.90	24.25	25.59	26.94
12"	20"	17.88	19.50	21.13	22.75	24.38	26.00	27.63	29.25	30.88	32.50
14"	20"	20.93	22.84	24.74	26.64	28.55	30.45	32.35	34.25	36.16	38.06
15"	20"	22.46	24.50	26.55	28.59	30.63	32.67	34.71	36.76	38.80	40.84

Depth of Steelform	Width of Steelform	10'	11'	12'	13'	14'	15'	16'	17'	18'	19'
8"	30"	16.25	17.88	19.50	21.13	22.75	24.38	26.00	27.63	29.25	30.88
10"	30"	20.42	22.46	24.50	26.55	28.59	30.63	32.67	34.71	36.76	38.80
12"	30"	24.58	27.04	29.50	31.95	34.41	36.87	39.33	41.79	44.24	46.70
14"	30"	28.75	31.63	34.50	37.38	40.25	43.13	46.00	48.88	51.75	54.63
15"	30"	30.83	33.91	37.00	40.08	43.16	46.25	49.33	52.41	55.49	58.58
		20'	21'	22'	23'	24'	25'	26'	27'	28'	29'
8"	30"	32.50	34.13	35.75	37.38	39.00	40.63	42.25	43.88	45.50	47.13
10"	30"	40.84	42.88	44.92	46.97	49.01	51.05	53.09	55.13	57.18	59.22
12"	30"	49.16	51.62	54.08	56.53	58.99	61.45	63.91	66.37	68.82	71.28
14"	30"	57.50	60.38	63.25	66.13	69.00	71.88	74.75	77.63	80.50	83.38
15"	30"	61.66	64.74	67.83	70.91	73.99	77.08	80.16	83.24	86.32	89.41

Sizes and Cu. Ft. Displacement of Ceco* Flange Type
Single Tapered Endforms

Endforms are 20" wide at one end tapering to 16" at the other end. 30" endforms are 30" wide at one end tapering to 25" wide at other end.

Sizes and Cu. Ft. Displacement of Ceco* Adjustable
Tapered Steel Endforms

Tapered endforms 3'-0" long are furnished for 30" wide steelforms only. Straight endforms 3'-0" long are furnished for the 30" standard widths and the 20", 15" and 10" special widths.

Depth of Steelform Inches	Width at Wide End	Width at Narrow End	Length in Feet	Cubic Foot Displacement per 3-Ft. Section
8"	30"	25"	3'-0"	4.317
10"	30"	25"	3'-0"	5.483
12"	30"	25"	3'-0"	6.629
14"	30"	25"	3'-0"	7.775
15"	30"	25"	3'-0"	8.347

*The Ceco Corporation.

Another Method of Estimating the Concrete Required for Ceco Steelform* Construction.—The following method of estimating the concrete requirements for Ceco steelform construction is that recommended by the manufacturers.

Figure the concrete as though it were a solid slab and then deduct for displacement of a certain number of lineal feet of straight metal steelforms and tapered endforms.

Ceco steelform construction is a combination of concrete joists and thin slabs. In computing the amount of concrete in the floor construction of a building designed for Ceco steelforms it is more convenient, and there is less chance for error, to multiply the floor area by the equivalent thickness of the floor construction than to separately figure the concrete in the joists by multiplying the lineal feet of joists by their unit of volume and adding thereto the concrete in the thin slabs.

The cubic feet of concrete per square foot of floor (or roof) area of Ceco steelform construction, is a function of the thickness of the thin slab, the depth of the joists, and the width and spacing of the joists. The additional concrete in the joists at their ends as formed by tapered endforms cannot be pro-rated on a square foot basis. The concrete added to the joists by the tapered endforms must be figured per lineal foot of beam or wall into which the joists frame, and this unit of volume must be multiplied by the number of lineal feet of beams and walls along which the tapered endforms are set, to determine the total amount of concrete added by the use of the tapered endforms.

For convenience in estimating, an arrangement in tabular form of the quantities of concrete required for Ceco flange type steelform construction, is given on the following pages. In this connection it is well to bear in mind that the strength of the metal of which Ceco steelforms are made permits sharp angles in the steelforms and an absolutely flat top, thus requiring a minimum of concrete. These tables should, therefore, not be used in computing the concrete for other types of metal form construction.

Typical Cross-section of Flangeforms

All Dimensions Are Outside To Outside

regular widths
FOR FORMING REGULARLY SPACED JOISTS

filler widths
FOR USE ONLY IN FILLING OUT ODD SPACES

Sizes of Ceco Flange Type Steelforms

---

*The Ceco Corporation.

Typical Cross-sections of Longforms

All Widths and Depths Are Outside Dimensions

regular widths
FOR FORMING REGULARLY SPACED JOISTS

filler widths
FOR USE ONLY IN FILLING OUT ODD SPACES

Sizes of Ceco Longform Steelforms

Typical Cross-sections of Adjustable Steelforms

All Dimensions Are Outside To Outside

regular widths
FOR FORMING REGULARLY SPACED JOISTS

filler widths
FOR USE ONLY IN FILLING OUT ODD SPACES

Sizes of Ceco Adjustable Type Steelforms

Concrete Required per Sq. Ft. of Floor Using Ceco† Flange** Type Steelforms 20 Inches Wide					
Depth of Steelform	Width of Joist	Cubic feet of concrete per sq. ft. of floor with slab thickness over steelforms of			Additional concrete required (in. cu. ft.) per lin. ft. of beam or bearing wall where standard tapered endforms are used*
		2"	2½"	3"	
6"	4"	.262	.304	.346	.13
	5"	.279	.321	.363	.12
	6"	.293	.335	.377	.12
8"	4"	.298	.340	.382	.17
	5"	.320	.362	.404	.16
	6"	.339	.381	.423	.16
10"	4"	.336	.378	.420	.21
	5"	.363	.405	.447	.20
	6"	.387	.429	.471	.19
12"	4"	.377	.419	.461	.25
	5"	.408	.450	.492	.24
	6"	.437	.479	.521	.23
14"	5"	.456	.498	.540	.28
	6"	.490	.532	.574	.27
	7"	.522	.564	.606	.26
16"	5"	.507	.549	.590	.32
	6"	.545	.587	.629	.31
	7"	.581	.622	.664	.30

*Amount of Concrete Given for Standard Tapered Endforms is for One Side of Beam Only.
**The above concrete requirements are also applicable to one piece longforms.
†The Ceco Corporation

Depth of Steelform	Width of Joist	Cubic feet of concrete per sq. ft. of floor with slab thickness over steelforms of			Additional concrete required (in. cu. ft.) per lin. ft. of beam or bearing wall where standard tapered endforms are used**
		2"	2½"	3"	
6"	3½"	.262	.304	.345	.13
	4½"	.279	.321	.362	.12
	5½"	.295	.337	.378	.12
8"	3½"	.289	.329	.370	.17
	4½"	.309	.350	.393	.16
	5½"	.331	.372	.414	.16
10"	3½"	.312	.353	.395	.21
	4½"	.340	.381	.424	.20
	5½"	.367	.408	.450	.19
12"	3½"	.336	.378	.420	.25
	4½"	.371	.412	.455	.24
	5½"	.403	.444	.486	.23
14"	3½"	.361	.403	.444	.29
	4½"	.402	.443	.485	.28
	5½"	.438	.480	.522	.27

Table title: Concrete required per Sq. Ft. of Floor Using Ceco† Adjustable Steelforms 20 Inches Wide

Depth of Steelform	Width of Joist	Cubic feet of concrete per sq. ft. of floor with slab thickness over steelforms of			Additional concrete required (in. cu. ft.) per lin. ft. of beam or bearing wall where standard tapered endforms are used*
		2½"	3"	3½"	
6"	5"	.293	.334	.376	.11
	6"	.304	.346	.387	.11
	7"	.315	.357	.398	.11
8"	5"	.322	.364	.405	.15
	6"	.338	.380	.421	.15
	7"	.351	.392	.434	.14
10"	5"	.353	.395	.436	.19
	6"	.372	.414	.455	.18
	7"	.389	.430	.472	.18
12"	5"	.386	.427	.469	.22
	6"	.408	.450	.491	.22
	7"	.430	.470	.512	.21
14"	5"	.420	.460	.503	.26
	6"	.447	.488	.530	.25
	7"	.470	.511	.553	.24
16"	5"	.451	.493	.535	.29
	6"	.482	.523	.565	.28
	7"	.510	.552	.594	.27

*Table title: Concrete Required per Sq. Ft. of Floor Using Ceco† Flange** Type Steelforms 30 Inches Wide*

*Amount of Concrete Given for Standard Tapered Endforms is for One Side of Beam Only.
**The above concrete requirements are also applicable to one piece longforms.
†The Ceco Corporation.

Concrete Required per Sq. Ft. of Floor Using Ceco* Adjustable Steelforms 30 Inches Wide					
Depth of Steelform	Width of Joist	Cubic feet of concrete per sq. ft. of floor with slab thickness over steelforms of			Additional concrete required (in. cu. ft.) per lin. ft. of beam or bearing wall where standard tapered endforms are used**
		2½"	3"	3½"	
6"	5½"	.300	.341	.383	.11
	6½"	.311	.353	.395	.11
	7½"	.322	.364	.406	.11
8"	5½"	.326	.368	.408	.15
	6½"	.341	.383	.424	.15
	7½"	.355	.397	.438	.14
10"	5½"	.352	.394	.434	.19
	6½"	.371	.413	.454	.18
	7½"	.389	.430	.472	.18
12"	5½"	.378	.420	.461	.22
	6½"	.401	.443	.484	.22
	7½"	.422	.464	.505	.21
14"	5½"	.404	.446	.486	.26
	6½"	.430	.472	.514	.25
	7½"	.456	.497	.539	.24

*The Ceco Corporation.
**Amount of concrete given for standard tapered endforms is for one side of beam only.

Ceco* One Piece Steeldomes for Two-Way Dome Slab Construction.—To estimate the quantity of concrete required for steeldome construction, figure the concrete required for a solid slab of the depth of the dome plus the top slab, and subtract the voids created for each steeldome as given in the table below.

Ceco Steeldomes

Steeldome Sizes				
	19" x 19" SYSTEM	30" x 30" SYSTEM		
Item	Standard Size	Standard Size	Filler Sizes	
Plan Size of Void	19" x 19"	30" x 30"	20" x 30" and 20" x 20"	
Overall Plan Size Including Flanges	24" x 24"	36" x 36"	26" x 36" and 26" x 26"	
Width of Flange	2½"	3"	3"	
Depths of Domes	4", 6", 8", 10" & 12"	8", 10", 12", 14" 16" & 20"	8", 10", 12", 14", 16" & 20"	

*The Ceco Corporation.

Table of Concrete Voids				
VOIDS CREATED BY STEELDOME, CUBIC FEET				
Depth of Steeldome	19" x 19" SYSTEM	30"x30" SYSTEM		
	Regular 19" x 19"	Regular 30" x 30"	Fillers	
			20" x 30"	20" x 20"
4"	0.77			
6"	1.09			
8"	1.41	3.85	2.54	1.65
10"	1.90	4.78	3.13	2.06
12"	2.14	5.53	3.63	2.41
14"		6.54	4.27	2.87
16"		7.44	4.85	3.14
20"		9.16	5.90	3.81

Lightweight Concrete, Insulating Concrete of Pumice, Perlite, Vermiculite and Other Lightweight Aggregates

Lightweight concrete of crushed slag, foamed slag, haydite, pumice, perlite and vermiculite are used for floor and roof fill, structural roofs, floors in cold storage rooms, etc.

A comparison of concrete aggregates is as follows:

Type of Aggregate	Aggregate Weight per cubic foot Pounds	Weight per Cu. Ft. of Concrete Using Aggregate, Pounds
Gravel	120	150
Sand	90-100	150
Crushed stone	100	145

	Aggregate Wt. per Cu. Ft. Lbs.	Concrete Wt. per Cu. Ft. Lbs.
Crushed Bank Slag	80	100-130
Haydite	40-60	100-120
Foamed slag	40-60	90-100
Cinders	40-50 (plus sand)	110-115
Pumice	30-60	60- 90
Diatomite	28-40	55- 70
Perlite	6-16	20- 50
Vermiculite	6-10	20- 40
Waylite	40-60	90-100
Expanded Shale	44-58	92- 98

Vermiculite Insulating Concrete Aggregate.—Vermiculite concrete aggregate is a type of mica which is mined, crushed and screened to size. When subjected to a temperature of about 2,000° F., it expands into a laminated granule containing millions of tiny dead air cells. This produces a lightweight aggregate weighing 6 to 10 lbs. per cu. ft., as compared to sand or crushed rock aggregate weighing about 100 lbs. per cu. ft. Vermiculite has high insulating value, is fireproof and will not rot or decay.

Vermiculite insulating concrete is a lightweight building material made like ordinary portland cement concrete, except vermiculite concrete aggregate is used instead of sand, gravel or crushed stone. Depending on the amount of vermiculite aggregate used, it is possible to make concrete weighing from 20 to 40 lbs. per cu. ft. Ordinary concrete weighs 145 to 150 lbs. per cu. ft.

Vermiculite concrete is used for poured concrete roof decks over a variety of forming materials; for roof insulation over such surfaces as concrete, steel, wood, etc.; and for poured concrete floor slabs.

Because of its high insulation value, vermiculite concrete is an excellent base for radiant heat floor installations and for grade level floors without radiant heat in homes, farm structures and all types of commercial buildings. At a density of approximately 25 lbs. per cu. ft., 1" of vermiculite concrete has an insulating value equal to approximately 16" of ordinary concrete. The density, insulating value and strength of vermiculite concrete can be varied to meet a wide range of design requirements.

Vermiculite concrete is incombustible and has earned the highest fire ratings attainable. It carries a 5-hour fire rating for spandrel wall construction and a 1½-hour rating on a steel deck covered with 1½" of vermiculite concrete and completely unprotected on the underside.

Mixing.—Vermiculite concrete should be mixed in a mechanical mixer. The required amount of water and cement shall be placed in the mixer, and then the aggregate. Mixing shall be limited to the minimum time required to obtain a thorough mix and proper fluidity. (Maximum time recommended—5 minutes)

Note—When transit-mixed vermiculite concrete is used, the operation shall be as follows:

(a) Introduce water and cement into mixer. (Fill auxiliary water tank before leaving plant)

(b) Rotate the mixer slowly until all aggregate has been added.

(c) Continue to rotate mixer for approximately one minute after aggregate is in the mixer.

(d) Do not rotate drum on way to job site. Mix concrete at job site at fastest speed until it is uniform and flows freely from the mixer.

VERMICULITE CONCRETE ROOF INSULATION
POURED OVER STRUCTURAL CONCRETE ROOF DECKS

VERMICULITE CONCRETE ROOF DECK
POURED-IN-PLACE OVER FORMBOARDS
CEMENT-ASBESTOS, GYPSUM, INSULATING FIBERBOARD, ACOUSTICAL FIBERBOARD
FIBROUS GLASS

VERMICULITE INSULATING CONCRETE
OVER VENTED METAL ROOF DECKS

Placing.—Vermiculite concrete shall be transported and placed immediately after mixing is completed, and the period between completion of the mixing and placing shall be of such short duration that the mixture does not appreciably change in consistency.

Vermiculite Concrete Insulation Over Vented Galvanized Steel Roof Decks, Precast Concrete Joists. (Two-hour fire rated.)

Vermiculite concrete shall not be placed when the temperature is less than 40°F. If the temperature expected after placing of the concrete is near or below 40°F., the mixing water shall be heated in the temperature range 75°–100°F.

Vermiculite concrete may be placed by the use of standard concrete equipment such as bucket hoists, buggies or by specially designed pumping equipment.

Machine Placement.—The use of high volume pumps to install cast-in-place vermiculite insulating concrete is an important development. Pumping or spraying the concrete is especially suited to roof decks with curved, sloped or irregular surfaces, and is just as efficient on flat roof decks.

A pump is fast and simple. It can be set up or taken down in about 30 minutes. The long hose lengths that are possible enable the crew to get over obstructions on the job easily.

The Vermiculite Institute annually approves a national roster of roof deck applicators appointed by institute members to assure the highest quality of finished work.

Vermiculite concrete is covered by the United States of America Standards Institute Specification A122.1-1965.

Vermiculite Concrete Insulation Over Precast Prestressed Concrete Double Tee Roof Slabs.

Vermiculite Insulating Concrete Floors-on-Grade.—A 1:4 mix of vermiculite concrete is recommended as an insulating base under a sand-gravel concrete topping for floors on grade. It is particularly effective in reducing heat loss where radiant heating coils or ducts are carried in the floor.

Minimum thickness of the slab should be 3". A reinforcing mesh of 6"x6"-10/10 ga. welded wire fabric, or equal, should be placed in the center of the vermiculite concrete, lapping the sides and ends one full mesh.

The sand-gravel topping should be designed to carry the maximum floor load. Minimum thickness should be 1½" except when radiant heating coils are used. In that case minimum thickness over the top of the coils should be 1". The radiant heating coils should be supported on chairs to assure a minimum of ½" of sand-gravel concrete below the coils.

Proportion Table for Vermiculite Concrete

Description of Work	Vermiculite Aggregate Cu. Ft.	Portland Cement Cu. Ft.
Poured concrete roofs	4	1
Poured concrete roof insulation	6 or 8	1
Poured concrete slabs		
On grade insulating slab under		
1½ ordinary cement topping	4	1
Panel or spandrel walls	4	1

Mixing Table For Vermiculite Concrete

Mixed by Volume Vermiculite Aggregate Cu. Ft.	Portland Cement Sacks	Approx. Density in Lbs. per Cu. Ft.	Compressive Strength in Lbs. per Sq. In.	Thermal Conductivity (K) in btu/sq/ft./inch/ hour/degree F
8	1	20 to 25	100 to 125	0.60
6	1	25 to 30	125 to 225	0.76
4	1	35 to 40	350 to 500	0.97

Approximate Quantities Required for One Cu. Yd. of Ready-Mixed Vermiculite Concrete

Mix	Bags (4 Cu. Ft.) Vermiculite Concrete Aggregate	Sacks (1 Cu. Ft.) Portland Cement	Water Gals.
8-1	7.5	3.75	85 to 95
6-1	7.5	5.0	85 to 95
4-1	7.5	7.5	85 to 95

Various Types of Roof Construction on Which Vermiculite Insulating Concrete May Be Used.

½" COMPRESSIBLE
EXPANSION JOINT IN
TOPPING AT PERIMETER

**SAND CONCRETE TOPPING
(1½" Min.)**

VERMICULITE MASONRY FILL

**VERMICULITE INSULATING CONCRETE
(3" Min.)**

Vermiculite Insulating Concrete Floor

— ROOF CONSTRUCTIONS —

3 HOURS:

Roof deck of 2" minimum thickness perlite concrete on 28 gauge galvanized steel form units supported by steel joists 4' on center. Ceiling of ⅞" perlite-gypsum plaster on expanded metal lath attached to ¾" furring channels wire-tied to lower chord of joists. (For details see Underwriters' Laboratories Class C-3, Design No. RC1-3.)

2" min Perlite concrete
28 gauge galvanized steel

7/8 Perlite-gypsum plaster Metal lath

2 HOURS:

Same as 3 hour rated system above except that the perlite-gypsum plaster ceiling shall be ¾" thick. (For details see Underwriters' Laboratories Class D-2, Design No. RC11-2.)

2" min Perlite concrete
28 gauge galvanized steel

3/4 Perlite-gypsum plaster Metal lath

3 HOURS:

Roof deck of 2½" thickness perlite concrete on paper-backed welded wire mesh supported by steel joists. Ceiling of ¾" perlite-gypsum plaster on high ribbed metal lath attached to lower chord of joists. (For details see Underwriters' Laboratories Class C-3, Design No. RC2-3.)

1 HOUR:

UNPROTECTED STEEL ROOF DECK

Corrugated galvanized steel deck topped with 2⅝" average thickness of perlite concrete reinforced with 48-1214 wire mesh. No ceiling protection. Beams protected with ⅞" perlite-gypsum plaster on metal lath. (For details see Underwriters' Laboratories Class E-I, Design RC2-1.)

— CURTAIN WALL CONSTRUCTIONS —

4 HOURS:

6" non-bearing wall (not including weatherproof facade). Exterior: 4" perlite-portland cement mix sprayed on paper-backed wire mesh. Interior: 1" furred perlite-gypsum plaster on paper-foil backed wire mesh. (For details see Underwriters' Laboratories Class B-4, Design No. 3.)

2 HOURS:

Non load-bearing wall consisting of 3" thickness perlite concrete spray-applied to paper-backed wire lath attached to 4" steel studs 16" on center. (For details see Underwriters' Laboratories Class D-2, Design No. 18.)

Perlite Concrete Aggregate

Perlite is a siliceous volcanic rock mined in western United States. When crushed and quickly heated to above 1500°F., it expands to form lightweight, non-combustible, glass-like particles of cellular structure. This material, white or light gray in color, weighs 7½ to 15 lbs. per cu. ft.

Perlite aggregate consists of expanded perlite sized for use in lightweight insulating concrete in place of sand or gravel. It is usually packed in 3 or 4 cu. ft. paper bags for easy handling.

The many, tiny glass sealed cells in each particle of expanded perlite make it highly insulating as well as comparatively nonabsorptive. Thus perlite mixes with about 30 percent less water than comparable lightweight aggregates.

Perlite concrete, in combination with portland cement forms a very lightweight concrete. Where ordinary concrete weighs 150 lbs. per cu. ft. the dry weight of perlite concrete can be designed from 20 lbs. to 40 lbs. per cu. ft. The extremely light weight and ease of handling make perlite concrete adaptable to practically any shape.

Perlite concrete has received up to 4-hour ratings in Underwriters' Laboratories fire tests. Its light weight makes it suitable for fireproof roof and floor fills, for thin concrete curtain walls (blocks, slabs or monolithic) and for many precast panel and block constructions.

For lightweight structural roof construction where perlite concrete is placed over galvanized steel forms, paper backed wire mesh, formboards, structural concrete, or other suitable form materials, use a 25 to 29 lb. per cu. ft. density perlite concrete. This offers an ideal balance of low dead weight, adequate compressive and indentation strengths, and good insulating value. For uses where higher strengths are more important than insulating value, i.e., floor fills and certain lightweight structural roof designs, use a 34 to 40 lb. per cu. ft. density perlite concrete.

Perlite concrete should be mixed in a paddle type plaster or a drum type concrete mixer. The required amount of water, air entraining admixture and portland cement should be placed in the mixer and mixed until a slurry is formed. The proper amount of perlite concrete aggregate should then be added to the slurry and all materials mixed until design wet density is reached. Perlite concrete may also be transit mixed.

Perlite concrete should be carefully deposited and screeded in a continuous operation until a panel or section is completed. Steel troweling should be avoided. Rodding, tamping and vibrating should not be used unless so specified by the architect.

It is recommended that a 1" air space or expansion joint be provided through the thickness of the perlite concrete at the juncture of all roof projections, such as skylights, penthouses, and parapet walls. Expansion joints are recommended in accordance with good concrete and steel construction practice and are based upon the provision that suitable consideration has been given to the use of adequate through-building expansion joints. Other types of insulating concrete may not require expansion joints because of high initial shrinkage.

Perlite concrete, during the curing period, should be protected for at least the first three days to keep it from drying out too rapidly or freezing. Freshly poured concrete should also be given adequate protection from heavy rain. Allow no traffic until concrete can sustain a man's weight without indentation.

Perlite concrete should not be placed in temperatures under 40°F. nor on frosted surfaces. In near-freezing weather the mixing water should be heated to between 75° and 100°F.

Perlite concrete is also used as a base for radiant heating coils. For grade level floors a moisture resistant barrier should be used and a suitable wearing surface applied over the perlite insulating concrete.

Considerable savings in dead weight are possible by using perlite concrete as a floor fill over cellular steel or pan-type floors in multiple story buildings. The surface may be covered with concrete topping and other surfacing materials after the perlite concrete has cured several days. Another floor fill material, using perlite blended with other aggregates, is perlite/sand or perlite/expanded shale concrete. These concretes have higher compressive strengths and require only resilient flooring or carpeting as a wearing surface. Following are the results of tests conducted at nationally recognized laboratories. Due to the varying characteristics of naturally occurring aggregates in different geographical areas, trial mixes of aggregate blends are recommended for determination of mix proportions.

Density Wet	Dry	Compression 28 day, psi	Cement (Sacks)	Perlite (cf)	Heavy Aggregate (cf)	AEA† (oz)	Water U.S. Gal/Sack	Ct. Factor cu. yd.
82	72	1000	1	3	2*	6	8.5	5.82
66	54	1200	1	3	2**	4	8.5	6.12

*Sand.   **Rotary Kiln Expanded Shale.   †Air Entraining Agent.

Perlite concrete can be nailed, sawed and worked with ordinary carpenter tools.

Mix Proportions By Volume for Lightweight Perlite Insulating Concrete

Cement (Sacks)	Perlite Cu. Ft.	Water-Gals. per Sack Cement	Air-Entraining Agent—Pints	Density Lbs. per Cu. Ft. Oven Dry	Comp. Str. psi-28 Days
1	4	9	1	36	400
1	5	11	1¼	30½	250
1	6	12	1½	27	170
1	7	14	1¾	24	120
1	8	16	2	22	95

Material Required for One Cu. Yd. of Placed Perlite Concrete

Proportions Cement Perlite	Cement Sacks	Perlite Cu. Ft.	Water Gallons	Air-Entraining Agent—Pints
1:4	6.75	27	61	6¾
1:5	5.40	27	59½	6¾
1:6	4.50	27	54	6¾
1:7	3.85	27	54	6¾
1:8	3.38	27	54	6¾

Pumice Concrete

Pumice is a natural white or gray glass foam. It is not a volcanic ash. It is mined extensively in the western part of the United States.

Uniform graded pumice is used as aggregate in monolithic concrete, concrete masonry units and precast products. It replaces sand and gravel in any concrete up to strengths of 3,000 lbs. per sq. in., where thermal insulation, acoustical and light weight properties are desirable.

All pumice aggregate shall be clean, free of all foreign matter and well graded so as to meet A.S.T.M. specifications C-330.

All water used in concrete work shall be free from oil, strong acids, alkali, or organic material.

Mixing.—Concrete for columns, walls above ground, floors, window sills and all self-supporting floor and roof slabs, canopies, and other parts of structure not coming in contact with ground shall be of a mixture of water, portland cement, and pumice aggregate, and shall develop at 28 days a strength of not less than 2,500 psi. and with a slump of not more than 2" by standard test.

Water contained in the aggregate shall be deducted from the amount of water used in the mix except that the water needed to saturate the pumice aggregate shall not be deducted. Pumice aggregate shall be thoroughly saturated either in the stockpile or in the mixer prior to the addition of cement and water for mixing.

SUGGESTED MIX DESIGN DATA FOR PYRAMID PUMICE CONCRETE

Compression lbs/ sq. in	Design Weight lbs/ cu ft	Sand lbs (dry)	Pyramid Pumice per Cubic Yard Yield						Cement		Water				Recommended slump
			Fine		Coarse		Blend		sax/ cu yd yield	lbs/ cu yd yield	to saturate		to mix		
			lbs (dry)	cu ft	lbs (dry)	cu ft	lbs (dry)	cu ft			gal	lbs	gal	lbs	
150	40	0	0	0	792	36	0	0	4	376	36.1	300	Determine by sight		0
250	46	0	914	24	264	12	1180	36	3	282	34.8	290	Determine by sight		0
500	55	0	914	24	264	12	1180	36	4	376	34.8	290	Determine by sight		0
1000	60	0	914	24	264	12	1180	36	5	470	34.8	290	50	417	½"
1500	65	0	914	24	264	12	1180	36	6.5	611	34.8	290	52	433	½"
2000	70	0	914	24	264	12	1180	36	7.5	705	34.8	290	53	442	1"
2500	75	0	914	24	264	12	1180	36	8	752	34.8	290	48	400	2"

Note: Pyramid Pumice Aggregate should be saturated prior to the addition of cement. This may be done in the stock pile or in the mixer. A 1" slump in pumice concrete will give a consistency equal to a 3" slump in hard rock concrete. Pumice concrete should be mixed with as little water as possible to get the required workability. It should be the consistency of thick mud and no free water should be in evidence. Air entraining agents are recommended to cut water-cement ratio and get greater workability. A combination of fine and coarse aggregate or the blended aggregate should be used. Volume measurements are based on loose volume noncompacted pumice.

Placing.—All pumice concrete shall be thoroughly vibrated with a mechanical vibrator. Vibration shall be conducted in such a manner as to assure that the vibrator passes through the pumice mass with not more than 6 inches of space between the successive positions of the vibrator. In all cases where pumice is placed in layers the vibrator must penetrate into the underlying concrete layer after passing through the newly placed layer. Ample time shall be allowed for the settlement of the concrete before final topping or screeding to finished floor level or grade. The pumice aggregate shall be thoroughly saturated prior to mixing with the cement and water used for mixing and ample time shall be allowed for the saturation to take place.

Application.—Pumice must be handled carefully if it is to be successful. Because it will float it must be pre-saturated before use. The stock pile should be sprayed for at least 48 hours or the pumice should be saturated in the mixer before use. Unless this is done, cement particles will be drawn into the cells of the pumice and excessive shrinkage may occur as a result of loss of water from the mix to the pumice particles. One half of the total water required should be added to the aggregate before the cement is introduced.

The concrete should be water cured for at least seven days after which time the outer fibers will have become strong and dense enough to protect against evaporation and resultant too rapid drying.

Gradations Available.—Pumice aggregate is graded to meet A.S.T.M. specifications for properly graded lightweight aggregrates for concrete. Designation C-330 in sizes No. 4 to 0 for fine aggregate; ⅜ and ½ to 0 for blended aggregate; ½" to No. 4 for coarse aggregate.

Shipping Weight.—Fine material or ¼" to 0 aggregate will have an average weight of about 1,200 lbs. per cu. yd. Blended aggregate or ⅜" to 0 material will weigh about 1,100 lbs. per cu. yd. The coarse aggregate or ½" to No. 4 material will have an average shipping weight of about 800 lbs. per cu. yd. These weights may increase from 100 to 200 lbs. per cu. yd. during wet seasons.

An open top, side dump, flat bottom (Caswell) car carries approximately 63 cu. yds. of material.

An open top, solid bottom, gondola car will carry approximately 63 cu. yds. of material.

An open top, hopper bottom car will vary in capacity from 60 to 100 cu. yds.

Waylite Lightweight Aggregate

A lightweight, cellular aggregate, refined and expanded by a process of agitating and rapidly cooling in an atmosphere of steam molten hot slag as it comes from the blast furnace at temperatures of 2500 to 3000°F. In this process the inherent gases expand and fill the material with minute air cells or bubbles completely sealed. These minute air cells are the active insulation and lightweight principle in all Waylite aggregates.

Waylite aggregate crushed and screened is available in three commercial sizes and average dry loose weights per cubic foot.

Aggregate	Grading	Average Weight
Coarse	7/16" to 3/16"	40 lbs.
Acoustical	3/16" to 3/32"	45 lbs.
Fine	1/8" to dust	60 lbs.

Waylite concrete requires a higher proportion of fine aggregate in relation to coarse than ordinary concrete. This is due to the lower top sizes of the coarse aggregate (7/16") and the additional angularity or harshness of the crushed fine and coarse aggregates.

The mixing of Waylite concrete requires a somewhat different manipulation procedure than for heavy aggregates. To obtain maximum workability with minimum weight, the aggregate and water should be mixed for one-half minute before the addition of the cement, and then mixed for at least an additional two minutes. A mix which appears harsh at first will improve in plasticity with continued mixing without additional water.

WAYLITE CONCRETE MIX DATA

Mix Loose Volume	Gals. Water per Sack	Cement Sacks	Fine Waylite Cu. Yds.	Coarse Waylite Cu. Yds.	Sand* Cu. Yds.
1-2¼-2½-¾*	9.4	6.4	0.53	0.60	0.17
1-2 -2¼-¾*	8.2	7.2	.53	.60	.20
1-1¾-2 -½*	7.3	8.1	.54	.60	.15
1-1½-2 -½*	6.5	8.8	.49	.65	.17
1-1½-1¾-½*	5.9	9.6	.53	.62	.18

Quantities Required per Cu. Yd., Loose Dry Volume

*Sand shall be a fine siliceous bank, lake, or mason sand.

The addition of approved concrete air entraining agents materially improve the workability of Waylite concrete mixes.

PLACING, PATCHING AND FINISHING ARCHITECTURAL CONCRETE

Architectural concrete demands the use of a plastic, workable mix that can be worked and molded readily in the forms. Surfaces of the hardened concrete should be free of defects such as sand streaking and honeycomb. It is important that sufficient fines be provided in the mixture and for these reasons the fine aggregate should have at least 15 percent passing through a 50 mesh sieve and 3 percent passing the 100-mesh sieve and the cement factor should not be less than 5½ sacks per cu. yd. of concrete. In other words, the concrete should not be richer than 1 :2 :3 by loose volumes and preferably 1 :2⅓, :3⅓, containing 6½ or 6 sacks of cement per cu. yd. of concrete respectively.

Not more than 6½ gals. of water per sack of cement should be allowed so that the concrete will have the durability required to withstand weathering.

Mixing and Placing Concrete.—Concreting of architectural concrete walls should proceed at a rather slow rate allowing sufficient time for thorough vibrating at the finished surfaces so that the pressure exerted by the concrete on the forms does not exceed 600 lbs. per sq. ft.

A reasonable rate of placing the concrete should be considered when locating joints. Generally a rate of not more than 2 feet an hour is most conductive to good workmanship and lifts of 6 to 8 feet between joints are desirable and 12 feet should be the absolute maximum.

The concrete should be spaded or vibrated into all corners and along all form surfaces, but even if vibrated, some spading is necessary in the corners and angles as vibrators will not fill these places at all times. On many jobs small electric hammers moved along the forms at the point of placement have been found helpful. Vibration should be sufficient only for complete compaction, not to the point where segregation occurs and excess water is forced to the form faces. Such overvibration may result in sand streaking and discoloration.

When concrete is splashed on the forms above the point of deposit and subsequently hardens, the wall in that area is likely to be rough and discolored. Splash boards are sometimes used but are not completely satisfactory as they do not prevent segregation of the falling concrete. Tremies of various lengths equipped with a hopper at the top through which the concrete is placed will avoid both splash and segregation. The tremies should be spaced not more than about 9 feet apart. Sometimes splashing is the result of mortar being thrown from vibrators as they are lifted out of the concrete and raised to the placing platform. This can be prevented by shutting off the power as the vibrator is withdrawn from the concrete.

Labor Placing Concrete.—The labor cost of placing architectural concrete will vary considerably, depending upon the size of job, size of sections, amount of ornamentation, quantity of reinforcing steel, etc., but on most work of this class, it will require 2 to 3 hours labor time per cu. yd. of concrete, plus the time for the hoisting engineer and foreman.

Based on a 64-cu. yd. pour per 8-hr. day, the labor cost should average as follows per cu. yd.:

	Hours	Rate	Total	Rate	Total
Foreman	0.13	$....	$....	$13.04	$ 1.70
Hoist engineer	0.13	....	....	17.95	2.33
Labor	2.50	....	....	12.54	31.35
Cost per cu. yd			$....		$35.38

Curing.—On this type of work one of the surest and best ways of curing the concrete is by leaving the forms in place. Whenever it is practical to do so, they should be left in place 4 or 5 days. By that time the concrete is hard enough so that it is not scuffed nor are corners so likely to be broken off during form removal. If the forms are removed earlier, other means must be provided to keep the concrete wet or damp for at least 5 days if a durable concrete is to be obtained.

Patching and Finishing.—When the forms are first stripped, the concrete does not always present a pleasing appearance as there may be pock marks, air holes, fins and some discoloration. Some of the latter will bleach out upon exposure but a simple clean down with cement grout will do wonders for the appearance. The projecting fins and other blemishes extending beyond the face of the wall are removed with a carborundum stone or abrasive hone and then a grout of 1 part portland cement to 1½ parts of very fine sand (sand passing through a fly screen will do) is applied. From ⅓ to ½ of the cement should be white cement.

The grout is mixed to the consistency of a heavy paste. After the wall is wet down, the grout is applied uniformly with a stiff fiber brush, completely filling all air voids and holes. Immediately after the grout has been applied, the wall is floated with a wood or cork float. A very fine abrasive hone can also be used. The grout is then allowed to harden partially, the exact time depending upon the weather. When it is hardened sufficiently so it will not be pulled out of the holes, all grout is scraped from the surface of the wall with the edge of a steel trowel. After the wall has dried thoroughly, it is rubbed with a dry piece of burlap to completely remove all of the dried grout or film. There should be no visible film left on the surface. Grout should not be left on the wall overnight. All cleaning of any one panel of concrete should be completed the same day.

Patching mortar used to fill tie holes and imperfections should be of the same mixture of gray and white cements, but should be a stiffer mixture and the cement and sand proportions should be the same as those used in the concrete. Patches should never be steel troweled, but may be finished with wood or cork floats. Allowing the mixed patching mortar to stand for an hour or two before using it reduces the amount of shrinkage but water must not be added in remixing it.

To correct blemishes appearing on the face of the walls, allow about 4 hrs. time for a cement finisher and a helper per 100 sq. ft. of wall. The labor cost per 100 sq. ft. of wall should average as follows:

	Hours	Rate	Total	Rate	Total
Cement mason	4	$....	$....	$16.62	$ 66.48
Helper	4	....	....	12.54	50.16
Cost per 100 sq. ft. wall			$....		$116.64
Cost per sq. ft. of wall			....		1.17

Cutting Concrete with Portable Concrete Saw

Many states now specify control joints in highway paving to be cut with a concrete saw in lieu of forming. Advantages of this method are as follows: Improved ridability of joint; less sealing material required to seal joint; better aggregate interlock at joint; no spalling or cracking at joint which in turn reduces maintenance costs.

This same technique is also widely used for cutting floors and paving for repair work, trenches for electric conduit, pipe lines, sewers, etc., and also for scoring before breaking out concrete for machinery bases and other foundation work.

Clipper* concrete saws, designed primarily for this service, are available in several electric and gasoline powered models.

The most popular models are the 18 hp. model C-188, the 30 hp. model C-305, the 37 hp. model C-375, and the 65 hp. model C-655, all gasoline powered. Approximate prices range from $1996.00 for the basic model C-188 manually propelled to $5595.00 for the model C-655 with self propelling unit and electric starter. The water pump and spotlight are optional features. The self propelling unit and other accessories are optional on the C-188 and C-305 models. The larger C-655 with 65 hp. engine is generally used for sawing contraction joints in airports, highways and turnpikes. The C-305 and C-188 are widely used in patching, trenching and plant maintenance work. Electrically powered models are available from 3 hp. and up.

Two types of blades are used in cutting concrete with concrete saws as follows: (1) Green-Con wet-dry abrasive blades for cutting green concrete (2) diamond blades, of various specifications, for cutting cured or old concrete.

The cost of cutting concrete with a concrete saw varies considerably, depending on age of concrete, type of aggregate, blade specification used, accessibility of work, skill of operator, etc.

Assuming proper blade specification is used for the cutting job, green concrete 24 to 48 hours old may be cut 1-inch deep at the rate of 12 to 14 lin. ft. per minute and 3-inches deep at the rate of 6 to 8 lin. ft. per minute. These rates cover only actual cutting time. Overall production will be reduced considerably depending on length of each individual cut, number of machine moves, changes in direction of cuts, etc.

---

*Norton Construction Products Division, Worchester, Mass.

Blade wear for a Green-Con blade of proper specification should cost from 1 to 2 cents per lin. ft. of cut 1-inch deep. For a 2-inch cut, blade cost should be 1½ to 3 cents per lin. ft.; 3-inch cut, 2 to 4 cents per lin. ft.

When cutting old or cured concrete, using a diamond blade of correct specification, cutting time may be double the amount given above for green concrete and cost of blade wear will probably run 5 to 6 times that given above.

Cutting Concrete Pavement with a Portable Concrete Saw

CEMENT FLOOR HARDENERS

There are many preparations on the market for waterproofing, dustproofing and wearproofing cement finish floors. These preparations fall into two classifications, integral treatments in which the entire slab or topping is treated and surface treatments in which the surface is specially prepared to become waterproof, dustproof and wear resistant.

Integral Treatments.—Integral treatments consist of powdered chemicals that are mixed with the cement before sand and water are added and become an integral part of the topping mix throughout its entire thickness. This type of treatment is said to densify and harden the topping and is generally recommended for areas subjected to light pedestrian traffic.

Surface Treatments.—There are two types of surface treatments for cement finish floors, dust coat applications which must take place as floors are being finished and liquid chemical applications which may be done at any time after the floor finish is set hard enough for foot traffic.

METALLIC FLOOR TREATMENTS (HARDENERS)

By far the most wide spread surface treatment is the metallic dust coat. The term "hardeners" usually applied to metallic floor treatments is somewhat misleading since iron has a much lower rating on the scale of hardness than even the most common sand or stone aggregate. In this sense iron treatments could be described as "softeners" since the basic technique is to provide a floor surface where-in the hard, brittle sand and stone found in plain concrete is overlaid with a coating of soft, ductile iron. It is reasoned that where hard, brittle aggregates crush under impact and abrasion, iron particles are malleable under these forces and will not fracture or crush out of the floor surface.

Metallic floor treatments consist of a dust coat application of pulverized iron, size graded to pass through mesh screens in the following proportions:

	Percent Passing
4 mesh screen	100 percent
8 mesh screen	Not less than 90 percent
14 mesh screen	Not less than 70 nor more than 85 percent
28 mesh screen	Not less than 35 nor more than 50 percent
48 mesh screen	Not more than 10 percent
100 mesh screen	Not more than 5 percent

Iron particles of the above graduation are mixed with cement and dusted onto freshly floated concrete and then floated and troweled to a smooth even finish. This optimum grading produces a more workable material while plastic and provides a dense surface having great resistance to wear upon hardening.

The iron material should be free from oil and should contain no non-ferrous metals. Oil will prevent the cement paste from bonding properly to the iron particles and non-ferrous metals may react with the cement to form gases and cause serious surface blistering.

Manufacturers, architects and engineers specify different methods and amounts of iron to be used, which is usually a certain number of pounds of iron mixed with one half that weight of cement. The material is dusted over the floor surface before the finish topping has taken its initial set. These quantities vary from 30 to 120 pounds of iron per 100 sq. ft. of floor, depending upon the traffic the floor will have to withstand. When the heavier dust coats are used, water-reducing and plasticizing agents are required to produce a dust coat that can readily be worked onto the surface without resorting to an overly wet concrete mix. In wet mixes there is danger that the iron aggregate will sink into the concrete where it serves no benefit as surface armoring.

It is not necessary to add anything to the labor costs of applying the regular cement finish top. It merely means that the finisher dusts the floor with the metallic powder while it is still too wet to trowel. By adding this dust coat it is possible to trowel the floor sooner than it would be without it, as it absorbs a portion of the surface water.

Durocon Metallic Floor Hardener (Castle Chemical).—A specially prepared metallic hardener for industrial and heavy duty concrete floors. Contains ocon, which provides concrete of increased strength, resisting acid, alkali, etc.

For average traffic, 30 to 40 lbs. hardener and 30 lbs. portland cement; for heavy traffic, 50 lbs. hardener and 30 lbs. portland cement; and for extra heavy traffic, 60 lbs. hardener and 30 lbs. portland cement per 100 sq. ft. floor.

Use 40 lbs. colored Durocon with 20 lbs. portland cement per 100 sq. ft.

Standard colors, natural, tile red, linoleum brown, French gray, Persian red, battleship gray, Nile green, black, russet and maroon.

Duroplate (Castle Chemical).—Specially prepared, size-graded iron particles combined with dispersing agent. Applied as a dust coat while concrete is still in a plastic condition and machine floated into the surface. Produces a ductile, highly impact and abrasion-resistant, non-absorbent floor that will resist oils, grease, etc., and is easily cleaned and maintained. Floor life is extended up to ten times that of normal concrete floors when Duroplate aggregate is used. Use 60 to 120 lbs. per 100 sq. ft. of floor depending on degree of duty required.

Durundum Non-Slip Hardener (Castle Chemical).—An aggregate floor hardener composed of powerful coloring mediums and hardening elements plus cementitious binders. Non-rusting and produces a non-slip surface. Entirely inert, very hard, tough and abrasive resistant. Apply as a dust coat from 30 to 60 lbs. per 100 sq. ft. as conditions require.

Ferro-Fax Metallic Floor Hardener (A. C. Horn, Inc.).—Ferro-Fax metallic floor hardener meets government specifications. Free from oil and foreign metals. Produces wearproof, waterproof, dustproof concrete floors. Apply as a dust coat, using 30 lbs. per 100 sq. ft. for light traffic; 40 lbs. per 100 sq. ft. for moderate traffic; heavy-duty floors, 70 to 100 lbs. per 100 sq. ft.

Comes in two types: Ferro-Fax Standard, to be mixed with portland cement, and Ferro-Fax Ready-Mixed.

Immediately following the leveling, deposit upon the surface ⅔ of a uniform dry mixture consisting of 2 sacks Ferro-Fax Standard and 1 sack portland cement.

Allow the Ferro-Fax to absorb the surface water. Float sufficiently to work the Ferro-Fax into the surface. Discontinue floating as soon as the surface becomes wet. Immediately after floating the first shake, apply the final ⅓ uniformly over the surface. Float just until moisture is brought to the surface. Trowel to the desired finish.

Mastercron Pre-Mixed (Master Builders).—A specially graded hard silica aggregate, combined with light-fast and alkali-fast coloring pigments, selected cement and a water-reducing agent which will produce a tough, wear-resistant floor surface.

Ready to use as it comes from the bag, the material is applied as a dust coat to freshly floated concrete floors at the rate of 50 lbs. per 100 sq. ft.

Available plain and in colors.

Masterplate Pre-Mixed (Master Builders).—Specially prepared metallic aggregate combined with a water reducing and cement dispersing agent. Makes possible the incorporation of 60 to 120 lbs. of iron on the surface to provide maximum resistance to abrasion under the heaviest of industrial traffic.

Average duty floors, 60 lbs. Masterplate plus 30 lbs. cement per 100 sq. ft.; heavy duty floors, 90 lbs. Masterplate plus 45 lbs. cement; extra heavy duty floors, 120 lbs. Masterplate plus 60 lbs. cement.

DPS Masterplate (Master Builders).— A complete ready to use product containing specially processed iron aggregate, a conductive cement binder and a water-reducing and plasticizing agent. Applied as a dust coat at the rate of 180 lbs. per 100 sq. ft. of floor surface, DPS Masterplate meets federal and industrial requirements for conductive flooring. It retains its conductivity at temperatures from — 20° to 150° F. and provides an extremely heavy duty floor that is spark-resistant and static-disseminating. Dark gray in color.

Sparkproof Duroplate Pre-Mixed (Castle Chemical).—Static-disseminating and spark-resistant, meets all requirements of the Navy BuDocks Type Specifications, TS-F-15, Metallic Type (Superseding 48Y) for use in hazardous areas. Will not spark from friction with metal as do ordinary concrete aggregates. A special conductive cement binder is interground in this ready-to-use product. Cure with Conductive Kuraseal (200 sq. ft. per gallon). Available in the following colors: French grey, battleship grey, tile red, maroon, seal brown, black, Nile green, tan, and terra cotta.

Harcol R. M. (Sonneborn).—A ready to use, rustproof, dry mix of abrasive resistant aggregates, specially selected for their flint-like hardness, portland cement and limeproof pigments of high and uniform tinctorial strength.

Applied as a dust coat at the rate of 30 to 45 lbs. per 100 sq. ft., floated into the surface and troweled. Harcol produces an integrally colored surface highly resistant to abrasion.

Colors: bright red, dark red, gray, terra cotta, green and natural.

Ferrolith H. (Sonneborn).—A finely ground metallic water absorbent hardener, free from oils and impurities.

Requires 60 to 180 lbs. per 100 sq. ft. for ordinary traffic, depending on planned use of floor.

Ferrocon Metallic (Castle Chemical).—Finely powdered iron mixed with cement and water. Prevents absorption of moisture and seepage of water. For use on old and new structures. 3 to 5 coats recommended for pressure. Use 10 to 15 lbs. per 100 sq. ft. per coat.

Kemox H (Sika).—A metallic hardener to be used for areas subject to heavy steel wheel traffic.

For monolithic metallic floors, mix 100 lbs. Kemox H, 2 sacks portland cement, 3 cu. ft. clean sand. Sprinkle evenly over 1,000 sq. ft. of freshly placed, concrete floor slab. Trowel with float.

For finish coat, mix 200 lbs. Kemox H with 1 sack portland cement, immediately after floating preliminary finish spread over 300 sq. ft. for heavy duty; 1,000 sq. ft. for light duty floors. Steel trowel again when set starts.

Liquid Chemical Floor Treatments (Hardeners)

As concrete wears down it is ground into dust and this condition has had an especially detrimental effect in mills and factories where the dust from the floors has been carried into the bearings of machinery, motors, engines, etc., causing considerable damage and expense in the upkeep of the equipment.

To overcome this condition, there have been a number of chemicals in liquid form placed on the market that create a chemical action in the cement and lime in the floors and cause a complete transformation of the lime so that it solidifies the entire concrete aggregate into a hard, flint-like, homogenous mass that prevents dusting and wearing of the floors.

These liquid chemical compounds are sprayed over the floor until all of the pores in the surface are filled. It is usually necessary to treat the floors two or three times with the liquid to secure satisfactory results. When using these preparations, it is necessary to treat the floors at stated intervals, as the chemical action they create on the cement and lime is not permanent, although some manufacturers claim their product is permanent.

When using these liquid chemicals care must be exercised to see that the floor is absolutely clean and free from all grease, dust, dirt, oil, or other foreign matter.

Different manufacturers specify varying numbers of applications of their materials but all of them must be applied just as often as the floor shows signs of dusting.

Prices given below are approximate and should be verified for each project.

Liquid Chemical Floor Hardeners

Kind of Hardener	No. Coats	Sq. Ft. per Gal.	Gals. Reqd. 100 Sq. Ft.	Approx. Price per Gal.	Labor Hours
Granitex, Devoe Paint	2	200	½	$7.80	0.8
Hornolith, A. C. Horn..........	3	150	⅔	5.00	0.8
Lapidolith, Sonneborn	3	150	⅔	4.25	1.0
Flintox, Toch......................	2	125	¾	3.75	0.8

Labor Cleaning Floors.—The labor cost of cleaning floors is a matter that is entirely up to the judgment of the contractor or estimator. Floors that are full of oil and grease will require much more time to clean than a floor free of foreign matter.

The labor applying the liquid is also a variable item, but a man should spray 700 sq. ft. of floor an hr. on the first application and 800 sq. ft. and hr. on subsequent applications.

The labor applying the first coat should cost as follows per 100 sq. ft. :

	Hours	Rate	Total	Rate	Total
Labor................................	0.15	$....	$....	$12.54	$1.88
Cost per sq. ft...................			....		.019

On the second and third applications, the labor per 100 sq. ft. should cost as follows:

	Hours	Rate	Total	Rate	Total
Labor................................	0.13	$....	$....	$12.54	$1.63
Cost per sq. ft...................			....		.016

Concrete Accelerators and Densifiers

There are numerous preparations on the market for controlling the set of concrete, increasing its early strength, densifying and waterproofing the concrete mass, and for preventing freezing during winter weather by lowering the freezing point of the mixing water.

Anti-Hydro Accelerator and Anti-Freeze. (Anti-Hydro Co., Newark, N. J.)—Can be mixed with the water or added to the wet mix. As a general rule, the standard proportion of 1.5 gals. of Anti-Hydro per cu. yd. of concrete, with reduced water content to compensate for increased slump, will give protection down to 23° F. on concrete deposited in forms except in the most exposed locations.

For cold weather work, be sure that sand and aggregate are free of ice.

To protect concrete or mortar against freezing, use the following table for anticipated outside air temperatures as follows:

32° to 25° F. use 1 part Anti-Hydro to 12 parts water.
25° to 15° F. use 1 part Anti-Hydro to 10 parts water.
15° or less, mechanical heat is necessary.

Calcium Chloride.—A white, dry, flaky chemical which, when added to the mixing water accelerates the initial hardening of concrete; at the same time densifies the concrete, making it more waterproof; reduces the freezing point of water, thereby aiding the fresh concrete to resist freezing in low temperatures.

Flake calcium chloride may be added to the mix in either dry form (as it comes from the package) or in the form of a solution. The solution is made up of flake calcium and water in proportions that will result in a liquid each quart of which will be equivalent to one lb. of 77–80% flake calcium chloride. One quart of standard solution can be substituted wherever one lb. of flake material is specified.

Calcium chloride is always added to the mix in terms of "pounds per sack of portland cement." The quantities recommended vary from one pound per sack of cement for use when the temperature is above 90° F. to as much as four pounds per sack of cement during the lower range of temperatures. Maximum efficiency is obtained in the use of up to two pounds the proportional reaction of quantities in excess of this amount being less marked. The use of quantities in excess of four pounds per sack of cement is not recommended at any time.

When dry flake calcium chloride is used it should be added to and with the aggregate and not the cement. When solution is used it should be introduced with the mixing water (which should in all cases be reduced by at least the amount of calcium chloride solution used).

The following quantities are recommended per sack of portland cement:

For temperatures above 90° F. .... 1 lb.
For temperatures 80° to 90° F. ....1.5 lbs.
For temperatures 32° to 80° F. ....2 lbs.
For temperatures below 32° F. ....2 to 4 lbs.,

POZZOLITH-HE (Master Builders).—An accelerator and water reducing agent recommended for use in concrete where accelerated set, increased early and ultimate strength and improved performance are required or desired: Concrete made with POZZOLITH-HE provides 3-day normal strength in 1 day, 7-day normal strength in 3 days, 28-day normal strength in 7 days and substantial increases in ultimate strength.

Also, non-chloride high early formulations are available for specialized concreting operations where calcium chloride is not permitted.

POZZOLITH-HE admixture is used at the dosage rate of anywhere from 16-64 fluid ounce per 100 lb (1040 ml to 4160 ml per 100 kg) of cement depending on formulation being used and the amount of set and strength acceleration needed or desired.

Plastiment (Sika).— Plastiment is a powder added in the proportion of ½ to 1–lb. per sack of cement. It will increase workability, density, surface hardness and decrease shrinkage. It may be used for concrete or mortar.

Also available in a concentrated liquid form for use with automatic dispensing equipment. Use 2 to 4 fl. oz. per sack of cement.

Dehydratine No. 80 (W. R. Grace & Co.).—Liquid integral cement floor hardener, accelerator and anti-freeze compound. Used in the proportions of 1 quart per sack of cement.

Sikacrete (Sika).— Accelerating and plasticizing liquid for concrete or mortar, used as anti-freeze or rapid hardening compound. Used in the dilution of 1 part Sikacrete to 3 to 10 parts water. A dilution of 1 to 7 will reduce the hardening rate of mortar or concrete to one-half or protect against freezing to a temperature of 22° F.

Trimix (W. R. Grace & Co.).—A chemical solution of uniform strength which when used according to directions, functions as an accelerator, integral hardener or anti-freeze compound, for use in mortar or concrete work. As an anti-freeze, use as follows:

25° to 32° F., 1½ qts. Trimix to each sack of cement.
20° to 24° F., 2 qts. Trimix to each sack of cement.
15° to 19° F., 2½ qts. Trimix to each sack of cement.
Below 15° F., heat is necessary.

BONDING OLD AND NEW CONCRETE

The problem of bonding new concrete to old so the new concrete will remain permanently in place, is one that has received considerable attention among users of concrete. Unless the utmost care is exercised in preparing the old surface, the new concrete will invariably come loose where it is joined to the old concrete.

Different methods have been used to overcome this fault, but the basic condition of all of them is that the old surface must be thoroughly cleaned and washed, and the old aggregate must be exposed, which means that the thin film of cement that covers the surface of the concrete must be removed. One of the methods of preparing the old surface consists of hacking or picking it, and then washing or turning a steam hose under pressure to remove all dust and dirt from the old floor so as to present a perfectly clean surface.

Another method is to wash the concrete surface with a solution of muriatic acid and water to remove the old film of cement and then thoroughly wash the surface with clear water. After this has been done, the old concrete surface should be given a slush coat of neat portland cement and water, and the new concrete applied.

Hacking and Chipping Old Concrete.—The cost of hacking and chipping old concrete surfaces in preparation for bonding new concrete to old, is a variable item, depending upon the hardness of the old concrete and the condition of the surface. Under average conditions, a laborer using a pick, should hack and roughen 175 to 200 sq. ft. of "green" concrete floor per 8-hr. day, at the following cost per 100 sq. ft.:

	Hours	Rate	Total	Rate	Total
Labor	4	$....	$....	$12.54	$50.16
Cost per sq. ft			....		.50

Where the concrete floors are of old concrete, a man using hand tools will do well to hack and roughen 8 to 10 sq. ft. an hour, and even when a compressor and jack-hammer or bush-hammering tool is used it is often possible to hack and roughen only 20 to 25 sq. ft. an hr. and the cost per 100 sq. ft. should average as follows:

	Hours	Rate	Total	Rate	Total
Compressor expense	2.25	$....	$....	$ 7.50	$16.88
Labor on jack-hammer	4.50	....	....	12.54	56.43
Cost per 100 sq. ft			$....		$73.31
Cost per sq. ft			....		.73

Anti-Hydro Bonding Coat. —Rough and clean the old concrete as previously described. Apply a coat of grout composed of ½ to ¾-sack of portland cement added gradually into a solution of 1-gal. of Anti-Hydro in 3 gals. of water until a thick, creamy consistency is obtained. After applying this grout to the prepared surface, the concrete or mortar should be placed while the grout is still wet.

Sufficient Anti-Hydro grout to cover 100 sq. ft. should cost as follows:

	Rate	Total	Rate	Total
0.25 gal. Anti-Hydro	$....	$....	$5.00	$1.25
0.25 sack portland cement	....	....	4.25	1.06
Cost per 100 sq. ft		$....		$2.31
Cost per sq. ft		....		.023

Daraweld-C (W. R. Grace & Co.)—Daraweld-C is an emulsion of special internally plasticized high polymer resins uniformly dispersed in water. Mixed with cement mortars or grouts, it forms a durable, highly water-resistant bond. A Daraweld-C bond withstands water immersion without softening or disintegrating, resists the action of heat, cold, oil and gasoline, most acids and other corrosive materials. Its coupling action is so great that grouts containing Daraweld-C will bond not only to concrete and masonry, but to tile, wood, steel—even glass. Ready-to-use, non-settling, Daraweld-C will not re-emulsify when subjected to moisture. It is completely compatible with mixes containing calcium chloride.

Daraweld-C is used as a bonding agent admixture in grouts and mortars made from portland or Lumnite cement— assures their lasting adhesion to existing concrete. Formulated for exterior or interior use, it is especially useful for patching concrete floors, repairing cracks in walls, securing an integral bond between successive concrete pours, for waterproofing exteriors below and above grade, in Gunite applications, in bonding concrete to asphalt surfaces. Diluted with water, Daraweld-C is an excellent brush-on sealer for dustproofing concrete floors, and as a primer for painting concrete masonry.

Tremco Floor Bond.—A chemically treated metallic powder to use in bonding new concrete toppings to set slabs, where a perfect bond must be secured to prevent cracking.

Old surface must be roughened and cleaned as described previously and the floor bond applied in 2 coats. Requires 25 lbs. of Metallic Floor Bond per 100 sq. ft. of surface.

When used for bonding vertical surfaces, 2 applications are required, using 15 lbs. of Floor Bond per 100 sq. ft.

Weld-Crete.*—A product that permanently bonds new concrete to old concrete. It also bonds new concrete to many other materials, such as brick, stucco, stone, etc.

For successful bonding of new concrete to old, the old concrete surface must be structurally sound and free from dust, dirt, loose material, grease, oil, wax, water soluble coatings, etc.

It is not necessary to chip, bush hammer or roughen the old surface in any way.

Weld-Crete may be used on either interior or exterior surfaces and may be applied over "green" concrete and damp or dry surfaces. May be applied with a brush, roller or spray. Spraying is most efficient and should be done with heavy industrial spray equipment.

Coverage, when sprayed, from 200 to 300 sq. ft. per gal. When brushed or rolled, coverage is somewhat less.

Repairing Leaks Against Hydrostatic Pressure

When repairing leaks in basement walls or floors, tunnels, pits, etc., where water pressure exists, the following methods and materials may be used.

Sika 2.—Sika 2, red, fast-setting, sealing liquid is mixed with standard portland cement in small quantities. It is used to plug small infiltrations of water against high pressure. Sika 2 mortar will have an initial set of 15 seconds and a final set of 30 seconds. It is used after surrounding leaking concrete has been sealed with Sika 4A. Quantity required varies with degree of leakage.

Sika 4A.—Sika 4A is a clear liquid mixed with neat portland cement in small quantities. The resulting mortar will have an initial set of less than 1 minute and a final set of approximately 5 minutes. Sika 4A may be diluted with water and the cement mixed with sand. This mortar is to be applied against damp surfaces or masonry over which water is running. If pressure is against damp surfaces or masonry over which water is running. If pressure is too heavy, bleeder pipes may be

---

*Larsen Products Corp., Rockville, Md.

installed and after smaller leaks are sealed with Sika 4A, the bleeder holes may be plugged with Sika 2, liquid. Quantity required depends upon the degree of leakage.

CONCRETE ADMIXTURES

Since the introduction of the water-cement ratio law governing the strength and durability of concrete, the necessity for using comparatively low water ratio mixes has been generally recognized. Of greatest importance, however, has been the necessity of producing such mixes with a degree of workability which would insure their easy and economical placing and compacting.

In a cubic yard of concrete, approximately 2½ gallons of water per sack of cement are required to hydrate the cement. Any water in excess of that amount must be regarded as "placing" water, merely providing sufficient workability to make the mix placeable. This excess water occupies about 10 per cent of the total space in concrete. As it evaporates, it causes the concrete to shrink. Excess water also reduces the strength of the concrete.

The purpose of admixtures is to improve the plasticity, workability and finish of the concrete, to prevent segregation of the aggregates and to a certain extent provide integral waterproofing qualities by filling the small voids in the concrete mixture.

Pozzolith (Master Builders).—A water reducing, set-controlling admixture available in several formulations to facilitate optimum performance benefits with the range of cements, sands and coarse aggregates used for making concrete.

In the plastic concrete, Pozzolith improves finishing characteristics for flat work and cast surfaces, reduces segregation of the mix, and enhances placement of low slump concrete. In the hardened concrete Pozzolith results in reduced cracking, increased compressive and flexural strengths, improved watertightness, and increased resistance of air entrained concrete to damage from freezing and thawing as well as scaling from de-icing salts.

Pozzolith Normal and Retarding formulations are generally used at the rate of 5 ± 2 fluid ounce per 100 pound (325 ± 130 ml per 100 kg) of cement.

Plastiment (Sika).— Plastiment, concrete densifier and retarding agent, is a powder to be added to concrete or mortar, using ½ to 1-lb. per sack of portland cement. For equal workability the water-cement ratio may be reduced approximately 10 percent. Results in better workability, density, adhesion to old concrete, surface hardness, and flexural and compressive strength. The delayed set results in elimination of cold joints and reduced shrinkage.

Also available in a concentrated liquid form for use with automatic dispensing equipment. Use 2 to 4 fl. oz. per sack of cement.

Plastocrete (Sika).—A water-reducing admixture for ready-mix and performance concrete to produce higher strengths and workability. Improves structural quality of all concrete. Can be used throughout the year.

Trimix (Sonneborn).—A multi-purpose concrete and mortar admixture which reduces water-cement ratio, produces higher early compression strengths, accelerates set and generally improves quality and workability of portland cement mixtures.

Used in the proportion of 1-qt. Trimix to each sack of cement in the mix.

Hydrated Lime as a Concrete Admixture.— Hydrated lime is often used as an admixture to increase the plasticity and improve the workability of concrete. For this purpose 5 to 8 lbs. of hydrated lime are usually added to each bag of portland cement used in the concrete mixture.

Air Entraining Admixtures

In air entrained concrete 3 to 5 or 6 percent of air is incorporated in the concrete in the form of minute separated air bubbles. Such concrete is more resistant to freezing and thawing and to salt action than normal concrete. Originally developed to prevent scaling of pavement where salts are used for ice removal, air entrained concrete is being widely used for all types of work and in all locations because of its better workability as well as its better resistance to weathering even where salts are not used.

MB-VR (Master Builders). —MB-VR is a neutralized Vinsol resin type air entraining agent for concrete. It is furnished in water solution form and is ready to use as it comes from the drum. MB-VR is compatible for use with other admixtures commonly used in concrete; however, if more than one admixture is used, each should be dispensed into the mix separately. MB-VR is available via bulk tank delivery or 55-gallon steel drums.

Sika Aer.—Air entraining resin solution of neutralized wood resin improves various properties of plastic and hardened concrete.

Add between 1 and 2 ounces of Sika Aer solution to entrain between 3 and 5 percent air as air content is in direct proportion to quantity of Sika Aer added.

Darex AEA (W. R. Grace & Co.)—Darex AEA is an aqueous solution of highly purified and modified salts of a sulfonated hydrocarbon. It contains a catalyst which promotes more rapid and complete hydration of portland cement. Darex AEA is specifically formulated for use as an air entraining admixture for concrete and is manufactured under rigid control which insures uniform, predictable performance. The addition of ¾ fluid ounce of Darex AEA per sack of cement will generally entrain 4 to 7% air in an average (5½ sack) concrete mix.

Concrete Curing Process

Efficient methods of curing concrete are of vital importance to every contractor because exposed surfaces such as concrete floors, roofs, road, pavements, airport runways, and other surfaces exposed to the sun, require protection during the curing period.

This was formerly accomplished by the use of damp burlap, sand, sawdust, dirt covering, paper or ponding.

A method that is being used to a large extent involves curing compounds, which provides an air-tight seal over the surface of the concrete and prevents the evaporation of the water from the concrete until it has cured normally.

These compounds are usually applied with a brush or spray as soon as the concrete has set sufficiently to prevent injury.

Servicised/Horn Concrete Curing Compounds. (A. C. Horn, Inc.)—Are scientifically formulated liquids for spray application to fresh concrete surfaces. The compound quickly forms a vapor-tight film which seals in 95 percent or more of the moisture in the concrete for 3 or more days, without changing color of the concrete. Can be applied with standard spray equipment. One application should control cure on any size surface area, from mass concrete to a few square feet.

Kure-N-Seal (Sonneborn). —Kure-N-Seal cures, seals and dustproofs in one application, and it increases surface hardness when applied as a curing membrane after final troweling. By locking in essential curing moisture the hydration process is refined and maximum hardness for the involved mix is assured.

Klearseal (Castle Chemical).—A clear curing compound for use on surfaces where moderate traffic or corrosive conditions are anticipated, and where a completely clear surface is desired. One gallon covers 300 to 450 sq. ft. for smooth trowelled concrete; 250 to 350 sq. ft. for broom finished concrete, and 200 to 300 sq. ft. for rough finished concrete.

03400   PRECAST CONCRETE

FLEXICORE* FLOOR AND ROOF SLABS

Prestressed Flexicore concrete floor and roof slabs designed in accordance with ACI 318 Building Code are precast in central plants with patented equipment. The cross sections and lengths of these units are indicated in Table I. Filler width slabs are also available.

	TABLE 1 SPAN-FT.	LBS. PER SQ. FT. (Based on 150 Lb. Concrete)
6"x24"	15-25	43
8"x24"	20-33	57
10"x20"	28-40	61
10"x24"	28-40	72
12"x24"	35-50	79

The amount of deflection allowed shall be consistent with the finish to be applied to the underside of the flexicore units, and shall be within the limits specified by ACI 318 Building Code.

These units are manufactured in a rigid steel form made sufficiently strong to resist the pretensioning force applied by the seven wire strand reinforcing steel. Specially constructed rubber tubes are inflated to form circular voids so as to constitute approximately 50% of the cross section. The concrete is thoroughly vibrated to assure maximum denseness and strength.

The slabs may be cast of concrete made of gravel, crushed stone or lightweight aggregate. When lightweight aggregate is used, the unit weight is approximately 25% less. They are cured in a heated kiln.

Since all the Flexicore plants are not equipped to make all the sections listed in Table I, the availability for each section and available length should be checked locally.

Erection.—Flexicore units are usually delivered to the job by truck. They are hoisted from the truck to the floor location and often placed directly into their final position by the crane. After the slabs are placed side by side, they are aligned and levelled. The keyways in the sides of the slab are then filled with a grout mixed in a ratio of one to three. The erection crew usually consists of six men, including a crane operator, an oiler, a foreman and three laborers. A crew will unload, erect and grout 2800 sq. ft. per day on smaller jobs to approximately 6000 sq. ft. on some larger jobs.

Cost of Flexicore Floor Slabs.—Flexicore slabs are usually quoted on an erected basis by the manufacturer and the price will usually include the cost of grouting, and caulking of the joints on the ceiling side. The price will vary according to the type of building, size, live load, span, section specified, location, etc. but usually runs from $3.75 to $5.00 per sq. ft. erected, grouted and caulked.

In localities distant from the manufacturing plant, the slabs are often sold on a "delivered only" basis and erection is performed by the contractor using his own men.

---

*Trademark of The Flexicore Co. Inc.

Flexicore on Lightweight Steel Frame.

Caulking of the joints on the underneath is undertaken after the building has been enclosed. A non-staining type caulking material is used and will average about 20 cents per lineal foot. An underlayment which is not included in the bid is generally used over the Flexicore slabs and is provided by the general contractor.

Flexicore Slabs Erected Directly from Truck.

CONCRETE PLANK

Cantilite* is a lightweight high-strength concrete plank that is nailable and easily cut in the field. In laying, it is clipped to steel or wood beams. Floor or roof materials are nailed to it. Planks are factory-made in steel forms to give a smooth, even surface on all sides and to assure proper positioning of the reinforcing steel. Side edges are tongued and grooved.

Concrete plank are used for floors and roofs. The plank are 2" and 2¾" thick, 16" wide and made in lengths up to 10'-0". The 2" plank are used on steel or concrete joists, 4'-0" on centers for floors. The 2¾" plank are placed on steel or concrete joists up to 5'-0" on centers for floors. A 1" or 2" cement finish top is placed over the plank. Roofs having a span up to 7'-0" will require 2" plank, while 2¾" plank will be required for a roof span up to 8'-0".

*Concrete Plank Co., Inc. North Arlington, N. J.

Concrete plank costs $1.90 to $2.60 cts. per sq. ft., f.o.b. factory, depending upon quantity in the job. The 2" plank weigh 13 lbs. per sq. ft.; the 2¾" plank, 18 lbs. per sq. ft.

Labor Erecting Concrete Plank.—A crew consisting of 3 carpenters and 1 laborer can handle and lay approximately 2,000 sq. ft. of plank per 8–hr. day, on a flat surface.

Labor Cost of 100 Sq. Ft. of Concrete Plank Laid on a Flat
Floor or Roof Surface.

	Hours	Rate	Total	Rate	Total
Carpenter	1.2	$....	$....	$16.47	$19.76
Labor	0.4	....	....	12.54	5.02
Hoisting engineer	0.4	....	....	17.95	7.18
Cost per 100 sq. ft.			$....		$31.96
Cost per sq. ft.			....		.32

PRECAST CONCRETE ROOF SLABS

Precast concrete roof slabs are of three general types: Rib, Flat and Channel. They are adapted to all roof decks, flat or sloping, are fireproof and with caulked joints, present a smooth surface for the application of built-up roofing. Where slate, ornamental tile or copper covering is specified, slabs with nailing surfaces can be furnished, permitting direct application.

Precast concrete roof slabs are furnished made with regular concrete aggregates or where a lightweight slab is desired, they are made of lightweight aggregate.

The rib tile is self-weathering and its attractive red color provides an architectural feature that is very desirable and so often lacking in industrial buildings. Auxiliary pieces are furnished to suit the particular design required, as ridges, saw-tooth ridges, gable end finishing tile, monitor flashing tile and other specials such as would be required in connection with a hip or valley.

Tile of the same standard stock sizes, but having a glass insert, may also be furnished where additional light may be required. On this type of roof slab it is necessary to obtain prices from the manufacturer due to the fact that the amount of trim varies with each project.

Example of Architectural Concrete

The other two types of precast concrete roof slabs, Flat and Channel, are used over all roof decks, whether flat or sloping, and present a smooth surface for the application of any type of built-up roofing. When used with a sloping roof and when slate or other types of ornamental covering are to be used, these slabs can have a concrete nailing surface superimposed and manufactured integrally with the structural slab.

The accompanying illustration gives the sizes and weights of the various types of tile.

TYPE	RIB TILE	FLAT SLAB	2½" CHANNEL TILE	3½" CHANNEL TILE	
WIDTH	24" } STOCK SIZE	24" } STOCK SIZE	24"	24"	18"
LENGTH	4'-4"	5'-0"	6'-8" MAX.	6'-0" MAX.	9'-0" MAX
THICKNESS	WEB - 1½"	1⅝"	LEG - 2½" - WEB-1"	LEG - 3½" - WEB-1"	
WEIGHT MAXDITL	16 LBS.	13.5 LBS.	10.5 LBS.	12 LBS.	13.5 LBS.
PER 22 FT. S∺ND	17 LBS.	19 LBS.	16 LBS.	17.5 LBS.	19 LBS.
REINFORCEMENT	WELDED GALV. MESH	WELDED GALV. MESH	MESH & 2 DEF. BARS	MESH & 2 DEF. BARS	
ROOF PITCH	⅓ PITCH OR GREATER	FLAT OR PITCHED	FLAT OR PITCHED	FLAT OR PITCHED	
ROOFING	SELF WEATHERING	COMP. ROOFING REQ'D	COMP. ROOFING REQ'D	COMP. ROOFING REQ'D	
PURLIN SPACING	4'-0½" MAX. - 3'-9"MIN.	5'-0" MAX.	6'-8" MAX.	6'-0" MAX.	9'-0" MAX.

Sizes and Weights of Precast Concrete Roof Tile

Details of Precast Concrete Roof Tile Construction

Labor Placing Precast Concrete Roof or Floor Slabs.—Because of their size and weight it requires 2 men to handle precast concrete roof or floor slabs, as each piece weighs from 125 to 200 lbs. Ordinarily it requires 2 men on the ground handling the slabs and getting them to the hoist and two men on the floor or roof placing them for the mason, and a mason and helper to lay and caulk the joints.

A carpenter and helper should lay and caulk 960 to 1,050 sq. ft. of lightweight concrete tile per 8-hr. day on straight roofs.

PRECAST CONCRETE JOIST FLOORS

Precast concrete joists are widely used to support concrete floors in homes, apartment buildings, office buildings, schools, hospitals and similar structures. Precast concrete joists are also extensively used for supporting concrete roof slabs for practically all types of buildings.

The design tables relate only to floors and roofs which carry relatively light loads. They are not intended for use in designing floors or roofs carrying heavy or concentrated loads, or where they are subject to heavy impact loads or to vibration of mounted machinery. These conditions impose special loading problems and to meet them floors must be designed accordingly.

The concrete joists are manufactured in a cement products plant and delivered to the job ready to set in place. The joists for each floor of a building should be laid out separately, showing size of joists, spacing, etc., which varies from 20 to 33 inches, depending upon depth of joists, span, floor load, etc.

To develop maximum strength, a reinforced concrete slab 2 to 2½" thick is placed over the tops of the joists with the joists embedded into the concrete slab to a depth of ½ to ¾-inch. Effective bond shall be obtained between the joists and slab. Concrete for both joists and slab shall have an average compressive strength of not less than 3,750 lbs. per sq. in. at age of 28 days.

Floors made of precast joists mortised or embedded into a monolithic floor placed or poured on the job are called "Precast joist cast-in-place concrete slab" floors.

Floors made of precast joist over which are laid precast slabs for flooring are called "Precast joist and slab" concrete floors.

Concrete joists are manufactured 6, 8, 10, 12 and 14 inches deep and various strength requirements of each size may be obtained by increasing the amount of reinforcing steel. For instance, 8-inch concrete joists are manufactured containing from .22 to .88 sq. in. of reinforcing bars at the bottom of the joists and ⅜" round reinforcing bars at the top, with ¼" round steel stirrups spaced according to the load requirements of the particular job. The strength of 10, 12, and 14-inch joists is determined in the same manner.

Notes for Table Nos. 1, 2 and 3 on following pages:

Superimposed loads shown in tables are the loads that the floor or roof is designed to support in addition to the weights shown for the precast joists and concrete slab.

Load values are limited by maximum moment or shear and will produce deflections less than L/360, except in Table No. 3 where load values shown with an asterisk are maximum loads for a deflection of L/360. In Table No. 3 load values in bold face type are limited only by maximum moment and shear and will produce deflections greater than L/360.

For Table Nos. 1 and 2, one row of bridging should be used at midspan in residential floor construction having spans over 20 ft. and in all other floor construction having spans over 16 ft. Bridging is not usually required for roof construction.

For Table No. 3, one row of bridging should be used at midspan for all floor construction having spans over 16 ft. Bridging is usually not required for roof construction.

For floor or roof construction shown in Table Nos. 1 and 3, no shoring is required during construction.

Table 1 • Joists and Slab Designed as T-Beams

No shoring used during construction

Superimposed loads = psf/Clear span = ft.

Joist size in.	Top	Bot	Wt. of joist and slab psf	Slab thickness in.	Joist spacing c. to c. in.	10	11	12	13	14	15	16	17	18	19	20	21	22	23	24
	#3	#5	34	2	20		113	88	69	53	40	30								
			32	2	24	118	90	69	53	40	29									
			32	2	27	101	76	57	42	31										
			37	2½	30	85	61	43	29											
			36	2½	33	74	52	36	23											
	#3	#6	34	2	20				113	91	73	59								
			32	2	24			113	90	72	57	45								
8			32	2	27		122	96	72	59	46	36								
			37	2½	30	139	106	81	61	46	33	22								
			36	2½	33	123	93	70	53	38	26									
	#3	#7	34	2	20					137	113	94								
			32	2	24					134	110	90	74							
			32	2	27					115	93	76	61							
			37	2½	30				125	99	78	61	48							
			36	2½	33			128	111	87	68	53	40							
	#3	#6	37	2	20					124	103	85	70	57	47	38				
			35	2	24					122	99	81	66	53	43	34	27			
			34	2	27				131	105	85	69	55	44	35	27				
			39	2½	30				113	89	70	54	41	31	22					
			38	2½	33			128	100	78	60	46	34	25						
	#3	#7	37	2	20							129	109	92	78	66				
			35	2	24						124	103	87	72	61	50				
10			34	2	27					129	107	89	74	61	51	42				
			39	2½	30				137	111	90	73	59	47	37	28				
			38	2½	33				121	98	79	63	50	39	30	22				
	#3	#8	37	2	20									134	115	99				
			35	2	24								125	107	91	78				
			34	2	27							127	108	92	78	66				
			39	2½	30					138	125	110	91	76	63	52				
			38	2½	33					122	111	97	80	66	54	44				
	#3	#6	44	2	20					126	104	86	71	59	48	39	31			
			41	2	24					123	100	83	68	55	44	35	27			
			39	2	27				131	106	87	70	57	46	36	28				
			43	2½	30				113	90	72	57	44	33	24					
			42	2½	33			126	99	79	62	48	36	27						
	#3	#7	44	2	20							135	115	97	83	70	59	50	42	
			41	2	24						128	108	91	77	64	54	45	37	30	
			39	2	27					133	111	93	78	65	54	45	37	30	23	
			43	2½	30					115	95	78	63	51	41	32	25			
12			42	2½	33				124	101	82	67	54	43	34	26				
	#3	#8	44	2	20											123	107	93	81	69
			41	2	24									133	114	98	85	73	62	53
			39	2	27								135	115	99	85	72	62	52	44
			43	2½	30								117	98	83	69	58	48	39	32
			42	2½	33							123	103	86	72	60	49	40	32	
	#3	#9	44	2	20													131	115	102
			41	2	24											137	120	105	92	80
			39	2	27										137	119	103	90	78	68
			43	2½	30									138	118	101	87	74	63	54
			42	2½	33									122	104	89	75	64	54	45

For floor and roof construction shown in Table No. 2, shoring at midspan is required during construction and must remain in place until the concrete has reached its required strength.

Bar sizes are given in numbers based on the number of eighths of an inch included in the nominal diameter of the bars as specified in ASTM A305.

Table 2 • Joists and Slab Designed as T-Beams

Joist size	Size of deformed reinforcing bars		Wt. of joist and slab	Slab thick-ness	Joist spac-ing c. to c.	Shoring used at midspan during construction														
						Superimposed loads = psf/Clear span = ft.														
in.	Top	Bot	psf	in.	in.	10	11	12	13	14	15	16	17	18	19	20	21	22	23	24
8	#3	#5	34	2	20		124	99	79	64	51	41								
			32	2	24	128	100	80	63	50	39	30								
			32	2	27	111	86	67	52	41	31									
			37	2½	30	100	76	58	44	33	24									
			36	2½	33	89	67	50	38	28										
	#3	#6	34	2	20					123	102	84	70							
			32	2	24				123	100	82	67	55							
			32	2	27			133	106	86	70	57	46							
			37	2½	30			121	96	76	61	48	38							
			36	2½	33	138	108	85	67	53	42	32								
	#3	#7	34	2	20						124	105								
			32	2	24						120	101	85							
			32	2	27					126	104	87	72							
			37	2½	30					115	94	77	63							
			36	2½	33				126	102	83	68	55							
10	#3	#6	37	2	20					133	112	94	79	66	55	46				
			35	2	24					130	108	89	75	62	51	42	35			
			34	2	27					113	93	77	63	52	43	35	28			
			39	2½	30				125	101	82	66	53	43	34	26				
			38	2½	33				112	90	72	58	46	37	29					
	#3	#7	37	2	20								118	101	87	75				
			35	2	24						132	112	95	81	69	59				
			34	2	27					137	115	97	82	69	59	50				
			39	2½	30					123	102	85	71	59	49	40				
			38	2½	33				134	110	91	75	62	52	42	35				
	#3	#8	37	2	20										124	109				
			35	2	24								133	115	100	87				
			34	2	27							136	116	100	86	75				
			39	2½	30						137	122	104	88	75	64				
			38	2½	33					134	123	109	92	78	66	56				
12	#3	#6	44	2	20						134	112	95	80	67	56	47	39	32	26
			41	2	24					130	109	90	75	63	52	43	35	29		
			39	2	27					114	94	78	65	54	44	36	29			
			43	2½	30				124	101	83	67	55	44	35	28				
			42	2½	33			136	110	89	72	58	47	37	29					
	#3	#7	44	2	20								123	106	92	79	68	59	50	
			41	2	24							136	116	99	84	72	62	53	45	38
			39	2	27							119	101	86	73	62	53	44	37	31
			43	2½	30						126	105	89	75	63	52	43	35	29	
			42	2½	33					135	112	93	78	65	54	44	36	29		
	#3	#8	44	2	20										133	116	102	90	79	
			41	2	24										123	106	93	81	700	61
			39	2	27									123	106	92	80	69	61	52
			43	2½	30								128	109	94	81	69	59	50	43
			42	2½	33							134	114	97	83	71	60	51	43	36
	#3	#9	44	2	20														125	111
			41	2	24												128	113	100	88
			39	2	27											127	111	98	86	76
			43	2½	30										129	113	98	85	75	65
			42	2½	33									133	115	99	86	75	65	56

Table 3 • Joists Designed as Independent Beams

No shoring used during construction

Superimposed loads = psf/Clear span = ft.

Joist size in.	Top	Bot	Wt. of joist and slab psf.	Joist spac -ing c. to c. in.	10	11	12	13	14	15	16	17	18	19	20	21	22	23	24
8	#3	#5	34	20	112	86	67	52	40	31									
			32	24	89	68	52	40	30	21* / 22									
			32	27	76	57	43	32	23										
	#4	#6	34	20		106	85	69	49* / 56	23* / 45									
			32	24	107	85	67	54	30* / 43	8* / 34									
			32	27	92	72	56	42* / 44	16* / 35										
	#6	#7	34	20					103	69* / 85	39* / 71								
			32	24				100	79* / 82	46* / 67	21* / 55								
			32	27			106	86	60* / 69	30* / 56	8* / 46								
10	#3	#6	37	20					100	82	68	56	46	26* / 37					
			35	24				98	79	64	52	42	29* / 34						
			34	27			104	84	68	54	44	35							
	#5	#7	37	20						104	88	74	70* / 63	43* / 53	21*				
			35	24							99	82	46* / 69	23* / 58	5* / 48	40			
			34	27					102	85	70	57* / 59	32* / 49	12* / 40	— / 33				
	#6	#8	37	20									85* / 107	56* / 93	32* / 80				
			35	24								89* / 100	59* / 85	34* / 73	14* / 63				
			34	27							102	70* / 86	43* / 73	22* / 62	4* / 53				
12	#3	#6	44	20						105	87	72	60	49	40	32	26		
			41	24					102	83	68	56	45	36	29	23			
			39	27				109	88	72	58	47	38	30	23				
			43	30					90	71	57	45	35	26					
			42	33				100	79	62	49	38	29						
	#3	#7	44	20								96	81	69	58	49	39* / 41	19* / 34	
			41	24							106	89	75	63	53	44	37	20* / 30	
			39	27							92	77	65	54	45	37	28* / 30	10* / 24	
			43	30						91	75	61	50	40	32	21* / 25			
			42	33					98	80	65	53	43	34	26				
	#4	#8	44	20										103	90	75* / 78	50* / 67	28* / 58	
			41	24									95	81	70	50* / 608	28* / 52	10* / 44	
			39	27									96	82	70	37* / 60	18* / 51	2* / 44	37
			43	30								93	78	66	55	30* / 46	9* / 38	— / 31	
			42	33							97	82	68	41* / 57	19* / 47	— / 39	32		

Concrete Joists Designed as Independent Beams.—Where concrete joists are designed to be used with precast concrete slabs, so that the slab does not form a part of the beam, it is necessary to use more reinforcing steel in the beams, which increases their cost considerably.

For instance 8" beams having 3/8" top bars and 5/8" bottom bars are used for spans 10 to 15 ft.; ½" top bars and ¾" bottom bars for spans 11 to 16 ft., also joists using 3/4" top bars and 7/8" bars for spans 12 to 16 ft. requiring heavier loads.

Ten inch joists having 3/8" top bars and 3/4" bottom bars are used for spans 12 to 19 ft. 5/8" top bars and 7/8" bottom bars for spans 14 to 20 ft.; and 3/4" top bars and 1" bottom bars for spans 16 to 20 ft. required to carry heavier loads.

Twelve inch joists having 3/8" top bars and 3/4" bottom bars are used for spans 12 to 22 ft.; 3/8" top bars and 7/8" bottom bars for spans 14 to 24 ft. and ½" top bars and 1" bottom bars for spans 16 to 24 ft. required to carry heavier loads.

These beams require from 1 to 2 lbs. of reinforcing steel per lin. ft. more than the T-beams and consequently cost 25 to 30 cts. per lin. ft. more than the T-beam joists.

Precast concrete joists are manufactured of Haydite, Waylite, Celocrete and other lightweight aggregate, also of ordinary aggregate consisting of gravel or crushed stone and sand.

Weight of Precast Concrete Joists, Ordinary Aggregate

Size of Joist In.	Length of Joist in Feet					
	10	12	14	16	18	20
8	190	228	266	304	...	...
10	230	276	322	368	414	460
12	350	420	490	560	630	700

Cost of Precast Concrete Joists.—The cost of precast concrete joists will vary with the size of joists, amount of reinforcing steel required, size of job, locality, etc.

Inasmuch as precast concrete joists are always manufactured in a concrete products plant, it is important to obtain definite prices for each particular job.

Approximate Prices per Lin. Ft. for Precast Concrete Joists.—

8-Inch Joists	10-Inch Joists	12-Inch Joists	14-Inch Joists
$4.25	$4.60	$6.25	$7.00

Where the concrete joists are designed as independent beams for use with precast concrete floor or roof slabs, add from 25 to 30 cts. per lin. ft. to the above prices to allow for the extra reinforcing steel required.

Labor Setting Precast Concrete Joists.—As long as the joist is not too heavy to handle by four men or in some cases by light derricks, the precast joist is more economical than one which is cast in place. The economical limits of largeness of precast joists has not been established, although at the present time they are obtainable to a depth of 14" and 36'-0" long. Such a joist weighs approximately 1,150 to 1,330 lbs. if made of lightweight aggregate—they fall within the weight where erection by machinery is necessary.

On every job there should be a joist setting plan, prepared by an architect or engineer, as this will expedite setting the joists on the job.

Where smaller size joists are used, two men can carry the joists from the stock pile to the building and set them in place, while with heavier joists it may require 4 to 6 men to carry the joists to the building and place them on the wall.

Where joists are used on the second floor, two men will probably be required to carry the joists into the building and 3 men above to hoist the joists to the second

	8" Joist	10" Joist	12" Joist
Sectional area sq.in.	18.2	22.1	33.4
Average weight (as used in Tables 1-3) lb. per lin. ft.	19	23	35

Sizes and Weights of Precast Concrete Floor Joists.

floor. If the joists are not too heavy for two men to handle, a rope can be slipped around each end of the joist and lifted to the second floor.

After the joists have been placed on the wall, it will require some additional time for exact spacing of the joists, placing headers, etc.

Where the joists can be handled and set by 2 men, it will require 6 to 8 min. to carry one joist from stock pile (not over 50 ft.) and set it on the wall, at the following labor cost per joist:

	Hours	Rate	Total	Rate	Total
Labor	0.23	$....	$....	$12.54	$2.88

If joists are hoisted to the second floor, double the above costs.

When handling and setting joists requiring 4 men to carry them to the building and set them on the wall, it will require 6 to 8 min. for 4 men, at the following labor cost per joist:

	Hours	Rate	Total	Rate	Total
Labor	0.47	$....	$....	$12.54	$5.89

If joists are hoisted to the second floor, double the above costs.

When handling and setting 12" joists up to 24-ft. long, requiring 6 men to handle and set same, it will require 6 to 8 min. for 6 men, at the following labor cost per joist:

	Hours	Rate	Total	Rate	Total
Labor	0.7	$....	$....	$12.54	$8.78

If joists are hoisted to the second floor, double the above costs.

On jobs having sufficient joists to warrant the use of a hydraulic crane, and using iron workers to set the joists instead of laborers, the cost of setting joists should average as follows:

Labor setting one concrete joist:

	Hours	Rate	Total	Rate	Total
Crane and operator	0.10	$....	$....	$52.00	$5.20
Iron workers	0.20	....	....	17.70	3.54
Cost per joist			$....		8.74

Wood Forms for Precast Concrete Joist Floors.—The reinforced concrete floor slab used with precast concrete joists is usually 2 to 2½ inches thick, and the joists are spaced from 20 to 33 inches apart, so that very light formwork is sufficient.

The formwork usually consists of short pieces of 2x4-inch wood spreaders as shown in the illustration, cut to bear on the bottom flange of the concrete joists. These spreaders are placed between the concrete joists, and the 5/8" plywood are placed on top of the spreaders to receive the concrete.

Method of Constructing Precast Concrete Joist Floors

Method of Constructing Precast Concrete Joist Floors Using Paper-Backed Floor Lath for Permanent Form Material

On forms of this kind, it will require about 1¼ BF of lumber per sq. ft. of floor area.

When placing forms for precast concrete joist floors, a carpenter should frame and place forms for 175 to 225 sq. ft. of floor per 8-hr. day, at the following cost per 100 sq. ft.:

	Hours	Rate	Total	Rate	Total
Carpenter	4	$. . . .	$. . . .	$16.47	$65.88
Cost per sq. ft			. . . .		.66

A laborer should remove forms from 600 to 750 sq. ft. of floor per 8-hr. day, at the following cost per 100 sq. ft. :

	Hours	Rate	Total	Rate	Total
Labor	1.2	$. . . .	$. . . .	$12.54	$15.05
Cost per sq. ft			. . . .		.15

Permanent Forms for Precast Concrete Joist Floors.—Permanent type forms, not requiring removal, are being widely used in precast concrete joist floor construction, especially for work such as floors over crawl spaces, tunnels, etc., where removal of wood forms would be difficult and expensive. Form materials usually used for this purpose are of two kinds—paper-backed floor lath, such as "Steeltex*" or corrugated steel sheets described in the first part of this chapter. With corrugated steel forms, reinforcing steel or wire mesh must be added to satisfy flexure and temperature steel requirements.

Steeltex* Forms for Precast Concrete Joist Floors.—Steeltex is a welded wire mesh with a waterproofed paper backing attached. The paper backing serves as the form and the mesh as the reinforcing for thin, short span concrete floor slabs. When concrete is placed, the paper backing sags away from the mesh to the limit of the stitch wires thereby placing the reinforcing mesh at the proper position for encasement in the slab.

Steeltex is furnished in rolls, 4'-0" wide and 125'-0" long, containing 500 sq. ft. Side laps should be at least 2" and end laps should be a minimum of 12" occurring only over a support with no adjacent splices across the same support. In estimating, compute area of slab and add about 5 percent for laps and waste.

For floor slabs at or near grade level, two ironworkers working together should install about 4000 sq. ft. of Steeltex over steel joists per 8-hr. day. For slabs at higher levels, additional handling labor and hoisting must be added.

Labor Cost of 100 Sq. Ft. Steeltex Forms for
Precast Concrete Joist Floors

	Hours	Rate	Total	Rate	Total
Ironworker	0.4	$. . . .	$. . . .	$17.70	$7.08
Cost per sq. ft			. . . .		.071

Corrugated Steel Forms for Precast Concrete Joist Floors.—There are two methods of using corrugated steel forms for precast concrete joist floors. Corrugated steel forms may be laid continuous over joist tops with end laps occurring only over supports and no adjacent laps over same joist.

Corrugated steel forms also may be installed between joists, so that tops of joist and protruding stirrups are encased in slab concrete, providing T-beam action. For this method, joists should be precast with ledges on the top edges to furnish support for the corrugated sheets.

---

*National Wire Products Corp. Baltimore, Md.

Side laps for corrugated steel forms should be one corrugation and end laps a minimum of 2". For estimating purposes, compute area of floor slab placed over corrugated steel sheets and add 5 to 10 percent for laps and waste, depending upon the method of installation.

Where corrugated steel forms are installed in continuous lengths over precast concrete joists for floors at or near grade level, with an area of 5,000 sq. ft. or more, a 4-man crew should place and secure 2,500 to 2,700 sq. ft. per 8-hr. day. For areas less than 5,000 sq. ft., the crew should install 2,100 to 2,300 sq. ft. per 8-hr. day. For slabs at higher levels, additional handling labor and hoisting must be added.

Labor Cost of 100 Sq. Ft. of Corrugated Steel Slab Forms
for Precase Concrete Joist Floors in
Residential Construction

	Hours	Rate	Total	Rate	Total
Ironworker	1.33	$....	$....	$17.70	$23.54
Cost per sq. ft.			....		.24

Where corrugated steel forms are installed between joists, the above costs should be increased about 10 percent.

Reinforcing Steel for Precast Concrete Joist Floors.—A thin slab of concrete 2 to 2½-in. thick is used with concrete joists and the reinforcing steel consists of ¼-inch round rods spaced 10 in. on centers at right angles to the joists. It requires about .33 lbs. of steel per sq. ft. or 33 lbs. per 100 sq. ft. of floor. Steel reinforcing mesh of equal effective steel area may be used instead of reinforcing rods, if desired.

A man should place about 400 lbs. of reinforcing steel per 8-hr. day and based on 33 lbs. per 100 sq. ft. the labor per 100 sq. ft. of floor should cost as follows:

	Hours	Rate	Total	Rate	Total
Ironworker	0.67	$....	$....	$17.70	$11.86
Cost per sq. ft.			....		.12

Where welded wire mesh is used for reinforcing concrete slabs for precast concrete joist floors, a man should place 1,400 to 1,600 sq. ft. of mesh per 8-hr. day, at the following labor cost per 100 sq. ft.:

	Hours	Rate	Total	Rate	Total
Ironworker	0.53	$....	$....	$17.70	$9.38
Cost per sq. ft.			....		.09

Precast Concrete Joist Floors with Wood Floor Screeds

Wood Sleepers For Use With Precast Concrete Joist Floors.—Where wood floors are used with precast concrete joist floors, it will be necessary to place 2x2 or 2x3-in. wood strips or "sleepers" in the concrete for nailing the wood floors. These strips should be spaced about 16 in. on centers and should rest on top of the steel stirrups that project above the concrete joist. The sleepers are secured to the joists by wiring around the wood strip and through the top of the steel stirrups. Care must be taken when placing the concrete to see that the strips are level to receive the finish wood floors.

The floor strips should be beveled or nails driven into the sides at short intervals to provide anchorage into the concrete, otherwise after the concrete hardens and the wood strips dry out and shrink they become loose and cause squeaking floors.

Where wood strips are placed over the tops of the joists without leveling, a carpenter should place 350 to 450 lin. ft. per 8-hr. day, at the following cost per 100 lin. ft.:

	Hours	Rate	Total	Rate	Total
Carpenter	2.0	$. . . .	$. . . .	$16.47	$32.94
Cost per lin. ft.			. . . .		.33
Cost per 100 sq. ft. (16-in. on centers):					
Carpenter	1.5	. . . .	. . . .	16.47	24.71
Cost per sq. ft.			. . . .		.25

Labor Handling and Placing Precast Concrete Slabs.—Precast concrete slabs are often used in conjunction with concrete joists instead of cast-in-place concrete slabs.

These slabs are ordinarily used for roofs or for floor slabs in unexcavated areas where it is either extremely difficult or impossible to remove the forms after the cast-in-place slab has been poured.

When precast slabs are used, it requires joists containing heavier reinforcing steel than where T-beam construction is used. This increases the cost of the concrete joists 10 to 15 cents per lineal foot.

Precast concrete slabs are furnished in several sizes, including, 2'-1"x 2'-6"x1¼"; 1'-2"x2'-1"x1¼"; 1'-0"x2'-6"x1¼"and 1'-6"x2'-1"x1¼". They are reinforced with 4"x4" No. 10 wire mesh.

When used as roof slabs, they are laid over the concrete joists with about a 1" space between the slabs. After the slabs are placed they are grouted in solid. Insulation board or roofing is applied directly over the slabs.

When used for floors, the precast slabs are placed over the joists and about 1½" of concrete is poured over the slab and either left rough or troweled to a smooth finish to receive the finish flooring.

This method eliminates form cost but it requires heavier reinforced joists and it is more difficult for the mechanical trades to place their pipes than where the cast-in-place floor is used.

On smaller jobs, the men usually work in groups of two carrying and placing the precast slabs over the concrete joists.

On work on the first floor, where the precast slabs are on the floor or convenient to the place where they are to be used, two men working together should handle and set 50 to 60 pcs. an hour, depending upon the size of the slabs used, at the following labor cost per 100 pcs.:

	Hours	Rate	Total	Rate	Total
Labor	3.6	$. . . .	$. . . .	$12.54	$45.15
Cost per piece			. . . .		.45

Add cost of carrying slabs from stock pile to floor.

If slabs are to be hoisted to the second floor, double the above costs.

On large jobs where it is possible to use a hydraulic crane to lift the slabs from the ground to the second floor or roof, a crane will hoist 7 slabs at one time and place them on the floor.

To the above costs it will be necessary to add cost of material and labor grouting joints between slabs.

PRESTRESSED CONCRETE

Within a relatively short period of time prestressed concrete has taken a very significant place in the building industry. Firms producing prestressed concrete structural members number about 320 and are located in every part of the United States and Canada.

The use of prestressed concrete began in Europe some 50 years ago. A French engineer, Eugene Freyssinet, is generally regarded as the developer of the prestressing concept. Many spectacular prestressed concrete structures have been designed by Mr. Freyssinet and others in Europe and in Latin America and more recently in this country. The first major application of the use of prestressed concrete in the United States was the Walnut Lane Bridge in Philadelphia, Pa. This structure was completed in 1950 and since that time thousands of bridges and buildings have been constructed with structural elements of prestressed concrete.

Prestressed concrete was developed partly out of necessity as scarcity and the consequent high price of conventional building materials was experienced in Europe during and after the war years. The concept of utilizing the available materials to their fullest capacities was experimented with until techniques were perfected which developed prestressing into a workable and practical method of construction.

The American Concrete Institute defines prestressed concrete as "concrete whose stresses resulting from external loadings are counterbalanced by prestressing reinforcement placed in the structure." This is accomplished by precompressing the concrete by means of internal high tensile strands stressed a predetermined amount. This prestressing introduces compressive stresses into the concrete which counteract the stresses induced when the external loading is applied. The high strength strands are restrained in the member by either bond between the concrete and the strands, or by end bearing devices. These two basic methods of restraining the strands are known as "pretensioning" and "post-tensioning."

In pretensioning, the strands are tensioned prior to casting the concrete and in post-tensioning the strands are tensioned after the concrete is cast. In general, the post-tensioning process is accomplished at the job site when casting large members. The member is cast with the strands properly positioned in the form and, after the concrete has reached sufficient strength, the tensioning members are stressed and secured by means of end bearing devices.

The pretensioning methods are particularly adaptable to the mass production of building elements such as double and single tees, channel sections, I joists and beams, T joists, hollow slabs and flat planks, among others. These members are usually cast in steel casting beds from 100 ft. to 600 ft. long. With the use of high quality concrete, low slumps and steam curing, a complete manufacturing cycle may be accomplished in a day.

The prestressed and precast industry is based upon the economics of prefabrication and maximum use of the casting facilities. The larger the project and the number of units of identical span and section, the lower the unit cost.

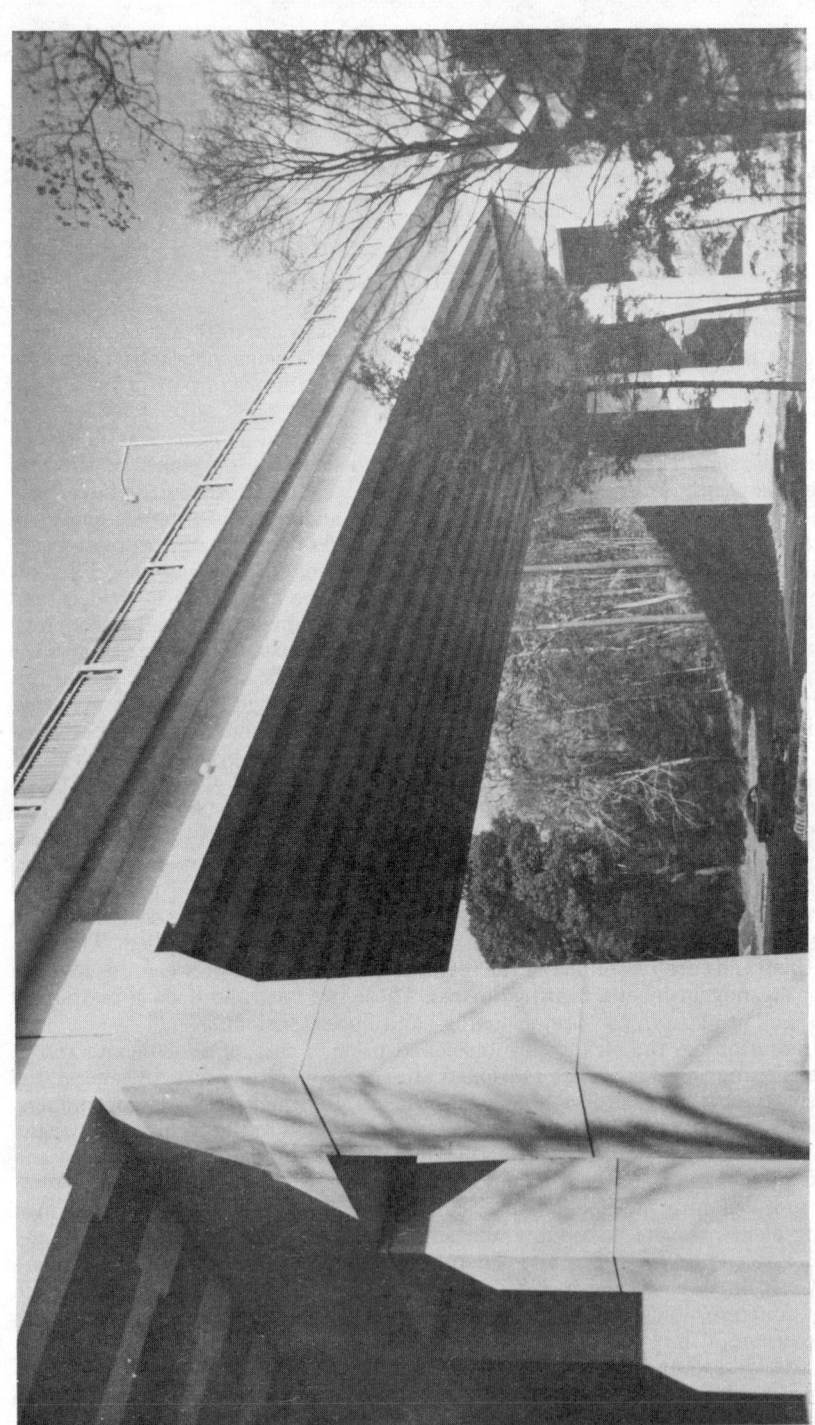

Walnut Lane Bridge, Philadelphia, Pa. The First Major Use of Prestressed Concrete Structural Members in the Western Hemisphere

Building Elements of Prestressed Concrete.—Precast/Schokbeton Inc. of Kalamazoo, Michigan has a modern prestressing plant for the manufacture of prestressed concrete building elements. The plant is equipped with a variety of casting beds capable of producing a full range of prestressed products, including both structural and architectural components.

Included in the structural items available in the United States are single and double tee slabs, "F" or Monowing slabs, channel slabs, and Flextee slabs for floor and roof construction. Columns, beams and girders are available in a wide range of sizes and shapes.

Architectural products made at the Precast/Schokbeton plant include precast concrete stadium seats, miscellaneous free form concrete made to order, and Schokbeton products produced under a license from N. V. Schokbeton of Zeist, Holland. These include wall panels, window facades and frames, column shells, stair treads, etc.

Cost of Prestressed Concrete Structural Elements.—As prestressed concrete building elements are now manufactured generally throughout the country, prices may usually be obtained from local sources. As stated previously, casting beds are in most cases long enough to cast any length member desired: length, width, height and weight must be acceptable to local trucking regulations.

There are, of course, many variables which can influence the cost of the installed product, such as the size of the job; the number of identical units in a job; the distance from the plant to the job site; the length of the units, as extremely long units will probably require special handling at the plant, in transit and at the job site; accessibility at the site; the size and number of cranes or other equipment required for erection; the condition at the ground, etc.

Courtesy Portland Cement Association

Precast concrete Tee bars are commonly trucked to site for placement

Double Tee being hoisted into position where it will be welded into place

Small Crew Places LongSpan Prestressed Double Tee Roof Member For Column-free Interior Space.

Typical Prestressed Concrete Products

Erection of Prestressed Concrete Building Units.—A typical erection setup for an average job would include a crane, crane operator, oiler, five iron workers, welding machines and other miscellaneous equipment and tools. With move-in and move-out time for the cranes, dead time due to the weather or breakdowns, contingencies, etc., the daily cost of an erection setup such as this will average about $2500 per day. On a large job and under good conditions an experienced crew can probably erect 6,000 to 10,000 sq. ft. or more per day. On smaller jobs or where

Erection of Roof Slabs

working conditions are below average, a crew might only erect 3,000 to 4,000 sq. ft. per day.

The Prestressed Concrete Institute, with offices in Chicago, Illinois, is a source of the latest reference material on prestressed concrete and can provide assistance to the design profession as well as to the purchaser of prestressed concrete products.

03500 CEMENTITIOUS DECKS
USG† METAL EDGE GYPSUM PLANK

A lightweight, noncombustible construction for roofs and floors in connection with steel framing.

USG† Metal Edge Gypsum Plank are made in 2"x15"x10'-0" units and designed for roofs on spans up to 7'-0". The plank has a water resistant core.

All plank are molded and have steel tongue and groove edging on sides and ends, firmly anchored into the gypsum. They also have an electrically welded galvanized steel fabric as reinforcement for the gypsum. Weight 13.0 lbs. per sq. ft.

On jobs requiring approximately 3,000 sq. ft. or more, the plank cost about $1.35 per sq. ft. in the eastern zone of the United States.

Labor Laying Metal Edge Gypsum Plank.—One plank (12½ sq. ft. in area) will weigh 160 lbs. requiring 2 men to handle a single plank. Ordinarily it will require 2 men on the ground, 2 men on the roof. This crew should lay approximately 1,200 sq. ft. per 8-hr. day at the following labor costs:

	Hours	Rate	Total	Rate	Total
Labor	2.7	$. . . .	$. . . .	$12.54	$33.86
Cost per sq. ft			. . . .		.34

The above costs contemplate ordinary flat slab work, on a job containing 3,000 to 4,000 sq. ft., with a minimum amount of cutting. Add for extra handling and cutting for pitched roof and double the above labor costs for pitched roofs having hips, valleys, dormers, etc. Add extra for hoisting engineer if required.

Monolithic or Poured-in-Place Gypsum Roof Construction

Monolithic or poured-in-place gypsum roof deck, with its continuous reinforcing and securely welded sub-purlins, becomes an integral part of the main steel construction of a building, adding rigidity to the entire structure and can be designed for any required loading condition.

Permanent formboards which remain as the undersurface of the completed deck may be ½" gypsum formboard; ¼" asbestos cement board; and 1" to 2" mineral fiber formboard.

Rolled steel sections designed to carry all loads between supports shall be laid on the top of walls, beams or roof supports at sufficient spacing to receive the formboards on the lower flanges; "reinforcing shall be 48-1214 galvanized welded wire fabric having 12 ga. longitudinal wire spaced 4" on centers and 14 ga. transverse wired 8" on centers or "Keydeck," a galvanized woven mesh as manufactured by Keystone Steel and Wire Co., having 16 ga. longitudinal wires and 19 ga. diagonal wires. The effective cross sectional area of reinforcing shall be not less than 0.026 sq. in. per ft. of slab width."

---

†United States Gypsum Co.

Erection of Single Tee Roof Slabs.

Standard (mill mixed) gypsum composition containing at least 87½ percent calcined gypsum is mixed with water to a medium quaking consistency and poured or pumped in place, forming a solid reinforced slab 2" thick (or more, if desired), plus the board undersurface. The top surface of the slab shall be screeded smooth and left ready to receive the specified roofing.

The density of poured gypsum concrete is approximately 50 lbs. per cu. ft. When cast in the minimum thickness of 2", the total dead weight of the slab is about 10 lbs. per sq. ft.

The weights (excluding sub purlins) and insulating values of poured gypsum roof decks are as follows:

Description	Weight per Sq. Ft.	"U" Value BTU
2" gypsum concrete on ½" gypsum form boards or plastic faced board	11.0 lbs.	.34
2¼" gypsum concrete on ¼" asbestos cement board	11.0 lbs.	.37
2" gypsum concrete on 1" mineral fiber formboard	11.2 lbs.	.16
2" gypsum concrete on 2" mineral fiber formboard	11.2 lbs.	.10

ERECTION PLAN

2" STEEL BOUND GYPSUM PLANK

CLIP AND EDGING DETAIL

SECTION "A-A" THRU PLANK — Galvanized Wire Mesh

Details of USG Metal Edge Gypsum Plank

Asbestos cement formboards should be used for poured gypsum roof installations in buildings where heat or humidity is higher than normal and on exterior overhangs.

The accompanying illustration shows the type of roof construction that is widely used for schools, auditoriums, field houses, gymnasiums, factory, industrial and many commercial buildings, in fact, for most any steel skeleton structure where a permanent, noncombustible highly insulated roof is desired.

Practically all gypsum concrete is now mixed mechanically and pumped from ground to roof level as a means of further increasing the speed of operation and uniformity of quality, with corresponding reduction in cost.

The cost of poured-in-place gypsum roof decks will vary with the size and type of structure, availability of materials, freight charges, labor rates, etc., but the simplicity of design, lightweight and speed of erection, noncombustibility and finished undersurface make the overall cost of this construction very economical.

Poured-in-place gypsum roof decks are applied by contractors approved by the manufacturers of the materials.

The following are approximate prices only on the various types of poured-in-place gypsum roof decks. These prices may vary locally due to labor and material costs, freight rates or size of specific job. For definite prices on any project it is always advisable to consult a roof deck contractor specializing in this type of construction.

Type of Roof	Approximate Price Range per Sq. Ft.
2" gypsum concrete on ½" gypsum formboards	$1.25 - $1.40
2" gypsum concrete on ¼" asbestos cement board	1.35 - 1.50
2" gypsum on 1" mineral fiber formboard	1.45 - 1.60
For each additional ½" in thickness of gypsum concrete add	0.12 - 0.15

Courtesy of U. S. Gypsum Co.

Two Hour Fire Rated Roof and Ceiling Assembly

The accompanying illustration shows a poured gypsum roof on steel joists used in combination with acoustic ceiling boards of mineral fiber to comply with Underwriters Laboratories design No. P207 for a two hour fire rating. This assembly has a "U" value of 0.19, a noise reduction coefficient of .55 to .65, and a light reflectance of 85%. Other 2 hr. U.L rated systems are also available.

USG* Floor Plank Systems

USG Floor Plank Systems are lightweight noncombustible assemblies for use with bar joists and floor loads of the magnitude generally found in residential buildings.

---

*United States Gypsum Co.

USG* Floor Plank System

USG* floor plank with tongue and grooved metal edges is laid dry without grout and welded to the joists. A topping of Mastical*, Flo-Fill, or ¼" plywood underlayment and a ceiling of ⅝" Sheetrock Firecode C* gypsum wallboard make up the complete floor system.

*United States Gypsum Co.

MASONRY

To prepare an accurate estimate on the cost of masonry requires a knowledge of the various factors entering into the work. This includes the type of building, thickness of walls, kind of mortar, style of mortar joint, class of workmanship, ability of workmen, and last but not least, the ability of the foreman in charge.

Estimating masonry is more than a matter of referring to a table or chart to obtain the prevailing wage scale and from that determine the labor cost of laying a thousand brick. If accuracy is important in the preparation of your estimates then all of the above items should be considered when determining the unit costs.

The type of building plays an important part in the costs because a mason can lay far more brick on a factory or warehouse building having long straight walls than on a school or office building, cut up with numerous windows, which require straight walls, plumb jambs, pilasters, etc.

A mason can lay far more brick on a 16" wall than on an 8" wall and he can lay more brick in smooth working lime mortar than in coarse portland cement mortar. He can also lay more brick by merely cutting the mortar joints than by striking them with the point of his trowel.

The class of workmanship plays an important part. Many of the cheaper con-structed speculative buildings have crooked walls, open mortar joints and present an all round slovenly appearance, while the best grade of workmanship requires full mortar joints, straight walls, plumb jambs, and corners, etc.—in other words, good workmanship.

Weather conditions also affect labor costs. A mason can lay more brick on a clear, dry day than when it is cold and wet, or when it is necessary to heat materials during freezing weather.

The demand for mechanics also affects their output. When there is an abundance of work, some (not all) mechanics take advantage of this condition and do just as little work as possible to hold their jobs. On the other hand, when work is scarce and there are two mechanics for every available job, they do a much better day's work when they know there is another man eagerly waiting to take his place.

The ability of the foreman in charge of the work has much to do with the final costs; his ability to plan, schedule and lay out his work in addition to obtaining a full day's work from the men under his supervision are equally important.

Masonry Specifications

The estimator in preparing his proposals from architectural plans and specifica-tions will most generally find units referred to by specifications published by the

American Society for Testing and Materials. These specifications set limitations on requirements and properties of the various materials and these are generally accepted by manufacturers.

ASTM specifications are available for all the following materials which the mason may encounter:

C 216*	Facing brick, clay or shale
C 62	Building brick, clay or shale
C 73	Building brick, sand-lime
C 126	Ceramic facing brick and tile
C 410	Industrial floor brick
C 279	Chemical resistant masonry units
C 7	Paving brick
C 32	Sewer brick, clay or shale
C 270	Mortar for unit masonry
C 476	Mortar and grout for reinforced masonry
C 466	Chemically setting silicate and silica types of chemical resistant mortars
C 395	Resin type chemical-resistant mortars
C 287	Sulphur mortar

*Check for latest edition

While every estimator should have a general acquaintance with all the above specifications, he will deal primarily with face brick, building (formerly called common) brick and mortar for unit masonry. Building brick is subdivided into 3 grades: SW, MW, and NW, indicating severe, medium and negligible weathering. Face brick is subdivided into only SE and MW grades, but these grades each have 3 types: FBX, FBS, and FBA, which set forth requirements as to appearance and size.

Mortar specifications cover both portland cement-lime and masonry cement mortars and are subdivided into types M, S, N, O and K, based on 28 day compressive strengths of 2,500, 1,800, 750, 350 and 75 lbs. per sq. in. ASTM specifications are also published for all the manufactured ingredients for the mortar.

An estimator should welcome a specification based on these ASTM designations as they enable him to figure within well defined limits and he might do well to adopt their terminology in proposals he submits.

04100   MORTAR

Cements.—Type I portland cement is the basic cement used in most mortars. Type II, for use where sulphate action is a problem, and Type III, a high early strength cement, are also included under ASTM specifications for mortar C 270-64T.

Air entraining cements should usually not be substituted, as bond strength is affected. Full information on tests and specifications are available in ASTM publications and in information published by the Portland Cement Association.

Masonry cements are included in ASTM specifications, but their chemical make-up is not regulated and varies widely among the manufacturers. It may therefore take some experimentation to find the product most suited to the job.

Masonry cements do, however, contain additives that increase workability and water retentivity, and the single product saves handling both portland cement and lime on the job, and mortar can be used as soon as mixed—all advantages. However, where a Type S mortar is selected for bonding strength, care should be taken not to select a masonry cement with air entrainment properties. Masonry cements are marketed under such names as Brixment, Atlas, Lehigh, Hy-Test, Medusa-Brisket, etc.

Lime.—Lime for construction purposes is available in both the quick (unslaked) and hydrated (slaked) forms, though in recent years hydrated lime has been preferred to quicklime because it is faster, easier, and safer to use. Building lime is supplied in bulk or packaged in 50-lb. multiple-wall, moisture resistant bags. Barrels, once widely used to package lime, are no longer in use as containers for domestic lime shipments. Lime products sold in bulk or bags are generally quoted on a ton (2000 lbs. net) basis.

Quicklime may be purchased in the following forms, the principal difference being in the size of the particles: pebble lime, crushed lime, ground lime, and pulverized lime. All quicklime must be slaked prior to use and manufacturer's directions for slaking should be followed to secure best results. Due to the greater ease with which complete slaking is secured, the finer divided types of quicklime are preferred. Since quicklime is very reactive with water, care should be exercised at all times during the slaking operation to prevent splattering of the lime and potentially serious burns of the eyes and skin. Slaked quicklime putty should be screened and permitted to age until all the lime particles are completely slaked. The aging period may vary from a few hours to several days according to the chemical and physical properties of the quicklime employed and the skill of the operator in slaking the lime.

Hydrated lime for structural use is furnished in the dry powdered form usually packaged in 50 lbs. net paper bags and sold by the ton (2000 lbs. net). It may also be purchased in bulk if proper facilities are available for handling and storage. In recent years a new special type of hydrated lime has been marketed known as Type "S" or pressure hydrated lime. This highly hydrated product contains a maximum of 8% combined unhydrated calcium and magnesium oxides; in addition putties made from this lime develop high plasticities instantly upon mixing. Type "S" hydrated limes meeting all ASTM requirements are commercially available in both the dolomitic and high calcium varieties. Type "N" (normal) hydrated limes should be soaked in a paste or putty form for several hours or overnight prior to use in order to enhance their plasticity and workability. The preferred method for soaking is to sift the hydrated lime evenly into a watertight box or vat which has been previously half filled with clean water. Do not stir, mix, or agitate the mass, but allow the lime to settle naturally; continue to add the hydrate slowly and evenly over the entire surface of the water until the thick paste is formed after which the mass is permitted to soak until required for use.

All building limes should comply with the requirements of the Standard Specifications for Hydrated Lime for Masonry Purposes (C-207) or Quicklime for Structural Purposes (C-5) as adopted by the American Society for Testing Materials.

Storage of Lime and Lime Putty.—Prior to slaking, or use, all quicklime and hydrated lime previously delivered to the site of the work should be stored in a clean, dry place.

Quicklime is shipped in bulk and waterproofed multi-wall paper bags. When furnished in bulk, it should be slaked immediately to assure maximum putty yield and avoid loss. Quicklime furnished in bags may be stored for considerable time depending upon the humidity and storage conditions.

Hydrated lime in bulk or packed in bags may be stored for relatively long periods of time provided it is placed in a clean, dry, properly ventilated warehouse.

Lime putty should be aged and stored in clean, tight vats and should be well protected from direct contact with the atmosphere by maintaining a thin film of water over the surface.

Commercial plants for the production of scientifically prepared, well aged putty, are in operation in many markets. The use of commercially prepared lime putty eliminates all necessity for slaking and aging or soaking of lime on the job, and has the further advantage of providing the estimator with reliable figures for estimating purposes since purchase order is based on net quantity of lime putty required for the operation. Other advantages of commercially produced lime putty include flexibility during construction, eliminating lost time awaiting preparation of lime putty and admitting of expansion in operations on short notice, reduction in storage space required for materials at the site of the work and improved quality of putty due to careful supervision during slaking and aging of lime putty.

Lime Putty and Its Preparation.—Job specifications generally require mortar to be mixed in definite predetermined proportions by volume and the lime proportion is usually determined in volumes of lime putty, therefore, some knowledge of putty yield of lime is necessary in preparing estimates of quantity of lime required. The putty yield of different limes varies over a relatively wide range according to chemical and physical properties of the lime, methods of processing, etc. However, the quantities given in the following table may be used as a sound, conservative average for estimating purposes.

Quantity of Lime Putty Obtainable from Various Types of Lime

Type of Lime	Average Cu. Ft. Putty per ton of lime	No. Lbs. Lime per Cu. Ft. Putty
Hydrated Lime	46 cu. ft.	44
Pebble or Pulverized Quicklime	80 cu. ft.	25

109 lbs. portland cement produces 1 cu. ft. of cement paste of normal working consistency.

9 cu. ft. of putty or paste with 27 cu. ft. of ordinary building sand will produce 1 cu. yd. of 1 :3 mortar

Quantity of Materials Required for Masonry Using
Pebble or Pulverized Quicklime at 80 cu. ft. Lime Putty per Ton

Lime Putty	Proportions by Volume Portland Cement	Sand	For one cu. yd. mortar Lime† Pounds	Cement Sacks	Sand Cu. Yd.	To lay 1000 brick (18 cu. ft.) Lime+ Pounds	Cement Sacks	Sand Cu. Yd.
1	0	3	225	0	1	150	0	.67
3	1	12	169	2.25	1	113	1.50	.67
2	1	9	150	3	1	100	2.00	.67
1½	1	7.5	135	3.6	1	90	2.40	.67
1	1	6	112.5	4.5	1	75	3.00	.67
½	1	4.5	75	6	1	50	4.00	.67
0	1	3	0	9	1	0	6.00	.67
*10%	1	3	22.5	9	1	15	6.00	.67
15%	1	3	34	9	1	23	6.00	.67

*Based on volume of cement required. Add 5 to 10 percent to the above quantities for waste.
†Dry unslaked lime.

Where hydrated lime is used in dry powdered form it is usually assumed that a given volume of dry hydrated lime will produce an equivalent volume of lime putty, thus a bag of dry hydrated lime (50 lbs. net) is equivalent to approximately 1.15 cu. ft. of lime putty.

Labor Slaking Lime and Making Mortar.—When mixing mortar by hand, a good mortar maker should slake and sand about a ton of quicklime per 8-hr. day, and should make about 4 cu. yds. of mortar per 8-hr. day.

Quantity of Materials Required for Masonry Mortar Using
Hydrated Lime at 46 cu. ft. Lime Putty per Ton

Lime Putty	Portland Cement	Sand	Lime Pounds	Cement Sacks	Sand Cu. Yd.	Lime Pounds	Cement Sacks	Sand Cu. Yd.
1	0	3	391	0	1	261	0	.67
3	1	12	293.5	2.25	1	196	1.50	.67
2	1	9	261	3	1	174	2.00	.67
1½	1	7.5	235	3.6	1	157	2.40	.67
1	1	6	195.5	4.5	1	130	3.00	.67
½	1	4.5	130.5	6	1	87	4.00	.67
0	1	3	0	9	1	0	6.00	.67
*10%	1	3	39	9	1	26	6.00	.67
*15%	1	3	59	9	1	39	6.00	.67

(Column groups: "Proportions by Volume" — Lime Putty, Portland Cement, Sand; "Quantity of Materials Required — For one cu. yd. mortar" — Lime Pounds, Cement Sacks, Sand Cu. Yd.; "To lay 1000 brick (18 cu. ft.)" — Lime Pounds, Cement Sacks, Sand Cu. Yd.)

*Based on volume of cement required. Add 5 to 10 percent to the above quantities for waste.

The actual labor cost of making 18 cu. ft. of mortar, sufficient for 1,000 common brick, should average as follows:

	Hours	Rate	Total	Rate	Total
Mortar maker	1.33	$. . . .	$. . . .	$14.95	$19.88

On all jobs of any size, the mortar is mixed in a mixer as it produces a more uniform mortar and one that is more easily spread. It not only enables a mason to lay more brick but also results in a saving of 10 to 20 percent in labor required mixing mortar.

Quik-Slak Masons Lime.—A lime furnished in pulverized form that slaks the instant it hits the water. Put up in 50 lb. paper sacks and the manufacturers say that 4 sacks are sufficient to lay, 1,000 brick under average conditions.

Aggregate.—Aggregate may be either natural sand that is clean and sharp or sand manufactured by crushing stone, gravel or air-cooled iron blast furnace slag. Sand should be properly graded as set forth in ASTM Specification C 144-66T, with all sand passing a No. 4 sieve and approximately 10% passing a No. 200 sieve. Too coarse a sand decreases workability while too fine a sand decreases water retentivity.

Mortar Colors.—Mortar coloring agents may be added, either natural, such as white sand, or pigments. While different color effects will require some experimentation, in general, 4 to 8 pounds of color to one bag of cement, hydrated lime or cu. ft. of putty and 3 cu. ft. of sand can be figured, or around one 50 pound bag per 1000 bricks with a ⅜" joint.

Waterproofing and Shrinkproofing Mortar

Considerable interest has been aroused as to the cause of leaky masonry walls and this has led to extensive investigation and tests by the National Bureau of Standards, Report BMS7, which investigated not only the materials such as brick, tile, concrete blocks, mortar, etc., but also the effects of different grades of workmanship.

After extensive tests, it was determined that workmanship affected the permeability of the walls more than any other factor. Walls with tooled joints were less permeable than similar walls with cut joints; but the quality of the workmanship inside the walls had a greater influence than the kind of surface finish on the joint. In other words, brick as laid on the average contract job, where mortar was spread and furrowed and the joints not completely filled with mortar leaked far more than bricks laid in a full bed of mortar and "shoved" into position with full bed and head joints and a complete filling of joints between all brick.

It was also determined that nearly all of the leakage was through the mortar joints regardless of the kind of brick used and that in practically every instance where the "commercial" grade of workmanship was used, leaky walls resulted.

Where brick of high absorptive properties were used, the brick absorbed the water from the mortar, making it difficult to work and resulting in small hair cracks between the mortar joints and the brick. This also caused leakage through the walls.

The use of lime with low plasticity producing a mortar with low water retentivity greatly increased the permeability of the walls. This effect was more pronounced when the mortar was used with high absorptive brick.

Water reduction in cement and lime mortars is a highly desirable objective because with less water there is less evaporation and mortar shrinkage is reduced below the critical point.

Since these investigations have been made, numerous admixtures have been placed on the market to be added to the mortar mixture which are said to give the mortar greater plasticity, retain the water in the mortar, prevent shrinkage and leakage in the mortar joints and make a smoother mortar that will work better under the trowel.

Some of these products are listed below:

Anti-Hydro.—Cement mortar for use with brick, block or stone shall be of 1 :3 mix, gauged and tempered with a solution of 1 gal. Anti-Hydro to 15 gals. of water, or use 1 qt. Anti-Hydro to each sack of cement.

Wet all brick. Push-lay and thoroughly grout all brick at every course. Parge with same mortar between the face brick and back-up masonry.

Castle Aquatite.—Mortar to be composed of 1 part portland cement, 3 parts clean and mixed with a solution of 1 part Aquatite to 15 parts water.

Hydratite Plus.—A powder in concentrated form which when mixed with the mortar is dissolved and miscible in the water. It anchors the particle of water intimately to the particle of the cement, sand and lime with a bond greater than the suction of the brick. With less water it increases the flow or workability of the mortar.

Use 2-lbs. per sack of cement in 1 :1 :6 mortar.

Hydrocide Powder.—When used for waterproofing masonry mortar, use 1-lb. of powder to each sack of cement or per cu. ft. of lime putty used in the mixture.

Omicron Mortarproofing.—A plasticizing and water reducing agent that increases the workability of mortar, increases water retentivity and minimizes shrinkage. Contains stearate for increased water repellency. Add 1 lb. of Omicron for each sack of cement, lime or putty in the mortar mix.

Requires 5 to 6 lbs. per 1,000 brick, depending upon mortar proportions and width of joints.

Plastiment (Sika).—A powder used in the proportion of up to 1-lb. per sack of cement. Increases the workability of mortar, allows a reduction in the water-cement ratio, reduces bleeding and increases the bond between brick and mortar. Of particular advantage in hot weather construction as it delays the setting time and preserves the workability of the mortar for a longer time.

Toxement IW.—Toxement IW powder, paste or concentrated paste may be used for waterproofing and shrinkproofing masonry mortar.

Sikacrete.—Accelerating and plasticizing liquid for use in brick mortar. Diluted with the gauging water 1 part Sikacrete to 3 to 7 parts water. Increases the bond and accelerates the hardening rate of mortar so that higher tiers of brick or glass block can be laid and the joints struck sooner.

Types of Mortar

The following mortars are based on proportion set forth in ASTM specifications. These mortars were designated Type A-1, A-2, B and C prior to 1954.

Type M mortar is a high strength mortar used primarily in foundation masonry, retaining walls, walks, sewers and manholes. Its proportions and costs for a cement-lime mixture are:

1 cu. ft. portland cement.....................................$3.75
¼ cu. ft. lime @ $2.40 per
    cu. ft... .60
3 cu. ft. sand @ $0.22 per
    cu. ft... .66
                       $5.01 or $1.67 per cu. ft.

Proportions and costs for a type M mortar using masonry cement are:
1 bag of portland cement.................................$3.75
1 bag Type II masonry cement........................ 3.25
6 cu. ft. sand @ $0.22 per
    cu. ft... 1.32
                       $8.82 or $1.39 per cu. ft.

Type S mortar also has a reasonably high compressive strength and develops maximum tensile bond strength between brick and cement-lime mortars. It is recommended for use in reinforced masonry and where flexural strengths are required, such as cavity walls exposed to strong winds, and for maximum bonding power, such as for ceramic veneers. Proportions and costs for cement-lime Type S mixtures are:

1 cu. ft. portland cement.................................$3.75
1/2 cu. ft. lime @ $2.40 per
    cu. ft... 1.20
41/2 cu. ft. sand @ .22
    cu. ft... .99
                       $4.94 or $1.10 per cu. ft.

Costs and proportions for cement masonry Type S Mortar are:
1/2 bag portland cement.....................................$3.75
1 bag masonry cement....................................... 3.25
41/2 cu. ft. sand @ $0.22
    cu. ft... .99
                       $7.99 or $1.78 per cu. ft.

Type N mortar is a medium strength mortar most generally used in exposed masonry above grade. The proportions and cost of Type N cement-lime mortar is:
1 cu. ft. portland cement.................................$3.75
1 cu. ft. lime 2.40
6 cu. ft. sand @ $0.22
    cu. ft... 1.32
                       $7.47 or $1.25 per cu. ft.

Proportions and cost of Type N masonry cement mortar is:
1 cu. ft. masonry cement.................................$3.25
3 cu. ft. sand @ $0.22
    cu. ft... .66
                       $3.91 or $1.30 per cu. ft.

Type O mortar is a low strength mixture for general interior use where compressive strengths do not exceed 100 psi. It may be used elsewhere where exposures are not severe and no freezing will be encountered. Proportions and costs of Type O cement-lime mortar is:

1 cu. ft. portland cement.....................................	$3.75
2 cu. ft. of lime @ per $2.40 cu. ft..................	4.80
9 cu. ft. sand @ per $0.22 cu. ft..................	1.98
	$10.53 or $1.17 per cu. ft.

Proportions and cost of Type O masonry cement mortar is:

1 cu. ft. masonry cement.....................................	$3.25
3 cu. ft. sand @ per $0.22 cu. ft..................	.66
	$3.91 or $1.30 per cu. ft.

In the above costs lime is figured in an exact proportion. As noted under "Quantity of Materials Required for Masonry Mortar Using Hydrated Lime . . .", a 50 lb. bag of hydrated lime will make 1.15 cu. ft. putty. The ASTM specifications for Types N and O mortar allow some leeway in lime proportions and a full bag of lime can be figured. Only with very high strength mortars might you have to figure splitting bags to achieve exact proportions.

Cu. Ft. of Mortar Required to Lay 1,000 Face Brick

Based on standard size brick having 1/4" to 3/8" end joints and "bed" joints as follows:

Width of Mortar Joints in Inches

1/8"	1/4"	3/8"	1/2"	5/8"	3/4"	7/8"	1"
4	7	9	12	14	16	18	20

For Dutch, English and Flemish Bond, add about 10 percent to the above quantities on account of additional head joints.

Cubic Feet of Mortar Required per Thousand Brick

For various thicknesses of walls and joints. No allowance for waste

Joint Thickness	4" Wall	8" Wall	12" Wall	16" Wall	20" Wall	24" Wall
1/8"	2.9	5.6	6.5	7.1	7.3	7.5
1/4"	5.7	8.7	9.7	10.2	10.5	10.7
3/8"	8.7	11.8	12.9	13.4	13.7	14.0
1/2"	11.7	15.0	16.2	16.8	17.1	17.3
5/8"	14.8	18.3	19.5	20.1	20.5	20.7
3/4"	17.9	21.7	23.0	23.6	24.0	24.2
7/8"	21.1	25.1	26.5	27.1	27.5	27.8
1"	24.4	28.6	30.1	30.8	31.2	31.5

Labor cost to mix mortars is usually figured as part of labor cost setting brickwork. However, if figured separately, it will run about $0.28 per cu. ft., machine mixed, $0.55 hand mixed.

Selecting Mortars.—Often the estimator will have no choice in the makeup of his mortar, the architect's specifications setting forth exactly what is required. As replacing masonry that is found not acceptable by an architect or owner is extremely expensive, it is false economy indeed to alter specifications without approval of the architect, not to mention the loss of reputation involved.

However, many jobs will leave considerable leeway in the selection of mortars and an informed estimator may even perform a valuable service in suggesting changes to the mortar proportions that will either save money or make a better job. All too often a mortar may be selected on the basis of high compressive strength when the job may require only a 100 psi. This is not only wasteful of cement, but such mortars do not have the adhesive and sealing powers of the weaker mortars, and these two characteristics are the primary functions required on most jobs.

Of further importance to the estimator is the workability of the mortar which will enable the bricklayer to lay more units. A complete discussion of mortars and recommendations is contained in Chapter 4 of "Brick and Tile Engineering" by Harry C. Plummer, published by Structural Clay Products Institute, Washington, D.C. It is highly recommended.

Mortar for Concrete Masonry.—For laying masonry walls subject to average conditions of exposure, use a mortar made in the proportions of 1 volume of masonry cement and between 2 and 3 volumes of damp, loose mortar sand; or 1 volume of portland cement and between 1 and 1¼ volumes of hydrated lime or lime putty and between 4 and 6 volumes of damp, loose mortar sand.

Walls which will be subjected to extremely heavy loads, severe winds, earthquakes, serious frost action, or other conditions requiring extra wall strength, should be laid with a mortar made of 1 volume masonry cement plus 1 volume portland cement and between 4 and 6 volumes of damp, loose mortar sand; or 1 volume of portland cement, to which may be added up to ¼ volume of hydrated lime or lime putty, and between 2 and 3 volumes of damp, loose mortar sand.

Cost of One Cubic Yard Cement Mortar for Concrete Masonry

1 volume masonry cement and 3 volumes loose mortar sand.

	Rate	Total	Rate	Total
9 sacks masonry cement	$....	$....	$3.25	$29.25
1.00 cu. yd. sand	....	....	5.75	5.75
Cost per cu. yd.		$....		$35.00
Cost per cu. ft.		....		1.28

1 volume portland cement: 1 volume hydrated lime: 6 volumes loose mortar sand.

	Rate	Total	Rate	Total
4.5 sacks portland cement	$....	$....	$3.75	$16.88
196 lbs. hydrated lime	....	....	.04	7.84
1.00 cu. yd. sand	....	....	5.75	5.75
Cost per cu. yd.		$....		$30.47
Cost per cu. ft.		....		1.13

1 volume portland cement: 2 volumes hydrated lime or lime putty; 9 volumes loose mortar sand.

	Rate	Total	Rate	Total
3 sacks portland cement	$....	$....	$3.75	$11.25
261 lbs. hydrated lime	....	....	.04	10.44
1.00 cu. yd. sand	....	....	5.75	5.75
Cost per cu. yd.		$....		$27.44
Cost per cu. ft.		....		1.02

1 volume masonry cement: 1 volume portland cement: 6 volume loose mortar sand.

	Rate	Total	Rate	Total
4.5 sacks masonry cement	$....	$....	$3.25	$14.63
4.5 sacks portland cement	....	....	3.75	16.88
1.00 cu. yd. sand	....	....	5.75	5.75
Cost per cu. yd.		$....		$37.06
Cost per cu. ft.		....		1.37

1 volume portland cement: ¼-volume hydrated lime: 3 volumes loose mortar sand.

	Rate	Total	Rate	Total
7.83 sacks portland cement	$....	$....	$3.75	$29.36
90 lbs. hydrated lime	....	....	.04	3.60
0.86 cu. yd. sand	....	....	5.75	4.95
Cost per cu. yd.		$....		37.91
Cost per cu. ft.		....		1.40

Quantity of Mortar Required to Lay 1,000 Concrete Brick of Various Sizes

Thickness of Wall

Width of Bed Joint	1 Brick or 4" Wall	2 Bricks or 8" Wall	2 Bricks or 8" Backup	3 Bricks or 12" Wall
	Modular Size Brick, 2¼"x3⅝"x7⅝"			
5/12"	9.64 cu. ft.	12.19 cu. ft.	14.73 cu. ft.	13.01 cu. ft.
	Jumbo Brick, 3⅝"x3⅝"x7⅝"			
3/8"	10.06 cu. ft.	13.88 cu. ft.	17.70 cu. ft.	15.15 cu. ft.
	Double Brick, 4⅞"x3⅝"x7⅝"			
11/24"	12.68 cu. ft.	17.77 cu. ft.	22.86 cu. ft.	19.47 cu. ft.
	Roman Brick, 1⅝"x3⅝"x11⅝"			
3/8"	11.79 cu. ft.	14.65 cu. ft.	....	....
	Roman Brick, 1⅝"x3⅝"x15⅝"			
3/8"	15.25 cu. ft.	19.07 cu. ft.	....	....

All end and back joints figured ⅜". Quantities include 10 percent for waste.

Mortar for glass blocks shall be composed of 1 part waterproof portland cement, 1 part lime and 4 parts well graded sand, all well mixed to a consistency as stiff and dry as possible.

Cost of Mortar for Laying Glass Blocks

	Rate	Total	Rate	Total
25 lb. quicklime	$....	$....	$0.04	$1.00
1 bag w'p'f portland cement	....	....	7.50	7.50
4 cu. ft. sand	....	....	.22	.88
Cost 4 cu. ft. mortar		$....		$9.38
Cost per cu. ft.		....		2.35

Mortar Required for Setting Stone.—The quantity of mortar required for stone setting will vary with the size of the stone, width of bed, etc., but as a general thing it will require 4 to 5 cu. ft. of mortar per 100 cu. ft. of stone.

If the stone is to be back-plastered, the quantity of mortar will vary with the thickness of the plaster coat but as most ashlar is 4" to 8" thick, it will require the following cu. ft. of mortar per 100 cu. ft. of stone:

Plaster Thickness	Thickness of Stone in Inches		
	4"	6"	8"
¼"	6½	4¼	3¼
½"	13	8½	6½
¾"	19	13	9½

It is customary to use white non-staining portland cement for setting and back-plastering limestone or other porous stone, as this prevents stains from appearing on the face of the stone. Never use ordinary portland cement for setting limestone.

Cost of One Cu. Yd. of 1:3 Non-Staining Cement Mortar With 1/5 Part of Hydrated Lime Based on the Cement Volume

	Rate	Total	Rate	Total
2 bbls. Medusa Stone Set cement	$....	$....	$10.25	$20.50
100 lbs. hydrated lime	....	....	2.00	4.00

1 cu. yd. sand	....	....	5.75	5.75
Cost per cu. yd.		$....		$30.25
Cost per cu. ft.			....	1.12

If white sand is used for mortar, figure $11.75 per ton (3,000 lbs. per cu. yd.)
If white cement is specified, figure $17.50 per bbl. If white cement (waterproofed) is specified, figure $18.75 per bbl.

Cost of One Cu. Yd. of Cement-Lime Mortar Consisting of ½ Part Lime, ½ Part Stone Set Cement and 3 Parts Sand

	Rate	Total	Rate	Total
1.25 bbls. Stone Set cement	$....	$....	$10.25	$12.81
225 lbs. hydrated lime	....	....	2.00	9.00
1 cu. yd. sand	....	....	5.75	5.75
Cost per cu. yd.		$....		$27.56
Cost per cu. ft.			....	1.02

If white sand is used for mortar, figure $11.75 per ton (3,000 lbs. per cu. yd.)
If white cement is specified, figure $17.50 per bbl. If white cement (waterproofed) is specified, figure $18.75 per bbl.

04200   UNIT MASONRY
04210   BRICK

The Structural Clay Products Institute separates brick walls into 3 general categories: conventional, bonded walls; drainage type walls; and barrier type walls.

The drainage type walls include brick veneer, the "SCR brick" wall, the cavity type wall, and the masonry bonded hollow wall. Drainage type walls are recommended for walls subjected to the most severe exposures. The success of this type of wall, of course, depends on the care taken to provide continuous means for water to escape.

The barrier type walls include metal tied walls and reinforced brick masonry. The success of these walls will depend on a solidly filled collar joint. The SCR presently recommends using an 8" metal tied wall in preference to an 8" bonded wall for areas subjected to moderate exposure. Metal tied walls adapt well to brick and block backup.

Bonded walls are still widely used and can be more interesting in design. Presently, there is renewed interest in interior masonry bearing walls, applied in some cases to high rise buildings. In Europe, buildings as high as 16 stories are being built with the full load of the structure carried on 8" walls. It is most often encountered in this country in apartment and dormitory type buildings of 3 to 4 stories high with many fixed interior partitions. In this type of construction, the compressive strength of the brick and mortar is of great importance.

When estimating the number of brick required for any job, obtain the length, height and thickness of each wall, stating the totals in cu. ft.

In present day estimating, all openings should be deducted in full, regardless of their size, as the estimate should show as accurately as possible the exact number of brick required for the job. The old method of counting corners twice, doubling the number of brick required for pilaster, chimney breasts, etc., is no longer used by progressive builders because they know they would never get a job using obsolete methods—and if they did get it, they would have no idea of the actual number of brick required to build it. Actual quantities and costs is the only sure method of getting your share of work at a profit.

Brick walls are usually designated on the plans as 4" or 4½"; 8" or 9"; 12" or 13"; 16" or 17", etc., increasing by 4" or 4½ in width. This variation in thickness need not be considered by the estimator because a 4" or 4½" wall is 1 brick thick; and 8" or 9" wall is 2 brick thick; a 12" or 13" wall is 3 brick thick; a 16" or 17" wall is 4 brick thick, etc. The easiest method for the estimator is to mark all walls, 4", 8", 12", or 16" and then reduce the totals to cu. ft. After the number of cu. ft. has been obtained, multiply by the number of brick required per cu. ft. of wall and the result will be the number of brick required for the job.

When brickwork is estimated on the above basis, it is advisable to add 1½ to 2 percent to the total to take care of salmon brick, broken brick, bats, etc., and the waste should not exceed this amount unless very poor brick or poor job handling methods are used.

Many years ago contractors figured their building brickwork on the basis of 7 or 7½ brick per sq. ft. of 4" or 4½" wall; 14 or 15 brick per sq. ft. of 8" or 9" wall; 21 or 22½ brick per sq. ft. of 12" or 13" wall, etc.

This method is no longer used by progressive contractors because modern business practice demands that actual quantities be figured. Work is being estimated too closely today to permit a 10 to 20 percent overrun on the brickwork, especially when many contractors are figuring on a 5 to 10 percent margin of profit.

After the number of cu. ft. of brickwork has been obtained it will be necessary to determine the number of brick required per cu. ft. of wall. This will vary with the size of the brick and the width of the mortar joints. The standard non-modular size is 8"x2¼"x3¾". Modular brick is 7½"x2²⁄₆"x3½". Sizes will vary somewhat depending upon the burning, the brick in the center of the kiln usually being more thoroughly burned.

When estimating the number of face brick required for any job, obtain the length and height of all walls to be faced with brick, and the total will be the number of sq. ft. of wall.

Always make deductions in full for all openings, regardless of their size, as the estimate should show as accurately as possible the exact number of brick required to complete the job.

When making deductions for door and window openings, always note the depth or width of the brick jambs or "reveals." If they are only 4" deep, deduct the full size of the opening as the 4" end of the brick forms the "reveal' or jamb.

If the jamb is more than 4" wide, the full depth of the jamb or "reveal" should be deducted. Example: Deduct for a 5'-0"x7'-0" opening in a 12" brick wall where the face brick jamb or "reveal" returns the full thickness of the wall. It will be necessary to deduct for the 12" jamb or "reveal" on each side of the opening, i.e., 2x12"=24" or 2'-0". Deduct this from the width of the opening, i.e., 5'-0"—2'-0" =3'- 0". The opening deducted should be 3'-0"x7'-0" instead of 5'-0"x7'-0", as shown on the plan.

Standard Brick

A standard brick is 8" long, 2¼" high and contains 18 sq. in. on the face. If laid dry without a mortar joint it would require 8 brick per sq. ft. of wall (144 ÷ 18=8). However, the thickness of the mortar joint must be added, and this will vary from 3/8" to 5/8", with 1/2" the average width. A brick is 8" long, plus 1/4" for the vertical or end mortar joint makes a total length of 8¼". A brick is 2¼" high, plus 1/2" for horizontal or bed mortar joint, makes the total height of each brick course 2¾", 8¼"x2¾"=22¹¹⁄₁₆ sq. in. on the face.

To obtain the number of brick required per sq. ft. of wall divide 144 by 22¹¹⁄₁₆" and the result is 6.35 or 6⅓ brick per sq. ft. of 4" or 4½" wall. If the wall is 8" or 9"

(2 brick) thick 2x6¹/₃=12²/₃ brick per sq. ft. of 8" or 9" wall. If 12" or 13" (3 brick) thick, 6¹/₃x3=19 brick per sq. ft. of wall, etc.

This same method should be used to obtain the number of brick of any size required for one sq. ft. of wall of any thickness.

Number of Standard Brick (8"x2¼"x3¾") Required for One Sq. Ft. of Brick Wall of any Thickness

Vertical or End Mortar Joints figured as ¼" wide

Thickness of wall	Number of Brick Thick	Width of Horizontal or "Bed" Mortar Joints					
		⅛"	¼"	⅜"	½"	⅝"	¾"
4" or 4½"	1	7.33	7	6.67	6.33	6.08	5.8
8" or 9"	2	14.67	14	13.33	12.67	12.17	11.6
*12" or 13"	3	22.00	21	20.00	19.00	18.25	17.4
16" or 17"	4	29.33	28	26.67	25.33	24.33	23.2
20" or 21"	5	36.67	35	33.33	31.67	30.42	29.0
24" or 25"	6	44.00	42	40.00	38.00	36.50	34.8

*Use this column for computing the number of brick required per cu. ft. of wall with any width mortar joint.

Variations in Face Brick Quantities.—If the brick are laid in running bond without headers, the above quantities will prove sufficient but if there is a full course of headers every 5th, 6th or 7th course, it will be necessary to allow for the extra brick required. Also, if the brick are laid in English, Flemish, English Cross or Dutch Bond, it will be necessary to make an allowance for extra brick where full header courses are required.

Number of Square Inches Occupied by One 8"x2¼" Face Brick With Various Width Mortar Joints

All Vertical Mortar Joints Figured ¼" Wide
Width of Horizontal or "Bed" Mortar Joints in Inches

⅛"	¼"	⅜"	½"	⅝"	¾"	⅞"	1"
19.30	20.625	21.656	22.69	23.72	24.75	25.78	26.81

Number of 8"x2¼"x3¾" Face Brick Required Per Sq. Ft. of Wall In Running Bond Without Headers

All Vertical Mortar Joints Figured ¼" Wide
Width of Horizontal or "Bed" Mortar Joints in Inches

⅛"	¼"	⅜"	½"	⅝"	¾"	⅞"	1"
7.46	7.0	6.65	6.35	6.08	5.82	5.60	5.38

The above table does not include any allowance for waste, breakage, nor for header courses extending into the building brick backing to form a bond, but is based on using all stretchers or blind headers with metal wall ties.

On face brick work, 3 to 5 percent should be added to the net quantity for waste.

For face brick laid with a row of full "headers" every 5th, 6th or 7th course add the following percentages to the above quantities:

Percentage to be Added for Various Brick Bonds

Common (full header course every 5th course)...20 % or ⅛
Common (full header course every 6th course).......................................16²/₃% or ⅛
Common (full header course every 7th course).......................................14¹/₃% or 1/7
English or English Cross* (full headers every other course)............... 50 % or ½
English or English Cross* (full headers every 6th course)................ 16²/₃% or 1/6
Dutch or Dutch Cross* (full headers every other course).............. 50 % or ½

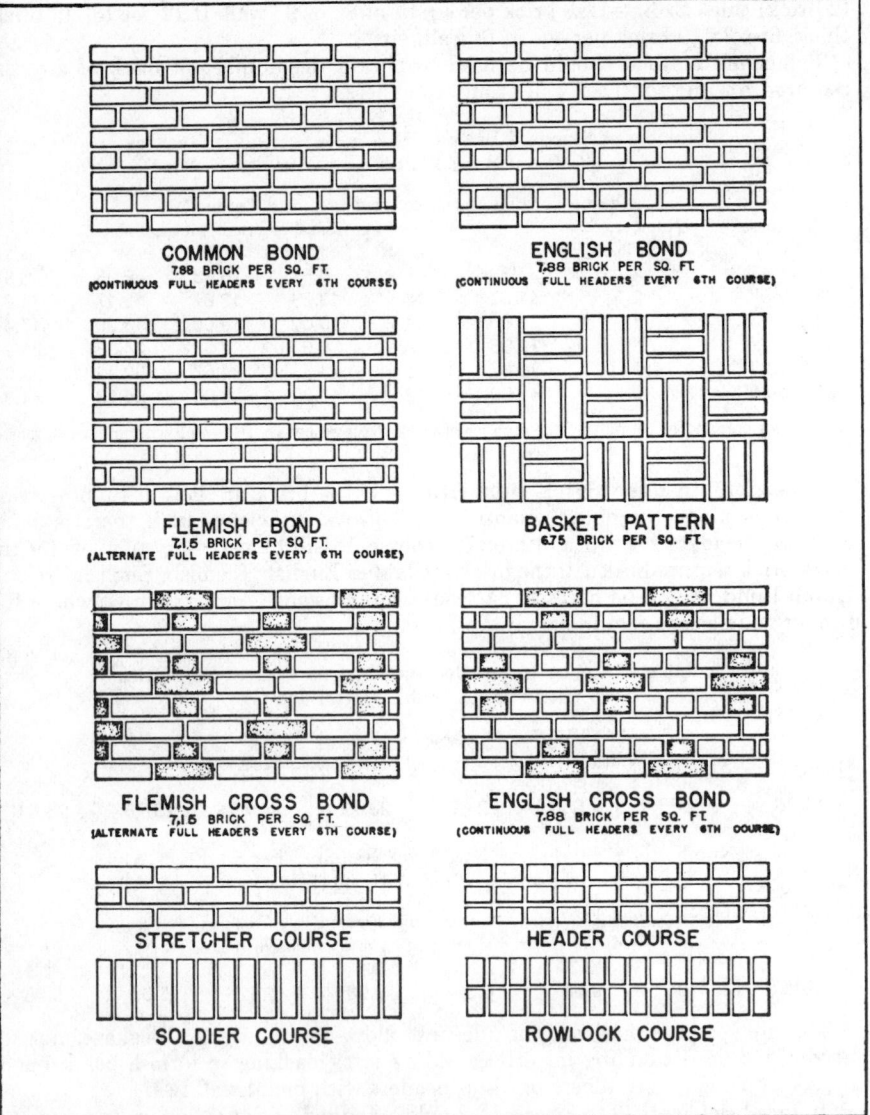

Elevations of Various Brick Bonds

Dutch or Dutch Cross* (full headers every 6th course)...................... 16 2/3 % or 1/6
Flemish (full headers every course)....................................... 33 1/3 % or 1/3
Flemish (full headers every 6th course)................................5.6 % or 1/18
Double Header (two headers and a stretcher every 6th course) .... 8 1/3 % or 1/12
Double Header (two headers and a stretcher every 5th course) .... 10 % or 1/10
Double Flemish (full headers every other course) ....10 % or 1/10

*Add 10 to 15 percent extra brick for waste in cutting unless a masonry saw is used.

Percentages To Be Added for Various Brick Bonds

Double Flemish (full headers every 3rd course).................................... 62/3% or 1/15
3 Stretcher Flemish (full headers every other course) .... 71/7% or 1/14
3 Stretcher Flemish (full headers every 3rd course) .... 4.8% or 1/21
4 Stretcher Flemish (full headers every other course) .... 5.6% or 1/18
4 Stretcher Flemish (full headers every third course)............................ 3.7% or 1/27

Face Brick Laid in
Common or American Bond.

For garden walls, porch walls, and other places where an 8" wall is used, with face brick on both sides of the wall, no additional brick are required for any type of bond.

For walks and floors with the brick laid on edge, in any pattern except diagonal, calculate as you would for face brick in common bond without headers. For herringbone or other diagonal work, additional brick will be required because of the waste in chipping or cutting the brick for the borders. The additional number of brick will vary with the width of the walk or floor, as the wider the surface, the smaller the average waste per sq. ft. Walks and floors having the brick laid flat require one-third less brick than where they are laid on edge.

Brick Bonds

Bond in brickwork is the overlapping of one brick upon the other, either along the length of the wall or through its thickness, in order to bind them together into a secure structural mass. Units are shifted back and forth so that the vertical joints in two successive layers or "courses" do not come into line; in other words, the brick are laid so as to break the joint, the whole forming a natural bond or a structural unit giving strength to the wall.

Running or Stretcher Bond.—The wall surface is made up of stretcher courses having a header at the corners which appears as a stretcher on the return side. This bond is very strong longitudinally, but lacks transverse strength, consequently it is modified into what is called Common or American Bond.

Face Brick Laid in English Bond.

Face Brick Laid in English Cross
or Dutch Cross Bond.

Common or American Bond.—This bond is obtained by laying a course of headers every fifth, sixth or seventh course. To maintain the effect of the running bond, a special double header bond is sometimes used. This method of using headers as in Common or American Bond, in order to secure transverse strength of wall, can be treated in a way to produce more pleasing effects, as may be seen in Flemish or English Bonds.

English Bond.—English Bond is made up of alternating courses of stretchers and headers. Ordinarily half brick are used for the header courses, except every sixth course a full header course is used to tie the face and common brick walls together. The allowance for headers is given on the previous page.

Dutch Bond or English Cross Bond.—Dutch or English Cross Bond is similar to English Bond, except starting courses differ. Each course starts off with ¼, ½, or ¾ of a brick, alternating as shown in the illustration, and each successive course consists of stretchers and headers. On account of the large amount of cutting necessary to obtain the ¼ and ¾ pieces of brick, it is necessary to add 10 to 15 percent additional brick to allow for waste, caused by breakage in cutting.

Face Brick Laid in Flemish Bond.

Face Brick Laid in Double Flemish
or Garden Wall Bond.

However, waste in cutting may be greatly reduced if a masonry saw is used for this purpose, as a brick may be cut in 10 to 15 seconds and the resulting waste and spoilage of face bricks greatly reduced.

If brick are laid with full header courses every second or sixth course, add the percentages given on the previous pages.

Flemish Bond.—Flemish Bond secures its effect by laying each course in alternate stretchers and headers, the header resting upon the middle of the stretcher in successive courses. When full brick are used for headers, it requires one-half more brick than where half brick or "blind" headers are used.

Double Flemish Bond.—The arrangement of bricks along the course so that each header is preceded and followed by two stretchers for the entire course. On consecutive courses, the position of the header is directly over the joint between the two stretchers. Each stretcher has a lap of three-fourths of its length over the stretcher beneath.

Face Brick Laid in Diamond Bond Patterns.

Face Brick Laid in Three
Stretcher Flemish Bond

Three Stretcher Flemish Bond.—An arrangement of bricks along the course so that a header is placed between each sequence of three stretchers. The header of one course is centered over the middle stretcher of the course below.

Four Stretcher Flemish Bond.—In this bond, a header occurs between each sequence of four stretchers. The header of one course is centered between the headers of the course below.

The allowances for header courses are given on the previous pages.

Garden Wall Bond.—Garden Wall Bond is merely a modification of Flemish Bond, secured by laying courses with two to four stretchers alternating with a header, and requires the same number of brick per square foot as Flemish Bond.

Face Brick Laid in Four
Stretcher Flemish Bond

Diamond Bond Patterns.—Diamond Bond Patterns are secured by a modification of the Garden Wall Bond. It is, however, only in case of large wall surfaces that patterns of an elaborate character could be recommended, as any departure from simple bonds adds to the cost of bricklaying.

Modular Size Brick

Economy is the prime objective of Modular Design. Its application has brought about economy of design, in layout and in construction.

While Modular Design involves the coordination of building materials, this basic principle is employed only as a means to the main objective—economy—and is not carried beyond the point where its use would defeat the main purpose.

It is not to be considered as a complex and revolutionary plan of coordinating building materials, but rather as a simplified and efficient system of design, layout and construction resulting in definite economies.

After an exhaustive study, a four inch module or increment was agreed upon as a unit that would afford the maximum practical standardization and simplification, and at the same time afford a sufficient flexibility in design.

On the basis of the four inch module or increment, certain sizes of brick and tile have been recommended. In this connection, it must be remembered that the four inch module, governing both the vertical and horizontal dimensions, is used to govern the layout of finished wall areas and to determine heights and widths of doors, windows, other openings.

Since the module is a unit of wall measurement, it cannot be applied to the individual brick or tile, but must be considered as governing the size of the brick or tile, plus the mortar joint.

In employing the modular system of measurement for masonry walls, the dimension is considered from center to center of the mortar joints. Therefore, to fit properly into the modular system, the actual sizes of the individual units of either brick or tile are determined by the modular dimension less the mortar joint.

The explanatory details, drawn with ½" mortar joints, present this matter very clearly. The first illustration indicates a wall section built of standard size brick measuring 2¼" x3¾" x8". It will be noted that a unit of this size does not comply

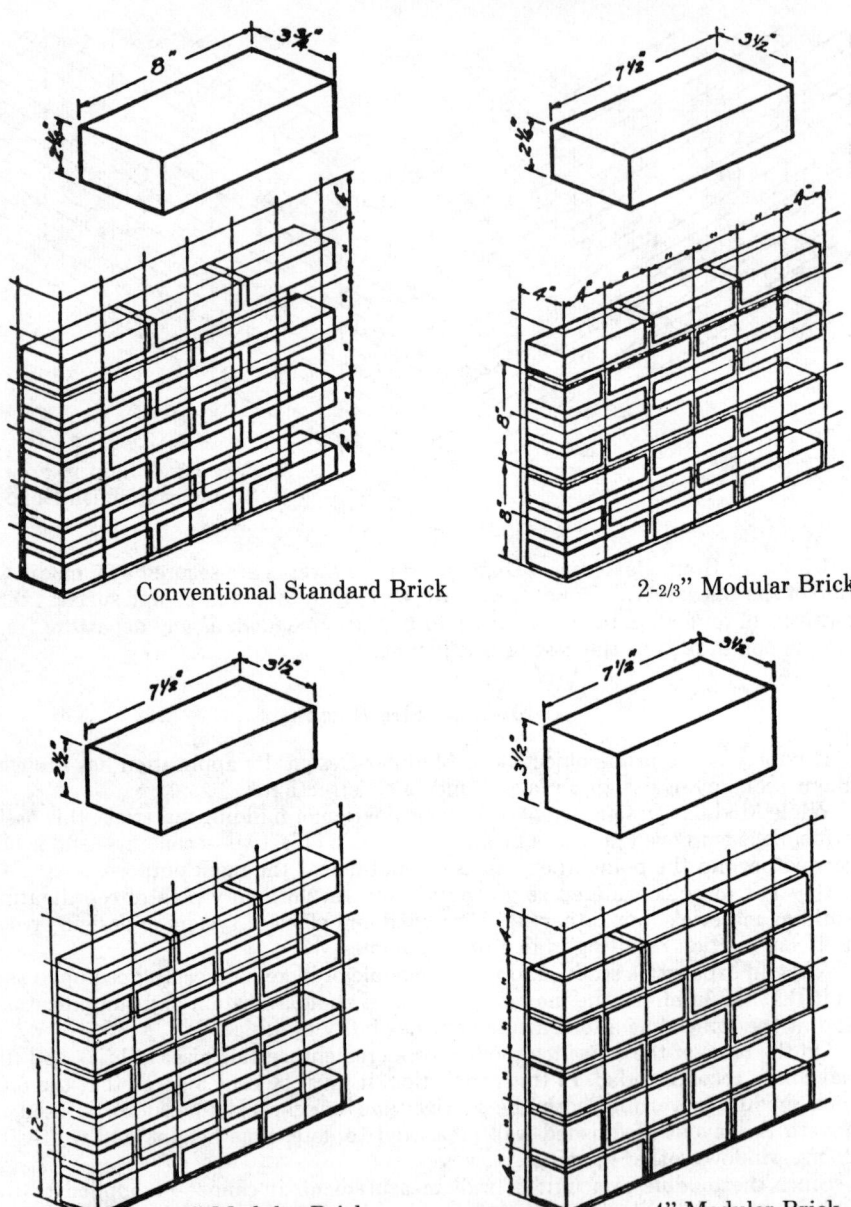

Conventional Standard Brick

2-2/3" Modular Brick

3" Modular Brick

4" Modular Brick

with the modular measurements as indicated by the four inch modular dimension lines, known as grid lines. Therefore, if standard size brick are used in the construction of a project designed according to the modular system, wasteful cutting and fitting of the brick would be necessary.

Note the difference in the second wall section. In this case a modular brick is employed, having an actual size of $2\frac{1}{6}$" x $3\frac{1}{2}$" x $7\frac{1}{2}$". Using a 1/2" mortar joint, three courses of brick lay up perfectly to every two horizontal grid lines, and one length of brick likewise fits two vertical grid lines. The same holds true of the other sizes of brick shown.

Exact heights of modular brick will vary, the difference being taken up in the bed joints. It is therefore important to know the exact size of brick. The following table lists standard joint widths.

Modular Face Brick

Three standard modular joint thicknesses are ¼", ⅜", and ½". The number of modular masonry units per square foot, of course, will remain constant for each nominal size, the difference being taken up in the joint.

Modular Brick Walls Without Headers

Nominal Brick Size in Inches	Brick Per	Cubic Feet of Mortar Per 100 Sq Ft of Wall			Cubic Feet of Mortar Per 1000 Brick		
	Sq Ft	Joint Thickness			Joint Thickness		
h t l	of Wall	¼ in.	⅜ in.	½ in.	¼ in.	⅜ in.	½ in.
2²/3x4x 8	6.750	3.81	5.47	6.95	5.65	8.10	10.30
3¹/5x4x 8	5.625	3.34	4.79	6.10	5.94	8.52	10.84
4 x4x 8	4.500	—	4.12	5.24	—	9.15	11.65
5¹/3x4x 8	3.375	—	3.44	4.34	—	10.19	12.87
2 x4x12	6.000	—	6.43	8.20	—	10.72	13.67
2²/3x4x12	4.500	3.52	5.06	6.46	7.82	11.24	14.35
3 x4x12	4.000	—	4.60	5.87	—	11.51	14.68
3¹/5x4x12	3.750	3.04	4.37	5.58	8.11	11.66	14.89
4 x4x12	3.000	2.56	3.69	4.71	8.54	12.29	15.70
5¹/3x6x12	2.250	—	3.00	3.84	—	13.34	17.05
2²/3x6x12	4.500	—	7.85	10.15	—	17.45	22.55
3¹/5x6x12	3.750	—	6.79	8.77	—	18.10	23.39
4 x6x12	3.000	—	5.72	7.40	—	19.07	24.67

The above quantities include mortar for bed and vertical joints only. The following table gives allowances for backing mortar or collar joints.

Cubic Feet of Mortar Per 100 Sq Ft of Wall

¼-in. Joint	⅜-in. Joint	½-in. Joint
2.08	3.13	4.17

Note: Cubic feet per 1000 units = $\dfrac{10 \times \text{cubic feet per 100 sq ft of wall}}{\text{number of units per square foot of wall}}$

Material Costs of Brick

Brick costs will vary widely around the country and within color ranges. Common brick averages around $115.00 per thousand nationally.

Face brick in standard size, nominally 2²/3"x4"x8", are quoted from $140.00 per 1000 to $195.00. Engineer units, nominally 3 1/5"x4"x8" so that 5 courses equal 16" in height, range from $160.00 to $185.00. King size units, nominally 2²/3"x4"x10", range from $170.00 to $200.00. Norwegian size, nominally 3¹/5"x4"x12" range from $220.00 to $250.00. Utility, nominally 4"x4"x12", range from $295.00 to $325.00.

In general red tones are cheapest, earth tones fall in the middle range and whites and grays are the most expensive except for glazed units which run from $225.00 up in the more brilliant shades.

Brick Cavity Walls

Brick cavity walls are distinguished from all other forms of hollow masonry by a continuous, vertical and horizontal air space, bridged by masonry ties. Prime purpose of this space is to provide an absolute barrier against the penetration of moisture to the inner side of the wall at any point, including heads and jambs of openings, joist bearing points, etc., thus assuring trouble-free masonry and permitting direct plastering without furring. It also increases resistance to heat-flow to about the same extent as the air space created by ordinary furring.

Since this construction affords no supporting material through which moisture penetrating the outer 4" wythe of the wall can soak to the inner wythe, it guarantees the inner wythe will remain perfectly dry, provided adequate precautions are taken to intercept and deflect moisture running down the inner side of the outer wythe at the heads of all openings, and provided the bottom of the wall is equipped with a suitable dampproof course to prevent absorption of soil water.

In place of the usual header courses, the inner and outer wythes of the cavity walls are connected by rust-proof metal ties—usually 3/16" diameter, square-end, Z bars—placed in every sixth course and 3'-0" on center horizontally.

Labels on figure:

WOOD PLATE
Anchor Bolt wired to double Z ties.

DAMPPROOFING
Over all window heads, one or two courses for inside leaf.

DAMPPROOFING
Under and back of all sills.

½" PLASTER
Applied direct to inside of wall.

Z TIES, 3" O.C.
every sixth course.

CONTINUOUS
2" AIR SPACE

DAMPPROOFING
ALTERNATE METHOD

Occasional vertical joints in first course may be left open as weep holes.

FILL GRADE

Std. Joist Anc.

Brick Cavity Wall Construction

Labor Laying Building Brick

There are so many factors entering into the labor cost of laying building brick that an explanation is necessary to use the following data intelligently.

There are no set quantities of brick that a mason will lay per hr. or per day, and with 10 men laying brick on the same wall and on exactly the same class of work, no two of them will lay the same number of brick. For this reason averages must be taken. There are good mechanics and poor mechanics—there are conscientious mechanics who desire to do a full day's work and there are unscrupulous ones who do only enough to "get-by". There are others who never "get-by" and are on one job only long enough for the foreman to see they are incompetent and off they go.

Solid brick walls are used much less frequently now than in the past, and where brick are used as a facing they are most often backed-up with cement blocks.

This is due to the lower labor costs in handling and setting the larger units of cement blocks in comparison with the smaller units of brick, as one 8"x8"x16" cement block usually displaces 12 building brick in the wall.

The class of work also has a great deal to do with the variation in labor costs, as it costs much more to lay brick in an 8" wall with struck joints than in a 16" wall having cut joints on one side of the wall and struck joints on the other. Also, it costs more to lay brick in portland cement mortar than in lime mortar because the lime mortar spreads far more easily than cement mortar.

Grid Paper Is Used by the Designer for All Layouts and Details in Modular Design. The Above Plan and Elevation of Brick between Two Window Openings Show How the Brick Fit the Intervening Space Perfectly with No Wasteful Cutting or Fitting.

Labor Production For Laying Building Brick

Class of Work	Mortar Joints Style	Kind of Mortar	Average Number Brick Laid per 8-hr. Day	Aver. Hrs. 1,000 Brick		
				Mason Hours	Labor Hours	Hoist Engr.*
8" walls, 1-story bungalows, garages, two-flat buildings, residences, etc.	Cut	Lime	750–800	10.5	9.0	
	Struck	Lime	700–750	11.0	9.0	
12" walls, ordinary construction, apartment buildings, houses, garages, factories, stores, store and apartment buildings, schools, etc.	Cut	Lime	950–1,050	8.0	8.0	
	Struck	Lime	925–975	8.5	8.0	
	Cut	Cement	825–900	9.3	8.0	
	Struck	Cement	750–850	10.0	8.0	
Hodding brick to second story, add					2.0	
16" walls, heavy warehouse, factory and industrial work. Straight walls	Cut	Lime	1,100–1,225	7.0	8.0	
	Struck	Lime	1,000–1,125	7.5	8.0	
	Cut	Cement	950–1,050	8.0	8.0	
	Struck	Cement	900–1,000	8.4	8.0	
Backing-up face brick, cut stone, terra cotta on ordinary wall bearing buildings, figure same as given above for 8" and 12" walls						

WINTER WORK. Heating brick, mortar, attending salamanders, removing snow and ice, add extra for this work.
*HOISTING. Add about ¼-hr. hoisting engineer time per 1,000 for reasonable size jobs using 35,000 or more brick per day; ½-hr. per 1,000 for jobs using 15,000 to 20,000 brick per day; and ⅓-hr. per 1,000 for jobs using 25,000 to 30,000 brick per day.
If automatic hoists are used, omit all time for hoisting engineer.

Labor Production For Laying Building Brick—Cont.

Class of Work	Mortar Joints Style	Kind of Mortar	Average Number Brick Laid per 8-hr. Day	Aver. Hrs. 1,000 Brick		
				Mason Hours	Labor Hours	Hoist Engr.*
Backing-up face brick, cut stone, terra cotta on steel or concrete skeleton frame buildings. First Grade workmanship. Walls 8" to 12" thick.	Cut	Lime	825–900	9.3	9.0	....
	Struck	Lime	750–850	10.0	9.0	....
	Cut	Cement	725–800	10.5	9.0	....
	Struck	Cement	700–775	10.8	9.0	....
Public buildings, First Grade workmanship. Schools, college and university bldgs. Courthouses, state capitols, public libraries, etc. 12" to 20" walls.	Cut	Lime	850–950	9.0	8.0	....
	Struck	Lime	800–900	9.4	8.0	....
	Cut	Cement	750–850	10.0	8.0	....
	Struck	Cement	700–800	10.6	8.0	....
Powerhouses and other structures, having high walls 12" to 20" thick, without intermediate floors.	Cut	Lime	1,100–1,200	7.0	9.0	....
	Struck	Lime	1,000–1,100	7.6	9.0	....
	Cut	Cement	925–1,050	8.0	9.0	....
	Struck	Cement	875–950	8.8	9.0	....
Shoved joints. Brick laid in full shoved joints with all vertical joints slushed full of mortar, add to all of the above				1.5		
Basement foundation walls, paving brick.	Cut	Cement	900–1,000	8.5	8.5	....
	Struck	Cement	850–950	9.0	9.0	....

WINTER WORK. Heating brick, mortar, attending salamanders, removing snow and ice, add extra for this work.
*HOISTING. Add about ¼-hr. hoisting engineer time per 1,000 for reasonable size jobs using 35,000 or more brick per day; ½-hr. per 1,000 for jobs using 15,000 to 20,000 brick per day; and ⅓-hr. per 1,000 for jobs using 25,000 to 30,000 brick per day.
If automatic hoists are used, omit all time for hoisting engineer.

Labor Production For Laying Building Brick—Cont.

Class of Work	Mortar Joints Style	Kind of Mortar	Average Number Brick Laid per 8-hr. Day	Aver. Hrs. 1,000 Brick		
				Mason Hours	Labor Hours	Hoist Engr.*
Building brick foundation walls 8" to 12" thick, ordinary workmanship.	Cut	Lime	1,000–1,100	7.6	8.0	
	Struck	Lime	900–975	8.5	8.0	
	Cut	Cement	900–975	8.5	8.0	
	Struck	Cement	825–900	9.3	8.0	
Chimneys and stacks, building brick, 1'-4" to 2'-0" sq. 15'-0" above roof, 4" to 8" walls. Hodding brick, extra.	Struck	Lime	500–550	16.0	16.0 3.0	
Large chimney and stacks, 3'-0" to 4'-0" Sq. 15' to 30' high above roof, 8" to 12" walls.	Struck	Lime	550–600	14.0	16.0	
Large brick stacks 100' to 150' high. walls from 1'-8" at base to 12" top. Inside put-log scaffold. Outside scaffold extra.	Struck	Lime	650–750	11.5	14.0	
Bricking-in boilers, fire-boxes, etc.	Struck		600–650	12.5	10.0	

WINTER WORK. Heating brick, mortar, attending salamanders, removing snow and ice, add extra for this work.

*HOISTING. Add about ¼-hr. hoisting engineer time per 1,000 for reasonable size jobs using 35,000 or more brick per day; ½-hr. per 1,000 for jobs using 15,000 to 20,000 brick per day; and ⅓-hr. per 1,000 for jobs using 25,000 to 30,000 brick per day.

If automatic hoists are used, omit all time for hoisting engineer.

It costs more to lay brick in cold, wet weather when the brick are slippery and the mason's hands cold than in fair weather with all conditions favorable.

It requires more labor to lay brick in the winter when the brick and mortar must be heated and salamanders placed along the floor or scaffold, requiring additional labor attending salamanders, removing ice, snow, etc., than in the summer when conditions are favorable.

The quantities and costs given on the following pages are based on a good grade of workmanship applicable to the type of building on which it is to be used.

The quantities given are based on the performances of experienced masons working regularly at their trade under present conditions. They will not hold good in small communities where the mason does the cement work and plastering as well as the bricklaying and stone setting of the community.

The costs are based on efficient management and supervision, the use of modern tools and appliances, machine mortar mixers, etc., on jobs large enough to warrant their use. On jobs of sufficient size to require a mason foreman, it will be necessary to add for this time on the basis of a certain price per 1,000 brick or at a certain price per week, based on the estimated duration of the job.

Labor Laying Face Brick

The labor cost of laying face brick will vary with the grade of workmanship, style of bond in which the brick are laid, width and kind of mortar joint, type of building, whether long, straight walls or walls cut up with pilasters, door and window openings, etc., and last but not least, the ability of the foreman in charge—a good foreman will probably have competent masons and vice versa.

Concave   Convex   V-Tooled   Stripped   Rodded

Raked Out   Weathered   Struck   Flush Cut

Styles of Mortar Joints

Common bond is the most economical method of laying face brick, and other styles, such as English, Flemish, Dutch and Garden Wall bond, require more labor on account of the various patterns and the additional cutting required. This is especially so on First Grade work where all vertical mortar joints must be plumb.

For instance, when laying Dutch, English and Flemish brick bonds, and the many variations of same, it is necessary to do a great deal of cutting, especially where half brick are used for headers, and it requires just as much time to cut the brick, spread mortar to lay one-half brick as for a full brick.

The labor cost of laying 1,000 face brick in Dutch, English or Flemish bond will be more if half-brick are used for headers than where full brick are used because it will take just as much time to cut, spread mortar bed, apply mortar to brick, lay same and cut the joints whether a full or half brick is used.

The style of mortar joint has considerable bearing on the labor costs. The flush cut joint is the most economical because the bricklayer cuts off the excess mortar with his trowel and the joint is finished. For concave, V-tooled, weathered and struck joints, it is necessary to cut off the excess mortar with the trowel and then another operation is required, using a trowel for the weathered or struck joints or a jointing tool for concave and V-tooled. For raked-out joints, a bricklayer must rake out the mortar ¼" to ⅜" back from the face of the brick, and for a rodded joint, the bricklayer must rake out the excess mortar and then strike the joint with a jointer. These operations all require time and cost money.

The stripped joint is the most expensive of any illustrated, as it is necessary to place narrow wood strips in the mortar joints between each course of brick, and these strips must remain until the mortar has set, so it will not squeeze out toward the face of the brick.

It is slow work laying brick where the joints are stripped.

If face brick are laid with ⅞" or 1" bed mortar joints, the work proceeds very slowly and requires a stiff mortar because the weight of the top brick courses squeeze the mortar out of the lower mortar joints, producing an uneven appearance. For this reason the work cannot be carried up many courses at a time. Where the joints are struck, it is necessary for the mortar to take its initial set before the joints can be struck to produce a workmanlike job. Brick work of this kind is very expensive.

The quantities and costs given on the following pages are intended to cover all kinds of face brick work, whether laid in veneer in frame buildings, in solid brick or brick and tile walls.

Labor Laying Face Brick in English, Flemish, Dutch, Garden Wall and Diamond Bonds

Face brick laid in any of the above bonds with the courses consisting of various arrangements of headers and stretchers requires considerable extra labor cutting brick on account of the large number of "bats" or headers required for each course. Ordinarily, a full header course extends through the wall every 5th or 6th course, and the other headers are only half brick. On First Grade workmanship it is necessary to plumb all vertical mortar joints in order that the various patterns will work out accurately.

Basing the labor laying brick in Common Bond as 100, it will require approximately the following additional time to lay face brick in Dutch, English and the different variations of Flemish bond, where half brick are used as headers instead of full brick.

Kind of Bond	Percentage
Common Bond using Full Headers	100
Dutch Bond using Full Headers every other Course	120
Dutch Bond using Full Headers every Sixth Course	129
English Bond using Full Headers every other Course	120
English Bond using Full Headers every Sixth Course	129
Flemish Bond using Full Headers every Course	120
Flemish Bond using Full Headers every Sixth Course	126
Double Flemish Bond using Full Headers every Course	100
Double Flemish Bond using Full Headers every Sixth Course	116
3-Stretcher Flemish Bond using Full Headers every Other Course	107
3-Stretcher Flemish Bond using Full Headers every Sixth Course	112
4-Stretcher Flemish Bond using Full Headers every Other Course	106
4-Stretcher Flemish Bond using Full Headers every Sixth Course	109

Labor Laying 1,000 Standard Size (8"x2¼"x3¾") or Modular Size (7⅝"x2¼"x3⅝") Face Brick in Common Bond

| 2:1:9 Lime-Cement Mortar | | | | Ordinary Workmanship | | Cement Mortar | | |
Mason Hours	Labor Hours	Hoist Engr.*	Number Laid per 8-Hr. Day	Style of Mortar Joint	Number Laid per 8-hr. Day	Mason Hours	Labor Hours	Hoist Engr.*
16.0	12	0.5	475-525	Flush Cut	425-475	17.8	13	0.5
17.8	12	0.5	425-475	V-Tooled or Concave	400-450	19.0	13	0.5
18.0	12	0.5	420-450	Struck or Weathered	390-435	19.5	13	0.5
17.8	12	0.5	425-475	Raked Out	400-450	19.0	13	0.5
19.0	12	0.5	400-450	Rodded	375-425	20.0	13	0.5
26.7	15	0.5	275-325	Stripped	275-300	27.8	16	0.5
				First Grade Workmanship				
19.0	13	0.5	400-450	Flush Cut	375-425	20.0	14	0.5
20.0	13	0.5	375-425	V-Tooled or Concave	360-400	21.0	14	0.5
21.0	13	0.5	360-400	Struck or Weathered	350-390	21.5	14	0.5
20.0	13	0.5	375-425	Raked Out	360-400	21.0	14	0.5
23.0	13	0.5	325-360	Rodded	300-340	25.0	14	0.5
32.0	16	0.5	225-275	Stripped	200-240	36.5	18	0.5
32.0	16	0.5	225-275	⅞" to 1" Flush Cut	200-240	36.5	18	0.5
36.0	16	0.5	200-250	⅞" to 1" Struck	190-230	38.0	18	0.5

*Add Hoisting engineer's time only if required.

Labor Laying 1,000 Standard Size (8"x2¼"x3¾") or Modular Size (7⅝"x2¼"x3⅝") Face Brick in English, Flemish**, Dutch and Garden Wall Bond, Laid with Full Headers Every 6th Course**

2:1:9 Lime-Cement Mortar				Ordinary Workmanship	Cement Mortar			
Mason Hours	Labor Hours	Hoist Engr.*	Number Laid per 8-Hr. Day	Style of Mortar Joint	Number Laid per 8-hr. Day	Mason Hours	Labor Hours	Hoist Engr.*
20.0	12	0.5	380-420	Flush Cut	360-400	21.0	13	0.5
21.3	12	0.5	350-400	V-Tooled or Concave	330-380	22.5	13	0.5
22.5	12	0.5	335-375	Struck or Weathered	320-360	23.5	13	0.5
21.3	12	0.5	350-400	Raked Out	330-380	22.5	13	0.5
22.5	12	0.5	335-375	Rodded	320-360	23.5	13	0.5
29.0	15	0.5	250-300	Stripped	240-285	31.0	15	0.5
				First Grade Workmanship				
22.5	13	0.5	335-375	Flush Cut	320-355	23.7	14	0.5
24.0	13	0.5	315-350	V-Tooled or Concave	300-335	25.0	14	0.5
25.0	13	0.5	300-340	Struck or Weathered	285-325	26.5	14	0.5
24.0	13	0.5	315-350	Raked Out	300-335	25.0	14	0.5
26.7	13	0.5	275-325	Rodded	260-310	29.0	14	0.5
35.5	16	0.5	200-250	⅞" to 1" Flush Cut	190-240	37.0	18	0.5
35.5	16	0.5	200-250	⅞" to 1" Struck	190-240	37.0	18	0.5
40.0	16	0.5	175-225		165-215	42.0	18	0.5

*Add hoisting engineer's time only if required.
**If laid with full headers every course or every 2nd course, increase quantities laid approximately 8 percent.

Labor Laying 1,000 English Size 8⅜"x2⅝"x4" Face Brick

2:1:9 Lime-Cement Mortar				Common Bond		Cement Mortar			
Mason Hours	Labor Hours	Hoist Engr.*	Number Laid per 8-Hr. Day	Style of Mortar Joint	Number Laid per 8-hr. Day	Mason Hours	Labor Hours	Hoist Engr.*	
22.5	17	0.6	335-375	Flush Cut	320-360	23.5	18	0.6	
23.5	17	0.6	325-360	V-Tooled or Concave	310-340	24.6	18	0.6	
26.3	17	0.6	285-325	Struck or Weathered	275-315	27.0	18	0.6	
23.5	17	0.6	325-360	Raked Out	310-340	24.6	18	0.6	
28.5	17	0.6	260-300	Rodded	250-290	29.5	18	0.6	
				English, Flemish or Dutch Bond					
24.7	17	0.6	300-350	Flush Cut	285-325	26.3	18	0.6	
26.3	17	0.6	285-325	V-Tooled or Concave	275-310	27.0	18	0.6	
28.5	17	0.6	260-300	Struck or Weathered	250-290	29.5	18	0.6	
26.3	17	0.6	285-325	Raked Out	275-310	27.0	18	0.6	
31.0	17	0.6	240-275	Rodded	230-265	32.2	18	0.6	

*Add hoisting engineer's time only if required.

Labor Laying 1,000 Standard Size (8"x2¼"x3¾") Vitrified Face Brick

| 2:1:9 Lime-Cement Mortar | | | | Common Bond | Cement Mortar | | | |
Mason Hours	Labor Hours	Hoist Engr.*	Number Laid per 8-Hr. Day	Style of Mortar Joint	Number Laid per 8-hr. Day	Mason Hours	Labor Hours	Hoist Engr.*
31.0	17	0.6	235-275	Buttered and Struck	225-265	32.7	18	0.6
22.5	16	0.6	335-375	Flush Cut	320-360	23.5	17	0.6
23.5	16	0.6	320-360	V-Tooled or Concave	300-340	25.0	17	0.6
26.3	16	0.6	285-325	Struck or Weathered	270-310	27.5	17	0.6
23.5	16	0.6	260-300	Rodded	250-285	30.0	17	0.6
28.5	16	0.6	320-360	Racked Out	300-340	25.0	17	0.6
				English, Flemish or Dutch Bond**				
34.8	17	0.6	210-250	Buttered and Struck	200-240	36.4	17	0.6
25.0	16	0.6	300-340	Flush Cut	285-325	26.3	17	0.6
26.3	16	0.6	285-325	V-Tooled or Concave	275-310	27.3	17	0.6
28.5	16	0.6	260-300	Struck or Weathered	250-290	29.7	17	0.6
26.3	16	0.6	285-325	Raked Out	275-310	27.3	17	0.6
31.0	17	0.6	240-275	Rodded	230-265	32.7	17	0.6

*Add hoisting engineer's time only if required.
**Full headers every sixth course. If full headers every other course, increase quantities laid approximately 8 percent.

Roman Face Brick

Roman size face brick, 12"x1½"x4" were extensively used some years ago for facing many types of buildings. Generally, they were laid with thin "buttered" joints and most of the original brick were made by the dry-pressed method and were relatively expensive.

Within the past few years, the modular Roman brick was introduced having a nominal size of 12"x2"x4", including mortar joint, while the actual brick size is 11⅝"x1⅝"x3⅝", requiring a ⅜" mortar joint to lay 2" high and 12" long. These brick are furnished with either standard brick textured faces or a split (rock-like) texture. The split face is generally preferred as it greatly accentuates the horizontal bonding effect unobtainable in any other type of construction. These units are now being produced in practically all sections of the United States and are in great demand, particularly for single story ranch homes, although they are also used for commercial structures, store fronts and for finished masonry interiors.

When laid with a ⅜" mortar joint, it requires 6 brick per sq. ft. of wall. Add for headers, breakage and waste.

When laid with a ⅜" mortar joint, it requires 10¾ cu. ft. of mortar per 1,000 brick, not including any back-up mortar or any allowance for waste.

Labor Laying 1,000 Roman Face Brick 3⅝"x1⅝"x11⅝" Face Brick

| Common Bond | | | | Style of Mortar Joint | Flemish Bond | | | |
Mason Hours	Labor Hours	Hoist Engr.*	Number Laid per 8-Hr. Day		Number Laid per 8-hr. Day	Mason Hours	Labor Hours	Hoist Engr.*
29.6	17	0.5	250-290	Buttered and Struck	210-240	35.5	17	0.5
20.5	17	0.5	370-410	Flush Cut	300-350	24.6	17	0.5
23.5	17	0.5	320-360	V-Tooled or Concave	265-300	28.3	17	0.5
25.0	17	0.5	300-340	Struck or Weathered	245-285	30.2	17	0.5
23.5	17	0.5	320-360	Raked Out	265-300	28.3	17	0.5
26.7	17	0.5	280-320	Rodded	230-270	32.0	17	0.5

*Add hoisting engineer's time only if required.

Double Size Building Brick

Double size building brick (3¾"x5"x8"), are used in different parts of the country. Their use effects a saving in both mortar and labor over standard size building brick (3¾"x2¼"x8").

Each brick contains 40 sq. in. on the face and when laid with a ½" bed or horizontal joint and a ¼" vertical or end joint, covers 45.375 or 45⅜ sq. in. and requires 3.2 brick per sq. ft. of 4" wall; 6.4 brick per sq. ft. of 8" wall and 9.6 brick per sq. ft. of 12" or 13" wall, not including any allowance for waste.

Mortar Required for Double Size Building Brick

Based on ½" bed or horizontal joints and ¼" end or vertical joints.

Mortar Required per 1,000 Double Brick			Mortar Required per 100 Sq. Ft. of Wall		
4" Walls*	8" Walls	12" Walls	4" Walls*	8" walls	12" Walls
12¾ Cu. Ft.	19¼ Cu. Ft.	21¼ Cu. Ft.	4 Cu. Ft.	12¼ Cu. Ft.	20½ Cu. Ft.

*Freestanding Walls.
No Allowance For Waste Included In Above Quantities.

Labor Laying Double Size Building Brick.—A mason will lay about 60 percent as many double brick (3¾"x5"x8") as standard size building brick (3¾"x2¼"x8"), i.e., if a mason lays 800 building brick on a certain class of work, he would lay 480 double size building brick on the same class of work. In other words, if it required 10 hrs. mason time per 1,000 building brick, it would require 1⅔ as many hours per 1,000 double brick, i.e., 10x1.667=16.67 or 16⅔ hrs. mason time per 1,000 double brick.

CENTER ½"x 6½"
8"
8"
2¼"
CORES 2½"x 3½"
WALLS ¾"
WALLS 2¼"
ROUND CORES ⅝"
WEIGHT 7½#

Twin Brick. Size 8"x8"x2¼"

Twin Brick

Twin brick is a double size brick that contains the face brick and a 4" backing brick in one unit, size 8"x8"x2¼".

It is made with a hollow core so that the twin brick weighs only 7½ lbs. and can be picked up and laid in a one-hand operation. It is designed so the wall has no continuous mortar joint from outer to inner wall faces, thereby tending to avoid moisture penetration.

It is intended primarily for residential work having 8" walls that are to be furred and plastered. Extra face brick must be used at openings and corners to insure proper bond.

It requires about 25 cu. ft. of mortar per 1,000 units.

On four jobs on which these brick were used, requiring 50,000 units, a mason laid 60 twin brick an hour or 480 per 8 hr. day, at the following labor cost per 1,000 units:

	Hours	Rate	Total	Rate	Total
Mason	17	$. . . .	$. . . .	$18.91	$321.47
Labor	11.5		. . . .	14.95	171.93
Cost per 1,000 units			$. . . .		$493.40

Skintled Brickwork

Skintled brickwork is enjoying a revival, particularly in storefront work and interiors of the "Olde English Pub" variety.

Skintled brickwork consists of laying common brick irregularly, with random projecting brick at intervals and using clinkers or irregular and other off-colored brick to produce a variegated wall.

These brick are laid in a number of different ways to produce the desired effect. In some of them the mortar projects beyond the face of the brick just as it has been squeezed out of the joint; in others the joints are cut flush or raked out to produce shadows, etc.

The labor cost will vary with the design and care used in laying up the brick.

When laying building brick without regard to color of wall and with uncut mortar joints or joints cut flush with the face of the wall, a mason should lay 350 to 400 brick per 8-hr. day, at the following labor cost per 1,000:

	Hours	Rate	Total	Rate	Total
Mason	21	$. . . .	$. . . .	$18.91	$397.11
Labor	15	. . . .	. . . .	14.95	224.25
Cost per 1,000 brick			$. . . .		$621.36

If brick joints are raked out, add 2.5 hrs. mason time per 1,000 brick.

On skintled brickwork using clinkers and other dark or irregularly shaped brick to produce a textured wall, more care is required in laying the brick, working out the designs and in raking out the joints to produce shadows, etc., and a mason should lay 250 to 300 brick per 8-hr. day, at the following labor cost per 1,000:

	Hours	Rate	Total	Rate	Total
Mason	29	$. . . .	$. . . .	$18.91	$548.39
Labor	20	. . . .	. . . .	14.95	299.00
Cost per 1,000 brick			$. . . .		$847.39

English Size Face Brick

English size face brick, 8⅞" x 2⅞" x4" in size, are seldom used in this country any more and while the size is not standard, they are still furnished by a few manufacturers.

Inasmuch as they are practically half again as large on the face as the standard brick, a mason is unable to lay as many as where the standard size brick are used. They also require more labor time handling, hoisting and wheeling to the masons.

Number of English Size, 8⅞"x2⅞"x4", Face Brick Required
Per Sq. Ft. of Wall

All Vertical Mortar Joints Figured ¼" Wide
Width of Horizontal or "Bed" Mortar Joints in Inches

Dry	⅛"	¼"	⅜"	½"	⅝"	¾"
5.65	5.34	5.06	4.88	4.67	4.52	4.37

Add for waste, headers and the different brick bonds as given on the previous pages under "Standard Brick."

Cu. Ft. of Mortar Required to Lay 1,000 English Size Face Brick

Based on ¼" End Joints and "Bed" Joints as Given Below

⅛"	¼"	⅜"	½"	⅝"	¾"
5	9	11	14	17	20

Vitrified Face Brick

There are a number of vitrified face brick on the market that are so hard it is almost impossible to cut them. On account of their denseness and brittleness they do not absorb water from the mortar and for that reason it requires more time for the mortar to take its initial set. It is impossible to lay up many courses of brick at one time because the weight of the upper brick courses forces the mortar from the joints of the lower courses, causing an unsightly wall. It is also difficult to hold the brick in the wall and keep the wall plumb, as the brick slide out of place very easily.

Brick of this kind are ordinarily laid with "buttered" mortar joints or with 1/4" mortar joints struck flush. They are furnished in the standard size, 8" x2 1/4" x 3 3/4", and require the same number of brick and the same amount of mortar as other standard size face brick.

SCR Face Brick

"SCR brick", developed by the Structural Clay Products Research Foundation, combines the beauty of the original Norman brick size with the utility and structural advantage of a single unit for nominal 6" masonry wall construction.

Designed primarily to meet the construction requirements of FHA and National Building Codes for one story, single family dwellings, where the wall height does not exceed 9' -0" to the eaves or 15' -0" to the peak of the gables.

The nominal face size of "SCR brick" is 2 2/3"x12", while the actual size of the brick is 2 1/6" x5 1/2" x11 1/2".

When laid with a ½" mortar joint, each brick lays up 32 sq. in. of wall, or it requires 4 1/2 brick per sq. ft. of wall, not including any allowance for breakage or waste.

When laid with a ½" mortar joint, it requires approximately 24 cu. ft. of mortar per 1,000 brick. This quantity includes an allowance for mortar entering the cores and also provides for 10 percent waste.

Depending upon the length of walls and the number of door and window openings, a mason should lay 400 and better brick per 8-hr. day, at the following labor cost per 1,000:

	Hours	Rate	Total	Rate	Total
Mason	20	$....	$....	$18.91	$378.20
Labor	20	....	....	14.95	299.00
Cost per 1,000 brick			$....		$677.20
Cost per sq. ft. wall			....		3.05

Brick Floors and Steps

Face Brick Floors Laid in Basket Weave Pattern.—Brick floors laid in square or basket weave pattern do not require any cutting and are much easier laid than herringbone designs.

On work of this kind, a mason should lay 180 to 225 brick per 8-hr. day, at the following labor cost per 1,000:

	Hours	Rate	Total	Rate	Total
Mason	40	$....	$....	$18.91	$ 756.40
Labor	20	....	....	14.95	299.00
Cost per 1,000 brick			$....		$1,055.40

BRICK FLOOR PATTERNS

Brick Laid in Basket Weave Pattern.

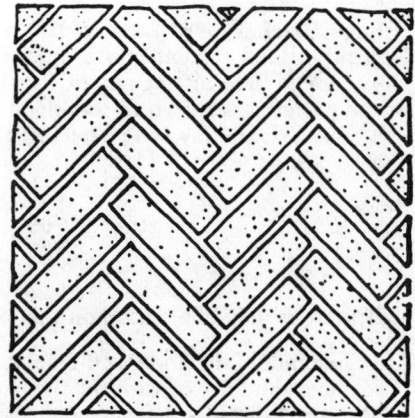

Brick Laid in Herringbone Pattern.

BRICK STAIR DETAILS

Brick Steps with End-Set Treads

Brick Steps with Edge-Set Treads.

Face Brick Floors Laid in Herringbone Pattern.—Brick floors are usually laid in portland cement mortar and when laid in herringbone pattern, it requires considerable time cutting and fitting the small pieces of brick at the edge or border of the pattern, as shown in the illustration.

A mason should lay 125 to 150 brick per 8-hr. day, at the following labor cost per 1,000:

	Hours	Rate	Total	Rate	Total
Mason	56	$....	$....	$18.91	$1,058.96
Labor	28	....	....	14.95	418.60
Cost per 1,000 brick			$....		$1,477.56

If there is only one mason on the job it will require a helper for each mason.

Promenade Tile floor and Decks. When laying 4"x8"x½" promenade tile on porch floors and decks, with the tile laid in portland cement mortar and 1/2" mortar joints, struck flush, a mason should lay and point 100 to 125 tile per 8-hr. day, at the following labor cost per 1,000:

	Hours	Rate	Total	Rate	Total
Mason	68	$....	$....	$18.91	$1,285.88
Labor	34	....	....	14.95	508.30
Cost per 1,000 tile			$....		$1,794.18

A tile is 4"x8"x 1/2" in size, or it requires 4 tile per sq. ft. not including mortar joints. Figuring a 1/2" mortar joint, it requires 3.5 tile per sq. ft. of floor.

A mason will lay 30 to 36 sq. ft. of floor per 8-hr. day, at the following labor cost per 100 sq. ft.:

	Hours	Rate	Total	Rate	Total
Mason	24	$....	$....	$18.91	$453.84
Labor	12	....	....	14.95	179.40
Cost per 100 sq. ft			$....		$633.24
Cost per sq. ft				....	6.33

Brick Steps.—When laying brick steps in portland cement mortar, similar to those illustrated, a mason should lay 125 to 150 brick per 8-hr. day, at the following labor cost per 1000:

	Hours	Rate	Total	Rate	Total
Mason	56	$....	$....	$18.91	$1,058.96
Labor	28	....	....	14.95	418.60
Cost per 1,000 brick			$....		$1,477.56

Turning Face Brick Segmental Arches Over Doors and Windows.—Where segmental brick arches are laid over wood centers for 3'-0" wide door and window openings, it will require about 4 hrs. mason time at the following labor cost per arch:

	Hours	Rate	Total	Rate	Total
Mason	4	$....	$....	$18.91	$75.64

For a 5'-0" arch, figure about 6.5 hrs. mason time, as follows:

	Hours	Rate	Total	Rate	Total
Mason	6.5	$....	$....	$18.91	$122.92

If the brick are cut or chipped to a radius, add cost of chipping as given below.

Laying Face Brick in Flat or "Jack" Arches.—When laying face brick in flat or "Jack" arches a mason should complete one arch up to 3'-6" wide in about 5 hrs. at the following labor cost:

	Hours	Rate	Total	Rate	Total
Mason	5	$....	$....	$18.91	$94.55

For a 5'-0" "Jack" arch, figure about 7 1/2 hrs. mason time, as follows:

	Hours	Rate	Total	Rate	Total
Mason	7.5	$....	$....	$18.91	$141.83

The above costs do not include chipping or rubbing brick to a radius. If this is necessary, add cost of chipping as given below.

Chipping Brick.—Where necessary to chip or cut brick for segmental or "Jack" arches, a mason should chip 400 to 500 brick per 8-hr. day, depending upon the texture and density of the brick, at the following labor cost per 1,000:

	Hours	Rate	Total	Rate	Total
Mason	17.5	$....	$....	$18.91	$330.93

There is considerable lost motion in cutting brick and unless there is a great deal of duplication in the cutting where a man can work continuously on a few special sizes, a mason will cut only 160 to 200 pieces per 8-hr. day, at the following labor cost, per 100 pieces:

	Hours	Rate	Total	Rate	Total
Mason	4.5	$. . . .	$. . . .	$18.91	$ 85.10
Labor	1.0	. . . .	. . . .	14.95	14.95
Cost per 100 cuts			$. . . .		$100.05
Cost per cut			. . . .		1.00

Cutting Masonry Products With Portable Masonry Saw

On practically every construction job there is a large amount of cutting necessary on masonry materials and with the high cost of labor this becomes an expensive item especially when the cutting is done by hammer and chisel.

This applies particularly to the cross and diamond bonds in brick masonry, arch brick, radial brick, glazed tile, rotary kiln blocks, concrete blocks, natural stone, roofing tile, conduits, etc.

By using a masonry saw, this work can be done at a fraction of the cost of hand cutting plus the saving of materials.

Masonry saws are obtainable with 12", 14", 18" and 20" blade capacities, and are designed for both wet (dustless) and dry cutting. The advantage of wet cutting is in the confined areas where dust would be a problem. An exhaust assembly is available for use with dry cutting masonry saws—this assembly consists of a high velocity fan that draws the dust away from the machine and disposes of it through the nearest opening.

A smaller compact model masonry saw has recently been developed. This model is light enough for a man to carry and may be set up on the scaffolding, operated from a pickup truck or the back of a station wagon.

Recently developed also is a small, highly portable concrete saw which can do the work of many of the heavier models.

There are four types of blades that can be used on masonry saws: (1) wet-dry abrasive blades; (2) dry "break-resistant" abrasive blades; (3) wet "break-resistant" abrasive blades; (4) diamond blades. The wet abrasive and diamond blades should be used only with a masonry saw equipped for wet cutting. There are many specifications of each type of blade; it is very important that the correct specification be used for the particular material to be cut.

Diamond blades for use on masonry saws vary in price from approximately $232.00 to $450.00 for 14" diameter blades and $402.00 to $854.00 for 18" diameter blades according to the specifications.

The cost of cutting masonry units varies greatly, depending upon the type of material, blade specifications and upon the operator.

Concrete products vary considerably in hardness, such as sand-gravel blocks compared with Haydite blocks. Therefore, blade wear and cutting time will vary accordingly. As an average, most 8"x8"x16" concrete blocks can be sliced in two in 14 to 16 seconds at a cost of less than one cent per unit.

Fire brick varies tremendously in hardness, as well as types. Ordinary dry press or silica can be cut in two very rapidly, while Super-duty refractories or Basic, Chrome and Magnesite requires 15 to 18 seconds for each cut completely through, at a cost varying from 3 to 7 cts. depending upon actual density.

All the above costs and cutting speeds are based on cross cuts. Longitudinal cuts will approximate the same for equivalent area.

The following is an example of the approximate time required making cuts in various masonry products:

Compact Masonry Saw

Material	Type of Cut	Length of Cut	Depth of Cut	Approximate Time Required
Acid Brick	Miter	4½"	2¼"	15 seconds
Concrete Block	Straight	8 "	8 "	19 seconds
Glazed Brick-Tile	Miter	6 "	4 "	15 seconds
Magnesite Brick	Miter	2¾"	4 "	15 seconds
Rotary Kiln Block	Straight	8 "	3 "	11 seconds
Silica Brick	Straight	8 "	2¼"	10 seconds
Roofing Tile	Miter	9 "	2 "	14 seconds
Clay Sewer Pipe	Cross	4 " diam.	. . .	18 seconds
Concrete Cylinder	Cross	4 " diam.	. . .	21 seconds

Cleaning Face Brick Work

The cost of cleaning face brick work will vary with the kind of brick, as it is much easier to clean a smooth face vitrified brick than one having a rough texture.

Where smooth face brick are used and it is not necessary to do any pointing but merely wash the wall with muriatic acid and water, an experienced mechanic should clean 5,000 to 5,500 brick (750 to 825 sq. ft. of wall) per 8-hr. day, at the following labor cost per 1,000 brick, or approximately 150 sq. ft. of wall:

	Hours	Rate	Total	Rate	Total
Mechanic	1.6	$....	$....	$18.91	$30.26
Helper	0.8	....	....	12.54	10.03
Cost per 1,000 brick			$....		$40.29
Cost per sq. ft. of wall			....		.27

The labor cost of cleaning and washing rough texture face brick will run higher than smooth brick on account of the cement and mortar getting into the surface of the brick.

When cleaning and washing rough textured brick, the same crew should clean and wash 3,500 to 4,000 brick (525 to 600 sq. ft. of wall) per 8-hr. day, at the following labor cost per 1,000 brick or per 150 sq. ft. of wall:

	Hour	Rate	Total	Rate	Total
Mechanic	2.2	$....	$....	$18.91	$41.60
Helper	1.1	....	....	12.54	13.80
Cost per 1,000 brick			$....		$55.40
Cost per sq. ft. of wall			....		.37

Brick Fireplaces, Mantels and Hearths

On account of the many sizes, shapes and styles of brick fireplaces, mantels and hearths, it is difficult to estimate the cost of the work at a certain price per 1,000 brick and the most satisfactory method is to figure a certain length of time to lay up each fireplace or mantel.

A mason should complete a plain brick fireplace 5'-0" to 6'-0" wide and 4'-0" to 5'-0" high, requiring 200 to 225 brick, not including hearth, in about 10 hrs. at the following labor cost:

	Hours	Rate	Total	Rate	Total
Mason	10	$....	$....	$18.91	$189.10
Labor	5	....	....	14.95	74.75
Cost per fireplace			$....		$263.85

On larger and more elaborate brick fireplaces and mantels having raked or rodded mortar joints, a mason should lay about 100 to 125 brick per 8-hr. day, at the following labor cost per 1,000:

	Hours	Rate	Total	Rate	Total
Mason	70	$....	$....	$18.91	$1,323.70
Labor	35	....	....	14.95	523.25
Cost per 1,000 brick			$....		$1,846.95

If there is only one mason working on the fireplace it will require one laborer with each mason.

Brick Fireplace Linings.—A mason should complete the lining for a brick fireplace having an opening 3'-0" to 4'-0" wide, 2'-6" to 3'-0" high, and 1'-4" to 1'-9" deep, in 6 to 8 hrs. all depending upon the class of work.

If the fireplace linings are laid in common bond and do not require an excessive amount of cutting and fitting, a mason should complete one fireplace lining in about 5 hrs. but where the back-hearths are of irregular shape and require considerable cutting and fitting at angles and at tops for dampers and smoke chambers, it will require about 8 hrs. mason time for each lining.

Fireplaces having an opening 3'-6" to 4'-0" wide, about 4'-0" high and 1'-9"deep, lined with firebrick laid in herringbone pattern, requiring cutting and fitting at all edges and corners, will require 20 to 24 hrs. mason time to line one fireplace.

Labor Laying Brick Hearths and Back-Hearths.—If brick hearths and back-hearths are laid in square or basket weave pattern, a mason should complete one back-hearth in 3 to 4 hrs. and the front hearth will require the same length of time.

If laid in herringbone pattern with cut brick at all edges, figure 4 to 5 hrs. mason time for both hearth and back-hearth.

Figure about ½-hr. laborer time to each hour of mason time, except where only one mason is on the job when it will figure hour for hour.

FIRE BRICK WORK

Fire brick and tile are furnished in innumerable sizes and shapes to meet the requirements of all classes of boilers, furnaces, stacks, etc., so that it is practically impossible to list them all in this book. The most common sizes and shapes are listed below. Standard fire brick size is 9"x4½"x2½" and will cost around $450.00 per 1,000 for low duty, $550.00 for high duty.

Fire Clay.—When a close thin joint is desired, the clay should be soaked and mixed thin, and the brick should be dipped and rubbed. For work of this kind 300 lbs. of finely ground fire clay is required per 1,000 brick.

Sizes and Shapes of Standard Fire Brick

Name	Size	Number Required per Sq. Ft.	
		Laid Flat	Laid on Edge
8" Straight	8"x3½"x2¼"	8.0	5.2
9" Straight	9"x4½"x2½"	6.5	3.5
Small 9" Brick	9"x3½"x2½"	6.5	4.5
Split Brick	9"x4½"x1¼"	13.0	3.5
2" Brick	9"x4½"x2 "	8.0	3.5
Soap	9"x2¼"x2½"	6.5	7.0
Checker	9"x2¾"x2¾"	6.0	6.0

Ordinary fire clay costs about $4.50 per 100 lbs. while high-grade refractory fire clay costs about $5.25 per 100 lbs.

Labor Lining Brick Chimneys and Stacks with Fire Brick.—When lining small brick chimneys and stacks from 2'-0" to 3'-0" square, laid in fire clay, a mason should lay 450 to 525 brick per 8-hr. day, at the following labor cost per 1,000:

	Hours	Rate	Total	Rate	Total
Mason	16	$....	$....	$18.91	$302.56
Labor	12	....	....	14.95	179.40
Cost per 1,000 brick			$....		$481.96

On large bricks stacks having an inside diameter of 4'-0" to 6'-0", a mason should lay 600 to 750 fire brick per 8-hr. day, at the following labor cost per 1,000:

	Hours	Rate	Total	Rate	Total
Mason	11.8	$....	$....	$18.91	$223.14
Labor	10.0	....	....	14.95	149.50
Cost per 1,000 brick			....		$372.64

Number of Wedge and Arch Brick Required for Various Circles

No. 1 Wedge Brick 9"x4½"x	No. 2 Wedge Brick 9"x4½"x
(2½"-1⅞")	(2½"-1½")
No. 1 Arch Brick 9"x4½"x	No. 2 Arch Brick 9"x4½"x
(2½"-2⅛")	(2½"-1¾")

Labor Laying Fire Brick in Fireboxes, Breechings, Etc.—When lining fireboxes, breechings, furnaces, etc., with fire brick, requiring arch and radial brick, special shapes, etc., a mason should lay 175 to 225 brick per 8-hr. day, at the following labor cost per 1,000:

	Hours	Rate	Total	Rate	Total
Mason	40	$....	$....	$18.91	$ 756.40
Labor	40	....	....	14.95	598.00
Cost per 1,000 brick			....		$1,354.40

When estimating brick arches, it is advisable to allow additional labor and it is doubtful if a mason thoroughly experienced in boiler work will average more than 25 brick an hour for the actual arch and skew-backs, due to adverse conditions, close working space, etc. One contractor specializing in this class of work, checks the penetrations through the wall and regardless of size, adds 4 hours mason time for each one.

Labor Bricking in Boilers.—When bricking in steel boilers, fireboxes, breechings, etc., with common or fire brick where the walls are 8" to 12" thick, a mason should lay 600 to 650 brick per 8-hr. day, at the following labor cost per 1,000:

	Hours	Rate	Total	Rate	Total
Mason	13	$....	$....	$18.91	$245.82
Labor	13	....	....	14.95	194.35
Cost per 1,000 brick			$....		$440.18

Table of Wedge Brick Required for Various Circles	Number of Arch Bricks Required for Various Circles

Diam. Ft.	In.	No. 2 Wedge	No. 1 Wedge	Square	Total	No. 2 Arch	No. 1 Arch	9-inch	Total
2	0	....	....	....	....	42	....	....	42
2	6	63.	....	....	63.	10	40	....	50
3	0	48.	20.	....	68.	....	57	....	57
3	6	36.	39.	....	75.	....	57	7	64
4	0	24.	57.	....	81.	....	57	15	72
4	6	12.	78.	....	90.	....	57	22	79
5	0	....	94.	....	94.	....	57	29	86
5	6	....	94.	7.5	102.	....	57	37	94
6	0	....	94.	15.	109.	....	57	44	101
6	6	....	94.	23.	117.	....	57	52	109
7	0	....	94.	30.	124.	....	57	59	116
7	6	....	94.	39.	133.	....	57	67	124
8	0	....	94.	45.	139.	....	57	74	131
8	6	....	94.	53.	147.	....	57	82	139
9	0	....	94.	60.	154.	....	57	89	146
9	6	....	94.	68.	162.	....	57	97	154
10	0	....	94.	75.	169	....	57	104	161

Fire Clay Tile Flue Lining

Flue lining is estimated by the lin. ft. It is furnished in 2'-0" lengths in the following sizes:

Size of Flue Lining	No. Lin. ft. Set per 8-Hr. Day	Hrs. Required to Set 100 Lin. Ft. Mason	Labor
4"x 8"	165–180	4.7	4.7
4"x12"	130–150	5.7	5.7
8"x 8"	130–150	5.7	5.7
8"x12"	105–125	7.0	7.0
12"x12"	85–100	8.7	8.7
8"x18"	75– 90	9.7	9.7
12"x18"	70– 85	10.2	10.2
18"x18"	60– 70	12.3	12.3
20"x20"	55– 65	13.3	13.3
20"x24"	50– 60	14.0	14.0
24"x24"	45– 55	16.0	16.0
20" round	55– 65	13.3	13.3

Approximate Prices of Tile Flue Lining

Size	Price Lin. Ft.	Size	Price Lin. Ft.
8"x8"	$ 1.50	6" Round	$ 1.40
8"x12"	2.40	8" Round	2.20
12"x12"	3.20	10" Round	3.10
12"x16"	5.90	12" Round	4.00
16"x16"	7.50	15" Round	6.50
20"x20"	12.50	18" Round	7.50
20"x24"	14.00	20" Round	11.00
24"x24"	17.00	24" Round	17.50

Labor Setting Terra Cotta Wall Coping

Size of Coping	No. Lin. Ft. Set per 8-Hr. Day	Hrs. Required to Set 100 Lin. Ft. Mason	Labor
9"	145–160	5.2	5.2
12"	115–130	6.5	6.5
18"	100–120	7.3	7.3

Approximate Lineal Foot Prices of Terra Cotta Wall Coping

Size	Double Slant	Single Slant
9"	2.50	2.75
13"	3.15	5.00
18"	6.30	12.00

Corners, ends and starters, four times price of straight coping; angles, six times the price of straight coping.

Illustrating Construction of One Type of Brick Incinerator

Brickwork for Incinerators

Incinerators were used in residences, apartments, apartment hotels, hotels, hospitals, schools, industrial plants, etc., for burning garbage, waste papers, etc.

With many types the refuse may be fed into the chimney at each floor, the deposited material dropping to the incinerator in the basement, or they may be fed direct to the incinerator in the basement.

Practically all incinerators require a separate flue or stack, which should be estimated in the regular way, figuring fire brick, flue lining and common brick as shown on the plans.

The number of fire and common brick required for the various sizes and types of incinerators will vary with the different manufacturers.

However, the following data will assist the estimator in computing the number of brick and labor hours required to construct the brickwork for the different size incinerators, where they are not detailed on the plans.

NOTE—If 4" outside walls are used, dimensions below will be reduced in length and height by 4" and in width by 8"

Model	Dimensions			Number Brick		Minimum Size Flue	Mason Hours	Labor Hours
	A	B	C	Fire	Common			
"A"	6' 0"	3' 5"	4' 0 "	300	900	12"x12"	20	14
"B"	5' 7"	3' 11"	4' 7½"	350	1100	14"x14"	24	18
"C"	6' 0"	4' 3"	4' 7½"	400	1300	16"x16"	28	20
"D"	6'10"	4' 3"	5' 11 "	525	1500	20"x20"	33	24

RADIAL BRICK CHIMNEYS

The construction of radial brick chimneys is a highly specialized business and they are usually constructed by contractors specializing in chimney construction.

They have in their employ workmen with years of experience in this class of work and can build them much more economically than the contractor who has just an occasional chimney to build. Also, most constructors of radial brick chimneys have their own designs and shapes of radial brick, manufactured especially for them, so they are not available to the average contractor.

However, some information on this important subject may be of value to many contractors. Radial brick chimneys are constructed of perforated radial brick formed to fit the circular and radial lines of the chimney. The brick are moulded with vertical perforations, and should be hard, well burned, acid proof, of necessary refractory powers and crushing strength and of maximum density.

The perforations serve to form a dead air space in the walls of the chimney which tends to prevent rapid heating and cooling of the walls by conserving the heat inside. Naturally, the higher the temperature of the gases the better the draft.

Four sizes of brick are ordinarily used in the construction of radial brick chimneys. All have the same face dimensions, approximately 6½" wide by 4½" high. The lengths of the blocks vary in order to make possible the breaking of the joints horizontally and vertically in the walls. The combination of bonds with this type of chimney provides for a lighter chimney than can be produced with ordinary building brick.

Radial chimney brick are generally either red or buff in color. Red radial brick are usually the most economical, as they are in greater demand and there are more sources of supply.

In laying the brick, the mortar is worked into the perforations, locking them together on the principle of a mortise and tenon joint. This produces the strongest bonded wall known to brick construction.

Radial Brick Chimney

Sizes and Weights of Radial Brick for Radial Brick Chimneys

Type Brick	Outside Face	Inside Face	Depth Inches	Height Inches	Weight Pounds
No. 4	6½"	5¾"	4"	4½"	7
No. 6	6½"	5½"	6"	4½"	9½
No. 7	6½"	5½"	7"	4½"	11
No. 8	6½"	5½"	8"	4½"	12
Pier	9"	9"	4"	4½"	9

Each radial brick having an outside face 6½" wide and 4½" high covers 29¼ sq. in. of surface, or it requires (4.923) 5 blocks per sq. ft. of wall, not including mortar joints. Allowing ½" for mortar joints, each brick covers 7"x5" or 35 sq. in. of surface, or it requires 4.12 blocks per sq. ft. of wall. Add about 5 percent for breakage and waste.

The walls of most radial chimneys taper about ¼" per 1'-0" in height, and the

Perforated Radial Chimney Brick and Details of Bonding and Jointing
of Perforated Radial Brick Construction

thickness of the walls vary from the bottom to the top, depending upon the height. The wall thickness must be made up of a number of different depth units to provide the proper bond and structural strength. For instance a wall 16¼" thick might be made up of 1 block 7" deep and 2 blocks 4" deep; a wall 8 5/8" thick of 2 blocks 4" deep, etc.

Estimating Quantities of Radial Brick.—Radial brick chimney work is frequently estimated by the ton, inasmuch as the wall thicknesses vary, depending upon the height. A chimney 125'-0" high might have walls of the following thicknesses:

First	25' height, 16¼" thick	Fourth	20' height, 10¼" thick
Second	20' height, 13¼" thick	Fifth	20' height, 8⅝" thick
Third	20' height, 11¾" thick	Top	20' height, 7⅛" thick

These wall thicknesses will be made up of radial brick of different depths to work out the correct wall thickness and proper bond, as follows:

Wall Thickness	Size of Units Used	Weight 29¼ Sq. In. Wall*	Weight per Sq. Ft. Wall*
20"–21"	2 pcs. 8"; 1 pc. 4"	31 lbs.	128 lbs.
18"–19"	2 pcs. 7"; 1 pc. 4"	29 lbs.	120 lbs.
16"–17"	2 pcs. 6"; 1 pc. 4" or 2 pcs. 8"	24–26 lbs.	99 –107 lbs.
14"–15"	1 pc. 6"; 2 pcs. 4" or 2 pcs. 7"	22–23½ lbs.	91 – 97 lbs.
12"–13"	1 pc. 7"; 1 pc. 4" or 1 pc. 8"; 1 pc. 4"	18–19 lbs.	75 – 79 lbs.
10"–11"	1 pc. 6"; 1 pc. 4"	16½ lbs.	68 lbs.
8"– 9"	2 pcs. 4" or 1 pc. 8"	12–14 lbs.	50 – 58 lbs.
7"– 8"	1 pc. 7" or 1 pc. 8"	11–12 lbs.	45½– 50 lbs.

*Net weight of brick. Does not include mortar.

Figure radial brick chimney lining at about 30 lbs. per sq. ft. of wall surface.

Mortar Required Laying Radial Brick.—It requires about two-thirds (1,334 lbs.) of a ton of mortar to lay one ton of radial brick. Mortar is usually mixed in the proportions of 1 volume portland cement; 1 volume lime putty and 4 to 6 volumes clean mortar sand.

Types and Sizes of Brick for Radial Brick Chimneys

UNIT	A	B	C	D	INSIDE RADIUS	APPROXIMATE SHIPPING WEIGHT	APPROXIMATE PERCENT VOIDS
4 (N)	4 1/2"	6 1/2"	4"	5 20/32"	3'-4"	5.5	32
6 (A)	4 1/2"	6 1/2"	6"	5 21/32"	3'-4"	8.3	32
7 (B)	4 1/2"	6 1/2"	7"	5 1/2"	3'-8"	9.7	32
7-S.R.(B-S.R.)	4 1/2"	6 1/2"	7"	4 15/16"	2'-0"	9.7	32
8 (C)	4 1/2"	6 1/2"	8"	5 1/2"	3'-9"	10.8	32
8-S.R.(C-S.R.)	4 1/2"	6 1/2"	8"	5 1/16"	2'-6"	10.8	32
S (S)	4 1/2"	8"	4"	8"	Straight	7.0	32

Radial chimney brick is priced by the ton and will vary with type of brick and locality. Prices are f.o.b. factory and estimator must add freight and trucking from factory to job site and cost of unloading and storing brick at job.

Labor Laying Brick in Radial Brick Chimneys.—The labor cost of laying radial brick chimneys will vary with the height of the chimney, thickness of walls, etc., but the following are actual production times on 6 radial brick chimneys constructed.

Description	Tons Radial Brick	Mason	Labor	Hoist Engr.
			Hours per Ton of Brick	
125'-0" high, Walls 16¼" to 7⅛" thick. 12'-9" diam. at bottom, 7'-0" at top. Lined 50'-0"	157	2.7	4.7	0.9
80'-0" high, walls 11¾" to 7⅛" thick. 8'-10" diam. at bottom, 5'-9" at top. Lined 35'-0"	66.5	3.1	5.3	1.0
60'-0" high, walls 10¼" to 7⅛" thick. 7'-2" diam. at bottom, 4'-4" at top. Lined 21'-0"	31.1	4.7	8.2	1.9
46'-0" high, walls 10¼" to 7⅛" thick. 6'-6" diam. at bottom, 4'-2" at top. Unlined	24.6	3.3	8.0	1.5
46'-0" high walls 10¼" to 7⅛" thick. 6'-6" diam. at bottom, 4'-2" at top. Unlined (2)	49.2	3.0	5.0	1.5
55'-0" high, walls 10¼" to 7⅛" thick. 6'-11" diam. at bottom, 4'-2" at top. Lined 20'-0"	31.6	3.0	5.0	1.5

The above does not include supervision time.

Labor time includes unloading materials from cars into trucks and unloading trucks at site. All mortar joints on the above stacks were struck flush with point of trowel.

Where stacks are lined, labor includes lining same with perforated radial fire clay brick, laid in high temperature cement. The lining is laid up from inside after stack is topped out.

Add for clean-out doors, band iron, ladders, lightning rods, cables and cable clips.

Miscellaneous Masonry Costs

Labor Costs on High Buildings.—On buildings over 10 to 12 stories high, it will be necessary to add an extra allowance for labor handling and hoisting the bricks and mortar, as the higher the building the longer it takes the hoist to make a trip from the street level to the floor on which the bricks are being laid and return: extra work raising and handling scaffolding, etc.

Taking a 12-story building or 120 ft. in height as normal, or 100 percent, add about 1 percent to the labor cost for each story or each 10 feet in height above 120 ft.

For instance, a 12-story building or 120 feet in height is 100, the 13th story would be 101; the 14th story, 102; the 15th story, 103, etc.

To find the average cost on a 15 story building, figure as follows:

The first 12 stories as 100 each or 1,200
The 13th story as 101 101
The 14th story as 102 102
The 15th story as 103 103
Total for 15 stories or 150 ft. in height 1,506

Dividing 1,506 by 15 gives 1.004 as the average cost of the masonry labor for a 15 story building. Other story heights would be computed in the same manner.

The following table applies to all classes of masonry work on high structures.

Height Stories	Normal Base	Add Above Normal %	Average for Job
1-12	100.00	None	100.00
13	100.00	1%	100.08
14	100.00	2%	100.22
15	100.00	3%	100.40
16	100.00	4%	100.63
18	100.00	6%	101.17
20	100.00	8%	101.80
22	100.00	10%	102.50
24	100.00	12%	103.25
26	100.00	14%	104.04
28	100.00	16%	104.86
30	100.00	18%	105.70
40	100.00	28%	110.15
50	100.00	38%	114.82

Brick Catch Basins and Manholes

When laying building or sewer brick in catch basins, manholes, etc., from 3'-0" to 5'- 0" in diameter and having 8" to 12" walls, a mason should lay 600 to 750 brick per 8-hr. day, at the following labor cost per 1,000 :

	Hours	Rate	Total	Rate	Total
Mason	11.5	$....	$....	$18.91	$217.47
Labor	11.5	....	....	14.95	171.93
Cost per 1,000 brick			$....		$389.40

Pointing Around Steel Sash With Cement Mortar.—All of the steel windows in a three story office building were pointed with portland cement mortar between the sash and the brick jambs after they had been set.

There were 39 windows 4'-0"x4'-10" and 96 windows 4'-0"x7'-0", making a total of 2,800 lin. ft. which were pointed at the following labor cost per 100 lin. ft.:

	Hours	Rate	Total	Rate	Total
Bricklayer	1.1	$....	$....	$18.91	$20.80
Labor	0.5	....	....	14.95	7.48
Cost per 100 lin. ft			$....		$28.28
Cost per lin. ft			....		.28

Reinforced Brick Masonry

Reinforced brick masonry has been widely used on the west coast because of its resistance to the lateral forces produced by earthquakes. It is now also being used throughout the country where high winds or blast conditions may occur, as well as for retaining walls.

These walls are comparable to reinforced concrete construction with the masonry units serving as a form. The exterior wythes are built up in a conventional manner, the reinforcing is set in the interior joints and the whole is made homogeneous by filling the interior joints with grout.

No special skills or techniques are needed on the part of the mason, but certain care must be taken, including seeing that there is adequate lime in the mortar; a sound base to start on with the aggregate in concrete exposed and wetted to saturation; that brick is wetted so grout will not dry out too quickly; that grout is sufficiently fluid so none will adhere to a trowel; that grout be poured in a layer of about three bricks high and be allowed to set for 15 minutes before another pour is made, and poured from the inside to avoid splashing the exposed surfaces; and that mortar cuts not be spilled into grout or grout allowed to accumulate on bed joints.

The amount of reinforcing will vary widely with the job. In lightly reinforced walls the reinforcing will fit within the widths of normal brick walls. The more heavily reinforced walls may require wider inner joints or the use of soap courses to accommodate the steel. As headers are not used in reinforced walls, the contractor should check the drawings to see if, for aesthetic reasons, header rows are shown as this will require cutting all these brick. This is especially true when figuring reinforced lintels in conventional walls so coursing can carry through.

Savings, as well as aesthetic advantages, can be realized often through the use of reinforced brick lintels. These are almost always poured in place with temporary shoring for the soffit. The economies result from savings in steel cost and elimination of painting. In better work it may call for the soffit brick joints to be filled temporarily with sand and then repointed later.

A variation on the above has been developed and is known as high lift grouted reinforced brick masonry. It varies in that it is essentially a cavity wall with a 2" or better inner joint. This allows the reinforcing to be preset and grouting delayed until a full 12' wall height is reached. Mechanical pumping can thus be used.

Precautions listed for standard reinforced walls also apply here with the following additions. Care must be taken to allow cleaning out of grout space by omitting every other unit on one side; brick must be allowed to set for at least 3 days before grouting; vibrating of grout is necessary; once grout pour is started, it must be continuous; and construction dams may be built every 20 ft. or 25 ft. to contain the grout laterally.

SCR Masonry Process

The Brick Institute of America at McLean, Va., has developed three aids for increasing the masonry wall area completed per day, as well as increasing the quality of the work, which should be given consideration. These aids consist of a marked line to show where each brick is to be laid; a scaffold easily adjustable in height and which, in addition to the platform on which the mason stands, incorporates a second platform at waist height for brick and mortar; and corner poles to support course guides which can be moved up course mark by course mark.

The foundation has its own patented hardware for the poles which can be mounted on locally bought lumber. The hardware is available through tool dealers. Adjustable scaffolds meeting the foundation's specifications are available from several manufacturers.

High lift booms for raising palleted brick and mortar hoppers can be used to advantage with this system on high rise buildings.

As with most new processes, savings cannot be expected until workmen become accustomed to taking advantage of its full utility. However, with union objections to the use of such auxiliary aids fast disappearing, any contractor expecting to remain competitive must utilize every possible means of increasing productivity.

Handling of Masonry By Palletization

For years builders have considered the use of palletization for handling masonry materials and many efforts have been made in this direction with indifferent success due to the lack of adequate handling equipment.

New systems have overcome this problem to the point where palletized masonry units, whether brick, block, or tile, can be transported from the manufacturing plant, through the job site, up the hoist and around the scaffold to the mason's station, eliminating all hand handling of units until the mason picks them up for laying. Two companies manufacturing this equipment are the PCM Division of Koehring Corp., Port Washington, Wis. and Prime-Mover Co., Muscatine, Iowa.

Time studies on over 300 jobs using the PCM system indicate that on jobs with ten or more masons the conventional ratio of 0.7 to 1.0 labor hour per mason hour can be reduced to 0.3 to 0.5 labor hours per mason hour—a saving of approximately 50 percent in labor time required for tending masons, or an overall saving on the total masonry labor cost of about 20 percent. These ratios include the labor time required to mix mortar, tend mortar, brick, tile, block, etc., and remove rubbish.

The success of these systems is mainly due to the equipment used which consists of properly sized wooden pallets, a tractor with special design fork lift mechanism, special brick buggies and special mortar buggies.

For efficient use of this system and equipment, certain job conditions must be established from the start and maintained throughout the life of the job.

All materials must be purchased on the basis of pallet delivery or palletized as soon as they arrive on the job. Up to the present time, contractors have been furnishing the wood pallets, but dealers and manufacturers are showing an increasing preference for this type of delivery and in some cases furnish their own pallets. Contractor should specify the quantity of material to be placed on each pallet and the arrangement of same. Weight should not exceed 1,000 lbs. per pallet to stay within the capacity of scaffolding and buggies.

The effect on material prices for pallet delivery has been variable. Some suppliers have maintained their same prices, claiming the extra cost of labor and material for palletization is offset by savings in loading and unloading. Other suppliers have increased brick prices $1.00 to $2.00 per thousand. Actual cost of a 24" x 32" pallet, if palletization is done on the job site, is around $3.50 per pallet. One laborer can handle 6 to 7 pallets per hour.

Access roads to job should be laid out and maintained to facilitate operation of the fork lift tractor. Area surrounding the building should be backfilled, consolidated and graded before masonry work starts and kept reasonably clear thereafter.

Exterior scaffolding must be wide enough to permit maneuvering brick and mortar buggies. Where tubular steel scaffolding is used, this may be accomplished with standard scaffolding components, using a 5-ft. width tubular scaffold with the addition of a 30" bracket on the inside for the masons and a 20" bracket on the outside for the extra width required for buggies and storage of pallets. No special planking is required, but plank laps should all run in the same direction and a wood easement strip placed at each lap to partially eliminate that irregularity.

Where suspended scaffolding is used, extra width can be obtained by using 8-ft wide Wheeling scaffold machines with the addition of 10-ft. pipe putlogs fastened directly under and through scaffold planking with U-bolts. This method moves the inside drum and cable 2'-0" out from the wall and creates a clear working space for the mason.

Scaffolding costs will be increased 5 to 15 percent, depending upon the type of scaffold used.

Standard Pallet 32" x 24".—A standard pallet size of 32" x 24" has been chosen because it accommodates all common sizes of brick, block and tile; loaded weight does not exceed safety regulations for scaffolding or light floor construction; accommodates just enough material to build 10 lineal feet of wall, 4 feet high, 4 inches thick, allowing proper spacing for stacking on the scaffold according to standard masonry practice; loaded weight can be handled by one man using hand powered equipment; pallet load can be split into two 16" x 24" pallet loads permitting passage through normal doorways, handling on occasional narrow scaffolds and maneuvering in confined spaces.

Pallets should be made to withstand hard use, preferably from a good grade of 1" lumber—1" x 8" boards for the tops and 2" x 4" for skids. Pallets made on the job, using power saws and a production line set-up, should cost about $1.75 per half pallet (16" x 24") each, including material and labor.

One standard full size pallet is composed of two 16 x 24s strapped together with metal banding. When required, the metal banding can be removed, breaking the package into two halves of 16" x 24" for use where operating space is limited.

Fork Lift Tractors.—Manufacturers provide various capacities and sizes of rubber tired fork lifts having lifting heights from 18'-6" up to 30'-6" with capacities from 2,500 lbs. to 4,000 lbs. The 3,000 lbs. unit can unload and stockpile as many as 4,000 brick in 10 minutes, and can be used to handle concrete planks and roofing materials as well as masonry.

Fork lift tractor unloads pallets of material from delivery trucks and places them in stockpiles. From stockpiles, pallets are then transported to masons' stations at ground level or on the scaffold up to maximum reach of tractor; to construction hoist for work on scaffolds or floors beyond tractor reach; to building floors within tractor reach for partition work or exterior masonry performed from inside the scaffold.

The fork lift can handle as much palletized masonry material as 20 laborers, using conventional hand methods and wheelbarrows. It can also be equipped with other attachments to perform many other tasks on the job.

The fork lift also delivers mortar to the bricklayers from the mixing point by use of mortar buggies, which are mortar containers with a capacity of 7 cu. ft., equipped with wheels and casters so that they can be readily pushed on floors or scaffolds. The tractor handles these mortar containers to and from the mixer the same as packaged materials.

Brick Buggies.—Pallet loads are handled on the scaffold and building floors with brick buggies. These come in a variety of sizes, either hand propelled or power driven, with capacities of 1,000 lbs. for half (16" x 24" pallets with 120 bricks) and 1,500 lbs. for full (24" x 32" pallets with 240 bricks) loads. Lifting forks can be had for 10", 7'-6" and 10'-0" lifts, the latter being able to feed double deck scaffolding. Buggies can be operated under average outdoor conditions as well as on up and down ramps.

Cost of lift trucks for 24" x 32" pallets is around $3,500. It can service 4 to 6 masons on a medium rise building instead of 1 laborer required to keep each mason supplied with brick. Two laborers, one on the ground and one delivering, should serve 5 masons.

The resulting savings in labor cost per day should amortize the lift and pallet costs in approximately 5 to 6 weeks under the best of conditions. Such handling also cuts down the time the hoist must be held for unloading by hand, which can be critical on larger jobs.

04220 CONCRETE UNIT MASONRY

Concrete masonry materials, commonly termed "Concrete Block" or "Concrete Brick" are extensively used in all parts of the country for exterior and interior bearing walls, interior partitions, floor and roof fillers, etc.

They are also widely used for backing up face brick as well as for backing up building brick when used as a wall facing. They are used both for bearing walls on wall bearing structures and for curtain walls in steel or reinforced concrete frame structures.

Prime Mover Mason Tender Transporting Brick From Stockpile to Hoist

The use of concrete masonry has increased greatly. This is partly due to the lower labor cost of handling and setting the larger concrete units in comparison with building brick, when used as a back-up material.

The term concrete masonry is applied to block, brick or tile building units molded from concrete and laid by masons into a wall.

The concrete is made by mixing portland cement with water and other suitable materials, such as sand, gravel, crushed stone, bituminous or anthracite cinders, burned clay or shale, blast furnace slag, pumice, perlite and vermiculite.

Concrete masonry units should always be manufactured in a cement products plant where facilities for manufacturing, curing, etc., are uniform so as to obtain high quality products.

The quality of all concrete masonry units should conform to the standards set by the American Society for Testing and Materials in their standard specifications for concrete masonry units and concrete brick.

Weight of Concrete Masonry.—A 7⅝"x7⅝"x15⅝" hollow load-bearing concrete block weighs from 40 to 50 lbs. when made from heavyweight aggregate and

Courtesy J. Emil Anderson & Son, Inc. Photographer Ed Nastek

Large mason station stocked with brick and mortar

from 25 to 35 lbs. when made from lightweight aggregate. Heavyweight aggregates are sand, gravel, crushed stone, air-cooled slag, etc. Lightweight aggregates are coal cinders, expanded shale, clay or slag, pumice, perlite, vermiculite, etc.

Estimating Quantities of Concrete Brick.—Concrete brick are manufactured, throughout the country generally, in the modular size of 2¼" x 3⅝" x 7⅝". In some localities, additional sizes are available as follows: Jumbo brick, 3⅝" x 3⅝" x 7⅝"; double brick, 4⅞" x 3⅝" x 7⅝"; and Roman brick, 1⅝" x 3⅝" x 11⅝" and 1⅝" x 3⅝" x 15⅝".

They should be estimated by the sq. ft. of wall of any thickness or by the cu. ft., if all wall thicknesses are combined. For instance, if there are only 8" walls in the job, it will be satisfactory to take the number of sq. ft. of walls but if the job contains 8", 12" and 16" walls, then it will be easier and more satisfactory to reduce the quantities to cu. ft. and multiply by a certain number of brick per cu. ft. of wall.

The number of concrete brick required per square or cubic foot of wall will vary with the size of brick and width of mortar joint used. Concrete brick sizes and widths of mortar joints have been standardized so that they will lay up to an even multiple of the 4-inch module.

All concrete brick units are 3⅝" thick, which together with a ⅜" mortar back joint equal the modular thickness of 4 inches. Since there is always one less mortar joint than brick, in walls of multiple brick thickness, the actual wall thickness will always be ⅜" less than the modular 4 inches ; i.e. 3⅝", 7⅝", 11⅝", 15⅝", etc.

In drawing plans, some architects give actual wall thickness dimensions, while others use nominal dimensions. For estimating purposes, nominal dimensions should be used, i.e. 4", 8", 12", 16", etc., to facilitate computations.

When measuring the plans and listing the quantities of brick required, always take exact measurements for length and height and deduct all openings in full. Do not count corners twice but have your estimate show as accurately as possible the actual number of brick required to complete the job. For unexposed backup work, add 1 to 2 percent for waste. For exposed face work, add 3 to 5 percent for waste.

Modular Size Concrete Brick

Modular size concrete brick, 2¼"x3⅝"x7⅝" are laid so that 3 courses of brick, with 5/12" mortar joints, builds 8" in height, and the 7⅝" length, plus ⅜" mortar joint, makes a total length of 8" laid in the wall. In other words, 3 brick lay 8"x8", or 64 sq. in. so it requires 6¾ or 6.75 brick per sq. ft. of wall of nominal 4" thickness.

Jumbo brick are 3⅝"x3⅝"x7⅝" in size and are laid so that 3 courses of brick, with ⅜" mortar joints, builds 12" in height, and the 7⅝" length, plus ⅜" mortar joint, makes a total length of 8" laid in the wall. One jumbo brick lays 32 sq. in. in the wall or it requires 4.5 brick per sq. ft. of wall of nominal 4" thickness.

Double brick are 4⅞"x3⅝"x7⅝" in size, so that 3 courses of brick, with 11/24" mortar joint builds 16" high, and the 7⅝" length, plus ⅜" mortar joint, makes a total length of 8" in the wall, or one brick lays 42.67 sq. in. in the wall or it requires 3.4 brick per sq. ft. of wall of nominal 4" thickness.

Roman size brick are 1⅝"x3⅝"x11⅝" and 1⅝"x3⅝"x15⅝" and are laid so that 6 courses of brick, with ⅜" mortar joints, builds 12" in height , and the 11⅝" or 15⅝" lengths, plus ⅜" mortar joints makes a total length of 12" or 16" respectively laid in the wall. One nominal 12" brick lays 24 sq. in. in the wall or it requires 6 brick per sq. ft. of wall of nominal 4" thickness. One nominal 16" brick lays 32 sq. in. in the wall or it requires 4.5 brick per sq. ft. of wall of nominal 4" thickness.

The number of modular size brick per sq. ft. of wall of various thicknesses is as follows :

	1-Brick or 4" Wall	2-Bricks or 8" Wall	3-Bricks or 12" Wall*
2¼"x3⅝"x7⅝"	6.75	13.50	20.25
3⅝"x3⅝"x7⅝"	4.50	9.00	13.50
4⅞"x3⅝"x7⅝"	3.40	6.80	10.20
1⅝"x3⅝"x11⅝"	6.00	12.00	18.00
1⅝"x3⅝"x15⅝"	4.50	9.00	13.50

*Use this column for number of brick required per cu. ft. of wall.
No allowance is included for waste in above quantities.

Labor Handling and Laying Concrete Brick

The labor cost of handling and laying concrete brick will vary with the size of brick, kind of mortar, class of work, thickness of walls, number of openings, etc. On buildings having long straight walls without many openings, such as basement walls, garages, factories, etc., a mason should lay 800 to 900 modular size concrete brick, 2¼"x3⅝"x7⅝", per 8-hr. day, while on dwellings and other structures having 8" walls, cut up with numerous openings, pilasters, etc., a mason should lay 650 to 700 brick per 8-hr. day.

Labor Cost of 1,000 Modular Size Concrete Brick, 2¼"x3⅝"x7⅝",
Laid in 8" Walls

	Hours	Rate	Total	Rate	Total
Mason	11.8	$....	$....	$18.91	$223.14
Labor	8.8	....	....	14.95	131.56
Cost per 1,000 brick			$....		$354.70

Above costs based on flush cut mortar joints. For struck joints, add ½-hr. mason time per 1,000 brick.

Estimate the labor cost of laying concrete face brick the same as given for face brick under "Unit Masonry/Brick."

On similar work, using jumbo brick, 3⅝"x3⅝"x7⅝", a mason should lay 675 to 775 brick per 8-hr. day in 12" walls and 550 to 600 brick per day in 8" walls.

Using double brick, 4⅞"x3⅝"x7⅝", a mason should lay 450 to 550 brick per 8-hr. day in 12" walls and 375 to 425 brick per day in 8" walls.

Roman size concrete brick, 1⅝"x3⅝"x11⅝" or 1⅝"x3⅝"x15⅝", are used only for facing or in 8" walls. The labor cost of handling and laying these brick will vary considerably, depending upon the type of joint treatment, flush cut, V-tooled or concave, struck or weathered, raked out, etc.

On average work with flush cut joints, a mason should lay 300 to 350 nominal 12" Roman brick per 8-hr. day or 225 to 265 nominal 16" Roman brick per 8-hr. day. For tooled, struck, weathered or raked out joints, reduce the above production quantities 8 to 12 percent.

Concrete Blocks and Partition Units

Concrete blocks and partition units are manufactured in two weights of concrete, one weighing approximately 145 lbs. per cu. ft. (sand, pebbles, crushed stone or slag aggregates) and another of lighter weight concrete weighing about 100 lbs. per cu. ft. of burned clay, cinder, processed slag, pumice, perlite or vermiculite aggregates.

These two types of units may be used interchangeably for all purposes, although the lightweight units afford a saving of weight to secure economy in design, and they afford increased heat and sound insulation value.

Estimating the Quantity of Concrete Blocks and Partition Units.—Concrete blocks and partition units should be estimated by the square foot of wall of any thickness and then multiplied by the number of blocks per 100 sq. ft. as given in the following tables.

When estimating quantities, always take exact measurements.

Make deductions in full for all openings, regardless of size. The result will be the actual number of sq. ft. or number of blocks required for the job.

Mortar.—The same kind of mortar should be used in laying concrete blocks as given for "Modular Size Concrete Brick."

When laying up exterior walls, face shell bedding shall be used with complete coverage of face shells. Furrowing of the mortar shall not be permitted. Extruded mortar shall be cut off flush with face of wall and the joints firmly compacted, after the mortar has stiffened somewhat.

Tooling is essential in producing tight mortar joints. Mortar has a tendency to shrink slightly and may pull away from the masonry units causing fine, almost invisible cracks at the junction of mortar and masonry units.

Sizes and Shapes of Concrete Building Units

Sizes and Weights of Concrete Masonry Units

Concrete masonry units are usually made with standard modular face dimensions 7⅝" high and 15⅝" long and are available in thicknesses ⅜" less than the nominal 3", 4", 5", 6", 8", 10" and 12" thickness.

When laid up with ⅜" mortar joints the units are 8" high and 16" long, requiring 112.5 units per 100 sq. ft. of wall, not including allowance for waste and breakage.

Hollow and solid partition and furring units have nominal thickness of 2", 3", 4", 5", 6" and 8".

Hollow and solid load bearing units have nominal thickness of 4", 6", 8", 10" and 12" and are also available in half lengths.

Standard specials such as steel and wood sash jambs, bullnose and closures are also available in full and half lengths.

Sizes and Shapes of Concrete Building Units

There are also many special sizes available on special order but are not usually stocked by the manufacturing plants.

The foregoing illustrations show many of the special sizes and shapes available upon special order.

Sizes, Weights and Quantities of Load-Bearing Concrete Blocks and Tile

Actual Size and Description of Units	Wall Thick- ness	Approx. Weight Lbs.		No. Units 100 Sq. Ft. of Wall	Cu. Ft. Mortar 100 Sq. Ft. of Wall*
		Hvy. Wt. Units	Light Wt. Units		
3⅝"x4⅞"x11⅝"	4"	11-13	8-10	240	8.0
5⅝"x4⅞"x11⅝"	6"	17-19	12-14	240	8.5
7⅝"x4⅞"x11⅝"	8"	22-24	14-16	240	9.0
3⅝"x7⅝"x11⅝"	4"	17-19	12-14	150	6.0

5⅝"x7⅝"x11⅝"	6"	26-28	17-19	150	6.5
7⅝"x7⅝"x11⅝"	8"	33-35	21-23	150	7.0
9⅝"x7⅝"x11⅝"	10"	42-45	27-29	150	7.5
11⅝"x7⅝"x11⅝"	12"	48-51	29-31	150	8.0
3⅝"x3⅝"x15⅝"	4"	12-14	9-10	225	9.0
5⅝"x3⅝"x15⅝"	6"	17-19	11-13	225	9.5
7⅝"x3⅝"x15⅝"	8"	22-24	14-16	225	10.0
3⅝"x7⅝"x15⅝"	4"	23-25	16-18	112.5	5.0
5⅝"x7⅝"x15⅝"	6"	35-37	24-26	112.5	5.0
7⅝"x7⅝"x15⅝"	8"	45-47	29-31	112.5	6.0
9⅝"x7⅝"x15½"	10"	57-60	36-38	112.5	6.5
11⅝"x7⅝"x15⅝"	12"	64-67	40-42	112.5	7.0

Sizes, Weights and Quantities of Concrete Partition Tile

Actual Size and Description of Units	Wall Thickness	Approx. Wt. Light Wt. Units Lbs.	No. Units. 100 Sq. Ft. Wall	Cu. Ft. Mortar 100 Sq. Ft. Wall*
3⅝"x7⅝"x11⅝"	4"	12-14	150	6.0
5⅝"x7⅝"x11⅝"	6"	17-19	150	6.5
7⅝"x7⅝"x11⅝"	8"	21-23	150	7.0
1⅝"x7⅝"x15⅝"	2"	11-12	112.5	3.0
3⅝"x7⅝"x15⅝"	4"	16-18	112.5	5.0
5⅝"x7⅝"x15⅝"	6"	24-26	112.5	5.0
7⅝"x7⅝"x15⅝"	8"	29-31	112.5	6.0

*Actual mortar quantities figure out about half those given in the above table but experience on the job shows that considerable more mortar is required, due to waste, droppings, etc.

Labor Laying Concrete Masonry

The labor cost of laying the various types and sizes of concrete blocks and tile will vary with the size and weight of the blocks, the class of work, whether long straight walls or walls cut up with numerous openings, etc.

Labor costs will also vary, depending upon whether or not the blocks are laid above or below grade, as basement walls usually proceed faster than exposed work above grade.

It is usually more economical to use lightweight units, even though they cost a few cents more per piece, as a mason can handle and lay them with less effort. In some localities union regulations require two masons to work together where the blocks weigh more than 35 pounds each.

Where concrete blocks are used for exterior or interior facing walls, with the blocks carefully laid to a line and in various patterns and with neatly tooled mortar joints, the labor costs will run considerably higher than on straight structural walls.

The quantities given in the following tables are based on average conditions with blocks laid in 1 :1 :6 cement-lime mortar. If portland cement mortar is used above grade, reduce daily output about 5 percent.

All quantities are based on the output of one mason and one laborer or hod carrier per 8-hr. day. If hoisting engineer time is required, add about ¼-hr. per 100 sq. ft.

Number of Concrete Partition Tile Laid Per 8-Hr. Day By One Mason

Actual Size of Units	Wall Thickness	Number Lightweight Units* Laid per 8-Hour Day	Hours per 100 Pieces Mason	Labor
3⅝"x7⅝"x11⅝"	4"	195-215	3.9	3.9
5⅝"x7⅝"x11⅝"	6"	175-195	4.3	4.3

Actual Size of Units	Wall Thickness	Number Lightweight Units* Laid per 8-Hour Day	Hours per 100 Pieces Mason	Labor
7⅝"x7⅝"x11⅝"	8"	155-175	4.8	4.8
1⅝"x7⅝"x15⅝"	2"	180-200	4.2	4.2
3⅝"x7⅝"x15⅝"	4"	190-210	4.0	4.0
5⅝"x7⅝"x15⅝"	6"	170-190	4.5	4.5
7⅝"x7⅝"x15⅝"	8"	150-170	5.0	5.0

Number of Concrete Building Units Laid Per 8-Hr. Day by One Mason

Actual Size of Units	Wall Thickness	Number Lightweight Units* Laid per 8-Hour Day	Hours Per 100 Pieces Mason	Labor
3⅝"x4⅞"x11⅝"	4"	215-235	3.5	3.5
5⅝"x4⅞"x11⅝"	6"	195-215	3.9	3.9
7⅝"x4⅞"x11⅝"	8"	175-195	4.3	4.3
3⅝"x7⅝"x11⅝"	4"	195-215	3.9	3.9
5⅝"x7⅝"x11⅝"	6"	175-195	4.3	4.3
7⅝"x7⅝"x11⅝"	8"	155-175	4.8	4.8
9⅝"x7⅝"x11⅝"	10"	135-155	5.5	5.5
11⅝"x7⅝"x11⅝"	12"	115-135	6.4	6.4
3⅝"x3⅝"x15⅝"	4"	215-235	3.5	3.5
5⅝"x3⅝"x15⅝"	6"	195-215	3.9	3.9
7⅝"x3⅝"x15⅝"	8"	175-195	4.3	4.3
3⅝"x7⅝"x15⅝"	4"	190-210	4.0	4.0
5⅝"x7⅝"x15⅝"	6"	170-190	4.5	4.5
7⅝"x7⅝"x15⅝"	8"	150-170	5.0	5.0
9⅝"x7⅝"x15⅝"	10"	130-150	5.7	5.7
11⅝"x7⅝"x15⅝"	12"	110-130	6.7	6.7
7⅝"x7⅝"x15⅝"	8"	225-250**	3.4	3.4
11⅝"x7⅝"x15⅝"	12"	120-160**	5.7	5.7

*For heavyweight concrete units decrease above quantities and increase labor 10 percent.
**For walls below grade.

Mortar Joints.—Hollow concrete blocks or tile should be laid with broken joints to secure the maximum resistance to moisture and heat penetration. Sufficient mortar is spread on the inner and outer shell bed and end joints to cover and to provide a joint of the required thickness when pressed into place. Solid units, brick, etc., are laid in solid beds.

Comparison of Concrete Masonry Sizes With Brick.—Allowing for the mortar joints in brickwork, the various sizes of concrete masonry units are equivalent to the following number of modular size brick, i.e., 2¼"x3⅝"x7⅝".

Size of Blocks	Brick Equivalent
4"x8"x16"	6
8"x8"x16"	12
12"x8"x16"	18

Cost of Concrete Masonry Units.—The prices quoted below are for index purposes only and are based on lightweight 8"x8"x16" block delivered to the job. As concrete masonry is widely manufactured, the estimator is always close to the source of supply and costs on different size blocks are readily obtainable.

City	Price Each	City	Price Each
Baltimore	$0.42	Cleveland	$0.43
Boston	.45	Dallas	.70
Chicago	.60	Denver	.55
Cincinnati	.40	Detroit	.48

City	Price Each	City	Price Each
Kansas City	.55	Pittsburgh	.50
Los Angeles	.58	San Francisco	.70
New Orleans	.48	St. Louis	.48
New York	.35	Seattle	.80
Philadelphia	.40		

Precast reinforced concrete lintels can usually be obtained from concrete masonry manufacturers.

Control Joints in Concrete Masonry

To control movements in concrete masonry walls from various kinds of stresses, increasing use is being made of control joints. Control joints are continuous vertical joints built into the walls in such locations and in such a manner as to permit slight wall movement without cracking the masonry.

The spacing and location of control joints will depend upon a number of factors, such as length of walls, architectural details and especially on the experience records as to the need for control joints in the particular locality where the structure is to be built. Control joints should be placed at junctions of bearing as well as nonbearing walls, at junctions of walls and columns or pilasters, in walls weakened by chases and openings, etc. In long walls, joints are ordinarily spaced at approximately 20-ft. intervals, again depending upon local experience. As a rule, control joint locations are determined by the architect or engineer and are usually indicated on the drawings.

Method of Constructing Control Joints in Concrete Masonry Walls.

Control joints may be built with regular full and half-length stretcher block or full and half-length offset jamb block. With this type of joint construction, a non-corroding metal Z-tiebar, placed in every other horizontal joint across the control joint will provide lateral support to wall sections on each side of the control joint.

In some localities, special control joint block are available. These block have tongue and groove ends which provide the required lateral support. These block are also made in full and half-length units.

To keep control joints as unnoticeable as possible, care must be taken to build them plumb and of the same thickness as the other mortar joints. If joint is to be exposed to weather or view, it should be raked out to a depth of at least ¾" and sealed with knife grade caulking compound.

The additional cost of building control joints, over and above the regular wall cost, will run 25 to 30 ct. per lin. ft. for material and 50 to 60 cts. per lin. ft. for labor.

Reinforced Concrete Masonry

Local conditions frequently demand special construction methods. For example, in earthquake regions it is necessary to provide more than ordinary stability for all types of structures. This is also true in areas where severe wind storms occur or where foundation soils are unstable. Moreover, in almost any locality there are structures which are subject to excessive vibrations, very heavy loads, extreme contraction or other stresses. Any of these examples of unusual stress conditions require special attention in designing and are often covered by local building codes.

In concrete masonry construction, additional strength is obtained by reinforcing the walls with steel reinforcing rods encased in poured grout or concrete. The walls may be reinforced horizontally, vertically or both.

Bond Beams.—To reinforce concrete masonry walls horizontally, bond beams are frequently used at each story height. Under extreme stress conditions, it may be necessary to use them in every second or third course.

Bond beams may be constructed by forming and pouring a continuous ribbon of reinforced concrete at the course height required, but this method breaks up the continuity of the block pattern and may be objectionable.

In many localities, special bond beam blocks are available which are trough-shaped. When laid open-side-up, they form a continuous trough in which reinforcing steel and concrete encasement may be placed. Thus, the need for wood side forms is eliminated and the block pattern of the wall is preserved.

Bond beam blocks are also used to construct lintels over openings. Lintels may be pre-constructed on the ground and set in place when they have attained sufficient strength or they may be built in place, using wood centers for support until mortar and concrete is strong enough to permit their removal.

Vertical Reinforcement in Concrete Masonry Walls.—Where vertical reinforcement is required in concrete masonry walls, it is usually located at building corners, jambs of wall openings and at regular intervals between wall openings. Size and spacing of reinforcement is usually covered by local building codes. When used in conjunction with bond beams, vertical and horizontal steel should be tied together.

In placing vertical reinforcement, advantage is taken of the vertical alignment of hollow block cores which form wells into which the reinforcing bars are placed and grouted solid with poured mortar or concrete. At locations where vertical reinforcement is to occur, the bottom block should be left out for a cleanout hole when wall is laid up. Just prior to setting steel bars in place, the wells should be rodded clean of extruded mortar and debris removed from the cleanout. After cleaning, setting bars and inspection, cleanouts are closed with side-forms and the wells are grouted solid.

Concrete Masonry Cavity Walls

A cavity wall consists of two walls separated by a continuous air space and securely tied together with non-corroding metal ties of adequate strength. For each 3 sq. ft. of wall surface a rectangular tie of No. 6 gauge wire can be used. The ties are embedded in the horizontal joints of both walls. Additional ties are necessary at

A- Continuous reinforced concrete bond beam. Lap bars at corners
B- Reinforced concrete studs tied to footing
C- Reinforced concrete footing
D- Reinforcement in horizontal mortar joints

Method of Constructing and Reinforcing Concrete Masonry Walls.

all openings, with ties spaced about 3 ft. apart around the perimeter and within 12 inches of the opening.

Typical codes require that 10 inch cavity walls should not exceed 25 ft. in height. In residential construction the overall thickness of concrete masonry cavity walls are nominally 10" or 12".

Neither the inner nor outer walls should be less than 4 inches thick (nominal dimension) and the space between them should not be less than 2 inches nor more than 3 inches wide. Usually the outer wall is nominally 4 inches thick and the remaining wall thickness made up by the air space and the inner wall. For example, a modular concrete masonry cavity wall nominally 12 inches thick is composed of a 4" outer wall, a 2" air space and a 6" inner wall.

A simple method of preventing the accumulation of mortar droppings between walls and maintaining a clear cavity is to lay a 1" x 2" wood strip across a level of ties to catch the droppings. As the masonry reaches the next level for placing ties the strip is raised, cleaned and laid on the ties placed at this level.

The practice of providing special flashing and weep holes in masonry cavity walls primarily presumes that water will enter the wall from the outside. If concrete masonry walls are properly designed and built with well-compacted mortar joints and are painted or stuccoed the walls should be weathertight. Thus, in the great majority of cases, there should be no need for special flashing and weep holes.

However, in limited areas subjected to severe driving rains, or where experience

Laying up inside concrete masonry wall which will be finished by facing with brick

has shown that sufficient water collects in the wall to make flashing and weep holes necessary, this practice is followed: The heads of windows, doors and other wall openings, and the bottom course of masonry immediately above any solid belt course or foundation, are flashed so that any moisture entering the wall cavity will be directed toward the outside walls. Only rust-resisting metal or approved materials treated with asphalt or pitch preparations should be used for flashing.

Weep holes are placed every 2 or 3 units apart in the vertical joints of the bottom course of the outside wall immediately above any solid belt course or foundation. In no case should the weep holes be located below grade. Weep holes can be formed by placing well-oiled rubber tubing in the mortar joint and then extracting it after mortar has become hard. The tubing should extend up into the cavity for several inches to provide a drainage channel through any mortar droppings that might have accumulated.

The labor cost of concrete masonry cavity walls varies the same as ordinary block walls according to the size of unit, pattern of laying, nature of job, etc.

Concrete Masonry Units Used for Exterior Facing

Concrete masonry walls are extensively used for exterior facing the same as face brick or stone. They may be laid up in any of the attractive designs shown in the accompanying illustrations, ranging from straight ashlar to the many variations of random or broken ashlar.

When used for exterior facing, the walls should be true and plumb and laid with full mortar coverage on vertical and horizontal face shells, (no furrowing permitted) with all vertical joints shoved tight. Mortar joints should be ⅜" thick.

Method of Constructing Cavity Walls of Concrete Masonry.

The mortar joints should be struck off flush with wall surface and when partially set shall be compressed and compacted with a rounded or V-shaped tool. This provides a more waterproof mortar joint. An attractive treatment is obtained by emphasizing the horizontal joints and obscuring the vertical joints in concrete walls. This is done by tooling the horizontal joints and striking the vertical joints flush with the wall surface and then rubbing with carpet or burlap to remove the sheen from the troweled mortar surface.

Lightweight concrete units should be used where obtainable as they provide better insulation than units made of ordinary concrete aggregates.

The labor costs given on the following pages are based on using lightweight units. If ordinary concrete units are used, add about 10 percent to labor costs given.

12-in. wall   8-in. wall

(Bonded every 7ᵗʰ course)

8-in. high block

8-in. high block

5-in. high block

12-in. wall
(Bonded every 6ᵗʰ course)

8-in. wall
(Bonded every 7ᵗʰ course)

Examples of Concrete Masonry Used as Backup for Brick.

Methods of Laying Concrete Masonry Units for Exterior Facing

Methods of Laying Concrete Masonry Units for Exterior Facing

Example of Horizontal and Vertical Face Shell Bedding

Labor Cost of 100 Sq. Ft. of Exterior Facing Concrete Masonry Walls
Using 8"x16" Units, Laid in Regular Ashlar or Running Bond
Walls 8" Thick

	Hours	Rate	Total	Rate	Total
Mason	7	$....	$....	$18.91	$132.37
Labor	7	....	....	14.95	104.65
Cost per 100 sq. ft.			$....		$237.02
Cost per sq. ft			....		2.37

Add 10 percent to labor costs for 10" thick walls and 20 percent for 12" thick walls. Deduct 25 percent from labor costs for 4" veneer facing.

Labor Cost of 100 Sq. Ft. of Exterior Facing Concrete Masonry Walls
Using 4"x16" or Half Height Units Laid in Regular Ashlar
or Running Bond
Walls 8" Thick

	Hours	Rate	Total	Rate	Total
Mason	10	$....	$....	$18.91	$189.10
Labor	7	....	....	14.95	104.65
Cost per 100 sq. ft.			$....		$293.75
Cost per sq. ft			....		2.94

Add 10 percent to labor costs for 10" thick walls and 20 percent for 12" thick walls. Deduct 25 percent from labor costs for 4" veneer facing.

Labor Cost of 100 Sq. Ft. of Exterior Facing Concrete Masonry Walls
Using Coursed Ashlar No. 1, Using Alternate
Courses of 8"x16" and 4"x16" Units
Walls 8" Thick

	Hours	Rate	Total	Rate	Total
Mason	8	$....	$....	$18.91	$151.28
Labor	6	....	....	14.95	89.70
Cost per 100 sq. ft			$....		$240.98
Cost per sq. ft			....		2.41

Add 10 percent to labor costs for 10" thick walls and 20 percent for 12" thick
walls. Deduct 25 percent from labor costs for 4" veneer facing.

Labor Cost of 100 Sq. Ft. of Exterior Facing Concrete Masonry Walls
Using Coursed Ashlar No. 2, Consisting of Two Courses
of 8"x16" Units and One Course of 4"x16" Units
Walls 8" Thick

	Hours	Rate	Total	Rate	Total
Mason	7.5	$....	$....	$18.91	$141.83
Labor	5.5	....	....	14.95	82.23
Cost per 100 sq. ft			$....		$224.06
Cost per sq. ft			....		2.24

Add 10 percent to labor costs for 10" thick walls and 20 percent for 12" thick
walls. Deduct 25 percent from labor costs for 4" veneer facing.

Labor Cost of 100 Sq. Ft. of Exterior Facing Concrete Masonry Walls
Using Coursed Ashlar No. 3, Consisting of
8"x16" and 4"x16" Units
Walls 8" Thick

	Hours	Rate	Total	Rate	Total
Mason	8.25	$....	$....	$18.91	$156.00
Labor	6.50	....	....	14.95	97.18
Cost per 100 sq. ft			$....		$253.18
Cost per sq. ft			....		2.53

Add 10 percent to labor costs for 10" thick walls and 20 percent for 12" thick
walls. Deduct 25 percent from labor costs for 4" veneer facing.

Labor Cost of 100 Sq. Ft. of Exterior Facing Concrete Masonry Walls
Using Coursed Ashlar No. 4, Consisting of
8"x16" and 4"x16" Units
Walls 8" Thick

	Hours	Rate	Total	Rate	Total
Mason	7.9	$....	$....	$18.91	$149.39
Labor	5.7	....	....	14.95	85.22
Cost per 100 sq. ft			$....		$234.61
Cost per sq. ft			....		2.35

Add 10 percent to labor costs for 10" thick walls and 20 percent for 12" thick
walls. Deduct 25 percent from labor costs for 4" veneer facing.

CONCRETE UNIT MASONRY

Labor Cost of 100 Sq. Ft. of Exterior Facing Concrete Masonry Walls
Using Vertical Stacking of 8"x16" Units
Walls 8" Thick

	Hours	Rate	Total	Rate	Total
Mason	9	$....	$....	$18.91	$170.19
Labor	7	....	....	14.95	104.65
Cost per 100 sq. ft.			$....		$274.84
Cost per sq. ft.			....		2.75

Add 10 percent to labor costs for 10" thick walls and 20 percent for 12" thick walls. Deduct 25 percent from labor costs for 4" veneer facing.

Labor Cost of 100 Sq. Ft. of Exterior Facing Concrete Masonry Walls
Using Horizontal Stacking of 8"x16" Units
Walls 8" Thick

	Hours	Rate	Total	Rate	Total
Mason	7.5	$....	$....	$18.91	$141.83
Labor	5.5	....	....	14.95	82.23
Cost per 100 sq. ft.			$....		$224.06
Cost per sq. ft.			....		2.24

Add 10 percent to labor costs for 10" thick walls and 20 percent for 12" thick walls. Deduct 25 percent from labor costs for 4" veneer facing.

Labor Cost of 100 Sq. Ft. of Exterior Facing Concrete Masonry Walls
Using Square Stacking of 8"x8" Units
Walls 8" Thick

	Hours	Rate	Total	Rate	Total
Mason	12.50	$....	$....	$18.91	$236.38
Labor	8.75	....	....	14.95	130.81
Cost per 100 sq. ft.			$....		$367.19
Cost per sq. ft.			....		3.67

Add 10 percent to labor costs for 10" thick walls and 20 percent for 12" thick walls. Deduct 25 percent from labor costs for 4" veneer facing.

Labor Cost of 100 Sq. Ft. of Exterior Facing Concrete Masonry Walls
Using Basket Weave Design Consisting of 8"x16" Units
Walls 8" Thick

	Hours	Rate	Total	Rate	Total
Mason	7.9	$....	$....	$18.91	$149.39
Labor	5.7	....	....	14.95	85.22
Cost per 100 sq. ft.			$....		$234.61
Cost per sq. ft.			....		2.35

Add 10 percent to labor costs for 10" thick walls and 20 percent for 12" thick walls. Deduct 25 percent from labor costs for 4" veneer facing.

Labor Cost of 100 Sq. Ft. of Exterior Facing Concrete Masonry Walls
Using Patterned Ashlar Consisting of 8"x16" and 8"x8" Units
Walls 8" Thick

	Hours	Rate	Total	Rate	Total
Mason	8.4	$....	$....	$18.91	$158.85
Labor	6.2	....	....	14.95	92.69
Cost per 100 sq. ft.			$....		$251.54
Cost per sq. ft.			....		2.52

Add 10 percent to labor costs for 10" thick walls and 20 percent for 12" thick walls. Deduct 25 percent from labor costs for 4" veneer facing.

Labor Cost of 100 Sq. Ft. of Exterior Facing Concrete Masonry Walls
Using Coursed Patterned Ashlar Consisting of 4"x8", 4"x12",
4"x16", 8"x12" and 8"x16" Units
Walls 8" Thick

	Hours	Rate	Total	Rate	Total
Mason	14	$. . . .	$. . . .	$18.91	$264.74
Labor	9	. . . .	. . . .	14.95	134.55
Cost per 100 sq. ft			$. . . .		$399.29
Cost per sq. ft				. . . .	3.99

Add 10 percent to labor costs for 10" thick walls and 20 percent for 12"thick walls. Deduct 25 percent from labor costs for 4" veneer facing.

Concrete block is often faced with brick. Three bricks equal the height of one block and every seventh course of brick can be a header course, or, in 12" walls every other block course can be a header block, that is a block notched to receive a header brick. This allows every sixth course to be a bond.

Labor Cost of 100 Sq. Ft. of 8" Brick Faced Block Wall Bonded Every Seventh
Course

	Hours	Rate	Total	Rate	Total
Mason	18.5	$. . . .	$. . . .	$18.91	$349.84
Labor	13.5	. . . .	. . . .	14.95	201.83
Cost per 100 sq. ft			$. . . .		$551.67
Cost per sq. ft				. . . .	5.52

Labor Cost of 100 Sq. Ft. of 12" Brick Faced Concrete Block
Wall Bonded Every Sixth Course

	Hours	Rate	Total	Rate	Total
Mason	22	$. . . .	$. . . .	$18.91	$416.02
Labor	17	. . . .	. . . .	14.95	254.15
Cost per 100 sq. ft			$. . . .		$670.17
Cost per sq. ft				. . . .	6.70

Insulating Fill for Concrete Masonry Walls.—For insulating masonry walls, a water-repellent vermiculite masonry fill is available. This is a free-flowing granular material, processed to provide a high degree of water-repellency. It is designed for the cores of masonry blocks and the voids of cavity walls. It does not bridge, and completely fills cores and cavities. Test installations show that the fill reduces heat loss up to 50 percent, and air-conditioning costs about 25 percent. Masonry fill is also effective sound-deadening in masonry block partitions.

Vermiculite masonry fill is marketed in 4 cu. ft. bags and costs about $4.75 per bag. Approximate coverage is :

	Wall Area Sq. Ft.
2" cavity	24
2½" cavity	20
4½" cavity	11
6" block, 2 or 3 core	21
8" block, 2 or 3 core	14½
12" block, 2 or 3 core	8

Silicone treated perlite loose fill insulation insulates masonry walls to give greater comfort at lower cost. Silicone treated perlite pours readily into the cores of concrete block or cavity type masonry walls. Tests prove it reduces winter and summer heat transfer 50 percent or more. Silicone treatment of perlite provides lasting water repellency, prevents moisture penetration, and assures constant insu-

Insulating the Cores of Masonry Units with Water-repellent Vermiculite Masonry Fill.

lating efficiency in the most severe weather, including wind driven rain. Performance is rated excellent in accordance with test procedures developed by the National Bureau of Standards.

Silicone treated perlite loose fill insulation is manufactured nationally and marketed in 4 cu. ft. bags. Coverage is approximately the same as shown for vermiculite. Installed costs will also be competitive with other loose fill materials.

Silicone Treated Perlite Fill in Masonry Walls

Special Facing Blocks

In many areas, special block are available for the purpose of creating special facing effects. These block range from ordinary cement block, scored across the face to give the appearance of additional joints, to block made or faced with special aggregates which produce various textures and colors.

Scored Block.—The simplest departure from ordinary block facing is obtained by scoring regular block one or more times across the face to give the illusion of shallower units, and more joints. Most block manufacturers can furnish scored units for about 5 cts. per score in addition to the regular block price.

Labor costs for laying up this type of facing is about the same as for a regular running bond facing except when scored grooves are pointed with mortar and tooled to match the actual joints.

Split-Block.—Split-block afford another variation in wall finish. Split-block are made by splitting a hardened concrete unit lengthwise. The units are laid in the wall with the fractured face exposed. Many interesting variations can be obtained by introducing mineral colors and by using aggregates of different gradings and colors in the concrete mixture.

Split-block can be laid in simple running bond or in any of the other patterns used in concrete masonry construction. The fractured faces of the units produce a wall of rugged appearance.

Prices for 4" thick split-block veneer units run from $0.80 to $1.00 per sq. ft. of wall depending upon the size of units and pattern used.

Labor costs will be about the same as given on previous pages for the various patterns illustrated.

Slump Block.—For a weathered stone effect, slump block are used. When manufacturing slump block, a special consistency mixture is used so that unit will sag or "slump" when released from the molds before complete "setting." This method produces many artistic irregularities with variations in height, texture, etc. Many are integrally colored in varying shades. Standard heights vary 1⅝" to 3⅝" and units are usually laid in a random ashlar pattern, although coursed patterns may be used if desired.

The cost of slump block varies with the locality and prices should be obtained locally when figuring this work. Prices range from 80 cts. to $1.00 per sq. ft. of wall facing.

Mortar requirements for slump block facing are high due to the irregularities of the material and approximate those of random ashlar stone work. As an average, it will require about 10 cu. ft. of mortar per 100 sq. ft. of 4" veneer facing.

When laying slump block in a random ashlar pattern, a mason and a helper should lay 45 to 55 sq. ft. of 4" veneer facing per 8-hr. day.

Glazed Concrete Block.—Concrete block may be ordered with a factory applied glazed surface on one or more sides in 8"x16", 8"x8", and 4"x16" face sizes and in even thicknesses of 2" through 12". Standard cap, base and finished end units are also furnished. This glaze is a satin finish, factory applied to an ⅛" thickness, and is available in a standard range of 48 colors and 11 scored patterns with special colors and shapes a possibility. This glazing process is available to local sources through special licensing rather than being centrally manufactured and warehoused, and the local firm should always be consulted for availability and cost.

These units are especially useful where it is desired to have a block finish on one side and a glazed on the other as one unit will provide both which results in considerable savings in both labor costs and floor space over building a 4" block partition and adding a 2" glazed furring block. The glazed surface meets both USDA and OSHA requirements for sanitary surfaces. The high performance standards of concrete block for fire resistance and insulation value are retained. Tolerance on face dimensions and face distortions is limited to 1/16".

Units are set in standard mortar and struck with a groover but may be pointed or grouted for more finished work. Acid cleaning solutions should not be used, so units must be kept clean as the work progresses. A final cleaning with a masonry cleaning compound should be figured.

Adding the glazed surface to a standard block will add around $1.25 to $1.50 per square foot of glazed surface. Some colors will be more expensive than others, and the amount of scoring within the block surface will also affect the price. Two masons and one laborer will set 240–250 square feet of 4" "partition in running bond per day.

Labor Cost of 100 Sq. Ft. of 4" Thick, 8"x16" Concrete
Block Glazed One Side Laid In Running Bond

	Hours	Rate	Total	Rate	Total
Mason	7	$. . . .	$. . . .	$18.91	$132.37
Labor	3.5	. . . .	. . . .	14.95	52.33
Cost per 100 sq. ft			$. . . .		$184.70
Cost per sq. ft			. . . .		1.85

Add to the above costs for pointing, grouting and cleaning as required.

CONCRETE STORAGE BINS AND SILOS

Concrete storage bins and silos are used for storing grain, coal, sand, gravel, etc., in addition to those used for the processing and preserving of forages for feed on the farm.

Silos are constructed of numerous types of materials such as concrete stave, cast-in-place concrete, concrete block, glazed tile, wood, and steel. The cost of storage silos varies with the storage capacity, the type of structure and the kind of materials to be stored.

Practically all silos are round or circular as this shape gives more capacity for the least amount of construction cost, and the stored material packs better.

Farm silos are built for storing corn, sorghum, hay crops and miscellaneous forages. Research studies reveal that lateral pressures increase with the moisture content of the silage and the size of the silo. All silages exert approximately equal lateral pressures when of the same moisture content. Therefore, the reinforcing requirements are identical. However, hay crop silages are usually stored at excessive moisture content and more reinforcing is recommended.

Excavating and Concrete Foundations for Silos

Footings should be ample to carry the weight of the silo without settlement or tipping. The width and depth, therefore, are varied according to the height of the silo and the load-carrying capacity of the soil. Recommended footing sizes are given in the accompanying table. These take into consideration the bearing capacities of the different soils as well as the weights of silos of different heights, including the weights of the concrete chutes, roofs, and the weight of the silage carried by the walls.

The depth of the foundation wall below ground level will vary according to the location of the silo. The footings should extend below frost penetration and to firm soil. The bottom of the trench for the footing should be flat and even in width to assure uniform distribution of the load to the soil. Forms will not be required for construction of the footings if care is taken to excavate the trench so that the walls stand vertically. Wall forms are used from the footing up.

Concrete Foundation and Floor for Cast-in-Place Concrete Silos.

Dimensions of Annular Footings for Silos with Walls 6 Inches Thick*

Height of Silo ft.	TYPE OF SOIL**					
	Sand and Gravel (Type 1)		Firm clay, wet sand or clay and sand mixture (Type 2)		Soft Clay (Type 3)	
	Width Inches	Depth Inches	Width Inches	Depth Inches	Width Inches	Depth Inches
20	12	8	12	8	17	8
25	12	8	12	8	24	8
30	12	8	16	8	32	11
35	12	8	21	10		
40	13	9	26	13	Silos Higher Than 30 ft. Not Recommended on Type 3 Soil	
45	16	11	32	15		
50	19	13	Silos Higher Than 45 ft. Not Recommended on Type 2 Soil			
55	23	16				

*A.C.I.—714.
**Soil Pressures in p.s.f. (Type 1) 8,000 (Type 2) 4,000 (Type 3) 2,000

Cubic Feet of Excavation or Concrete Footing Required For Silo Foundations of Various Sizes Per Inch of Depth or Thickness

Footing Width Inches	Diameter in Feet							
	10	12	14	16	18	20	22	24
12	2.67	3.17	3.67	4.17	4.75	5.25	5.75	6.33
13	2.89	3.43	3.97	4.51	5.14	5.69	6.23	6.86
16	3.55	4.22	4.88	5.55	6.33	7.00	7.67	8.44
17	3.77	4.48	5.19	5.90	6.72	7.44	8.14	8.97
19	4.21	5.01	5.80	6.59	7.51	8.31	9.10	10.03
21	4.66	5.53	6.40	7.28	8.30	9.19	10.06	11.08
23	5.10	6.06	7.01	7.97	9.10	10.06	11.02	12.15
24	5.33	6.33	7.33	8.33	9.50	10.50	11.50	12.67
26	5.55	6.86	7.94	9.03	10.29	11.37	12.45	13.72
27	5.77	7.12	8.25	9.37	10.68	11.81	12.95	14.25
32	7.10	8.44	9.77	11.10	12.65	14.00	15.33	16.90

Diameter and Circumference of Standard Size Silos

Diam. Silo in Ft.	Circum. Silo in Ft.	Nominal Circum. in Ft.	Diam. Silo in Ft.	Circum. Silo in Ft.	Nominal Circum. in Ft.
10	31.42	32	18	56.55	57
12	37.70	38	20	62.83	63
14	44.00	44	22	69.10	69
16	50.27	50	24	75.41	76

To obtain number of cubic feet or cubic yards of excavation or footings required for any job, proceed as follows: Find number of cubic feet of concrete required for a silo 16 feet in diameter, requiring a footing 19" wide and 13" thick. Refer to above table: a 16-ft. diameter silo requiring a footing 19" wide, requires 6.59 cu. ft. of concrete per inch in thickness. Multiply 6.59 by 13" (thickness of footing) and the result is 85.67 cu. ft. of concrete for the footing.

Silo Wall Forms.—The use of commercial steel forms is recommended for the construction of cast-in-place silos. Such forms are quickly set and removed and result in true, smooth walls. Several types of commercial steel forms are available. They differ from each other mainly in the size of the sections or panels and in the manner they are fitted together. In all types, the sections are curved and joined together snugly with clamps to form a complete circle or course. One course of forms is referred to as a lift. Each course or circle is 2 or more feet high, according to the forms used. From 6 to 8 feet of wall can be built in a day, depending upon the size of the crew and the outdoor temperature. In summer weather, forms can usually be removed in 24 hours. In building 6 or 8-ft. of wall per day, three or four complete courses of forms, using 2-ft. high sections, will be required. The usual practice is to set and fill one lift at a time. Where three or more courses of forms are in use it is the usual practice to remove the lower course and set it on top. This procedure is continued until the required height of wall is built. Form faces are cleaned and re-oiled for each resetting.

A plumb wall is assured when using commercial forms provided the first course is carefully leveled. Each manufacturer usually furnishes instructions covering the use of his forms. These companies also make forms for casting chutes and roofs and can furnish erection derricks, hoists and other construction accessories.

Courtesy A. O. Smith Harvestore Products, Inc.

Concrete foundation being poured for A. O. Smith Harvestore silo

Constructing the Walls.—Walls of well-built cast-in-place silos are smooth, plumb, air and watertight. The walls are 4 to 6-in. thick. A 5 to 6-in. thickness is recommended for large silos.

Concrete for durable silo walls should be of high quality, using clean sharp sand and gravel. Sand should be from very fine up to particles which pass a No. 4 screen (4 openings per lin. in.). Gravel should be clean, hard and range in size from ¼ " up to 1½ ".

A workable concrete mix is one which is "mushy" but not "soupy." It should be somewhat sticky when worked with a shovel or trowel and should be stiff enough to require some spading in the forms. For machine mixing allow about 2 minutes mixing after all materials are in the mixer. Concrete for footing should be mixed in approximately the proportions of 1 bag air entrained portland cement; 2¾ cu. ft. sand and 4 cu. ft. gravel. Use about 5 gals. water per bag of cement. Concrete for walls, floor, chute and roof should be mixed in the proportions of 1 bag of air entrained portland cement; 2¼ cu. ft. sand and 3 cu. ft. gravel. Use about 5 gals. water per bag of cement.

Total Weight of Horizontal and Vertical Bars Required
For Cast-in-Place Concrete Silos

Height of silo, ft.	Diameter of silo					
	10 ft.	12 ft.	14 ft.	16 ft.	18 ft.	20 ft.
	Lb.	Lb.	Lb.	Lb.	Lb.	Lb.
20	547	652	....	....	....	....
24	675	796	968	1,263	....	....
28	787	935	1,174	1,492	....	....
32	910	1,047	1,381	1,721	2,014	2,609
36	1,034	1,213	1,587	2,017	2,351	3,112
40	....	1,400	1,816	2,339	2,836	3,649
44	....	....	2,044	2,702	3,291	4,317
48	....	....	2,351	3,066	3,780	4,986
52	....	....	....	3,430	4,386	5,654

No allowance made for lapping of bars. Add approximately 10 percent when three lenghts of bars are used per circle.

Number and Length of Horizontal Reinforcing Bars Required
For Cast-in-Place Concrete Silos with Continuous Doorway

Height of silo, ft.	Diameter of Silo					
	10 ft.	12 ft.	14 ft.	16 ft.	18 ft.	20 ft.
	Length of bars*					
	31 ft. 10 in.	38 ft. 0 in.	44 ft. 4 in.	50 ft. 6 in.	56 ft. 10 in.	63 ft. 4 in.
	Number of bars required					
20	14	17	....	16	....	....
24	17	21	14	20	....	....
28	20	25	18	24	....	....
32	24	29	22	28	20	26
36	28	33	26	34	24	32
40	....	39	30	40	30	38
44	....	....	34	48	36	46
48	....	....	38	56	42	54
52	....	....	....	64	50	62
	No. 3 Bars		No. 4 Bars		No. 5 Bars	

*Bar lengths do not allow for laps. Add approximately 10% when three lengths of bar are used per circle.

Number and Length of Vertical Reinforcing Bars Required
For Cast-in-Place Concrete Silos

Height of silo, ft.	Length of bars	Diameter of Silo							
		10'	12'	14'	16'	18'	20'	22'	24'
		Number of bars required							
20	20 ft. 0 in.	22	26	....	....	....	....	....	....
24	13 ft. 0 in.	44	52	60	68	....	....	....	....
28	15 ft. 0 in.	44	52	60	68	....	....	....	....
32	17 ft. 0 in.	44	52	60	68	76	86	92	102
36	19 ft. 0 in.	44	52	60	68	76	86	92	102
40	14 ft. 8 in.	....	78	90	102	114	129	138	153
44	16 ft. 0 in.	....	78	90	102	114	129	139	153
48	17 ft. 4 in.	....	....	90	102	114	129	138	153
52	18 ft. 8 in.	....	....	90	102	114	129	138	153
60	16 ft. 6 in.	....	....	120	136	152	168	184	204

All bars are No. 3, spaced 18" apart. Bars are lapped 24 in. at splices.

Basis for Estimating Labor for Cast-in-Place Concrete Silos.—The following estimating data is applicable to and based on circular type cast-in-place concrete silos.

The labor excavating and placing concrete footings for silos should be estimated separately, as they will vary with the depth of excavation and size of silo.

After the footings are in place, an experienced crew of three men should build 8 feet of 10'-0", 12'-0" or 14'-0" diameter silo per 8-hr. day, including the labor on feed room. It will also require about a day to place the roof, and after the concrete is placed, a day will be required to tear down and remove the forms and equipment.

Labor Required to Construct Cast-in-Place Concrete Silos

Diam. of Silo	Thickness of Wall	Number of Men in Crew	Number Hours for Crew	Labor Hours for Each 8'-0" Height
		Labor Required to Construct Each 8'-0" of Wall Height		
10'-0"	5"-6"	3	8	24
12'-0"	5"-6"	3	8	24
14'-0"	5"-6"	3	8	24
Roof	....	3	8	24
16'-0"	5"-6"	4	8	32
18'-0"	5"-6"	4	8	32
20'-0"	5"-6"	4	8	32
22'-0"*	5"-6"	4	8	32
24'-0"*	5"-6"	4	8	32
Roof	....	4	8	32

*Labor required to construct each 6 ft. of wall height.

A set of metal forms usually includes 4 sets of forms 2'-0" high.

Reinforcing steel may be either rods or mesh but many silo builders prefer mesh reinforcing as it may be placed at lower cost than rods.

To the above, add labor cost excavating for footings, also labor placing concrete footings.

To the above, add time for 1 day for crew bringing tools and equipment to job and setting up; also 1 day tearing down and removing equipment at completion.

Silo Walls Using Slip Forms.—For casting monolithic concrete structures of great height the method of slip forms is often used and this type of form is particularly adaptable to the construction of silos and storage bins. Despite the fact that the sliding forms must be built with more precision and of heavier material than fixed forms, slip forms have proven to be more economical in the long run for high structures. Naturally the higher the construction the less the cost per square foot of wall.

The forms are prefabricated of wood or metal and are usually raised by using screw jacks or a hydraulic jacking system. The yokes which are usually made of wood and tied together with bolts are suspended with a screw or jack arrangement on a steel jackrod. The raising is done by a crew of men, each man operating from 1 to 5 jacks depending on the size of the job. This operation must be done on a shift basis, continuing day and night until the structure is completed to prevent the forms from adhering to the concrete.

Concrete Stave Silos

Footings for concrete stave silos can be made smaller than those for cast-in-place silos, because staves are only 2½" thick. American Concrete Institute Bulletin 714 gives the following information for footings for concrete stave silos.

Concrete Staves.—Concrete stave silos are built of cast concrete staves usually 10" wide, 30" long and 2½" thick. Every other stave in the bottom circle is 24" high with a 6" stave at the top of the corresponding column.

The number of staves required may be computed by figuring the number of staves for each circle, with the height of the structure governing the number of circles required.

Labor Setting Staves.—The setting of 1,000 10"x30" staves may be considered a fair day's work under average conditions for a 4-man crew, consisting of 2 men on the ground, 1 man setting staves, and a foreman to follow up, leveling staves and inspecting the tension of hoops. This includes the raising of the necessary scaffolding, placing and tightening hoops, etc.

Dimensions of Annular Footings for Silos with Walls 2½ Inches Thick

	TYPE OF SOIL					
Height of Silo ft.	Sand and Gravel (Type 1)		Firm clay, wet sand or clay and sand mixture (Type 2)		Soft Clay (Type 3)	
	Width Inches	Depth Inches	Width Inches	Depth Inches	Width Inches	Depth Inches
20	12	8	12	8	12	8
25	12	8	12	8	18	8
30	12	8	12	8	24	8
35	12	8	16	8	32	11
40	12	8	21	10		
45	13	9	26	13	Silos Higher Than 35 ft. Not Recommended on Type 3 Soil	
50	16	11	32	15		
55	19	13				
60	23	16	Silos Higher Than 50 ft Not Recommended on Type 2 Soil			

Number of Staves Required Per Circle for Concrete Stave Silos

The number of staves required for one complete circle of various size silos. Use last (*) column where door frames are included.

Diameter of Silo	Outside Diameter of Silo	Number of Staves per Circle	Number of Staves per Circle*
10'-0"	10'-4"	38	35
12'-0"	12'-6"	46	43
14'-0"	14'-4"	53	50
16'-0"	16'-5"	61	58
18'-0"	18'-6¾"	69	66
20'-0"	20'-8"	77	74

Hoops for Concrete Stave Silos.—It will be necessary to place steel rods or "hoops" around the concrete stave silo to reinforce the staves. The hoops spacing for either grass or corn silage of moisture content not over 75 percent should not exceed that given in the table of "Hoop Spacing for Concrete Stave Silos."

Lugs (connections for hoops of round cross-section) should be of malleable cast iron or steel. The strength of the lug should be such that the ultimate strength of the hooping to be used with it can be developed by the lug in place. For silos up to and including 10'-0" in diameter, at least two lugs per hoop should be used; from 10'-0" to 16'-0", at least 3 lugs per hoop should be used; and for diameters from 16'-0" to 22'-0", at least 4 lugs per hoop should be used.

On a new silo a final retightening of the reinforcing rods should be made 5 to 7 days after the silo is erected.

Tile Storage Bins

Glen-Gery storage bin and silo tile are clay, burned at a temperature of over 2,000 degrees and are salt glazed. They are made with a channel space at the top of each tile which permits the use of steel bands or reinforcing bars in each row of tile, amount of steel being proportioned to the amount of pressure, which is greatest at the bottom of the bin and decreases at the top.

These tile are furnished 12" high, 12" long and 6" and 8" thick, and it requires one tile per sq. ft. of wall, excluding the mortar joints which generally average ½" thick.

Mortar Required for Tile Storage Bins.—Pure portland cement mortar should be used for laying all tile to make the strongest possible wall. Including mortar required to fill the channel space at the top of each tile, it will require approximately 9 cu. ft. of mortar for 6" thick tile and 12 cu. ft. for 8" thick tile per 100 sq. ft. of wall.

Labor Laying Silo Tile.—When laying-up circular walls of 6"x12"x12" salt glazed tile, a mason should set 110 to 125 sq. ft. of tile per 8-hr. day, and will require one laborer to each mason.

When circular bin walls are laid up with 8"x12"x12" salt glazed tile, a mason should lay 90 to 100 sq. ft. of tile per 8-hr. day with a laborer helping full time.

Hoop Spacing for Concrete Stave Silos*

Based upon round rods of the sizes indicated of intermediate grade steel, with rolled threads. If cut threads are used the ends must be upset, or the spacing be closer. Unit tensile stress 20,000 psi (°)

Distance from top of silo in feet (a)	Lateral Pressure of silo in psf (c)	Hoop Spacing in Inches for 9/16 Inch Diameter Rods — Silo Diameter in feet						Hoop Spacing in Inches for 1/2 or 9/16 Inch Diameter Rds — Silo Diameter in feet					
		10	12	14	16	18	20	10	12	14	16	18	20
2½	12	30	30	30	30	30	30	30	30	30	30	30	30
5	33	30	30	30	30	30	30	30	30	30	30	30	30
7½	60	30	30	30	30	30	30	30	30	30	30	30	30
10	91	30	30	30	30	30	30	30	30	30	30	30	30
12½	125	30	30	30	30	30	30	30	30	30	30	30	30
15	162	30	30	30	30	30	30	30	30	30	30	30	30
17½	203	30	30	30	30	30	30	30	30	30	30	15	15
20	246	30	30	30	30	15	15	30	30	15	15	15	15
22½	294	30	30	15	15	15	15	30	15	15	15	15	15
25	340	30	30	15	15	15	15	30	15	15	15	15	15
27½	388	30	15	15	15	15	15	15	15	15	15	15	10
30	440	15	15	15	15	15	10	15	15	15	15	10	10
32½	495	15	15	15	15	15	10	15	15	15	10	10	10
35	550	15	15	15	10	10	10	15	15	10	10	10	10
37½	609	15	15	15	10	10	10	15	15	10	10	10	10
40	668	15	15	10	10	10	7½	15	15	10	10	10	7½
42½	725	15	10	10	10	7½	7½	15	15	10	10	7½	7½
45	793	15	10	10	7½	7½	7½	15	10	10	7½	7½	7½
47½	855	10	10	10	7½	7½	6	10	10	7½	7½	7½	6
50	925	10	10	7½	7½	6	6	10	10	7½	7½	7½	6
52½	990	10	10	7½	7½	6	6	10	10	7½	7½	7½	6
55	1055	10	7½	7½	6	6	6	10	7½	7½	7½	6	6
57½	1110	10	7½	7½	6	6	5	10	7½	7½	7½	6	6
60	1200	7½	7½	6	6	5	5	7½	7½	6	6	5	5

For spacings above zig-zag line use 1/2" rods. For spacings below zig-zag line use 9/16" rods. For spacings above zig-zag line use 9/16" rods.
Psi denotes pounds per sq. in., psf denotes pounds per sq. ft. * A.C.I. 714

Concrete Stave Silo

Capacities of Storage Bins of Various Diameters—Per Foot of Height

DIAMETERS		10' Dia.	12' Dia.	14' Dia.	16' Dia.	18' Dia.	20' Dia.	22' Dia.	24' Dia.
Cubical Contents—	Cu. Ft.	78.54	113.10	153.94	201.06	254.47	314.16	380.13	452.39
Sand, Crushed Stone, etc	Cu. Yds.	2.91	4.18	5.70	7.44	9.42	11.63	14.08	16.75
Crushed Stone or Sand (100 lbs. per cu. ft.)	Tons	3.93	5.65	7.70	10.05	12.72	15.71	19.01	22.62
Coal—Anthracite (52 lbs. per cu. ft.)	Tons	2.04	2.94	4.00	5.23	6.62	8.17	9.88	11.76
Coal—Bituminous (50 lbs. per cu. ft.)	Tons	1.96	2.82	3.85	5.02	6.36	7.85	9.50	11.31
Cinders (45 lbs. per cu. ft.)	Tons	1.77	2.54	3.46	4.52	5.72	7.07	8.55	10.18
Salt (48 lbs. per cu. ft.)	Tons	1.88	2.71	3.69	4.82	6.11	7.54	9.12	10.86
Lime—Unslaked (53 lbs. per cu. ft.)	Tons	2.08	3.00	4.08	5.33	6.74	8.32	10.07	11.99
Lime—Hydrated (40 lbs. per cu. ft.)	Tons	1.57	2.62	3.08	4.02	5.09	6.28	7.60	9.05
Cement in Bulk (94 lbs. per cu. ft.)	Tons	3.69	5.32	7.23	9.45	11.96	14.76	17.87	21.26
Grain (1 bu.=1.24445 cu. ft.)	Bushels	63.11	90.88	123.70	161.56	204.48	252.45	305.46	363.52
Liquid (231 cu. in.)	Gallons	587.52	846.05	1151.55	1504.03	1903.57	2350.08	2843.57	3384.11
Interspace between 4 bins	Cu. Ft.	46.97	61.27	77.28	95.02	114.47	135.64	158.52	183.12
Interspace between 4 bins	Bushels	37.38	49.28	62.14	76.36	92.04	109.00	127.38	147.15

TO USE TABLE:—Example:

How many tons of hard coal will a bin 20' in diameter and 25' high hold—From table, 8.17 x 25'=204 tons capacity.

Reinforcing Steel.—Reinforcing steel will have to be added for each row of tile, according to size and weight of contents.

It is impossible to give the amount of reinforcing steel required, as this will vary with the height of the storage bins and the pressure exerted by the contents and should be figured separately for each job.

Number of 6"x7¾"x6¾" Double Grooved Concrete Manhole Blocks and Quantity of Mortar Required to Lay Up Manholes and Catch Basins of Various Sizes					
Depth of Manhole		Internal Diameter of Manhole			
		30"	36"	42"	48"
	No Block (Cone)	8	16	24	32
4 Ft.	No. Block (Bbl.)	35	32	27	20
	Total	43	48	51	52
	Cu. Ft. Mortar	1½	2	2½	3

6 Ft.	No. Block (Cone)	8	16	24	32
	No. Block (Bbl.)	56	56	54	50
	Total	64	72	78	82
	Cu. Ft. Mortar	2	3	4	5
8 Ft.	No. Block (Cone)	8	16	24	32
	No. Block (Bbl.)	77	80	81	80
	Total	85	96	105	112
	Cu. Ft. Mortar	2½	3½	4½	5½
10 Ft.	No Block (Cone)	8	16	24	32
	No. Block (Bbl.)	99	104	108	110
	Total	107	120	132	142
	Cu. Ft. Mortar	3	4	5	6
12 Ft.	No. Block (Cone)	8	16	24	32
	No. Block (Bbl.)	119	128	135	140
	Total	127	144	159	172
	Cu. Ft. Mortar	3½	5	6½	8
14 Ft.	No. Block (Cone)	8	16	24	32
	No. Block (Bbl.)	140	152	162	179
	Total	148	168	186	202
	Cu. Ft. Mortar	4	5½	7½	9½

Concrete Manhole and Catch Basin Block

Concrete manhole and catch basin block are radial concrete units used for building circular manhole or catch basins in storm or sanitary sewers. The best quality units meet or exceed the specifications of the American Concrete Institute.

Size and Weight of Units.—For manholes or catch basins having 48" inside diameter, 10 units are usually required for each course; for 42" 9 units; 36" 8 units, and 30" require 7 units. However, one popular type of block requires 12 units per course in 48" structures and 9 units in 36". A block 7¾" high and 6" thick weighs approximately 50 lbs. and a block 5¾" high and 6" thick weighs approximately 35 lbs.

Approximate Prices of Manhole Covers and Frames

Steel 18" Diameter	$85.00
Steel 24" Diameter	$140.00
Steel 36" Diameter	$250.00

Method of Construction—The concrete base is placed and then the first course of blocks is embedded at least 3" in the plastic concrete, leveling the units at once, as this construction insures a watertight joint at the base.

Labor Costs for Concrete Block Manholes and Catch Basins. —The estimated cost of labor for a concrete block manhole, 48" inside diameter and 8'-0" deep, is as follows :

	Hours	Rate	Total	Rate	Total
Mason	7.5	$....	$....	$18.91	$141.83
Labor	7.5	....	....	14.95	112.13
Cost per manhole			$....		$253.96

Add for excavation, backfill, concrete base and cover.

Note:"Cone" construction consists of four courses of special batter block.

CONCRETE BLOCK MANHOLE

Note:"Cone" construction consists of one course of bevel block and subsequent courses of standard units, using one less in each succeeding ring.

CONCRETE BLOCK MANHOLE

UNIT MASONRY/CLAY TILE

Tile units are furnished in many styles, sizes and shapes, for both interior and exterior, and load and non-load bearing conditions.

Much tile used today has a finished surface, others are scored for plaster and stucco finish applied by other trades. Tile is used as backup on exterior walls, as it provides better insulation than brick (a "U" factor of .26 as against .35 for face brick walls backed 8" of tile and common brick respectively), and provides a drier surface for plastering direct.

Clay and gypsum tile are used as non-bearing partitions and serve as a base for plastering direct. Clay tile partitions are preferable to gypsum units in damp locations such as around toilet rooms, whether plastered or used with finished surfaces.

Tile was long used as fireproofing for beams and columns, but is largely replaced today by plaster, although gypsum tile is often used to enclose columns.

Tile arch and book tile roof construction was once quite common but is seldom encountered today except in remodeling older structures.

LOAD BEARING TILE

Load bearing tile are used to carry part of the structural load of the building and are used for both interior and exterior walls. Some units have finished surfaces, some are scored for plaster, or other applied finishes.

The American Society for Testing and Materials (A.S.T.M.) has set up three specifications covering load bearing tile. Specification C-34-62 covers structural clay load-bearing wall tile in general and sets two grades: LBX for exposed locations; LB for units not exposed to frost or where protected with at least 3" of other masonry facing.

Tile are manufactured as end construction tile, to be placed in walls with axes of cells vertical and as side construction tile, to be placed with axes of cells horizontal. Average compressive strength of end construction tile, Grade LBX, is 1,400 p.s.i., and for Grade LB, 1,000 p.s.i. Average compressive strength for side construction tile is 700 p.s.i., for both grades, based on gross area.

A.S.T.M. Specification C-212 covers structural facing tile designed for either interior or exterior exposure, including salt glazed and unglazed units. The two grades are FTX for smoothest faces, low absorption and high degree of mechanical perfection, and FTS, which provides moderate absorption and mechanical perfection.

The compressive strengths given above for LBX grade tile apply here except that a special duty tile is available with 2,500 p.s.i. and 1,200 p.s.i. average compressive strength for end and side construction types respectively. The Facing Tile Institute's specifications list considerably higher strengths of 3,000 and 2,000 p.s.i. respectively, for "select" tiles.

A.S.T.M. Specifications C-126-62 covers structural tile with ceramic glazed surfaces. Two grades are given: "S" (select) for use with comparatively narrow joints, and "G" (Facing Tile Institute lists this as "SS") for ground edge tile. Grade S is available in all standard sizes except 7¾" x15¾" and "G" (or "SS") is available in 7¾"x15¾" units. These units also carry the designation, Types I and II, which indicates single or double finished faces. The Facing Tile Institute also lists a "B" quality ceramic glaze structural tile. These are not produced to meet a specification, but are units that may be available at the yard which fail to meet one or more of the specifications for "S" grade tile. The estimator should be aware, however, that even when "B" quality is acceptable on a job, it may be impossible to get all the special shapes required to complete the job in anything other than select quality.

INTERIOR STRUCTURAL CLAY FACING TILE

Facing tile discussed below includes those units at least one of whose faces, ceramic glazed, salt glazed or unglazed, are designed to be left exposed either as a partition, as backup to another type of facing, or as the inner wythe of a cavity wall.

These units are manufactured in accordance with sizes and specifications adopted by the Facing Tile Institute. The four sizes adopted are modular.

Facing tile are available in such a wide variety of colors and finishes as to be adaptable to most any design requirement, and the estimator must be knowledgeable about the product and grade and the desired end result.

While most facing tile is furnished in 4" thicknesses, the estimator should always check the cost of using 2" soaps in combination with a locally produced backing material against the cost of the full width tile, particularly if the wall carries no superimposed loads. Often, structural tile is used where its load bearing capacity is never utilized.

Mortar design mixes for tile follow those listed in the previous chapter for brick, except interior non-load-bearing partitions may utilize a low strength type "0" mortar which calls for .111 cu. ft. of cement and .222 cu. ft. of lime for each cubic foot of sand. Note, however, that in much tile work white portland cement and white sand may be called for. Most facing tile is figured for ¼" joints which limits sand to that passing a No. 16 sieve.

In the figures given below, mortar is figured as two separated 1" wide joints except for the bed joints of horizontal celled units which are figured as full.

Where walls are built up of two units in thickness, metal ties should be figured 16" vertically and 36" horizontally, which will work out about 25 per 100 sq. ft. in all. Also if a solid filled collar joint is desired add 2.6 cu. ft. of mortar per 100 sq. ft. of wall.

Material Requirements for Facing Tile

8W Series.
   Size: 8"x16"x2" or 4" thick
      (7¾"x15¾" actual face size)
   Tile per sq. ft. wall - 1.125
   Mortar in cu. ft.:
     For vertical cell tile:
      .77 per 100 sq. ft.
      6.87 per 1000 units
     For horizontal cell tile:
      1.23 per 100 sq. ft.
      10.92 per 1,000 units
   Cost: 2" thick - $1,980 per M, or $2.23 per sq. ft., glazed 1 side
       4" thick - $2,180 per M, or $2.45 per sq. ft., glazed 1 side
       4" thick - $3,330 per M, or $3.71 per sq. ft., glazed 2 sides
       6" thick - $3,090 per M, or $3.48 per sq. ft., glazed 1 side
       8" thick - $3,700 per M, or $4.16 per sq. ft., glazed 1 side
6T Series.
   Size: 5 1/3 x 12"x2", 4", 6" and 8" thick.
      (5 1/16" x11 3/4" actual face size)
   Tile per sq. ft. wall - 2.250
   Mortar in cu. ft.:
     For vertical cell tile:
      1.11 per 100 sq. ft.
      4.94 per 1,000 units
     For horizontal cell tile:
      1.80 per 100 sq. ft. for 4" wall
      7.98 per 1,000 units for 4" wall
      2.58 per 100 sq. ft. for 6" wall
      11.45 per 1,000 units for 6" wall
      3.36 per 100 sq. ft. for 8" wall
      14.93 per 1,000 units for 8" wall
   Cost: 2" thick - $850 per M, or $1.91 per sq. ft., glazed 1 side
       4" thick - $1,080 per M, or $2.43 per sq. ft., glazed 1 side
       4" thick - $1,500 per M, or $3.38 per sq. ft., glazed 2 sides,
       6" thick - $1,580 per M, or $3.56 per sq. ft., glazed 1 side,
       8" thick - $1,890 per M, or $4.25 per sq. ft., glazed 1 side,

6T Series Structural Facing Tile, 5-⅓"x12" Face

8W Series Structural Facing Tile,
8"x16" Face

4S Series
2-2/3"x8" Face

Special Shapes.—In addition to stretcher, or field units, these series also include many special shapes. These are given as belonging to groups; the higher the number, the more expensive the unit.

In general, the various groups include the following, although not all units will be available in all tile.

Group I :
  4" bed return, bullnose or square
  2" and 4" soap return, bullnose or square
  coved internal corner
  octagonal external corner
  octagonal interior corner
  cost: 8W—$3,150.00 per M
    6T —$1,390.00 per M

Group II :
  2" and 4" bed, bullnose cap
  2" bed, cove base
  cost: 8W—$3,620.00 per M
    6T —$1,610.00 per M

Group III :
  6" and 8" bed, bullnose cap
  4" bed, cove base
  2" bed, round top cove base
  4" and 8" bed, full and half ends glazed both sides
  4" bed, bullnose cap, glazed both sides
  4" bed, sloped sill
  Standard radial external or internal field units:
  2" and 4" bed, 6 3/8" high, bullnose cap unit
  2" and 4" bed, 7¾" high, bullnose cap unit
  cost: 8W—$5,130 per M
    6T —$2,540 per M

Group IV :
  4" bed, miter unit
  4" bed, miter unit with top return
  4" bed, miter unit with finished end
  2" and 4" bed, bullnose return
  4" bed, bullnose return with starter

2" and 4" bed, cove base with return
4" bed, round top cove base
6" bed, bullnose glazed both sides
6" and 8" bed, sloped sill
4" bed, bullnose starter
2" bed, bullnose reveal
2" bed, bullnose coved internal corner
2" bed, bullnose octagonal internal and external corners
4" bed, 6 3/8" high slope sill
4" bed, 7¾" high slope sill
cost: 8W—$8,180.00 per M
     6T —$5,340.00 per M
Group V :
  4" bed, bullnose with 4" return
  4" bed, miter 7¾" high with face return
  4" bed, round top cove base with 4" return
  2" bed, round top cove base with 4" return
  2" bed, bullnose radial
  2" bed, bullnose cove base radial
  4" bed, bullnose internal corner with 4" reveal
  6" bed, 6 3/8" high slope sill
  2" bed, round top cove base with square corner
  2" bed, round top cove base with coved interior corner
  2" bed, round top cove base with octagonal exterior corner
  2" bed, round top cove base with octagonal internal corner
  4" bed, field radial starter
  cost: 8W—$11,790 per M
       6T —$5,730 per M
Group VI :
  4" bed, bullnose full end, or half end glazed both sides
  8" bed, bullnose half end, glazed both sides
    8W—$14,580 per M
    6T —$7,070 per M
Group VII :
  4" bed, bullnose radial starter
  4" bed, cove base radial starter

SCR "Acoustile"*

In addition to standard facing tile, a glazed facing tile unit with a perforated face shell and the adjacent cell filled at the factory with fiberglass has been developed by the Brick Institute of America to combine high sound absorption with the advantages of ceramic finish and a structural wall material. These units are produced by several manufacturers with a wide selection of colors, types of slots and in both 6T and 8W series, of facing tile. Thickness is 4" and, upon special order, both sides can be perforated.

Not only does this tile allow acoustical control in areas where ceramic veneer is desired, but it opens possibilities of adding acoustical control to other wall areas where less durable materials would be damaged by wear and tear or would lose much of their value because of the need of constant repainting.

*Reg. TM, US Pat. Off., Patent No. 3,001,602

Acoustile walls provide mass so as to reduce sound transmission as well as provide sound absorption, and the combination of fiberglass and open cells provide a barrier for heat transmission. These units are available as field or stretcher units only, and are used in connection with special shapes of standard series ceramic facing block. These are horizontal celled units and requirements given above will apply here also. Cost is $3,510 per M in 8W series, $1,990 per M in 6T series.

SCR "Acoustile"*

Sculptured Ceramic Tile

Certain manufacturers produce units with cast in designs and profiles. These are usually on special order, but may have some application in lobbies or meeting rooms. Costs run around $1.75 per unit in the 6T series.

Estimating Facing Tile Quantities

The most widely used estimating procedure is the wall area method, as follows :
1. Compute gross wall area (height times length). Where coved base and/or bullnosed caps are specified, do not figure these in overall height.
2. Determine net area of wall; deduct wall openings. These are computed by using the actual width multiplied by the actual height plus the height of special sill and lintel units. Where openings overlap base or cap units, do not include this in the opening height as it was not included in the gross area. Also deduct for other special shapes by multiplying their height by the following factors :
   For exterior bullnose or square corners—1.167 per ft. of height.
   For Interior coved corners—.833 per ft. of height.
   For bullnose or square jambs—.75 per ft. of height.
3. Determine number of stretchers required by multiplying net area by number of stretchers required per square foot.
   30,58W : 1.12
      6T : 2.25
      4D : 3.37
4. Compute number of standard fittings such as corners and jambs by the number of courses. Cove bases and caps, which should now be included as well as sills and lintels, are computed by the linear foot. Miters, cove base corners, etc., should be taken off individually.
5. Summarize quantities and separate fittings in particular group.

---

*Reg. TM, US Pat. Off., Patent No. 3,001,602

6.   Multiply by the unit cost figures.
7.   Add 2% (or more, depending on job conditions) to cover breakage.
    Example:

Step 1
Wall length = 29'-0"
Wall height, exclusive of the base course = 8'-0"
Gross area = (8.00) (29) =232 sq. ft.

Step 2
Deductions:

Window area (4) (4.89)	= 19.56 sq. ft.
Door Area (2) (6.67)	= 13.33 sq. ft.
Exterior Corner (1.167) (8.00)	= 9.33 sq. ft.
Interior Coved Corner (0.833) (8.00)	= 6.66 sq. ft.
Jambs (4+4+6.22+6.22) (0.75)	= 15.33 sq. ft.
Total	= 64.21 sq. ft.
Net wall area = 232 minus 64.21	= 167.79 sq. ft.

Step 3
Number of "6T" stretchers required (167.79) (2.25) = 378

Step 4

	Group				
	I	II	III	IV	V
Exterior Bullnose corner					
18 courses of 5T4	18				
Interior coved corner					
18 courses of 4T8	18				
Jambs					
24 courses of 6T4	24				
24 courses of 3T4	24				
Lintels					
5 lin. ft. of 6T20		5			
1 miter 6T30R (door)				1	
1 miter 6T30L (door)				1	
1 miter 8T31R (window)					1
1 miter 8T31L (window)					1
Sill					
3 lin. ft. of 6T20		3			
1 miter 8T31R					1
1 miter 8T31L					1
Cove Base					
25 lin. ft. of 6T50			25		
1 outside corner 5T54L				1	
1 inside cove corner 4T58L				1	
1 bullnose jamb 6T504R				1	
1 bullnose jamb 6T504L				1	
Totals	84	8	25	6	4

Step 5— Summary

Group	Series	Quantity
Stretcher	6T	378
I	6T	84
II	6T	8
III	6T	25
IV	6T	6
V	6T	4

Example of Wall Area Method for Determining Facing Tile Quantites

Add an arbitrary percentage to the quantities taken off for breakage, to vary between 2 and 5% according to job conditions as well as distance of job from the plant. Chippage can frequently be used for special short lengths or other cuts.

A short cut method, devised by Stark Ceramics, is based on the following formula:

(Length x height minus actual openings) x (cost per sq. ft.) + (lineal feet x lineal foot cost factor) = (total material cost).

In using this method, actual total heights are used to figure cross areas and openings are figured using actual height and width with no allowances for sills, lintels, caps or bases. Lineal footage includes all special shapes : cove bases, internal coved corners, jambs, sills, lintels, caps, etc. The lineal foot cost factor is an average additional cost to cover these items. It is determined by taking an average of the group costs and deducting the stretcher unit cost already included in the net area figure.

Labor Laying Glazed Structural Facing Tile.—To insure the most economical construction, as well as the most attractive appearance in the finished work, all cutting of tile on the job should be done with a power saw using a carborundum blade.

The following labor quantities are based on the average job requiring not more than 20 percent of special shapes, such as bullnose corners, cove base, cap, etc. If the percentage of special pieces vary from the above, increase or decrease labor time proportionately.

Labor Laying Glazed Structural Facing Tile

Nominal Size of Tile*	Class of Work	No. Pcs. Set per 8-Hr. Day	Labor Hrs. per 100 Pcs. Mason	Labor
2"x5⅓"x 8"	Glazed one side	155-175	4.8	4.8
4"x5⅓"x 8"	Glazed one side	155-175	4.8	4.8
4"x5⅓"x 8"	Glazed two sides	125-155	5.7	5.7
6"x5⅓"x 8"	Glazed one side	155-175	4.8	4.8
6"x5⅓"x 8"	Glazed two sides	125-155	5.7	5.7
8"x5⅓"x 8"	Glazed one side	125-155	5.7	5.7
8"x5⅓"x 8"	Glazed two sides	105-120	7.0	7.0
2"x5⅓"x12"	Glazed one side	115-135	6.4	6.4
4"x5⅓"x12"	Glazed one side	115-135	6.4	6.4
4"x5⅓"x12"	Glazed two sides	90-110	8.0	8.0
6"x5⅓"x12"	Glazed one side	105-125	7.0	7.0
6"x5⅓"x12"	Glazed two sides	80-100	8.9	8.9
8"x5⅓"x12"	Glazed one side	85-100	8.7	8.7
8"x5⅓"x12"	Glazed two sides	65-85	10.6	10.6
2"x8 "x16"	Glazed one side	65-80	11.0	11.0
4"x8 "x16"	Glazed one side	55-65	13.3	13.3

Add for hoisting engineer if required.
*Nominal size includes thickness of standard mortar joint (¼" for glazed tile) in length, height and thickness.

Tile 1¾" thick is usually used for wall furring or partition wainscoting. Tile glazed one side is usually used for partitions that are glazed on one side and plastered on the other side. Tile glazed two sides are usually used for partition walls glazed on both sides of wall.

Unglazed Structural Clay Tile

Backup tile are sized for backing up brickwork, but are also used with stone and other facing. These units are available with nominal face sizes of 5"x12", 8"x12", and 12"x12", in thickness of 4", 6", 8" and 12", and are usually furnished scored one side, smooth on the other. These units are often available, especially in larger sizes, in special designs with cast in handles for greater speed and ease in handling and non-continuous mortar joints to protect against moisture penetration. An example of this is the Speed-A-Backer tile, Kwiklay and Denison backup tile, patented units produced by many local manufacturers. These all come with special header and closure units. Generally, backup tile is sized to have ½" joints, which is the mortar amount given below, but certain units will be found to allow for less. Dimensional tolerances are wider in this type than in facing tile. The mortar figures below assume joints consist of two separated 1" wide joints and bed joints are two separated 2" wide joints.

Nominal 5"x12" tile :
4⅞"x11½"x3½", 5½", 7½" and 11½" actual
Tile per sq. ft. wall—2.25
Mortar—per 100 sq. ft. wall—3.75 cu. ft.
—per 1,000 units —16.69 cu. ft.

Nominal 8"x12" tile :
7½"x11½"x3½", 5½", 7½" and 11½" actual
Tile per sq. ft. wall—1.5
Mortar—per 100 sq. ft. wall—2.73
—per 1,000 units —18.23

8" S.A.B. Stretcher   12" S.A.B. Stretcher   12" Full Closure

12" Wall Section   16" Wall Section   TYPICAL JAMB DETAIL—12" WALL

Speed-A-Backer Tile

Labor Laying Backup Tile.—The following quantities are based on the average job, having the usual number of openings, pilasters, etc.

Size Tile	No. Pcs. Laid per 8-Hr. Day	Labor Hrs. 100 Pcs. Mason	Labor	Engr.*	No. Sq. Ft. Laid per 8-Hr. Day	Labor Hrs. 100 Sq. Ft. Mason	Labor	Engr.*
4"x5"x12"	330-370	2.25	2.25	0.3	150-170	5.0	5.0	0.6
8"x5"x12"	250-290	3.00	3.00	0.3	115-130	6.5	6.5	0.7
4"x8"x12"	200-225	3.80	3.80	0.3	135-150	5.7	5.7	0.6
8"x8"x12"	155-185	4.70	4.70	0.5	110-132	6.6	6.6	0.7

*Add for hoisting engineer only if required.

Unit Wall Tile

Some tile units are designed to be used for a complete single unit, load bearing, nominal 8" wall construction, with finished surfaces inside and out. These are usually manufactured under special patents and include such products as Uniwall, Dri-Speed wall and Tex Dri-wall, all manufactured by Glen-Gery Corporation, and certain units produced by local manufacturers under patents and known as Kwiklay. Some of these units may have one finished surface only and be used in combination with an inside backup.

Faces are available in buff unglazed, either smooth or Ruggtex finish; salt glazed, red textured (to give the appearance of high quality face brick); and ceramic glaze. Certain units are offered with supplemental full and half jamb corners, and lintel and pilaster units along with interior stretcher units to match the interior face of the unit tile.

DENISON BACK UP TILE

8816 SERIES — 8" x 16" face — Scored one side, smooth one side

128-16-BU   8816-BU   6816-BU   4816-BU

88 SERIES — 8" x 12" face — Scored one side, smooth one side

128-BU   88-BU   68-BU   48-BU

HEADER UNITS

128-16-HBU or   8816-HBU   6816-HBU   Closure unit
128-HBU   or 88-HBU   or 68-HBU   88-CBU
                                                    Smooth both sides,
                                                    both ends

85 SERIES — 5" x 12" face — Scored one side, smooth one side

105-BU   85-BU   65-BU   45-BU

85-BU FLATBED   Closure Unit
                            85-CBU
                            Smooth both sides

ESTIMATING DATA

8816 Series — 1.1 units/sq. ft.
88   Series — 1.5 units/sq. ft.
85   Series — 2.25 units/sq. ft.

Units may be scored two sides or smooth two sides if desired.

DENISON STRUCTURAL CLAY TILE
MASON CITY BRICK AND TILE COMPANY, MASON CITY, IOWA

Units are also available in 4" widths for use as exterior wythes in cavity wall construction.

A wall of this type is highly impervious to water, thoroughly fire, termite and vermin proof, and cannot rot or decay. Textures offer a broad range including brick-like textures and colors suitable for homes and schools; glazed for situations where a sanitary finish is desired, and smooth faced units in natural range shades suitable for larger structures and areas.

SS–8512	SI25DI2	S45KI2	S45DI2
SS–8512 (GAUGED)	8" DRIP-BOND	4" BOND	4" DRIP JOIST
8" STRETCHER	STRETCHER	STRETCHER	STRETCHER

UNIVERSAL FITTING   INTERIOR WALL STRETCHER UNITS

| | 4" BED | 2" BED | 6" BED |

| S45 | 6TC SU | 6TCA | 6TC60 SU |
| | 6TCD (SELECT 2 FACES) | | |

SUPPLEMENTAL SQUARE AND BULLNOSE JAMBS, CORNER, SILL, CAP, LINTEL SHAPES AND FITTINGS, ALSO BULLNOSE STARTERS, CAP AND COPING CORNERS AVAILABLE FOR COMPLETE INSTALLATION AS REQUIRED.

12" BRICK & TILE WALL   12" TILE WALL   8" TILE WALL

Glen-Gery Dri-Speedwall Tile

GLEN-GERY TEX DRI-WALL TILE
(NOMINAL 4 X 12 FACE SIZE)

TEX 80

TEX 40

TEX 45

WEEP HOLE

The estimator must know exactly the quality of interior finish desired. While it is possible to obtain high grade finishes and workmanship both sides, such high quality is not always required, especially in larger warehouse and manufacturing buildings, and large economies can result where "select one face" units can be used. Often the lower wall may be constructed "select two faces" while "select one face" is used above. In general, though, the estimator must expect a variation of ⅛" in thickness and, if this is not acceptable, he should plan to use a built-up wall with a separate interior facing.

Unit wall tile comes in two nominal face sizes : 4"x12", and 5⅓"x12", in 4" and 8" widths.

4"x12" units, which include Uniwall and Tex Dri-wall, have an actual face dimension of 3⅝" x 11⅝" with 3⅝" and 7⅝" thicknesses. These units will lay up 300 per 100 sq. ft. of wall and require 7.5 cu. ft. of mortar for 4" walls; 9.75 cu. ft. for 8" walls.

Labor laying 4"x12" unit wall tile :

		Labor Hours per 100 Pieces			Labor Hours per 100 Sq. Ft.	
Size of Tile	No. Laid per 8-Hr. Day	Mason	Labor	No. Sq. Ft. 8-Hr. Day	Mason	Labor
3⅝"x3⅝"x11⅝"	170-200	4.33	4.33	57-67	13.0	13.0
7¾"x3⅝"x11⅝"	150-170	5.0	5.0	50-57	15.0	15.0

5½"x12" units, which include Kwiklay and Dri-Speedwall, have actual face dimensions of 5"x11⅝" with 3⅝" and 7⅝" thicknesses, although Kwiklay may vary somewhat in various parts of the country from 4⅞" to 5 1/16" in height and 11½" to 11¾" in length. Units can be figured to lay up 225 tile per 100 sq. ft. of wall and require 8.5 cu. ft. of mortar for 4" walls and 9.25 for 8" walls, the latter varying as to actual face size.

Labor laying 5 1/3"x12" unit tile :

Size of Tile	No. Laid per 8-hr. Day	Labor Hours per 100 Pieces Mason	Labor	No. Sq. Ft. per 8-Hr. Day	Labor Hours per 100 Sq. Ft. Mason	Labor
3⅝"x5"x11⅝"	175-200	4.25	1.90	80-90	9.50	4.12
7⅝"x5"x11⅝"	130-155	5.50	2.50	60-70	12.25	5.50

Clay Masonry Shading Devices

Architects have been experimenting with sun controls beyond the usual shades, blinds, and drapes. An effective way of sun control is the egg crate screen wall design which baffles both the low east and west rays of the sun as well as the high direct south rays.

Clay masonry units are available to construct such baffles in both glazed and unglazed finishes and in a large range of sizes, shapes and designs. Selection will depend on the degree of shading desired, the wall thickness necessary to provide structural stability and aesthetic considerations. Glazed solar screen units 4" thick will cost around $5.50 per unit.

C-105   C-107

C-106   C-108

FOUR PATTERNS OF SOLAR SCREEN TILE

Size: 7 5/8 x 7 5/8 for 3/8" mortar joints   Thickness: 3 5/8", 5 1/16" and 7 3/4"

UNIT MASONRY/CERAMIC VENEER

Terra cotta is a building material made of a high grade clay (usually a mixture of several different kinds), to which is added a percentage of calcined clay to uniformly control the shrinkage and prevent undue warping.

Terra cotta is used for the same purposes as other building materials, like sandstone, limestone, granite, press brick, etc.; i.e., for interior or exterior trimmings or for entire fronts. It is also used extensively as glazed terra cotta for light courts, for the interior walls of swimming pools, operating rooms in hospitals, clinics and patients' rooms, corridors, vestibules and stair halls (full height of wainscot) and in large kitchens where a ceramic material is required for sanitary purposes as well as for fire resisting advantages. It is used for service stations, power houses and rooms housing machinery on account of the ease with which it can be kept free from dust, dirt and grease. Its unlimited possibilities in color effects make it an ideal material for store fronts. It is used as a lining for the interior of restaurants, palm courts, hotels, residences and railroad stations. In stations and public buildings it is also used in the construction of ceilings and domes. In fact, it is a universal building material and has the advantage over other materials in that it may be produced in almost any color and may be glazed or unglazed with any surface treatment desired. It furnishes unlimited possibilities to the architect not only in color effect, but in the beauty of the original modeling in the plastic clay, an effect that cannot be obtained in carved stone except at tremendous cost.

Method of Manufacturing Architectural Terra Cotta.—After shop drawings have been approved, full size models are made of plaster to a shrinkage scale; to an increased size to allow for the shrinkage in drying and burning. A model is made for each different shape on the shop drawings. If a piece is ornamental, such ornament as may be required is modeled in clay and attached to the plaster model. From the models, sectional molds of plaster are cast, from which the required number of pieces of terra cotta are produced.

The mixture of clays and fusible materials used in forming terra cotta is carefully selected and proportioned to give the desired degree of plasticity when mixed with water. The composition must be such that when fired at high temperatures, it will produce a homogeneous body amply strong to support the required loads.

Forming terra cotta is a manual operation and consists of pressing the soft clay mixture into the molds. The walls of the pieces should be not less than one inch thick following the contour of the mold and the partitions should be properly sized and spaced to produce the strength required. The pressed piece remains in the mold until the clay stiffens, after which it is removed, retouched and placed in the drier where most of the moisture in the clay is removed.

Following the drying process the exposed surfaces of each piece are sprayed with the ceramic mixture which, when fired, develops into the desired glaze color. The pieces are then placed in kilns and fired by gradually increasing the temperature to 2,000°F or more, depending on the temperature of maturity of the clay and glaze. After proper firing the kiln is allowed to cool slowly to normal temperature.

After firing, the terra cotta is fitted for alignment by cutting or grinding and marked to correspond with the piece numbers on the shop drawings.

Method of Manufacturing Ceramic Veneer.—Ceramic veneer is the usual term applied to machine made architectural terra cotta and the process of manufacturing is the same as for handmade pieces except that the material is extruded instead of hand pressed into molds.

A machine mixes the proportioned clay with water and forces the plastic mixture through a steel die which forms it into the desired shape. In most cases, before entering the die, the plastic mixture passes through a de-airing chamber where it is subjected to a partial vacuum and most of the entrapped air is removed. The pieces are then dried and surface finished to a true plane.

Subsequent operations are the same as for handmade pieces and include application of ceramic mixture, firing, fitting and marking.

Cost of Architectural Terra Cotta and Ceramic Veneer.—In estimating the cost of architectural terra cotta and ceramic veneer there is no unit that may be taken as a definite basis for estimating. All estimates are usually made by the manufacturer in a lump sum price for the entire job.

To be able to accurately arrive at the cost of these materials, an estimator must have an intimate knowledge of the manufacturing process. Each job is a custom job, there are no standard sizes, colors or shapes which are stocked.

One manufacturer suggests that the cost of architectural terra cotta will vary from $350.00 to $500.00 and higher per ton and that the cost of ceramic veneer will run from $2.50 to $4.00 and higher per sq. ft. The extreme range of these costs serves to emphasize the necessity for obtaining firm quotations from the manufacturers.

Some of the conditions governing the costs of both architectural terra cotta and ceramic veneer are design of work, character and amount of ornamentation, amount of duplication, sizes of pieces, color of glaze, weight, supply and demand, shipping conditions, etc.

Estimating Quantities of Architectural Terra Cotta and Ceramic Veneer.—When estimating quantities of terra cotta from the plans or details, all measurements should be squared ; i.e., all molded courses, cornices, round columns, column caps, bases, coursed ashlar, sills, copings, etc., should be figured from the extreme dimensions and reduced to cubic feet.

The weight of terra cotta varies from about 80 lbs. per cu. ft. for plain 4" thick ashlar to 60 lbs. per cu ft. for large molded courses. For an average job about 72 lbs. per cu. ft., equivalent to 28 cu. ft. per ton of 2,000 lbs. may be used.

After the quantities have been taken off the plans and reduced to cu. ft., they may either be priced in that manner, or multiplying the number of cu. ft. by 72 lbs. will give the total weight of the terra cotta required for the job.

Ceramic veneer quantities are measured by the square foot area and should include all returns at doors, windows, soffits, etc.

Cost of Handling and Setting Architectural Terra Cotta.—The labor cost of handling and setting architectural terra cotta will vary with the sizes of the different pieces, the amount of duplication and the method used in handling and setting same.

Terra cotta weighs about one-half as much as stone, so that it is seldom necessary to use derricks in handling and setting, as it is usually furnished in sizes that can be conveniently handled by 2 or 3 men. Considerable space is usually required on the job for storing the terra cotta, as it is sent to the job each piece is marked to show its exact position in the building and must be carefully handled and placed in convenient piles for use as required. When handling the terra cotta on the job it is customary to load it into wheelbarrows or stone barrows and wheel it to the location where it is to be used. However, if it is necessary to hoist it to the upper floors or roof of the building the barrows can be placed on the material hoist, raised to the proper floor and then wheeled to the location where it is to be set.

Inasmuch as a great deal of terra cotta consists of trimmings or 4" ashlar, it requires considerable handling to set a cubic foot. For instance, most ashlar being 4" thick, it is necessary to set terra cotta in 3 sq. ft. of wall to set one cu. ft. This also applies to sills, lintels, etc.

Based on the average job consisting of ashlar, sills, lintels, ornamental trimmings, etc., a mason will set about 35 to 40 cu. ft. (105 to 120 sq. ft. wall) per 8-hr. day, at the following labor cost per 100 cu. ft.:

	Hours	Rate	Total	Rate	Total
Terra cotta setter	21	$. . . .	$. . . .	$17.87	$ 375.27
Helper	21	. . . .	. . . .	14.95	313.95
Labor sorting, handling wheeling, etc.	44	. . . .	. . . .	14.95	657.80
Cost per 100 cu. ft.			$. . . .		$1,347.02
Cost per cu. ft.				. . . .	13.47
Cost per sq. ft (300)				. . . .	4.49
Cost per ton (28 cu. ft.)				. . . .	377.17

For labor cost of setting plain terra cotta wall ashlar, use same quantities and costs as given on the previous pages under "Labor Laying Glazed Structural Facing Tile."

On more complicated work consisting of gothic architecture, balusters and balustrade, glazed and enameled terra cotta, and other intricate work having a great deal of detail and requiring the utmost care in setting, a terra cotta setter will set only 20 to 25 cu. ft. (60 to 75 sq. ft. wall) per 8-hr. day, at the following labor cost per 100 cu. ft.:

	Hours	Rate	Total	Rate	Total
Terra cotta setter	36	$. . . .	$. . . .	$17.87	$ 643.32
Helper	36	. . . .	. . . .	14.95	538.20
Labor sorting, handling, wheeling, etc.	66	. . . .	. . . .	14.95	986.70
Cost per 100 cu. ft.			$. . . .		$2,168.22
Cost per cu. ft.				. . . .	21.68
Cost per sq. ft. (300)				. . . .	7.23
Cost per ton (28 cu. ft.)				. . . .	607.10

Small lintels and similar courses that have to be hung, may cost 25 to 50 percent higher than given above but these items are exceptional.

In general, the voids in architectural terra cotta should be completely filled with setting mortar or grout to prevent the possible accumulation of moisture in the cavities.

All supporting iron should be firmly embedded in masonry and completely covered with mortar to prevent rusting. Bronze or other non-corrosive metal or cadmium coated steel anchors should invariably be used for suspended features or for anchoring or bracing freestanding work.

Mortar Required for Setting Architectural Terra Cotta.—Under average conditions it will require about 4 cu. ft. of mortar to set one ton of terra cotta, or about ½-cu. yd. of mortar per 100 cu. ft.

The following proportions are recommended as being suitable for making satisfactory mortar for setting terra cotta : One part standard portland cement, one part lime putty and six parts sand.

Do not under any circumstances use plaster of Paris or salt in the mortar.

All exposed joints in projecting and overhanging features, parapet work, and work standing free, should be raked out and pointed with a good asphaltic elastic cement.

Never use a cement that swells in setting for filling the voids in terra cotta. Use only the best quality of cement.

Handling and Setting Ceramic Veneer.—Ceramic veneer is available in two types, namely, adhesion type, commonly called "thin" ceramic veneer, and anchored type.

Adhesion Type Ceramic Veneer.—Adhesion type ceramic veneer is not over 1⅛" in thickness and the maximum face areas of individual slabs do not exceed 540 sq. in. Maximum overall face dimensions are 18"x30" or 20"x27". Actual face dimensions will vary with the manufacturer.

This type of ceramic veneer requires no metal anchorage. It is held in place by the adhesion of the mortar to the veneer body and the backing wall. The overall thickness from the face of the veneer to the face of the backing wall is 1¾" to 2".

Adhesion type ceramic veneer may be applied to a variety of backings, such as concrete, masonry, wood and metal lath. Where the backing is wood, a sheet of waterproof building paper should be installed first, followed by a layer of wire mesh. In the case of metal lath backing or wire mesh on wood backing, 9¼" scratch coat of mortar should be applied in advance of the setting of veneer slabs.

Mortar Required for Adhesion Type Ceramic Veneer.—Under average conditions, it will require 8 to 10 cu. ft. of mortar to set 100 sq. ft. of adhesion type ceramic veneer.

The following proportions are recommended as being suitable for making satisfactory mortar for setting adhesion type ceramic veneer: Two parts portland cement, one part lime putty and eight parts sand. A stearate type admix may be added to the mix in the proportions recommended by the manufacturer.

Setting Adhesion Type Ceramic Veneer.—Strict compliance with the manufacturer's recommendations for setting adhesion type ceramic veneer is necessary to assure the development of full shearing strength of veneer to backing. Step-by-step setting directions, recommended by manufacturers are as follows:

BRICK   CONCRETE

METAL   WOOD

All units must be soaked in clean water for at least one hour before setting. The backing wall surface should also be damp at the time of setting. Immediately before setting, the backing wall and the backs of veneer units should both receive a brush coat of portland cement and water. One-half of the setting mortar should be spread on the back of the unit to be set and the other half on the wall surface to be covered. The unit is then set in place, using wood wedges to maintain correct bed joint thickness. Sufficient mortar should be used to create a slight excess which will be forced out of the joints from the back when the unit is tapped into place, thus

Method of Setting Adhesion Type Ceramic Veneer

eliminating all air pockets and filling all voids. Face joints may be pointed as soon as mortar has set sufficiently to support the veneer slabs. As a final step, the surface of the veneer should be washed down with clean water.

Cost of Handling and Setting Adhesion Type Ceramic Veneer.—On most jobs adhesion type ceramic veneer is delivered to the job in trucks and must be carefully unloaded, stored and protected in advance of the setting. Two laborers should do this work at the rate of 360 to 400 sq. ft. of veneer area per 8-hr. day.

Labor Cost Handling and Setting 100 Sq. Ft. Adhesion Type
Ceramic Veneer on Concrete or Masonry Backing

Unloading and Storing	Hours	Rate	Total	Rate	Total
Labor	4.2	$. . . .	$. . . .	$14.95	$ 62.79
Setting					
Mason	11.5	. . . .	. . . .	17.87	205.51
Helper	11.5	. . . .	. . . .	14.95	171.93
Cost per 100 sq. ft			$. . . .		$440.23
Cost per sq. ft			. . . .		4.40

On the average job, where most of the veneer consists of typical panels, a mason and a helper should set 65 to 75 sq.ft. of veneer per 8-hr. day.

If veneer is to be applied to a backing of metal lath, add cost of ¼" scratch coat of mortar.

If veneer is to be applied to a wood backing, add cost of one layer of waterproof building paper, one layer of wire mesh and ¼" scratch coat of mortar.

Above costs do not include any allowance for scaffolding or hoisting. If hoisting is required, add 1-hr. hoisting engineer and 2-hrs. labor per 100 sq. ft. of veneer.

Anchored Type Ceramic Veneer.—Anchored type ceramic veneer is recommended where a larger size slab is desired. Anchored type slabs are available in face sizes considerably larger than the adhesion type, again depending upon the manufacturer. Ribs or scoring are provided on the backs of such units and the overall thickness of the slabs range from 2" to 2½". Depending on slab thickness, a total of 3" to 4½" is required from rough wall to finished veneer surface to provide adequate grout space between the veneer and the backing. Anchor holes are provided in the bed edges of the slabs for the installation of loose wire anchors which, in turn, are fastened to pencil rods anchored to the backing. Once the wire anchoring is in place, the units are bonded to the wall by a poured grout core.

Grout Required for Anchored Type Ceramic Veneer.—For ordinary work, it will require 20 to 25 cu. ft. of grout to fill the space back of 100 sq. ft. of anchored type ceramic veneer.

The following proportions are recommended for a suitable grout mix: One part portland cement, one part sand and five parts graded pea gravel passing a ¼" sieve. Sufficient water should be added to cause the mixture to flow readily.

In addition to the above grout, it may require 1 to 2 cu. ft. of mortar per 100 sq. ft. of veneer for buttering the joints in setting to prevent leakage of grout. Any workable mortar mix may be used.

Setting Anchored Type Ceramic Veneer.—The following directions for setting anchored type ceramic veneer are recommended by the manufacturers. As in the case of adhesion type veneer, both units and backing wall should be thoroughly soaked with clean water prior to setting.

Assuming the backing wall has had loop tie inserts or eye bolts properly located, the first step in veneer slab erection is the insertion of ¼" diameter pencil rods into the loops projecting from the backing wall. The veneer slabs are then set in place and secured to the pencil rods with wire anchors inserted in the top of each unit. Wood wedges should be used to maintain the proper joint widths and the joints

Illustrating Steps in Setting Anchored Type Ceramic Veneer

may be buttered with mortar or caulked with rope yarn or other material to prevent grout leakage. Temporary wood wedges should also be used to hold each piece in place at the required distance from the backing wall.

When the setting of each horizontal row of units is completed, grout is poured into the space between the veneer and the backing, puddled thoroughly and allowed to set. Wedges may be removed and joints pointed when grout has set sufficiently to hold slabs in place. To complete the operation the finished veneer surface should be washed down with clean water.

In structural concrete walls, wire loop dowel anchors are cast in place 1'-0" on vertical centers and not more than 1'-6" on horizontal centers.

1"x6" horizontal groove formed in concrete walls and piers not more than 2'-6" on center. (1'-0" on center in spandrel beams)

1/4" diameter pencil rods placed through eye bolts or loop dowels so as to be held out from wall at least 1"

No. 6 galvanized wire anchors hooked around pencil rods and into holes in edges of CERAMIC VENEER.

Space between wall and CERAMIC VENEER filled with grout.

In BRICK, TILE or other types of unit masonry walls, eye bolts are set in place by the mason, not more than 1'-0" on vertical centers or 1'-6" on horizontal centers.

Method of Setting and Securing Anchored Type Ceramic Veneer Terra Cotta to Brick and Concrete Backing

Cost of Handling and Setting Anchored Type Ceramic Veneer.—Unloading, storing and protecting anchored type veneer slabs is carried on in the same manner as for adhesion type slabs except that these units, being larger and heavier, are somewhat more difficult to handle. Two laborers should unload and store 320 to 360 sq. ft. anchored type slabs per 8-hr. day.

Setting anchored type veneer slabs is dependent on the size and weight of slabs, nature of design, etc., but on an average facing job a mason and a helper should set, grout and point 60 to 70 sq. ft. of veneer per 8-hr. day.

Labor Cost Handling and Setting 100 Sq. Ft. Anchored Type
Ceramic Veneer

Unloading and Storing	Hours	Rate	Total	Rate	Total
Labor	4.7	$....	$....	$14.95	$ 70.27
Setting					
Mason	12.3	....	....	17.87	219.80
Helper	12.3	....	....	14.95	183.89
Cost per 100 sq. ft			$....		$473.96
Cost per sq. ft				....	4.74

The above costs do not include any allowance for scaffolding or hoisting. If hoisting is required, add 1.5 hrs. hoisting engineer and 3 hrs. labor per 100 sq. ft. of veneer.

Cleaning Down or Washing Terra Cotta.—Broadly speaking, there are three different surface finishes in terra cotta; lustrous glazed, matt glazed and unglazed. The glazed surfaces are the easiest to clean.

For both lustrous and matt glazed finishes a good abrasive soap or washing powder is best. Where required, there can be added to the latter a slight proportion of sharp sand. A stiff lather should be made and the surface scrubbed hard, allowing the lather to remain long enough to soften the dirt and then rinse off with clean water. Acid is not necessary for cleaning glazed surfaces.

In cleaning unglazed terra cotta the addition of a slight proportion of commercial muriatic acid is recommended. The proportion of acid to water may be varied. One quart of acid to 4 gallons of water is sufficient in ordinary cases. In no event should it exceed 1½ pints of acid to one gallon of water and care should be taken to rinse off shortly after applying.

Where acid is employed use only acid-resistant pails and fiber brushes. Metal pails cause a chemical reaction under which the solution becomes a yellow stain instead of a cleaning fluid. Do not permit hydrofluoric acid to be used. Many cleaners in order to save hand labor employ strong solutions of this which are highly injurious.

Pointing Terra Cotta.—Recommended practice today is to strike the joints as the work progresses. Pointing mortar applied afterward is usually of different composition and accordingly of questionable permanency.

Cold Weather Protection.—Architectural terra cotta or ceramic veneer should not be set in freezing temperatures or when freezing can be expected within 30 days after setting, without providing enclosures and temporary heat.

GLASS BUILDING BLOCKS

Glass blocks are hollow, partially evacuated units of clear or colored glass. They have been engineered to serve several functions, all providing good thermal and sound insulating values.

Clear and other non-light directing units are available in 5¾", 8" and 12" square units, all 3⅞" thick. These units give high light transmission and are available in clear glass for visibility, wide fluted patterns, and wavy, translucent surfaces. These units are often used in partitions and do not screen out sunlight.

Light directing units are engineered for use on sun exposures of rooms having finished ceilings to provide efficient reflection where such indirect daylighting is desired, such as in schools and offices. The basic size is 7¾" x 7¾" x 3⅞". These units can also be had with a white or green fibrous insert to reduce both glare and solar heat gain and are also available in 11¾" square size. The insert reduces the overall light transmission.

A	B	C
36"	4' 4"	14¾"
29¾"	4' 3"	14¾"
27¾"	4' 3"	11¾"
25¾"	4' 3"	11¾"
23¾"	4' 3"	11¾"
21¾"	3' 6"	7¾"
20¾"	3' 6"	7¾"
19¾"	3' 6"	7¾"
18¾"	3' 0"	7¾"
17¾"	3' 0"	7¾"

MAXIMUM SIZES
AVAILABLE IN
ANCHORED TYPE
CERAMIC VENEER

A	B	C
5¾"	2' 0"	3¾"
7¾"	2' 0"	3¾"
11¾"	2' 0"	3¾"
1' 2¾"	2' 2"	3¾"
1' 3¾"	2' 2"	3¾"

MAXIMUM SIZES
AVAILABLE IN
ADHESION TYPE
CERAMIC VENEER

Light diffusing units diffuse light in all directions and give maximum light transmission. Size is 7¾" square. These units are also available with sun control inserts.

Color glass units have a ceramic coating fused to a single face of each block. They are used for decoration, often as accents with other blocks, and are available in 5¾" and 7¾" square units and 3¾" x 11¾" rectangles.

Sculptured glass modules have various geometric shapes pressed into the glass to a depth of approximately 1½" on both sides of the unit. They are 11¾" square, and are available in clear glass and the lighter hues of the color glass units.

Grille wall units are patterned glass with a smooth exterior with an outline of grey colored ceramic frit fused at the edges leaving ovals, circles, dots and hourglass designs exposed in the center. They are also available in solid grey frit. Sizes are 3¾" x 7¾" x 3⅞" and 7¾" square. The finish is scratch, abrasive and chemical resistant and will not fade from exposure. They make handsome and practical screen walls.

Mortar for Glass Blocks.—Materials used in making mortar for laying glass blocks shall be measured by volume. For this purpose 25 lbs. of quicklime or 40 lbs. of hydrated lime shall equal one cu. ft.

Lime shall be high-calcium or dolomitic hydrated lime. Dolomitic type lime must be pressure hydrated so that it does not contain more than 8 percent by weight of unhydrated oxides, in accordance with A.S.T.M. Specification C207-49 Type S.

The mortar shall be composed of 1 part waterproof portland cement, 1 part lime and 4 parts well graded sand. It shall be mixed to a consistency as stiff and dry as possible and still retain good working characteristics. Prepared masonry mortars of high strength and low volume change may be substituted, when approved.

Setting accelerators or anti-freeze compounds are not to be used.

Expansion Joints.—Expansion joints must be provided at the head and jambs of each glass block panel. This expansion area shall contain expansion strips as designated by the details or specifications. Caulk head and jamb expansion joints and at intermediate reinforcing members with sponge plastic strip or rope and a good grade of non-hardening mastic so that all expansion joints are waterproof.

Panels in residential construction which are not over 5'-0" in width or 7'- 0" in height, or 25 sq. ft. in area, may be mortared in solid without the use of expansion joints at the jambs. The expansion joint at the head must still be used to prevent loads from being carried by the panels. Reinforcing wall ties may be omitted in residential panels which do not exceed the above size limits.

Reinforcement.—Reinforcing wall ties shall consist of two No. 9 wires spaced 2" apart to which are welded No. 14 gauge cross wires. Ties are to be 8'-0" long and not more than .20" thick at the weld. They shall be galvanized or treated with some other approved corrosion resisting coating. Ties are to run continuously with ends lapped 6" and are to be installed in horizontal mortar joints approximately 24" on centers.

Ties shall not be laid directly on the block, but shall be completely embedded in the center of the mortar joint.

Glass Building Units—Installation

1. Sill area to be covered by mortar shall first have a heavy coat of PC Asphalt Emulsion and allowed to dry.
2. Adhere PC Expansion Strips to jambs and head with PC Asphalt Emulsion. Expansion strip must extend to sill.
3. When emulsion on sill is dry, place full mortar bed joint—do not furrow.
4. Set lower course of block. All mortar joints must be full and not furrowed. Steel tools must not be used to tap blocks into position. Mortar shall not bridge expansion joints. Visible width mortar joint shall be ¼" or as specified.
5. Install PC Panel Reinforcing in horizontal joints where required as follows:
   (a) Place lower half of mortar bed joint. Do not furrow.
   (b) Press panel reinforcing into place.
   (c) Cover panel reinforcing with upper half of mortar bed and trowel smooth. Do not furrow.
   (d) Panel reinforcing must run from end to end of panels and where used continuously must lap 6 inches. Reinforcing must not bridge expansion joints.
6. Place full mortar bed for joints not requiring panel reinforcing. Do not furrow.
7. Follow above instructions for succeeding courses. The number of blocks in successive lifts shall be limited to prevent squeezing out of mortar or movement of blocks.
8. Strike joints smoothly while mortar is still plastic and before final set. At this time rake out all spaces requiring caulking to a depth equal to the width of the spaces. Remove surplus mortar from faces of glass blocks and wipe dry. Tool joints smooth and concave, before mortar sets so that exposed edges of blocks have sharp clean lines.
9. After final mortar set pack PC Oakum tightly between glass block panel and jamb and head construction. Leave space for caulking.
10. Caulk panels as indicated on details.
11. Final cleaning of glass block faces shall not be done until after final mortar set.

CLEANING—Surplus mortar shall be removed and block faces wiped dry as joints are tooled. Final cleaning shall be done by others after mortar has attained final set.

Courtesy Pittsburgh Corning

Labor Laying Glass Blocks.—In laying glass blocks, all mortar joints shall be completely filled with mortar. The joints shall not be less than 3/16" thick or more than 3/8" thick. Blocks shall not be hit with a metal tool but laid by the method known as "shoving," (working into place with the hands) thereby compressing the vertical joints.

It has been found when bricklayers' trowel handles are capped with a common type of rubber ferule, such as used on the bottom of crutches, they prevent damage to the glass.

After the mortar has passed its initial set, the exposed edges of the joints shall be tooled and thoroughly compressed with a round jointer. The finished surface of the joint shall be slightly concave, smooth and nonporous. The final cleaning shall not be done until after the mortar has reached its final set.

If blocks are in any way disturbed after laying, during the plumbing, tooling, or cleaning process, such blocks shall be removed and relaid.

Inasmuch as glass blocks do not absorb water, the mortar must be stiff as possible and still be workable, otherwise the blocks will be inclined to slide and it will be difficult to lay up a straight wall and very few courses can be laid at a time as the weight of the glass blocks will squeeze the mortar out of the joints.

Basis for Estimating Labor Costs.—The labor costs on the following pages are based on one laborer to two masons, which includes time mixing and supplying mortar, glass blocks, etc. It does not include time building and removing scaffolding, placing expansion joint strips, packing joints with sponge plastic rope and caulking, nor does it include time required cleaning glass blocks after erection. These are given as separate items.

Panels containing more than 144 sq. ft. and not more than 25'-0" in length or 20'-0" in height should be braced by and anchored to structural stiffeners, so a fair average for all classes of work would be panels 10'- 0"x10'-0" in size, containing 100 sq. ft. Panels of this size have been used as a fair example for determining average costs.

Expansion joint strips ⅜" thick, 4⅛" wide and 25" long are used at the heads, jambs and mullions of all openings. These strips are secured in place by an adhesive such as asphalt emulsion, and are butted together end to end to form a continuous cushion around the edges of the panels.

After the panels have been laid and the mortar has set, sponge plastic rope should be packed tightly between the side of the block and the sides of the "chase." Sponge plastic rope should be kept back ⅜" from the finished surface. The recess thus formed should then be caulked with a non-hardening waterproof caulking material to a depth of not less than ⅜".

In a 10'-0" x10'-0" opening, caulked on both sides of the blocks, there will be 40 lin. ft. of caulking on the inside of the wall and 30 lin. ft. of caulking on outside of the wall or 70 lin. ft. for each opening. A man should pack rope into joints and caulk 30 to 40 lin. ft. (on one side) an hour.

To clean both sides of 100 sq. ft. of smooth face glass block wall will require 2 to 2½ hrs. labor time.

To clean both sides of 100 sq. ft. of ribbed face glass block (light directional type) will require 3 to 3½ hrs. labor time.

Erecting and removing scaffolding on the average job will require 1½ to 2 hrs. labor time per 100 sq. ft. of glass area.

Glass blocks must be carefully handled in unloading and handling on the job, as they require the same care as fine enameled brick.

Cost of Glass Blocks.—The cost of glass blocks varies with the kind, size and quantity required. Always obtain delivered prices on blocks, expansion strips and wall ties for each job figured.

Approximate Prices of Standard Glass Blocks Per Unit

Size Blocks	Up to 1,000 Sq. Ft.	From 1,000 Sq. Ft.
5¾"x5¾"x3⅞"	$1.75	$1.70
7¾"x7¾"x7⅞"	2.40	2.30
11¾"x11¾"x3⅞"	5.60	5.40

Special block runs considerably more and should be checked locally for both price and availability.

Estimating Data on Laying Glass Blocks Using ¼-Inch Mortar Joints

	Size of Glass Blocks in Inches		
Cost per 100 Sq. Ft. of Wall	5¾x5¾x3⅞	7¾x7¾x7⅞	11¾x11¾x3⅞
Number of Blocks per Sq. Ft. of Wall*	4.00	2.25	1.00
Number of Blocks per 100 Sq. Ft. of Wall*	400	225	100
Cu. Ft. of Mortar Required**	5	3.6	2.33
Mason Hours	21 to 23	14 to 16	10 to 12
Labor Hours	10½ to 11½	7 to 8	5 to 6
Labor on Scaffolding, Hours	2	2	2
Mason Ramming Oakum and Caulking, Hours	1½	1½	1½
Mason Cleaning Blocks, Hours	2½	2½	2½
Expansion Strips, ⅜"x4⅛", Lin. Ft	30	30	30
Wall Ties, Lin. Ft.	60	60	60
Sponge Plastic Rope Joints, Lin. Ft.	460	60	60
Caulking Joints, Lin. Ft.	70	70	70
Asphalt Emulsion, 40'-0"x4½"	½ pt.	½ pt.	½ pt.
Cost per 1,000 Glass Blocks (Based on 10x10 Panels)			
Cu. Ft. of Mortar Required**	12.5	16	23.3
Mason Hours	53 to 57	62 to 71	100 to 120
Labor Hours	27 to 29	31 to 36	50 to 60
Labor on Scaffolding, Hours	5	9	20
Mason Ramming Oakum and Caulking, Hours	3¾	6¾	15
Mason Cleaning Blocks, Hours	6¼	11	25
Expansion Strips, ⅜"x4⅛", Lin. Ft	75	135	300
Wall Ties—Reinforcement, Lin. Ft.	110	200	440
Sponge Plastic Rope Joints, Lin. Ft.	150	266	600
Caulking Joints, Lin. Ft.	175	310	700
Asphalt Emulsion, 3½" Wide	1¼ pt.	2¼ pt.	2½ qts.
Number pcs. laid per 8-hr. day	140 to 150	115 to 125	70 to 80

*Does not include allowance for breakage.

**Includes 10 percent for waste.

TWO STYLES OF DECORATIVE GLASS BLOCK

Two styles of 8"x8"x4" block showing three possible patterns formed by each style when arranged into panels of sixteen units.

GYPSUM TILE PARTITIONS AND FIREPROOFING

Gypsum tile are used for partitions and column fireproofing. They are 12"x30" in face dimensions. The large units are light in weight and require a minimum of labor for erection.

Estimating Quantities of Gypsum Tile.—Gypsum tile should be estimated by the sq. ft. taking the area of all walls or partitions and making deductions in full for all openings. Gypsum tile are easily handled and may be cut with an ordinary hand saw, reducing waste to a minimum.

The average mortar joint is ⅜" to ½" thick.

Sizes and Weights of Gypsum Partition and Furring Tile

	Size of Tile	For Ceiling Heights Up To	Weight per Sq. Ft. Lbs.
2" Solid	2"x12"x30"	10'-0"	10.0
3" Hollow	3"x12"x30"	13'-0"	10.0
4" Hollow	4"x12"x30"	17'-0"	13.0
6" Hollow	6"x12"x30"	30'-0"	19.0

Cu. Ft. of Mortar Required to Lay 100 Sq. Ft. of Gypsum Partition Tile.

Width of Mortar Joint in Inches	Thickness of Gypsum Tile in Inches			
	2-in.	3-in.	4-in.	6-in.
¼	1	1½	2	3
⅜	1½	2	2½	3½
½	2 *	2½*	3 *	4 *

*Quantities ordinarily used.

Note: Mortar quantities determined through actual experience and includes waste due to dropping, filling around pipe, etc.

Gypsum partition tile cement should always be used for laying up gypsum tile to obtain a secure bond and consequently a rigid wall.

Labor Setting Gypsum Tile.—When laying gypsum tile a man sets 2½ sq. ft. at a time but on account of the greater footage of tile set, it requires more labor time handling and getting them to the mason.

Gypsum tile are set by bricklayers or tile setters (bricklayers who specialize in tile setting) and the labor costs will vary with the experience of the men setting them. There are concerns which specialize in the erection of gypsum tile partitions and fireproofing, and their workmen are skilled in this branch of work. Naturally they set more tile per day than a bricklayer, who lays brick today, clay tile tomorrow and gypsum tile the day after.

Experienced tile setters should set 15 to 25 percent more tile than inexperienced men.

Labor Setting Gypsum Partition Tile

Size of Tile in Inches	No. Sq. Ft. Set per 8-Hr. Day	Labor Hrs. per 100 Sq. Ft.		
		B'layer	Labor	Engr.*
2" Partition	250-300	2.91	2.75	0.25
3" Partition	220-260	3.34	3.00	0.25
4" Partition	195-235	3.72	3.50	0.25
6" Partition	175-215	4.10	3.75	0.50
2" Column	200-230	3.72	3.00	0.25
3" Column	160-190	4.57	4.00	0.25

* Add hoisting engineer time only if required.

04400   STONE

Stone, being a natural product, varies widely in type and in geographical distribution. Being a very heavy material, except for the lava type stones which have very limited use, the availability of stone locally is of utmost importance to the estimator as transportation costs can become a high percentage of the total cost.

Geologically, stones are classified as follows :

a) Igneous: those formed by cooling molten material inside the earth including granite, trap rock and lava.

b) Sedimentary: those consolidated from particles of decayed rocks deposited along streams and including limestone and sandstone.

c) Metamorphic: those either igneous or sedimentary which have been altered
    by deep-seated heat and pressure or intrusion of rock materials and includ-
    ing gneiss, marble, quartzite and slate.

AN INDIANA LIMESTONE QUARRY

Within the industry, the Building Stone Institute divides stones as follows:

a) Bluestone: a hard sandstone, blue gray and buff in color, quarried in New
    York and Pennsylvania, used for paving, treads, copings, sills and stools,
    hearth and in random ashlar patterns for walls.

b) Granite: a fine to coarse grained igneous rock, grey, red, green, black or
    yellow incolor, quarried in New England, New York, Virginia, Michigan,
    Minnesota and Colorado, used for paving, curbing and rubble, mosaic, ashlar
    and veneered wall construction for almost all building conditions.

c) Greenstone: a metamorphosed stone, a distinctive green in color, quarried in Virginia, and used inside and out for flooring, treads, spandrels and curtain walls, sills, copings, hearths.
d) Limestone: a sedimentary rock, oolitic, dolomitic and crystalline, grey, buff and pinkish in color, quarried in Indiana, Kentucky, Massachusetts, Minnesota, New York, Alabama, Kansas and Texas, used for wall facings in all patterns, from rubble through the various ashlars to smooth facings in large sheets and for all trim pieces including carved moldings, panels and sculpture.
e) Marble: a metamorphic rock, generally recrystallized limestone, available in a wide range of colors and variegated patterns, quarried in Georgia, Alabama and Vermont, used generally as interior facings and much of which is imported, but are increasingly used as exterior veneers on precast concrete panels and as brick sized and random ashlar units for exterior walls.
f) Quartzite: a compact granular rock composed of quartz crystals in a wide variety of colors, many of which are variegated, quarried in Colorado, Tennessee and Utah, and used for paving, ashlar and mosaic wall veneers, copings, sills, hearths and treads. Quartzite is also highly popular as aggregate for precast panels.
g) Sandstone: a sedimentary rock, in warm earth tones from near white to dark browns and reds, quarried in Colorado, Minnesota, Michigan, Indiana, Ohio, New York and Massachusetts, used for wall facings of all types rubble and mosaics, ashlars and smooth sheet facings.
h) Slate: a fine grained metamorphic rock derived from sedimentary rock shale, predominantly gray and blue-black in color but also available in red, green, mottled green, purple, mottled purple and weathering green, quarried in Pennsylvania, Virginia, New York and Vermont, and used in slabs for exterior facings, flooring, paving, treads, bases, sills and stools, copings, fireplace trim and, of course, roofing.

The stone pattern and the type of finish also are of utmost importance to the estimator.

Patterns used in laying up stone work in walls include:
a) Rubble: including uncoursed field stone, polygonal or mosaic and random coursed.
b) Ashlar: including rubble field mixed with coursed ashlar, random range coursed ashlar, random range interrupted coursed, broken range ashlar and coursed ashlar.
c) Panelling: including spandrel and curtain wall units of both natural stone and stone aggregate precast with concrete.

Finishes for stone will vary as to the hardness and evenness of the material, and may be generally classified as natural, rough worked, and smooth.

Natural finishes may be weathered as in field stone, or seam or split faced, when the natural quarry texture is exposed.

Rough finishes are obtained mechanically by cutting with pitching chisels; gang sawing with steel shot or rough chat; plain sawn which still leaves saw marks; planed finish including those where small particles are plucked after planing to roughen the texture; hammered, hand and machine tooled, where 2 to 10 parallel concave grooved cuts per inch are cut into the exposed surface. A newly devised rough finish for granite is obtained by exposing it to flames.

Smooth finishes include those rubbed wet by hand with carborundum block and water; machine rubbed with carborundum disc; and honed, obtained with a polishing machine and which leaves the stone almost without texture. As this last finish is expensive, it should be limited to interior work or the finest exterior within 30' of eye range. Granite and marble may be polished smooth.

Rough faced stone is used in the construction of churches, college and university buildings, chapels, residences, and retaining walls. It is available in granite, quartzite and sandstone.

RUBBLE STONE WORK

Rubble stone consists of irregularly shaped pieces, only partly trimmed or squared if at all. Finish is usually split faced or natural. Some specifications will restrict the size range. Strip rubble is ledge stone with beds of the natural cleft.

Rubble stone work may be estimated by the cu. yd. containing 27 cu. ft. or by the perch containing 24¾ cu. ft. A perch of stone is nominally 16½ ft. long, 1 ft. high and 1½ ft. wide, but is often computed differently in different locations. In many states, especially west of the Mississippi River, rubble work is figured by the perch containing 16½ cu. ft. Before submitting prices on rubble work by the perch, be sure to find out the common practice in your locality.

Cost of Rubble Stone.—The cost of rubble stone will vary with the locality, kind of stone used, distance of quarry from the job, etc.

The method of buying and selling rubble stone also varies with the locality, as in some places it is sold by the ton containing 2,000 lbs. while in others it is sold by the perch containing 24¾ cu. ft. or by the cord containing 100 cu. ft. and weighing 13,000 lbs. which is about 130 lbs. per cu. ft. In some localities it is sold by the cu. yd. containing 27 cu. ft.

Mortar for Rubble Stone Work.—The quantity of mortar required for rubble stone will vary with the size and shape of the stone and the thickness of the walls, but a fair average is 7 to 9 cu. ft. of mortar per cu. yd. of wall. Mortar costs are given under the mortar section early in this chapter.

Cost of Setting Rubble Stone.—When handling and setting rubble stone in walls up to 1'-6" thick, a mason and laborer should handle and set 70 to 85 cu. ft. per 8-hr. day.

On heavy walls, 2'-0" to 3'-0" thick, a mason should set 100 cu. ft. per 8-hr. day, and will require about 1½ hrs. labor time to each hr. mason time.

Cost of One Cu. Yd. of Rubble Stone for Walls Over 2'-0" Thick

	Hours	Rate	Total	Rate	Total
Mason	3.0	$. . . .	$. . . .	$17.87	$ 53.61
Labor	4.5	. . . .	. . . .	14.95	67.28
Cost per cu. yd			$. . . .		$120.89
Cost per cu. ft			. . . .		1.21

ASHLAR STONE WORK

Ashlar Stone is defined as stone having a flat faced surface generally square or rectangular with sawed or dressed beds and joints. Wall patterns are referred to as "coursed" when set to form continuous horizontal joints; "stacked" when set to form continuous vertical joints; and "random" when set with stones of various lengths and heights so that neither horizontal or vertical joints are continuous. The usual restrictions for random ashlar design are that no vertical joint should be higher than the highest course height being used; that no horizontal joint should be more than three stones long; and that no two stones the same height should be placed end to end. Random work will take more planning on the part of the mason than coarsed work.

Finish for ashlar work is usually split face which is the least expensive and brings out the character of the stone. It may also be ordered chat, sand or shot sawn. The grade of stone for ashlar work is usually a mixture of all the grades available as a full range is part of the desired effect.

Many types of stone are furnished in ashlar stock including granite; limestone in the greys and buffs from Indiana, the creamy pinks from Minnesota and the popular lannon from Wisconsin; both Vermont and Georgia marble; Tennessee quartzite; and the local sandstones, those from Ohio and Pennsylvania being especially popular.

The common heights for ashlar are 2¼", 5", 7¾" and sometimes 10½", 13¼"and 16". Lengths are shipped from 1' to 6' for cutting on the job. Most ashlar work today is set as veneer and comes in thicknesses of 3½" to 4". Some ashlar comes in varying thicknesses to be set to a uniform back and to create deep shadow lines on the face. Some jobs will still specify that the ashlar be built as part of the back-up wall, not veneered to it. Ashlar must then be ordered with bond blocks. These units will be a nominal 8" deep and usually constitute around 10% of the stone furnished, or one stone in each 10 sq. ft. of wall.

Many producers offer stock patterns under their own trade names. These are less expensive than stone cut to specification, but are limited to veneer thickness and a fixed percentage of heights, usually around 15% of 2¼", 40% of 5" and 45% of 7¾".

Veneer ashlar must be tied to the back-up wall with noncorrosive ties not less than 18" apart horizontally and 24" vertically. The stock heights when used with ½" mortar joints level well with brick and block back-up, but adjustable anchors and stud anchors are also available.

Costs of ashlar stone will vary widely depending on the stone selected and the distance from the quarry as well as whether stock supplies or special cutting and finishes are required.

The cost of setting this stone also varies widely, depending upon the size of the stone and the amount of cutting required. On some of the more intricate jobs it requires as much mason's time trimming and cutting the stone as is required to set it. The entire success of the use of this random ashlar depends upon the artistic manner in which the work is performed, and the labor costs will be governed accordingly.

Labor time handling stone and helping masons varies widely, depending upon method of handling stone at job, access of material piles, size of stone, etc.

To show the variations in the cost of setting, six different classes of work are described.

Estimating Quantities of Random Ashlar.—Split faced or other finishes of ashlar stone is usually sold by the ton and is furnished with rough beds, backs and joints. The stone weighs 160 to 165 lbs. per cu. ft. and there are 12 to 12½ cu. ft. of stone to the ton.

Stone 4-in. thick produces 36 to 38 sq. ft. of wall to the ton and when laid with a ½-in. mortar joint should lay 40 to 42 sq. ft. of wall.

Stone 5-in. thick produces 29 to 30 sq. ft. of wall to the ton and when laid with a ½-in. mortar joint should lay 32 to 33 sq. ft. of wall.

Where the stone is backed with brick or block, the stone is usually furnished 4-in. to 8-in. thick to bond with the brick backing. Based on an average of 6-in. thick for the job, one ton is sufficient for 24 to 25 sq. ft. of wall and when laid with a ½-in. mortar joint should lay 26 to 28 sq. ft. of wall.

RUBBLE
Thickness: 2" and 2½"
Coverage: 70 sq. ft. per ton

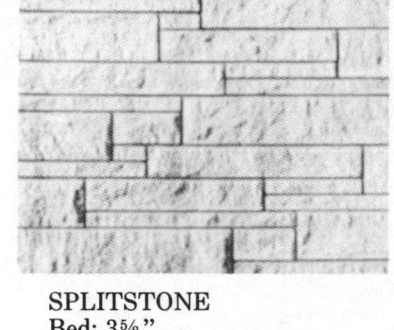

SPLITSTONE
Bed: 3⅝"
Height: 2¼", 5", 7¾"
Coverage: 48 sq. ft. per ton

WEBALL
Coverage: 30 sq. ft. per ton

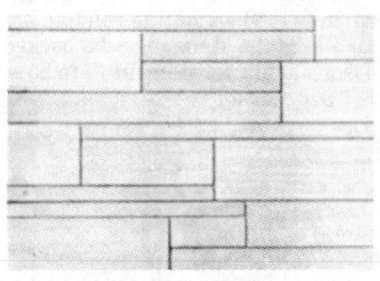

CHAT SAWED ASHLAR
Bed: 3" and 4"
Height: 2¼", 5", 7¾", 10½"
Coverage: 3" bed, 60 sq. ft. per ton
4" bed, 40 sq. ft. per ton

HAND PITCHED ROCKFACE
Bed: 3"
Height: 3"
Coverage: 60 sq. ft. per ton

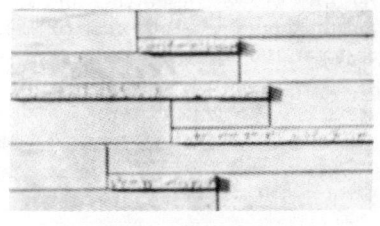

**SPLITSTONE WITH
CHAT SAWED**
Height: Splitstone–2¼"
Chat Sawed–5" and 7¾"
Coverage: 55 sq. ft. per ton

LIMESTONE VENEER PATTERNS

Courtesy of Victor Oolitic Stone Co., Bloomington, Ind.

Cost of 100 Sq. Ft. of Random Ashlar Used as a Veneer Over
Wood Framing and Sheathing, Stone Veneer 4-In. Thick

	Hours	Rate	Total	Rate	Total
Mason	24	$. . . .	$. . . .	$17.87	$428.88
Labor	12	. . . .	. . . .	14.95	179.40
Cost per 100 sq. ft			$. . . .		$608.28
Cost per sq. ft			. . . .		6.08

Cost Per Ton of Stone

	Hours	Rate	Total	Rate	Total
Mason	10	$. . . .	$. . . .	$17.87	$178.70
Labor	5	. . . .	. . . .	14.95	74.75
Cost per ton			$. . . .		$253.45

On work above the second story or where large stones are used, it may require an extra laborer bringing stone to the mason.

Setting Random Ashlar in Straight Work Requiring Backing, Cutting and Fitting.—When setting random ashlar in fairly straight walls as described above, but where the stone must be backed, and end joints cut on the job, an experienced mason should set stone in 17 to 25 sq. ft. of wall per 8-hr. day, at the following labor cost per 100 sq. ft. :

	Hours	Rate	Total	Rate	Total
Mason	38	$. . . .	$. . . .	$17.87	$ 679.06
Labor	38	. . . .	. . . .	14.95	568.10
Cost per 100 sq. ft			$. . . .		$1,247.16
Cost per sq. ft			. . . .		12.47

The labor time will vary widely according to job conditions, as some jobs may require only one-half as much labor time as mason time, while others may require twice as much.

Setting Random Ashlar in Churches, Chapels, Etc., Where Stone is Cut to Size in Shops.—On churches, chapels, and similar structures having numerous corners, offsets, piers, pilasters, etc., where the stone is all backed in the shop and requires very little cutting and fitting by the masons on the job, an experienced mason should set stone in 25 to 30 sq. ft. of wall per 8-hr. day, at the following labor cost per 100 sq. ft. :

	Hours	Rate	Total	Rate	Total
Mason	29	$. . . .	$. . . .	$17.87	$518.23
Labor	29	. . . .	. . . .	14.95	433.55
Cost per 100 sq. ft			$. . . .		$951.78
Cost per sq. ft			. . . .		9.52

Setting Random Ashlar in Churches, Chapels, Etc., Where Cutting and Fitting is Done on the Job.—On churches, chapels, and similar structures having numerous corners, offsets, piers, pilasters, etc., where the rough stone is sent to the job and all backing, jointing and fitting is done by the masons on the job, an experienced mason should set stone in 14 to 18 sq. ft of wall per 8-hr. day, at the following labor cost per 100 sq. ft.:

	Hours	Rate	Total	Rate	Total
Mason	50	$. . . .	$. . . .	$17.87	$ 893.50
Labor	50	. . . .	. . . .	14.95	747.50
Cost per 100 sq. ft			$. . . .		$1,641.00
Cost per sq. ft			. . . .		16.41

Random Ashlar in Residence Construction—Mr. Walker made an actual study of a residence faced with 4500 square feet of 5" Lannon stone veneer. The stone varied from 1" to 10" high and consisted of one third weathered edge, one third bed face stone and one third rock face stone which was only one to two inches high. Care was taken to produce a natural looking pattern with very little cutting and fitting.

Detail of Random Ashlar Stonework

A mason set 34 sq. ft. of stone per 8-hr. day, at the following labor cost per 100 sq. ft. :

	Hours	Rate	Total	Rate	Total
Mason	24	$....	$....	$17.87	$428.88
Labor	14	....	....	14.95	209.30
Cost per 100 sq. ft			$....		$638.18
Cost per sq. ft			....		6.38

Random Ashlar in Memorial Chapel.—The memorial chapel illustrated is also built of Lannon (Wisconsin) stone, but on account of the numerous corners, pilasters, piers, etc., and the fact that all backing, jointing and fitting of stone was done on the job, the labor costs ran rather high, as on work of this kind the time required cutting and fitting the stone is almost as much as the setting time.

On this job, a mason averaged 17 sq. ft. of 8" ashlar per 8-hr. day, at the following labor cost per 100 sq. ft. :

	Hours	Rate	Total	Rate	Total
Mason	47	$....	$....	$17.87	$ 839.89
Labor	40	....	....	14.95	598.00
Cost per 100 sq. ft			$....		$1,437.89
Cost per sq. ft			....		14.38

CUT STONE

The term "cut stone" refers here to stone that is cut to an exact dimension from selected quality stone, given a uniform texture and delivered to the site ready to set in place. In order to provide such quality control shop drawings must be made and approved in advance. Cut stone is used on the highest quality work and thus demands the highest quality of installation procedures.

Cut stone may be built up with and bonded to a masonry back-up, but today it is most often applied as a veneer or a facing panel sometimes preassembled with an insulating backing and frame.

In the past cut stone usually referred to granite or limestone or sometimes sandstone which was used to face the many "brownstones" which line the older streets of our northern cities. With the shift from slabs to thin veneers the denser marbles and slate are now often specified.

Shop Drawings.—All cut stone work should be cut and set in accordance with approved shop drawings. These should show the bedding, bonding, jointing and anchoring details. Each stone is dimensioned and given a setting number. The producer will not cut the stone until all details and dimensions are agreed upon. As jointing in veneer work is generally ¼", extremely accurate measurements are required. Also, the contractor must plan ahead and have cutouts and holes required for lifting large slabs (called Lewis holes) as well as cutouts for posts, door checks, electrical fixtures, etc., as may be required for the work of other trades.

Holes for anchorage are also cut by the contractor furnishing the stone.

Mortar for Cut Stone.—The following three types of mortar are recommended for setting cut stone. The best type adapted to the particular work should be selected and included.

Non-Staining Cement Mortar.—The cement mortar shall be composed of one (1) part of non-staining cement to three (3) parts of sand with the addition of one-fifth (1/5) part of hydrated lime based on the cement volume.

Non-Staining Cement-Lime Mortar. —Cement-lime mortar shall be composed of one-half (½) part of non-staining cement, one-half (½) part of hydrated lime or quick lime paste and not to exceed three (3) parts of sand.

If quick lime paste is used, first thoroughly mix the quick lime paste and sand and stack to age; the cement shall be added and thoroughly worked into the lime and sand mixture in small batches just prior to use.

Quick lime shall be reduced to a paste by thorough and complete slaking with cold water and screening through mesh screen into a settling box. Lime putty shall stand not less than one week before use.

Sand shall be clean and sharp, free from loam, silt, vegetable matter, salts, and all other injurious substances with grains graded from fine to coarse. Screen through a 6-mesh screen.

Non-Staining Waterproof Cement Mortar.—Recommended for use wherever parging is required or when any of the construction materials, such as water, sand, brick, tile, etc., are known to contain salts which may cause staining or efflorescence; and to walls exposed to excessive amounts of moisture.

Non-staining waterproof cement mortar shall be composed of one (1) part of non-staining waterproof cement to three (3) parts of sand with the addition of one-fifth (1/5) part of hydrated lime based on the cement volume.

For quantity and cost of mortar required for setting limestone, refer to mortar section of this chapter.

Job Conditions.—The cut stone contractor must check job conditions carefully. His product is expensive: defective or damaged work is almost always refused, and replacements can cause costly job delays.

Carefully unpack and inspect all material to assure it matches samples, is cut to dimensions of shop drawings and tolerances of specifications, and to check for shipping damages.

Carefully store all stone on non-staining platforms at least 4" off the ground. Plug all holes during freezing weather.

Check hoist and scaffolding conditions if they are involved. Dirty hoists and scaffolding can stain stone.

Check backup construction. Backup must be laid to a true line. Concrete and masonry backup should be dampproofed. If this is not in other contracts, this contractor should provide for it.

If work is to continue alongside or above stone after it is set, add an allowance for protection of exposed surfaces.

Estimating Cut Stone.—The cost of cut stone will depend upon the kind of stone used (whether granite, limestone or marble), the amount of cutting and hand work necessary, and freight from mill to destination.

Cut stone for exterior use is estimated by the cu. ft. (taking the cube of the largest dimensions), and any special work such as hand cutting, carving, etc., is estimated separately and added to the cost of the plain stone.

When estimating the cost of plain door and window sills, lintels, steps, platforms, copings, etc., the quantities are sometimes estimated by the lin. ft. for sills and coping and by the sq. ft. for platforms, but the cu. ft. method is more generally used.

Setting Cut Stone.—When estimating the labor cost of handling and setting cut stone, the size of the job should be considered, average size of stone, class of workmanship required and the method of setting, i.e., whether by hand, hand operated derricks or by using a mobile crane, stiff leg, guy or other types of derricks.

Setting Cut Stone by Hand.—On small cut stone jobs consisting of door and window sills, lintels, wall coping, steps and other light stone that can be handled and set by two men without the use of derricks, a stone setter should set 50 to 60 cu. ft. of stone per 8-hr. day, at the following labor cost per 100 cu. ft. :

	Hours	Rate	Total	Rate	Total
Stone setter	14	$. . . .	$. . . .	$17.87	$250.18
Labor	28	. . . .	. . . .	14.95	418.60
Cost per 100 cu. ft			$. . . .		$668.78
Cost per cu. ft			. . . .		6.69

Setting Cut Stone Using Hand Derricks.—Where the stone consists of coursed ashlar, sills, lintels, lightweight cornices, copings, etc., which can be handled on stone barrows or rollers and set with a breast derrick, gin pole, or other hand operated derricks, a stone setter should set 80 to 100 cu. ft. of stone per 8-hr. day, at the following labor cost per 100 cu. ft. :

	Hours	Rate	Total	Rate	Total
Stone setter	9	$. . . .	$. . . .	$17.87	$160.83
Helper	9	. . . .	. . . .	14.95	134.55
2 derrickmen	18	. . . .	. . . .	17.87	321.66
2 men handling stone	18	. . . .	. . . .	14.95	269.10
Cost per 100 cu. ft			$. . . .		$886.14
Cost per cu. ft			. . . .		8.86

Setting Heavy Cut Stone Using Hand Derricks.—On jobs requiring heavy stone, such as thick ashlar, heavy molded courses, cornices, copings, etc., where the stone is hoisted and set using hand operated derricks, a stone setter should set 115 to 140 cu. ft. of stone per 8-hr. day, at the following labor cost per 100 cu. ft. :

	Hours	Rate	Total	Rate	Total
Stone setter	7	$. . . .	$. . . .	$17.87	$125.09
Helper	7	. . . .	. . . .	14.95	104.65
2 derrickmen	14	. . . .	. . . .	17.87	250.18
3 men handling stone	21	. . . .	. . . .	14.95	313.95
Cost per 100 cu. ft			$. . . .		$793.87
Cost per cu. ft			. . . .		7.94

Setting Cut Stone Using a Power Crane.—On many large jobs it will be possible to set the stone, using a mobile power crane to lift the stone from the ground to the wall.

This method is usually less expensive than setting up and dismantling stiff-leg or guy derricks, as the crane can cover the entire perimeter of the building without any setup or dismantling to speak of.

Where the stone can be set in this manner, it will require one operator for the crane, 2 laborers on the scaffold helping to land the stone and set it, 4 or 5 laborers on the ground sorting and handling stone from stock pile to derricks, plus the stone setter.

On a job of this kind, a crew should handle and set 200 to 240 cu. ft. of stone per 8-hr. day, depending upon size of stone being set, at the following labor cost per 100 cu. ft. (excluding crane and operator costs):

	Hours	Rate	Total	Rate	Total
Stone setter	4	$. . . .	$. . . .	$17.87	$ 71.48
2 helpers on scaffold	8	. . . .	. . . .	14.95	119.60
5 men handling stone	20	. . . .	. . . .	14.95	299.00
Cost per 100 cu. ft			$. . . .		$490.08
Cost per cu. ft			. . . .		4.90

The above does not include any allowance for scaffold, if a separate scaffold is required for stone setting, add for same.

Power Crane Used for Setting Stone

Setting Cut Stone Using Power Derricks.—On large jobs where it is necessary to use stiff leg or guy derricks for hoisting and handling the stone, the costs will vary with the size and weight of the stone, quantity, and with the method of handling the stone on the ground.

On work of this kind it is usually necessary to use stiff leg or guy derricks, or by having the boom of the derrick placed in a boom seat and bolted to the steel frame of the building. On jobs covering a large area, it will be necessary to use a number of derricks, so regardless of the method used in setting the stone, the cost of setting and removing derricks, hoisting engines, etc., should be estimated separately, as this charge will remain the same whether there are 5,000 or 50,000 cu. ft. of stone to be set.

After the derricks and equipment are in place, a stone setter should set 200 to 240 cu. ft. of stone per 8-hr. day. It will require about 2 laborers helping to land the stone and set it, 4 or 5 laborers on the ground sorting and handling the stone from stock piles to derricks, 3 derrickmen and one hoisting engineer.

In connection with work of this kind there is always a certain amount of lost time caused by delays in delivery of stone, repairs to derricks and hoisting engines, oiling and attending derricks, moving and raising derricks, inspecting cables, tightening guy lines and many other items that require attention and which must be charged against the cost of stone setting.

Pointing Cut Stone.—Cut stone is often pointed at the time the stone is set, as recommended practice is to strike the joints as the work progresses. Pointing mortar applied afterward is usually of different composition and accordingly of questionable permanency.

However, on some jobs it may be necessary to point all mortar joints with white or other non-staining cement after all stone has been set. The mortar joints may be struck flush or slightly rounded with a jointer, as preferred.

The cost of this work will vary with the size of the individual pieces of stone, as the fewer joints the more surface a tuckpointer will be able to point per day.

On an average job, a tuckpointer should point 500 to 600 sq. ft. of surface per 8-hr. day, at the following labor cost per 100 sq. ft.:

	Hours	Rate	Total	Rate	Total
Tuckpointer	1.50	$. . . .	$. . . .	$17.87	$26.81
Helper	0.75	. . . .	. . . .	14.95	11.21
Cost per 100 sq. ft			$. . . .		$38.02
Cost per sq. ft			. . . .		.38

Allow one helper on the ground to each 2 tuckpointers or to each scaffold. Add for mortar and scaffold.

Washing and Cleaning Cut Stone.— After the stone work has been pointed it will be necessary to wash and clean it, using a stiff fiber brush and plenty of clean water or soapy water, if necessary, and then rinsed with clean water, which will remove the construction and mortar stains. A wire brush should never be used, as it tends to produce rust marks in the stone.

On new work or work that is fairly clean, an experienced tuckpointer should wash and clean 600 to 800 sq. ft. of surface per 8-hr. day, at the following labor cost per 100 sq. ft.:

	Hours	Rate	Total	Rate	Total
Tuckpointer	1.1	$. . . .	$. . . .	$17.87	$19.66
Helper	0.6	. . . .	. . . .	14.95	8.97
Cost per 100 sq. ft			$. . . .		$28.63
Cost per sq. ft			. . . .		.29

Allow one helper on the ground to each 2 tuckpointers or to each scaffold. Add for scaffold.

Carving Cut Stone.—The cost of carving cut stone is usually figured from full-size details or models furnished by the architect. This cost will vary with the amount of detail and the method used in carving, i.e., whether performed in the shop before setting or after the stone has been set on the job.

Where you are sure of a first class job of setting, it is more economical to have the carving done in the shop as all carved work consisting of a number of pieces is set up with wood strips forming the thickness of mortar joints and accurately fitted before shipping to the job.

Carving Stone by Hand in the Shop. —When carving cut stone for medium priced buildings requiring just an average grade of workmanship and no fine detail work, a carver should complete 2½ to 3½ sq. ft. per 8-hr. day, at the following labor cost per sq. ft.:

	Hours	Rate	Total	Rate	Total
Stone carver	2.7	$. . . .	$. . . .	$20.00	$54.00

On jobs having fine detail work and requiring first class workmanship, a carver should complete 1 to 1½ sq. ft. per 8-hr. day, at the following cost per sq. ft.:

	Hours	Rate	Total	Rate	Total
Stone carver	7	$. . . .	$. . . .	$20.00	$140.00

Carving Stone in the Shop Using Pneumatic Tools.—The use of air tools will greatly increase the daily output of a stone carver, regardless of the class of workmanship required.

On jobs requiring just an average grade of workmanship and no fine detail work, a carver should complete 5 to 8 sq. ft. per 8-hr. day, at the following labor cost per sq. ft.:

	Hours	Rate	Total	Rate	Total
Stone carver	1.5	$. . . .	$. . . .	$20.00	$30.00

Where first class workmanship and fine detail is required, a carver should complete 2 to 2½ sq. ft. per 8-hr. day, at the following labor cost per sq. ft.:

	Hours	Rate	Total	Rate	Total
Stone carver	4.0	$. . . .	$. . . .	$20.00	$80.00

Carving Stone by Hand After It Is Set in the Building.—Where the stone is carved on the job after it has been set in place and where first class workmanship is required, a carver should complete 1 to 1¼ sq. ft. per 8-hr. day, at the following labor cost per sq. ft.:

	Hours	Rate	Total	Rate	Total
Stone carver	7	$. . . .	$. . . .	$20.00	$140.00

If the carving contains considerable fine detail, add ¾ to 1-hr. per sq. ft.
If outside scaffold is required, add labor time to above.

Carving Stone on the Job Using Pneumatic Tools.—If air tools may be used for carving the stone on the job, a carver should complete 1½ to 2 sq. ft. per 8-hr. day, at the following labor cost per sq. ft.:

	Hours	Rate	Total	Rate	Total
Stone carver	4.5	$. . . .	$. . . .	$20.00	$90.00

If the carving contains considerable fine detail, add ½ to ¾-hr. per sq. ft.
If outside scaffold is required, add labor time to above.

Cost of Models for Stone Carving.—On the better class of work, clay or plaster models for all carving are usually furnished by the architect or a stipulated sum is allowed for models.

Figuring on a sq. ft. basis, models in clay will run around $25.00 to $75.00 per sq. ft. while plaster models will cost about $40.00 to $100.00 per sq. ft. A small rosette or head cast in plaster will cost about $75.00. The models quite generally cost more than the carving, and often twice as much where models are made for an individual piece. It will often require a week's work for a skilled artisan to produce a carefully made model for a carved plaque or panel, 3'-6" to 4'-0" in diameter.

The sq. ft. price of models should be based on the sq. ft. of model required and not on the sq. ft. surface of carved stone, as one model may often be used for carving 10 or 15 pieces of stone, or one lin. ft. model of carved molding in plaster may often be used for carving hundreds of lin. ft. of stone.

Indiana Limestone

Indiana Limestone is probably more favorably known and more generally used in all parts of the United States than any other stone used for building purposes. This is doubtless due to its appearance, as well as the ease with which it is worked, its freedom from stratification and cleavage plane, and while it is a sedimentary stone, it is to all intents and purposes a free stone, having about equal strength in all directions.

It is used for the exterior facings of buildings, as well as all kinds of trimmings, such as sills, lintels, copings, steps, platforms, etc.

Indiana limestone can be produced in any reasonable quantity and size and is usually cut ready to place in the building by the mills located adjacent to the quarries.

The creamy-pink limestone quarried in Minnesota is another high quality material.

Limestone is furnished in a variety of different grades and color-tones, a brief summary of which is given below:

Select stock is uniform in color and texture, more so than is required for ordinary building construction. It is recommended for entrance work and those portions of a building within ready range of vision, carving, interior and other exceptional uses.

Standard stock is sound stone with a range of variation in color shades and texture not found in Select but necessarily confined within limits that make it impossible to determine at a distance of a few feet whether it is Standard or Select stock.

Rustic stock is of more or less coarse grain. Rustic stock is used to a considerable extent for sawed ashlar and various rough textured finishes.

Variegated stock in an irregular mixture of buff and gray produced from the blocks that are quarried where the buff and gray color-tones adjoin in the quarry. It shows variation in texture as well as color. Variegated stone when cut into building units will exhibit, in these, units of each color-tone and a small percentage of units with both color-tones in one piece. Variegated stone is unusually effective in giving variety to plain surfaces and is a desirable class of material for trim as well as for facing.

There are certain grades of Indiana Oolitic limestone especially adapted for grade courses, steps, buttresses, floor tiling, etc., or any position in the building exposed to abrasion. Some are available in different color-tones and may be identified by the following names: Special Hard Gray and Special Hard Buff Indiana Limestone.

The following are the trade terms by which the different grades of Indiana limestone are classified:

Select Buff	Standard Buff	
Select Gray	Standard Gray	Variegated

Some of the machine finishes suitable for limestone are sand sawed, chat or shotsawed finish, smooth machine finish, wet rubbed finish, machine tooled finishes, carborundum finish and plucked finish. There are also some rough textured finishes. The class of tooling in common practice is understood to cover what is properly termed fluting, that is, concave depressions, four and six parallel, concave grooves to the inch. Convex tooling is sometimes used effectively, but it is never applied unless specifically designated. Tooling and machinery has been developed to produce a variety of rough textured finishes.

It is generally recommended that all molded work be finished smooth. If it is tooled at all, the tooling should run in the direction of the mold and not across it, unless largely increased expense is of no consequence. As to tooled ashlar, some architects contend vertical tooling is preferable, others prefer the horizontal.

Indiana Oolitic limestone will take most hand finishes which under certain conditions are considered preferable to machine finish. Notable examples of this class of work are crandalling and tooth chiselling, but a number of the hand finishes can now be simulated by machine work.

A smooth finish is not commonly furnished other than in the form it comes from the planers. If sand rubbing is desired, it should be specified, and in such event the sand rubbing should be by the wet process or carborundum-and-water rubbed finish.

Indiana limestone weighs 200 lbs. per cu. ft. when furnished in rough quarry blocks and 180 lbs. per cu. ft. in scabbled quarry blocks. These figures include the weight of rough excess stone.

Sawed slabs or other squared material weighs 170 lbs. per cu. ft. The weight of the finished cut stone ready to set in the building is figured at 150 lbs. per cu. ft. If there are any large round columns in the job, they may be figured at 140 lbs. per cu. ft. of stock. The actual net weight of a cu. ft. of dry seasoned stone is 144 lbs.

Where the cut stone for any job consisting of ashlar, sills, steps, coping, plain machine moldings, etc., is furnished by any of the mills in the Bedford or Indiana limestone district, the price will vary from $9.00 to $16.00 per cu. ft. f.o.b. cars at the mills.

Considerable variation in price is due to the following reasons: Stone consisting of lightweight pieces, such as 4" or 6" ashlar, small molded courses, small turned columns, bases, sills, coping, etc., require a comparatively small amount of stone but involves considerable cutting and finishing expense, and even higher costs will prevail.

On other jobs consisting of fairly large stone, 8" or more thick, large molded pieces, bases, columns, etc., the cubical contents are a deciding factor, and the cu. ft. price of this stone, even for more elaborate work will usually average less in cost than the lighter stone described above.

Jobs containing many large pieces of stone will take the minimum price.

Always remember that the cube of the largest dimensions must always be used when computing the quantity, cost and weights of stone of any kind.

Indiana Limestone Sill Stock

Sill stock is available as standard items and sold by the lineal foot. The designation S4S means stock will be sent in random lengths of from 5' to 12'. S6S means stock will be sawed to exact lengths. Sill stock is sold in rectangular shapes. If slopes, lugs, drips or moldings are required, profile cuts will be made at additional cost. Sand and other special finishes are also extra.

Sill stock will run around $7.00 per cu. ft. If specific lengths are ordered add $1.50 per cu. ft. If custom cut to shop drawing details sills, copings, steps and other simple trim will run around $20.00 per cu. ft. Belt courses, simple cornice molds and the like will run around $28.00 per cu. ft. Elaborately carved detail work will run $30.00 and up per cu. ft.

Cutting Beds and Joints.—When cutting beds and joints the cost will vary with the size of the stone. The smaller the stone the higher the price per cu. ft. and the larger the stone the less the price per cu. ft. For instance, a piece of ashlar 8" thick contains twice as many cu.ft. of stone as a piece 4" thick, while the cost of cutting the joints will not be much more on the larger piece than on the smaller.

Where the joints are cut by hand, the cost will vary from $1.00 to $1.50 per cu. ft., depending upon the conditions stated above.

STANDARD ITEMS

On the second sawing, where sawed slabs are used and it is necessary to run them through the saw the second time for top and bottom beds, it will cost 80 cts. per cu. ft. for this work.

Where the jointing is done on the saw, figure 65 cts. per cu. ft. for cutting joints.

Smooth or Planer Finish.—When Indiana limestone is wanted with a smooth finish as it comes from the planer, the cost will vary from 45 to 50 cts. per sq. ft. of surface planed.

The slabs are usually furnished sawed to the correct thickness and then the stone is run through the planer to remove the saw marks, which is not a very expensive process on straight work.

On the cheaper classes of work, and on some of the higher grade work where a sawed finish is desired, the sawed slabs are jointed and the stone is sent to the building without having been run through the planer. Work of this class is a little cheaper than stone having a first class finish.

Tooled Finish.—If the stone has a tooled finish, having 4, 6, 8 or 10 bats to the inch, the cost of horizontal tooling on plain surfaces will cost 45 to 50 cts. per sq. ft. (not cu. ft.) of surface tooled, or about the same as planer finish.

Hand vertical tooling in connection with machine molded work will add 50 to 55 cts. per sq. ft. of surface finished.

Cross or vertical tooling on curved surfaces will add $1.00 per sq.ft. of surface finished.

Two bat tooling, which is always used vertically, can be done for almost the same cost as given above if it is not made to register. This tooling, however, usually has to be made to register and for that reason is generally more costly than 4, 6, 8 or 10 bat work.

Sand Rubbed Finish.—If the stone is to have a sand rubbed finish, making it necessary to rub the surface with sand and water, add 45 to 50 cts. per sq. ft. for all surface to be rubbed.

Rustication.—If the stone is to be furnished with rustication or a "rabbet" along the edges of the stone, the cost will vary according to the depth of the rabbet. This is all planer work, so the width of the rabbet does not make as much difference as the depth to which it must be cut.

The following will prove a fair average of the cost of rustication:

Depth in inches	Price Per Lin. Ft.	Depth in inches	Price Per Lin. Ft.
½	$0.46	2	$0.77
1	.54	2½	.90
1½	.65	3	1.04

Rounded Edges on Stone.—Add about 22 cts. per lin. ft. for stone with edges slightly rounded.

Stone Moldings.—The cost of stone moldings will vary greatly according to the size of the stone, the type of mold, breaks, heads, jointing, etc. The two kinds of moldings most commonly used are plain or Classic moldings and Gothic moldings.

When cutting stone moldings on a planer, the smaller the piece of stone the higher the price per cu. ft., on account of the additional labor involved and the small amount of stone required.

Tools for cutting moldings vary from 3" to 8" wide, so when the moldings are over 8" in width, it is necessary to run them through the planer a second time.

The moldings that are the cheapest to run are similar to the illustration of plain moldings, where a large amount of stone is required on which no special work is necessary.

When estimating plain moldings, take the largest dimensions of the stone to obtain the cubical contents and figure at the average price as given on the previous pages, as this will cover most jobs where the molded courses do not form too large a proportion of the total stone in the job.

Plain Moldings.   Gothic Moldings.

Gothic Moldings.—Gothic moldings are much more expensive than plain or Classic moldings, as in almost every instance it is necessary to run the stone through the planer twice or perform a certain amount of hand labor in connection with same. Gothic moldings cost at least twice as much to run as plain or Classic moldings. The cutting of heads for Gothic moldings will run much higher than plain moldings, or from $2.50 to $10.00 each, depending upon the size and character of the molding.

When entire jobs consist of elaborate Gothic stone, they frequently will run from $12.00 to $30.00 and more per cu. ft. On work of this kind it is advisable to obtain figures from a reputable concern covering the stone delivered ready to set.

Round Stone Columns.—When estimating the cost of plain turned columns, the cubical contents should be obtained by taking the cube of the largest dimension (measured square) and price as follows:

Obtain the cu. ft. cost of the grade of stock to be used, whether select buff, standard buff, select gray, etc., and add seven times the cu. ft. cost of the rough quarry blocks. This will give the cu. ft. price at which the columns should be estimated. For instance, a turned column 3'-0" in diameter and 36'-0" long would contain 3'x3'x36' or 324 cu. ft. of stone. Select buff is worth $12.00 per cu. ft. in rough blocks. Multiply $12.00 x 7 gives $84.00 per cu. ft. 324 cu. ft. @ $84.00 equals $27,216.00, the cost of one column.

A stone column 38'-0" long is the usual limit of present day machine capacity, so when the column shaft is over 36'-0" long, it is best to get them out in three sections.

Fluted Stone Columns.—Fluted stone columns should be figured on the same basis as plain turned columns described above. After the cost of the plain column has been obtained, the cost of the flutes should be estimated at 50 cts. per lin. ft. of each flute in the column.

Cutting Raised Letters in Indiana Limestone.—The cost of cutting raised letters in Indiana limestone will vary according to the size and style of the letters.

On average work, a stone carver should cut one 6" letter an hr. at the following labor cost per letter:

	Hours	Rate	Total	Rate	Total
Stone carver	1	$. . .	$. . .	$20.00	$20.00

If the letters are 12" high, it will require about 1½ hrs. time per letter, at the following labor cost per letter:

	Hours	Rate	Total	Rate	Total
Stone carver	1.5	$. . . .	$. . . .	$20.00	$30.00

Cutting Sunk Letters in Indiana Limestone.—The cost of cutting sunk letters is considerably less than raised ones.

When cutting letters up to 6" high, a cutter should cut one letter in about 20 minutes at the following labor cost per letter:

	Hours	Rate	Total	Rate	Total
Stone cutter	.33	$. . . .	$. . . .	$20.00	$6.60

If the letters are 12" high, a cutter will cut 1 letter an hr. at the following labor cost per letter:

	Hours	Rate	Total	Rate	Total
Stone cutter	1	$. . . .	$. . . .	$20.00	$20.00

TYPICAL WALL SECTION

Custom Fabricated Limestone Wall Panels*
*Harding & Cogswell, Corp. Bedford, Indiana

Textured Indiana Limestone Facing Panels.

Limestone panels are available as "curtain wall" units from the Harding and Cogswell Corp., Bedford, Indiana. These can be had as wall units spanning a full story height, or as spandrels spanning a full bay. Lengths of up to 15' are possible but the most practical panels will fall in the range of 40 to 80 sq. ft., 120 sq. ft. being the maximum. Thicknesses will depend on panel size and depth of texture but usually fall in the 4" to 5" range. Nine textures are available. The price range for panels in the 40 to 60 sq. ft. size will run from $8.00 to 11.00 per sq. ft. plus shipping and erection costs.

SECTION OF PREASSEMBLED SPANDREL
(3-pcs. stone 14'-0" length epoxied together and assembled on steel frames at our plant)

PLAN OF PREASSEMBLED COLUMN COVER
(3-pcs. plant assembled)

Custom Fabricated Built-Up
Limestone Wall Panels*
*Courtesy Harding & Cogswell, Corp.

Erection costs are estimated to run from $3.50 to $5.00 per sq. ft. including cranes or hoists. On a job in Indianapolis where 346 deep textured, full story height panels were used on construction four stories high as continuous piers a crew of seven using a crain and dual C-clamps erected 23,000 sq. ft. in 23 days. However 750 sq. ft. per day is probably a safer average unless the crew is experienced in this type of construction or the job very simple.

Panels are shipped with no back-up applied. Conventional masonry units can, of course, be used or rigid sheets of insulation board can be field applied to the back of the panels. Or the backing may be metal studs and plasterboard with either batt or insulating lath attached to the studs. Any "U" factor may be designed for.

Panels may also be pre-assembled at the plant from several stones with epoxy bond so that deep sills and returns or special pier or spandrel shapes can be delivered ready for erection in one piece at the job site.

Granite

Granite is an igneous stone. It is extremely hard and weighs 160 lbs. per cu. ft., or 15 to 20 percent more than cut stone. It is one of the most durable materials available for construction but, because of the high costs of quarrying, fabrication, shipping, handling, and setting, considerable time and care must be taken in planning the job and estimating in order to make the best use of the material.

Granite is usually thought of as a facing material for the lower stories of buildings where durability is paramount. However, several tall buildings, such as the 38-story CBS Building in New York City and the Canadian National Bank in Montreal, have been faced with granite full height. It is also now popular as a sidewalk material for the large plazas around new skyscrapers such as the John Hancock Center in Chicago.

Granite can be classified into three basic groups, each having its own design and specification requirements.

Standard building granite is used structurally either self-supporting or as bonded masonry. Granite has a compressive strength of 20,000 p.s.i. and is well suited for bearing walls.

Granite veneer is a facing used for decoration or protection, rather than for its structural value and is applied to masonry backup with stainless steel anchors engineered for the job. Veneers will vary from 7/8" in thickness up to 8", the majority being in the 2" to 4" range. Also being developed today are precast "curtain wall" panels using granite facing. These are supported from the building framing rather than from the backup and present different construction problems altogether.

Masonry granite is granite used as an integral part of a masonry wall in which one of the more rugged finishes are used such as in churches, rambling residences, bridge abutments, and the like.

Granite is available in a wide range of finishes from split faced to mirror finished slabs.

Polished face is a mirror gloss; honed face, a dull gloss; and fine rubbed, a smooth finish with no gloss. Allowable tolerance for these three types is 3/64".

Next are the rubbed finishes. Sand finish is slightly pebbled with a tolerance of 1/10"; spot finish has circular markings with an indefinite pattern and 3/32 surface tolerance.

Hammered surfaced granite comes in three grades: fine or 8-cut with parallel markings not over 3/32"; medium or 6-cut with parallel markings not over 1/8"; and coarse with parallel markings not over 3/32". Tolerances are limited to 7/32" for the first two, 1/8" for the third. These are corrugated finishes which become smoother near arris lines, that is, near the intersections of two dressed faces.

Sawn faces may be had with vertical scorings (V.S.M.), horizontal scorings (H.S.M.), or circular scorings (R.S.M.). The first two are produced by gang saws, the second, by rotary or circular saws. Tolerance is 1/8" with scorings 3/32" deep.

Thermal finishes are produced by flame texturing a sawn surface with torches to flake granite to a dull, uniform, regular rough surface of coarse texture. It is popular for large paving blocks. Tolerance is 1/8".

Jet honing is produced by honing previously cut surfaces with a jet of water and compressed air fed with tiny glass spheres under pressure. Tolerance is 1/8".

Granite is a natural material and colors are varied, often two to three per material. Color may be selected from samples or specified by quarry. As shipping costs can be a major factor in a material as heavy as granite, local quarries with similar colors can effect major savings.

However, when the architect specifies a color by quarry, any substitution should be made only with his written approval or as an alternate, lest the contractor be held for the difference in shipping costs.

Granites with feldspar produce the red, pink, brown, buff, grey and off-white varieties; those with horn blende or mica produce the dark greens and blacks.

Graining as well as color is a consideration; a uniformity of texture, whether fine, medium or coarse, being desirable.

The National Building Granite Quarries Assoc., Inc., P.O. Box 444, Concord, N.H. 03302, publishes a list of granite suppliers with colors and graining available from each. Canada and Sweden are also major sources.

Material Cost of Granite.—As discussed above, several factors determine the price of granite, including color, graining and transportation to the site.

The lower priced polished veneers will include the solid blacks and mottled greys. These will run around $8.00 to $10.00 per sq. ft. in 1" thickness, $10.00 to $12.00 in 2" thickness and $12.00 to $14.00 in 4" thickness. The variegated reds, pinks, and bluish and greenish blacks will cost $10.00 to $15.00 per sq. ft. Sawn finish will cost around $1.75 less per sq. ft. and thermal finish about $1.00 less.

Random ashlar granites are considerably cheaper, running from $4.00 to $7.50 in 4" thicknesses. This is also often priced as so much per cu. ft., the range being $12.00 to $25.00.

Dressed granite, machine cut in structural thicknesses, will run $12.50 to $40.00 a cu. ft. depending upon the number of openings, washes and corners. Hand cut granite, as one might find on a mausoleum, can run to $100.00 or more a sq. ft.

Hand cutting, except on the most elaborate work, has been replaced by precision machines. Even surface carving and lettering is done today by sandblasting. Carved letters 6" to 10" high will cost around $15.00 per letter if cut into the granite, $25.00 if raised from the granite.

Cutting of granite into shapes other than square blocks adds quickly to the cost. One source suggests that if a simple square step, which might run $15.00 per cu. ft., were to have a wash or 1/8" slope on the top surface, the cost would increase 23%; a wash on two surfaces, 30%; a wash on two surfaces plus a drip, 42%; and an ogee molding plus a drip, 55%.

Labor Setting Granite Veneer.—One mason and one helper should set from 30 to 40 sq. ft. of veneer per day, polished work with 1/4" joints taking more time than straight sawn work. Column enclosures can be figured at around 35 sq. ft. per day. Straight run work with no openings and no hoisting involved can be figured at 50 sq. ft. per day.

The following figures are based on one mason and one helper. On large jobs this ratio may be cut down with 3 laborers serving 4 masons providing scaffolding is arranged so men can move from position to position. In addition, hoisting costs may have to be added.

On a low-rise building, the square foot labor cost for 100 sq. ft. of polished granite veneer would be figured as follows:

	Hours	Rate	Total	Rate	Total
1 mason	20	$....	$....	$17.87	$357.40
1 helper	20	....	....	14.95	299.00
Cost per 100 sq. ft.			$....		$656.40
Cost per sq. ft.					6.56

Work involving special trim at openings, copings and moldings would require more time:

	Hours	Rate	Total	Rate	Total
1 mason	26.5	$....	$....	$17.87	$473.56
1 helper	26.5	....	....	14.95	396.18
Cost per 100 sq. ft.			$....		$869.74
Cost per sq. ft.				....	8.70

Artificial Granite.—Cast granite is manufactured in slabs 1½" to 2½" thick as a less expensive substitute for the natural material. It is not as durable and will not take or hold quite the high polish of the true stone. Still, it has been popular on remodeling and store front work. It is set and handled on the job like true granite. The material costs less than half, or in the $4.00 to $6.00 range. One mason and one helper should set 50 to 60 sq. ft. per day.

A Residence Constructed of Sawed Bed Rock Face Granite Ashlar

Labor Setting Granite Ashlar.—The labor cost of setting granite ashlar will vary considerably, depending upon type of wall, height of rises and the skill of the mason. Being thin, sawed bed ashlar is light in weight and easy to handle and when finished or semi-finished, a mason should set 45 to 55 sq. ft. of wall consisting predominantly of low rises per 8-hr. day, and up to 70 sq. ft. of wall, where the ashlar consists principally of high rises.

Marble is classified according to soundness into A, B, C and D grades. The A grade is in general the cheapest and also the best for exterior use. A more detailed discussion of marble as a material is given under the chapter on interior finishes.

Exterior marble veneer will vary in thickness from 7/8" to 2". Curtain wall material is usually 1¼" thick, although if set in a metal frame or precast with a concrete back-up 7/8" material is used.

The finish for exterior marble is either sawn, sand or honed. A polished surface is not recommended.

Marble will vary widely in cost both because of the type selected and the distance it must be shipped. In Chicago 7/8" Class A marble will run from $7.00 to $10.00 per sq. ft.

Jointing for marble is 1/16" to 1/8" for thin veneer, 3/16" for the thicker stock. All bed and face joints must be solidly filled with either a non-staining mastic or mortar. Most work is either pointed or caulked.

Exterior veneer should be anchored to the back-up as follows: two anchors for up to 2 sq. ft.; four for up to 20 sq. ft.; and two for each additional 10 sq. ft. Anchors may be dowels, straps, wire ties or dovetails, but all must be non-corrosive.

In addition to anchors marble veneer must be supported over all openings and at each story height or not more than 20' apart vertically.

Marble may be set against a dampproofed, prebuilt wall in which case it is set out 1" to 1½", anchored and spot mortared 18" on centers; or set along with the back-up in which case it is completely parged on the back with the anchors built into both facing and back-up as the work progresses; or it may be solidly grouted to the back-up with the grouting carefully paced so as not to dislodge the marble. In all cases the mortar must be non-staining. Soffit marble must have liners.

Marble is cleaned with clear water and fibre brushes. Acids will stain the surface.

Two masons and a laborer will set around 8 sq. ft. of 1" veneer per hour in straight runs. Pointing and caulking and hoisting costs must be added.

Labor Cost of 100 Sq. Ft. of 1¼" Marble Veneer Set Against a Prebuilt, Dampproofed Masonry Wall

	Hours	Rate	Total	Rate	Total
Mason	25	$....	$....	$17.87	$446.75
Labor	12.5	....	....	14.95	186.88
Cost per 100 sq. ft.			$....		$633.63
Cost per sq. ft.			....		6.34

Slate is sometimes used as an exterior veneer, the U. S. Gypsum building in Chicago being an outstanding example. When used in this way it is usually 1" to 1¼" thick and furnished with a natural cleft finish. Panel sizes should not exceed 9-6" in length, 5' in width or around 25 sq. ft. in area.

Slate veneer is held in place with anchors and spot mortar. Relief angles should be provided at each story height. All angles and anchors must be non-corrosive. Anchors should be provided four to a slab or every 3 sq. ft. Round anchor holes cut in the edge of the slate to receive wire anchors is the recommended procedure. Mortar should be applied in 6" spots, 18" on center. All joints should be full and are usually 3/8" wide.

Slate will run around $8.00 per sq. ft. for material. Two masons and one laborer should set around 7 sq. ft. per hour.

Labor Cost of 100 Sq. Ft. Slate Veneer

	Hours	Rate	Total	Rate	Total
Mason	28	$....	$....	$17.87	$500.36
Labor	14	....	....	14.95	209.30
Cost per 100 sq. ft			$....		$709.66
Cost per sq. ft				....	7.10

04500 MASONRY RESTORATION
Cutting out Mortar Joints in Old Brickwork Using an Electric Grinder.—On old buildings, where it is necessary to cut out the mortar joints in the old brickwork to a depth of ½" and repoint same, the most efficient method of performing this work is to use an electrically operated carborundum wheel about 5" in diameter and ¼" thick, depending upon the width of the mortar joints. The motor is hung above the scaffolding with a flexible shaft connected to the carborundum wheel.

This method is nearly twice as fast as using an ordinary electric hand saw on account of the weight of the saw (approximately 17 lbs.) which is difficult and awkward to handle by men working on a swinging scaffold.

Either two or four men work off of each scaffold, depending upon the size of the job, and it will require one helper on the ground to each scaffold.

An experienced tuckpointer or operator should grind out lime or lime-cement mortar joints in 350 to 450 sq. ft. of brick wall per 8-hr. day, at the following labor cost per 100 sq. ft.:

	Hours	Rate	Total	Rate	Total
Tuckpointer	2.0	$....	$....	$18.91	$37.82
Helper	1.0	....	....	14.95	14.95
Cost per 100 sq. ft			$....		$52.77
Cost per sq. ft				....	.53

Using ordinary methods, with an electric saw or ginder, a mason will cut out about 25 sq. ft. of joints an hr. at the following labor cost per 100 sq. ft.:

	Hours	Rate	Total	Rate	Total
Mason	4	$....	$....	$18.91	$75.64
Cost per sq. ft				....	.76

Repointing Mortar Joints in Old Brickwork.—After the mortar joints have been cut out, two tuckpointers working on the scaffold, with one helper on the ground should point 400 to 500 sq. ft. of brick wall per 8-hr. day, at the following labor cost per 100 sq. ft.:

	Hours	Rate	Total	Rate	Total
Tuckpointers	3.5	$....	$....	$18.91	$66.19
Helper	1.5	....	....	14.95	22.43
Cost per 100 sq. ft			$....		$88.61
Cost per sq. ft				....	.89

The ordinary hanging or swinging scaffold is 25 ft. long and the men working off of each scaffold can cover approximately 6 feet in height, so that each lift of the scaffold covers 150 sq. ft.

Cleaning Paving Brick Using Pneumatic Chipping Hammers.—A satisfactory method of cleaning brick with pneumatic chipping hammers is to use a 4"x12" plank about 10'-0" long, placed on 2 carpenters' horses and raised at one end 15 to 18 degrees. This is used to hold the brick which may be placed lengthwise across the plank, so that one side and one end can be cleaned without moving. A plank will hold about 56 brick. Using a pneumatic chipper with a hexagon nozzle and a 2" chisel bit, a man should clean 150 to 180 brick an hr. or 1,200 to 1,440 per 8-hr. day.

SANDBLAST CLEANING OF BUILDINGS

The portable compressor and the sandblast are widely used for cleaning buildings, bridges and other structures, either preparatory to repainting or for the purpose of brightening up the surface.

The cost of this work, however, is contingent entirely upon the condition of the old buildings, as some of them have nothing more than many years' accumulation of dirt while others have been painted numerous times, making it necessary to remove several coats of paint before the surface of the brick or stone can be touched. Also, it is often necessary to repoint all of the old mortar joints. The estimator should consider all of these items when preparing the estimate.

The process of sandblasting is to force sand against the surface to be cleaned through a small hose connected with the air compressor, and the force of the sand striking the object to be cleaned removes the old surface. Where it requires a maximum pressure of sand and air to remove the old dirt, a ¾" hose is used but where volume is desired more than high pressure a 1" hose is used.

It usually requires four men to operate a sandblasting machine, as follows: one man attending the machine, one man at the nozzle, and two men on the ground assisting with hose, scaffold, sand, etc.

When estimating quantities, take the entire area of the surface to be cleaned, make no deductions for door or window openings.

On court houses, state capitols, post offices, public libraries, and other ornamental structures, the quantities are computed as above and in addition, balustrades, balusters, etc., are measured on both sides, as it is necessary to go around each baluster separately to clean it.

With a portable compressor supplying air, the average operator can cover a strip 25'-0" wide and 60'-0" to 75'-0" high per 8-hr day. The exact area varies with the quality of the stone encountered but the following table shows the speed at which various stones can be cleaned.

	Sq. Ft. per Min.		Sq. Ft. per Min.
Limestone	7 to 9	Terra Cotta	8 to 10
Marble	3 to 5	Brick or Brownstone	8 to 10
Granite	6 to 8	Sand stone	10 to 12

The quantity of sand required for sandblasting will vary with the method used but where a canvas screen or covering surrounding the scaffold catches most of the sand, it is then carried back to the tank on the ground. If this is done about 75% of the sand is saved for use again. This, however, is governed somewhat by the conditions of the work.

The following costs are for removing and cleaning only the natural accumulation of dirt, and do not include removing paint, oil stains, etc., unless otherwise mentioned.

Maximum quantities are not figured in the tables as average job conditions usually include delays, moving from one job to another, placing and removing scaffold, etc.

Labor Cost of Sandblasting 100 Sq. Ft. of Old Brick or Limestone
Buildings.

	Hours	Rate	Total	Rate	Total
Compressor Operator	0.33	$....	$....	$17.95	$ 5.92
Nozzleman	0.33	....	....	13.04	4.30
Labor	0.67	....	....	12.54	8.40
Cost per 100 sq. ft			$....		$18.62
Cost per sq. ft				....	.19

Add for Equipment rental, sand, and scaffold costs.

Sandblasting Old Brick Work That Has Been Painted.—If the old brick work has been painted and the paint can be blasted off without the necessity of repointing the mortar joints, the costs will run about the same as given above for plain work.

Sandblasting Ornamental Buildings, Such as Post Offices, Courthouses, State Capitols, Public Libraries, Etc., Constructed of Granite or Limestone.—The cost of sandblasting buildings as enumerated above, having exterior elaborations, such as balustrades, gables, pediments, plain or fluted columns, etc., will be higher than plain buildings on account of the extra labor required for scaffolding, sandblasting balusters, columns, caps, etc., that are not found on the average building.

Repointing Mortar Joints After Sandblasting.—On many old buildings that are unusually dirty or that have been painted, it often happens that the mortar is blasted from the joints when removing the paint from the brick. After the job has been cleaned the tuckpointers must go over the entire surface and repoint all of the brick joints.

CHAPTER 10

METALS

CSI DIVISION 5
05100 STRUCTURAL METAL
FRAMING
05200 METAL JOISTS
05300 METAL DECKING
05400 LIGHTGAGE FRAMING
05500 METAL FABRICATIONS

05100 STRUCTURAL METAL FRAMING

The preparation of a structural steel estimate involves a great deal of work because of the large amount of detail required in listing the quantities from the plans, computing the weights of the various sections, making of shop drawings and detailing the steel, estimating the cost of shop fabrication, freight, trucking, and the labor cost of erecting the steel on the job.

The structural steel framework of a building usually consists of anchor bolts, setting plates, base plates, columns, girders, beams, lintels, roof trusses, etc., which are fabricated from standard shapes, such as angles, "S" beams, channels, "W" beams and columns, plates, rods, etc., in combinations designed to give the required strength.

Estimating Quantities of Structural Steel

When estimating the quantity of structural steel required for any job, each class of work should be estimated separately, i.e., column bases, columns, girders, beams, lintels, trusses, etc., as they all involve different labor operations in fabrication and erection.

All standard connections can be taken from the A.I.S.C. Manual of Steel Construction. Be sure to include rivets and bolts for connecting steel to steel.

In addition to the structural framing of the building, it is often necessary for the structural steel contractor to furnish numerous miscellaneous items, such as bearing plates, loose lintels, anchor bolts, etc., although they are set in place by other contractors.

Items to be Included in a Structural Steel Estimate.—When preparing an estimate on structural steel, the following items should be included to obtain the erected price:

1. Cost of structural steel shapes at the mill or warehouse.
2. Freight or trucking of structural steel shapes from rolling mill or warehouse to fabricating shop.
3. Cost of making shop drawings and details for shop fabrication.
4. Cost of assembling and fabricating shapes into columns, girders, beams, trusses, etc. including templates in connection with same.
5. Cost of applying one or more coats of shop paint.
6. Shop overhead expense and profit.
7. Freight on fabricated steel from shop to destination.
8. Cost of erecting steel, including eqiupment rental, setting up and removing cranes, hoisting equipment, unloading steel at site, erecting steel, and field riveting, welded or bolted connections.

Estimating Cost of Structural Shapes at Mill or Warehouse.—After the quantities have been listed from the plans, it will be necessary to compute the cost of the steel sections required.

All prices on structural steel are based on mill shipment from rolling mills in Pittsburgh, Birmingham, Chicago, etc., depending upon the location of the mill where the steel is purchased, plus freight or trucking from rolling mill to fabricating shop.

Where steel is wanted in a hurry, it is sometimes necessary to purchase it from warehouse stocks, which usually costs from 5 to 15 cents per lb. more than when purchased from mills, depending upon quantity.

Another item that must be considered when pricing steel work is the base price on standard sections and the additions thereto. Example: Assuming the base price of structural shapes is $23.00 per 100 lbs. or $460.00 per ton, find the cost of 4"x4"x½" steel angles. The standard classification of extras on the following pages shows that angles of this size cost $1.40 per 100 lbs. more than the base price or $24.40 per 100 lbs.

Warehouse prices vary from warehouse base depending upon size and quantity ordered. On orders of 30,000 lbs. or more, the decrease from base price is 70 cts. per 100 lbs. while on orders from 100 to 399 lbs. the increase over base price is about $8.00 per 100 lbs.

Structural steel prices change so often, depending upon market conditions, that latest quotations should always be obtained before figuring jobs of any size.

Cost of Making Working Drawings and Detailing Structural Steel.—The cost of making shop or working drawings for steel structures is usually figured by the number of sheets estimated to be required for layout, details, erection diagrams, etc. Drawings, 24"x36" average about $300.00 to $500.00 per sheet, including drafting room overhead.

If drawing costs are reduced to a cost per ton, the unit costs will vary considerably, depending upon the type of building, weight of steel, amount of duplication, etc. The ton cost will run considerably less where fairly heavy shapes are used than on light structures using tube columns, large lightweight trusses, etc., as the latter type of construction requires considerable detailing without involving much tonnage.

Steel skeleton construction, such as office buildings, hotels, etc., without unusual features can usually be detailed for $35.00 to $50.00 per ton.

Ordinary mill and factory buildings consisting of columns, beams, light trusses and crane runs, requiring considerable detailing without much duplication, will cost from $50.00 to $65.00 per ton.

Theaters, churches and other structures of this class will cost from $60.00 to $80.00 per ton for details.

Shop Painting of Structural Steel.—After the steel has been fabricated it is customary to give it a coat of paint before it leaves the shop, except where steel is to be encased in concrete.

The cost of painting is usually estimated at a certain price per ton and the quantity of paint required and the surface area to be covered will vary with the class of steel. Ordinary structural beams contain 225 to 275 sq. ft. of surface per ton; plate girders 125 to 175 sq. ft. of surface per ton, while trusses contain 275 to 325 sq. ft. of surface per ton. The average for a job will vary from 175 to 225 sq. ft. per ton of steel.

A gallon of paint will cover 400 to 500 sq. ft. of surface or about 2 tons of steel, at a material cost of $10 per gallon, or $5 per ton.

The labor necessary to paint one ton of structural steel (200 sq. ft.) with one shop coat of paint should cost as follows:

	Hours	Rate	Total	Rate	Total
Labor (spray)	.75	$. . . .	$. . . .	$17.70	$13.28

Basis for Estimating All Classes of Structural Steel Work

Every structural shop has constant or fixed costs that must be considered before estimating the detailed fabricating operations. The following items should be computed before adding the various shop operations required for fabricating the different members.

Approximate Constant or Fixed Costs	Price per 100 lbs.
Base price of steel, f.o.b. Pittsburgh or Chicago	$23.00
Freight or trucking on steel from rolling mill	2.25*
Extras	2.00*
Cost of structural steel, f.o.b. fabrication shop	$27.25
Shop handling charge which includes unloading steel from cars or trucks upon arrival and loading cars or trucks after fabrication	1.25
Tonnage Charges—Painting structural steel in shop	.92
Handling structural steel in shop	.75
Cutting structural shapes to lengths	1.10
Shop Drawings (varies from $30.00 to $60.00 per ton) see explanation on previous pages, average cost	2.25**
Total Constant or Fixed Costs	$33.52

*Varies. **Varies according to type of building.

The above constant or fixed costs must be determined in every estimate before adding special fabrications necessary.

Cost of Fabricating Structural Steel

The cost of fabricating structural steel varies widely, depending upon the type of building, amount of duplication, weight of structural members, number of fabricating operations necessary, etc., as the cost of fabricating light steel members usually runs considerably higher than heavier members on a per ton basis.

The following gives approximate costs, including shop overhead, of the various shop operations and classifications:

Beam and Channel Punching Only.—Where beams and channels are punched only, without any other operation, estimate as follows:

	Price per 100 lbs.
6" to 10" beams and channels	$5.50
12" to 18" beams and channels	4.25
Over 18" beams and channels	4.00

Beams and Channels Framed with Connections.—When beam and channel sections are punched and fabricated with the necessary connections, estimate as follows:

	Price per 100 lbs.
Up to 10" beams and channels	$9.00
12" to 18" beams and channels	7.50
Over 18" beams and channels	6.00

Beam and Angle Framing with Shelf Angles.—Where S-beams are fabricated with shelf angles as illustrated, estimate as follows:

	Price per 100 lbs.
Up to 10" beams with shelf angles	$8.50
12" to 18" beams with shelf angles	6.50
Over 18" beams with shelf angles	6.00

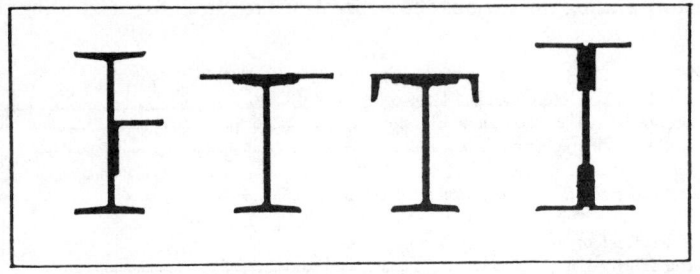

Beam Framing with Shelf Angles	Beam Framing with Plates or Channels	Built-up Plate Girder

Beam Framing with Plates or Channels.—Where beams are fabricated with a plate or channel bolted or welded to same, as illustrated, estimate as follows:

	Price per 100 lbs.
Up to 10" beams	$12.50
12" to 18" beams	9.50
Over 18" beams	8.00

Built-Up Plate Girders.—When estimating the cost of built-up plate girders consisting of plates and angles as illustrated, estimate as follows :............ $12.50

Built-Up Plate Girders with Cover Plates.—Built-up plate girders consisting of plates, angles and cover plates, as illustrated, should be estimated, per 100 lbs. $14.00

Plate and Angle Columns.—Structural steel columns built-up of plates and angles as illustrated, should be estimated as follows:

	Price per 100 lbs.
Columns weighing up to 50 lbs. per lin. ft.	$16.50
Columns weighing from 50 lbs. to 100 lbs. per lin. ft.	13.50
Columns weighing over 100 lbs. per lin. ft.	12.50

Built-up Plate Girder with Cover Plates	Built-up Plate and Angle Columns	Rolled W Sections	Built-up Plate and Angle Columns with Cover Plates

Plate and Angle Columns with Cover Plates.—Structural steel columns built-up of plates and angles with cover plates as illustrated, should be estimated as follows:

Not including Weight of Cover Plates.   Price per 100 lbs.

Columns weighing up to 50 lbs. per lin. ft.	$18.00
Columns weighing 50 lbs. to 100 lbs. per lin. ft.	15.00
Columns weighing 100 lbs. per lin. ft.	14.00

Rolled W Columns.—Structural steel columns consisting of rolled wide flange and similar sections as illustrated, should be estimated as follows:

               Price per 100 lbs.

WF Columns, up to 6"	$12.50
WF Columns, 8"	11.50
WF Columns, over 8"	10.50

Crane Columns.—Structural steel columns for supporting crane runs, etc........ $14.00

Steel Roof Trusses.—When estimating the cost of structural steel roof trusses, the fabricating cost will vary with the weight of the trusses, as follows:

               Price per 100 lbs.

Steel roof trusses, weighing up to 1,000 lbs. each	$24.00
Steel roof trusses, weighing from 1,000 to 2,000 lbs. each	22.50
Steel roof trusses, weighing from 2,000 to 3,000 lbs. each	20.50
Steel roof trusses, weighing from 3,000 to 4,000 lbs. each	18.00
Steel roof trusses, weighing over 4,000 lbs. each	17.00

Knee Bracing.—Fabricating cost per 100 lbs. $23.00
Angle Struts.—Fabricating cost per 100 lbs. 22.50
"X" Framing and Bracing.—Fabricating cost per 100 lbs. 13.00
Tie Rods.—Tie rods ⅝" rounds, fabricating cost per 100 lbs. 45.00
Bracing Rods.—Bracing rods of ¾" to 1" rounds, fabricating cost per 100 lbs. 27.00

Fabricating Steel Bins, Hoppers and Similar Plate Work.— When estimating the cost of fabricating steel plate work, such as steel bins, hoppers and work of this class, figure the fabricating cost, per 100 lbs. 28.00

Floor Plates, Plain or Checkered.—Estimate the weight of the material in the usual manner.

The material usually costs 50 to 60 cts. per lb. depending upon market conditions and the cost of fabricating should be figured at 23 to 30 cts. per lb. additional.

Make Up of a Complete Structural Steel Price

The following items should be included to obtain the total price for structural steel delivered f.o.b. cars or job:

               Price per 100 lbs.

Total steel price, f.o.b. fabricating shop, including constant or fixed costs, as given on previous pages	$33.52*
Shop labor and fabrication cost (varies according to operations required, as given on previous pages), varies	12.00**
Total direct shop costs	$45.52
Business administration, overhead and sales expense, 15%	6.83
	$52.35

	Price per 100 lbs.
Profit on total cost, including overhead, 10%	5.24
	$57.59
Transportation from shop to job—varies...	1.25
Price of structural steel, delivered at job...	$58.84

*This item varies as described on previous pages.
**Fabricating costs vary according to amount of shop labor required as given under cost of the various fabricating operations.

Mill and Shop Inspection of Structural Steel.—Where the structural steel is inspected by one of the testing laboratories, the usual charges for this work are as follows:

Mill inspection is usually figured at $3.00 to $3.50 per ton and shop inspection from $4.00 to $4.50 per ton. Where both mill and shop inspection are let to one firm, a common price is $6.00 to $7.00 per ton for both.

Field Inspection of Structural Steel.—Where field inspection of the erection of the structural steel is required, figure $1,000.00 per week for salary and expenses of a field inspector, plus transportation to and from job.

Erection of Structural Steel

The field cost of handling and erecting structural steel is difficult to estimate. This is due to a number of reasons, such as weather conditions, strikes, delays in receipt of materials, storage facilities, size and type of building, equipment available, labor conditions, number and type of connections, etc.

Any one of the above items may cause a considerable variation in the estimated labor costs and for that reason should be carefully considered when preparing the estimate.

Hauling Structural Steel From Shops to Job.—The cost of hauling structural steel from the shop to the job will vary with the facilities for loading trucks, distance of haul, weight of steel members, etc.

The average cost of hauling structural steel is $25 per ton.

Plant and Equipment for Steel Erection.—The equipment required for steel erection will vary considerably with the size and type of building.

Skeleton frame buildings up to 100'-0"x100'-0" in size will require one derrick to unload the steel from the trucks and erect it in place, while buildings of irregular shape, such as "L" shape, "U" shape, etc., will probably require additional derricks, as one derrick can handle a job 100'-0"x100'-0" but if the job is 75'-0"x125'-0" it will probably require two derricks to set the job, while a "U" shape building may require 3 derricks.

Where the buildings have setbacks at certain story heights, as required by many cities, it will often require 2 derricks on the job, one to unload the steel and hoist it from the street and another to set the steel.

It costs $5,000.00 to $10,000.00 to set up and remove a hoisting engine, derrick, compressor outfit, etc., for a one derrick job, with additional derricks at approximately the same cost.

On buildings covering considerable ground area, such as high schools, office buildings, auditoriums, mill and factory buildings, etc., up to 8 to 10 stories high, it is often more economical to use a crane, mounted on a truck or crawler, depending upon location of the job. Where a movable crane can be used to advantage, that is the most economical method to use, as the crane can be moved about at slight expense.

Steel Erection Contractors.—Most structural steel is erected by contractors making a specialty of steel erection. These contractors can erect structural steel much cheaper than the average general contractor because they have well organized crews of workmen specializing in this branch of work, they also have on hand the most modern equipment, such as crawler cranes, truck cranes, lightweight and modern air tools, such as grinders, drills, rivet busters, impact wrenches, welding machines, and other special tools which are furnished to the job as needed and seldom figured in the job cost; to say nothing of the background of experience of its organization and men.

Erecting Structural Steel in Skeleton Frame Buildings.—When figuring erection on a job of this type, consider carefully the amount of equipment, cranes, etc., required, as described previously.

On a one crane job, an erection gang should unload, handle, erect and connect 9 to 10 tons of steel per 8-hr. day, at the following labor cost per ton:

Erection gang	Hours	Rate	Total	Rate	Total
1 foreman	0.85	$....	$....	$18.20	$ 15.47
2 ir. wkrs. "hooking on"	1.70	....	....	17.70	30.09
1 ir. wkr. giving signals	0.85	....	....	17.70	15.05
4 ir. wkrs. "connecting"	3.40	....	....	17.70	60.18
Cost per ton			$....		$120.79

Add for workmen's compensation and liability insurance.

If the structural connections are welded instead of bolted, add $25.00 to $30.00 per ton.

Erecting Structural Steel in Wall Bearing Buildings.—On wall bearing buildings the ends of all beams and channels usually rest on the exterior masonry walls while the interior framing is carried on masonry walls or structural steel columns. Field connections are usually bolted.

Eight men working together should handle and erect (including bolted connections) about 8 tons of steel per 8-hr. day, at the following labor cost per ton:

	Hours	Rate	Total	Rate	Total
1 Foreman	1	$....	$....	$18.20	$ 18.20
7 Iron Workers	7	....	....	17.70	123.90
Cost per ton			$....		$142.10

Add for workmen's compensation and liability insurance.

Erecting Lightweight Steel for Mill Buildings, Foundries, Auditoriums, Etc.—On light constructed steel buildings, such as machine shops, foundries, mill buildings, garages, auditoriums, etc., where the structural steel consists of steel columns and lightweight roof trusses, requiring considerable handling to set only a small tonnage, the labor costs run higher than on average skeleton frame construction.

A crawler crane or a truck crane should be used for setting the steel.

After the crane has been delivered at the job, the labor unloading, handling, erecting and connecting one ton of steel should cost as follows:

Erection gang	Hours	Rate	Total	Rate	Total
1 Foreman	1.1	$....	$....	$18.20	$ 20.02
7 iron workers	7.7	....	....	17.70	136.29
Cost per ton			$....		$156.31

To the above costs, add workmen's compensation and liability insurance.

Erecting Steel Bins, Hoppers and Similar Plate Work.—When erecting steel bins, material hoppers and similar plate work, an erection crew consisting of 5 iron workers, and an engineer should erect in place and bolt, 4 to 5 tons of steel a day at the following labor cost per ton:

	Hours	Rate	Total	Rate	Total
1 Foreman.......................................	2	$....	$....	$18.20	$ 36.40
5 iron workers..................................	10	....	....	17.70	177.00
Cost per ton...................................			$....		$213.40

Riveting or welding the joints or seams in hopper work will run from $50.00 to $60.00 per ton, additional.

Add for workmen's compensation and liability insurance.

Erecting Structural Steel Using Crawler Mounted Cranes

Riveting Steel After Erection.—The labor cost of riveting steel after erection will vary with the type of construction and the weight of the structural members. Light constructed jobs usually require fewer rivets at each connection than heavy constructed jobs and require more moving of scaffold to drive the same number of rivets. The more rivets that can be driven from one scaffold, the greater the number driven per day.

On light constructed buildings, a riveting gang should drive 275 to 325 rivets per 8-hr. day, at the following labor cost per 100:

	Hours	Rate	Total	Rate	Total
Ir. wkr. heating rivets	2.6	$....	$....	$17.70	$ 46.02
Ir. wkr. catching rivets	2.6	....	....	17.70	46.02
Ir. wkr. driving rivets	5.2	....	....	17.70	92.04
Compressor engineer	1.3	....	....	17.05	22.17
Cost per 100 rivets			$....		$206.25
Cost per rivet				....	2.06

The cost of rivets will vary with their size and length, as steel of different thicknesses requires different size rivets. Example: when riveting steel ½" or less in thickness, it is customary to use ¾" rivets by the required length; ¾" steel requires ⅞" rivets, etc.

Bolting Field Connections.—Most of the structural steel jobs have the sections bolted together instead of riveted.

On work of this kind an iron worker should place 150 to 200 bolts per 8-hr. day, at the following labor cost per 100 :

	Hours	Rate	Total	Rate	Total
Iron worker	5	$....	$....	$17.70	$88.50
Cost per bolt				....	.89

ARC WELDING IN BUILDING CONSTRUCTION

The following information on Arc Welding Structural Steel has been furnished through the courtesy of The Lincoln Electric Company, Cleveland, Ohio.

Arc welding structural steel in the construction of buildings and bridges has become an established practice. To aid the structural designer in its use, the American Welding Society, the American Institute of Steel Construction and the American Association of State Highway Officials have established recommended codes and specifications governing design, fabrication and erection of welded steel structures and applicable codes of most cities, counties and states are in accordance with them. In addition, most fabricating shops now have the equipment, personnel and experience required to fabricate steel members for welded construction and in the field, an increasing number of qualified operators are available for welding field joints during erection.

Advantages of Arc Welded Steel Design.—Advantages of arc welded steel design in building construction are lower overall construction costs; less construction noise, particularly desirable in hospital zones, residential districts, etc., and smooth, clean appearance of erected members which permits leaving structural steel exposed where desired or necessary.

Overall lower costs are realized through reduced steel material requirements and savings in foundation and exterior wall construction. When a structure with welded connections is properly designed, savings in steel material are gained from the following :

Gross section, rather than effective net section, is used for design, permitting lighter members.

Rigid connections provide continuous beam action, permitting a further reduction in member sections.

Connection steel, such as splice plates, gussets, etc., are greatly reduced—sometimes eliminated.

With beams of less depth, height of building may be reduced, saving column steel and exterior wall construction.

In addition, overall costs for detailing, fabrication, freight, trucking, handling and erection are lower due to reduced weight, fewer pieces and simplified connections.

On many structures, where designs and estimates were made for both riveted and welded construction, savings up to 15 percent on structural steel costs were obtained by using the welded method.

Conditions Affecting Arc Welding Costs

Best results and lowest costs are obtainable only when proper plate preparation and welding procedure is followed. The following conditions have considerable bearing on the cost and quality of welded joints.

Fit-Up.—Care in cutting, forming and handling shapes to be welded, to avoid poor fit-up, is a major factor in the cost and performance of welded joints. However, a gap of 1/64" to 1/32" is useful in preventing angular distortion and weld cracking. The accompanying graph shows how various sizes of gaps affect welding speeds for square or grooved butt welds. Fillet welds are affected by oversize gaps in a similar manner.

Position of Joints.—The position of the joint has considerable effect on the speed and case of welding. Wherever practical, welds should be made in the downhand position with the joint level. Vertical or overhead welds require much more time and skill.

Foreign Matter in Joint.—Excessive scale, paint, oil, rust, etc., all tend to interfere with welding and should be removed to obtain best speeds and results.

Build-Up or Overwelding.—Any amount of weld metal in addition to that actually needed for the specified strength is useless, costly, wasteful and, in some instances, actually harmful. For butt welds, there should be just enough build-up (no more than 1/16") to make sure weld is flush with the plate. Excessive build-up not only wastes weld metal but increases welding time.

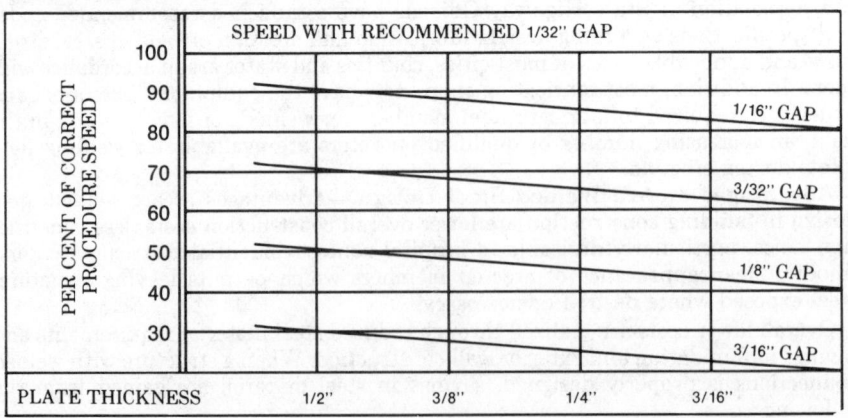

Effect of Gap Size on Welding Speeds for Flat Butt Joints
Welded on Both Sides

WELDED JOINTS
Standard symbols

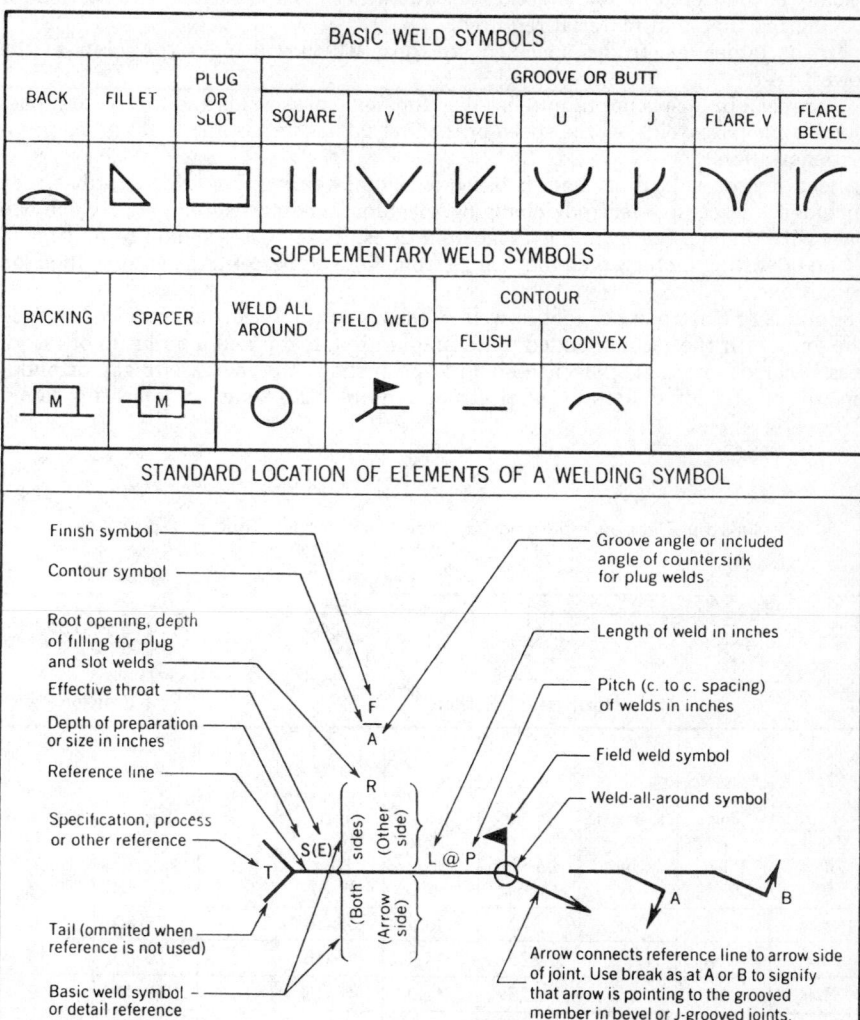

BASIC WELD SYMBOLS									
BACK	FILLET	PLUG OR SLOT	GROOVE OR BUTT						
			SQUARE	V	BEVEL	U	J	FLARE V	FLARE BEVEL

SUPPLEMENTARY WELD SYMBOLS					
BACKING	SPACER	WELD ALL AROUND	FIELD WELD	CONTOUR	
				FLUSH	CONVEX

STANDARD LOCATION OF ELEMENTS OF A WELDING SYMBOL

Finish symbol

Contour symbol

Root opening, depth of filling for plug and slot welds

Effective throat

Depth of preparation or size in inches

Reference line

Specification, process or other reference

Tail (ommited when reference is not used)

Basic weld symbol or detail reference

Groove angle or included angle of countersink for plug welds

Length of weld in inches

Pitch (c. to c. spacing) of welds in inches

Field weld symbol

Weld-all-around symbol

Arrow connects reference line to arrow side of joint. Use break as at A or B to signify that arrow is pointing to the grooved member in bevel or J-grooved joints.

Note:

Size, weld symbol, length of weld and spacing must read in that order from left to right along the reference line. Neither orientation of reference line nor location of the arrow alter this rule.

The perpendicular leg of △, V, ⊬, ⟨ weld symbols must be at left.

Arrow and Other Side welds are of the same size unless otherwise shown. Dimensions of fillet welds must be shown on both the Arrow Side and the Other Side Symbol.

The point of the field weld symbol must point toward the tail.

Symbols apply between abrupt changes in direction of welding unless governed by the "all around" symbol or otherwise dimensioned.

These symbols do not explicitly provide for the case that frequently occurs in structural work, where duplicate material (such as stiffeners) occurs on the far side of a web or gusset plate. The fabricating industry has adopted this convention: that when the billing of the detail material discloses the existence of a member on the far side as well as on the near side, the welding shown for the near side shall be duplicated on the far side.

Courtesy of American Institute of Steel Construction, Inc.

Procedures for Making Various Types of Welded Joints

The following tables give procedures for making various types of welded joints usually encountered in field erection of structural steel and furnish information for calculating time and material required.

In the tables, electrodes specified are those which will make the joints at the lowest cost.

Arc speed in inches per minute is given for single pass welds and for the first pass in multiple pass welds, as the speed of the first pass is important in obtaining proper penetration.

Feet of joint welded per hour is based on actual welding time. No factor has been included for set-up, electrode changing, cleaning, moving scaffold, etc., which will vary with the job. For overall average quantities, these figures should be multiplied by an operating factor which may range from 25 to 50 percent, depending upon job conditions.

Pounds of electrode per foot of weld is the amount of electrode required to weld the joint with the recommended plate preparation, fit-up and a build-up of 1/16" or less—includes normal spatter loss and 2" stub ends. Any increase in gap or build-up will greatly affect amount of electrode required and time for depositing same.

Estimating Data on Fillet and Lap Welds in Horizontal and Flat Positions

Horizontal Position   Flat Position

Plate Thick-ness (In.)	Gauge Size of Fillet (In.)	Electrode Size (In.)	Current (Amps.)	Electrode Melt-off Rate (In. per Min.)	Arc. Speed (In. per Min. for First Pass)	Passes or Beads	Ft. of Joint Welded Per Hr. (100% Operating Factor)	Lbs. of Electrode Per Ft. of Weld
3/16	5/32	1/8	170	15.0	15-16	1	75-80	.10
1/4	3/16	5/32	225	14.0	15-16	1	75-80	.15
5/16	1/4	3/16	275	12.0	14-15	1	70-75	.19
3/8	5/16	1/4	350-375	9.6-10.2	12-13	1	60-65	.30
1/2	7/16	1/4	350-375	9.6-10.2	10-11	2	28-31	.57
3/4	5/8	1/4	350-375	9.6-10.2	10-11	3-4	16-14	1.34
1	3/4	1/4	350-375	9.6-10.2	10-11	5	9-10	1.66

Preparation: square edge.

Fit-Up: recommended gap, 1/32".

Electrode: iron powder E6024 AC. Use E6014 AC electrodes when weldment cannot be positioned close enough to flat for E6024 AC electrodes.

Estimating Data on Fillet and Lap Welds in Vertical Position, Welded Up*

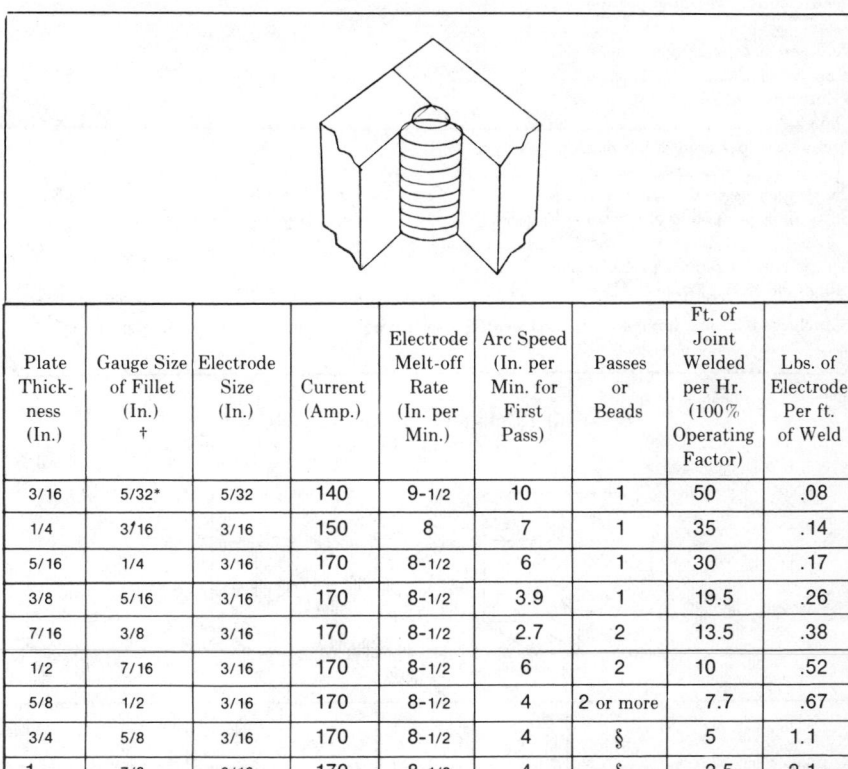

Plate Thickness (In.)	Gauge Size of Fillet (In.) †	Electrode Size (In.)	Current (Amp.)	Electrode Melt-off Rate (In. per Min.)	Arc Speed (In. per Min. for First Pass)	Passes or Beads	Ft. of Joint Welded per Hr. (100% Operating Factor)	Lbs. of Electrode Per ft. of Weld
3/16	5/32*	5/32	140	9-1/2	10	1	50	.08
1/4	3/16	3/16	150	8	7	1	35	.14
5/16	1/4	3/16	170	8-1/2	6	1	30	.17
3/8	5/16	3/16	170	8-1/2	3.9	1	19.5	.26
7/16	3/8	3/16	170	8-1/2	2.7	2	13.5	.38
1/2	7/16	3/16	170	8-1/2	6	2	10	.52
5/8	1/2	3/16	170	8-1/2	4	2 or more	7.7	.67
3/4	5/8	3/16	170	8-1/2	4	§	5	1.1
1	7/8	3/16	170	8-1/2	4	§	2.5	2.1

*Weld 5/32" fillet vertically down. †T-joint welded both sides with this fillet will have plate strength. §Total no. of passes varies with operator.

Preparation: square edge.
Fit-up: recommended gap, 1/32 "; maximum gap 1/16". If greater gaps must be welded, use same procedure but add width of gap to fillet size.
Electrode: E6010
Polarity: electrode positive.

The following are actual production quantities of field welding on two different jobs located in different parts of the country and will give the estimator an idea of the amount of welding and labor required.

Description	19 Story Off Bldg.	18 Story Hospital
Size of building on the ground	100'x100'	140'x200'
Total tonnage of steel welded, tons	1,170	2,274
Fillet per ton of steel, lineal inches	63.5	50.68
Electrode used per ton of steel, lbs.	4.0	1.7

Description	19 Story Off Bldg.	18 Story Hospital
Electric current required per ton of steel, kw. hrs.	8.1	6.05
Weld per lb. of electrode bought, lin. inches..	16.00	21.34
Welding man hours per ton of steel, hours..	1.51	1.80*
Steel welded per month, per machine, tons...	128.00	. . . .
Lin. ft. fillet per 8-hr. day, per welder	28.0	36.0
Lin ft. fillet per pound of electrode bought..	1.33	1.78
Electric current required to melt one lb. of electrode, kw. hrs.	2.90	3.63

*Includes 0.26 hrs. foreman, 0.91 hrs. welder, 0.26 hrs. helpers and 0.37 hrs. hoist. engr. time.

Estimating Data on Fillet and Lap Welds in Overhead Position

Plate Thickness (In.)	Gauge Size of Fillet (In.) †	Electrode Size (In.)	Current (Amp.)	Electrode Melt-off Rate (In. per Min.)	Arc Speed (In. per Min. for First Pass)	Passes or Beads	Ft. of Joint Welded per Hr. (100% Operating Factor)	Lbs. of Electrode per ft. of Weld
3/16	5/32	5/32	140	9-1/2	9	1	45	.09
1/4	3/16	3/16	160	8-1/4	7	1	35	.14
5/16	1/4	3/16	160	8-1/4	5.7	1	28.5	.18
3/8	5/16	3/16	160	8-1/4	3.7	2	18.5	.27
7/16	3/8	3/16	160	8-1/4	7-3/4	3	12.5	.40
1/2	7/16	3/16	160	8-1/4	7	§	9.2	.55
5/8	1/2	3/16	160	8-1/4	6	§	7.0	.72
3/4	5/8	3/16	160	8-1/4	6	§	4.5	1.2
1	7/8	3/16	160	8-1/4	6	§	2.2	2.3

†T-joint welded both sides with this fillet will have plate strength.
§Total number of passes varies with operators.

Preparation: square edge.

Fit-up: recommended gap, 1/32", maximum gap, 1/16". If greater gaps must be welded, use same procedure but add width of gap to fillet size.

Approximate Prices of Iron and Steel Products

All prices for structural steel are based on carload shipments from mills, Pittsburgh, Pa., Chicago, Ill., or Birmingham, Ala. Freight must be added from nearest mill point to destination.

Prices on steel should always be obtained from mills or nearest jobber, as steel prices are subject to frequent change.

Structural steel shapes are currently quoted at the basic price of $23.00 per 100 pounds for ASTM A36 steel.

Extras Over Base Price on Structural Shapes

S Shape Beams

Size Inches	Extra 100 Lbs.	Size Inches	Extra 100 Lbs.
3"	$6.25	10"	$2.25
4"	5.25	12" 31.8 and 35 lbs.	2.25
5"	5.00	12" 40.8 and 50 lbs.	2.25
6"	2.75	15"	2.25
7"	2.75	18"	2.25
8"	2.35	20"	2.25
		24"	2.25

W Shape Beams

Size Inches	Weight Pounds	Extra 100 lbs.	Size Inches	Weight Pounds	Extra 100 Lbs.
6"	8.5	$4.00	10"	11.5	$3.50
6"	12 and 16	4.00	10"	15 to 19	3.50
6"	15.5, 20 and 25	2.75	10"	22 to 30	2.75
8"	10	3.50	10"	33 to 45	2.00
8"	13 and 15	3.50	10"	49 to 112	1.25
8"	18 and 21	2.75	12"	14	3.50
8"	24 to 28	2.25	12"	16.5 to 22	3.50
8"	31 to 67	1.50	12"	26 to 35	2.75

W Shape Beams

Size Inches	Weight Pounds	Extra 100 Lbs.	Size Inches	Weight Pounds	Extra 100 Lbs.
12"	40 to 50	$1.75	21"	62 to 147	$1.75
12"	65 to 190	1.25	24"	68 to 162	1.75
14"	30 to 38	2.00	27"	84 to 114	1.75
14"	43 to 53	1.75	30"	99 to 132	2.00
14"	61 to 136	1.75	30"	172 to 210	2.00
16"	36 to 57	1.50	33"	118 to 152	2.25
16"	67 to 100	1.00	33"	200 to 240	2.25
18"	50 to 71	1.75	36"	135 to 194	2.60
18"	76 to 119	1.75	36"	230 to 300	2.60
21"	62 to 93	1.75			

M Shape Beams

Size Inches	Weight Pounds	Extra 100 lbs.	Size Inches	Weight Pounds	Extra 100 lbs.
4"	13	$3.50	6"	20	$2.75
5"	18.9	3.50	8"	6.5	3.50

Angles

Size Inches		Extra 100 Lbs.	Size Inches		Extra 100 Lbs.
3 x2	X3/16	$2.30	5 x3	X1/4	$1.90
3 x2	X1/4 to 1/2	2.10	5 x3	X5/16	1.90
3 x21/2	X1/4 to 1/2	2.10	5 x31/2	X3/8 to 3/4	1.75
3 x3	X3/16	2.30	5 x5	X5/16	2.70
3 x3	X1/4 to 1/2	2.10	5 x5	X3/8 to 3/4	2.50
31/2x21/2	X1/4 to 5/16	2.70	6 x31/2	X5/16 to 1/2	2.50
31/2x3	X1/4 to 5/16	2.10	6 x4	X5/16	2.80
31/2x31/2	X1/4 to 5/16	1.90	6 x4	X3/8 to 7/8	2.50
4 x3	X1/4	2.05	6 x6	X3/8 to 1	2.50
4 x3	X5/16	2.05	7 x4	all thicknesses	3.00
4 x31/2	X1/4 to 5/16	1.95	8 x4	all thicknesses	3.00
4 x31/2	X3/8 to 1/2	1.80	8 x6	all thicknesses	3.00
4 x4	X1/4 to 5/16	1.55	8 x8	all thicknesses	2.75
4 x4	X3/8 to 3/4	1.40	9 x4	all thicknesses	3.00

Channels—Standard

Size Inches	Extra 100 Lbs.	Size Inches	Extra 100 Lbs.
3"	$5.00	8"	$3.00
4"	4.00	9"	3.00
5"	3.60	10"	2.50
6"	3.00	12"	2.00
7"	3.00	15"	2.00

Zees

Size	Extra 100 lbs.
Z3 x 6.7 to 12.6 lb.	$5.00
Z4 x 8.2 and 10.3 lb.	5.00

Estimating the Weight of Wrought Iron, Steel, or Cast Iron

When tables of weights are not handy, the following rules will prove of value to the estimator when computing the weights of wrought iron, steel, or cast iron.

Weight of Wrought Iron.—One cubic foot of wrought iron weighs 480 pounds. One square foot of wrought iron 1-inch thick weighs 40 pounds. One square inch of wrought iron one foot long weighs 3 1/3 lbs.

To find the weight of one square foot of flat iron of any thickness, multiply the thickness in inches by 40, and the result will be the weight of the iron in pounds.

To find the weight of one lineal foot of wrought iron bar of any size, multiply the cross sectional area in square inches by 3 1/3, and the result will be the weight per lineal foot.

Weight of Steel.—One cubic foot of steel weighs 489.6 pounds, or 2 percent more than wrought iron. One square foot of steel 1-inch thick weighs 40.8 pounds. A piece of steel 1-inch square and one foot long weighs 3.4 pounds.

To find the weight of one lineal foot of steel bar of any size, multiply the cross sectional area in square inches by 3.4, and the result will be the weight of the steel in pounds. If the weight per lineal foot is known, the exact sectional area in square inches may be obtained by dividing the weight by 3.4.

Weight of Cast Iron.—One cubic foot of cast iron weighs 450 pounds. One square foot of cast iron 1-inch thick weighs 37½ pounds. A piece of cast iron 1-inch square and one foot long weighs 3 1/8 pounds. One cubic inch of cast iron weighs .26 pound.

Weights of Square and Round Bars

Thickness or Diameter in Inches	Weight of Square Bar 1 foot long	Weight of Round Bar 1 foot long	Thickness or Diameter in Inches	Weight of Square Bar 1 foot long	Weight of Round Bar 1 foot long
1/4	.213	.167	1 11/16	....	7.60
5/16	.332	.261	3/4	10.41	8.18
3/8	.478	.376	13/16	....	8.773
7/16	.651	.511	7/8	11.95	9.39
1/2	.850	.668	2	13.60	10.68
9/16	1.076	.845	1/8	15.35	12.06
5/8	1.382	1.04	1/4	17.21	13.52
11/16	1.607	1.26	3/8	....	15.06
3/4	1.913	1.50	1/2	21.25	16.69
13/16	2.240	1.76	5/8	23.43	18.40
7/8	2.603	2.04	3/4	25.71	20.20
15/16	....	2.35	7/8	....	22.07
1	3.400	2.67	3	30.60	24.03
1/16	....	3.01	1/8	....	26.08
1/8	4.303	3.38	1/4	35.91	28.21
1/4	5.313	4.17	3/8	....	30.42
5/16	....	4.6	1/2	41.65	32.71
3/8	6.428	5.05	5/8	....	35.09
7/16	....	5.52	3/4	47.81	37.55
1/2	7.650	6.01	4	54.40	42.73
5/8	8.928	7.05			

Weights of Flat Steel Bars

Thickness In Inches	Width of Bars 1"	2"	3"	4"	5"	6"
3/16	.638	1.28	1.91	2.55	3.19	3.83
1/4	.850	1.70	2.55	3.40	4.25	5.10
5/16	1.06	2.12	3.19	4.25	5.31	6.38
3/8	1.28	2.55	3.83	5.10	6.38	7.65
7/16	1.49	2.98	4.46	5.95	7.44	8.95
1/2	1.70	3.40	5.10	6.80	8.50	10.20
9/16	1.92	3.83	5.74	7.65	9.57	11.48
5/8	2.12	4.25	6.38	8.50	10.63	12.75
11/16	2.34	4.67	7.02	9.35	11.69	14.03
3/4	2.55	5.10	7.65	10.20	12.75	15.30
13/16	2.76	5.53	8.29	11.05	13.81	16.58
7/8	2.98	5.95	8.93	11.90	14.87	17.85
15/16	3.19	6.38	9.57	12.75	15.94	19.13
1	3.40	6.80	10.20	13.60	17.00	20.40
1 1/16	3.61	7.22	10.84	14.45	18.06	21.68
1 1/8	3.83	7.65	11.48	15.30	19.13	22.95
1 3/16	4.04	8.08	12.12	16.15	20.19	24.23
1 1/4	4.25	8.50	12.75	17.00	21.25	25.50
1 5/16	4.46	8.93	13.39	17.85	22.32	26.78
1 3/8	4.67	9.35	14.03	18.70	23.38	28.05

Thickness In Inches	Width of Bars					
	1"	2"	3"	4"	5"	6"
1 7/16	4.89	9.78	14.66	19.55	24.44	29.33
1 1/2	5.10	10.20	15.30	20.40	25.50	30.60
1 9/16	5.32	10.63	15.94	21/25	26.57	31.88
1 5/8	5.52	11.05	16.58	22.10	27.63	33.15
1 11/16	5.74	11.47	17.22	22.95	28.69	34.43
1 3/4	5.95	11.90	17.85	23.80	29.75	35.70
1 13/16	6.16	12.33	18.49	24.65	30.81	36.98
1 7/8	6.38	12.75	19.13	25.50	31.87	38.25
1 15/16	6.59	13.18	19.77	26.35	32.94	39.53
2	6.80	13.60	20.40	27.20	34.00	40.80

Weights and Dimensions of S Shape Beams

Size in Inches	Weight Per Ft. Lbs.	Thickness of Web Inches	Width of Flange Inches	Size in Inches	Weight Per Ft. Lbs.	Thickness of Web Inches	Width of Flange Inches
3	5.7	0.170	2.330	5	14.75	0.494	3.284
	7.5	0.349	2.509	6	12.5	0.230	3.330
4	7.7	0.190	2.660		17.25	0.465	3.565
	9.5	0.326	2.796	7	15.3	0.250	3.660
5	10	0.210	3.000		20	0.450	3.860
8	18.4	0.270	4.000	18	70	0.711	6.251
	23	0.441	4.171	20	66	0.505	6.255
10	25.4	0.310	4.660				
	35	0.594	4.944		75.0	0.641	6.391
12	31.8	0.350	5.000		86	0.660	7.053
	35	0.428	5.078		96	0.800	7.200
	40.8	0.460	5.250	24	80	0.500	7.000
	50	0.687	5.477		90.0	0.625	7.124
15	42.9	0.410	5.500		100	0.745	7.247
	50	0.550	5.640		106	0.620	7.875
18	54.7	0.460	6.000		121	0.800	8.048

Weights of American Standard Channels

Depth of Channels, Inches	Weight Per Ft., Pounds	Thickness of Web, Inches	Width of Flange, Inches	Depth of Channels, Inches	Weight Per Ft., Pounds	Thickness of Web, Inches	Width of Flange, Inches
3	6.00	0.362	1.602	9	20.00	0.452	2.652
	5.00	0.264	1.504		15.00	0.288	2.488
	4.10	0.170	1.410		13.40	0.230	2.430
4	7.25	0.325	1.725	10	30.00	0.676	3.036
	5.40	0.180	1.580		25.00	0.529	2.889
5	9.00	0.330	1.890		20.00	0.382	2.742
	6.70	0.190	1.750		15.30	0.240	2.600
6	13.00	0.440	2.160	12	30.00	0.513	3.173
	10.50	0.318	2.038		25.00	0.390	3.050
	8.20	0.200	1.920		20.70	0.280	2.940
7	14.75	0.423	2.303	15	50.00	0.720	3.720
	12.25	0.318	2.198		40.00	0.524	3.524
	9.80	0.210	2.090		33.90	0.400	3.400
8	18.75	0.490	2.530				
	13.75	0.307	2.347				
	11.50	0.220	2.260				

Weights of Steel Angles

Size in Inches	Weight per Ft. Lbs.	Size In Inches	Weight per Ft. Lbs.
1/2X 1/2X 1/8	.38	2 x2 X 1/8	1.65
5/8X 5/8X 1/8	.48	3/16	2.44
3/4X 3/4X 1/8	.59	1/4	3.19
3/4X 3/4X3/16	.84	5/16	3.92
7/8X 7/8X 1/8	.70	3/8	3.70
1 X 5/8X 1/8	.64	1/2	6.00
1 x1 X 1/8	.80	21/4X11/2X3/16	2.28
3/16	1.16	21/4X21/4X 1/8	1.86
1/4	1.49	3/16	2.75
11/8X11/8X 1/8	.91	1/4	3.62
11/4X11/4X 1/8	1.01	5/16	4.50
3/16	1.48	21/2X11/2X3/16	2.44
1/2	1.92	1/4	3.19
13/8X 7/8X 1/8	.91	21/2X2 X3/16	2.75
11/2X11/2X 1/8	1.23	1/4	3.62
3/16	1.80	5/16	4.50
1/4	2.34	3/8	5.30
5/16	2.86	1/2	6.74
3/8	3.35	21/2X21/2X 1/8	2.08
13/4X13/4X 1/8	1.44	3/16	3.07
3/16	2.12	1/4	4.10
1/4	2.77	5/16	5.00
5/16	3.39	3/8	5.90
2 X11/4X3/16	1.96	1/2	7.70
1/4	2.55	2 x2 X3/16	3.07
2 X11/2X 1/8	1.44	3 x3 X 1/8	2.50
3/16	2.12	3/16	3.71
1/4	2.77		
5/16	3.39		

Weights of Structural Steel Angles

Size in Inches	Weight per Ft. Lbs.	Size in Inches	Weight per Ft. Lbs.	Size in Inches	Weight per Ft. Lbs.
3 x2 X 1/4	4.1	4 x3 X 5/8	13.6	6x31/2X 3/8	11.7
5/16	5.0	3/4	16.0	7/16	13.5
3/8	5.9	4 X31/2X 1/4	6.2	1/2	15.3
3 x21/2X 1/4	4.5	5/16	7.7	5/8	18.9
5/16	5.6	3/8	9.1	6x4 X5/16	10.3
3/8	6.6	7/16	10.6	3/8	12.3
1/2	8.5	1/2	11.9	7/16	14.3
3 x3 X 1/4	4.9	4 x4 X 1/4	6.6	1/2	16.2
5/16	6.1	5/16	8.2	9/16	18.1
3/8	7.2	3/8	9.8	5/8	20.0
7/16	8.3	7/16	11.3	3/4	23.6
1/2	9.4	1/2	12.8	7/8	27.2
5/8	11.5	5/8	15.7	6x6 X 3/8	14.9
31/2X21/2X1/4	4.9	3/4	18.5	7/16	17.2
5/16	6.1	41/2X3 X 3/8	9.1	1/2	19.6
3/8	7.2	5 x3 X5/16	8.2	9/16	21.9
1/2	9.4	3/8	9.8	5/8	24.2
5/8	11.5	7/16	11.3	3/4	28.7
31/2X3 X 1/4	5.4	1/2	12.8	7/8	33.1
5/16	6.6	3/4	18.5	1	37.4

Size in Inches	Weight per Ft. Lbs.	Size in Inches	Weight per Ft. Lbs.	Size in Inches	Weight per Ft. Lbs.
3/8	7.9	5 x31/2x5/16	8.7	7x31/2x 3/8	13.0
1/2	10.2	3/8	10.4	7/16	15.0
31/2x31/2x 1/4	5.8	7/16	12.0	1/2	17.0
5/16	7.2	1/2	13.6	5/8	21.0
3/8	8.5	5/8	16.8	8x31/2x 1/2	18.7
7/16	9.8	5 x4 x 3/8	11.0	8x6 x 1/2	23.0
1/2	11.1	1/2	14.5	3/4	33.8
5/8	13.6	5 x5 x 3/8	12.3	8x8 x 1/2	26.4
4 x3 x 1/4	5.8	7/16	14.3	5/8	32.7
5/16	7.2	1/2	16.2	3/4	38.9
3/8	8.5	5/8	20.0	7/8	45.0
7/16	9.8	3/4	23.6	1	51.0
1/2	11.1	6 x31/2x5/16	9.8	11/8	56.9

Weight of Steel Tees

Size Flange by Stem in Inches	Pounds Per Lineal Foot — Thickness in Inches					
	1/8	3/16	1/4	5/16	3/8	1/2
3/4X 3/4	0.60	....	....	....	....	....
7/8X 7/8	.73	....	....	....	....	....
1 x1	.90	1.20	....	....	....	....
11/4x11/4	....	1.55	1.98	....	....	....
11/2x11/2	....	1.88	2.43	....	....	....
13/4x13/4	....	2.30	2.90	....	....	....
2 x11/2	....	....	3.10	....	....	....
2 x2	....	....	3.60	4.40	....	....
21/4x21/4	....	....	4.10	4.90	....	....
21/2x21/2	....	....	4.60	5.50	6.4	....
21/2x23/4	....	....	....	5.9	6.8	....
21/2x3	....	....	....	6.1	7.2	....
3 x21/2	....	....	....	6.1	7.2	....
3 x3	....	....	....	6.7	7.8	....
31/2x3	....	....	....	....	8.5	....
3 x4	....	....	....	....	9.3	....
3 x31/2	....	....	....	....	8.5	....
31/2x31/2	....	....	....	....	9.2	11.7
31/2x4	....	....	....	....	9.8	....
4 x21/2	....	....	....	....	8.5	....
4 x3	....	....	....	....	9.2	....
4 x4	....	....	....	....	10.5	13.5
4 x5	....	....	....	....	11.9	15.3
41/2x21/2	....	....	....	....	9.2	....
41/2x3	....	....	....	8.6	10.0	....
5 x3	....	....	....	....	11.0	13.6
61/2x61/2	....	....	....	....	....	19.8

Weights of Standard Diamond Steel Floor Plates

Thickness Inches	Weight Per Sq. Ft. Pounds	Thickness Inches	Weight Per Sq. Ft. Pounds
1/8	8.00	5/16	13.75
3/16	8.75	3/8	16.25
1/4	11.25	1/2	21.5

Weights of Steel M Shapes

Size	Weight Per Ft., Pounds	Size	Weight Per Ft., Pounds
M4x13	13.0	M 8x 6.5	6.5
M5x18.9	18.9	M10x 9	9.0
M6x 4.4	4.4	M12x11.8	11.8
M6x20	20.0	M14x18	18.0

Weights of Round Head Structural Rivets

Diam. and Length Inches	Weight per 100 Rivets Lbs.	Diam. and Length Inches	Weight per 100 Rivets Lbs.	Diam. and Length Inches	Weight per 100 Rivets Lbs.	Diam. and Length Inches	Weight per 100 Rivets Lbs.
½x1	9.49	⅝x4½	46.76	¾x6	87.61	⅞x6	122.54
1½	12.24	5	51.10	6½	93.80	6½	130.97
2	14.99	5½	55.45	7	99.97	7	139.40
2½	17.74	6	59.80	7½	106.15	7½	147.83
3	20.49	¾x1	25.81	8	112.33	8	156.26
3½	23.24	1½	31.99	⅞x1½	46.67	1x2½	86.82
4	25.99	2	38.17	2	55.10	3	97.80
⅝x1	16.38	2½	44.35	2½	63.53	3½	108.79
1½	20.68	3	50.53	3	71.96	4	119.77
2	24.98	3½	56.71	3½	80.39	4½	130.76
2½	29.28	4	62.89	4	88.82	5	141.74
3	33.58	4½	69.07	4½	97.25	5½	152.73
3½	37.88	5	75.25	5	105.68	6	163.71
4	42.18	¾x5½	82.05	5½	114.11		

Weights of Steel Plates in Pounds per Square Foot

Thickness	lb./ sq. ft.	Thickness	lb./ sq. ft.
3/16	7.65	3/4	30.60
1/4	10.20	7/8	35.70
5/16	12.75	1	40.80
3/8	15.30	1 1/4	51.00
7/16	17.85	1 1/2	61.20
1/2	20.40	1 3/4	71.40
5/8	25.50	2	81.60

Weight of Corrugated Sheets in Pounds Per 100 Square Feet

Gauge Nos.	Black and Painted 2, 2½ and 3-Inch Corrug.	1¼-Inch Corrug.	Galvanized 2, 2½ and 8-Inch Corrug.	1¼-Inch Corrug.
16	271	...	286	...
18	217	...	232	...
20	163	170	178	185
22	136	142	151	157
24	110	114	125	129
26	83	86	98	101
27	76	79	91	94
28	68	72	85	87

JUNIOR STEEL BEAMS

Junior beams, (M shapes), are lightweight hot rolled structural beams which are used as secondary floor and roof beams in schools, stores, apartments, hospitals and other types of light occupancy buildings. They are well adapted for purlins in mill buildings.

They are rolled to ASTM specifications, and are made in 6", 8", 10", 12", and 14" sizes.

Sizes and properties are given below:

Designation	Depth in Inches	Wt. per Ft. Lbs.	Web Thick. Inches	Flange Wdth. Inches	Area Inches
M 6 X 4.4	6	4.4	.114	1.844	1.29
M 6 X 20	6	20.0	.250	5.938	5.89
M 8 X 6.5	8	6.5	.135	2.281	1.92
M10 X 9	10	9.0	.157	2.690	2.65
M12 X 11.8	12	11.8	.177	3.065	3.47
M14 X 18	14	18.0	.215	4.000	5.10

The cost of cutting, punching holes, and coping junior beams will run approximately as given below although when much fabricating is required it is advisable to refer plans to a local fabricator or warehouse for a subbid.

Top cope, per end	$ 4.00
Bottom cope and seat angles, per end	10.00
Cutting end on diagonal, per end	5.00
Anchor holes in web, per hole	1.00
Conduit holes in web, per hole	1.00

Junior steel beams can be used in conjunction with monolithically finished concrete floors or precast gypsum slabs laid over the top flange of the junior beams.

By encasing the top flanges in concrete, the floor is made rigid and does not require bridging in light occupancy buildings.

Where the slab is poured on metal lath or where gypsum plank is used, adjustable steel rigid bridging is furnished.

When a poured-in-place concrete floor is used, the forms are easily constructed by cutting short pieces of lumber to rest on the lower flanges of the beams, which in turn support the floor sheathing.

The reinforcement of the concrete slab usually consists of welded wire fabric, the size and spacing of the wire depending upon the span of the slabs and the loads to be carried.

Where joists rest on walls or bearing partitions, anchors are provided. Where the tops of junior beams are flush with tops of supporting beams either clip angles or shelf angles are provided. Where the tops of beams are 2½" above the tops of supporting beams, part of the beam is coped out and an angle seat is welded to the junior beam which rests on the top flange of the supporting beam.

Illustrating Method of Constructing Wood Forms for Concrete Floors
Supported by Junior Beams

050200 METAL JOISTS

Standard and longspan steel joists were developed primarily to take the place of wood construction. The Underwriters and Building Departments recognize the merits of this construction from the point of fire resistive and structural qualities with the result that this type of floor and roof construction has not only replaced wood to a great extent but other types of fire resistive floor construction to an equal extent. Steel joist construction is recognized and accepted by all building codes.

Standard open web and longspan steel joists are manufactured by welding chord members made up of hot rolled structural or cold formed sections to round bar or angle web members to form a truss. Information on the various types of steel joists can be obtained from the latest catalogs published by the Steel Joist Institute and the joist manufacturers.

Typical Shortspan Steel Joist

When steel joists are used in fireproof construction, it is necessary to place a 2" to 2½" reinforced concrete slab over the joists, and a fire resistant ceiling of gypsum board or plaster beneath the joists to obtain the required fire rating.

Where a wood floor is required, wooden nailing screeds supported on screed clips are embedded in the concrete slab and the wood flooring is then nailed to the screeds.

Steel joists may also be used in semi-fireproof construction by nailing the wood sheathing directly to nailers attached to the joists without the use of a protective concrete slab. Some manufacturers of steel joists have met the demand for this type of construction by fabricating a nailer joist with the wood nailer attached in the shops by the use of lag screws or bolts.

Open web joists may be used to great advantage in all types of light occupancy buildings. The open web construction greatly facilitates the placing of plumbing, conduits, etc., making unnecessary the need of a furred ceiling or heavy fills on top of the structural slab.

Steel joists are also used in industrial construction to replace structural steel roof purlins with a resulting economy of materials. Metal roof deck is then usually welded to the joists and light weight insulating material either poured or rigid board is applied to the metal deck and the roofing is installed.

Estimating Steel Joists.—Steel joists and related products are generally sold by the manufacturer directly to the contractor. When preparing general contract estimates the most satisfactory method is to send plans or sketches to the manufacturer or his agent thus assuring a sub-bid based on an economical layout and design and one which includes the necessary accessories.

In his quotation the manufacturer will note the tonnage of joists which will enable the contractor to figure the necessary handling and labor of erection. The cost of joist erection including placing of accessories varies from $150.00 to $200.00 per ton (excluding material costs) depending on labor conditions, location and type of building.

Labor Setting Steel Joists.—The labor cost of handling and setting steel joists is usually estimated by the ton, based on the total weight of the joists to be placed.

This cost will vary considerably, depending upon the type of building, size of joists, length of spans, amount of handling and hoisting necessary, etc.

It is much cheaper to set joists on long straight spans and large floor areas, such as garage and factory buildings, office buildings, etc., than on small or other irregularly constructed buildings.

All steel joists must be bridged and connected according to the Steel Joist Institute standards.

On an ordinary job of steel floor joists of regular construction and fairly long spans, five men working together should handle and place about 4½ to 5 tons of joists (including bridging), per 8-hr. day, at the following labor cost per ton:

	Hours	Rate	Total	Rate	Total
Ironworker	40	$....	$....	$17.70	$708.00
Cost per ton			....		149.05

On jobs of irregular construction, five men working together should handle and place about 3½ to 4 tons of joists (including bridging), per 8-hr. day, at the following labor cost per ton:

	Hours	Rate	Total	Rate	Total
Ironworker	40	$....	$....	$17.70	$708.00
Cost per ton			....		188.80

Prices of Steel Joists.—Costs of standard steel joists will vary with several factors among which are the size of the job, the number of joists of one particular size and length, and the location of the job. For an average job the cost of standard joists will be in the range of $600.00 to $700.00 per ton, which would include nominal bridging members and other accessories.

Costs of longspan steel joists will vary with the same factors as given above for standard joists. For an average job the cost of the longspan joists will range from $580.00 to $750.00 per ton including accessories.

05300 METAL DECKING

Steel roof deck is formed from steel sheets in 18, 20, or 22 gauge. Most commercial decks have a standardized cross section with longitudinal ribs spaced 6" on centers. Sections are 30" in width and are manufactured in various lengths to suit job conditions, usually in the 14-ft. to 31-ft. range, although some can be supplied in greater lengths.

Deck sections are usually 1½" deep, but some manufacturers supply a 1¾", 2" or 2½" deep deck. Sections have interlocking or nesting side laps and telescoping or nesting end laps.

Under ordinary roof live loads—30 to 40 lbs. per sq. ft.—20 gauge deck is normally used for spans from 6'-0" to 7'-6" centers of purlins and 18 gauge deck for spans from 7'-0" to 8'-0". Specific information on allowable loads can be obtained from the various manufacturers. By extending a single sheet over two or more purlin spaces, structural continuity, increasing load carrying capacity, can be obtained. Dead load of deck, insulation and built-up roofing is approximately 7 lbs. per sq. ft.

Approximate cost of steel roof deck, delivered to job site in the eastern or northern states, on an average size job (10,000 sq. ft. or more) is as follows:

Gauge of Roof Deck	Price per Sq. Ft. 1½" Deck
22 Ga.	$ .70
20 Ga.	.80
18 Ga.	1.00

Placing Metal Deck on Top of Steel Joists.—When metal deck is placed over the top of steel joists to receive the concrete slab or lightweight insulating materials, the metal deck is usually spot welded to the joists.

On work of this type two men working together should place and weld about 200 sq. ft. of deck per hour at the following labor cost per square of 100 sq. ft.:

	Hours	Rate	Total	Rate	Total
Iron Worker	1.0	$....	$....	$17.70	$17.70
Cost per sq. ft.			....		.18

Steel roof deck can be erected in almost any kind of weather, permitting other trades to work under cover much more rapidly. Deck must be welded to steel support members in accordance with the Steel Deck Institute recommendations. Two tack welds per 18" width of deck section at end laps and one weld on the outside rib on intermediate supports are recommended.

Longspan Deck

Longspan deck is formed from a single 14, 16, 18 or 20 gauge steel sheet into 4½", 6" or 7½" deep pans 12" wide. These are rolled sections with integral stiffening ribs formed into the top flange. Interlocking longitudinal side laps are provided on opposite edges for positive side lap attachment and are fastened by welding. Maximum spans allowable vary from 20'-0" to 30'-0".

Longspan deck is easily erected—on an average job, a 5 man crew should erect 2,200 to 2,400 sq. ft. per 8-hr. day.

Longspan Roof Deck

Approximate cost of longspan steel roof deck, delivered to job sites in the eastern or northern states, is $1.50 to $2.80 per sq. ft., depending upon gauge and length of sections.

Concrete Slab With Steel Forms Used In Conjunction With
Composite Beam Design

Reinforcing Floor Forms

Steel deck, erected in an inverted position (with ribs up) can be used as a reinforcing form for concrete construction. It acts as a form to support the concrete and permits the ribs to act as reinforcing.

Steel deck reinforcing forms can be erected for an entire structure as soon as the steel framework is placed, providing an immediate working platform and protective staging for all trades. Concrete can be placed at any portion of the building without regard for removing, cleaning and re-setting temporary formwork. Steel deck forms usually provide all necessary reinforcing to satisfy flexure requirements—the only additional reinforcing ordinarily required is temperature mesh to minimize shrinkage cracking. Tight form joints prevent concrete from dripping to lower floors, saving clean-up time.

Steel deck reinforcing forms can also be used in conjunction with composite beam design. Ample area for concrete around stud shear connectors and between deck ribs permits full effectiveness of the connectors. Standard A.I.S.C. Composite Design Procedure may be followed using the total slab depth in beam property calculations.

Reinforcing floor forms are installed by welding, the same as ordinary steel roof deck, and costs of materials and erection are about the same as given for "Metal Decking."

CELLULAR STEEL DECK AND FLOOR PANELS

Light gauge cellular steel structural panels are used for longspan structural floors and roofs. Some panels are composed of two identically formed beam sections and a flat plate—others are formed from two ribbed sections and a flat plate. All panels have interlocking side joints. Components are factory assembled by spot welding. Panels are manufactured in gauges from 20 to 13 and are normally available in 24" widths—some panels are fabricated in 12" widths. Lengths are available up to 40'- 0", while depths vary from 1½" to 7½".

Cellular Steel Deck and Floor Panel

This type of panel is easily handled and erected, as it is deck and joist combined. Concrete floor forms are eliminated and panels provide a working platform for other trades.

A flat, plate down, installation is often used to provide an attractive ceiling which needs only a finish coat of paint to complete the job. Voids between beam sections may be filled and covered with 2½" of concrete for floors or covered with rigid board insulation and built-up roofing for roofs. Installations with beam section forming the ceiling can be merely painted to give a finished fluted ceiling or a suspended ceiling can be erected to provide a flush type ceiling.

Where panels are covered with a concrete fill, the use of temperature mesh, to reduce shrinkage cracks, is recommended.

The cost of cellular panels, delivered to job site, ranges from $2.25 per sq. ft. for 20 ga. 2" to 3" thick panels, $2.70 to $3.10 for 3" to 4" 18 ga. units on up to $3.70 for 16 ga. deck.

Acoustically Treated Deck and Floor Panel

A crew of four structural iron workers plus a foreman should install around 1600 sq. ft. of light deck per 8 hour day, 1200 sq. ft. of 18 ga. deck, and around 1000 sq. ft. of the heavy gauge deck under normal job conditions. For example, 100 sq. ft. of 18 ga. labor costs would be as follows:

	Hours	Rate	Total	Rate	Total
Foreman	.65	$....	$....	$18.20	$11.83
Str'l Ironworkers	2.60	....	....	17.70	46.02
Cost per 100 sq. ft.					$57.85
Cost per sq. ft.					.58

05400 LIGHTGAGE FRAMING

The components involved are a complete range of studs, joists and accessories for the steel framing of buildings. Sections are fabricated from structural-grade, high tensile strip steel by cold forming and are designed specifically for strength, light weight and low cost. Yet, structural framing carries all the benefits of conventional steel framing.

A nailing groove, developed for easy, economical attachment of other materials, is a feature of all double studs and joists. It is obtained by welding two cold-formed steel elements together, so designed that a nail driven into this space is not only held by friction, but is also deformed to provide maximum holding power.

The usual limitations imposed by prefabrication are avoided. Using nailable framing sections architects and engineers have unrestricted freedom in design. All sections are painted at the plant with a coat of oven-dried, rust-resisting red zinc chromate paint, or are galvanized.

Lightgage framing systems can supply complete wall, floor and roof construction for buildings up to four stories in height, or can be used in combination with other framing systems for interior, load bearing partitions, exterior curtain walls, fire separation walls, parapets, penthouses, trusses, suspended ceilings and mansard roofs.

Advantages of using these systems include reduced dead loads, uniform fabrication, less on-the-job storage space, no warping, shrinkage or swelling, incombustibility, punched slots for passage of other trades and secure nailing to eliminate nail popping. When properly engineered and utilized this system may be cheaper than conventional wood or steel construction; but even when comparable in price the advantages offered to other trades may yield savings on the overall job.

The components can be completely detailed, cut and assembled in the shop and delivered for erection to the job site. Where on-the-job cutting and assembly is preferred, the material can be cut with a radial saw fitted with a ⅛" high speed circular blade.

Fastening may be by bolts, sheet metal screws or welding.

Joists come in 6", 8", 10", and 12" depths and in 12, 14, 16, and 18 gauge material. There are generally three styles: double nailable, "C" joists with 1⅝" flanges and "C" joists with 2½" flanges. Lengths from 6' to 40' are available. Joist webs may be solid, selectively punched or continuously punched for maximum raceway flexibility. Unnecessary punching should be avoided as the slotting lowers the structural value of the joist.

Joist bridging, which may be by stock "V" units or solid channels, must be supplied in the center of all spans up to 14'; at third points on spans from 14' to 20'; at quarter points on spans from 26' to 32'; and at eight foot centers on all spans over 32'.

Studs are available in 2½", 3⅝", 4" and 6" depths and in 14, 16, 18 and 20 gauges. 25 gauge studs are also available but are suitable for non-load bearing

Erecting Lightgauge, Preassembled Wall.

partitions only and are discussed under the drywall section of Chapter 14 "Finishes". Lengths from 7' to 40' are available with either slotted or solid webs. Studs may be of the double-nailable type, channels, or "C" type with knurled flanges.

Studs mount into standard tracks at floors, ceilings, sills and facias. They should be stiffened with lateral bridging at the midpoint of all walls to 10' in height; at third points on walls from 10' to 14'; and every 4' on walls above 14'. Stock "V" bars, channels or special clipping systems may be ordered for this.

Costs will vary because of the various options available and each job should be priced with the supplier when all the design criteria are known. For general budget purposes joists can be figured at around 75 cts per pound galvanized. A punched, galvanized "C" joist will vary from 2.24 lbs. per ft. 6" deep to 5.85 lbs. 12" deep or from around $1.70 to $4.40 per foot. Carload lots will run less.

DIMENSIONAL DATA — TABLE 1

12", 10",   8" and 6" NAILABLE JOISTS

4" NAILABLE DOUBLE STUDS

3⅝" NAILABLE DOUBLE STUDS

V-BAR BRIDGING*
(Weight = 0.26 lbs. ft.; gage = 16; t = 0.0568 ; net area = 0.075 sq. in.)

UNPUNCHED CHANNEL AND TRACK

2½" NAILABLE DOUBLE STUDS

Studs 3⅝" deep, punched and galvanized will run 40 cts per foot in 20 gauge material in less than carload lots; 65 cts. per foot in 18 gauge; and 75 cts per foot in 16 gauge. Tracks will approximate stud costs.

DIMENSIONAL DATA – TABLE 2

6" AND 8" CEE STUDS

4", 3 5/8", 3 1/4" SCREW CEE STUDS

2 1/2" SCREW CEE STUDS

Dimension A: 0.4" for all gages.

6" PUNCHED SCREW STUDS AND CHANNEL STUDS

4", 3 5/8", 3 1/4" PUNCHED SCREW STUDS AND CHANNEL STUDS

2 1/2" PUNCHED SCREW STUDS AND CHANNEL STUDS

NOTE: If holes are to line up, this specification must accompany order.

Attachment of Other Materials.—Steel or aluminum roof decking, metal siding or corrugated sheets can be quickly and economically attached to sections with self-tapping screws, or by other conventional methods.

When self-tapping screws are used, a hole slightly smaller than the screw is drilled through the material and the lightgage section. Then, the screw can be placed and tightened by one man. This is faster and more economical than most conventional methods of attachment which require two men. ⅝" hex-head self-tapping, cadmium-plated screws are recommended.

Attachment of materials is generally with standard nails when the nailable sections are used. When channel studs or other non-nailable sections are used, the

Solid Bridging   "V" Bridging

"V" Bridging of Studs   Channel Bridging of Studs

self-drilling screws are recommended for metal siding or plywood. When drywall is applied to non-nailable sections, self-drilling screws can be ordered from the drywall manufacturer.

Other types of materials can be readily attached to framing members by the use of bolts, clips, welds or other methods consistent with the finish desired.

05500   METAL FABRICATIONS

Metal Stairs.—Metal stairs are usually figured at so much per riser with stringers, treads, nosings, and railings included. A standard, three foot wide simple run metal pan stair can be budgeted at around $85.00 per riser, material cost. For each additional foot of width up to five feet add 10%. Custom stairs will run from $95.00 per riser to twice that amount where any of the stair components vary from the norm. Simple landings will run around $7.25 per square foot without an allowance for railings.

A four man crew should erect around 40 to 50 risers or 150 sq. ft. of landing per 8 hr. day.

The labor erection cost of a three foot wide steel stair consisting of three flights of sixteen risers and one 3' by 6' landing each run would figure as follows:

	Hours	Rate	Total	Rate	Total
Ironworkers - stairs 48 risers	34.0	$....	$....	$17.70	$601.80
landing 54 sq. ft.	11.1	....	....	17.70	196.47
Total Cost					$798.27
Cost per stair run (3)					266.09
Cost per riser incl. landings					16.63

Pipe Railings.—When estimating plain pipe railing, 3'-6" high, consisting of two horizontal runs of pipe with uprights 6 ft. to 8 ft. on center, figure as follows for the various sizes, erected in place:

1¼" pipe railing, complete as described above,
per lineal foot $16.50
1½" pipe railing, complete as described above,
per lineal foot $18.00
2" pipe railing, complete as described above,
per lineal foot $20.00

For each curved section or termination point that is formed to radius, add $25.00 to $30.00.

For following the rake of a stair, add $7.50 per upright.

For each foot of wall railing $7.50

Aluminum pipe railings can also be ordered from stock and will cost around $30.00 per foot, anodized. Aluminum wall rails will cost around $8.50 per foot.

A two man crew should erect some 100 lineal feet of straight run railing per day, and 130 feet of wall railing.

Steel Ladders.—Straight steel ladders cost from $30.00 to $35.00 per lin. ft., erected in place. Add $30.00 to $40.00 for curved handles and platform over coping walls. Add $30.00 per lin. ft. if cage is required.

Steel Gratings.—Gratings may be welded or of expanded construction and made of steel, aluminum or stainless steel.

Welded steel gratings weighing 4# will run around $3.50 per sq. ft.; weighing 5#, $4.50; 9#, $5.50; 12#, $8.00; and 16#, $8.75 to $9.75, all depending on the spacing and size of the bars. If the material is galvanized add 90 cts. per sq. ft.

Aluminum grating will run from $9.00 per sq. ft. ¾" thick and weighing 1.5# to $20.00 per sq. ft. 2¼" thick and weighing 5.9# in amounts of 75 sq. ft. or more.

Checkered plate is sometimes specified in place of or in combination with gratings and can be figured at about $9.50 per sq. ft. for ¼" thickness.

A typical crew laying grating will consist of four men who can install about 100 square feet per hour at the following cost per 100 sq. ft.:

	Hours	Rate	Total	Rate	Total
Ironworkers	4.0	$....	$....	$17.70	$70.80
Cost per sq. ft.					.71

Steel Window Guards.—Window guards made with ⅜"x2" horizontal bars, top and bottom and round or square upright bars spaced 4" on centers. Delivered to job.

	Approx. Price per Sq. Ft.
Window guards with ½" round upright bars	$ 6.50
Window guards with ½" square upright bars	7.00
Window guards with ⅝" round upright bars	9.00
Window guards with ⅝" square upright bars	9.50

Approx.
Price per
Sq. Ft.

Window guards with ¾" round upright bars **11.00**
Window guards with ¾" square upright bars **11.50**

Do not figure any opening as containing less than 15 sq. ft.

Window Guards Made of Crimped Diamond Mesh.—Window guards made with 1-inch channel iron frames and crimped diamond mesh wire. Per sq. ft., erected in place .... $9.00 to $10.00
Do not figure any opening as containing less than 20 sq. ft.

Curb Angles, Floor Frames, Trench Frames and Covers.—For miscellaneous items, such as curb angles, floor frames, trench frames and covers, etc., figure $1.00 per lb. delivered to job.

Ornamental Iron Entrance Rails.—Fabricated from 1¼"x¼" or 1¼"x⅜" bar stock or 1½" bar size channels for rails and ½" square bar pickets. Stock designs, delivered to job, cost about $30.00 per lin. ft. of rail depending on height and amount of ornamentation. For rails following the rake of stairs, add 25 per cent. Rails made to special design may cost 2 to 3 times the above prices.

Ornamental Porch Columns.—Stock design, scrolled, wrought iron columns, with 3/4"x3/4" bar stock or 15/16" square tubing frames and 3/8"x1/4" or 1/2"x3/16" bar or strip stock scrolls. Flat columns, 8½" to 12" wide x 8'-0" high, $50.00 to $60.00 each. Corner columns to match flat columns $70.00 to $90.00 each.

Stock design columns, with ¾"x¾" bar stock frames and cast iron panels, delivered to job, cost as follows:

Flat columns, 9½" to 11" wide x 8'-0" high, $70.00 to $80.00 each.
Corner columns to match flat columns, $85.00 to $100.00 each.
Brackets to match columns, $20.00 each.

Steel Stacks

Steel stack costs vary widely and are dependent on height, diameter, gauge of material, number of joints, transportation, etc. Steel stacks are usually fabricated by companies specializing in steel boiler and tank manufacturing.

When estimating a job that has a steel stack, always obtain a firm figure from a steel stack fabricator before submitting a bid on the work.

Labor Erecting Steel Stacks.—When estimating the labor cost of erecting steel stacks for office buildings or other buildings of this type, where the stack extends from the ground to the top of the building and is either built into the building or placed on the outside of one of the rear walls, there are several items to be taken into consideration; viz., the length and weight of each section of stack, the facilities for hoisting and placing, and the amount of rigging necessary.

Where each section of the stack weighs 500 to 1,000 pounds, it is customary to place a small gin pole on the roof of the building and lift each section in place.

Four iron workers should set up all necessary rigging in one day, ready to place the sections. After the derricks and rigging have been placed to start hoisting, the gang should handle, place and bolt one 15 to 20 foot section of stack in 4 hours, or 2 sections (30 to 40 lin. ft.), per 8-hour day.

The labor cost placing and removing derricks for erecting a steel stack should average as follows:

	Hours	Rate	Total	Rate	Total
Ir. wkrs. placing equip	32	$. . . .	$. . . .	$17.70	$ 566.40
Ir. wkrs. remove equip	32	. . . .	. . . .	17.70	566.40
Labor on equipment			$. . . .		$1,132.80

The labor cost erecting a steel stack up to 30" in diam. and 60 feet long, should average as follows:

	Hours	Rate	Total	Rate	Total
Ir. wkrs. erecting stack	48	$. . . .	$. . . .	$17.70	$849.60
Cost per lin. ft.			. . . .		14.16

To the above costs, add equipment rental and move charges to and from job for equipment.

CHAPTER 11

CARPENTRY

06100   ROUGH CARPENTRY

To estimate carpentry quantities and costs accurately, an estimator should possess a working knowledge of the trade and be familiar with job conditions. He should know how the various classes of work are constructed, where joists and studs should be doubled, where to place wood furring strips for walls and floors, openings requiring wood bucks, and where it is necessary to place wood grounds. These may appear to be small items but they are important, nevertheless.

For example, an inexperienced estimator taking off carpentry quantities might figure wood bucks for the exterior window frames because he is unfamiliar with how work is conducted in the field. He must know how a job goes together and what is required for each item of work. He is much like an insurance policy—an expense until it is needed—at which time it becomes the most valuable document in a contractor's possession.

When taking off lumber quantities from the plans, always estimate as closely as possible the exact quantity of each kind of lumber required. Some contractors estimate the cost of wood floors at a certain price per sq. ft. including joists, bridging, subfloor, deadening felt, furring strips and the finish wood flooring. Entering into the construction of such a floor are materials and labor operations of widely varying costs, yet certain contractors are content to estimate their work in this haphazard manner, and at the completion of the job have no more idea of their labor costs than before they took the job.

When estimating labor costs of rough carpentry, bear in mind the costs will vary with the class of work performed, with the ability of the carpenters employed and with the experience and executive ability of the foreman or superintendent in charge of the work.

Modern tools, such as electric hand saws, electric drills, sanders, etc., play a big part in reducing labor costs. For instance an electric handsaw will cut off a 2"x12" plank in 4 or 5 seconds while it takes about a minute to do it by hand; it takes 10 to 12 minutes to rip a 12'-0" plank 2" thick when sawing by hand while an electric handsaw will do the work in less than a minute. In other words, an electric saw will cut 10 to 20 times as fast as can be done by hand. The handling cost remains the same in either instance but when sawing for any length of time, a man can turn out three or four times as much work with an electric saw as by hand.

Another thing that should be considered is the cost of filing saws where there are a number of carpenters working. This will vary with the condition of the saw but on an average it will require ½-hr. to file a saw by hand, while an automatic saw filer will do the same work in ¼-hr. The same applies to circular saws used on saw rigs or electric hand saws, as these can be filed in ½ to ¾-hr. by machine, while hand filing requires nearly twice as much time.

The same thing applies to the use of electric drills, electric sanders, etc., so the contractor who is going to keep up with the procession must use modern, labor-saving tools and equipment wherever possible.

Here is another thing that should be remembered by every contractor. No matter how carefully the plans have been measured and how much consideration they have been given before pricing, if the work is not executed in an efficient manner, careful estimating will not avert losses.

Estimating Lumber Quantities

When estimating the quantity of lumber required for any job, the only safe method is to take off every piece of lumber required to complete that portion of the work. This is seldom done, however, and the following tables are given to simplify the work as much as possible and at the same time provide accurate material quantities.

Estimating Wood Joists.—When estimating wood joists, always allow 4" to 6" on each end of the joist for bearing on the wall.

To obtain the number of joists required for any floor, take the length of the floor in feet, divide by the distance the joists are spaced and add 1 to allow for the extra joist required at end of span.

Example: If the floor is 28 ft. long and 15 ft. wide, it will require 16-ft. joists to allow for wall bearing at each end. Assuming the joists are spaced 16" on centers, one joist will be required every 16" or every 1 1/3 ft. In other words it will require 3/4 as many joists as the length of the span, plus one. Three-quarters of 28 equal 21, plus 1 extra joist at end, makes 22 joists 16 ft. long for this space.

The following table gives the number of joists required for any spacing:

Number of Wood Floor Joists Required for any Spacing

Distance Joists are Placed on Centers	Multiply Length of Floor Span by	Add Joists	Distance Joists are Placed on Centers	Multiply Length of Floor Span by	Add Joists
12 inches	1	1	36 inches	1/3 or .33	1
16 inches	3/4 or .75	1	42 inches	2/7 or .29	1
20 inches	3/5 or .60	1	48 inches	1/4 or .25	1
24 inches	1/2 or .50	1	54 inches	2/9 or .22	1
30 inches	2/5 or .40	1	60 inches	1/5 or .20	1

Number of Feet of Lumber B.M. Required per 100 Sq. Ft. of Surface When Used for Studs, Joists, Rafters, Wall and Floor Furring Strips, etc.

The following table does not include any allowance for waste in cutting, doubling joists under partitions or around stair wells, extra joists at end of each span, top or bottom plates, etc. These items vary with each job. Add as required.

Size of Lumber	12-Inch Centers	16-Inch Centers	20-Inch Centers	24-Inch Centers
1"x2"	16⅔	12½	10	8⅓
2"x2"	33⅓	25	20	16⅔
2"x4"	66⅔	50	40	33⅓
2"x5"	83⅓	62½	50	41⅔
2"x6"	100	75	60	50
2"x8"	133⅓	100	80	66⅔

2"x10"	166⅔	125	100	83⅓
2"x12"	200	150	120	100
2"x14"	233⅓	175	140	116⅔
3"x6"	150	112½	90	75
3"x8"	200	133⅓	120	100
3"x10"	250	187½	150	125
3"x12"	300	225	180	150
3"x14"	350	262½	210	175

Number of Wood Joists Required for any Floor and Spacing

Length of Floor	\(Spacing of Joists\) 12"	16"	20"	24"	30"	36"	42"	48"	54"	60"
6	7	6	5	4	3	3	3	3	2	2
7	8	6	5	5	4	4	3	3	3	2
8	9	7	6	5	4	4	3	3	3	3
9	10	8	6	6	5	4	4	3	3	3
10	11	9	7	6	5	4	4	4	3	3
11	12	9	8	7	5	5	4	4	3	3
12	13	10	8	7	6	5	4	4	4	3
13	14	11	9	8	6	5	5	4	4	4
14	15	12	9	8	7	6	5	5	4	4
15	16	12	10	9	7	6	5	5	4	4
16	17	13	11	9	7	6	6	5	5	4
17	18	14	11	10	8	7	6	5	5	4
18	19	15	12	10	8	7	6	6	5	4
19	20	15	12	11	9	7	6	6	5	5
20	21	16	13	11	9	8	7	6	5	5
21	22	17	14	12	9	8	7	6	6	5
22	23	18	14	12	10	8	7	7	6	5
23	24	18	15	13	10	9	8	7	6	6
24	25	19	15	13	11	9	8	7	6	6
25	26	20	16	14	11	9	8	7	7	6
26	27	21	17	14	11	10	8	8	7	6
27	28	21	17	15	12	10	9	8	7	6
28	29	22	18	15	12	10	9	8	7	7
29	30	23	18	16	13	11	9	8	7	7
30	31	24	19	16	13	11	10	9	8	7
31	32	24	20	17	13	11	10	9	8	7
32	33	25	20	17	14	12	10	9	8	7
33	34	26	21	18	14	12	10	9	8	8
34	35	27	21	18	15	12	11	10	9	8
35	36	27	22	19	15	13	11	10	9	8
36	37	28	23	19	15	13	11	10	9	8
37	38	29	23	20	16	13	12	10	9	8
38	39	30	24	20	16	14	12	11	9	9
39	40	30	24	21	17	14	12	11	10	9
40	41	31	25	21	17	14	12	11	10	9

One joist has been added to each of the above quantities to take care of extra joist required at end of span.

Add for doubling joists under all partitions.

Estimating Quantity of Bridging.—It is customary to place a double row of bridging between joists about 6'-0" to 8'-0" on centers. Joists 10'-0" to 12'-0" long will require one double row of bridging or 2 pcs. to each joist.

Joists 14'-0" to 20'-0" long will require 2 double rows of bridging or 4 pcs. to each joist.

Bridging is usually cut from 1"x3", l"x4", 2"x2", or 2"x4" lumber.

The following table gives the approximate number of pcs. and the lin. ft. of bridging required per 100 sq. ft. of floor.

Joists Up to 12 Feet Long				Joists Up to 20 Feet Long			
12-Inch Centers		16-Inch Centers		12-Inch Centers		16-Inch Centers	
No. Pcs.	Lin. Ft.	No. Pcs.	Lin. Ft.	No. Pcs.	Lin. Ft.	No. Pcs.	Lin. Ft.
20	30	16	24	40	60	32	48

Steel Bridging for Wood Joists.—Several types of steel bridging are available, fabricated from de-formed strip steel varying from 16 to 20 gauge in thickness. Some are galvanized while others are painted with black asphaltum or have a baked enamel finish. Stock sizes fit all joist depths and spacings encountered in average construction work—special sizes are available made to order.

Prices vary according to type, size, finish and quantity, but a fair average for galvanized bridging is 30 to 40 cts. per set.

Installation time is about 50 percent of that required for wood bridging and with many types can be done after subflooring is laid.

Estimating Number of Wood Studs.—When estimating the number of wood partition studs, take the length of each partition and the total length of all partitions.

If a top and bottom plate is required, take the length of the wood partition and multiply by 2. The result will be the number of lin. ft. of plates required.

If a double plate consisting of 2 top members and a single bottom plate is used, multiply the length of the wood partitions by 3.

Example: Find the quantity of lumber required to build a stud partition 16'-0" long, 8'-0" high, with studs spaced 16" on centers and having single top and bottom plates. 16'-0"=192". 192" ÷ 16" =12 studs, plus 1 extra at the end equals 13 studs 8'-0" long. Two top and bottom plates 16'-0" long equal 32 lin. ft.

$$13 \text{ pcs. } 2\text{x}4 @ \quad 8'\text{-}0 = 104$$
$$\underline{2 \text{ pcs. } 2\text{x}4 @ \ 16'\text{-}0 = \ \ 32}$$
Total = 136 lin. ft.
$$136 \text{ lin. ft.} \times \tfrac{2}{3} = 90.67 \text{ bd. ft.}$$

After the total number of lin. ft. of lumber has been obtained, reduce to board measure, as above.

Feet of Lumber, B.M. Required for Wood Stud Partitions

2"x4" studs 16" on centers, with single top and bottom plates.

| Length of Partition | Height of Partition | | | | |
	8'-0"	8'-6"	9'-0"	10'-0"	12'-0"
3'-0"	20	22	22	24	28
4'-0"	27	29	29	32	37
5'-0"	33	37	37	40	47
6'-0"	40	44	44	48	56
7'-0"	41	45	45	49	57
8'-0"	48	53	53	57	67
9'-0"	55	60	60	65	76
10'-0"	61	67	67	73	85
11'-0"	63	69	69	75	87
12'-0"	69	76	76	83	96
13'-0"	76	83	83	91	105
14'-0"	83	91	91	99	115
15'-0"	84	92	92	100	116
16'-0"	91	99	99	108	125
17'-0"	97	107	107	116	135
18'-0"	104	114	114	124	144

19'-0"	105	115	115	125	145
20'-0"	112	123	123	133	155
21'-0"	119	130	130	141	164
22'-0"	125	137	137	149	173
23'-0"	127	139	139	151	175
24'-0"	133	146	146	159	184
25'-0"	140	153	153	167	193
26'-0"	147	161	161	175	203
27'-0"	148	162	162	176	204
28'-0"	155	169	169	184	213
29'-0"	161	177	177	192	223
30'-0"	168	184	184	200	232
31'-0"	169	185	185	201	233
32'-0"	176	193	193	209	243
33'-0"	183	200	200	217	252
34'-0"	189	207	207	225	261
35'-0"	191	209	209	227	263
36'-0"	197	216	216	235	272
37'-0"	204	223	223	243	281
38'-0"	211	231	231	251	291
39'-0"	212	232	232	252	292
40'-0"	219	239	239	260	301

Add ⅔ ft. of lumber, b.m. for each lin. ft. of double top or bottom plate.

Number of Partition Studs Required for Any Spacing

Distance Apart Studs	Multiply Length of Partition by	Add Wood Studs
12 inches	1.0	1
16 inches	0.75	1
20 inches	0.60	1
24 inches	0.50	1

Add for top and bottom plates.

Number of Feet of Lumber Required Per Sq. Ft. of Wood Stud Partition Using 2"x4" Studs.

Studs spaced 16" on centers, with single top and bottom plates.

Length Partition in Feet	No. Studs Req'd	8'-0"	Ceiling Heights in Feet 9'-0"	10'-0"	12'-0"
2	3	1.25	1.167	1.13	1.13
3	3	0.833	.812	.80	.80
4	4	0.833	.812	.80	.80
5	5	0.833	.812	.80	.80
6	6	0.833	.812	.80	.80
7	6	0.833	.75	.75	.80
8	7	0.75	.75	.75	.70
9	8	0.75	.75	.75	.70
10	9	0.75	.75	.75	.70
11	9	0.75	.70	.70	.67
12	10	0.75	.70	.70	.67
13	11	0.75	.70	.70	.67
14	12	0.75	.70	.70	.67
15	12	0.70	.70	.70	.67
16	13	0.70	.70	.70	.67
17	14	0.70	.70	.70	.67
18	15	0.70	.70	.67	.67

Length Partition in Feet	No. Studs Req'd	8'-0"	Ceiling Heights in Feet 9'-0"	10'-0"	12'-0"
19..................	15	0.70	.70	.67	.67
20..................	16	0.70	.70	.67	.67
For dbl. plate, add per sq. ft.		0.13	.11	.10	.083

For 2"x8" studs, double above quantities.

For 2"x6" studs, increase above quantities 50%.

Example: Find the number of feet of lumber, b.m. required for a stud partition 18'-0" long and 9'-0" high. This partition would contain 18x9=162 sq. ft.

The table gives 0.70 ft. of lumber, b.m. per sq. ft. of partition. Multiply 162 by 0.70 equals 113.4 ft. b.m.

Quantity of Plain End Softwood Flooring Required Per 100 Sq. Ft. of Floor

Measured Size Inches	Actual Size Inches	Add for Width	Ft. B.M. Req. per 100 Sq. Ft. Surface	Weight per 1000 Ft.
1x3..............	¾x2⅜	27%	132	1800
1x4..............	¾x3¼	23%	128	1900

The above quantities include 5% for end cutting and waste.

Quantity of End Matched Softwood Flooring Required Per 100 Sq. Ft. of Floor

Measured Size Inches	Actual Size Inches	Add for Width	Ft. B.M. Req. per 100 Sq. Ft. Surface	Weight per 1000 Ft.
1x3..............	13/16x2⅜	27%	130	1800
1x4..............	13/16x3 1/4	23%	126	1900

The above quantities include 3% for end cutting and waste.

Quantity of Square Edged (S4S) Boards Required Per 100 Sq. Ft. of Surface

Measured Size Inches	Actual Size Inches	Add for Width	Ft. B.M. Req. per 100 Sq. Ft. Surface	Weight per 1000 Ft.
1x 4..............	¾x 3½	14%	119	2300
1x 6..............	¾x 5½	9%	114	2300
1x 8..............	¾x 7¼	10%	115	2300
1x10..............	¾x 9¼	8%	113	2300
1x12..............	¾x11¼	7%	112	2400

The above quantities include 5% for end cutting and waste.

Lineal Foot Table of Board Measure
Number of Feet of Lumber, B.M., Per Lineal Foot of any Size.

2" x 4"=0.667	4"x 4"=1.333	8"x14"= 9.333
2" x 6"=1.	4"x 6"=2.	8"x16"=10.667
2" x 8"=1.333	4"x 8"=2.667	10"x10"= 8.333
2" x10"=1.667	4"x10"=3.333	10"x12"=10.
2" x12"=2.	4"x12"=4.	10"x14"=11.667
2" x14"=2.333	4"x14"=4.667	10"x16"=13.333
2" x16"=2.667	4"x16"=5.333	10"x18"=15.
2½"x12"=2.5	6"x 6"=3.	12"x12"=12.

2½"x14"=2.917	6"x 8"=4.	12"x14"=14.
2½"x16"=3.333	6"x10"=5.	12"x16"=16.
3" x 6"=1.5	6"x12"=6.	12"x18"=18.
3" x 8"=2.	6"x14"=7.	14"x14"=16.333
3" x10"=2.5	6"x16"=8.	14"x16"=18.667
3" x12"=3.	8"x 8"=5.333	14"x18"=21.
3" x14"=3.5	8"x10"=6.667	16"x16"=21.333
3" x16"=4.	8"x12"=8.	16"x18"=24.

Lengths of Common, Hip, and Valley Rafters Per 12 Inches of Run

1	2	3	4	5*	6†
Pitch of Roof	Rise and Run or Cut	Length in Inches Common Rafter per 12" of Run	Percent Increase in Length of Com. Rafter over Run		Length in Inches Hip or Valley Rafters
1/12	2 and 12	12.165	.014	1.014	17.088
1/8	3 and 12	12.369	.031	1.031	17.233
1/6	4 and 12	12.649	.054	1.054	17.433
5/24	5 and 12	13.000	.083	1.083	17.692
1/4	6 and 12	13.417	.118	1.118	18.000
7/24	7 and 12	13.892	.158	1.158	18.358
1/3	8 and 12	14.422	.202	1.202	18.762
3/8	9 and 12	15.000	.250	1.250	19.209
5/12	10 and 12	15.620	.302	1.302	19.698
11/24	11 and 12	16.279	.357	1.357	20.224
1/2	12 and 12	16.971	.413	1.413	20.785
13/24	13 and 12	17.692	.474	1.474	21.378
7/12	14 and 12	18.439	.537	1.537	22.000
5/8	15 and 12	19.210	.601	1.601	22.649
2/3	16 and 12	20.000	.667	1.667	23.324
17/24	17 and 12	20.809	.734	1.734	24.021
3/4	18 and 12	21.633	.803	1.803	24.739
19/24	19 and 12	22.500	.875	1.875	25.475
5/6	20 and 12	23.375	.948	1.948	26.230
7/8	21 and 12	24.125	1.010	2.010	27.000
11/24	22 and 12	25.000	1.083	2.083	27.785
11/12	23 and 12	26.000	1.167	2.167	28.583
Full	24 and 12	26.875	1.240	2.240	29.394

*Use figures in this column to obtain area of roof for any pitch. See explanation below.
†Figures in last column are length of hip and valley rafters in inches for each 12 inches of common rafter run.

To Obtain Area of Roofs For Any Pitch.—To obtain the number of square feet of roof area for roofs of any pitch, take the entire flat or horizontal area of the roof and multiply by the figure given in the fifth column (*), and the result will be the area of the roof. Always bear in mind that the width of any overhanging cornice must be added to the building area to obtain the total area to be covered. Example: Find the area of a roof 26'-0"x42'-0", having a 12" or 1'-0" overhanging cornice. Roof having a ¼ pitch. To obtain roof area, 26'-0"+1'-0"+1'-0"=28'-0" width. 42'-0"+1'-0"+1'-0"=44'-0" length. 28 x 44=1232 or 1,232 sq. ft. of flat or horizontal area.

To obtain area at ¼ pitch: multiply 1,232 by 1.12*=1379.84 or 1,380 sq. ft. roof surface.

Add allowance for overhang on dormer roofs and sides.

*See column 5 on this page.

Table of Board Measure
Giving Contents in Feet of Joists, Scantlings and Timbers

Size in Inches	10	12	14	16	18	20	22	24	26	28	30
1 x 2	1 2/3	2	2 1/3	2 2/3	3	3 1/3	....	....	....	....	....
1 x 3	2 1/2	3	3 1/2	4	4 1/2	5	....	....	....	....	....
1 x 4	3 1/3	4	4 2/3	5 1/3	6	6 2/3	....	....	....	....	....
1 x 6	5	6	7	8	9	10	....	....	....	....	....
1 x 8	6 2/3	8	9 1/3	10 2/3	12	13 1/3	....	....	....	....	....
1 x10	8 1/3	10	11 2/3	13 1/3	15	16 2/3	....	....	....	....	....
1 x12	10	12	14	16	18	20	....	....	....	....	....
1 1/4x 4	4 1/6	5	5 5/6	6 2/3	7 1/2	8 1/3	....	....	....	....	....
1 1/4x 6	6 1/4	7 1/2	8 3/4	10	11 1/4	12 1/2	....	....	....	....	....
1 1/4x 8	8 1/3	10	11 2/3	13 1/3	15	16 2/3	....	....	....	....	....
1 1/4x10	10 5/12	12 1/2	14 7/12	16 2/3	18 3/4	20 5/6	....	....	....	....	....
1 1/4x12	12 1/2	15	17 1/2	20	22 1/2	25	....	....	....	....	....
1 1/2x 4	5	6	7	8	9	10	....	....	....	....	....
1 1/2x 6	7 1/2	9	10 1/2	12	13 1/2	15	....	....	....	....	....
1 1/2x 8	10	12	14	16	18	20	....	....	....	....	....
1 1/2x10	12 1/2	15	17 1/2	20	22 1/2	25	....	....	....	....	....
1 1/2x12	15	18	21	24	27	30	....	....	....	....	....
2 x 2	3 1/3	4	4 2/3	5 1/3	6	6 2/3	....	....	....	....	....
2 x 3	5	6	7	8	9	10	11	12	13	14	15
2 x 4	6 2/3	8	9 1/3	10 2/3	12	13 1/3	14 2/3	16	17 1/3	18 2/3	20
2 x 6	10	12	14	16	18	20	22	24	26	28	30
2 x 8	13 1/3	16	18 2/3	21 1/3	24	26 2/3	29 1/3	32	34 2/3	37 1/3	40
2 x10	16 2/3	20	23 1/3	26 2/3	30	33 1/3	36 2/3	40	43 1/3	46 2/3	50
2 x12	20	24	28	32	36	40	44	48	52	56	60
2 x14	23 1/3	28	32 2/3	37 1/3	42	46 2/3	51 1/3	56	60 2/3	65 1/3	70
3 x 4	10	12	14	16	18	20	22	24	26	28	30
3 x 6	15	18	21	24	27	30	33	36	39	42	65
3 x 8	20	24	28	32	36	40	44	48	52	56	60
3 x10	25	30	35	40	45	50	55	60	65	70	75
3 x12	30	36	42	48	54	60	66	72	78	84	90
3 x14	35	42	49	56	63	70	77	84	19	98	105

Table of Board Measure
Giving Contents in Feet of Joists, Scantlings and Timbers

Size in Inches	10	12	14	16	18	20	22	24	26	28	30
4x 4	13	16	19	21	24	27	29	32	35	37	40
4x 6	20	24	28	32	36	40	44	48	52	56	60
4x 8	27	32	37	43	48	53	59	64	69	75	80
4x10	33	40	47	53	60	67	73	80	87	93	100
4x12	40	48	56	64	72	80	88	96	104	112	120
4x14	47	56	65	75	84	93	103	112	121	131	140
6x 6	30	36	42	48	54	60	66	72	78	84	90
6x 8	40	48	56	64	72	80	88	96	104	112	120
6x10	50	60	70	80	90	100	110	120	130	140	150
6x12	60	72	84	96	108	120	132	144	156	168	180
6x14	70	84	98	112	126	140	154	168	182	196	210
6x16	80	96	112	128	144	160	176	192	208	224	240
8x 8	53	64	75	85	96	107	117	128	139	149	160
8x10	67	80	93	107	120	133	147	160	173	187	200
8x12	80	96	112	128	144	160	176	192	208	224	240

8x14	93	112	131	149	168	187	205	224	243	261	280
8x16	107	128	149	171	192	213	235	256	277	298	320
10x10	83	100	117	133	150	167	183	200	217	233	250
10x12	100	120	140	160	180	200	220	240	260	280	300
10x14	117	140	163	187	210	233	257	280	303	327	350
10x16	133	160	187	218	240	267	293	320	347	373	400
12x12	120	144	168	192	216	240	264	288	312	336	360
12x14	140	168	196	224	252	280	308	336	364	392	420
12x16	160	192	224	256	288	320	352	384	416	448	480
14x14	163	196	229	261	294	327	359	392	425	457	490
14x16	187	224	261	299	336	373	411	448	485	523	560
14x18	210	252	294	336	378	420	462	504	546	588	630
14x20	233	280	327	373	420	467	513	560	607	653	700
16x16	213	256	299	341	384	427	469	512	555	597	640
16x18	240	288	336	384	432	480	528	576	624	672	720
16x20	267	320	373	425	480	533	587	640	693	747	800
18x18	270	324	378	432	486	540	594	648	702	756	810
18x20	300	360	420	480	540	600	660	720	780	840	900
20x20	333	400	467	533	600	667	733	800	867	933	1000

Nails Required for Carpenter Work

The following table gives the number of wire nails in pounds for the various kinds of lumber per 1,000 ft., board measure, or per 1,000 shingles and lath or per square (100 sq. ft.) of asphalt slate surfaced shingle, with the number of nails added for loss of material on account of lap or matching of shiplap, flooring, ceiling and siding of the various widths. The table gives the sizes generally used for certain purposes with the nailing space 16" on centers, and 1 or 2 nails per board for each nailing space.

Description of Material	Unit of Measure	Size and Kind of Nail	Number of Nails Required	Pounds of Nails Required
Wood Shingles	1,000	3d Common	2,560	4 lbs.
Individual Asphalt Shingles	100 sq. ft.	7/8" Roofing	848	4 lbs.
Three in One Asphalt Shingles	100 sq. ft.	7/8" Roofing	320	1 lb.
Wood Lath	1,000'	3d Fine	4,000	6 lbs.
Wood Lath	1,000'	2d Fine	4,000	4 lbs.
Bevel or Lap Siding, 1/2"x4"	1,000'	6d Coated	2,250	* 15 lbs.
Bevel or Lap Siding, 1/2"x6"	1,000'	6d Coated	1,500	* 10 lbs.
Byrkit Lath, 1"x6"	1,000'	6d Common	2,400	15 lbs.
Drop Siding, 1"x6"	1,000'	8d Common	3,000	25 lbs.
3/8" Hardwood Flooring	1,000'	4d Common	9,300	16 lbs.
25/32" Hardwood Flooring	1,000'	8d Casing	9,300	64 lbs.
Subflooring, 1"x3"	1,000'	8d Casing	3,350	23 lbs.
Subflooring, 1"x4"	1,000'	8d Casing	2,500	17 lbs.
Subflooring, 1"x6"	1,000'	8d Casing	2,600	18 lbs.

Description of Material	Unit of Measure	Size and Kind of Nail	Number of Nails Required	Pounds of Nails Required
Ceiling, 5/8"x4"	1,000'	6d Casing	2,250	10 lbs.
Sheathing Boards, 1"x4".....	1,000'	8d Common	4,500	40 lbs.
Sheathing Boards, 1"x6".....	1,000'	8d Common	3,000	25 lbs.
Sheathing Boards, 1"x8".....	1,000'	8d Common	2,250	20 lbs.
Sheathing Boards, 1"x10"...	1,000'	8d Common	1,800	15 lbs.
Sheathing Boards, 1"x12"...	1,000'	8d Common	1,500	12 1/2 lbs.
Studding, 2"x4"	1,000'	16d Common	500	10 lbs.
Joist, 2"x6"............................	1,000'	16d Common	332	7 lbs.
Joist, 2"x8"............................	1,000'	16d Common	252	5 lbs.
Joist, 2"x10"..........................	1,000'	16d Common	200	4 lbs.
Joist, 2"x12"..........................	1,000'	16d Common	168	3 1/2 lbs.
Interior Trim, 5/8" thick......	1,000'	6d Finish	2,250	7 lbs.
Interior Trim, 3/4" thick......	1,000'	8d Finish	3,000	14 lbs.
5/8" Trim where nailed to jamb.................................	1,000'	4d Finish	2,250	3 lbs.
1"x2" Furring or Bridging ..	1,000'	6d Common	2,400	15 lbs.
1"x1" Grounds	1,000'	6d Common	4,800	30 lbs.

*NOTE—Cement coated nails sold as two-thirds of pound equals 1 pound of common nails.

Nails Required for Subflooring.—On wood floors up to 4" wide, quantities are based on 8d flooring nails. For flooring 6" and wider, 10d nails have been figured. The quantities given below are sufficient to lay 1,000 ft. of flooring, b.m.

Width Flooring	Joist Spacing 12" on Centers	Joist Spacing 16" on Centers
2"	40 lbs. 8d flg.	30 lbs. 8d flg.
3"	30 lbs. 8d flg.	23 lbs. 8d flg.
4"	22 lbs. 8d flg.	17 lbs. 8d flg.
6"	24 lbs. 10d com.	18 lbs. 10d com.
8"	17 lbs. 10d com.	13 lbs. 10d com.

Data on Common Wire Nails

Size of Nails	Length of Nails Inches	Gauge Number	Approximate Number to Pound	Approx. Price Per 100 lbs.
4d.................................	1½	12½	316	$36.00
5d.................................	1¾	12½	271	36.00
6d.................................	2	11½	181	36.00
8d.................................	2½	10¼	106	34.00
10d...............................	3	9	69	34.00
12d...............................	3¼	9	63	34.00
16d...............................	3½	8	49	34.00
20d...............................	4	6	31	34.00
30d...............................	4½	5	24	34.00
40d...............................	5	4	18	34.00
50d...............................	5½	3	14	34.00
60d...............................	6	2	11	34.00

Check local market as these prices are subject to rapid change.

The following table gives the recommended nailing schedule for special (helically threaded) nails that are being used today throughout the building industry.

Recommended Nailing Schedule for Common Applications in Building Construction

Application	Nailed Into	Nail Size Inches	Nail Type	Head Diameter Type	Point Size	Point Type	Nailing	Spacing o.c.	Nails Per Joint
Mudsill, partition plate, 2"	Concrete	2½-2¾x0.148	Sc-1z	5/16" Checkered	Long	Dia.	Face	12"-24"	—
Ditto, in earthquake regions	Concrete	(3¼-)3½x0.250	Sc-1z	9/16" Checkered	Med.	Ndl.	Face	24"-48"	—
Ditto, 3"	Concrete	4½x0.250	Sc-1z	9/16" Checkered	Med.	Ndl.	Face	24"-48"	—
Furring strips	Concrete	1½-1¾x0.148	Sc-1z	5/16" Checkered	Long	Dia.	Face	12"-24"	—
Mudsill	Mudsill	2½x0.120	Sc-2	9/32" Flat	Med.	Dia.	Toe	—	2
Sleepers	Mudsill	2½x0.120	Sc-2	9/32" Flat	Med.	Dia.	Toe	—	2
Joists	Mudsill	3¼x0.135	Sc-2	5/16" Flat	Med.	Dia.	Toe	—	2-3
Subflrg, 1" lumber, plywood	Mudsill, sleeper, joist	2⅛x0.105	St-14	1/4" Flat, Csk.	Med.	Dia.	Face	6" & 12"	2(3)
Subflrg, 2" lumber, plywood	Mudsill, sleeper, joist	2⅞x0.120	St-14	9/32" Flat, Csk.	Med.	Dia.	Face	—	2(3)
Subflrg, 3/8"-1/2" plywood (dph.)	Mudsill, sleeper, joist	1½x0.135	Hi-28	5/16" Flat, Csk.	Med.	Dia.	Face	6" & 12"	—
Subflrg, 5/8" plywood (dph.)	Mudsill, sleeper, joist	1¾x0.135	Hi-28	5/16" Flat, Csk.	Med.	Dia.	Face	6" & 12"	—
Subflrg, 3/4" plywd. (dph.), part bd.	Mudsill, sleeper, joist	2 x0.148	Hi-28	5/16" Flat, Csk.	Med.	Dia.	Face	6" & 12"	—
Subflrg, 1"-11/8" plywood (dph.) part. bd.	Mudsill, sleeper, joist	2½x0.148	Hi-28	5/16" Flat, Csk.	Med.	Dia.	Face	6" & 12"	—
Underlayment, 1/4"-5/16" plywood	Subfloor	1 x0.083	St-16	3/16" Flat, Csk.	Med.	Dia.	Face	6" & 6"-12"	—
Underlayment, 3/8"-1/2" plywood	Subfloor	1¼x0.083	St-16	3/16" Flat, Csk.	Med.	Dia.	Face	6" & 6"-12"	—
Underlayment, 5/8" plywood	Subfloor	1⅜x0.098	St-16	1/4" Flat, Csk.	Med.	Dia.	Face	6" & 6"-12"	—
Underlayment, 3/4" plywood	Subfloor	1½x0.098	St-16	1/4" Flat, Csk.	Med.	Dia.	Face	6" & 6"-12"	—
Underlayment, 7/8" plywood	Subfloor	1⅝x0.098	St-16	1/4" Flat, Csk.	Med.	Dia.	Face	6" & 6"-12"	—
Underlayment, 3/16"-5/8" hardboard	Subfloor	1-13/8x0.083	St-15	3/16" Casing	Med.	Dia.	Face	6" & 12"	—
Flooring, T & G hardwood	Subfloor, joist, sleeper	2-2½x0.115	Sc-4	13/64" Casing	Blunt	Dia.	Toe	10"-18"	—
Flooring, T & G softwood	Subfloor, joist, sleeper	2-2½x0.115	Sc-4	13/64" Casing	Blunt	Dia.	Toe	10"-18"	—
Flooring, T & G hardwood, 3/8" and 1/2"	Subfloor, joist, sleeper	1-1¼x0.072	Sc-4	9/64" Casing	Blunt	Dia.	Toe	10"-18"	—
Flooring, T & G parquet	Subfloor	1½x0.105	Sc-4	9/64" Casing	Blunt	Dia.	Face	—	—
Framing plates	Stud	3¼x0.135	Sc-2	5/16" Flat	Med.	Dia.	Face	16"-24"	2
Framing studs	Stud, cripple, lintel, sill	2½x0.120	Sc-2	9/32" Flat	Med.	Dia.	Face	—	—
Framing studs	Plate, cripple, lintel, sill	2½x0.120	Sc-2	9/32" Flat	Med.	Dia.	Toe	16"	3
Framing sole plate	Mudsill	3¼x0.135	Sc-2	5/16" Flat	Med.	Dia.	Toe	16"	—
Framing top plate	Lower top plate	3¼x0.135	Sc-2	5/16" Flat	Med.	Dia.	Face	24"	—
Trussed rafter assembly		3¼x0.135	Sc-5	5/16" Flat	Med.	Dia.	Face	2½"-3"	Given
Trussed rafter assembly		2½x0.120	Sc-5	9/32" Flat	Med.	Dia.	Face	2½"-3"	Given

See Notes, Key to Nail Types and Abbreviations on later page.

Application	Nailed Into	Nail Size Inches	Nail Type	Head Diameter, Type	Point Size	Point Type	Nailing	Spacing o.c.	Nails Per Joint
Rafter, 4"	Top plate	31/4x0.135	Sc-2	5/16" Flat	Med.	Dia.	Toe	—	3
Rafter, 4"	Top plate	6 x0.177	St-34	7/16" Flat	Med.	Dia.	Face	—	2
Rafter, 4"	Top plate	7 x0.207	St-34	1/2" Flat	Med.	Dia.	Face	—	2
Rafter, 6", 8", 10"	Top plate	{4-6 x0.177, 7-9 x0.203}	St-34	7/16" Flat	Med.	Dia.	Toe	—	2-3
Sheathing, 1" lumber	Framing, rafter	2 x0.120	St-3	9/32" Flat	Med.	Dia.	Face	6" & 12"	2
Sheathing, 3/8"-1/2" plywood	Framing, rafter	13/4x0.120	St-17	9/32" Flat	Med.	Dia.	Face	6" & 12"	—
Sheathing, 5/16"-1/2" plywood (dph.)	Framing, rafter	11/2x0.135	Hi-17	5/16" Flat, Csk.	Med.	Dia.	Face	6" & 12"	—
Sheathing, 5/8" plywood (dph.)	Framing, rafter	13/4x0.135	Hi-17	5/16" Flat, Csk.	Med.	Dia.	Face	6" & 12"	—
Sheathing, 3/4" plywood (dph.)	Framing, rafter	2 x0.148	Hi-17	5/16" Flat, Csk.	Med.	Dia.	Face	6" & 12"	—
Sheathing, 1"-11/8" plywood (dph.)	Framing, rafter	21/2x0.148	Hi-17	5/16" Flat, Csk.	Med.	Dia.	Face	6" & 12"	—
Sheathing, insulation board, gypsumboard	Framing, rafter	11/2-2x0.120	St-10g	3/8", 7/16" Flat	Blunt	Dia.	Face	3-4" & 6-8"	—
Sheathing, asbestosboard, 1/8"	Framing, rafter	11/4x0.083	St-6g	3/16" Flat, Csk.	Blunt	Dia.	Face	3-4" & 6-8"	—
Sheathing, asbestosboard, 1/4"	Framing, rafter	11/4x0.120	St-or Sc-6g	5/16" Flat, Csk.	Blunt	Dia.	Face	3-4" & 6-8"	—
Sheathing, hardboard, 3/8"-5/8"	Framing, rafter	2 x0.115	Sc-7g	13/64" Flat, Csk.	Med.	Ndl.	Face	3-4" & 6-8"	—
Building paper	Sheathing	1/2-3/4x0.105	Sq-30	15/16" Square	Med.	Dia.	Face	6"-12"	—
Stripping, 3/8"x35/8"	Framing, joist, rafter	2 x0.120	St-3	9/32" Flat	Med.	Dia.	Face		2
Stripping, 1"x4"	Framing, joist, rafter	21/2x0.135	St-3	5/16" Flat	Med.	Dia.	Face		2
Stripping, 2"x3"	Framing, joist, rafter	31/2x0.165	St-3	5/8" Flat	Med.	Dia.	Face		2
Siding, wood, 1"	Sheathing and framing	21/8x0.101-0.115	St-14g	1/4" Flat, Csk.	Med.	Dia.	Face		—
Siding, wood, 1"	Sheathing and framing	2 x0.120	Dr-14	5/32" Flat, Csk.	Med.	Ndl.	Face		1
Siding, wood, 2"	Sheathing and framing	3 x0.135	Dr-14	5/32" Flat, Csk.	Med.	Ndl.	Face	6" & 12"	1
Siding, plywood	Sheathing and framing	17/8x0.109	Dr-8	5/32" Casing	Med.	Ndl.	Face	6" & 12"	1
Siding, T & G wood	Sheathing and framing	13/4x0.105	Sc-8	5/32" Casing	Med.	Dia.	Toe		Given
Siding, asbestos shingle	Sheathing and framing	11/2-13/4x0.105	Dr-33	3/16" Flat Button	Med.	Dia.	Face		Given
Siding, asbestos shingle	Sheathing and framing	11/2-13/4x0.083	St-19t	3/16" Flat	Med.	Dia.	Face		Given
Siding, asbestos shingle	Sheathing and framing	11/2-13/4x0.076	St-20	3/16" Flat	Med.	Dia.	Face		2
Siding, insulated brick, wood shingle	Sheathing and framing	13/4x0.095	St-18ge	3/16" Flat	Med.	Dia.	Face	8"-12"	2
Siding, wood shingle	Insulating sheathing	13/4-2 x0.083	St-18ge	5/32" Finishing	Blunt	Dia.	Face		2
Siding, wood shingle	Insulating sheathing	13/4-2 x0.105	Dr-18	5/32" Finishing	Blunt	Dia.	Face		2
Siding, wood shingle	Plywood	11/8x0.102	Dr-18	3/16" Flat	Long	Dia.	Face		—
Siding, hardboard	Framing	2-21/2x0.115	Sc-7z	13/64" Casing	Long	Ndl.	Face	12"	—
Siding, hardboard battenboard	Framing	11/2x0.083	Sc-7z	9/64" Casing	Long	Ndl.	Face	12"	—

See Notes, Key to Nail Types and Abbreviations on later page.

Application	Nailed Into	Nail Size Inches	Nail Type	Head Diameter Type	Point Size	Point Type	Nailing	Spacing o.c.	Nails Per Joint
Fascia, 1"	Framing, rafter	2 1/2x0.120	Sc-2g	9/32" Flat	Med.	Dia.	Face	—	2
Fascia, 2" lumber	Framing, rafter	3 1/4x0.135	Sc-2g	5/16" Flat	Med.	Dia.	Face	—	2
Roofing, built-up	Sheathing	3/4-1 1/4x0.105	Sq-30	15/16" Square	Med.	Dia.	Face	10"	—
Roofing, built-up	Poured gypsum	1 1/2-1 3/4x0.120	Sq-31	15/16" Square	Med.	Dia.	Face	10"	—
Roofing, asphalt shingle	Sheathing	3/4-2x0.120	St-10g	3/8" Flat	Blunt	Dia.	Face	—	2-3
Roofing, asphalt shingle	Sheathing	3/4-2x.120-.135	Dr-10	3/8" Flat	Blunt	Dia.	Face	—	2-3
Roofing, wood shingle	Sheathing	3/4-2x.105-.120	Dr-18	3/16" Flat	Blunt	Dia.	Face	—	2
Roofing, asbestos shingle	Sheathing	1 3/4-2x0.083	St-18g	1/8" Flat, Csk.	Blunt	Dia.	Face	—	2
Roofing, aluminum (corr. and flat)	Rafter, purlin	As for siding 1 1/2-1 3/4x0.145	Dr-10	3/32" Flat ★	Long	Dia.	Face	12"	—
Roofing, sheet metal (corr. and flat)	Rafter, purlin	1-3x0.135	St-or	7/16" Flat ★	Long	Dia.	Face	12"	—
Roofing, glass fiber (corr. and flat)	Rafter, purlin	1 1/2-3x0.135	Sc-9g	7/16" Flat ★	Long	Dia.	Face	12"	—
Roofing, glass fiber (corr. and flat)	Rafter, purlin	1 1/2-3x0.148	Dr-9 or 10	7/16" Flat ★	Long	Dia.	Face	12"	—
				★ With Neoprene washer attached					
Lath, expanded metal, K-lath	Framing, joist	1 1/2x0.148	St-22g	L-Shaped	Med.	Dia.	Face	6" & 12"	—
Lath, gypsum plasterboard	Framing, joist	1 1/4x0.101	St-23b	19/64" Flat, Csk.	Long	Dia.	Face	5"	—
Gypsumboard, 3/8"	Framing, joist	1 1/4x0.098	St-24	1/4"-19/64" Flat, Csk.	Long	Dia.	Face	5"-8"	—
Gypsumboard, 1/2"-5/8"	Framing, joist	1 3/8x0.098	St-24	1/4"-19/64" Flat, Csk.	Long	Dia.	Face	5"-8"	—
Gypsumboard, prefinished	Framing, joist	1 3/8x0.083	K-32e	3/16" Flat, Csk.	Long	Dia.	Face	5"-8"	—
Paneling, trim		1-1 1/4x0.054	K-32e	3/32" Casing	Blunt	Dia.	Face	—	—
Paneling, trim, exterior		1-1 1/2x0.072	St-13	3/32" Casing	Blunt	Dia.	Face	—	—
Paneling, trim, exterior		1 x0.065	St-12	3/32" Casing	Blunt	Dia.	Face	—	—
Paneling, trim, exterior		1 1/2x0.076	Sc-12	3/32" Oval	Blunt	Dia.	Face	—	—
Paneling, trim		1 x0.072	Sc-11	3/32" Casing	Blunt	Dia.	Face	—	—
Paneling, trim		1 1/2-1 3/4x0.083	Sc-11	1/8" Casing	Blunt	Dia.	Face	—	—
Paneling, trim		2 1/2x0.105	Sc-11	9/64" Casing	Blunt	Dia.	Face	—	—
Acoustic tile		1-1 3/4x0.062	St-25z	—	Blunt	Dia.	Face	—	—
Electric conduit	Wood	1 1/2-2 x0.162	St-26z	1" Hook	Blunt	Ndl.	Face	—	—
Electric conduit	Masonry	1 1/2-2 x0.162	St-27z	1" Hook	Blunt	Ndl.	Face	—	—
Fencing wire	Softwood (treated)	1 1/2x0.148	St-22g	L-Shaped	Med.	Dia.	Face	—	—
Fencing wire	Hardwood	1 1/2x0.148	St-21g	L-Shaped	Med.	Dia.	Face	—	—

See Notes, Key to Nail Types and Abbreviations on next page.

Key to Nail Types

Code	Description
Sc-iz	Screw-Tite Masonry Nail, hardened HCS, zinc plated
Sc-2	Screw-Tite Framing Nail, hardened HCS
Sc-2g	Screw-Tite Framing Nail, hardened HCS, galvanized
Sc-3	Screw-Tite Framing Nail, bright LCS
Sc-3g	Screw-Tite Framing Nail, bright LCS, galvanized
Sc-4	Screw-Tite Flooring Nail, hardened HCS
Sc-5	Screw-Tite Trussed Rafter Nail, hardened HCS
Sc-6g	Screw-Tite Asbestosboard Nail, hardened HCS, galvanized
Sc-7g	Screw-Tite Exterior Hardboard Nail, hardened HCS, galvanized
Sc-7z	Screw-Tite Exterior Hardboard Nail, hardened HCS, zinc plated
Sc-8	Screw-Tite Casing Nail, silver bronze
Sc-9g	Screw-Tite Roofing Nail, hardened HCS, galvanized
Sc-10g	Screw-Tite Roofing Nail, bright LCS, galvanized
Sc-11	Screw-Tite Finishing Nail, bright LCS,
Sc-12	Screw-Tite Finishing Nail, stainless steel
St-3	Stronghold Framing Nail, bright LCS
St-4	Stronghold Parquet Flooring Nail, hardened HCS
St-6g	Stronghold Asbestosboard Nail, hardened HCS, galvanized
St-9g	Stronghold Roofing Nail, hardened HCS, galvanized
St-10g	Stronghold Roofing Nail, bright LCS, galvanized
St-12	Stronghold Finishing Nail, stainless steel
St-13	Stronghold Finishing Nail, monel metal
St-14	Stronghold Sinker Nail, bright LCS
St-14g	Stronghold Sinker Nail, bright LCS, galvanized
St-15	Stronghold Underlay Nail, Hardened HCS
St-16	Stronghold Underlay Nail, bright LCS
St-17	Stronghold Sheathing Nail, bright LCS
St-18g	Stronghold Shingle Nail, bright LCS, galvanized
St-18ge	Stronghold Shingle Nail, bright LCS, galvanized and enameled
St-19t	Stronghold Shingle Nail, bronze, tin plated
St-20	Stronghold Shingle Nail, stainless steel
St-21g	Stronghold Fence Staple, hardened HCS, galvanized
St-22g	Stronghold Fence Staple, bright LCS, galvanized
St-22z	Stronghold Fence Staple, bright LCS, zinc plated
St-23b	Stronghold Lath Nail, bright LCS, blued
St-24	Stronghold Drywall Nail, bright LCS
St-25z	Stronghold Kollarnail, hardened HCS, zinc plated
St-26z	Stronghold Conduit Staple, bright LCS, zinc plated
St-27z	Stronghold Knurled Conduit Staple, hardened HCS, zinc plated
St-34	Stronghold Spike, hardened HCS
Hi-28	"Hi-Load" Shear-Resistant Nail, bright LCS
Hi-17	"Hi-Load" Sheathing Nail, bright LCS
Sq-30	Squarehed Annular Thread Cap Nail, bright LCS
Sq-31	Squarehed Spiral Thread Cap Nail, bright LCS
K-32e	Annular Thread Kolorpin, bright LCS, enameled
Dr-8	Drive-Rite Spiral Thread Casing Nail, aluminum
Dr-9	Drive-Rite Spiral Thread Roofing Nail, aluminum
Dr-10	Drive-Rite Spiral Thread Roofing Nail, aluminum
Dr-14	Drive-Rite Spiral Thread Sinker Nail, aluminum
Dr-18	Drive-Rite Screw Thread Shingle Nail, aluminum
Dr-20	Drive-Rite Spiral Thread Shingle Nail, aluminum
Dr-33	Drive-Rite Knurled Asbestos-Cement Shingle Face, aluminum

The above chart is based on a table appearing in Bulletin No. 38 (Revised Edition). "Better Utilization of Wood Through Assembly with Improved Fasteners," a study undertaken at Wood Research Laboratory, Virginia Polytechnic Institute, under the sponsorship of Independent Nail & Packing Company, Bridgewater, Mass., manufacturers of Stronghold® Annular Thread and Screw-Tite® Spiral Thread Nails and other improved fasteners.

NOTES: For fastening redwood, use only aluminum or stainless steel nails.

Local conditions, customs and popular usage may dictate minor variations in length and gauge of nails. Consult the Technical Service Department of Independent Nail & Packing Company, Bridgewater, Mass.

STRONGHOLD ® ANNULAR THREAD NAIL

SCREW-TITE ® SPIRAL THREAD NAIL

STRONGHOLD ® SCREW THREAD NAIL

SCREW-TITE ® KNURLED MASONRY NAIL

ABBREVIATIONS USED IN THIS TABLE:

Corr.—Corrugated
Dph.—Diaphragm
Part. bd—Particle board
LCS—Low Carbon Steel
HCS—High Carbon Steel
Plywd—Plywood
Csk.—Countersunk   Med.—Medium
Flrg.—Flooring   Dia.—Diamond
Shgl.—Shingle   Ndl.—Needle

Aluminum Nails

Aluminum nails are excellent for use whenever the nailhead is exposed to the atmosphere or corrosive conditions. In other words, use aluminum nails wherever there is the possibility of rust or nail stain.

The manufacturers do not ordinarily recommend aluminum common nails for ordinary framing because there is no economic advantage over steel wire nails for most common purposes and the bending resistance of aluminum nails is less than that of steel nails despite the larger diameter of the aluminum nails.

Aluminum nails weigh about one-third as much as steel wire nails but they are more expensive nail for nail—however they do save labor and painting where rust or nail stain is a factor.

Common 3d to 20d nails will run $2.30 per lb., 30d to 60d around $2.05 per lb.

Hardware Accessories Used for Wood Framing

The following items may be used to advantage in all types of wood construction. Developed primarily to provide better joints between wood framing members, in many cases their use has resulted in lower overall costs—the additional material cost being more than offset by increased labor efficiency.

Steel Joist Hangers.—Used for framing joists to beams and around openings for stair wells, chimneys, hearths, ducts, etc. Made of galv. steel, varying from 12 ga. to 3/16" in thickness, with square supporting arms, holes punched for nails and with bearing surfaces proportioned to size of lumber. Approximate prices for sizes most commonly used are as follows:

Joist Size	Gauge	Depth of seat	Opening in Hanger	Price Each
2"x 6"	12	2"	15/8"x5"	$0.70
2"x 8"	12	2"	15/8"x5"	.78
2"x10"	12	2"	15/8"x81/2"	.81
2"x12"	10	21/2"	15/8"x81/2"	1.00
4"x 8"	11	2"	35/8"x51/4"	.94
4"x10"	9	2"	35/8"x51/4"	1.12
4"x12"	3/16	21/2"	35/8"x81/2"	1.73

Teco* Framing Accessories.—Used in light wood construction to provide face-nailed connections for framing members. Adaptable to most framing connections, they eliminate the uncertainties and weaknesses of toe-nailing. Manufactured of zinc-coated, sheet steel in various gauges and styles. Framing accessories are designed to provide nailing on various surfaces. Special nails, approximately equal to 8d common nails, but only 1¼" long, to prevent complete penetration of standard nominal 2" lumber, are furnished with anchors. Approximate prices are as follows:

Type	Price Per 100	Type	Price Per 100
Trip-L-Grips	$25.00	Post Caps	$66.00
Du-Al-Clip	20.00	H Clips	4.00
Truss Plates	30.00	Angles	45.00

*Timber Engineering Company, Inc., Washington, D.C.

Teco* Timber Connectors.—Timber connectors are devices for increasing the strength of bolted joints in timber construction. They are placed between adjacent faces of overlapping members, embedded to half-depth in each, to increase the bearing area at the connection. Several types of timber connectors are available, each for a particular purpose, as follows:

Teco Timber Connectors

Teco* Wedge-Fit Split Rings.—Split ring connectors are hot-rolled carbon steel bands, bent to form a ring with a tongue and groove meeting joint. Cross-section of steel band is tapered both ways from center to edges, providing a wedge-fit when connectors are inserted into tapered grooves, conforming to ring section, and precut to half-depth in each contact face of overlapping members.

Grooves in wood members must be cut with a special grooving tool, using a power drill press or portable drill. Grooves and bolt holes may be cut in a single operation, using a combination grooving tool and drill bit. If bolt holes are bored first, a smooth shank pilot is used to guide the grooving tool.

Split rings are available in two sizes—2½" diameter rings are used for wood framing involving lumber of 2" nominal dimension and are used in trussed rafter

*Timber Engineering Company, Inc., Washington, D.C.

construction for spans up to 50 ft.—4" diameter rings are used for medium to heavy construction, generally with members 3" or more in nominal thickness. Approximate prices of split rings and grooving tools are as follows:

Description	Approximate Prices	
	2 1/2" Dia.	4" Dia.
Split rings, per 100 pcs.	$ 42.00	$ 68.00
Grooving tool cutter head, complete	120.00	175.00
Replacement cutters,		
Set of 4 for 2 1/2" dia.	69.00	. . . .
Set of 6 for 4" dia.	. . . .	93.00
Smooth shank pilot,		
9/16" dia.	10.00	. . . .
11/16" dia.	10.00	. . . .
13/16" dia.	. . . .	12.00
15/16" dia.	. . . .	15.00

Teco* Shear Plates.—Shear plates are used for wood-to-steel connections and for wood structures that are frequently disassembled—used singly for wood-to-steel joints and in pairs for wood-to-wood assemblies. They are also often used, in place of split rings, for field connected joints. Shear plates are placed in daps precut in the contact faces of wood members with a special grooving tool. Load is transferred from wood member to shear plate, from shear plate to bolt and thence to opposing member, which may be another shear plate in wood, a steel gusset plate or rolled steel shape. Drilling, grooving and dapping of wood members are accomplished in the same manner as described for split ring connectors.

Shear plates are available in two sizes—2 5/8" diameter, made of pressed steel, and 4" diameter made of malleable iron. Approximate prices of shear plates and grooving tools are as follows:

Description	Approximate Prices	
	2 5/8" Dia.	4" Dia.
Shear plates, per 100 pcs.	$ 62.00	$220.00
Grooving tool cutter head, complete	147.00	235.00
Replacement cutters,		
Set of 3 for 2 5/8" dia.	79.00	. . . .
Set of 5 for 4" dia.	. . . .	98.00
Smooth shank pilot,		
13/16" dia.	12.00	11.00
15/16" dia.	. . . .	15.00

Teco* Toothed Rings.—Toothed ring connectors are toothed metal bands with each tooth corrugated or curved along its cross-section for greater rigidity. Used for wood-to-wood connections between lighter structural members. Also used for strengthening structures in place and structural repairs.

The function of toothed rings is similar to that of split rings, but the method of installation is not as efficient and requires more labor. Toothed rings are embedded in wood members by applying pressure. Where a large number of joints are to be assembled, a hydraulic jack set-up can be used to advantage. It requires a 5- to 7 1/2-ton jack for single bolt connections and a 10-ton jack for joints with two bolts. Where fewer joints are involved, the high strength rod assembly is recommended. The assembly consists of a high strength rod with Acme threads on one end, double depth nuts, ball bearing thrust washer, heavy plate washers, lock washer and nut. Metal sleeve adaptors permit the use of 1/2", 5/8" or 3/4" diameter rods with a standard ball bearing washer. The length of rod required equals the total thickness of

---

*Timber Engineering Company, Inc., Washington, D.C.

wood members plus one inch for each layer of toothed rings plus 5 inches for nuts and washers. Ratchet wrenches help speed assembly—impact wrenches have also been used with success. After joint is drawn closed, high strength rod assembly is replaced with ordinary bolt and washers.

Teco* Spike Grids.—Spike grid connectors are malleable iron castings of a grid-like structure with teeth protruding from both faces. Available in Three styles—square flat type, 4⅛"x4⅛" and circular type, 3¼" dia., are used between sawn timbers—single curve type, 4⅛"x4⅛", is used between curved faces of poles or piling and sawn timbers. Spike grids are embedded under pressure, using the high strength rod assembly, as previously described for toothed rings. Approximate prices are as follows:

Type of spike grid	Square Flat	Single Curve	Circular
Price per 100 pcs.	$245.00	$245.00	$190.00

Installation tool for spiked grids, complete with adaptor -$185.00.

WOOD ROOF TRUSSES

Wood roof trusses are used in buildings where clear floor space is a requirement and the width of the building exceeds the economical span of roof joists. Some of the advantages of trussed roof construction are fast erection, unobstructed weather protected space available sooner, simplified installation of ceiling, floor, mechanical and electrical systems, use of non-load-bearing movable partitions, etc.

Wood roof trusses are used for spans as short as 25'-0", and may be used up to 200'-0".

Wood roof trusses are widely used in garages, factories, hangars, gymnasiums, auditoriums, bowling alleys, dance halls, supermarkets, and other structures of this type.

When using wood roof trusses, it is common practice to span the shorter dimension of the building. This may be accomplished with any of several truss designs available, such as bowstring, crescent, Belgian, flattop, super-bow, scissors and Gothic. The type of truss is dependent upon the degree in which economy is emphasized, the use of the building and the general architectural effect desired.

Prices on roof trusses are governed by the following conditions:

1. Cost of material and labor.
2. Loading conditions and spacing.
3. Difficulties of erection.
4. Requirements of local building ordinances.

The fourth condition causes the price of trusses to vary considerably in different parts of the country since some cities require only a 25 lb. live load while others require a 50 lb. live load, thus necessitating heavier construction.

Spacing of wood roof trusses is usually 16'-0" to 20'-0" on centers, but the 16'-0" spacing is more generally used. The spacing of trussed rafters which support the roof sheathing directly (instead of with purlins as is the case where trusses are used) is usually 2'-0". Where roof loads are light and the installation of a ceiling is not required, spacings of 4'-0" and 5'-0" can be advantageous. Where snow loads are especially heavy, spacings of 16" and even 12" have been used.

The prices given on the following pages are approximate only and it is advisable to obtain definite prices from the manufacturers for either trusses delivered at the

---

*Timber Engineering Company, Inc., Washington, D.C.

building site or installed in place, as they are familiar with conditions in practically all parts of the country and are in position to quote definite prices.

Bowstring Roof Trusses.—For industrial and many commercial structures, the bowstring truss is the most popular. This type of truss has a curved top chord which starts from a point approximately 10 inches high at the end of the truss, rising to a maximum height at center-line of about ⅛ of the total span distance. This type of truss gives a curved roof shape. Because of its low height at the ends of truss, a minimum of masonry is required for parapet or fire walls.

Bowstring Roof Truss

The truss may be left exposed, carrying roof load only, as in garages or industrial plants, or it can be designed to carry monorail, floor systems, or other concentrated loads. The bowstring truss is also designed to carry ceiling loads for use in stores, automobile agency salesrooms, and other commercial structures. Bowstring trusses provide wind bracing for walls through diaphragm action, and can be used with "knee braces" to develop portal action in mill type buildings. Bowstring trusses are efficiently fabricated with bolts, thereby causing the least reduction in cross-sectional areas of the pieces connected.

Approximate Prices of Wood Bowstring Trusses

Designed with glued laminated or nailed 2"x3" or 2"x4" top chords—built to conform to parabolic curve.

Furnished in Washington, D.C. in lots of 5 or more, spaced 16'-0" on centers. 25 lb. live load. 40 lb. total load.

Span	Height	Price	Span	Height	Price
20'-0"	2'-6"	$ 296.00	90'-0"	11'-3"	$1275.00
25'-0"	3'-1"	310.00	95'-0"	11'-10"	1426.00
30'-0"	3'-9"	316.00	100'-0"	12'-6"	1518.00
35'-0"	4'-4"	322.00	105'-0"	13'-1"	1610.00
40'-0"	5'-0"	356.00	110'-0"	13'-9"	1679.00
45'-0"	5'-7"	380.00	115'-0"	14'-4"	1782.00
50'-0"	6'-3"	402.00	120'-0"	15'-0"	1863.00
55'-0"	6'-10"	506.00	125'-0"	15'-7"	2000.00
60'-0"	7'-6"	575.00	130'-0"	16'-3"	2150.00
65'-0"	8'-1"	661.00	135'-0"	16'-10"	2277.00
70'-0"	8'-9"	793.00	140'-0"	17'-6"	2415.00
75'-0"	9'-4"	874.00	145'-0"	18'-2"	2484.00
80'-0"	10'-0"	1023.00	150'-0"	18'-9"	2610.00
85'-0"	10'-7"	1150.00			

Crescent Type Roof Trusses.—The crescent type roof truss is becoming increasingly popular and is identified by its curved lower chord which affords a higher ceiling at the mid- span of the truss. The roof presents the curved roof, similar to the bowstring truss.

Crescent Design Roof Truss

This truss presents a curved ceiling effect and is used in gymnasiums, auditoriums, low-cost churches, more elaborate stores and restaurants, etc. Recommended spans 20'-0" to 85'-0".

Approximate Prices of Crescent Type Wood Roof Trusses

Furnished in Washington, D.C. in lots of 5 or more, spaced 16'-0" on centers. 25 lb. live load. 40 lb. total load.

Span	Height	Camber	Price	Span	Height	Camber	Price
20'-0"	5'-0"	1'-0"	$356.00	55'-0"	8'-6"	2'-9"	$ 615.00
25'-0"	5'-6"	1'-3"	375.00	60'-0"	9'-0"	3'-0"	680.00
30'-0"	6'-0"	1'-6"	400.00	65'-0"	9'-6"	3'-3"	805.00
35'-0"	6'-6"	1'-9"	415.00	70'-0"	10'-0"	3'-6"	870.00
40'-0"	7'-0"	2'-0"	442.00	75'-0"	10'-6"	3'-9"	1025.00
45'-0"	7'-6"	2'-3"	477.00	80'-0"	11'-0"	4'-0"	1185.00
50'-0"	8'-0"	2'-6"	506.00	85'-0"	11'-6"	4'-3"	1305.00

Belgian Design Roof Truss

Belgian Roof Trusses—Belgian roof trusses are used where it is desirable to give a conventional building a more pleasing appearance through the use of a peaked roof. It is recognized by its sloping top chord and horizontal lower chord. In most instances, a ceiling is applied to the lower chord of these trusses, while the top chord may have shingles or slate as a covering. Used on some higher class store buildings and low-cost churches. Recommended for spans 20'-0" to 85'-0".

Belgian roof trusses are less efficient than the bowstring type because the connections generally govern the member sizes and therefore cost approximately 50 percent more than bowstring type trusses.

Flat Top or Howe Design Roof Truss

Howe or Flattop Roof Trusses.—Industrial plants, warehouses, sheds, etc., when built of frame construction, often find use for a truss which absorbs wind loads in addition to carrying roof and concentrated loads. This truss is ideal for this purpose as it has parallel top and bottom chords and presents the over-all appearance of a large rectangle. Spans should not exceed 65 feet where costs are an important factor.

Howe roof trusses are generally more expensive than a bowstring type truss for reasons similar to those noted for the Belgian type trusses.

Labor Erecting Wood Roof Trusses.—Erection of wood roof trusses in confined areas, or where the work is not heavy enough to support new loads, is usually accomplished with a gin pole. However, the bulk of wood truss erection is now accomplished by the use of special hydraulic cranes mounted on truck chassis and with lifting capacity up to 12 tons.

Buildings should be clear before the crane is brought in and all overhead electrical wires removed. Door opening should be of predetermined size to permit crane entrance.

Where a gin pole is used, the labor cost of erecting will vary with job conditions, as deep excavations, piles of brick, mortar boxes, lumber piles, etc., in the way of the derricks, will prevent the erecting crew from making as good time as where a clear unobstructed space is available.

Under average working conditions on jobs requiring 5 or more trusses, a crew consisting of 5 men should erect one truss in the following time:

Length of Truss	No. Hrs. Crew	Labor Hrs. per Truss	Hours	Labor Cost per Truss			
				Rate	Total	Rate	Total
50'-0"	1	5	5	$....	$....	$16.47	$ 82.35
60'-0"	1	5	5	....	....	16.47	82.35
75'-0"	2	10	10	....	....	16.47	164.70
90'-0"	3	15	15	....	....	16.47	247.05
100'-0"	4	20	20	....	....	16.47	329.40
125'-0"	5	25	25	....	....	16.47	411.75

TRUSSED RAFTERS

Designers and builders of multiple unit housing projects and typical plan residence developments have utilized truss principles in wood roof framing in an effort to reduce costs and to shorten construction time. Most small home constructors can benefit from their experience and research by considering this type of framing.

Trussed rafter framing is most effective on roofs of simple gable design without dormers, hips or valleys, using trussed rafter framing, the following advantages may be realized:

1. Roof framing and ceiling framing are accomplished at the same time.
2. Members can be made of lighter stock, saving material and reducing weight.
3. Trusses may be pre-assembled.
4. Trusses can be erected rapidly and the job put under cover quickly.
5. Trusses permit the use of non-bearing partitions for all interior walls.

Types of Trussed Rafters.—Trussed rafters used in house building and similar construction, vary as to truss pattern and method of construction, depending upon design requirements and individual preference of the builder.

Several forms of truss patterns are as follows:

The W-type truss is the most popular type and is adaptable for spans from 18 ft. up to 40 ft.; roof slope from 2 in 12 to 6 in 12 and higher.

The kingpost truss is economical for spans 18 ft. up to 40 ft. with roof slopes from 2 in 12 to 6 in 12 and higher.

The scissors truss is used where sloping ceilings are desired for architetural effect. Economical spans are from 18 ft. to 36 ft.; roof slopes 5 in 12 to 6 in 12; ceiling slopes 2 in 12 to 3 in 12.

Trussed rafter construction varies principally as to the method of assembly, whether members are fastened together at joints by bolting or nailing, with split ring connectors and bolts, or flat, bent, pronged or toothed metal plates, or by gluing.

Using the bolting method, the members at all major joints, i.e., heel joints, apex, splices, etc., are connected by one or more bolts, or by using split rings to transfer the loads. In some cases, one or both ends of diagonal members may be fastened by nailing.

For nailed trussed rafters, all joint connections use metal truss plates, plywood gussets and nails, or ordinary nailed joints.

Fabrication of Trussed Rafter Members.—Fabrication of trussed rafter members is ideally suited for mass production methods, due to the large amount of repetition possible. In most cases, trussed rafters are symmetrical about their center lines so that corresponding members of each half are identical.

Since every member of a truss must do its share of the work, each one must be fabricated to exacting specifications. Design dimensions must be rigidly followed and all joints must have a good mechanical fit.

For efficient fabrication, a full-size layout should be made, either in the shop or on the subfloor of the building, after which one trussed rafter is carefully patterned and constructed with temporary connections. The unit may then be disassembled and the members used as templates for the balance of the trusses. On large projects, templates may be made of plywood or sheet metal.

To insure smooth roof and ceiling lines after erection, layout measurements should be made to upper edges of top chords and lower edges of bottom chords to eliminate variations in stock lumber widths.

A word of caution—attention is called to the fact that the joints and fabrication of various sized trusses are similar so it is important to exercise caution and see that the appropriate drawings are consulted and followed when trusses are fabricated.

Split-Ring Connected Trussed Rafters

As previously mentioned, members are connected by means of bolts and split rings at all major joints as the split rings transfer the loads and the bolts hold the members in contact.

In order that each member may function properly, the bolt holes must be located exactly in accordance with design dimensions and grooves for split rings must be of correct diameter, width and depth to provide a snug fit. Grooves must be cut with a grooving tool designed especially for this purpose.

Proper procedure for cutting grooves is as follows: after bolt locations have been established, drill bolt holes 1/16" larger in diameter than bolt; grooving tool is then fitted with a pilot mandrel of same diameter as drill and cutters are set to cut to a depth equal to half the width of the split ring; using bolt holes as pilot holes, concentric grooves are then cut in the faces of members which will contact each other. The grooving tool may also be fitted with the correct size drill bit, instead of a pilot mandrel, permitting both drilling and grooving operations to be done at one setup. For best results, drilling and grooving should be done with a bench drill or portable drill mounted in a drill stand.

Material Requirements for Split-Ring Connected Trussed Rafters.—The material requirements for split-ring connected trussed rafters are furnished through the courtesy of Timber Engineering Company, Washington, D.C., and are based on the following design conditions:

Design for Split-Ring Connected Trussed Rafter

Trussed rafters are designed to support a dead load plus live load on roof of 35 lbs. per sq. ft. and a ceiling load of 10 lbs. per sq. ft. with rafters spaced at 2'-0" centers.

Lumber shall be good grade of sufficient quality to permit the following unit stresses:

c=900 lbs. per sq. in. Compression parallel to grain.

f=900 lbs. per sq. in. Extreme fiber in bending.

E=1,600,000 lbs. per sq. in. Modulus of elasticity.

All split rings to be 2½" diameter. All bolts to be ½" diameter machine bolts. Washers may be 2"x2"x1/8" plate washers, 2 1/8" diameter cast or malleable iron washers or cut washers.

No allowances have been included for roof overhangs.

Quantities listed for diagonals are most economical lengths to produce a long and short member from a single length.

Material Requirements for One Split-Ring Connected Trussed Rafter of Various Spans for a 4 in 12 Roof Slope

Span	20'-0"	24'-0"	26'-0"	28'-0"	30'-0"	32'-0"
Top Chords	2-2x6	2-2x6	2-2x6	2-2x8	2-2x8	2-2x8
	12'-0"	14'-0"	16'-0"	16'-0"	18'-0"	18'-0"
Bottom Chords	2-2x4	2-2x4	2-2x4	2-2x4	2-2x4	2-2x4
	12'-0"	14'-0"	16'-0"	16'-0"	18'-0"	18'-0"
Diagonals	2-2x4	2-2x4	2-2x4	2-2x4	2-2x4	2-2x4
	8'-0"	10'-0"	10'-0"	12'-0"	12'-0"	12'-0"
Joint Scabs	1-1x4	1-1x4	1-1x4	1-1x4	1-1x4	1-1x4
	2'-0"	2'-0"	2'-0"	3'-0"	3'-0"	3'-0"
Total Ft. B.M.	51 1/3	60 2/3	67 1/3	81	89	89
Split Rings	11	11	11	13	13	13

Bolts, ½"x4"	4	4	4	6	6	6
½"x6"	2	2	2	2	2	2
½"x7½"	1	1	1	1	1	1
Washers	14	14	14	18	18	18
Nails, 8d	16	16	16	16	16	16

Material Requirements for One Split-Ring Connected Trussed Rafter of Various Spans for a 5 in 12 Roof Slope

Span	20'-0"	24'-0"	26'-0"	28'-0"	30'-0"	32'-0"
Top Chords	2-2x6	2-2x6	2-2x6	2-2x6	2-2x6	2-2x6
	12'-0"	14'-0"	16'-0"	18'-0"	18'-0"	20'-0"
Bottom Chords	2-2x4	2-2x4	2-2x4	2-2x4	2-2x4	2-2x4
	12'-0"	14'-0"	16'-0"	16'-0"	18'-0"	18'-0"
Diagonals	2-2x4	2-2x4	2-2x4	2-2x4	2-2x4	2-2x4
	10'-0"	12'-0"	12'-0"	12'-0"	14'-0"	14'-0"
Joint Scabs	1-1x4	1-1x4	1-1x4	1-1x4	1-1x4	1-1x4
	2'-0"	2'-0"	2'-0"	2-0"	2'-0"	2'-0"
Total Ft. B.M.	54	63 1/3	70	74	79 1/3	83 1/3
Split Rings	11	11	11	11	11	11
Bolts, ½"x4"	4	4	4	4	4	4
½"x6"	2	2	2	2	2	2
½"x7½"	1	1	1	1	1	1
Washers	14	14	14	14	14	14
Nails, 8d	16	16	16	16	16	16

Material Requirements for One Split-Ring Connected Trussed Rafter of Various Spans for a 6 in 12 Roof Slope

Span	20-0"	24'-0"	26'-0"	28'-0"	30'-0"	32'-0"
Top Chords	2-2x6	2-2x6	2-2x6	2-2x6	2-2x6	2-2x6
	14'-0"	16'-0"	16'-0"	18'-0"	18'-0"	20'-0"
Bottom Chords	2-2x4	2-2x4	2-2x4	2-2x4	2-2x4	2-2x4
	12'-0"	14'-0"	16'-0"	18'-0"	18'-0"	18'-0"
Diagonals	2-2x4	2-2x4	2-2x4	2-2x4	2-2x4	2-2x4
	10'-0"	12'-0"	14'-0"	14'-0"	16'-0"	16'-0"
Joint Scabs	1-1x4	1-1x4	1-1x4	1-1x4	1-1x4	1-1x4
	2'-0"	2'-0"	2'-0"	2'-0"	2'-0"	2'-0"
Total Ft. B.M.	58	67 1/3	72 2/3	76 2/3	82	86
Bolts, ½" x 4"	4	4	4	4	4	4
½" x 6"	2	2	2	2	2	2
½" x 7½"	1	1	1	1	1	1
Washers	14	14	14	14	14	14
Nails, 8d	16	16	16	16	16	16

Material Requirements for One Split-Ring Connected Trussed Rafter of Various Spans for a 7 in 12 Roof Slope

Span	20'-0"	24'-0"	26'-0"	28'-0"-	30'-0"	32'-0"
Top Chords	2-2x6	2-2x6	2-2x6	2-2x6	2-2x6	2-2x6
	14'-0"	16'-0"	16'-0"	18'-0"	20'-0"	20'-0"
Bottom Chords	2-2x4	2-2x4	2-2x4	2-2x4	2-2x4	2-2x4
	12'-0"	14'-0"	16'-0"	16'-0"	18'-0"	18'-0"
Diagonals	2-2x4	2-2x4	2-2x4	2-2x4	2-2x4	2-2x4
	12'-0"	14'-0"	14'-0"	16'-0"	16'-0"	18'-0"
Joint Scabs	1-1x4	1-1x4	1-1x4	1-1x4	1-1x4	1-1x4
	2'-0"	2'-0"	2'-0"	2'-0"	2'-0"	2'-0"
Total Ft. B.M.	60 2/3	70	72 2/3	79 1/3	86	88 2/3
Split Rings	11	11	11	11	11	11

Bolts, ½" x 4"	4	4	4	4	4	4
½" x 6"	2	2	2	2	2	2
½" x 7½"	1	1	1	1	1	1
Washers	14	14	14	14	15	14
Nails, 8d	16	16	16	16	16	16

Nailed Trussed Rafters

Many contractors prefer nailed trussed rafters for the simplicity of their construction. For small one-house jobs, containing only 20 to 30 rafter units, they may be easily fabricated and assembled on the job, using simple hand tools—hand saw and hammer. They are also adaptable to mass production methods using portable power tools or shop equipment, for projects involving thousands of trussed rafters.

Another advantage of these all-nailed trussed rafters is the fact that only standard dimension lumber and nails are used in the case of "W" and scissor-type rafters, and standard plywood for gusset and splice plates. In the case of kingpost trussed rafters, no steel plates, bolts, special fasteners, glue, etc. are required.

Design and Details for a Low Rise Nailed Trussed Rafter

Design and Details for a Scissors Type Nailed Trussed Rafter

To function properly, nailed trussed rafters must be fabricated and assembled with the same degree of accuracy as bolted and other types of trussed rafters. Fabrication and assembly of rafter units must adhere closely to design dimensions and joint details. It is important to have good bearing fits for top chord and compression diagonal member joints. Location, number and size of nails must be in accordance with joint details.

For fast, accurate marking of nail locations, use nailing templates and spray colored paint or shellac through template openings with a paint gun or an aerosol-type container. To permit nailing from one side without nail clinching, use helically threaded nails. For dry, dense lumber, use hardened nails, which are more slender and less likely to split material.

While the use of seasoned lumber is not absolutely essential in the construction of nailed truss rafters, better results are obtained with dry lumber and its use is recommended. This is particularly true for splice plates over 4" wide and is a design requirement for 8" wide material. 1"x8" splice plates made from green lumber tend to split along nail rows, when undergoing shrinkage, causing the joint to fail.

Material Requirements for Nailed Trussed Rafters.—The material requirements for nailed trussed rafters are furnished through the courtesy of Dr. E. George Stern, Research Professor of Wood Construction at the Virginia Polytechnic Institute, Blacksburg, Va. Truss designs for the various spans and roof slopes are based on experimental data obtained in the Wood Research Laboratory of the Virginia Engineering Experimental Station under the direction of Dr. Stern. Trussed rafters, constructed in accordance with these designs, have been subjected to all types of tests and have proved conclusively to be of ample strength for the use intended.

Material requirements for trussed rafters are based on the following design conditions:

Trussed rafters are designed to support a dead load plus live load on roof of 35 lbs. per sq. ft. and a ceiling load of 10 lbs. per sq. ft. with rafters spaced at 2'-0" centers.

Lumber shall be stress-graded No. 1 southern pine or Douglas fir, preferably seasoned 12 to 15 percent moisture content, permitting the following unit stresses:

$c=1350$ lbs. per sq. in. Compression parallel to grain.

$f=1500$ lbs. per sq. in. Extreme fiber stress.

$E=1,760,000$ lbs. per sq. in. Modulus of elasticity.

Quantities listed for diagonals are most economical lengths to produce a long and short member from a single length.

Material Requirements for One Nailed Trussed Rafter
of Various Spans for a 2 in 12 Roof Slope

Span	18'-0"	20'-0"	24'-0"	26'-0"	28'-0"	30'-0"
Approx. Overhang	3'-0"	2'-0"	4'-0"	3'-0"	4'-0"	3'-0"
Top Chords	2-2x6	2-2x6	2-2x6	2-2x6	2-2x6	2-2x6
	12'-0"	12'-0"	16'-0"	16'-0"	18'-0"	18'-0"
Bottom Chords	2-2x4	2-2x4	2-2x6	2-2x6	2-2x6	2-2x6
	8'-0"	10'-0"	12'-0"	12'-0"	14'-0"	14'-0"
Diagonals	2-1x6	2-1x6	2-1x6	2-1x6	2-1x6	2-1x6
	8'-0"	10'-0"	10'-0"	10'-0"	12'-0"	12'-0"
	1-2x4	1-2x4	1-2x4	1-2x4	1-2x4	1-2x4
	4'-0"	4'-0"	4'-0"	4'-0"	6'-0"	6'-0"
Splice Plates	4-1x4	4-1x4	4-1x6	4-1x6	4-1x6	4-1x6
	3'-0"	3'-0"	3'-6"	3'-6"	3'-6"	3'-6"
	2-1x6	2-1x6	2-1x6	2-1x6	2-1x6	2-1x6
	3'-0"	3'-0"	2'-0"	2'-0"	3'-0"	3'-0"
	1-1x8	1-1x8	1-1x8	1-1x8	1-1x8	1-1x8
	4'-0"	4'-0"	4'-0"	4'-0"	4'-0"	4'-0"
Total Ft. B.M.	55	59 2/3	80 1/3	80 1/3	92 2/3	92 2/3
Nails, 12d	144	144	198	198	242	242
8d	40	40	58	58	72	72

Material Requirements for One Nailed Trussed Rafter
of Various Spans for a 3 in 12 Roof Slope

Span	24'-0"	26'-0"	28'-0"	30'-0"	32'-0"	34'-0"
Approx. Overhang	3'-6"	4'-0"	3'-6"	4'-0"	3'-6"	2'-6"
Top Chords	2-2x6	2-2x6	2-2x6	2-2x6	2-2x6	2-2x6
	16'-0"	18'-0"	18'-0"	20'-0"	20'-0"	20'-0"
Bottom Chords	2-2x4	2-2x4	2-2x4	2-2x4	2-2x4	2-2x4
	12'-0"	14'-0"	14'-0"	16'-0"	16'-0"	18'-0"

Diagonals	2-1x6	2-1x6	2-1x6	2-1x6	2-1x6	2-1x6
	10'-0"	10'-0"	12'-0"	12'-0"	12'-0"	14'-0"
	1-2x4	1-2x4	1-2x4	1-2x4	1-2x4	1-2x4
	6'-0"	6'-0"	6'-0"	6'-0"	8'-0"	8'-0"
Splice Plates	4-1x4	4-1x4	4-1x4	4-1x4	4-1x4	4-1x4
	3'-0"	3'-0"	4'-0"	4'-0"	4'-0"	4'-0"
	3-1x6	2-1x6	2-1x6	2-1x6	2-1x6	2-1x6
	3'-0"	3'-0"	3'-0"	3'-0"	3'-0"	3'-0"
	. . . .	1-1x6	1-1x6	1-1x6	1-1x8	1-1x8
	. . . .	4'-0"	4'-0"	4'-0"	4'-0"	4'-0"
Total Ft. B.M.	70 1/2	77 2/3	81	87 2/3	89 2/3	94 1/3
Nails, 12d	138	138	146	146	168	168
8d	38	46	58	60	56	56

Material Requirements for One Nailed Trussed Rafter of Various Spans for a 5 in 12 Roof Slope

Span	24'-0"	26'-0"	28'-0"	30'-0"	32'-0"	34'-0"
Approx. Overhang	3 1/4	3 1/4	3 1/4	3 1/4	3 1/4	3 1/4
Top Chords	2-2x4	2-2x4	2-2x6	2-2x6	2-2x6	2-2x6
	14'-0"	16'-0"	16'-0"	18'-0"	18'-0"	20'-0"
Bottom Chords	2-2x4	2-2x4	2-2x4	2-2x4	2-2x4	2-2x4
	12'-0"	14'-0"	14'-0"	16'-0"	16'-0"	18'-0"
Diagonals	2-1x6	2-1x6	2-1x6	2-1x6	2-1x6	2-1x6
	12'-0"	12'-0"	14'-0"	14'-0"	14'-0"	16'-0"
	1-2x4	1-2x4	1-2x4	1-2x4	1-2x4	1-2x4
	6'-0"	8'-0"	8'-0"	8'-0"	8-0"	10'-0"
Splice Plates	6-1x6	6-1x6	4-1x4	4-1x4	4-1x4	4-1x4
	2'-0"	2'-0"	3'-0"	3'-0"	3'-0"	3'-0"
	1-1x6	1-1x6	2-1x6	2-1x6	2-1x6	2-1x6
	3'-0"	3'-0"	2'-0"	2'-0"	2'-0"	2'-0"
	. . . .	. . . .	1-1x6	1-1x6	1-1x8	1-1x8
	. . . .	. . . .	3'-0"	3'-0"	3'-0"	3'-0"
Total Ft. B.M.	57 5/6	64 5/6	77 1/2	84 1/6	84 2/3	94 2/3
Nails, 12d	100	112	110	110	124	124
8d	26	18	42	44	34	34

Material Requirements for One Scissors Typed Nailed Trussed Rafter of Various Spans for a 5 in 12 Roof Slope and 2 in 12 Ceiling Slope

Span	18'-0"	20'-0"	24'-0"	26'-0"	28'-0"	30'-0"
Overhang	2"	2"	3 3/8"	3 3/8"	3 3/8"	3 3/8"
Top Chords	2-2x4	2-2x4	2-2x4	2-2x6	2-2x6	2-2x6
	10'-0"	12'-0"	14'-0"	16'-0"	16'-0"	18'-0"
Bottom Chords	2-2x4	2-2x4	2-2x4	2-2x4	2-2x6	2-2x6
& Compr. Diag.	12'-0"	14'-0"	18'-0"	18'-0"	20'-0"	22'-0"
Vert. Member	1-1x8	1-1x8	1-1x8	1-1x8	1-1x8	1-1x8
	3'-0"	3'-0"	4'-0"	4'-0"	4'-0"	5'-0"
Splice Plates	4-1x4	4-1x4	4-1x4	4-1x4	4-1x6	4-1x6
	2'-6"	2'-6"	3'-0"	3'-0"	3'-0"	3'-0"
	2-1x4	2-1x4	2-1x4	2-1x4	3-1x6	3-1x6
	1'-6"	1'-6"	1'-6"	1'-6"	2'-0"	2'-0"
	1-1x6	1-1x6	1-1x6	1-1x6	1-1x6	1-1x6
	2'-0"	2'-0"	2'-0"	2'-0"	4'-0"	4'-0"
	1-1x6	1-1x6	1-1x6	1-1x6	. . . .	. . . .
	3'-0"	3'-0"	4'-0"	4'-0"	. . . .	. . . .
Total Ft. B.M.	38 1/6	43 1/2	53 1/3	66 2/3	85 2/3	94 1/3
Nails, 12d	63	63	78	80	104	109
8d	106	106	115	120	14	147

All-Nailed Kingpost Trussed Rafters.—All nailed kingpost trussed rafters are a result of research conducted at Virginia Polytechnic Institute under the sponsorship of Independent Nail & Packing Company of Bridgewater, Mass. The studies, which have been supported by this largest manufacturer of improved fasteners, have brought about a new era in nail technology and wood construction. Indebtedness to Independent Nail for their continuing cooperation and foresight is acknowledged.

Test data are available on request from the V.P.I. Wood Research Laboratory at Blacksburg, Va. Samples of the hardened-steel spiral-thread Hi-Load nails, used for the assembly of the trussed rafters, can be obtained from the manufacturer, the Independent Nail & Packing Company of Bridgewater, Mass.

Labor Required to Fabricate and Assemble Trussed Rafters.—The labor required to fabricate and assemble trussed rafters will vary with the scope of the job, number of rafter units and production methods used, also upon the size of the rafter units.

On an average one-house job, two carpenters will require about 6 hours to make a layout on the subfloor, place jig stops to guide assembly work and construct a pilot truss with temporary connections, at the following labor cost:

	Hours	Rate	Total	Rate	Total
Carpenter	12	$. . .	$. . .	$16.47	$197.64

For the balance of the rafter units, a 2-man crew, using portable power equipment for cutting, drilling, etc., should fabricate, assemble and stockpile 10 to 12, 24 ft. to 28 ft. span trussed rafters per 8-hr. day, at the following labor cost per truss:

	Hours	Rate	Total	Rate	Total
Carpenter	1.5	$. . .	$. . .	$16.47	$24.71

For larger spans, up to 36 ft., add about 5 percent to above costs. For smaller spans, reduce costs about 5 percent.

On large projects, containing several thousand trussed rafters, where the expense of setting up a shop for fabrication and assembly is warranted, a 2-man crew should complete 20 to 24 trussed rafters per 8-hr. day, at the following labor cost per truss:

	Hours	Rate	Total	Rate	Total
Carpenter	.75	$. . .	$. . .	$16.47	$12.35

To the above, add equipment charges and cost of setting up and dismantling fabrication shop.

The above costs may be applied to either bolted trussed rafters or nailed trussed rafters as any differential in their costs will vary with the individual builders.

Labor Erecting Trussed Rafters.—Erection of trussed rafters, for one-story structures, can usually be accomplished with a 4-man crew — 2 laborers carrying rafter units from stockpile to setting location, placing units in an inverted position on the bearing wall plates and tilting up into an erect position ready for 2 carpenters to anchor and brace same in final position. For efficient operation, the layout of rafter locations should be marked on the wall plates and metal framing anchors set in place before erection is started. The 4-man crew should lay out, set anchors and erect into place 22 to 26 trussed rafters, up to 28'-0" span, per 8-hr. day, at the following labor cost per truss:

	Hours	Rate	Total	Rate	Total
Carpenter	0.67	$. . .	$. . .	$16.47	$11.03
Labor	0.67	. . .	. . .	12.54	8.40
Cost per truss			$. . .		$19.43

For spans greater than 28'-0" it may be necessary to add an extra man to the crew.

Material Requirements for All-Nailed Kingpost Trussed Rafters with ½" Plywood Gusset and Splice Plates

Span	20'-8"	24'-8"	28'-8"	32'-8"	36'-5 1/2"	40'-0"
Roof Slope	2/12	2/12	3/12	2/12	5/12	4/12
Overhang	19"	18"	16"	18"	51/2	12"
Top Chord, 2 ea.	2x4x12'	2x4x14'	2x6x16'	2x6x18'	2x6x20'	2x8x22'
Bottom Chord, 2 ea.	2x4x10'	2x4x12'	2x4x14'	2x6x16'	2x6x18'	2x6x20'
Post, 1 (or 2) ea.	2x4x2	2x4x2'-35/8"	2x4x3'-101/2"	2x4x3'-17/8"	2x6x8'	2-2x4x6'
Heel Gusset, 4 ea.	2'-9"x1'3/4"	4'x1'-31/4"	3'x1'-61/4"	4'x1'-7"	4'x2'-7"	4'x2'-31/2"
Ridge Gusset, 2 ea.	2'-9"x7"	4'x75/8"	3'x101/8"	5'x1'-0"	4'x1'-4"	4'x1'-51/4"
Center Splice, 2 ea.	3'x4'	4'x4'	4'x4'	4'x6"	4'x53/4"	4'x9"
Lumber, Bd. ft.	311/3	36	55	71	100	109
Plywood, Sq. Ft. (2½"x0.135")	16	24	22	36	43	44
Hardened Steel Nos.	140	168	218	304	308	342
Spiral-Thread Hi-Load Nails, lb.	1.39	1.66	2.16	3.01	3.05	3.40

Labor Framing Lumber in Building Construction

The labor quantities and costs given in the following pages are intended to include all classes of frame construction, such as residences, ranch houses, barns, stables, apartment buildings, combined store and apartment buildings, country clubs, schools, and in fact, all buildings that are constructed entirely of wood or have brick, stone, tile or cement block walls and wood joists, rafters, stud walls, partitions, wood subfloors, wall and roof sheathing, finish wood floors, etc.

Costs are given on two classes of workmanship. The classification "Ordinary Workmanship" is intended to cover the grade of workmanship encountered in most of our present day buildings where price is a factor. First Grade Workmanship designates the grade of workmanship demanded where quality is a factor as well as price, and is usually found in high-grade residences, apartments, hotel and office buildings, public buildings, high school, college and university buildings, etc.

The estimator must use his own judgment as to the grade of workmanship required, depending upon the type of building, the reputation of the architect, and the requirements of the specifications.

Another thing that is going to govern labor costs is the kind of equipment used, as the contractor who uses electric handsaws, drills, etc., will not have any trouble in outdistancing his competitors who are still working on the old handsaw basis.

All costs given on the following pages are based on using an electric saw for cutting all joists, studs and rafters to length, cutting off ends of subflooring, roof sheathing, etc. If an ordinary handsaw is used for this purpose, add 1 to 1½ hrs. carpenter time per 1,000 ft. b.m.

All of the labor quantities and costs on the following pages are based on laborers or helpers handling and carrying the lumber from the stock piles or benches where the lumber is being cut to length to the building. If the lumber is handled and carried by carpenters, figure labor time at carpenter wages.

Framing and Placing Foundation Wall Plates.—Where 2"x4" or 2"x6" wood plates are placed on top of foundation walls or concrete slab to receive the floor joists and exterior wall studs, it is customary to place anchor bolts in the walls and floors and the plates are then bored to receive the bolts, the plates are placed on the wall (and usually wedged with shingles) ready to receive the joists and exterior wall studs. Special purpose nails may be used for anchoring wall plates to concrete.

Where just an ordinary grade of workmanship is required, two carpenters working together should handle, frame and place 225 to 275 lin. ft. of 2"x4" or 2"x6" plates per 8-hr. day, at the following cost per 100 lin. ft.:

	Hours	Rate	Total	Rate	Total
Carpenter	6.4	$....	$....	$16.47	$105.41
Cost per lin. ft.			....		1.05

First Grade Workmanship

On jobs where first grade workmanship is required, with the foundation wall plates drilled for bolts, plates set and bedded absolutely level in a bed of cement mortar, two carpenters working together should handle, frame and place 175 to 225 lin. ft. of 2"x4" or 2"x6" wall plates per 8-hr. day, at the following labor cost per 100 lin. ft. :

	Hours	Rate	Total	Rate	Total
Carpenter	8	$....	$....	$16.47	$131.75
Cost per lin. ft.			....		1.32

Framing and Placing Box Sills and Plates.—On platform framing, where a wood box sill and plate is formed by using a 2"x4" or 2"x6" plate and a 2"x8" or

2"x10" on the end of the joists to form a box sill, the wall plate should be drilled, leveled and set the same as described above, and the side or end piece is nailed after the joists are set.

This work should be estimated as given above for "Foundation Wall Plates" and the 2"x8" or 2"x10" end pieces should be figured in with the floor joists.

Framing and Erecting Exterior Stud Walls for Frame Buildings.—The labor cost of framing and erecting stud walls is subject to wide variation, depending upon the type of building, height, regularity of the walls, etc.

The framing on square or rectangular buildings, such as Cape Cod and colonial type houses and buildings of similar shape and construction will cost much less to frame than English type houses, having walls of irregular shape and height.

When you consider the average house requires only 1,500 to 2,500 ft. of lumber for outside stud walls, it shows just how much labor is required to frame it.

On square or rectangular shaped buildings, such as colonial, Georgian houses, etc., a carpenter should frame and erect 350 to 400 ft. of lumber, b.m. per 8-hr. day, at the following labor cost per 1,000 ft. :

	Hours	Rate	Total	Rate	Total
Carpenter	21.4	$. ...	$. ...	$16.47	$353.46
Labor	6	....	....	12.54	75.24
Cost 1,000 ft. b.m.			$. ...		$428.70

On English type and other buildings having irregular wall construction, a carpenter should frame and erect 250 to 300 ft. of lumber, b.m. per 8-hr. day, at the following labor cost per 1,000 ft. :

	Hours	Rate	Total	Rate	Total
Carpenter	29.1	$. ...	$. ...	$16.47	$479.28
Labor	6	....	....	12.54	75.24
Cost 1,000 ft. b.m.			$. ...		$554.52

First Grade Workmanship

In high class wood constructed buildings where every precaution is taken to prevent settling due to shrinkage, where the wood studs rest on masonry walls or steel I beams, and are not set on top of the floor joists and where it is necessary to bridge or truss between all studs, truss over door and window openings, etc., a carpenter should frame and erect 225 to 275 ft. of lumber, b.m. per 8-hr. day, on square or rectangular type buildings, at the following labor cost per 1,000 ft. :

	Hours	Rate	Total	Rate	Total
Carpenter	32	$. ...	$. ...	$16.47	$527.04
Labor	10	....	....	12.54	125.40
Cost 1,000 ft. b.m.			$. ...		$652.44

On English type and other buildings having irregular wall construction, a carpenter should frame and erect 150 to 200 ft. of lumber, b.m. per 8-hr day, at the following labor cost per 1,000 ft. :

	Hours	Rate	Total	Rate	Total
Carpenter	45.7	$. ...	$. ...	$16.47	$752.68
Labor	10	....	....	12.54	125.40
Cost 1,000 ft. b.m.			$. ...		$878.08

Framing Interior Stud Partitions.—When framing and setting interior stud partitions set on top of rough wood floors that require just the ordinary amount of framing for door openings, a carpenter should frame and erect 375 to 425 ft. of lumber, b.m. per 8-hr. day, at the following labor cost per 1,000 ft. :

	Hours	Rate	Total	Rate	Total
Carpenter	20	$....	$....	$16.47	$329.40
Labor	6	....	....	12.54	75.24
Cost 1,000 ft. b.m.			$....		$404.64

First Grade Workmanship

Where the wood partition studs are set on masonry walls or steel I beams instead of the wood subfloor and where it is necessary to brace between studs and truss over all door openings, a carpenter should frame and erect 250 to 300 ft. of lumber, b.m. per 8-hr. day, at the following labor cost per 1,000 ft. :

	Hours	Rate	Total	Rate	Total
Carpenter	29.1	$....	$....	$16.47	$479.28
Labor	8.0	....	....	12.54	100.32
Costs 1,000 ft. b.m.			$....		$579.60

Framing and Setting Floor Joists.—When framing and placing wood floor joists up to 2"x8" in buildings of regular construction, a carpenter should frame and place 550 to 600 ft. b.m. per 8-hr. day, at the following labor cost per 1,000 ft. :

	Hours	Rate	Total	Rate	Total
Carpenter	13.9	$....	$....	$16.47	$228.93
Labor	6	....	....	12.54	75.24
Cost 1,000 ft. b.m.			$....		$304.17

If 2"x10"or 2"x12" joists are used, a carpenter should frame and erect 600 to 650 ft. b.m. per 8-hr. day, at the following labor cost per 1,000 ft. :

	Hours	Rate	Total	Rate	Total
Carpenter	12.8	$....	$....	$16.47	$210.82
Labor	6	....	....	12.54	75.24
Cost 1,000 ft. b.m.			$....		$286.06

First Grade Workmanship

On jobs where the wood joists must be set with the crowning edge up and it is not permissible to block up under the joists with shingles or wood wedges, a carpenter should frame and erect 500 to 550 ft. b.m. (sizes up to 2"x8") per 8-hr. day, at the following labor cost per 1,000 ft. :

	Hours	Rate	Total	Rate	Total
Carpenter	15.2	$....	$....	$16.47	$250.34
Labor	4.5	....	....	12.54	56.43
Cost 1,000 ft. b.m.			$....		$306.77

Using 2"x10" or 2"x12" joists, a carpenter should frame and erect 550 to 650 ft. of lumber, b.m. per 8-hr. day, at the following labor cost per 1,000 ft. :

	Hours	Rate	Total	Rate	Total
Carpenter	13.9	$....	$....	$16.47	$228.93
Labor	4.0	....	....	12.54	50.16
Cost 1,000 ft. b.m.			$....		$279.09

Framing Panel and Girder Floor Systems.—When framing and placing 4"x6" wood beams and 2"x4" spacers to form 4-ft. square grids for panel and girder floor systems, 2 carpenters and a helper should frame and place 1,500 to 1,700 ft. of lumber, b.m. per 8-hr. day, at the following labor cost per 1,000 ft. :

	Hours	Rate	Total	Rate	Total
Carpenter	10	$....	$....	$16.47	$164.70
Labor	5	....	....	12.54	62.70
Cost per 1,000 ft. b.m.			$....		$227.40

Method of Framing Panel and Girder Floor Systems

Installing Cross-Bridging.—Under average conditions a carpenter should cut and place 90 to 110 sets (2 pieces) of crossbridging per 8-hr. day at the following labor cost per 100 sets:

	Hours	Rate	Total	Rate	Total
Carpenter	8	$....	$....	$16.47	$131.76
Cost per set			....		1.32

Framing and Erecting Rafters for Gable Roofs.—When framing and erecting rafters for plain double pitch or gable roofs, without dormers or gables, and where 2"x6" or 2"x8" lumber is used, a carpenter should frame and erect 285 to 335 ft. of lumber, b.m. per 8-hr. day, at the following labor cost per 1,000 ft. :

	Hours	Rate	Total	Rate	Total
Carpenter	24.6	$....	$....	$16.47	$405.16
Labor	7.0	....	....	12.54	87.78
Cost 1,000 ft. b.m.			$....		$492.94

On buildings having double pitch or gable roofs, cut up with dormers and gables framing into the main roof, a carpenter should frame and erect 250 to 300 ft. of lumber, b.m. per 8-hr. day, at the following labor cost per 1,000 ft. :

	Hours	Rate	Total	Rate	Total
Carpenter	29.1	$....	$....	$16.47	$479.28
Labor	8.0	....	....	12.54	100.32
Cost 1,000 ft. b.m.			$....		$579.60

Framing and Erecting Rafters for Hip Roofs.—When framing and erecting plain hip roofs, without dormers or gables framing into the main roof, a carpenter should frame and erect 250 to 300 ft. of lumber, b.m. per 8-hr. day, at the following labor cost per 1,000 ft. :

	Hours	Rate	Total	Rate	Total
Carpenter	29.1	$....	$....	$16.47	$479.28
Labor	8.0	....	....	12.54	100.32
Cost 1,000 ft. b.m.			$....		$579.60

On difficult constructed hip roofs, where it is necessary to frame for dormers, gables, valleys, etc., a carpenter should frame and erect 180 to 220 ft. of lumber, b.m. per 8-hr. day, at the following labor cost per 1,000 ft. :

	Hours	Rate	Total	Rate	Total
Carpenter	40	$....	$....	$16.47	$658.80
Labor	8	....	....	12.54	100.32
Cost 1,000 ft. b.m.			$....		$759.12

Framing Light Timbers for Exposed Roof Beam Construction.—Light timbers, varying from 4"x6" to 4"x14" and larger, depending upon span and spacing, are used for exposed beam roof construction. This type of construction is now in wide use for contemporary design houses, stores, medical groups, small churches, etc., with flat or low rise sloping roofs.

On work of this type, two carpenters and a helper should frame and place 750 to 850 ft. of lumber, b.m. per 8-hr. day, at the following labor cost per 1,000 ft. :

	Hours	Rate	Total	Rate	Total
Carpenter	20	$....	$....	$16.47	$329.40
Labor	10	....	....	12.54	125.40
Cost per 1,000 ft. b.m.			$....		$454.80

Framing for Roof Saddles on Flat Roofs.—On flat roofs where it is necessary to frame saddles to pitch the water toward the drains, and where the saddles are built up of 2"x4" framing covered with 1" sheathing, a carpenter should frame and erect 400 to 450 ft. of lumber, b.m. per 8-hr. day, at the following labor cost per 1,000 ft. :

	Hours	Rate	Total	Rate	Total
Carpenter	19	$....	$....	$16.47	$312.93
Labor	6	....	....	12.54	75.24
Cost 1,000 ft. b.m.			$....		$388.17

Placing Wood Cant Strips.—When cant strips, diagonally cut from 4"x4" or 6"x6" lumber, occur in long runs with few breaks a carpenter should place 475 to 525 lin. ft. per 8-hr. day at the following labor cost per 100 lin. ft:

	Hours	Rate	Total	Rate	Total
Carpenter	1.6	$....	$....	$16.47	$26.35
Cost per lin. ft.				....	.26

If cant strips occur in short runs, such as around roof opening curbs, or if the runs are cut up with pilasters and other breaks, a carpenter should place 300 to 325 lin. ft. per 8-hr. day at the following labor cost per 100 lin. ft. :

	Hours	Rate	Total	Rate	Total
Carpenter	2.6	$....	$....	$16.47	$42.82
Cost per lin. ft.				....	.43

Laying Rough Wood Floors.—When laying 1"x6" or 1"x8" rough wood subflooring, a carpenter should lay 825 to 925 ft. b.m. per 8-hr. day, at the following labor cost per 1,000 ft. :

	Hours	Rate	Total	Rate	Total
Carpenter	9.2	$....	$....	$16.47	$151.52
Labor	5.3	....	....	12.54	66.46
Cost 1,000 ft. b.m.			$....		$217.98

Laying Rough Wood Floors Diagonally.—When laying rough wood subflooring diagonally, making it necessary to cut both ends of the flooring on a bevel, a carpenter should lay 650 to 750 ft. of flooring b.m. per 8-hr. day, at the following labor cost per 1,000 ft. :

	Hours	Rate	Total	Rate	Total
Carpenter	11.5	$....	$....	$16.47	$189.41
Labor	5.5	....	....	12.54	68.97
Cost 1,000 ft. b.m.			$....		$248.38

Laying Wood Sheathing on Flat Roofs.—Figure same as given above for Rough Wood Floors.

Labor Laying Plywood Subflooring.—The labor cost of handling and laying plywood subflooring can vary widely, depending upon the size of the floor area, the regularity of the joist spacing and upon the method of laying.

When plywood is used as subflooring, nailed directly to the joists and with the finish floor applied directly to the plywood, all edges (both longitudinal and cross-wise) should be nailed about 6" on centers, otherwise the finish floor may develop squeaks caused by deflection of the long edges of the sheets when walked upon.

In other words, with the joists spaced 16" on centers, it is necessary to cut pieces of 2"x4"s approximately 14 3/8" long and the ends of these 2"x4"s must be nailed into the sides of the joists to provide nailing for the long edges of the plywood sheets. As the plywood is usually 4'-0" wide, it means a row of 2"x4" blocking must be run at right angles to the floor joists every 4'-0" apart to provide nailing for the long edges of the plywood sheets. This requires a lot of extra work—and at the present rate of wages—costs a lot of money.

In buildings with regular spans, where an underlayment is to be provided, and where the long edges of the plywood sheets are not nailed, two carpenters working together should handle, fit, lay and nail 52 to 60 sheets (1,664 to 1,920 sq. ft.) of 4'-0"x8'-0" plywood 5/8" to 3/4" thick per 8-hr. day, at the following labor cost per 100 sq. ft. :

	Hours	Rate	Total	Rate	Total
Carpenter	0.9	$....	$....	$16.47	$14.83
Labor	0.3	....	....	12.54	3.76
Cost per 100 sq. ft.			$....		$18.59
Cost per sq. ft.			....		.19

If 2"x4" wood blocking is required between the joists, as described above, and the plywood sheets must be nailed both longitudinally and crosswise, two carpenters working together should cut and nail blocks, and handle, fit, lay and nail 26 to 30 sheets of 4'-0"x8'-0" plywood (832 to 960 sq. ft.) 5/8" to 3/4" thick per 8-hr. day, at the following labor cost per 100 sq. ft. :

	Hours	Rate	Total	Rate	Total
Carpenter	1.8	$....	$....	$16.47	$29.65
Labor	0.5	....	....	12.54	6.27
Cost per 100 sq. ft.			$....		$35.92
Cost per sq. ft.			....		.36

Laying Plywood Decking for Panel and Girder Floor Systems.—Plywood decking for panel and girder floor construction should be 1" or 1¼" thick and laid with staggered joints. To minimize waste, inside dimensions of building should be laid out on a 4-ft. module, which also reduces cutting of panels to starters in every other row. In some localities, 4'x4' plywood panels are available as a stock size.

On large projects, where the work may be highly organized, a carpenter and 2 helpers should lay 4,500 to 4,700 sq. ft. of plywood decking per 8-hr. day, or enough for 3 to 4 average one-story houses, at the following labor cost per 1,000 sq. ft. :

	Hours	Rate	Total	Rate	Total
Carpenter	1.75	$....	$....	$16.47	$28.82
Labor	3.50	....	....	12.54	43.89
Cost per 1,000 sq. ft.			$....		$72.71
Cost per sq. ft.			....		.072

For smaller jobs, where only 1 or 2 houses are involved, increase above costs about 25 percent to absorb lost motion in laying out work and getting started.

Where plywood decking is to be covered with linoleum or other resilient flooring, requiring a fairly smooth surface, a carpenter should spackle and sand joints, surface cracks and other imperfections at the rate of 200 sq. ft. per hr.

Laying Wood Sheathing on Pitch or Gable Roofs.—When laying roof sheathing on plain hip or gable roofs, a carpenter should lay 575 to 615 ft. of lumber, b.m. per 8-hr. day, at the following labor cost per 1,000 ft. :

	Hours	Rate	Total	Rate	Total
Carpenter	13.5	$....	$....	$16.47	$222.35
Labor	6.5	....	....	12.54	81.51
Cost 1,000 ft. b.m.			$....		$303.86

On very steep roofs and roofs cut up with dormers, hips, valleys, etc., such as English type houses, etc., a carpenter will lay only 275 to 325 ft. of sheathing, b.m. per 8-hr. day, at the following labor cost per 1,000 ft. :

	Hours	Rate	Total	Rate	Total
Carpenter	26.5	$....	$....	$16.47	$436.46
Labor	7.5	....	....	12.54	94.05
Cost 1,000 ft. b.m.			$....		$530.51

Labor Placing Sidewall Sheathing.—When sheathing sidewalls of frame buildings with 1"x6" or 1"x8" boards laid horizontally, a carpenter should handle and place 625 to 700 ft. b.m. per 8-hr. day, at the following labor cost per 1,000 ft. :

	Hours	Rate	Total	Rate	Total
Carpenter	12	$....	$....	$16.47	$197.64
Labor	5	....	....	12.54	62.70
Cost 1,000 ft. b.m.			$....		$260.34

Labor Placing Diagonal Sidewall Sheathing.—When sidewall sheathing is placed diagonally, it usually requires two carpenters working together—one at each end of the board. It also usually requires scaffolding the outside walls before the sheathing is placed, as the man at the top must have a scaffold from which to work.

On work of this class, a carpenter should place 450 to 525 ft. b.m. of 1"x6" or 1"x8" sheathing per 8 hr. day, at the following labor cost per 1,000 ft. :

	Hours	Rate	Total	Rate	Total
Carpenter	16.5	$....	$....	$16.47	$271.76
Labor	5.5	....	....	12.54	68.97
Cost 1,000 ft. b.m.			$....		$340.73

WOOD BLOCKING, FURRING AND GROUNDS

Placing Wood Furring Strips on Masonry Walls.—Where it is necessary to place wood furring strips on brick or tile walls before lathing and plastering, allowing the strips to follow the line of the walls without wedging or blocking out to make them straight or plumb, with the nails driven into dry joints in the brickwork, a carpenter should place 500 to 550 lin. ft. of furring per 8-hr. day, at the following labor cost per 100 lin. ft. :

	Hours	Rate	Total	Rate	Total
Carpenter	1.5	$. . . .	$. . . .	$16.47	$24.71
Cost per lin. ft.			. . . .		.25
		Cost per Square (100 Sq. Ft.)			
Strips 12" centers	1.7	$. . . .	$. . . .	$16.47	$27.99
Strips 16" centers	1.3	. . . .	. . . .	16.47	21.41

First Grade Workmanship

Where it is necessary to plug the masonry walls and place all furring strips absolutely straight and plumb to produce a level surface to receive lath and plaster, a carpenter should plug walls and place 200 to 250 lin. ft. of strips per 8-hr. day, at the following labor cost per 100 lin. ft. :

	Hours	Rate	Total	Rate	Total
Carpenter	3.5	$. . . .	$. . . .	$16.47	$57.65
Cost per lin. ft.			. . . .		.58
		Cost per Square (100 Sq. Ft.)			
Strips 12" centers	3.9	$. . . .	$. . . .	$16.47	$64.23
Strips 16" centers	2.9	. . . .	. . . .	16.47	48.43

Placing Wood Grounds.—Where wood grounds are nailed direct to wood furring strips, door and window openings, etc., allowing them to follow the rough furring, partitions and wood bucks, without wedging or blocking to make them absolutely straight, a carpenter should place 540 to 590 lin. ft. per 8-hr. day, at the following labor cost per 100 lin. ft. :

	Hours	Rate	Total	Rate	Total
Carpenter	1.4	$. . . .	$. . . .	$16.47	$23.06
Cost per lin. ft.			. . . .		.23

First Grade Workmanship

Where it is necessary to keep all wood grounds absolutely straight, plumb and level for the finish plaster and to receive the interior wood finish, and where it is necessary to plug the masonry walls with wood plugs to hold the nails, a carpenter should place 200 to 250 lin. ft. of grounds per 8-hr. day, at the following labor cost per 100 lin. ft. :

	Hours	Rate	Total	Rate	Total
Carpenter	3.5	$. . . .	$. . . .	$16.47	$57.65
Cost per lin. ft.			. . . .		.58

Wood Blocking and Grounds Around Window Openings.—Where the masonry walls are 12" or more in thickness, requiring a wood jamb lining, a carpenter should place all necessary wood blocking and grounds for a window up to 4'-0"x 6'-0" in size in ¾ to 1¼ hr. at the following labor cost. :

	Hours	Rate	Total	Rate	Total
Carpenter	1	$. . . .	$. . . .	$16.47	$16.47

Windows 4'-0"x7'-0" to 5'-0"x8'-0" in size will require 1 1/8 to 1 3/8 hrs. to place blocking and grounds, at the following labor cost per window:

Carpenter	1.25	$. . . .	$. . . .	$16.47	$20.58

First Grade Workmanship

In the better class of buildings where all grounds must be absolutely straight and plumb and it is necessary to plug the brick and block walls for securing the grounds and blocking, a carpenter should complete one window up to 4'-0"x6'-0" in size in 1¼ to 1¾ hrs. at the following labor cost:

	Hours	Rate	Total	Rate	Total
Carpenter	1.5	$. . . .	$. . . .	$16.47	$24.71

A carpenter should plug walls, place blocking and grounds for one window 4'-0"x7'-0" to 5'-0"x8'-0" in 1½ to 2 hrs. at the following labor cost per window:

	Hours	Rate	Total	Rate	Total
Carpenter	1.75	$. . . .	$. . . .	$16.47	$28.82

Wood Blocking for Cabinet and Case Bases.—A carpenter should locate, frame and set 115 to 135 ft. b.m. of 2"x4" or 2"x6" blocking for cabinet or case bases per 8-hr. day at the following labor cost per 1,000 ft. b.m. :

	Hours	Rate	Total	Rate	Total
Carpenter	64	$. . . .	$. . . .	$16.47	$1,054.08

Wood Furring Strips Over Wood Subfloors.—When placing 1"x2" or 2"x2" wood furring strips over wood subfloors to receive finish flooring, a carpenter should place 550 to 600 lin. ft. (without wedging or leveling) per 8-hr. day, at the following labor cost per 100 lin. ft. :

	Hours	Rate	Total	Rate	Total
Carpenter	1.4	$. . . .	$. . . .	$16.47	$23.06
Labor	0.3	. . . .	. . . .	12.54	3.76
Cost per 100 lin. ft.			$. . . .		$26.82
Cost per lin. ft.				. . . .	.27
Cost per Square (100 Sq. Ft.)					
Strips 12" centers	1.9	$. . . .	$. . . .	$16.47	$31.29
Strips 16" centers	1.4	. . . .	. . . .	16.47	23.06

First Grade Workmanship

In the better class of buildings where the wood furring strips must be wedged up and blocked to produce an absolutely level surface to receive finish flooring, a carpenter should place 250 to 300 lin. ft. of furring strips per 8-hr. day, at the following cost per 100 lin. ft. :

	Hours	Rate	Total	Rate	Total
Carpenter	3.0	$. . . .	$. . . .	$16.47	$49.41
Labor	0.3	. . . .	. . . .	12.54	3.76
Cost per 100 lin. ft.			$. . . .		$53.17
Cost per lin. ft.				. . . .	.53
Cost per Square (100 Sq. Ft.)					
Strips 12" centers	3.6	$. . . .	$. . . .	$16.47	$59.29
Strips 16" centers	2.8	. . . .	. . . .	16.47	46.12

Placing Strip Deadening Felt and Wood Furring Strips Over Wood Subfloors.—Frequently a narrow strip of deadening felt ½"x3" is placed under the wood furring strips and both are nailed to the wood subfloor, as this provides better insulation and deadening than where the furring strips are nailed directly to the subfloor.

On work of this kind, a carpenter should place 425 to 475 lin. ft. of deadening felt and furring strips per 8-hr. day, at the following labor cost per 100 lin. ft. :

	Hours	Rate	Total	Rate	Total
Carpenter	1.8	$. . . .	$. . . .	$16.47	$29.65
Labor	0.4	. . . .	. . . .	12.54	5.02
Cost 100 lin. ft.			$. . . .		$34.67
Cost per lin. ft.				. . . .	.35
Cost per Square (100 Sq. Ft.)					
Strips 12" centers	2.4	$. . . .	$. . . .	$16.47	$39.53
Strips 16" centers	1.9	. . . .	. . . .	16.47	31.29

First Grade Workmanship

Where it is necessary to wedge and block up the strips to produce an absolutely level surface to receive the finish flooring, a carpenter should place 220 to 250 lin. ft. of deadening and furring strips per 8-hr. day, at the following labor cost per 100 lin. ft. :

	Hours	Rate	Total	Rate	Total
Carpenter	3.5	$....	$....	$16.47	$57.65
Labor	0.4	....	....	12.54	5.02
Cost 100 lin. ft.			$....		$62.67
Cost per lin. ft.			....		.63
	Cost per Square (100 Sq. Ft.)				
Strips 12" centers	4.3	$....	$....	$16.47	$70.82
Strips 16" centers	3.3	....	....	16.47	54.35

Placing Deadening Quilt Over Rough Wood Floors.—Where deadening felt or quilt is laid over wood subfloors, a man should handle and lay 200 to 250 sq. ft. of felt an hr. at the following labor cost per 100 sq. ft. :

	Hours	Rate	Total	Rate	Total
Labor	0.5	$....	$....	$12.54	$6.27
Cost per sq. ft.			....		.063

The above costs are based on deadening felts or quilts in rolls or sheets that can be laid on the rough floor and the furring strips placed directly over them.

Cost of Placing Deadening Felt and Wood Furring Strips Over 100 Sq. Ft. of Floor
Strips 12" on Centers—Ordinary Workmanship

	Hours	Rate	Total	Rate	Total
Carpenter	2.0	$....	$....	$16.47	$32.94
Labor	0.3	....	....	12.54	3.76
Cost 100 sq. ft.			$....		36.70
Cost per sq. ft.			....		.37

Strips 16" on Centers—Ordinary Workmanship

	Hours	Rate	Total	Rate	Total
Carpenter	1.6	$....	$....	$16.47	$26.35
Labor	0.3	....	....	12.54	3.76
Cost 100 sq. ft.			$....		$30.11
Cost per sq. ft.			....		.30

Strips 12" on Centers—First Grade Workmanship

	Hours	Rate	Total	Rate	Total
Carpenter	3.8	$....	$....	$16.47	$62.59
Labor	0.3	....	....	12.54	3.76
Cost 100 sq. ft.			$....		$66.35
Cost per sq. ft.			....		.66

Strips 16" on Center—First Grade Workmanship

	Hours	Rate	Total	Rate	Total
Carpenter	3.0	$....	$....	$16.47	$49.41
Labor	0.3	....	....	12.54	3.76
Cost 100 sq. ft.			$....		$53.17
Cost per sq. ft.			....		.53

Placing Wood Floor Sleepers.—When placing 2"x3" or 2"x4" wood floor screeds or sleepers over rough tile or concrete floors, to receive finish flooring, a carpenter should place 225 to 275 lin. ft. per 8-hr. day, at the following labor cost per 100 lin. ft. :

	Hours	Rate	Total	Rate	Total
Carpenter	3.2	$....	$....	$16.47	$52.70
Labor	0.8	....	....	12.54	10.03
Cost 100 lin. ft.			$....		$62.73
Cost per lin. ft.			....		.63

First Grade Workmanship

In the better class of buildings, 2"x3" or 2"x4" beveled floor sleepers are placed over the rough concrete floors and wedged or blocked up to provide a perfectly level surface to receive the finish flooring.

The screeds are usually held in place by metal clips placed in the rough concrete, or anchored with special purpose nails.

On work of this class, a carpenter should place, wedge-up and level 130 to 170 lin. ft. of sleepers per 8-hr. day, including setting sleeper clips, at the following labor cost per 100 lin. ft. :

	Hours	Rate	Total	Rate	Total
Carpenter	5.3	$....	$....	$16.47	$87.29
Labor	0.8	....	....	12.54	10.03
Cost per 100 lin. ft.			$....		$97.32
Cost per lin. ft.			....		.97

Lumber Required for Wood Door Bucks of 2"x4" Lumber

Size of Opening	Lin. Ft. of Lumber Req'd.	Board Feet of Lumber Req'd.	With Transom Add Lineal Ft.	With Transom Add Board Ft.
2'-0"x4'-0"	12	8	6'-0"	4
2'-0"x4'-6"	12	8	6'-0"	4
2'-0"x5'-0"	13½	9	6'-0"	4
2'-6"x4'-0"	12	8	6'-6"	4⅓
2'-6"x4'-6"	13	8⅔	6'-6"	4⅓
2'-6"x5'-0"	14	9⅓	6'-6"	4⅓
2'-6"x5'-6"	15	10	6'-6"	4⅓
2'-6"x6'-0"	16	10⅔	6'-6"	4⅓
2'-6"x6'-6"	17	11⅓	6'-6"	4⅓
2'-6"x7'-0"	18	12	6'-6"	4⅓
2'-8"x6'-8"	17	11⅓	7'-0"	4⅔
2'-8"x7'-0"	18	12	7'-0"	4⅔
3'-0"x7'-0"	19	12⅔	7'-0"	4⅔
3'-6"x7'-0"	19	12⅔	7'-6"	5
4'-0"x7'-0"	20	13⅓	8'-0"	5⅓
5'-0"x7'-0"	21	14	9'-0"	6
6'-0"x7'-0"	22	14⅔	10'-0"	6⅔
7'-0"x7'-0"	24	16	11'-0"	7⅓
3'-6"x7'-6"	20	13⅓	7'-6"	5
4'-0"x7'-6"	20	13⅓	8'-0"	5⅓
4'-6"x7'-6"	21	14	8'-6"	5⅔
5'-0"x7'-6"	21	14	9'-0"	6
6'-0"x7'-6"	22	14⅔	10'-0"	6⅔
7'-0"x7'-6"	24	16	11'-0"	7⅓
3'-6"x8'-0"	21	14	7'-6"	5
4'-0"x8'-0"	22	14⅔	8'-0"	5⅓
5'-0"x8'-0"	22	14⅔	9'-0"	6
6'-0"x8'-0"	23	15⅔	10'-0"	6⅓

Size of Opening	Lin. Ft. of Lumber Req'd.	Board Feet of Lumber Req'd.	With Transom Add Lineal Ft.	With Transom Add Board Ft.
7'-0"x8'-0"	24	16	11'-0"	7⅓
8'-0"x8'-0"	25	16⅔	12'-0"	8
9'-0"x9'-0"	28	18⅔	9'-0"	6

The lumber quantities given in the table are estimated in lengths that cut with the least waste.

Lumber Required for Wood Door Bucks of 2"x6" Lumber

Size of Opening	Lin. Ft. of Lumber Req'd.	Board Ft. of Lumber Req'd.	With Transom Add Lineal Ft.	With Transom Add Board Ft.
2'-0"x4'-0"	12	12	6'-0"	6
2'-0"x4'-6"	12	12	6'-0"	6
2'-0"x5'-0"	13½	13½	6'-0"	6
2'-6"x4'-0"	12	12	6'-6"	6½
2'-6"x4'-6"	13	13	6'-6"	6½
2'-6"x5'-0"	14	14	6'-6"	6½
2'-6"x5'-6"	15	15	6'-6"	6½
2'-6"x6'-0"	16	16	6'-6"	6½
2'-6"x6'-6"	17	17	6'-6"	6½
2'-6"x7'-0"	18	18	6'-6"	6½
2'-8"x6'-8"	17	17	7'-0"	7
3'-0"x7'-0"	19	19	7'-0"	7
3'-6"x7'-0"	19	19	7'-6"	7½
4'-0"x7'-0"	20	20	8'-0"	8
5'-0"x7'-0"	21	21	9'-0"	9
6'-0"x7'-0"	22	22	10'-0"	10
7'-0"x7'-0"	24	24	11'-0"	11
3'-6"x7'-6"	20	20	7'-6"	7½
4'-0"x7'-6"	20	20	8'-0"	8
4'-6"x7'-6"	21	21	8'-6"	8½
5'-0"x7'-6"	21	21	9'-0"	9
6'-0"x7'-6"	22	22	10'-0"	10
7'-0"x7'-6"	24	24	11'-0"	11
3'-6"x8'-0"	21	21	7'-6"	7½
4'-0"x8'-0"	22	22	8'-0"	8
5'-0"x8'-0"	22	22	9'-0"	9
6'-0"x8'-0"	23	23	10'-0"	10
7'-0"x8'-0"	24	24	11'-0"	11
8'-0"x8'-0"	25	25	12'-0"	12
9'-0"x9'-0"	28	28	9'-0"	9

Wood Door Bucks

Wood door bucks are made from 2"x4" or 2"x 6" lumber, depending upon the thickness of the partition in which they are to be used.

Considerable saving may be affected by using bucks made from 2"x6" lumber, having the back grooved out the thickness of the partition and ½" to ¾" deep. For a 5½" finished partition, this allows for a 4" tile or gypsum partition block and 1½" for plaster on both sides of the partition. The partition tile fit into the groove in the wood buck and the ¾" extension on each side provides a ground for nailing the trim.

The mills furnishing the lumber usually charge $12.00 to $15.00 per 1,000 ft. b.m. for grooving the backs of 2"x6"s.

The lumber quantities given above are estimated in lengths that cut with the least waste.

Labor Making Rough Door Bucks.—A carpenter should make and brace a rough door buck for an opening up to 3'-0"x7'-0" in about 1-hr. If the door has a transom, add ½-hr. carpenter time.

When making rough door bucks for door openings 5'-0"x7'-0" to 6'-0"x9'-0", a carpenter should make and brace one door buck in 1¼ to 1½ hrs. If the door has a transom, add ½ to ¾-hr. carpenter time.

Rough Wood Bucks for Borrowed Lights.—When making rough wood bucks for borrowed lights, a carpenter should make and brace one buck for an average size opening in 1 to 1½ hrs.

Setting Rough Door Bucks.—After the rough door buck has been made, a carpenter should set, plumb and brace on door buck for an opening up to 3'-0"x 9'-0" in about 1 to 1¼ hr.

For double doors requiring an opening from 5'-0"x7'-0" to 6'-0"x9'-0", a carpenter should set, plumb and brace one large door buck in 1¼ to 1½ hrs.

Fireproofing Wood.—Where framing lumber and timbers are chemically treated to fireproof the wood, the cost of this treatment practically doubles the cost of the untreated wood.

BUILDING AND INSULATING SHEATHING

There are any number of insulating sheathing boards on the market, consisting of felted wood fiber or vegetable fiber products treated to make them moisture resistant.

Some of these sheets are coated with high melting point asphalt or are asphalt impregnated to form a moisture resistant surface which retards moisture penetration.

Insulating sheathing is furnished in sheets 4 ft. wide and 6, 7, 8, 9, 10 and 12 ft. long, and ½-in. and 25/32-in. thick, the same thickness as wood sheathing.

Most manufacturers furnish this type of sheathing 25/32-in. thick, 2 ft. wide and 8 ft. long, with the long edges V-jointed or shiplapped.

Insulating sheathing is usually applied to wood studs under wood siding, shingles, stucco or brick veneer and because of the large size sheets used and the asphalt treatment, building paper is not ordinarily used with it, except under stucco.

Use 2 in. galvanized nails with 3/8 or ½-in. heads for insulating sheathing.

Place nails 3 inches apart on all outside edges of board and 6 inches apart for all intermediate nailing.

Insulating sheathing 25/32-in thick costs $160 to $180 per 1,000 sq. ft.

Labor Placing Insulating Sheathing.—When placing insulating sheathing on square or rectangular houses of regular construction, a carpenter should place 700 to 900 sq. ft. per 8-hr. day, at the following labor cost per 100 sq. ft. :

	Hour	Rate	Total	Rate	Total
Carpenter	1.0	$....	$....	$16.47	$16.47
Labor unloading and carrying sheets	0.3	....	....	12.54	3.76
Cost per 100 sq. ft			$....		$20.23
Cost per sq. ft			....		.20

On buildings of irregular construction, requiring a great deal of cutting and fitting, it is not possible to take advantage of the large size sheets as too much cutting and fitting is required. A carpenter will place only 350 to 450 sq. ft. of sheathing per 8-hr. day, at the following labor cost per 100 sq. ft. :

	Hours	Rate	Total	Rate	Total
Carpenter	2.0	$. . . .	$. . . .	$16.47	$32.94
Labor unloading and carrying sheets	0.3	. . . .	. . . .	12.54	3.76
Cost per 100 sq. ft			$. . . .		$36.70
Cost per sq. ft			. . . .		.37

Insulating Roof Decking

Several manufacturers produce insulating roof decking made of multiple layers of ½-in. insulating board laminated together with vapor resistant cement and fabricated into tongue and groove planks, 2'-0" wide, 8'-0" long and 1½", 2" or 3" thick, with a finish painted undersurface. It provides a roof deck, insulation, vapor barrier and interior ceiling finish using only one material, and is especially adaptable for use in contemporary design ranch houses with exposed beam ceilings.

Insulating roof decking can be used for flat, pitched or monosloped roofs. Flat roofs and surfaces with a slope of 3" in 12" or less are usually covered with built-up roofing. Steeper roofs may be covered with rigid shingles, slate or tile roofing, providing wood nailing strips are fastened through decking to supporting beams below.

Decking should be laid so that cross joints are staggered and occur only over supports. Decking should be face-nailed to all framing members, spacing nails 4" to 6" apart and keeping back ¾" to 1" from edges of plank. Nails should be galvanized common of sufficient length to pass through decking and penetrate supports at least 1½" and should be driven flush but not countersunk. Where underside of decking will be exposed, planks must be handled and laid with care to prevent finished ceiling surface from being marred or damaged— avoid excessive sliding of plank on roof beams.

Estimating Data on Insulating Roof Decking

Thickness	Max. Distance Between Supports	Size & Type of Nails	Nails Req'd. Per 100 Sq. Ft.	Approx. Price Per Sq. Ft.
1½"	24"	10d Com. Galv.	6.0 lbs.	$0.45
2"	32"	16d Com. Galv.	3.5 lbs.	.55
3"	48"	30d Com. Galv.	3.0 lbs.	.68

Labor Placing Insulating Roof Decking.—The labor cost of handling and placing insulating roof decking will be about the same for the various thicknesses available, as the saving in handling the lighter material is offset by the increased nailing required due to closer allowable spacing of supports.

On simple, rectangular, flat or low-rise roofs, two carpenters and a helper should place 750 to 850 sq. ft. of decking per 8-hr. day, including stapling paper strips on roof beams for added protection against marring finished surface, at the following labor cost per 100 sq. ft. :

	Hours	Rate	Total	Rate	Total
Carpenter	2	$. . . .	$. . . .	$16.47	$32.94
Labor	1	. . . .	. . . .	12.54	12.54
Cost per 100 sq. ft.			$. . . .		$45.48
Cost per sq. ft			. . . .		.45

For steep roofs, with a slope greater than 6" in 12", add 15 to 25 percent.

Method of Installing Insulating Roof Decking.

PLYWOOD ROOF AND WALL SHEATHING, SUBFLOORING AND UNDERLAYMENT

Roof and Wall Sheathing—Plywood for use in roof and wall sheathing is readily available, easily worked with ordinary tools and skills, and is adaptable to almost any light construction application. Because of the large size sheets, installation time is less than for other types of sheathing, and waste is minimal. It may be used under any type of shingle or roofing material, or any type of siding.

Standard size sheets are 4"x8" but other sizes are available by special order. Most common sheathing thicknesses are 5/16", 3/8", 1/2", 5/8" and 3/4", which are unsanded, and are available with interior, intermediate or exterior glue.

For most roof and wall sheathing installations, where sheathing is to be covered, interior type plywood is used. Where installations are to be exposed to the weather, such as for roof overhangs or service building siding, Exterior type plywood should be used.

Nailing of plywood sheathing should be at 6" o.c. along panel edges and 12" o.c. at intermediate supports. 6d common nails should be used for panels ½" or less in thickness, and 8d for greater thickness.

Where plywood is used for wall sheathing, corner bracing is not required because of the rigidity of the plywood. Where other types of sheathing are used, plywood of the same thickness as the sheathing is sometimes used at corners to take the place of corner diagonal bracing as shown in the photograph.

Method of Applying Plywood Sheathing

The American Plywood Association has prepared maximum load and span tables for the various thicknesses of plywood for use in roof decking and wall sheathing installations as follows:

TABLE E   Plywood Wall Sheathing (a)   (Plywood continuous over 2 or more spans)

Panel Identification Index	Panel Thickness (inch)	Maximum Stud Spacing (inches) Exterior Covering Nailed to:		Nail Size (b)	Nail Spacing (inches) Panel Edges (when over framing)	Intermediate (each stud)
		Stud	Sheathing			
12/0, 16/0, 20/0	5/16	16	16 (c)	6d	6	12
16/0, 20/0, 24/0	3/8	24	16 24 (c)	6d	6	12
24/0, 30/12, 32/16	1/2, 5/8	24	24	6d	6	12

NOTES:

(a) When plywood sheathing is used, building paper and diagonal wall bracing can be omitted.

(b) Common smooth, annular, spiral-thread, or galvanized box, or T-nails of the same diameter as common nails (0.113" dia. for 6d) may be used. Staples also permitted at reduced spacing.

(c) When sidings such as shingles are nailed only to the plywood sheathing, apply plywood with face grain across studs.

TABLE A Plywood roof decking (a) (b) (c)

(Plywood continuous over 2 or more spans; grain of face plys across supports)

Panel Ident. Index	Plywood Thickness (inch)	Max. Span (inches) (d)	Unsupported Edge—Max. Length (inches) (e)	Allowable Roof Loads (psf) (f) (g) (Spacing of Supports [Inches] Center to Center)										
				12	16	20	24	30	32	36	42	48	60	72
12/0	5/16	12	12	100 (130)										
16/0	5/16, 3/8	16	16	130 (170)	55 (75)									
20/0	5/16, 3/8	20	20		85 (110)	45 (55)								
24/0	3/8, 1/2	24	24		150 (160)	75 (100)	45 (60)							
30/12	5/8	30	26				145 (165)	85 (110)	40 (55)					
32/16	1/2, 5/8	32	28					90 (105)	45 (60)	40 (50)				
36/16	3/4	36	30				125 (145)	65 (85)	55 (70)	35 (50)				
42/20	5/8, 3/4, 7/8	42	32						80 (105)	65 (90)	45 (60)	35 (40)		
48/24	3/4, 7/8	48	36						105 (115)	75 (90)	55 (55)	40 (40)		
2-4-1	1-1/8	72	48							175 (175)	105 (105)	80 (80)	50 (50)	30 (35)
1-1/8" Grp 1&2	1-1/8	72	48							145 (145)	85 (85)	65 (65)	40 (40)	30 (30)
1-1/4" Grp 3&4	1-1/4	72	48							160 (165)	95 (95)	75 (75)	45 (45)	25 (35)

NOTES:

(a) These values apply for STANDARD C-D INT-DFPA, STRUCTURAL I AND II C-D INT-DFPA, C-C EXT-DFPA and STRUCTURAL I C-C EXT-DFPA grades only.

(b) For application where the roofing is to be guaranteed by a performance bond, recommendations may differ somewhat from these values. See Table D.

(c) Use 6d common, smooth, ring-shank or spiral-thread nails for 1/2" thick or less, and 8d common, smooth, ring-shank or spiral-thread for plywood 1" thick or less. Use 8d ring-shank or spiral-thread or 10d common smooth shank nails for 2-4-1, 1-1/8" and 1-1/4" panels. Space nails 6" at panel edges and 12" at intermediate supports, except that where spans are 48" or more, nails shall be 6" at all supports.

(d) These spans shall not be exceeded for any load conditions.

(e) Provide adequate blocking, tongue and grooved edges or other suitable edge support such as Plyclips when spans exceed indicated value. Use two Plyclips for 48" or greater spans and one for lesser spans.

(f) Uniform load deflection limitation: 1/180th of the span under live load plus dead load, 1/240th under live load only. Allowable live load shown in boldface type and allowable total load shown within parentheses. The allowable live load should in no case exceed the total load less the dead load supported by the plywood.

(g) Allowable roof loads were established by laboratory test and calculations assuming uniformly distributed loads.

Subflooring and Underlayment.—Plywood subflooring, underlayment or combined subfloor-underlayment, are products available for use in most areas of the country. Some particular panels may not be stocked in all areas so it is advisable to check with local plywood sources as to their availability.

Subflooring panels are engineered grades of plywood marked by the American Plywood Association for maximum spans for the various thicknesses. See table below. Deflection of the panels under concentrated loads at panel edges is the limiting factor in the design of plywood floors. Panel edges should be supported by solid blocking, or tongue and grooved panels should be used, unless a separate layer of underlayment is installed with its joints offset from the joints in the subfloor.

Nailing of subflooring should be at 6" o.c. along the edges and 10" o.c. at intermediate supports. 6d common nails may be used for ½" plywood, and 8d for ⅝" to ⅞" thick plywood. For 1⅛" thick panels, use 8d ring shank or 10d common nails spaced 6" o.c. at both panel edges and intermediate supports.

Underlayment grades of plywood are touch sanded panels of ¼" and ⅜" thickness with a smooth, solid surface for application of nonstructural flooring finished directly to them. This grade or plywood is made with special inner-ply construction which resists punch-through by concentrated point loading.

Joints in the underlayment should be staggered with respect to the joints in the subfloor. Nailing should be done with 3d ring shanked underlayment nails at 6" o.c. along the edges and 8" each way in the interior of the panels. Nails should be

TABLE G Plywood subflooring (a) (c) (d)   For direct application of T&G wood strip and block flooring and lightweight concrete.
(Plywood continuous over two or more spans, face grain across supports)

Panel Identification Index (b)	Plywood Thickness (inch)	Maximum Span (e) (inches)	Nail Size & Type	Nail Spacing (inches) Panel Edges	Intermediate
30/12	5/8	12 (f)	8d common	6	10
32/16	1/2, 5/8	16 (g)	8d common (h)	6	10
36/16	3/4	16 (g)	8d common	6	10
42/20	5/8, 3/4, 7/8	20 (g)	8d common	6	10
48/24	3/4, 7/8	24	8d common	6	10
1-1/8" Groups 1 & 2	1-1/8 (i)	48	10d common	6	6
1-1/4" Groups 3 & 4	1-1/4	48	10d common	6	6

NOTES:

(a) These values apply for STANDARD C-D INT-DFPA, STRUCTURAL I AND II C-D INT-DFPA, C-C EXT-DFPA and STRUCTURAL I C-C EXT-DFPA grades only.

(b) Identification Index appears on all panels except 1-1/8" and 1-1/4" panels.

(c) In some nonresidential buildings, special conditions may impose heavy concentrated loads and heavy traffic requiring subfloor constructions in excess of these minimums.

(d) Edges shall be tongue and grooved or supported with blocking for square edge wood flooring, unless separate underlayment layer (1/4" minimum thickness) is installed.

(e) Spans limited to values shown because of possible effect of concentrated loads. At indicated maximum spans, floor panels carrying Identification Index numbers will support uniform loads of more than 100 psf.

(f) May be 16" if 25/32" wood strip flooring is installed at right angles to joists.

(g) May be 24" if 25/32" wood strip flooring is installed at right angles to joists.

(h) 6d common nail permitted if plywood is 1/2".

(i) For 2•4•1 application details see page 12.

countersunk if subfloor and joists are not completely dry, to prevent nail popping. See table below for American Plywood Association recommendations on Plywood Underlayment, and fastening.

TABLE H Plywood underlayment / For application of tile, carpeting, linoleum or other non-structural flooring.

Plywood Grades and Species Group	Application	Minimum Plywood Thickness	Fastener size (approx.) and Type (set nails 1/16")	Fastener Spacing (inches) Panel Edges	Inter-mediate
Groups 1, 2, 3, 4 UNDERLAYMENT INT-DFPA (with interior, intermediate or exterior glue) UNDERLAYMENT INT-DFPA C-C Plugged EXT-DFPA	over plywood subfloor	1/4"	18 Ga. staples or 3d ring-shank nails (a) (b)	3	6 each way
	over lumber subfloor or other uneven surfaces	3/8"	16 Ga. staples (a)	3	6 each way
			3d ring-shank nails (b)	6	8 each way
Same Grades as above, but Group 1 only.	over lumber floor up to 4" wide. Face grain must be perpendicular to boards.	1/4"	18 Ga. staples or 3d ring-shank nails	3	6 each way

(a) Crown 3/8" for 16 ga., 3/16" for 18 ga. staples; length sufficient to penetrate completely through, or at least 5/8" into, subflooring.

(b) Use 3d ring-shank nail also for 1/2" plywood and 4d ring-shank nail for 5/8" or 3/4" plywood.

Installation: Apply UNDERLAYMENT just prior to laying finish floor or protect against water or physical damage. Stagger panel end joints with respect to

each other and offset all joints with respect to the joints in the subfloor. Space panel ends and edges about 1/32". For maximum stiffness, place face grain of panel across supports and end joints over framing. Unless subfloor and joists are thoroughly seasoned and dry, countersink nails 1/16" just prior to laying finish floor to avoid nail popping. Countersink staples 1/32". Fill any damaged, split or open areas exceeding 1/16". Do not fill nail holes. Lightly sand any rough areas, particularly around joints or nail holes.

Combined subfloor-underlayment plywood—is a single application installation combining both subfloor and underlayment, thus eliminating the expense of installing a separate underlayment. Panel edges perpendicular to the joists must be supported by solid blocking if tongue and grooved panels are not used.

Panels should be nailed at 6" o.c. at edges and 10" at intermediate supports with ringshank or spiral-thread nails. 6d deformed-shank nails may be used for panel thickness up to ¾" and 8d for thicker panels.

Approximate costs of plywood sheathing, subflooring and underlayment per 1000 sq. ft. are as follows:

Interior Type Plywood STANDARD (C-D INT)		Interior Type Plywood Structural 1 C-D		Exterior Type Plywood C-C EXT	
5/16"	$240.00	1/2"	$310.00	5/16"	$290.00
3/8	260.00	5/8	390.00	3/8	350.00
1/2	310.00	3/4	470.00	1/2	390.00
5/8	360.00			5/8	430.00
3/4	420.00			3/4	500.00

Plywood Underlayment Group 1, Interior	
1/2"	$300.00
5/8	390.00
3/4	450.00

Following is a table showing recommended support spacing for combined subfloor-underlayment.

TABLE I Combined subfloor-underlayment / For direct application of tile, carpeting, linoleum or other non-structural flooring. (Plywood continuous over two or more spans; grain of face plys across supports. Seasoned lumber is recommended.)

Plywood Grade (c)	Plywood Species Group	16" o.c. Panel Thickness	16" o.c. Deformed Shank Nail Size (d)	20" o.c. Panel Thickness	20" o.c. Deformed Shank Nail Size (d)	24" o.c. Panel Thickness	24" o.c. Deformed Shank Nail Size (d)	32" (e) or 48" o.c. Panel Thickness	32" (e) or 48" o.c. Deformed Shank Nail Size (d)	Nail Spacing (inches) Panel Edges	Nail Spacing (inches) Inter-mediate
UNDERLAYMENT INT-DFPA (with interior, intermediate or exterior glue)	1	1/2"	6d	5/8" (f)	6d	3/4"(g)	6d	—	—	6	10
UNDERLAYMENT EXT-DFPA	2 & 3	5/8" (f)	6d	3/4"(g)	6d	7/8"	8d	—	—	6	10
C-C Plugged EXT-DFPA	4	3/4"(g)	6d	7/8"	8d	1"	8d	—	—	6	10
2-4-1	1, 2 & 3	(2-4-1 specifications are so written that panels from all groups have equal properties.)						1-1/8"	8d (or 10d common smooth shank if supports well-seasoned)	6	(h)

(a) Edges shall be tongue & grooved, or supported with framing.
(b) In some non-residential buildings, special conditions may impose heavy concentrated loads and heavy traffic requiring subfloor-underlayment constructions in excess of these minimums.
(c) For certain types of flooring such as wood block or terrazzo, sheathing grades of plywood may be used.
(d) Set nails 1/16" (1/8" for 2-4-1 panels) and lightly sand subfloor at joints if resilient flooring is to be applied. Don't fill nail holes.
(e) 2" wide supports for 32" o.c.; 4" wide supports for 48" centers.
(f) May be 19/32".
(g) May be 23/32".
(h) 10" for 32" o.c.; 6" for 48" o.c. supports.

Prices on all grades and types of plywood should be verified with local suppliers prior to submitting a proposal for work which includes these materials.

Approximate costs of various materials and labor required for sidewall sheathing are given below.

Labor Cost of 100 Sq. Ft. of Structural Insulating Sidewall Sheathing
Applied to Square or Rectanglar Buildings of
Regular Construction

	Hours	Rate	Total	Rate	Total
Carpenter sheathing	1.0	$....	$....	$16.47	$16.47
Helper unloading and carrying sheets	0.3	....	....	12.54	3.76
Cost per 100 sq. ft			$....		$20.23
Cost per sq. ft			....		.20

No paper required as insulating sheathing is treated to resist moisture.

If used on buildings of irregular construction, as described above, add 1-hr. carpenter time.

Labor Cost of 100 Sq. Ft. of 1x6-In. Wood
Sidewall Sheathing Applied Horizontally to Square or
Rectangular Buildings of Regular Construction

	Hours	Rate	Total	Rate	Total
Carpenter sheathing	1.5	$....	$....	$16.47	$24.71
Labor unloading and carrying lumber	0.5	....	....	12.54	6.27
Carpenter applying building paper.	0.4	....	....	16.47	6.59
Cost per 100 sq. ft			$....		$37.57
Cost per sq. ft			....		.38

If 1x8-in. shiplap is used instead of 1x6-in. D&M lumber, deduct 5 ft. lumber b.m. and 0.1-hr. carpenter time per 100 sq. ft. of wall.

If used on buildings of irregular construction, as described above, add 1-hr. carpenter time.

Labor Cost of 100 Sq. Ft. of 1x6-In. Wood
Sidewall Sheathing Applied Diagonally to Square or
Rectangular Buildings of Regular Construction

	Hours	Rate	Total	Rate	Total
Carpenter sheathing............................	2.0	$. . . .	$. . . .	$16.47	$32.94
Labor unloading and car-					
rying lumber.......................................	0.5	. . . .	. . . .	12.54	6.27
Carpenter applying paper	0.4	. . . .	. . . .	16.47	6.59
Cost per 100 sq. ft.........................			$. . . .		$45.80
Cost per sq. ft..................................			. . . .		.46

If 1x8-in. shiplap is used instead of 1x6-in. D&M lumber, deduct 5 ft. lumber
b.m. and 0.1-hr. carpenter time per 100 sq. ft. of wall.

If used on buildings of irregular construction, as described above, add 1-hr. carpenter time.

Labor Cost of 100 Sq. Ft. of ½"-in. Plywood Sidewall
Sheathing Applied to Square or Rectangular Buildings
of Regular Construction

	Hours	Rate	Total	Rate	Total
Carpenter sheathing............................	1.0	$. . . .	$. . . .	$16.47	$16.47
Helper unloading and car-					
rying sheets.....................................	0.3	. . . .	. . . .	12.54	3.76
Carpenter applying paper	0.4	. . . .	. . . .	16.47	6.59
Cost per 100 sq. ft.........................			$. . . .		$26.82
Cost per sq. ft..................................			. . . .		.27

Paper optional as Plywood comes in large sheets 48"x96" but is not treated with asphalt.

If used on buildings of irregular construction, as described above, add 1-hr.
carpenter time.

Gypsum Sheathing.—Gypsum sheathing is a fireproof solid sheet of gypsum
encased in a tough, fibrous, water-resisting covering. The sides and ends are treated
to resist moisture. Some manufacturers blend asphalt emulsion into the wet gypsum core mix to provide additional resistance to moisture. It is used for sheathing
on frame structures under siding, shingles, stucco and brick veneer.

Gypsum sheathing has V-joint edges on the long dimension to provide positive
assurance of a tight fit at the unsupported joints. It is applied with its length at
right angles to the studs.

Gypsum sheathing is ½-in. thick, 2 ft. wide and 8 and 10 ft. long, and 4 ft. wide
and 8 and 9 ft. long to fit supports 16 in. on centers. Nails should be 1¾-in. long,
No. 10½ gauge, galvanized flat head roofing nails, spaced 4-in. on centers, except
under wood siding and stucco, 8". Requires 14 to 21 lbs. per 1,000 sq. ft.

Price $75.00 to $90.00 per 1,000 sq. ft.

Labor cost same as given for insulating or plywood sheathing.

Labor Cost of 100 Sq. Ft. of Gypsum Sidewall Sheathing Applied to
Square or Rectangular Buildings of Regular Construction

	Hours	Rate	Total	Rate	Total
Carpenter sheathing............................	1.0	$. . . .	$. . . .	$16.47	$16.47
Helper unloading and carrying					
sheets..	0.3	. . . .	. . . .	12.54	3.76
Carpenter applying paper	0.4	. . . .	. . . .	16.47	6.59
Cost per 100 sq. ft.........................			$. . . .		$26.82
Cost per sq. ft..................................			. . . .		.27

*Paper optional as gypsum sheathing is now waterproofed sides and edges. Some manufacturers mix
asphalt emulsion with wet gypsum to provide a waterproof board.

If used on buildings of irregular construction, as described above, add 1-hr. carpenter time.

Insulating Shingle Backer Strips

A number of manufacturers produce a shingle backer strip, of the same material as insulating sheathing board, to be used as the undercourse in double course shingle work. These strips are 4'-0" long and either 5/16" or 3/8"thick. They are available in two widths, 13½" for 12" shingle exposure and 15½" for 14" shingle exposure. They provide added insulation, improve the shadow line appearance of the finished wall, eliminate the necessity of building paper over wood sheathing and are more economical to use than second grade wood shingle undercoursing. Insulating shingle backer strips cost approximately 20 cts. per sq. ft.

Labor Placing Insulating Shingle Backer Strips.—On straight run jobs, with an average number of openings, a carpenter should place 500 to 600 sq. ft. of shingle backer strips per 8- hr. day at the following labor cost per 100 sq. ft. :

	Hours	Rate	Total	Rate	Total
Carpenter	1.5	$. . . .	$. . . .	$16.47	$24.71
Labor	0.5	. . . .	. . . .	12.54	6.27
Cost 100 sq. ft			$. . . .		$30.98
Cost per sq. ft				. . . .	.31

On complicated jobs, requiring a large amount of cutting and fitting, a carpenter should place 400 to 500 sq. ft. per 8-hr. day at the following labor cost per 100 sq. ft. :

	Hours	Rate	Total	Rate	Total
Carpenter	1.8	$. . . .	$. . . .	$16.47	$29.65
Labor	0.5	. . . .	. . . .	12.54	6.27
Cost 100 sq. ft			$. . . .		$35.92
Cost per sq. ft				. . . .	.36

Insulating Lath

Insulating lath are made by several manufacturers of insulating building sheathing. They are made of the same materials as the building boards previously described.

Insulating lath is usually furnished in the 18"x48" size, ½",¾" and 1" thick, although they are obtainable from some sources 16"x48".

Some manufacturers furnish lath both plain and with the back coated with asphalt to form a vapor barrier.

Most insulating lath are manufactured with the long edges shiplapped and all edges beveled to reinforce the plaster against cracking and eliminate unsightly lath marks. Some of these insulating lath are metal reinforced along the edges to reinforce the unsupported horizontal joint.

Insulating gypsum lath consists of regular gypsum board lath with a thin sheet of aluminum attached to one side to provide insulating value. They are furnished in sheets 16x32-in., 16x48-in., and ⅜ and ½-in. thick.

Estimating Quantities of Insulating Lath

Size of Lath in Inches	No Sq. Ft. per Lath	No. Lath per 100 Sq. Ft.
16x48	5.33	18.76
18x48	6.00	16.67

Approximate prices of insulating lath are as follows per 1,000 sq. ft.: ½", $187.00; ⅝", $220.00.

Labor Applying Insulating Lath.—The labor cost of applying insulating lath will vary with the locality and labor restrictions. Where there are no labor restrictions, an experienced man should place 125 sq. yds. per 8-hr. day; while in the larger cities a lather will place only 100 sq. yds. per 8-hr. day, at the following labor cost per 100 sq. yds. :

	Hours	Rate	Total	Rate	Total
Lather	8	$....	$....	$16.38	$131.04
Cost per sq. yd				....	1.31

WOOD AND WIRE FENCE

Setting Wood Fence Posts.—When placing wire fence around the lots of 16 houses, it required 218 cedar posts, 5" in diameter and 7' -0" long, also 32 clothes posts, 7" in diameter and 10' -0" long.

After the holes had been dug, a man set 7 posts an hr. at the following labor cost per post :

	Hours	Rate	Total	Rate	Total
Carpenter	0.5	$....	$....	$16.47	$8.23

Treated pine posts and locust posts may also be used for fencing. Locust and cedar posts should be all heartwood.

Placing 2"x4" Top Bottom Fence Rail Ready to Receive Wire.—The posts for these fences were spaced about 6'-0" apart and a 2"x4" top rail was nailed flatwise to the top of the post and a 2"x4" bottom rail was notched into the side of the post so that the wire could be stapled to it.

There was a total of 1,400 lin. ft. of fence or 2,800 lin. ft. of top and bottom rail. A carpenter placed rail for 110 lin. ft. of fence or 220 lin. ft. of single rail per 8-hr. day, at the following labor cost per 100 lin. ft. :

	Hours	Rate	Total	Rate	Total
Carpenter	7.3	$....	$....	$16.47	$120.23
Labor	0.9	....	....	12.54	11.29
Cost per 100 lin. ft. fence			$....		$131.52
Cost per lin. ft. fence				....	1.32
Cost per lin. ft. rail				....	.66

Stringing Fence Wire.—Cyclone wire fencing in 42" rolls was used for these fences and was stapled to the top and bottom rails and posts about 4" on centers.

There were 1,400 lin. ft. of 42" wire fencing in the job and a carpenter would string and staple about 280 lin. ft. per 8-hr. day, at the following labor cost per 100 lin. ft. :

	Hours	Rate	Total	Rate	Total
Carpenter	2.9	$....	$....	$16.47	$47.76
Cost per lin. ft.				....	.48

Setting Lally Columns in Basements.—In a group of brick houses, 3" iron columns were used to support the 6"x8" wood girders carrying the floor joists.

These columns were 3" in diameter and 7'-0" long, and a carpenter set, plumbed and braced 1 column an hr. at the following labor cost per column :

	Hours	Rate	Total	Rate	Total
Carpenter	1.0	$....	$....	$16.47	$16.47
Labor	1.0	....	....	12.54	12.54
Cost per column			$....		$29.01

MISCELLANEOUS HARDWARE AND ACCESSORIES

Steel Area Walls and Gratings

Used around basement windows when first floor is at grade level. Made of 16-ga. and 20-ga. galvanized steel, with stiffening ribs and rounded tops. Attached to masonry walls by screws or bolts.

Size Width Depth	For Use with Basement Windows	Price 16 ga.	Each 20 ga.
3'-2"x0'-11½"	15"x12"-2-Lt.	$ 9.50	$ 6.00
3'-2"x1'- 5½"	15"x16"-2-Lt.	12.00	8.00
3'-2"x1'-11½"	15"x20"-2-Lt.	16.00	12.00

Steel gratings for use with above walls. Frames made of 1¼" x¼" steel bars with cross bars 1" x 13/16" welded into one piece. Supports on the grating hold it flush with the top of the area wall. Price each painted...$22.00

Metal Foundation Wall Ventilators

Brick Ventilators.—Made of aluminum exactly the size of a brick (2½" x4" x8"). The bottom edge is flanged both front and back to insure a positive mortar lock. Louvered face allows 13 sq. in. of free area. Screened back. Price each ..$8.75

Frame Ventilators.—Nailed to sheathing just above the foundation or to ends of joists spaced 16" on centers. Made of sheet aluminum, size 16"x8", providing 89 sq. in. of free area. Screened back. Price each $22.00

Concrete Block Ventilators.—Made of cast aluminum 16" wide x 8" high, providing 89 sq. in. of free area. Screened back. Price each.................................$23.50

Metal Roof Ventilators.

Metal Roof Ventilators

Metal roof ventilators provide air in the attic space under the roof. They are made of steel or aluminum and usually provide 18 to 75 sq. in. of free area.

Roof Opg. Inches	Free Area Sq. Inches	Price Each Aluminum
12x18	30	$ 7.00
15x20	50	9.50
16x16	65	15.00

Steel Clothes Chute Doors

Furnished in steel, white enameled, opening 9" x12".
Price each $10.00

Steel Access Doors

For access to piping or wiring installations. Furnished with a flanged frame that may be nailed to studs 16" on centers. Size of door 16" x 24", overall size 17½" x25½".

Price each $24.00

ROUGH HARDWARE

While it is possible to include the cost of rough hardware, such as nails, screws, etc., in the material unit for each carpentry item figured, most contractors and estimators find it more convenient to make an allowance for this item at the end of the carpentry estimate based on the amount of lumber and other material involved. The usual method of arriving at a rough hardware allowance is to total all the material in the carpentry estimate taking ft. b.m. of dimension lumber and boards, sq. ft. area of plywood, paneling, etc., and lin. ft. of grounds, furring strips, trim, etc., and then figure from 25 to 35 lbs. of nails per 1,000 ft. b.m. The weight of nails multiplied by the cost per lb. will give the rough hardware allowance required.

06130   HEAVY TIMBER CONSTRUCTION

The quantities and costs included under this classification cover all types of heavy constructed mill, warehouse and factory buildings, etc.

Framing and Erecting Heavy Wood Columns or Posts.—The labor cost of erecting and framing heavy wood columns or posts, 10" x10" or larger, will vary according to the number of operations necessary, such as chamfering corners, cutting out and framing for post caps and bases, boring holes through the center, and upon the distance the posts must be hoisted before placing.

On the average mill constructed job requiring 10"x10" or 12"x12" posts 10'-0" to 16'-0" long, it will require 3½ to 4 hrs. time to chamfer corners, frame for bases and caps, and erect each column, at the following labor cost:

	Hours	Rate	Total	Rate	Total
Carpenter	3.7	$. . . .	$. . . .	$16.47	$60.94
Labor	0.5	. . . .	. . . .	12.54	6.27
Cost per column			$. . . .		$67.21

If hoisted above the second floor, add ½-hr. labor time per column.

It is more difficult to estimate the cost of this work by the 1,000 ft. of lumber, b.m. as it costs just as much to chamfer the 4 corners and frame for post caps and bases for an 8"x8" as a 12"x12" column, while the former contains less than half as much lumber per lin. ft. as the latter. The following costs are based on an average 10"x10"-12'-0" timber and should be increased or reduced accordingly for larger or smaller timbers.

Based on a 10"x10"-12'-0" column, a carpenter should chamfer corners, frame for post caps and bases, handle and erect 200 to 250 ft. of lumber, b.m. per 8-hr. day, at the following labor cost per 1,000 ft. :

	Hours	Rate	Total	Rate	Total
Carpenter	35	$. . . .	$. . . .	$16.47	$576.45
Labor	8	. . . .	. . . .	12.54	100.32
Cost 1,000 ft. b.m.			$. . . .		$676.77

Rounding Corners on Large Timbers.—Where the corners of wood columns, beams or girders must be rounded to a small radius, the yards or mills furnishing the lumber usually charge $15.00 to $20.00 per 1,000 ft. b.m. for performing this work.

Boring Holes Through Center of Large Timbers.—Where the wood columns or posts have a hole bored through the center for their entire length, the yard or mill furnishing the lumber usually charges $30.00 to $35.00 per 1,000 ft. b.m. for boring these holes.

Framing and Erecting Heavy Wood Beams and Girders for First Floor.—The labor cost of framing and erecting heavy wood beams and girders will vary according to the amount of framing necessary and the method used in placing them; i.e., whether they set on top of the wood columns or whether it is necessary to frame for post caps, stirrups, etc.

Placing heavy beams and girders for the first floor of a building does not require as much labor as for the upper floors because the timbers can be rolled to the place they are to be used without hoisting.

Where heavy wood girders are set on concrete or masonry piers without framing for post caps, joist hangers, etc., a carpenter should frame and erect 1,000 to 1,250 ft. per 8-hr. day, at the following labor cost per 1,000 ft. :

	Hours	Rate	Total	Rate	Total
Carpenter	7	$. . . .	$. . . .	$16.47	$115.29
Labor	7	. . . .	. . . .	12.54	87.78
Cost 1,000 ft. b.m.			$. . . .		$203.07

Where necessary to frame for post caps and bases, stirrups, etc., a carpenter should frame and erect 500 to 600 ft. of lumber, b.m. per 8-hr. day, at the following labor cost per 1,000 ft. :

	Hours	Rate	Total	Rate	Total
Carpenter	14.5	$. . . .	$. . . .	$16.47	$238.82
Labor	7.0	. . . .	. . . .	12.54	87.78
Cost 1,000 ft. b.m.			$. . . .		$326.60

Framing and Erecting Heavy Wood Girders Above First Floor.—Where heavy wood girders must be hoisted above the first floor, using a timber hoist or derrick, the carpenter time framing and placing the timbers will remain practically the same but the labor time handling, hoisting and skidding the timbers into place will be increased.

On work of this kind, a carpenter should frame and erect 450 to 550 ft. of lumber, b.m. per 8-hr. day, at the following labor cost per 1,000 ft.:

	Hours	Rate	Total	Rate	Total
Carpenter	16	$. . . .	$. . . .	$16.47	$262.53
Labor	10	. . . .	. . . .	12.54	125.40
Cost 1,000 ft. b.m.			$. . . .		$387.93

Framing and Erecting Heavy Wood Joists.—When placing heavy wood joists 4"x12" or 4"x16" in size, where the joists set on top of wood girders or in iron stirrups, a carpenter should frame and erect 1,000 to 1,200 ft. b.m. per 8-hr. day, at the following labor cost per 1,000 ft. :

	Hours	Rate	Total	Rate	Total
Carpenter	7.3	$. . . .	$. . . .	$16.47	$120.23
Labor	3.5	. . . .	. . . .	12.54	43.89
Cost 1,000 ft. b.m.			$. . . .		$164.12

Where smaller joists are used, ranging from 3"x10" to 4"x10" in size, a carpenter should frame and erect 800 to 1,000 ft. b.m. per 8-hr. day, at the following labor cost per 1,000 ft. :

	Hours	Rate	Total	Rate	Total
Carpenter	9	$....	$....	$16.47	$148.23
Labor	4	....	....	12.54	50.16
Cost 1,000 ft. b.m.			$....		$198.39

Framing and Placing Oak Bumpers at Edges of Loading Docks.—Oak timbers, varying from 4"x6" to 12"x12", are usually called for at loading dock edges. The timbers generally are secured to the edge of the dock slab by bolts set in the concrete or with cinch anchors.

With bolts already set in place, a carpenter should lay out, drill and countersink holes and secure in place 150 to 175 ft. b.m. of oak bumper per 8-hr. day at the following labor cost per 1,000 ft. b.m. :

	Hours	Rate	Total	Rate	Total
Carpenter	48	$....	$....	$16.47	$ 790.56
Labor	24	....	....	12.54	300.96
Cost 1,000 ft. b.m.			$....		$1,091.52

A carpenter should locate, set and secure in place in the slab edge or wall form about 25 anchor bolts per 8-hr. day at the following labor cost per bolt:

	Hours	Rate	Total	Rate	Total
Carpenter	0.3	$....	$....	$16.47	$4.94

Laying 2"x6" or 3"x6" Tongue and Groove Timber Subfloors or Roof Sheathing.—When laying 2" x6" or 3" x6" tongue and groove flooring or roof decking in large areas without much cutting and fitting, a carpenter should lay 800 to 1,000 ft. b.m. per 8-hr. day at the following labor cost per 1,000 ft.

	Hours	Rate	Total	Rate	Total
Carpenter	8.9	$....	$....	$16.47	$146.58
Labor	6.0	....	....	12.54	75.24
Cost per 1,000 ft. b.m.			$....		$221.82

If tongue and groove flooring or roof sheathing is laid in small spaces requiring considerable cutting and fitting around openings, stair wells, elevator shafts, skylights, etc., a carpenter should lay 700 to 800 ft. b.m. per 8-hr. day at the following labor cost per 1,000 ft.

	Hours	Rate	Total	Rate	Total
Carpenter	10.6	$....	$....	$16.47	$174.58
Labor	6.0	$....	....	12.54	75.24
Cost per 1,000 ft. b.m.			$....		$249.82

LAMINATED WOOD FLOORS

Laminated wood floors consisting of 2" or 3" plank set on edge and spiked together, are used where a floor is desired capable of carrying heavy loads or supporting heavy machinery with a minimum amount of vibration.

Sometimes the floors are designed to use certain length plank reaching from bearing to bearing while on others they are spiked together with broken joints, regardless of bearing.

Quantity of Lumber Required for Laminated Wood Floors

Measured Size Inches	Finished Thickness Inches	Add for Waste Per Cent	To Obtain Quantity of Lumber Required Multiply Area by	No. Feet Lumber Per 100 Square Feet of Surface
2x 4	1⅝	23	4.90	490 ft. or 74 pcs.
2x 6	1⅝	23	7.40	740 ft. or 74 pcs.

2x 8	1⅝	23	9.84	984 ft. or 74 pcs.
2x10	1⅝	23	12.30	1230 ft. or 74 pcs.
2x12	1⅝	23	14.76	1476 ft. or 74 pcs.
2x14	1⅝	23	17.22	1722 ft. or 74 pcs.
3x 6	2¾	10	6.60	660 ft. or 44 pcs.
3x 8	2¾	10	8.80	880 ft. or 44 pcs.
3x10	2¾	10	11.10	1110 ft. or 44 pcs.
3x12	2¾	10	13.20	1320 ft. or 44 pcs.
3x14	2¾	10	15.40	1540 ft. or 44 pcs.
4x 6	3¾	7	6.42	642 ft. or 32 pcs.
4x 8	3¾	7	8.56	856 ft. or 32 pcs.
4x10	3¾	7	10.70	1070 ft. or 32 pcs.
4x12	3¾	7	12.84	1284 ft. or 32 pcs.

The above quantities do not include any allowance for waste due to end matching, but are based on spans taking lumber of even lengths or laying flooring with broken joints.

Nails Required for 100 Sq. Ft. of 2" Laminated Flooring

	Distance Spikes Are Spaced Apart									
	12 Inches		16 Inches		20 Inches		24 Inches		30 Inches	
Size	No.	No.	No.	No.	No.	No.	No.	No.	No.	No.
Spikes	Spikes	Lbs.	Spikes	Lbs.	Spikes	Lbs.	Spikes	Lbs.	Spikes	Lbs.
16d	740	15	600	12¼	518	10½	450	9¼	370	7½
20d	740	24	600	19¼	518	16¾	450	14½	370	12
30d	740	31	600	25	518	21½	450	18¾	370	15½
40d	740	41	600	33¼	518	28¾	450	25	370	20½
50d	740	53	600	43	518	37	450	32	370	26½
60d	740	67	600	55	518	47	450	41	370	33¾

Nails Required for 100 Sq. Ft. of 3" Laminated Flooring

	Distance Spikes Are Spaced Apart									
	12 Inches		16 Inches		20 Inches		24 Inches		30 Inches	
Size	No.	No.	No.	No.	No.	No.	No.	No.	No.	No.
Spikes	Spikes	Lbs.	Spikes	Lbs.	Spikes	Lbs.	Spikes	Lbs.	Spikes	Lbs.
16d	440	9	352	7¼	308	0¼	264	5½	220	4½
20d	440	14¼	352	11½	308	10	264	8½	220	7
30d	440	18½	352	14¾	308	12¾	264	11	220	9¼
40d	440	24½	352	19½	308	17	264	14¾	220	12¼
50d	440	31½	352	25¼	308	22	264	18¾	220	15¾
60d	440	40	352	32	308	28	264	24	220	20

Nails Required for 100 Sq. Ft. of 4" Laminated Flooring

	Distance Spikes are Spaced Apart									
	12 Inches		16 Inches		20 Inches		24 Inches		30 Inches	
Size	No.	No.	No.	No.	No.	No.	No.	No.	No.	No.
Spikes	Spikes	Lbs.	Spikes	Lbs.	Spikes	Lbs.	Spikes	Lbs.	Spikes	Lbs.
30d	320	13½	256	10¾	224	9½	192	8	160	6¾
40d	320	17¾	256	14¼	224	12½	192	10¾	160	9
50d	320	23	256	18¼	224	16	192	13¾	160	11½
60d	320	29	256	23¼	224	20½	192	17½	160	14½

06170 PREFABRICATED STRUCTURAL WOOD

Glued Laminated Beam Construction.—Beams of glued laminated construction are popular where price is not the controlling factor.

They are made in uniform or tapered shapes. In the beam shape, similar to illustration, they can be used in spans from 20'-0" to 80'-0".

They are used in schools, auditoriums, churches, stores and ranch-style homes.

They are made of kiln-dried structural woods bonded together by glue, applied under controlled conditions of temperature and pressure. They are planed and sanded to a very smooth surface and later treated with a transparent preservative, making their overall appearance very pleasing. With this type of construction the ceilings are usually omitted, leaving the beams exposed to the room below. Roof insulation is accomplished through the use of standard insulating boards placed upon the roof deck and covered with roofing material. Because of the purlin construction generally used, two or four inch decking is recommended.

Span Feet	Center Height	End Height	Camber	Price
20'-0"	2'- 4"	0'- 9½"	0'-9"	$ 270.00
30'-0"	2'-10"	0'-11"	1'-0"	670.00
40'-0"	3'-11"	1'- 2½"	1'-6"	965.00
50'-0"	5'- 1"	1'- 6½"	2'-0"	1530.00
60'-0"	6'- 1"	1'- 9½"	2'-6"	2200.00

Glued Laminated Three Hinged Arch

Another type of glued laminated construction is the three-hinged arch, which gains its support from floor level, incorporating column and beam in one compact design. Fabricated and finished in the manner of the glued beam, they are used where good appearance and function are the major factors and are used quite extensively in churches and schools.

Purlins are generally used to span the resulting bays and are covered with two inch decking and suitable insulating material.

This is the most expensive of the trusses discussed and costs from 5 to 6 times that of the bowstring type truss.

General Notes Regarding Trusses.—Where rafters are used between trusses, one-inch roof sheathing or ½ inch plywood is generally used. Where purlins are spaced 4'-0" to 6'-0" apart, two-inch sheathing is common practice. Purlins may be eliminated and wood decking of 4 to 6 inches may be used to span up to 18'-0" between trusses.

The use of ceilings supported by the lower chord of the truss should be called to the attention of the truss designer, as it requires a heavier load.

Trusses may bear upon pilasters, steel columns, wood posts and plates, grillages or steel beams. A suitable steel bearing plate should be placed under each end of the truss to distribute the load throughout the masonry. Trusses may afford lateral support through angle irons fastened to truss and column.

Labor Framing Wood Roof Trusses.—When framing wood roof trusses, a carpenter experienced in truss construction will handle and frame approximately the following amounts of lumber per hour on the various types of trusses :

Type of Truss	Ft. B.M. per Hr.
Bowstring Type Trusses	40
Scissors and Gothic Type Trusses	30
Glued Laminated Type Trusses	15

PREFABRICATED STRUCTURAL TIMBER

Structural timber decking in thickness of 2" to 4", generally 6" wide and double tongue and grooved is available in hemlock cedar, Douglas fir and spruce. It can be used for roof decks, structural flooring and for planked walls. Approximate costs are as follows :

	Hemlock PER SQ. FT.	Cedar or Fir PER SQ. FT.	Spruce PER SQ. FT.
4"x6" Select	$2.15	$2.45	$3.05
3"x6" Select	1.80	2.05	2.25

Glued Laminated Beams in a Church.

All prices should be verified with local distributers prior to submitting proposals including these materials.

06200 FINISH CARPENTRY

The estimator figuring finish carpentry work for a job whose specifications are arranged in accordance with the Construction Specification Institute Uniform Index must check at least five divisions to make certain that he has included in his bid all the work that is expected of him. Division 6, the subject of this chapter, covers rough and finish carpentry. But wood and related sidings will be in Division 7, Thermal and Moisture Protection. Wood doors and windows will be in a section of their own, Division 8. Division 10, Specialties, will include such items as tackboards, laminated plastic toilet partitions, signs, folding partitions, telephone booths and toilet and wardrobe accessories which may be furnished and/or set by the carpenter. Division 11, Equipment, may cover kitchen and laundry cabinets and counters for residential work and such unrelated work as wood bank counters, church pews, bars, lab furniture, dock bumpers and library shelving. Finally Division 12, Furnishings, could include wood seating, class room furniture and similar semi-attached items. In addition it is prudent to check all the allowances set forth in the specifications as often the furnishing of some of the items in Divisions 10, 11 and 12 will be covered under a cash allowance but the installation is to be included in the bid submitted by the carpenter.

EXTERIOR FINISH CARPENTRY

As noted above, wood siding and wood doors and windows are discussed in Chapters 12 and 13 respectively. The following items cover other adjacent exterior work.

Placing Corner Boards, Fascia Boards, Etc.—When placing wood fascia boards, corner boards, etc., on houses, cottages, etc., a carpenter should place 175 to 225 lin. ft. per 8-hr. day, at the following labor cost per 100 lin. ft.:

	Hours	Rate	Total	Rate	Total
Carpenter	4	$. . . .	$. . . .	$16.47	$65.88
Cost per lin. ft.			. . . .		.66

Placing Exterior Wood Cornices, Verge Boards, Etc.—When placing exterior wood cornices, verge boards, fascia, soffits, etc., consisting of 2 members, two carpenters working together should place 150 to 175 lin. ft. per 8-hr. day, at the following labor cost per 100 lin ft. :

	Hours	Rate	Total	Rate	Total
Carpenter	10	$. . . .	$. . . .	$16.47	$164.70
Cost per lin. ft.			. . . .		1.65

When the exterior cornices consist of 3 members (crown mold, bed mold, fascia, etc.), two carpenters working together should place 100 to 125 lin. ft. per 8-hr. day, at the following labor cost per 100 lin ft. :

	Hours	Rate	Total	Rate	Total
Carpenter	14.2	$. . . .	$. . . .	$16.47	$233.87
Cost per lin. ft.			. . . .		2.34

If a 4-member wood cornice is used, two carpenters working together should place 60 to 75 lin. ft. per 8-hr. day, at the following labor cost per 100 lin. ft.:

	Hours	Rate	Total	Rate	Total
Carpenter	23.7	$. . . .	$. . . .	$16.47	$390.34
Cost per lin. ft.			. . . .		3.90

The above quantities and costs do not include time blocking out for fascia boards, cornices, etc. An extra allowance should be made for all blocking required.

Placing Brick Mouldings.—A carpenter should fit and set around 32 lin. ft. of brick moulding per hour at the following cost per 100 lin. ft.:

	Hours	Rate	Total	Rate	Total
Carpenter	3.2	$. . . .	$. . . .	$16.47	$52.70
Cost per lin. ft.			. . . .		.53

Brick moulding runs around 20¢ per foot in random lengths and 25¢ in specified lengths.

Placing Wood Cupolas.—One carpenter should set a prefabricated pine cupola in around two hours.

	Hours	Rate	Total	Rate	Total
Carpenter	2	$....	$....	$16.47	$32.94
Labor	2	....	....	12.54	25.08
Cost per each			$....		$58.02

Porch Work.—The labor placing porch work is a variable item owing to the vast difference in the style and construction of porches and the amount of detail involved.

Placing Plain Porch Columns.—When placing plain square or turned porch columns, such as commonly used for rear porches and other inexpensive work, a carpenter should place one post in about 1 hr., at the following labor cost:

	Hours	Rate	Total	Rate	Total
Carpenter	1.0	$....	$....	$16.47	$16.47

Placing Porch Top and Bottom Rail and Balusters.—When placing wood top and bottom rail and wood balusters, such as used on front porches, a carpenter should complete 15 to 20 lin. ft. of rail per 8-hr. day, at the following labor cost per lin. ft.:

	Hours	Rate	Total	Rate	Total
Carpenter	0.46	$....	$....	$16.47	$7.58

When placing top and bottom rails with open balusters or using matched and beaded ceiling, such as is often used for the cheaper grades of work, a carpenter should complete 35 to 45 lin. ft. of rail per 8-hr. day, at the following labor cost per lin. ft.:

	Hours	Rate	Total	Rate	Total
Carpenter	0.2	$....	$....	$16.47	$3.29

Framing and Erecting Exterior Wood Stairs for Rear Porches.—When framing and erecting outside wood stairs for rear porches on apartment buildings, etc., where the stringers are 2"x10" or 2"x12" lumber, with treads and risers nailed on the face of the stringers, it will require 18 to 22 hrs. carpenter time per flight of stairs.

This is for an ordinary stair having 14 to 18 risers extending from story to story, and the labor per flight should cost as follows:

	Hours	Rate	Total	Rate	Total
Carpenter	20	$....	$....	$16.47	$329.40

If the stair consists of 2 short flights with an intermediate landing platform between stories it will require 12 to 13 hrs. carpenter time per flight or 24 to 26 hrs. per story, including platform, at the following labor cost:

	Hours	Rate	Total	Rate	Total
Carpenter	25	$....	$....	$16.47	$411.75

Framing and Erecting Wood Stairs Having Winders.—Where the wood stairs to rear porches have 4 to 6 winders in each story height, figure 24 to 28 hrs. carpenter time, at the following labor cost per story:

	Hours	Rate	Total	Rate	Total
Carpenter	26	$....	$....	$16.47	$428.22

Porch Top Rail and Balustrade Cap

Rabbeted Porch Jamb, Plowed Shoe and Bottom Rail

Clear Ponderosa Pine Porch Material

Kind of Molding	Size in Inches	Price per 100 Lin. Ft.
Square Baluster Stock	1⅛"x1⅛"	$ 48.00
Square Baluster Stock	1⅜"x1⅜"	61.00
Square Baluster Stock	1⅝"x1⅝"	70.00
Balustrade Cap	3⅝"x1⅝"	125.00
Balustrade Shoe	3⅝"x1⅝"	125.00
Rabbeted Porch Jamb	1½"x3½"	125.00
Plowed Porch Shoe	1½"x3½"	125.00
Top Rail	3 "x2¼"	140.00
Top Rail	3¾"x1¾"	125.00
Bottom Rail	3½"x1¾"	125.00

Cornice Boards
Select Grade
Prices are per 100 lin. ft. in random lengths 6'-0" to 16'-0".

Measured Size	S4S to	Price
1"x 3"	¾"x 2⅝"	$ 30.00
1"x 4"	¾"x 3½"	38.00
1"x 6"	¾"x 5½"	70.00
1"x 8"	¾"x 7½"	85.00
1"x10"	¾"x 9½"	115.00
1"x12"	¾"x11½"	160.00

Wood Columns

The finest columns for exterior work are made from clear heart redwood staves glued together under clamp pressure with cold water waterproof glue. Shafts are turned in the lathe with the correct entasis and, to insure uniform stave thickness after turning, the staves are made straight for the lower third of the shaft and swelltapered for the upper two-thirds thus giving the correct entasis in the rough. Any style of fluting or reeding may be had.

The columns receive a prime coat of lead and oil on the exterior surface. Columns over 14" in diameter receive a coat of waterproof asphaltum paint on the hollow core surface.

All columns are furnished with a guarantee as to their workmanship and durability.

No printed price lists are available for these items and each must be priced at the factory. In order of price, the Roman is cheapest followed by the unfluted Ionic, the fluted Attic, the fluted Doric and finally the fluted Corinthian. A 12" diameter Doric style fluted column 10' high with a plinth and cap is currently quoted at around $300.00 at the factory.

Stock, tapered column shafts in pine can be figured at around $25.00 per vertical linear foot in 12" diameter ; $35.00 in 14" ; $47.00 in 18" ; and $67.00 in 24" .

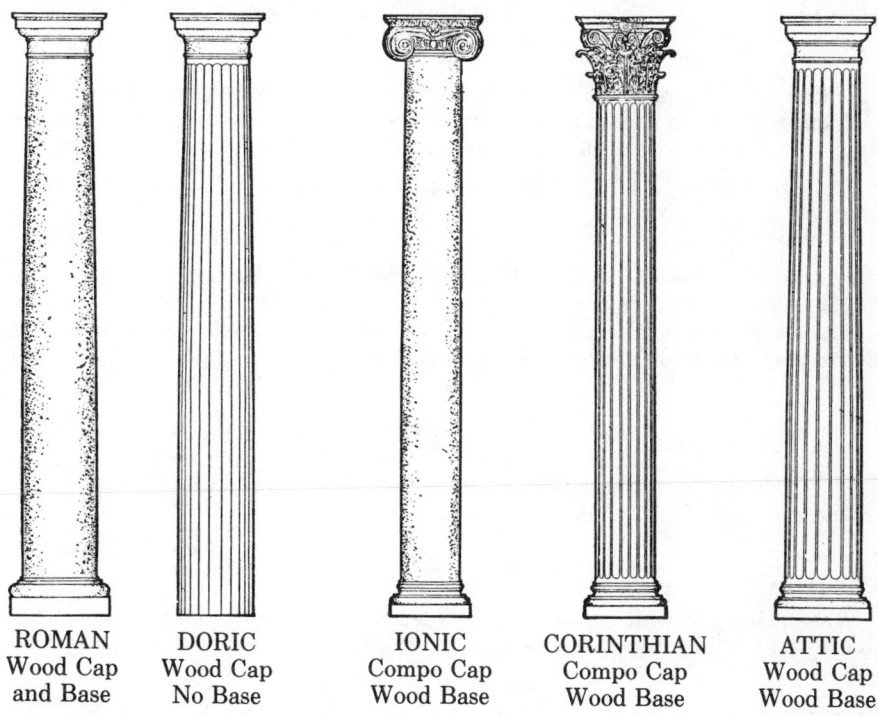

ROMAN	DORIC	IONIC	CORINTHIAN	ATTIC
Wood Cap	Wood Cap	Compo Cap	Compo Cap	Wood Cap
and Base	No Base	Wood Base	Wood Base	Wood Base

Fiberglass columns cast in a colonial design will run $5.50 per linear foot in 6" diameter; $6.75 in 8"; and $8.00 in 10".

Aluminum columns extruded to a fluted design are stocked in assorted sizes from 6" diameter in 8' lengths to 12" diameter in 18' lengths. They may be used as load-bearing members and come factory finished in a high-gloss, white, baked enamel. An 8" diameter shaft 12' long will run around $46.00. If no base or cap is used it may be attached to the slab and ceiling by cutting 2" lumber to fit the column diameter and bolting it to the adjacent construction. If a base and cap is ordered they will be furnished in aluminum castings and will run around $10.00 each.

Square porch columns will run around $1.75 per foot in 4x4 fir. Turned colonial posts in 8' lengths will run $14.00 in 4x4 stock, $21.00 in 5x5 stock.

It will take two carpenters to set columns, and they should erect a medium sized column in an hour.

Cast Iron Column Bases.—Cast iron column bases keep columns off the floor and permit air to circulate on the inside of the column.

Prices of Cast Iron Column Bases

Base Size	Height	For Col. Diam.*	Wt. Lbs.	Approximate Price
11 "x11 "	2 "	8"	18	$ 30.00
13½"x13½"	2½"	10"	24	45.00
16 "x16 "	2½"	12"	32	60.00
18½"x18½"	2½"	14"	35	85.00
21¼"x21¼"	2½"	16"	54	115.00
24 "x24 "	3¼"	18"	87	135.00
26½"x26½"	3¼"	20"	96	150.00

*For Doric columns without base use one size smaller.

Cast iron bases for pilasters are 2/3 the price of bases for corresponding columns.

Placing Plywood Soffits.—Two carpenters should set 75 sq. ft. of ¼" and 64 sq. ft. of ½" plywood on soffits per hour at the following labor cost per 100 sq. ft.:

	Hours	Rate	Total	Rate	Total
Carpenter	2.8	$....	$....	$16.47	$45.92
Cost per sq. ft.			....		.46

Placing Matched and Beaded Ceiling.—Matched and beaded ceiling is usually furnished 4" and 6" wide and ⅜", ½" and ¾" thick.

Where the 4" material is used for porch ceilings, overhang soffits, etc., a carpenter should place 285 to 325 ft. b.m., per 8-hr. day, at the following labor cost per 1,000 ft.:

	Hours	Rate	Total	Rate	Total
Carpenter	26	$....	$....	$16.47	$428.22

The labor cost per square (100 sq. ft. surface) requiring 128 ft. of ceiling, b.m. should run as follows:

	Hours	Rate	Total	Rate	Total
Carpenter	3.33	$....	$....	$16.47	$54.85

When placing 6" wide matched and beaded ceiling, a carpenter should place 400 to 450 ft. of ceiling, b.m. per 8-hr. day, at the following labor cost per 1,000 ft.:

	Hours	Rate	Total	Rate	Total
Carpenter	18.8	$....	$....	$16.47	$309.64

The labor cost per square (100 sq. ft. surface) requiring 120 ft. of ceiling, b.m. should run as follows:

	Hours	Rate	Total	Rate	Total
Carpenter	2.25	$....	$....	$16.47	$37.06

Placing Exterior Door Trim.—Much of the architectural type door trim today is of molded material rather than milled wood. It is pre-engineered to fit together with little or no cutting or fitting providing the door frame itself is set plumb and true. Stock trim can be ordered for widths of 3', 5' and 6' and for either brick or wood wall openings. Side trim is delivered for 7' door height, but designed so that 4" may be cut off to fit a 6-8" door without altering the design. Pilasters are nominally 8" in width. Cornices are 6" to 8" high and available in several styles. In addition pediments may be added. A basic cornice piece costs around $50.00 for a 3' opening, $75.00 for the larger ones. Pediments will run from $50.00 to $75.00 extra, for 3'-0" openings and up to $120.00 for the double door openings. Pilaster assemblies run $65.00 per pair. Two carpenters should set a cornice in ½ hour; the pediment in ½ hour; and the pilasters in one hour.

Outside Blinds or Shutters

Shutters are available in wood, aluminum and molded plastic. Because of the popularity of the latter it is not always easy to find wood stocked in all sizes. Custom made units will run considerably more than the prices given below.

All shutters have stiles, top and middle rails 2¼" wide; bottom rail 4½" wide. All slats are stationary.

Shutters with raised panels can be fitted with decorative cutouts in the upper panel at an extra cost of $1.45 per pair. Designs include pine trees, cloverleafs, squirrels, crescents and roosters.

A carpenter should hang 8–10 pair a day depending on size and height off the ground.

Price Per Pair, Hemlock, Preservative Treated

Window Size	With Fixed Slats	With Raised Panels
2'-0"x3'- 3"	$13.00	$17.00
3'-11"	15.50	19.00
4'- 7"	18.00	24.00
5'- 3"	20.00	28.00
2'-4"x3'- 3"	13.50	16.25
3'- 7"	14.25	17.50
3'-11"	15.50	18.75
4'- 3"	16.50	20.25
4'- 7"	19.00	21.50
4'-11"	20.00	23.00
2'-8"x3'- 3"	14.00	16.75
3'- 7"	14.50	18.00
3'-11"	16.00	19.00
4'- 3"	17.50	21.00
4'- 7"	19.00	22.00
4'-11"	20.00	23.50
5'- 3"	22.00	25.00
3'-0"x3'- 3"	19.00	21.00
3'- 7"	20.00	22.75
3'-11"	21.00	24.00
4'x 3"	22.00	26.00
4'- 7"	23.00	27.00
4'-11"	24.00	28.50
5'- 3"	25.00	29.00
3'-4"x3'- 3"	20.00	22.50
3'-11"	22.00	25.00
4'- 7"	24.00	28.00
4'-11"	25.00	29.00
5'- 3"	27.00	31.00
3'-8"x4'- 7"	22.00	29.50
4'-11"	24.00	31.00
5'- 3"	26.00	32.75

Door Blinds

Same as above, except 3 panels of slats in height.

Size	Price
2'-8"x6'-9", per pair	$28.00
3'-0"x6'-9", per pair	29.00

| Style "A" | Outside Blinds or Shutters Style "B" | Style "C" |

Aluminum Shutters

Metal shutters of aluminum are designed to be screwed on the wall permanently. The louver section is stamped out of one piece of metal. They are furnished in white, black, green or brown and come packaged with mounting screws.

Size	Price per Pair
2'-4"x34"	$14.50
2'-4"x36"	15.00
2'-4"x39"	15.75
2'-4"x43"	16.50
2'-4"x48"	17.00
2'-4"x55"	18.00
2'-4"x64"	21.00
2'-4"x80"	25.00
2'-8"x34"	15.50
2'-8"x36"	16.00
2'-8"x39"	16.75
2'-8"x43"	17.50
2'-8"x48"	18.00
2'-8"x55"	19.50
2'-8"x64"	22.00
2'-8"x80"	27.00

One carpenter should hang one pair in .8 hour.

Molded Shutters

Shutters are now molded of thermo-formed polymer. They are prefinished in either black or white, but may be painted with an exterior latex paint. They are screwed to siding with 2" aluminum screws, 4 per unit to 55", 6 per unit over 55".

Size Per Panel	Price Per Pair
14"x35"	$21.80
39	22.75
43	23.82
47	25.24
51	26.07
55	27.26
59	28.44
63	32.00
67	32.94
71	33.89
75	35.31
16"x80"	36.26

Labor Costs on Half Timber Work.—The following costs on half timber work are taken from a college building.

The timber was white oak 3-in thick, 10-in. and 11-in. wide and various lengths to meet the requirements of the job.

Nothing was done to the face of the timber and it was left just as received. The backs of all timbers were beveled, the thickness remaining 3-in. A shop was set up on the job where the timbers were cut, beveled and fit ready for erection.

There were 14,500 ft. b.m. of lumber used in this half-timber work and the labor cost per 1,000 ft. b.m. was as follows:

	Hours	Rate	Total	Rate	Total
Carpenter	66.3	$....	$....	$16.47	$1,091.96
Labor helping	12.0	....	....	12.54	150.48
Cost per 1,000 ft. b.m			$....		$1,242.44

Half timber work on commercial building

After the timbers were cut, beveled and fit in the shop, the labor cost of placing and bracing the timbers in the building was as follows per 1,000 ft. b.m.:

	Hours	Rate	Total	Rate	Total
Carpenter	41.5	$....	$....	$16.47	$683.51
Labor helping	2.2	....	....	12.54	27.59
Cost per 1,000 ft b.m			$....		$711.10

INTERIOR FINISH CARPENTRY

When estimating the labor cost of erecting interior finish, it is advisable for the estimator to ascertain the grade of workmanship required by the architect or owner, as the quantity of work a man will perform per hour or per day will vary considerably with the grade of workmanship required.

The costs given on the following pages embody two distinct grades of workmanship. The grade termed Ordinary Work is by far the most common and is usually found in medium priced residences, cottages, apartment buildings, apartment buildings with stores underneath, factory and warehouse buildings, non-fireproof store, office and school buildings, and other buildings of this class.

The other grade, known as First Grade Workmanship, is usually required in finishing fine residences, high-class fireproof apartments, hotel, bank and office buildings, high grade store buildings, fireproof school and university buildings, courthouses, city halls, state capitols, post offices, and other buildings of this type. The interior finish in buildings of this class is usually selected birch, selected gumwood, plain or quarter-sawed oak, mahogany, walnut or other first-class hardwoods, and is discussed later in this chapter under Architectural Woodwork.

Placing Wood Base.—This cost will vary with the size of the rooms and whether a single member or 2 or 3-member base is specified.

Where there are 55 to 60 lin. ft. of 2 member base in each room without an unusually large amount of cutting and fitting, a carpenter should place 125 to 150 lin. ft. per 8-hr day, at the following labor cost per 100 lin. ft.:

	Hours	Rate	Total	Rate	Total
Carpenter	5.8	$....	$....	$16.47	$ 95.53
Labor	1.0	....	....	12.54	12.54
Cost per 100 lin. ft			$....		$108.07
Cost per lin. ft				....	1.08

If there are an unusually large number of miters to make, such as required in closets and other small rooms, increase the above costs accordingly.

A carpenter should place almost as many lin. ft. of 3-member base (consisting of 2 base members and a carpet strip), as 2-member base (consisting of one member and carpet strip), as it is much easier to fit a small top member against the plastered wall than it is to nail a wide piece of base so that it will fit snug against the wall and follow the irregularities in the plaster.

Where there are 50 to 60 lin. ft. of base in a room, a carpenter should place 110 to 130 lin. ft. per 8-hr. day, at the following labor cost per 100 lin. ft.:

	Hours	Rate	Total	Rate	Total
Carpenter	6.7	$....	$....	$16.47	$110.35
Labor	1.0	....	....	12.54	12.54
Cost per 100 lin. ft				....	$122.89
Cost per lin. ft				....	1.23

First Grade Workmanship

In average size rooms, a carpenter should place 100 to 115 lin. ft. of 2-member hardwood base per 8-hr. day, at the following labor cost per 100 lin. ft.:

	Hours	Rate	Total	Rate	Total
Carpenter	7.4	$....	$....	$16.47	$121.88
Labor	1.0	....	....	12.54	12.54
Cost per 100 lin. ft			$....		$134.42
Cost per lin. ft				....	1.34

Where 3-member hardwood base is used in average size rooms, a carpenter should place 85 to 100 lin. ft. (2 ordinary rooms) per 8-hr. day, at the following labor cost per 100 lin. ft.:

	Hours	Rate	Total	Rate	Total
Carpenter	8.7	$....	$....	$16.47	$143.29
Labor	1.0	....	....	12.54	12.54
Cost per 100 lin. ft			$....		$155.83
Cost per lin. ft				....	1.56

On work of this class, the wood grounds should be straight, so that it will not be necessary to "force" the wood base to make it fit tight against the finished wall. Where a single 1x4 pine base is to be fitted to straight runs, a single carpenter should set around 200' per day.

	Hours	Rate	Total	Rate	Total
Carpenter	4	$....	$....	$16.47	$65.88
Cost per lin. ft					.66

Placing Wood Picture Molding.—Where just an ordinary grade of workmanship is required, a carpenter should place picture molding in 5 or 6 ordinary sized rooms per 8-hr. day. This is equivalent to 250 to 275 lin. ft. at the following labor cost per 100 lin. ft.:

	Hours	Rate	Total	Rate	Total
Carpenter	3.0	$....	$....	$16.47	$49.41
Labor	0.5	....	....	12.54	6.27
Cost per 100 lin. ft			$....		$55.68
Cost per lin. ft				....	.56

First Grade Workmanship
Where the wood picture molding must fit close to the plastered walls with perfect fitting miters, a carpenter should place molding in 4 to 5 ordinary sized rooms per 8-hr. day. This is equivalent to 175 to 200 lin. ft., at the following labor cost per 100 lin.ft.:

	Hours	Rate	Total	Rate	Total
Carpenter	4.4	$....	$....	$16.47	$72.47
Labor	0.5	....	....	12.54	6.27
Cost per 100 lin. ft			$....		$78.74
Cost per lin. ft				....	.79

If the picture molding is placed in fireproof buildings having tile or brick partitions, it will be necessary to place wood grounds for nailing the picture molding but in non-fireproof buildings the nails may be driven into the plaster, as the nails will obtain a bearing in the wood studs or wall furring.

Placing Wood Chair or Dado Rail.—In large rooms, long, straight corridors, etc., a carpenter should fit and place 275 to 300 lin. ft. of wood chair rail per 8-hr. day, at the following labor cost per 100 lin. ft.:

	Hours	Rate	Total	Rate	Total
Carpenter	2.8	$....	$....	$16.47	$46.12
Labor	0.5	....	....	12.54	6.27
Cost per 100 lin. ft			$....		$52.39
Cost per lin. ft				....	.52

In small kitchens, pantries, closets, bathrooms, etc., a carpenter will place only 160 to 180 lin. ft. of chair rail per 8-hr. day, at the following labor cost per 100 lin. ft.:

	Hours	Rate	Total	Rate	Total
Carpenter	4.7	$....	$....	$16.47	$77.41
Labor	0.5	....	....	12.54	6.27
Cost per 100 lin. ft			$....		$83.68
Cost per lin. ft.				....	.84

First Grade Workmanship
Where first grade workmanship is required, a carpenter should place 200 to 225 lin. ft. of chair rail per 8-hr. day, at the following labor cost per 100 lin. ft.:

	Hours	Rate	Total	Rate	Total
Carpenter	3.8	$. . .	$. . .	$16.47	$62.59
Labor	0.5	. . .	. . .	12.54	6.27
Cost per 100 lin. ft			$. . .		$68.86
Cost per lin. ft			. . .		.69

In small rooms, such as kitchens, pantries, bathrooms, etc., requiring considerable cutting and fitting around medicine cabinets, wardrobes, kitchen cases, etc., a carpenter should place 120 to 135 lin. ft. of chair rail per 8-hr. day, at the following labor cost per 100 lin. ft.:

	Hours	Rate	Total	Rate	Total
Carpenter	6.3	$. . .	$. . .	$16.47	$103.76
Labor	0.5	. . .	. . .	12.54	6.27
Cost per 100 lin. ft			$. . .		$110.03
Cost per lin. ft			. . .		1.10

Placing Wood Cornices.—Where 3 or 4-member wood cornices are placed in living rooms, reception rooms, dining rooms, etc., a carpenter should place cornice in one average sized room per 8-hr. day, which is equivalent to 50 to 60 lin. ft., and the labor cost per 100 lin. ft. would be as follows:

	Hours	Rate	Total	Rate	Total
Carpenter	14.5	$. . .	$. . .	$16.47	$238.82
Labor	2.0	. . .	. . .	12.54	25.08
Cost per 100 lin. ft			$. . .		$263.90
Cost per lin. ft			. . .		2.64

First Grade Workmanship

Where it is necessary that the wood members fit the plastered walls and ceilings closely, with all miters true and even, two carpenters working together should complete one to one and a quarter rooms per day.

This is at the rate of 35 to 40 lin. ft. per 8-hr. day for one carpenter, at the following labor cost per 100 lin. ft. :

	Hours	Rate	Total	Rate	Total
Carpenter	21	$. . .	$. . .	$16.47	$345.87
Labor	2	. . .	. . .	12.54	25.08
Cost per 100 lin. ft			$. . .		$370.95
Cost per lin. ft			. . .		3.71

Placing Vertical Wood Panel Strips.—When vertical wood panel strips or "battens" are nailed to plastered walls to produce a paneled effect, a carpenter should place 22 to 28 pcs. (175 to 225 lin. ft.) per 8-hr. day, at the following labor cost per 100 lin. ft. :

	Hours	Rate	Total	Rate	Total
Carpenter	4.0	$. . .	$. . .	$16.47	$65.88
Labor	0.5	. . .	. . .	12.54	6.27
Cost per 100 lin. ft			$. . .		$72.15
Cost per lin. ft			. . .		.72

Placing Wood Strip Paneling.—Where panels are formed of wood molding 1½" to 2½" wide, making it necessary to cut and miter both ends of each panel strip, the lin. ft. cost will vary with the size of the panels and the amount of cutting and fitting necessary, as there is almost as much labor required on a panel 2'-0"x3'-0" as one 3'-0"x6'-0", although there are only half as many lin. ft. in the former as in the latter.

On small sized panels up to 2'-0"x4'-0", requiring 12 lin. ft. of molding, a carpenter should complete 9 to 11 panels, containing 110 to 135 lin. ft. of molding per 8-hr. day, at the following labor cost per 100 lin. ft. :

	Hours	Rate	Total	Rate	Total
Carpenter	6.5	$....	$....	$16.47	$107.06
Labor	0.5	....	....	12.54	6.27
Cost per 100 lin. ft			$....		$113.33
Cost per lin. ft			....		1.13

On larger panels 3'-0"x5'-0" to 4'-0"x6'-0" in size, where each panel contains 16 to 20 lin. ft. of molding, a carpenter should complete 7 to 9 panels, containing 140 to 180 lin. ft. of molding per 8-hr. day, at the following labor cost per 100 lin. ft. :

	Hours	Rate	Total	Rate	Total
Carpenter	5.0	$....	$....	$16.47	$82.35
Labor	0.5	....	....	12.54	6.27
Cost per 100 lin. ft			$....		$88.62
Cost per lin. ft.			....		.89

First Grade Workmanship

Where wood panel moldings are used over canvassed or burlap walls, with all strips plumb and level, fitting closely to the plastered walls with perfect fitting miters, a carpenter should complete 7 to 9 small panels, requiring 90 to 115 lin. ft. of molding per 8-hr. day, at the following labor cost per 100 lin. ft. :

	Hours	Rate	Total	Rate	Total
Carpenter	7.8	$....	$....	$16.47	$128.47
Labor	0.5	....	....	12.54	6.27
Cost per 100 lin. ft			$....		$134.74
Cost per lin. ft			....		1.35

On larger panels 3'-0"x5'-0" to 4'-0"x6'-0" in size, where each panel contains 16 to 20 lin. ft. of panel molding, a carpenter should complete about 6 to 8 panels, containing 120 to 150 lin. ft. of molding per 8-hr. day, at the following labor cost per 100 lin. ft. :

	Hours	Rate	Total	Rate	Total
Carpenter	6.0	$....	$....	$16.47	98.82
Labor	0.5	....	....	12.54	6.27
Cost per 100 lin. ft			$....		$105.09
Cost per lin. ft			....		1.05

Placing Wood Ceiling Beams.—In buildings where built-up ceiling beams are used, the labor costs will vary according to the number of intersections of beams in each room and the length of the beams, as it is just as easy to erect a 12'-0" built-up beam as an 8'-0" one.

On average work, a carpenter should place 35 to 45 lin. ft. of built-up wood beams per 8-hr. day, at the following labor cost per 100 lin. ft. :

	Hours	Rate	Total	Rate	Total
Carpenter	20	$....	$....	$16.47	$329.40
Labor	3	....	....	12.54	37.62
Cost per 100 lin. ft			$....		$367.02
Cost per lin. ft			....		3.67

First Grade Workmanship

In the better class of buildings using wood ceiling beams, a carpenter should place 30 to 35 lin. ft. per 8-hr. day, at the following labor cost per 100 lin. ft. :

	Hours	Rate	Total	Rate	Total
Carpenter	25	$....	$....	$16.47	$411.75
Labor	3	....	....	12.54	37.62
Cost per 100 lin. ft			$....		$449.37
Cost per lin. ft			....		4.49

Ponderosa Pine Moldings
Prices are per 100 lin. ft. unless noted otherwise.

Aprons
11/16"x21/4"	$46.00
11/16"x21/4" ogee	46.00
9/16"x31/2"	85.00

Astragals
13/8"x7'-0"	3.15 ea.
11/4"x7'-0"	3.50 ea.

Bases
9/16"x3"	40.00
1/2"x31/2"	42.00
1/2"x41/4"	48.00

Base Cap
11/16"x13/8"	20.00

Backband
11/16"x11/16"	16.00

Balusters
3/4"x3/4"	25.00
11/16"x11/16"	40.00
15/16"x15/16"	65.00
15/8"x15/8"	70.00

Base Shoe Combination
3/4"x25/8"	30.00

Bed Molding
11/16"x13/4"	35.00
11/16"x21/4"	50.00

Base Shoe
1/2"x3/4"	20.00

Blind Stop
3/4"x13/8"	40.00

Brick Molding
15/16"x2"	60.00

Casings
11/16"x21/4"	37.00
11/16"x31/2"	68.00

Chair Rails
11/16"x21/8"	50.00
7/16"x21/2"	56.00

Corner Guards
3/4"x3/4"	21.00
11/8"x11/8"	42.00
15/16"x15/16"	46.00

Coves
11/16"x31/4"	70.00
11/16"x21/4"	48.00
11/16"x13/4"	34.00
3/4"x7/8"	26.00
3/4"x3/4"	24.00
1/2"x1/2"	20.00
11/16"x11/8"	27.00

Crowns
11/16"x41/4"	102.00
11/16"x31/4"	68.00
11/16"x21.4"	48.00

Drip Caps
11/16"x15/8"	48.00
3/4"x15/8"	42.00

Full Rounds
15/8"	75.00
15/16"	50.00

Glazing Beads
1/2"x9/16"	18.00
3/8"x3/8"	15.00

Half Rounds
5/16"x5/8"	20.00

Hand Rails
11/2"x13/4"	58.00

Hook Strips
9/16"x27/16"	42.00

Jamb Extensions
3/4"x115/16"	37.00
3/4"x7/8"	28.00

Lattice
1/4"x11/8"	13.00
1/4"x13/8"	15.00
1/4"x13/4"	19.00
1/4"x21/4"	26.00

Mullion Casings
9/16"x51/2"	85.00
3/16"x2	42.00

Parting Strip
1/2"x3/4"	16.00

Picture Molding
3/4"x13/4"	39.00

Quarter Rounds
1/4"x1/4"	6.00
1/2"x1/2"	12.00
3/4"x3/4"	20.00
11/16"x11/16"	38.00

Screen Moldings
5/16"x5/8"	12.00
1/4"x3/4"	10.00

Screen Stock
3/4"x13/4"	40.00
3/4"x23/4"	61.00
11/16"x13/4"	63.00
11/16"x23/4"	75.00

Shelf Cleat
11/16"x11/2"	20.00

Stools

11/16"x31/4"	170.00
11/16"x35/8"	185.00
11/16"x31/4"	95.00
11/16"x23/4"	70.00
11/16"x21/4"	65.00
11/16"x21/4"	90.00

Stops

7/16"x21/8"	48.00
7/16"x15/8"	42.00
7/16"x13/8"	37.00
7/16"x11/8"	31.00
7/16"x7/8"	30.00
7/16"x15/16"	31.00
7/16"X5/8"	24.00

Wood Interior Trim, Casings, Back-band, Stops, Stool, etc.

Wood Ceiling Cornice and Picture Mold

PLASTIC LAMINATE COUNTERTOPS

Plastic laminate for countertops will vary in price depending on the number of cutouts, corners, and whether the fabricator must include the cost of taking job measurements. A typical L-shaped kitchen countertop covering 18 ft. of inside wall, all edges plastic, laminated to ⅞" particle board, with 4" splashes back and ends, will cost around $16.00 per lineal foot in a standard pattern. Rounded edges will cost considerably more; metal banded ones somewhat less. For each cut out add $8.00. For each end splash add $6.00.

Setting Cabinets and Cases.—When setting mill assembled cabinets and cases, such as used in the average residence or apartment building, it will require about 1/6 hr. carpenter time for each sq. ft. of cabinet face area at the following labor cost per sq. ft. :

	Hours	Rate	Total	Rate	Total
Carpenter	0.16	$....	$....	$16.47	$2.64

A carpenter should fit and hang, complete with pulls and catches, 15 to 17 flush case doors per 8-hr. day, at the following labor cost per door :

	Hours	Rate	Total	Rate	Total
Carpenter	0.5	$....	$....	$16.47	$8.24

For lip case doors, not requiring fitting, a carpenter should install, complete with pulls and catches, 30 to 34 doors per 8-hr. day, at the following cost per door :

	Hours	Rate	Total	Rate	Total
Carpenter	0.25	$....	$....	$16.47	$4.12

In the better class buildings where cases, wardrobes, cabinets, etc., are used and it is necessary to scribe and fit them against the plastered walls or run a small molding to conceal joints between cases and plastered walls, it will require about 1/3 hr. carpenter time for each sq. ft. of cabinet face area at the following labor cost per sq. ft. :

	Hours	Rate	Total	Rate	Total
Carpenter	0.33	$....	$....	$16.47	$5.44

A carpenter should fit and hang, complete with pulls and catches, 11 to 13 flush case doors per 8-hr. day, at the following labor cost per door :

	Hours	Rate	Total	Rate	Total
Carpenter	0.67	$....	$....	$16.47	$11.04

Often, in restricted areas, such as adding cabinets in narrow lavatory areas, it is necessary to have cabinet work delivered in sections so that it may be turned and properly set within the space. Seldom is a job so organized that cabinet work is delivered before wall finishes are up, so the fit must be preplanned and exact.

Setting Factory Assembled and Finished Kitchen Cabinets.—When setting factory assembled and finished kitchen cabinets, such as used in the average apartment or residence, it will require about 0.1-hr. carpenter time for each sq. ft. of cabinet face area at the following labor cost per sq. ft. :

	Hours	Rate	Total	Rate	Total
Carpenter	0.1	$....	$....	$16.47	$1.65

First Grade Workmanship

In better class apartments and residences where a high class job of installation is required it will take about 1/6-hr. carpenter time for each sq. ft. of cabinet face area at the following labor cost per sq. ft. :

	Hours	Rate	Total	Rate	Total
Carpenter	0.16	$....	$....	$16.47	$2.64

Setting Wood Fireplace Mantels.—If wood fireplace mantels are factory assembled, merely requiring fitting and setting, a carpenter should install 2 to 3 mantels per 8-hr. day at the following labor cost per mantel :

	Hours	Rate	Total	Rate	Total
Carpenter	3.2	$. . . .	$. . . .	$16.47	$52.70
Labor	1.0	. . . .	. . . .	12.54	12.54
Cost per mantel		. . . .	$. . . .		$65.24

Fitting and Placing Closet Shelving.—A carpenter should fit and place 115 to 135 sq. ft. of closet shelving per 8-hr. day including setting of shelf cleats at the following labor cost per 100 sq. ft. :

	Hours	Rate	Total	Rate	Total
Carpenter	6.4	$. . . .	$. . . .	$16.47	$105.41
Cost per sq. ft			. . . .		1.05

Installing Hanging Rods in Closets.—A carpenter should install about three hanging rods, including supports, per hr. at the following labor cost per rod :

	Hours	Rate	Total	Rate	Total
Carpenter	0.33	$. . . .	$. . . .	$16.47	$5.44

Setting Metal Medicine Cabinets.—A carpenter should set 10 to 12 average size medicine cabinets per 8-hr. day at the following labor cost per cabinet :

	Hours	Rate	Total	Rate	Total
Carpenter	0.75	$. . . .	$. . . .	$16.47	$12.35

Setting Bathroom Accessories.—In an 8-hr. day a carpenter should locate and set 30 to 34 bathroom accessories, such as towel bars, soap dishes, paper holders, etc. at the follwoing labor cost per accessory :

	Hours	Rate	Total	Rate	Total
Carpenter	0.25	$. . . .	$. . . .	$16.47	$4.12

Installing Chalkboards, Corkboards, and Accessories.—Installation of chalkboard, according to manufacturer's instructions, is as follows :

Before starting, all grounds should be in place and final coat of finish should be thoroughly dry.

Chalkboard should be set so that all edges will be adequately covered by trim moldings. The writing surface should be plumb and in a true plane with ¼-in. clearance on all sides. Chalkboard panels should be nailed at the top edges only and the balance secured to the wall with an even coat of spotting cement.

Two carpenters working together should install 125 to 150 sq. ft. of chalkboard per 8-hr. day at the following labor cost per 100 sq. ft. :

	Hours	Rate	Total	Rate	Total
Carpenter	11	$. . . .	$. . . .	$16.47	$181.17
Cost per sq. ft			. . . .		1.81

Installation of corkboard is similar to that of chalkboard except nailing is eliminated. The labor cost of installing corkboard is approximately the same as for chalkboard.

Installation of aluminum trim moldings in conjunction with chalkboards and corkboards may be accomplished either with exposed wood screws or with concealed snap-on clips. A carpenter should install 90 to 110 lin. ft. of molding per 8-hr. day at the following labor cost per 100 lin. ft. :

	Hours	Rate	Total	Rate	Total
Carpenter	8	$. . . .	$. . . .	$16.47	$131.75
Cost per lin. ft			. . . .		1.32

Labor Placing Finish Hardware

Description of Work	No. Set per 8-Hr. Day	Carp. Hrs. Each
Rim locks or night latches, cheap work	20-24	0.3-0.4
Mortised locks in soft or hardwood doors, Ordinary Workmanship	12-16	0.5-0.7
Mortised locks in hardwood doors, First Grade Workmanship	6-8	1.0-1.3*
Cylindrical locks	17-23	0.3-0.5
Front entrance cylinder locks in hardware doors, Ordinary Workmanship	6-8	1.0-1.3**
Front entrance cylinder locks in hardwood doors, First Grade Workmanship	4-5	1.6-2.0***
Panic bolts, First Grade Workmanship	2-3	2.7-4.0
Door closers, exposed	7-9	0.9-1.1
Door closers, concealed	2-3	2.7-4.0
Door holders	20-24	0.3-0.4
Sash lifts and locks, No. of windows	20-24	0.3-0.4
Cremone bolts for casement doors and windows	20-24	0.3-0.4
Kickplates, Ordinary Workmanship	8-10	0.8-1.0
Kickplates, First Grade Workmanship	6-8	1.0-1.3

*On this grade of work the lock must be flush with the face of the door and must operate perfectly.
**On this class of work the architect or owner must expect to have more or less trouble with hardware because the time given is not sufficient to adjust lock and see that it operates properly, etc.
***This includes front door locks having fancy escutcheon plates, knobs, knockers, etc.

Labor Placing Cabinet Hardware

Description of Work	No. Set per 8-Hr. Day	Carp Hrs. Each
Surface butts	180-200	.040-.045
Offset butts	90-100	.080-.090
Elbow catches	90-100	.080-.090
Friction catches	90-100	.080-.090
Drawer and door pulls	140-150	.053-.057
Knobs	180-200	.040-.045
Rim latches	90-100	.080-.090
Mortised locks	20-24	.3-.4

Mantels

There are many stock mantel designs from which to choose. A simple molding 6" wide extending out from the wall 2" on 3 sides will run around $55 for an outside dimension of 4'x7'. A full mantel with 7" shelf and of colonial design will cost from $90.00 to $125.00. More elaborate designs in Georgian or French provincial and with end returns to fit over the brick projection will run from $150.00 to $350.00. These costs are for wood members only. Brick or marble trim around the fireplace opening are, of course, extra. It will take a carpenter 3 to 4 hours to set a prefabricated unit in place.

China Cabinets

Corner cabinets can be purchased from stock to add architectural interest to dining, breakfast or recreation rooms. Mostly of Ponderosa pine with either Colonial or Georgian design, they will vary from $95 for a unit 2'-8" wide with single doors top and bottom to $225.00 for a unit 3'-6" wide with double doors. Cabinets are from 7' to 7'-6" high. Bottom doors are usually paneled, tops factory glazed. Shipped with front assembled and body knocked down, the unit includes 3 shelves for the upper portion, 1 for the lower. Hardware is not included. Allow three hours of one carpenter's time to fit and set each unit, and apply the hardware.

Ironing Board Cabinet

A built-in feature which appeals to most housewives is an ironing board cabinet unit complete with adjustable board plus sleeve board, all operating hardware, and flush door which is sized to fit between wall studs 16" on centers. It is shipped in a carton and costs around $75.00. The door is available in either birch or Philippine mahogany. No casing is included, but if the unit is set in a matching paneled wall, it is almost inconspicuous.

Louvered Pine Panels

Shuttered panels are available in stock sizes, complete with hardware, or customized to fit exact openings. They are not only used inside window openings, but for cabinet doors and room separators.

Cost of Stock Single Panel Horizontal Slat Unit

			WIDTHS		
HEIGHT	6"	7"	8"	10"	12"
20	$2.70	$3.00	$3.40	$4.20	$5.00
23	3.25	3.50	3.90	4.60	5.50
29	3.70	4.00	4.60	5.30	6.30
32	4.20	4.50	5.10	6.00	7.20
36	4.80	5.00	5.70	6.75	8.30

A four panel hardware kit costs $1.50. If no trimming is required one carpenter should set two panels per hour or 2 window openings consisting of 8 panel units each in a day.

Custom shutters in lengths of 6" to 18', and for openings from 12" to 24", 2 panels will cost $23.90 unfinished and $32.60 prefinished; for an opening 24" to 30", four panels same height, $40.20 and $56.60; for 30" to 42" opening, $37.00 and $60.50; in lengths of 18"–24", openings 12" to 24", 2 panels will run $27.80 and $39.10; openings 24"–30", 4 panels, $48.20 and $69.60; openings 30" to 42", 4 panels, $52.10 and $73.60. Custom door panels to 7' will run $92.50 unfinished, $129.20 finished for 2 panel for an opening to 24" wide. For openings to 30", 2 panels, $95.40 and $132; openings 30"–42", 4 panels, $186.50 and $258.70.

06300   WOOD TREATMENT

Wood and plywood are often given special treatments at the mill to give additional protection against fire, rot and termites.

Fire protected or fire retardant wood is pressure-impregnated with mineral salts that will react chemically at a temperature below the ignition point of the wood. An Underwriter's Laboratory Flame Spread Rating of 25 can be obtained, and no periodic maintenance is required to keep the rating. However the lumber must also be kiln dried to 12% moisture content if it is to be painted. Also, wood is not suitable for direct exposure to weather and ground conditions unless further treatment is given. A fire retardant treatment will add 12 cts. to 15 cts. per board foot to lumber costs. If it is necessary to kiln dry the lumber add 6 cts for soft woods, 8 cts. for hard woods.

Rot and termite resistance can be obtained by several methods. Creosote, a coal tar product, is impregnated at the rate of around 8# per cubic foot at a cost of 12 cts. to 15 cts. per board foot. This finish is not readily paintable and can stain adjacent finishes such as plaster and wallboard. Pentachlorophenol is a popular oil bourne treatment, and costs 11 cts. to 13 cts. per sq. ft. and leaves the wood with an oil residue. This residue can be eliminated and the wood left clean and paintable if the pantachlorophenal is impregnated into the wood by liquid petroleum. This will add 5 cts. to the cost.

06400   ARCHITECTURAL WOODWORK

Architectural woodwork, insofar as this section is concerned, covers the assembly and installation of cabinets, wood and wallboard paneling, wood stairs and railing from components made up in a shop or factory off the job premises.

If the items under this heading are not stock items, it will be necessary to include the cost of making on-the-job measurements and preparing shop drawings. Most cabinet makers are equipped to do this and will include such costs in their proposal. Often they will also install their work, but most usually the general contractor will handle this.

There is considerable difference between the grades of workmanship for architectural woodwork and the estimator must be able to determine the grade required.

These same differences apply to the erection and installation of all classes of millwork and interior finish.

The Architectural Woodwork Institute publishes a manual of "Illustrated Quality Standards" in which the work is divided into three grades: premium, custom and economy. Premium is the highest grade available in both material and workmanship, intended for the finest work. Sometimes it is used throughout an entire building, but most often only in selected spaces or items within a building. Custom is the middle or normal grade in both material and workmanship, and is intended for high quality regular work. Economy is the lowest grade and intended for work where price outweighs quality considerations.

An estimator may find references to these three grades in an architect's specification and, as this manual sets forth very exactly what is to be expected with both drawings and notes, he should have a copy of this booklet in his library. It is obtainable from the Architectural Woodwork Institute, 2310 S. Walter Reed Drive, Arlington, Va. 22206.

Methods of Estimating the Labor Cost of Erecting Mill Work and Interior Finish.—Different contractors use different methods of estimating the labor cost of erecting and setting millwork and interior finish, but the system generally used is to add a certain percentage of the cost of the millwork to cover the labor cost of erection. While this method is the easiest one known it does not produce the best estimates. Example: suppose you are estimating the cost of a building on which you receive 4 millwork bids ranging from $4,000.00 to $5,000.00. After receiving the bids you add the customary percentage to the lowest acceptable bid to take care of the labor cost of handling and erecting the millwork. This percentage usually runs from 40 to 50% of the cost of the millwork. Taking 40% of $4,000.00 gives $1,600.00 for labor to handle and erect all millwork and interior finish. Suppose a competitor received a bid for $3,500.00 from a mill which did not quote you. If your competitor added 40% to his bid, he would have only $1,400.00 to erect the same amount of millwork on which you figured $1,600.00.

Why this difference of 12½ % on such a small item?

While this method saves a great deal of time and is almost universally used, it does not produce the most accurate estimates.

The only accurate method is to list each item separately; i.e., the number of lin. ft. of wood base, chair rail, picture mold, etc., the number of door jambs to set, the number of sash to fit and hang, together with the number of window openings to be cased or trimmed; the number of wardrobes, linen cases, vanity and kitchen units to set, etc. Estimating labor costs by this method produces more uniform and accurate estimates, reduces risk to a minimum and enables the contractor to check actual and estimated costs during construction—but it does require a lot more work.

INTERIOR PANELING

A few years ago interior paneling meant solid wood with stiles and rails and elaborate mouldings for the mansion or bank job; Knotty pine for the suburban retreat; and beaver board for the cottage. Today the selection is much wider.

Solid wood paneling is still with us but it is more apt to be in the form of planks with mouldings face applied than as elaborate paneling. Plywood veneered with hard wood is often favored over solid wood not only for cost savings but because the veneer exposes the beauty of the wood and offers a more easily cleaned surface.

Most paneling today though is sheet board faced to look like wood grain, marble, tile, stucco, leather or with a decorative pattern all its own. The designs come with a selection of moldings that blend into the overall wall.

Insulating wallboards and tiles are still sometimes used as exposed finishes but are selected more for their structural or insulating value rather than their decorative contribution.

Estimating Paneling Quantities. Estimating paneling is more of a task than merely computing the number of sq. ft. of surface to be covered. Each room should be laid out to obtain certain panel effects if the design of the panels will govern to any extent the amount of cutting and fitting necessary. Most paneling will come in stock lengths and widths. The job conditions must be carefully studied to minimize cutting, to make maximum use of studs for attachment, and to eliminate any awkward jointing.

The labor cost of applying wallboard depends upon three things: the cost of placing headers and nailing strips between wood studs to which wallboard and decorative strips are nailed; the cost of installing the wallboard and the cost of placing decorative panel strips or other means of concealing joints. When installing a ceiling, furring must be applied to provide nailing pieces for all edges of the decorative strips.

Wood decorative or panel strips are usually ¼" to ½" thick and of widths to meet architectural requirements.

There are several ways of covering joints. Today much paneling is grooved so joints are butted and appear as an overall pattern. Most flush panels have cover strips, caps, etc. available to match the basic material or to contrast with it and make a decorative feature of the joint.

A sketch of each room, should be made giving an elevation of each of the four walls and a plan showing the ceiling, similar to that illustrated. From this it is possible to obtain the number of lin. ft. of "headers" or nailing strips required, the size of each piece of paneling as well as the number of lin. ft. of panel strips and moldings necessary to complete the job.

Placing "Headers" or Nailing Strips.—When placing "headers" or nailing strips to receive paneling it is necessary to cut them to length and place them flush with the face of the studs, as the paneling must be nailed top and bottom to the "headers" and at the sides to the wood studs.

If the studs are not wide enough to form nailing pieces for the panel strips, a piece of wood, usually 1"x1", must be nailed to either side of the stud along the entire length and flush with the face of the stud.

When estimating this class of work, care should be taken to obtain the exact number of lin. ft. of "headers" required as this is an important item in the cost of the work.

With studs spaced 16" on centers, a carpenter should cut and place 225 to 275 lin. ft. per 8-hr. day, at the following labor cost per 100 lin. ft. :

	Hours	Rate	Total	Rate	Total
Carpenter	3.2	$....	$....	$16.47	$52.70
Cost per lin. ft			....		.53

Placing Furring Strips for Ceilings.—Furring strips should be nailed at right angles to joists. Any convenient width of furring may be used but it must be wide enough to provide a nailing base to nail all edges of the decorative strips at least every 6". Furring may be any No. 2 material, soft wood being preferable. Make sure that all furring strips are level.

With joists 16" on centers, a carpenter should cut and place 300 to 350 lin. ft. of furring strips per 8-hr. day, at the following labor cost per 100 lin. ft. :

	Hours	Rate	Total	Rate	Total
Carpenter	2.5	$....	$....	$16.47	$41.18
Cost per lin. ft			....		.41

Placing Sheet Paneling.—The actual cost of placing sheet paneling will vary with the size and shape of the room, whether full size sheets may be used or considerable cutting and fitting necessary. It requires practically as much time to place a sheet 16"x96" as one 48"x96" on account of the cutting and fitting, while there are three times as many sq. ft. in the latter as in the former.

Rough Sketch of Room Showing Location of Doors and Windows.

Finished Drawing Showing Location of Doors and Windows.

On straight work, in large rooms, a carpenter should fit and place 450 to 550 sq. ft. per 8-hr. day, at the following labor cost per 100 sq. ft. :

	Hours	Rate	Total	Rate	Total
Carpenter	1.6	$. . . .	$. . . .	$16.47	$26.35
Cost per sq. ft			. . . .		.26

When applying sheets in smaller rooms requiring considerable cutting and fitting around doors, windows, etc., a carpenter should fit and place 275 to 300 sq. ft. per 8-hr. day, at the following labor cost per 100 sq. ft. :

	Hours	Rate	Total	Rate	Total
Carpenter	2.7	$. . . .	$. . . .	$16.47	$44.47
Cost per lin. ft			. . . .		.44

As previously mentioned, it usually costs more to place a small piece than a large one, as this generally calls for cutting and fitting above and below windows, around door openings, etc., which requires considerably more work than placing a large sheet where no cutting and fitting is necessary. For this reason, estimating the labor cost by the sq. ft. is not the safest method to use, although it is the most common.

Where time in estimating is not a factor, the safest method is to lay out the work as illustrated on the previous pages, dividing it into panels and then allow 20 to 30 minutes to place each piece of wallboard or plywood depending upon the class of work.

06421 Hardwood Plywood Panels

During recent years hardwood plywood panels are being increasingly used for the interior walls of modern buildings.

These plywood panels of birch, oak, mahogany, etc., are usually furnished in sheets 4'-0" wide and 8'-0" to 10'-0" high and are used instead of plaster or other wall finishes.

When using ¼" plywood panels in large sizes for this class of work it is necessary to have a substantial backing for the thin plywood panels. This may be obtained by applying the plywood direct to a brown or finish plaster coat wall; a ½" gypsum board backing, or better still, a ½" thick plywood backing with the sheets running in the opposite direction to that of the finish plywood panels.

These plywood panels are usually applied with the panel running from floor to ceiling in one length, i.e., 4'-0"x8'-0" or 4'-0"x10'-0", with the vertical joints butted flush or having a very slight bevel at the edges.

To obtain a first class job it is usually necessary to apply adhesive to the back of the plywood panels and also use small brads to secure the plywood in place until the adhesive sets.

This is particular work as the adjoining sheets must be accurately fitted together if a first class job is to result.

If wood panel strips are to be used to conceal the joints, it is not necessary to use so much care in placing the panels.

Hardwood laminated veneer panels cover a wide variety of products, from the ¼" thick "grooved" veneers so popular now to ¾" thick flush panels with or without sequence matching of the flitches on up to special burl veneers suitable for the tops of the finest furniture.

Plywood may have a veneer core with a center and at least two intermediate cross banded veneers, a particleboard core, or a lumber core with strips of lumber edge glued into a solid slab with cross banding veneers either side. The back and face of each sheet then receive additional veneers at right angles to the cross banding.

The face veneers may be rotary cut with the log's annular rings exposing a bold variegated grain; flat sliced, which produces a variegated figure similar to that of plain sawn lumber; quarter sliced with growth rings of the log at right angles to the knife, producing a series of stripes in the veneer; and rift cut, which is used with oak logs to produce fairly uniform grains with a combed look caused by the medullary rays.

Then there are various grades of face veneers: premium, or architectural panels which have veneers sequence matched and no patching; custom, which allows no patching on face veneers but are not sequence matched; and economy, which allows "good" face veneers, but no matching of grain or colors.

And, finally, there is the type of tree itself which affects the price. The range from natural yellow birch to Brazilian rosewood is around 800%.

Costs should always be determined locally, since not all veneers are available at all times. The following will serve as a general outline of costs.

Natural yellow birch	4'x8'x¼"	custom	$ 30.00
Natural hard maple	"	"	30.00
Natural gum	"	"	30.00
Select white birch	"	"	38.00
Popular	"	"	34.00
Select birch heartwood	"	"	35.00
Select white maple	"	"	38.00
Plain sawn red oak	"	"	45.00
African mahogany	"	"	45.00
White ash	"	"	48.00
Limba	"	"	50.00
Rift sawn white oak	"	"	60.00
Pecan	"	"	53.00

Cherry	"	"	67.00
Butternut	"	"	58.00
Quarter sawn white oak	"	"	65.00
Walnut	"	"	65.00
Primavera	"	"	75.00
Avodire	"	"	80.00
Teak	"	"	120.00
Brazilian rosewood	"	"	240.00

If matched sequence panels are ordered, the cost will be about 50% more. Add around $10.00 per sheet for ¾" thickness.

The stock V-grooved prefinished 4'x8'x¼" panels are considerably cheaper. Although the range of woods is not so extensive, the finishes vary widely and approximate the colors of more expensive woods still using the grains of the more common species. A random sampling of costs indicate the following range.

			Clear	Colonial
Birch	4'x8'x¼"	V-grooved	$18.00	$13.00
Oak	"	"	28.00	16.00
Walnut	"	"	32.00	19.00
Pecan	"	"	27.00	25.00

Inlaid paneling in 4'x8'x¼" sheets consisting of pecan panels striped with 1¼" wide walnut stripes runs $30.00 per sheet; with walnut striped in pecan the cost is $35.00 per sheet.

For lower budget jobs, simulated wood grain finishes on ¼" plywood will run $7.00 per 4'x8' sheet.

For the lowest budget jobs, simulated wood finishes printed on hardboard are available in the $6.00 to $7.00 range per 4'x8'x¼" sheet.

Labor Placing Flush Plywood Wall Panels.— When applying plywood panels 4'-0"x8'-0" to 4'-0"x10'-0", with the abutting joints fitted tightly together, using adhesive and small brads to secure the plywood to the backing, two carpenters working together should handle, fit and place one plywood panel in 1½ to 2 hours, or they should place 4 to 5 panels per 8-hr. day.

Many of the panels will have to be cut to half or three-quarter size, and in some instances cut out for doors and windows, so a certain length of time should be allowed for each piece of plywood, rather than figuring on a square foot basis.

06422 SOFTWOOD PLYWOOD PANELING

Softwood plywood consists of an odd number of thin sheets or veneers of Douglas fir, consisting of 3, 5, 7 or 9 standard thicknesses—these being laminated with alternating grain direction. Because the grain of each ply composing the material is at right angles to the grain of adjacent plies, the strength of Douglas fir plywood panels is approximately equal in both directions.

All Douglas fir plywood from Association mills is made to conform to the moisture-resistance requirements set forth in the revised standards established by the National Bureau of Standards, as U. S. Commercial Standards CS45-55.

Plywall wallboard is the grade of plywood commonly used for interior walls and ceilings instead of lath and plaster or other types of wallboard. It is furnished in sheets 24 to 48-in. wide and 5 to 12 ft. long.

Plywall wallboard is furnished ¼, ⅜, ½, ⅝, ¾, 1 and 1⅛-in. thick. The ¼-in. thickness is most widely used for walls and ceilings, although for more substantial construction ⅜ or ½-in. thick is often preferred. Thicker material is used for shelving, wardrobes, cases, etc.

Practically any design may be obtained by the use of plywood as with other types of wallboard. Four penny casing or finishing nails are recommended for paneling up to ½-in. thick with the nails spaced 6 to 10 in. apart. One horizontal or fire stop should be cut in between top and bottom plates as a nailing piece.

Plywood panels may be finished by staining, painting, wall papering or mechanical surfacing.

Labor Applying Plywood Wallboard.—The labor cost of applying plywood wallboard will run about the same as given for fiber wallboard covering the same class of work.

Where special jointing of the panels is required to work out wall and ceiling panels, allow extra for grooving, beveling and any other designs worked into the wallboard.

Also add extra for applying panel strips as given on previous pages.

Prices of Douglas Fir Plywood

Made with water resisting glue. All panels sanded 2 sides to net thickness shown. As plywood prices have been very volatile recently, check local sources.

Size of Panel	Inches Thick	Number of Plies	Price per 1,000 Sq. Ft. Good (A-D) 1 Side	Good (A-A) 2 Sides
24"x60" to 48"x96"	¼	3	$280.00	$380.00
24"x60" to 48"x96"	⅜	3	350.00	450.00
24"x60" to 48"x96"	½	5	390.00	490.00
24"x60" to 48"x96"	⅝	5	430.00	530.00
24"x60" to 48"x96"	¾	5	525.00	580.00

Types of Joints Used with Plywood Wallboard.

Saw textured plywood siding is popular and relatively inexpensive. It is readily available in Douglas fir and cedar.

Saw Textured Douglas Fir Plywood

Size	Per 1000 Sq. Ft.
⅜" Natural	$520.00
⅜" Grooved	580.00
⅝" Natural	650.00
⅝" Grooved	750.00

Saw Textured Cedar

Size	Per 1,000 Sq. Ft.
⅜" Natural	$ 720.00
⅜" Grooved	760.00
⅝" Natural	1100.00
⅝" Grooved	1150.00

06423   Solid Wood Paneling

While most paneling used today is of veneered plywood, nothing matches the beauty of solid wood paneling; and its ability to withstand years of usage and often even improve with age.

Solid paneling is usually quoted in random widths of 4" to 8" and random lengths of 6' to 16'. Specified widths and lengths will run 10% or more additional. Traditional thickness is ¾", but much paneling today is available at a considerable savings in ½" thickness.

There is no nationally established grading rules for solid wall paneling, so it is necessary to know the product of the mill quoting to compare prices. Generally, woods are available in two grades: "select," "luxury" or "clear," and "knotty," "wormy" or "pecky," often designated as "Colonial" grade.

In figuring paneling, add 15% to actual areas for waste and matching.

Paneling is available with V-joint tongue and groove, square joint, shiplap and Colonial beaded.

The most commonly encountered paneling is knotty pine. Cost is around 85¢ per sq. ft.

Clear pine will run around $1.25.

Clear birch in ¾" thickness will run around $1.75 per board foot in lots over 1000 B.F., random width and length. Birch is an excellent cabinet wood with a firm texture, but has a rather strong grain pattern.

White ash is an open grained wood, light in color with contrasting light brown heartwood. It is easy to work and somewhat resembles flat grained white oak. It will cost $1.85 for ¾" select grade.

Butternut, often called white walnut, is a soft-textured, easy-to-nail wood which takes a beautiful natural finish. It will run $2.10 per B.F. select, $1.05 character marked, ¾" thick.

Cherry, cut from the wild black cherry, is a traditional Colonial paneling. It is a firm textured cabinet wood, but is available only in a rather narrow range of sizes. It runs around $2.00 for ¾" select down to $1.50 for ¾" Colonial.

Oak is available from either white or red varieties cut on the quarter grain. Oak is a very hard wood and subject to shrinkage and splitting in adverse conditions, but well worth the extra effort for the right place. Oak runs $1.60 for the red and $1.85 for the white in ¾" select, 1.00 and .85¢ in ¾" Colonial.

Walnut has become so popular and the sources for it so restricted that it is now a high priced wood and will run $2.81 or better a square foot for ¾" select. As it is a hard wood, ½" thickness is practical and will run $1.75 for select and $1.30 for Colonial.

Other domestic woods available are maple, around $1.40 for ¾" select; wormy chestnut, at $1.50 for ¾"; and cypress at 90¢ for ¾" select.

Imported woods are also available. The so-called "exotic" woods are more often found in the architectural grade, book-matched plywood where their veneers can be shown off to better advantage.

Some foreign woods stocked in the Chicago area include paldao, a firm textured striped wood for medium brown from the Philippines; African mahogany and the less costly Philippine mahogany which is more difficult to finish; avodire, a light creamy-colored wood which comes from Africa with a roll figure somewhat on the diagonal; limba, or korina, African wood which is cream or tan in color with a darker heartwood, and tigerwood or African lovea, a dullish brown to golden colored wood with a wide or ribbon stripe.

These are all special woods for special jobs. Prices and availability should always be checked as they vary considerably, but all are expensive woods.

Placing Wood Wainscot or Paneling.—The labor cost of placing wood paneling or wainscoting will vary with the height of the wainscot and the manner in which it is delivered, i.e., whether assembled or knocked down.

It takes nearly as long to place wood wainscoting 4'-0" high as 7'-0" high, as the top and bottom members are identical and the only difference is the height of the panels.

If the wood wainscot is put together in the mill and the carpenters on the job do what little fitting is necessary, set it against the walls and nail it to the grounds, a carpenter should place 12 to 18 lin. ft. of wainscot per 8-hr. day, at the following labor costs for the different heights :

Height of Wainscot	No. Sq. Ft. 8-Hr. Day	Carp. Hrs. 100 Sq. Ft.
4'-0"	60-70	11-13
5'-0"	70-85	9-11
6'-0"	80-100	8-10
7'-0"	90-110	7- 9
8'-0"	95-115	7- 8

Placing Wood Wainscot or Paneling (Knocked Down).—If wood wainscot or paneling is sent to the job knocked down, making it necessary to assemble the pieces and put them together on the job, a carpenter should place top and bottom rails, cap and intermediate panels for 10 to 15 lin. ft. of wainscoting per 8-hr. day, at the following costs for the different heights :

Height of Wainscot	No. Sq. Ft. 8-Hr. Day	Carp. Hrs. 100 Sq. Ft.
4'-0"	50-60	13-16
5'-0"	60-70	11-13
6'-0"	70-80	10-11
7'-0"	80-90	9-10
8'-0"	90-100	8- 9

06424   SHEET BOARD PANELING

An inexpensive prefinished paneling is plastic coated hardboard. This comes in ⅛" and ¼" thick, 4'x8' panels, and 16" wide, ¼" thick grooved planks. The thinner boards should have solid backing for first class work. These panels have a wide range of printed designs from wood grains, solid colors, scenic designs, and marble textures. A full line of aluminum and presdwood moldings is available colored to match or compliment the selected material.

A recent innovation is vinyl surfaced gypsum wallboard. This comes in ½" thickness 8', 9' and 10' lengths. It can be applied with direct nailing using colored nails; glued to furring with a combination adhesive nail-on system, or laminated to a base layer of gypsum backed board. An advantage of this finish is that it can be patched and eventually painted.

Open Work Panels.—A popular addition to stock partitioning available today are panels with open work. These vary from ⅛" thick prefinished masonite to solid walnut grilles ranging as high as $20.00 a square foot. The amount of openings can vary from 15% to 50%. These materials make good room dividers, help control sun and glare at windows and cut air conditioning and drapery costs.

The masonite product is known as Filigree and is available in a series of sizes up to 4'x8' sheets. Prefinished frame sections ¾" thick and 1½" wide, factory grooved to receive the panel are available.

Sculptured hardwood panels ¾" thick are available in many designs and woods including walnut, maple, ash, oak and poplar, in stock sizes from 2'x4' to 4'x8' from Customwood Mfg. Co., Albuquerque, New Mexico. These may be ordered sanded for finishing on the job or factory finished. Woods available listed in order of ascending cost include sycamore or gumwood; poplar, maple, red oak, ash, and walnut. Some designs, nine are available, are also made in ⅝" clear acrylic.

The least expensive design in the least expensive wood, sycamore, will run $54.00 for a 2'x6' unfinished, unframed panel. Framed and finished it will cost $174.00, plus shipping charges. The most expensive designs will run $102.00 unfinished and unframed, $222.00 framed and finished.

Wallboard—Kinds, Sizes and Costs.—There are so many kinds and sizes of wall and insulating board, it is impossible to list all of them.

The thinnest boards are made of wood fiber and are usually furnished in sheets 4 ft. wide, 4 to 12 ft. long, and ⅛", 3/16", ¼" and ⅜" thick. Prices vary from 13 cts. per sq. ft. for the ⅛-in. up to 20 cts. per sq. ft. for the ⅜" board.

Insulating and building board is usually furnished ⅜ and ½-in. thick, 4 ft. wide and 6, 7, 8, 9, 10 and 12 ft. long, all having square edges. Board of this type will cost 22 cts. per sq. ft. for the ⅜-in. and 28 cts. per sq. ft. for the ½-in. thickness.

Insulation Board Finish Plank.—Used for interior finish to combine insulation and decoration in one material.

Insulating plank are ½-in. thick and furnished in random widths 8, 10, 12 and 16 in. wide and 8, 10 and 12 ft. long. The long edges may be shiplapped with either a plain bevel or bead and bevel design, or have a blind nailing flange.

Plank ½-in. thick cost about 50 cts. per sq. ft. Wall plank with acoustical properties is also now available in 8' and 10' lengths. One such product is Cushiontone as manufactured by the Armstrong Cork Co.

Plank Effect Board.—The plank effect board is ½" thick, 4'-0" wide and 8'-0" high. A series of 5 score lines on each edge and through the center of the board gives the effect of 3 planks 16" wide. It may be applied vertically or horizontally either over old plaster or directly to studs. All edges are flush for butt jointing.

The board is painted ivory color on one side. Cost 55 cts. per sq. ft.

Concealed Nailing Clips.—Provide an invisible means for attaching tile or plank to nailing base—clips fit into tongue and groove. Usually packed 1,000 to a carton. Four pieces are required for each 16"x16" tile. Prices per 1,000 pcs..... $6.00

Use 2d common coated nails with clips. One pound required per 1,000 clips.

Cement or Mastic.—Waterproof cement or mastic for installing insulating plank. One gallon required for 80 to 100 sq. ft.

Price per gal. $3.50

Nails Required for Applying Fiberboard, Tile, Plank, Wainscot and Moldings

For Material ½"-in. Thick	Gauge of Nails	Length of Nails	Spacing Directions
Board	No. 17	1¼"	3"
Tile	No. 17	1¼"	6"—10"
Plank	No. 17	1¼"	6"
Wainscot	No. 17	1¼"	6"
Moldings	No. 15	1¾"	6"

As a general rule, board, plank and wainscot require about 6 brads per sq. ft. or 4 to 5 lbs. of 1¼" brads per 1,000 sq. ft. Galvanized wire brads are recommended. When they are not available, bright wire brads may be used.

Placing Insulating Plank.—The labor cost of placing insulating plank will vary with the surface to which it is applied, length of walls, amount of cutting and fitting necessary, etc. Ceiling work costs more than wall work on account of working on a scaffold and placing the plank overhead.

The most economical method of application is to a solid backing of plywood, wood boards or to other satisfactory nailing surface, as it is necessary to have nailing facilities every 8, 10, 12 or 16 inches, depending upon width of the plank. In the event a solid backing is not available, it is necessary to place wood furring strips to provide for nailing the boards at edges and at ends, also for chair rails, wainscot cap and other ornamental strips.

For plank without blind nailing flange nail through face or bead of the plank, never through bevel. Bury nail heads below surface with properly gauged hammer blow or with a nail set.

The cost of wood blocking and furring is given on previous pages of this chapter.

Number of Sq. Ft. of Insulating Plank of Various Widths
Placed Per 8-Hr. Day by One Carpenter, When Nailed In Place

Add for wood furring strips or other backing.

Width of Plank In.	4'-0"	Height of Wainscoting in Feet 6'-0"	8'-0"	10'-0"
8	210-260	275-315	315-365	325-390
10	240-300	300-360	360-420	375-450
12	275-340	350-410	410-475	425-500
16	300-360	375-440	450-525	465-525

When applied to ceilings, reduce above quantites approximately 25 per cent.

Carpenter Hours Required to Place 100 Sq. Ft. of Insulating
Plank On Walls When Nailed In Place

Add for wood furring strips or other backing.

Width of Plank In.	4'-0"	Height of Wainscoting in Feet 6'-0"	8'-0"	10'0"
8	3.5	2.8	2.4	2.3
10	3.0	2.5	2.0	1.9
12	2.6	2.1	1.8	1.8
16	2.5	2.0	1.7	1.7

Panel Strips and Moldings For Use With Sheet Paneling.—Strips for working out decorative effects on walls or ceilings and for concealing the joints may be of the same material as the sheet or they may be wood or metal, depending upon the effect desired.

The labor cost of placing wood or composition decorative strips over the joints will vary with the design of panel strip used, whether it consists of one, two or three members, the size of the panels and the design of the walls and ceilings, as the more cutting and fitting required the fewer lin. ft. of strips a man will place per day.

On straight work, using an ordinary panel strip ¼"x2" to ½"x3", a carpenter should fit and place 350 to 450 lin. ft. per 8-hr. day, at the following labor cost per 100 lin. ft. :

	Hours	Rate	Total	Rate	Total
Carpenter	2	$....	$....	$16.47	$32.94
Cost per lin. ft			....		.33

On more complicated work, using a 1-member panel or decorative strip, a carpenter should place 275 to 325 lin. ft. per 8-hr. day, at the following labor cost per 100 lin. ft. :

	Hours	Rate	Total	Rate	Total
Carpenter	2.7	$....	$....	$16.47	$44.47
Cost per lin. ft			....		.44

Modern Method of Applying Wallboard and Treating Joints.

Placing 3-Member Decorative or Panel Strips.—When the panel strips form a plain wall and ceiling design but consist of 3 members (a strip 3" or 4" wide with a small molding on each side) a carpenter should fit and place 150 to 175 lin. ft. per 8-hr. day, at the following labor cost per 100 lin. ft. :

	Hours	Rate	Total	Rate	Total
Carpenter	5	$....	$....	$16.47	$82.35
Cost per lin. ft			....		.82

Placing Wood Ceiling Beams, Cornices, Etc. —In rooms having beam ceiling effects, wood ceiling cornices, etc., a carpenter should fit and place 175 to 200 lin. ft. of each member per 8-hr. day, at the following labor cost per 100 lin. ft. :

	Hours	Rate	Total	Rate	Total
Carpenter	4.3	$....	$....	$16.47	$70.82
Cost per lin. ft			....		.71

Three member cornices would cost 3 times the above price, 4 member, 4 times, etc.

Applying Wallboard Over Existing Plastered Walls and Ceilings.—In remodeling work where the wallboard is applied over existing walls and ceilings, it is advisable to remove the existing plaster wherever possible. To obtain a first class job it is recommended that all interior trim, such as door and window casings, base, etc., be removed before applying the wallboard. It is also recommended that furring strips be placed over all existing plastered walls, as this will straighten the walls and provide a much better base for securing the wallboard.

After the furring strips have been placed, the cost of applying the wallboard will run about the same as given on the previous pages.

The labor cost of placing wood furring strips is given on the previous pages under "Placing Furring Strips for Ceilings."

Prefinished Panels.—New developments in wallboards have been most noticeable in prefinishing and the consequent care needed in concealing nailing.

One of the most popular items is V-grooved hardwood, prefinished with a baked-on finish. Grooves are generally random but with a groove every 16" on center to hit a stud. Most boards are ¼" thick in sheets 4' wide by 8', 10' or 12' long. These sheets should be applied over a backing board for first class work. Colored nails are available as well as prefinished moldings for cornices, trim and base. Similar paneling is also available in 7/16" thickness for application directly to studs. A wider range of finishes is generally available in the heavier material.

Method of Placing Wallboard and Horizontal Scoring Marks and
Division Strips.

Insulating Tile.—Tile made of Celotex, Fir-Tex, Insulite, Nu-Wood, Temlok and other fiber composition are still sometimes used for wall and ceiling finish. They are furnished with beveled edges either square or with a blind nailing or stapling flange. They come in sizes from 12"x12" to 16"x32" and may be used for obtaining ashlar effects on walls and tile effects on ceilings.

They may be applied to wood backing or to furring strips with adhesive, nails, staples or special clips on the walls or ceilings. When furring strips are used, the strips must be spaced to work out with the tile sizes.

In applying square edged tiles, nails should be driven through the tile at an angle to the surface of approximately 60 degrees. Reverse the angle of half the nails and nail through the face of the tile, never through the bevel. Bury nail heads below the surface with properly gauged hammer blow or with nail set.

When applying tiles with blind nailing or stapling flanges, drive the nails or staples in the thick part of the flange perpendicular to the surface.

Insulating fiber tile ½-in. thick costs $7.50 per carton of 64.

Nailing Recommendations for Square Edged Fiber Tile.—Where nailing is to be exposed, galvanized wire brads are recommended. When they are not available, bright wire brads may be used. 1¼" brads or finishing nails should be used for exposed nailing of ½" thick tile.

Nailing Recommendations for Square Edged Fiber Tile

Tile Size in Inches	Location of Brad	No. Brads per Tile	Approx. Lbs. Brads 1,000 Sq. Ft. Tile
12x12	1 in each corner	4	3½
12x24	4 in each long edge	8	3½
16x16	{ 3 on 2 opposite edges { 3 on center line	9	4
16x32	{ 5 on each long edge { 5 on center line	15	3½

Nailing Recommendations for Flanged Tile. —Where attachment of flanged tile is to be by nails, 3d blued lath nails 1⅛" long should be used. Where application is by means of staples, rust resistant staples ½" or 9/16" in length should be used in conjunction with a gun type stapler.

Tile Size in Inches	No. of Nails or Staples per Tile	Approximate No. of Staples per 1,000 Sq. Ft.	Approximate Lbs. of Nails per 1,000 Sq. Ft.
12x12	4	4,000	6.0
12x24	6	3,000	4.5
16x16	4	2,250	3.4
16x32	6	1,690	2.6

Labor Placing Fiber Tile When Nailed In Place.—The labor cost of handling and placing fiber tile will vary considerably with the size of the tile and the skill of the mechanic placing same, the amount of nailing necessary and the surface to which the tile is applied. A man will place more tile on a good solid backing than on furring strips.

To the quantities given in the following table it will be necessary to add for blocking and furring, where necessary.

Quantities and Labor Costs of Applying Fiber Tile to Walls
and Ceilings Where Tile are Nailed In Place

Add for wood furring strips or other necessary backing.

Size of Tile in Inches	No. Sq. Ft. per Tile	No. Tile Req'd per 100 Sq. Ft.	No. Placed per 8-Hr. Day	No. Sq. Ft. per 8-Hr. Day	Carp. Hrs. 100 Sq. Ft.
12x12*	1.0	100	180-200	180-200	4.2
12x24*	2.0	50	95-105	190-210	4.0
16x16*	1.78	56¼	120-130	215-235	3.6
16x32*	3.56	28¼	83- 90	295-315	2.7

*Stock sizes.

Applying Insulating Tile Using Adhesives.— Special adhesives known as Adhestik, Stickum, Armstrong's Brush-on Ceiling Cement, Celotex Adhesive, etc., are available for applying fiber tile.

Apply the adhesive in spots or dabs about 2" in diameter, ¼" to ⅜" thick, one on each corner of the square tileboard and two or more additional spots on the rectangular sizes. Keep adhesive about 1" from edges of fiberboard units and use approximately 4 spots per sq. ft. Adhesive spots should not be spaced over 10" on centers in any direction.

After dabs of adhesive have been applied, place the tile in its final position. Slide back and forth for a distance of about ¾ inch under uniform pressure of the hands to obtain good contact. When working units larger than 16"x16", two men should exert pressure simultaneously for best results. If the surface of the unit is below the level of adjoining pieces, do not pull the unit away from the base to bring surface level but remove it and add more adhesive. Again work into position as described above to obtain correct surface level.

Brad or shore the first units on either walls or ceilings to prevent sliding out of position during application of surrounding units. Do not bend units over high spots.

Tile and plank may be applied to plasterboard, sound plaster, plywood or lumber. In each case the use of adhesive or a combination of adhesive and brads is recommended.

Coverage Chart of Adhesive

Size of Tile in Inches	No. Spots per Tile	Sq. Ft. Per Gal.	Gals. Per 1,000 Sq. Ft.
12x12	4	40	13
12x24	6	40	13
16x16	4	60	10
16x32	6	60	10

Price, $3.50 per gal.

Labor Placing Fiber Tile Using Adhesive.— The labor cost of placing fiber tile using adhesive or adhesive and nails will vary greatly with the skill and experience of the workman applying same, size of tile, etc. For instance, where an experienced man is placing the tile, and has a helper to build scaffold, unpack tile and apply the adhesive to the backs of the tile and the mechanic merely slides them in place, an experienced mechanic and helper should place 40 to 45 sq. ft. of 12x12-in. tile an hour but if one man is working alone and he must stop to apply the adhesive to each tile, set it in place and then drive several brads through them, 15 to 20 sq. ft. an hour would probably be the average. For this reason, conditions under which the tile are to be installed must be carefully considered before arriving at the labor cost.

The design in which the tile are laid also affects the labor costs. For instance when tile are laid in herringbone pattern, it requires about 30 to 40 percent more time than regular square or common bond. Diagonal patterns require about 20 to 25 percent more time than regular squares or common bond while tile laid in mixed ashlar designs require about 50 percent more time than straight squares or common bond.

The quantities and costs given in the following table are based on using two men consisting of an experienced mechanic to place the tile and one man to build scaffold, unpack the tile and place them on the scaffold and to place the adhesive on the back of the tile ready for the mechanic.

If the work is performed by inexperienced men, the labor costs as given, may be doubled.

Quantities and Labor Costs of Applying Fiber Wall Tile to Walls and Ceilings Using Adhesive and Nails Where Necessary

Add for cleaning plaster or other work necessary for applying tile.

The quantities and costs given in the following table are based on using a 4 or 5-ft. scaffold consisting of horses and planks and laying the tile in plain squares or common bond.

Size of Tile in Inches	No. Sq. Ft. per Tile	No. Tile Req'd per 100 Sq. Ft.	No. Placed per 8-Hr. Day	No. Sq. Ft. per 8-Hr. Day	Hrs. per 100 Sq. Ft.
12x12*	1.0	100	165–185	165–185	4.5
12x24*	2.0	50	88– 98	175–195	4.4
16x16*	1.78	56¼	112–125	200–220	3.9
16x32*	3.56	28¼	65– 70	230–250	3.3

*Standard sizes

If tile are laid in herringbone pattern, add 30 to 40 percent to the above labor costs; if laid in diagonal patterns, increase the above labor costs 20 to 25 percent, and if laid in mixed ashlar designs, add about 50 percent to the above labor costs.

When placing tile in auditoriums and other spaces requiring high scaffolding, add cost of scaffolding.

Large jobs are applied faster proportionately than small jobs.

Presdwood, Hardboard, Panelboard, Temwood.—There are a number of makes of board made of wood fiber and pressed into board form under flat bed hydraulic presses that are much thinner and possess much greater strength than ordinary wallboard and are used for both interior and exterior work.

These hardboards are used for interior finish of walls and ceilings, cupboards, lining of closets, clothes chutes and similar purposes.

They are usually furnished 1/8", 3/16", 1/4" and 5/16" thick, 4 ft. wide and 4 to 12 ft. long, although they vary somewhat among different manufacturers.

Tempered hardboard, Temwood or Presdwood are the same materials as described above, except they have been subjected to a tempering treatment which further increase the strength, resistance to abrasion, durability and ease of painting. Tempered hardboards, ¼" and 5/16" thick are used for the exterior of homes and other small buildings. It is also used quite extensively for concrete form work.

These hardboards or tempered boards are in no way comparable with ordinary wallboard, as it is a far more dense material, having greater tensile and transverse strength, greater resistance to abrasion, lower moisture absorption, and when properly applied, no warping or buckling.

The labor cost of placing hardboard will run approximately the same as placing other types of board under the same conditions.

Method of Placing Tileboard in Bathrooms, Kitchens, etc.

Approximate Square Foot Prices of Hardboard.

Untempered,	Plain	⅛" Thick	$0.12
Untempered,	Plain	¼" Thick	.17
Untempered,	Pegboard	⅛" Thick	.15
Untempered,	Pegboard	¼" Thick	.20
Tempered	Plain	⅛" Thick	.14
Tempered	Plain	¼" Thick	.22
Tempered	Pegboard	⅛" Thick	.16
Tempered	Pegboard	¼" Thick	.23
Tempered	Plastic Face	⅛" Thick	.70
Tempered	Plastic Face	¼" Thick	.78
Tempered	Pl. F. Pegb'd	⅛" Thick	.70
Tempered	PL. F. Pegb'd	¼" Thick	.80

Tileboard

Tileboard has score lines impressed in one surface forming 4"x4" squares. The board is denser than Tempered Hardboard and has a lower rate of water absorption.

This material is used for walls in bathrooms, kitchens, lavatories, restaurants, barber shops and similar places. It is furnished in sheets 4'-0" wide and 2'-0" to 12'-0" long.

Tempered board or tileboard should be cemented to the walls, using Armstrong's Panelboard Cement or a cement of equal quality. The adhesive should be applied to the back of the boards with a saw-tooth trowel, the boards having first been cut to fit. Use supplementary nailing or shoring to hold the boards firmly until the adhesive takes a partial set. Where it is necessary to nail the board to the walls, small brads should be used where scored lines intersect.

Figure about 1 gal. waterproof cement per 50 sq. ft. of wall.

Presdwood Temprtile ⅛" thick costs about 20 cts. per sq. ft. unpainted.

Prefinished Tileboard.—Tileboard 5/32" thick, made of masonite tempered Presdwood, is furnished scored 4"x4" to represent tile and finished with multiple coats of enamel or lacquer. Tileboard is also furnished without scoring, if desired.

These boards are furnished with dull or velvet finish, or a high gloss finish in a variety of colors with contrasting joints.

Tileboard is furnished in sheets 4 ft. wide and 4, 5, 6, 7, 8, 9, 10 and 12 ft. long.

Tileboard having a semi-gloss or high gloss finish costs about 55 to 65 cts. per sq. ft. depending upon size of board.

Cap molding 1¾-in. wide for use at the top of the wainscoting is furnished in strips 4 and 8-ft. long, scored to match tile or unscored, costs 30 cts. per lin. ft.

Base molding 3¾-in. high, scored or unscored, in colors to match tileboard costs 36 cts. per lin. ft.

Placing Tempered Tileboard.—Where tempered tileboard is used in kitchens, for sink splashboards, bathrooms, shower stalls, and other small places, there is usually considerable cutting and fitting required to place a comparatively small square footage of board.

The safest method of estimating is to lay out each wall, showing number and sizes of pieces required, as it will cost almost as much to place a small panel as a large one.

Where the tileboard is secured by nails (without the use of adhesive), figure ¾-hr. to 1 hr. to fit and place each panel. Taking an average of 12 to 16 sq. ft. per panel, this will equal 125 to 150 sq. ft. per 8-hr. day.

However, if the tileboard is applied to walls, using an adhesive, a man should fit and place one panel in 1 to 1¼-hrs. or 100 to 125 sq. ft. per 8-hr. day.

This applies to both plain and prefinished tileboard.

One gallon of adhesive will cover about 100 sq. ft. of surface if "spot" glued or about 65 sq. ft. of surface if "spread" glued.

Price per Gallon...$6.50

Labor Cost of 100 Sq. Ft. Tempered Tileboard, Secured
to Walls With Nails

	Hours	Rate	Total	Rate	Total
Carpenter	4	$. . . .	$. . . .	$16.47	$65.88
Cost per sq. ft			. . . .		.66

Labor Cost of 100 Sq. Ft. Tempered Tileboard, Secured
to Walls Using Adhesive

	Hours	Rate	Total	Rate	Total
Carpenter	3.5	$. . . .	$. . . .	$16.47	$57.65
Cost per sq. ft			. . . .		.58

Wainscot Trim for Wallboard and Wall Tile

Moldings made of stainless steel, polished finish, in plain designs similar to illustrations. Punched for nails. Furnished in 4'-0", 6'-0" and 8'-0" lengths.

Description	For Wall-board	Per Ft.	Description	For Wall-board	Per Ft.
Cap Mold	1/8" thick	$0.20	Outside Corner	1/8" thick	$0.30
Joint Mold	1/8" thick	.25	Inside Corner	1/8" thick	.30
Tub edging or cove base				1/8" thick	.30

Cap Mold   Joint Mold   Outside Corner   Inside Corner

Metal covered wood cap, base or trim, size 5/16" x 11/2", chrome zinc, polished finish. Price per lin. ft $0.30

Labor Placing Metal Moldings.—The labor cost of placing metal moldings will vary with the design and size of the panels, cutting and fitting necessary, etc., but on the average job, a carpenter should place 100 to 150 lin. ft. per 8-hr. day, at the following labor cost per 100 lin. ft.:

	Hours	Rate	Total	Rate	Total
Carpenter	6	$....	$....	$16.47	$98.82
Cost per lin. ft			....		.99

Items to Include When Making Up Wallboard Estimates.—When making up estimates on wallboard construction the following items should be included to insure a complete and dependable estimate:

1. Headers or nailing strips to receive wallboard.
2. Furring strips on walls and ceilings where necessary.
3. Wallboard and labor placing same.
4. Wood or metal panel strips and labor placing same.
5. Labor filling nail holes with wood putty or crevice filler.
6. Beam ceilings, ceiling cornices, metal moldings, etc.
7. Painting and decorating wallboard and panel strips.
8. Nails and adhesive.
9. Overhead expense and profit.

06430   STAIR WORK

Framing and Erecting Wood Stairs of Unassembled Stock Material.—When framing and erecting ordinary wood stairs having closed stringers where the treads and risers are not housed out, it will require 3 to 4 hrs. carpenter time to lay out the work, cut out for treads and risers and place wood stringers for a flight of ordinary wood stairs up to 4'-0" wide and 11'-0" story height, at the following labor cost:

	Hours	Rate	Total	Rate	Total
Carpenter	4	$....	$....	$16.47	$65.88
Labor	2	....	....	12.54	25.08
			$....		$90.96

After the wood stringers are in place, it will require 4 hrs. carpenter time to place treads and risers for each flight of stairs (providing it is a plain box stair without nosings to form and without newels, balusters or handrail), at the following labor cost per flight:

	Hours	Rate	Total	Rate	Total
Carpenter	4	$....	$....	$16.47	$65.88

Figure 3 hrs. carpenter time to place 4"x4" newels, 2"x4" rails and 1"x2" balusters for one side of this type stairs, at the following labor cost per flight:

	Hours	Rate	Total	Rate	Total
Carpenter	3	$....	$....	$16.47	$49.41

To lay out the work and erect the stairs complete, including plain wall hand rail secured by brackets (instead of newels, hand rails and balusters), figure 8 hrs. carpenter time, at the following labor cost per flight:

	Hours	Rate	Total	Rate	Total
Carpenter	8	$....	$....	$16.47	$131.76

For the above stair complete, including newels, handrail and balusters, one side, figure 10 hrs. carpenter time, at the following labor cost per flight:

	Hours	Rate	Total	Rate	Total
Carpenter	10	$....	$....	$16.47	$164.70

Erecting Wood Stairs Made in Shop.—Where wood stairs are made in the shop with stringers housed out to receive treads and risers, it will require 8 hrs. stair builder's time to lay out the work, set stringers, treads and risers (with plain wall rail attached to brackets), at the following labor cost per flight:

	Hours	Rate	Total	Rate	Total
Stair builder	8	$....	$....	$16.47	$131.76
Labor	2	....	....	12.54	25.08
			$....		$156.84

If the stair consists of two short flights, having an intermediate landing platform between floors, it will require 10 to 12 hrs, stair builder's time to lay out work, place stringers, treads, risers and plain wall handrail, at the following labor cost:

	Hours	Rate	Total	Rate	Total
Stair builder	11	$....	$....	$16.47	$181.17
Labor	2	....	....	12.54	25.08
			$....		$206.25

Where a plain box stair (3'-0" to 4'-0" wide and 9'-0" to 10'-0" story height) with open stringers is used, it will require 10 to 12 hrs. stair builder's time to lay out work, place stringers, treads and risers (not including newels, balusters or handrail) for each flight of stairs, at the following labor cost:

	Hours	Rate	Total	Rate	Total
Stair builder	11	$....	$....	$16.47	$181.17
Labor	2	....	....	12.54	25.08
			$....		$206.25

If the stairs have newels, wood handrail and balusters, add 6 to 8 hrs. stair builder's time, at the following labor cost per flight:

	Hours	Rate	Total	Rate	Total
Star builder	7	$....	$....	$16.47	$115.29

Plain box stairs consisting of two short flights having an intermediate landing platform between stories will require 6 to 8 hrs. stair builder's time to lay out the work, place stringers, treads and risers for each short flight of stairs at the following labor cost per flight:

	Hours	Rate	Total	Rate	Total
Stair builder...	7	$....	$....	$16.47	$115.29
Labor...	2	....	....	12.54	25.08
			$....		$140.37

To place newels, balusters and handrail on each short flight of stairs will require about 4 hrs. stair builder's time, at the following labor cost per flight:

	Hours	Rate	Total	Rate	Total
Stair builder.....................................	4	$....	$....	$16.47	$65.88

The labor cost complete for two short flights of stairs, including newels, handrails and balusters should cost as follows per story height:

	Hours	Rate	Total	Rate	Total
Stair builder.....................................	22	$....	$....	$16.47	$362.34

Erecting Wood Stairs Having Open Stringers.—Wood stairs of the open stringer type having treads with a return nosing projecting beyond the face of the stringer, will require about 12 to 14 hrs. stair builder's time to lay out the work and place stringers, treads and risers for one flight of stairs containing 16 to 18 risers, at the following labor cost per flight:

	Hours	Rate	Total	Rate	Total
Stair builder.....................................	13	$....	$....	$16.47	$214.11
Labor...	2	....	....	12.54	25.08
			$....		$239.19

If necessary to place newels, handrails and baluster, add about 8 to 9 hrs. stair builder's time, at the following labor cost per flight:

	Hours	Rate	Total	Rate	Total
Stair builder.....................................	8.5	$....	$....	$16.47	$139.99

If the stair consists of two short flights having an intermediate landing platform between stories (each short flight contains 8 to 10 risers), it will require about 8 to 9 hrs. stair builder's time per flight or 16 to 18 hrs. per story, at the following labor cost:

	Hours	Rate	Total	Rate	Total
Stair builder.....................................	17.0	$....	$....	$16.47	$279.99
Cost per short flight.........................	8.5		....		139.99

To set starting and landing newels, handrails and wood balusters will require 7 to 9 hrs. stair builder's time per flight or 14 to 18 hrs. per story, at the following labor cost:

	Hours	Rate	Total	Rate	Total
Stair builder.....................................	16	$....	$....	$16.47	$263.52
Cost per short flight.........................	8		....		131.76

The above does not include carpenter time framing and erecting intermediate landing platform.

Placing Wood Handrail on Metal Balustrades.—A carpenter should place 45 to 55 lin. ft. of wood handrail on metal balustrades per 8-hr. day at the following labor cost per 100 lin. ft.:

	Hours	Rate	Total	Rate	Total
Carpenter..............................	16	$....	$....	$16.47	$263.52
Cost per lin. ft			....		2.64

Placing Wood Handrail on Wall Brackets.—A carpenter should set brackets and install 65 to 70 lin. ft. of wall hung wood handrail per 8-hr. day at the following labor cost per 100 lin. ft.:

	Hours	Rate	Total	Rate	Total
Carpenter	12	$. ...	$. ...	$16.47	$197.64
Cost per lin. ft			. . . .		1.98

Inside Stair Material

Starting Steps

Quarter circle	Red Oak	4' long	$32.00
Half circle	Red Oak	4'-6" long	40.00
Bull nose	Red Oak	4'-6" long	36.00
Scroll end	Red Oak	4'-6" long	40.00

Stringers (not housed)

3/4"x11 1/4" x 8'-0" long	Red Oak	$26.00
x10'-0" long	Red Oak	32.00
x12'-0" long	Red Oak	48.00
x14'-0" long	Red Oak	54.00
x16'-0" long	Red Oak	62.00

Treads (not returned; for returns add $1.75)

1 1/6"x10 1/2" x3'-0" long	Red Oak	$10.00
x3'-6" long	Red Oak	11.50
x4'-0" long	Red Oak	13.00
1 1/16"x11 1/2" x3'-0" long	Red Oak	13.00
x3'-6" long	Red Oak	15.00
x4'-0" long	Red Oak	17.50
x5'-0" long	Red Oak	22.00
x6'-0" long	Red Oak	27.00

Return Nosing

11/8"x13/4"x1'-2" with 2" return	Red Oak	$ 4.50

Tread Nosing

11/8"x13/4"x1'-2" long	Red Oak	1.80
11/16"x11/8"x1'-2" long	Red Oak	1.65

Landing Tread Nosing

11/16"x31/2"x3'-0" long	Red Oak	2.40
x3'-6" long	Red Oak	2.70
x4'-0" long	Red Oak	3.00

Tread Brackets

12"x73/4"x1/4" 3 ply	Nat. Birch	1.75

Newel Posts

	Birch	Red Oak
31/4"x31/4"x3'- 2" tapered starting—	$14.00	$13.00
x3'-10" tapered, starting—	16.00	16.00
x5'- 3" tapered, landing—	22.00	20.00
x7'- 0" tapered, angle—	26.00	24.00

Balusters

		Birch	Red Oak
15/16" round	30" L.	$ 2.15	$ 1.90
or square	33"	2.25	2.00
	36"	2.35	2.10
	39" or 42"	2.65	2.40

Railings—Natural birch, random length

13/4"x15/8"	$ 1.70
13/4"x111/16"	1.95
25/8"x111/16"	2.10
21/4"x23/8"	4.25

Rail Bolts and Plug   .35

Inside Stair Materials

Starting Steps and Colonial Newels

Prefabricated Folding Stairways

For 9' finished floors to finished ceiling heights:

Premium grade: for 2'-6"x6'-0" opng $250.00
Medium grade: for 2'-2"x4'-6" opng... 75.00
Economy grade: for 2'-2"x4'-6" opng 50.00

Railing Fittings

| Starting | Volute | Turned Balusters |
| Newels | Newels | |

Turned Colonial Stair Newels and Balusters

Prefabricated Spiral Stairs.—Wood stairs are available in 3'6", 4', 4'6" and 5' diameters with heights made to fit the actual job conditions. The balusters and center column are steel, but the treads, platforms and railings can be wood to give a finished appearance. Several styles are available and the cost runs from $850.00 for the smallest diameter and the simplest design in an 8' rise to $1250.00 for the 5' diameter. The more elaborate designs will run from $1125.00 to $1500.00 for the same sizes. The stairs are factory assembled and finished so that two carpenters should be able to install the unit in one day.

CHAPTER **12**

THERMAL AND MOISTURE PROTECTION

CSI   DIVISION 7
07100 WATERPROOFING
07150 DAMPPROOFING
07200 INSULATION
07300 SHINGLES AND ROOFING TILES
07400 PREFORMED ROOFING AND SIDING
07500 MEMBRANE ROOFING
07600 FLASHING AND SHEET METAL
07800 ROOF ACCESSORIES
07900 SEALANTS

07100   WATERPROOFING

Nearly every building requires water and damp resisting material, plaster bonds or floor hardeners of one kind or another, and with the great number of preparations on the market, it is impossible for the best informed contractor or estimator to familiarize himself with all of them.

No attempt has been made here to pass on the relative merits of the various preparations, but this data deals strictly with information that is of interest to the estimator, namely,—use of the material, covering capacity, and cost of applying same. The prices quoted are approximate, and may vary considerably in different locations. Always check with local distributors for current prices.

Water and Damp Resisting Methods.—There are several methods used by the different manufacturers for mixing and applying their products to produce a water resisting wall or surface. In some instances, the water repelling compounds are incorporated with the cement, sand, and gravel while the concrete is being mixed, and by filling the voids in the cement and sand produce a waterproof mixture. Where the various powders or pastes are mixed in the concrete mass it is called the integral method of waterproofing.

Some water and dampness resisting preparations are furnished in mastic form, and are applied with a trowel. Others are mixed with portland cement and sand and applied as a plaster coat to the surfaces to be treated.

There are also numerous water and dampness resisting compounds in the form of heavy paints, which are applied to the concrete or masonry walls, floors, and ceilings with a brush or mop. One or more applications of these materials are supposed to penetrate and fill the pores in the concrete or masonry surfaces to such an extent that the treated surfaces will be impervious to dampness or moisture.

Where there is considerable pressure or a large head of water, one of the most satisfactory methods is known as membrane waterproofing. This method consists of applying 2 to 7 hot moppings of pitch or asphalt on 1 to 7 plies of saturated felt or fabric. Waterproofing of this kind is applied in much the same manner as built-up roofing. After the last ply of saturated felt or fabric is applied, the entire surface is mopped with a heavy application of hot pitch or asphalt.

Estimating Quantities of Water and Damp Resisting Preparations.—When estimating quantities of the various plaster coats, paints and membrane waterproofing compounds required for any job, the entire area of the walls or surfaces to be treated should be measured and the quantities stated in sq. ft., sq. yds. or by the square (100 sq. ft.).

Integral water resisting compounds are usually estimated by the cu. ft. or cu. yd. of concrete to which the liquid, paste, or powder is added.

Water Required for Mixing Concrete.—When mixing concrete sufficient water should be used to produce a plastic mix—no more and no less. A plastic mix is concrete which can be readily molded and when the mold is removed will flow sluggishly without segregation of the water or the fine materials from the coarse. Concrete should be mixed at least 2 minutes after all materials have been placed into the mixer.

Remember there is nothing more injurious to good concrete than too much mixing water.

Refer to water required for mixing concrete as given in Step 4. "Selection of water-cement ratio", under Cast-in Place Concrete.

WATERPROOFING BY THE INTEGRAL METHOD

When using the integral method, the admixtures are usually mixed with the cement and aggregate and form a part of the concrete mass. The quantities are based on the specifications of the various manufacturers.

Material prices given are approximate, as it is impossible to quote prices that will apply to every locality. The contractor or estimator should obtain local prices on the materials specified. Prices are based on material being purchased in drum or barrel quantities. When purchased in one gallon or one pound quantities the cost will be increased considerably.

Integral Liquid Admixtures

Integral liquid admixtures consist of (1) calcium chloride solutions, (2) oil water repellent preparations and (3) other integral mixtures which flow freely from a drum.

The cost of hauling the barrel to the mixer, setting it up on blocks and drawing off the liquid by spigot into a measuring container will cost from 10 to 15¢ per gallon of waterproofing on small jobs. On large jobs this cost may be halved.

Concrete mixed in the proportions of 1 :2 :4 or 1 :2½ :3½ requires 5½ to 6 sacks of cement per cu. yd. of concrete.

Trade Name	Quan. per Sack Cement	Concrete Proportion	Quan. per Cu. Yd. Concrete	Approx. Price per Gallon
Anti-Hydro*†	1 qt.	1 :2 :4	1½ gal.	$3.00
Aquatite	2/3 qt.	1 :2 :4	1 gal.	1.50
Aquatite*	1 qt.	1 :2 :4	1½ gal.	1.50
Hydratite	1 qt.	1 :2 :4	1½ gal.	2.25
Toxement IW	1 qt.	1 :2 :4	1½ gal.	2.20
Sikacrete	1 qt.	1 :2 :4	1½ gal.	3.00
Trimix	1 qt.	1 :2 :4	1½ gal.	1.30

* Use where there is water pressure.
† A 5-year maintenance guarantee furnished by manufacturer where supervision is supplied at additional cost.

Powdered Integral Admixtures

This material should be added to the portland cement before the addition of aggregates and then dry mixed thoroughly before adding gauging water. Add about 10 to 15¢ per lb. to the cost of material for labor handling and mixing.

Trade Name	Quan. per Sack Cement	Concrete Proportion	Quan. per Cu. Yd. Concrete	Approx. Price per Lb.
Hydratite Powder	1 lb.	1 :2 :4	6 lbs.	$ .25
Hydrocide Powder	1 lb.	1 :2 :4	6 lbs.	.25
Plastiment	½ to 1 lb.	1 :2 :4	3 to 6 lbs.	.48
Toxement IW	1½ lbs.	1 :2 :4	7½ lbs.	.35

WATER AND DAMP-RESISTING PLASTER COATS

There are numerous preparations on the market to be mixed with portland cement and sand and applied as a plaster coat for waterproofing concrete and masonry walls. There are other preparations in mastic form that are applied with a trowel, the same as ordinary cement mortar.

The materials described on the following pages include many of the better known preparations used for this purpose.

Estimating Quantities of Water or Damp-Resisting Plaster Coats.—Where water or damp-resisting plaster coats are applied to brick or concrete walls, the water resisting ingredient is usually in liquid, paste or powder form.

This compound is usually added to the mortar on the basis of a certain quantity to each sack of cement used and practically all manufacturers specify a mortar mixed in the proportions of 1 part portland cement to 2 parts sand, adding only sufficient water to make a workable mortar.

Quantity of Cement and Sand Required for One Cu. Yd. of Cement Mortar

Based on damp loose sand containing 5 percent of moisture and weighing 2,565 lbs. per cu. yd.

Proportions by Volume Cement Sand	Packed Cement Sacks	Loose Sand Cu. Yds.	Proportions by Volume Cement Sand	Packed Cement Sacks	Loose Sand Cu. Yds.
1 :1	18.0	0.70	1 :2½	11.0	1.03
1 :1½	15.2	0.84	1 :3	9.5	1.10
1 :2	12.5	0.95			

Above quantities of sand include 5 percent for waste.

Number of Sq. Ft. of Cement Plaster Coat Obtainable from One Cu. Yd. of Cement Mortar

Thickness in Inches	Square Feet	Thickness in Inches	Square Feet
¼	1,296	¾	432
½	648	1	324

Cleaning and Washing Old Masonry or Concrete Surfaces.—Where old masonry or concrete surfaces are to be given a waterproof plaster coat, it will first be necessary to prepare the old surface to receive the cement plaster.

On some jobs it will be necessary to rake out the old brick joints to form a key for the plaster, and on others it will be necessary to thoroughly wire brush the old surface to remove all surface dirt and then wash the surface with clean water or a mixture of chemical cleaner and water.

Where the surfaces are to be brushed and washed with water, the labor cost of cleaning 100 sq. ft. of surface should average as follows :

	Hours	Rate	Total	Rate	Total
Labor	1.5	$....	$....	$12.54	$18.81
Cost per sq. ft.			....		.19

Roughing and Hacking Old Concrete Walls and Other Vertical Surfaces.— Where it is necessary to roughen old, hard concrete surfaces to prepare a suitable bond for the new waterproof plaster coat, this can prove to be very expensive especially where the old concrete surfaces are very hard and must be hacked and roughened using hand tools, a jackhammer, or bush-hammering tool as the men often work in cramped quarters, which prevents the best daily output.

On work of this kind, a man will roughen about 20 sq. ft. of surface per hour. The following table indicates the labor cost per 100 sq. ft. to roughen and clean the surface:

	Hrs.	Rate	Total	Rate	Total
Labor on Jackhammer	5	$....	$....	$12.54	$62.70
Labor for cleaning	2	....	....	12.54	25.08
Cost per 100 sq. ft.			$....		$87.78
Cost per sq. ft.			....		.88

Cost of tools (purchase or rental), fuels, etc. will have to be estimated, in addition to the labor requirements and will vary largely on the accessibility, depth to be roughened, and the quantity of work. For example: Light work or small areas can be accomplished with an electric bush-hammer, while heavier work will require an air compressor, hose and light jackhammer.

Equipment rental and fuel costs for an air compressor and tools to roughen the 100 sq. ft. of surface (as given in the preceeding table) will be as follows:

	Hrs.	Rate	Total	Rate	Total
Air Compressor and Tool Rental	5	$....	$....	$7.00	$35.00
Fuel Costs 10 gal.		....	....	1.35	13.50
Cost per 100 sq. ft.			$....		$48.50
Cost per sq. ft.			....		.49

Applying Slush or Grout Coat of Neat Cement.—After the old surfaces have been cleaned, it will be necessary to give them a slush coat of neat portland cement thoroughly brushed into the pores of the wall, which acts as a bond for the subsequent plaster coats.

The labor cost per 100 sq. ft. of surface should run as follows :

	Hours	Rate	Total	Rate	Total
Labor	2	$....	$....	$12.54	$25.08
Cost per sq. ft.			....		.25

The material needed for a slush coat of neat portland cement will require one bag of portland cement, mixed with water, per 100 sq. ft. of surface to be coated. A bag of cement costs about $4.00 each which amounts to the following material costs:

Per 100 sq. ft. - $4.00 or per sq. ft. - $.04

Labor Applying Waterproof Plaster Coats to Vertical Surfaces.—When applying waterproof plaster coats to vertical surfaces, the costs run high for the reasons listed below:

1. Interference of other trades.
2. Difficulty of mixing and handling.

3. Delay in setting of coats.
4. Difficulty in finishing vertical surfaces of this type with plaster or cement finish.
5. Contingency cost of return to job for breaks and cracks.
6. Waste and spillage of materials.

This work is usually performed by cement masons, plasterers, or bricklayers.

When applying waterproof plaster coats it is not necessary to give the surface a float or brush finish, but it must be reasonably smooth and without uneven projections that would prevent the water from running down the sides of the walls.

Most manufacturers specify the mortar to be mixed in the proportions of 1 :1½ or 1 :2, and the material is usually applied in one or two coats.

A man should apply about 350 sq. ft. of first coat per 8-hr. day, at the following labor cost per 100 sq. ft. :

	Hours	Rate	Total	Rate	Total
Plasterer	2.3	$....	$....	$16.35	$37.61
Helper	2.3	....	....	13.77	31.67
Cost per 100 sq. ft.			$....		$69.29
Cost per sq. ft.			....		.69

On the second or finish coat, a man should apply about 400 sq. ft. per 8-hr. day, at the following labor cost per 100 sq. ft.:

	Hours	Rate	Total	Rate	Total
Plasterer	2.0	$....	$....	$16.35	$32.70
Helper	2.0	....	....	13.77	27.54
Cost per 100 sq. ft.			$....		$60.24
Cost per sq. ft.			....		.60

If necessary to clean the old masonry or concrete surfaces, or to roughen the old concrete, add for this work as given on previous page.

Water Resisting Plaster Coats Using Powdered Intregal Water Repellents, Per 100 Sq. Ft.

Trade Name	Inches Thick	Mix Prop.	Number Coats	Sack Cement	Cu. Ft. Sand	Quantity Waterp'g	Approx. Price per lb.	Mechanic Hours	Labor Hours
Hydratite	1	1 :2	2	4	8	8 lbs.	$0.23	8	8
Hydrocide Powder	1	1 :2	2	4	8	4 lbs.	.22	8	8
Toxement IW	¾	1 :2	2	3	6	4½ lbs.	.30	7	7

Water Resisting Plaster Coats Using Liquid Intregal Admixtures, Per 100 Sq. Ft.

Trade Name	Inches Thick	Mix Prop.	Number Coats	Sack Cement	Cu. Ft. Sand	Quantity Waterp'g	Approx. Price per gal.	Mechanic Hours	Labor Hours
Anti-Hydro	¾	1 :2	2	3	6	¾ gal.	$3.00	7	7
Anti-Hydro	1	1 :2	2	4	8	1 gal.	3.00	8	8
Aquatite	¾	1 :2	2	3	6	¾ gal.	1.50	7	7
Aquatite	1	1 :2	2	4	8	1 gal.	1.50	8	8
Hydratite	¾	1 :2	2	3	6	¾ gal.	2.25	7	7
Hydratite	1	1 :2	2	4	8	1 gal.	2.25	8	8

Trade Name	Inches Thick	Mix Prop.	Number Coats	Sack Cement	Cu. Ft. Sand	Quantity Waterp'g	Approx. Price per gal.	Mechanic Hours	Labor Hours
Toxement IW.........	¾	1 :2	2	3	6	¾ gal.	2.20	7	7
Toxement IW.........	1	1 :2	2	4	8	1 gal.	2.20	8	8
Sikacrete.................	¾	1 :2	2	3	6	¾ gal.	3.00	7	7
Trimix......................	¾	1 :2	2	3	6	¾ gal.	1.30	7	7

MEMBRANE WATERPROOFING

Membrane waterproofing is constructed in place by building up a strong, water-proof and impermeable blanket with overlapping plies of tar or asphalt saturated open mesh cotton fabric or rag felt. The plies are coated and cemented together with hot coal tar pitch or waterproofing asphalt. There is always one or more applications of pitch or asphalt than plies of felt or fabric.

Estimating the Quantity of Felt or Fabric Required for Membrane Water-proofing.—When estimating the quantity of felt or fabric required for waterproof-ing masonry or concrete walls, pits, floors, etc., it will be necessary to add a certain percentage for lap and waste. If the felt is applied to walls, it will be necessary to allow about 6" to lap over the footings at the bottom, also a small allowance at the top of the walls above grade. This will average from 7½ to 10 percent of the total area to be waterproofed.

If the walls or floors are covered with a single thickness of felt or membrane, it will be necessary to lap each strip of felt 4" to 6", depending upon the specifica-tions. As most felts are put up in 36" rolls it will be necessary to add about 12 percent for waste for a 4" lap, and 17 percent for a 6" lap.

If the lap is 4" wide and only a single thickness of felt is used, it will be necessary to add about 20 percent to the actual area to allow for laps and the additional felt required at the top and bottom of the wall. On floors it is customary to run the felt or membrane up the sides of the walls 4" to 6".

If the lap is 6" and only a single thickness of felt is used, add about 25 percent to the actual wall or floor area for laps, etc.

When 2, 3, 4 or more plies of felt are used, it will not be necessary to allow for laps except at the ends of the wall or floor, as one thickness of felt overlaps the next without waste. However, extra material will be required at the tops and bottoms of the walls and where the paper runs up the sides of the walls 4" to 6". On work of this kind, add about 10 percent to the actual floor or wall area.

Fabric or Felt Required to Cover 100 Sq. Ft. of Surface

Class of Work	1 Ply 4" lap	1 Ply 6" lap	2-Ply	3-Ply	4-Ply	5-Ply	Ea. Add. Ply
Add percentage	20	25	10	10	10	10	10
Sq. Ft. Felt......................	120	125	220	330	440	550	110

WATERPROOFING

Weight of Tar or Asphalt Felt for Membrane Waterproofing.—Tar or as-phalt felt for waterproofing is currently furnished in 4 sq. rolls of 432 sq. ft. weigh-ing 60 lbs. per roll.

Double thickness asphalt felt is also furnished in 60 lb. per roll containing 216 sq. ft. and this felt is known as No. 30.

When specifying the grade or weight of felt to be used, it is customary to state that "felt shall weigh not less than 15 lbs. per 108 sq. ft." This is known as No. 15 felt. Inasmuch as felt is furnished in 4 sq. rolls of 432 sq. ft. there are 32 sq. ft. per roll or 8 sq. ft. per 100 sq. ft. allowed for laps.

Tar or asphalt saturated fabric is usually sold by the roll containing 50 sq. yds. or by the sq. yd.

Estimating the Quantity of Pitch or Asphalt Required for Membrane Waterproofing.—The quantity of pitch or asphalt required for mopping 100 sq. ft. of surface with one application is approximately 30 to 35 lbs.

Many special coatings are sold under trade names. Some are furnished solid, making it necessary to heat and melt them before using, while others are furnished in the form of a heavy black paint that can be either mopped or brushed on cold. The manufacturers usually specify how their materials shall be applied.

Engineers and architects usually specify the quantity of asphalt or compound to be used for each 100 sq. ft. of surface.

Weight of Pitch or Asphalt Per 100 Sq. Ft. of Surface

No. Plies of Saturated Fabric or Tarred Felt	Alternate Moppings of Pitch or Asphalt Required	Lbs. of Pitch or Asphalt Required 100 Sq. Ft. Waterproofing
2	3	90 to 105
3	4	120 to 140
4	5	150 to 175
5	6	180 to 210
6	7	210 to 245

Applying Membrane Waterproofing.—Whether to use saturated fabric or saturated felt in constructing membrane waterproofing depends entirely upon local conditions. Saturated fabric weighing 12 oz. per sq. yd. has approximately three times the tensile strength of 15-lb. saturated felt, but it costs considerably more per sq. yd. than saturated felt. On flat surfaces where the waterproofing will be covered with a protection course immediately after it is applied and where it will not be subjected to any unusual strain, saturated felts can be used. On any surface where greater strength is required, fabric should be used. It is always desirable to provide a protection course over the waterproofing construction. While the labor operations involved in applying membrane waterproofing are similar to those followed in applying built-in roofing, the work is usually more expensive than roofing because of the limited working space and the necessity for using scaffolds on vertical surfaces over 8 feet high.

On vertical surfaces the work can be expedited if the plies of felt or fabric are installed up and down the surface like wallpaper rather than across. The pieces of felt or fabric can be cut to the required length and folded so that after the surface has been mopped the top edges can be pushed into the hot bitumen and the remainder of the piece will then fall into place where it can be rubbed into the hot pitch or asphalt. After the required number of plies have been embedded in alternate moppings of hot bitumen all angles at corners, walls, etc., should be reinforced with at least two additional plies of felt or fabric and alternate moppings of bitumen. A protection course of beadboard, fiberboard, or other material should immediately be placed over the waterproofing for best results.

Prior to the application of the first hot bitumen mopping coat, the wall surface is primed, uniformly and completely, with one gallon of primer (asphalt primer for asphalt specifications or tarbase primer for pitch specifications) per 100 sq. ft. of surface area.

For the application of the primer ot the surface, one man should coat about 100 sq. ft. per hour, at the following labor cost:

	Hours	Rate	Total	Rate	Total
Labor	1.0	$. . . .	$. . . .	$16.46	$16.46
Cost per sq. ft.					.17

Labor Applying Membrane Waterproofing.—When applying membrane waterproofing to walls and floors, three or four men usually work together, as it requires one man to attend the fire and heat the bitumen, while two or three men are mopping the walls and applying the felt or fabric.

If only one application of hot bitumen is applied, two men working together should heat materials and mop 2,000 to 2,200 sq. ft. per 8-hr. day, at the following labor cost per 100 sq. ft. :

	Hours	Rate	Total	Rate	Total
Labor	0.8	$. . . .	$. . . .	$16.46	$13.17
Cost per sq. ft.			. . . .		.13

Water Pressure Table and Waterproofing Required for Varying Heads of Water

Hydrostatic Head in Feet	Pressure per Sq. In. in Pounds	Lifting Pressure per Sq. Ft., Lbs.	Wall Pressure per Sq. Ft., Lbs.	Plies of Saturated Felt or Fabric	Asphalt Mastic Thickness
0.5	0.21	39.2	15.6	2-Ply	⅛ inch
1.0	0.43	62.5	31.2	2-Ply	⅛ inch
2.0	0.86	125.0	62.5	2-Ply	¼ inch
3.0	1.30	187.5	93.7	2-Ply	¼ inch
4.0	1.73	250.0	125.0	3-Ply	⅝ inch
5.0	2.17	312.5	156.2	3-Ply	⅝ inch
6.0	2.60	375.0	187.5	4-Ply	⅝ inch
8.0	3.47	500.0	250.0	4-Ply	⅝ inch
10.0	4.34	625.0	312.5	5-Ply	⅝ inch
12.0	5.21	750.0	375.0	6-Ply	⅝ inch
15.0	6.51	937.5	468.7	6-Ply	¾ inch
20.0	8.68	1250.0	625.0	7-Ply	¾ inch
25.0	10.85	1562.5	781.2	8-Ply	¾ inch
30.0	13.02	1875.0	937.5	10-Ply	¾ inch
40.0	17.36	2500.0	1250.0	11-Ply	¾ inch

After the surface has been mopped with hot bitumen, it will be necessary for two men to cut the felt to lengths and place it on the wall. Usually it is cut long enough to lap over the top of the wall a few inches, so that it can be held in place by brick or stone until it is self-supporting.

Two men placing felt or fabric have to work along with the men mopping, so they will place felt or fabric on 2,000 to 2,200 sq. ft. of surface per 8-hr. day, at the following labor cost per 100 sq. ft. :

	Hours	Rate	Total	Rate	Total
Labor	0.8	$. . . .	$. . . .	$16.46	$13.17
Cost per sq. ft.			. . . .		.13

Labor Applying 100 Sq. Ft. of Membrane Waterproofing

Number Plies	Description of Work	Mopping Hours	Felt Hours	Total Hours
1	1-ply fabric, 2 moppings	1.6	0.8	2.4

2	2-ply fabric, 3 moppings	2.4	1.6	4.0
3	3-ply fabric, 4 moppings	3.2	2.4	5.6
4	4-ply fabric, 5 moppings	4.0	3.2	7.2
1	Each additional ply of felt and mopping	0.8	0.8	1.6

Prices of Waterproofing Materials

Prices on waterproofing materials must be checked by the estimator, as prices vary considerably due to market conditions, geographical location and with the various manufacturers.

Asphalt primer; $2.00 per gal. in 55-gal. drums; $2.10 per gal. in 30-gal. drums and $2.20 per gal. in 5-gal cans.

Waterproofing pitch or asphalt $250.00 per ton in carload lots; $260.00 per ton in less than carloads.

No. 15 tar or asphalt saturated felt, $12.00 per roll of 432 sq. ft. in carloads and $12.50 per roll in less than carloads.

Tar or asphalt saturated cotton fabrics, $1.00 to $1.25 per sq. yd. in carloads and $1.25 to $1.30 per sq. yd. in less than carloads.

Material Cost of 100 Sq. Ft. of 1-Ply Membrane Waterproofing
Consisting of 1-ply 15-lb. felt and 2 moppings of hot bitumen.

	Rate	Total	Rate	Total
1 gal. asphalt primer	$....	$....	$ 2.00	$ 2.00
70 lbs. asphalt or pitch*	....	....	.13	9.10
.28 rolls No. 15 felt (120 sq. ft.)	....	....	12.00	3.36
Fuel, mops, etc.	....	....	1.00	1.00
Cost per 100 sq. ft.		$....		$15.46
Cost per sq. ft.		....		.16

Labor Cost of 100 Sq. Ft. of 1-Ply Membrane Waterproofing
Consisting of 1 ply of felt and 2 moppings of hot bitumen.

	Hours	Rate	Total	Rate	Total
Roofer inc. primer coat	2.6	$....	$....	$16.46	$42.80
Roofer placing felt	0.8	....	....	16.46	13.17
Cost per 100 sq. ft.			$....		$55.97
Cost per sq. ft.			....		.56

Material Cost of 100 Sq. Ft. of Each Additional Ply of Felt and Hot Bitumen

	Rate	Total	Rate	Total
35 lbs. asphalt or pitch	$....	$....	$ .13	$4.55
.24 rolls No. 15 felt (100 sq. ft.)	....	....	12.00	2.88
Fuel, mops, etc.	....	....	.50	.50
Cost per 100 sq. ft.		$....		$7.93
Cost per sq. ft.		....		.08

Labor Cost per 100 Sq. Ft. for Each Additional Ply of Felt and Hot Bitumen

	Hours	Rate	Total	Rate	Total
Labor	1.6	$....	$....	$16.46	$26.34
Cost per sq. ft.			....		.26

Material Cost of 100 Sq. Ft. of 1-Ply Membrane Waterproofing Using Tar or Asphalt
Saturated Fabric and Hot Pitch or Asphalt
Consisting of 1-ply of saturated fabric and 2 moppings of hot bitumen.

	Rate	Total	Rate	Total
1 gal. asphalt primer	$. . . .	$. . . .	$2.00	$ 2.00
70 lbs. asphalt or pitch	. . . .	. . . .	.13	9.10
13.4 sq. yds. satur. fabric	. . . .	. . . .	1.25	16.75
Fuel, mops, etc.	. . . .	. . . .	.85	.85
Cost per 100 sq. ft.		$. . . .		$28.70
Cost per sq. ft.			. . . .	.29

Material Cost of 100 Sq. Ft. Additional Ply of Saturated
Fabric and Hot Pitch or Asphalt

	Rate	Total	Rate	Total
35 lbs. asphalt or pitch	$. . . .	$. . . .	$ .13	$ 4.55
12 sq. yds. satur. fabric	. . . .	. . . .	1.25	15.00
Fuel, mops, etc.	. . . .	. . . .	.50	.50
Cost per 100 sq. ft.		$. . . .		$20.05
Cost per sq. ft.			. . . .	.20

The labor costs for placing fabric and moppings of hot bitumen are the same as used in placing felts and moppings of hot bitumen.

THE IRON METHOD OF WATERPROOFING

The iron method of waterproofing uses an exceedingly fine metallic powder, containing no grease, asphalt, oil or other substances subject to disintegration.

The waterproofing is applied to the inside or outside surfaces of walls and to the tops of rough footings and floor slabs, in the form of brush coats or in a combination of brush and plaster coats.

Applied either upon the inside or outside of walls, this type of waterproofing will resist hydrostatic pressure to a considerable degree.

When preparing old concrete vertical surfaces for iron waterproofing, the vertical surfaces should be given an entirely new bonding exposure by cutting not less than 1/16" with bush-hammers or other suitable tools. See cost of roughing old surfaces on previous pages. The new surface shall then be thoroughly cleaned by brushing with wire brushes. Surfaces of brick and stone need only thorough cleaning with the wire brush and washing.

Holes, cracks, and soft or porous spots in vertical or horizontal surfaces shall be cut out and pointed with one part iron floor bond, 2 parts portland cement and 3 parts sand, all by weight. These materials shall be mixed dry, screened and enough water added to make a stiff mix. Particular care must be taken at the intersections of all walls and floors, in corners, around pipes and other projections through the waterproofing and at construction joints in the concrete.

After the wall surfaces have been cleaned and pointed, they shall be thoroughly cleaned again by hosing. Excess water shall then be removed.

It usually requires 2 to 6 coats of iron waterproofing, depending upon the density of the surface and the conditions encountered.

Some of these materials are mixed with sand and water and applied as a brush coat while others are furnished already mixed ready to apply by the addition of water.

1-Coat Waterproofing Using the
Iron Method, Brushed On, Per 100 Sq. Ft.

Trade Name	Number Coats	Quan. Req'd 100 Sq. Ft.	Approx. Cost Lb.	Labor Hrs. Applying
Ferrocon (Castle)	1	10–15 lbs.	$0.13	2
Ferrolith W	1	13–15 lbs.	.25	2
Kemox H (Sika)	1	60 lbs.–100 lbs.	.14	2
Irontox*	1	10–12 lbs.	.19	2

For each additional application, figure same as given above.

Three to five applications are required where there is water pressure to be overcome.

Liquid Membrane Waterproofing

Liquid membrane (elastomeric) waterproofing is a cold, fluid applied, synthetic rubber, seamless coating system. It is available in single or two-component form and can be applied by trowel or spray. The material forms a continuous, flexible, water impervious film that bonds tightly to a wide range of surfaces, including concrete, stone, masonry, wood, and metal. On a troweled surface, an application rate of five gallons per 125 sq. ft. of surface will produce a cured coating thickness of approximately 55 mils.

Two workmen should coat about 650 sq. ft. per day at the following labor cost:

	Hours	Rate	Total	Rate	Total
Labor	16.0	$. . . .	$. . . .	$16.46	$263.36
Cost per sq. ft.			. . . .		.41

Cost of 100 Sq. Ft. Mastic Waterproof Coatings

The following materials are furnished in the form of a mastic or plastic cement and are applied with a trowel.

Kind of Waterproofing	Inches Thick	Number Coats	Cover Cap. Sq. Ft. Gal.	Quan. Req'd 100 Sq. Ft.	Approx. Price per gal.	Labor Hours
Dehydratine No. 10	1/16	1	26	4 Gal.	$1.65	3
Hydrocide Mastic	1/16	1	20–25	5 Gal.	1.13	3
Hydrocide 700	1/16	1	20–25	5 Gal.	1.00	3
Tremco 103 Mastic						
Asphalt base	1/8	1	20	5 Gal.	1.00	3
Tar base	1/8	1	20	5 Gal.	1.70	3

07150 DAMPPROOFING

Dampproof Paints for Exterior Concrete or Masonry Surfaces.

Where a heavy head of water is to be overcome, the concrete or masonry should preferably be waterproofed by the membrane method. Where only a dampproofing material is required, there are a number of paints on the market for this purpose. The application of one, two or more coats of paint is intended to make the walls impervious to dampness.

Number of coats required and covering capacity will vary with porosity and surface of wall.

Heavy Dampproof Paints, Per 100 Sq. Ft.

Name of Material	Recommended for use on	No. Cts.	Sq. Ft. per Gal.	Gals. Reqd. 100 Sq. Ft.	Price per Gal.	Lab. Hrs. Apply-ing
Dehydratine 4	Fndtn. Walls	2	33	3	$1.43	2.2
Dehydratine 10	Mason Walls	1	30	3⅓	1.65	1.2
Hydrocide Semi-Mastic	Fndtn. Walls	1	30–50	3	1.05	1.2
	Fndtn. Walls	2	15–18	6	1.05	2.2
Hydrocide 600	Fndtn. Walls	1	75–100	1⅓	1.13	1.2
Hydrocide 648	Fndtn. Walls	2	50	2	1.13	2.2
Marine Liquid	Below Grade	1	50–75	1¾	1.24	1.2
Sikaseal	Fndtn. Walls	2	60–80	1½	2.20	2.2
Tremco 110	Fndtn. Walls	1	100	1	.95	1.2

Labor Applying Water and Dampproofing Paint to Exterior Surfaces Below Grade.—Where the water or dampproofing consists of applying one or more coats of heavy paint to exterior concrete or masonry surfaces below grade, it is customary to use an ordinary roofer's brush or a mop. On foundation walls below grade, the men are usually obliged to work in cramped quarters which reduces output.

To make these paints effective, they must be thoroughly brushed into the surface and all voids and pin holes filled, otherwise water and moisture will seep through.

On work of this kind, a man should apply 600 to 700 sq. ft. per 8-hr. day, at the following labor cost per sq. ft. :

	Hours	Rate	Total	Rate	Total
Roofer..	1.3	$. . . .	$. . . .	$16.46	$21.40
Cost per sq. ft.			. . . .		.21

On the second and third coats the labor cost will be slightly less than the first coat on account of the smoother surface to work on.

Dampproof and Plaster Bond Paints

Labor Applying Dampproofing and Plaster Bond Paints.—Before applying the first coat of plaster to the interior of exterior brick or masonry walls, the walls are often painted with one or more coats of heavy black paint, which acts as a dampproofing and plaster bond.

The labor cost will vary with the consistency of the paint and the condition of the walls, as the paint must be well brushed in and all pin holes and porous places thoroughly covered with paint, otherwise moisture will seep through the walls.

Where the paint is applied by hand, a man should cover 700 to 800 sq. ft. per 8-hr. day, at the following labor cost per 100 sq. ft. :

	Hours	Rate	Total	Rate	Total
Roofer..	1.1	$. . . .	$. . . .	$16.46	$18.11
Cost per sq. ft.			. . . .		.18

When the dampproof paints are applied with an air spray, a man should cover 300 to 350 sq. ft. per hr. at the following labor cost per 100 sq. ft. :

	Hours	Rate	Total	Rate	Total
Roofer	0.33	$. . . .	$. . . .	$16.46	$5.43
Cost per sq. ft.			. . . .		.05

Dampproof and Plaster Bond Paints, Per 100 Sq. Ft.

Heavy black paints used for dampproofing interior of exterior walls and to form a bond for plaster. Number of coats required will depend upon porosity of wall on which it is used.

Labor costs in the following table are based on hand application. If an air spray is used, deduct as given above.

Name of Material	No. Coats	Sq. Ft. per Gal.	Gals. Reqd. 100 Sq. Ft.	Price per Gal.	Labor Hrs. Applying
Dehydratine 4	1	70–80	1⅓	$1.43	1.1
Dehydratine 4	2	45–55	2⅓	1.43	2.0
Plasterbond 232	1	75–80	1⅓	1.47	1.1
Plasterbond 232	2	35–40	2½	1.47	2.0

Transparent Dampproofing of Exterior Masonry Walls

During the past few years there has been a big change in the use of transparent dampproofings and the method of application due to an investigation by the National Bureau of Standards, Publications BMS7.

From the results of this investigation it was determined that "leaky" walls were mainly caused by poor workmanship. In other words, unless the bricks were laid in a full bed of mortar (not furrowed), with all bed, end and vertical joints completely filled with mortar, the walls were likely to leak within a short period after the completion of the structure. A hairline crack between the brick and mortar joints was also given as a cause of leakage.

From these tests it was determined that only painting the surfaces of masonry walls with transparent "paints" or "waxes" did not stop these leaks and one method that was really efficient was to cut out the old mortar joints to a depth of ½ to ¾-in., and repoint the joints with new mortar.

This class of work is quite expensive as the cost depends upon the hardness and tenacity of the mortar joints that must be cut out.

The basis for the use of exterior waterproofing, labor and material, is calculated on the ordinary wall constructed with common brick of high absorption with ⅜ to ½-in. lime-cement mortar joints.

On lime-cement mortar joints, two men can rake out and repoint about 12 sq. ft. of brick masonry surface an hour, removing joints to a depth of ¾-in. Allow about ¼-hr. helper's time for moving scaffolds, mixing, etc.

On the above basis, the labor cost of cutting out old mortar joints and repointing mortar joints to 100 sq. ft. of wall surface will cost as follows :

	Hours	Rate	Total	Rate	Total
Mason	16.0	$. . . .	$. . . .	$18.91	$302.56
Helper	4.0	. . . .	. . . .	14.95	59.80
Cost per 100 sq. ft.			$. . . .		$362.36
Cost per sq. ft.			. . . .		3.62

On hard portland cement mortar joints, two men using a power saw can cut out and repoint joints in about 8 sq. ft. of wall per hour and requires ¼ -hr. helper time handling tools and mixing mortar.

On the above basis, the labor cost of cutting out old portland cement mortar joints, and repointing joints to 100 sq. ft. of wall will cost as follows :

	Hours	Rate	Total	Rate	Total
Mason	24	$. ...	$. ...	$18.91	$453.84
Helper	6	....	....	14.95	89.70
Cost per 100 sq. ft.			$. ...		$543.54
Cost per sq. ft.				....	5.44

Add cost of scaffold.

Colorless or Transparent Liquid Water Repellent Treatments

There are a number of colorless or transparent liquid preparations on the market used on brick, stone, stucco, cement and concrete surfaces to render them water repellent and help to prevent rain absorption.

These liquids are colorless and do not affect the original color of the surface to which they are applied.

The covering capacities and quantities given in the following table were furnished by the various manufacturers.

It will be noted that some manufacturers specify one coat and others two coats. Regardless of the material used, sufficient liquid must be applied to completely seal the pores of the surface, otherwise the results will be unsatisfactory.

Transparent Liquid Water Repellents, Per 100 Sq. Ft.

Transparent liquids used for dampproofing brick, stone, stucco, concrete and cement surfaces.

Name of Material	No. Coats	Sq. Ft. per Gal.	Gals. Reqd. 100 Sq. Ft.	Price per Gal.	Labor Hrs. Applying
Daracone	1	75–100	1¼	$4.05	1.6
Supertox	1	125–150	¾	3.50	0.8
Tremco 141 Invisible	2	75–100	1¼	3.25	1.6

Colorless or Transparent Silicone Base Liquid Water Repellent Treatments

Recent developments in liquid water repellent treatments are the new silicone base products which are truly colorless and permit the masonry to "breathe" while rendering the surface water repellent. They penetrate deeply into the cement or masonry surfaces coating pores, cracks, and fissures with an insoluble, nonoxidizing film of silicone which effectively stops capillary action by which water is absorbed.

Usually one flooding coat is sufficient and may be applied by either brushing or spraying.

The covering capacities and quantities given in the following table were furnished by the various manufacturers.

Silicone Base, Transparent Liquid Water Repellents, Per 100 Sq. Ft.

Name of Material	No. Coats	Sq. Ft. Per Gal.	Gals. Req'd. 100 Sq. Ft.	Price Per Gal.	Labor Hours Applying
Dehydratine 22	1	75–100	1¼	$3.40	0.8
Hydrocide Colorless SX	1	75–100	1	4.50	0.8
Sika Transparent	1	80–200	⅔	2.70	0.8
Tremco 147-3%	1	100–200	⅔	4.70	0.8
Tremco 147-5%	1	100–200	⅔	6.40	0.8
Devoe Super-Por-Seal	1	150	⅔	5.29	0.8

Labor Applying Transparent Liquid Waterproofing.—When applying transparent liquid dampproof paints by hand, a man should cover 900 to 1,100 sq. ft. of surface per 8-hr. day, at the following labor cost per 100 sq. ft. :

	Hours	Rate	Total	Rate	Total
Roofer	0.8	$. . . .	$. . . .	$16.46	$13.17
Cost per sq. ft.			. . . .		.13

On the second and third coats, a man should apply 1,000 to 1,200 sq. ft. per 8-hr. day, at the following labor cost per 100 sq. ft. :

	Hours	Rate	Total	Rate	Total
Roofer	0.75	$. . . .	$. . . .	$16.46	$12.35
Cost per sq. ft.			. . . .		.12

When applied by spray, a man should cover 300 to 350 sq. ft. per hr. at the following labor cost per 100 sq. ft. :

	Hours	Rate	Total	Rate	Total
Roofer	0.33	$. . . .	$. . . .	$16.46	$5.43
Cost per sq. ft.			. . . .		.05

Add cost of compressor and spray equipment.

Back-Plastering and Back-Painting Cut Stone and Marble to Prevent Staining

When marble, limestone, brownstone or sandstone are used in building work, considerable difficulty is encountered in preventing the moisture in the brick or concrete backing from penetrating the surface of the stone or marble and staining or discoloring the face. This is due to the open grain or porous nature of the stone or marble.

To overcome this condition it is customary to paint the back of the stone, beds and end joints up to within 1" of the face to prevent staining and discoloration. White or non-staining cement is also used for this purpose. The stone is set in non-staining cement mortar and the back of the stone is plastered or parged with the same mortar.

In many instances the backs, beds, and joints of the stone are painted and then back-plastered with non-staining cement which acts as a double preventative.

Back-painting of Indiana limestone is not recommended by quarrymen and producers because they state it tends to create conditions it is supposed to counteract. A full explanation is given under Cut Stone.

Where cut stone comes in contact with concrete, it is recommended that the concrete be painted instead of the stone and a one inch space left between the concrete and cut stone.

Estimating the Cost of Back-Painting Cut Stone and Marble.

Inasmuch as it is customary to estimate the cost of cut stone by the cu. ft., the cost of back-painting should be estimated in the same manner.

Plain stone ashlar is usually 4" to 8" thick or in alternating courses of 4" and 8" stone so as to provide a bond with the brick or masonry backing. Where just a veneer of cut stone is required, the stone is usually 4" thick and is bonded into the masonry backing by metal anchors.

If cut stone is furnished in alternating courses of 4" and 8", the stone for the entire job will average 6" thick. It will therefore be necessary to back-paint 2 sq. ft. of surface to each cu. ft. of stone. Including beds and joints, it will double the above quantities or it will be necessary to back-paint 4⅓ sq. ft. of surface to one cu. ft. of stone.

If the stone averages 4" thick, it will be necessary to back-paint about 5 sq. ft. of surface to each cu. ft. of stone, including the backs, beds, and joints.

If the stone averages 8" thick, it will be necessary to back-paint about 4 sq. ft. of surface to each cu. ft. of stone.

Back-Painting Cut Stone and Marble.—When back-painting limestone, sandstone, marble, etc., the stone is often painted in the yard before delivery, although on many jobs it is painted after delivery. It is customary to paint the stone a few days before setting, although the backs are frequently painted after the stone has been set.

Care must be exercised to thoroughly brush the paint into all the pores and not drop any paint on the face of the stone.

When painted at the building site, a man should paint 600 to 700 sq. ft. of surface per 8-hr. day, at the following labor cost per 100 sq. ft. :

	Hours	Rate	Total	Rate	Total
Mason	1.3	$. . . .	$. . . .	$18.91	$24.58
Cost per sq. ft.			. . . .		.25

Back-Painting 100 Sq. Ft. Cut Stone or Marble to Prevent Staining

Heavy black paints applied to backs and sides of stone to prevent cement and mortar stains. Quantities based on 1 coat.

Name of Material	Sq. Ft. per Gal.	Gals. Reqd. 100 Sq. ft.	Price per Gal.	Labor Hours
Dehydratine 4	100	1.00	$1.43	1.3
Hydrocide No. 648	125–175	0.75	1.13	1.3
Sikaseal	120	0.80	2.20	1.3

Number of Cu. Ft. of Stone to be Painted Per 100 Sq. Ft. of Surface
Includes back, beds and end joints
Average Thickness of Stone

4"	6"	8"	10"	12"
20	23	25	28.5	31

Number of Sq. Ft. of Painted Surface in One Cu. Ft. of Stone
Includes back, beds and end joints
Average Thickness of Stone

4"	6"	8"	10"	12"
5	4.3	4	3.5	3.2

Back-Plastering Cut Stone and Marble.—After the stone or marble has been set, it is customary to plaster the back of the stone with non-staining cement, using an ordinary trowel and applying it 3/16" to 1/4" thick.

Inasmuch as the stone must be back-plastered as the setting progresses, so as not to delay the bricklayers in backing-up the stone, the time required is rather indefinite. Working steadily, however, a mason should back-plaster 65 to 75 sq. ft. an hr. at the following labor cost per 100 sq. ft. :

	Hours	Rate	Total	Rate	Total
Mason or stone setter	1.5	$. . . .	$. . . .	17.87	$26.81
Cost per sq. ft.			. . . .		.27

07190   VAPOR BARRIERS/RETARDANTS

The combination of high indoor relative humidities and low outside temperatures causes condensation to form within the structure. Serious damage can result from this moisture in the way of rotting framing members, paint deterioration, wet walls and ceilings, etc.

Vapor barriers are recommended where the above conditions exist.

Vapor seal paper should be installed on the warm side of wall, floor, ceiling, roof, etc. Paper should be installed with joints running parallel to and over framing members. All joints should be lapped about 2 inches. Where paper is to be exposed, nail wood lath strips over paper along framing members to provide a neat and permanent job.

A satisfactory vapor proof paper consists of a 50 lb. continuous asphalt film, faced on both sides with 30 lb. basis kraft paper. Other vapor proof papers consist of two 30 lb. basis sheets of kraft paper cemented together with asphalt and reinforced with strong jute cords, spaced ½ to 1-in. on centers and running in both directions. This same reinforced sheet is furnished with two sheets of asphalt coated paper cemented together and reinforced as described above. Another very satisfactory vapor proof paper consists of a heavy kraft paper coated on one side with a thin sheet of aluminum or copper, which prevents the penetration of moisture.

Approximate prices on the various types of vapor seal paper are as follows :

Description	Price per Roll 500 Sq. Ft.
2 sheets 30-lb basis kraft paper cemented together with asphalt	$ 5.00
2 sheets asphalt coated 30-lb kraft paper cemented together with asphalt and reinforced with jute cords	$ 9.00
1 sheet heavy kraft paper coated two sides with a reflective surface	$12.00

Sisalkraft.—A strong, waterproof, windproof building paper consisting of 2 sheets of pre-treated kraft paper cemented together with 2 layers of special asphalt and reinforced with 2 layers of crossed sisal fibers. Costs 2 cts. per sq. ft.

Copper-Armored Sisalkraft is a combination of Sisalkraft and Anaconda electrosheet copper bonded under heat and pressure. It is available in 3 weights, 1, 2 or 3 oz. copper per sq. ft.

It may be used for waterproofing, flashings, ridge roll flashing, etc. Approximate prices are as follows: 1-oz. costs 30 cts. per sq. ft.; 2-oz. costs 50 cts. per sq. ft. and 3-oz. costs 70 cts. per sq. ft. Above prices subject to discount on large orders.

Polyethylene film comes in 100' rolls of various widths in thicknesses of .002", .004", .006", and .008". Costs vary from 1½¢ to 4¢ per sq. ft. Tape for sealing joints costs $3.25 for a roll 2" wide by 100' long in the .004" thickness.

Labor Placing Vapor Seal Paper.—Where vapor seal paper is applied to the interior of exterior stud walls, using a stapling machine, to prevent the penetration of moisture and condensation, a carpenter should handle and place 2,000 to 2,500 sq. ft. per 8-hr. day, at the following labor cost per 100 sq. ft. :

	Hours	Rate	Total	Rate	Total
Carpenter	0.4	$. . . .	$. . . .	$16.47	$6.59
Cost per sq. ft.			. . . .		.07

07200   INSULATION

To insulate is defined as "to make an island of". This "island" not only excludes the outside elements but retains those created within.

Today with energy costs soaring, and traditional energy sources drying up, the use of insulation becomes part of a larger, more inclusive concern—energy conservation. Building design must go beyond the casual approach of letting esthetic and use-function considerations dictate the design and then turning to technology to provide a satisfactory interior climate regardless of the initial and operating cost. The building envelope itself must be modified to contribute in every possible way to interior conditions of comfort—excluding the extremes of weather but welcoming those that pleasantly modify; retaining our manufactured interior weather but also providing means to allow it naturally to renew and maintain itself without resorting exclusively to mechanical means.

There are three stages where crucial decisions are made concerning the building and its environment. First in the preliminary planning; second in material selection; and finally in the actual construction.

Preliminary planning considers the basics. First the building form—should approximate a cube and have a minimum exposure to the elements or sprawl and take maximum advantage of natural light, solar heat and breezes. Buildings should be sited to work with nature, not fight it. Fenestration should be concentrated to the south and tempered with overhangs engineered to exclude the sun when it is high in the summer sky, but to allow the rays to penetrate and add their warmth when the sun is low in the winter. Overhangs need not always be eaves but can be louvres, screens, balconies, porches or a few well spaced deciduous trees. Roof forms should not be chosen solely to exclude the elements but also to shade and vent the space above the top habitable floor. Wind conditions should be considered, doors protected, and windows placed to provide the optimum natural ventilation supplimenting them with clear stories and skylights if need be.

The earth itself is both a very cheap and a very good insulator. While few may opt to live in some of the "cave" houses one sees in the Sunday supplements, intelligent use of banking can pay large dividends.

Once the general arrangement has been determined the second consideration is to select the best wall, ceiling and floor materials for coping with the conditions they will face, and the best methods of installation to assure that the selected materials will perform to their maximum capabilities.

This discussion has concerned itself with new construction. One of the largest opportunities developing for the contractor is bringing existing buildings up to the energy efficient standards acceptable today. Most of the materials and procedures discussed in the following text apply to both new and remodeling work. However the latter presents its own problems and for that type of work the reader is also referred to Chapter 19, which deals with work on existing buildings.

The exterior and interior faces of a building will be chosen for appearance and wearability, and will by nature be hard and dense and therefore good conductors of heat. Also by nature insulating materials will be full of air pockets and be soft and easily damaged and need protection. Therefore most exterior envelopes will consist of three plies; an outside wearing surface, and intermediate space designed to interrupt the heat flow, and an interior wearing surface.

Even the exterior and interior plies will contribute some insulating quality. Materials are rated for their thermal resistance or their "R" value, the temperature difference between two exposed faces required to cause one BTU to flow through one square foot of the material per hour. "R" values can be added, one to the other, to arrive at a rating for the total wall, ceiling, or floor construction. The following list gives the "R" ratings for some of the more common building materials. Ratings for insulating materials will be given as each is discussed in the following text.

Material	R Value
4" Face brick	.44
4" Common brick	.80
8" Poured concrete	.64
8" Cinder block	1.11
8" Light weight block	2.00
4" Concrete slab	.32
½" Plywood	.47
½" Gypsum board	.45
½" Gypsum plaster	.32
1x8 Drop siding	.79

Even from the above small sampling one can see that facing materials can be chosen one over the other to contribute to the insulating quality of the envelope. In addition, air spaces of over ¾" should also be added at an average of R = .91, this depending on the position of the air space and whether the air movement is up or down. Also in choosing exterior, facings colors should be considered for their ability to reflect; the darker the color, the more heat will be absorbed. And finally the jointing of all materials must be carefully studied to eliminate infiltration. A study of one apartment tower found 50% of the heat loss was due to infiltration around the windows!

Heat transfer through the building enclosure is by three means; convection, conduction and radiation.

Convection is the thermally produced upward and downward movement of air. Warm air rises, cold air falls. The concern here is to construct a total building envelope that blocks air currents. Hollow walls should be solidly blocked top and bottom at each story, and floor and roof construction should be continuously sealed from wall construction so air conditions in the exterior walls will not be able to spread out over the interior. This can be done either by extending the interior facing or by adding insulation or a combination of both.

Conduction is the transmission of heat through a material. The rate of conductance of a material or combination of materials, known as its "U" factor, is the BTU per hour per inch of thickness per square foot per degree temperature difference. The "R" ratings discussed previously are the reciprocals of conductance or 1 divided by the "U" factor. Most common insulating materials are manufactured to be as poor a conductor of heat as possible. However their positioning within the wall can also make them block convection and when enclosed in reflective coverings and separated from adjacent construction by at least ¾" they also can cut down on the third means of heat transfer, radiation.

Radiation is the emission of energy from a surface. Bright surfaces, such as aluminum foil, are good reflectors and have low emissive coefficients and are therefore poor absorbers of heat. The interior faces of the two outer plies of building envelope tend to be dark and the heat will radiate from one surface to the other constantly unless interrupted. By inserting one or more layers of reflective surfaces, the heat will be reflected back to the surface from which it escaped. Installation is crucial to the success of a radiant barrier as the reflective surface will be highly conductive and if allowed to come into contact with either side of the enclosing construction it will speed heat transfer rather than retard it.

Infiltration, as discussed above, will generally be solved by good detailing at the juncture of one material to another and by caulking. Strips of insulation, such as under sill plates, and the stuffing in of loose wools can also be used to plug air leaks.

There is no simple answer as to how much insulation should be added. Both comfort and the additional costs must be considered. Minimum accepted "R" values in the northern regions have been R19 for ceilings, R13 for floors and R11 for walls. These drop to R13, R9 and R8 in milder areas but should never be less than R9 for ceilings and R7 elsewhere. With the recent increase in fuel costs many consider these figures far too low and recommend R24 for ceilings, R19 for floors and R13 for walls. If fuel costs keep advancing at recent rates some project an "R" value of 49 for ceilings in the Chicago area, over two and a half times todays standard. If it is a matter of fuel supplies actually running out perhaps such increases will come about. Based on whether it would be a good financial investment, it is doubtful, as each additional inch of insulation beyond a certain point reduces the heat loss less than the one before it. There is a point of sharply diminishing returns.

To determine the best "R" values for a particular project, one must consider the following:

a) Determine all wall, ceiling, and floor construction types and give each its proper "R" value without insulation.
b) Determine the heat loss (or heat gain for cooling) without insulation through each of the above.
c) Choose an "R" rating for walls, ceiling and floors based on averages for existing climate, and consistant with the type of construction contemplated.
d) Refigure heat loss with new "R" values.
e) Figure fuel savings of "d)" over "b)". Several alternate fuels might be considered and local utilities can be helpful in determining these costs.
f) Figure the difference in cost of the insulated construction over the uninsulated, remembering to deduct for savings in size of the mechanical equipment, if any.
g) Determine how long it will take savings in fuel to pay for insulation costs. This is the payback period and will have to take many things into consideration such as future fuel costs, and the interest on the additional mortgage to finance the insulation. If the payback period looks attractive, the whole proceedure can be repeated with a higher "R" value assigned, and this can be repeated until a point of diminishing returns is evident.

Once the desired "R" value is established, the selection of the best type of insulation can begin. The possibilities include:

a) Blankets and batts, with or without reflective and/or vapor barriers.
b) Rigid board type which may also serve as sheathing or lathing.
c) Sprayed on or foamed in place, which may also serve as sound or fire retardants.
d) Poured fill
e) Reflective barriers

Many of these may be used in combination with the other.

Method of Placing Insulating
Wool Batts.

Method of Placing and Stapling
Paper-Backed Insulating
Wool Batts.

07210 BUILDING INSULATION BLANKET AND BATT INSULATION

Insulation formed into batts or rolls are made in a number of materials such as stone, slag, glass wools, vegetable and cotton fibers and suspended pulps. They are all light in weight, easily cut and handled and need support to stay in place. They are usually encased in either kraft paper or foil and can be used as a vapor barrier. The enclosing material is usually made with flapped edges for easy attachment to studs and joists. Units may be ordered unfaced and fitted between studs by friction. Unfaced units are most often used as a second layer of insulation in attic spaces as they do not form a second vapor barrier if the original layer has one. A narrow form of unfaced blanket 1" thick is made for inserting under sill plates. It will compress to 1/32".

Method of Placing Strip Insulating Wool.

Most batt and blanket materials have "R" values of around 3.5 to 3.7 per inch of thickness. 3" units will provide an "R" of 11, 3½" of 13 and 6" of 19. The 6" units are commonly used in attics and fitted between the joists; but more and more they are used in walls with 2x6 studs placed 24" o.c. in place of the standard 2x4s 16" o.c. If foil faced units are used and an air space of at least ¾" maintained between the foil and the warm side of the room the "R" rating can be increased some 15%.

Most glass fiber insulation is distributed on a nationwide basis, but other materials are often produced regionally and prices may vary widely. Widths are 15" and 23" to fit normal 16" and 24" stud and joist spacing. Batts come 48" and 96" long, blankets in rolls of from 24' to 40', often two rolls per package. The square foot cost for foil faced units 3½" thick is 16¢; 6" thick 26¢. Units bound in kraft paper may run a couple of cents less and unfaced units even less. Sill sealer rolls come in 50' rolls 1" thick and in 3⅝" and 6" widths and cost 5¢ and 7¢ per lineal foot.

Because these materials are light and precut to fit standard construction one carpenter can handle the installation. Friction fit type, with no attachments, will install at the rate of 2000 sq. ft. per day. Flanged type bound in kraft paper or foil will install at 1800 sq. ft. per day.

Labor Cost of 100 sq. ft. Faced Insulation

	Hours	Rate	Total	Rate	Total
Carpenter	.45	$. . . .	$. . . .	$16.47	7.41
Cost per sq. ft.			. . . .		.075

07211   LOOSE FILL INSULATION

Loose fill insulation includes mineral wool, which is molten rock extruded by air and steam into fibers, known as blowing wool for machine applications, or nodules for pouring or spreading by hand; and expanded volcanic rocks such as vermiculite or perlite. The latter are most often specified for filling concrete block or cavity walls.

Loose Insulating Wool.—Is suitable for any purpose where insulation can be packed by hand between ceiling joists or side wall studding. For maximum results it is recommended that wool be applied full thickness of side wall studding and approximately 4" to 16" over ceiling areas.

The covering capacity varies considerably with the density to which it is packed.

The National Bureau of Standards conductivity figure for glass or rock wool is 0.27 Btu at a density of 10 lbs. per cu. ft.

The covering capacity of bulk wool as given by most manufacturers is based on a density of 6 to 8 lbs. per cu. ft.

Number of Sq. Ft. of Surface Covered By One Bag of Loose
Insulating Wool Weighing 40 Lbs. and Containing 4 Cu. Ft.
Not Including Area Covered By Studs or Joists

Density per Cu. Ft. Lbs.	Actual Cu. Ft.	1	2	3	3½	3⅝	4
6	6.67	80.0	40.0	26.6	22.8	22.0	20.0
7	5.72	68.5	34.4	22.9	19.7	18.8	17.2
8	5.00	60.0	30.0	20.0	17.2	16.6	15.0
9	4.45	53.4	26.7	17.8	15.2	14.7	13.4
10	4.00	48.0	24.0	16.0	13.7	13.3	12.0

Number of Sq. Ft. of Surface Covered By One Bag of Loose Insulating
Wool Weighing 40 Lbs. and Containing 4 Cu. Ft.
Including Area Covered By Studs or Joists

Density per Cu. Ft. Lbs.	Actual Cu. Ft.	1	2	3	3½	3⅝	4
6	6.67	85.0	42.5	28.4	24.3	23.5	21.2
7	5.72	73.0	36.5	24.3	20.8	20.0	18.3
8	5.00	63.8	31.9	21.3	18.3	17.6	16.0
9	4.45	56.7	28.4	18.9	16.2	15.7	14.2
10	4.00	51.0	25.5	17.0	14.5	14.1	12.7

Loose or bulk rock wool is usually furnished in bags weighing 40 lbs. and containing 4 cu. ft. Average price $1.50 to $1.70 per bag.

Granule or Pellet Type Insulating Wool.—Granule or pellet type insulating wool is furnished in particles sufficiently large that the material will not sift or dust through wall or ceiling cracks and is used for insulating ceiling areas between supporting joists, or void wall spaces where accessible and which permit pouring the pellets into place, particularly in old house construction.

Granule or pellet type insulating wool is usually furnished in bags weighing 40 lbs. and is placed at a density of 7 to 8 lbs. per cu. ft.

The tables for loose or granular insulating wool, based on 7 to 8 lbs. per cu. ft. may be used for estimating quantities obtainable per bag.

Granule or pellet type insulating wool in bags or cartons weighing 40 lbs. cost about $2.00 each.

Labor Placing Loose or Bulk Insulating Wool.—When placing loose or bulk insulating wool between wood studs, a man should place about 350 to 450 sq. ft. per 8-hr. day, at the following labor cost per 100 sq. ft. :

	Hours	Rate	Total	Rate	Total
Labor	2	$. . . .	$. . . .	$16.47	$32.94
Cost per sq. ft.			. . . .		.33

Labor Placing Granule or Pellet Type Insulating Wool Between Ceiling Joists.—Where granule or pellet type insulating wool is poured between ceiling joists and other open spaces, to a thickness of 3½ or 3⅝-in. a man should place 700 to 900 sq. ft. per 8-hr. day, at the following labor cost per 100 sq. ft. :

	Hours	Rate	Total	Rate	Total
Carpenter	1	$. . . .	$. . . .	$16.47	$16.47
Cost per sq. ft.			. . . .		.17

Vermiculite Or Perlite Loose-fill Building Insulation.—Vermiculite loose-fill insulation is used to insulate attics, lofts, and side walls. It is fireproof, not merely "fire resistant." The fusion point is 2200° to 2400° F.

It is completely mineral and does not decompose, decay, or rot. Because of its granular structure, vermiculite is free-flowing, assuring a complete insulation job without joints or seams. Rodents cannot tunnel into it, and it does not attract termites or other vermin. Vermiculite is a non-conductor and is excellent protection around electrical wiring.

Vermiculite loose-fill is marketed in 4 cu. ft. bags. Approximate coverage per bag, based on joists spaced 16" o.c., is 14 sq. ft. for a 3⅝" thickness; 9 sq. ft. for 5½" thickness. A bag will cost about $3.60.

Approximate Coverage of Vermiculite
Fill in Cavity and Block Walls

Sq. ft. wall area	1" cavity	2" cavity	2½" cavity	8" block	12" block
100	2 bags*	4 bags*	5 bags*	7 bags*	13 bags*
500	10 "	20 "	25 "	34 "	63 "
1,000	21 "	42 "	50 "	69 "	125 "
5,000	104 "	208 "	250 "	545 "	625 "
10,000	208 "	416 "	500 "	1,090 "	1,250 "

*1 bag=4 cu. ft.

Courtesy Johns-Manville Corporation
Applying fiberglass batts over cellulose in attics

One mason can pour around 50 bags or 200 cu. ft. per 8 hr. day, at the following labor cost:

	Hours	Rate	Total	Rate	Total
Mason	8	$. . . .	$. . . .	$18.91	$151.28
Cost per cu. ft.			. . . .		.76

When adequate ventilation is provided in attics or similar spaces, it is not necessary to use a vapor barrier in the ceiling construction below vermiculite fill insulation. A vapor barrier is recommended on the warm side of insulated exterior walls, except when the insulated space is ventilated to permit movement of water vapor and air from the space.

Reinsulation: Where an existing thickness of blanket or other insulation is inadequate, vermiculite can be poured over it to obtain the thickness desired.

07212   RIGID INSULATION

Rigid insulation boards are made of expanded polystyrene, often called "beadboard"; extruded polystyrene sold as Styrofoam; urethane; glass fibers; glass foams; and wood and vegetable fibers. Many rigid boards are made to serve dual purposes such as sheathing, lath and even interior finish. Many are also used for roof decks and perimeter insulation which are discussed separately below.

Urethane boards have the highest "R" value per inch of thickness, 7.14, but are flammable and must be covered over, or treated. The cost is around 25¢ per sq. ft.

plain, 33¢ treated. The sizes available are many including the standard 4' x 8' sheet and in thicknesses from ½" to 24". Because of its high insulating value per thickness it is ideal for use in cavity walls and for applying to the inside face of an exterior masonry wall. In cavity walls it is applied against the outside face of the inside wythe leaving a clear air space for any penetrating water to drain out.

Where urethane board is used against exterior masonry walls the surface must be clean and even. Nailers must be applied wherever wood trim will be added, such as at the base and around doors and windows. The board is then applied to the wall with sufficient mastic to cover at least 50% of the back when shoved in place. Some manufacturers supply special channel systems for mechanical attachment as an alternative. The urethane board can then be covered with plasterboard or may be plastered direct. Because of the fire hazard, which ever method is chosen, the interior finish coat must be run up above the finished ceiling to cover all portions of the urethane as it is the fireproofing for it.

Extruded polystyrene, or Styrofoam under which name it is marketed by Dow Chemical Company, also has a high "R" value, 5.4 per inch of thickness. It costs about 29¢ per sq. ft. in 1" thickness. It is installed in the same locations and by the same methods as the urethane board above, and like urethane board, it should have a continous interior finish that will protect it from fire.

Polystyrene which is molded is known as beadboard. It has a lower "R" rating, 3.85, and is absorbent, but it is much cheaper, about 12¢ per sq. ft. in 1" thickness. It too must be covered with a fireproofing finish.

Glass fiber boards come in various densities. 1" board in a medium weight gives an "R" of 4.35 and costs around 30¢ per sq. ft.

The stronger rigid boards may also be used for sheathing on wood frame construction. But a 1" board in combination with standard 3½" batts will give a wall with the highly desirable rating of R19 or better and may be preferrable to using 6" batts and 6" studs to gain such a rating.

Rigid board is light, easily cut and can be installed by one man who should erect about 1,500 sq. ft. per day depending on the amount of fitting.

Labor Cost of 100 sq. ft. of 1" Rigid Board
Applied to a Masonry Wall

	Hours	Rate	Total	Rate	Total
Mechanic	.54	$. . . .	$. . . .	$16.47	$8.90
Cost per sq. ft.			. . . .		.09

07213 REFLECTIVE INSULATION

Some reflective insulation can be incorporated as the enclosing sheet on batt and blanket insulation and as a backing on rigid insulation. Also available are sheets of insulating material depending entirely on the reflective principle and consisting of alternate layers of trapped air and foil, the outer layers of which are usually laminated to a backing sheet of kraft paper for extra strength and which are extended at the edges to form flaps for stapling. Another type depends on an inner layer or layers of accordian foils. Such insulation must be held away from the inner and outer wall finishes at least ¾" so that there will be an uninterrupted air space which serves the dual purpose of allowing the foil surface to reflect escaping heat (or entering cold) back to where it came from and keeping the highly conductive foil from transmitting the heat by contact.

A typical lamination would consist of three layers; the two outer sheets of foil laminated to a backing sheet and separated from a center sheet of pure foil, usually aluminum, with sealed air spaces both sides. The unit is held together with flanges extended to allow stapling to studs and joists. The "R" value will depend on the placement of the material as reflective insulation is more effective with the heat flow down than with heat flow up. Where the heat flows down from a hot roof to a ceiling below 93% of the heat transfer is by radiation and only 7% by conduction. As heated air always travels up no convection is directed toward the ceiling. Considering the reverse travel, heated interior air travelling through the ceiling to the cold roof above, 50% will be transmitted by radiation, the same 7% by conduction and the remainder by convection. Heat transfer through the side walls will be 65–80% by radiation, 15–28% by convection and 5–7% by conduction. The "R" rating then on a three ply, two inch thick reflective sheet will be 26.18 down, 9.52 up and 14.65 through the wall for a mean "R" of 17.85. When we speak of this as 2" thick that is the distance from foil to foil. Another ¾" minimum must be added to each side to isolate the foil from other wall construction so that a full 3½" must be reserved for the reflective insulation.

One inch thick units with just one inner and one outer layer of reinforced foil separated by a 1" air space will give an "R" value of 21.79 down, 7 up and 10.72 through the wall for a mean "R" of 14.4.

Single sheets with aluminum foil either side of a reinforcing sheet with no integral air space if properly installed to preserve air spaces to adjacent construction will give "R" values of 14.28 down, 3.45 up and 7 through the wall, for a mean "R" of 9.

The cost of two ply, material is 12¢ per sq. ft.; three ply, 14¢; and five ply, 20¢.

The material is preformed to fit average stud and joist spacing and is very light; one man can install 2200 sq. ft. per day if the wall or ceiling space is clear, at the following labor cost per 100 sq. ft.:

	Hours	Rate	Total	Rate	Total
Carpenter	.66	$. . . .	$. . . .	$16.47	$6.00
Cost per sq. ft.			. . . .		.06

07214   FOAMED INSULATION

Foamed in place insulations are very efficient, and are usually sublet to an experienced applicator. Foamed insulation will adhere well to most surfaces, will conform to irregular forms and will set up quickly. The plastic foams flow better into restricted areas than the sprayed on fibers which have longer fibers and tend to clog in small spaces.

Application will require special equipment and the space must be well ventilated. Some foams expand and will set up stresses in restricted areas if care is not taken. Despite these drawbacks the resulting "R" can be as much as 7.7 per inch and will apply at the rate of 800 sq. ft. per day with a three man crew. Material costs for urethane will run around 22¢ per inch per sq. ft. If subcontracted, a budget figure of 85¢ per sq. ft. for 1" and $1.40 for 2" should be adequate.

07215   SPRAYED INSULATION

Sprayed fibers of rock wool, glass, gypsum, perlite and cellulose will have "R" values in the 4 to 5 range. Special binders are required as well as special equipment, so this work, too, will be subcontracted. A two man crew will apply some 1200 sq. ft. of 1" thickness per day, possibly more on wide open ceilings. Material costs will be around 15¢ to 20¢ per sq. ft. per 1" of material. If subcontracted, a budget figure of $1.10 per sq. ft. should be adequate.

07230 HIGH AND LOW TEMPERATURE INSULATION

Rigid foam insulation, composed of expanded polystyrene, glass or polyurethane bound together into rigid board has largely replaced corkboard for low temperature insulation. It is also used in general building for roofs, as it is quite strong, tough and resilient; for curtain wall back up as it is lightweight; for core walls as it is self-supporting; for perimeter and crawl space insulation as it has a low water absorption and a zero rating for capillary action. In addition, it forms a vapor barrier, is dimensionally stable, incombustible, and easily cut.

Rigid foam insulation can be placed on concrete and masonry surfaces by adhering units, with hot asphalt; or with gobs of cement.

Units come in 1", 1½", 2", 3", 4" and 6" thicknesses and sizes from 12"x18" to 36"x36" in packaged forms and up to 9' lengths when shipped in carload lots. There are no standard sizes within the industy and sizes, costs, and insulating values must be carefully checked with the various manufacturers.

Back Plaster.—Where insulation is to be applied against masonry walls in hot asphalt, the walls must be made true and even with portland cement plaster. The parge coat or back plaster may be omitted if the wall is of poured concrete.

Back plaster should be made by mixing 1 part portland cement and 3 parts clean sharp sand. Back plaster ½-in. thick requires 1.5 bags portland cement and 4.6 cu. ft. of sand per 100 sq. ft.

Material Cost of One Cu. Yd. of 1:3 Portland Cement Back Plaster

	Rate	Total	Rate	Total
9.5 sacks portland cement	$....	$....	$4.00	$38.00
1.10 cu. yds. sand	....	....	6.50	7.15
Cost per cu. yd.		$....		$45.15
Cost per cu. ft.		....		1.67

Asphaltic Priming Coat.—All concrete or plaster surfaces, to which insulation is to be applied in hot asphalt, must be primed with asphaltic priming coat, made especially for this purpose.

Asphalt.—The proper grade of asphalt to be used is very important. It must be odorless, as any material with an odor will taint foodstuffs that may be stored in insulated rooms. It should be an oxidized asphalt of 180° F. to 200° F. melting point and should preferably be made from a Mexican base oil. The hot asphalt in which corkboard is dipped should be held between 350° and 375° F. Do not heat asphalt to higher temperatures than these. Most manufacturers recommend an asphalt for use with their product.

On dipping floor, wall, and ceiling insulation allow .4 lbs. of asphalt per sq. ft. of insulation per layer. Mop coating on floors or top of ceiling requires .8 lbs. per sq. ft.

Cold Erection Plastic.—Where the use of hot asphalt for the erection of foamboard is not practical use cold erection plastic on floors, walls and ceilings. A 1/16–in thick coat requires 1-gal. of cold erection plastic per 25 sq. ft.

Nails or Impregnated Skewers.—Use nails for the attachment of the first layer of insulation to wood and frame construction, for the attachment of the second layer to wood stripped concrete ceilings, or for toenailing the first layer of solid cork walls. All nails should be of galvanized wire and should be 1-in longer than the thickness of the insulation. Where two courses of insulation are erected in hot asphalt or cold erection plastic, impregnated hardwood skewers should be used to additionally secure the second course in position. Do not use either nails or skewers when erecting the first course against brick, stone or concrete walls. Hot asphalt or cold erection plastic is all that is required.

Erection.—The requirements (material and labor) for installing one layer of 2" thick insulation, using the hot asphalt dip method of setting, onto a masonry wall surface are given in the following tables. Bear in mind that the labor times given are for straight, flat surfaces; and that any irregularities such as columns, pilasters, or beams will require additional labor time due to cutting and setting smaller pieces. If the insulation is thicker than 2", then adjust for the difference in material costs and add about ¼ hour to both mechanic and helper time per 1" thickness increase over 2".

2" Thick Single Insulation Layer, Hot Dip Method
Requirements Per 100 Sq. Ft.

Operation	Material Required	Mechanic Time	Helper Time
Masonry Wall Parge Coat-½"	5.5 cu. ft. mortar	3.0 hrs.	3.0 hrs.
Asphalt Prime Coat	1 gallon primer	1.0	...
Hot Asphalt	40 lbs.	...	...
Insulation Application	100 sq. ft.	1.5 hrs.	1.5 hrs.

Note: For applications against concrete walls, the parge coat is not required.

Low Temperature Insulation Using Two Layers Of Pittsburg Corning Foamglas

Material Cost of 100 Sq. Ft. of Single Layer Insulation,
Set In Hot Asphalt on Masonry Walls

	Rate	Total	Rate	Total
5.5 cu. ft. cement plaster	$....	$....	$1.67	$ 9.19
1 gal. asphalt primer	....	....	2.00	2.00
40 lbs. asphalt	....	....	.13	5.20
100 sq. ft. 2" corkboard	....	....	.97	97.00
Cost 100 sq. ft.		$....		$113.39
Cost per sq. ft.		....		1.13

Labor Cost of 100 Sq. Ft. of Single Layer Insulation,
Set in Hot Asphalt on Masonry Walls

	Hours	Rate	Total	Rate	Total
Plaster Coat					
Mechanic	3.0	$....	$....	$16.35	$ 49.05
Helper	3.0	....	....	13.77	41.31
Prime Coat					
Mechanic	1.0	....	....	16.46	16.46
Insulation					
Mechanic	1.5	....	....	16.47	24.71
Helper	1.5	....	....	13.77	18.81
Cost 100 sq. ft.			$....		$150.34
Cost per sq. ft.				....	1.50

Finishes.—Corkboard insulation should receive a finish which will give long service under the temperature and moisture conditions common to cold storage rooms. The finish should bond securely to the corkboard and be reasonably resistant to bumps and abrasion. Cost, good appearance, and availability should also be considered. These requirements are best met by asphalt emulsion finishes or portland cement plaster. Asphalt emulsion finishes are generally used on ceilings and portland cement plaster on walls. Rigid foam insulations may be sprayed or brushed with special white vinyl emulsion paints, aluminum paint, tile set with adhesive or cement plaster over mesh. Direct plastering should not be considered.

Asphalt emulsion finishes shall be applied directly to corkboard in two coats, each ⅛" thick. Before applying the first coat, all voids, open joints, or broken corners must be pointed with emulsion finish. The first coat must be hand dry before the second coat is applied. Trowel the second coat smooth. The final thickness of the finish shall be approximately ⅛" when dry. Finish all corners true and clean.

The emulsion finishes must be thoroughly dry before any paint is applied, to guard against cracking.

Approximately 5 gallons of asphalt emulsion, at a material cost of about $7.00, will be required for the application of the two coats on 100 sq. ft. of surface, at the following labor costs:

	Hours	Rate	Total	Rate	Total
First Coat					
Mechanic	2	$....	$....	16.46	$32.92
Second Coat					
Mechanic	2	....	....	16.46	32.92
Cost 100 sq. ft.			$....		$65.84
Cost per sq. ft.				....	.66

The labor cost of 100 sq. ft. of ½" of 1:3 portland cement plaster should average as follows:

	Hours	Rate	Total	Rate	Total
Mechanic	3	$....	$....	$16.35	$49.05
Helper	3	....	....	13.77	41.31
Cost per 100 sq. ft.			$....		$90.36
Cost per sq. ft.				....	.90

Portland cement plaster should be mixed in the proportions of 1 part portland cement, 3 parts clean, screened sand, and 5 percent hydrated lime.

To reduce cracking to a minimum, finish coat of plaster may be scored each way approximately 4'-0" on centers.

Some manufacturers state that plaster can be applied directly to insulation; others specify wire mesh must be applied, particularly on ceiling work. As rigid foam insulation does not absorb water, plaster coats will dry very slowly. It is also necessary to add the cost of nailer strips for casings, trim, and bases, if any occur. Light trim such as beads may be stapled on.

07240　ROOF AND DECK INSULATION

Roof and deck insulation is often of the same material as has been discussed under rigid board insulation but the units are generally furnished in smaller sizes more suitable for handling in exposed conditions. In addition to glass fiber, perlite, urethane and polystyrene boards, wood and mineral fiber boards costing as little as 25¢ per sq. ft. per 1" of thickness, and foam glass blocks are often specified. Poured decks are also common, but these have been covered at the end of Chapter 8, "Concrete". Foamglas* is a unique material which is inorganic, incombustible, dimensionally stable, water and vapor proof and weighs only 8.5# per cubic foot but is strong enough to be used under parking and promonade decks. It can also be ordered with an 1/8 " per foot slope. Minimum thickness is 1½" and is available in ½" increments up to 4". 1½" material has an "R" value of 4.2 and costs about 60¢ per sq. ft. in 1½" thickness. One roofer will install some 800 sq. ft. per day at the following labor cost:

Labor Cost of 100 sq. ft. of 1" Deck Insulation

	Hours	Rate	Total	Rate	Total
Roofer	1.0	$....	$....	$17.18	$17.18
Cost per sq. ft.			....		.17

07250　PERIMETER INSULATION

Board forms of polystyrene, urethane, fiberglass and cellular glass are commonly employed as insulation on foundation walls and under slab edges to insulate an on-grade floor construction at the outer wall. Material costs are from 15¢ to 25¢ per sq. ft. in 1" thickness.

As perimeter insulation must fit the outline of the interior face of the outside wall and must be run continuously around offsets, perimeter ducts, column foundations and other irregularities, unit costs will vary widely. In general, insulation is run down 24" below outside grade. In addition, it may be run back 24" under the outer edge of the slab. One carpenter should apply around 800 sq. ft. per day on straight run work, at the following labor cost per 100 sq. ft.:

	Hours	Rate	Total	Rate	Total
Carpenter	1.14	$....	$....	$16.47	$16.47
Cost per sq. ft.			....		.17

07300 ROOF SHINGLES AND ROOFING TILES

Roofing is estimated by the square, containing 100 sq. ft. The method used in computing the quantities will vary with the kind of roofing and the shape of the roof.

The labor cost of applying any type of roofing will be governed by the pitch or slope of the roof, size, plan of same (whether cut up with openings, such as sky-

*Pittsburg Corning

lights, penthouses, gables, dormers, etc.), and upon the distance of the roof above the ground, etc.

Rules for Measuring Plain Double Pitch or Gable Roofs.—To obtain the area of a plain double pitch or gable roof as shown in Figure 1, multiply the length of the ridge (A to B), by the length of the rafter (A to C). This will give the area of one-half the roof. Multiply this by 2 to obtain the total sq. ft. of roof surface.

Example: Assume the length of the ridge (A to B), is 30'-0" and the length of the rafter (A to C), 20'-0". By multiplying (A to B), 30'-0" by (A to C), 20'-0" equals 600. The area of one-half the roof is 600 sq. ft. 600x2=1,200 sq. ft. of roof.

Rules for Measuring Hip Roofs.—To obtain the area of a hip roof as shown in Figure 3, multiply the length of the eaves (C to D) by ½ the length of the rafter (A to E). This will give the number of sq. ft. of one end of the roof, which multiplied by 2 gives the area of both ends. To obtain the area of the sides of the roof, add the length of the ridge (A to B) to the length of the eaves (D to H). Divide this sum by 2 and multiply by the length of the rafter (F to G). This gives the area of one side of the roof and when multiplied by 2 gives the number of sq. ft. on both sides of the roof.

Fig 1. Plain Double Pitch or Gable Roof   Fig 2. Conical Building and Roof

To obtain the total number of sq. ft. of roof surface, add the area of the two ends to the area of the two sides. This total divided by 100 equals the number of squares in the roof.

Example: Assume the length of the eaves (C to D) is 20'-0" and the length of the rafter (A to E) is 20'-0". Multiply (C to D) 20'-0" by ½ the length of the rafter (A to E), or 10'-0" equals 200 sq. ft., the area of one end of the roof. To obtain the area of both ends, 200x2=400 sq. ft.

To obtain the area of the sides of the roof, the length of the ridge (A to B), is 10'-0" and the length of the eaves (D to H), is 30'-0". Add (A to B) 10'-0" to (D to H) 30'-0", and the result is 40'-0". By taking ½ the combined length of the ridge and eaves, ½ of 40'-0"=20'-0", the average length of the roof. Assuming the length of the rafter (F to G) as 20'-0", 20'-0"x20'-0"=400 sq. ft., the area of one side of the roof, which multiplied by 2 equals 800 sq. ft., the area of both sides of the roof.

Fig 3. Hip Roof

Adding the area of the two ends to the area of the two sides gives 1,200 sq. ft. of roof area.

The area of a plain hip roof (Fig. 4) running to a point at the top is obtained by multiplying the length of the eaves (B to E) by ½ the length of the rafter (A to F). This gives the area of one end of the roof. To obtain the area of all four sides, multiply by 4.

Example: Multiply the length of the eaves (B to E), which is 30'-0", by ½ the length of the rafter (A to F), 10'-0", and the result is 300 sq. ft., the area of one end of the roof. 300x4=1,200 sq. ft., the area of the 4 sides.

ROOF PLAN   ELEVATION

Fig 4. Hip Roof

Rules for Measuring Conical Tower Roofs and Circular Buildings.—To obtain the area of a conical tower roof as shown in Figure 2, multiply ½ the length of the rafter (A to B) by the distance around the eaves at B. As an example, assume the length of the rafter (distance from A to B) is 15'-0" and the diameter of the building at B is 20'-0".

To obtain the distance around the building, multiply the diameter by 3.1416. If the eaves project beyond the outside walls, and the diameter is given only to the outside walls of the building, add the length of the roof projection on both sides of the building to obtain the correct diameter. Example: If the diameter of the building (C to D) is 20'-0" and the eaves project 2'-0" on each side, the diameter of the building at the eaves would be 24'-0".

Multiplying 24'-0" (B to E), the diameter of the building at eaves, by 3.1416, gives 75.3984, or approximately 75'-5" around the eaves at projection (B to E).

To obtain the area of the roof, multiply ½ the length of the rafter (A to B), which is 15'-0", (½ of 15 is 7½) or 7'-6", by the distance around the eaves at B and E, which is 75'-5" or 75.4 feet, and the result is 565.5, or 565½ sq. ft., the area of the roof.

To obtain the wall area of a cylindrical or circular building, multiply the height by the circumference, or the distance around the building, and the result will be the number of sq. ft. to be covered. The circumference is obtained by multiplying the diameter (C to D) of the building by 3.1416. To obtain the area of the outside walls of a cylindrical building whose diameter is 20'-0" and the height 15'-0", 20x3.1416=62.832, or 62'-10". Multiply 62.832 ft. (the distance around the building) by 15 ft. (the height of the building)=942.48, or 942½ sq. ft., the area of the outside walls.

A Short Method of Figuring Roof Areas

To obtain the number of square feet of roof area, where the pitch (rise and run) of the roof is known, take the entire flat or horizontal area of the roof and multiply by the factor given below for the roof slope applicable and the result will be the area of the roof.

Always bear in mind, the width of any overhanging cornice must be added to the building area to obtain the total area to be covered.

Example: Find the area of a roof 26'-0"x42'-0, having a 12" or 1'-0" overhanging cornice. Roof having a ¼ pitch or a 6 in 12 rise and run.

To obtain the roof area, 26'-0"+1'-0"+1'-0"=28'-0" width. 42'-0"+1'-0"+1'-0"=44'-0" length. 28x44=1232 or 1,232 sq. ft. flat or horizontal area.

To obtain area at ¼ pitch or 6 in 12 rise and run: multiply 1,232 by 1.118=1377.376 or 1,378 sq. ft. of roof surface.

Add allowance for overhang on dormer roofs and sides.

Pitch of Roof	Rise and Run	Multiply Flat Area by	Lin. Ft. of Hips or Valleys per Lin. Ft. of Common Run
1/12	2 in 12	1.014	1.424
1/8	3 in 12	1.031	1.436
1/6	4 in 12	1.054	1.453
5/24	5 in 12	1.083	1.474
1/4	6 in 12	1.118	1.500
7/24	7 in 12	1.158	1.530
1/3	8 in 12	1.202	1.564
3/8	9 in 12	1.250	1.600
5/12	10 in 12	1.302	1.641
11/24	11 in 12	1.357	1.685
1/2	12 in 12	1.413	1.732

Hips and Valleys.—The length of hips and valleys, formed by intersecting roof surfaces, running perpendicular to each other and having the same slope is also a function of the roof rise and run. For full hips or valleys, i.e. where both roofs intersect for their full width, the length may be determined by taking the square root of the sum of the rise squared plus twice the run squared.

Using the factors given in the last column of the above table, the length of full hips or valleys may be obtained by multiplying the total roof run from eave to ridge, (not the hip or valley run), by the factor listed for the roof slope involved.

07311   ASPHALT SHINGLES

Asphalt shingles are furnished in several styles and the labor cost of laying them will vary with the style of shingle used and the type of roof to which they are applied.

Individual asphalt shingles are furnished and laid in exactly the same manner as wood shingles, the only difference being their size.

Strip asphalt shingles are furnished in strips, each containing 3 shingles, so when laying them a man handles and lays 3 shingles at a time.

Hexagon asphalt shingles are furnished in strips 11⅓"x36".

Strip Asphalt Shingles

Estimating Quantities of Asphalt Shingles.—Asphalt shingles are sold by the square containing sufficient shingles to cover 100 sq. ft. of roof.

When measuring roofs of any shape, always allow one extra course of shingles for the "starters" at the eaves, as the first or "starting" course of shingles must always be doubled.

Obtain the number of lin. ft. of hips, valleys and ridges to be covered with asphalt shingles and compute same as 1'-0" wide.

Many roofing contractors do not measure hips, valleys and ridges, preferring to add a percentage of the roof area to cover these items and waste. The following percentages are commonly used :

Gable roofs, 10 percent; hip roofs, 15 percent; hip roofs with dormers and valleys, 20 percent.

Asphalt shingles must be properly nailed—6 nails to a strip, and nailed low enough on the shingle (right at the cut-out), otherwise they will blow off the roof.

Most manufacturers produce a shingle, designed for high wind areas, that interlock in such a manner that all of the shingles are integrated into a single unit. Interlocking shingles are available in single coverage for re-roofing and double coverage for new construction.

Self-sealing shingles with adhesive tabs are produced by most manufacturers.

When using asphalt shingles for roofs, roll asphalt roofing of the same materials as the shingles is often used for forming valleys, hips and roof ridges.

Nails Required for Asphalt Shingles.—When laying individual asphalt shingles, use a 12 ga. aluminum nail, 1½" long, with a 7/16" head. For laying over old roofs, use nails 1¾" long.

When laying square butt strip shingles, use 11 ga. aluminum nails, 1" long, with a 7/16" head. For laying over old roofs, use nails 1¼" long.

It will require about 1 lb. of nails per sq. of shingles laid.

Sizes and Estimating Data on Asphalt Shingles

Kind of Shingle	Size	No. Shingles per Sq.	Expos. Inches	Length Nails	No. Nails per Shingle	Lbs. Nails per Sq.
3-in.-1 strip	12"x36"	80	5	1	4	1

Approximate Prices on Mineral Surfaced Asphalt Shingles, Seal Tab

Kind of Shingle	Size	Weight per Square	Price per Square
3-in-1 strip	12"x36"	235 lbs.	$26.00
3-in-1 strip	12"x36"	340 lbs.	55.00
3-in-1 strip	12"x36"	350 lbs.	65.00

Most asphalt shingles carry an Underwriters Class "C" rating. By incorporating a glass fiber layer an "A" rating can be had. Class "A" shingles weighing 225 lbs. will run $30.00 per square and weighing 300 lbs., $59.00 per square.

A roofer will lay one square (100 sq. ft.) of asphalt shingles in 1 hour on double pitched roofs with no hips, valleys or dormers. Figure about 35 lin. feet per hour for fitting hips, valleys and ridges. Hip and ridge roll material will run around 16 cts. per lin. ft.; valley, 30¢ per lin. ft.

Labor Cost of 100 Sq. Ft. (1 Sq.) Strip Asphalt Shingles on Plain Double Pitch or Gable Roofs

	Hours	Rate	Total	Rate	Total
Carpenter	1.0	$. . . .	$. . . .	$16.47	$16.47

On roofs having gables, dormers, etc. add 0.2-hr. time per sq.
On difficult constructed hip or English type roofs, add 0.5-hr. time per sq.

07312   ASBESTOS CEMENT SHINGLES

Asbestos cement shingles were widely used as a roofing and wall siding material through the 1950's. With the advent of improved construction materials technology, materials that are less brittle, lighter in weight, easier to install, and cheaper to manufacture have replaced the asbestos cement materials. Additionally, with Governmental O.S.H.A. requirements to reduce the potential of air-borne asbestos fibers, this material is near "unavailable" in the current market. Johns Manville, once the largest manufacturer of this product, has ceased making the material.

What current demand there is, is for repairs to existing buildings and, in particular, historical preservation work; and this demand is met by a handful of manufacturers on a special order basis.

The following text on this product is written in order that those estimators requiring information on this material may be familiar with the patterns and labor requirements; however, material prices are not given and must be obtained by the estimator.

Asbestos shingles are made of asbestos fiber and portland cement, united under hydraulic pressure and cut to single shapes. They are furnished in a variety of sizes, shapes and colors, so the cost will depend upon the style and color of shingles used, and method of laying.

When estimating quantities of asbestos shingles, the measurements are taken in the same manner as for shingle or slate roofs. The length of the eaves must be measured to obtain the number of lin. ft. of starters required, also the length of the hips and ridges, as either shingles or a special ridge roll will be required. Asbestos shingles are sold by the square containing sufficient shingles to cover 100 sq. ft. of surface.

They are made in hexagonal and Scotch or Dutch Lap styles, individual shingles, and slates and strip shingles, among others, which are described on the following pages.

Preparation of Roof.—Asbestos shingles are for use on pitched roofs only. For strip shingles the minimum allowable slope of roof is 4" per foot; for the hexagonal and Dutch Lap Method shingles, 5" per foot. A slope of 3" per foot is permissible with adequate underlayment procedure.

The roofing boards should be well seasoned, of narrow width and laid in the usual manner breaking joints and nailing securely in place with at least two nails at each rafter or purlin, leaving no loose ends. Plywood may be used for sheathing. A cant for the eaves shingles should be provided by the projection of the cornice molding ¼" above the roof boarding or by a lath (¼"x1½") nailed parallel and flush with lower edge of roof boarding. Over the roof boarding lay one thickness of 15-lb. or heavier, asbestos roofing felt, laying horizontally with a 4" lap and with a 12" lap on hips, ridges and valleys. Ridge shingles should be laid in the direction away from prevailing winds.

Strip Asbestos Shingles.—When laying strip shingles, apply over the roofing felt, one course of starter (9"x16") shingles at eaves lengthwise and parallel with same, overhanging the eaves about 1". Apply the second course, using main body shingles, entirely covering the first course, breaking joints, after which proceed in the regular manner as with wooden shingles or slate exposing 7" to the weather, with a minimum headlap of 2". Fasten each shingle in place with at least 2 galvanized iron or copper needle pointed shingle nails. Never drive nails down tight as it is only necessary to drive them firmly. Over the ridges and hips apply asbestos hip and ridge shingles 4¾" to 5⅜" wide by 14" long.

All chimneys, valleys, etc., should be flashed with copper or other approved material.

American Method shingles are 8"x16" in size, ¼" thick, and require about 260 shingles per square, or 16x16 in size, ¼" thick and requires 130 shingles per square.

For new work figure 3 lbs. 1½" nails, and for re-roofing figure 4 lbs. 2" galvanized nails per square.

Hexagonal Method Asbestos Shingles.—The Hexagonal or "Honeycomb" Method of applying asbestos shingles appears to show 6 sides of the shingle, thus overcoming the objections to severely straight lines and producing a pleasing effect.

Apply one course of starter (4"x16") shingles end to end, parallel with and over-hanging the eaves 1", over which apply one course of half shingles, entirely covering the starter and breaking all joints.

Cover balance of roof with body shingles 16"x16", exposing 13"x13" to the weather. Securely fasten all shingles in place with galvanized or copper needle pointed nails, and fasten the points of the main body shingles with special copper storm nails. Never drive nails down tight as it is only necessary to drive them firmly. All the main body shingles should be laid with the diagonal lines on a 45-degree angle with the eaves. Over the ridges and hips apply asbestos ridge and hip rolls, with not less than 3" lap, fastening in place with special roll fasteners furnished for the purpose.

Scotch or Dutch Lap Method of Laying Asbestos Shingles.—The Scotch or Dutch Lap Method of applying asbestos shingles closely resembles the American Method in appearance yet approximates the cost of the Hexagonal Method.

Apply over the felt at eaves, one course of eave starter shingles (4"x16") parallel with and overhanging eaves 1", with each shingle overlapping preceding shingle approximately ⅓ its length so that nail holes will occur one over the other. Cut the first starter to ⅔ its original length and with a new hole punched 2 13/16" from the cut end, apply at the end of the roof to overhang the gable 1". Fasten with a nail through the new hole. Then apply a full length starter to overlap the first starter one-half its length. Fasten both with a nail through coinciding holes. Proceed in a similar manner to end of roof.

Completely cover starters and balance of roof with body shingles (16"x16"), exposing 10⅔"x13" to the weather, securing each shingle with 2 shingle nails and 1 storm anchor. The shingles are aligned on nail and storm anchor holes and therefore automatically establish a one-third side lap and 3" head lap. If the shingles are laid with one-quarter side lap, 12"x13" is exposed. Ridges, hips and exterior corners are finished with asbestos ridge and hip shingles or asbestos ridge roll in the same manner as with other types of asbestos shingles.

Data on Scotch or Dutch Lap Method Asbestos Shingles

Size	Thickness	Laid with 4" or ¼ Side Lap			Laid with 5½" or ⅓ Side Lap		
			No.	Wt.		No.	Wt.
	Inches	Expos.	per Sq.	per Sq.	Expos.	per Sq.	per Sq.
16"x16"	5/32"	12"x13"	92	265 lbs.	10⅔"x13"	104	300 lbs.

When laid with 4" or ¼ side-lap, it requires (184) 0.8-lb. of 1¼" needlepoint nails and 92 storm buttons per square.

When laid with 5⅓" or ⅓ side-lap, it requires 104 shingles, (216) 1 lb. of 1¼" needlepoint nails and 104 storm buttons per square.

Starter shingles for the above may be 4"x16" laid lengthwise.

Data on Hexagonal Method Asbestos Shingles

Size	Thickness Inches	Expos. Inches	No. Singles per Sq.	Wt. Lbs. per Sq.	Nails Reqd. Lgth.	Lbs.
16"x16"	5/32"	13"x13"	86	240	1¼"	1*

*Also requires 86 storm anchors per square.

Requires 172 needlepoint nails and 86 storm buttons per square.
Starter shingles for above may be 4"x16" laid lengthwise.
Eaves shingle for use with hexagonal shingles 21" long requires 58 pcs. per 100 lin. ft.

Laying Asbestos Shingles by the Hexagonal Method.—When laying 16"x16" asbestos shingles by the hexagonal method on fairly plain roofs, a crew consisting of two roofers and one helper should handle and lay 600 to 700 sq. ft. (6 to 7 squares) per 8-hr. day.

On more complicated roofs, the labor cost will be higher due to the necessity of cutting and fitting shingles. This additional labor will vary, depending upon how badly the roof is broken up, but a crew consisting of two roofers and a helper should handle and lay 400 to 500 sq. ft. (4 to 5 sqs.) per 8-hr. day, unless very unusual conditions exist.

Laying Asbestos Shingles by the Dutch Lap Method.—When laying 16"x16" asbestos shingles by the Dutch Lap Method on either plain or complicated roofs, figure approximately 15% more than labor costs given for the Hexagonal Method.

Labor Cost of 100 Sq. Ft. (1 Sq.) 8"x16" Roof Shingles Laid on Plain Roofs

	Hours	Rate	Total	Rate	Total
Roofer	4	$....	$....	$16.47	$65.88
Helper	2	....	....	12.54	25.08
Cost per 100 sq. ft.			$....		$90.96

If laid on a complicated or "cut-up" roof, add 2.5-hrs. roofer and 1.25-hrs. helper time per 100 sq. ft.

Labor Cost of 100 Sq. Ft. (1 Sq.) 16"x16" Asbestos Shingles Laid
Hexagonal Method on Plain Roofs

	Hours	Rate	Total	Rate	Total
Roofer	2.4	$....	$....	$16.47	$39.53
Helper	1.2	....	....	12.54	15.05
Cost per 100 sq. ft.			$....		$54.58

If laid on a complicated or "cut-up" roof, add 1.2-hrs. roofer and 0.6-hr. helper time per 100 sq. ft.
Add for starters, eaves starters and hip and ridge shingles.

Labor Cost of 100 Sq. Ft. (1 Sq.) 16"x16" Asbestos Shingles Laid Dutch
Lap Method with 4" or ¼ Side Lap on Plain Roofs

	Hours	Rate	Total	Rate	Total
Roofer	2.8	$....	$....	$16.47	$46.12
Helper	1.4	....	....	12.54	17.56
Cost per 100 sq. ft.			$....		$63.68

If laid on a complicated or "cut-up" roof, add 1.4-hrs. roofer and 0.7-hr. helper time per 100 sq. ft.
Add for starters and hip and ridge shingles.

Labor Cost of 100 Sq. Ft. (1 Sq.) 16"x16" Asbestos Shingles Laid Dutch
Lap Method with 5⅓" or ⅓ Side Lap on Plain Roofs

	Hours	Rate	Total	Rate	Total
Roofer	3.2	$....	$....	$16.47	$52.71
Helper	1.6	....	....	12.54	20.06
Cost per 100 sq. ft.			$....		$72.77

If laid on a complicated or "cut-up" roof, add 1.6-hrs. roofer and 0.8-hr. helper time per 100 sq. ft.

Applying Asbestos Shingles over Old Roofs.—When applying asbestos shingles over old roofs, it will be necessary to place wood strips or laths over the old roof. Figure about 300 lin. ft. of strips per 100 sq. ft. of roof, and 0.5 to 1-hr. time to place same.

Asbestos Cement Sidewall Shingles and Siding are manufactured from asbestos fibers and portland cement, which are formed into various shingle shapes under hydraulic pressure.

They are furnished in various colors and white, and are wood grained to represent wood shingles and siding, and are used on both new and old work.

Asbestos cement siding shingles are furnished in strips 12"x24, 12"x27", or 32"x14⅝", and may be obtained with either staggered or thatched butts or with a wavy edge.

Estimating Data on Asbestos Cement Shingle Siding

Description Textured or Smooth	Dimensions	No. Req'd Per Sq.	Exposure Inches
Rock Shakes	24x12	54	11
Rock Shakes	32x14⅝	33	13⅝
Rock Shakes	32x 9	57	8
Textured Only			
Wavy Butt	24x12	54	11
Smooth Only			
Sheet	32x96	4.7	*

*Use 3½" backer strips.
Nails are furnished with shingles.

Special alloy nails for exposed nailing and asphalt joint strips are usually furnished by the manufacturers and are included in the price of the siding.

Labor Placing Asbestos Shingles on Sidewalls.—The labor cost of placing asbestos shingle siding will vary with the size of the shingle and the type of building, whether one having fairly long straight walls or one cut up with numerous openings requiring considerable cutting and fitting.

Where the old walls are crooked and it is necessary to straighten the old walls and place furring strips on same ready to receive the shingles, an extra allowance should be made for this work. Figure about 1.25 lin. ft. of furring to each square foot of siding.

Before placing asbestos shingles on sidewalls, the walls should first be covered with a heavy insulating and sheathing felt to keep out air and wind and prevent the penetration of water.

After the walls have been covered with paper, the courses of shingles should be marked with a chalk line, with the tops of the shingles following the chalk line. This will result in straight courses and a satisfactory job.

An asbestos siding cutter should be used for cutting siding as it will result in lower labor costs and prevent breakage.

On straight sidewalls using 24"x12" shingles, a carpenter should place 250 to 300 sq. ft. per 8-hr. day. On complicated work requiring considerable cutting, figure 150 to 200 sq. ft. per 8-hr. day. Where 32"x14⅝" shingles are used, figure 375 to 425 sq. ft. per 8-hr. day for straight walls and 225 to 275 sq. ft. per 8-hr. day for cut-up work.

If wood stripping is required for nailing siding, add extra for strips and labor applying same.

Labor Cost of 100 Sq. Ft. (1 Sq.) 24"x12" Asbestos Sidewall Siding on
Straight Walls

	Hours	Rate	Total	Rate	Total
Carpenter	3	$....	$....	$16.47	$49.41
Cost per sq. ft.			....		.49

On jobs requiring considerable cutting and fitting, add 1.5-hrs. carpenter time.

Colorbestos*.—Colorbestos is an asbestos-cement product manufactured in sheets 32" wide, 96" long and 3/16" thick and is intended primarily for use as an exterior wall finish to be applied vertically from sill to plate. It is equally adaptable for soffits and exterior ceilings.

The face of this material is textured with striations running the long dimension of the sheet accomplished by embedding colored granules in the asbestos-cement sheet. Although the sheet appears to have an ingrained texture, the surface is smooth to the touch.

Labor Placing Colorbestos.—Colorbestos is intended primarily for use on new construction and the large sheets are applied over a continuous backing of wood or non-wood sheathing. The sheets are applied with the vertical joints butted. Horizontal joints are lapped or may be butted when flashed with a "Z" type, non-corroding metal flashing. Exterior corners may be finished with either wood or metal trim.

Each sheet has 33 nail holes, ⅛" in diameter, pre-punched at the factory, along the sides and down the center. Where non-wood sheathing is used, studs must be accurately spaced 16" on center to permit nails to be driven through the sheathing into the studs; studs should be centered behind all vertical joints. Both wood and non-wood sheathing should first be covered with No. 15 asphalt saturated felt lapped 2" at horizontal joints and 6" at vertical joints. Nails, for fastening sheets, should be of non-corroding metal such as aluminum, etc.

Field cutting of Colorbestos is usually done by the "score and snap" method. Using a carbide tipped blade, a score mark is made in the surface of the sheet at the location of the cut and, with the sheet supported so that the edge of the support occurs under the cut-off line, the piece is snapped off cleanly. Cutting may also be accomplished by using a power saw equipped with a carbide tipped blade. To keep field cutting to a minimum, shorter factory-cut sizes may be ordered along with full size sheets.

If additional holes for nailing are required, they may be field punched using a sharpened nail set. However, holes close to the edge should be drilled.

Handling and placing full sheets of Colorbestos at ground level may be done by one man as a 32"x96" sheet weighs only 36 lbs. Each full sheet covers 21⅓ sq. ft. or 4.7 sheets provide 100 sq. ft. of finished surface. On ground story wall with a minimum of cutting, a carpenter should place 400 to 450 sq. ft. per 8-hr. day. Where the walls are cut up with door and window openings, a man should place 250 to 300 sq. ft. per 8-hr. day.

---

*Johns-Manville, Denver, Colo. - No longer makes this product.

For gable ends, where each piece must be cut to fit the rake of the roof and placing must be done from a scaffold, two carpenters working together, should place 375 to 425 sq. ft. per 8-hr. day. For 1 story soffit and exterior ceiling work, a carpenter should place 300 to 350 sq. ft. per 8-hr. day.

Labor Cost of 100 Sq. Ft. (1 Sq.) Colorbestos Siding on Ground
Story Walls with a Minimum of Cutting

	Hours	Rate	Total	Rate	Total
Carpenter	1.9	$. . . .	$. . . .	$16.47	$31.29
Cost per sq. ft.			. . . .		.31

For walls requiring cutting and fitting around door and window openings, add 1-hr. carpenter time.

Add cost of metal or wood corner trim.

Labor Cost of 100 Sq. Ft. (1 Sq. Colorbestos Siding on Gable Ends)

	Hours	Rate	Total	Rate	Total
Carpenter	4	$. . . .	$. . . .	$16.47	$65.88
Cost per sq. ft.			. . . .		.66

Add cost of scaffolding.

Labor Cost of 100 Sq. Ft. (1 Sq.) Colorbestos on 1-Story Soffits or Ceilings

	Hours	Rate	Total	Rate	Total
Carpenter	2.5	$. . . .	$. . . .	$16.47	$41.18
Cost per sq. ft.			. . . .		.41

07313 WOOD SHINGLES AND SHAKES

The labor cost of laying wood shingles will vary with the type of roof, whether a plain gable roof, a steep roof, or one cut up with gables, dormers, etc. Also with the manner in which they are laid, whether with regular butts, irregular or staggered butts, or thatched butts.

The costs given on the following pages are based on the actual number of shingles a man will lay per day and not upon the number of squares covered, as this will vary with the spacing of the shingles. It does not make any particular difference to the shingler whether the shingles are laid 4", 4½", or 5" to the weather, as he will lay practically the same number either way. It does make considerable difference in the number of sq. ft. of surface covered, which will vary from 10 to 40%.

The number of shingles laid will vary with the ability of the workman and the class of work, as ordinary carpenters will not lay as many shingles as carpenters who specialize in shingle laying. On the other hand, experienced shinglers usually demand a higher wage rate than ordinary carpenters.

Some carpenters claim to be able to lay 16 bundles, (3,200 shingles) per 8-hr. day, but this is unusual and is generally found only on the cheapest grade of work where only one nail is driven into each shingle instead of two that are necessary to secure a workmanlike job.

Estimating the Quantity of Wood Shingles.—Ordinary wood shingles are furnished in random widths, but 1,000 shingles are equivalent to 1,000 shingles 4" wide.

Dimension shingles are sawed to a uniform width, being either 4", 5" or 6" wide.

Wood shingles are usually sold by the square based on sufficient shingles to lay 100 sq. ft. of surface, when laid 5" to the weather, 4-bundles to the square.

Labor Laying One Square (100 Sq. Ft.) Wood Shingles

Class of Work	Mechanic	Number Laid per 8-Hr. Day	Distance Shingles are Laid to Weather					
			4"†	4¼"	4½"	5"	5½"	6"
Plain Gable or Hip Roofs	Carpenter	2,000–2,200	3.8 hrs.	3.6 hrs.	3.4 hrs.	3.0 hrs.	2.8 hrs.	2.5 hrs.
	Shingler	2,750–3,000	2.8	2.6	2.5	2.3	2.0	1.9
Difficult Gable Roofs, cut up with gables, dormers, hips, valleys, etc.	Carpenter	1,700–1,900	4.5	4.2	4.0	3.6	3.3	3.0
	Shingler	2,200–2,500	3.4	3.2	3.0	2.7	2.5	2.3
Difficult Hip Roofs, Steep English Roofs, hips, valleys, etc.	Carpenter	1,300–1,500	5.7	5.4	5.1	4.6	4.2	3.8
	Shingler	2,000–2,200	3.8	3.6	3.4	3.0	2.8	2.5
Shingles Laid Irregularly or with Staggered Butts on Plain Roofs.	Carpenter	1,700–1,900	4.5	4.2	4.0	3.6	3.3	3.0
	Shingler	2,400–2,700	3.1	3.0	2.8	2.5	2.3	2.1
Shingles Laid Irregularly or with Staggered Butts on Difficult Constructed Roofs.	Carpenter	1,100–1,300	6.7	6.3	6.0	5.4	4.9	4.5
	Shingler	1,600–1,800	4.7	4.5	4.2	3.8	3.4	3.1
Shingles with Thatched Butts.*	Shingler	800–1,000	8.9	8.4	8.0	7.1	6.5	6.0
Plain Sidewalls.	Carpenter	1,300–1,500	5.7	5.4	5.1	4.6	4.2	3.8
	Shingler	1,700–1,900	4.5	4.3	4.0	3.6	3.3	3.0
Difficult Sidewalls, having bays, windows, breaks, etc.	Carpenter	1,100–1,250	6.8	6.4	6.0	5.4	5.0	4.5
	Shingler	1,400–1,650	5.2	5.0	4.6	4.2	3.8	3.5

To obtain number of bundles of shingles required, divide number of shingles as given above by 200, and the result will be the number of bundles required, i.e, 2,200 shingles ÷ 200 equals 11 bundles; 1,000 ÷ 200=5 bundles; 800 ÷ 200=4 bundles, etc.

*Shingles with thatched butts require 25% more shingles than when laid regularly.

†Use 4" column for carpenter or shingler time per 1,000 shingles. (5 bundles).

The above table is based on using 2 nails to each shingle and 10 per cent waste.

When estimating the quantity of ordinary wood shingles required to cover any roof, bear in mind that the distance the shingles are laid to the weather makes considerable difference in the actual quantity required.

There are 144 sq. in. in 1 sq. ft. and an ordinary shingle is 4" wide. When laid with 4" exposed to the weather, each shingle covers 16 sq. in. or it requires 9 shingles per sq. ft. of surface. There are 100 sq. ft. in a square. 100x9=900, and allowing 10% to cover the double row of shingles at the eaves, waste in cutting, narrow shingles, etc., it will require 990 shingles (5 bundles) per 100 sq. ft. of surface.

Number of Shingles and Quantity of Nails Required Per 100 Sq. Ft. of Surface

Distance Laid to Weather	Area Covered by One Shingle Sq. In.	Add for Waste Per Cent	Actual No. per Square Without Waste	Number per Square With Waste	No. of 4-Square Bundles Required	Pounds 3d Nails Req'd
4 "	16	10	900	990	5.0	3.2
4½"	17	10	850	935	4.7	2.8
4½"	18	10	800	880	4.4	2.5
5 "	20	10	720	792	4.0	2.0
5½"	22	10	655	720	3.6	1.6
6 "	24	10	600	660	3.3	1.5

How to Apply Shingles for Different Roof Slopes.—Roof pitches are computed in fractions, such as ⅛, ⅓, ½ pitch. In the following cross-section, the steepness of distances AB and BC constitutes pitch. Distance AC, extending from one eave-line to the other, is known as the span. One-half of this span, distance AD or DC, is called the run, and distance BD is called the rise. The relationship of the rise to run obviously affects the slope of AB or BD; in fact, roof pitches are computed from the ratio of rise to run. Therefore, the first step is to determine length of the run (AD or DC) and the rise (BD).

Wood shingles are manufactured in three lengths—16-inch, 18-inch, and 24-inch. The standard weather exposure (portion of shingle exposed to weather on roof) for 16 inch shingles is 5", for 18-inch shingles it is 5½", and for 24-inch shingles it is 7½". These standard exposures are recommended on all roofs of ¼ pitch and steeper (6" rise in 12" run). On flatter roof slopes, the weather exposure should be reduced to 3¾" for 16-inch shingles, 4¼" for 18-inch shingles, and 5¾" for 24-inch.

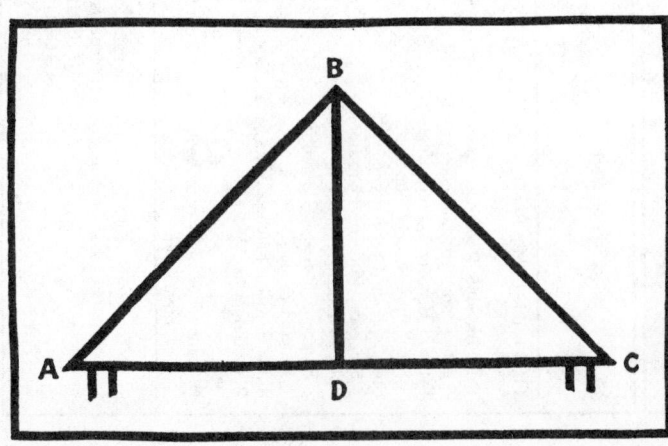

The following diagram shows at a glance, the weather exposure to be used for various roof pitches. For example, if a roof has a rise of 8" in a run of 12", it can be seen that this is ⅓ pitch and that an exposure of either 5", 5½" or 7½" should be employed, depending upon the length of the shingles used.

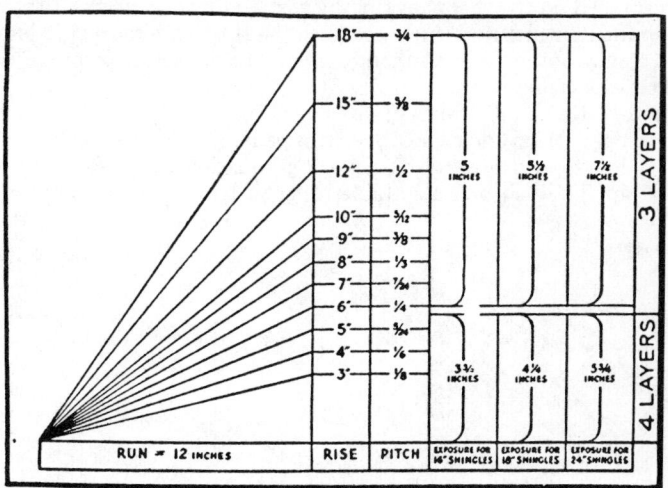

Wood shingles and shakes may be laid over open or solid sheathing, with solid preferred where wind driven snow conditions prevail. If open sheathing is selected, use 1x4s at centers equal to the weather exposure at which the shakes are to be laid, but not over 10". Building paper must be added for additional weather protection.

Western Red Cedar Shingles Will Cost as Follows:

144# 16" #1-5X Shingles 4 Bdl. per Sq.. $85.00
144# 16" #2-5X Shingles 4 Bdl. per Sq.. 75.00
60# 16" Undercourse 14/14 2 Bdl. per Sq....................................... 17.00
158# 18" #1 Perfection 4 Bdl. per Sq. 85.00
350# 24" #1 3/4" to 5/4 Handsplit
    Shakes, 5 Bdl. per Sq.. 95.00
    For fire retardant shingles add 100%

07314 SLATE ROOFING

Roofing slate is furnished in a number of sizes, thicknesses and finishes to meet the architectural requirements of the buildings on which they are to be used. The four principal classifications are: standard slate; textural slate; graduated slate and flat slate roofs.

Standard Roofs.—Standard slate roofs are composed of slate approximately 3/16" thick (commercial standard slate), of one uniform standard length and width; also one length and random widths or random lengths and widths. Any width less than one-half the length is not recommended. Standard roofs are suitable for any building where a permanent roofing material is desired at a minimum cost. If desired, the butts or corners may be trimmed to give a hexagonal, diamond or "Gothic" pattern for all or part of the roof.

Textural Roofs.—The term "textural" is used to designate slate of rougher texture than the standard, with uneven tails or butts, and with variations of thickness or size. In general, this term is not applied to slate over ⅜ " in thickness.

Graduated Roofs.—The graduated roof combines the artistic features of the textural slate roof with additional variations in thickness, size, and exposure. The slate is so arranged on the roof that the thickest and longest occur at the eaves and gradually diminish in size and thickness until the ridges are reached. Slate for roofs of this type can be obtained in any combination of thicknesses from 3/16" to 3/4" and heavier when especially desired.

In addition to the usual standard sizes, slate above ½ " thick are produced in lengths up to 24". The graduations in lengths generally range from 24" to 12". The variations in length will at once provide a graduation in exposure by using the standard 3" lap. To illustrate, a suitable range of lengths might be :

Under Eaves Course	3/8" thick	14" long	No exposure
First Course	3/4" "	24" "	10 1/2" "
1 "	3/4" "	24" "	10 1/2" "
2 Courses	1/2" "	22" "	9 1/2" "
2 "	1/2" "	20" "	8 1/2" "
2 "	3/8" "	20" "	8 1/2" "
4 "	3/8" "	18" "	7 1/2" "
5 "	1/4" "	16" "	6 1/2" "
3 "	1/4" "	14" "	5 1/2" "
3 "	3/16" "	14" "	5 1/2" "
8 "	3/16" "	12" "	4 1/2" "

Random widths should be used and so laid that the vertical joints in each course are broken and covered by the slate of the course above.

Flat Roofs.—Flat roofs offer a wide field for roofing slate and are designated whether or not they are used for "promenade" purposes.

Slate of any thickness may be used instead of slag or gravel as a surfacing material for the usual built-up type of roof. For ordinary roofs the standard 3/16" slate are used but for promenade or extraordinary service, the slate should be 1/4" or 3/8" thick.

Slate for flat roofs are furnished in the following standard sizes: 6"x6", 6"x8", 6"x9", 10"x6", 10"x7", 10"x8", 12"x6", 12"x7" and 12"x8".

Estimating Quantities of Roofing Slate.—Obtain the net area of the roof in sq. ft. as described earlier in this chapter. Add 6" to rafter length to allow for waste of normal roof.

Deduct one-half the area of chimneys and dormers if over 20 sq. ft. and less than 80 sq. ft. Make no deductions if less than 20 sq. ft. and deduct 20 sq. ft. less than actual area if more than 80 sq. ft.

Include area of dormer sides, sides of dormers if slated, slate saddles, and other places where slate is used in addition to the main roof area. Include overhanging parts of dormers, etc.

Add 1 sq. ft. for each lin. ft. of hips and valleys, for loss in cutting and fitting.

Allow 2 to 15 percent additional slate, depending upon the extent to which the roof is intersected by other roofs, dormers, walls, etc.

Divide the total of the above by 100, which will give the number of "squares" of roofing required.

Slate is always sold at the quarry on the basis of the quantity required to cover 100 sq. ft. or a "square" of roof when slate is laid with a 3" head lap. If the roof is flat or other than 3" lap is used, the quantity must be corrected to the equivalent amount required as though the 3" lap were used.

Items to be Included in an Estimate for Slate Roofing.—The following items should be included to make a complete and accurate estimate on slate roofing:

1. Cost of slate (punched) on cars or trucks at the quarry.
2. Freight or trucking from quarry to destination.
3. Unloading from cars and hauling to job, also unloading and piling at job, unless delivered to the job by truck. Trucking from quarry direct to job within a radius of 500 miles is now general practice in the Vermont-New York slate district.
4. Placing on roof and laying, (a) roofing felt, (b) elastic cement, (c) nails, (d) snow guards or snow rails, (e) sheet metal, (f) labor, (g) waste in handling, cutting and fitting.

Size of Slate.—The following table gives the different sizes in which roofing slate is furnished, number required per square when laid with 3" lap, and number of nails required per square based on 2 nails to each piece of slate.

Size of Slate Inches	Number per Square	Number of Nails Reqd. per Square	Size of Slate Inches	Number per Square	Number of Nails Reqd. per Square
10x 6	686	1,372	16x12	185	370
10x 7	588	1,176	16x14	159	318
10x 8	515	1,030	18x 9	213	426
10x10	414	828	18x10	192	384
12x 6	533	1,066	18x11	175	350
12x 7	457	914	18x12	160	320
12x 8	400	800	18x14	137	274
12x 9	355	710	20x10	169	338
12x10	320	640	20x11	154	308
12x12	267	534	20x12	141	282
14x 7	374	748	20x14	121	242
14x 8	327	654	20x16	106	212
14x 9	290	580	22x11	138	276
14x10	261	522	22x12	126	252
14x12	218	436	22x14	109	218
14x14	188	376	22x16	95	190
16x 8	277	554	24x12	115	230
16x 9	246	492	24x14	98	196
16x10	221	442	24x16	86	172

Slate ¾" and thicker and 24" or over in length should have 4 nail holes.

Weight of Slate Roofing.—A square of slate roofing, i.e., sufficient slate to cover 100 sq. ft. of roof surface with a standard 3" lap, will vary from 650 to 8,000 lbs. for thicknesses from the commercial standard 3/16" to 2".

The weight of slate varies with the size of the slate, color and quarry, and even sometimes in the same quarry. The variation may be from 10% above to 15% below the weights given in the following table :

Average Weight of Slate per Square (100 sq. ft.)

	Slate Thickness in Inches	Sloping Roof Allowing for 3" Lap	Flat Roof Without Lap
Standard	3/16	700	240
Selected full	3/16	750	250
	1/4	900	335
	3/8	1,400	500
	1/2	1,800	675
	3/4	2,700	1,000
	1	4,000	1,330

Slate Thickness in Inches	Sloping Roof Allowing for 3" Lap	Flat Roof Without Lap
1 1/4	5,000	1,670
1 1/2	6,000	2,000
1 3/4	7,000	
2	8,000	

Nails Required for Slate Roofing.—The quantity of nails required for slate roofing will vary with size, kind, etc., but the following table gives quantities of those commonly used :

Approximate Number of Nails to the Pound

Length in Inches	"Copperweld" Slating Nails	Copper Wire Slat. Nails	Cut Copper Slat. Nails	Cut Brass Nails	Cut Yellow-Metal Slat. Nails
1	386	270	. . .	. . .	. . .
1¼	211	144	190	164	154
1½	176	134	135	140	140
1¾	133	112	100	108	. . .
2	87	104	. . .	88	. . .
2¼	. . .	46	. . .	80	. . .
2½	. . .	. . .	. . .	64	. . .
2¾	. . .	. . .	. . .	52	. . .
3	. . .	. . .	. . .	48	. . .

Labor Punching Slate.—Because of high labor costs on the job, it is much cheaper to have slate punched at the quarry. The labor cost of punching slate on the job will vary with the size of the slate, as the larger the slate the fewer pieces required per square.

If one slater is working on the job with a helper, the helper should punch the slate and carry them to the slater as fast as required. In instances of this kind, there should be no additional charge for punching. If, however, there is only one man working on the job, who must punch the slate and lay them, a man should punch one square of slate in 30 to 40 min. or 12 to 16 sqs. per 8-hr. day, at the following labor cost per 100 sq. ft. (sq.) :

	Hours	Rate	Total	Rate	Total
Slater	0.6	$. . . .	$. . . .	$17.18	$10.31

Labor Cost of Laying Roofing Slate.—The labor cost of laying roofing slate will vary considerably with the size of the slate, thickness, type of roofs, etc., but when large thick slate are used for graduated roofs, the labor costs will run higher than when the same size slate are used on standard roofs.

The size of slate best adapted for plain roofs are large, wide slates, such as 16"x12", 18"x12", 20"x12", or 24"x12". Slate from 16"x8" to 20"x10" are also popular sizes. For roofs cut up into smaller sections, the 14"x7" and 16"x8" are very popular.

A roofing crew consisting of two slaters and one helper can work to advantage and perform the work more economically than one man working alone. In some localities the helper is considered an apprentice and is permitted to lay slate, in others he can only punch and carry the slate to the slaters.

Laying 16"x8" Roofing Slate.—When working on plain roofs that do not require much cutting, 2 slaters and a helper should lay felt, handle and lay 350 to 450 sq. ft. (3½ to 4½ sqs.) of 16"x8" slate per 8-hr. day. On more complicated hip or gable roofs, requiring considerable cutting and fitting for hips, valleys, dormers, etc., 2 slaters and a helper should handle and lay 175 to 225 sq. ft. (1¾ to 2¼ sqs.) of 16"x8" slate per 8-hr. day.

Laying 18"x9" Roofing Slate.—When laying 18"x9" slate on plain roofs, 2 slaters and a helper should lay felt, handle and lay 400 to 500 sq. ft. (4 to 5 sqs.) per 8-hr. day.

On more complicated hip or gable roofs, requiring considerable cutting and fitting for hips, valleys, dormers, etc., 2 slaters and a helper should lay 250 to 300 sq. ft. (2½ to 3 sqs.) per 8-hr. day.

Laying 20"x10" Roofing Slate.—When laying 20"x10" slate on plain roofs, 2 slaters and a helper should lay felt, handle and lay 500 to 600 sq. ft. (5 to 6 sqs.) of roof per 8-hr. day, but on more complicated roofs, requiring considerable cutting and fitting for hips, valleys, dormers, etc., the same crew would lay only 300 to 350 sq. ft. (3 to 3½ sqs.) per 8-hr. day.

Laying 22"x12" Roofing Slate.—On straight roofs, 2 slaters and a helper should lay felt, handle and lay 550 to 650 sq. ft. (5½ to 6½ sqs.) of 22"x12" slate per 8-hr. day, but on the more complicated roofs, requiring cutting and fitting for hips, valleys, dormers, etc., the same crew would lay only 350 to 400 sq. ft. (3½ to 4 sqs.) per 8-hr. day.

Laying Graduated Roofing Slate.—When laying graduated roofing slate from 3/16" to 3/4" thick and 12" to 24" long, considerable care is required in selecting the right slate for each course and laying them to obtain the desired effect. On work of this kind 2 slaters and a helper should lay felt, handle and lay 150 to 200 sq. ft. (1½ to 2 sqs.) of roof per 8-hr. day.

Labor Cost of 100 Sq. Ft. (1 Sq.) 16"x8" Standard Slate Roofing

	Hours	Rate	Total	Rate	Total
Slater	4	$....	$....	$17.18	$68.72
Helper	2	....	....	12.54	25.08
Cost per 100 sq. ft.			$....		$93.80

For complicated hip or gable roofs, double the time given above.

Labor Cost of 100 Sq. Ft. (1 Sq.) 18"x9" Standard Roofing Slate

	Hours	Rate	Total	Rate	Total
Slater	3.50	$....	$....	$17.18	$60.13
Helper	1.75	....	....	12.54	21.95
Cost per 100 sq. ft.			$....		$82.08

For complicated hip or gable roofs, add 2.5-hrs. slater and 1.25-hrs. helper time.

Labor Cost of 100 Sq. Ft. (1 Sq.) 20"x10" Standard Roofing Slate

	Hours	Rate	Total	Rate	Total
Slater	3.0	$....	$....	$17.18	$51.54
Helper	1.5	....	....	12.54	18.81
Cost per 100 sq. ft.			$....		$70.35

For complicated hip or gable roofs, add 2-hrs. slater and 1-hr. helper time.

Labor Cost of 100 Sq. Ft. (1 Sq.) 22"x12" Standard Roofing Slate

	Hours	Rate	Total	Rate	Total
Slater	2.6	$....	$....	$17.18	$44.67
Helper	1.3	....	....	12.54	16.30
Cost per 100 sq. ft.			$....		$60.97

For complicated hip or gable roofs, add 1.6-hrs. slater and 0.8-hr. helper time.

Labor Cost of 100 Sq. Ft. (1 Sq.) Rough Texture Slate
Roof Random Widths and 3/16" to 3/8" Thick

	Hours	Rate	Total	Rate	Total
Slater	7.50	$....	$....	$17.18	$128.85
Helper	3.75	....	....	12.54	47.03
Cost per 100 sq. ft.			....		$175.88

Labor Cost of 100 Sq. Ft. (1 Sq.) Graduated Slate Roof, Slate 12" to
24" Long and 3/16" to 3/4" Thick

	Hours	Rate	Total	Rate	Total
Slater	9.0	$....	$....	$17.18	$154.62
Helper	4.5	....	....	12.54	56.43
Cost per 100 sq. ft.			$....		$211.05

Prices on Slate Roofing Material

Slate shingles will come from various parts of the country; and the estimator must obtain a quotation from the quarries in that geographical location, not only for price and shipping costs, but for availability of certain sizes and colors.

Vermont slate shingles in 3/16" thicknesses will cost as follows: black or gray - $230.00 per square, green and gray - $240.00, and purples - $300.00. The Pennsylvania black #1 clear and the Buckingham, Virginia black slates will cost about $300.00 per square. Slates thicker than 3/16" will cost appreciably more.

Copper nails are usually used for nailing the slates. Use 1 1/4" long nails for the 3/16" thick slates and 1 3/4" long nails for the rough texture and graduated slate roofs where thicker slates are encountered.

07315 PORCELAIN ENAMEL SHINGLES

Porcelain enamel shingles are manufactured to have an exposed surface of 10"x10" with 144 shingles per square. Weight is 225 lbs. per 100 sq. ft. Finish is fused on at 1500° F. and provides a long lasting, self-cleaning finish that will not peel or blister.

In ordering these units, it is important to have complete shop drawings based on actual job conditions so that field cutting can be held to a very minimum. Shingles are nailed to the roof deck over 2 layers of 30 lb. felt. Special nails, ridge and hip sections are furnished by the manufacturer. Valley flashings and drips should be of non-corrosive metal and furnished by the sheet metal contractor.

Material costs will run in the vicinity of $225.00 per square, depending on the number of roof irregularities which run up the cost considerably, as all special fittings should be made up in the shop. One roofer can lay a square in about 6½ hours.

07316 METAL SHINGLES

Metal shingles formed with a wood grain texture and available in a wide range of colors can be used for a lightweight, decorative roof or false mansard. They can be applied in a conventional way on a solid nailable roof deck over 30 lb. felt underlayment or supported and clipped to sub-purlins on steel frames. Weights are 36 lbs. per square for .020 aluminum, 54 lbs. for .030 aluminum, and 88 lbs. for 30 ga. steel.

The shingles are not individual units, but made in strips either 10"x60", 12"x36", or 12"x48".

The shingle manufacturer can provide anchor clips, drip caps, starter strips, trim, ridge, and hip laps.

Material Costs will run as follows:

Aluminum, .020", mill finish	$ 58.00 per square
Aluminum, .030", mill finish	70.00 per square
Aluminum, .020", anodized color	105.00 per square
Aluminum, .030", anodized color	120.00 per square
Aluminum, .020", bonderized	160.00 per square
Aluminum, .030", bonderized	170.00 per square
Steel, 26 ga., galvanized	27.00 per square
Steel, 24 ga., galvanized	29.00 per square
Steel, 26 ga., color galvanized	60.00 per square
Steel, 24 ga., color galvanized	63.00 per square

Ridge and cap sections will run $1.50 per lin ft. for .020" aluminum, $1.75 for 030". Valley sections will run $1.10 and $1.40. One carpenter will take about 4–5 hours to lay one square.

07321   CLAY ROOFING TILE

The following information on clay roofing tile has been furnished through the courtesy of the Ludowici-Celadon Company, Chicago, Illinois.

When estimating the quantities of clay roofing tile, the roof areas are obtained in the same manner as described at the beginning of this chapter.

After the total number of sq. ft. of surface has been obtained, add about 3 per cent to the net measurement to take care of cutting and waste. Material is sold only in full pallets which vary from 3 to 8.7 squares.

Hip and valley fittings should be figured by the piece. Closure and finishing pieces, such as hip starters, closed ridge ends, terminals, etc., should also be figured by the piece. Fittings are sold only in full boxes.

Where interlocking shingle tiles are used, detached gable rake pieces are required for all gable roofs not intersecting with vertical surfaces. These pieces are measured by the total lineal feet of gable rake, keeping separate quantities for lefts and rights as determined by the roof eave to be covered.

For all shingle tile roofs, end bands or half-tile are required on each course, alternating from one end to the other to break joints. End bands are figured by the lineal feet of gable rake, taking half the length at each gable rake.

Laying Interlocking Shingle Tile.—On a roof of average difficulty, a crew consisting of 2 roofers, and an apprentice or helper should handle and lay 500 to 600 sq. ft. of interlocking shingle tile per 8-hr. day, at the following labor cost per 100 sq. ft. (1 sq.) :

	Hours	Rate	Total	Rate	Total
Roofer	3.0	$. . . .	$. . . .	$17.18	$51.54
Helper	1.5	. . . .	. . . .	12.54	18.81
Cost per 100 sq. ft.			$. . . .		$70.35

On steep or cut-up roofs, add 15 to 25 percent. For roofs over 3 stories high to eaves, add 1.5-hrs. helper time.

Laying Tile Shingles (Architectural Patterns.)—On a roof of average difficulty, 2 roofers and a helper should handle and lay 325 to 375 sq. ft. of tile shingles per 8-hr. day, at the following labor cost per 100 sq. ft. (1 sq.):

	Hours	Rate	Total	Rate	Total
Roofer	4.6	$. . . .	$. . . .	$17.18	$ 79.03
Helper	2.3	. . . .	. . . .	12.54	28.84
Cost per 100 sq. ft.			$. . . .		$107.87

On steep or cut-up roofs, add 15 to 25 percent. For roofs over 3 stories high to eaves, add 2.3-hrs. helper time.

Laying Miscellaneous Tile Pattern Roofs.—Where Spanish or French pattern tile is used on new work, figure same labor costs as for interlocking shingle tile.

Labor laying Mission tile will vary according to size and exposure of units. On new work of average difficulty, using 15"x8" tile with 12" exposure, 2 roofers and a helper should handle and lay 350 to 400 sq. ft. per 8-hr. day.

Items to Be Included When Estimating Tile Roofs.—The following items should be included when estimating the cost of a clay shingle tile roof.

One roll coated felt or roll roofing weighing 40 lbs. per sq. and 3-lbs. roofing nails, 1¾" long.

Allow 1-lb. colored elastic cement for pointing joints for each 20 lin. ft. of hips and ridges.

Cost of Clay Tile Roofing Materials.—The Luduwici-Celadon Company offers a number of different patterns and colors in each of their "Standard" and "Special" series product line. They will also produce custom shapes and colors, if given enough lead time.

Each of the patterns are produced with field tiles and special shapes for the ridge, hip, rake, etc.

Approximate material prices for different patterns and colors are as follows:

Per 100 Sq. Ft.

Spanish Red	$180.00
Spanish Buff	350.00
Spanish Blue	550.00
Mission Granada Red	350.00
Americana Gray or Green	225.00
Lanai Black or Brown	225.00
Classic Red	225.00
French Red	300.00
French Blue	650.00

Details of Clay Interlocking Shingle Tile Roofing

Details of Clay Tile Shingle Roofing

07400 PREFORMED ROOFING & SIDING

07411 PREFORMED METAL SIDING

Corrugated steel is used extensively for roofing and siding steel mills, manufacturing plants, sheds, grain elevators and other industrial structures.

It is made with various corrugations, varying in width and depth, but the 2½" corrugation width is the most commonly used.

The sheets are usually furnished 26" wide and 6'-0" to 30'-0" long. Some manufacturers offer siding protected with vinyl coatings.

When estimating quantities of corrugated siding or roofing, always select lengths that work to best advantage and be sure and allow for both end and side laps.

Corrugated siding or roofing is nailed to the wood framework or siding when used on wood constructed buildings. When used as a wall and roof covering on structural steel framing, the corrugated sheets are fastened to the steel framework, using clip and bolts, which are passed around or under the purlins (which usually consist of channels, angles or Z bars). When angles are used for purlins, clinch nails are sometimes used for fastening the corrugated sheets.

When used for siding one corrugation lap is usually sufficient, but for roofing two corrugations should be used and if the roof has only a slight pitch, the lap should be three corrugations.

When used for siding, a 1" to 2" end lap is sufficient, but when laid on roofs it should have an end lap of 3" to 6" depending upon the pitch of the roof. For a 1/3 pitch a 3" lap is sufficient; for a 1/4 pitch a 4" lap should be used; and for a 1/8 pitch a 5" end lap.

When applying to wood sheathing or strips, the nails should be spaced about 8" apart at the sides. When applied to steel purlins, the side laps should extend over at least 1½ corrugations, and the sheets should be riveted together every 8" on the sides and at every alternate corrugation on the ends.

Number of Sq. Ft. of Corrugated Sheets Required to Cover
100 Sq. Ft. of Surface, Using 26"x96" Sheets

2½-Inch Corrugation Width	Length of End Lap in Inches					
	1	2	3	4	5	6
	Square Feet of Corrugated Metal					
Side lap, 1 corrugation	110	111	112	113	114	115
Side lap, 1½ corrugations	116	117	118	119	120	121
Side lap, 2 corrugations	123	124	125	126	127	128
Side lap, 2½ corrugations	130	131	132	133	134	135
Side lap, 3 corrugations	138	139	140	141	142	143

If shorter sheets are used the allowance should be slightly increased.

Material Prices.—Corrugated steel siding and roofing panels are manufactured in a number of metal gauges, widths and depths of corrugations, and in galvanized and pre-finished painted surfaces. Average material prices, per 100 sq. ft., for sheeting with a 2½" corrugation width are as follows:

Metal Gauge	Galvanized	Painted
28	$39.00	$ 53.00
26	48.00	60.00
24	71.00	88.00
22	78.00	94.00
20	86.00	100.00

Placing Corrugated Steel Roofing or Siding on Wood Framing.—If corrugated sheets used for roofing or siding are nailed to wood strips or framing, a man and helper should place 100 sq. ft. (1 sq.) of 26"x96" or larger, in ⅞ to 1⅛ hr. or 7 to 9 sqs. per 8-hr. day, at the following labor cost per 100 sq. ft. (1 sq.) :

Labor Cost of 100 Sq. Ft. (1 Sq.) Corrugated Metal Roofing on
Wood Framing

	Hours	Rate	Total	Rate	Total
Carpenter	1	$....	$....	$16.47	$16.47
Labor	1	....	....	12.54	12.54
Cost per 100 sq. ft.			$....		$29.01

Labor Cost of 100 Sq. Ft. (1 Sq.) Corrugated Metal Siding on
Wood Framing

	Hours	Rate	Total	Rate	Total
Carpenter	1.2	$....	$....	$16.47	$19.76
Labor	1.2	....	....	12.54	15.05
Cost per 100 sq. ft.			$....		$34.81

Placing Corrugated Steel Roofing on Steel Framing.—The labor cost of applying corrugated steel roofing to structural steel framing where the sheets are fastened with clips or bolts, will vary with the size of the roof area, regularity of roof, height above ground, etc., but on a straight job, 2 sheeters and 2 laborers should place 550 to 650 sq. ft. per 8-hr. day, at the following labor cost per 100 sq. ft. :

	Hours	Rate	Total	Rate	Total
Labor handling, hoisting	2.67	$....	$....	$12.54	$33.48
Sheeters	2.67	....	....	17.70	47.26
Cost per 100 sq. ft.			$....		$80.74

On irregular roofs, above costs may be increased 25 to 50%.

Placing Corrugated Steel Siding on Steel Framing.—The labor cost of applying corrugated steel siding to walls will vary greatly according to length and height of walls, quantity and regularity of openings, etc., but on a straight job, 2 sheeters and 2 laborers should place 450 to 550 sq. ft. per 8-hr. day, at the following labor cost per 100 sq. ft. :

	Hours	Rate	Total	Rate	Total
Labor handling, hoisting	3.2	$....	$	$12.54	$40.13
Sheeters	3.2	....	....	12.70	56.64
Cost per 100 sq. ft.			$....		$96.77

On coal tipples, grain elevators and other structures having high and irregular walls, the above costs may be doubled.

Protected Metal Roofing and Siding

Protected metal panels consists of a steel core sheet to which asbestos felt is bonded by hot molten zinc. When the zinc has cooled, the asbestos felt and the steel core are metallurgically bound together. The felt is then impregnated with asphalt saturant. Finally, a tough, thick, waterproof outer coating is applied. All material is made with protective coating both sides.

There are several colors to choose from. Protected metal panels are manufactured in 4 gauges : 18, 20, 22 and 24, and are available in lengths up to 30 feet.

Estimating and Cost Data.—Panels are available in several profiles. The H.H. Robertson Company, Pittsburg, PA., a major supplier, has the following:

Huski-Rib has a deep fluted profile, designed specifically for high strength long span roofing and siding installations. The Huski-Rib sheet is 31 5/32" wide x 1½" deep, and provides a coverage of 28". It is manufactured in 18, 20, 22 and 24 gauge metal, a maximum of 30' long, and is suitable for spans up to 15'-9" for siding, and up to 14'-14" for roofing depending upon local design load requirements.

Box-Rib has a strong, precise vertical texture, best suited for sidings only, either in standard or inverted positions. The standard sheet is 29¼" wide x 1½" deep and provides a coverage of 28". It is manufactured in 18, 20, 22 and 24 gauge metal and is available in lengths up to 30 ft.

Sturdi-Rib sheets come 34" wide x 9/16" deep providing a coverage of 29 7/16", and are made in lengths up to 30 feet. They are made of 18, 20, 22, and 24 gauge metal and are capable of spanning up to 10'-1" for siding and 8'-9" for roofing installations, depending upon local design load requirements.

Profiles of Robertson Galbestos Roofing and Siding

The Magna-Rib profile sheets are 26⅝" wide x 4" deep providing a coverage of 24" installed. They come in 18, 20, and 22 gauge metal in lengths up to 38 feet. Magna-Rib sheets are capable of spanning up to 22'-8" for siding in one span.

Corrugated Galbestos is the original Robertson protected metal profile and comes in sheets 33" wide x 9/16" deep. Coverage is 29¾". Corrugated sheets are available in 18, 20, 22 and 24 gauge metal in lengths up to 30 feet. They are capable of spanning up to 9'-3" for siding and 7'-11" for roofing installations.

Approximate Price per Sq. Ft. of Corrugated Protected Metal Panels

	Gauge of Sheets			
Description	18	20	22	24
Galbestos C2S (Asphalt) Black	$2.20	$1.80	$1.65	$1.45
Galbestos C2S (Resin) Color	2.60	2.15	1.90	1.70

After figuring the price of sheets from the above, approximately 25 percent should be added to cover cost of flashings, fastenings, closures, corners, etc.

True cost comparisons must include the cost of supporting steel girts or purlins. Longer spans may save up to $0.15 per sq. ft. or more.

Matching profiled translucent fiberglass reinforced plastic daylight panels are available.

07415   CORRUGATED ALUMINUM ROOFING AND SIDING

Types of Fasteners Available for Corrugated Aluminum Roofing and Siding.

When using aluminum roofing and siding, it is very important when applying the sheets that they be insulated against electro-galvanic action. Ordinarily aluminum has a high resistance to corrosion but when it is contiguous to steel or copper in the presence of moisture, an electrolytic cell is formed and the resulting flow of current dissolves the aluminum, causing severe pitting. This condition can be avoided by preventing metal-to-metal contact.

Another significant requirement in using aluminum sheet is to seal the openings made for the fastening devices. Methods for achieving this objective effectively include the use of special washers and waterproof roofing compounds. Washers should be used with fasteners to distribute stresses and prevent tearing of the sheet.

To provide adequate drainage, the roof surface should never have a slope less than 2½ in. per ft. and preferably not less than 3 in. per ft.

For roofing, sheets should have a side lap of 1½ corrugations. For siding, sheets should be lapped 1 corrugation.

A 6 in. overlap at the ends is recommended for roofing and 4 in. for siding.

When corrugated aluminum sheets are used without sheathing on a steel frame, the method shown in (a) is suggested for fastening to purlins and girts. This method utilizes aluminum straps, 12 in. apart, with aluminum bolts and nuts, aluminum rivets or cadmium-plated steel bolts.

To prevent galvanic action, direct contact between the top of the purlin and roofing should be prevented preferably by insertion of an aluminum saddle or by painting the flange of the purlin with aluminum or bitumastic paint. Roofing paper may also be used to separate the metals.

Purlin clips (b) are also widely used as a fastening device. The U bend is slipped over the purlin flange and a leg hammered down against the web to lock the clip in place. The roofing is bolted to the clip. If standard steel purlin clips are used, they should be hot-dip galvanized.

Purlin nails, either single (c) or double (d) are also used for anchoring sheeting. A variation of this method employing aluminum washers, nuts and base plate under the purlin is shown in (e).

The best method of nailing to sheathing is to use galvanized roofing nails about 1¾-in. long with a head about ⅜-in. in diameter, spaced at 8-in. intervals. These nails should have a hot-dip zinc coating.

Corrugated sheets should always be riveted, bolted or nailed through the top of a corrugation. The reason is that rain running down the roof tends to collect in the bottom of corrugations and would penetrate imperfectly sealed holes there. At the top, water runs away from the holes.

A washer of non-metallic material, such as zinc chromate impregnated fabric, neoprene or other rubber should be employed under a nailhead. The purpose of this washer is to prevent metal-to-metal contact between the underside of the head and the roofing sheet, to seal the opening made in the sheet and to permit thermal expansion and contraction of the sheeting.

Hot-dip galvanized nails without washers may be used when washers are not available. Roofing nails with cast-on lead heads can be used in other than industrial or seacoast corrosive atmospheres.

Sizes, Weights and Coverage Data on Corrugated Aluminum Sheets Used for Roofing

Lengths 5'-0" to 12'-0" in 6" increments. Widths, 35" and 48⅜". Thicknesses, .024" and .032". Corrugation, 2.67" pitch and ⅞" deep. Coverage, 32" for 35" width and 45⅜" for 48⅜".

Data on .032" Corrugated Aluminum Roofing Sheets

Sheet Length in Feet	Square Feet per Sheet 35"	48⅜"	Weight per Sheet in Lbs. 35"	48⅜"	Approx. No. Sheets* per 100 Sq. Ft. 35"	48⅜"
5	14.58	20.16	8.04	11.03	6.86	4.96
5½	16.04	22.17	8.85	12.13	6.23	4.51
6	17.50	24.19	9.65	13.24	5.71	4.13
6½	18.96	26.20	10.46	14.34	5.27	3.82
7	20.42	28.22	11.26	15.44	4.90	3.54
7½	21.88	30.23	12.07	16.55	4.57	3.31
8	23.33	32.25	12.87	17.65	4.29	3.10
8½	24.79	34.27	13.68	18.75	4.03	2.92
9	26.25	36.28	14.48	19.86	3.81	2.76
9½	27.71	38.30	15.29	20.96	3.61	2.61
10	29.17	40.31	16.09	22.06	3.43	2.48
10½	30.63	42.33	16.90	23.16	3.26	2.36
11	32.08	44.34	17.70	24.27	3.12	2.26
11½	33.54	46.36	18.51	25.37	2.98	2.16
12	35.00	48.37	19.31	26.47	2.86	2.07

For .024" thickness use data given above except reduce weights 25 percent.

*For side and end lap allowance, add approximately 16% for 35" width and 12% for 48⅜" width.

Sizes, Weight and Coverage Data on Corrugated Aluminum Sheets Used for Siding

Lengths, 5'-0" to 12'-0" in 6" increments. Widths, 33¾" and 47⅛". Thicknesses, .024" and .032". Corrugation, 2.67" pitch and ⅞" deep. Coverage, 32" for 33¾" width and 45⅜ for 47⅛" width.

Data on .032" Corrugated Aluminum Siding Sheets

Sheet Length in Feet	Square Feet per Sheet		Weight per Sheet in Lbs.		Approx. No. Sheets* per 100 Sq. Ft.	
	33¾"	47⅛"	33¾"	47⅛"	33¾"	47⅛"
5	14.06	19.64	7.76	10.75	7.11	5.09
5½	15.47	21.60	8.54	11.82	6.46	4.63
6	16.87	23.56	9.31	12.90	5.93	4.24
6½	18.28	25.53	10.09	13.97	5.47	3.92
7	19.69	27.49	10.87	15.05	5.08	3.64
7½	21.09	29.45	11.65	16.12	4.74	3.40
8	22.50	31.42	12.42	17.20	4.44	3.18
8½	23.91	33.38	13.20	18.27	4.18	3.00
9	25.31	35.34	13.97	19.35	3.95	2.83
9½	26.72	37.31	14.75	20.42	3.74	2.68
10	28.12	39.27	15.52	21.50	3.56	2.55
10½	29.53	41.23	16.30	22.57	3.39	2.43
11	30.94	43.20	17.08	23.65	3.23	2.31
11½	32.34	45.16	17.86	24.72	3.09	2.21
12	33.75	47.13	18.63	25.80	2.96	2.12

For .024" thickness use data given above, except reduce weights 25 percent.

*For side and end lap allowance, add approximately 9% for 33¾" width and 6% for 47⅛" width.

Data on Nails Required for Roofing

Kind of Nail	Length Inches	Number per Lb.	Price per Lb.
Galvanized Needle Point..	1¾"	83	$ .75
Galvanized Needle Point..	2 "	77	.75
Aluminum Nails—Neoprene Washers	1¾"	318	2.80
Aluminum Nails—Neoprene Washers	2 "	285	2.80

Approximate Prices Per Sq. Ft. of Corrugated Aluminum Roofing and Siding Material

Thickness	Finish	Price
.0175"	Mill	$0.35
.0175"	Painted	.40
.024"	Mill	.65
	Painted	.85
.032"	Mill	.85
	Painted	1.05

After figuring the price of sheets from the above, approximately 20 percent should be added to cover cost of closures, flashings, fillers, corners, etc.

Labor Placing Corrugated Aluminum Roofing and Siding.—Estimate about the same labor cost for erecting aluminum corrugated roofing and siding as given on the previous pages under "Corrugated Steel Roofing and Siding."

Aluminum is much lighter than steel but the labor operations are practically the same.

Labor Cost of 100 Sq. Ft. (1 Sq.) Corrugated Aluminum Roofing on Wood Framing.

Based on using sheets 35" wide, .032" thick, weighing .552 lbs. per sq. ft.

	Hours	Rate	Total	Rate	Total
Carpenter..	1	$. . . .	$. . . .	$16.47	$16.47
Labor	1	. . . .	. . . .	12.54	12.54
Cost per 100 sq. ft.			$. . . .		$29.01

Cost of 100 Sq. Ft. (1 Sq.) Corrugated Aluminum Siding on Wood Framing
Based on using sheets 33¾" wide, .032" thick, weighing .552 lbs. per sq. ft.

	Hours	Rate	Total	Rate	Total
Carpenter	1.2	$....	$....	$16.47	$19.76
Labor	1.2	....	....	12.54	15.05
Cost per 100 sq. ft.			$....		$34.81

Cost of 100 Sq. Ft. (1 Sq.) Corrugated Aluminum Siding on Steel Framing, Straight Work
Based on using sheets 33¾" wide, .032" thick, weighing .552 lbs. per sq. ft.

	Hours	Rate	Total	Rate	Total
Labor handling-hoisting	3.2	$....	$....	$12.54	$40.13
Sheeters	3.2	....	....	17.70	56.64
Cost per 100 sq. ft.			$....		$96.77

Add extra labor for high and irregular walls such as grain elevators, conveyor housing, etc. Add for closures, flashings, etc.

Cost of 100 Sq. Ft. (1 Sq.) Corrugated Aluminum Roofing on Steel Framing, Straight Work
Based on using sheets 35" wide, .032" thick, weighing .552 lbs. per sq. ft.

	Hours	Rate	Total	Rate	Total
Labor handling-hoisting	2.67	$....	$....	$12.54	$33.48
Sheeter	2.67	....	....	17.70	47.26
Cost per 100 sq. ft.			$....		$80.74

Add extra labor for high or irregular roof areas. Add for closures, flashings, etc.

07460   CLADING/SIDING

Wood Siding is available in plain bevel, plain shiplap, and tongue and groove patterns to be used horizontally in heights from 4" to 12". Bevel siding is either ½" or ¾" thick, T&G, and shiplap nominal 1". Cedar is the most common wood for siding but redwood, fir, hemlock and spruce are also stocked. Wood may also be installed vertically with wood battens at the joints. Bevel siding ½" thick in clear "A" grade cedar will run around $1.10 a board foot; in redwood around $1.25. T&G red cedar 1x8s "D" and better will run 80¢ a board foot; T&G redwood, clear and better, runs $1.40.

Hardboard Siding, 7/16" thick and factory primed, will run 35¢ a square foot.

Plywood Siding in 4x8 sheets, rough textured and grooved to look like individual boards, ⅝" thick, will run around 75¢ a square foot in select grade Douglas fir, $1.10 in redwood.

Aluminum Siding factory finished white and in 8" widths will run 50¢ uninsulated, 85¢ insulated per sq. ft. Flashing strips will cost 20¢ a foot; inside corners 45¢; outside corners 90¢ a lineal foot. Colored nails cost $2.25 a pound.

Vinyl Siding, factory finished in 8" wide strips will cost about 60¢ a square foot, uninsulated, 80¢ with insulating backup. Corner boards will run 80¢ for outside positions, 45¢ for inside. Trim moldings will run 23¢ per lin. ft., while soffits will run 90¢ a sq. ft.

Placing Bevel and Drop Siding.—The labor cost of placing bevel or drop siding will vary with the class of work and the method of placing same.

On the less expensive type of construction it is customary to square only one end of the siding and the ends at the corners are left rough and are later covered with metal corner pieces. This is the cheapest method of placing siding. Metal corners are extensively used in the South because the long summers and intense heat cause mitered wood corners to open up, making a very unsightly appearance.

On more expensive buildings, the door and window casings and corner boards are placed and then it is necessary to cut and fit each piece of siding between the corner boards or between the casings and corner boards. On work of this class two carpenters usually work together as it is necessary to square one end of the siding and then measure and cut each board separately to insure a snug fit. This increases labor costs considerably but makes a nicer appearing job than the metal corners. This also applies where the siding is mitered at the corners. Two carpenters working together must measure and miter each piece of siding so that it will fit the adjoining corner.

The following quantities and production time are based on using ordinary brackets with plank scaffolding. If a more elaborate scaffold is required for placing siding, add extra for same.

Labor Placing Drop Siding

Measured Size Inches	Actual Size Inches	Class of Workmanship	Feet B.M. Placed per 8-Hr. Day	Carpenter Hours per 1000 Ft. B.M.
6	5¼	Rough Ends	525–575	14.5*
6	5¼	Fitted Ends	350–400	21.4
6	5¼	Mitered Corners	285–325	26.3
8	7¼	Rough Ends	600–650	12.8*
8	7¼	Fitted Ends	415–460	18.2
8	7¼	Mitered Corners	325–375	23.0

*Where an electric saw is used to square both ends of the siding before placing, leaving the exposed corners rough to be covered with metal corner pieces, deduct 1 to 1½ hours time per 1,000 ft. b.m. from the time given above.

Quantity of Bevel Siding Required Per 100 Sq. Ft. of Wall

Measured Size Inches	Actual Size Inches	Exposed to Weather	Pattern	Add for Lap	Ft. B.M. Req.- per 100 Sq. Ft. Surface
½x 4	½x 3¼	2¾	Regular	46%	151
½x 5	½x 4¼	3¾	Regular	33%	138
½x 6	½x 5¼	4¾	Regular	26%	131
½x 8	½x 7¼	6¾	Regular	18%	123
⅝x 8	⅝x 7¼	6¾	Regular	18%	123
¾x 8	¾x 7¼	6¾	Rabbetted	18%	123
⅝x10	⅝x 9¼	8¾	Rabbetted	14%	119
¾x10	¾x 9¼	8¾	Rabbetted	14%	119
¾x12	¾x11¼	10¾	Rabbetted	12%	117

The above quantities include 5% for end cutting and waste.

Quantity of Drop Siding Required Per 100 Sq. Ft. of Wall

Measured Size Inches	Actual Size Inches	Exposed to Weather	Add for Lap	Ft. B.M. Req. per 100 Sq. Ft. Surface
1x6	¾x5¼	5¼	14%	119
1x8	¾x7¼	7¼	10%	115

The above quantities include 5% for end cutting and waste.

Quantity of Shiplap Required Per 100 Sq. Ft. of Surface

Measured Size Inches	Actual Size Inches	Add for Lap	Ft. B.M. Req. per 100 Sq. Ft. Surface
1x 8	¾x7¼	10%	115
1x10	¾x9¼	8%	113

The above quantities include 5% for end cutting and waste.

Quantity of Bead Ceiling and Partition Required Per 100 Sq. Ft. of Surface

Measured Size Inches	Actual Size Inches	Add for Width	Ft. B.M. Req. per 100 Sq. Ft. Surface
1x4	⅝x3¼	23%	128
1x4	¾x3¼	23%	128

The above quantities include 5% for end cutting and waste.

Quantity of Dressed and Matched (D & M) or Tongued and Grooved (T & G) Boards Required Per 100 Sq. Ft. of Surface

Measured Size Inches	Actual Size Inches	Add for Width	Ft. B.M. Req. per 100 Sq. Ft. Surface
1x6	¾x5¼	14%	119
2x6	1⅝x5¼	14%	238

The above quantities include 5% for end cutting and waste.

Labor Placing Bevel Siding

Measured Size Inches	Actual Size Inches	Exposed to Weather Inches	Class of Workmanship	Feet B.M. Placed per 8-Hr. Day	Carpenter Hours per 1000 Ft. B.M.
4	3¼	2¾	Rough Ends	350–400	21.3**
4	3¼	2¾	Fitted Ends	240–285	30.5
4	3¼	2¾	Mitered Corners	200–240	36.3
5	4¼	3¾	Rough Ends	415–460	18.2**
5	4¼	3¾	Fitted Ends	285–330	25
5	4¼	3¾	Mitered Corners	240–285	30.5
6	5¼	4¾	Rough Ends	475–525	16*
6	5¼	4¾	Fitted Ends	325–375	23
6	5¼	4¾	Mitered Corners	265–310	28
8	7¼	6¾	Rough Ends	570–620	13.4**
8	7¼	6¾	Fitted Ends	375–415	20
8	7¼	6¾	Mitered Corners	300–350	24
10	9¼	8¾	Rough Ends	650–700	12**
10	9¼	8¾	Fitted Ends	440–480	17.5
10	9¼	8¾	Mitered Corners	375–425	20
12	11¼	10¾	Fitted Ends	475–525	16
12	11¼	10¾	Mitered Corners	400–450	19

**Where an electric handsaw is used to square both ends of the siding before placing, leaving the exposed corners rough to be covered with metal corner pieces, deduct 1 to 1½ hours time per 1,000 ft. b.m.

07463 ASBESTOS-CEMENT ROOFING AND SIDING

As previously stated in the Asbestos Cement Shingle Section (07312), this material is less in demand than it used to be. However, it is readily available at local material suppliers in flat sheets only; it will require a "Special Order" to obtain the corrugated sheets or the insulating core sandwich panels discussed in this section.

Bear in mind, that if this material needs cutting-to-size at the jobsite, that OSHA will require an enclosed, controlled environment to reduce the escape of air-borne asbestos fibers to the atmosphere and all workmen in that area will wear breathing apparatus (respirators) and disposable clothing. The asbestos/cement residue resulting from the cutting will have to be collected, bagged, and disposed of properly.

Corrugated asbestos roofing and siding is used over skeleton frame construction that can be applied with the minimum of time and labor. It is composed of asbestos fiber and portland cement combined under hydraulic pressure into homogeneous sheets of a light gray color, structurally strong, comparatively light in weight, and possessing the highest qualities of resistance to all of the forces of rot and decay.

Types of Corrugated Asbestos Roofing and Siding.

The same materials that are used in forming the corrugated sheets are molded at the factory into special standard shapes to be used as ridge roll, corner roll, louvers and ventilators.

Standard sheets have corrugations with a 4.2" pitch and a depth of 1 1/2". The thickness is approximately 7/16" at ridge and valley of corrugations and approximately 5/16" on tangent, an average thickness of 3/8". Sheets are furnished 42", or ten corrugations, wide. Standard lengths are 0'-6" to 12'-0" in 6" increments. Sheets can be furnished on special order cut to any desired length possible from the standard sizes. Special curved sheets can also be furnished with the following limitations: sheets may be curved longitudinally, the arc being formed along the direction of the corrugation; minimum radius when so curved 60". Sheets may be curved crosswise, the arc being formed across the corrugations, minimum radius when so curved 24". Sheets may be curved either way but not in combination. The sheets are furnished in three types, as illustrated. The cut corner construction eliminates a thickness of material in the corner lap, thus allowing the sheets to nest perfectly at these points and permit them to be laid with side laps forming vertical lines. The inside and outside radii of the corrugations are the same, thereby allowing the sheets to nest perfectly at all laps.

Corrugated asbestos roofing should not be laid on roofs having a pitch of less than 3" per foot; a pitch of 4" or more is preferred. Plastic cement or caulking should be used in all laps.

Supporting purlins or girts should be spaced not to exceed 54" center to center for roofing, and 66" for siding, depending upon the design loading.

Method of Applying Corrugated Roofing and Siding

Corrugated asbestos roofing and siding is applied directly over purlins and girts of skeleton frame construction or may be applied over solid wood sheathing.

When used as either roofing or siding, all sheets shall be lapped one corrugation at the sides and 6" at the ends. The size of the end lap is governed by the cut corner construction. The width of weather exposure of sheets for estimating purposes is 37.8".

Sheets shall be of proper length so that all end laps will occur over a purlin or girt and so that fasteners at ends of sheets will pass through both upper and underlying sheets.

Fasteners especially designed for use in connection with this material can be furnished by the manufacturer. The style of fastener to be used will be governed by the type, shape and position of purlins and girts. Sheets shall be secured at all purlins and girts, spacing fasteners as follows:

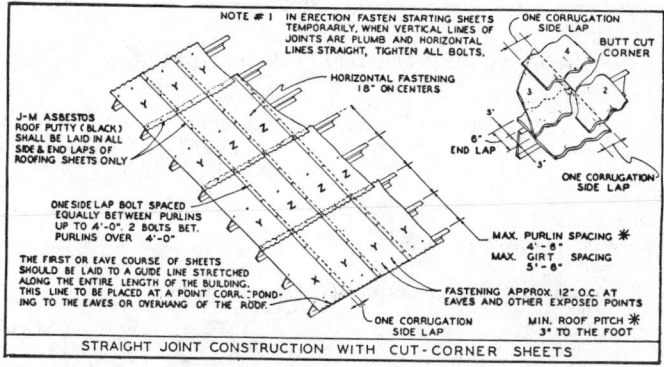

Erecting Corrugated Asbestos Roofing and Siding Using
Straight Joint Construction

Erecting Corrugated Asbestos Roofing and Siding Using
Staggered Joint Construction.

The percentage of the area to be covered which is absorbed by the end and side laps of the corrugated asbestos sheets will vary with the length of that area and the number of courses of sheets required. Therefore, a rough general layout of sheets should be made and area of laps determined before any estimating figures are made. See illustration.

Representative Diagram of Roofing or Siding

Table of Representative Diagram

Area of 10 side laps	(8'-6"+4'-0") .35' =	43.75 sq. ft.
Area of 1 end lap	(35'-0" x.5) =	17.5 sq. ft.
		61.25 sq. ft.
Area to be covered 12'-0"x35'-0"	=	420.00 sq. ft.
Sq. ft. of corrugated asbestos roofing or siding required		481.25 sq. ft.

Approximate Prices of Corrugated Asbestos Roofing and Siding

Weight per Square Foot	Price per 100 Sq. Ft. of Roof or Siding Area
4.1 lb.	$200.00

Prices of Ridge, Corner Roll and Louver Blades

	Approx. Wt. Lin. Ft.	Price per Lin. Ft.
Ridge and corner roll, including battens and fasteners	4 lbs.	$1.75
Louver blades	3.3 lbs.	.60

Fasteners for Asbestos Roofing and Siding

	Approx. Wt. 100 Sq. Ft.	Price per 100 Sq. Ft.
Lead headed fasteners (galv. iron) for use over wood framing	4 lbs.	$3.20
Lead headed fasteners (galv. iron) for use over steel framing	6 lbs.	4.80

Labor Handling and Placing Corrugated Asbestos Roofing and Siding.— The labor cost of handling and placing asbestos siding and roofing will vary according to the class of work, whether placed over wood or steel purlins or girts, size of sheets used and conditions under which the work is performed.

The most economical organization for applying asbestos roofing and siding consists of 2 mechanics and 2 laborers; and the labor costs given on the following pages are computed on this basis. Local labor regulations may require all crew members to be mechanics, in which case apply journeyman's wage rate to all labor hours.

The labor costs include all time necessary for cementing joints, and placing ridge roll, corner roll, louver blades, etc. Add for scaffolding, if necessary.

Placing Corrugated Asbestos Siding Over Wood Framing.—When placing corrugated asbestos siding on wood framing, a crew should handle, place and fasten 800 sq. ft. per 8-hr day.

Placing Corrugated Asbestos Siding on Steel Framing.—When placing corrugated asbestos siding on steel framing, a crew should handle, place and fasten 600 sq. ft., per 8-hr. day.

Placing Corrugated Asbestos Roofing Over Wood Purlins.—When placing corrugated sheets on roofs over wood purlins, an erection crew should handle, place and fasten 900 sq. ft., per 8-hr. day.

Placing Corrugated Asbestos Roofing Over Steel Purlins.—When placing corrugated asbestos roofing over steel purlins, an erection crew should handle, place and fasten 700 sq. ft. per 8-hr. day.

Labor Cost of 100 Sq. Ft. (1 Sq.) ⅜" Corrugated Asbestos
Siding Over Wood Framing

	Hours	Rate	Total	Rate	Total
Carpenter	2.0	$....	$....	$16.47	$32.94
Labor	2.0	....	....	12.54	25.08
Cost per 100 sq. ft.			$....		$58.02

Labor Cost of 100 Sq. Ft. (1 Sq.) ⅜" Corrugated Asbestos
Siding Over Steel Framing

	Hours	Rate	Total	Rate	Total
Sht. Met. wkr.	2.7	$....	$....	$19.75	$53.33
Labor	2.7	....	....	12.54	33.88
Cost per 100 sq. ft.			$....		$87.21

Labor Cost of 100 Sq. Ft. (1 Sq.) ⅜" Corrugated Asbestos
Roofing Over Wood Framing

	Hours	Rate	Total	Rate	Total
Carpenter	1.8	$....	$....	$16.47	$29.66
Labor	1.8	....	....	12.54	22.57
Cost per 100 sq. ft.			$....		$52.23

Labor Cost of 100 Sq. Ft. (1 Sq.) ⅜" Corrugated Asbestos
Roofing Over Steel Framing

	Hours	Rate	Total	Rate	Total
Sht. met. wkr.	2.3	$....	$....	$19.75	$45.43
Labor	2.3	....	....	12.54	28.84
Cost per 100 sq. ft.			$....		$74.27

Asbestos Cement Insulating Panels.—An insulating unit consisting of a core of insulating board surfaced on both sides with asbestos-cement board using a moisture-proof bituminous adhesive as a bond between the asbestos-cement and core material. The panel provides good strength, insulation and maintenance-free exterior and interior finish in a single thickness. This material is used in all types of buildings for exterior walls, roof decks, partitions and as a lining material. It may be applied to steel or wood framing using bolts and clips, screws, or nails. Nailing will not fracture the surface when reasonable care is used in driving. It is workable with ordinary carpenter tools but, for large volume cutting and drilling, carboloy tipped tools are more efficient.

Panels are furnished in sheets 4'-0" wide and from 4'-0" to 12'-0" long, in 2'-0" increments. Panel thickness varies depending on the insulating board core as follows:

Wood fiber core	- 11/16", 11/8", 19/16", 2"
Polystyrene Core	- 1", 11/8", 19/16", 2"
Perlite Core	- 1", 11/8", 19/16", 2"

Material costs increase with the thickness and the type of insulating core used, e.g., wood fiber core in 11/16" thick will cost about $1.50 per sq. ft., wood fiber core in 2" thick will cost about $2.50 per sq. ft., polystyrene core in 2" thick will cost about $2.60 per sq. ft., and Perlite core in 2" thick will cost about $3.10 per sq. ft.

Labor Installing Insulating Board.—When this material is used, it will require more labor time than the corrugated sheets previously discussed, due to the increased weight per sq. ft. and the fastening time (this material is laid with butt joints, thereby doubling the number of fasteners required around the edges). Using the same size crew as on the corrugated panels, add about 40% more time to the corrugated panel installation time, as previously given, to install the insulating board sandwich panels.

Labor Installing Insulating Board as Roof Decking.—When 19/16" or 2" board is used for roof decking over wood framing, 2 carpenters and 2 laborers working together should place 600 to 650 sq. ft. of decking per 8-hr. day at the following labor cost per 100 sq. ft.:

	Hours	Rate	Total	Rate	Total
Carpenter	2.5	$....	$....	$16.47	$41.18
Labor	2.5	....	....	12.54	31.35
Cost 100 sq. ft.			$....		$72.53
Cost per sq. ft.			....		.73

Add for caulking joints with plastic roofing cement and covering joints with 6" wide strips of saturated fabric or 30-lb. roofing felt.

Labor Installing Asbestos Insulating Board as Siding. —When 11/8", 1 9/16" or 2" board is applied as curtain walls on wood framing, 2 carpenters and 2 laborers working together should place 550 sq. ft. of curtain wall per 8-hr. day at the following labor cost per 100 sq. ft.:

	Hours	Rate	Total	Rate	Total
Carpenter	2.8	$....	$....	$16.47	$46.12
Labor	2.8	....	....	12.54	35.11
Cost 100 sq. ft.			$....		$81.23
Cost per sq. ft.			....		.81

Add for caulking joints and covering joints with batten strips or metal plates.

Labor Installing Asbestos Insulating Board as Partitions.—When boards are used for partitions, several methods of framing may be employed, however, very little difference in labor placing the panels should be experienced.

In general, floor and ceiling plates, either rabbeted or built up to form a retaining slot, are installed and panels are set in place. Joints between panels may be covered with battens, concealed with metal flush panel mouldings or splined together. In some instances 2"x4" studs and plates are used, the studs being spaced to allow one full panel to be installed between them and stops being used to hold the panels in place.

After framing is in place, 2 carpenters an 1 laborer working together should place 1,024 sq. ft. of partition panels per 8-hr. day at the following labor cost per 100 sq. ft. :

	Hours	Rate	Total	Rate	Total
Carpenter	1.6	$. ...	$. ...	$16.47	$26.35
Labor	0.8	....	....	12.54	10.03
Cost 100 sq. ft.			$. ...		$36.38
Cost per sq. ft.			....		.37

07464 PLASTIC SIDING

Fiberglass panels are made to most of the usual configurations and widths and in lengths up to 30', although stock lengths are in the 3' to 12' range. Fiberglass panels may be used to enclose an entire building, walls and roof, or in combination with similarly configured panels of steel, aluminum, protected metal or asbestos. The standard panels are translucent and can be inserted as skylights or windows. They can be either clear or colored, smooth or embossed. Fiberglass has an ignition point of 850-900 degrees. Where greater fire protection is required they may be ordered to be rated "fire retardant" at around 60¢ per square foot extra. The panels can be further treated to resist erosion or corrosion, or reinforced to be resistant to breakage. The weight of the sheet determines the cost. Standard sheets weighing 4 oz. per sq. ft. will cost around 70¢ per sq. ft.; 5 oz., 80¢; 6 oz., 95¢; and 8 oz., $1.25. The material is light and one carpenter should set around 100 sq. ft. per hour at the following cost per 100 sq. ft. ;

	Hours	Rate	total	Rate	Total
Carpenter	1.0	$. ...	$. ...	$16.47	$16.47
Cost per sq. ft.			....		.17

07500 MEMBRANE ROOFING

Built-up roofing consists of alternate plies of saturated felt and moppings of pitch with tar saturated felt or asphalt with asphalt saturated felt, covered with a top pouring of pitch or asphalt into which slag or gravel is embedded. On flat roofs with slopes of less than ¼" per foot, on which water may collect and stand, coal tar pitch and felt or a low melting point asphalt bitumen and asphalt felt are generally used. Coal tar pitch is not recommended for roofs having an incline in excess of 1" per foot.

On built-up roofs, where slope is over 2" per foot up to 4" per foot, steep asphalt (180°-200° melting point) is more suitably used and slag is embedded in preference to gravel because it remains embedded better than well rounded gravel. Where slag is not available and where a light gray or white surface is desired, hard limestone chips, angular pieces ¼" to ¾" in size, are embedded in the top pouring of bitumen. On slopes over 2" per foot, double coverage mineral surfaced roll roofing with a 19" selvage edge may be used as the top finish.

Smooth surface built-up roofing consists of alternate plies of asbestos felt cemented solid to the base sheet and to each other with asphalt, and a top coating of hot or cold asphalt, uniformly distributed.

The specifications on the various types of roofing vary widely, depending upon the surface to which the built-up roofing is applied and the service required.

The cost of built-up roofing is governed by the incline of the roof, size, plan (whether cut up with openings, skylights, penthouses, irregular roof levels, etc.), and the distance of the roof above the ground; the higher the roof, the higher materials must be hoisted, which increases the labor costs.

The term, "built-up roof" will be commonly heard in the trade, but is more properly "built-up roofing". Technically, the "roof" is the supporting structure over which the "roofing" is applied. Although the term "built-up roof" may be loosely used in the trade, and seems to flow more easily in conversation, the specification writer, designer, or their attorneys will be careful to distinguish between "roof" and "roofing".

Rules for Measuring Flat Roofs.— When measuring flat roof surfaces that are to be covered with composition, tar and gravel, tin, metal or prepared roofing, the measurements should be taken from the outside of the walls on all four sides to allow for flashing up the side of each wall. The flashing is usually 8" to 1'-0" high. This applies particularly to brick, stone or tile buildings having parapet walls above the roof level.

Fig. 1. Flat Roof with Parapet Walls

On flat roofs projecting or overhanging beyond the walls of the building, the measurements should cover the outside dimensions of the roof and not merely to the outside of the walls.

When estimating the area of flat roof surfaces, do not make deductions for openings containing less than 100 sq. ft. and then deductions should be made for just one-half the size of the opening.

Make deductions in full for all openings having an area of 500 sq. ft. or more.

Example : The length of the building (as shown in Figure 1), is 50'-0", which is the distance from A to B. The width of the building from B to C is 20'-0". Multiplying the length by the width, 50x20=1,000 sq. ft. By referring to Figure 1, you will not the building is only 48'-0" long and 18'-8" wide between walls. The method of measuring the full length and width of the building is to allow for flashing up the side of the walls, which usually extends 8" to 12" high.

Figure 2 illustrates a building of the same size with projecting roof. The size of the building proper is 20'-0"x50'-0", while the roof projects beyond the walls 2'-0" on each side, making the size of the roof 24'-0"x54'-0". The roof area is obtained by multiplying the length of the roof (the distance from E to F), 54'-0", by the width of the roof (the distance from F to G), 24'-0", making the total roof area, 54'-0"x24'-0"=1,296 sq. ft.

Fig. 2. Flat Roof Overhanging Walls

The skylights on both roofs measure 6'-0"x9'-0" and contain only 54 sq. ft. each, so no deduction should be made for these openings, as they contain less than 100 sq. ft. each and the extra labor flashing up the sides of the walls more than offsets the cost of the small openings.

Quantity of Pitch or Asphalt Required for Built-Up Roofs.—The quantity of roofing pitch or asphalt required for any built-up roof will vary with the number of thicknesses or plies of felt used. All manufacturers of roofing materials indicate in their specifications the amount of materials required for first class roof. On every first class roof plenty of pitch or asphalt is used in each mopping so that each ply of felt is well cemented to the next and in no instance does felt touch felt.

Approximately 25 lbs. of coal tar pitch or approximately 20 lbs. of asphalt should be used for each mopping per 100 sq. ft. of surface and the last pouring, in which the gravel or slag is imbedded, should use 60 lbs. of asphalt or 75 lbs. of pitch per 100 sq. ft.

Quantity of Asphalt Required For 100 Sq. Ft of Roof.

Number	Number of Plies		Surface to Which	Lbs. per
Plies	Dry	Mopped	Roof is Applied	100 Sq. Ft.
5	2	3	Wood, plywood, structural wood fiber	120
4	1	3	Poured sypsum, lightweight concrete	120
4		4	Concrete, precast concrete or gypsum	140

For roofing pitch add approximately 5 lbs. per mopping each ply per 100 sq. ft.—add 15 lbs. for top pouring per 100 sq. ft.

Estimating the Quantity of Roofing Felt.—Asphalt or tarred felt for built-up roofing is furnished in 4 sq. rolls containing 432 sq. ft. and weighing from 56 to 62 lbs. per roll for No. 15 felt. No. 30 felt which is generally used as the base sheet over wood construction when an asphalt specification is used is furnished in 2 sq. rolls containing 216 sq. ft. and weighing 60 lbs. These weights are those used on a bonded job in the best type of building construction.

When specifying the grade or weight of felt to be used it is customary to state that "felt shall weigh not less than 15 lbs. per 108 sq. ft." 15 lbs. per 108 sq. ft. is standard practice of all roofing manufacturers. Inasmuch as felt is furnished in 4 sq. rolls of 432 sq. ft., 32 sq. ft. per roll or 8 sq. ft. per 100 sq. ft. is allowed for laps.

The following table gives the quantity of roofing felt required to cover 100 sq. ft. of surface of various thicknesses :

Number of Plies	Add for Waste	No. Sq. Ft. Req'd. 100 Sq. Ft. Roof	Weight per Square of Roof Using No. 15 Felt
1	8%	108	15
2	8%	216	30
3	8%	324	45
4	8%	432	60
5	8%	540	75

The general use of No. 30 felt is as a base sheet over wood decks in 1-ply thickness over which subsequent layers of No. 15 felts are mopped in.

Quantity of Roofing Gravel Required for Built-Up Roofs.—Roofing gravel should be uniformly embedded into a heavy top pouring of asphalt or pitch so that approximately 400 lbs. of gravel or 300 lbs. of slag is used per 100 sq. ft. of roof area.

Labor Applying Built-Up Roofing.—The labor cost of handling materials and applying built-up roofing will vary with the specification used and the type of building to which the roofing is applied. On the low type of building, 1, 2 or 3 stories high, where there is no great distance from the ground to the roof deck, the labor for hoisting materials, and supplying hot asphalt or pitch from the ground to the roof deck is considerably less than on a high building, where there is a greater distance from the kettle to the roof.

On low type buildings, a 5-man crew composed of 1 man at the kettle, 1 man carrying 'hot stuff,' 1 man rolling in felts, 1 man nailing in and 1 man mopping in, can lay the following areas per 8-hr. day :

18 sqs. of 5-ply asphalt and gravel or pitch and gravel over wood roof deck, consisting of 1 sheathing paper, 2 dry felts and 3 plies of felt mopped in with a top pouring of pitch or asphalt and gravel surfacing.

18 sqs. of 4-ply asphalt and gravel or pitch and gravel over a concrete deck, all mopped including a top pouring of pitch or asphalt and gravel or slag surfacing.

22 sqs. of 4-ply asphalt and gravel or pitch and gravel over wood roof deck, consisting of 1 sheathing paper, 2 dry felts and 2 plies of felt mopped in with a top pouring of pitch or asphalt and gravel or slag surfacing.

22 sqs. of 3-ply asphalt and gravel or pitch and gravel over concrete roof deck, all mopped and including a top pouring of pitch or asphalt and gravel or slag surfacing.

24 sqs. of 3-ply asphalt and gravel, or pitch and gravel, over wood roof deck, consisting of 1 dry felt, and 2 plies of felt mopped in, with a top pouring of pitch or asphalt, and gravel or slag surfacing.

The above figures apply to buildings up to 3 stories in height.

If the roof surfaces are broken up with skylights, irregular roof level, etc., the above quantities should be reduced about 15 to 20 per cent.

A 5-man crew should apply the following number of squares of roof per 8-hr. day, based on first class workmanship throughout:

Type of Roofing	Type "A" Structures	Type "B" Structures
3-ply roof over wood roof deck	24 Sqs.	18 Sqs.
3-ply roof over concrete roof deck or insulation	22	17
4-ply roof over wood roof deck	22	17
4-ply roof over concrete roof deck or insulation	18	14
5-ply roof over wood roof deck	18	14

Type "A" buildings are the most convenient type of structure for application of built-up roofing. A low type of building from 1 to 3 stories in height, straight, practically flat area not broken up by very many skylights or variations in the deck elevation.

Type "B" buildings are not convenient type of structures for the application of built-up roofing. These would include high buildings which require considerable handling of materials, and buildings with roofs broken up by skylights or penthouses, sawtooth construction, monitors, or considerable variation in deck levels.

An easy and accurate method of computing the labor cost on any type of roof is to take the cost of a 5-man crew for an 8-hr. day, and divide this by the number of squares of roof applied per day, as follows:

	Hours	Rate	Total	Rate	Total
Foreman (working)....................	8	$. ...	$. ...	$17.58	$140.64
Roofers (4)............................	32	....	....	17.18	549.76
Crew cost per 8-hr. day................			$. ...		$690.40
Cost per sq. 14 sqs. per day............				....	49.31
Cost per sq. 17 sqs. per day............				....	40.61
Cost per sq. 18 sqs. per day............				....	38.36
Cost per sq. 22 sqs. per day............				.... .	31.38
Cost per sq. 24 sqs. per day............				....	28.77

On the average roof where the mopping between layers consists of mopping the width of the lap, a roofer should mop 100 sq. ft. of surface in 6 to 7 minutes.

On first grade work where the entire roof surface is mopped between each layer of felt, a man should mop 100 sq. ft. of roof in about 10 minutes.

After the roof surface has been mopped, a roofer should place roofing felt over 100 sq. ft. in about 6 minutes.

Where first grade workmanship is required, a roofer should handle pitch and gravel, pour hot pitch or asphalt and spread gravel over one square of roof in 35 to 40 minutes.

Prices on Roofing Materials.—Prices on roofing materials such as coal tar pitch, asphalt, felt, etc., vary according to market fluctuations, location of job, etc. Current material prices should always be obtained, before submitting bid.

Material Cost of 100 Sq. Ft. (1 Sq.) 3-Ply Tar and Gravel Built-Up Roof Applied Over Wood Roof Deck

Maximum incline 2-in. per ft. One ply No. 30 felt, 2 plies No. 15 felt (1 dry, nailed and 2 mopped), coal tar pitch surfaced with gravel or slag.

	Rate	Total	Rate	Total
1.1 sqs. No. 30 felt..........................	$. ...	$. ...	$6.00	$ 6.60
2.2 sqs. No. 15 felt..........................	....	....	3.00	6.60
125 lbs. coal tar pitch......................	....	....	.13	16.25
400 lbs. roofing gravel.....................	....	....	.50	2.00
Nails, fuel, mops, etc.	....	....	2.50	2.50
Cost per 100 sq. ft......................		$. ...		$33.95

Material Cost of 100 Sq. Ft. (1 Sq.) 3-Ply Tar and Gravel Built-Up Roof Over Poured Concrete, Poured Gypsum Roof Decks, or Insulation Board

Maximum incline 2-in. per ft. Roof consisting of 3 plies of felt, coal tar pitch surfaced with gravel or slag.

	Rate	Total	Rate	Total
3.3 sqs. No. 15 felt..........................	$. ...	$. ...	$3.00	$ 9.90
150 lbs. coal tar pitch	....	....	.13	19.50
400 lbs. roofing gravel.....................	....	....	.50	2.00
Fuel, mops, etc............................	....	....	3.00	3.00
Cost per 100 sq. ft......................		$. ...		$34.40

Material Cost of 100 Sq. Ft. (1 Sq.) of Each Additional Ply of
No. 15 Felt and Hot Mopping of Pitch

	Rate	Total	Rate	Total
1.1 Sqs. No. 15 Felt...............................	$. . . .	$. . . .	$3.00	$3.30
25 lbs. Pitch	. . . .	. . . .	.13	3.25
Fuel, Mops, etc.	. . . .	. . . .	.50	.50
Cost per 100 sq. ft.		$. . . .		$7.05

07520 PREPARED ROLL ROOFING

Ready roofing can be used on small buildings where an inexpensive, yet satisfactory, roofing is required. It is furnished in various grades and weights and is sold by the roll or in flat sheets packed with the nails and cement necessary for application.
Laying Ready-to-Lay Roll Roofing.—When laying ready-to-lay roll roofing, a man should handle, lay and cement or nail in place about 800 sq. ft. per 8-hr. day, at the following cost per 100 sq. ft. (1 sq.) :

	Hours	Rate	Total	Rate	Total
Carpenter...	1.0	$. . . .	$. . . .	$16.47	$16.47

Asphalt Slate Surfaced Ready-to-Lay Roofing

A ready-to-lay roofing consisting of several plies of asphalt impregnated felt, surfaced with a thick layer of slate granules, either red or green. For roofs having a pitch of 1½ " or more per foot.
Furnished in rolls 36" wide and 36'-0" long, containing 108 sq. ft., sufficient for 100 sq. ft. of roof, at the following price per 90 lb. roll............................. $13.00

Smooth Surfaced Ready-to-Lay Roofing

Mica surface asphalt roofing consisting of good quality felt saturated with pure asphalt and both sides covered with flake mica. Rolls 36" wide and 36'-0" long.

Description	Weight per roll	No. Sq. Ft. in Roll	Price per Roll
Mica surfaced asphalt roofing	55 lbs.	108	$11.50
Mica surfaced asphalt roofing	65 lbs.	108	13.25

7530-40   ELASTIC SHEET AND FLUID APPLIED ROOFING

These types of elastomeric coatings include neoprene, Hypalon, urethane, butyl and silicone. Some are applied by spraying, others in sheet form, and all are subcontracted to firms licensed by the manufacturer.
Since these coatings are thin, the roof surface to which they are to be applied must be firm, continuous, smooth, clean, and dry. New concrete should be sealed with a primer.
Because of their thinness and ability to conform to any shape and the fact that they can be had in most any colors including white they are often chosen as the roofing surface for decorative and fluid roof forms. Some materials are also used as traffic decks.

A 1/16" thick butyl sheet will cost around 65¢ a sq. ft. and one roofer can install some 225 sq. ft. per day for a total cost of around $1.25 a sq. ft.

A 1/16" neoprene sheet will have a material cost of some $1.25 a sq. ft. and will install at the same rate as the butyl for an installed price of around $1.80 a sq. ft.

Fluid applied Hypalon-neoprene .02" thick will be applied at the rate of some 100–110 sq. ft. per day and have a material cost of 85¢ for a total figure of around $2.00 a sq. ft.

07600   FLASHING AND SHEET METAL
07610   SHEET METAL ROOFING

Metal roofing includes galvanized steel, copper, lead, stainless steel and aluminum plus the many combinations and alloys of these metals such as lead-coated copper, terne (80% lead, 20% tin over copper-bearing carbon steel) microzinc and terne coated stainless. Terne and aluminum are the least expensive and copper and lead the costliest. Metal prices tend to fluctuate broadly as many of the ores are imported and at the mercy of the value of the dollar and the political climate of the country they are mined in.

Metal roofs may be applied in many ways. The simplest is the flat seam roof which may be used on slopes as low as ¼" to the foot. Standing seams and battened seams generally need a slope of at least 2½". They are more decorative, and in fact batten designs are often selected solely on their decorative value. Some materials may be ordered with prefabricated battens, others are formed in the traditional way over wood strips. Another decorative roof is the "Bermuda" type, where the metal is applied over wood "steps" provided by the carpenter. The step is based on the width of the metal roll to be used and is sloped at least 2½" to the foot. The roofer interlocks the rolls at each step edge giving a sharp shadowline.

For custom work terne and copper, with or without special coatings, are the usual choices. Many batten roofs today have the batten stamped into the metal and are usually factory finished aluminum or steel.

When estimating quantities of metal roofing, the measurements are taken in the same manner as for other types of roofs.

Where the metal roof is applied over flat roofs having parapet walls, an allowance of 1'-0" on each side and end of the wall should be allowed for flashing up the sides of the walls.

If metal roofing is applied to a flat surface or pitch roof with overhanging eaves the measurement is taken to cover the entire roof surface.

Do not make deductions for openings containing less than 25 sq. ft. as the extra flashing costs as much as the roofing omitted.

Deduct in full for all openings containing over 50 sq. ft. and add for flashing.

Flat Seam Metal Roofing.—The common sizes of tin plates are 10"x14" or multiples of that size. The sizes generally used for roofing are 14"x20" and 20"x28". The larger sizes are more economical to lay, and for flat roofs the 20"x28" size is preferable.

For a flat seam roof, the edges of the sheets are turned in about ½", locked together and well soaked with solder. The sheets should be fastened to the roof boards by cleats locked in the seams. These cleats are usually spaced about 8" apart and 2 nails are used in each cleat.

For flat seam roofing, a 14"x20" sheet of tin, with edges turned up ½" on each side and end, measures 13"x19" and contains 247 sq. in.; but the covering capacity when locked to other sheets is only 12½"x18½" or 231¼ sq. in. It requires 62½ sheets of 14"x20" tin to cover 100 sq. ft. of roof.

Method of Laying Flat Seam Metal Roofing

Sheets measuring 20"x28" with the edges turned for flat seam roofing measure 19"x27", and when locked to other sheets, have a covering capacity of 18½"x26½" or 490¼ sq. in. It requires 20½ sheets to cover 100 sq. ft. of roof.

Tin for roofs may also be obtained in rolls 14", 20", 24" and 28" wide. When used for flat seam roofing the loss due to turning edges amounts to 1½" on the width of any sheet.

Tin in 14" rolls loses 1½" or 11.2 percent of its width due to turning edges and laps. Another method is to take the length of the roof and divide by the net width of each sheet and the result will be the number of strips of tin required. Multiply the number of strips by the length of each strip to obtain the quantity of tin required for any roof.

The following table gives the covering capacity and allowances for edging and laps to be figured when using tin roofing in rolls for flat seam roofs:

Width of Sheets Inches	Allowance for Edge and Laps	Actual Width Inches	Add for Loss Due to Edges and Laps	No. Sq. Ft. Required Per Sq.
14	1½	12½	12.0%	112
20	1½	18½	8.0%	108
24	1½	22½	7.0%	107
28	1½	26½	6.0%	106

The quanitites of galvanized iron or steel sheets required for flat seam roofing will depend upon the size of the sheets, as they may be obtained in sheets from 24"x96" to 48"x120" in size.

The actual measurements of the sheets will be 1½" less in width and length than the size of the sheet used. For instance, the actual covering capacity of a 24"x96" sheet will be only 22½"x94=12,126 sq. in. or 14.76 sq. ft.

The following table gives the approximate covering capacity of one sheet of the different sizes, allowing 1½" in width and length for turning edge and lapping, together with the number of sheets of the different sizes required to cover one sq. (100 sq. ft.), of surface:

Size of Sheet Inches	Allowance for Edge of Lap		Actual Size Sheet Ins.	Cover Cap. One Sheet	No. Sheets Per Sq.
	Length	Width			
24x 96	1½	1½	22½x 94½	14.76	6.8
24x120	1½	1½	22½x118½	18.51	5.4
26x 96	1½	1½	24½x 94½	16.08	6.2
26x120	1½	1½	24½x118½	20.16	5.0
28x 72	1½	1½	26½x 70½	13.00	7.7
28x 84	1½	1½	26½x 82½	15.18	6.6
28x 96	1½	1½	26½x 94½	17.40	5.8
28x108	1½	1½	26½x106½	19.60	5.1
28x120	1½	1½	26½x118½	21.80	4.6
30x 96	1½	1½	28½x 94½	18.70	5.4
30x120	1½	1½	28½x118½	23.45	4.3

Standing Seam Metal Roofing.—Standing seam roofing requires a larger allowance for waste than flat seam roofing. The standing seam, edged 1¼" and 1½" takes 2¾" off the width and the flat cross seams edged ⅜", take 1⅛" off the length.

If 14"x20" tin is used, each sheet covers 11¼x18⅞" or 212.34 sq. in. It requires 68 sheets of 14"x20" tin per 100 sq. ft. of roof.

If 20"x28" tin is used, each sheet covers 17¼x26⅞" or 463.59 sq. in. It requires 32 sheets of 20"x28 tin per 100 sq. ft.

If roll tin is used, 2¾" should be allowed for the two standing seams. The end waste will vary with the length of the sheets.

Tin in 14" rolls loses 2¾" off the width due to the standing seams, making it necessary to add 20 percent for waste. Another method is to take the width of the roof and divide by 11¼" (the net width of the sheet), and the result is the number of strips of tin required. Multiply the number of strips by the length of the roof to obtain the quantity of tin required.

Method of Laying Standing Seam Metal Roofing

The following table gives the covering capacity and allowances for standing seams, when estimating roll roofing:

Width of Sheets Inches	Allowance for Standing Seams	Actual Width Inches	Add for Loss Due to Standing Seams	No. Sq. Ft. Required Per Square
14	2¾	11¼	25%	125
20	2¾	17¼	16%	116
24	2¾	21¼	13%	113
28	2¾	25¼	11%	111

If galvanized steel or any of the special process metals, such as Armco, Toncan, etc., are used for standing seam roofs, an allowance of 1¼" and 1½" or a total of 2¾" should be taken from the width of the sheets to allow for standing seams. The cross seams edged ⅜" take 1⅛" off the length of each sheet. For instance, where 24"x96" sheets are used, the actual size of the sheets is 21¼"x94⅞", or each sheet will cover 2,016 sq. in. or 14 sq. ft.

The following table gives the covering capacity of painted or galvanized steel sheets of the different sizes:

Size of Sheet Inches	Standing Seam and End Lap Inches Width	Length	Actual Size Sheet Inches	Cover Cap. One Sheet Sq. Ft.	No. Sheets Per Square (100 Sq. Ft.)
24x 96	2¾	1⅛	21¼x 94⅞	14.00	7.2
24x120	2¾	1⅛	21¼x118⅞	17.54	5.7
26x 96	2¾	1⅛	23¼x 94⅞	15.33	6.6
26x120	2¾	1⅛	23¼x118⅞	19.20	5.3
30x 96	2¾	1⅛	27¼x 94⅞	17.92	5.6
30x120	2¾	1⅛	27¼x118⅞	22.50	4.5

Size of Sheet Inches	Standing Seam and End Lap Inches Width	Length	Actual Size Sheet Inches	Cover Cap. One Sheet Sq. Ft.	No. Sheets Per Square (100 Sq. Ft.)
36x 96	2¾	1⅛	33¼x 94⅞	22.00	4.6
36x120	2¾	1⅛	33¼x118⅞	27.45	3.7
42x 96	2¾	1⅛	39¼x 94⅞	25.85	3.9
42x120	2¾	1⅛	39¼x118⅞	32.40	3.1
48x 96	2¾	1⅛	45¼x 94⅞	29.80	3.4
48x120	2¾	1⅛	45¼x118⅞	37.40	2.7

V-Crimped Roofing.—V-crimped roofing is used the same as standing seam roofing, the V-crimp answering the same purpose as the standing seam.

This roofing is usually furnished in sheets covering 24" in width and 6'-0" to 10'-0" long.

When estimating quantities of V-crimped roofing, allow for the end lap but there is no waste in the width as only the actual covering capacity is charged for by the manufacturers. For instance, a sheet 26" wide. before crimping is 24" wide after crimping, but the sheet is sold as 24" wide.

The following table gives the quantity of V-crimp roofing required to cover 100 sq. ft. of roof with end laps 1" to 6":

End Laps in Inches

Length of Sheet Feet	1	2	3	4	5	6
			Square Feet of V-Crimp Roofing Required			
6	102	103	105	106	108	109
7	102	103	104	105	106	108
8	101	102	103	104	106	107
9	101	102	103	104	105	105
10	101	102	103	104	105	105

Material Cost of 100 Sq. Ft. (1 Sq.) Flat Seam Tin Roofing Using 14"x20" Plates

	Rate	Total	Rate	Total
1.1 sq. No. 15 felt	$....	$....	$3.00	$ 3.30
63 shts. 14"x20" 40 lb. coated	....	....	1.50	94.50
6 lbs. solder	....	....	1.00	6.00
252 cleats	....	....	.03	7.56
2 lbs. nails	....	....	.80	1.60
Flux and charcoal	....	....	1.30	1.30
Cost per 100 sq. ft.		$....		$114.26*

*Add for ridges, hips, valleys, flashing etc.

Material Cost of 100 Sq. Ft. (1 Sq.) Flat Seam Tin Roofing Using 20"x28" Plates

	Rate	Total	Rate	Total
1.1 sq. No. 15 felt	$....	$....	$3.00	$ 3.30
29 shts. 20"x28" 40 lb. coated	....	....	3.00	87.00
5 lbs. solder	....	....	1.00	5.00
145 cleats	....	....	.03	4.35
1 lb. nails	....	....	.80	.80
Flux and charcoal	....	....	1.30	1.30
Cost per 100 sq. ft.		$....		$101.75*

*Add for ridges, hips, valleys, flashing etc.

Labor Placing 100 Sq. Ft. Flat and Standing Seam Tin and Metal Roofing

Description of Roof	Number Sq. Ft. per 8-Hr. Day	Hours per 100 Sq. Ft.		*Combined Labor Cost per 100 Sq. Ft.
		Tinner	Helper	
Flat seam metal roofing, using 24"x96" sheets or larger	300–350	2.5	2.5	$ 80.73
Flat seam tin roofing, using 14"x20" plates	175–225	4.0	4.0	129.16
Flat seam tin roofing, using 20"x28" plates	225–275	3.3	3.3	106.56
Flat seam tin roofing, using 20" tin in rolls	325–375	2.3	2.3	74.27
Flat seam tin roofing, using 14" tin in rolls	275–325	2.8	2.8	90.41
Flat seam tin roofing, using 28" tin in rolls	425–475	1.8	1.8	58.12
Standing seam metal roofing, using 24"x96" sheets	250–300	3.0	3.0	96.87
Standing seam tin roofing, using 14"x20" plates	125–175	5.3	5.3	171.14
Standing seam tin roofing, using 20"x28" plates	175–225	4.0	4.0	129.16
Standing seam tin roofing, using 14" tin in rolls	200–225	3.8	3.8	122.70
Standing seam tin roofing, using 20" tin in rolls	250–300	3.0	3.0	96.87
Standing seam tin roofing, using 24" tin in rolls	275–325	2.8	2.8	90.41
Standing seam tin roofing, using 28" tin in rolls	300–350	2.4	2.4	77.50
V-Crimped metal roofing, using 24"x96" sheets or larger	425–475	1.8	1.8	58.12

*Tinner - $19.75 per hr. and Helper - $12.54 per hr.

Material Cost of 100 Sq. Ft. (1 Sq.) Flat Seam Tin Roofing Using Roll Roofing

	Rate	Total	Rate	Total
1.1 sq. No. 15 felt	$. . . .	$. . . .	$3.00	$ 3.30
108 sq. ft. 20" roll rfg.	. . . .	. . . .	1.20	129.60
4 lbs. solder	. . . .	. . . .	1.00	4.00
110 cleats	. . . .	. . . .	.03	3.30
1 lb. nails	. . . .	. . . .	.80	.80
Flux and charcoal	. . . .	. . . .	1.00	1.00
Cost per 100 sq. ft.		$. . . .		$142.00*

*Add for ridges, hips, valleys, flashing, etc.

Material Cost of 100 Sq. Ft. (1 Sq.) Flat Seam Roofing Using 24"x96" Galvanized Steel Sheets

	Rate	Total	Rate	Total
1.1 sq. No. 15 felt	$. . . .	$. . . .	$3.00	$ 3.30
7 shts. 26 ga. steel, 105 lbs.	. . . .	. . . .	.80	89.60
3 lbs. solder	. . . .	. . . .	1.00	3.00
100 cleats	. . . .	. . . .	.03	3.00
1 lb. nails	. . . .	. . . .	.80	.80
Flux and charcoal	. . . .	. . . .	.90	.90
Cost per 100 sq. ft.		$. . . .		$100.60*

*Add for ridges, hips, valleys, flashing, etc.

Material Cost of 100 Sq. Ft. (1 Sq.) Standing Seam Roofing Using 14"x20" Plates

	Rate	Total	Rate	Total
1.1 sq. No. 15 felt	$. . . .	$. . . .	$3.00	$ 3.30
68 shts. 14"x20" 40-lb. coated	. . . .	. . . .	1.50	102.00
100 cleats	. . . .	. . . .	.03	3.00
2.5 lbs. solder	. . . .	. . . .	1.00	2.50
1 lb. nails	. . . .	. . . .	.80	.80
Flux and charcoal	. . . .	. . . .	.90	.90
Cost per 100 sq. ft.		$. . . .		$112.50*

*Add for ridges, hips, valleys, flashing, etc.

Material Cost of 100 Sq. Ft. (1 Sq.) Standing Seam Roofing Using 20"x28" Plates

	Rate	Total	Rate	Total
1.1 sq. No. 15 felt	$. . . .	$. . . .	$3.00	$ 3.30
32 shts. 20"x28" 40-lb. coated	. . . .	. . . .	3.00	96.00
80 cleats	. . . .	. . . .	.03	2.40
2 lbs. solder	. . . .	. . . .	1.00	2.00
1 lb. nails	. . . .	. . . .	.80	.80
Flux and charcoal	. . . .	. . . .	.90	.90
Cost per 100 sq. ft.		$. . . .		$105.40*

*Add for ridges, hips, valleys, flashings, etc.

Material Cost of 100 Sq. Ft. (1 Sq.) Standing Seam Roofing Using Roll Roofing

	Rate	Total	Rate	Total
1.1 sq. No. 15 felt	$. . . .	$. . . .	$3.00	$ 3.30
116 sq. ft. 20" roll rfg 40 lb.	. . . .	. . . .	1.20	139.20
80 cleats	. . . .	. . . .	.03	2.40
1 lb. solder	. . . .	. . . .	1.00	1.00

1 lb. nails	. . .	. . . .	.80	.80
Flux and charcoal	. . .	. . . .	.70	.70
Cost per 100 sq. ft.		$. . . .		$147.40*

*Add for ridges, hips, valleys, flashing, etc.

Material Cost of 100 Sq. Ft. (1 Sq.) Standing Seam Roofing Using 24"x96" Galvanized Steel Sheets

	Rate	Total	Rate	Total
1.1 sq. No. 15 felt	$. . .	$. . .	$3.00	$ 3.30
8 shts. 26 ga. steel, 120 lbs.	. . .	. . .	.80	102.40
60 cleats	. . .	. . .	.03	1.80
1 lb. solder	. . .	. . .	1.00	1.00
1 lb. nails	. . .	. . .	.80	.80
Flux and charcoal	. . .	. . .	.70	.70
Cost per 100 sq. ft.		$. . . .		$110.00*

*Add for ridges, hips, valleys, flashing, etc.

Material Cost of 100 Sq. Ft. (1 Sq.) V-Crimped Metal Roofing

	Rate	Total	Rate	Total
1.1 sq. No. 15 Felt	$. . .	$. . .	$ 3.00	$ 3.30
1 sq. 26 ga. V-crimped rfg	. . .	. . .	95.00	95.00
5 lbs. lead headed nails	. . .	. . .	1.50	7.50
60 lin. ft. wood strips	. . .	. . .	.20	12.00
Cost per 100 sq. ft.		$. . . .		$117.80*

*Add for ridges, hips, valleys, flashing, etc.

COPPER ROOFING

Copper may be used as roofing for any type of roof. There are various methods of applying copper for roofing, but most frequently used are the flat seam and standing seam methods. The batten seam method is also used, where durability is of more importance than cost, e.g., cathedrals, government buildings, monumental structures, etc.

Flat Seam Roofing.—Flat seam roofing is commonly used on dead level or flat surfaces, but is adaptable to any slope. This method is also adaptable for the construction of water cooling roof panels, permitting water to stand up to a depth of 3 inches.

Method of Placing Flat Seam Copper Roofing.

For flat seam roofing, the use of small sheets of 20-oz, cold-rolled copper, not larger than 16"x18", with ¾" seams flat-locked and soldered, is recommended.

The accompanying illustration shows the method of laying flat seam roofing. The sheets are tinned on all edges, for soldering, and are then formed to lock ¾" with adjacent sheet. Corners must be clipped to permit folding as shown. Opposite sides of the sheets are folded in opposite directions so they will hook into adjoining sheets.

The sheets are held down by 2" wide copper cleats, two on each of two adjacent sides. The other two sides are held by the edges of adjacent sheets already cleated. For small areas, it is general practice to use only three cleats, two on a long side and one on a short side.

The cross seams are folded in the direction of flow. All seams should be flattened with a mallet and solder sweated to completely fill the seam.

Each 16"x18" sheet when folded covers 14½"x16½" or 239¼ sq. in. It requires 61 sheets of 16"x18" copper per 100 sq. ft. of roof.

Standing Seam Roofing.—Standing seam copper roofing may be used for surfaces with a minumum slope of 3" per foot. Standing seams are usually made to finish 1" high. The spacing of the seams is a matter of scale and architectural effect, but for economy, a choice should be made that will use sheets of stock sizes. The use of 20"x96" sheets is recommended as maximum for ordinary purposes.

Method of Placing Standing Seam Copper Roofing.

For standing seams to finish 1" high, the vertical bends are made on the long edges of the sheets—1¾" on one side and 1½" on the other; the short bend on the one sheet always adjoins the long bend on the adjacent sheet.

Recommended weights of material are: 16-oz. for widths between seams up to 20"; 20-oz. for widths over 20".

Formed pans are held to the roof deck with 2"x3" cleats spaced 12" apart and locked into the standing seam as shown in the accompanying illustration. Transverse joints should be staggered and each secured with one cleat locked into the flat seam.

Standing seams should not be riveted or soldered—cross seams should be left unsoldered whenever conditions permit.

A sheet loses 3¼" in width on account of the standing seam, i.e. a sheet 20" wide would cover only 16¾". Where 20" sheets are used, add 20 percent to roof area for standing seams.

The loss at the end of the sheet will depend upon the length of the roof and the length of copper sheet used, but allow 1½" for each flat seam.

Sizes and Weights of Copper Sheets Commonly Used in Building Work

Size of Sheet Copper	Weight in Pounds per Sheet of Various Weight Copper			
	16-Oz.	20-Oz.	24-Oz.	32-Oz.
20"x96"	13.33	16.67	20.00	26.67
24"x96"	16.00	20.00	24.00	32.00
30"x96"	20.00	25.00	30.00	40.00
36"x96"	24.00	30.00	36.00	48.00
20"x120"	16.67	20.83	25.00	33.33
24"x120"	20.00	25.00	30.00	40.00
30"x120"	25.00	31.25	37.50	50.00
36"x120"	30.00	37.50	45.00	60.00

Labor Placing Flat Seam Copper Roofing.—When using 16"x18" copper sheets for flat seam roofing, a tinner and helper should place 175 to 225 sq. ft. of roof per 8-hr. day, on average size roofs, at a labor cost of $129.16 per 100 sq. ft.

Labor Placing Standing Seam Copper Roofing.—When using copper sheets 20" wide and up to 8'-0" long for standing seam roofing, a tinner and helper should place 200 to 250 sq. ft. per 8-hr. day, on average size roofs, at a labor cost of $113.02 per 100 sq. ft.

Material Cost of 100 Sq. Ft. (1 Sq.) Flat Seam Copper Roofing Using 16"x18" Sheets

	Rate	Total	Rate	Total
1.1 sq. No. 15 felt	$....	$....	$3.00	$ 3.30
61 shts. 16"x18" (20-oz.) 153 lbs.	....	....	3.20	195.20
250 copper cleats	....	....	.25	62.50
2 lbs. copper nails	....	....	2.85	5.70
6 lbs. solder	....	....	1.00	6.00
Flux and charcoal	....	....	1.25	1.25
Cost per 100 sq. ft.		$....		$273.95*

*Add for ridges, hips, valleys, flashing, etc.

Material Cost of 100 Sq. Ft. (1 Sq.) Standing Seam Copper Roofing Using 20" Copper

	Rate	Total	Rate	Total
1.1 sq. No. 15 felt	$....	$....	$3.00	$ 3.30
120 sq. ft. (16-oz.) 120 lbs.	....	....	1.60	192.00
80 copper cleats	....	....	.25	20.00
1 lb. copper nails	....	....	2.85	2.85
Cost per 100 sq. ft.		$....		$218.15*

*Add for ridges, hips, valleys, flashing, etc.

Stainless steel, type 304, monel and lead are also used for flat, standing and batten seams.

07620   SHEET METAL FLASHING AND TRIM

Sheet metal work in its many branches is a highly specialized business, because the greater part of the work is performed in the shop and the fabricated materials are sent to the job ready to erect. For that reason an estimate on sheet metal work must take into consideration the cost of the finished materials at the shop or delivered to the building site, plus the labor cost of erection.

Estimating Quantities of Sheet Metal Work

The data given on the following pages will assist the estimator in measuring quantites from the plans and listing them on the estimate sheet.

When preparing the quantity survey for the different types, shapes, and sizes of metal items, be sure to identify the type and weight, or gauge, of metal that is to be used, e.g., galvanized steel, aluminum, stainless steel, copper, lead, terne, etc. Each one will have its own waste factor, difficulty in shaping, method of jointing, etc.; all affecting the production and installation costs, in addition to the sheet goods cost delivered to the shop.

Metal Gutters and Eave Trough.—Metal gutters and eave trough are furnished in numerous designs, from 2" to 12" in size. When estimating quantities, obtain the number of pieces of each size gutter and the length of same, as the estimate should state the lin. ft. of each size.

Metal Conductor Pipe or Downspout.—Metal conductor pipe is furnished both round and square and from 2" to 6" in size.

When estimating quantities, measure the distance from the roof to the ground, which will be the length of each downspout. After the quantities have been obtained in this manner, list the total number of lin. ft. of each size.

Metal Conductor Heads.—Metal conductor heads are made in various styles, shapes and sizes.

Conductor heads are generally used on buildings having flat roofs where a cast iron conductor pipe is placed inside the building and installed by the plumbing contractor. When placed on the outside of a wall at the roof level, an opening is usually left in the masonry wall at the low point of the roof to allow the water to pass through the wall opening into the conductor head and down the conductor pipe.

When estimating conductor heads always note style and size of each head and the material specified.

Metal Flashing and Counterflashing.—Flashing and counterflashing is furnished in tin, stainless steel, galvanized steel, aluminum or copper. They are measured by the lin. ft. when less than 12" wide and by the sq. ft. when over 12" wide.

Metal flashing used in connection with a flat roof deck covered with felt or built-up roofing usually extends 6" to 12" up the side of the brick fire wall and projects under the roofing the same distance. A counterflashing is then placed over and above this flashing. The upper edge of the counterflashing is inserted in a reglet (slot) in the brick or stone wall. After the counterflashing has been placed, it is caulked with caulking compound to prevent water from getting in back of the flashing and running under the roofing.

On many jobs using felt or built-up roofing, the felt is run up the wall about 12" and mopped. The metal counterflashing is then placed to extend down over the felt, which prevents water from getting in back of the flashing and running under the roofing.

Where metal counterflashing is used it is necessary to cut an open joint or reglet in the mortar joint of the masonry wall.

The cost of cutting reglets should be estimated by the lin. ft.

Metal Valleys and Hips.—Metal valleys and hips are frequently used with certain types of shingle and slate roofs. They are either continuous metal strips of a certain size or small shingles of tin, galvanized steel, aluminum, stainless steel or copper.

They should be estimated by the lin. ft. stating the width of the valley or hip and the kind of metal used.

Metal Ridge Roll, Hip Roll and Cap.—Metal ridge roll, hip roll and cap should be estimated by the lin. ft. giving the width, gauge and kind of metal used.

Metal Ventilators.—Metal ventilators for roofs and skylights should be estimated at a certain price for each ventilator, f.o.b. factory or erected. Always state size and type of ventilators required, as there are numerous kinds on the market at varying prices.

Metal Skylights.—When estimating metal skylights, consider the following: size, type of skylight, whether single or double pitch or hip skylight; with or without ventilators or side sash; kind of glass, and whether a metal curb flashing is required.

Labor Placing Sheet Metal Work

The costs given on the following pages are for the erection of various kinds of sheet metal work on the job. Add cost of the fabricated material, contractor's overhead expense and profit.

Placing Hanging Gutter or Eave Trough.—This cost will vary with the slope of the roof, distance above the ground and the method used in hanging the gutter.

On wood constructed buildings having hip or gable roofs, metal hangers or supports are placed over the gutters or eave trough and fastened to the eaves under the shingles or other roofing. The eave trough or gutter must always slope in the direction of the downspout, in order to drain satisfactorily.

On hip or gable roofs having eaves 20'-0" to 25'-0" above the ground, a tinner and helper should place 180 to 220 lin. ft. of gutter per 8-hr. day, at the following labor cost per 100 lin. ft.:

	Hours	Rate	Total	Rate	Total
Sht. met. wkr	4	$....	$....	$19.75	$ 79.00
Labor	4	....	....	12.54	50.16
Cost 100 lin. ft.			$....		$129.16
Cost per lin. ft.				....	1.29

Placing Metal Conductor Pipes or Downspouts.—Metal conductor pipes or downspouts are furnished in both round and square and from 1½" to 6" diameter. The conductor pipe extends down the side of the building from the eave trough to grade, where it is connected to an elbow that throws the water away from the building or is connected with the drainage system. Metal conductor pipe is usually held in place by hooks fastened to the walls.

On one or two story buildings where the conductor pipe is 12'-0" to 25'-0" long, a tinner and helper should place 200 to 250 lin. ft. per 8-hr. day, at the following labor cost per 100 lin. ft.:

	Hours	Rate	Total	Rate	Total
Sht. met. wkr	3.5	$....	$....	$19.75	$ 69.13
Labor	3.5	....	....	12.54	43.89
Cost 100 lin. ft.			$....		$113.02
Cost per lin. ft.				....	1.13

Where outside conductor pipe is installed on buildings 3 to 4 stories high, a tinner and helper should place 160 to 200 lin. ft. per 8-hr. day, at the following labor cost per 100 lin. ft.:

	Hours	Rate	Total	Rate	Total
Sht. met. wkr.	4.5	$....	$....	$19.75	$ 88.88
Labor	4.5	....	....	12.54	56.43
Cost 100 lin. ft.			$....		$145.31
Cost per lin. ft.				....	1.45

Placing Metal Flashing.—When placing metal roof flashing around parapet walls, etc., it is customary to have the flashing extend under the roofing and up the side of the wall 6" to 12", and then place the counterflashing above to lap over the flashing.

Where it is not necessary to cut reglets in the masonry wall, a tinner should place 140 to 160 lin. ft. of flashing per 8-hr. day, at the following cost per 100 lin. ft.:

	Hours	Rate	Total	Rate	Total
Sht. met. wkr.	5.4	$. . . .	$. . . .	$19.75	$106.65
Cost per lin. ft.			. . . .		1.07

On frame buildings where the flashing extends under the wood wall siding to protect porch roofs, etc., it is only necessary to place the flashing or metal shingles on the roof and tack them against the sheathing. A tinner should place 165 to 185 lin. ft. per 8-hr. day, at the following labor cost per 100 lin. ft.:

	Hours	Rate	Total	Rate	Total
Sht. Met Wkr.	4.6	$. . . .	$. . . .	$19.75	$90.85
Cost per lin. ft.			. . . .		.91

Placing Metal Counterflashing.—When placing metal counterflashing over felt or metal flashing that extends up on the side of the wall, it will be necessary to place the counterflashing in a reglet or open joint in the masonry that can be caulked or sealed to prevent the water getting in back of the counterflashing and under the roof.

Where reglets have previously been cut, as is usually the case on stone copings, requiring only the sealing of the joint, a tinner should place 140 to 160 lin. ft. per 8-hr. day, at the following labor cost per 100 lin. ft.:

	Hours	Rate	Total	Rate	Total
Sht. Met. Wkr.	5.4	$. . . .	$. . . .	$19.75	$106.65
Cost per lin. ft.			. . . .		1.07

Cutting Reglets in Masonry Walls.—The labor cost of cutting reglets in brick walls will depend entirely upon the kind of mortar used and the condition of the mortar at the time of cutting the reglets.

If the brick are laid in portland cement mortar, it will cost more to cut the reglet than when laid in lime mortar. Also, if the reglets are cut within 24 hrs. after the brick are laid, the labor cost will be much less than cutting them two or three weeks later.

If reglets are cut before the mortar has had sufficient time to set, a man should cut 175 to 225 lin. ft. per 8-hr. day, at the following labor cost per 100 lin. ft.:

	Hours	Rate	Total	Rate	Total
Labor	4	$. . . .	$. . . .	$12.54	$50.16
Cost per lin. ft.			. . . .		.50

If the reglets are cut several days after the brick have been laid, giving the mortar sufficient time to harden, it will require a hammer and chisel to cut out the joints. Under these conditions a man should cut 65 to 80 lin. ft. per 8-hr. day, at the following labor cost per 100 lin. ft.:

	Hours	Rate	Total	Rate	Total
Labor	11	$. . . .	$. . . .	$12.54	$137.94
Cost per lin. ft.			. . . .		1.38

If a portable electric saw, having an abrasive blade or cutting wheel is used for cutting the reglets, a man should cut 10 to 15 lin. ft. of reglet an hour.

Cutting reglets in stone coping or balustrade is usually performed by stone cutters at about the same cost as given above.

Placing Metal Valleys.—When placing metal shingles or valleys in connection with gable or hip roofs, dormers, etc., a tinner should place 115 to 135 lin ft. per 8-hr. day, at the following labor cost per 100 lin. ft.:

	Hours	Rate	Total	Rate	Total
Sht. Met. Wkr.	6.4	$. . . .	$. . . .	$19.75	$126.40
Cost per lin. ft.			. . . .		1.26

Placing Metal Ridge Roll of Cap.—If metal ridge roll is placed on the ridge of a hip or gable roof, a tinner should place 90 to 110 lin. ft. per 8-hr. day, at the following labor cost per 100 lin. ft.:

	Hours	Rate	Total	Rate	Total
Sht. Met. Wkr.	8	$. . . .	$. . . .	$19.75	$158.00
Cost per lin. ft.			. . . .		1.58

Placing Metal Conductor Heads.—On buildings having flat roofs where the conductor pipe or downspout is placed inside of the building, requiring a conductor head at the low point of the roof which is connected to the downspout, a tinner should place one conductor head (made complete in the shop) in 1½ to 2 hrs. at the following labor cost:

	Hours	Rate	Total	Rate	Total
Sht. Met. Wkr.	1.8	$. . . .	$. . . .	$19.75	$35.55

07630   ROOFING SPECIALTIES

Flashings can be formed from a wide variety of materials of varying permanence, as follows:

MATERIAL	SQ. FT. COST	SQ. FT. INSTALLED PER DAY
Aluminum .013" thick	$0.24	150
(Mill \ .016 "	.28	"
(Finish) .019 "	.38	"
.032 "	.72	145
.040 "	.86	"
.050 "	1.05	"
Copper 16 oz.	2.20	140
20 oz.	2.70	"
24 oz.	3.10	"
Stainless Steel 26 ga.	2.00	150
Asph. C't'd Cotton 17 oz.	.15	350
40 oz.	.22	"
Lead 2.5 lb.	1.65	140
Zinc/Copper Alloy	1.35	"
Butyl 1/32"	.52	275
1/16"	.75	"
Neoprene 1/16"	1.25	"
Polyvinyl .02"	.14	"
.03	.22	"

Gravel stops may be formed on the job or ordered out in extruded form. If extruded be sure to check if any special finish is specified as this will raise costs considerably. A few costs and rates for installation are given below:

MATERIAL	LIN. FT. COST		LIN. FT. INSTALLED
	4" high	6" High	PER DAY
Aluminum .050" mill finish	$2.50	$2.90	135
Duranodic	3.00	3.40	"

Preformed gutters can be installed by one sheet metal worker at the rate of from 110' to 120' per day. Material costs are:

MATERIAL	COST PER LIN. FT.		
	4"	5"	6"
Aluminum .027" plain	$0.60	$0.65	$0.80
enam.	.65	.73	.90
Copper 16 oz.	3.40	3.80	4.70
Stainless 22 ga.	3.40	3.70	4.00
Vinyl	1.00	1.10	1.25
Galvanized 28 ga.	.55	.70	.90

Downspouts can be installed by one sheet metal worker at the rate of 175 lineal feet for small sizes, 150 for medium and 125 for the larger sizes per day. The material costs will run as follows:

MATERIAL	COST PER LIN. FT.				
	3" dia.	4" dia.	5" dia.	2"x3"	3"x4"
Aluminum .025" plain	$0.50	$0.80	$ 1.10	$0.75	$0.95
enam.	.58	.90	1.15	0.85	1.10
Copper 16 oz.	3.00	3.50	4.30	3.30	4.05
Stainless	7.00	9.00	14.00	7.00	9.00
Vinyl	.60	—	—	.65	—
Galv. Steel 28 ga.	.45	.60	1.05	.50	.65

For elbows for the above work add the following:

3" diam. or 2"x 3" in plain aluminum-	$1.00
" " " " enam. " "	1.25
" " " " 16 oz. copper	5.00
4" diam. or 3"x 4" in plain aluminum	1.25
" " " " enam. " "	1.50
" " " " 16 oz. copper	8.00

07800　ROOF ACCESSORIES

Erecting and Glazing Metal Skylights.—Metal skylights are manufactured in the shop and sent to the job "knocked down" or in sections ready to erect and glaze.

An average skylight (single pitch, double pitch or hip) up to 8'-0"x12'-0" in size, containing 100 sq. ft. of area, should be erected and glazed complete by a sheet metal worker and glazier in 8 to 10 hrs. time, at the following labor cost:

	Hours	Rate	Total	Rate	Total
Sheet metal worker	5	$. . . .	$. . . .	$19.75	$98.75
Glazier	4	. . . .	. . . .	15.56	62.24
Cost per skylight			$. . . .		$160.99

When erecting skylights 10'-0"x10'-0" to 10'-0"x15'-0", containing 100 to 150 sq. ft., a sheet metal worker and glazier should erect and glaze one skylight in 14 to 16 hrs. at the following labor cost:

	Hours	Rate	Total	Rate	Total
Sheet metal worker	9	$....	$....	$19.75	$177.75
Glazier	6	....	....	15.56	93.36
Cost per skylight			$....		$271.11

When erecting skylights 10'-0"x16'-0" to 10'-0"x20'-0" in size, containing 160 to 200 sq. ft. it will require about 24 hrs. for a sheet metal worker and glazier to erect the skylight and glaze it complete, at the following labor cost:

	Hours	Rate	Total	Rate	Total
Sheet metal worker	16	$....	$....	$19.75	$316.00
Glazier	8	....	....	15.56	124.48
Cost per skylight			$....		$440.48

The above costs do not include overhead and profit or the cost of placing a metal curb flashing for skylight. For costs on this work, see below.

Placing Metal Skylight Curb Flashing.—Metal skylights are often placed on top of a wood curb projecting 6" or more above the roof level. The space from the top of the curb to the roof level is covered with metal flashing.

When erecting metal skylight curb flashing, a tinner should place 190 to 210 sq. ft. of metal per 8-hr day, at the following labor cost per 100 sq. ft.:

	Hours	Rate	Total	Rate	Total
Sheet metal worker	4	$....	$....	$19.75	$79.00
Cost per sq. ft.				....	.79

Erecting Skylights With Side Sash.—If the skylights have side sash, the erection cost will vary with the number of sash in the skylight and whether stationary or pivoted. On an average it will require 1 to 1½ hrs. labor time for each sash in the skylight, at the following labor cost per sash:

	Hours	Rate	Total	Rate	Total
Sheet metal worker	1.25	$....	$....	$19.75	$24.69

Worm and Gear Sash Operating Device

Erecting Sash Operating Devices for Movable Sash.—Skylights furnished with hinged or pivoted sidewall sash require an operating device to open and close them.

An operating device usually consists of a mechanism similar to that illustrated, which is operated from the floor by a worm and gear or a chain attachment, from which all sash on each side of the skylight can be opened and closed.

The labor cost of erecting sash operating devices will vary with the length of each run and the number of runs in each skylight but on an average, a man should erect 50 to 75 lin. ft. per 8-hr. day, at the following labor cost per 100 lin. ft.:

	Hours	Rate	Total	Rate	Total
Sheet metal worker	13	$....	$....	$19.75	$256.75
Cost per lin. ft.			....		2.57

Ventilating Skylights.—Ventilating skylights are used in factories, foundries, machine shops, garages, dairies, laundries and in any other style of building requiring economical light and ventilation. They are furnished in any type of metal desired, galvanized copper-bearing steel, aluminum or copper, depending upon the nature of the building on which they are to be used. Normally, for building of ordinary construction, the standard 18-gauge galvanized copper-bearing steel is furnished. Should the building be of reinforced concrete or steel and concrete construction, the non-corrosive aluminum or copper construction should be furnished.

Section Through Ventilating Skylight

These skylights are of gable type as illustrated and are furnished in standard widths of 4'-0" to 20'-0" and any desired length in multiples of 2'-0". It is standard practice to interpret the size of a skylight as being outside of curb dimensions.

The length of the ventilating sections depend upon the width of the skylights; 4'-0", 6'-0" and 8'-0" widths having maximum ventilating sections of 40'-0" each controlled from one operating station; 10'-0" and 12'-0" wide skylights have maximum ventilation sections of 30'-0" in length; 14'-0" and 16'-0" widths have 40'-0" maximum lengths, and 18'-0" and 20'-0" widths have 30'-0" maximum ventilating sections.

A mechanical operating device consisting of either loose rod or chain control is furnished as required, so that the skylight can be readily controlled from the floor. All skylights under 30'-0" in length are arranged for full length ventilation. Longer skylights are arranged with ventilating sections as required.

The cost of ventilating skylights vary with their size and number of ventilating sections in each skylight. Prices should be secured from the manufacturer.

In estimating the cost of ventilating skylights, the principal thing to remember is : The smaller the area of the skylight, the higher the price per sq. ft.

Plastic Dome Skylights.—This type skylight consists of a thermo-formed acrylic plastic sheet free blown into square, rectangular and circular domes.

This type skylight is furnished as a factory assembled unit; the plastic is mounted in an extruded aluminum frame, and prefabricated curbs are available. The aluminum frame is generally designed to form counterflashing.

Plastic Domes

Size	Cost Per Roof Dome
16"x16"	$ 40.00
24"x24"	50.00
36"x36"	65.00
48"x48"	115.00
24" Round	85.00
48" Round	150.00

Erection cost will vary widely as to the number involved and the preparatory work done by others. As units are vulnerable to breakage and scratching, it takes two men to handle all but the smallest unit. On a job with several units involved two men should set 2 units of up to 10 sq. ft. in a day if the curbing is already in place.

The dome unit is available as an insulating unit which has a secondary acrylic dome mounted under the primary (exterior) dome. This double dome unit provides a dead air space between the domes that reduces heat loss and condensation formation.

ROOF HATCH

A standard size 2'-6"x3'-0" steel roof hatch with insulated curb and cover and hardware will run around $275.00. A mechanic and helper should set this in 2 hrs.

GRAVITY VENTILATORS

12" stationary ventilators of the syphon type with a 24" base will cost around $40 and take two hours to set. Stationary ventilators of the mushroom type with a 24" base will cost $130.00 with the same labor cost.

07900   SEALANTS

With modern construction materials and designs, came the need for good sealants and caulking compounds. No one sealant can solve the requirements across the full range of construction applications; therefore, sealants are manufactured for interior use, exterior use, gun or pour grade, ability to expand and contract, service temperature range, paintability and for compatability with the material to be sealed.

An oil base caulking compound has a life expectancy of about 4 or 5 years. Its restrictions list a joint maximum size of ½" wide x ¾" deep and is incapable of withstanding any joint movement without rupture. Therefore, using this material for exterior applications may not be prudent or cost effective.

Acrylic latex caulk is probably a good starting point for exterior use. It is considered to be the best value among sealants in the middle performance range due to; relatively inexpensive, life expectancy of 8 to 12 years, cures quickly, excellent paintability, non-staining, and has reasonable elongation and recovery for movement of joints. However, this material has a maximum joint size of ½" wide and ½" deep and should not be used for expansion joints.

Polysulfide base sealants, available in one or two part compounds, are among the premiere compounds on the market. It is reasonably expensive and requires a little more labor time to install, however its movement capabilities, bonding strength, wide use range (including expansion joints) and a life expectancy of 15± years makes it cost effective while providing excellent joint sealing characteristics.

It is important to have a properly designed joint, both in width and depth. Good sealants will expand and contract with the movement of the joint. As the joint moves, the sealant changes shape while the volume remains constant. It is critical that the width-to-depth ratio be designed to withstand the constant elongation and compression cycles over long periods of time.

Deep beads of sealant wastes material and are more prone to failure than shallow sealant beads. Another important factor to consider is that sealant works best when adhering to the two opposing faces and isolated from the third (back) side of the joint to be sealed. This is accomplished by filling the joint with a material such as oakum or polyethylene foam rod to within ⅜" or ½" of the face of the joint. In shallow depth joints, use a non-adhering tape at the back of the joint to prevent the sealant from bonding.

Sealants are manufactured in colors from white to black and some are clear (silicones). This is necessary to be architecturally compatible with the adjoining surfaces, in addition paint does not adhere at all to some sealants.

The following chart will provide a guide for the amount of sealant required to fill various size joints; expressed in linear feet of joint obtained per gallon of material.

		WIDTH, INCHES							
		1/16	1/8	1/4	3/8	1/2	5/8	3/4	1
DEPTH, INCHES	1/16	4928	2464	1232	821	616	493	411	307
	1/8	—	1232	616	411	307	246	205	154
	3/16	—	—	411	275	205	164	137	103
	1/4	—	—	307	205	154	123	103	77
	3/8	—	—	—	137	103	82	68	51

In using the above chart, if the size of the joint to be sealed is not given on the drawings; use the principle that the depth of the sealant is ½ of the width of the joint to be sealed.

When estimating the labor required to caulk a joint, it will depend upon the width of the joint, the material used, and the location or accessibility of the joint. One man should caulk about 600 lin. ft. per 8 hr. day of joints around doors or window frames when they are close to the ground or accessible without the use of scaffolding. This time may be reduced in half if working off ladders or scaffolding. This same man, working on sealing wide expansion joints, will only finish about 200 lin. ft. per 8 hr. day.

CHAPTER 13

DOORS AND WINDOWS

CSI   DIVISION 8
08100 METAL DOORS AND FRAMES
08200 WOOD AND PLASTIC DOORS
08300 SPECIAL DOORS
08400 ENTRANCES AND STOREFRONTS
08500 METAL WINDOWS
08600 WOOD AND PLASTIC WINDOWS
08700 HARDWARE AND SPECIALTIES
08800 GLAZING
08900 WINDOW WALLS/CURTAIN WALLS

Most of the items covered in this chapter are furnished by material distributors, but set by the carpenter on the job; in the estimate, the cost of labor would be included under the proper trade.

Glazing is an exception to this, and both material and labor will be included under a single item. However, it is more and more prevalent these days for sash and doors to come to the job factory glazed. The contractor must always check this out so as to eliminate any duplication in the estimate and to make certain the pre-glazing complies with both the specifications and local union rules.

Curtainwall material and labor are usually let to a single subcontractor. It is advantageous to have the metal erection, sealants and glazing all in one contract so that one source will be responsible for the weather tightness of the wall.

For the estimator who will be required to perform the quantity survey for materials covered by this Chapter, it is suggested that he become very familiar with the various "Schedules" that the project Architect will include on the plans.

These Schedules are usually laid out in block column form consisting of a listing of the material involved and all of the particulars for each item, and may show architectural details for each type of listed item.

Door Schedules will normally list the door openings by "mark" numbers or letters, e.g., #1, #2, #3 or A, B, C. After the "mark" designation will be the size, material of the door and the frame, a code designation for the type and size of the frame, fire rating requirements (if any), a code designation for any louver or glass panels required, finish hardware set designations, and a "remarks" column for any special instructions. Bear in mind that the "mark" designation refers to that particular design type, and may have nothing to do with the total quantity of doors and frames of that design type required for the project.

Window schedules are laid out in much the same fashion, showing: mark designation, size, material, glass, type, etc.

Finish hardware schedules are normally set up in the hardware section of the specifications, rather than on the drawings. The hardware schedule is a listing of the hardware items required to fit a particular door opening, e.g., hinges, latchset or lockset, closer, kickplates, etc.

Each type of opening, such as for toilets, closets, offices, etc., will have its own "H W Set #_____"designation; and each will vary from the other in some way. For example, a door that leads to a stairway in an office building could list under its HW Set #_____: 1½ pair hinges, latchset, and closer. A door that leads to an office in the same building could list under its HW Set #_____: 1½ pair hinges, lockset, and closer. As you can see from the two prior examples, they are very similar, but, because of the lock requirements for the office door opening, a separate HW Set #_____will be given.

Here again, keep in mind that the HW Set #_____has nothing to do with the total quantity of finish hardware required, i.e., one set number may be applied to 10 door openings and the next set number applies to only one or two door openings.

08100   METAL DOORS AND FRAMES

Each hollow metal manufacturer produces their stock design doors and frames as well as a custom line to suit the project Architect's requirements. Bear in mind that the "stock" line will vary between manufacturers, e.g., sizes, width of frames, hardware preparation, etc. For example, a hollow metal door from one manufacturer usually can not be hung in the hollow metal frame from another manufacturer, due to unmatched hinge or lockset locations.

"Stock" will also refer to the cut-out preparation for hinges and, in particular, locksets. Most manufacturers prepare the frames for 3 hinge locations each sized 3½" or 4½" in height and for the standard cylindrical lockset. If the finish hardware specified for the project varies from the manufacturers' standards, the hollow metal manufacturer will consider the order "custom" and increase the price accordingly as well as increase the delivery time to the project.

Material prices for hollow metal items are quoted inclusive of all mortising, reinforcing, drilling and tapping for mounting the door into the frame and preparation for items of finish hardware. Also included are frame anchors, and shop prime coat of paint.

Hollow Metal Frames.—Frames are made of 18, 16, and 14 gauge metal and formed to receive a 1⅜" or 1¾" thick door, depending on which side of the stop the hinge and strike cut-outs are placed. See the following section through a standard hollow metal frame.

TYPICAL FRAME PROFILE

Therefore, when the manufacturer varies the overall width of the frame to suit the different wall conditions that may be encountered, he simply increases the width of the stop. This allows them to set up any width frame for either thickness of door in lieu of manufacturing frames for 1⅜" doors and 1¾" doors.

As stated, frames are made in various widths (profile) to suit the thicknesses of the walls to which they are mounted. The standard widths with most manufacturers are 4¾", 5¾", 6¾" and 8¾".

In addition to the frame width dimensions, frames are ordered according to the door opening size, the standard being 2'-0", 2'-4", 2'-6", 2'-8", 3'-0", 3'-4", 3'-6",

3'-8" and 4'-0" wide for single swing and 4'-0", 4'-8", 5'-0", 5'-4", 6'-0", 6'-8", 7'-0", 7'-4" and 8'-0" wide for double swing by 6'-8", 7'-0", 7'-2" and 8'-0" in height.

Frames are available from the factory in welded joint construction (all 3 sides put together to form a unit) or in knocked down (K.D.) form for use on drywall partitions. The K.D. frame is considered to be a pressure-fit frame that once in place (capping the wall) it exerts a gripping action onto the wall. The welded frame is stronger and must be erected prior to the wall erection, whereas, the K.D. frame is erected after the wall is built.

Special frames can be ordered which will incorporate a transom panel above the door and/or sidelites of various heights and widths beside the door.

Average material prices for hollow metal frames of various sizes are as follows:

18 Ga. Welded Hollow Metal Frames

Size	4¾"	5¾"	6¾"	8¾"
2'-6"x7'-0"	57.00	59.00	61.00	63.00
3'-0"x7'-0"	58.00	60.00	62.00	64.00
5'-0"x7'-0"	73.00	75.00	77.00	79.00
6'-0"x7'-0"	75.00	77.00	79.00	81.00

Note: Deduct $10.00 each for knocked down (K.D.) frames.
Add $10.00 for "B" Label, $15.00 for "A" Label.

Labor Erecting Hollow Metal Frames.—Two carpenters should erect about 16 welded frames per 8 hour day at the following labor cost per frame:

	Hours	Rate	Total	Rate	Total
Carpenter	1.0	$....	$....	$16.47	$16.47

One carpenter should erect about 12 K.D. frames per 8 hour day at the following labor cost per frame:

	Hours	Rate	Total	Rate	Total
Carpenter	0.67	$....	$....	$16.47	$11.04

Hollow Metal Doors.—Doors are made with 20, 18, or 16 gauge face sheets and in 1⅜" and 1¾" thicknesses. Door sizes will correspond to the frame opening sizes stated previously in the Hollow Metal Frame data.

Doors may be ordered in a variety of styles including: small vision lites, half glass, full glass, dutch, louvered, and flush. Actually, when a door is ordered with a vision lite, what is received from the manufacturer is a flush door that has been cut out and prepared for future glass installation at the jobsite; the glass is not supplied by the door manufacturer.

The physical construction of the door itself is where the manufacturers really differ. They each have their own techniques for assembly i.e., location of seams, inverted channel or flush top and bottom edges, gauge of hinge and closer reinforcements, leading edge reinforcement, core materials, etc. The contractor is obligated to determine the compliance of the manufacturers product with the requirements contained in the specification. Therefore, be sure of compliance with specification requirements when pricing hollow metal doors, because prices vary widely.

Average material prices for hollow metal doors are as follows:

20 Ga. Hollow Metal Doors

Size	1⅜"	1¾"
2'-6"x7'-0"	125.00	135.00

Size	1⅜"	1¾"
3'-0"x7'-0"	135.00	145.00
3'-6"x7'-0"	150.00	160.00

Note: Add $20.00 for "B" Label, $45.00 for "A" Label
    Add $60.00 for louver, $50.00 for lite cut-out
    Add $40.00 for mineral core (for temperature rise rating)

Labor Erecting Hollow Metal Doors.—Two carpenters should hang, on hinges, about 12 doors per 8 hour day at the following labor cost per door:

	Hours	Rate	Total	Rate	Total
Carpenter	1.34	$. . . .	$. . . .	$16.47	$22.07
Laborer (for distribution)	.5	. . . .	. . . .	12.54	6.27
			$. . . .		$28.34

Fire Label Requirements.—According to building codes, certain door locations within a building; such as at stairs, trash rooms, furnace rooms, garage entries, storage rooms, etc., will require a fire rating for the door and frame. This rating is expressed as "C Label" for ¾ hour rating, "B Label" for 1½ - 2 hour rating, and "A Label" for 3 hour rating.

The rating is obtained by the manufacturer subjecting their product to actual fire testing by either Underwriters Laboratory or a Factory Mutual testing laboratory. The door and frame is then approved based on the physical construction of the unit. If the manufacturer changes the design after the approval, they must submit for a new test based on the re-designed unit. The approved designed and tested door and frame then qualifies to have a label affixed to it, which bears proof that it is a fire rated assembly.

When a door is to be fire rated, it also carries other restrictions such as; (1) the latch or lockset must be installed with a UL throw bolt, (2) glass is either not allowed (as in "A Label" rating) or restricted in size (as in "B or C Label" ratings), (3) if a louver is needed in the door, it must have a fusible link that will melt and allow the louver blades to close when the fire reaches the door, and (4) the door must have a closer device or spring hinges.

Metal Covered Jambs and Trim

Wood jambs and trim covered with No. 24 gauge zinc coated sheet steel, for use with kalamein doors.

Jambs are 1¾"x5½" rabbeted for 1¾" doors. Plain casing is ⅞"x3½". Round edge and brick mold is ⅞"x⅞".

		Prices Jamb Width To:		
Size Up to		6¾	8¾	12
3'-0"x7'-0"	....	$70.00	$75.00	$80.00
4'-0"x7'-0"	....	75.00	80.00	85.00
5'-0"x7'-0"	....	80.00	85.00	90.00
6'-0"x7'-0"	....	85.00	90.00	95.00

For frames over 7'-0" add $5.00 for each 6" or less.

Exterior Metal Residential Door Assemblies

Prehung metal doors complete with moldings of the styles usually found on traditional wood doors and with a frame, threshold, weatherstripping and a variety of locks are now available in over fifty stock designs. Prices for a 3' wide unit will vary from $180 to $685.00 depending on the type of lock and the type of glazing. A plain, flush unit with a double cylinder lock and no glazing runs $200.00; a unit with raised panels runs $230.00; if the upper section is glazed with insulating glass the cost will raise to $260.00. Units similar to the above but in pairs for a 5' opening will run $375, $440 and $493 respectively. If moldings are ordered for the inside face of the door add $17.00.

08200   WOOD AND PLASTIC DOORS

Flush Veneer Slab Doors for Exterior Use

Furnished with water and weather resistant birch face veneers both sides of the door. All doors 1³/₄" thick.

Size	Hol. Core	Sol. Core
2'-6"x6'-8"	$38.00	$ 58.00
2'-8"x6'-8"	40.00	60.00
3'-0"x6'-8"	42.00	65.00
3'-0"x7'-0"	47.00	68.00
3'-0"x8'-0"	55.00	$100.00

Extras for Altering Solid Core Flush Doors

For V-grooves, add per groove	$ 7.00
For cutting in one square light, add	17.00
For one circle light, approx. 12" diam., add	21.00
For one circle light, approx. 18" diam., add	25.00
For one peek light, add	17.00
For one diamond light, add	17.00

White Pine Entrance Doors

Type	2'-8" x 6'-8" x 1³/₄"	3'-0" x 6'-8" x 1³/₄"
6 Panel Colonial	$90.00	$ 95.00
4 Panel & Light	92.00	97.00
X Buck & 9 Light Rect.	105.00	110.00
X Buck & 12 Light Diag.	120.00	125.00
12 Panel Spanish	130.00	135.00
21 Panel Spanish	—	150.00

These doors can usually be purchased prehung in rabetted frames of standard widths with 1½ pair of butts, without sill, lock or weatherstripping, for about $75.00 extra.

Exterior trim sets for front entrances are available; the price range is great, depending on depth of members, elaborateness and sharpness of carvings and flutings. One manufacturer offers trim resembling entrance "A" above at $150.00 in the deluxe model, $85.00 for a modified design; trim such as doorways "C," "D" and "E," around $85.00. These may also be purchased as complete entrance sets with frames, thresholds and sidelights if desired.

Western Pine Colonial Front Entrances

Colonial Front Entrances.

Exterior wood doors are usually fitted with combination storm and screen units. These are usually of 1⅛" thick Ponderosa pine with aluminum wire. They run as follows for 6'-9" height:

	2'-6"	2'-8"	3'-0"
4 Light	$40.00	$42.00	$46.00
X Buck & 1 Light	—	52.00	55.00

Combination doors are also available in aluminum, 1" thick for either 6'-8" or 7'-0" openings in various qualities.

Standard quality, 2'-6", 2'-8" & 3'-0" wide...	$40.00
Deluxe quality, 2'-6", 2'-8" & 3'-0" wide...	50.00
Deluxe quality with white finish, 3'-0" wide...	56.00
X-Buck type with white finish, 3'-0" wide...	65.00

Soft Wood French Doors

Made of clear Ponderosa pine. Stiles and top rails are 4¾" and bottom rails are 9⅝" overall with bead and cove sticking. Muntins are ½" between glass. Wood stops or glazing beads are mitred and tacked in place.

Type	2'-6" x 6'-8"		2'-8" x 6'-8"		3'-0" x 6'-8"	
	1⅜"	1¾"	1⅜"	1¾"	1⅜"	1¾"
1 Light French	$66.00	$70.00	$66.00	$70.00	$84.00	$90.00
5 Light French	67.00	71.00	67.00	71.00	85.00	92.00
15 Light French	68.00	72.00	68.00	72.00	86.00	95.00

French Doors.

Stud Wall
Drip Cap Frame.

Brick Wall
Frame.

Brick Veneer
Wall Frame.

Prices of Western Pine Exterior Door Frames

	Drip Cap		Masonry	
	4⅝"	5¼"	4⅝"	5¼"
2'-6"x6'-8"	$32.00	$33.00	$27.50	$28.50
2'-8"x6'-8"	33.00	34.00	28.00	29.00
2'-8"x7'-0"	34.00	35.00	28.50	29.50
3'-0"x6'-8"	35.00	35.50	28.50	30.50
3'-0"x7'-0"	35.50	36.00	29.50	30.50
4'-0"x6'-8"	38.00	40.00	30.50	31.50
5'-0"x6'-8"	41.00	43.00	31.50	32.50
6'-0"x6'-8"	45.50	47.00	33.50	34.50

Door Frame Prices with Preglazed
Sidelight One Side for 3'-0"x6'-8" Doors

Glass	With 1-1'-0" x 6'-8" S.L.		With 1-1'-4" x 6'-8" S.L.		With 1-1'-8" x 6'-8" S.L.	
	Drip Cap	Masonry	Drip Cap	Masonry	Drip Cap	Masonry
Saf. Gl.-1 Lite	$113.00	$108.00	$125.50	$121.00	$139.00	$134.00
3 Light Saf. Gl.	121.00	116.00	128.50	123.50	136.00	131.00
3 Light & Flat Pan.	134.50	130.50	142.00	137.00	149.50	144.50
3 Light & Raised Pan.	145.00	140.00	152.50	147.50	160.00	155.00

Also available in 4 and 5 light with or without flat or raised panel and with fluted, latticed rondel or loxenged, or insulating glass.

Door Frame Prices with Preglazed
Sidelights Both Sides for 3'-0"x6'-8" Doors

Glass	With 2-1'-0" x 6'-8" S.L.		With 2-1'-4" x 6'-8" S.L.		With 2-1'-8" x 6'-8" S.L.	
	Drip Cap	Masonry	Drip Cap	Masonry	Drip Cap	Masonry
Saf. Gl.-1 Light	$202.00	$199.00	$228.00	$225.00	$255.00	$252.00
3 Light Saf. Gl.	206.00	203.00	221.00	217.50	236.00	232.50
3 Light & Flat Pan.	227.00	223.50	241.50	240.00	256.50	253.00
3 Light & Raised Pan.	248.00	245.00	263.00	259.50	278.00	274.50

Setting Exterior Wood Door Frames

Size of Door Opening	No. Frames Set 8-Hr. Day	Carp. Hrs. per Frame	Add Hrs. for Transom
3'-0"x7'-0"	7–9	0.9–1.1	0.3
6'-0"x7'-0"	5–6	1.3–1.6	0.3
8'-0"x8'-0"	5–6	1.3–1.6	0.3
	First Grade Workmanship		
3'-0"x7'-0"	5–7	1.1–1.6	0.3
6'-0"x7'-0"	4–5	1.6–2.0	0.3
8'-0"x8'-0"	4–5	1.6–2.0	0.3

Interior Flush Doors

Interior hollow core (H.C.) flush doors are manufactured with face veneers of tempered hardboard or plywood of various species, e.g., luan mahogany, birch, or oak, walnut, etc. The core, which is not hollow, will have some material such as honeycomb fiber board or wood stripping to give lateral support to the face veneers.

Interior Hollow Core Flush Doors

Doors 1⅜" Thick Size		Plain Birch	Luan
2'-0"x6'-8"	....	$27.00	$15.00
2'-4"x6'-8"	....	28.00	17.00
2'-6"x6'-8"	....	30.00	20.00
2'-8"x6'-8"	....	33.00	21.00
3'-0"x6'-8"	....	34.00	22.00

For doors 7'-0" in height, add $6.00 per door.
For doors 1¾" thick, add $5.00 per door.

Interior solid core (S.C.) doors have face veneers of plywood in various species, the same as hollow core doors. The cores of the solid door are usually constructed of solid wood blocks (stave) or particleboard.

Interior Solid Core Flush Doors, Particleboard Core

Size	Rotary Birch	Red Oak	Select Birch
2'-8"x6'-8"x1⅜"	$48.00	$52.00	$58.00

3'-0"x6'-8"x1⅜"	50.00	54.00	65.00
2'-8"x6'-8"x1¾"	52.00	56.00	68.00
3'-0"x6'-8"x1¾"	54.00	58.00	72.00

Pre-Hung Interior Door Units

Pre-hung door units, for interior use, consist of the door (usually 1⅜" H.C.), frame, stops, face trim, and hinges all assembled together as a unit. The door is bored and mortised for the lockset and bolt and the frame is mortised for the strike plate.

The frame, which is referred to as "split jamb", parts into two pieces at the stop for installation in the wall opening.

Prices are for units with adjustable jambs, from 4⅜" to 5½" jambs.

Door Size	Luan Doors	Birch Doors
1'-6"x6'-8"	$60.00	$72.00
2'-0"x6'-8"	62.00	74.00
2'-4"x6'-8"	64.00	76.00
2'-6"x6'-8"	65.00	78.00
2'-8"x6'-8"	67.00	80.00

Interior Paneled Doors

Pine panel doors built up of stiles and rails have largely been supplanted by molded doors consisting of a single piece of resin-impregnated wood fibre. The doors are factory primed for finish painting on the job site and are joint-free and shrink and swell resistant.

Colonial style 6 panel doors will cost as follows:

Size	
2'-0"x6'-8"x1⅜"	$30.00
2'-4"x6'-8"x1⅜"	32.00
2'-6"x6'-8"x1⅜"	33.00
2'-8"x6'-8"x1⅜"	35.00
3'-0"x6'-8"x1⅜"	38.00

Sliding and Bi-Folding Closet Doors

With the popularity of sliding doors for closets, many different combinations of doors, materials and sizes are on the market.

With some of them, the doors are the same height as the other doors in the room or building, i.e. 6'-8" or 7'-0", while in others the doors extend to the ceiling, so that when opened, they provide access to the storage shelves at the top.

While there are many different types of doors, hardware, etc., the ones described below are of the type usually found in moderate class housing and apartments.

Frames for sliding doors may be obtained in kits containing two side jambs, head assembly with track attached and necessary hardware for hanging doors. Available with 5⅜" jambs for lath and plaster construction, or with 4⅝" jambs for dry wall construction. Openings may be arranged for two, three, or four by-passing doors. Any 1⅜" thick doors can be used.

The following table lists by-passing units, doors and hardware only for a variety of 1⅜" thick doors.

2 Panel Opng.	H. C. Luan	H. C. Birch	W. P. Louver	W. P. Half Louver	6 Pan. W. P. Col.
4'-0"x6'-8"	$45.00	$55.00	$ 95.00	$120.00	$170.00
5'-0"x6'-8"	50.00	60.00	100.00	125.00	175.00
6'-0"x6'-8"	55.00	65.00	105.00	130.00	180.00

Bi-folding doors offer the advantage of full access to the closet behind. They are available in kits with doors, hinges, pivots, knobs, and guide assembly and track. Jambs and casing not included. Openings to 3'-2" panels hinged one side. 4'-0" and over 4 panels hinged two sides.

Panels Per Opng.	Opng.	1⅜" Hardboard	1⅜" H. C. Birch	1⅛" Louver	1⅛" Half Louver	1⅜" 3 Pan. W. P. Col.
2	2'-0"x6'-8"	$25.00	$35.00	$45.00	$50.00	$ 55.00
2	2'-6"x6'-8"	27.00	40.00	50.00	55.00	60.00
2	3'-0"x6'-8"	30.00	42.00	54.00	59.00	65.00
4	4'-0"x6'-8"	36.00	67.00	75.00	85.00	110.00
4	5'-0"x6'-8"	43.00	73.00	80.00	90.00	115.00
4	6'-0"x6'-8"	48.00	76.00	85.00	95.00	120.00

Sliding Closet Doors.

Louver Sliding Closet Doors.

PLASTIC FACED WOOD DOORS

For institutional use some doors are veneered on all surfaces with plastic laminate. The laminate can be used on both solid and hollow core doors as well as fire, acoustical, x-ray and other special units. Any standard plastic finish may be applied.

Unless otherwise specified, doors are edged as well as faced with plastic. The doors will be shipped in a carton prefinished and pre-mortised for all hardware and cannot be trimmed on the job. Therefore, special care must be taken in checking all details on the shop drawings to see that dimensions are accurately followed on the job.

Since all mortising and fitting is eliminated, hanging of the doors is simplified. One carpenter and one laborer should hang two doors an hour.

These doors are fabricated by manufacturers on special order. The veneers are not applied by contact adhesives, but are bonded under pressure. Costs should be obtained from the manufacturer or his distributor.

For budget purposes a door 2'-8"x6'-8" plastic faced both sides of a 1¾" solid core will run around $135.00. For one face plastic, one face birch around $100.00.

Interior Door Jambs

The following prices are for plain door jambs ¾"x4½". Side jambs are dadoed at top for head jambs. The prices are for jambs only, not including stops or trim which must be added. Furnished knocked down.

Price Per Set, Dadoed and Sanded, Not Including Stops

Size	Pond. Pine	Nat. Birch
2'-8"x6'-8" or smaller	$15.00	$24.00
3'-0"x7'-0" or smaller	16.00	25.00
6'-0"x6'-8" or smaller	19.00	28.00

Add for Door Stops 7/16"x13/8", Cut to Length

	Pond. Pine	Nat. Birch
3'-0"x7'-0" or smaller	$1.10	$3.00
6'-0"x7'-0" or smaller	1.25	3.75

If door jambs are wanted put together in the mill, add $3.00 per set to the above prices.

Setting Interior Wood Door Jambs

Size of Door Opening	Description of Jamb	No. Set 8-Hr. Day	Carp. Hrs. per Jamb	Add for Transom
3'-0"x7'-0"	Plain door jambs	8–10	0.8–1.0	0.6
3'-0"x7'-0"	Paneled door jambs	5– 6	1.3–1.6	0.6
6'-0"x7'-0"	Plain door jambs	5– 6	1.3–1.6	0.6
	First Grade Workmanship			
3'-0"x7'-0"	Plain door jambs	7– 9	0.9–1.1	0.6
3'-0"x7'-0"	Paneled door jambs	4– 5	1.6–2.0	0.6
6'-0"x7'-0"	Plain door jambs	4– 5	1.6–2.0	0.6

Door Jambs

Stock door casings per 100 lineal feet will cost $38.00 in Ponderosa pine, and $70.00 in birch. Stops will run $12.00 and $24.00 respectively.

Oak thresholds, 3⅝"x⅝" will run $105.00 per 100 ft. and oak sills, 1₅/₁₆"x7¼" will run $300.00 per 100 ft. Pre-cut sills will run $3.25 for 2'-8" opening, $3.75 for 3' opening; in the narrow size, $9.50 and $11.50 for the wider.

Interior Door Trim

The labor cost of placing interior door trim will vary with the type of trim or casings used and the class of workmanship. Production time is given on 4 different types of interior door trim, as follows :

Style "A" consists of 1-member casing with either a mitered or square cut head or "cap." This is the simplest trim to use.

Style "B" consists of 1-member trim for sides and a built-up cap trim for heads. Caps may either be put together in the mill or assembled and put together on the job.

Style "C" consists of 2-member back-band trim, consisting of a one-piece casing with a back-band of thicker material.

Style "D" consists of any type of door trim that is assembled and glued-up in the mill, ready to set in place as a unit.

Labor Erecting Interior Door Casings and Trim

Style Trim	Kind of Trim	Size of Opening	No. Sides Trim per 8-Hr. Day	Carp. Hrs. per Side	Carp. Hrs. per Opening	Add for Transom†
"A"	Single Casing	3'-0"x7'-0"	14–16	0.5–0.6	1.0–1.2	0.6
"B"	Cap Trim	3'-0"x7'-0"	14–16	0.5–0.6	1.0–1.2	0.6
"B"	Cap Trim*	3'-0"x7'-0"	7– 8	1.0–1.1	2.0–2.2	0.6
"C"	Back-band Trim	3'-0"x7'-0"	9–10	0.8–0.9	1.6–1.8	0.6
"D"	Mill Assembled**	3'-0"x7'-0"	14–16	0.5–0.6	1.0–1.2	0.6
"A"	Single Casing	6'-0"x7'-0"	9–10	0.8–0.9	1.6–1.8	0.8
"B"	Cap Trim	6'-0"x7'-0"	9–10	0.8–0.9	1.6–1.8	0.8
"B"	Cap Trim*	6'-0"x7'-0"	5– 7	1.1–1.6	2.2–3.2	0.8
"C"	Back-band Trim	6'-0"x7'-0"	7– 8	1.0–1.1	2.0–2.2	0.8
"D"	Mill Assembled**	6'-0"x7'-0"	9–10	0.8–0.9	1.6–1.8	0.8
		First Grade Workmanship				
"A"	Single Casing	3'-0"x7'-0"	9–10	0.8–0.9	1.6–1.8	0.8
"B"	Cap Trim	3'-0"x7'-0"	9–10	0.8–0.9	1.6–1.8	0.8
"B"	Cap Trim*	3'-0"x7'-0"	5– 7	1.1–1.6	2.2–3.2	0.8
"C"	Back-band Trim	3'-0"x7'-0"	8– 9	0.9–1.0	1.8–2.0	0.8
"D"	Mill Assembled**	3'-0"x7'-0"	11–12	0.7–0.8	1.4–1.6	0.8
"A"	Single Casing	6'-0"x7'-0"	7– 8	1.0–1.1	2.0–2.2	1.0
"B"	Cap Trim	6'-0"x7'-0"	7– 8	1.0–1.1	2.0–2.2	1.0
"B"	Cap Trim*	6'-0"x7'-0"	5– 6	1.3–1.6	2.6–3.2	1.0
"C"	Back-band Trim	6'-0"x7'-0"	6– 7	1.1–1.3	2.2–2.6	1.0
"D"	Mill Assembled**	6'-0"x7'-0"	7– 8	1.0–1.1	2.0–2.2	1.0

*Cap trim put together by carpenters on the job.
**Trim assembled and glued together in the mill.
†Includes transom jambs, stops and casings but no transom sash.

The class of workmanship will have considerable bearing on the labor costs, as in ordinary workmanship the casings are nailed to the rough bucks or grounds and the trim may show some hammer marks but where First Grade Workmanship is required, the utmost care must be used in cutting and fitting all casings, miters, etc. All nails must be carefully driven and set and the casings must show the same margin on all sides of the jamb and trim must fit close to the walls. In other words, it must be a first class job in every respect.

Fitting and Placing Stationary Transom Sash.—Where just an ordinary grade of workmanship is required, a carpenter should fit and place about 12 to 14 stationary transom sash per 8-hr. day, at the following labor cost per sash :

	Hours	Rate	Total	Rate	Total
Carpenter	0.6	$. . .	$. . .	$16.47	$9.88

First Grade Workmanship

Figure about 9 to 11 transom sash per 8-hr. day, at the following labor cost per sash:

	Hours	Rate	Total	Rate	Total
Carpenter	0.8	$. . .	$. . .	$16.47	$13.18

Fitting and Hanging Hinged or Pivoted Transom Sash.—If the transom sash are hinged or pivoted, a carpenter should fit and hang 7 to 9 sash per 8-hr. day, at the following labor cost per sash :

	Hours	Rate	Total	Rate	Total
Carpenter	1.0	$. . .	$. . .	$16.47	$16.47

First Grade Workmanship

Figure about 5 to 7 hinged or pivoted transom sash per 8-hr. day, at the following labor cost per sash :

	Hours	Rate	Total	Rate	Total
Carpenter	1.3	$. . .	$. . .	$16.47	$21.41

Fitting and Hanging Wood Doors.—The labor cost of fitting and hanging doors will vary with the class of workmanship, weight of door, whether soft or hard wood, and upon the tools and equipment used. There are power planes, hinge butt routers and lock mortisers that enable a carpenter to prepare about 3 times as much work per day as where the doors are fitted by hand. This must be considered when making up the estimate.

Where just an ordinary grade of workmanship is required, the doors do not always show the same margin at the top and sides, the screws are frequently driven instead of placed with a screw driver and in many instances the doors are "hingebound" so they will not open and close freely.

Ordinary Workmanship.—Where doors are fit and hung by hand, a carpenter should fit and hang 8 doors per 8-hr. day, at the following labor cost per door :

	Hours	Rate	Total	Rate	Total
Carpenter	1.0	$. . .	$. . .	$16.47	$16.47

On large production line jobs where power tools can be used to advantage and where the same class of work is performed using a power plane, an electric hinge butt router for door and jambs and an electric lock mortiser, a carpenter should fit and hang 12 to 16 doors per 8-hr. day, at the following labor cost per door :

	Hours	Rate	Total	Rate	Total
Carpenter	0.6	$. . .	$. . .	$16.47	$9.88

The above costs will vary with the ability of the carpenter and the job setup, as a power planer should plane off a door in 5 to 10 minutes; a hinge butt router should do all the way from 60 to 75 openings per 8-hr. day, while a lock router should mortise out for a lock in about ½-minute, after it has been set on the door. The labor handling the doors is the same for both hand and machine work.

First Grade Workmanship

In the better class of buildings, the doors must show the same margin between the door and jamb at the top and sides, the hinges must be set even and flush with the jambs and the edge of the door and the doors must be free from "hinge binding" so they will fit well and work easily at the same time.

On this grade of work a carpenter should fit and hang about 4 doors per 8-hr. day, at the following labor cost per door :

	Hours	Rate	Total	Rate	Total
Carpenter	2.0	$....	$....	$16.47	$32.94

Where power planers, hinge butt routers and lock mortisers are used, a carpenter should fit and hang 7 doors per 8-hr. day, at the following labor cost per door :

	Hours	Rate	Total	Rate	Total
Carpenter	1.1	$....	$....	$16.47	18.12

Labor Setting Jambs, Fitting and Hanging Doors, Placing Stops,
Hardware and Casing One Door Opening Complete

Size of Door Opening	Style Trim	Kind of Trim	Carp. Hrs. per Opening	Add for Transom†
		Ordinary Workmanship		
3'-0"x7'-0"	"A"	Single Casing	3.1– 3.7	2.2
3'-0"x7'-0"	"B"	Cap Trim	3.1– 3.7	2.2
3'-0"x7'-0"	"B"	Cap Trim*	4.1– 4.7	2.2
3'-0"x7'-0"	"C"	Back-band Trim	3.7– 4.3	2.2
3'-0"x7'-0"	"D"	Mill Assembled**	3.1– 3.7	2.2
6'-0"x7'-0"	"A"	Single Casing	5.6– 6.5	2.8
6'-0"x7'-0"	"B"	Cap Trim	5.6– 6.5	2.8
6'-0"x7'-0"	"B"	Cap Trim*	6.2– 7.9	2.8
6'-0"x7'-0"	"C"	Back-band Trim	6.0– 6.9	2.8
6'-0"x7'-0"	"D"	Mill Assembled**	5.6– 6.5	2.8
		First Grade Workmanship		
3'-0"x7'-0"	"A"	Single Casing	5.5– 6.2	2.7
3'-0"x7'-0"	"B"	Cap Trim	5.5– 6.2	2.7
3'-0"x7'-0"	"B"	Cap Trim*	6.1– 7.6	2.7
3'-0"x7'-0"	"C"	Back-band Trim	5.7– 6.4	2.7
3'-0"x7'-0"	"D"	Mill Assembled**	5.3– 6.0	2.7
6'-0"x7'-0"	"A"	Single Casing	9.2–10.3	3.4
6'-0"x7'-0"	"B"	Cap Trim	9.2–10.3	3.4
6'-0"x7'-0"	"B"	Cap Trim*	9.8–11.3	3.4
6'-0"x7'-0"	"C"	Back-band Trim	9.4–10.7	3.4
6'-0"x7'-0"	"D"	Mill Assembled**	9.2–10.3	3.4

*Cap trim put together by carpenters on the job.
**Trim assembled and glued together in the mill.
†Above is for hinged or pivoted transom. For stationary transom deduct 0.4-hr.

For front door hardware, add 0.5-hr. for ordinary work and 1 to 2 hrs. for First Grade.

Where power planers, hinge butt routers and lock mortisers are used, deduct 0.4-hr. for single door and 0.8-hr. for double door openings on Ordinary Workmanship; for First Grade Workmanship, deduct 0.9-hrs. for single door and 1.8-hrs. for double door openings.

Labor Installing Pre-Hung Door Units.—Pre-hung door units are used extensively in the construction of houses and apartments ranging from "economy" grade to middle class.

In the completely assembled package, jambs are assembled, door is hung in place with hinges applied and stops are mitered and nailed into place.

Where assembled pre-hung door units are used, a carpenter should install 16 units per 8-hr. day, at the following labor cost per unit :

	Hours	Rate	Total	Rate	Total
Carpenter	0.5	$. . . .	$. . . .	$16.47	$8.24

Fitting and Hanging Hardwood Acoustical Doors.—The labor cost of fitting and hanging hardwood acoustical doors is always figured on a first grade workmanship basis as they must be perfectly installed to insure their efficient operation.

Assuming a carpenter, experienced in hanging acoustical doors, is employed to do the work, the following production should be obtained :

Type of Door	Approximate Weight per Sq. Ft.	Carpenter Hours per Opening	Helper Hours per Opening
Hardwood Class 36	4 lbs.	6	3
Hardwood Class 41	7 lbs.	8	4

Fitting and Hanging Toilet Stall Doors.—A carpenter should fit and hang 9 to 11, ⅞" or 1⅛" toilet doors per 8-hr. day, at the following labor cost per door :

	Hours	Rate	Total	Rate	Total
Carpenter	0.8	$. . . .	$. . . .	$16.47	$13.18

First Grade Workmanship

Where hardwood stall doors are used and carefully hung, a carpenter should fit and hang 7 to 9 doors per 8-hr. day, at the following labor cost per door :

	Hours	Rate	Total	Rate	Total
Carpenter	1	$. . . .	$. . . .	$16.47	$16.47

Fitting and Hanging Heavy Wood Swinging Doors Up to 4'-0"x8'-0".—When hanging heavy wood swinging doors, such as used in garages, stables, mill and factory buildings, etc., where the doors are 3'-0" to 4'-0" wide and 7'-0" to 8'-0" high, it will require about 4 to 4½-hrs. time to fit and hang one door or 8 to 9-hrs. per pair, at the following labor cost per door :

	Hours	Rate	Total	Rate	Total
Carpenter	4.3	$. . . .	$. . . .	$16.47	$70.82

Fitting and Hanging Heavy Wood Swinging Doors Up to 5'-0"x10'-0".—To fit and hang extra heavy wood swinging doors up to 5'-0"x10'-0"x2½" or 3" thick, will require about 4 to 5-hrs. time for 2 carpenters, at the following labor cost per door :

	Hours	Rate	Total	Rate	Total
Carpenter	9	$. . . .	$. . . .	$16.47	$148.23

Fitting and Hanging Wood Sliding Doors Up to 4'-0"x8'-0".—When fitting and hanging heavy wood sliding doors, such as used in garages, factories, etc., two carpenters working together should place tracks and hang one door complete in 3 to 3½ hrs. at the following labor cost per door :

	Hours	Rate	Total	Rate	Total
Carpenter	6.5	$. . . .	$. . . .	$16.47	$107.06

To place track and hang a pair of double sliding doors for and opening up to 8'-0"x8'-0", will require 10 to 12 hrs. carpenter time, at the following labor cost per opening :

	Hours	Rate	Total	Rate	Total
Carpenter	11	$. . . .	$. . . .	$16.47	$181.17

The above does not include cutting or drilling in masonry walls to place bolts or anchors but is based on this work being done when the walls are built.

Fitting and Hanging Heavy Wood Sliding Doors Up to 10'-0"x10'-0".—To place track, fit and hang one sliding door up to 10'-0"x10'-0" in size and 2½" to 3" thick, will require about 12 hrs. carpenter time, at the following labor cost per opening :

	Hours	Rate	Total	Rate	Total
Carpenter	12	$. . . .	$. . . .	$16.47	$197.64

The above does not include cutting or drilling masonry walls for bolts or anchors.

Fitting and Hanging Heavy Wood Sliding Doors Up to 12'-0"x18'-0".—To place track and hang one large door for an opening up to 12'-0"x18'-0", will require about 24 to 28 hrs. carpenter time, at the following labor cost per door :

	Hours	Rate	Total	Rate	Total
Carpenter	26	$. . . .	$. . . .	$16.47	$428.22

It will require 4 to 6 men to handle a door of this size.

Making and Setting Sliding Door Pockets.—Where sliding door pockets are made on the job for door openings up to 3'-0"x7'-0", a carpenter should make the pocket, fit track and set the pocket in about 4 hrs. at the following labor cost :

	Hours	Rate	Total	Rate	Total
Carpenter	4	$. . . .	$. . . .	$16.47	$65.88

Where double doors are used and it is necessary to make and set 2 pockets, double the above costs.

Setting Sliding Door Pockets.—If sliding door pockets are sent to the job put together with track in place or steel pocket door T-frames are used, and the carpenters on the job set the pocket or steel frame ready to receive the door, it will require about 1 hr. carpenter time, at the following labor cost per opening :

	Hours	Rate	Total	Rate	Total
Carpenter	1	$. . . .	$. . . .	$16.47	$16.47

If double sliding doors are used and it is necessary to set 2 pockets for each opening, double the above costs.

Fitting and Hanging Sliding Doors.—To fit an ordinary sliding door up to 3'-0"x7'-0", place hangers and hang the door complete, not including finish hardware, will require about 1 hr. carpenter time, at the following labor cost per door :

	Hours	Rate	Total	Rate	Total
Carpenter	1	$. . . .	$. . . .	$16.47	$16.47

For double sliding doors, double the above costs.

Labor on Sliding Door Openings Complete with Pockets.—Where necessary to make the sliding door pocket, hang the track, set the pocket and fit and hang the door complete, exclusive of finish hardware, figure about 5 hrs. carpenter time, at the following labor cost per opening :

	Hours	Rate	Total	Rate	Total
Carpenter	5	$. . . .	$. . . .	$16.47	$82.35

For double sliding door openings, double the above costs.

If pockets are sent to the job with track fitted in place or steel pocket door T-frames are used, and the job carpenters set the box or steel frame and fit and hang the sliding door, it will require about 2 hrs. carpenter time per opening, exclusive of finish hardware, at the following labor cost per opening :

	Hours	Rate	Total	Rate	Total
Carpenter	2	$. . . .	$. . . .	$16.47	$32.94

For double sliding door openings, double the above costs.

Labor Fitting and Hanging Sliding Wood Closet Doors.—When hanging and fitting wood closet doors that slide past each other without the necessity for pockets, a carpenter should place track, attach hangers, fit and hang two doors complete for an opening 3'-0" to 5'-0" wide and 6'-8" to 8'-0" high in about 2 hours.

This does not include time setting door jambs and casing openings on two sides, as this should be added according to time given on previous pages.

The labor cost per opening should average as follows :

	Hours	Rate	Total	Rate	Total
Carpenter	2	$. . . .	$. . . .	$16.47	$32.94

08300 SPECIAL DOORS

TIN CLAD FIRE DOORS

Tin clad fire doors are used in factory, mill and warehouse buildings and may be either sliding or swinging, having fusible links and automatic closing devices.

Sliding doors and lap type swinging doors in brick walls do not ordinarily require a steel frame but they may be necessary with tile walls. Flush type swinging doors require a frame.

Standard tin clad fire doors are usually manufactured in two thicknesses depending on the fire rating required. For Class "A" rating, the thickness is 2⅝" consisting of 3-plies of tongue-and-groove lumber clad with tin sheeting on all sides. For Class "B" or "C" rating, the thickness is 1¾" consisting of 2-plies of tongue-and-groove lumber with cladding.

Method of Calculating the Area of Fire Doors.—For flush doors, multiply the width of the opening in feet by the height of the opening in feet.

Square top lap doors. Obtain size of door by adding 8" to width and 4" to height of opening, then multiply width of door in feet by height of door in feet.

Inclined top lap doors. Obtain width of door by adding 8" to width of opening and obtain height by adding 8" to height of door for doors up to 4'-0" wide and 12" for doors up to 6'-0" wide. Multiply width of door in feet by height of door in feet.

Prices of Tin Clad Fire Doors and Shutters

Price includes door only, covered both sides with tin of a quality and in a manner specified by the National Board of Fire Underwriters.

	Price per Sq. Ft.
Standard 3-ply (2⅝" thick), Wt. 13 lbs. per sq. ft.	$9.00
Standard 2-ply (1¾" thick), Wt. 10 lbs. per sq. ft.	8.00
Standard 2-ply (1¾" thick), Shutters, Wt. 10 lbs. per sq. ft.	8.00

Above prices include doors only, no hardware or fixtures of any kind or delivery costs.

Hardware for Swinging, Automatic Closing Tin Clad Doors

Fixture sets include hinges, pintles, weights, chains, catches, latches, complete automatic device with bolts and screws for attaching. Installation not included. Door frames not included.

Door Width Up to and Including	Door Height		Single Swinging Doors		Double Swinging Doors	
			Flush Type	Lap Type	Flush Type	Lap Type
3'-0"x 6'-6"		....	$ 62.00	$ 81.00	$138.00	$159.00
3'-0"x 8'-6"		....	76.00	103.00	157.00	187.00
3'-0"x10'-6"		....	91.00	123.00	177.00	211.00
4'-0"x 6'-6"		....	64.00	84.00	140.00	161.00
4'-0"x 8'-6"		....	79.00	106.00	160.00	189.00
4'-0"x12'-0"		....	95.00	128.00	180.00	215.00
5'-0"x 6'-6"		....	67.00	86.00	142.00	164.00
5'-0"x 8'-6"		....	82.00	110.00	163.00	193.00
5'-0"x12'-0"		....	99.00	132.00	184.00	222.00
6'-0"x 6'-6"		....	71.00	89.00	145.00	166.00
6'-0"x 8'-6"		....	86.00	113.00	166.00	196.00
6'-0"x12'-0"		....	104.00	136.00	189.00	223.00

If automatic controls are not wanted with swinging tin clad doors, deduct from the above prices, $8.00 for single doors and $12.00 per opening for double swinging doors.

Hardware for Sliding Tin Clad Doors

Prices include non-adjustable hangers, track, weights and bolts for attaching hardware to doors. Through wall bolts not included.

Hardware for single sliding door, 3'-0" wide opening $180.00
For each 6" additional width of opening, add.. 15.00
Openings over 6'-0" wide require 3 hangers on each door, add for each additional hanger.. 10.00
Hardware for double sliding doors, 6'-0" wide opening................................. 310.00
For each 6" additional width of opening, add.. 15.00

Labor Erecting Tin Clad Doors

The following schedules, furnished by the Richmond Fireproof Door Co., are guides for estimating labor for erection.

The hours of labor are based on brick walls 13" thick. For walls greater in thickness, add ¼ hour for one man for each 4" in thickness per bolt involved. Concrete and stone walls require slightly more time drilling; terra cotta, slightly less. Where "doors both sides" are noted, this assumes doors back to back and tracks bolted together. When tracks cannot be bolted together use "one side" schedule multiplied by two.

		Labor			
		Door One Side		Door Both Sides	
Type	Width	Mech.	Help.	Mech.	Help.
Single Slide	2'-0" to 4'-0"	7	7	10	10

	4'-0" to 5'-2"	8	8	12	12
	5'-3" to 7'-8"	9	9	14	14
	7'-9" to 8'-8"	11	11	17	17
	8'-9" to 12'-0"	12	12	18	18
Slide in Pairs	4'-0" to 4'-5"	9	9	14	14
	4'-6" to 9'-8"	11	11	16	16
	9'-9" to 10'-6"	13	13	18	18
	10'-7" to 12'-0"	15	15	20	20
	Width				
Vertical	to 4'-10"	13	13	21	21
Sliding	4'-11" to 7'-6"	15	15	24	24
	7'-7" to 10'-2"	17	17	26	26
	10'-3" to 12'-0"	19	19	28	28
	Door Height				
Single Swing	to 5'-9"	5	5	7	7
(Flush or	5'-10" to 8'-9"	5½	5½	8	8
Lap)	8'-10" to 12'	6	6	9	9
	Door Height				
Swing in Pairs	to 5'-9"	8	8	12	12
(Flush or Lap)	5'-10" to 8'-9"	9	9	14	14
	8'-10" to 12'-0"	10	10	16	16

Swinging Type Tin Clad Door   Gravity Sliding Type Tin Clad Door

Labor Cost of Installing One Single 3 Ply Sliding Tin Clad Fire Door For 4'-0"x7'-0" Wall Opening

The following estimate includes door, track, hangers and necessary hardware installation.

	Hours	Rate	Total	Rate	Total
Mechanic	7	$. . . .	$. . . .	$16.47	$111.29
Helper	7	. . . .	. . . .	12.54	87.78
Cost per opening			$. . . .		$203.07

Labor Cost of Installing One Pair Double Sliding 3 Ply Tin Clad Fire Doors for 6'-0"x7'-0" Wall Opening

The following estimate includes doors, track, hangers and necessary hardware installation.

	Hours	Rate	Total	Rate	Total
Mechanic	11	$. . . .	$. . . .	$16.47	$181.17
Helper	11	. . . .	. . . .	12.54	137.94
Cost per opening			$. . . .		$319.11

Labor Cost of Installing One Lap Type Single Swinging 2 Ply Tin Clad Fire Door for 3'-0"x7'-0" Wall Opening

The following estimate includes door, hardware, and automatic controls installation.

	Hours	Rate	Total	Rate	Total
Mechanic	5	$....	$....	$16.47	$ 82.35
Helper	5	....	....	12.54	62.70
Cost per opening			$....		$145.05

Labor Cost of Installing One Pair Lap Type Double Swinging 2 Ply Tin Clad Fire Doors for 6'-0"x7'-0" Wall Opening

The following estimate includes doors, hardware, and automatic controls installation.

	Hours	Rate	Total	Rate	Total
Mechanic	9	$....	$....	$16.47	$148.23
Helper	9	....	....	12.54	112.86
Cost per opening			$....		$261.09

Labor Cost of Installing One Flush Type Single Swinging 2 Ply Tin Clad Fire Door for 3'-0"x7'-0" Wall Opening

The following estimate includes door, hardware, and automatic controls installation.

	Hours	Rate	Total	Rate	Total
Mechanic	5	$....	$....	$16.47	$ 82.35
Helper	5	....	....	12.54	62.70
Cost per opening			$....		$145.05

Labor Cost of Installing One Pair Flush Type Double 3 Ply Swinging Tin Clad Fire Doors for 6'-0"x7'-0" Wall Opening

The following estimate includes doors, hardware and automatic controls installation.

	Hours	Rate	Total	Rate	Total
Mechanic	9	$....	$....	$16.47	$148.23
Helper	9	....	....	12.54	112.86
Cost per opening			$....		$261.09

Frames for tin clad doors are made of angles and channels usually furnished under metal fabrications and set by the mason. A 3x3 angle frame will run around $5.50 a foot installed. An 8" channel frame will run around $9.00 per foot installed.

08330   COILING DOORS

Coiling doors are usually called rolling shutters and are fully factory assembled units made of steel or aluminum. Often, they are labeled fire doors and installation is governed by Underwriter's specifications. Steel mounting jambs and heads are usually furnished by the metal fabrications contractor.

Costs for a manually operated, 20 gauge, door up to 12'x12' will cost around $5.50 to $6.50 a sq. ft., including hood, guides and operating hardware.

Where class A fire ratings are required, doors will cost $5.00 a square foot more. 18 gauge construction will add another $0.80 a sq. ft. Motor operation will cost $500.00 for smaller sizes $600 for medium. Each pass door will add another $500.00. Two mechanics should hang and adjust a rolling shutter in size up to 12'x12' in one 8 hour day, at the following cost per shutter:

	Hours	Rate	Total	Rate	Total
2 Mechanics	16.0	$. . . .	$. . . .	$17.70	$283.20

Coiling Grilles

Rolling grilles are especially popular now for protection of store fronts, for open fronted stores in enclosed shopping areas and for closing off counter areas.

Stock commercial grilles will cost around $12.00 per sq. ft. with all hardware. As these are often part of a decorative front, they are often made of anodized aluminum, and cost around $15.00 a sq. ft. Exact prices can be obtained from the manufacturer who will also furnish shop drawings upon request to show all the necessary anchoring involved. Steel grilles run around $6.50 a sq. ft. Motorizing and installation costs are the same as for rolling shutters.

08350 FOLDING DOORS

Steel bi-fold closet doors may be installed in openings with drywall, wood, or steel jambs and head frame. Each bi-fold door package comes complete with door panels, track, guides, and pulls for a complete installation.

Doors are available in a variety of styles including flush, panel, half louver, full louver, etc. The doors are shipped "factory finished" with a baked-on paint finish.

A two panel unit is used for openings up to 3'-0" wide and will cost about $45.00 each. A four panel unit is used for openings from 4'-0" to 6'-0" wide, with the 6'-0" unit costing about $90.00 each. Units are sold in 6'-8" and 8'-0" heights, with the 8'-0" heights costing an additional $10–$20 more per unit.

One carpenter should install a complete bi-fold unit and all hardware in about 45 minutes at the following labor cost per unit:

	Hours	Rate	Total	Rate	Total
Carpenter	0.75	$. . . .	$. . . .	$16.47	$12.35

Folding doors, which are often folding partitions, can be ordered completely furnished and with all hardware in wood or vinyl. Folding doors for a 3'x6'-8" opening will run around $60.00 in vinyl, $90.00 in wood slats. Doors for 6'x6'-8" openings will run $140 for vinyl, $170 for wood. Deluxe and custom models will run somewhat more. All the above prices include hardware but the frame must be furnished by others. When the opening is over 150 sq. ft. the units are considered "partitions" rather than doors and will run around $4.25 a sq. ft. for light models to $6.50 a sq. ft. for heavy duty units.

08360 OVERHEAD DOORS

Overhead Garage Doors.—Overhead garage doors are furnished in several different styles, including the one-piece over-head door; self-balancing overhead door and the spring balanced over-head doors.

For an average opening up to 8'-0"x7'-0", two carpenters should place tracks, fit and hang doors and apply the necessary hardware in 4 hours, at the following labor cost per door :

	Hours	Rate	Total	Rate	Total
Carpenter	8	$. . . .	$. . . .	16.47	$131.76

For larger openings up to 16'-0" wide and 7'-0" high, where one wide overhead folding door is used, two carpenters should place tracks, fit and hang doors and apply the necessary hardware complete in 5 hours, at the following labor cost per door :

	Hours	Rate	Total	Rate	Total
Carpenter	10	$. . . .	$. . . .	$16.47	$164.70

GARAGE, FACTORY AND WAREHOUSE DOORS

Overhead Type Garage Doors

Overhead type garage doors are counter-balanced and controlled by oil tempered torsion springs. Cables are secured at the bottom of the door on each side and are directly connected with the pair of balancing springs located immediately over the door.

Overhead Type Garage Doors

Ball bearing rollers carry all sections of door over a continuous channel track. The track has a flange on it which keeps the rollers in the track.

Doors have kiln dried fir or spruce sections, with hardboard panels. All joints are mortised, tenoned or doweled with steel pins and glued. All sections are fitted to size, squared for opening and drilled for hardware.

Doors over 12'-0" wide have steel struts in intermediate sections to prevent sagging. Doors over 15'-0" wide have steel struts in all sections.

Standard Size, Sectional Overhead Type, Garage Doors

Furnished complete with hardware, including cylinder lock, springs and track. Doors may be furnished with all wood panels or top section open for glass. Glass not included. Approximate prices :

Size Opening	1⅜" Thick	1¾" Thick
9'-0"x 7'-0"	225.00	250.00
14'-0"x 7'-0"	475.00	500.00
15'-0"x 7'-0"	500.00	550.00

16'-0"x 7'-0"	525.00	600.00
10'-0"x10'-0"	....	400.00
12'-0"x10'-0"	....	480.00
10'-0"x12'-0"	....	490.00
12'-0"x12'-0"	....	575.00

Electric Door Operators

Operators may be used with any type of overhead doors. Motors are suitable for single phase, 60 cycle, 110–120 Volt AC power supply, and are of the reversing type which permit stopping and changing direction at any point of the door travel. Adjustable limit switches stop the door automatically at top and bottom of door openings.

Safety mechanism automatically reverses the door if it meets an obstruction while in operation, or the door will stop and the operator will shut off.

Doors are driven by means of a roller chain fastened to a shoe sliding on 2 steel tracks. The shoe is connected to the door by a detachable arm.

Single door residential operator with ¼-hp. motor for door up to 22 ft. wide, 7 ft. high, complete with one toggle type control station...... $175.00

For commercial door operator, with ¼-hp. motor for door not over 140 sq. ft. nor over 10 ft. high............ 250.00

Commercial door operator, with ⅓-hp. motor for door not over 168 sq. ft. or 12 ft. high 275.00

Commercial door operator, with ½-hp. motor for door not over 224 sq. ft. or 14 ft. high 330.00

Extra toggle type control station for commercial operator...................... 10.00

Extra key switch (for residential operator only)..................... 20.00

An experienced mechanic should install an electric door operator in 1 to 2 hours after the necessary wiring is in place.

Electronic Door Operators

For use where a transmitter is located in each automobile using the garage. A push of the button operates the door from the car.

For residential doors up to 16'-0"x8'-0"
approx. price installed $250.00

An experienced mechanic should install an electric door operator in 2 hours after the necessary wiring is in place.

Special Hardware—Add per Door

Low headroom, 6" minimum, for doors not over 132 sq. ft........................ $15.00

Low headroom, 6" minimum, for doors not over 133 to 175 sq. ft............ $26.00

Low headroom, 6" minimum, for doors over 16'-0" wide or over 12'-0" high.. $38.00

High lift (for lubritoriums) up to 2'-0" above head jamb $22.00

High lift (for lubritoriums) up to 5'-0" above head jamb $45.00

On commercial jobs steel doors are often specified. A 10'x10' door will cost about $400.00 with manual operator, $700.00 with electric operator.

Fiberglass and aluminum doors are light in weight and let in daylight. A stock door will cost about $225.00 for a single car size, $375.00 for a double car size.

Doors of special designs

| Colonial | Fan Top | Vertical Battens | Diagonal Battens |

Colonial doors, heavily molded outside with raised panels, add per sq. ft. $1.00.
For fan top on door, add $125.00 extra.
Vertical battens, 2-ply V-jointed or beaded, add per sq. ft. $1.00.
Diagonal battens, 2-ply V-jointed or beaded, add per sq. ft. $1.20.

08370   SLIDING GLASS DOORS

Type XO   Type OXO   Type OXXO

Sliding Glass Doors

Aluminum sliding glass doors are available in several types, as illustrated, for glazing with crystal glass or ⅝" insulated glass. "X" indicates sliding panel; "O" indicates a fixed panel.

Frame, glazing bead and door members are heavy aluminum extrusions. Door sections move upon bottom mounted, ball bearing, grooved brass rollers. Frames are designed with provisions for installing bottom roller, horizontally sliding type screens. Weatherstripping is mohair pile, factory mounted in channels to form a continuous double seal. Hardware includes full-grip lucite pulls and touch latch. Cylinder lock sets are available at extra cost.

Screens are made from aluminum extrusions of tubular box type section with bottom rollers and fitted with fiberglass screen.

Prices are for door units set up. Factory assembly includes attachment of all accessory parts, such as weatherstripping, roller mechanisms, latch, corner connecting brackets, bumpers, etc. Screen units are assembled. Glass and glazing, is included.

Material Prices of Aluminum Sliding Doors

		Prices	
		Economy	Deluxe
Type	Frame Size	Crystal Gl.	Insulating Gl.
XO	6'-0"x7'-0"	$250.00	$500.00
OXO	9'-0"x7'-0"	300.00	600.00
OXXO	12'-0"x7'-0"	400.00	850.00

For cylinder lock set in lieu of latch, add $10.00.

Labor Erecting Aluminum Sliding Glass Doors.—Assembly and erection of aluminum sliding glass doors is simple and does not require special tools. All aluminum members are prefit, with holes drilled and connecting brackets attached. The following table gives average labor hours required for installing aluminum sliding glass doors in ordinary residential openings.

Type	Number of Doors	Number of Fixed Panels	Mechanic Hours Required for Assembly	for Erection
XO	1	1	2	6
OXO	1	2	3	8
OXXO	2	2	4	10

Performance Standards.—There are many suppliers of aluminum windows and sliding doors and it is often difficult to compare the quality of one manufacturer with that of another. To aid one's selection, the Architectural Aluminum Manufacturers Association, 35 East Wacker Drive, Chicago 60601, publishes quality specifications and manufacturers belonging to that association affix the AAMA "Quality-Certified" label to their products attesting that it complies with the designated specification.

These specifications include the following for sliding glass doors:

SGD-B1	For residential and limited commercial applications
SGD-A2	For commercial and residential application requiring increased size or performance characteristics.
HP Series	Either of the above designations followed by "HP" indicates that the door complies with High Performance requirements.

The specifications for aluminum windows are as follows:

DH-B1	Double hung residential
DH-A2	" " commercial
DH-A3 & A4	" " monumental
C-B1	Casement residential
C-A2	" commercial
C-A3	" monumental
P-B1	Projected residential
P-A2	" commercial
P-A3	" monumental
A-B1	Awning residential
A-A2	" commercial
HS-B1 & B2	Horizontal sliding residential
HS-A2	" " commercial
HS-A3	" " monumental
J-B1	Jalousie residential
JA-B1	Jalousie-awning residential
VS-B1	Vertical sliding residential
TH-A2	Top hinged commercial
TH-A3	" " monumental
VP-A2	Vertically pivoted commercial
VP-A3	" " monumental
HP Series	Any of the above followed by "HP" indicates the window complies with High Performance requirements.

The AAMA also publishes a certified products directory with a current listing of all certified windows and sliding glass doors by company, model and maximum size approved.

Wood Sliding Or Patio Doors

Sliding doors are now available from most of the nationally distributed millwork houses, and follow the general designs established for aluminum sliding units. These doors can usually be purchased in two basic widths per panel and in various arrangements including two panels, one fixed, one operating; three panels, two sides fixed and center panel operable; and four panels, two end panels fixed and two center panels operable.

The following price schedule is based on units preglazed with ⅝" safety glass and complete with operating hardware, weatherstripping, sill, and screens for operable sash.

Glass Size	Nominal Door Size	Type	Material Cost
28" wide	5'-0"x6'-8"	Double XO	$ 470.00
34" wide	6'-0"x6'-8"	Double XO	508.00
46" wide	8'-0"x6'-8"	Double XO	668.00
34" wide	9'-0"x6'-8"	Triple OXO	800.00
46" wide	12'-0"x6'-8"	Triple OXXO	1,010.00

Add for keyed lock, $10.00

Grilles which may be removed for cleaning are available in either diamond or rectangular patterns. Costs are $41.00 each for the 28" width, $42.00 each for the 34" width, $50.00 for the 46" width in the rectangular pattern; and $42.50, $44.00 and $62.00 each for the diamond patterns.

08380   SOUND RETARDANT DOORS

Hardwood Sound Insulating Doors

Acoustical doors are for use in consultation rooms, conference rooms, hospitals, clinics, music rooms, television and radio studios, and other areas where acoustical levels must be closely maintained.

Dependent on the desired decibel reduction, or acoustical efficiency, doors are identified and furnished in several models.

Units 1¾" thick are rated in sound transmission class 36. Units 2½" thick are rated in sound transmission class 41. A door 3" thick is also made which has superior accoustical qualities.

Acoustical doors are made in a wide selection of face woods. Custom finishing of doors is available. These are veneered doors and, as such, can be furnished with veneers to match those of the balance of the doors in the building. Any standard hardware may be used for door erection.

The following prices include bottom closers, stops, gaskets and stop adjusters and are for standard rotary cut veneers of natural birch and red oak.

Material Prices of Hardwood Sound Insulating Doors*

Size	1¾" 35 Decibel	2½" 40 Decibel	3" 50 Decibel
2'-6"x6'-8"	$250.00	$325.00	$430.00
2'-8"x6'-8"	260.00	335.00	440.00
3'-0"x6'-8"	270.00	345.00	450.00
2'-6"x7'-0"	260.00	335.00	440.00
2'-8"x7'-0"	270.00	345.00	450.00
3'-0"x7'-0"	280.00	355.00	460.00

*f.o.b. factory, uncrated.

	1¾"	2½"	3"
Light openings: Rectangular vision light not over 18" square	$35.00	$37.00	$40.00

DETAILS OF HPC ACCOUSTICAL DOORS

Astragals.—Astragals for pairs of doors, 2½"x1" in any wood to match face veneer with 1"x½" rubber gasket, $25.00 per pair.

For rounding meeting edges and grooving one edge with Eveleth bumper strip, add $30.00.

Other woods.—For other woods add to the total price the following percentages:

Sliced natural birch and sliced red oak.. 8%
Rotary cut red or white birch and rotary cut plain white oak............................ 15%
Sliced red or white birch and sliced plain white oak... 20%
Rift sliced white oak and sliced African mahogany... 30%
Sliced American black walnut.. 50%

08400 ENTRANCES AND STORE FRONTS

METAL STORE FRONT CONSTRUCTION

Aluminum storefront and entrance doors are manufactured by a number of firms, in a wide variety of styles, to suit the architectural aesthetics desired or a particular design function for most any type of building.

The glass and glazing subcontractor is the firm to contact for this type of work, and they typically represent more than one manufacturer of these materials. This subcontractor will handle all the necessary functions of bidding, shop drawing preparation, ordering, and erection of the storefront materials, as well as the glass, for a complete installation.

Materials are made from an aluminum alloy and extruded into many different shapes and sizes of various thicknesses. The extrusions are offered in stock lengths of 24' long, however, longer lengths are available upon special order.

To prevent oxidation of the aluminum surfaces, the extrusions are anodized in a number of colors as well as the clear anodized finish. In recent years this material has become available with a paint coating that offers exceptional durability in addition to a controlled color gradation that is difficult to obtain with the color anodized finishes.

Entrance doors are available in numerous sizes and designs (narrow stile, medium stile, and wide or monumental stile) with various styles of applied panels and finish hardware.

The doors can be hung into the frame by using butt hinges, offset pivots, or center pivots. The controlled automatic closing of the doors can be achieved by installing an exposed overhead closer, a closer concealed in the horizontal head frame member, or by a floor closer that is set into a pocket in the floor under the hinge point of the door.

Installation—Once the subcontractor has a contract agreement to provide the storefront construction, he will prepare shop drawings based on the architectural drawings for the project. The shop drawings clearly define all dimensions and jointing details necessary for the installation; and, the materials are ordered based on these drawings.

Prior to fabrication of the exact lengths necessary for erection, the subcontractor will verify the shop drawing dimensions with actual field dimensions taken at the jobsite.

After the extrusions are cut to the proper length for field erection, they are delivered to the jobsite, distributed to the proper location, and installation begins.

The extrusions are fastened to each other with concealed, extruded aluminum clips at all joints. The pre-fabricated frames are then set into the opening and fastened, through the glazing pocket, to the substrate material with screws or expansion anchors. Proper assembly allows the vertical extrusions to run top to bottom, with all horizontal extrusions fitted between the verticals.

Method Of Estimating Materials.—Measure the length of each type and size extrusion and list into proper groupings. Since this material is sold in stock lengths, the estimator will have to combine the measured lengths together to arrive at the total number of stock lengths needed for the job with the least amount of waste. Count the joints to ascertain the required number of concealed clips (one per joint). It will also be necessary to measure the total length of vinyl gasket material needed for glazing, since this material is ordered separately from the aluminum extrusions. After all of the material quantities are measured, grouped, and totaled, the appropriate material prices can be applied for the extension of the material costs.

Doors are listed by size and style and depending upon the architectural requirements, they may be stock units or custom units. The pricing manual, supplied by the aluminum manufacturer, will list the material pricing strategy for the stock units as well as for the component parts necessary for the custom units.

Installed Costs Of Storefront Construction.—Since there are so many different sizes and shapes of extrusions available for the installation of this type of construction, it would be inappropriate to try to list the material prices and labor time required for the installation, as the labor hours are directly related to the size and shapes of the extrusions to be installed. For instance, a simple storefront and entrance door installation, 40'-0" wide x 10'-0" high, in one system might take 72 man hours, while a more sophisticated system would require 100 man hours.

For budget purposes only, the following installed costs of aluminum storefront framing (glass not included) are given on a per square foot basis for the opening to be enclosed:

1). Windows without intermediate vertical mullions up to 5'-0"x5'-0" in size, using 1¾"x4½" tube - $7.50 per sq. ft.
2). Same as above, but with one vertical mullion - $8.50 per sq. ft.
3). Windows 40'-0"x10'-0" in size with vertical mullions spaced 5'-0" on center, using 1¾"x4½" tube - $6.00 per sq. ft.
4). Same as above, but with 3"x6" tube - $9.50 per sq. ft.
5). For a narrow stile single door and frame, 3'-0"x7'-0" - $550.00 each.
6). For a pair of narrow stile doors and frame, 6'-0"x7'-0" - $850.00 each.

Note: Add 15% for bronze anodized finish
      Add 20% for black anodized finish
      Add for transom over single door, $80.00; over pair of doors, $140.00

Revolving Doors.—Revolving doors are made in a variety of styles and sizes in aluminum, bronze, and stainless steel. The doors are fitted with a mechanism to control the travel speed and are pivoted so that they may collapse together to provide an unobstructed opening. The semi-circular wings at each side of the four leaf revolving door provides a surface for the door sweep to ride against, therefore, at no point during normal operation, is there a clear path to the exterior for the escape of the building heat or air conditioning to the outside atmosphere.

Average installed prices for revolving doors are as follows: Aluminum, $14,000; bronze, $25,000; stainless steel, $20,000.

Storefront Members.—The following detail drawings will show some typical cut sections through aluminum storefront extrusions:

08500   METAL WINDOWS

Steel and aluminum windows are available in a great variety of types and sizes to meet almost all building requirements. These products have been standardized by the manufacturers to reduce costs and promote quicker deliveries.

The illustrations included in this chapter are for reference when selecting requirements. For full description and specifications of available products refer to manufacturers' catalogs.

All metal window and door products are either classified as "Standard," "Commodity" or "Special."

"Commodity Units" are those selected from "Standards" which are subject to the greatest demand and are available from manufacturers' warehouse stocks or through dealers engaged in the distribution of these products. "Standard Units" and "Special Units" are not available from stock and must be fabricated and shipped from factory. The trend is for distributors to send all orders to the factory for both pricing and fabrication.

The following information will deal with "Commodity Units" only. For "Standard Units" and "Special Units" obtain quotations from manufacturers or their dealer representatives.

Labor Installing Steel Windows and Doors.—Installation costs will vary with the class of labor used, type and size of individual units and job working conditions. However, a reasonably accurate estimate of installation labor costs can be made by multiplying the number of man hours required to install each unit by the prevailing hourly rate of pay, plus rates for pension, welfare, etc., in effect in the area in which job is located.

ELEVATIONS VIEWED FROM OUTSIDE

DETAILS SHOWN QUARTER SIZE

NOTE: Details are shown as 2¾" x 5½".
2¾" x 7½" system similar.

ENTRANCE DOORS

CENTER HUNG   OFFSET HUNG

ENTRANCE DOOR

CENTER HUNG

ENTRANCE DOOR

OFFSET HUNG

1 3/4" x 3"   **1 3/4" x 4"**   **1 3/4" x 4 1/2"**   **1 3/4" x 5"**

Product	Window Opening	Man Hours Each Unit
Pivoted, Commercial	1. Single Unit*...	2.25
Projected and Architectural	2. Two or more Units*	
Projected Windows	(a) Precast Sills...............................	1.25
	(b) Poured Sills...............................	1.50
Fixed Industrial	1. Single Unit....................................	1.75
	2. Two or More Units	
	(a) Precast Sills...............................	.75
	(b) Poured Sills...............................	1.00
Fixed Architectural	1. Single Unit....................................	2.00
	2. Two or More Units	
	(a) Precast Sills...............................	1.00
	(b) Poured Sills...............................	1.25
Intermediate Windows	1. In Wood	
(Casement, Combination	(a) Single Unit	2.75
and Projected)	(b) Two or More	
	Units...	2.25
	2. In Stone	
	(a) Single Unit	3.00
	(b) Two or More	
	Units...	2.50
	3. In Masonry	
	(a) Single Unit	2.25
	(b) Two or More	
	Units...	1.25
Mullions Only		.50
Mullions with Covers..............................		.75
Residence Casements (Roto)	1. Single Unit	
	(a) In Wood.............................	1.75
	(b) In Masonry or	
	Brick Veneer.......................	1.25
	2. Two or More Units	
	(a) In Wood.............................	1.50
	(b) In Masonry or	
	Brick Veneer.......................	1.00
Residence Casements (Fixed)	1. Single Unit	
	(a) In Wood.............................	1.50
	(b) In Masonry or	
	Brick Veneer.......................	1.00
	2. Two or More Units	
	(a) In Wood.............................	1.25
	(b) In Masonry or	
	Brick Veneer.......................	.75
Casements in Casings	Single or Multiple	
	Units...	1.50
Continuous Top Hung, Fixed	Per Lineal Foot............................	.50
Continuous Top Hung, Swing	Per Lineal Foot............................	.75
Mechanical Operators,		
Rack and Pinion	Per Lineal Foot............................	.75
Mechanical Operator,		
Lever Arm	Per Lineal Foot............................	.50
Mechanical Operators,		
Tension	Per Lineal Foot............................	1.50
Industrial Doors and Frames	1. Up to 35 Sq. Ft.	
	Per Sq. Ft.	.125
	2. Over 35 Sq. Ft.	
	Per Sq. Ft.	.10

METAL WINDOWS

Double Hung Windows	1. Single Units	1.50
	2. Two or More Units	1.25

Screen Installation		
For Projected Windows	Net per Screen	1.10
For Basement and Utility Windows	Net per Screen	.75
For Security Windows	Net per Screen	1.10
For Casement Windows	Net per Screen	.75
For Pivoted Windows	Net per Vent	1.50

*Single ventilator units. For each additional ventilator, ADD .25

Prices of Commercial Projected Steel Windows

VENTILATED			FIXED	
Window Type	Window Price	Screen Price	Window Type	Window Price
*A12120	$ 55.00	$10.00	B-22	$ 29.00
*A13121	56.00	10.00	B-23	38.00
*B22140	67.00	14.00	B-24	43.00
*B23141	74.00	"	B-25	55.00
*B24141	82.00	"	B-26	67.00
*B25141	91.00	"	A-32	38.00
B2522402	96.00	†22.00	A-33	45.00
*B26141	101.00	14.00	A-34	57.00
B2622403	110.00	†22.00	A-35	68.00
*A32160	82.00	17.00	A-36	80.00
*A33161	90.00	"	A-42	84.00
*A34161	101.00	"	A-43	100.00
*A35161	112.00	"	A-44	112.00
*A36161	123.00	"	A-45	129.00
A3523602	112.00	†26.00	A-46	145.00
A3623603	123.00	"		
A43141	98.00	15.00		
A44141	112.00	"		
A45141	129.00	"		
A46141	135.00	"		

*These types available with either project-out or project-in vents.
†Price includes wicket screen for project-out vent and fixed screen for project-in vent.

Prices include cam handle and pole ring for project out ventilators and cam handle with strike or spring latch with strike for project in ventilators. Hardware is cadmium plated.

Prices also include standard installation fittings, provisions for standard screens, bonderizing and one factory coat of paint. Glass, putty, glazing and glazing clips not included.

Prices of Mullions and Covers

Lights High	2	3	4	5	6
Mullions	$6.00	$8.00	$11.00	$14.00	$16.00
Covers	7.00	9.00	12.00	15.00	18.00

Prices of Architectural Projected Steel Windows

Window Type	Window Price	Screen Price	Window Type	Window Price	Screen Price
252A	$ 65.00	$11.00	215	$129.00	$20.00
252B	65.00	8.00	216	132.00	20.00
253	81.00	11.00	218	138.00	20.00

Window Type	Window Price	Screen Price	Window Type	Window Price	Screen Price
261	$ 69.00	$ 8.00	221	$ 77.00	$ 9.00
262A	71.00	12.00	222A	86.00	14.00
262B	73.00	9.00	222B	90.00	12.00
263	84.00	12.00	223	92.00	14.00
264	100.00	12.00	224	104.00	14.00
265	119.00	18.00	225	124.00	23.00
266	170.00	18.00	226	134.00	23.00
268	184.00	18.00	228	139.00	23.00
211	77.00	8.00	273	107.00	17.00
212A	88.00	13.00	274	116.00	17.00
212B	89.00	11.00	275	158.00	25.00
213	93.00	13.00	276	173.00	25.00
214	97.00	13.00	278	190.00	25.00

Prices include bronze operating hardware for ventilated units. Cam handle, strike and pole ring for project out vents and cam handle with strike or spring latch with strike for project in vents. Also included are standard installation fittings, bonderizing and one factory coat of paint. Glass, putty, glazing and glazing clips not included.

Screen prices include screens for all vents in unit. Wicket screens are furnished for project out vents and outside fixed screens for project in vents. Prices also include clips for attachment.

Polished bronze hardware in place of standard bronze hardware, add per vent..$2.50

Types and Sizes of Commercial Projected Windows

Prices of Security Steel Windows

Window Type	Window Price	Screen Price	Mullion Price
32130	$ 75.00	$ 8.00	$3.00
33160	93.00	11.00	4.00
62 Fixed	72.00	....	3.00
62160	104.00	11.00	3.00
631120	107.00	15.00	4.00
641121	135.00	15.00	5.00
931180	172.00	19.00	4.00
941181	193.00	19.00	5.00

Prices include standard operating hardware, installation, fittings, bonderizing and one factory coat of paint. Glass, putty, glazing and glazing clips are not included.

Anchors

Anchors for securing frames to wood or masonry walls are included with window units.

Four anchors per opening are furnished for heights up to and including 4'-5½". Six anchors per opening to and including 5'-9½". Eight anchors per opening to and including 8'-1½".

Prices of Residential Steel Casements Ventilated Casements

Window Type	Window Price	Head and Jamb Fins Per Set	Inside Casing Price	In/Out Trim Price	Screen Price
1212	$ 36.00	$3.00	$ 7.00	$11.00	$4.00
1313	41.00	3.00	9.00	12.00	5.00
1413	48.00	4.00	10.00	13.00	5.00
1414	46.00	4.00	10.00	13.00	6.00
1514	54.00	5.00	11.00	15.00	6.00
2212	41.00	3.00	9.00	13.00	4.00
2313	51.00	4.00	10.00	15.00	5.00
2414	63.00	4.00	11.00	16.00	6.00
2413	60.00	4.00	11.00	16.00	5.00
2514	76.00	5.00	12.00	18.00	5.00
2222	41.00	4.00	9.00	13.00	6.00
2323	48.00	4.00	10.00	15.00	8.00
2424	51.00	5.00	11.00	16.00	9.00
2423	50.00	5.00	11.00	16.00	8.00
2524	54.00	6.00	12.00	18.00	9.00
3222	50.00	4.00	10.00	16.00	6.00
3323	70.00	5.00	12.00	17.00	8.00
3424	94.00	6.00	13.00	19.00	9.00
3423	90.00	6.00	13.00	19.00	8.00
3524	118.00	6.00	14.00	21.00	9.00
4222	54.00	5.00	12.00	18.00	6.00
4323	82.00	6.00	13.00	20.00	8.00
4424	107.00	7.00	14.00	21.00	9.00
4423	106.00	7.00	14.00	21.00	8.00
4524	140.00	8.00	15.00	24.00	9.00
5222	80.00	5.00	13.00	21.00	6.00
5323	112.00	6.00	14.00	22.00	8.00
5424	156.00	8.00	15.00	24.00	9.00
5423	154.00	8.00	15.00	24.00	8.00
5524	190.00	9.00	16.00	26.00	9.00

Types and Sizes of Architectural Projected Steel Windows

Types and Sizes of Security Steel Windows

Fixed Casements

Window Type	Window Price	Window Type	Window Price
12	$13.00	34	$ 60.00
13	19.00	35	75.00
14	24.00	42	35.00
15	30.00	43	54.00
22	19.00	44	74.00
23	27.00	45	90.00
24	35.00	52	48.00
25	41.00	53	74.00
32	31.00	54	96.00
33	42.00	55	118.00

Prices of ventilated units include locking handles, Roto underscreen operators (lever underscreen operators also available at reduced price) and standard installation fittings, bonderizing and one factory coat of paint for both ventilated and fixed units. Glass, putty, glazing, glazing clips and mastic not included.

Screen prices include screens for all vents in unit and clips for attachment.

Prices of Extras

Bronze hardware for roto casements in lieu of bronze
lacquer finish hardware, add per side hinged vent......................................$ 6.00
Mastic, use 1 quart for each 40 lin. ft. of application
Per quart (4 lbs.) 3.50
Per gallon (16½ lbs.) 13.00
5 gallons (85 lbs.)...60.00

Prices of Mullions

Vertical Mullions		Horizontal Mullions		Pipe Corners	
2-Lt..................	$3.00	1-Lt..................	$3.00	2-Lt..................	$ 6.00
3-Lt..................	4.00	2-Lt..................	4.00	3-Lt..................	8.00
4-Lt..................	5.00	3-Lt..................	5.00	4-Lt..................	9.00
5-Lt..................	6.00	4-Lt..................	6.00	5-Lt..................	11.00
		5-Lt..................	7.00		

Types and Sizes of Residential Steel Casements

Prices of Residential Steel Casement Picture Windows

For Standard Glazing		For Insulating Glazing	
Window Type	Window Price	Window Type	Window Price
PW33	$19.00	DG34	$23.00
PW34	21.00	DG35	26.00
PW35	23.00	DG44	27.00
PW43	22.00	DG45	29.00
PW44	25.00		
PW45	27.00		

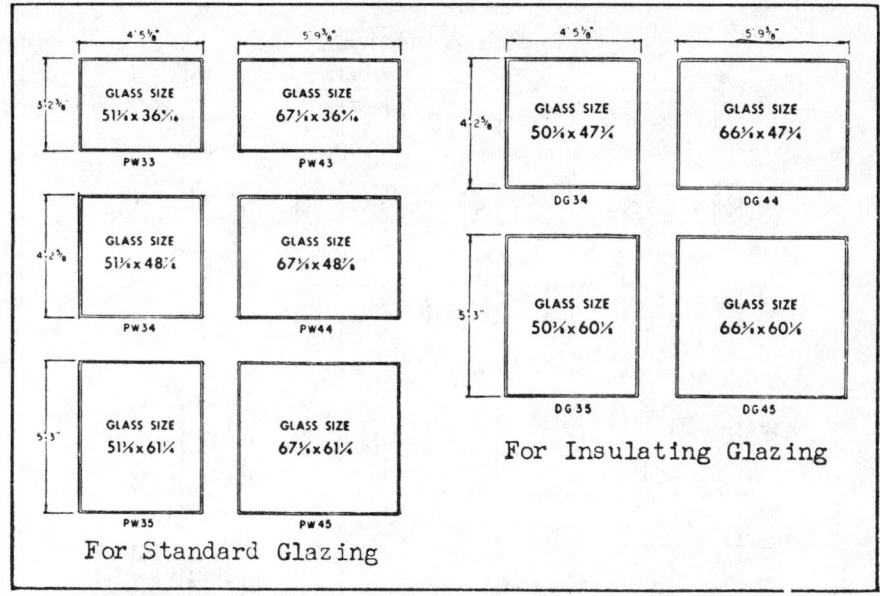

For Insulating Glazing

For Standard Glazing

Picture Window Units for Use with Residential Steel Casements

Prices are for four sided frames, shipped assembled, with two standard vertical mullions attached. Units for insulating glazing have glazing strips attached. Units are bonderized and painted one prime coat of oven-baked paint. Glass, putty and glazing are not included.

Prices of Mullions and Fins

	Horizontal Widths			Vertical Heights		
	3'-1¾"	6'-2½"	9'-2½"	2'-1¾"	4'-2⅛"	6'-2½"
Mullions				$2.00	$3.00	$4.00
Fins	$2.00	$3.00	$4.00	2.00	3.00	3.00

Prices of Basement and Utility Steel Windows

Window prices include steel latch with ring and keeper, bonderizing and one factory coat of paint. Glass, putty, glazing and glazing clips not included.

Screen and storm window prices include clips for attachment. Storm panels are factory glazed.

Basement Windows

Window Type	Opening Dimensions	Window Price	Mullion Price	Screen Price	Storm Sash Price
2-Lt.	2'-8½"x1'- 2⅝"	$26.00	$2.00	$4.00	$ 9.00
2-Lt.	2'-8½"x1'- 6⅝"	30.00	2.00	4.00	10.00
2-Lt.	2'-8½"x1'-10⅝"	38.00	2.00	5.00	12.00

Utility Windows

Window Type	Opening Dim. Width	Height	Window List Price	Mullion List Price	Screen List Price
22121	2'-8½"	3'-6⅝"	$25.00	$3.00	$6.00

Vented Windows

RA1212 L/R   RA2212 L/R   RA2222   RA3222   RA4222   RA5222

RA1313 L/R   RA2313 L/R   RA2323   RA3323   RA4323   RA5323

RA1413 L/R   RA2413 L/R   RA2423   RA3423   RA4423

RA1414 L/R   RA2414 L/R   RA2424   RA3424   RA4424   RA5424

RA1514 L/R   RA2514 L/R   RA2524   RA3524   RA4524   RA5524

Aluminum Residential Casements

PICTURE WINDOWS - SINGLE GLASS
PICTURE WINDOWS - DOUBLE GLASS
FOR STOCK SIZE DOUBLE INSULATING GLASS

- DIMENSIONS SHOWN ARE WINDOW DIMENSIONS.
- ELEVATIONS ARE SHOWN AS VIEWED FROM THE OUTSIDE. HAND OF CASEMENTS IS DETERMINED BY LOCATION OF HINGES AS VIEWED FROM THE OUTSIDE.
- SINGLE GLASS PICTURE WINDOWS ARE SHIPPED ASSEMBLED.
- DOUBLE GLASS ("DG") UNITS ARE SHIPPED KNOCKED DOWN FOR FIELD ASSEMBLY.
- "DG" PICTURE WINDOW UNITS ACCOMODATE 1" THICK STANDARD SIZE INSULATING GLASS.
- FOR STUD OPENING OR ROUGH OPENING SIZES USE WINDOW DIMENSIONS SHOWN AND ADD 1½" IN WIDTH AND 1½" IN HEIGHT.

Picture Window Units for Use with Aluminum Residential Casements

08520 ALUMINUM CASEMENT WINDOWS

Prices include aluminum die cast roto operator with aluminum arm. Operator case and crank are brushed to a satin finish and given one coat of clear Duranite phenolic resin baked-on plastic coating. Casements are etched in a hot acid bath and receive a dip coat of methacrylate lacquer. Also included are wood screws and sill anchors when specified. Prices include ½" insulating glass and screens.

Prices of Aluminum Residential Casements

———— VENTILATED ————

Window Type	Window Price
1212	$ 36.00
1313	38.00
1413	46.00
1414	46.00
1514	56.00
2212	42.00
2313	50.00
2413	66.00
2414	66.00
2514	80.00
2222	53.00
2323	63.00
2423	70.00
2424	70.00
2524	86.00
3222	85.00
3323	72.00
3423	96.00
3424	96.00
3524	119.00
4222	58.00
4323	105.00
4423	112.00
4424	115.00
4524	144.00
5222	77.00
5323	115.00
5424	154.00
5524	192.00

Prices of Picture Windows for Use with Aluminum Residential Casements

Glazed With ½" Insulating Glass.

Window Type	Window Price
33MO	$ 65.00
34MO	76.00
35MO	86.00
43MO	75.00
44MO	90.00
45MO	102.00

Price of Aluminum Awning Windows

Window Type	Window Price
1722	$ 90.00
1732	144.00
1742	180.00
1753	210.00
3122	113.00
3132	144.00
3142	210.00
3153	250.00
4522	168.00
4532	210.00
4542	240.00
4553	300.00

Prices of Mullions and Fins

Lts. Wide or High	Vertical Mullions	Head Fins Each Head	Jamb Fins Each Jamb
1	$...	$2.00	$...
2	5.00	2.00	1.50
3	8.00	3.00	2.00
4	10.00	4.00	3.00
5	13.00	5.00	3.00

Prices include a mechanical operator with crank handle. Windows are glazed by using snap-in aluminum beads. Prices include integral fin, and glazing with ½" insulating glass.

Windows are weatherstripped and receive a dip coat of methacrylate lacquer.
Screen prices include screens for all vents in unit. Screens have aluminum frames
and 18x14 mesh aluminum cloth.
Storm sash are factory glazed.

Aluminum Awning Windows

Prices of Aluminum Double Hung Windows

Prices of windows include silicone-treated wool-pile weatherstripping, spiral,
sash balances, glazing clips, standard anchors, mullion screws and window hardware consisting of sash lock, keeper and pull down for upper sash. Prices include
½" insulating glass and screens.

All windows are finished with a coat of water clear lacquer.

Screens have extruded or rolled aluminum frame sections and are wired with
18x14 mesh aluminum cloth. Screens consist of full or half panels to allow for full

or half-length screening as desired. Prices include all necessary installation hardware.

Storm sash are furnished glazed with single strength glass set in a plastic seal. All points of contact with window frame are insulated. Prices include all necessary hardware for installation.

Picture windows are shipped knocked down. Picture windows are designed to receive ¼" single glass or ½", ¾" or 1" insulating glass if specified. Prices include necessary assembly and installation hardware.

Prices of Aluminum Double Hung Windows

Window Type	2-Lt. Top 2-Lt. Bot.	6-Lt. Top 6-Lt. Bot.	Two Light
2030	$ 75.00	$ 80.00	$ 73.00
2830	80.00	85.00	78.00
3030	83.00	88.00	81.00
3430	86.00	91.00	84.00
3830	92.50	98.00	90.00
2038	78.00	83.00	76.00
2838	82.00	87.00	80.00
3038	90.00	95.00	88.00
3438	93.00	98.00	90.00
3838	112.00	117.00	110.00
2044	80.00	85.00	78.00
2844	83.00	88.00	81.00
3044	91.00	96.00	89.00
3444	93.00	98.00	91.00
3844	96.00	101.00	94.00
2050	85.00	90.00	83.00
2850	87.00	92.00	85.00
3050	90.00	95.00	88.00
3450	105.00	110.00	103.00
3850	110.00	115.00	108.00
2060	95.00	100.00	93.00
3060	100.00	105.00	98.00
3860	105.00	110.00	103.00

Prices of Aluminum Fins and Trim or Brick Molds

	Height of Window				
	3'-1⅝"	3'-9⅝"	4'-5⅝"	5'-1⅝"	5'-9⅝"
Jamb fin and trim or brick mold, per pair	$4.00	$5.00	$5.50	$6.00	$6.50

	Width of Window				
	2'-0½"	2'-4½"	2'-8½"	3'-0½"	3'-4½"	3'-8½"
Head Section	$0.60	$0.65	$0.75	$0.80	$0.85	$0.90
Sill Section	.50	.55	.60	.65	.75	.80

Prices of Aluminum Screens, Storm Sash and Inside Casings for Aluminum Double Hung Windows

Window Type	Half Screen	Full Screen	Storm Sash 2-Piece	Inside Casing
2031	$ 5.50	$ 9.00	$38.50	$ 7.50
2431	6.00	10.00	40.00	7.70
2831	6.50	11.00	42.00	8.10
3031	7.75	13.50	44.00	8.50
3431	8.25	14.50	45.00	8.70

Window Type	Half Screen	Full Screen	Storm Sash 2-Piece	Inside Casing
3831	$ 9.00	$15.00	$49.00	$ 9.10
2039	5.75	9.50	40.00	8.20
2439	7.00	12.00	42.00	8.50
2839	8.00	14.00	44.50	8.80
3039	9.00	16.00	45.50	9.20
3439	10.00	18.00	49.00	9.40
3839	11.00	20.00	53.00	9.80
2045	7.00	12.00	42.00	8.60
2445	8.00	14.00	44.50	8.80
2845	9.00	16.00	45.00	9.20
3045	10.00	18.00	49.00	9.50
3445	11.50	21.00	52.00	9.80
3845	12.50	23.00	54.00	10.20
2051	8.00	14.00	44.50	9.10
2451	9.00	16.00	45.50	9.30
2851	10.00	18.00	49.00	9.70
3051	12.00	22.00	52.00	10.00
3451	13.00	24.00	54.00	10.20
3851	14.00	26.00	63.00	10.60
2059	9.00	16.00	45.50	9.70
2459	10.00	18.00	49.00	9.90
2859	11.00	20.00	52.00	10.30
3059	13.00	24.00	54.00	10.60
3459	14.00	26.00	63.00	10.90
3859	15.00	28.00	64.00	11.30

Prices of Aluminum Picture Windows for Use with
Aluminum Double Hung Windows, Glass Not Included

Window Type	Price	Window Type	Price
40P31	$36.50	44P45	$44.50
44P31	37.50	48P45	47.00
40P39	38.00	*410P45	47.50
44P39	40.00	50P45	48.00
48P39	40.75	*56P45	49.00
50P39	42.00	*62P45	50.00
40P45	43.00	*610P45	51.00
*42P45	44.00		

*These sizes are set up (standard) for insulated glazing.

Prices are for units for single glazing. For double glazing, add $3.00 per unit.

Prices of Aluminum Horizontally Sliding Windows

Prices of windows include silicone-treated wool-pile weatherstripping, and integral fin. Prices are for frames and sash assembled and glazed with ½" insulating glass.

All frame and sash members are cleaned, caustic etched and coated with methacrylate-type lacquer at the factory.

Screens are aluminum, covering half of window opening, and are easily installed and removed from inside.

Storm sash are installed same as screens and are two piece SSB glazed.

Window Type	Window Price	Half Screen	Inside Casing
2020	$32.00	$5.00	$ 8.60
3020	34.00	6.00	9.50
4020	37.00	5.00	9.10
5020	39.00	6.00	10.20
6020	42.00	6.00	11.20
3030	46.00	6.00	8.80
4030	48.00	6.00	9.90
5030	52.00	7.00	11.00
6030	60.00	7.00	11.90
4040	58.00	7.00	11.00
5040	70.00	8.00	12.00
6040	81.00	8.00	13.50

Prices of Aluminum Picture Windows with Horizontally Sliding Flankers

Picture window prices are for frames and sash assembled and glazed.
Screen prices include two per unit.
Storm sash prices are for three piece glazed.

Window Type	Window Price	2-Piece Screens	3-Piece Storm Sash	Inside Casing
7020	$ 66.00	$10.00	$52.00	$15.40
8020	69.00	10.00	54.00	16.00
9020	76.00	10.00	60.00	18.40
7030	82.00	11.00	64.00	16.90
8030	86.00	12.00	68.00	18.20
9030	93.00	12.00	73.00	19.60
7040	97.00	12.00	74.00	18.40
8040	100.00	13.00	78.00	19.50
9040	110.00	13.00	83.00	21.00

08600 WOOD AND PLASTIC WINDOWS

Wood windows are most generally sold as complete units with frame, sash, operating hardware, weatherstripping and glazing assembled at the factory. Frames are treated. The following sizes are based on units manufactured in the Chicago area by ROW Window Company. Sizes for other manufacturers will vary slightly. Prices as given should, of course, be checked locally and will vary considerably geographically and as to the size of order and even as to delivery to local dealer or to job site. The prices quoted will be helpful, however, for comparing the various window and glazing types.

Double Hung Units Preglazed SS*

Glass Size	Stud Opening Width	Stud Opening Height	Two	Four	Six	Eight
16x16	1'11"	3'- 6"	$ 56.72	$65.45	$. . . .	$. . . .
20		4'- 2"	63.13	72.17		
24		4'-10"	69.33	78.84		
28		5'- 6"	75.84	84.67		
32		6'- 2"	84.37			
36		6'-10"	91.75			
20x16	2'- 3"	3'- 6"	61.22	70.53		

5¼" Stud Wall		9" or 13"	10"
No Wind Break	With Wind Break	Brick Wall	Brick Ven. Wall

Stock Western Pine Windows

One Pane   Two Pane   Six Pane

Double Hung Units Preglazed SS*

Glass Size	Stud Opening Width	Height	Two	Four	Six	Eight
20x20		4'- 2"	$ 68.72	$77.62		
24	2'- 3"	4'-10"	75.26	84.74		
28		5'- 6"	80.92	91.31		
32		6'- 2"	91.18			
36		6'-10"	98.24			
24x16	2'- 7"	3'- 6"	64.45	75.13	77.38	
20		4'- 2"	71.98	81.87	84.00	
24		4'-10"	79.38	89.36	92.03	
28		5'- 6"	85.80	98.15	97.54	
32		6'- 2"	96.33			
36		6'-10"	104.14			
28x16	2'-11"	3'- 6"	67.55		81.72	
20		4'- 2"	74.50		89.86	
24		4'-10"	82.53		98.39	
28		5'- 6"	90.01		105.70	
32		6'- 2"	101.58		120.13	
36		6'-10"	109.95		128.81	

*For double hung units in pairs, triple or quadruple arrangements, multiply costs by two, three and four respectively.

Glass Size	Stud Opening Width	Height	Two	Four	Six	Eight
32x16	3'- 3"	3'- 6"	$ 72.24		$ 88.38	$ 95.20
20		4'- 2"	80.75		97.30	99.91

Glass Size	Width	Height	Cost	Diamond*	Colonial
24		4'-10"	$ 88.95	$104.27	$108.07
28		5'- 6"	97.56	111.95	115.75
32		6'- 2"	106.72	126.42	130.20
36		6'-10"	116.10	135.86	141.72
36x16	3'- 7"	3'- 6"	76.10	94.13	97.91
20		4'- 2"	84.33	101.56	106.26
24		4'-10"	93.18	106.85	113.10
28		5'- 6"	102.34	116.98	121.22
32		6'- 2"	112.58	128.57	132.30
36		6'-10"	122.06	137.99	144.96
40x16	3'-11"	3'- 6"	81.48	100.10	104.25
20		4'- 2"	88.15	105.22	110.28
24		4'-10"	97.13	112.97	119.57
28		5'- 6"	107.81	123.19	128.10
32		6'- 2"	118.40	138.64	142.35
36		6'-10"	129.35	149.01	155.98

Double Hung Sash Glazed With ½" Insulating Glass

Glass Size	Stud Opening Width	Height	Cost	Add for Rem. Grilles Diamond*	Colonial
16x20	1'-11"	4'- 2"	$ 92.64	$. . . .	$ 7.46
24		4'-10"	102.29	. . . .	7.76
28		5'- 6"	113.12	. . . .	8.12
20x16	2'- 3"	3'- 6"	90.16	34.76	7.79
20		4'- 2"	98.89	35.46	7.99
24		4'-10"	111.34	36.34	8.18
28		5'- 6"	122.08	37.14	8.76
32		6'- 2"	138.23		9.15
36		6'-10"	149.88		9.68
24x16	2'- 7"	3'- 6"	96.67	35.46	11.98
20		4'- 2"	108.22	36.34	12.48
24		4'-10"	118.72	37.14	13.00
28		5'- 6"	130.09	37.88	13.48
32		6'- 2"	146.65	53.81	13.95
36		6'-10"	158.32		14.39
28x16	2'-11"	3'- 6"	103.92	40.70	12.53
20		4'- 2"	115.36	41.62	12.87
24		4'-10"	126.49	42.46	13.39
28		5'- 6"	135.19	43.34	13.89
32		6'- 2"	154.85	63.10	14.34
36		6'-10"	168.39	64.33	14.72
32x16	3'- 3"	3'- 6"	112.45	41.62	13.31
20		4'- 2"	124.36	42.46	13.67
24		4'-10"	136.17	43.34	14.14
28		5'- 6"	144.52	44.31	14.70
32		6'- 2"	165.79	64.33	15.19
36		6'-10"	178.22	65.52	15.72
36x16	3'- 7"	3'- 6"	119.83	42.46	14.17
20		4'- 2"	132.30	43.34	14.53
24		4'-10"	144.13	44.31	14.92
28		5'- 6"	157.82	45.19	15.80
32		6'- 2"	174.38	65.52	16.25
36		6'-10"	189.09	66.84	16.75
40x16	3'-11"	3'- 6"	125.53	48.49	14.83
20		4'- 2"	137.40	49.46	15.22
24		4'-10"	150.10	50.47	15.72
28		5'- 6"	162.47	51.44	16.39

Glass Size	Stud Opening Width	Height	Cost	Add for Rem. Grilles Diamond*	Colonial
40x32		6'- 2"	$186.16	$78.19	$17.08
36		6'-10"	201.90		17.86
24x24/36	2'- 7"	5'-10"	137.73	40.72	15.97
28x24/36	2'-11"	5'-10"	148.19	54.78	16.55
32x24/36	3'- 3"	5'-10"	157.72	55.77	17.30
36x24/36	3'- 7"	5'-10"	169.61	56.80	18.08
40x24/36	3'-11"	5'-10"	176.12	62.48	18.88

*Price Per 3 Or More Sets

Panorama Window Units

Picture Window Units
For Use Alone or With Double Hung Units

Glass Size	Glazed Insulating 1"
36x32½	$154.28
48	189.14
36x40½	170.34
48	214.78
36x48½	229.98
48	253.92
56	297.24
64	359.69
72	443.42
80	500.25
96	582.97
36x56½	226.33
48	321.59
56	362.03
64	462.15
72	521.42
80	573.18
96	645.29
48x60½	413.89
56	469.21
64	524.44
72	585.76
48x64½	434.18
56	487.06
64	549.14
72	603.54

Gliding Window Units

Gliding window units may be used in any room of the house, but are more frequently used in enclosed porches, as high windows in bedrooms, baths and other similar areas where interior wall space is a consideration.

By using the high narrow windows, furniture may be placed under them making much better interior arrangements possible.

Gliding Window Units

The gliding windows described below are furnished complete with frames, sash, weatherstrips, so that each sash slides to the opposite side of the opening for ventilation, but when fully closed is weatherstripped on all four sides.

Prices for complete gliding window units, assembled and carton packed:

Gliding Window Units

Glass Size	Stud Opening Width	Stud Opening Height	Cost per unit glazed ss	Combinations 3 track
16x16	3'- 3⅝"	1'-11"	$ 56.15	$51.30
20		2'- 3"	59.35	"
24		2'- 7"	62.46	"
28		2'-11"	66.09	"
32		3'- 3"	69.76	"
36		3'- 7"	73.17	"
40		3'-11"	78.77	"
48		4'- 7"	92.84	"
20x16	3'-11⅝"	1'-11"	63.36	"
20		2'- 3"	67.22	"
24		2'- 7"	70.54	"
28		2'-11"	73.74	"
32		3'- 3"	78.39	"
36		3'- 7"	81.70	"
40		3'-11"	85.92	"
48		4'- 7"	100.63	"
24x16	4'- 7⅝"	1'-11"	70.38	"
20		2'- 3"	74.28	"
24		2'- 7"	78.08	"
28		2'-11"	81.72	"
32		3'- 3"	86.54	"
36		3'- 7"	90.28	"
40	4'- 7¼"	3'-11½"	94.65	$53.38
48		4'- 7"	108.30	55.46
28x16	5'- 3¼"	1'-11½"	77.50	59.62
20		2'- 3½"	81.13	51.30
24		2'- 7½"	85.39	51.30
28		2'-11½"	89.74	53.38
32		3'- 3½"	95.32	55.46

Glass Size	Stud Opening Width	Height	Cost per unit glazed ss	Combinations 3 track
28x36		3'- 7½"	$ 99.42	$57.54
40		3'-11½"	104.89	59.62
48		4'- 7"	119.78	63.78
32x16	5'-11¼"	1'-11½"	85.34	51.30
20		2'- 3½"	89.95	53.38
24		2'- 7½"	94.36	55.54
28		2'-11½"	99.44	57.54
32		3'- 3½"	103.69	59.62
36		3'- 7½"	108.48	61.70
40		3'-11½"	114.24	63.78
48		4'- 7"	128.10	67.94

Screens and operating hardware are included in the above prices

Gliding type window units may be ordered with a fixed glass in the center and two gliding sash on either end which, when open, pass in front of the fixed panel.

Frame Extras

Liners for 4⅞" or 5¼" jamb..$ 1.50
Liners for 5" or 5⅛" jamb.. 2.80
Mullion centers, wider than standard (1½"), for each 1", add....................... 4.80
Sill horn only, up to 12", one side only, add.. 3.20
Sill horn only, up to 12", both sides, add 6.50
Sill horn and 10" S4S casing, one side only, add.. 26.50
Sill horn and 10" S4S casing, both sides, add... 53.00
Sill horn and 10" S4S casing, over 10", per inch, add...................................... 5.30
Drip cap add per single opening.. 1.50

Sash Extras

Glazing with obscure glass, add to SS list, per square foot $1.70
Rectangular grilles same as double hung.

Glass Panel Wall Units.

Stationary   Glass Panel Wall Units.   Ventilating

Awning Panel Wall Units

Awning units are a system of window panel arrangement ranging from a single individual panel to a multiple 12 panel unit, as illustrated, which is used as a picture frame unit.

Awning units are designed to give a selection of either stationary glass panels, ventilating sash panels or a combination of both. Both come glazed with ½" insulated glass, ¼" plate or DS glass, bedded in glazing compound.

Ventilating units have 1¾" sash, hinged at the top which can be opened about 35 degrees, closed or held in intermediate position by rotary operators. Operators have worm gears, twin arms of cadmium plated steel, removable brass plated handles and are self-locking.

Screens for all ventilating units have aluminum frames and 14 x 18 mesh aluminum screen wire.

Glass panel wall units come completely set up, glazed and carton packed. The ventilating units are weatherstripped, operators are applied and screens installed in frames.

7'- 10 ½"   2'-3¼"
2 wide and 1 high can be used in bedrooms, providing extra wall space for furniture placement.

11'- 9 ¾"   4'-6½"
3 wide and 2 high are effectively used in dining rooms or dining alcoves.

15'- 9"   6'- 9¾"
Awning units, room width, ceiling height, add charm to living rooms and solariums.

Awning Window Units

Glass Size	Stud Opening		Cost per Unit Glazed ½" in.*
	Width	Height	
1 High 27x14	2'-9⅞"	1'-8½"	$106.00
18		2'-0½"	115.00

Glass Size	Stud Opening Width	Height	Cost per Unit Glazed ½" in.*
36x14	3'-7¼"	1'-8½"	$126.00
18		2'-0½"	132.00
44x14	4'-3¼"	1'-8½"	136.00
18		2'-0½"	147.00
2 High 27x14	2'-9⅞"	3'-1⅜"	185.00
18		3'-9⅜"	192.00

Awning Window Units

Glass Size	Stud Opening Width	Height	Cost per Unit Glazed ½" in.*
36x14	3'-7 1/4"	3'-1 3/8 "	$198.00
18		3'-9 3/8 "	228.00
44x14	4'-3 1/4"	3'-1 3/8 "	230.00
18		3'-9 3/8 "	155.00
3 High 27x14	2'-9 7/8"	4'-6 3/16"	156.00
18		5'-6 3/16"	275.00
36x14	3'-7 1/4"	4'-6 3/16"	280.00
18		5'-6 3/16"	302.00
44x14	4'-3 1/4"	4'-6 3/16"	300.00
18		5'-6 3/16"	324.00

*Screens included

Combination unit with fixed sash above, awning unit below.

Vista-Vent Units

Vista-vents combine a bottom panel of an awning window unit with a fixed panel above. In the following table the first glass size is the upper fixed light; the second, the vent height below. Prices include storms and screens for SSB glazed, screens for ½" insulating glass.

Glass Size	Stud Opng. Width Height	Cost
36x30 Over 16	3'-5⅝"x4'-9⅝"	$196.00
20x36 " " 16	2'-1⅝"x5'-3⅝"	164.00
28x36 " " 16	2'-9⅝"x5'-3⅝"	186.00
36x36 " " 16	3'-5⅝"x5'-3⅝"	207.50

44x36	Over 16	4'-2⅝"x5'-3⅝"	$227.00
20x48	" " 16	2'-1⅝"x6'-3½"	187.50
28x48	" " 16	2'-9⅝"x6'-3½"	208.50
36x48	" " 16	3'-5⅝"x6'-3½"	243.00
44x48	" " 16	4'-2⅝"x6'-3½"	264.00

Casement Units

Glass Size	Stud Opening Width	Height	Cost Glazed: 1/2" insulating	Rect. Grille	Diam. Grille
16x24	1'-9⅞"	2'-6 9/16"	$ 91.58	$ 6.92	$20.08
30		3'-0 3/4 "	97.98	8.98	21.20
36		3'-6 3/4 "	108.10	8.98	21.76
49		4'-615/16"	123.54	13.62	22.90
61		5'-7 1/8 "	142.56	17.90	23.88
21x24	2'-2⅜"	2'-6 9/16"	102.74	6.92	20.64
30		3'-0 3/4 "	110.10	8.98	21.76
36		3'-6 3/4 "	121.04	8.98	22.32
49		4'-615/16"	138.96	13.62	23.88
61		5'-7 1/8 "	159.96	17.90	24.92
25x36	2'-6⅞"	3'-6 3/4 "	128.30	8.98	22.92
49		4'-615/16"	144.70	13.62	24.92
61		5'-7 1/8 "	168.94	17.90	26.02

Casement sash may be ordered in gangs of two, three, four or five. Prices for each additional unit in a gang arrangement decline approximately 10%. It is thus cheaper to use casements in groups than as individual units. Additional savings will also be realized in gang arrangements in fewer jambs to form and caulk providing the lintel construction does not become more costly due to the longer span. If some of the sash are ordered as stationary units a further savings of $30.00 on the basic price per opening can be made. The above prices include screens.

Casements are also often used in combination with larger fixed units or "picture windows".

Picture Window Units For Casements

Glass Size	1" insul.	Cost glazed with: 11/16" insul.
35x36	$165.00	$135.00
44	186.00	156.00
55	213.00	183.00
68	243.00	
75	283.00	
93	325.00	
35x48	190.00	160.00
44	214.00	179.00
55	248.00	218.00
68	286.00	
75	365.00	
93	480.00	
35x60	220.00	
44	251.00	
55	290.00	
68	400.00	
75	490.00	
93	560.00	

A panelled study with diamond-grilled casements used either side of a fixed sash.

Prefabricated Bow Window Units

Prices include all casements venting and fitted with screens.

Glass Size	No. of Units	Cost ½" Insul.*	For Head & Seat Board Add
16x36	3	$ 382.82	$107.18
	4	509.42	126.52
	5	636.78	148.62
	6	768.32	241.70
	7	899.86	324.24
16x49	3	430.94	107.18
	4	573.40	126.52
	5	716.74	148.62
	6	865.24	241.70
	7	1,013.76	324.24
16x61	3	487.64	107.18
	4	648.82	126.52
	5	811.02	148.62
	6	976.05	241.70
	7	1,141.98	324.24
21x36	3	419.52	129.56
	4	558.24	148.74
	5	702.34	178.84
	6	833.88	286.88
	7	965.42	391.74
21x49	3	475.18	129.56
	4	632.26	148.74
	5	795.62	178.84
	6	944.14	286.88
	7	1,092.60	391.74
21x61	3	537.86	129.56
	4	715.62	148.74
	5	900.66	178.84
	6	1,066.16	286.88
	7	1,231.64	391.74

*For bronze insulating glass add 25%

Another type of prefabricated bow window is made up of fixed 22" x 18" glass units to form a shallow arc. These run in cost as follows, glazed with ½" insulating glass and with head and seat board a part of the basic unit.

Stud Opng. Height	Width	Cost ½" Insul.
4'-9¾"	x 7'-9⅝"	$395.00
	9'-7 "	506.43
	11'-3¾"	632.01
6'-4⅝"	x 7'-9 "	491.70
	9'-7 "	619.98
	11'-3¾"	766.67

A further variation of this is a similar unit prefabricated in a flat pattern. It is available glazed, SSB, or ½" insulating.

Stud Opng. Height	Width	Cost ½" Insul.
4'-10⅜"	x 7'-8¾"	$343.31
	9'-7⅜"	430.75
	11'-6 "	507.03
6'-5¼"	x 7'-8¾"	436.09
	9'-7⅜"	532.55
	11'-6 "	634.81

Vented sash can be made to replace a fixed sash in both the flat and bow units. Add, including operator and screen, $54.00.

WINDOW SCREENS

Wood window screens have been largely replaced by aluminum units. Most prefabricated wood sash have aluminum screens available to fit them and can be ordered as part of the unit or individually later. The following list gives a sampling of prices.

Glass Size	Full Frame Screen Size	Full Frame
16x16	19¾x38¾	$15.68
20	46¾	16.48
24	54¾	17.30
28	62¾	18.16
32	70¾	18.94
36	78¾	19.70
20x16	23¾x38¾	16.44
20	46¾	17.30
24	54¾	18.12
28	62¾	19.08
32	70¾	19.94
36	78¾	20.74
24x16	27¾x38¾	17.16
20	46¾	18.08
24	54¾	19.08
28	62¾	19.98
32	70¾	20.92
36	78¾	21.84
28x16	31¾x38¾	17.94
20	46¾	18.90
24	54¾	19.94

Full Screen (order by screen sizes).
Length: Measure the distance from the bottom of the head casing to the sill of the window.
Width: Measure the distance between the side casing of each window.

Half-Screen (order by screen sizes).
Length: Measure the distance from the top of the bottom rail of the top sash to the sill of the window.
Width: Measure the distance between the blind stop of each window.

Glass Size	Full Frame Screen Size	Full Frame
28x28	31¾x62¾	$20.92
32	70¾	21.88
36	78¾	22.88
32x16	35¾x38¾	18.66
20	46¾	19.76
24	54¾	20.80
28	62¾	21.84
32	70¾	22.92
36	78¾	24.00
36x16	39¾x38¾	19.34
20	46¾	20.52
24	54¾	21.66
28	62¾	22.74
32	70¾	23.88
36	78¾	25.06
40x16	43¾x38¾	20.16
20	46¾	21.34
24	54¾	22.52
28	62¾	23.70
32	70¾	24.92
36	78¾	26.10
44x16	47¾x38¾	20.88
20	46¾	22.16
24	54¾	23.38

44x28	47¾x62¾	$24.60
32	70¾	25.92
36	78¾	27.18

Nailing Wood Window Frames Together.—When wood window frames are sent to the job knocked down, it will require 1 to 1¼ -hr. carpenter time to assemble and nail each frame together, at the following labor cost per frame:

	Hours	Rate	Total	Rate	Total
Carpenter	1.2	$. . . .	$. . . .	$16.47	$19.76

It will require 1¾ to 2¼ hrs. carpenter time to assemble and nail together, a double window frame up to 6'-0" wide and 5'-6" or 6'-0" high, at the following labor cost per frame:

	Hours	Rate	Total	Rate	Total
Carpenter	2	$. . . .	$. . . .	$16.47	$32.94

Setting Single Wood Window Frames.—The labor cost of setting wood window frames will vary with the size of the frame, amount of bracing necessary and the distance they have to be carried or hoisted.

After a frame has been delivered at the job, it will require 10 to 15 minutes sorting frames and carrying them to the building and it will then require 20 to 30 minutes for 2 carpenters to set, plumb and brace each frame, which is at the rate of 8 to 12 frames per 8-hr. day.

The labor handling and setting each frame should cost as follows:

	Hours	Rate	Total	Rate	Total
Carpenter	0.8	$. . . .	$. . . .	$16.47	$13.18
Labor	0.2	. . . .	. . . .	12.54	2.51
Cost per frame			$. . . .		$15.69

Setting Double or Triple Window Frames.—It will ordinarily require 10 to 15 minutes for 2 men to carry a double or triple window frame from the stock pile to place where it is to be used and it will then require 30 to 35 minutes for 2 carpenters to set, plumb and brace the frame on the wall, at the following labor cost per frame:

	Hours	Rate	Total	Rate	Total
Carpenter	1.1	$. . . .	$. . . .	$16.47	$18.12
Labor	0.4	. . . .	. . . .	12.54	5.02
Cost per frame			$. . . .		$23.24

Labor Handling and Setting Complete Window Units.—When handling and setting complete window units consisting of frame, sash, balances, weatherstrips, etc., all installed complete, figure about ¼ -hr. more than for handling and setting window frames of the same size.

Fitting Wood Sash.—When fitting wood sash for double hung windows, a carpenter should fit 24 to 28 single sash per 8-hr. day, at the folowing labor cost per pair:

	Hours	Rate	Total	Rate	Total
Carpenter	0.6	$. . . .	$. . . .	$16.47	$9.88

First Grade Workmanship

Where the sash must fit perfectly, with neither too much nor too little play, a carpenter should fit 18 to 22 single sash per 8-hr. day, at the following labor cost per pair:

	Hours	Rate	Total	Rate	Total
Carpenter	0.8	$....	$....	$16.47	$13.18

Hanging Wood Sash.—Including the time required placing sash cord, chain and weights or counterbalances, a carpenter should hang 24 to 28 single sash per 8-hr. day, at the following labor cost per pair:

	Hours	Rate	Total	Rate	Total
Carpenter	0.6	$....	$....	$16.47	$9.88

First Grade Workmanship

Where care must be exercised to see that all sash work fits perfectly and the cord or chain is the exact length with the correct weights or counterbalances, a carpenter should hang 22 to 26 single sash per 8-hr. day, at the following labor cost per pair:

	Hours	Rate	Total	Rate	Total
Carpenter	0.7	$....	$....	$16.47	$11.53

Fitting and Hanging Wood Sash.—A carpenter should fit and hang, complete with sash cord, chain and weights or counterbalances, 12 to 14 single sash per 8-hr. day, at the following labor cost per pair:

	Hours	Rate	Total	Rate	Total
Carpenter	1.2	$....	$....	$16.47	$19.76

First Grade Workmanship

Where the sash must be fitted to have neither too much nor too little play and hung so they will work with ease and not be too loose, a carpenter should fit and hang (including sash cord, chain and weights or counterbalances), 10 to 12 single sash per 8-hr. day, at the following labor cost per pair:

	Hours	Rate	Total	Rate	Total
Carpenter	1.5	$....	$....	$16.47	$24.71

Fitting Casement Sash.—A carpenter should fit 22 to 26 average size wood casement sash per 8-hr. day, at the following labor cost per pair:

	Hours	Rate	Total	Rate	Total
Carpenter	0.7	$....	$....	$16.47	$11.53

First Grade Workmanship

Where first class workmanship is required with all sash fitted to work easily without binding and hung so that each sash shows the same top, bottom and side margin, a carpenter should fit 14 to 18 single sash per 8-hr. day, at the following labor cost per pair:

	Hours	Rate	Total	Rate	Total
Carpenter	1.0	$....	$....	$16.47	$16.47

Hanging Casement Sash.—The labor cost of hanging casement sash will vary, depending on whether they are hung with 2 or 3 butts, style of butts used, etc.

Where the sash are hung using 2 butts per sash, a carpenter should hang 20 to 24 single sash per 8-hr. day, at the following labor cost per pair:

	Hours	Rate	Total	Rate	Total
Carpenter	0.75	$....	$....	$16.47	$12.35

First Grade Workmanship

In fine residences, apartments and other buildings where the casement sash must be hung without binding and to show the same margin at the top, bottom and sides, a carpenter should hang 12 to 14 single sash per 8-hr. day, at the following labor cost per pair:

	Hours	Rate	Total	Rate	Total
Carpenter	1.2	$....	$....	$16.47	$19.76

Fitting and Hanging Casement Sash.—Under average conditions, a carpenter should fit and hang 10 to 13 single casement sash per 8-hr. day, at the following labor cost per pair:

	Hours	Rate	Total	Rate	Total
Carpenter	1.4	$....	$....	$16.47	$23.06

First Grade Workmanship

Where casement sash must be fitted and hung to be neither too loose nor so tight as to cause binding when opening and closing and show the same margin between the sash and frame on all sides, a carpenter should fit and hang 7 to 9 single sash per 8-hr. day, at the following labor cost per pair:

	Hours	Rate	Total	Rate	Total
Carpenter	2.2	$....	$....	$16.47	$36.23

Fitting and Hanging Outside Window Shutters.—When fitting and hanging outside window shutters on medium priced residences, cottages, etc., a carpenter should fit and hang 7 to 8 pairs per 8-hr. day, at the following labor cost per pair:

	Hours	Rate	Total	Rate	Total
Carpenter	1.1	$....	$....	$16.47	$18.12
Cost per single shutter	0.55		....		9.06

First Grade Workmanship

On the better class of buildings where more care must be used in fitting and hanging outside window shutters, a carpenter should fit and hang 5 to 6 pairs per 8-hr. day, at the following labor cost per pair:

	Hours	Rate	Total	Rate	Total
Carpenter	1.4	$....	$....	$16.47	$23.06
Cost per single shutter	0.7		....		11.53

Interior Window Trim

The labor cost of placing interior window trim will vary considerably with the type of building, thickness of walls, style of trim or casing, etc. Frame buildings, stucco buildings, brick veneer buildings, etc., having walls 6" to 8" thick will not require inside wood jamb linings, as the wood window frames usually extend to the inside face of the wall and the wood casings, window stools, aprons, stops, etc., are nailed directly to the wood frame.

Buildings having walls 10", 12", 16" or more in thickness will require a wood jamb lining to extend from the inside of the wood window frame to the face of the finished wall. It will also require wood blocking and grounds to nail the jamb linings, window stools, aprons, etc.

The style of trim will affect labor costs considerably and for that reason costs are given on 4 distinct types of interior casings, as follows:

Style "A" consists of 1-member trim with either a mitered or square cut head or "cap". This is the simplest trim it is possible to use.

Style "B" consists of 1-member trim for sides and a built-up cap trim for window heads. Caps may be put together in the mill or assembled and put together on the job.

Style "C" consists of 2-member back-band trim, consisting of a one-piece casing with a back-band of thicker material.

Style "D" consists of any type of window trim that are assembled and glued together in the mill and set in the building as a unit.

The class of workmanship will also have considerable bearing on the labor costs, as in ordinary structures the casings are nailed to the frames or jamb linings and the trim may show some hammer marks but where First Grade Workmanship is required, the utmost care must be used in cutting and fitting all casings, miters, etc. All nails must be carefully driven and set, casings must show the same margin on all sides and must fit closely to the walls. In other words, it must be a first class job in every respect.

Labor Time Erecting Interior Window Trim

Style Trim	Kind of Trim	No. Windows 8-Hr. Day	Carp. Hrs. per Window	Add for Wood Jamb Lining
"A"	Single Casing	7–8	1.0–1.1	0.6
"B"	Cap Trim*	5–7	1.1–1.6	0.6
"C"	Back-bank Trim	5–6	1.3–1.6	0.6
"D"	Mill Assembled**	8–10	0.8–1.0	0.6
	First Grade Workmanship			
"A"	Single Casing	5–7	1.1–1.6	0.8
"B"	Cap Trim*	4–6	1.3–2.0*	0.8
"C"	Back-band Trim	4–5	1.6–2.0	0.8
"D"	Mill Assembled**	7–9	0.9–1.1	0.8

*For cap trim put together on job.
**Trim assembled and glued up in the mill.

On complicated window trim having recessed jamb linings and window trim that has both a casing and a subcasing and a subcasing or panel under the stools, a carpenter will complete only about 1 window per 8-hr. day. This class of work is exceptional and is found only in high grade work.

Labor Setting Window Frames, Fitting and Hanging Wood Sash and Trimming Interior Window Openings Complete

The following table includes all labor required to set window frame, fit and hang wood sash, place jamb linings, stops, window stool, apron, interior casings, hardware, etc., for one complete single window opening.

Window Frames, Sash and Trim

Style Trim	Kind of Trim	No. Windows 8-Hr. Day	Hrs. per Window Carpenter	Labor	Add for Wood Jamb Lining
"A"	Single Casing	2½–3	3.0	0.3	0.6
"B"	Cap Trim	2½–3	3.0	0.3	0.6
"B"	Cap Trim*	2¼	3.5	0.3	0.6
"C"	Back-band Trim	2¼	3.5	0.3	0.6
"D"	Mill Assembled**	2½–3	3.0	0.3	0.6
	First Grade Workmanship				
"A"	Single Casing	2	3.5	0.3	0.8
"B"	Cap Trim	2	3.5	0.3	0.8

"B"	Cap Trim*	2	41.	0.3	0.8
"C"	Back-band Trim	2	4.4	0.3	0.8
"D"	Mill Assembled**	2¼–2½	3.3	0.3	0.8

*For cap trim put together on job.
**Trim assembled and glued up in the mill.

08700   HARDWARE AND SPECIALTIES

08710   FINISH HARDWARE

Finish hardware refers to any item that is usually fitted to a door and frame to perform a specific function, e.g., hinges, latch or locksets, closers, stops, bolts, etc.

Most items of finish hardware can be obtained in various finishes and each finish has a U.S. Code Symbol Designation as follows:

US P   - Prime Paint Coat
US 3   - Polished Brass
US 4   - Satin (dull) Brass
US 9   - Polished Bronze
US 10 - Satin (dull) Bronze
US 10B - Satin Bronze - Oil Rubbed
US 14 - Polished Nickel
US 15 - Satin (dull) Nickel
US 20 - Statuary Bronze
US 26 - Polished Chrome
US 26D - Satin (dull) Chrome
US 28 - Satin Aluminum - Anodized
US 32 - Polished Stainless Steel
US 32A - Satin (dull) Stainless Steel

Keep in mind that the finish will affect two areas of concern for the estimator: (1) cost and (2) delivery time. The US 28 finish on finish hardware is one of the cheapest to buy and readily available from most suppliers, compared to US 32 for example, that will be more expensive and a special order from the manufacturer.

Another item that will affect the cost and the delivery time will be the knob and rose trim-ring specified for latch and locksets. There are numerous styles that are made, however most suppliers only stock the most popular 3 or 4 patterns.

The specification writer will refer to a specific Federal Specification Series Designation when specifying the type of lockset required for the project, and each of these series relates to the construction and design function of the lockset. For example:

ANSI* Series 1000 - Mortise type latchset or lockset and keyed deadbolt housed in one casing.

ANSI* Series 2000 - Mortise type latchset or lockset with integral deadbolt operated by the key in the knob.

ANSI* Series 4000 - Cylindrical latch and locksets
Grade #1 - Heavy Duty
Grade #2 - Medium Duty
Grade #3 - Light Duty

*American National Standards Institute

The following listing of finish hardware material prices should be used as a guide only; always obtain a quotation from a material supplier for the specific type, function, finish, and style of finish hardware specified for the project.

Material Prices For Finish Hardware

Item	Price Range
Cylindrical Locksets	
Std. Duty	$ 15.00–$ 45.00
Heavy Duty	50.00– 80.00
Mortise Locksets	
Heavy Duty	55.00– 95.00
Push-Pull Bars	50.00– 150.00
Anti-Panic Device, Single	150.00– 250.00
Hinges (per pair)	
Steel Plain	10.00– 15.00
Steel B.B.	25.00– 50.00
Bronze B.B.	75.00– 100.00
Door Closers	
III	45.00– 75.00
IV	55.00– 80.00
V	65.00– 100.00
Surface Bolt	6.00– 15.00
Kick Plate: Alum	5.00– 15.00
Bronze	15.00– 35.00
Door Stops:	
Floor Bumper	3.50– 4.50
Wall Bumper	7.50– 9.00
Bumper Plus Holder	16.00– 20.00
Door Plunger	15.00– 18.00
Thresholds	
Alum.	4.00– 16.00
Bronze	25.00– 65.00

The following table will give average production time, in carpenter hours, necessary for the installation of finish hardware items:

Items of Finish Hardware	Hours to Install
Cylindrical Locksets (door prepared)	0.5
Mortise Locksets (door prepared)	0.7
Push-Pull Bars	0.4
Anti-Panic Device	1.5
Door Closers	1.0
Surface Bolts	0.3
Flush Bolts	1.0
Door Stops - Wall Type	0.3
Floor Type	0.5
Thresholds	0.5

08730   WEATHERSTRIPPING AND SEALS

The cost of metal weather strips will vary with the type of window, whether double hung or casement, size of opening, thickness of sash, etc., as they all affect the material and labor costs. It will make considerable difference in the labor costs whether the weather strips are installed in an occupied or unoccupied building. Ordinarily a workman must use more care when working in an occupied building and for that reason it will require ¼ to ½ hour more time per opening than when working in new or unoccupied buildings.

The following quantities and costs are based on using men experienced in placing weather strips.

Metal Weather Strips for Double Hung Windows.—When weatherstripping double hung windows, standard specifications require a track strip at least ¾ " wide at the top and bottom of the window, side strips for the lower sash not less than 1/16" wider than the width of the sash runway and side strips for the upper sash not less than 1/16" narrower than the width of the sash runway. Meeting rails to receive two interlocking hook strips or interlocking hook and flat strips.

When estimating the quantities of weather stripping required take the distance around the window, for the sides, head and sill strips plus the width, to take care of the meeting rail between the top and bottom sash. A 4'-0"x6'-0" double hung window will require 20 lin. ft. of weather strips for the sides, head and sill and 4 lin. ft. of interlocking material for the meeting rails.

The lineal foot cost of metal weather strip for double hung windows will average as follows in zinc:

Description	Sash Thickness		
	1⅜ "	1¾ "	2¼ "
Side, head and sill strips	$0.30	$0.35	$0.45
Meeting rail strips	.70	.75	1.15

The weather strip for a 4'-0"x6'-0"x1⅜ " double hung window would cost 20 x .30 plus 4 x .70 or $8.80. Bronze will run about twice as much.

Labor Installing Metal Weather Strips for Double Hung Windows.—When installing metal weather strips on 1⅜ " sash for windows up to 3'-0"x5'-6", an experienced carpenter should weatherstrip 1 window in ¾ to ⅞-hr. or 9 to 10 windows per 8-hr. day, at the following labor cost per window:

	Hours	Rate	Total	Rate	Total
Carpenter	0.9	$....	$....	$16.47	$14.82

When installing metal weather strips on 1¾ " sash for windows up to 4'-0"x7'-0", an experienced carpenter should complete one window in ⅞ to 1-hr. or 8 to 9 windows per 8-hr. day, at the following labor cost per window:

	Hours	Rate	Total	Rate	Total
Carpenter	1	$....	$....	$16.47	$16.47

If installed in an occupied building, add ¼-hr. per window.

Installing Metal Weather Strips for Double Hung Windows Glazed with Plate Glass.—When installing metal weather strips on 1¾ " or 2¼ " sash, glazed with plate glass, for openings up to 4'-0"x7'-0", a carpenter should weatherstrip one window in 1⅛ to 1¼ hrs. or 6 to 7 windows per 8-hr. day, at the following labor cost per window:

	Hours	Rate	Total	Rate	Total
Carpenter	1.25	$....	$....	$16.47	$20.59

If installed in occupied buildings, add ¼ to ½-hr. per window.

Weatherstripping In-Swinging Casement Windows.—Single or double casement windows that swing in require a ¾ " wide track strip on the hinged side, running in a groove in the sash. At the top, on the latch side, and where the sash meet (in the case of double casement windows) interlocking hook and flat strips are generally used. A trough strip and and a hook strip is generally used at the bottom of the sash.

Material for jambs, head and center stiles, in-swinging casement windows costs 40 cts. per lin. ft. Brass trough section and interlocking sillstrip, $1.20 per lin. ft.

When estimating the number of lin. ft. of strip required for single casement windows, measure the entire distance around the window, but price the sill section separately.

For double casement windows, take the distance around the window, plus the height, which allows for the strip where the two sash meet. A 4'-0"x6'-0" double casement window will require 22 lin. ft. of weather strips for jambs, head and center stiles and 4'-0" of sill section.

Labor Weatherstripping In-Swinging Single Casement Windows.—An experienced carpenter should weatherstrip one single casement window in 1 to 1 1/3, hrs. or 6 to 8 windows per 8-hr. day, depending upon the size, at the following labor cost per window:

	Hours	Rate	Total	Rate	Total
Carpenter	1.1	$. . . .	$. . . .	$16.47	$18.12

If installed in an occupied building, add ¼-hr. per window

Labor Weatherstripping In-Swinging Double Casement Windows.—When weatherstripping in-swinging double casement windows, an experienced carpenter should weatherstrip one window in 1 3/4 to 2 hrs. or 4 to 5 windows per 8-hr. day, depending upon the size, at the following labor cost per window:

	Hours	Rate	Total	Rate	Total
Carpenter	2	$. . . .	$. . . .	$16.47	$32.94

If installed in an occupied building, add ¼-hr. per window.

Weatherstripping Out-Swinging Casement Windows.—Out-swinging single casement windows should have ¾" wide track strip on the hinged side, running in a groove in the sash. At the top, on the latch side and where the sash meet (in the case of double casement sash), interlocking hook and flat strips are required. A heavy sill section with interlocking hook strip is generally used at the bottom of the sash. The material for jambs, heads and center stiles will cost about 40 cts. per lin. ft., while the sill section will cost about $1.00 per lin. ft.

Labor Weatherstripping Out-Swinging Single Casement Sash.—An experienced carpenter should place weather strips on one single casement sash in ¾ 1 hr. or 8 to 10 windows per 8-hr. day, at the following labor cost per window:

	Hours	Rate	Total	Rate	Total
Carpenter	1	$. . . .	$. . . .	$16.47	$16.47

If installed in an occupied building, add ¼-hr. per window.

Labor Weatherstripping Out-Swinging Double Casement Sash.—An experienced carpenter should weatherstrip one double casement window in 1 1/4 to 1 1/2 hrs. or 5 to 6 windows per 8-hr. day, at the following labor cost per window:

	Hours	Rate	Total	Rate	Total
Carpenter	1.4	$. . . .	$. . . .	$16.47	$23.06

If installed in an occupied building, add ¼-hr. per window.

Weather Strips for Single Wood Doors.—Consisting of spring bronze strips for jambs and heads and 1½" interlocking brass threshold for sill.

Door Size Up To	Thick	Price	Door Size Up To	Thick	Price
2'-8"x6'-8"	1⅜"	$11.20	2'-8"x6'-8"	1¾"	$12.00
3'-0"x7'-0"	1⅜"	11.90	3'-0"x7'-0"	1¾"	12.80

If aluminum interlocking thresholds are used instead of brass, deduct $1.75 for 2'-8" openings and $2.00 for 3'-0"x7'-0" openings.

If interlocking bronze is used instead of spring bronze, add $7.00 per door to the above prices.

Extruded aluminum and felt or neoprene sponge 11/8" wide and 1/4" thick to be set on stops at heads and jambs runs $1.10 per foot. Extruded aluminum with neoprene gasket 11/4" wide by 25/32" thick to be set on stops runs $2.20 per foot. A similar gasket for mounting on bottom rail of door runs $1.20 per foot.

A complete extruded aluminum channel with integral drip and vinyl insert at still runs $1.00 per foot. 6" wide extruded aluminum thresholds for use with above rail strips runs $2.25 per foot. A complete extruded aluminum threshold 3¾" wide by ¾" thick with integral latch track, interlock and hook stop for mounting on outside of door runs $3.00 per foot complete. A two-piece extruded aluminum with neoprene gasket astragal for surface mounting on double doors runs $2.60 per lineal foot.

A one-piece extruded aluminum and neoprene gasket designed for recessing in meeting rail will run $1.40 per lineal foot. Adjustable two-piece astragals with pile will run $3.50 per foot per each door leaf in extruded aluminum, $4.00 in dull bronze, and $4.80 in polished bronze or dull chrome.

Labor Weatherstripping Single Wood Doors.—An experienced carpenter should weatherstrip one 3'-0"x7'-0"x1¾" door in 1 hour or about 8 doors per 8-hr. day, at the following labor cost per door:

	Hours	Rate	Total	Rate	Total
Carpenter	1.0	$....	$....	$16.47	$16.47

Where interlocking weather strip is used for head and jambs of doors instead of spring bronze, add ½ to ¾-hr. carpenter time to the time given above.

Metal Weatherstripping for French Doors.—The material for weatherstripping double doors or French doors consists of spring bronze strips for jambs and head, spring bronze interlocking strip and threshold for sill. For doors up to 5'-0"x7'-0"-1⅜" thick, the material should cost about $13.00; for a 5'-0"x7'-0"-1¾" door, about $14.00.

Labor Weatherstripping French Doors.—An experienced carpenter should weatherstrip one pair of 1¾" French doors in 2 hrs. or 3 to 4 pair per 8-hr. day, at the following labor cost per pair:

	Hours	Rate	Total	Rate	Total
Carpenter	2.0	$....	$....	$16.47	$32.94

Weatherstripping Double Doors.—Where double exterior doors are used for openings up to 5'-0"x7'-0", using spring bronze weather strip for the head, jambs and between doors, with an interlocking threshold at the bottom of the door, figure 3 to 4 hrs. carpenter time per opening, as follows:

	Hours	Rate	Total	Rate	Total
Carpenter	2.0	$....	$....	$16.47	$32.94

If interlocking weather strip is used for the head, jambs and meeting rail, add 1 hour per opening.

Metal Thresholds

Price per lin. ft. or fraction thereof. Drilled and countersunk, with screws, ready for installation.

Metal	Size	Price per Lin. Ft.	Metal	Size	Price per Lin. Ft.
Brass	1½"x¼"	$ 5.00	Aluminum	1½"x¼"	$1.50
Brass	3½"x¾"	9.00	Aluminum	3½"x¾"	2.80
Brass	4½"x¾"	12.00	Aluminum	4½"x¾"	3.90
Brass	5 "x⅞"	13.50	Aluminum	5 "x⅞"	5.50

All the above thresholds are furnished with interlocking strip.

08800   GLAZING

During the late 1970's, the glass manufacturers changed the process for making certain types of flat glass.

Prior to that time, vertically drawn sheet glass was manufactured in various qualities (AA, A and B) and thicknesses (S.S., D.S., 3/16" and 7/32"). The 3/16" and 7/32" glass was further known as crystal sheet. Plate glass was made by grinding and polishing both surfaces to a level, parallel plane, thereby removing any distortion lines from the glass.

The new process of making glass consists of floating molten glass over large pools of molten tin. Through automated production stages, the floated glass is fire polished to remove all distortion, allowed to cool, and fed onto the tables in a continuous ribbon where it is cut to sheet size for packing into shipping crates. The glass is classified as "float glass" and is made in glazing quality and mirror quality of various thicknesses.

Clear Float Glass - Glazing Quality

Product	Thickness	Wt./Sq. Ft.	Standard Max. Size	Material Cost/Sq. Ft.
Float	3/32"	1.22 lbs.	40"x100"	$1.10
	1/8"	1.62	80"x120"	1.25
	3/16"	2.43	120"x212"	1.35*
Float/Plate	1/4"	3.24	130"x212"	1.50*
	3/8"	4.92	124"x204"	2.50*
	1/2"	6.56	124"x204"	3.75
	3/4"	9.85	124"x204"	5.50

*Note: Add $0.75 per sq. ft. for gray or bronze tint

Float glass is also available in glare and heat reducing tints of gray and bronze, heat absorbing (blue-green), tempered, coated with thin metallic coverings for reflective qualities, laminated, and in insulating glass units consisting of two sheets of glass separated by an internal air space.

Tempered Glass.—Tempered glass is a heat treated annealed glass which has high mechanical strength. It is 4 to 5 times as strong as annealed glass of the same thickness. When it is broken, it disrupts into innumerable small fragments of a cubical shape. In recent years, tempered glass has become much more reasonable in cost and should be given more consideration in general construction. However, the nature of its manufacture does not allow cutting or drilling on the job and all fabricating must be done prior to heat treatment. It is available in clear, tinted and heat absorbing. One-quarter inch thick units are available in sizes to 72"x120". As this glass is all custom ordered, costs should be checked locally, but tempering will add around $1.50 a sq. ft. premium to base glass prices.

Laminated Glass.—Laminated or "safety" glass is composed of two or more lights of glass with a layer or layers of tough, transparent vinyl plastic sandwiched between glass under heat and pressure to form a single unit. Laminated glass will crack but almost always will hold together. It is available in a wide range of thicknesses and sizes from 7/32" thick to 1" thick in sheets up to 80"x120". One-quarter inch laminated glass will cost about $3.90 a square foot, 1/2" will cost about $9.00 per sq. ft.

Insulating Glass.—Glass units made from two sheets of glass with a sealed air space between are used to decrease heat loss through windows. The seal can be either metal or glass, and air space will vary from 3/16" to 1/2" in thickness.

Insulating glass units must be made to order in the exact size to be used, although standard sizes are manufactured.

They require a slightly larger allowance for clearance in setting them into the rabbet than single lights of glass.

Each piece weighs about 2½ times what a single lite of the same thickness would weigh and the glazing cost is about double that of single lites.

The price per square foot varies widely, depending upon kind of glass, size, etc., so it is always advisable to obtain definite prices on the sizes required.

For quick estimates two sheets of 1/8" for 1/2" thick unit will cost $3.50 a square foot.

Two sheets of 1/4" for 1" thick unit will cost about $4.50 per sq. ft.

For tinted glass one side, add 15%.

Spandrel Glass.—Heat-strengthened glass is used for spandrels and other opaque panels of curtain walls. The glass is coated on the back with a ceramic frit to give opacity and color.

It is manufactured in several standard colors, plus black and white, but can be ordered in almost any color desired.

Spandrel glass is factory cut to size (job cutting is not possible due to the heat-strengthening process in manufacturing). It is usually 1/4" in thickness but may be ordered up to 3/4" thick. Sealing of spandrel glass is usually accomplished by the use of liquid polysulfide sealants and neoprene gaskets.

Material prices of 1/4" spandrel glass are about $3.50 per sq. ft. for standard colors, and about $4.00 per sq. ft. for non-standard colors.

Wired Glass.—Wire glass is a protective type glass for use in fire doors and windows, skylights, and other places where breakage is a problem. It is available in 1/4" thicknesses in sheets 60"x144" maximum size in hexagonal, square, diamond and pin stripe patterns, and comes in clear, hammered and many other patterned glass. Patterned wire glass will cost around $3.00 a sq. ft. Clear wire glass will cost $4.00 a sq. ft. in diamond pattern, $6.00 in pin stripe.

Patterned Glass.—Glass can be rolled or figured for decoration and light diffusion. Patterns vary from manufacturer to manufacturer but usually include sandblasted, hammered, ribbed, fluted, pebbled, and rough hammered designs. It is usually available in two thicknesses:

1/8" in 48"x132" maximum size
7/32" in 60"x132" maximum size

Some patterns are illustrated on the accompanying pages. Cost of material are about $1.30 per sq. ft. for 1/8" thicknesses, $1.50 for 7/32".

Mirrors.—Mirrors are manufactured by taking a sheet of "mirror quality" float glass, in 1/8" or 1/4 thickness, and hermetically sealing a silver coating with a film of electrolytic copper plating onto one side of the glass. The coating is then given a protective paint coat to seal out moisture and reduce the potential of damaging the silver coating.

Crossweld Mottled Wired
¼" thick in sheets to 72"x130"
86% light transmission

Hex Mottled Wired
¼" thick in sheets to 72"x130"
86% light transmission

Crossweld Polished Wired
¼" thick in sheets to 72"x98"
86% light transmission

Hex Polished Wired
¼" thick in sheets to 72"x98"
86% light transmission

Hammered
⅛"and 7/32" thick in sheets to 96"x132"
90% light transmission

Surf
⅛" and 7/32" thick in sheets to 96"x132"
90% light transmission

Average material prices for mirrors will be about $2.75 per sq. ft.

Any edge work associated with mirrors will cost extra, as follows: plain polished edges at $0.15 per linear inch, bevelled edges at $0.60 per linear inch.

Estimating The Quantity Of Glass Required.—When measuring and listing the quantities of glass for a project, list the glass by:

1). Type of glass and thickness
2). Size, measure to the next even inch dimension listing the width first, then the height

Crossnet
⅛" and 7/32" thick in sheets to 96"x132"
90% light transmission

Bamboo
7/32" thick in sheets to 96"x132"
90% light transmission

Textured Colonnade
7/32" thick in sheets to 96"x132"
90% light transmission

Colonnade
7/32" thick in sheets to 96"x132"
90% light transmission

Textured Doric
⅛" and 7/32" thick in sheets to 96"x132"
90% light transmission

Doric
⅛" and 7/32" thick in sheets to 96"x132"
90% light transmission

3). Type of frame the glass is placed in
4). Note if required to glaze from the inside or outside of the building. If from the outside, ladders or scaffolding may be necessary.
5). Height from the ground (as in multi-story buildings) for additional labor requirements of distributing the materials.

Be sure not to list any glass for products that will be delivered to the job pre-glazed. It is common practice for sliding door units, aluminum sliding and single or double hung sash, and some wood sash to be glazed at the factory.

Labor Glazing Window Glass.—The labor cost of glazing will vary with the size and kind of glass, and whether set in putty or glazing compound, a Thiokol or Silicone sealant, or with wood or metal stops. Putty is now seldom recommended for glazing except on a very small light under 50 united inches (length plus width).

The estimator should have in his reference library the "Glazing Manual of the Flat Glass Jobbers Association" which reviews in detail recommended glazing procedures.

Labor Setting Window or Plate Glass in Wood Sash,
Using Putty or Glazing Compound

Approximate Size of Glass	Number Lights Set 8-Hr. Day	Glazier Hours per 100 Lights
12"x14"	65	12.5
20"x28"	40	20.0
30"x40"	30	26.7
40"x48"	20	40.0

Labor Setting Window or Plate Glass Using Wood Stops

Approximate Size of Glass	Number Lights Set 8-Hr. Day	Glazier Hours per 100 Lights
12"x14"	40	20.0
20"x28"	20	40.0
30"x40"	15	53.3
40"x48"	12	66.7

Putty Required Setting Glass in Wood Sash.—Window glass set in 1⅜" wood sash requires 1-lb. of putty for each 8 to 8½ lin. ft. of glass or sash rabbet; when set in 1¾" sash it requires 1-lb. of putty to each 7¼ to 7½ lin. ft. of glass or sash rabbet.

The following table gives the quantity of putty required for glazing various size glass :

Sash Thick	Size of Glass Inches				
	12"x14"	14"x20"	20"x28"	30"x36"	40"x48"
13/8"	1/2-lb.	2/3-lb.	1 lb.	13/8-lb.	13/4-lb.
13/4"	3/5-lb.	3/4-lb.	11/8 lb.	11/2-lb.	2 -lbs.

A good grade of wood sash putty will cost about 50 cts. per lb.

Labor Setting Glass in Metal Doors.—When glazing metal doors and enclosures, the stops are usually steel and are fastened to the sash with small screws, and a glazier must remove all the screws from the stops, place the glass, and then replace the screws for each small light of glass. On work of this kind a glazier should set about 15 lights per 8-hr. day, at the following labor cost per 100 lights :

	Hours	Rate	Total	Rate	Total
Glazier	8.0	$. . . .	$. . . .	$15.56	$124.48
Cost per light			. . . .		8.30

Labor Glazing Steel Sash.—The labor cost of glazing steel sash will vary with the size of the glass, amount of putty required per light, and the class of work; i.e., whether sidewall sash, monitor sash, etc., also whether it is necessary to hoist the glass and erect special scaffolding to set it.

The time of year in which the work is performed will also influence the labor costs, as a glazier can set more glass in warm weather than during the winter months. The labor costs given below are based on summer glazing. For winter work add up to 25% to costs given.

Putty Required for Glazing Steel Sash.—Putty used for glazing steel sash is a special mixture, part of which is red or whitelead. One lb. should be sufficient to glaze 5 to 5½ lin. ft. of sash rabbet. For instance, a 16"x20" light would contain 72" or 6'-0" to be puttied.

Elastic glazing compound is displacing the regular steel sash putty to a great extent for use in steel sash and must be used in aluminum sash. This compound costs about 60¢ per lb.

Sufficient putty to glaze 100 lin. ft. of steel sash should cost as follows :

	Pounds	Rate	Total	Rate	Total
Putty	20	$. . . .	$. . . .	$0.50	$10.00
Cost per lin. ft.			. . . .		.10

	Pounds	Rate	Total	Rate	Total
Elastic compound	20	$. . . .	$. . . .	$0.60	$12.00
Cost per lin. ft.			. . . .		.12

Labor Glazing Steel Sash.—On jobs where there are no unusual window heights or other abnormal conditions, a glazier should set the following number of lights of glass in the various types of steel sash per 8-hr. day.

In clerestory, monitor or outside setting (from scaffold or swing stage), it will require one man on the ground for each two men on the scaffold to keep them supplied with glass, putty, etc. This adds approximately 50 percent to the costs as given below.

Add up to 25 percent to the following costs for winter work.

Pivoted Steel or Commercial Projected Steel Windows*

Glass Size Inches	Number Lights per 8-Hr. Day	Glazier Hrs. per 100 Lts.	Lbs. Putty per Light
22"x16"	50	26.7	1.25
32"x16"	40	26.7	1.50
32"x22"	30	32.0	1.67

Housing and Residence Steel Casements**

9"x12"	60	20.0	0.75
16"x12"	60	23.0	1.0
18"x12"	60	23.0	1.0

Architectural Projected Steel Windows**

18"x16"	30	23.0	1.25
20"x16"	30	23.0	1.25
38"x16"	25	40.0	1.75
40"x16"	25	50.0	2.00
46"x16"	25	50.0	2.00
48"x16"	25	50.0	2.25

*Based on setting glass from inside and working from floor.
**Based on setting glass from outside, using glazing compound or steel sash putty.

Labor Setting Plate Glass in Store Fronts.—The labor cost of handling and setting plate glass in store fronts will vary with the size of the glass, the length of the haul, etc.

The average size plate glass used in modern store fronts requires 4 to 6 men to handle and set each lite of glass. No matter how many lites of glass there are in the job, it will be necessary to allow time for all the men going to the job and returning to the warehouse after completion, and time for the truck hauling glass from the warehouse to the job and return.

Unless the job is a considerable distance from the warehouse, an allowance of 1 hour time for each man going to the job and returning to the warehouse should prove sufficient.

It is customary for the truck to remain on the job until the glass is set and return to the warehouse with the men, so that truck time should be figured on this basis.

Number of Men Required for Various Sizes of Plate Glass in Store Fronts

From 100 united inches up to 140 united inches.................................. Two (2) men
From 140 united inches up to 160 united inches................................Three (3) men
From 160 united inches up to 180 united inches..................................Four (4) men
Not over 210 united inches.. Five (5) men
Not over 240 united inches.. Six (6) men

On sizes larger than the above, some thought should be given to a mechanical means of hoisting the glass, in addition to the labor requirement.

It is further required that (1) when glass is set six feet or more over street level, (2) when weather conditions require it, or (3) when insulated glass (such as Thermopane or Twindow) is installed, additional men shall be used on the following basis :

First size enumerated in the schedule .... 1 additional man
Next two sizes enumerated in the schedule.................................... 2 additional men
Next two sizes enumerated in the schedule.................................... 3 additional men

LEADED AND ART GLASS

When estimating leaded or art glass, odd or fractional parts of inches are charged as even inches of the next larger size; for example, a 12¼"x13¼" lite will be charged as 14"x14".

Circles, ovals or irregular patterns will be charged at the rate of a square or rectangle which will contain them.

The following prices are based on leading, using zinc cames not over ⅜" wide :

	Price per Sq. Ft.
Rectangles and squares, Clear sheet glass	$ 3.75
Rectangles and squares, Cathedral glass, white or colored	$4.00 to 15.00
Rectangles or squares, Plate glass	5.25
Diamonds, Clear sheet glass	4.25
Diamonds, Cathedral glass, white or colored	$5.00 to 20.00
Diamonds, Plate glass	6.30

Colored Art Glass Leaded.—Leaded art glass furnished in colors may be had in many designs at $6.00 to $10.00 a square foot, which includes only straight work.

Landscape and water scenes in colored art glass will run from $15.00 to $25.00 a square foot.

Very elaborate designs of art glass in colors with flowers, fruits, etc., may be purchased at from $35.00 up per sq. ft.

The architect usually specifies an allowance of so much per square foot to cover the cost of all art glass in the building.

08900   WINDOW WALLS/CURTAIN WALLS

INSULATED METAL WALL PANELS

Insulated steel, aluminum, and aluminum-steel wall panels were developed to provide the answer to dry wall construction, enabling the builder to enclose large areas in a minimum of time. They are designed to serve as exterior walls and partitions.

One typical wall panel consists of two members pressed together at the side laps to form a structural unit. This type has a strip of felt inserted the full length of the joint to prevent metal-to-metal contact between the two members. These panels are factory filled with borosilicate glass fiber insulation, 2½-lb. density. An end closure is attached to both ends of the panel to insure the proper retention of insulation and to seal the panels. Panels are manufactured in standard widths of 12", 16", 24" or 32" and thicknesses vary from 1 9/16" to 3⅜", depending upon whether surfaces are flat or fluted and thickness of insulation used.

Insulated Metal Wall Panel

Lengths also vary from 6'-0" to 30'-0". Maximum allowable clear span for an all galvanized steel panel, based on a wind load of 20 lbs. per sq. ft., is 14'-0"; for an all aluminum panel, 10'-6" and for an aluminum-galvanized steel panel, 12'-0".

The interlocking tongue and groove between adjacent panels offers a joint that unites the series of units into a continuous wall surface. When using the interlocking tongue and groove connection the panels may be erected either vertically or horizontally in walls and partitions. Also, this double acting tongue and groove joints has three positive bearing surfaces which permit a series of panels to act as a structural unit in resisting loads from either side of a wall. With panels in the horizontal position, the double tongue and groove joint forms a shiplap construction, adding to the weather-resisting properties of the complete wall unit. Where panels are used as semi-permanent constructions enclosing the face of a building to which a future extension is contemplated, details can allow for 100 percent salvaging of the panels.

Steel Wall Panels.—Steel type panels are manufactured of 18 to 26 gauge sheet steel, with either a baked-on coat of shop primer or galvanized finish. The all-steel panel has the advantage of high structural strength and low initial cost. The panels are caulked during erection by buttering a mastic in the female joints prior to putting the panels in place. Welding of panels to supporting steel is the most common method of attachment. The material cost of shop assembled steel wall panels is approximately $2.50 per sq. ft. For two sheets of 26 ga. metal and 1" insulation to $2.75 per sq. ft. for two sheets of 22 ga. If the outer sheet is to have a porcelain enamel finish, add $1.25 to the sq. ft. cost. An asbestos coating will add $0.75 to the sq. ft. cost.

Aluminum Wall Panels.—The extreme light weight of the long length aluminum panels allows easier and faster handling and erecting in the field. A typical aluminum panel, 12'-0" long, 24" wide weighs only 60 lbs. The erection procedure is the same as for steel panels except that direct contact between the aluminum and steel must be prevented. A coating of bituminous paint is recommended for this purpose. Maintenance costs become negligible with an all-aluminum construction. The original mill finish will eventually darken with weathering. Bolting of panels to supporting steel and use of pre-formed sections are the most common methods of attachments. Aluminum wall panels cost about $2.75 per sq. ft.

Acoustically Treated Wall Panel

Aluminum-Steel Wall Panels.—To supply the need for an economical panel which would have an aluminum exterior surface for high resistance to corrosion and weather, but did not require this type surface on the interior, the aluminum-steel panel was developed.

For particular use in conjunction with aluminum-steel panels and all aluminum panels, extrusions have been developed which enable the panel to be placed in position and held through the clamping action of the extrusion at the base of the panel. These extrusions eliminate bolting and accelerate erection. They also provide 100 percent salvageability of the panel walls when placed in buildings which are to be expanded later. At the head of the panels there is usually no extrusion, but rather a toggle bolt is used to fasten at this position. The material cost of aluminum-steel wall panels is about $2.50 per sq. ft., depending on the gauge of metals used.

Labor Erecting Metal Wall Panels.—A normal wall panel job with panel lengths ranging between 6 and 8 ft. should be constructed at the rate of 12 man-hours per 100 sq. ft., which includes handling all work from unloading to actual fastening of the panels in place. A difficult job, involving the use of toggle bolts, aluminum panels, short runs, either long or short panels, will take somewhat longer—up to 20 manhours per 100 sq. ft. A typical crew would consist of one man to caulk and hoist, a man to receive and distribute the panel, a man to place and weld the bottom, one to plumb and weld the top, also a man to do the cutting, attach closures, place flashings, bitumastic, etc. The erection costs for a normal job are about $1.10 per sq. ft. and $2.00 per sq. ft. for a difficult job. On large jobs with no openings, erection costs of $0.55 per sq. ft. can be achieved, with 180–190 sq. ft. erected per man per 8 hour day.

Field Assembled Wall Panels.—Steel, aluminum and combination aluminum-steel wall panels are also available for field assembly—to be assembled and erected simultaneously. Manufactured with a deeply fluted or ribbed exterior surface, they are particularly suited to modern architectural designs.

Panel lengths generally range from 6'-0" to 40'-0"—some are available up to 60'-0". Widths are 12", 18" and 24" and depth varies from 3" to 3½".

Steel panels—prime coat, galvanized or stainless—are available in 18, 20 or 22 gauge. Aluminum panels—mill or grained finish—are available in 14, 16, 18 or 20 B&S gauge. A combination of aluminum exterior face and galvanized steel interior face is also manufactured. Interior faces of panels may be flat or have small V-shaped ribs.

Maximum girt spacings for all-steel panels is 11'-0" for simple spans and 13'-0" for double spans; for all-aluminum or aluminum-steel panels, 7'-6" and 8'-6".

Material and erection costs for field assembled wall panels total practically the same as given on previous pages for shop assembled panels of similar materials.

Movable Fire Partitions.—Movable fire partitions, long needed in industrial plants, are available for installation where permanent fire walls are not desired. One such partition has a 2½ hour fire rating.

These lightweight partitions are formed of two 18 gauge galvanized steel, 1½" deep fluted sheets of the same type as used for exterior face of field assembled wall panels. Four ½" laminations of gypsum board, containing glass fibers and unexpanded vermiculite, are inserted between the sheets and the build-up is held together by bolts.

Depth of the fire partition is 5" and coverage width per section is 24". Length of section varies with ceiling to floor dimension.

Fire partitions are lightweight and easily erected. Portable attachment to structural steel is by bolting—permanent attachment is by welding.

Cost in place of movable fire partitions is approximately $3.25 per sq. ft.

Curtain Wall Construction.—Buildings of all types are now using the curtain wall method of exterior wall construction, in preference to the conventional masonry, for the following reasons :

Curtain walls are lightweight, permitting savings in structural framing and foundations.

Units are thin—varying from 1¼" to 4½"—increasing ratio of usable floor area to total building area.

They provide good insulation with the use of efficient insulating filler materials and smooth reflective exterior and interior surfaces.

The large prefabricated and shop assembled panels, containing window units, spandrels, mullions, etc., are erected rapidly, reducing labor costs and permitting earlier completion and occupancy of building.

At the present time, curtain walls must be designed and fabricated to order to fit a particular job. Each job is custom built and costs vary considerably depending upon design, materials used, size of units, amount of duplication, type of windows, secondary framing requirements, etc.

Materials commonly used for curtain wall panels are painted, galvanized or stainless steel, aluminum and porcelain enameled steel or aluminum. Various combinations of these materials are also used.

Various types of windows may be used including intermediate projected, casement or combination types, awning type, reversible windows and others.

Panel sizes vary considerably, but most economical designs for multiple story structures have a coverage width of about 4'-0" and a story to story height of 10'-0" to 12'-0". For 2 and 3 story structures, full height panels have been used successfully.

The following average curtain wall costs are given for preliminary estimating purposes only. Always obtain firm figures for curtain walls before submitting bid, unless an allowance has been set up for this work.

	Approx. Cost Per Sq. Ft.
Description	
1½" Steel frames and sash, insul. panels	$10.50–14.00
Same in aluminum	13.50–20.00
Same in stainless steel	23.00–30.00
Same in bronze	30.00–40.00

The curtain wall can be priced out after four elements are agreed upon: the framing system, the window system; the spandrel material and the type of glazing.

CHAPTER 14

FINISHES

09100 LATH AND PLASTER

Plastering costs vary widely depending upon the kind of materials used and the class of workmanship performed, and the estimator should consider these items carefully before pricing any plastering.

There are two distinct grades of plastering, and the estimator should possess a fair knowledge of the class of work required and estimate accordingly.

The first grade of plastering and the one commonly used on most types of ordinary buildings have the walls rodded and angles and corners fairly straight and true. Work of this class permits variations and "waves" in the finish up to ⅛" to 3/16". This is the regular commercial grade of plastering and on the following pages is headed "Ordinary Work."

The highest grade of work is found only in high grade residences, hotels, schools, courthouses, federal government buildings, state capitol buildings, public libraries, etc., where all walls and ceilings together with all interior and exterior angles must be absolutely straight, and plumb and not permit variations or "waves" exceeding 1/32" to 1/16". This is termed "First Grade Workmanship."

Estimating Quantities of Lathing and Plastering.—Lathing and plastering are estimated by the square yard, but the method of deducting openings, etc., varies with the class of work and the individual contractor.

Many plastering contractors make no deductions for door or window openings when estimating new work, while others deduct for one-half of all openings over a certain size.

Both lathing and plastering are estimated by the sq. yd. which is obtained by multiplying the girth of the room by the height, plus the ceiling area. Example : Obtain the number of sq. yds. of plastering in a room 12'-0"x18'-0", having 9'-0" ceilings. 18+18+12+12=60 lin. ft. of wall.

$$60 \times 9 = 540 \text{ Sq. Ft. Wall Area}$$
$$12 \times 18 = \underline{216} \text{ Sq. Ft. Ceiling Area}$$
$$756 \text{ Sq. Ft. Area of Walls and Ceiling}$$
$$756 \div 9 \text{ Sq. Ft.} = 84 \text{ Sq. Yds.}$$

Openings should be measured and no deductions should be made for openings less than 2'-0" wide, and then only one-half the area of each opening should be deducted.

The complete contents of large openings should be deducted, and the contractor

Courtesy Wm. A. Duiguid Co.

Machine application of fine stone finish to stucco wall

should figure 1'-6" wide for each jamb by the height, when the return or "reveal" is less than 1'-0".

For all beams or girders projecting below ceiling line, allow 1'-0" in width by the total length as an extra for each internal or external angle, in addition to the actual area.

Corner Beads, Moldings, Etc.—All corner beads, quirks, rule joints, moldings, casings, cornerites, screeds and other lathing trim accessories necessary to complete a plastering job are measured by the lin. ft., from their longest extension with 1'-0" extra added for each stop or miter.

Plaster Cornices.—Plaster cornices should be measured by the total length of the walls. If the cornices are 1'-0" or less in girth, measurement should be made by the lin. ft., stating the girth of each cornice. If the cornices are over 1'-0" in girth, measurement may be made either by the sq. ft. of exposed face or by the lin. ft., stating the girth of each cornice.

Allow one lin. ft. for each internal miter and 3 lin. ft. for each external miter.

Enriched cornices (cast work) should be measured by the lin. ft. for each enrichment.

Arches, Corbels, Brackets, Ceiling Frieze Plates, Etc.—Arches, corbels, brackets, rings, center pieces, pilasters, columns, capitals, bases, rosettes, pendants, and niches are estimated by the piece. Ceiling frieze plates should be estimated by the sq. ft.

Columns.—Plain plastered columns should be measured by the lin. ft. or by the sq. yd., multiplying the girth by the height.

Cement Wainscot.—Cement wainscot should be measured by the sq. ft. or sq. yd.

Circular or Elliptical Work.—Circular or elliptical work on a small radius should be estimated at double the rate of straight work.

GYPSUM LATH

Gypsum lath is manufactured in 3 types—plain, perforated, and insulating. Each type is made of a core of gypsum with a tough, fibrous surfacing material. For the usual surfaces, either plain or perforated gypsum lath is used. Where maximum fire protection is required or desired, perforated gypsum lath should be employed. For insulation against heat and cold, and as a vapor barrier, insulating gypsum lath having an aluminum foil back should be used on the inside face of exterior walls and for top floor ceilings.

Gypsum lath is furnished ⅜" thick and face dimension is approximately 16"x48", 6 laths per bundle. Each bundle of laths contains 32 sq. ft. and weighs about 50 lbs. These sizes fit the usual stud or joist spacings without cutting.

Sizes and Weights of Gypsum Lath

Size and Thickness	Wt. Per Sq. Yd.	Approx. Price 1,000 Sq. Ft.
16"x48"x⅜"	14 Lbs.	$105.00
16"x48"x½"	19 Lbs.	$110.00

Insulating or aluminum foil backed gypsum lath costs about 8 cents per sq. ft. more than plain lath.

Gypsum lath is also made ½"x16"x48" for framing over 16" on center but not exceeding 24" on center ⅜"x24"x job length for exterior wall furring with or without foil backing; and ⅜"x16"x96" plain or perforated.

For use in 2" solid studless partitions, gypsum lath is made ½" thick, 24" wide, and in lengths up to 12 ft. These laths are erected with the long dimension vertical.

Nails Required for Gypsum Lath.—Nails shall be 1⅛" long for ⅜" lath and 1¼" long for ½" lath, 13 gauge with 19/64" flathead and blued. For ⅜" lath on 16" spacing, nails shall be spaced approximately 5" apart using 4 nails per stud and for ½" lath on 24" spacing, nails shall be spaced approximately 4" apart, using 5 nails per framing member. Lath should be butted lightly together. Gaps over ½" wide should be reinforced with metal lath.

With studs spaced 16" on centers it will require 6 to 7 lbs. of nails per 100 sq. yds. of gypsum lath and with studs spaced 24" on centers it will also require 6 to 7 lbs. per 100 sq. yds. of ½" lath. Lath may also be attached by staples, screws or clips.

Labor Placing Gypsum Lath.—A lather should place 90 to 105 sq. yds. of gypsum lath per 8-hr. day on average work, at the following labor cost:

Courtesy Wm. A. Duguid Co.

Mechanically fastening metal lath

Labor Cost of 100 Sq. Yds. of ⅜" Gypsum Lath Applied to Wood
Studs or Joists with Nails

	Hours	Rate	Total	Rate	Total
Lather	8	$. . . .	$. . . .	$16.35	$130.80
Cost per sq. yd.			. . . .		1.31

Mechanical Stapling of Gypsum Lath.—Increased production in placing gypsum lath can be gained where labor conditions permit the use of air-driven staplers for attaching lath to wood studs or joists.

Air-driven stapler heads, using galvanized frozen staples ⅞" long with a ⅜" crown, cost about $150.00 each. Staples cost 80 to 90 cts. per thousand depending upon quantity purchased and are packed 12,000 to the carton. An air compressor able to handle one stapler head will cost about $350.00 and a 2-stapler head capacity machine will cost about $700.00.

For this type of attachment, it is general practice to tack lath into place, using 2 nails per piece for wall lath and 4 nails per piece for ceiling lath, until enough area is covered to permit continuous stapling. Attachment is completed by stapling 4" on center to all wood studs or joists. Approximately 3,500 staples and 1-lb. of nails are required per 100 sq. yds. of gypsum lath.

Labor Cost of 100 Sq. Yds. of ⅜" Gypsum Lath Applied to Wood Studs
or Joists Using Air-Driven Staples

	Hours	Rate	Total	Rate	Total
Lather	5.3	$. . . .	$. . . .	$16.35	$86.66
Cost per sq. yd.			. . . .		.87

Two (2") Inch Solid Gypsum Long Length Lath and Plaster Partition

A studless, non-load-bearing partition consists of ½" thick, 24" wide, long length (up to 12'-0") gypsum lath with V'd tongue and groove edges, held vertically in floor and ceiling runners and plastered to ¾" grounds each side.

Temporary bracing consists of ¾" cold rolled channels horizontally (one channel for heights up to 9'-0" and two channels at the third points for heights up to 12'-0") is required with adequate vertical braces on partitions over 6'-0" long, and every 6'-0" or fraction thereof during initial plastering stages.

A metal ceiling runner is used for attachment at the ceiling and either a 2½" flush metal base or a routed out wood floor runner may be used at the floor to secure the long sheets of gypsum board lath.

On work of this class it will require 20 to 24 hours lather time to set and place floor and ceiling runners and gypsum lath for 100 sq. yds. of partition, after the material has been hoisted and stocked by laborers, which will average about 4 hrs. per 100 sq. yds.

Material Cost of 100 Sq. Yds. 2" Solid Gypsum Long Length Lath and Plaster Partition
(Partition only, not including plastering)

	Rate	Total	Rate	Total
115 lin. ft. ceiling runner...........................	$....	$....	$.24	$ 27.60
115 lin. ft. 2½" flush metal base*..	....	....	.45	51.75
945 sq. ft. ½" long length gypsum lath..	....	....	.11	103.95
50 lin. ft. 3" strip lath and paper backing...................................	....	....	.12	6.00
155 lin. ft. temp. bracing channels**..	....	....	.08	12.40
100 temp. bracing clips...............................	....	....	.04	4.00
Cost per 100 sq. yds....................................		$....		$205.70
Cost per sq. yd.		....		2.06
Cost per sq. ft.		....		.23

*Floor runner may be a special wood or steel runner channel at correspondingly lower material and installation cost in carpentry labor, but will require wood base at additional cost.

**115 lin. ft. of horizontal temporary bracing channels, plus five 8-ft. verticals figured on a minimum of 3 re-uses.

Labor Cost of 100 Sq. Yds. 2" Solid Gypsum Long Length Lath and Plaster Partition
(Partition only, not including plastering)

	Hours	Rate	Total	Rate	Total
Lather..	22	$....	$....	$16.35	$359.70
Labor..	4	....	....	13.77	55.08
Cost per 100 sq. yds......................			$....		$414.78
Cost per sq. yd..................................			....		4.15
Cost per sq. ft.			....		.46

Add for corner beads, cornerites and other accessories required to complete lathing work. Also available are 1" thick gypsum lath either solid or laminated which reduce the requirements for bracing and grounds.

Floating Systems of Attachment of Gypsum Lath.—In order to eliminate the sources of sound transmission in walls and ceilings, the principle of resiliently isolating the interior walls and ceilings from their structural elements (studs and

Laying Out and Placing Floor Runners for Long Length
Gypsum Lath Partitions

joists) has proven extremely effective as a means of confining the sound at its source. It interposes a shock absorber that reduces the energy transmitted to the structural elements of the building. Interior wall and ceiling finish should be completely and thoroughly isolated from all structural members surrounding the source of sound.

Attachment devices for gypsum lath are designed to isolate the lath from framing members. They are made from sheet metal or wire, and generally include field and corner clips for the attachment's to wood, metal furring and masonry.

In addition to reducing sound transmission, the source of plaster cracking is also often eliminated when occasioned by movement of the structural elements which are stronger than the plastered interior. Floating systems of attachment permit a limited movement of the structural members without imposing undue stresses in the lath and plaster interior.

The spacing of framing members should not exceed 16" o.c. for clip attachments. Plain gypsum lath should be used on ceilings, and either plain or perforated lath may be used on walls.

Suspended Gypsum Lath Ceilings.—Suspended gypsum lath ceilings consist of metal hangers, runners and furring channels, which in turn support gypsum lath as a base for plaster, or the adhesive application of acoustical tile.

The hangers, runners and furring channels for ceilings are approximately the same as required for metal lath. Furring channel spacing must not exceed 16". Some fire resistive ratings require 12" spacing. Starter channels along walls should be within 2" of wall. A ½" space should be provided between ends of channels and masonry walls. Each lath is supported transversely under the channels by a long wire clip, with not less than 3 such clips per full length of lath.

Stagger the end joints by starting alternative rows with half lengths of lath, allowing end joints to fall between channels.

Material Cost of 100 Sq. Yds. Suspended Gypsum Lath Ceilings Using
16"x48"x⅜" Gypsum Lath with Main Runners 3'-0" on
Centers and Cross Furring Channels 16" on Centers

	Rate	Total	Rate	Total
77 pcs. 3/16"x3'-0" hangers, 231'-0"	$....	$....	$.03	$ 6.93
364 lin. ft. 11/2" channels	....	....	.24	87.36
700 lin. ft. 3/4" channels	....	....	.16	112.00
945 sq. ft. 16"x48"x3/8" gypsum lath	....	....	.105	99.23
500 Brace-tite field clips	....	....	.08	40.00
340 Bridjoint clips, B-1	....	....	.05	17.00
Cost 100 sq. yds.		$....		$362.52
Cost per sq. yd.			....	3.63

Labor Cost of 100 Sq. Yds. Suspended Gypsum Lath Ceilings Using
16"x48"x⅜" Gypsum Lath with Main Runners 3'-0"on
Centers and Cross Furring Channels 16" on Centers

	Hours	Rate	Total	Rate	Total
Lather	20	$....	$....	$16.35	$327.00
Labor	4	....	....	13.77	55.08
Cost 100 sq. yds.			$....		$382.08
Cost per sq. yd.				....	3.82

The above ceiling has a fire rating of 1 hour. If the job specification calls for a fire resistance rating of two hours, 14 ga. diagonal wire reinforcement must be added below the lath, to obtain a 4 hour rating, 1" 20 ga. galvanized 1" hex wire mesh must be added. All the above ratings are dependent on the type of floor construction above.

Hollow Steel Stud Partitions Using Gypsum Lath as a Plaster Base.—A non-load-bearing partition of trussed steel studs, track and shoes to which gypsum lath is applied to both sides by means of metal clips, nailing or stapling.

The construction is similar to hollow steel stud non-bearing partitions except that gypsum lath is applied to the studs by means of special metal clips and the studs are spaced 16" on centers to receive the 16"x48" gypsum lath.

Material Cost of 100 Sq. Yds. of Hollow Steel Stud Non-Bearing Plastered
Partitions Using Gypsum Lath on Both Sides of Partitions,
2½" Studs 16" on Centers, Ready to Receive Plaster

	Rate	Total	Rate	Total
105 lin. ft. top track	$....	$....	$.24	$ 25.20
105 lin. ft. bottom track	....	....	.24	25.20
76 9'-0" studs, use 750 lin. ft.*	....	....	.32	240.00
1,900 sq. ft. gypsum lath. ⅜"	....	....	.105	199.50

Perforated Gypsum Lath on Steel Studs

	Rate	Total	Rate	Total
1,000 Trus-Lok field clips	....	....	.07	70.00
680 Bridjoint clips	....	....	.05	34.00
Cost 100 sq. yds.		$....		$593.90
Cost per sq. yd.			....	5.94

*Allowance made for opening framing.

Labor Cost of 100 Sq. Yds. of Hollow Steel Stud Non-Bearing Plastered
Partitions Using Gypsum Lath on Both Sides of Partitions,
2½" Studs 16" on Centers, Ready to Receive Plaster

	Hours	Rate	Total	Rate	Total
Lather	28	$....	$....	$16.35	$457.80
Labor	4	....	....	13.77	55.08
Cost 100 sq. yds.			$....		$512.88
Cost per sq. yd.			....		5.13

Add for corner beads, cornerites and other accessories required to complete lathing work.

Gypsum lath is also used for exterior wall furring and column fireproofing. Special framing support systems are manufactured for this. Foil backed lath may be used for exterior walls.

WOOD LATH

Wood lath for plastering is a thing of the past. Gypsum lath and metal lath are commonly used.

The following data is included for that one in a thousand job where wood lath may be used, for patching in remodeling work.

Wood lath should be estimated by the sq. yd. Inasmuch as it is customary to deduct only one-half the area of openings, it is not necessary to add any allowance for waste. When placing wood lath it is customary to allow ⅜" between each lath for a plaster key.

A standard wood lath is 1½" wide and 48" long, as follows:

Size of Lath	No. Req'd. 100 Sq. Yds.
1½"x48"	1,425–1,450

Labor Placing Wood Lath.—A lather should place 1,400 to 1,600 1½"x48" wood lath, (95 to 110 sq. yds.) per 8-hr. day.

METAL LATH AND FURRING

Where metal lath is used as a plaster base in partition work, it may be nailed or stapled to wood or steel studs, nailed to masonry, or wire-tied to steel channels and prefabricated steel studs. It may also be used without studs or back-up when attached to floor and ceiling runners and temporarily braced during application and curing of first coat of plaster. Where used as a plaster base in ceiling work, it may be nailed to wood joists, wire-tied to various metal support systems either in contact with or suspended from the building structure, or wire-tied directly to the bottom chords of steel joists. It may also be attached directly to the underside of reinforced concrete floors or joists by means of attachments previously embedded in the concrete.

In addition to its utility as a plaster base, metal lath is also used to reinforce plaster applied to other surfaces. Prefabricated, right-angled strips of metal lath are attached to masonry or other rigid plaster bases, prior to plastering, at the interior angels formed by intersecting walls and along the junction of walls and ceilings. Flat strips of metal lath are attached to solid plaster bases to reinforce the joints in such bases, or to bridge potential crack lines at the corners of door and window openings.

Metal lath is used in the individual fireproofing of beams and columns. It may be adapted to decorative shapes for cornices, coves, curved walls, convoluted ceilings and various acoustic shapes such as inverted pyramids. In specialty classifications, it is used to form the membrane of hyperbolic paraboloids, or the web section of concrete beams and girders.

Metal Lath Construction.—The costs of metal lathing will vary according to the complexities of the work in which it is used, the size of the rooms, the height above floors, and the method of attachment. In the sections which follow, attempt is made to cover only the type of lathing which is usual and normal in building construction.

Materials

A number of metal and wire laths, and welded wire fabrics are available to suit the various conditions which might be encountered in building construction. In general, there are sheets of slitted and expanded metal, or perforated metal, commonly called metal laths; rolls of netting formed of 19 ga. wire woven into a pattern which provides 2½ strands per inch, commonly called wire laths; and sheets of open mesh laths formed of strands of 16 gauge wire welded at right angles to one another and spaced 2" apart, commonly called welded wire fabric. Laths may be of galvanized metal or painted with a rust-inhibitive paint. Some metal laths and all welded wire fabrics are available with a paper backing. Also, some laths may be obtained with a backing of aluminum foil laminated with paper. The backings sometimes perform useful functions insofar as utility and plastering are concerned, but many such laths are often more difficult to install.

Expanded metal lath can be either flat dimpled (self-furring), or ribbed. The flat expanded type is classed as small mesh, having approximately 1000 openings per square foot. The perforated metal lath, commonly called sheet lath, is fabricated either with furring indentations or with channel-shaped ribs. The openings in sheet lath and in rib lath are classified as small. Although they are somewhat larger than those of the expanded metal laths, they are separated by larger areas of unperforated metal. Stucco mesh has diamond-shaped openings which are approximately 1½"x3". Wire laths are flat when installed and the openings are classified as of medium size. The welded wire fabrics are either flat or, in the crimped, self-furring variety, all the wires laying in one direction are bent approximately ¼" out of a flat plane, at their intersection with every third wire lying in the other direction.

Installation Procedures

Rather rigid industry standards govern the installation of metal laths, particularly with respect to ceiling work. Complete specifications are prepared by associations of the manufacturers, by individual manufacturers, or by sectional committees of the American National Standards Institute, New York City. The building construction estimator will require a brief outline of the critical factors of installation in order to derive a realistic estimate of costs.

Spans of Lath.—In all work except where laths are supported by solid surfaces, and except in the case of studless solid partitions, the weight of metal lath must be suitable to the distance between studs, joists or other supporting members. The laths are attached to these members by nails, wire-ties, staples or clips, at intervals not exceeding 6". The following chart gives maximum spans, or distance between supports for various weights of metal and wire laths and welded wire fabric.

Laps of Lath.—Industry standards require overlapping at ends and sides of sheets of lath. Most sheets are manufactured slightly larger than nominal size and

Type of Lath	Min. Wt. Lath Lbs. per Sq. Yd.	Maximum Allowable Spacing of Supports in Inches				
		Vertical Supports			Horizontal	Supports
		Wood	Metal Solid Partit.	Others	Wood or Concrete	Metal
Flat Expanded	2.5	16	16	12	0	0
Metal Lath	3.4	16	16	16	16	13½
Flat Rib	2.75	16	16	16	16	12
Metal Lath	3.4	19	24	19	19	19
⅜" Rib	3.4	24	..	24	24	24
Metal Lath	4.0	24	..	24	24	24
Sheet Lath*	4.5	24	..	24	24	24
Wire Lath	2.48	16	16	16	13½	13½
V-Stiffened						
Wire Lath	3.3	24	24	24	19	19
Wire Fabric	..	16	0	16	16	16

*Used in studless solid partitions.

the estimator need not allow extra quantities on this account. A few extra ties are required in all laps: one between each supporting member and, where laps of the ends of sheets fall between supports, such laps should be laced with tie wire. These extra ties are considered part of normal lathing work and no extra allowance need be made for them.

Attachment of Lath.—Nail-on attachment of metal lath to horizontal or sloping surfaces (ceilings and soffits) requires the use of No. 11 ga. barbed, galvanized roofing nails with a head diameter of 7/16". The length of such nails must permit a minimum penetration into the wood of 1⅜". In nail-on attachment of metal lath to vertical surfaces (walls and partitions) 4d common nails are permitted. They should penetrate the wood at least ¾" however, and the remainder should be bent over the strands of lath. Also, in vertical work, 1" long roofing nails with a head diameter of 7/16" are permitted. They are driven home without crushing the strands of metal. In direct attachment to masonry, concrete stub nails with ⅜" flat heads are generally used.

Tie-on attachment of metal lath is generally done with 18 ga. galvanized annealed wire, a strand of wire passing through the mesh, around the support member, and back through the mesh with the two ends twisted together approximately three turns. This operation is more difficult for the lather when paper or foilbacked lath is being used, unless adequate openings in the backing have been provided. Therefore, when estimating hand tying of lath having heavy backing, a reduction in labor productivity should be allowed. In certain cases 18 ga. tie wire is not permitted. Where metal lath is attached directly to structural elements of the building, such as concrete slabs or joists or the bottom chord of open web steel joists, wire of heavier gauge, or more than one loop of 18 ga. wire, is required (16 ga. or 2-18 ga. for steel joist work; 14 ga. twisted or 10 ga. bent over for concrete work). The extra

---

*Wire sizes specified are U. S. Steel wire gauge.

difficulty of twisting and cutting the heavier gauge wires has been considered in the productivity rates suggested below.

Special devices for attachment of laths are permitted, where they develop a fastening of strength equal to the accepted standards. Such devices often result in noticeable labor savings but the estimator should not neglect to adjust his material costs as necessary.

Horizontal Support Systems.—Horizontal supports for metal lath fall under the general classifications of contact systems, where the lath is fastened in direct contact with the structural elements of the building, furred systems, or suspended systems. In both of the latter systems, the laths are attached in the usual manner to a framework of pencil rods or channel shapes which have been fastened directly to, or suspended from, structural elements of the building. In a furred system, the channels or pencil rod furring are attached to the structural elements of the building, by means of 16 ga. wire saddle ties—in open web steel joist construction, or by 18 ga. wire saddle ties supported by nails driven in or through wood joists. The forming of saddle ties is considered normal lathing work and the productivity rates suggested below contemplate this type of tying.

In a suspended system, the channels or pencil rods to which the lath is attached, are called cross-furring. These members are saddle-tied with 16 ga. wire to the underside of an intersecting framework of heavier channels, which have been hung from the structural elements of the building by means of wire, rod or flat steel hangers. Wire hangers and rods are saddle-tied around these support channels, which are called main runners, and flats may be welded or bolted to them.

A combination of the furred and suspended systems, is usually found where furring members are installed in contact with concrete joists. The furring members which support the lath are in contact with the bottoms of the joists, but they are supported by transverse channels, or runners, which are suspended between and run parallel to the joists. The furring members are saddle-tied with 16 ga. wire to the runners, and the runners are suspended by means of 10 ga. wire hangers embedded in the slab above. The wire hangers are saddle-tied to the runners.

In contact, furred or suspended systems, the lath must be supported at spans not exceeding those given in the table of maximum spans, above. In furred and suspended systems, all pencil rods and channels are limited as to distance between supports and center to center spacing, and—in suspended systems—the size of each hanger is dependent upon the area of ceiling which it supports. The following tables list the usual elements and arrangements to be found in furred and suspended ceilings.

Hanger Sizes for Suspended Ceilings

Ceiling Area, SF Max.	Hanger Size Min.	Hangers for Supporting up to 25 SF Ceiling	
8	12 ga. wire	6 ga. wire	1 "x3/16" flat
12	10 ga. wire	5 ga. wire	11/4"x 1/8" flat
12½	9 ga. wire	3/16" rod	11/4"x3/16" flat
16	8 ga. wire	7/32" rod	11/2"x 1/8" flat
17½	7 ga. wire	1/4" rod	11/2"x3/16" flat

All wire hangers should be galvanized steel, all flat and rod hangers should be coated with rust-inhibitive paint.

Spans and Spacing for Main Runners in Suspended Ceilings

Min. Size and Type	Max. Span Between Hangers or Supports	Max. CtoC Spacing of Runners
¾" channel, .3 lb.	3'-0"	2'-4"*
¾" channel, .3 lb.	2'-6"	2'-6"

¾" channel, .3 lb.	2'-0"	3'-0"
1½" channel, .475 lb.	5'-0"	2'-0"
1½" channel, .475 lb.	4'-0"	3'-0"
1½" channel, .475 lb.	3'-6"	3'-6"
1½" channel, .475 lb.	3'-0"	4'-0"
2 " channel, .59 lb.	7'-0"	2'-0"
2 " channel, .59 lb.	5'-0"	3'-6"
2 " channel, .59 lb.	4'-6"	4'-0"

All weights of channel are for cold-rolled members.
*This spacing for runner channels supporting furring members against concrete joists, only.

Spans and Spacing for Cross Furring

Min. Size and Type	Max. Span Between Runners or Supports	Max. CtoC Spacing of Cross Furring
⅜" pencil rod	2'-6"	12"
⅜" pencil rod	2'-0"	19"
¾" channel, .3 lb.	4'-0"	16"
¾" channel, .3 lb.	3'-6"	19"
¾" channel, .3 lb.	3'-0"	24"

All weights of channel are for cold-rolled members.

Details of Suspended Ceiling*

In most horizontal work, the laths, framing members, hangers and inserts are installed in the usual manner. The estimator should look for the presence of large ducts which might require additional framing or hangers, not detailed on the drawings. He should allow for hung or tied supports at the perimeter of rooms and not plan to set channels into wall pockets. The estimator's check list of materials should usually contain allowances for lath, furring channels, carrying channels, tie wire, hanger wire, nails, special inserts if any, corner reinforcing, and accessory items, if required, such as expansion bead or casing bead.

Vertical Support Systems.—Metal lath may be used in a number of wall and partition framing systems. A generalized classification would include hollow walls and partitions, solid plaster partitions, vertical furring and wall cladding. Hollow assemblies may have either metal or wood studs. The metal studs are generally used where non-combustible construction is required, where concealed horizontal pipe must be accommodated, or where sound resistive qualities are needed. Although the majority of such studs are rated as non-load-bearing, because they are

---

*Metal Lath Association.

fabricated from light-gauge steel, or rod, tests of completed walls have indicated surprising load carrying capacity and resistance to horizontal impact.

Thin solid plaster partitions may be formed over a core of metal lath alone, or an assembly of metal lath and metal studs. The studless solid partitions are generally 2" thick and may be as much as 10' in height. The solid partitions with ¾" cold-rolled channels may range between 1½" thickness and 8½" height to 2½" thickness and 16' height. Heavier channels may be used for thicker and higher solid partitions.

Vertical, or wall, furring systems may be braced from exterior walls, pilasters, or interior columns, or may be free-standing. Their utility is sometimes decorative, at other times functional. Functional aspects include concealment, fireproofing, acoustic treatment, condensation and temperature control. The lathing installation is essentially that of a partition with studs, lathed on one side only. However, horizontally-placed ¾" channels are required for stiffening steel studs, and above certain heights—dependent upon size of these studs—the framing must be braced from the structure behind.

Exterior and interior wall claddings include overcoating (remodeling work), decorative surfacing, acoustic surfacing, and reinforcement.

In most vertical work, the laths are applied in the usual manner and the estimator need not look for hidden items of extra material and labor. However, a few unusual operations are worthy of note. In erection of studless solid partitions, the laths must be braced temporarily until first, or scratch coat of plaster has set. In erection of sound resistive partitions, various systems for separating the components of the partition may be encountered. These might include resilient clips and pencil rod furring between the metal studs and metal lath, staggered stud arrangements which require double the normal number of studs, or double rows of small studs with short pieces of cross-tie channels between them. Note also that most soundproofing partitions extend up to the structure itself, and are not terminated at the level of a suspended ceiling or the bottoms of joists. In wall cladding work, some of the more dense back-up materials, such as reinforced concrete or old brick, may be difficult to penetrate with nails, thus slowing the progress of the lather. In certain areas of the country, the exterior walls of stuccoed dwellings may be constructed without sheathing. In this style of erection, known as line wire construction, the lathing work sometimes includes installation of 18 ga. wires, stretched taut completely around the building, and fastened securely to the studs at 6" intervals vertically. This operation is generally performed only when woven wire or stucco mesh is the plaster base.

The estimator's check list of materials for vertical work should usually contain allowances for lath, studs if any, four shoes for each prefabricated metal stud (2 top and 2 bottom), ceiling and floor tracks for holding studs and shoes, nails for attachment of either the tracks or bent-over ends (of small studs) to ceiling and floor, nails for attachment of lath to wood or solid back-up, wire for tying lath and shoes to studs, and shoes to tracks, if required, metal base or screeds, if any, and accessory items such as corner, casing or expansion beads, and picture molds, if required. In estimating requirements for studs, allow for double-studding at door frames where necessary and, where splicing is required, allow for an 8" lap and double wire-tying.

Installation—Labor

Nail-On Attachment of Metal Lath.—For either horizontal or vertical work, the lather should apply 90 to 100 sq. yds. of metal lath per 8-hr. day, depending on size of rooms.

(1) Stiffeners placed at mid-height of partitions which are more than 10′ long or 9′ high, not to exceed 6′ on centers vertically.

Partition Framing At Door Openings*

Tie-On Attachment of Metal Lath.—Usually the actual tying of metal lath is not isolated as a separate operation, for the lather must also install the studs or channels to which the lath is tied. However, in contact ceiling work where the lath is sometimes tied directly to open web joists or to hangers previously embedded in concrete, the lather should apply 100 sq. yds. of metal lath per 8-hr. day.

Furred and Suspended Horizontal Systems.—In furred ceiling work, including installation of furring members and lath but not including drilling for and placement of inserts in concrete, the lather should apply 50 sq. yds. per 8-hr. day. In suspended ceiling work, including installation of wire and rod hangers (but not concrete inserts for hangers), main runner channels, cross furring channels, and lath, the lather should apply 40 sq. yds. per 8-hr. day. Where strap hangers are used (bolted construction), the lather should complete 30 sq. yds. per 8-hr day. In drilling concrete and placing hanger inserts, the lather should cover 25 sq. yds. of area per 8-hr. day. In closets, toilet rooms and other small spaces, a lather may complete only 30 sq. yds. of suspended ceiling per 8-hr. day.

Beam Fireproofing.—In furring and lathing around structural beams, the lather should complete 9 lin. ft. of beam per hour.

Vertical Systems Other Than Nail-On Work: Hollow Partitions.—In hollow partition work, not including sound-resistive construction, a lather should install top and bottom tracks, tie shoes on studs, tie studs and shoes to tracks, apply lath, and complete 30 sq. yds. of partition per 8-hr. day. In 100 sq. yds. of such partitions, approximately 16 man-hours are required to install runners and studs and the attachment of two surfaces of lath should require approximately 11 man-hours.

---

*Metal Lath Association.

Another version of the steel stud hollow partition (using a framework of ¾" cold-rolled channels) requires slightly more labor for installation. In this work the lather must erect two walls and tie them together. Productivity to be expected is between 24 and 28 sq. yds. per 8-hr. day. For estimating sound-resistive hollow partitions with prefabricated steel studs, add 5 man-hours per 100 sq. yds. to cover installation of resilient clips and pencil rods. No extra allowance need be made in the case of hollow partitions with double rows of ¾" channels.

Hollow Partitions: Attachments to Suspended Ceilings

Hollow Partitions: Attachments to Floor*

Vertical Systems, Solid Partitions and Vertical Furring.—In erecting solid partitions with ¾" cold rolled channels, a lather should complete 22 sq. yds. per 8-hr. day, depending on spacing of channels and method of attaching channels at floor and ceiling. Studless solid partitions should be erected at the rate 27 sq. yds. per 8-hr. day. Labor savings are realized in the studless system due to elimination of wire tying the lath to channel studs. Vertical furring systems should generally be erected at the same rate as solid partitions with channel studs. However, where wall braces are required, add 4 hrs. per 100 sq. yrds.

Vertical Systems, Wall Cladding.—In overcoating work, and in other direct-applied lath attachments, a lather should nail metal lath to surfaces of wood and concrete block at the same rate as for nail-on work to wood studs and joists. Lathing should proceed at the rate of 160 sq. yds. per 8-hr. day. In attachment to old brick or reinforced concrete, however, a lather should be expected to complete only 125 sq. yds. per 8-hr. day.

Accessories.—Many metal lathing accessories are available for a variety of functional or trimming purposes. Only a few of the more common items are covered herein. The labor rates given below for erection of some of the functional accessories are inclusive of items of work covered as part of the total assembly. The estimator should, therefore, use the highest productivity rate for erection of an entire

*Metal Lath Association.

assembly, when he lists these functional accessories as separate items. For example: installation of metal base sometimes includes layout of partition and attachment of clips to floor—items of work which are also contemplated in erection of some partition types.

Flush Metal Base.—Double for solid partitions—including layout of partition, attachment of clips to floor, cutting and fitting, attachment of side plates: 5 hrs. per 100 lin. ft. of partition.

Details At Ceiling

METAL BASE   COMPOSITION BASE   WOOD BASE

Floor Attachments*

Single for Masonry Trim.—Including drilling for and setting masonry clips, cutting and fitting, attachment of side plates : 3 hrs. per 100 lin. ft.

Single for Wall Furring.—Including layout of furring, attachment of clips to floor, cutting and fitting, attachment of side plates : 4 hrs. per 100 lin. ft.

Flush or Applied Metal Base.—Single for snap-on to metal studs or for screwing to wood grounds—including cutting and fitting, attaching side plates, and attaching clips to studs, but not including wood grounds : 3 hrs. per 100 lin. ft.

Beads and Molding.—Including corner and bullnose beads, picture molds, base screeds, casing bead (except when used for door and window trim), and expansion bead (except when lath and/or framing is severed beneath line of bead):

wired or stapled to lath bases ....2½ hrs. per 100 lin. ft.
nailed or stapled to wood construction...................................2 hrs. per 100 lin. ft.
nailed to brick, tile or block construction ....3 hrs. per 100 lin. ft.
stub-nailed to concrete construction ....3½ hrs. per 100 lin. ft.

If on-the-job mitering is required, add up to ½ hr. per 100 lin. ft. to above rates.

When windows and doors are being trimmed with casing bead, and the operation is separate from the remainder of the lathing work, the lather should complete 1 door or window per hour. When expansion bead is to be installed to a lath base and the base is to be severed beneath the line of bead, add 3 hrs. per 100 lin. ft. for cutting lath.

---

*Metal Lath Association

Cornerite and Stripite.—Placed in interior wall angles and at junctions of walls and ceilings where adjoining constructions are of rigid lath or masonry, and installed along the joints in nonmetallic plaster bases, should be placed at the rate of 500 lin. ft. per 8-hr. day, depending upon whether material is applied in isolated, or repetitious patterns.

Material Cost of 100 Sq. Yds. of 3.4 Lb. Diamond Mesh Metal Lath Applied
to Wood Studs or Joists @ 16" O.C., Ready to Receive Plaster

	Rate	Total	Rate	Total
105 sq. yds. diamond mesh	$. . . .	$. . . .	$1.50	$157.50
9 lbs. 1½" roofing nails	. . . .	. . . .	1.00	9.00
Cost 100 sq. yds.		$. . . .		$166.50
Cost per sq. yd.		. . . .		1.67

Labor Cost of 100 Sq. Yds. of 3.4 Lb. Diamond Mesh Metal Lath Applied
to Wood Studs or Joists @ 16" O.C., Ready to Receive Plaster

	Hours	Rate	Total	Rate	Total
Lather	7.5	$. . . .	$. . . .	$16.35	$122.63
Cost per sq. yd.			. . . .		1.23

Material Cost of 100 Sq. Yds. of 3.4 Lb. ⅜" Rib Metal Lath Applied to
Wood Studs or Joists @ 24" O.C., Ready to Receive Plaster

	Rate	Total	Rate	Total
105 sq. yds. ⅜" rib metal lath	$. . . .	$. . . .	$2.20	$231.00
6 lbs. 1½" roofing nails	. . . .	. . . .	1.00	6.00
Cost 100 sq. yds.		$. . . .		$237.00
Cost per sq. yd.		. . . .		2.37

Labor Cost of 100 Sq. Yds. of 3.4 Lb. ⅜" Rib Metal Lath Applied to
Wood Studs or Joists @ 24" O.C., Ready to Receive Plaster

	Hours	Rate	Total	Rate	Total
Lather	7.0	$. . . .	$. . . .	$16.35	$114.45
Cost per sq. yd.			. . . .		1.15

Material Cost of 100 Sq. Yds. of 3.4 Lb. ⅜" Rib Metal Lath Applied Directly
to Ties Embedded in Concrete or to Steel Joists at 24" on Centers

	Rate	Total	Rate	Total
105 sq. yds. ⅜" rib metal lath	$. . . .	$. . . .	$2.20	$231.00
15 lbs. 18 ga. tie wire	. . . .	. . . .	.90	13.50
Cost 100 sq. yds.		$. . . .		$244.50
Cost per sq. yd.		. . . .		2.45

Labor Cost of 100 Sq. Yds. of 3.4 Lb. ⅜" Rib Metal Lath Applied Directly
to Ties Embedded in Concrete or to Steel Joists at 24" on Centers

	Hours	Rate	Total	Rate	Total
Lather*	9	$. . . .	$. . . .	$16.35	$147.15
Cost per sq. yd.			. . . .		1.47

Material Cost of 100 Sq. Yds. of Furred Ceiling Using 3.4 Lb. Flat Rib
Metal Lath Applied to ⅜" Pencil Rods 19" on Centers
Tied to Structure at 2 Ft. Intervals

	Rate	Total	Rate	Total
105 sq. yds. flat rib metal lath	$. . . .	$. . . .	$2.20	$231.00
660 lin. ft. ⅜" pencil rods	. . . .	. . . .	.075	49.50
6 lbs. 18 ga. tie wire (lath to rods)	. . . .	. . . .	.90	5.40
2 lbs. 16 ga. wire for splicing rods	. . . .	. . . .	.80	1.60
4 lbs. 16 ga. wire for attaching rods	. . . .	. . . .	.80	3.20
Cost 100 sq. yds.		$. . . .		$290.70
Cost per sq. yd.		. . . .		2.91

Labor Cost of 100 Sq. Yds. of Furred Ceiling Using 3.4 Lb. Flat Rib
Metal Lath Applied to ⅜" Pencil Rods 19" on Centers
Tied to Structure at 2 Ft. Intervals

	Hours	Rate	Total	Rate	Total
Lather*	16	$. . . .	$. . . .	$16.35	$261.60
Cost per sq. yd.			. . . .		2.62

*Lather hours listed are for attachment of lath only. For concrete joist construction, add for placing wire in forms before pour: or, if necessary to drill holes in concrete.

Material Cost of 100 Sq. Yds. Suspended Metal Lath Ceiling Using 3.4 Lb.
⅜" Rib Metal Lath With Furring Channels 24" on Centers
Runner Channels 36" on Centers, Hangers 48" on Centers

	Rate	Total	Rate	Total
100-9 ga. x 3'-0" hangers 300'-0" or 19 lbs.	$. . . .	$. . . .	$ .70	$ 13.30
345 lin. ft. 1½" c.r. channels	. . . .	. . . .	.24	82.80
490 lin. ft. ¾" c.r. channels	. . . .	. . . .	.16	78.40
3 lbs. 16 ga. wire	. . . .	. . . .	.80	2.40
105 sq. yds. ⅜" rib metal lath	. . . .	. . . .	2.20	231.00
5 lbs. 18 ga. wire	. . . .	. . . .	.90	4.50
Cost per 100 sq. yds.		$. . . .		$412.40
Cost per sq. yd.		. . . .		4.13

Labor Cost of 100 Sq. Yds. Suspended Metal Lath Ceiling Using 3.4 Lb.
⅜" Rib Metal Lath With Furring Channels 24" on Centers
Runner Channels 36" on Centers, Hangers 48" on Centers

	Hours	Rate	Total	Rate	Total
Lather*	50	$. . . .	$. . . .	$16.35	$817.50
Cost per sq. yd.			. . . .		8.18

*For erection under concrete, add for placing hangers in forms before pour; or, if necessary to drill holes in concrete or tile, and place inserts.

Material Cost of 100 Sq. Yds. Suspended Metal Lath Ceiling Using 3.4 Lb.
Diamond Mesh Metal Lath With Furring Channels 13½" on Centers,
Runner Channels 48" on Centers, Hangers 48" on Centers

	Rate	Total	Rate	Total
85-8 ga. x 3'-0" hangers 255'-0" or 19 lbs.	$. . . .	$. . . .	$0.70	$ 13.30
280 lin. ft. 1½" c.r. channels	. . . .	. . . .	.24	67.20

	Rate	Total	Rate	Total
850 lin. ft. ¾" c.r. channels	....	....	.16	136.00
4 lbs. 16 ga. wire	....	....	.80	3.20
105 sq. yds. diamond mesh	....	....	1.50	157.50
8 lbs. 18 ga. wire	....	....	.90	7.20
Cost per 100 sq. yds.		$....		$384.40
Cost per sq. yd.			....	3.85

*See above for erection under concrete.

Labor Cost of 100 Sq. Yds. Suspended Metal Lath Ceiling Using 3.4 Lb. Diamond Mesh Metal Lath With Furring Channels 13½" on Centers, Runner Channels 48" on Centers, Hangers 48" on Centers

	Hours	Rate	Total	Rate	Total
Lather*	56	$....	$....	16.35	$915.60
Cost per sq. yd.				....	9.16

*See above for erection under concrete.

Material Cost of 100 Lin. Ft. Beam or Girder Fireproofing Using 3.4 Lb. Self-Furring Diamond Mesh Metal Lath Tied to Adjacent Ceiling Lath and Wrapped Downwards Around 12" WF Beam

	Rate	Total	Rate	Total
33 sq. yds. 3.4 lb. self-furring diamond mesh	$....	$....	$1.60	$52.80
2.5 lbs. 18 ga. wire	....	....	.90	2.25
Cost per 100 lin. ft.		$....		$55.05
Cost per lin. ft.			....	.55

Labor Cost of 100 Lin. Ft. Beam or Girder Fireproofing Using 3.4 Lb. Self-Furring Diamond Mesh Metal Lath Tied to Adjacent Ceiling Lath and Wrapped Downwards Around 12" WF Beam

	Hours	Rate	Total	Rate	Total
Lather	12	$....	$....	$16.35	$196.20
Cost per lin. ft.				....	1.96

Material Cost of 100 Sq. Yds. of Hollow Steel Stud Partitions; Lathed Both Sides, Studs 12" on Centers, Using 2.5 Lb. Diamond Mesh Metal Lath

	Rate	Total	Rate	Total
200 lin. ft. snap-in-runner track	$....	$....	$ .24	$ 48.00
Nails or expansion drives-allowance	....	....	25.00	25.00
101—9'-0" 3⅝" galv. steel studs (909 lin. ft.)*	....	....	.35	318.15
210 sq. yds. diamond mesh	....	....	1.20	252.00
17 lbs. 18 ga. wire	....	....	.90	15.30
Cost per 100 sq. yds.		$....		$658.45
Cost per sq. yd.			....	6.59

*Add necessary extras for framing around openings, returns, etc.

Labor Cost of 100 Sq. Yds. of Hollow Steel Stud Partitions; Lathed Both Sides, Studs 12" on Centers, Using 2.5 Lb. Diamond Mesh Metal Lath

	Hours	Rate	Total	Rate	Total
Lather	27	....	....	$16.35	$441.45
Cost per sq. yd.			....		4.42

Material Cost of 100 Sq. Yds. of Sound Insulation Partition Using Prefabricated Steel Studs 24" on Centers, Resilient Clips and Pencil Rods, Lathed Both Sides With 3.4 Lb. ⅜" Rib Metal Lath

	Rate	Total	Rate	Total
200 lin. ft. plain runner track	$....	$....	$ .24	$ 48.00
Nails or expansion drives-allowance	....	....	25.00	25.00
51—9'-0" 3⅝" galv. steel studs (459 lin. ft.)*	....	....	.35	160.65
204 shoes	....	....	.075	15.30
210 sq. yds. ⅜" rib metal lath	....	....	2.20	462.00
11 lbs. 18 ga. wire	....	....	.90	9.90
Add for sound treatment:				
100 lin. ft. cork, 3½" wide x 1" thick	....	....	.50	50.00
1020 resilient clips	....	....	.07	71.40
918 lin. ft. ¼" rod	....	....	.06	55.08
Cost per 100 sq. yds.		$....		$897.33
Cost per sq. yd.			....	8.97

*Add necessary extras for framing around openings, etc.

Labor Cost of 100 Sq. Yds. of Sound Insulation Partition Using Prefabricated Steel Studs 24" on Centers, Resilient Clips and Pencil Rods, Lathed Both Sides With 3.4 Lb. ⅜" Rib Metal Lath

	Hours	Rate	Total	Rate	Total
Lather	27	$....	$....	$16.35	$441.45
Add for sound treatment:					
Lather	5	....	....	16.35	81.75
Cost per 100 sq. yds.					$523.20
Cost per sq. yd.				....	5.23

Material Cost of 100 Sq. Yds. Metal Lath Solid Partitions, Lathed One Side Only, Using Channels 16'" on Centers, Metal Floor and Ceiling Runners and 2.5 Lb. Diamond Mesh Metal Lath

	Rate	Total	Rate	Total
690 lin. ft. ¾" cold rolled channel*	$....	$....	$ .16	$110.40
200 lin. ft. Z-type runner, slotted	....	....	.18	36.00
Nails or screws—allowance	....	....	25.00	25.00
105 sq. yds. diamond mesh	....	....	1.20	126.00
8 lbs. 18 ga. wire	....	....	.90	7.20
Cost per 100 sq. yds.		$....		$304.60
Cost per sq. yd.			....	3.05

Labor Cost of 100 Sq. Yds. Metal Lath Solid Partitions, Lathed One Side Only, Using Channels 16" on Centers, Metal Floor and Ceiling Runners and 2.5 Lb. Diamond Mesh Metal Lath

	Hours	Rate	Total	Rate	Total
Lather	37	$...	$...	$16.35	$604.95
Cost per sq. yd.			....		6.05

Material Cost of 100 Sq. Yds. of Hollow Partitions (Concealment) Using Double Rows of ¾" Channel Studs at 16" on Centers and 3.4 Lb. Diamond Mesh Metal Lath

	Rate	Total	Rate	Total
152 pcs. ¾" studs x 9'-2" 1390 lin. ft. : (tops of studs tied to ceiling furring channels)*	$...	$...	$ .16	$222.40
208 lin. ft. ¾" stiffener	....	....	.16	33.28
152 pcs. ¾" spacer 512 lin. ft. : (based on 2' be-tween rows of studs	....	....	.16	81.92
200 lin. ft. snap-on metal base	....	....	.45	90.00
152 single base clips (furring clip)	....	....	.07	10.64
Nails or expansion drive-allowance	....	....	25.00	25.00
210 sq. yds. diamond mesh	....	....	1.50	315.00
17 lbs. 18 ga. wire	....	....	.90	15.30
Cost per 100 sq. yds.		$....		$793.54
Cost per sq. yd.		....		7.94

*Add necessary extras for framing around opening, returns, etc.

Labor Cost of 100 Sq. Yds. of Hollow Partitions (Concealment) Using Double Rows of ¾" Channel Studs at 16" on Centers and 3.4 Lb. Diamond Mesh Metal Lath

	Hours	Rate	Total	Rate	Total
Lather	32	$...	$...	$16.35	$523.20
Cost per sq. yd.			....		5.23

Material Cost of 100 Sq. Yds. of Vertical Furring Using 2.75 Lb. Flat Rib Metal Lath and ¾" Channels 16" on Centers

	Rate	Total	Rate	Total
76 pcs. ¾" studs x 9'-0" 684 lin. ft.*	$...	$...	$ .16	$109.44
104 lin. ft: ¾" stiffener	....	....	.16	16.64
51 pcs. ¾" braces x 2'-0" 102 lin. ft.	....	....	.16	16.32
200 lin. ft. double-pronged ceiling-floor runner	....	....	.24	48.00
Nails or expansion drives-allowance	....	....	25.00	25.00
105 sq. yds. flat rib metal lath	....	....	1.50	157.50

8 lbs. 18 ga. wire	....	....	.90	7.20
Cost per 100 sq. yds.		$....		$380.10
Cost per sq. yd.			....	3.80

*Add necessary extras for framing around opening, for soffits, returns, etc.

Labor Cost of 100 Sq. Yds. of Vertical Furring Using 2.75 Lb. Flat Rib Metal Lath and ¾" Channels 16" on Centers

	Hours	Rate	Total	Rate	Total
Lather*	22	$....	$....	$16.35	$359.70
Cost per sq. yd.				....	3.60

Material Cost of 100 Lin. Ft. of Metal Corner Bead Attached to Lath Bases

	Rate	Total	Rate	Total
100 lin. ft. corner bead	$....	$....	$.15	$15.00
Staples or tie wire	....	....		1.00
Cost per 100 lin. ft.		$....		$16.00
Cost per lin. ft.		....		.16

Labor Cost of 100 Lin. Ft. of Metal Corner Bead Attached to Lath Bases

	Hours	Rate	Total	Rate	Total
Lather	2.5	$....	$....	$16.35	$40.88
Cost per lin. ft.				....	.41

In estimating for other beads, molding or screeds, substitute proper material price, and for other bases consult discussion of labor on beads and molding for necessary additions to hours required. Also, if beads or molding are to serve as plaster grounds and all plumbing or straightening work is assigned to the installation of the accessory, add ½ hrs. per 100 lin. ft.

ESTIMATING PLASTER QUANTITIES

The following pages contain covering capacities and quantities of the different kinds of plastering materials required and are based on manufacturer's* specifications.

Gypsum plasters are sold by the ton and are usually packed in 100 lb. sacks. When sand is mixed with plaster in the proportion of 1 to 1, 1 to 2, or 1 to 3, it means 100, 200, or 300 lbs. of sand respectively are to be added to each 100 lb. sack of plaster.

Certain fire-resistant ratings may require an increase in the ratio of plaster to aggregate (such as 1 :1; 1 :2 in lieu of 1 :2, 1 :3) or an increase in the thickness of the plaster over normal grounds, or both, and the estimator must take this into consideration in the use of the following data.

Covering Capacity.—The covering capacity of plaster necessarily depends upon conditions under which the plaster is applied—upon the mechanic, thickness of grounds, kind and trueness of the surface to be plastered, type of lath, quality of sand, etc.

The covering capacity given in the following paragraphs for each type of material and for the various surfaces is based on average conditions, and may be accepted as a safe basis for calculating the quantity of each class of material required when applied according to the manufacturer's specifications.

---

*U.S. Gypsum Co.

Hollow Steel Stud Non-Bearing Partition

Thickness of grounds will have considerable bearing on the covering capacity of plaster and the quantities given are based on the following: ¾" grounds for metal lath, or about ⅝" of plaster over the face of the lath; ⅞" grounds for ⅜" gypsum lath or ½" of plaster; 1" for ½" insulating board lath or ½" of plaster and ⅝" for unit masonry.

Number of 100 Lb. Sacks of Gypsum Cement Plaster Required Per 100 Sq. Yds.

Metal Lath	Kind of Plastering Surface Gypsum Lath	Unit Masonry
18 to 20	9 to 11	10 to 12
(Sanded	(Sanded	(Sanded
1 :2;1: 3)	1 :2;1 :3)	1 :3)

For each ⅛" in Thickness, Proportions 1 :3; Add or Deduct Sacks

| 2.5 | 2.5 | 2.5 |

For insulation board lath, use same plaster quantities as given for gypsum lath.
Standard specifications specify that the first (scratch) coat on all types of lath shall be mixed in the proportions of 1 part gypsum neat plaster to not more than 2 parts of sand by weight (or 100 lbs. of gypsum neat plaster to 2 cu. ft. of perlite or vermiculite aggregate).

First coat on masonry surfaces (except monolithic concrete) and second (brown) coat on all lath bases shall be mixed in the proportions of one (1) part gypsum neat plaster to not more than three (3) parts of sand by weight (or 100 lbs. of gypsum neat plaster to 2 cu. ft. of perlite or vermiculite aggregate).

The sacks of gypsum cement plaster shown in the table above per 100 sq. yds. for various bases are based on these proportions when used with sand aggregate.

Perlite and Vermiculite Plaster

Vermiculite is a mineral that expands when heated at about 2000° F. Each ore particle expands to about 12 times its original size. It will not rot, decay, burn or be attacked by vermin or termites.

Perlite is a siliceous volcanic rock mined in western United States. When crushed and quickly heated to above 1500° F., it expands to form lightweight, noncombustible, inert glass-like particles of cellular structure. This material, white in color is about 1/10 the weight of sand.

Perlite or vermiculite plaster aggregate is a lightweight material. Compared with 100 lbs. per cu. ft. for sand aggregate, perlite ranges from 7½ to 15 pounds per cu. ft., and vermiculite 6 to 10 pounds per cu. ft. Plaster using this aggregate will weigh approximately 30 to 45 lbs. per cu. ft. (depending upon proportions used), against 100 lbs. per cu. ft. where sand aggregate is used.

When perlite or vermiculite is expanded each granule has countless numbers of small dead air spaces which, in addition to making the material very light, gives it a high insulating value. It has approximately three and one half times as much insulating value as ordinary plaster.

It is this insulating quality that provides good fire resistance when combined with gypsum plaster. These aggregates are frequently called for when specific fire ratings are required for the protection of steel framing. As such fire ratings usually require greater plaster thicknesses than normal grounds provide, the estimator is cautioned to determine the proper quantities required per 100 sq. yds.

Perlite or vermiculite plaster aggregate is usually packed in multi-wall paper bags containing 3 or 4 cu. ft. Since there may be some compression or packing in shipping or storage, to insure correct proportioning do not use the original package but a cubic foot measuring box.

Machine-applied vermiculite portland cement plaster is used for panel or spandrel walls and exterior columns. The contractor can apply it to exterior columns simultaneously with the wall, since the paper-backed wire fabric used in wall construction is also wrapped around the columns. The protection is also used for interior columns where more than ordinary wear and tear is expected due to the type of occupancy.

The following table gives the approximate quantities of gypsum plaster and perlite or vermiculite for 100 sq. yds. of various mixtures.

Recommended Proportions and Quantities

In the following table giving recommended proportions for gypsum plaster and perlite or vermiculite, the proportions 1-2, means 100 lbs. of gypsum plaster to 2 cu. ft. of perlite or vermiculite; 1-3 means 100 lbs. of gypsum plaster to 3 cu. ft. of perlite or vermiculite.

Type of Construction	Grounds (See Note)	Actual Plaster Thickness	Coat	Recommended Volume Proportions	Required per 100 Sq. Yds.	
					Bags Plaster	Cu. Ft. Aggregate
Metal Lath	3/4"	1/4"	Scratch	1—2	10	20
		7/16"	Brown	1—2	12½	45
Gypsum Lath	7/8"	3/16"	Scratch	1—2	5	10
		1/4"	Brown	1—2	10	20
2" Solid Partition	2"	3/8"	Scratch (One side only)	1—2	15	30
		3/8"	Brown (One side only)	1—2	15	30
		1 1/8"	Brown 2nd Side	1—2	45	90
Unit Masonry	5/8"	3/16"	Scratch	1—3	4	12
		3/8"	Brown	1—3	8	24

Note: Grounds attached to studs, joists or channels—not the lath. 1/16" allowed for finish coat thickness.

Number of 100 Lb. Sacks or Wood Fiber Plaster Required Per 100 Sq. Yds.

Kind of Plastering Surface

Metal Lath	Gypsum Lath	Unit Masonry
32 to 34 unsanded	17 to 18 unsanded	12 to 14 Equal parts of sand by weight to be added

For each ⅛" in Thickness, Add or Deduct Sacks

5	4½	3

For insulation board lath, use same plaster quantities as given for gypsum lath.

For two and three-coat work on all types of lath, gypsum wood fiber plaster may be used without the addition of aggregate.

For two and three-coat work on masonry surfaces and as scratch coat on lath (except monolithic concrete), gypsum wood fiber plaster shall be mixed in the proportions of 1 part of plaster to 1 part of sand, or perlite by weight.

Gypsum Ready-Mix Plasters Made with Lightweight Aggregate

Gypsum ready-mix plasters are generally made with perlite and the regular formula for lath bases is packed in 80-lb. sacks and the masonry formula for application over all masonry bases is packed in 67-lb. sacks.

The following table gives the number of sacks required per 100 sq. yds. applied over various surfaces:

Kind of Plastering Surface

Metal Lath	Gypsum Lath	Unit Masonry
32 to 34*	17 to 19*	24 to 26**

For Each ⅛" in Thickness, Add or Deduct Sacks

5	5	5

*80-lb. sacks.   **67-lb. sacks.

Quantity of Putty Obtained from Various Plastering Materials

The quantity of lime putty obtainable from one ton of lime will vary with the kind of lime used, whether pebble quicklime, pulverized quicklime or hydrated lime, also upon the quality of the lime, as lime produced in different parts of the country will produce varying amounts of lime putty.

Under average conditions the following quantities should be obtained.

Plastering Material	Cu. Ft. of Putty per Ton of Material	Lbs. Req'd for One Cu. Ft. Putty
Pebble quicklime	80	25
Pulverized quicklime	80	25
Hydrated lime	46	43½
Gypsum plaster	37	54
Keene's cement	30	66⅔
Portland cement	18½	109

BASE COAT

Covering Capacity of Lime Plaster.—There has been some confusion in the past regarding the covering capacity of lime plaster due to the lack of definite specifications covering the method of mixing and proportioning the materials.

Lime plaster is usually proportioned by volume, using 1 volume of stiff, well aged lime putty to 2 volumes of dry sand, plus necessary fiber for scratch coat.

For the brown coat, use 1 volume of lime putty to 3 volumes of dry sand, plus fiber as needed.

One cu. ft. of sand is considered 100 lbs. or 2,700 lbs. per cu. yd.

Covering Capacity of One Cu. Yd. of Lime Plaster

Based on same thickness of grounds as given on previous pages.

	Plastering Surface Metal Lath	Unit Masonry ⅝-in. Grounds
Scratch Coat	50-60 sq. yds.	. . . .
Brown Coat	70-80 sq. yds.	60-75 sq. yds.

Bondcrete For Interior Concrete Surfaces.—Used as a bond coat on rough concrete surfaces only. If the concrete surfaces are smooth, they must be roughened by bush-hammering or hacking or a dash coat of portland cement grout, composed of one part portland cement and one and one-half parts of fine sand mixed to a mushy consistency, shall be applied. Use a stiff fiber brush forcibly dashing the mix onto the surface and damp cure for two days and then allow to dry.

Bondcrete adheres to dry concrete surfaces because of its cementitious qualities and because it maintains a practically unchanging volume during the process of setting and hardening.

On properly prepared rough concrete ceilings, apply a coat of bond plaster scratched in thoroughly, doubled backed, and filled out to a true, even surface and left rough, ready to receive the finish coat. Total thickness of the plaster not to exceed ⅜".

On properly prepared rough concrete columns or walls, scratch in a minimum ¼" coat of bond plaster and then follow with a brown coat of gypsum plaster (one part gypsum neat plaster to not more than three parts sand, by weight) troweled into the scratch coat before it sets. Brown coat shall be brought out to ground, straightened to a true surface and left rough ready to receive the finish coat. Total thickness of plaster not to exceed ⅝".

It usually requires 1,700 to 2,200 lbs. of Bondcrete per 100 sq. yds. depending upon the regularity of the concrete surface and thickness applied.

Plaster-Weld*.—Plaster-Weld permanently bonds new gypsum plaster (finish or base coat), or acoustical plaster to most surfaces, but cannot be used over calcimine or other water soluble coatings.

No chipping or roughening of surfaces is required. The surface must be structurally sound and free from dust, dirt, grease, oil or wax.

It may be applied by brush, roller or spray equipment. After allowing Plaster-Weld to become dry to the touch (one hour on porous surfaces and two hours on non-porous surfaces), plaster can be applied either immediately or later. Too much time lag in excess of a week or ten days may require recoating.

Plaster-Weld covers up to 500 sq. ft. per gal. when sprayed on, which gives the most satisfactory results. When applied by brush or roller, coverage is reduced.

When sprayed on the labor cost per 100 sq. yds. should average as follows:

	Hours	Rate	Total	Rate	Total
Labor	2.25	$. . . .	$. . . .	$16.35	$36.79
Cost per sq. yd.			. . . .		.37
Cost per sq. ft.			. . . .		.04

Plaster-Weld costs approximately $18.00 per gal. in 1-gal. cans.

Finishing Lime.—Finishing lime is essential to two common types of job mixed finish coats—the white putty coat trowel finish and the sand float finish.

For the trowel finish coat, the proportion is generally 1 part of gypsum gauging plaster to 2 parts of hydrated lime on a dry weight basis, or 100 lbs. of gypsum gauging plaster to 4½ cu. ft. of lime putty. One ton of gypsum gauging plaster and 2 tons of hydrated finishing lime, or 90 cu. ft. of lime putty will finish approximately 1,200 sq. yds.

While a sand float finish may be mixed with 1 part of lime putty and 2½ to 3 parts of sand, it is generally necessary to add Keene's cement or gypsum gauging plaster to give the finish early hardness and strength. A standard mix for a Keene's cement-lime putty sand float finish is 2 parts of lime putty to 1½ parts of Keene's cement and 4½ parts of sand by volume. This is equivalent to 1 ton of hydrated lime, or 45 cu. ft. of putty, to 1¼ tons of Keene's cement and 4¾ tons of sand and this quantity will cover approximately 1,600 sq. yds.

If type "S" (fully) hydrated limes are used, they must be added to the mix in dry form, as they form immediate plasticity on wetting. If quicklimes are used, they must be slaked and properly aged and added to the mix in the volume indicated above for putty.

Prepared Gypsum Trowel Finishes.—There are a number of prepared gypsum trowel finishes on the market which require only the addition of water to make them ready for use. A few of these are Universal White Trowel Finish, Red Top* Trowel, and Red Top* Imperial.

Under average conditions, if mixed and applied in accordance with standard specifications, one ton will cover 350 to 400 sq. yds. of surface or it requires 500 to 575 lbs. per 100 sq. yds.

Prepared Gypsum Sand Float Finishes.—There are a number of gypsum sand float finishes, such as Red Top* Sand Float Finish.

These finishes are ready for use when water is added and under average conditions one ton is sufficient for 250 to 275 sq. yds. or it requires 725 to 800 lbs., per 100 sq. yds.

---

*Larson Products Corp., Bethesda, Md.
*U.S. Gypsum Co.

Estimating Labor Costs of Plastering

Before it is possible to prepare an accurate estimate on the cost of any plastering job, the quality of work must be considered, as labor costs may vary up to 25 or 30 percent.

The costs given on the following pages cover two grades of workmanship, which are fully described at the beginning of this chapter.

Applying Scratch Coat of Plaster.—A plasterer should apply 140 to 165 sq. yds. of scratch coat per 8-hr. day, where ordinary grade of workmanship is required, and 135 to 155 sq. yds. per 8-hr. day where first grade workmanship is required.

Applying Brown Coat.—Where an ordinary grade of workmanship is required, a plasterer should apply 90 to 110 sq. yds. of brown coat per 8-hr. day.

Where first grade workmanship is required, a plasterer should brown out 70 to 85 sq. yds. per 8-hr. day.

Applying "Scratch—Double-Back" Basecoat.—Over masonry bases and in many markets over gypsum lath base it is common practice to apply a single plaster mix in a "scratch-in" and "double-back" operation to fill out to grounds.

For ordinary work, a plasterer should apply 90 to 110 sq. yds. per 8-hr. day, and for first grade workmanship, a plasterer should apply 70 to 85 sq. yds. per 8-hr. day in this manner.

Applying White Finish or "Putty" Coat.—On an ordinary job of white coating, a plasterer should apply 90 to 110 sq. yds. per 8-hr. day.

Where first grade workmanship is specified, a plasterer will apply 60 to 80 sq. yds. per 8-hr. day.

General Notes Applying to Finish Plaster Coats

All of the tables on the following pages give costs on a finish white coat. If wanted with sand float finish, deduct quantities given in tables for Finish White Coat and add cost of Sand Float Finish as given on the following pages under "Sand Float Finish."

Applying Sand Float Finish.—A plasterer should apply 100 to 120 sq. yds. of simple sand float finish per 8-hr. day, where an ordinary grade of workmanship is required or 70 to 90 sq. yds. when first grade workmanship is required.

Labor Applying 100 Sq. Yds. of Plaster to Various Plaster Bases
Ordinary Workmanship

Description of Work	Sq. Yds. per 8-Hour Day	Plasterer Hours	Tender Hours
Gypsum Mortar on Gypsum lath—½" Plaster—2-Coat Work			
1 :2½ Brown Coat	90–110	8	5
Gypsum Mortar on Unit Masonry—⅝" Plaster—2-Coat Work—1 :3 Brown Coat	80–100	9	6
Lime Mortar on Unit Masonry—⅝" Plaster—2-Coat Work			
Brown Coat 1 :3	90–100	8.5	5.5
Lime Mortar on Metal Lath—¾" Grounds—⅝" Plaster—3-Coat Work			
Scratch Coat 1 :2	150–160	5	5
Brown Coat 1 :3	90–110	8	5
Lime Mortar on Unit Masonry—⅝" Plaster—3-Coat Work			
Scratch Coat 1 :2	150–160	5	5
Brown Coat 1 :3	90–110	8	5

Description of Work	Sq. Yds. per 8-Hour Day	Plasterer Hours	Tender Hours
Gypsum Mortar on Unit Masonry Tile—⅝" Plaster—			
3-Coat Work			
Scratch Coat 1 :3	150–160	5	5
Brown Coat 1 :3	90–110	8	5
Gypsum Mortar on Gypsum Lath—½" Plaster—3-Coat Work			
Scratch Coat 1 :2	150–160	5	5
Brown Coat 1 :3	90–110	8	5
Gypsum Mortar on Metal Lath—¾" Grounds—⅝" Plaster—			
3-Coat Work			
Scratch Coat 1 :2	150–160	5	5
Brown Coat 1 :3	85–100	9	6
Two (2") Inch Solid Plaster Partitions—Metal Lath One			
Side—Back Up—Finished 2 Sides—Gypsum Mortar—¾"			
Grounds—2" Plaster			
Scratch Coat 1 :2	150–160	5	5
Back Up Coat 1 :2	150–160	5	5
2 Brown Coats 1 :2	40– 50	18	12

Labor Applying 100 Sq. Yds. of Plaster to Various Plaster Bases
Ordinary Workmanship

Description of Work	Sq. Yds. per 8-Hour Day	Plasterer Hours	Tender Hours
Hollow Stud Partitions—Lathed 2 Sides—Plastered to ¾"			
Grounds			
Scratch Coat 1 :2—2 Sides	75– 80	10	10
Brown Coat—2 Sides	40– 50	18	12
Finish White Coat—All Classes of Undercoat	90–110	8	4
Sand Finish Coat—All Classes of Undercoat	100–120	7.25	5
First Grade Workmanship—Where First Grade Workmanship is			
Required as described on the previous pages, add to Hours			
Given above			
Brown Coat	. . . .	2.25	1
Finish White Coat	. . . .	3.50	1
Sand Finish	. . . .	2.75	1

Applying Finish Coat of Cement Wainscoting.—Where cement plaster finishes are used in kitchens, bathrooms, laundries, corridors, etc., using Keene's cement, or similar finishes applied over a brown coat of gypsum cement plaster or portland cement, a plasterer should apply 50 to 60 sq. yds. per 8-hr. day.

If the wainscoting is blocked off into squares, 3"x6", 4"x4" or 6"x6" to represent tile, a plasterer should block off 45 to 55 sq. yds. per 8-hr. day.

To apply the finish coat of plaster and block same off into squares to represent tile, a plasterer should complete 25 to 30 sq. yds. per 8-hr. day.

Where first grade workmanship is required, a plasterer should apply 35 to 45 sq. yds. of cement wainscoting per 8-hr. day, and block off 40 to 45 sq. yds. or he will apply cement wainscoting and block off 18 to 22 sq. yds. per 8-hr. day.

ITEMIZED PLASTERING COSTS

The itemized cost given on the following pages are intended to furnish the estimator and contractor detailed information regarding material quantities and labor costs of all kinds of plastering.

All plaster quantities are based on thickness of grounds as given and if the work is "skimped" and the plaster does not fill out the grounds or is applied thinner than specified, then the quantities in the tables will not prove correct. Also, if sand is not added in accordance with manufacturer's specifications, the quantities may be more or less than given in the tables.

Labor hours include time required for building and removing ordinary scaffolding, but will not cover scaffolding for high work in churches, auditoriums, etc.

2-Coat Plastering

Material Cost of 100 Sq. Yds. 2-Coat Gypsum Cement Plaster, White
Finish on Gypsum Lath

Add for gypsum lath.

Ordinary Workmanship—⅞" Grounds—½" Plaster

Brown Coat—1 :2.5	Rate	Total	Rate	Total
1,000 lbs. gypsum plaster	$....	$....	$6.00*	$ 60.00
25 cu. ft. sand	....	....	.45	11.25
Finish White Coat				
340 lbs. fin. hydrated lime	....	....	8.00*	27.20
170 lbs. gauging plaster	....	....	9.00*	15.30
Cost 100 sq. yds.		$....		$113.75
Cost per sq. yd.		....		1.14

*100 lb. bag.

Labor Cost of 100 Sq. Yds. 2-Coat Gypsum Cement Plaster, White
Finish on Gypsum Lath

Add for gypsum lath.

Ordinary Workmanship—⅞" Grounds—½" Plaster

Brown Coat—1 :2.5	Hours	Rate	Total	Rate	Total
Plasterer	8	$....	$....	$16.35	$130.80
Labor	5	....	....	13.77	68.85
Finish White Coat					
Plasterer	8	....	....	16.35	130.80
Labor	4	....	....	13.77	55.08
Cost 100 sq. yds.			$....		$385.53
Cost per sq. yd.			....		3.86

Material Cost of 100 Sq. Yds. 2-Coat Gypsum Cement Plaster, White
Finish on Unit Masonry
Ordinary Workmanship—⅝" Grounds

Brown Coat—1 :3	Rate	Total	Rate	Total
1,100 lbs. gypsum plaster	$....	$....	$6.00	$ 66.00
33 cu. ft. sand	....	....	.45	14.85
Finish White Coat				
340 lbs. fin. hydrated lime	....	....	8.00	27.20
170 lbs. gauging plaster	....	....	9.00	15.30
Cost 100 sq. yds.		$....		$123.35
Cost per sq. yd.		....		1.23

Labor Cost of 100 Sq. Yds. 2-Coat Gypsum Cement Plaster, White
Finish on Unit Masonry
Ordinary Workmanship—⅝" Grounds

Brown Coat—1 :3	Hours	Rate	Total	Rate	Total
Plasterer	9	$....	$....	$16.35	$147.15
Labor	6	....	....	13.77	82.62

Finish White Coat	Hours	Rate	Total	Rate	Total
Plasterer	8	....	....	16.35	130.80
Labor	4	....	....	13.77	55.08
Cost 100 sq. yds.			$....		$415.65
Cost per sq. yd.				....	4.16

Material Cost of 100 Sq. Yds. 2-Coat Wood Fiber Plaster, White Finish on Gypsum Lath

Add for gypsum lath.

Ordinary Workmanship—⅞" Grounds—½" Plaster

Brown Coat	Rate	Total	Rate	Total
1,700 lbs. wood fiber	$....	$....	$6.00	$102.00
Finish White Coat				
340 lbs. fin. hydrated lime	....	....	8.00	27.20
170 lbs. gauging plaster	....	....	9.00	15.30
Cost 100 sq. yds.		$....		$144.50
Cost per sq. yd.			....	1.45

Labor Cost of 100 Sq. Yds. 2-Coat Wood Fiber Plaster, White Finish on Gypsum Lath

Add for gypsum lath.

Ordinary Workmanship—⅞" Grounds—½" Plaster

Brown Coat	Hours	Rate	Total	Rate	Total
Plasterer	9	$....	$....	$16.35	$147.15
Labor	5	....	....	13.77	68.85
Finish White Coat					
Plasterer	8	....	....	16.35	130.80
Labor	4	....	....	13.77	55.08
Cost 100 sq. yds.			$....		$401.88
Cost per sq. yd.				....	4.02

Bondcrete plaster can only be applied to rough concrete surfaces. If the concrete surfaces are smooth, they must be roughened by bush-hammering or hacking or a dash coat of portland cement grout, composed of one part portland cement and one and one-half parts fine sand, mixed to a mushy consistency. Use a stiff fiber brush forcibly dashing the mix onto the surface and damp cure for two days and then allow to dry.

For costs on this work refer to section of "Waterproofing."

Material Cost of 100 Sq. Yds. 2-Coat Bondcrete Plaster, White Finish on Concrete Ceilings

Ordinary Workmanship—Plaster ⅜" Thick

Brown Coat	Rate	Total	Rate	Total
2,000 lbs. Bondcrete	$....	$....	$7.00	$140.00
Finish White Coat				
340 lbs. fin. hydrated lime	....	....	8.00	27.20
170 lbs. gauging plaster	....	....	9.00	15.30
Cost 100 sq. yds.		$....		$182.50
Cost per sq. yd.			....	1.83

Labor Cost of 100 Sq. Yds. 2-Coat Bondcrete Plaster, White Finish on
Concrete Ceilings

Ordinary Workmanship—Plaster ⅜" Thick

Brown Coat	Hours	Rate	Total	Rate	Total
Plasterer	8	$....	$....	$16.35	$130.80
Labor	5	....	....	13.77	68.85
Finish White Coat					
Plasterer	8	....	....	16.35	130.80
Labor	4	....	....	13.77	55.08
Cost 100 sq. yds.			$....		$385.53
Cost per sq. yd.				....	3.86

3-COAT PLASTERING

Material Cost of 100 Sq. Yds. 3-Coat Gypsum Cement Plaster, White
Finish on Gypsum Lath

Add for gypsum lath.

Ordinary Workmanship—⅞" Grounds—½" Plaster

Scratch Coat—1 :2	Rate	Total	Rate	Total
400 lbs. gypsum plaster	$....	$....	$6.00	$24.00
8 cu. ft. sand	....	....	.45	3.60
Brown Coat—1 :3				
700 lbs. gypsum plaster	....	....	6.00	42.00
21 cu. ft. sand	....	....	.45	9.45
Finish White Coat				
340 lbs. fin. hydrated lime	....	....	8.00	27.20
170 lbs. gauging plaster	....	....	9.00	15.30
Cost 100 sq. yds.		$....		$121.55
Cost per sq. yd.			....	1.22

Labor Cost of 100 Sq. Yds. 3-Coat Gypsum Cement Plaster, White
Finish on Gypsum Lath

Add for gypsum lath.

Ordinary Workmanship—⅞" Grounds—½" Plaster

Scratch Coat—1 :2	Hours	Rate	Total	Rate	Total
Plasterer	5	$....	$....	$16.35	$ 81.75
Labor	5	....	....	13.77	68.85
Brown Coat—1 :3					
Plasterer	8	....	....	16.35	130.80
Labor	5	....	....	13.77	68.85
Finish White Coat					
Plasterer	8	....	....	16.35	130.80
Labor	4	....	....	13.77	55.08
Cost 100 sq. yds.			$....		$536.13
Cost per sq. yd.				....	5.36

If first grade workmanship is required, add 5.75 hrs. plasterer time and 2 hrs.
labor time per 100 sq. yds.

Material Cost of 100 Sq. Yds. 3-Coat Gypsum Cement Plaster, White
Finish on Unit Masonry

Ordinary Workmanship—⅝" Grounds

Scratch Coat—1 :3	Rate	Total	Rate	Total
500 lbs. gypsum plaster	$....	$....	$6.00	$ 30.00
15 cu. ft. sand	....	....	.45	6.75

Brown Coat—1 :3	Rate	Total	Rate	Total
700 lbs. gypsum plaster	....	....	6.00	42.00
21 cu. ft. sand	....	....	.45	9.45
Finish White Coat				
340 lbs. fin. hydrated lime	....	....	8.00	27.20
170 lbs. gauging plaster	....	....	9.00	15.30
Cost 100 sq. yds.		$....		$130.70
Cost per sq. yd.			....	1.31

Labor Cost of 100 Sq. Yds. 3-Coat Gypsum Cement Plaster, White Finish on Unit Masonry
Ordinary Workmanship—⅝" Grounds

Scratch Coat—1 :3	Hours	Rate	Total	Rate	Total
Plasterer	5	$....	$....	$16.35	$ 81.75
Labor	5	....	....	13.77	68.85
Brown Coat—1 :3					
Plasterer	8	....	....	16.35	130.80
Labor	5	....	....	13.77	68.85
Finish White Coat					
Plasterer	8	....	....	16.35	130.80
Labor	4	....	....	13.77	55.08
Cost 100 sq. yds.			$....		$536.13
Cost per sq. yd.				....	5.36

If first grade workmanship is required, add 5.75 hrs. plasterer time and 2 hrs. labor time per 100 sq. yds.

Applying Plaster Scratch Coat to Metal Lath Walls and Ceilings

Material Cost of 100 Sq. Yds. 3-Coat Gypsum Cement Plaster, White Finish on Metal Lath Walls and Ceilings

Add for metal lath and furring.

Ordinary Workmanship—¾" Grounds—⅝" Plaster

Scratch Coat—1 :2	Rate	Total	Rate	Total
800 lbs. gypsum plaster	$. . . .	$. . . .	$6.00	$ 48.00
16 cu. ft. sand	. . . .	. . . .	.45	7.20
Brown Coat—1 :3				
1,200 lbs. gypsum plaster	. . . .	. . . .	6.00	72.00
36 cu. ft. sand	. . . .	. . . .	.45	16.20
Finish White Coat				
340 lbs. fin. hydrated lime	. . . .	. . . .	8.00	27.20
170 lbs. gauging plaster	. . . .	. . . .	9.00	15.30
Cost 100 sq. yds.		$. . . .		$185.90
Cost per sq. yd.		. . . .		1.86

Labor Cost of 100 Sq. Yds. 3-Coat Gypsum Cement Plaster, White Finish on Metal Lath Walls and Ceilings

Add for metal lath and furring.

Ordinary Workmanship—¾" Grounds—⅝" Plaster

Scratch Coat—1 :2	Hours	Rate	Total	Rate	Total
Plasterer	5	$. . . .	$. . . .	$16.35	$ 81.75
Labor	5	. . . .	. . . .	13.77	68.85
Brown Coat—1 :3					
Plasterer	9	. . . .	. . . .	16.35	147.15
Labor	6	. . . .	. . . .	13.77	82.62
Finish White Coat					
Plasterer	8	. . . .	. . . .	16.35	130.80
Labor	4	. . . .	. . . .	13.77	55.08
Cost 100 sq. yds.			$. . . .		$566.25
Cost per sq. yd.			. . . .		5.66

If first grade workmanship is required, add 5.75 hrs. plasterer time and 2 hrs. labor time per 100 sq. yds.

Material Cost of 100 Sq. Yds. 2" Solid Plaster Partition on Metal Lath.

Measurement taken on one side only. Add for metal lath and furring

Ordinary Workmanship

Scratch Coat—1 :2	Rate	Total	Rate	Total
2,000 lbs. gypsum plaster	$. . . .	$. . . .	$6.00	$120.00
40 cu. ft. sand	. . . .	. . . .	.45	18.00
Brown Coat—1 :3				
2,800 lbs. gypsum plaster	. . . .	. . . .	6.00	168.00
84 cu. ft. sand	. . . .	. . . .	.45	37.80
Finish White Coat				
680 lbs. fin. hydrated lime	. . . .	. . . .	8.00	54.40
340 lbs. gauging plaster	. . . .	. . . .	9.00	30.60
Cost 100 sq. yds.		$. . . .		$428.80
Cost per sq. yd.		. . . .		4.29

Labor Cost of 100 Sq. Yds. 2" Solid Plaster Partition on Metal Lath.

Measurement taken on one side only. Add for metal lath and furring

Ordinary Workmanship

Scratch Coat—1 :2	Hours	Rate	Total	Rate	Total
Plasterer	10	$. . . .	$. . . .	$16.35	$ 163.50
Labor	10	. . . .	. . . .	13.77	137.70

Brown Coat—1 :3	Hours	Rate	Total	Rate	Total
Plasterer	18	....	....	16.35	294.30
Labor	12	....	....	13.77	165.24
Finish White Coat					
Plasterer	16	....	....	16.35	261.60
Labor	8	....	....	13.77	110.16
Cost 100 sq. yds.			$....		$1,132.50
Cost per sq. yd.				....	11.33

If first grade workmanship is required, add 11.5 hrs. plasterer time and 4 hrs. labor time per 100 sq. yds.

Material Cost of 100 Sq. Yds. 2" Solid Gypsum Long Length Lath and Plaster Partition

Add for gypsum lath.

Ordinary Workmanship
Partition 2" Thick; ½" Gypsum Lath; ¾" of Plaster Each Side of Lath

Scratch Coat—1 :2	Rate	Total	Rate	Total
800 lbs. gypsum plaster	$....	$....	$6.00	$ 48.00
16 cu. ft. sand	....	....	.45	7.20
Brown Coat—1 :3				
2,400 lbs. gypsum plaster	....	....	6.00	144.00
72 cu. ft. sand	....	....	.45	32.40
Finish White Coat				
680 lbs. fin. hydrated lime	....	....	8.00	54.40
340 lbs. gauging plaster	....	....	9.00	30.60
Cost 100 sq. yds.		$....		$316.60
Cost per sq. yd.		....		3.17

Labor Cost of 100 Sq. Yds. 2" Solid Gypsum Long Length Lath and Plaster Partition

Add for gypsum lath.

Ordinary Workmanship
Partition 2" Thick; ½" Gypsum Lath; ¾" of Plaster Each Side of Lath

Scratch Coat—1 :2	Hours	Rate	Total	Rate	Total
Plasterer	10	$....	$....	$16.35	$ 163.50
Labor	10	....	....	13.77	137.70
Brown Coat—1 :3					
Plasterer	16	....	....	16.35	261.60
Labor	10	....	....	13.77	137.70
Finish White Coat					
Plasterer	16	....	....	16.35	261.60
Labor	8	....	....	13.77	110.16
Cost 100 sq. yds.			$....		$1,072.26
Cost per sq. yd.				....	10.72

Material Cost of 100 Sq. Yds. 2" Solid Studless Metal Lath and Plaster Partition

Add for metal lath and furring.

Ordinary Workmanship
Partition 2" Thick; 3.4-lb. Metal Lath; Plaster 2" Thick

Scratch Coat—1 :2	Rate	Total	Rate	Total
2,000 lbs. gypsum plaster	$....	$....	$6.00	$120.00
40 cu. ft. sand	....	....	.45	18.00
Brown Coat—1 :3				
2,800 lbs. gypsum plaster	....	....	6.00	168.00
84 cu. ft. sand	....	....	.45	37.80

Finish White Coat

			Rate	Total
680 lbs. fin. hydrated lime	....	....	8.00	54.40
340 lbs. gauging plaster	....	....	9.00	30.60
Cost 100 sq. yds.		$....		$428.80
Cost per sq. yd.			....	4.29

Labor Cost of 100 Sq. Yds. 2" Solid Studless Metal Lath and Plaster Partition

Add for metal lath and furring.

Ordinary Workmanship
Partition 2" Thick; 3.4-lb. Metal Lath; Plaster 2" Thick

Scratch Coat—1 :2	Hours	Rate	Total	Rate	Total
Plasterer	10	$....	$....	$16.35	$ 163.50
Labor	10	....	....	13.77	137.70
Brown Coat—1 :3					
Plasterer	18	....	....	16.35	294.30
Labor	12	....	....	13.77	165.24
Finish White Coat					
Plasterer	16	....	....	16.35	261.60
Labor	8	....	....	13.77	110.16
Cost 100 sq. yds.			$....		$1,132.50
Cost per sq. yd.				....	11.33

Material Cost of 100 Sq. Yds. 3-Coat Wood Fiber Plaster, White Finish on Metal Lath Walls and Ceilings

Add for metal lath and furring.

Ordinary Workmanship—¾" Grounds—⅝" Plaster

Scratch Coat	Rate	Total	Rate	Total
1,100 lbs. wood fiber	$....	$....	$6.00	$ 66.00
Brown Coat				
2,300 lbs. wood fiber	....	....	6.00	138.00
Finish White Coat				
340 lbs. fin. hydrated lime	....	....	8.00	27.20
170 lbs. gauging plaster	....	....	9.00	15.30
Cost 100 sq. yds.		$....		$246.50
Cost per sq. yd.			....	2.47

Labor Cost of 100 Sq. Yds. 3-Coat Wood Fiber Plaster, White Finish on Metal Lath Walls and Ceilings

Add for metal lath and furring.

Ordinary Workmanship—¾" Grounds—⅝" Plaster

Scratch Coat	Hours	Rate	Total	Rate	Total
Plasterer	5.5	$....	$....	$16.35	$ 89.93
Labor	5.0	....	....	13.77	68.85
Brown Coat					
Plasterer	10.0	....	....	16.35	163.50
Labor	6.0	....	....	13.77	82.62
Finish White Coat					
Plasterer	8.0	....	....	16.35	130.80
Labor	4.0	....	....	13.77	55.08
Cost 100 sq. yds.			$....		$590.78
Cost per sq. yd.				....	5.91

If first grade workmanship is required, add 5.75 hrs. plasterer time and 2 hrs. labor time per 100 sq. yds.

On properly prepared rough concrete columns or walls, scratch in a minimum ¼" coat of bond plaster and then follow with a brown coat of gypsum plaster (one part gypsum neat plaster to not more than three parts sand, by weight) troweled into the scratch coat before it sets. Brown coat shall be brought out to grounds, straightened to a true surface and left rough ready to receive the finish coat. Total thickness of plaster not to exceed ⅝".

Add for preparing concrete surfaces, if required.

Material Cost of 100 Sq. Yds. 3-Coat Bondcrete Plaster, White Finish on
Concrete Walls and Columns
Ordinary Workmanship—Plaster ⅝" Thick

	Rate	Total	Rate	Total
Scratch Coat—¼"				
1,500 lbs. Bondcrete	$. . . .	$. . . .	$7.00	$105.00
Brown Coat—1 :3				
800 lbs. gypsum plaster	. . . .	. . . .	6.00	48.00
24 cu. ft. sand	. . . .	. . . .	.45	10.80
Finish White Coat				
340 lbs. fin. hydrated lime	. . . .	. . . .	8.00	27.20
170 lbs. gauging plaster	. . . .	. . . .	9.00	15.30
Cost 100 sq. yds.		$. . . .		$206.30
Cost per sq. yd.		. . . .		2.06

Labor Cost of 100 Sq. Yds. 3-Coat Bondcrete Plaster, White Finish on
Concrete Walls and Columns
Ordinary Workmanship—Plaster ⅝" Thick

	Hours	Rate	Total	Rate	Total
Scratch Coat—¼"					
Plasterer	5	$. . . .	$. . . .	$16.35	$ 81.75
Labor	5	. . . .	. . . .	13.77	68.85
Brown Coat—1 :3					
Plasterer	8	. . . .	. . . .	16.35	130.80
Labor	5	. . . .	. . . .	13.77	68.85
Finish White Coat					
Plasterer	8	. . . .	. . . .	16.35	130.80
Labor	4	. . . .	. . . .	13.77	55.08
Cost 100 sq. yds.			$. . . .		$536.13
Cost per sq. yd.			. . . .		5.36

If first grade workmanship is required, add 5.75 hrs. plasterer time and 2 hrs. labor time per 100 sq. yds.

Plastering Radiant Panel Heat Ceilings.—Where radiant panel heating coils are embedded in ceiling plaster, base coat plastering costs will be higher than normal for both material and labor.

For this type of heating, ⅜" I.D. (½" O.D.) copper tubing or electric cable is secured to ceiling framing after lath is installed, contacting underside of same. For proper operation of the heating system, all ceiling coils must be completely embedded in plaster, in perfect contact with plaster (no air pockets permitted) and have a plaster cover of ⅜" or more. This means at least ⅞" of plaster below the face of the lath for copper tubing and ½" for electric cable.

Scratch coat application is more difficult, due to the care which must be taken to obtain proper contact with tubing. Brown coat application will require more labor, due to the increased thickness of material required and the difficulty in making it stick to the ceiling. In some cases, a second scratch coat may be necessary to obtain the required thickness.

There should be no change in the method or cost of applying the various plaster finish coats.

Use sand aggregate only for radiant heat ceilings and walls. Lightweight aggregate is an insulator.

Material Cost of 100 Sq. Yds. 3-Coat Gypsum Plaster ⅞" Thick, Applied to Gypsum Lath Ceilings with Radiant Heating Coils Embedded in the Plaster

Add for gypsum lath.

Ordinary Workmanship—⅞" Plaster

	Rate	Total	Rate	Total
Scratch Coat 1 :2				
400 lbs. gypsum plaster	$....	$....	$6.00	$ 24.00
8 cu. ft. sand	....	....	.45	3.60
Brown Coat 1 :3				
1,400 lbs. gypsum plaster	....	....	6.00	84.00
42 cu. ft. sand	....	....	.45	18.90
Finish White Coat				
340 lbs. fin. hydrated lime	....	....	8.00	27.20
170 lbs. gauging plaster	....	....	9.00	15.30
Cost 100 sq. yds.		$....		$173.00
Cost per sq. yd.			....	1.73

Labor Cost of 100 Sq. Yds. 3-Coat Gypsum Plaster ⅞" Thick, Applied to Gypsum Lath Ceilings with Radiant Heating Coils Embedded in the Plaster

Add for gypsum lath.

Ordinary Workmanship—⅞" Plaster

	Hours	Rate	Total	Rate	Total
Scratch Coat 1 :2					
Plasterer	6	$....	$....	$16.35	$ 98.10
Labor	6	....	....	13.77	82.62
Brown Coat 1 :3					
Plasterer	10	....	....	16.35	163.50
Labor	6	....	....	13.77	82.62
Finish White Coat					
Plasterer	8	....	....	16.35	130.80
Labor	4	....	....	13.77	55.08
Cost 100 sq. yds.			$....		$612.72
Cost per sq. yd.				....	6.13

Material Cost of 100 Sq. Yds. of 3-Coat Gypsum Plaster ⅝" Thick, Applied to Metal Lath Ceilings With Radiant Electric Heating Cables Embedded in the Plaster

Add for metal lath.

Ordinary Workmanship—⅝" Plaster

	Rate	Total	Rate	Total
Scratch Coat—1 :2				
800 lbs. gypsum plaster	$....	$....	$6.00	$ 48.00
16 cu. ft. sand	....	....	.45	7.20
Brown Coat—1 :3				
1,200 lbs. gypsum plaster	....	....	6.00	72.00
36 cu. ft. sand	....	....	.45	16.20
Finish White Coat				
340 lbs. fin. hydrated lime	....	....	8.00	27.20
170 lbs. gauging plaster	....	....	9.00	15.30
Cost 100 sq. yds.		$....		$185.90
Cost per sq. yd.			....	1.86

Labor Cost of 100 Sq. Yds. of 3-Coat Gypsum Plaster ⅝" Thick, Applied
to Metal Lath Ceilings With Radiant Electric Heating Cables
Embedded in the Plaster

Add for metal lath.

Ordinary Workmanship—⅝" Plaster

Scratch Coat—1 :2	Hours	Rate	Total	Rate	Total
Plasterer	6	$. . . .	$. . . .	$16.35	$ 98.10
Labor	6	. . . .	. . . .	13.77	82.62
Brown Coat—1 :3					
Plasterer	9	. . . .	. . . .	16.35	147.15
Labor	6	. . . .	. . . .	13.77	82.62
Finish White Coat					
Plasterer	8	. . . .	. . . .	16.35	130.80
Labor	4	. . . .	. . . .	13.77	55.08
Cost 100 sq. yds.			$. . . .		$596.37
Cost per sq. yd.				. . . .	5.96

If it is necessary to apply the plaster in four coats instead of three, add labor for a
second scratch coat, as follows:

	Hours	Rate	Total	Rate	Total
Plasterer	6	$. . . .	$. . . .	$16.35	$ 98.10
Labor	6	. . . .	. . . .	13.77	82.62
Cost 100 sq. yds			$. . . .		$180.72
Cost per sq. yd.				. . . .	1.81

Labor Applying Perlite or Vermiculite Plaster

When using perlite or vermiculite plaster aggregate, a plasterer is applying material weighing 30 to 45 lbs. per cu. ft. against 100 lbs. per cu. ft. for gypsum-sand plaster, so that a plasterer should apply a greater yardage of light aggregate plaster.

Material Cost of 100 Sq. Yds. 3-Coat Plaster Applied to Gypsum
Lath, Using Perlite or Vermiculite Aggregate

Add for gypsum lath.

Ordinary Workmanship—⅞" Grounds—1½" of Plaster

Scratch Coat—1 :2	Rate	Total	Rate	Total
500 lbs. gypsum plaster	$. . . .	$. . . .	$6.00	$ 30.00
10 cu. ft. aggregate	. . . .	. . . .	3.00	30.00
Brown Coat—1 :2				
1,000 lbs. gypsum plaster	. . . .	. . . .	6.00	60.00
20 cu. ft. aggregate	. . . .	. . . .	3.00	60.00
Finish White Coat				
340 lbs. fin. hydrated lime	. . . .	. . . .	8.00	27.20
170 lbs. gauging plaster	. . . .	. . . .	9.00	15.30
Cost 100 sq. yds.		$. . . .		$222.50
Cost per sq. yd.			. . . .	2.23

Labor Cost of 100 Sq. Yds. 3-Coat Plaster Applied to Gypsum
Lath, Using Perlite or Vermiculite Aggregate

Add for gypsum lath.

Ordinary Workmanship—⅞" Grounds—1½" of Plaster

Scratch Coat—1 :2	Hours	Rate	Total	Rate	Total
Plasterer	4.5	$. . . .	$. . . .	$16.35	$ 73.58
Labor	4.5	. . . .	. . . .	13.77	61.97

Courtesy Wm. A. Duguid Co.

Applying finish plaster coat

				Rate	Total
Brown Coat—1 :2					
Plasterer	7.5	. . . .	. . . .	16.35	122.63
Labor	5.0	. . . .	. . . .	13.77	68.85
Finish White Coat					
Plaster	8.0	. . . .	. . . .	16.35	130.80
Labor	4.0	. . . .	. . . .	13.77	55.08
Cost 100 sq. yds.			$. . . .		$512.91
Cost per sq. yd.		. . . .	. . . .		5.13

Material Cost of 100 Sq. Yds. 3-Coat Plaster Applied to Unit Masonry Walls, Using Perlite or Vermiculite Aggregate
Ordinary Workmanship—⅝" Grounds—⅝" of Plaster

	Rate	Total	Rate	Total
Scratch Coat—1 :3				
400 lbs. gypsum plaster	$. . . .	$. . . .	$6.00	$ 24.00
12 cu. ft. aggregate	. . . .	. . . .	3.00	36.00

Brown Coat—1 :3	Rate	Total	Rate	Total
800 lbs. gypsum plaster	....	....	6.00	48.00
24 cu. ft. aggregate	....	....	3.00	72.00
Finish White Coat				
340 lbs. fin. gypsum plaster	....	....	8.00	27.20
170 lbs. gauging plaster	....	....	9.00	15.30
Cost 100 Sq. yds.		$....		$222.50
Cost per sq. yd.		....		2.23

Labor Cost of 100 Sq. Yds. 3-Coat Plaster Applied to Unit Masonry Walls, Using Perlite or Vermiculite Aggregate
Ordinary Workmanship—⅝" Grounds—⅝" of Plaster

Scratch Coat—1 :3	Hours	Rate	Total	Rate	Total
Plasterer	4.5	$....	$....	$16.35	$ 73.58
Labor	4.5	....	....	13.77	61.97
Brown Coat—1 :3					
Plasterer	7.5	....	....	16.35	122.63
Labor	5.0	....	....	13.77	68.85
Finish White Coat					
Plasterer	8.0	....	....	16.35	130.80
Labor	4.0	....	....	13.77	55.08
Cost 100 Sq. yds.			$....		$512.91
Cost per sq. yd.			....		5.13

Material Cost of 100 Sq. Yds. 3-Coat Plaster Applied to Metal Lath, Using Perlite or Vermiculite Aggregate
Add for metal lath and furring.

Ordinary Workmanship—¾" Grounds—¾" Plaster

Scratch—1 :2	Rate	Total	Rate	Total
1,000 lbs. gypsum plaster	$....	$....	$6.00	$ 60.00
20 cu. ft. aggregate	....	....	3.00	60.00
Brown Coat—1 :2				
1,250 lbs. gypsum plaster	....	....	6.00	75.00
40 cu. ft. aggregate	....	....	3.00	120.00
Finish White Coat				
340 lbs. fin. hydrated lime	....	....	8.00	27.20
170 lbs. gauging plaster	....	....	9.00	15.30
Cost per 100 sq. yds.		$....		$357.50
Cost per sq. yd.		....		3.58

Labor Cost of 100 Sq. Yds. 3-Coat Plaster Applied to Metal Lath, Using Perlite or Vermiculite Aggregate
Add for metal lath and furring.

Ordinary Workmanship—¾" Grounds—¾" Plaster

Scratch Coat—1 :2	Hours	Rate	Total	Rate	Total
Plasterer	4.5	$....	$....	$16.35	$ 73.58
Labor	4.5	....	....	13.77	61.97
Brown Coat—1 :2					
Plasterer	8.0	....	....	16.35	130.80
Labor	6.0	....	....	13.77	82.62
Finish White Coat					
Plasterer	8.0	....	....	16.35	130.80
Labor	4.0	....	....	13.77	55.08
Cost per 100 sq. yds.			$....		$534.85
Cost per sq. yd.			....		5.35

Material Cost of 100 Sq. Yds. 2" Solid Plaster Partition Applied to Metal Lath, Using Perlite or Vermiculite Aggregate
(Measurement One Side Only)
Add for metal lath and furring.

Ordinary Workmanship—2" Partition—2" of Plaster

	Rate	Total	Rate	Total
Scratch Coat—1 :2				
1,500 lbs. gypsum plaster	$. . . .	$. . . .	$6.00	$ 90.00
30 cu. ft. aggregate	. . . .	. . . .	3.00	90.00
Brown Coat—1 :2				
6,000 lbs. gypsum plaster	. . . .	. . . .	6.00	360.00
120 cu. ft. aggregate	. . . .	. . . .	3.00	360.00
Finish White Coat				
680 lbs. fin. hydrated lime	. . . .	. . . .	8.00	54.40
340 lbs. gauging plaster	. . . .	. . . .	9.00	30.60
Cost 100 sq. yds.		$. . . .		$985.00
Cost per sq. yd.			. . . .	9.85

Labor Cost of 100 Sq. Yds. 2" Solid Plaster Partition Applied to Metal Lath, Using Perlite or Vermiculite Aggregate
(Measurement One Side Only)
Add for metal lath and furring.

Ordinary Workmanship—2" Partition—2" of Plaster

	Hours	Rate	Total	Rate	Total
Scratch Coat—1 :2					
Plasterer	5	$. . . .	$. . . .	$16.35	$ 81.75
Labor	5	. . . .	. . . .	13.77	68.85
Brown Coat—1 :2					
Plasterer	23	. . . .	. . . .	16.35	376.05
Labor	17	. . . .	. . . .	13.77	234.09
Finish White Coat					
Plasterer	16	. . . .	. . . .	16.35	261.60
Labor	8	. . . .	. . . .	13.77	110.16
Cost 100 sq. yds.			$. . . .		$1,132.50
Cost per sq. yd.				. . . .	11.33

Material Cost of 100 Sq. Yds. 3-Coat Plaster (Scratch-Double-Back Basecoat) Applied to Gypsum Lath Using Pre-Mixed Perlite Plaster
Add for gypsum lath.

Ordinary Workmanship—⅞" Grounds—½" of Plaster

	Rate	Total	Rate	Total
Scratch Coat				
8–80-lb. sacks P. plaster	$. . . .	$. . . .	$6.50	$ 52.00
Double-Back Brown Coat				
11–80-lb. sacks P. plaster	. . . .	. . . .	6.50	71.50
Finish White Coat				
340 lbs. fin. hydrated lime	. . . .	. . . .	8.00	27.20
170 lbs. gauging plaster	. . . .	. . . .	9.00	15.30
Cost 100 sq. yds.		$. . . .		$166.00
Cost per sq. yd.			. . . .	1.66

Labor Cost of 100 Sq. Yds. 3-Coat Plaster (Scratch-Double-Back Basecoat) Applied to Gypsum Lath Using Pre-Mixed Perlite Plaster
Add for gypsum lath.

Ordinary Workmanship—⅞" Grounds—½" of Plaster

	Hours	Rate	Total	Rate	Total
Scratch Coat					
Plasterer	4.5	$. . . .	$. . . .	$16.35	$ 73.58
Labor	4.5	. . . .	. . . .	13.77	61.97

Double-Back Brown Coat	Hours	Rate	Total	Rate	Total
Plasterer	7.5	....	....	16.35	122.63
Labor	5.0	....	....	13.77	68.85
Finish White Coat					
Plasterer	8.0	....	....	16.35	130.80
Labor	4.0	....	....	13.77	55.08
Cost 100 sq. yds.			$....		$512.91
Cost per sq. yd.				....	5.13

Material Cost of 100 Sq. Yds. 3-Coat Plaster (Scratch-Double-Back Basecoat) Applied to Unit Masonry Walls Using Pre-Mixed Perlite Plaster
Ordinary Workmanship—⅝" Grounds—⅝" of Plaster

Scratch Coat	Rate	Total	Rate	Total
10–80-lb. sacks P. Plaster	$....	$....	$6.50	$ 65.00
Double-Back Brown Coat				
16–80-lb. sacks P. Plaster	....	....	6.50	104.00
Finish White Coat				
340 lbs. fin. hydrated lime	....	....	8.00	27.20
170 lbs. gauging plaster	....	....	9.00	15.30
Cost 100 sq. yds.		$....		$211.50
Cost per sq. yd.			....	2.12

Labor Cost of 100 Sq. Yds. 3-Coat Plaster (Scratch-Double-Back Basecoat) Applied to Unit Masonry Walls Using Pre-Mixed Perlite Plaster
Ordinary Workmanship—⅝" Grounds—⅝" of Plaster

Scratch Coat	Hours	Rate	Total	Rate	Total
Plasterer	4.5	$....	$....	$16.35	$ 73.58
Labor	4.5	....	....	13.77	61.97
Double-Back Brown Coat					
Plasterer	7.5	....	....	16.35	122.63
Labor	5.0	....	....	13.77	68.85
Finish White Coat					
Plasterer	8.0	....	....	16.35	130.80
Labor	4.0	....	....	13.77	55.08
Cost 100 sq. yds.			$....		$512.91
Cost per sq. yd.				....	5.13

Material Cost of 100 Sq. Yds. 3-Coat Plaster Applied to Metal Lath Using Pre-Mixed Perlite Plaster
Add for metal lath and furring.
Ordinary Workmanship—¾" Grounds—¾" Plaster

Scratch Coat	Rate	Total	Rate	Total
15–80-lb. sacks P. Plaster	$....	$....	$6.50	$ 97.50
Brown Coat				
18–80-lb. sacks P. Plaster	....	....	6.50	117.00
Finish White Coat				
340 lbs. fin. hydrated lime	....	....	8.00	27.20
170 lbs. gauging plaster	....	....	9.00	15.30
Cost 100 sq. yds.		$....		$257.00
Cost per sq. yd.			....	2.57

Labor Cost of 100 Sq. Yds. 3-Coat Plaster Applied to Metal Lath Using Pre-Mixed Perlite Plaster

Add for metal lath and furring.

Ordinary Workmanship—¾" Grounds—¾" Plaster

Scratch Coat	Hours	Rate	Total	Rate	Total
Plasterer	4.5	$....	$....	$16.35	$ 73.58
Labor	4.5	....	....	13.77	61.97
Brown Coat					
Plasterer	8.0	....	....	16.35	130.80
Labor	6.0	....	....	13.77	82.62
Finish White Coat					
Plasterer	8.0	....	....	16.35	130.80
Labor	4.0	....	....	13.77	55.08
Cost 100 sq. yds.			$....		$534.85
Cost per sq. yd.				....	5.35

Material Cost of 100 Sq. Yds. 2" Solid Plaster Partition Applied to Metal Lath Using Pre-Mixed Perlite Plaster
(Measurement One Side Only)

Add for metal lath and furring.

Ordinary Workmanship—2" Partition—2" of Plaster

	Rate	Total	Rate	Total
15–80-lb. sacks P. Plaster	$....	$....	$6.50	$ 97.50
Brown Coat				
70–80-lb. sacks P. Plaster	....	....	6.50	455.00
Finish White Coat				
680 lbs. fin. hydrated lime	....	....	8.00	54.40
340 lbs. gauging plaster	....	....	9.00	30.60
Cost 100 sq. yds.		$....		$637.50
Cost per sq. yd.			....	6.38

Labor Cost of 100 Sq. Yds. 2" Solid Plaster Partition Applied to Metal Lath Using Pre-Mixed Perlite Plaster
(Measurement One Side Only)

Add for metal lath and furring.

Ordinary Workmanship—2" Partition—2" of Plaster

	Hours	Rate	Total	Rate	Total
Plasterer	5	$....	$....	$16.35	$ 81.75
Labor	5	....	....	13.77	68.85
Brown Coat					
Plasterer	23	....	....	16.35	376.05
Labor	17	....	....	13.77	234.09
Finish White Coat					
Plasterer	16	....	....	16.35	261.60
Labor	8	....	....	13.77	110.16
Cost 100 sq. yds.			$....		$1,132.50
Cost per sq. yd.				....	11.33

Machine Application of Plaster.—Various types of machines are available for the spray application of base coat plaster mortar to walls and ceilings where it is then worked in the normal manner. They vary in size from large machines capable of spraying any plaster mortars, including base coat materials for ordinary plastering and membrane fireproofing, to small machines which handle only the finish coat of acoustical plaster or a sand float finish. Most models have self-contained pump units with gasoline or electric power air compressors. Several small hand gun-hopper units are available for attachment to separate air compressors.

When applying plaster mortars by machine, approximately 20 to 25 percent additional material will be required to provide equal coverage to hawk and trowel application. This is caused by material densification due to machine action and increased material loss from spattering.

Using the machine method, the labor cost of applying the various plaster coats will be considerably less than for the hawk and trowel method, but an allowance must be made for machine cost and maintenance and additional protection and clean-up labor will be required.

Where scratch coat is machine applied over metal lath, an experienced operator can apply 750 to 850 sq. yds. per 8-hr. day. For this area, an additional 4-hrs. plasterer time will be required for touching up and scratching surface.

Material Cost of 100 Sq. Yds. of Machine Applied Scratch Coat on Metal Lath
Using Perlite or Vermiculite Aggregate

Add for metal lath and furring.

	Rate	Total	Rate	Total
1,200 lbs. gypsum plaster	$....	$....	$6.00	$ 72.00
24 cu. ft. aggregate	....	....	3.00	72.00
Machine cost	....	....	5.00	5.00
Cost 100 sq. yds.		$....		$149.00
Cost per sq. yd.		....		1.49

Labor Cost of 100 Sq. Yds. of Machine Applied Scratch Coat on Metal Lath
Using Perlite or Vermiculite Aggregate

Add for metal lath and furring.

	Hours	Rate	Total	Rate	Total
Plasterer	1.5	$....	$....	$16.35	$24.53
Labor	2.0	....	....	13.77	27.54
Cost 100 sq. yds.			$....		$52.07
Cost per sq. yd.			....		.52

Where brown coat is machine applied over scratch coat or gypsum lath, an experienced operator, with another plasterer rodding and leveling, can apply 375 to 425 sq. yds. per 8-hr. day.

Material Cost of 100 Sq. Yds. of Machine Applied Brown Coat,
7/16" Thick over Scratch Coat or Gypsum Lath,
Using Perlite or Vermiculite Aggregate

	Rate	Total	Rate	Total
1,550 lbs. gypsum plaster	$....	$....	$ 6.00	$ 93.00
50 cu. ft. aggregate	....	....	3.00	150.00
Machine cost	....	....	10.00	10.00
Cost 100 sq. yds.		$....		$253.00
Cost per sq. yd.		....		2.53

Labor Cost of 100 Sq. Yds. of Machine Applied Brown Coat,
7/16" Thick over Scratch Coat or Gypsum Lath,
Using Perlite or Vermiculite Aggregate

	Hours	Rate	Total	Rate	Total
Plasterer	4	$....	$....	$16.35	$ 65.40
Labor	4	....	....	13.77	55.08
Cost 100 sq. yds.			$....		$120.48
Cost per sq. yd.			....		1.21

Special Prepared Finishes

There are a number of prepared plaster finishes on the market that are ready for use by the addition of water. These may be applied over a brown coat of gypsum cement plaster.

Smooth trowel finishes such as gypsum gauging-lime putty require a much higher degree of rigidity and strength in the lath and basecoat plaster for good visual performance. When the use of high strength basecoat plasters is impractical, consideration should be given to the elimination of restraint at the perimeter angles or the selection of a more rigid lath, or both. They help to compensate for the lower strength of some basecoat plasters.

Where a special finish is specified, deduct the cost of the finish white coat as given under the ordinary plastering classification and add the cost of the special finish specified.

Material Cost of 100 Sq. Yds. Gypsum Trowel Finish
Ordinary Workmanship

	Rate	Total	Rate	Total
550 lbs. gyp. trowel finish	$. . . .	$. . . .	$15.00	$82.50
Cost per sq. yd.		. . . .		.83

Labor Cost of 100 Sq. Yds. Gypsum Trowel Finish
Ordinary Workmanship

	Hours	Rate	Total	Rate	Total
Plasterer	8	$. . . .	$. . . .	$16.35	$130.80
Labor	4	. . . .	. . . .	13.77	55.08
Cost 100 sq. yds.			$. . . .		$185.88
Cost per sq. yd.			. . . .		1.86

For first grade workmanship, add 3.5 hrs. plasterer time and 1-hr. labor time per 100 sq. yds.

If plaster is wanted with a sand float finish, deduct quantities of plaster and labor given under finish white coat and add the following:

Material Cost of 100 Sq. Yds. Sand Float Finish
Ordinary Workmanship

	Rate	Total	Rate	Total
200 lbs. unfibered plaster	$. . . .	$. . . .	$6.50	$13.00
4 cu. ft. sand	. . . .	. . . .	.45	1.80
Cost per sq. yds.		$. . . .		$14.80
Cost per sq. yd.		. . . .		.15

Labor Cost of 100 Sq. Yds. Sand Float Finish
Ordinary Workmanship

	Hours	Rate	Total	Rate	Total
Plasterer	7.25	$. . . .	$. . . .	$16.35	$118.54
Labor	5.00	. . . .	. . . .	13.77	68.85
Cost per sq. yds.			$. . . .		$187.39
Cost per sq. yd.			. . . .		1.88

If first grade workmanship is required, add 3 hours plasterer time and 1 hour labor time per 100 sq. yds.

Material Cost of 100 Sq. Yds. Keene's Cement Smooth Finish, (Medium
Hard) Applied Over a Brown Coat of Gypsum Cement Plaster
Ordinary Workmanship

	Rate	Total	Rate	Total
530 lbs. Keene's cement	$....	$....	$16.00	$ 84.80
270 lbs. fin. hydrated lime	....	....	8.00	21.60
Cost 100 sq. yds.		$....		$106.40
Cost per sq. yd.		....		1.07

Labor Cost of 100 Sq. Yds. Keene's Cement Smooth Finish, (Medium
Hard) Applied Over a Brown Coat of Gypsum Cement Plaster
Ordinary Workmanship

	Hours	Rate	Total	Rate	Total
Plasterer	14.5	$....	$....	$16.35	$237.08
Labor	5.0	....	....	13.77	68.85
Cost 100 sq. yds.			$....		$305.93
Cost per sq. yd.			....		3.06

If blocked off into squares to represent tile, add about 16 hrs. plasterer time per 100 sq. yds.

If first grade workmanship is required, add 5.5 hrs. plasterer time and 1-hr. labor time per 100 sq. yds.

Keene's cement sand float finish coat may be applied over a brown coat of gypsum cement plaster, either plain or blocked off into ashlar effects. If blocked, add extra for this work.

Material Cost of 100 Sq. Yds. Keene's Cement Sand Float Finish
Ordinary Workmanship

	Rate	Total	Rate	Total
160 lbs. Keene's cement	$....	$....	$16.00	$25.60
125 lbs. fin. hydrated lime	....	....	8.00	10.00
6 cu. ft. clean sand	....	....	.45	2.70
Cost 100 sq. yds.		$....		$38.30
Cost per sq. yd.		....		.38

Labor Cost of 100 Sq. Yds. Keene's Cement Sand Float Finish
Ordinary Workmanship

	Hours	Rate	Total	Rate	Total
Plasterer	13	$....	$....	$16.35	$212.55
Labor	6	....	....	13.77	82.62
Cost 100 sq. yds.			$....		$295.17
Cost per sq. yd.			....		2.95

If first grade workmanship is required, add 4 hrs. plasterer time and 1-hr. labor time per 100 sq. yds.

The resistance of gypsum gauging-lime putty trowel finishes to cracking, particularly to check cracking, can be increased by the addition of fine aggregate. The use of not less than ½ cubic foot of fine silica sand or perlite to each 100 pounds of gauging plaster or Keene's cement, increases the factor of safety in trowel finishes.

Acoustical Plaster.—Acoustical plaster is occasionally specified, in lieu of trowel putty coat or sand float finishes, for noise quieting in classrooms, corridors, etc., and for acoustical correction of churches and auditoriums. A number of prepared acoustical plasters are available having sound absorption values of 40 to 45 percent at 512 cycles and a noise reduction coefficient of 55 to 60 percent.

There are two general types of acoustical plaster. Those with a gypsum or lime binder, such as Sabinite, Hi-Lite, Kilnoise, Gold Bond, Perfolite, etc., are for trowel application only with either a troweled, floated or stippled and perforated finish. Other acoustical plasters, such as Audicote, Sprayolite, Zonolite Acoustical Plaster, etc., are made with an adhesive type binder that shrinks on drying, providing the porosity for sound absorption, and may be applied by machine or trowel.

In general, acoustical plasters are for application over gypsum base coats (scratch and brown) and should be applied in two coats to a total thickness of ½"—usually a ⅜" first coat and a ⅛" finish coat.

Because of restrictions due to the asbestos fiber content, acoustical plaster is temporarily out of supply but will probably be back on the market with a new mixture. The prices below are for the old formula material.

Where applied by trowel, a plasterer should complete 65 to 75 sq. yds. of acoustical plaster, two coats to ½" total thickness, per 8-hr. day.

Material Cost of 100 Sq. Yds. of Acoustical Plaster
Applied by Trowel
Add for lath and base coats required.

	Rate	Total	Rate	Total
20 bags (5 sq. yds., ½" thick), acoustical plaster..............................	$....	$....	$7.50	$150.00
Cost per sq. yd..		....		1.50

Labor Cost of 100 Sq. Yds. of Acoustical Plaster
Applied by Trowel
Add for lath and base coats required.

	Hours	Rate	Total	Rate	Total
Plasterer.............................	11.5	$....	$....	$16.35	$188.03
Labor...................................	6.0	....	....	13.77	82.62
Cost 100 sq. yds.............................			$....		$270.65
Cost per sq. yd.................................			....		2.71

Where application of the two coat acoustical finish is by machine, a plasterer should complete 100 to 110 sq. yds. per 8-hr. day, but one extra hour of labor time must be figured for the additional cleanup and protection required, and an allowance included for cost and maintenance of equipment.

Material Cost of 100 Sq. Yds. of Acoustical Plaster Applied by Machine
Add for lath and base coats required.

	Rate	Total	Rate	Total
14 bags (7.2 sq. yds. ½" thick) acoustical plaster	$....	$....	$ 7.50	$105.00
Machine cost..............................	....	....	37.50	37.50
Cost per 100 sq. yds....................................		$....		$142.50
Cost per sq. yd.................................		....		1.43

Labor Cost of 100 Sq. Yds. of Acoustical Plaster Applied by Machine
Add for lath and base coats required.

	Hours	Rate	Total	Rate	Total
Plasterer.............................	8.0	$....	$....	$16.35	$130.80
Labor...................................	7.0	....	....	13.77	96.39
Labor (masking)	7.0	....	....	13.77	96.39
Cost per 100 sq. yds.....................			$....		$323.58
Cost per sq. yd.................................			....		3.24

Vermiculite Direct-to-Steel Fireproofing.—Vermiculite Type-MK direct-to-steel fireproofing is a new and economical method of protecting steel columns, beams, girders, and trusses, and steel floor and roof sections. It is applied by machine directly to the metal, without a primer, lath, or reinforcing. It presents a tough, fire-protective and insulating coating free of fissures, and can be applied in the early stages of construction when the building is free of other trades.

This is a mill-mixed material that requires only the addition of water. It is packaged in 25- and 50-pound bags.

Coverage varies with specified thickness.

Colored Interior Finish Plaster

There are a number of colored finish plasters on the market used for textured or float plaster finish.

These special finishes are applied over a brown coat of gypsum cement plaster and may be finished in a number of different ways, such as Spanish Palm Finish, Brush Swirl, Float Finish, English Trowel finish, etc., as shown in the accompanying illustrations, in addition to a simple float finish.

Labor Applying Special Finish Coats.—For colored finish plaster with a simple float finish, a plasterer should apply 100 to 120 sq. yds. per 8-hr. day, where an ordinary grade of workmanship is permitted.

Oriental* Interior Color Finish Plaster.—Used for textured or floated plaster finish. It is made in 13 standard shades and white and is ready for use with the addition of water only.

It requires approximately 8 lbs. of Oriental finish per sq. yd. depending upon the texture used. Price about $12.00 per 100 lbs.

Material Cost of 100 Sq. Yds. of Oriental* Interior Colored Finish
Plaster, with Simple Float Finish, Applied Over a Brown
Coat of Gypsum Cement Plaster
Ordinary Workmanship

Finish Coat	Rate	Total	Rate	Total
800 lbs. Oriental finish	$. . . .	$. . . .	$12.00	$96.00
Cost per sq. yd.		. . . .		.96

*U.S. Gypsum Co.

Labor Cost of 100 Sq. Yds. of Oriental* Interior Colored Finish
Plaster, with Simple Float Finish, Applied Over a Brown
Coat of Gypsum Cement Plaster
Ordinary Workmanship

Finish Coat	Hours	Rate	Total	Rate	Total
Plasterer	7.25	$. . . .	$. . . .	$16.35	$118.54
Labor	4.50	. . . .	. . . .	13.77	61.97
Cost 100 sq. yds.			$. . . .		$180.51
Cost per sq. yd.			. . . .		1.81

*U.S. Gypsum Co.

If first grade workmanship is required, add 2.75 hrs. plasterer time and 1-hr. labor time per 100 sq. yds.

---

*U.S. Gypsum Co.

Material Costs
Colored Plaster with Smooth or Textured Finish

	Rate	Total	Rate	Total
350 lbs. Keene's cement	$. . . .	$. . . .	$16.00	$ 56.00
450 lbs. fin. hydrated lime	. . . .	. . . .	8.00	36.00
16 lbs. color	. . . .	. . . .	1.00	16.00
Cost 100 sq. yds		$. . . .		$108.00
Cost per sq. yd. (material only)		. . . .		1.08

Material Costs
Colored Plaster with Sand Float Finish

	Rate	Total	Rate	Total
100 lbs. Keene's cement	$. . . .	$. . . .	$16.00	$ 16.00
500 lbs. fin. hydrated lime	. . . .	. . . .	8.00	40.00
1,200 lbs. white silica sand	. . . .	. . . .	1.00	12.00
36 lbs. color	. . . .	. . . .	1.00	36.00
Cost 100 sq. yds		$. . . .		$104.00
Cost per sq. yd. (material only)		. . . .		1.04

The above formulas are standard mixture for permanent colors. Best results are obtained by soaking lime overnight or 24 hours.

ORNAMENTAL PLASTERING

Ornamental plastering is not used as extensively as it once was, and mechanics who can do this type of work are not always available. It is best to contact a plastering subcontractor for firm prices on intricate ornamental plaster work.

Material Cost of 100 Lin. Ft. of Plaster Cornice Under 12" in Girth

	Rate	Total	Rate	Total
200 lbs. gypsum plaster	$. . . .	$. . . .	$ 6.00	$12.00
4 cu. ft. sand	. . . .	. . . .	.45	1.80
50 lbs. molding plaster	. . . .	. . . .	10.00	5.00
Cost 100 lin. ft.		$. . . .		$18.80
Cost per lin. ft.		. . . .		.19

Add for metal lath and furring as given on previous pages.

Labor Cost of 100 Lin. Ft. of Plaster Cornice Under 12" in Girth

	Hours	Rate	Total	Rate	Total
Plasterer	18	$. . . .	$. . . .	$16.35	$294.30
Labor	6	. . . .	. . . .	13.77	82.62
Cost 100 lin. ft.			$. . . .		$376.92
Cost per lin. ft.			. . . .		3.77

Add for metal lath and furring as given on previous pages.

Material Cost of 100 Sq. Ft. of Plaster Cornice Over 1'-0" in Girth

	Rate	Total	Rate	Total
200 lbs. gypsum plaster	$. . . .	$. . . .	$ 6.00	$12.00
4 cu. ft. sand	. . . .	. . . .	.45	1.80
50 lbs. molding plaster	. . . .	. . . .	10.00	5.00
Cost 100 sq. ft.		$. . . .		$18.80
Cost per sq. ft.		. . . .		.19

Add for metal lath and furring as given on previous pages.

Labor Cost of 100 Sq. Ft. of Plaster Cornice Over 1'-0" Girth

	Hours	Rate	Total	Rate	Total
Plasterer	17	$....	$....	$16.35	$277.95
Labor	6	....	....	13.77	82.62
Cost 100 sq. ft.			$....		$360.57
Cost per sq. ft.				....	3.61

Add for metal lath and furring as given on previous pages.

Material Cost of 100 Sq. Ft. of Portland Cement Cornice

	Rate	Total	Rate	Total
4 sacks portland cement	$....	$....	$4.00	$16.00
40 lbs. hydrated lime	....	....	8.00	3.20
10 cu. ft. sand	....	....	.45	4.50
Cost 100 sq. ft.		$....		$23.70
Cost per sq. ft.			....	.24

Add for metal lath and furring as given on previous pages.

Labor Cost of 100 Sq. Ft. of Portland Cement Cornice

	Hours	Rate	Total	Rate	Total
Plasterer	46	$....	$....	$16.35	$752.10
Labor	10	....	....	13.77	137.70
Cost 100 sq. ft.			$....		$889.80
Cost per sq. ft.				....	8.90

Add for metal lath and furring as given on previous pages.

Running Bull Nose Corners, Ceiling Coves, Etc.—When running bull nose corners on plaster columns, pilasters, window jambs, etc., and plain ceiling coves and wall angles, a plasterer should run 80 to 100 lin. ft. per 8-hr. day, at the following labor cost per 100 lin. ft.:

	Hours	Rate	Total	Rate	Total
Plasterer	9	$....	$....	$16.35	$147.15
Labor	2	....	....	13.77	27.54
Cost 100 lin. ft.			$....		$174.69
Cost per lin. ft.				....	1.75

PORTLAND CEMENT PLASTER

Portland cement plaster is used for both interior and exterior work. It should not be mixed too rich, as too rich a mixture has a tendency to check and shrink.

Proper attention to curing is important. After each coat has set, and before the following coat is applied the mortar should be kept moist for at least three days, and then permitted to dry out gradually.

When using portland cement mortar it is customary to add a small percentage of lime (usually about 10%) to make the mortar work easier, as a straight portland cement mortar is very difficult to apply.

For the cost of portland cement plaster applied over metal lath and the cost of metal lath and furring see previous pages of this chapter.

Labor Applying 100 Sq. Yds. Portland Cement Plaster to Various Plaster Bases

Description of Work Ordinary Work*	No. Sq. Yds. per 8-Hr. Day	Hrs. 100 Sq. Yds. Plasterer	Tender
Interior Work on Brick, Clay Tile or			
Cement Block—¾" Thick			
Scratch Coat	120-140	6	4
Brown Coat	75- 85	10	5
Float Finish Coat	60- 70	12	4
Trowel Finish Coat	45- 55	16	4
Interior Portland Cement Plaster on Metal Lath			
Scratch Coat	120-140	6	5
Brown Coat	60- 70	12	6
Float Finish Coat	60- 70	12	4
Trowel Finish Coat	45- 55	16	4
Exterior Portland Cement Stucco on			
Brick, Clay Tile or Cement Block			
Scratch Coat	120-140	6	5
Brown Coat	75- 85	10	6
Scaffold	....	..	12
Exterior Portland Cement Stucco on Metal Lath			
on Frame Construction			
Scaffold	....	..	12
Scratch Coat	120-140	6	6
Brown Coat	60- 70	12	8
Floated Finish Coat	60- 70	12	6
Troweled Finish Coat	45- 55	16	6
Textured Finish Coat	50- 60	14	6
Blocking Off Portland Cement Plaster			
into Squares to Represent Tile	45- 55	16	4
Portland Cement Straight Base	45- 55**	16	4
Portland Cement Coved Base	30- 35**	24	6

*For first grade workmanship as described at the beginning of this chapter, add 30 to 40 percent to plasterer time for brown and finish coats.
**Lineal feet.

Material Cost of 100 Sq. Yds. 2-Coat Portland Cement Plaster on Brick, Clay Tile or Cement Block
Ordinary Workmanship—⅝" Grounds

Brown Coat	Rate	Total	Rate	Total
14 sacks portland cement	$....	$....	$4.00	$ 56.00
140 lbs. hydrated lime	....	....	8.00	11.20
42 cu. ft. sand	....	....	.45	18.90
Finish Coat				
4 sacks portland cement	....	....	4.00	16.00
40 lbs. hydrated lime	....	....	8.00	3.20
12 cu. ft. sand	....	....	.45	5.40
Cost 100 sq. yds.		$....		$110.70
Cost per sq. yd.		....		1.11

Labor Cost of 100 Sq. Yds. 2-Coat Portland Cement Plaster on Brick, Clay Tile or Cement Block
Ordinary Workmanship—⅝" Grounds

Brown Coat	Hours	Rate	Total	Rate	Total
Plasterer	13	$....	$....	$16.35	$212.55
Labor	6	....	....	13.77	82.62
Finish Coat					
Plasterer	16	....	....	16.35	261.60

Finish Coat	Hours	Rate	Total	Rate	Total
Labor	4	....	....	13.77	55.08
Cost 100 sq. yds			$....		$611.85
Cost per sq. yd				....	6.12

Material Cost of 100 Sq. Yds. 3-Coat Portland Cement Plaster on Brick, Clay Tile or Cement Block
Ordinary Workmanship—⅝" Grounds

Scratch Coat	Rate	Total	Rate	Total
6 sacks portland cement	$....	$....	$4.00	$ 24.00
60 lbs. hydrated lime	....	....	8.00	4.80
18 cu. ft. sand	....	....	.45	8.10
Brown Coat				
8 sacks portland cement	....	....	4.00	32.00
80 lbs. hydrated lime	....	....	8.00	6.40
24 cu. ft. sand	....	....	.45	10.80
Finish Coat				
4 sacks portland cement	....	....	4.00	16.00
40 lbs. hydrated lime	....	....	8.00	3.20
12 cu. ft. sand	....	....	.45	5.40
Cost 100 sq. yds		$....		$110.70
Cost per sq. yd		....		1.11

Labor Cost of 100 Sq. Yds. 3-Coat Portland Cement Plaster on Brick, Clay Tile or Cement Block
Ordinary Workmanship—⅝" Grounds

Scratch Coat	Hours	Rate	Total	Rate	Total
Plasterer	6	$....	$....	$16.35	$ 98.10
Labor	4	....	....	13.77	55.08
Brown Coat					
Plasterer	10	....	....	16.35	163.50
Labor	5	....	....	13.77	68.85
Finish Coat					
Plasterer	16	....	....	16.35	261.60
Labor	4	....	....	13.77	55.08
Cost 100 sq. yds			$....		$702.21
Cost per sq. yd				....	7.02

If first grade workmanship is required, add 8 hrs. plasterer time and 2 hrs. labor time per 100 sq. yds.

If blocked off into squares to represent tile, add 16 hrs. plasterer time for ordinary work and 18 hrs. plasterer time for first grade workmanship per 100 sq. yds.

Material Cost of 100 Sq. Yds. 3-Coat Portland Cement Plaster on Metal Lath

Add for metal lath and furring.
Ordinary Workmanship—¾" Grounds

Scratch Coat	Rate	Total	Rate	Total
7 sacks portland cement	$....	$....	$4.00	$ 28.00
70 lbs. hydrated lime	....	....	8.00	5.60
21 cu. ft. sand	....	....	.45	9.45
Brown Coat				
11 sacks portland cement	....	....	4.00	44.00
110 lbs. hydrated lime	....	....	8.00	8.80
33 cu. ft. sand	....	....	.45	14.85
Finish Coat				
4 sacks portland cement	....	....	4.00	16.00

40 lbs. hydrated lime	....	....	8.00	3.20
12 cu. ft. sand	....	....	.45	5.40
Cost 100 sq. yds		$....		$135.30
Cost per sq. yd.		....		1.35

Labor Cost of 100 Sq. Yds. 3-Coat Portland Cement Plaster on Metal Lath

Add for metal lath and furring.

Ordinary Workmanship—¾" Grounds

Scratch Coat	Hours	Rate	Total	Rate	Total
Plasterer	6	$....	$....	$16.35	$ 98.10
Labor	5	....	....	13.77	68.85
Brown Coat					
Plasterer	12	....	....	16.35	196.20
Labor	6	....	....	13.77	82.62
Finish Coat					
Plasterer	16	....	....	16.35	261.60
Labor	4	....	....	13.77	55.08
Cost 100 sq. yds			$....		$762.45
Cost per sq. yd.			....		7.63

If blocked off into squares to represent tile, add 16 hrs. plasterer time per 100 sq. yds.

If ⅝" grounds are used instead of ¾", deduct 2 sacks portland cement, 20 lbs. lime and 6 cu. ft. sand.

If first grade workmanship is required, add 8 hrs. plasterer time and 2 hrs. labor time per 100 sq. yds.

Material Cost of 100 Lin. Ft. of 6" Portland Cement Straight Base

	Rate	Total	Rate	Total
2 sacks portland cement	$....	$....	$4.00	$ 8.00
20 lbs. hydrated lime	....	....	8.00	1.60
6 cu. ft. sand	....	....	.45	2.70
Cost 100 lin. ft.		$....		$12.30
Cost per lin. ft.		....		.12

Labor Cost of 100 Lin. Ft. of 6" Portland Cement Straight Base

	Hours	Rate	Total	Rate	Total
Plasterer	16	$....	$....	$16.35	$261.60
Labor	4	....	....	13.77	55.08
Cost 100 lin. ft.			$....		$316.68
Cost per lin. ft.			....		3.17

For coved base, add 50 percent to material and labor.

EXTERIOR STUCCO

Exterior stucco is generally composed of a portland cement base. The scratch and brown coats should be mixed in the proportions of 1 part of portland cement by weight to 3 parts of clean sand by weight. Do not add more than 8 lbs. of lime to each 100 lbs. of portland cement used in the mixture.

Portland cement plaster should not be applied when the outside temperature is below 32° F. as freezing is injurious to portland cement.

The cost of portland cement stucco will vary according to the materials used and the grade of workmanship required.

Applying Scratch Coat of Portland Cement Stucco.—Where just an ordinary grade of workmanship is required, a plasterer should apply 120 to 140 sq. yds. of scratch coat per 8-hr. day on metal lath.

Applying Brown Coat of Portland Cement Stucco.—A plasterer should apply 75 to 85 sq. yds. of brown coat per 8-hr. day, on brick, tile or concrete block walls and 60 to 70 sq. yds. per 8-hr. day on metal lath.

Applying Trowel Finish Coat of Portland Cement.—A plasterer should apply 45 to 55 sq. yds. of portland cement trowel finish per 8-hr. day, where just an ordinary grade of workmanship is required and 35 to 40 sq. yds. per 8-hr. day, where first grade workmanship is required.

Applying Wet Rough Cast Finish to Portland Cement Stucco.—If a wet rough cast finish is used and the mortar and aggregate are thrown against the wall with a paddle or similar tool, a plasterer should complete 30 to 40 sq. yds. per 8-hr. day.

Applying Pebble Dash or Dry Rough Cast.—When applying a pebble dash or dry rough cast finish where the aggregate is thrown against the wet cement or "butter" coat, a plasterer should apply 35 to 40 sq. yds. per 8-hr. day.

Washing Exterior Stucco With Acid to Expose Crystals.—Where the finish coat of stucco contains granite, marble or crystal screenings, it is washed off with a solution of muriatic acid to expose the crystals. A man should wash 475 to 525 sq. ft. (53 to 58 sq. yds.) per 8-hr. day.

On some jobs it will be necessary to wash the walls two or three times to bring out the crystals satisfactorily, and in such instances the labor cost should be increased accordingly.

Material Cost of 100 Sq. Yds. 3-Coat 1:3 Portland Cement Stucco, Float
Finish, Applied to Metal Lath Over Wood Framing

Add for metal lath.

Ordinary Workmanship

	Rate	Total	Rate	Total
Scratch Coat—1:3				
8 sacks portland cement	$. . . .	$. . . .	$4.00	$ 32.00
60 lbs. hydrated lime	. . . .	. . . .	8.00	4.80
24 cu. ft. sand	. . . .	. . . .	.45	10.80
Brown Coat—1:3				
16 sacks portland cement	. . . .	. . . .	4.00	64.00
120 lbs. hydrated lime	. . . .	. . . .	8.00	9.60
48 cu. ft. sand	. . . .	. . . .	.45	21.60
Float Finish—1:3				
5 sacks w'p'f. portland cement	. . . .	. . . .	5.00	25.00
15 cu. ft. sand	. . . .	. . . .	.45	6.75
Cost 100 sq. yds.		$. . . .		$174.55
Cost per sq. yd.		. . . .		1.75

Labor Cost of 100 Sq. Yds 3-Coat 1:3 Portland Cement Stucco, Float
Finish, Applied to Metal Lath Over Wood Framing

Add for metal lath.

Ordinary Workmanship

	Hours	Rate	Total	Rate	Total
Scratch Coat—1:3					
Plasterer	6	$. . . .	$. . . .	$16.35	$ 98.10
Labor	6	. . . .	. . . .	13.77	82.62
Brown Coat—1:3					
Plasterer	12	. . . .	. . . .	16.35	196.20
Labor	8	. . . .	. . . .	13.77	110.16
Float Finish—1:3					
Plasterer	12	. . . .	. . . .	16.35	196.20

Labor	6	....	....	13.77	82.62
Cost 100 sq. yds.			$....		$765.90
Cost per sq. yd.				....	7.66

If a smooth troweled finish is desired, add 4 hrs. plasterer time.
If a textured finish coat is desired, add 2 hrs. plasterer time.
If first grade workmanship is required, add 8 hrs. plasterer time to the above.
Add for scaffold.

Material Cost of 100 Sq. Yds. 3-Coat 1:3 Portland Cement Stucco, Float Finish, Applied Over Brick, Clay Tile or Concrete Block Surfaces
Ordinary Workmanship

	Rate	Total	Rate	Total
Scratch Coat—1:3				
6 sacks portland cement	$....	$....	$4.00	$ 24.00
45 lbs. hydrated lime	....	....	8.00	3.60
18 cu. ft. sand	....	....	.45	8.10
Brown Coat 1:3				
10 sacks portland cement	....	....	4.00	40.00
75 lbs. hydrated lime	....	....	8.00	6.00
30 cu. ft. sand	....	....	.45	13.50
Float Finish 1:3				
5 sacks w'p'f. portland cement	....	....	5.00	25.00
15 cu. ft. sand	....	....	.45	6.75
Cost 100 sq. yds.		$....		$126.95
Cost per sq. yd.		....		1.27

Labor Cost of 100 Sq. Yds. 3-Coat 1:3 Portland Cement Stucco, Float Finish, Applied Over Brick, Clay Tile or Concrete Block Surfaces
Ordinary Workmanship

	Hours	Rate	Total	Rate	Total
Scratch Coat—1:3					
Plasterer	6	$....	$....	$16.35	$ 98.10
Labor	5	....	....	13.77	68.85
Brown Coat 1:3					
Plasterer	10	....	....	16.35	163.50
Labor	6	....	....	13.77	82.62
Float Finish 1:3					
Plasterer	12	....	....	16.35	196.20
Labor	6	....	....	13.77	82.62
Cost 100 sq. yds.			$....		$691.89
Cost per sq. yd.			....		6.92

For smooth troweled finish, add 4 hrs. plasterer time.
If a textured finish coat is desired, add 2 hrs. plasterer time.
If first grade workmanship is required, add 8 hrs. plasterer time to the above.
Add for scaffold.

Special Finishes for Portland Cement Stucco

If any of the following special finishes are wanted with portland cement stucco, deduct the finish coat given above and add cost of the finish coat desired.

Material Cost of 100 Sq. Yds. of White Cement Float Finish

	Rate	Total	Rate	Total
Ordinary Workmanship	$....	$....	$8.00	$40.00
5 sacks w'p'f. white cement				
1,500 lbs. white silica sand	....	....	1.00	15.00
Cost 100 sq. yds.		$....		$55.00
Cost per sq. yd.		....		.55

Labor Cost of 100 Sq. Yds. of White Cement Float Finish

Ordinary Workmanship	Hours	Rate	Total	Rate	Total
Plasterer	12	$....	$....	$16.35	$196.20
Labor	6	....	....	13.77	82.62
Cost 100 sq. yds.			$....		$278.82
Cost per sq. yd.			....		2.79

Material Cost of 100 Sq. Yds. Finish Coat Colored Stucco, Float Finish

Ordinary Workmanship	Rate	Total	Rate	Total
2 sacks w'p'f. white cement	$....	$....	$8.00	$16.00
100 lbs. fin. hydrated lime	....	....	8.00	8.00
600 lbs. white silica sand	....	....	1.00	6.00
18 lbs. color	....	....	1.00	18.00
Cost 100 sq. yds.		$....		$48.00
Cost per sq. yd.		....		.48

Labor Cost of 100 Sq. Yds. Finish Coat Colored Stucco, Float Finish

Ordinary Workmanship	Hours	Rate	Total	Rate	Total
Plasterer	12	$....	$....	$16.35	$196.20
Labor	6	....	....	13.77	82.62
Cost 100 sq. yds.			$....		$278.82
Cost per sq. yd.			....		2.79

If a smooth troweled finish is desired, add 4 hrs. plasterer time.
If a textured finish is desired, add 2 hrs. plasterer time.
If first grade workmanship is required, add 8 hrs. plasterer time to the above.

Material Cost of 100 Sq. Yds. Wet Rough Cast Finish Coat

	Rate	Total	Rate	Total
4 sacks w'p'f. portland cement	$....	$....	$5.00	$20.00
12 cu. ft. sand	....	....	.45	5.40
500 lbs. aggregate	....	....	6.00	30.00
Cost 100 sq. yds.		$....		$55.40
Cost per sq. yd.		....		.56

Labor Cost of 100 Sq. Yds. Wet Rough Cast Finish Coat

	Hours	Rate	Total	Rate	Total
Plasterer	20	$....	$....	$16.35	$327.00
Labor	10	....	....	13.77	137.70
Cost 100 sq. yds.			$....		$464.70
Cost per sq. yd.			....		4.65

Material Cost of 100 Sq. Yds. of Dry Rough Cast Finish

	Rate	Total	Rate	Total
4 sacks w'p'f. portland cement	$....	$....	$5.00	$20.00
12 cu. ft. sand	....	....	.45	5.40
750 lbs. aggregate	....	....	6.00	45.00
Cost 100 sq. yds.		$....		$70.40
Cost per sq. yd.		....		.71

Labor Cost of 100 Sq. Yds. of Dry Rough Cast Finish

	Hours	Rate	Total	Rate	Total
Plasterer	20	$....	$....	$16.35	$327.00
Labor	10	....	....	13.77	137.70
Cost 100 sq. yds.			$....		$464.70
Cost per sq. yd.			....		4.65

A dry rough cast finish requires a finish or "butter" coat ¼" thick applied directly over the brown coat. This is brought to a straight, smooth finish and then the aggregate is thrown onto the "butter" coat dry.

If coarse aggregate is used, it requires about 1,000 lbs. per 100 sq. yds.

If medium size aggregate is used, it requires 750 lbs. per 100 sq. yds.

If fine aggregate is used, it requires 500 lbs. per 100 sq. yds.

Approximate Prices on Lathing and Plastering Materials
Metal Lath

Description	Finish	Wt. Lbs. Sq. Yd.	Price Sq. Yd.
Expanded diamond mesh (flat)	Painted	2.5 lbs.	$1.20
Expanded diamond mesh (flat)	Painted	3.4 lbs.	1.50
Expanded diamond mesh (flat)	Galvanized	3.4 lbs.	1.75

Self-furring lath, add 10¢ per sq. yd. to the above prices.

Rib Metal Lath

Description	Finish	Height of Rib	Wt. Lbs. Sq. Yd.	Price Sq. Yd.
Flat rib lath	Painted	⅛"	2.75	$1.50
Flat rib lath	Painted	⅛"	3.4	1.60
Rib or rod ribbed lath	Painted	⅜"	3.4	2.20
Rib or rod ribbed lath	Painted	⅜"	4.0	2.40

Hot Rolled Channels

Width	Leg	Weight per 1,000 Ft.	Price per 1,000 Lin. Ft.
3/4" Standard	15/32"	300	$165.00
1 " Standard	3/8"	410	200.00
11/2" Standard	19/32"	650	250.00
2 " Standard	7/16"	1,260	310.00

Peforating, add $10.00 per 1,000 ft.

Cold Rolled Channels

Width	Leg	Weight per 1,000 Ft.	Price per 1,000 Lin. Ft.
3/4"	1/2"	300 lbs.	$160.00
11/2"	19/32"	475 lbs.	240.00
2"	19/32"	590 lbs.	300.00

Wire

16 Ga	$80.00 per cwt.
18 Ga	$90.00 per cwt.

Studs & Tracks
20 Ga.

Size	Per 1,000 Ln. Ft.
1⅝"	$300.00
2½"	320.00
3¼"	350.00
4 "	370.00
6 "	470.00
Shoes	$75.00 per 1,000 pcs.

Corner Bead

Type	Per 1,000 Ln. Ft.
Regular	$150.00
Expanded	175.00
Bull Nose	350.00
Cornerite	100.00

Rods

	Price per 1,000 Lin Ft.
3/16" plain	$45.00
3/16" galvanized	55.00
1/4" plain	60.00
1/4" galvanized	70.00
3/8" plain	75.00

Gypsum Lath

Description	Price per 1,000 sq. ft.
Gypsum lath, plain or perforated, 16"x48"x⅜"	$105.00
Gypsum lath, foilback, 16"x48"x⅜"	185.00
Gypsum lath, plain, 16"x48"x½"	110.00
Gypsum lath, foilback, 16"x48"x½"	190.00

Nails and Staples

Description	Price per 100 Lbs.
Gypsum lath, 1⅛" blued	$ 90.00
Gypsum lath, 1¼" blued	90.00
Roofing, barbed, 1½" galv	100.00
Concrete stub	115.00
Annular ring nail (blued)	90.00
Brick—plain 2" and 2½"	100.00

Lime and Finishing Plaster

	Per 100 Lbs.
Pulverized Quicklime	$ 7.00
Plasterer's Hydrated Lime	8.00
Gypsum Gauging Plaster	9.00
No. 1 Molding Plaster	10.00
Keene's Cement, Regular or Fast	16.00
Portland Cement	4.00

Plastering Items

Liquid Bonding Agent	$25.00 Per Gal.

Prepared Finishes

Hair (4 Lb. to Bu.)	$8.00 Per Bu.
Fiber (2 Lb. to Bu.)	$8.00 Per Bu.

Gypsum Plaster

	Per 100 Lbs.		Per 100 Lbs.
Gypsum Cement Plaster	$6.00	Perlite Plaster	$12.00
Wood Fiber (prepared)	6.00	Gypsum Trowel Finish	7.00
Bondcrete (for concrete)	7.00	Silica Sand Float	9.00

09250   GYPSUM WALLBOARD

Gypsum wallboard, commonly referred to as drywall, is manufactured from a gray-white colored rock called gypsum, which is a nonmetallic mineral composed of calcium sulphate chemically combined with crystallized water.

After the gypsum ore is mined, it is crushed, dried, and ground to a fine powder. It is then heated to remove most of the chemically combined water. The calcined gypsum is then mixed with other ingredients and water and sandwiched between two sheets of treated paper to form a smooth gypsum wallboard panel. After the gypsum core has achieved a set, it is cut to length, dried, finished, and packaged for shipment.

Advantages.—Gypsum wallboard is fire resistent and will not support combustion. When one surface is exposed to a flame, the opposite panel surface remains cool until the gypsum core is calcinized.

The panel is easily cut and quickly hung in large sheets. This tends to promote an earlier completion and occupancy of the building.

Since the wallboard is a dry panel, less moisture is introduced into the building and cold weather is less of a factor for a completed finished surface. Only the joint finishing must be kept from freezing.

The strong, smooth face paper of the gypsum wallboard panel is suitable for any decorative treatment such as paint, wallpaper and textured coatings.

Size and Thickness.—Gypsum wallboard panels are manufactured in a number of thicknesses and sizes with various longitudinal edge designs.

Thickness	Size	Edge Design
¼"	4'-0"x8'-12'-0"	S.E., T.E.
⅜"	4'-0"x8'-12'-0"	S.E., T.E.
½"	4'-0"x8'-12'-0"	S.E., T.E.
⅝"	4'-0"x8'-12'-0"	S.E., T.E.

S.E. - Square Edge
T.E. - Tapered Edge

Square edge panels may be used as the first layer of a two layer system or where the joints will be covered by a batten strip. The tapered edge panel should always be used on the finished surface layer, where the joints are to be finished with tape and compound. The tapered edge is a depression along the longitudinal edge that receives the tape and compound build-up to smooth and hide the butting edges of the gypsum panels.

One-quarter inch gypsum wallboard is used as a lightweight, low cost, utility wallboard. It is also used for curved surfaces or to cover existing wall surfaces. Three-eighths inch gypsum wallboard is used principally for repair or remodeling work or in double wall construction. One-half inch gypsum wallboard is used in single layer new construction. Five-eighths inch gypsum wallboard is used where a one-hour fire rating is required or where the framing is spaced in excess of ½-in. wallboard limitations.

Perforated Tape Joint System for Gypsum Wallboard.—This method of concealing joints in gypsum wallboard is backed by over 20 years of joint reinforcing experience. It conceals the joints between boards and bonds the gypsum wallboard units together into a single, smooth, even wall and ceiling surface unit.

First, the hollow or channel at the edges of the wallboard is filled with compound, using a 4" or 5" joint finishing broadknife.

Apply the perforated tape immediately, directly over the compound and press it into place with the broadknife, squeezing excess compound out from under the tape, but leaving enough compound under tape for proper bond.

Apply thin covering coat of compound immediately and let dry.

Vertical Application of Gypsum Wallboard

Horizontal Application of Gypsum Wallboard

Next, apply another thin coating of compound so that the tape will be completely hidden. Feather out edges beyond previous coat as smoothly as possible and let dry thoroughly. Intermediate nail heads should also be carefully filled with compound and brought flush with the surface of the board.

Apply third and finish coat of compound, of thinner consistency to even up surface and edges of joint.

Sand, as necessary, between and after coats to insure smooth, inconspicuous joint.

Foil-Back Gypsum Panels.—These panels are made by laminating a sheet of aluminum foil to the back surface of the gypsum wallboard. The foil reduces outward heat flow in winter and inward heat flow in summer, has a significant thermal insulating value if facing an air space of ¾" or more, and is effective as a vapor barrier.

Moisture Resistant Gypsum Panels.—MR board has a special asphalt composition gypsum core and is covered with chemically treated face papers to prevent moisture penetration. This board is a little harder to cut, has a brownish color core, and is usually covered with a light green finish face paper. These panels were developed for application in bathrooms, kitchens, utility rooms, and other high moisture areas.

Fire Rated Gypsum Panels.—A specially formulated mineral gypsum core is used to make panels in ½" and ⅝" thicknesses for application to walls, ceilings, and columns where a fire rated assembly is needed. Based on tests by Underwriters' Laboratories, Inc., certain wall, floor/ceiling, and column assemblies give 45 minute to 4 hour fire resistance ratings.

Drywall Accessories.—The following items will be necessary to provide a complete finish for most any drywall job:

1). Fasteners: Nails or screws may be used for attachment of gypsum panels to wood framing, however screws must be used for attachment to metal framing members.

Wood Frame Fastening Applications

Fastener Description	Nail Spacing c. to c.	Approx. Lbs. Nails Req'd per MSF Gypsum Panels
1¼" GWB-54 Annular Ring Nail 12½ ga.; ¼" dia. head with a slight taper to a small fillet at shank; bright finish; medium diamond point; meets ASTM C380	7" ceiling 8" walls	5¼
1⅜" Annular Ring Nail (Same as GWB-54 except for length)	7" ceiling 8" walls	5¼

Fastening Application	Fastener Used

GYPSUM PANELS TO STANDARD METAL FRAMING

½" single-layer panels to standard studs, runners, channels

⅞" Type S Bugle Head

⅝" single-layer panels to standard studs, runners, channels

1" Type S Bugle Head

½" double-layer panels to standard studs, runners, channels

1⁵⁄₁₆" Type S Bugle Head

⅝" double-layer panels to standard studs, runners, channels

1⅝" Type S Bugle Head

1" coreboard to metal angle runners in solid partitions

1¼" Type S Bugle Head

½" panels through coreboard to metal angle runners in solid partitions

1⅞" Type S Bugle Head

⅝" panels through coreboard to metal angle runners in solid partitions

2¼" Type S Bugle Head

GYPSUM PANELS TO 12-GA. (MAX.) METAL FRAMING

½" and ⅝" panels and gypsum sheathing to 20-ga. studs and runners

1" Type S-12 Bugle Head

USG Self-Furring Metal Lath through gypsum sheathing to 20-ga. studs and runners

1¼" Type S-12 Bugle Head

½" and ⅝" double-layer gypsum panels to 20-ga. studs and runners

1⅝" Type S-12 Bugle Head

Multi-layer gypsum panels to 20-ga. studs and runners

1⅞" Type S-12 Bugle Head

2.) Corner Bead: Metal corner bead, furnished in lengths of 8' and 10', are used at all external corners in drywall construction to protect and prevent damage to the gypsum wallboard. This metal bead also provides a true and straight finish line to the corner.

3). **Stop Bead:** Metal stop beads are used to cap the edge of a gypsum panel at its termination point, whenever that edge is exposed and not to be covered by either a tape/compound finish or other trim members.

4). **Joint Tape:** A strong fiber paper tape designed with feathered edges, random pin-hole perforations, and lightly pre-creased for internal corners is used with the joint compound for finishing the butt joints of gypsum panels. This tape is packaged and sold in rolls of 250 and 500 linear feet per roll.

5). Joint Compound: This material is available in many different forms, e.g., powdered, ready-mixed, all-purpose, fast setting, exterior application, etc. The ready-mixed all-purpose joint compound is used extensively for all interior applications of taping and joint finishing, and is readily available in 5 gallon cans weighing about 68 pounds each.

Estimating Material Quantities.—The estimator would first measure the wall and ceiling surfaces to ascertain the required square footage of surface to be covered. Do not deduct for openings smaller than 50 sq. ft.

Separate the square foot quantities into the proper categories of types, thicknesses, and sizes of panels required.

Fasteners (nails or screws) should be estimated as requiring about 1,000 each per 1,000 sq. ft. of board to be installed (or 1 fastener per sq. ft.).

It will require about 400 L.F. of joint tape and about 1.3 5-gal. cans of joint compound to finish 1,000 sq. ft. of drywall surface. The amount of compound may be reduced to 1 5-gal. can per 1,000 sq. ft. if there are very few external corners to be finished.

Drywall Material Prices

Item	Price
¼" Gypsum Panels - Regular	$102. per M sq. ft.
⅜" Gypsum Panels - Regular	107. per M sq. ft.
½" Gypsum Panels - Regular	110. per M sq. ft.
⅝" Gypsum Panels - Regular	125. per M sq. ft.
½" Gypsum Panels - Fire Rated	124. per M sq. ft.
⅝" Gypsum Panels - Fire Rated	135. per M sq. ft.
½" Gypsum Panels - Moisture Resistant	160. per M sq. ft.
⅝" Gypsum Panels - Moisture Resistant	172. per M sq. ft.
Add: For Foil Back Panels	75. per M sq. ft.
1¼" GWB Ring Nails	2.00 per M pcs.
1" Type S Bugle Head Screws	5.50 per M pcs.
Corner Bead	.07 per lin. ft.
Stop Bead	.10 per lin. ft.
Joint Tape - 250' roll	.90 per roll
Joint Compound - 5 gal. can	8.00 per can

Example: The following list of material quantities may be required for installation and finishing of 300 lin. ft. of drywall partition, 8'-0" high, covering both sides of partition (300 L.F. x 8' Ht. x 2 sides = 4,800 sq. ft.).

"Material Only"

Item	Rate	Total	Rate	Total
½" Drywall Reg. 4,800 sq. ft.	$....	$....	$ .11	$528.00
1¼" Nails 4.8 M Pcs.	....	....	2.00	9.60
Corner Bead 96 L.F.	....	....	.07	6.72
Stop Bead 40 L.F.	....	....	.10	4.00
Joint Tape-8 rolls	....	....	.90	7.20
Joint Compound - 6 5-gal cans	....	....	8.00	48.00
Cost for 4,800 sq. ft.		$....		$603.52
Cost per sq. ft		....		.13

Labor Placing Gypsum Wallboard.—When placing ½" gypsum wallboard in average size rooms, 2 carpenters experienced in wallboard erection should place about 3,000 sq. ft. of board per 8-hr. day, at the following labor cost per sq. ft.:

	Hours	Rate	Total	Rate	Total
Carpenter	16	$. . . .	$. . . .	$16.47	$263.52
Cost per sq. ft.			. . . .		.09

When placing ½" gypsum wallboard in small rooms and spaces requiring considerable cutting and fitting in proportion to the number of square feet of board placed, 2 carpenters should place about 2,000 sq. ft. per 8-hr. day, at the following labor cost per sq. ft.:

	Hours	Rate	Total	Rate	Total
Carpenter	16	$. . . .	$. . . .	$16.47	$263.52
Cost per sq. ft.			. . . .		.13

Where fire resistant ⅝" material is specified, 2 carpenters should place about 2,500 sq. ft. per 8 hr. day on straight wall and ceiling construction at the following labor cost per sq. ft.

	Hours	Rate	Total	Rate	Total
Carpenter	16	$. . . .	$. . . .	$16.47	$263.52
Cost per sq. ft.			. . . .		.11

Where gypsum wallboard is used to fireproof beams and columns, 2 carpenters should place about 1,200 sq. ft. per 8 hr. day at the following labor cost:

	Hours	Rate	Total	Rate	Total
Carpenter	16	$. . . .	$. . . .	$16.47	$263.52
Cost per sq. ft.			. . . .		.22

Labor Finishing Gypsum Wallboard.—The finishing of gypsum wallboard is a multi-step process of tape coat, block coat, skim coat, and point-up coat. The necessity of sanding between the block and skim coats is not required if the finisher is careful and a first class mechanic.

The complete finishing labor for 1,000 sq. ft. of wallboard is as follows:

	Hours	Rate	Total	Rate	Total
Tape Coat	2.0	$. . . .	$. . . .	$16.47	$ 32.94
Block Coat	1.6	. . . .	. . . .	16.47	26.35
Skim Coat	1.3	. . . .	. . . .	16.47	21.41
Sand	0.4	. . . .	. . . .	16.47	6.59
Point-Up	0.8	. . . .	. . . .	16.47	13.18
Cost per 1,000 sq. ft			$. . . .		$100.47
Cost per sq. ft			. . . .		.10

METAL STUDS AND FURRING

Stud framing and wall or ceiling furring on commercial projects is normally accomplished with light gauge metal studs, runner channels and furring members in lieu of wood framing.

This material is manufactured from cold rolled galvanized metal and is available in a number of sizes, shapes and gauges.

Metal Stud Partitions.—Runner channels are channel shaped members that are positioned at the top and bottom of the studs, as in top and bottom plates for a wood stud partition. The runner channels are attached to the floor and overhead structure with nails, screws, or powder actuated drive pins.

The metal studs, which are channel shaped with a backbend to give stiffness, are placed within the web of the runner channels, located for "on-center" spacing, plumbed, and screwed into place with a ⅜" pan head screw, top and bottom. The web of the metal studs have cutouts for the passage of conduit and piping.

Runner channels and studs are made in 1⅝", 2½", 3⅝", 4", 6" widths in 25 ga., 2½", 3⅝", 4", 6" widths in 20 ga., and 4", 6" widths in 18 ga. and 16 ga. metals.

The 25 ga. runners and studs are usually specified for non-load bearing partitions up to 14'-16' in height. Over this height and up to 22', the 20 ga. material is used. The 18 ga. and 16 ga. materials are used for greater heights, load bearing walls, and for exterior wall construction where lateral pressure (wind loads) will be imposed, therefore, requiring more stiffness and less deflection.

Metal Furring.—This material is made from 25 gauge metal in the following shapes:

1). Z furring channel is used in conjunction with rigid insulation board when furring exterior wall surfaces.
2). The hat shaped furring channel is used for furring walls and ceilings when insulation board is not required.
3). The resilient channel is used primarily on ceilings to provide a separation between the gypsum panels and the framing members for the floor above. It is used extensively in wood framed garden apartments for noise dampening between apartment units.

Metal Stud

Metal Runner

1¼"
1"
¼"
Metal
Z-Furring Channel

2¾"
1⅜"
⅞"
Metal
Furring Channel

Resilient Channel

Estimating Material Quantities.—To determine the quantity of runner channel, studs, and fasteners for a partition, the estimator would measure the total length of partition to be erected. Note the thickness and gauge of materials specified, ascertain the height of the partition for length of studs, ascertain the required spacing of studs and add extras for door or window openings and at corners, and determine what type and spacing of fasteners is required for attachment of the runner channels to the floor and overhead structure. Horizontal stud bridging or fire stops are usually not required as in wood frame partition, but be sure to check, as this item in metal, is expensive to install.

To determine the quantity of furring channel required, ascertain the on-center spacing of the members for quantity of pieces by the length of the members (height of wall, as furring members are usually vertically installed) to give total linear footage to be installed. Ascertain the fastener type and spacing for attachment to the furring channel.

Material Prices

25 ga. - 1⅝" Runner Channel	130. MLF
25 ga. - 1⅝" Stud	140. MLF
25 ga. - 2½" Runner Channel	155. MLF
25 ga. - 2½" Stud	170. MLF
25 ga. - 3⅝" Runner Channel	190. MLF
25 ga. - 3⅝" Stud	200. MLF
25 ga. - 4" Runner Channel	205. MLF
25 ga. - 4" Stud	220. MLF
25 ga. - 6" Runner Channel	265. MLF
25 ga. - 6" Stud	285. MLF
25 ga. - ⅞" Hat Furring Channel	140. MLF
25 ga. - 1" Z Furring Channel	165. MLF
25 ga. - Resilient Channel	145. MLF
20 ga. - 2½" Runner Channel	305. MLF
20 ga. - 2½" Stud	320. MLF
20 ga. - 3⅝" Runner Channel	340. MLF
20 ga. - 3⅝" Stud	365. MLF
20 ga. - 4" Runner Channel	360. MLF
20 ga. - 4" Stud	380. MLF
20 ga. - 6" Runner Channel	460. MLF
20 ga. - 6" Stud	485. MLF

⅜" Pan Head Screws	4.50 M
Powder Actuated Drive Pins (Average cost inc. shot)	.25 ea.

Example.—The following list of material quantities would be required for a metal stud partition 200 feet long, 10 feet high, stud at 24" on-center, 25 ga. metal, and installed with drive pins at 24" on-center, top and bottom to concrete slabs:

"Material Only"

Item	Rate	Total	Rate	Total
2½" 25 ga. Runner - 400 L.F.	$. . . .	$. . . .	$ .155	$ 62.00
2½" 25 ga. Stud - 1,010 L.F.	. . . .	. . . .	.17	171.70
1" Drive Pins - 202 ea.	. . . .	. . . .	.25	50.50
⅜" Pan Screws - 0.4 M	. . . .	. . . .	4.50	1.80
Cost for 200 Lin. Ft		$. . . .		$286.00
Cost per Lin. Ft.		. . . .		1.43

Add extra studs for door and window openings and corners.

Labor to Install Metal Stud Partitions.—Using the above example given for materials, two (2) carpenters should erect the 200 lin. ft. of partition at the following labor cost:

	Hours	Rate	Total	Rate	Total
Layout Wall Line	1.0	$. . . .	$. . . .	$16.47	$ 16.47
Fasten Bottom Runner	4.0	. . . .	. . . .	16.47	65.88
Fasten Top Runner	4.0	. . . .	. . . .	16.47	65.88
Install Studs	3.0	. . . .	. . . .	16.47	49.41
Plumb & Screw Studs	4.0	. . . .	. . . .	16.47	65.88
Cost for 200 lin. ft			$. . . .		$263.52
Cost Per Lin. Ft			. . . .		1.32

09300   TILE
CERAMIC WALL AND FLOOR TILE

Ceramic tile provides a durable, colorful surface that is virtually maintenance-free. Its applications include interior and exterior finishes for functional and decorative purposes in all types of structures. Special uses include acid resistant and electrically conductive installations.

Ceramic tile is available in many sizes, shapes and finishes. Terms of reference for the most commonly used ceramic tile are :

Glazed Ceramic Wall Tile—Ceramic tile that has an impervious facial finish fused onto the body of the tile. The glazed surface comes in a wide variety of colors.

Ceramic Mosaic Tile—Ceramic tile, either glazed or unglazed, having a facial area of less than six square inches and which is usually mounted on sheets or mesh approximately 2 feet by 1 foot to facilitate setting.

Quarry Tile—A rugged ceramic tile used primarily as a finish flooring (interior and exterior) where a long wearing, easily cleanable surface is desired.

Recent developments in the ceramic tile industry make it necessary to stress the importance of relating tile costs to each individual job specification. Glazed ceramic wall tile, for example, can be backmounted or unmounted, and can be installed using conventional portland cement mortar, various types of adhesives or the more recently developed 'dry-set' portland cement mortar. All of these variations can affect costs, both material and labor, and, consequently, the estimator should review his job requirements carefully prior to making the estimate.

The above notwithstanding, the estimator's job has been simplified in recent years. Virtually the entire ceramic tile industry has adopted a 'simplified practice', prepared under the auspices of the Tile Council of America, which has reduced the number of sizes and shapes and which has established generally recognized standards for the Industry.

Estimating Quantities.—Ceramic tile is estimated by the square foot, with trim pieces such as base, cap, etc. being estimated by the lineal foot. The estimator should deduct door and window openings, but the trim pieces necessary to finish the openings must be added. The quantities should be related to the type and size of ceramic tile, since these items will affect the cost of the ceramic tile when priced.

Estimate Composition.—The finished ceramic tile estimate will include the following items :

1. Cost of ceramic tile delivered to the job site.
2. Cost of accessory materials such as wire mesh, sand and cement for floor fill under ceramic tile.
3. Cost of mixing and placing floor fill.
4. Direct labor cost of laying and cleaning the ceramic tile.
5. The ceramic tile contractor's overhead and profit.

Variable Factors Influencing Costs.—As in the other construction trades, estimating the labor costs of setting ceramic tile requires an intimate knowledge of the labor market in which the work is to be performed. Wage rates vary throughout the country and it is important the estimator determine the rate in his locality.

Ceramic tile is unique in that it is often necessary to install it in very small quantities (for example, in 1 or 2 bathrooms in a house) and, therefore, the costs can vary greatly. It is felt, however, that such small installations are the exception, rather than the norm, and, therefore, the costs given on the following pages are based on the assumption that the areas involved are large enough to permit the tile contractor to operate efficiently.

The costs developed on the following pages are based on 4¼"x4¼" glazed wall tile, 1"x1" ceramic mosaic tile and 6"x6"x½" quarry tile, since these are the sizes most commonly employed. If other sizes are specified, substitute the material price only. For all practical purposes, the labor cost remains constant regardless of size. The developed costs do not include areas where an unusually large amount of trim is required. Therefore, the estimator should make an allowance for extra trim pieces in such special cases.

Setting Methods.—Recent years have seen the development of many methods of adhering ceramic tile to a subsurface. There are, however, three methods which are most commonly used and accepted. These are :

Conventional Portland Cement Mortar Method. This method is to bond each ceramic tile with a layer of pure portland cement paste to a portland cement set-

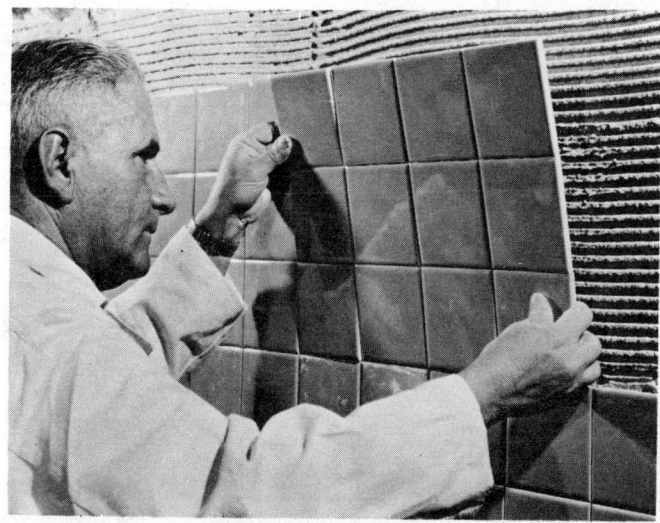

Dry-set Mortar and Back Mounted Tile Sheets.

ting bed. This is done while the setting bed is still plastic. Wall tile must be soaked in water so that the water needed for curing is not absorbed from the paste. While being the traditionally accepted method, it is also more costly than the other methods.

Dry-Set Portland Cement Mortar Method. This method utilizes a dry curing portland cement mortar (accomplished through the use of water retaining additives) and has made ceramic tile installation cheaper and simpler. "Dry-set" is ideally suitable for use with concrete masonry, brick, poured concrete and portland cement plaster. It should not be used over wood or gypsum plaster. Labor costs are appreciably reduced when this method is used.

Water-Resistant Organic Adhesive. Organic adhesives can be used over smooth base materials, such as wallboards, plywood and metal. Labor productivity is comparable to that of 'dry-set' mortar.

Labor Productivity
(sq. ft. per team day; team is 1 tile setter and 1 helper)

	Description	Face-mounted	Back-mounted	Unmounted
Conventional Mortar	Glazed wall tile	....	75- 90 SF	55- 65 SF
	Ceramic mosaic tile—walls	45- 50 SF	55- 65 SF	....
	floors	100-125 SF	100-125 SF	....
	Quarry tile floors	....	....	100-125 SF
'Dry-Set' Mortar	Glazed wall tile	....	125-150 SF	100-125 SF
	Ceramic mosaic tile—walls	100-125 SF	120-140 SF	....
	floors	125-150 SF	125-150 SF	....
	Quarry tile floors	....	....	125-150 SF
Organic Adhesive	Glazed wall tile	....	150-175 SF	120-140 SF
	Ceramic mosaic tile—walls	120-140 SF	125-150 SF	....
	floors	150-175 SF	175-200 SF	....
	Quarry tile floors	....	....	....

Cove Or Base	65-75 lin. ft.
Cap	80-90 lin. ft.

Approximate Prices
of Ceramic Tile Materials
Ceramic Wall Tile

Description	Size	Price Per sq. ft.
Flat tile, all colors, including white	4¼"x4½"	$1.25
	6 "x6 "	1.35
	6 "x4¼"	1.40
	6 "x9 "	1.55
	8½"x4¼"	1.45

		Price per piece[1]
for use with 4¼"x4¼" tile		
Cove	6 "x3¾"	$ .80
Cove	4¼"x4¼"	.80
Bullnose	2 "x6 "	.80

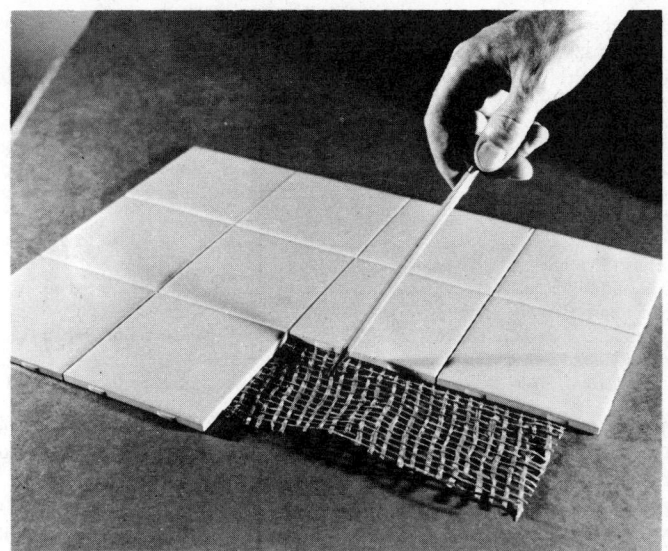

Glazed Wall Tile of the 4¼ by 4¼ Inch Size Mounted on Mesh Backing

Bullnose	4¼"x4¼"	.85
Double bullnose	4¼"x4¼"	1.20
for use with 6"x6" tile		
Cove	6 "x6 "	$1.10
Bullnose	6 "x6 "	1.15
Double bullnose	6 "x6 "	1.40
miscellaneous trim shapes		
Bead	6 "x ¾"	$ .80
Bullnose	6 "x3¾"	1.10
Base	6 "x4 "	1.20
Base	6 "x6 "	1.25
Sink Trim	6 "x2 "	1.20
Window Sill	9 "x5⅛"	1.40

NOTES
1) Prices are for stretchers; for angles, multiply prices by two

Ceramic Mosaic Tile

Description (price per sq. ft.)
Unglazed, Modular, Solid Color

	White & Grays	Other Colors
1"x1"	$1.20	$1.40
2"x1"	1.30	1.50
2"x2"	1.40	1.60
3/4"x3/4"	1.20	1.40
19/16"x 3/4 "	1.20	1.40
19/16"x19/16"	1.20	1.40
Trim Shapes (price per lineal foot)		
1"x1" Bullnose		$1.30
2"x1" Bullnose		1.50

3/4" Series trim pieces same price as 1" Series

Various manufacturers furnish, as standard items, patterns made from combinations of modular tile sizes and colors. Prices vary depending on number of different sizes, number of colors and depth of colors used. Patterns such as these are too numerous to itemize, but the general range is from $1.25 to $1.60 per sq. ft. There are exceptions to this range however, and when a specific pattern price is needed, the estimator should check with the specified supplier.

Abrasive Ceramic Mosaic Tile	from $1.35 to $1.70 per sq. ft.
Conductive Floor Tile	from $2.40 to $3.00 per sq. ft.
Granitized Ceramic Mosaic Tile	from $1.25 to $1.45 per sq. ft.
Abrasive Granitized Ceramic Mosaic Tile	from $1.35 to $1.60 per sq. ft.
Glazed Satin Finish Ceramic Mosaic Tile	$1.30 per sq. ft.
Glazed Textured Finish Ceramic Mosaic Tile	$1.50 per sq ft.
Special Decorative Glazed Ceramic Mosaic Tile for Accent Effects	from $3.00 to $13.00 per sq. ft.

Venetian glass mosaics mounted on sheets 12¼"x12¼" square run $2.50 a sq. ft. For ¾"x¾" $2.70 for 1"x1" and $2.90 for 1½"x1½". Metallic finishes run considerably more. Beads and coves for glass mosaics will run $2.00 a lin. ft.

Sculptured tiles which can be used for wall surfacing, murals, screens and sculpture are available in a wide range of ceramic designs and treatments. Design-Technics of New York have tiles in sizes 4¼"x8½"x¼" at approximately $4 to $5 a sq. ft., 12"x12"x⅜" at $5 to $6.25 a sq. ft. and 12"x12"x1¾" at $6.50 to $7.75 a sq. ft.

Quarry Tile

Description	Size	Price per sq. ft.
Deep red, plain surface	6"x6 "x½"	$1.20
	6"x2¾"x½"	1.25
	6"x6 "x¾"	2.00
	9"x9 "x¾"	3.25
Deep red, abrasive	6"x6 "x¾"	2.25
	6"x6 "x¾"	2.75

Note: Other colors, with the exception of green, are virtually the same price as the red shown above.
Green in the 6"x6"x½" size costs $2.00 per sq. ft.
Extras, when specified, are:

Tile ground square after burning	$0.08 per sq. ft.
Wax coating	.20 per sq. ft.

Floor brick, deep red, plain		
tray packed	8"x4"x1⅜"	$165.00 per M pieces
in sealed cartons		195.00 per M pieces
Floor brick, deep red, abrasive		
tray packed	8"x4"x1⅜"	180.00 per M pieces
in sealed cartons		210.00 per M pieces

There are many specially cast heavy floor and wall tiles specified today that compliment the provincial and country modes popular now. Many of these are quite expensive and are not regularly stocked. Always check locally for availability and prices, and for similar tiles that might be allowed for substitution should this become necessary.

Ceramic Tile Bathroom Accessories

Numerous ceramic tile bathroom accessories are available in a variety of sizes and qualities. They may be recessed or surface mounted and they come in the full range of tile colors. Some of the more commonly used items are as follows:

Ceramic Mosaic Tile of 1 by 1 Inch Tile, Back Mesh-Mounted.

Description	Price per Piece
Recessed soap holder	$12.00
Recessed glass holder	12.00
Roll paper holder	15.00
Robe hook	4.00
Double robe hook	4.50
Toothbrush holder	6.00
Tile bar brackets (pair)	8.00

Decorative Ceramic Wall Tiles

Glazed ceramic wall tile containing designs as an integral part of each piece of tile is also available. Normally supplied in 4¼"x4¼" and 6"x6" sizes, the price of this material will vary with the design selected and number of colors in the design. If the design is known, the estimator should obtain a firm price from a supplier. The general range is as follows:

One color	4¼"x4¼"	$2.50 per sq. ft.
	6 "x6 "	2.35 per sq. ft.
Two colors	4¼"x4¼"	3.00 per sq. ft.
	6 "x6 "	2.80 per sq. ft.
Three colors	4¼"x4¼"	3.50 per sq. ft.
	6 "x6 "	3.25 per sq. ft.

Ceramic Tile Adhesives and Accessory Materials

Item	Coverage	Price
Organic adhesive	40-50 sf/gal	$8.00/gal.
Dry-set mortar mix	35 lbs./100 sf	.45/lb.
Primer (for use in damp areas)	125 sf/gal.	6.00/gal.
Plastic underlayment	100 sf/unit*	8.00/unit*
Wet tile grout	15-20 sf/lb**	.25/lb.
Dry tile grout	12-15/sf/lb.**	.35/lb.

*Unit consists of 1 gallon liquid & 34 lbs. aggregate.
**When used with 4¼"x4¼" tile.

Material Cost of 100 sq. ft. of 4¼"x4¼" Glazed Ceramic Wall Tile using conventional mortar method of installation (unmounted tile).

	Rate	Total	Rate	Total
2 sacks portland cement	$....	$....	$4.00	$ 8.00
4 cu. ft. sand	....	....	.45	1.80
100 sq. ft. glazed wall tile	....	....	1.25	125.00
7 lbs. wet tile grout mix	....	....	.25	1.75
Cost per 100 sq. ft		$....		$136.55
Cost per sq. ft		....		1.37

Labor Cost of 100 sq. ft. of 4¼"x4¼" Glazed Ceramic Wall Tile using conventional mortar method of installation (unmounted tile).

	Hours	Rate	Total	Rate	Total
Tile Setter	13.5	$....	$....	$14.95	$201.83
Cost per sq. ft			....		2.02

Material Cost of 100 sq. ft. of 4¼"x4¼" Glazed Ceramic Wall Tile using water resistant organic adhesive (unmounted tile).

	Rate	Total	Rate	Total
2.5 gallons adhesive	$....	$....	$8.00	$ 20.00
100 sq. ft. glazed wall tile	....	....	1.25	125.00
9 lbs. dry tile grout mix	....	....	.35	3.15
Cost per 100 sq. ft		$....		$148.15
Cost per sq. ft		....		1.48

Labor Cost of 100 sq. ft. of 4¼"x4¼" Glazed Ceramic Wall Tile using water resistant organic adhesive (unmounted tile).

	Hours	Rate	Total	Rate	Total
Tile Setter	8.0	$....	$....	$14.95	$119.60
Cost per sq. ft			....		1.20

Material Cost of 100 sq. ft. of 4¼"x4¼" Glazed Ceramic Wall Tile using 'dry-set' portland cement mortar (unmounted tile).

	Rate	Total	Rate	Total
35 lbs. 'dry-set' mortar mix	$....	$....	$ .45	$ 15.75
100 sq. ft. glazed wall tile	....	....	1.25	125.00
9 lbs. dry tile grout mix	....	....	.35	3.15
Cost per 100 sq. ft		$....		$143.90
Cost per sq. ft		....		1.44

Labor Cost of 100 sq. ft. of 4¼"x4¼" Glazed Ceramic Wall Tile using 'dry-set' portland cement mortar (unmounted tile).

	Hours	Rate	Total	Rate	Total
Tile Setter	10.0	$....	$....	$14.95	$149.50
Cost per sq. ft			....		1.50

Material Cost of 100 sq. ft. 1"x1" Ceramic Mosaic Tile floors using conventional mortar method of installation (face-mounted tile).

	Rate	Total	Rate	Total
2 sacks portland cement	$....	$....	$4.00	$ 8.00
4 cu. ft. sand	....	....	.45	1.80

100 sq. ft. ceramic				
mosaic tile	....	....	1.20	120.00
17 lbs. wet tile grout mix	....	....	.25	4.25
Cost per 100 sq. ft		$....		$134.05
Cost per sq. ft			....	1.34

Labor Cost of 100 sq. ft. 1"x1" Ceramic Mosaic Tile floors using conventional mortar method of installation (face-mounted tile).

	Hours	Rate	Total	Rate	Total
Tile setter	10.5	$....	$....	$14.95	$156.98
Cost per sq. ft				....	1.57

Material Cost of 100 sq. ft. of 1"x1" Ceramic Mosaic Tile floors using water resistant organic adhesive (face-mounted tile).

	Rate	Total	Rate	Total
2.5 gallons adhesive	$....	$....	$8.00	$ 20.00
100 sq. ft. ceramic				
mosaic tile	....	....	1.20	120.00
23 lbs. dry tile grout mix	....	....	.35	8.05
Cost per 100 sq. ft		$....		$148.05
Cost per sq. ft			....	1.48

Labor Cost of 100 sq. ft. of 1"x1" Ceramic Mosaic Tile floors using water resistant organic adhesive (face-mounted tile).

	Hours	Rate	Total	Rate	Total
Tile setter	8.0	$....	$....	$14.95	$119.60
Cost per sq. ft				....	1.20

Material Cost of 100 sq. ft. of 1"x1" Ceramic Mosaic Tile floor using 'dry-set' portland cement mortar (face-mounted tile).

	Rate	Total	Rate	Total
35 lbs. 'dry-set' mortar				
mix	$....	$....	$ .45	$ 15.75
1 cu. ft. of sand	....	....	.45	.45
100 sq. ft. ceramic				
mosaic tile	....	....	1.20	120.00
23 lbs. dry tile grout mix	....	....	.35	8.05
Cost per 100 sq. ft		$....		$144.25
Cost per sq. ft			....	1.44

Labor Cost of 100 sq. ft. of 1"x1" Ceramic Mosaic Tile floors using 'dry-set' portland cement mortar (face-mounted tile).

	Hours	Rate	Total	Rate	Total
Tile setter	9.0	$....	$....	$14.95	$134.55
Cost per sq. ft				....	1.35

Material Cost of 100 sq. ft. of 1"x1" Ceramic Mosaic Tile on walls using conventional mortar method of installation (face-mounted tile).

	Rate	Total	Rate	Total
2 sacks portland cement	$....	$....	$4.00	$ 8.00
4 cu. ft. sand	....	....	.45	1.80

	Rate	Total	Rate	Total
100 sq. ft. ceramic mosaic tile	....	....	1.20	120.00
17 lbs. wet tile grout mix	....	....	.25	4.25
Cost per 100 sq. ft.		$....		$134.05
Cost per sq. ft.		....		1.34

Labor Cost of 100 sq. ft. of 1"x1" Ceramic Mosaic Tile on walls using conventional mortar method of installation (face-mounted tile).

	Hours	Rate	Total	Rate	Total
Tile setter	11.0	$....	$....	$14.95	$164.45
Cost per sq. ft.			....		1.65

Material Cost of 100 sq. ft. of 1"x1" Ceramic Mosaic Tile on walls using water resistant organic adhesive (face-mounted tile).

	Rate	Total	Rate	Total
2.5 gallons adhesive	$....	$....	$8.00	$ 20.00
100 sq. ft. ceramic mosaic tile	....	....	$1.20	120.00
23 lbs. dry tile grout mix	....	....	.35	8.05
Cost per 100 sq. ft.		$....		$148.05
Cost per sq. ft.		....		1.48

Labor Cost of 100 sq. ft. of 1"x1" Ceramic Mosaic Tile on walls using water resistant organic adhesive (face-mounted tile).

	Hours	Rate	Total	Rate	Total
Tile setter	9.0	$....	$....	$14.95	$134.55
Cost per sq. ft.			....		1.35

Material Cost of 100 sq. ft. of 1"x1" Ceramic Mosaic Tile on walls using 'dry-set' portland cement mortar (face-mounted tile).

	Rate	Total	Rate	Total
35 lbs. 'dry-set' mortar mix	$....	$....	$ .45	$ 15.75
1 cu. ft. sand	....	....	.45	.45
100 sq. ft. ceramic mosaic tile	....	....	1.20	120.00
23 lbs. dry tile grout mix	....	....	.35	8.05
Cost per 100 sq. ft.		$....		$144.25
Cost per sq. ft.		....		1.44

Labor Cost of 100 sq. ft. of 1"x1" Ceramic Mosaic Tile on walls using 'dry-set' portland cement mortar (face-mounted tile).

	Hours	Rate	Total	Rate	Total
Tile setter	9.0	$....	$....	$14.95	$134.55
Cost per sq. ft.			....		1.35

Material Cost of 100 sq. ft. of 6"x6"x½" Quarry Tile floors using conventional mortar method of installation

	Rate	Total	Rate	Total
2 sacks portland cement	$....	$....	$4.00	$ 8.00
6 cu. ft. sand	....	....	.45	2.70

100 sq. ft. quarry tile plain surface	. . . .	. . . .	1.20	120.00
35 lbs. portland cement grout	. . . .	. . . .	.50	17.50
Cost per 100 sq. ft.		$. . . .		$148.20
Cost per sq. ft.		. . . .		1.48

Labor Cost of 100 sq. ft. of 6"x6"x½" Quarry Tile floors using conventional mortar method of installation

	Hours	Rate	Total	Rate	Total
Tile setter	16.0	$. . . .	$. . . .	$14.95	$239.20
Cost per sq. ft.			. . . .		2.39

Material Cost of 100 sq. ft. of 6"x6"x½" Quarry Tile floors using 'dry-set' portland cement mortar

	Rate	Total	Rate	Total
35 lbs. 'dry-set' mortar mix	$. . . .	$. . . .	$ .45	$ 15.75
2 cu. ft. of sand	. . . .	. . . .	.45	.90
100 sq. ft. quarry tile plain surface	. . . .	. . . .	1.20	120.00
35 lbs. portland cement grout	. . . .	. . . .	.50	17.50
Cost per 100 sq. ft.		$. . . .		$154.15
Cost per sq. ft.		. . . .		1.54

Labor Cost of 100 sq. ft. 6"x6"x½" Quarry Tile floors using 'dry-set' portland cement mortar

	Hours	Rate	Total	Rate	Total
Tile setter	12.0	$. . . .	$. . . .	$14.95	$179.40
Cost per sq. ft.			. . . .		1.80

Material Cost of 100 lin. ft. of cove or base (Ceramic or Quarry Tile)

	Rate	Total	Rate	Total
100 lin. ft. cove	$. . . .	$. . . .	$1.60	$160.00
Adhesive or mortar—allow	. . . .	. . . .	6.00	6.00
Cost per 100 lin. ft.		$. . . .		$166.00
Cost per lin. ft.		. . . .		1.66

Labor Cost of 100 lin. ft. of cove or base (Ceramic or Quarry Tile)

	Hours	Rate	Total	Rate	Total
Tile setter	20.0	$. . . .	$. . . .	$14.95	$299.00
Cost per lin. ft.			. . . .		2.99

Material Cost of 100 lin. ft. of wainscot cap (Glazed Ceramic Wall Tile)

	Rate	Total	Rate	Total
100 lin. ft. bullnose	$. . . .	$. . . .	$1.60	$160.00
Adhesive or mortar—allow	. . . .	. . . .	6.00	6.00
Cost per 100 lin. ft.		$. . . .		$166.00
Cost per lin. ft.		. . . .		1.66

Labor Cost of 100 lin. ft. of wainscot cap (Glazed Ceramic Wall Tile)

	Hours	Rate	Total	Rate	Total
Tile setter	20.0	$. . . .	$. . . .	$14.95	$299.00
Cost per lin. ft.			. . . .		2.99

Placing Cement Floor Fill.—Cement floor fill under ceramic tile floors is usually placed by tile setters and helpers (one tile setter and one or two helpers working together). The fill is placed one or two days in advance of the tile if the overall thickness from rough floor to finished tile surface is over 3". For 3" thickness and under, fill and setting bed may be placed in one operation. When such tile floors are to be placed over wood subfloors it is necessary to first place a layer of waterproof building paper and a layer of wire mesh reinforcing before placing the fill.

A tile setter and two helpers should place 450 to 500 sq. ft. of fill per 8-hour day in areas large enough to permit efficient operations, at the following cost per 100 sq. ft.:

Material Costs

	Rate	Total	Rate	Total
6 bags portland cement	$....	$....	$ 4.00	$24.00
1 cu. yd. sand	....	....	12.00	12.00
Cost per 100 sq. ft.		$....		$36.00
Cost per sq. ft.		....		.36

Labor Costs

	Hours	Rate	Total	Rate	Total
Tile setter	1.5	$....	$....	$14.95	$22.43
Helper	3.0	....	....	12.54	37.62
Cost per 100 sq. ft.			$....		$60.05
Cost per sq. ft.			....		.60

Metal Wall Tile

Wall tile of aluminum, having a baked enamel finish in various colors, or tile of copper or stainless steel are used for the same purposes.

They require the same type of base; are applied with adhesive and a man will apply about 100 square feet per day.

Metal tile mastic weighs about 16 lbs. per gal. and it requires ½-lb. per sq. ft. or one gal. is sufficient for 32 sq. ft. of tile. Price about $8.00 per gal. Aluminum metal tile cost about $1.50 per sq. ft. Copper or stainless steel tile cost about $3.70 per sq. ft.

MARBLE TILE

To prepare an accurate estimate of cost on marble work, the estimator should understand the correct method of measuring and listing quantities from the plans; the manner in which the marble should be handled and set on the job, as well as the most economical and practical methods of handling the work.

When estimating marble wainscoting, it is incorrect to take the height of the wainscot from the floor to the cap and multiply by the number of lin. ft. because in all probability the wainscot is made up of three members, (base, die and cap) and each member should be estimated separately; the base and cap by the lin. ft. and the die by the square foot.

The length and size of the pieces of marble has considerable bearing on the labor cost, because it usually costs as much to set a 6" piece as one 2'-0" long, so that a job made up of short pieces will cost more per foot than a job having a larger percentage of long pieces of marble.

Trade Practice.—Trade practice in the marble industry recognizes all marbles and stones commonly used for interior building purposes, as falling within one of four groups or Classifications, according to their respective characteristics and working qualities. Therefore, for purposes of standardization and clearer definition, the Marble Institute of America has officially adopted these four classifications and defines them as follows:

Group "A". Sound marbles and stones which require no "sticking", "waxing," or "filling;" characteristically uniform and favorable working qualities.

Group "B". Marbles and stones similar in character to the preceding group but somewhat less favorable working qualities; occasional natural faults and requiring a limited amount of "waxing" and "sticking."

Group "C". Marbles of uncertain variation in working qualities, geological flaws, voids, veins and lines of separation common; common shop practice to repair nature's shortcomings by sticking, waxing and filling; "liners" and other forms of reinforcement freely employed when necessary.

Group "D". Marbles and stones similar to the preceding group and subject to the same methods of finishing and manufacture but embracing those materials which contain a larger proportion of natural faults and a maximum variation in working qualities, etc. This group comprises many of the highly colored marbles prized for their decorative qualities.

Below is a list of some of the more commonly used varieties of marbles and stones and the groups to which they belong:

Group "A"
Alabama—usual grades
Blanco P
Brocadillo, Vermont
Carthage
Georgia
Italian White
Italian English Vein
Napoleon Gray (Vt.)
Tennessee, Gray and Pink
Vermont, Black
Vermont, White Grades
Vermont Pavonazzo
Westland Green Vein Cream

Group "B"
Alabama Cream Veined A
Belgian Black
Champville
Cremo Italian
Imperial Black, Tenn.
Mankato Buff
Travertine (Italian)

Group "C"
Belgian Grand Antique
Bois Jourdan
Botticino
Breche Opal

Group "C"
Escalette
Hauteville
Mankato Pink
Ozark Rouge
Red Levanto
Red Verona
Rosato
Tavernelle, all types
Verdona
Verona, Yellow
Vermont Verde Antique
Westfield Green

Group "D"
Alps Green
Black and Gold
Bleu Belge
Breche Oriental
Breche Rosora
Forest Green
Grand Antique, French
Grand Antique, Italian
Onyx, Pedrara
Rouge Antique
Rouge Jascolin
Sienna
Verde Antico (Italian)
Vert St. Denis

Estimating Quantities of Marble Work.—When measuring and listing quantities of marble work from the plans, there are several general rules that should be followed to insure accuracy.

To estimate accurately the quantity of marble base, obtain the number of lineal feet of base, listing the various heights separately. All base under 1'-0" high is extended and priced on a lineal foot basis. All base 1'-0" or over is extended and priced on a square foot basis. Always bear in mind that no piece of marble should be figured as being less than 1'-0" long, as it requires just as much time to set a 6" piece as one 2'-0" long. Short pieces are usually found around pilasters, door and window returns, etc.

Marble die is estimated by the square foot, and the labor costs computed in the same manner. When listing quantities, obtain the length of each run and multiply by the height, and the result will be the number of sq. ft. of die. If marble pilasters project beyond the face of the wall, list each projection according to its actual dimensions and if less than a 6" projection, it should be priced as 6" wide, but if the pilasters are over 6" wide, then use actual dimensions.

Marble wainscot cap under 1'-0" wide or any marble under 1'-0" wide, should be measured by the lineal foot.

Marble stair treads should be measured by the square foot, as they are ordinarily 12" or more in width. Be sure to mention length, width and thickness of all treads, type of nosing, etc.

Marble stair risers should be estimated by the lin. ft. giving length, height and thickness of each riser, as it costs practically as much to set a short riser as a long one.

When estimating toilet stall work consisting of toilet backs, partitions, stall fronts, stiles, etc., the fronts should be estimated by the square foot if over 12" wide and by the lin. ft. if less than 12" wide. Partitions and backs should be estimated by the square foot. Marble cap is usually estimated by the lin. ft. Some marble contractors figure toilet work at a certain price per stall, including all labor in connection with same, although most contractors recommend pricing on a square or lineal foot basis.

Marble column bases and caps formed of solid stock over 2" thick, should be estimated at a certain price for each base or cap, mentioning the size and number of cu. ft. in each.

Marble floor tile and border are usually estimated by the square foot, giving size of tile, such as 6"x6", 9"x9", 12"x12", etc. Small or irregular sized tile, such as dots, etc., are estimated in the same manner, except that a notation should be made of the class of work, as it costs more to lay small and irregular shaped tile.

Marble handrail, moldings, etc., should be estimated by the lin. ft., giving the size of each, in order that the quantities may also be stated in cu. ft. if over 2" thick.

All classes of circular work, such as base, wainscot or die, should be estimated in the same units as straight work, except that circular work should be estimated separately, as it usually requires considerable more cutting and fitting than straight work.

All marble work cut and set "on the rake," such as stair wainscoting, wainscoting under stairs, etc., should be estimated by the square foot, using the largest dimensions. Work of this kind should be estimated separately as it is more expensive to set than straight work.

Handling and Setting Interior Marble.—Estimating the labor cost of handling and setting interior marble will depend as much upon the conditions under which the work is performed as any other class of construction work. The amount of labor handling, sorting and distributing the marble on the job before it is set should be considered, also whether it is necessary to hoist the marble above the first floor, as this will require considerably more handling. The average size of the pieces should be considered, such as the average length of base, size and thickness of wainscot, length of thresholds, treads, risers, caps, etc., as it costs almost as much to handle and set a short piece of marble as a long one. This is especially true on stock that can be handled by one man.

The labor costs given on the following pages are based on the performance of the regular crew of any well organized marble company but if it is necessary to hire "extra" men or "floaters", the daily production may be decreased 25 percent and the labor costs increased accordingly.

On practically all jobs it is necessary to have laborers handle the marble and distribute it ready for the setters, although in the larger cities all marble must be handled by marble setters and helpers.

Hoisting Marble.—The proposal sheet of practically all marble contractors states that the marble contractor shall have free use of the general contractor's hoisting facilities, runways, etc., so for this reason no allowance has been made for hoisting in the following estimates of cost.

However, if it is necessary for the marble contractor to pay for the use of hoist for hoisting marble to the upper floors of a building, figure about ½-hr. hoist time per 100 sq. ft. of marble.

Trucking on Marble From Mill or Cars to Job.—If the marble is furnished by an out of town mill and it is necessary to haul it in trucks from cars to the job, the cost will vary with the length of the haul. If the marble is furnished by a mill located in the same town as the job, the price of the marble will probably include delivery to the job.

Marble Contractor's Foreman Expense.—It is necessary to have a foreman on all jobs of any size to supervise the arrival, unloading and distribution of the stock, as well as to lay out the work and supervise the setting. On an average allow 1/16 to ⅛-hr. foreman time to each hour setter's time, depending upon the size of the job and the number of setters employed.

Most small jobs will not require a foreman and no allowance should be made for same.

Average Labor Cost Setting Interior Marble.—While the labor cost of marble setting will vary with the kind of material handled, some of the large producing marble companies estimate the setting cost of the entire job at a certain price per foot.

The quantities for the entire job are taken, which include base, cap, wainscot or die, treads, risers, etc., and the labor priced per sq. ft., or 100 sq. ft. should cost as follows :

	Hours	Rate	Total	Rate	Total
Foreman	1.5	$....	$....	$15.50	$ 23.25
Marble setter	12.0	....	....	14.95	179.40
Helper	12.0	....	....	12.54	150.48
Labor or helper	8.0	....	....	12.54	100.32
Cost 100 sq. ft			$....		$453.45
Cost per sq. ft				....	4.54

The above price includes setting on jobs consisting principally of ⅞" and 1¼" stock.

If 2" marble is used, the handling and setting cost will be increased considerably, and the labor per 100 sq. ft. should cost as follows :

	Hours	Rate	Total	Rate	Total
Foreman	1.5	$....	$....	$15.50	$ 23.25
Marble setter	16.0	....	....	14.95	239.20
Helper	16.0	....	....	12.54	200.64
Labor or helper	12.0	....	....	12.54	150.48
Cost 100 sq. ft			$....		$613.57
Cost per sq. ft				....	6.14

On jobs having an average amount of base, wainscot or die, cap, toilet stalls, treads and risers, etc., the above prices will prove close enough, but where there are large quantities of any one class of work, it is advisable to refer to that particular class of work to obtain accurate quantities and costs.

Setting Marble Base.—The cost of setting marble base will vary with the size of the rooms, whether straight walls or broken up with pilasters, piers, etc., and whether the job consists of long or short pieces of base.

In small corridors and other spaces having numerous pilasters, piers, etc., requiring short pieces of base, a setter and helper should set 60 to 75 lin. ft. per 8-hr. day, at the following labor cost per 100 lin. ft. :

	Hours	Rate	Total	Rate	Total
Foreman	1.5	$. . . .	$. . . .	$15.50	$ 23.25
Marble setter	12.0	. . . .	. . . .	14.95	179.40
Helper	12.0	. . . .	. . . .	12.54	150.48
Labor or helper	8.0	. . . .	. . . .	12.54	100.32
Cost 100 lin. ft			$. . . .		$453.45
Cost per lin. ft				. . . .	4.54

If the base can be set in reasonably long pieces, a marble setter and helper should set 80 to 100 lin. ft. per 8-hr. day, at the following labor cost per 100 lin. ft. :

	Hours	Rate	Total	Rate	Total
Foreman	1	$. . . .	$. . . .	$15.50	$ 15.50
Marble setter	9	. . . .	. . . .	14.95	134.55
Helper	9	. . . .	. . . .	12.54	112.86
Labor or helper	8	. . . .	. . . .	12.54	100.32
Cost 100 lin. ft			$. . . .		$363.23
Cost per lin. ft				. . . .	3.63

Setting Circular Base.—Setting circular base will run about double the cost of straight work of the same kind and set under similar conditions. There is usually considerable cutting and fitting on all circular work.

Setting Marble Wainscot or Die.—The labor cost of setting marble wainscot or die will vary with the size of the pieces, the height of the wainscot or die, etc., as a marble setter will set almost as many lin. ft. of 5'-0" wainscot as 3'-0" wainscot. The labor cost handling stock will be increased somewhat, as it will require 2 to 4 laborers or helpers to handle each piece of marble and place it on the floor ready for the setter, while lighter stock, such as base, risers, treads, etc., can be handled by one man.

Setting Marble Wainscot Up to 3'-0" High.—When setting marble wainscot up to 3'-0" high, a marble setter and helper should set 22 to 26 lin. ft. containing 66 to 78 sq. ft. per 8-hr. day, at the following labor cost per 100 sq. ft. :

	Hours	Rate	Total	Rate	Total
Foreman	1.5	$. . . .	$. . . .	$15.50	$ 23.25
Marble setter	11.0	. . . .	. . . .	14.95	164.45
Helper	11.0	. . . .	. . . .	12.54	137.94
Labor or helper	8.0	. . . .	. . . .	12.54	100.32
Cost 100 sq. ft			$. . . .		$425.96
Cost per sq. ft				. . . .	4.26

The cost of setting circular wainscot or die will run about double the cost of straight work on account of the extra cutting and fitting.

Setting Marble Wainscot 3'-0" to 4'-0" High.—A marble setter and helper should set 20 to 25 lin. ft. containing 75 to 90 sq. ft. of wainscot per 8-hr. day, at the following labor cost per 100 sq. ft. :

	Hours	Rate	Total	Rate	Total
Foreman	1.2	$. ...	$. ...	$15.50	$ 18.60
Marble setter	10.0	....	....	14.95	149.50
Helper	10.0	....	....	12.54	125.40
Labor or helper	8.0	....	....	12.54	100.32
Cost 100 sq. ft			$. ...		$393.82
Cost per sq. ft				....	3.94

The cost of setting circular wainscot or die will run about double the cost of straight work on account of the extra cutting and fitting.

Setting Marble Wainscot 4'-0" to 5'-0" High.—Where the marble wainscot varies from 4'-0" to 5'-0" high, a marble setter and helper should set 20 to 23 lin. ft. containing 85 to 100 sq. ft. per 8-hr. day, at the following labor cost per 100 sq. ft. :

	Hours	Rate	Total	Rate	Total
Foreman	1	$. ...	$. ...	$15.50	$ 15.50
Marble setter	9	....	....	14.95	134.55
Helper	9	....	....	12.54	112.86
Labor or helper	8	....	....	12.54	100.32
Cost 100 sq. ft			$. ...		$363.23
Cost per sq. ft				....	3.63

The cost of setting circular wainscot or die will run about double the cost of straight work on account of the extra cutting and fitting.

Setting Marble Wainscot Over 5'-0" High.—When setting marble wainscot 5'-0" to 7'-0" high, a marble setter and helper should set 90 to 110 sq. ft. per 8-hr. day, at the following labor cost per 100 sq. ft. :

	Hours	Rate	Total	Rate	Total
Foreman	1	$. ...	$. ...	$15.50	$ 15.50
Marble setter	8	....	....	14.95	119.60
Helper	8	....	....	12.54	100.32
Labor or helper	8	....	....	12.54	100.32
Cost 100 sq. ft			$. ...		$335.74
Cost per sq. ft				....	3.36

The cost of setting circular wainscot or die will run about double the cost of straight work on account of the extra cutting and fitting.

Wainscot over 7'-0" high will require working from a scaffold which decreases the production considerably. In such cases the labor cost of that portion of wainscot over 7'-0" high should be increased 50 to 100 percent and the cost of erecting and removing steel tubular scaffold should be added.

Setting Marble Wainscot Cap.—When setting marble wainscot cap, a marble setter and helper should set 60 to 75 lin. ft. per 8-hr. day, at the following labor cost per 100 lin. ft. :

	Hours	Rate	Total	Rate	Total
Foreman	1.5	$. ...	$. ...	$15.50	$ 23.25
Marble setter	12.0	....	....	14.95	179.40
Helper	12.0	....	....	12.54	150.48
Labor or helper	8.0	....	....	12.54	100.32
Cost 100 lin. ft			$. ...		$453.45
Cost per lin. ft				....	4.54

*Add for overhead and profit.

Setting Marble Stair Treads.—The cost of setting marble stair treads will vary with the thickness and length of the treads because the shorter each piece of marble, the higher the cost per lin. ft.

If the treads are 3'-0" to 3'-6" long and 1¼" thick, a marble setter and helper should set 20 to 22 treads containing 60 to 77 lin. ft. per 8-hr. day, at the following labor cost per 100 lin. ft. :

	Hours	Rate	Total	Rate	Total
Foreman	1.5	$....	$....	$15.50	$ 23.25
Marble setter	11.0	....	....	14.95	164.45
Helper	11.0	....	....	12.54	137.94
Labor or helper	8.0	....	....	12.54	100.32
Cost 100 lin. ft			$....		$425.96
Cost per lin. ft			....		4.26

If the treads vary from 4'-0" to 6'-0" long, a marble setter and helper should set 80 to 90 lin. ft. per 8-hr. day, at the following labor cost per 100 lin. ft. :

	Hours	Rate	Total	Rate	Total
Foreman	1.1	$....	$....	$15.50	$ 17.05
Marble setter	9.5	....	....	14.95	142.03
Helper	9.5	....	....	12.54	119.13
Labor or helper	8.0	....	....	12.54	100.32
Cost 100 lin. ft			$....		$378.53
Cost per lin. ft			....		3.79

Setting Marble Stair Risers.—The cost of setting marble stair risers will vary with the length of the pieces of marble, width of stairs, etc.

Where the stairs are 3'-0" to 3'-6" wide, a marble setter and helper should set 60 to 75 lin. ft. of risers per 8-hr. day, at the following labor cost per 100 lin. ft. :

	Hours	Rate	Total	Rate	Total
Foreman	1.5	$....	$....	$15.50	$ 23.25
Marble setter	11.0	....	....	14.95	164.45
Helper	11.0	....	....	12.54	137.94
Labor or helper	8.0	....	....	12.54	100.32
Cost 100 lin. ft			$....		$425.96
Cost per lin. ft			....		4.26

If the marble stairs are over 4'-0" wide, a marble setter and helper should set 80 to 90 lin. ft. of risers per 8-hr. day, at the following labor cost per 100 lin. ft. :

	Hours	Rate	Total	Rate	Total
Foreman	1.1	$....	$....	$15.50	$ 17.05
Marble setter	9.5	....	....	14.95	142.03
Helper	9.5	....	....	12.54	119.13
Labor or helper	8.0	....	....	12.54	100.32
Cost 100 lin. ft			$....		$378.53
Cost per lin. ft			....		3.79

Setting Marble Stair Wainscot on the Rake.—Where marble wainscoting is set on the rake of the stairs, the labor cost will run considerably higher than straight work on account of the additional cutting and fitting necessary.

If the wainscoting is 3'-0" to 3'-6" high, a marble setter and helper should set 10 to 15 lin. ft. containing 35 to 45 sq. ft. per 8-hr. day, at the following labor cost per 100 sq. ft. :

	Hours	Rate	Total	Rate	Total
Foreman	2	$....	$....	$15.50	$ 31.00
Marble setter	20	....	....	14.95	299.00
Helper	20	....	....	12.54	250.80
Labor or helper	8	....	....	12.54	100.32
Cost 100 sq. ft			$....		$681.12
Cost per sq. ft			....		6.81

Where the wainscot is 5'-0" to 7'-0" high and follows the rake of the stairs, a marble setter and helper should set 50 to 65 sq. ft. per 8-hr. day, at the following labor cost per 100 sq. ft. :

	Hours	Rate	Total	Rate	Total
Foreman	1.5	$. . . .	$. . . .	15.50	$ 23.25
Marble setter	14.0	. . . .	. . . .	14.95	209.30
Helper	14.0	. . . .	. . . .	12.54	175.56
Labor or Helper	8.0	. . . .	. . . .	12.54	100.32
Cost 100 sq. ft			$. . . .		$508.43
Cost per sq. ft				. . . .	5.09

Setting Marble Toilet Stalls and Partitions.—The cost of handling and setting marble toilet stalls, backs, partitions, stiles, etc., will vary greatly with the individual job. On work of this kind the marble partition slabs are seldom less than 5'-0" in height and width and each slab contains 25 to 30 sq. ft. of marble. The backs are usually the same height as the partitions but not so wide and each back contains 16 to 20 sq. ft. of marble. The stall fronts or stiles are usually 6" to 8" wide and 6'-0" to 7'-0" high and extend to the floor to support the dividing partitions where metal standards are not used.

As an average on toilet stall work, a good marble setter should complete a stall in 3 to 3½ hours, after the marble backs are in place. Since the average partition contains about 25 sq. ft. and each stile and cap about 7 sq. ft., a setter and helper should set 75 to 80 sq. ft. of marble per 8-hr. day, at the following labor cost for 1 complete stall exclusive of back:

	Hours	Rate	Total	Rate	Total
Foreman	0.4	$. . . .	$. . . .	$15.50	$ 6.20
Marble setter	3.3	. . . .	. . . .	14.95	49.34
Helper	3.3	. . . .	. . . .	12.54	41.38
Labor or helper	2.6	. . . .	. . . .	12.54	32.61
Setting cost 1 stall			$. . . .		$129.53

Toilet partition work is estimated by the sq. ft. by most marble contractors.

Setting Marble Toilet Backs.—The marble backs for toilet stalls are usually 6'-0" to 7'-0" high. A marble setter and helper should set 75 to 85 sq. ft. per 8-hr. day, at the following labor cost per 100 sq. ft.:

	Hours	Rate	Total	Rate	Total
Foreman	1.2	$. . . .	$. . . .	$15.50	$ 18.60
Marble setter	10.0	. . . .	. . . .	14.95	149.50
Helper	10.0	. . . .	. . . .	12.54	125.40
Labor or helper	8.0	. . . .	. . . .	12.54	100.32
Cost 100 sq. ft			$. . . .		$393.82
Cost per sq. ft				. . . .	3.94

Setting Marble Door and Window Trim.—A marble setter and helper should set 45 to 55 lin. ft. of marble door and window trim or casings per 8-hr. day, at the following labor cost per 100 lin. ft.:

	Hours	Rate	Total	Rate	Total
Foreman	2	$. . . .	$. . . .	$15.50	$ 31.00
Marble setter	16	. . . .	. . . .	14.95	239.20
Helper	16	. . . .	. . . .	12.54	200.64
Labor or helper	8	. . . .	. . . .	12.54	100.32
Cost 100 lin. ft			$. . . .		$571.16
Cost per lin. ft				. . . .	5.71

*Add for overhead and profit.

Setting Marble Countertops.—When setting marble countertops, a marble setter and helper should set 75 to 85 sq. ft. per 8-hr. day, at the following labor cost per 100 sq. ft.:

	Hours	Rate	Total	Rate	Total
Foreman	1.2	$....	$....	$15.50	$ 18.60
Marble setter	10.0	....	....	14.95	149.50
Helper	10.0	....	....	12.54	125.40
Labor or helper	8.0	....	....	12.54	100.32
Cost 100 sq. ft			$....		$393.82
Cost per sq. ft				....	3.94

Setting Marble Thresholds.—When setting marble threshholds 3'-0" to 3'-6" long, a marble setter and helper should set 12 to 14 thresholds per 8-hr. day, at the following labor cost each:

	Hours	Rate	Total	Rate	Total
Marble setter	.60	$....	$....	$14.95	$ 8.97
Helper	.60	....	....	12.54	7.53
Cost per threshold			$....		$16.50

Setting Marble Plinths.—When setting marble plinths at door openings, a marble setter and helper should set 20 to 25 plinths per 8-hr. day, at the following labor cost per 100 plinths:

	Hours	Rate	Total	Rate	Total
Foreman	4.5	$....	$....	$15.50	$ 69.75
Marble setter	36.0	....	....	14.95	538.20
Helper	36.0	....	....	12.54	451.44
Labor or helper	8.0	....	....	12.54	100.32
Cost 100 plinths			$....		$1,159.71
Cost per plinth				....	11.60

Setting Marble Stair Railings and Balusters.—On jobs having marble balusters up to 6" in diameter and 2'-6" long, a marble setter and helper should set about 20 balusters per 8-hr. day, at the following labor cost per 100 balusters:

	Hours	Rate	Total	Rate	Total
Foreman	5	$....	$....	$15.50	$ 77.50
Marble setter	40	....	....	14.95	598.00
Helper	40	....	....	12.54	501.60
Labor or helper	16	....	....	12.54	200.64
Cost 100 balusters			$....		$1,377.74
Cost per baluster				....	13.78

Setting Marble Ashlar.—Where marble ashlar is used for the interior walls of buildings, it is customary to use ⅞" or 1¼" marble and the method of handling and setting is similar to that used for setting exterior work, as it is often necessary to use a light breast derrick to set the marble.

On work of this kind, a marble setter and helper should handle and set 60 to 80 sq. ft. of ashlar per 8 hr. day, at the following labor cost per 100 sq. ft.:

	Hours	Rate	Total	Rate	Total
Foreman	1.5	$....	$....	$15.50	$ 23.25
Marble setter	12.0	....	....	14.95	179.40
Helper	12.0	....	....	12.54	150.48
Labor or helper	12.0	....	....	12.54	150.48
Cost 100 sq. ft			$....		$503.61
Cost per sq. ft				....	5.04

Setting Marble Moldings, Handrail, Etc.—On jobs having light marble moldings, stair handrail, well hole rail, etc., where the marble is about 6" square and furnished in reasonably long pieces, a marble setter and helper should set 75 to 85 lin. ft. per 8-hr. day, at the following labor cost per 100 lin. ft.:

	Hours	Rate	Total	Rate	Total
Foreman	1.2	$....	$....	$15.50	$ 18.60
Marble setter	10.0	....	....	14.95	149.50
Helper	10.0	....	....	12.54	125.40
Labor or helper	8.0	....	....	12.54	100.32
Cost 100 lin. ft			$....		$393.82
Cost per lin. ft				....	3.94

Setting Marble Column Bases.—When setting marble column bases 2'-6" to 3'-0" square and 0'-9" to 1'-3" high, a marble setter and helper should handle and set about one base an hr. at the following labor cost per base:

	Hours	Rate	Total	Rate	Total
Foreman	0.1	$....	$....	$15.50	$ 1.55
Marble setter	1.0	....	....	14.95	14.95
Helper	1.0	....	....	12.54	12.54
Labor or helper	0.5	....	....	12.54	6.27
Cost per base			$....		$35.31

Laying Marble Floor Tile.—There are three labor operations to be considered when estimating the labor cost of setting marble floor tile, viz., the cement bed under the marble floor, labor handling and setting floor tile and the cost of rubbing or smoothing the floors after they have been laid.

As a general rule, the cement bed under the marble floor is 2½" to 3" thick, composed of cement and sand mixed rather dry. It requires about 1½ bbls. of portland cement and one cu. yd. of sand to each 100 sq. ft. of floor.

Material and labor per 100 sq. ft. of floor fill should cost as follows:

Material Costs

	Rate	Total	Rate	Total
6 sacks portland cement	$....	$....	$ 4.00	$24.00
1 cu. yd. sand	....	....	12.00	12.00
Cost 100 sq. ft		$....		$36.00
Cost per sq. ft			....	.36

Labor Costs

	Hours	Rate	Total	Rate	Total
Foreman	0.2	$....	$....	$15.50	$ 3.10
Marble setter	1.5	....	....	14.95	22.43
Helper	3.0	....	....	12.54	37.62
Cost 100 sq. ft			$....		$63.15
Cost per sq. ft				....	.63

If the marble floor tile vary from 6"x6" to 12"x12" in size, a marble setter and helper should handle and lay 90 to 110 sq. ft. of floor per 8-hr. day, at the following labor cost per 100 sq. ft.:

	Hours	Rate	Total	Rate	Total
Foreman	1	$....	$....	$15.50	$ 15.50
Marble setter	8	....	....	14.95	119.60
Helper	8	....	....	12.54	100.32
Labor or helper	8	....	....	12.54	100.32
Cost 100 sq. ft			$....		$335.74
Cost per sq. ft				....	3.36

When laying marble floors it is customary to place the cement bed just ahead of the tile, and the entire work is performed by the setter and helper laying the floor. If the subfloor is sound and even, marble in residential and light commercial work is only ⅜" thick and set in mastic. Setting costs would average about 15¢ per sq. ft. less, unless the area was small and irregular, or set with an intricate pattern.

Smoothing Marble Floors.—After the marble floors have been laid, it is necessary to rub them with carborundum stone to remove the uneven spots at the joints. It is customary to use a rubbing machine powered by gasoline or electricity, which is operated by one man and runs back and forth over the floors and grinds them to an even surface. A machine rents for about $6.00 per hour.

The quantity of floor that can be surfaced per day will vary with the kind of marble used, as it is possible to surface considerably more soft marble than hard marble.

When rubbing and surfacing soft marble, a machine should surface 125 to 150 sq. ft. per 8-hr. day, at the following labor cost per 100 sq. ft.:

	Hours	Rate	Total	Rate	Total
Machine operator	6.0	$. . . .	$. . . .	$14.95	$ 89.70
Foreman	0.7	. . . .	. . . .	15.50	10.85
Cost 100 sq. ft			$. . . .		$100.55
Cost per sq. ft				. . . .	1.01

When rubbing and surfacing hard marble, a rubbing machine should surface 100 to 125 sq. ft. of floor per 8-hr. day, at the following labor cost per 100 sq. ft.:

	Hours	Rate	Total	Rate	Total
Machine operator	7.0	$. . . .	$. . . .	$14.95	$104.65
Foreman	0.8	. . . .	. . . .	15.50	12.40
Cost 100 sq. ft			$. . . .		$117.05
Cost per sq. ft				. . . .	1.17

Setting Marble Floor Border.—Where the marble floor has a different colored marble border, the labor setting the border will cost about the same as the balance of the floor.

Setting Structural Slate.—The labor cost of setting structural slate work, such as base, treads and risers, toilet stalls and partitions, etc., will run about the same as for the same class of marble work and should be estimated accordingly.

If, however, field cutting should be required, as is often the case with window stools and door thresholds, the labor costs should be increased 20 to 25 percent as slate is more difficult to work than marble.

Estimating Quantities of Interior Marble

When estimating quantities of interior marble, any fraction of an inch will be treated as a whole inch when figuring.

No piece of marble will be considered as being less than 6" wide and 1'-0" long.

Thickness.—Thicknesses given are approximate but not exact. For slabs gauged to a given thickness, an extra charge of 10 percent to the list price will be made.

Extra Lengths.—On slabs 10'-0" to 12'-0" long, add 10 percent to list price; on slabs over 12'-0" long, add 20 percent.

Extra Widths.—On slabs over 5'-6" wide, add 10 percent to list price.

Polished Both Sides.—Prices given are based on marble having one side polished unless otherwise noted. If both sides are to be polished, add 85 cents per sq.

ft. for marbles in Group "A" and "B" and $1.00 per sq. ft. for marbles in Group "C" and "D."

Hone Finish.—Hone finish takes the same price as polished work.

Sand Finish.—If marble is finished with sand finish on one side only, deduct 15 cents per sq. ft. If sand finish on both sides, add 15 cents per sq. ft.

Polished Edges.—For each polished edge of ⅞" and 1¼" thickness, add 50 cts. per lin. ft. of edge; on 1½" and 2" thickness, add 60 cts. per lin. ft.

Beveled or Rounded Edges.—For each beveled or rounded edge on ⅞" and 1¼" thickness, add 50 cts. per lin. ft.; on 1½" and 2" thickness, add 60 cts. per lin. ft.

Ogee Molding.—For single member ogee molding $2.50 per lin. ft. of edge on ⅞" thick; $3.00 per lin. ft. on 1¼", and $3.50 per lin. ft. on 1½".

For each additional member, add $1.75 per lin. ft. of edge on ⅞" and 1¼", and $2.00 on 1½".

Countersinking.—For countersinking ¼" deep or less, $1.50 per sq. ft. of surface. For each additional ¼" deeper, or fraction thereof, add extra $1.20 per sq. ft. of surface.

Grooving.—For grooving ¼" deep or less $2.00 per lin. ft. for each groove. For each additional ¼" deeper, or fraction thereof, add extra $1.20 per lin. ft. of each groove.

Rabbeting.—For rabbeting ¼" deep or less, $1.60 per lin. ft. for each rabbet. For each additional ¼" deeper, or fraction thereof, add extra $1.00 cents per lin. ft. of each rabbet.

Special Cutting, Polishing, Drilling, Etc.—All special work other than listed above will be charged for extra.

Shipping Weights of Marble.—The approximate shipping weights of marble, boxed for shipment, are as follows per square foot: ⅞" thick 15 lbs.; 1¼" thick 20 lbs.; 1½" thick 24 lbs.; 2" thick 32 lbs.; per cu. ft. 192 lbs.

Approximate Material Prices per Square Foot of ⅞" Marble, Polished One Side

Always check local sources for price and availability. Foreign marbles are always affected by exchange rates. The trend has been to marbles of even color. Many of the quarries for the more colorful marbles have been shut down, and when a block of such marble is found the price may be out of line because of its scarcity.

Kind of Marble		Kind of Marble	
Alabama Clouded "A" or		Tennessee Marbles	$ 8.00
Cream "A"	$ 8.00	Verde Antique	10.00
Brocadillo	10.00	Westfield Green	11.00
Carthage Gray	7.50	York Fossil	9.00
French Gray	8.00	Foreign Marbles	
Georgia Creole, Cherokee		Alps Green	13.50
or Mezzotint	8.00	Belgian Black	12.00
Gravina	7.50	Black and Gold	16.00
Kasota Pink	7.50	Botticino	12.00
Kasota Yellow	9.00	Bois Jordan	12.00
Light Cloud Vermont	8.00	Breche Oriental	17.00
Napoleon Gray	8.00	Grand Antique, French	20.00
Radio Black	9.00	Grand Antique, Italian	15.00

The above prices are for marble in less than carload lots, uncrated, f.o.b. mill. Crating 30 to 40 cts. per sq. ft. for marble up to 2" thick.

Approximate Material Prices on Sand Finish Marble Floor Tile and
Border ⅞" Thick

Tile sizes 6"x6" to 1'-0"x2'-0": Prices per Square Foot

Alabama Clouded "A"	$5.50	Tennessee Pink	$4.50
Carthage Gray	3.75	Tennessee Gray	4.50
Georgia Creole	5.50	Tennessee Dark	4.00
Georgia Cherokee	3.00	Verde Antique	7.00
Italian White	3.50	Travernelle	5.50
Italian English Vein	3.50	Travertine	5.00

Approximate Material Prices on ⅜" Thinset Marble Tiles Per Square Foot

	Polished	Honed
Cremo	$2.55	$2.50
Negro-Marquina	3.00	2.95
Rose Aurora	2.40	2.35
Statuary Vein	2.10	2.05
Travertine Filled	2.00	1.95
Travertine Unfilled	1.95	1.90
White Italian	2.00	1.95

12"x12" tiles, ⅜" thick, 10 tiles per box.

SLATE TILE

Slate is a natural mica granular crystalline stone. Its main sources are Vermont, Pennsylvania and Virginia. Vermont produces colored slates, including greens; purples; mottled green, purple and red; as well as the standard dark grays.

Slate is usually graded into clear stock and ribbon stock. Ribbon stock has bands of a darker color running through it and is cheaper than clear stock, which has no ribbons but will have some spots and veining in it.

Finishes available are as follows:

Natural Cleft—Natural split or cleaved face, moderately rough with some textural variations.

Sand Rubbed Finish—Wet sand on a rubbing bed is used to eliminate any natural cleft, and leave slate in an even plane.

Honed Finish—This is a semi-polished surface.

Slate is used for spandrels, sills, stools, treads and risers, toilet stalls, floors and walks, fireplace facings and hearths.

Spandrels are generally clear slate with either cleft, rubbed or honed finish. Thickness is generally 1", 1¼" or 1½". Sizes are somewhat limited, with a length of no more than 6 ft. and width of 2'-6" to 4' being standard recommendations, although larger sizes may be available for certain conditions. Many spandrels today are set in metal grid frames. Otherwise, they must be anchored with bronze or stainless steel anchors in each corner.

Sills and stools may be either ribbon or clear stock. Exterior sills are generally furnished with a sand rubbed finish in lengths up to 4' and 1" to 2" in thickness. Sills may have drip grooves cut in at additional cost.

Interior stools and fireplace surrounds are usually 1" thick with honed finish, a maximum length of 4' being most economical.

Treads and risers on the exterior are generally natural cleft finish. Interior work is usually sand rubbed rather than honed as it is both cheaper and gives a good non-slip surface, although the front edges may be honed. Thickness is generally 1" or 1¼" for stairs, 1½" to 2" for landings.

The cost of slate is determined by the grade, the size involved and the finish.

Ribbon grade, often used on stairs, will run around $4.50 per sq. ft. in 1" thickness and $7.00 for 2" thickness, sand finished. Honed finishes will add $0.50 per sq. ft. to the cost.

Clear grade, 1" thick and sand finished, will run $5.00 per sq. ft. in sizes to 3 sq. ft., $6.50 to $8.00 in sizes 3 sq. ft. to 6 sq. ft., and $10.00 in sizes over 6 sq. ft. Lineal foot costs for sills and stools will vary with both length and width. Up to 6" wide stock will run around $4.00 per lineal foot for 1" thickness; $6.00 for 2" thickness. 10" wide stock will cost $6.25 per lineal foot for 1" thickness to $8.50 per lineal foot for 2" thickness.

MORTAR BED METHOD
SLATE FLOORING OVER METAL LATH

LATH NAILED TO FELT.
COVERED ¾" THICK PLYWOOD

NATURAL
CLEFT
FACE
FINISH
SLATE
FLOORING

WOOD FLOOR   JOIST

THINSET MASTIC METHOD
OVER PLYWOOD SUB-FLOOR

PECORA NO. A-135 MASTIC OR
OTHER THINSET MASTIC.
USE A ¼" NOTCHED TROWEL.
BE SURE PLYWOOD NAILED DOWN TIGHT.

½" JOINT OR TIGHT,
NO GROUT JOINT

WOOD
FLOOR   JOIST

Slate for flooring is available for either mortar bed or thin set application. This slate is ¼" thick and available in 3" multiples 6"x6" to 12"x12". It weighs 3¾ lbs. per sq. ft. It is set in ¼" mastic applied with a notched trowel over concrete or plywood. It can have grouted joints or be ordered for butt joints. Cost will run $1.50 per sq. ft. for material and $2.00 for labor.

Slate for heavier duty floors and for outside applications is generally ½", ¾", or 1" thick, weighing 7½ lbs., 11¼ lbs. and 15 lbs. respectively. It is set in 1" mortar beds on concrete slabs or plywood. Plywood should be covered with felt and metal lath, nailed on. Minimum thickness is 1½". If plywood is set over joists without a subfloor, it should be at least ¾" thick. Sizes are 6"x6" to 24"x18", in multiples of 3". Cost runs from $3.00 per sq. ft. for ½" to $4.00 for 1".

09400 TERRAZZO

The cost of terrazzo work varies with the size of the job, size of rooms or spaces where the floors are to be installed, floor designs, strips of brass or other non-rusting material, and method of laying the floors.

Two methods are used in laying terrazzo floors over concrete construction. One is to bond it to the concrete and the other to separate it from the structural slab.

When the first method is used, the concrete fill is provided for by another contractor and should be left 2" below the finished floor. Before the terrazzo contractor installs his underbed, he must see that this concrete fill is thoroughly cleaned of plaster droppings, wood chips and other debris, and wetted to insure cohesion.

The second method is used in buildings where cracking is anticipated either from settlement, expansion and contraction or vibration. In this case the terrazzo contractor begins his work from the structural floor slab up. This method requires a thickness of at least 3". The concrete slab is covered with a thin bed of dry sand over which a sheet of tar paper is laid. Over the paper the underbed is installed as in the first method except that coarser aggregate can be used in the underbed, such as cinders or fine gravel where its thickness exceeds 2½". When this method is used, the cracks originating in the structural slab are not likely to appear on the surface but terminate at the sand bed.

When terrazzo is laid over wood floors a thickness of not less than 2" is required. The floor should first be covered with tarred paper. Over this paper, nail galvanized wire netting NO. 14 gauge, 2" mesh. Then lay the concrete underbed as specified under the second method.

The underbed for terrazzo consisting of 1 part portland cement and 4 parts sharp, screened sand shall be spread and brought to a level not less than ½" nor more than ¾" below the finished floor.

Into this underbed, while still in a semi-plastic state, install metal strips or preassembled decorative units, having proper bonding features.

The terrazzo topping shall be not less than ½" nor more than ¾" in thickness and shall be of granulated marble of colors selected. The topping shall be uniform in composition and the marble granule that appears on the surface shall be used for its entire thickness. The granule to be composed of such proportions of Nos. 1-2-3 sizes as required.

Composition for terrazzo mixture shall be in the proportion of 200 lbs. of marble granules to 100 lbs. of gray or white portland cement, mixed dry, and water added afterward to make the mix plastic, but, not too wet. The mix shall then be placed in the spaces formed by the metal dividing strips and rolled into a compact mass by means of heavy stone or metal rollers until all superfluous cement and water is extracted, after which, it must be hand troweled to an even surface, disclosing the lines of the metal strips on a level with the terrazzo filling.

When the floors have set sufficiently hard, they shall be machine rubbed, using coarse carborundum grit stones for the initial rubbing after which a light grouting of pure gray white portland cement shall be applied to the surface, filling all voids, and allowed to remain until the time of final cleaning.

Floors shall have the grouting coat removed by machine, using a fine carborundum grit, after which it must be thoroughly washed. Do not use acids in cleaning terrazzo floors.

Ramped or other surfaces in terrazzo floors so specified, shall be made anti-slip by the addition of an abrasive aggregate. For heavy duty floors, the proportion shall be 2 parts of abrasive to 3 parts of marble granule and the abrasive shall be mixed in the terrazzo topping for its entire thickness. For light traffic floors, the abrasive shall be sprinkled on the surface only, and show a proportion of 5 parts marble granule and 1 part abrasive aggregate.

Terrazzo Base.—Terrazzo base is usually coved at the floor and may be any height desired. If the base is to finish flush with the finish plaster, a metal base bead shall be furnished and set by the plasterer. The base shall be divided approximately every 4 lin. ft. using metal base dividers. The walls back of the base shall have a scratch coat of cement and sand mortar brought to a line ⅜" back of the finished face of the base, into which the base dividers shall be set. Base shall be finished with a very fine stone, so as to leave the surface at a hone finish.

Estimating the Cost of Terrazzo Work.— In addition to the direct cost of labor and materials entering into terrazzo work, an allowance should be made to provide for wood grounds, carborundum stone for rubbing, wiring for electric power, depreciation on machines, use of hoist, cleaning up rubbish, freight and trucking on machines and tools. This will vary with size and location of job.

The costs on the following pages are based on an average job containing about 3,000 sq. ft. The labor on a small job containing 100 to 200 sq. ft. will run twice as much per sq. ft. as on a job containing 5,000 sq. ft.

Metal Strips in Terrazzo Floors and Base. —Brass strips or other non-rusting metals are used in practically all terrazzo floors and add to the cost in proportion to the quantity used.

Standard brass strips B & S gauge No 20 cost 75 cts. per lin. ft. including waste. Metal strips B & S gauge No. 14, cost 1.45 per lin. ft. including waste in brass and 50 cts. per lin. ft. in zinc.

The cost of metal strips per sq. ft. of floor will vary with the size of the squares or design and should be estimated accordingly.

Terrazzo base is jointed every 4'-0" with vertical metal joint dividers.

Estimating Quantities of Terrazzo Work. —Terrazzo floors are estimated by the sq. ft. and will vary from 2" to 3" thick, depending upon the subfloors, as described in the previous paragraphs.

Terrazzo floor borders are estimated by the sq. ft. when over 12" wide and by the lin. ft. when less than 12" wide.

Terrazzo base is estimated by the lin. ft. stating height. An allowance of 1'-0" should be made for each corner or miter.

Double cove terrazzo base having 2 finished faces, such as ordinarily used under narrow partitions, shower stalls, etc. are estimated by the lin. ft. and the cost is double that of single cove base.

Labor Placing Terrazzo, Floor and Base. —The labor cost of placing terrazzo work will vary with the size of the rooms, design of the floor, size of the squares or pattern requiring brass strips, etc., as the smaller the squares, the more metal strip required and the higher the cost.

The following are approximate quantities of various classes of terrazzo work a crew of 2 mechanics and 3 helpers should install per 8-hr. day :

Description of Work	No. Sq. ft. per 8-Hr. Day	Hrs. per 100 Sq. Ft. Mechanic	Helpers
Floor blocked off into 5'-0" squares	400–450	3.75	5.65
Floor blocked off into 4'-0" squares	375–425	4.00	6.00
Floor blocked off into 3'-0" squares	350–400	4.25	6.40
Floor blocked off into 2'-0" squares	325–375	4.60	6.90
Floor blocked off into 1'-0" squares	275–325	5.33	8.00
Floor border 12" to 24" wide	275–325	5.33	8.00

Terrazzo Cove Base	No. Lin. Ft.	Hours per 100 Lin. Ft.	
Terrazzo cove base, 3" high	60–65	13	13
Terrazzo cove base, 6" high	50–55	16	16

Double cove base, double time given for single base.

Rubbing and Finishing Terrazzo Work.— Terrazzo floors are rubbed by machine and by hand only in corners where the machine cannot reach.

On jobs consisting of large and small rooms, a man will rub and complete about 100 sq. ft. of terrazzo floor per 8-hr. day.

A man should rub and finish about 80 lin. ft. of terrazzo base per 8-hr. day.

Cost of Concrete Subbase Under Terrazzo Floors.—As described in the previous paragraphs, there are three methods of applying the concrete underbed for terrazzo floors; whether bonded to the structural concrete slab, separate from the structural slab or applied over wood floors.

Inasmuch as terrazzo topping is ½" thick for all types of floors, cost of finished floor varies with type of underbed.

Material Cost of 100 Sq. Ft. of Concrete Underbed 1¼ " Thick Bonded
To Structural Concrete Slab

3 sacks portland cement	$. . .	$. . .	$4.00	$12.00
11 cu. ft. sand	. . . .	. . . .	.45	4.95
Cost 100 sq. ft		$. . .		$16.95
Cost per sq. ft		. . . .		.17

Material Cost of 100 Sq. Ft. of Concrete Underbed, Consisting of ¼" Dry Sand, 1 Thickness Tarred Felt and 2" of Concrete, Underbed Separated From Structural Slab

	Rate	Total	Rate	Total
2.5 cu. ft. dry sand	$. . . .	$. . . .	$ .45	$ 1.13
1 sq. No. 15 felt	. . . .	. . . .	3.00	3.00
4.5 sacks portland cement	. . . .	. . . .	4.00	18.00
18 cu. ft. sand	. . . .	. . . .	.45	8.10
Cost 100 sq. ft		$. . . .		$30.23
Cost per sq. ft			. . . .	.30

Material Cost of 100 Sq. Ft. of Concrete Underbed 2" Thick, Applied Over Wood Floor

	Rate	Total	Rate	Total
1 sq. No. 15 felt	$. . . .	$. . . .	$3.00	$ 3.00
105 sq. ft. No. 14 ga. galv. wire netting	. . . .	. . . .	.12	12.60
4.5 sacks portland cement	. . . .	. . . .	4.00	18.00
18 cu. ft. sand	. . . .	. . . .	.45	8.10
Cost 100 sq. ft		$. . . .		$41.70
Cost per sq. ft			. . . .	.42

Material Cost of 100 Sq. Ft. of Terrazzo Floor ½" Thick, Applied Over Concrete Underbed

	Rate	Total	Rate	Total
3 sacks portable cement	$. . . .	$. . . .	$4.00	$12.00
600 lbs. marble granules	. . . .	. . . .	.12*	72.00
0.5 sack port. cement (grout)	. . . .	. . . .	4.00	2.00
Cost 100 sq. ft		$. . . .		$86.00
Cost per sq. ft			. . . .	.86

*Varies according to kind of granules used.
Add for metal strips.

Material Cost of 100 Sq. Ft. Terrazzo Floors Blocked Into 5'-0" Squares, Including 1¼" Concrete Underbed

Underbed 1¼" Thick	Rate	Total	Rate	Total
3 sacks portland cement	$. . . .	$. . . .	$4.00	$ 12.00
11 cu. ft. sand	. . . .	. . . .	.45	4.95
Terrazzo Floor				
3 sacks portland cement	. . . .	. . . .	4.00	12.00
600 lbs. marble granules	. . . .	. . . .	.12*	72.00
0.5 sack cement (grout)	. . . .	. . . .	4.00	2.00
50 lin. ft. metal strips	. . . .	. . . .	.50**	25.00
Cost 100 sq. ft		$. . . .		$127.95
Cost per sq. ft			. . . .	1.28

Labor Cost of 100 Sq. Ft. Terrazzo Floors Blocked Into 5'-0" Squares, Including 1¼" Concrete Underbed

	Hours	Rate	Total	Rate	Total
Mechanic	3.75	$. . . .	$. . . .	$14.95	$ 56.06
Helper	5.65	. . . .	. . . .	12.54	70.85
Labor rub and finish	8.00	. . . .	. . . .	14.95	119.60
Cost 100 sq. ft			$. . . .		$246.51
Cost per sq. ft				. . . .	2.47

*Varies according to kind of granules used. **Varies according to kind and gauge of metal.

If terrazzo is placed over wood subfloor or separated from concrete floor, add difference in cost of underbed.

Material Cost of 100 Sq. Ft. Terrazzo Floors Blocked Into 4'-0" Squares, Including 1¼" Concrete Underbed

Underbed 1¼" Thick	Rate	Total	Rate	Total
3 sacks portland cement	$....	$....	$4.00	$ 12.00
11 cu. ft. sand	....	....	.45	4.95
Terrazzo Floor				
3 sacks portland cement	....	....	4.00	12.00
600 lbs. marble granules	....	....	.12*	72.00
0.5 sack cement (grout)	....	....	4.00	2.00
60 lin. ft. metal strips	....	....	.50**	30.00
Cost 100 sq. ft		$....		$132.95
Cost per sq. ft			....	1.33

Labor Cost of 100 Sq. Ft. Terrazzo Floors Blocked Into 4'-0" Squares, Including 1¼" Concrete Underbed

	Hours	Rate	Total	Rate	Total
Mechanic	4	$....	$....	$14.95	$ 59.80
Helper	6	....	....	12.54	75.24
Labor rub and finish	8	....	....	14.95	119.60
Cost 100 sq. ft			$....		$254.64
Cost per sq. ft				....	2.55

If terrazzo is placed over wood subfloor or separated from concrete floor, add difference in cost of underbed.

Material Cost of 100 Sq. Ft. Terrazzo Floors Blocked Into 3'-0" Squares, Including 1¼" Concrete Underbed

Underbed 1¼" Thick	Rate	Total	Rate	Total
3 sacks portland cement	$....	$....	$4.00	$ 12.00
11 cu. ft. sand	....	....	.45	4.95
Terrazzo Floor				
3 sacks portland cement	....	....	4.00	12.00
600 lbs. marble granules	....	....	.12*	72.00
0.5 sacks cement (grout)	....	....	4.00	2.00
80 lin. ft. metal strips	....	....	.50**	40.00
Cost 100 sq. ft		$....		$142.95
Cost per sq. ft			....	1.43

Labor Cost of 100 Sq. Ft. Terrazzo Floors Blocked Into 3'-0" Squares, Including 1¼" Concrete Underbed

	Hours	Rate	Total	Rate	Total
Mechanic	4.25	$....	$....	$14.95	$ 63.54
Helper	6.40	....	....	12.54	80.26
Labor rub and finish	8.00	....	....	14.95	119.60
Cost 100 sq. ft			$....		$263.40
Cost per sq. ft				....	2.64

If terrazzo is placed over wood subfloor or separated from concrete floor, add difference in cost of underbed.

*Varies according to kind of granules used. **Varies according to kind and gauge of metal.

Material Cost of 100 Sq. Ft. Terrazzo Floors Blocked Into 2'-0" Squares,
Including 1¼" Concrete Underbed

	Rate	Total	Rate	Total
Underbed 1¼" Thick				
3 sacks portland cement	$. ...	$. ...	$4.00	$ 12.00
11 cu. ft. sand	....	....	.45	4.95
Terrazzo Floor				
3 sacks portland cement	....	....	4.00	12.00
600 lbs. marble granules	....	....	.12*	72.00
0.5 sack cement (grout)	....	....	4.00	2.00
120 lin. ft. metal strips	....	....	.50**	60.00
Cost 100 sq. ft		$. ...		$162.95
Cost per sq. ft			....	1.63

Labor Cost of 100 Sq. Ft. Terrazzo Floors Blocked Into 2'-0" Squares,
Including 1¼" Concrete Underbed

	Hours	Rate	Total	Rate	Total
Mechanic	4.6	$. ...	$. ...	$14.95	$ 68.77
Helper	6.9	....	....	12.54	86.53
Labor rub and finish	8.0	....	....	14.95	119.60
Cost 100 sq. ft			$. ...		$274.90
Cost per sq. ft			....		2.75

If terrazzo is placed over wood subfloor or separated from concrete floor, add difference in cost of underbed.

Material Cost of 100 Sq. Ft. Terrazzo Floors Blocked Into 1'-0" Squares,
Including 1¼" Concrete Underbed

	Rate	Total	Rate	Total
Underbed 1¼" Thick				
3 sacks portland cement	$. ...	$. ...	$4.00	$ 12.00
11 cu. ft. sand	....	....	.45	4.95
Terrazzo Floor				
3 sacks portland cement	....	....	4.00	12.00
600 lbs. marble granules	....	....	.12*	72.00
0.5 sacks cement (grout)	....	....	4.00	2.00
210 lin. ft. metal strips	....	....	.50	105.00
Cost 100 sq. ft		$. ...		$207.95
Cost per sq. ft			....	2.08

Labor Cost of 100 Sq. Ft. Terrazzo Floors Blocked Into 1'-0" Squares,
Including 1¼" Concrete Underbed

	Hours	Rate	Total	Rate	Total
Mechanic	5.35	$. ...	$. ...	$14.95	$ 79.98
Helper	8.00	....	....	12.54	100.32
Labor rub and finish	8.00	....	....	14.95	119.60
Cost 100 sq. ft			$. ...		$299.90
Cost per sq. ft			....		3.00

If terrazzo is placed over wood subfloor or separated from concrete floor, add difference in cost of underbed.

Material Cost of 100 Lin. Ft. of 6" Terrazzo Cove Base

	Rate	Total	Rate	Total
Scratch Coat				
1.5 sacks portland cement	$. ...	$. ...	$4.00	$ 6.00
3 cu. ft. sand	....	....	.45	1.35

*Varies according to kind of granules used. **Varies according to kind and gauge of metal.

Terazzo Base

1 sack portland cement	. . .	. . .	4.00	4.00
200 lbs. marble granules	. . .	. . .	.12*	24.00
25 metal dividers	. . .	. . .	.50**	12.50
Cost 100 lin. ft		$. . .		$47.85
Cost per lin. ft			. . .	.48

Labor Cost of 100 Lin. Ft. of 6" Terrazzo Cove Base

	Hours	Rate	Total	Rate	Total
Mechanic	16	$. . .	$. . .	$14.95	$239.20
Helper	16	. . .	. . .	12.54	200.64
Labor rub and finish	10	. . .	. . .	14.95	149.50
Cost 100 lin. ft			$. . .		$589.34
Cost per lin. ft			. . .		5.89

Double cove base, double cost of single cove base.

Abrasive Materials for Terrazzo Floors and Stair Treads.—Where an anti-slip surface is required on terrazzo floors, platforms, landings and treads, mix 2 parts abrasive aggregates with 3 parts marble granule. For heavy duty floors the abrasive aggregate shall be mixed in the terrazzo topping for its entire thickness. For light traffic floors, the abrasive shall be sprinkled on the surface only, to show in the proportion of 3 parts marble granule and 1 part abrasive aggregate.

Abrasive aggregates for 100 sq. ft. of heavy duty floors should cost as follows :

	Rate	Total	Rate	Total
240 lbs. abrasive aggregates	$. . .	$. . .	$0.50	$120.00
Cost per sq. ft		. . .		1.20

Deduct cost of 240 lbs. of marble granules included in itemized estimates.

Abrasive aggregates for 100 sq. ft. of light duty floors should cost as follows:

	Rate	Total	Rate	Total
10 lbs. abrasive aggregates	$. . .	$. . .	$0.50	$5.00
Cost per sq. ft		. . .		.05

For prices on special terrazzo floors such as venetian, mosaic and conductive, it is advisable to contact a local installation contractor.

There are several products on the market that resemble terrazzo floors and are installed by terrazzo mechanics. These are magnisite, latex, epoxy, and polyester resin type terrazzo floors. These floors can be installed over any sound existing floor in ½" thickness, with the exception of the epoxy and polyester resin types which can be installed as thin as the marble granules will allow. The costs of these floors are similar to that of regular terrazzo.

Precast terrazzo is available for base, floor, wainscots and stair treads. Straight 6" base will run around $3.50 per l. f., and one mechanic will set 125 lin. ft. per 8 hr. day. Floor tiles, 1" thick will run $3.70 per sq. ft. and one man will set 60-70 sq. ft. per 8 hr. day. Stair treads will cost around $9.00 per lineal foot, as these will be heavy, on a stair of any width it will take a mechanic and helper to set 100 lineal ft. per 8 hr. day.

09500 ACOUSTICAL TREATMENT

A large variety of prefabricated acoustical products is available on today's market. In general, they fall into three categories : acoustical tiles which are mounted on ceilings, suspended boards, clipped or splined to suspension systems; acoustical

*Varies according to kind of granules used. **Varies according to kind and gauge of metal.

panels for lay-in systems on exposed runners; and metal pan systems for mechanical suspension. Many systems today are also engineered to distribute air and light as well as to control sound; also, some systems are rated as fire resistive assemblies.

Acoustical tiles are applied to ceilings and walls for the purpose of absorbing and deadening sound in offices, banks and other spaces where the utmost in quiet is desired, and in auditoriums to obtain optimum reverberation time.

Acoustical tiles are made of many different materials, such as wood fiber, cane fiber, mineral wool, cork, specially processed mineral filaments, perforated metal units, and other insulating materials.

They are made in a variety of sizes, such as 12"x12", 12"x24", 24"x24", 24"x48", etc., and vary from ½" to 2" thick, depending upon the materials used.

Acoustical tile may be installed over sound, dry plaster or concrete that is thoroughly seasoned, over gypsum board, or nailed or screwed to 1"x3" wood furring strips or by means of mechanical suspension systems.

Acoustical tile work is simple to estimate, but as it is almost always factory cut and prefinished, it is extremely important to see that job conditions are of the very best, as once installed, there is very little opportunity to "touch up" work. Call backs can quickly eat up profits as well as ruin reputations.

One of the most important conditions to check is temperature and humidity. These should always approximate the interior conditions that will exist when the building is occupied.

All plastering, poured roof decks, concrete and floor work should be complete and dry; all windows and doors in place and glazed. If weather demands it, the building heating, rather than space heating, should be turned on and kept on as though the building were occupied.

The trend today is to square edge tile which minimizes joint visibility. However, tile work will always have slight unevennesses in it and if the job conditions include cove lights or high windows where light strikes the tile surface at a small angle, tile with beveled edges will help conceal any unevenness.

Budget Prices on Acoustical Tile.—Most acoustical tile jobs are sublet to concerns specializing in this work as they have mechanics experienced in applying their particular brands of tile and can complete the job at less cost than a general contractor who purchases the tile and attempts to install them.

The following are approximate prices on the various classes of tile erected in place on fair size jobs as they might be quoted by a subcontractor including his overhead and profit.

Type of Installation	Price per Sq. Ft.
½" wood fibre tile surface applied	$1.10
⅝" mineral tile suspended*	1.40
¾" mineral tile suspended*	1.55
1" pad on steel pan suspended*	2.40
1" pad on alum. pan suspended*	3.50
⅝" mineral fibreboard suspended*	.95
Air distributing ceilings-board* 2'x2'	1.35
Air distributing ceilings tile* 12"x12"	1.65

*Including hangers but not additional channel supports.

Acoustical Tile Directly Applied.—When cemented, four spots of cement or adhesive should be applied to each tile on each corner of a 12"x12" tile. The spots should be placed as uniformly as possible with relation to the corners of the tile. A sufficient area on each corner of the tile should be primed prior to the application of the spot to receive the entire spot when it has been pressed to its final diameter. The spots of cement should be applied to these primed areas using a putty knife or trowel. The spot should be approximately the size and shape of a walnut, and should be nearly uniform in size and shape except where thicker spots are required to level surfaces. They should stand out from the tile about ½" to ¾" so as to contact the surface at about the same time. The diameter of the spot when pressed in place should be approximately 2½" to 2¾" and about ⅛" thick.

In placing, the tile should be held with both hands so that when the pressure is applied it will be between the spots rather than outside them. The tile should be held as nearly parallel to its final position as possible and pressed to position with a firm pressure. A slight lateral sliding motion of the hands while setting is necessary to settle the cement into the ceiling surface. It is important that the tile be pressed to level when setting.

A crew of at least two men is required to handle the work of installation efficiently, one man on the scaffold applying the adhesive to the tile and the other man placing the tiles on the ceiling.

One gallon of acoustical cement is sufficient for approximately 60 sq. ft. of tile under average conditions, although this varies somewhat with the surface to which it is applied. Adhesive costs about $7.50 per gallon.

The labor cost of placing acoustical tile on ceilings will vary with the size of the rooms or spaces where the tile are to be applied, pattern in which the tile are laid, height of scaffolding, etc.

The following quantities are based on 2 mechanics working together from a 4 to 6-ft. scaffold and laying tile in plain square or ashlar designs :

Size of Acoustical Tile	No. Pcs. Placed per 8-Hr. Day	No. Sq. Ft. Placed per 8-Hr. Day	Mechanic Hrs. per 100 Sq. Ft.
12"x12"..........................	600-700	600-700	2.5
12"x24"..........................	400-450	800-900	2.0

When acoustical tile are laid in herringbone pattern, it requires 30 to 40 percent more time than regular squares or common bond. Diagonal patterns require 20 to 25 percent more time than regular squares or common bond, while mixed ashlar designs require about 50 percent more time than straight squares or common bond.

When placing tile in auditoriums and other spaces requiring high scaffolding, the additional cost of the scaffolding will have to be added to the cost of installing the tile.

Large jobs are applied faster proportionately than small jobs.

Material Cost of 100 Sq. Ft. 12"x12" Acoustical Tile on Ceilings

	Rate	Total	Rate	Total
100 sq. ft. mineral tile ¾".............................	$. . . .	$. . . .	$ .65*	$65.00
1.75-gal. adhesive...	. . . .	. . . .	7.50	13.13
Cost 100 sq. ft..		$. . . .		$78.13
Cost per sq. ft...		. . . .		.78

*Cost of tile varies according to kind and thickness.

Labor Cost of 100 Sq. Ft. 12"x12" Acoustical Tile on Ceilings

	Hours	Rate	Total	Rate	Total
Mechanics...	2.5	$. . . .	$. . . .	$16.47	$41.18
Cost per sq. ft...................................			. . . .		.41

Acoustical Tile On Suspended Systems

Lay-in type systems for 2'x2' and 2'x4' grids will have a material cost of around

28¢ and 24¢ per sq. ft. respectively not including any channel supports if they are needed. One man should install 700 sq. ft. per eight hour day of 2'x4' grid, 600 sq. ft. for 2'x2' grid.

Concealed spline systems are more time consuming. Material costs will run around 30¢ a sq. ft. and one man will install 500 sq. ft. per day.

If channel carriers are required, the material cost will be about 8¢ per sq. ft. and one man should install about 500 sq. ft. per 8 hr. day.

Labor Cost To Install 100 Sq. Ft. 2'x2' Lay-in Grid

	Hours	Rate	Total	Rate	Total
Mechanic	1.3	$. . . .	$. . . .	$16.47	$21.41
Cost per sq. ft					.22

Labor Cost To Install 100 Sq. Ft. 2'x4' Lay-in Grid

	Hours	Rate	Total	Rate	Total
Mechanic	1.2	$. . . .	$. . . .	$16.47	$19.77
Cost per sq. ft					.20

Labor Cost To Install 100 Sq. Ft. 12 x 12 Z Bar System Grid

	Hours	Rate	Total	Rate	Total
Mechanic	1.6	$. . . .	$. . . .	$16.47	$26.35
Cost per sq. ft					.27

Labor Cost To Install 100 Sq. Ft. Channel Ceiling Supports

	Hours	Rate	Total	Rate	Total
Mechanic	1.6	$. . . .	$. . . .	$16.47	$26.35
Cost per sq. ft					.27

Acoustical Tile For Suspended Grid Systems.—This material is made "cut to size" ready to install into the grid system. 24"x24" and 48" boards will cost about 30¢ to 40¢ per sq. ft. depending on the type of tile, and one man should install about 100 sq. ft. per hour. The 12"x12" tiles for the Z bar system will cost about 50¢ to 65¢ per sq. ft. and one man should install about 100 sq. ft. in two (2) hours.

Metal Pan Type Acoustical Tile.—This type of acoustical treatment consists of metal units 12"x24" in size, having center grooves which result in the appearance of 12"x12" units when applied. The edges, which are beveled and returned vertically, are firmly held in place on the 12" sides by special tee runners. Within the metal units and supported on crimped, galvanized wire mesh, or on miniature channels, are mineral wool absorbent pads.

These systems will install at the rate of around 250 sq. ft. per day, including sound absorbing pads, but without channel supports if they are required. Material costs will run around $1.50 for steel, $2.50 for aluminum and $3.00 for stainless.

Labor Cost To Install 100 Sq. Ft. Suspended Steel Pan System

	Hours	Rate	Total	Rate	Total
Mechanic	3.2	$. . . .	$. . . .	$16.47	$52.71
Cost per sq. ft					.53

09550   WOOD FLOORING

Estimating Quantities of Wood Flooring. —When estimating the quantity of board feet (b.f.) of wood flooring required for any job, take the actual number of sq. ft. in any room or space to be floored, and add allowances as given in the following tables :

Measured Size Inches	Finished Size Inches	No. of Pieces in Bundle	Add for Waste Percent	To Obtain Quantity of Flooring Repaired Multiply Area by	No. B.F. Flg. Required for 100 Sq. Ft. Floor
1x1	3/8 X 7/8	24	162/3	11/6 or 1.17	117
1x2	3/8 x11/2	24	331/3	11/3 or 1.33	133
1x21/2	3/8 x2	24	25	11/4 or 1.25	125
1x21/4	25/32x11/2	12	501/3	11/2 or 1.50	150
1x23/4	25/32x2	12	371/2	13/8 or 1.375	1371/2
1x3	25/32x21/4	12	331/3	11/3 or 1.33	133
1x4	25/32x31/4	8	25	11/4 or 1.25	125

Amount of Surface 1,000 Board Feet of Flooring Will Cover and
Quantity of Nails Required to Lay it

How Measured	Size Flooring	Will Cover Sq. Ft. Fl.	Nailed Every	Cut Nails		Lbs. Nails Req'd. Helically threaded nails	
1x2	3/8x11/2	750	8 in.	20 lbs.	4 "d" cut		
1x21/2	3/8x2	800	8 in	17	4 "d" cut		
1x21/4	25/32x11/2	667	12 in.	70	8 "d" cut	60 lbs	7 "d" screw
1x23/4	25/32x2	727	12 in.	56	8 "d" cut	47	7 "d" screw
1x3	25/32x21/4	750	10 in.	64	8 "d" cut	54	7 "d" screw
1x4	25/32x31/4	800	10 in.	29	8 "d" cut	24	7 "d" screw

Drill flooring for nails, if possible, for best results. No predrilling is required for helically threaded nails.

Note : The above figures are based on laying the flooring straight across a rectangular room without producing any design whatever.

Rules for Grading Hardwood Flooring.— In many instances the specifications of the architects and owners are confusing and not in accordance with the standard rules for grading as adopted by the manufacturers. The rules adopted by the flooring manufacturers are given below, so that they may be compared with the architect's specifications, and the proper grade of flooring figured.

FIRST GRADE

First Grade—25/32 inch and thicker, shall have the face practically free of all defects, but the varying natural color of the wood shall not be considered a defect. Standard lengths in all widths in this grade shall be in 2 foot bundles and longer as the stock will produce. Not over 30% of the total footage shall be in bundles under 4 feet, and not more than 17% of the total footage shall be in 2 foot bundles.

SECOND GRADE

Second Grade—25/32 inch and thicker, will admit tight, sound knots and slight imperfections in dressing, but must lay without waste. Standard lengths in all widths in this grade shall be in 2 foot bundles and longer as the stock will produce. Not over 45% of the total footage shall be in bundles under 4 feet, and not more than 25% of the total footage shall be in 2-foot bundles.

THIRD GRADE

Third Grade—25/32 inch and thicker must be of such character as will lay and give a good serviceable floor. Standard lengths in all widths of this grade shall be in 1¼ foot bundles and longer as the stock will produce. Not over 65% of the total footage shall be in bundles under 4 feet, and not more than 27% of the total footage shall be in 1¼ -foot bundles.

COMBINATION GRADES

Second and Better Grade—In all thicknesses and widths is a combination of First and Second Grades developing in the strip without crosscutting for each grade. The lowest grade pieces admissible shall not be less than standard Second Grade. Standard lengths in all widths in this grade shall be in 2 foot bundles and longer as the stock will produce. Not over 40% of the total footage shall be in bundles under 4 feet, and not more than 20% shall be in 2-foot bundles.

Third and Better Grade—In all thicknesses and widths is a combination of First, Second and Third Grades developing in a strip without crosscutting. The lowest grade pieces admissible shall not be less than standard Third Grade. Standard lengths in all widths of this grade shall be in 1¼ foot bundles and longer as the stock will produce. Not over 50% of the total footage shall be in bundles under 4 feet.

SPECIAL COLOR GRADES
The Finest Grades Produced

SELECTED FIRST GRADE LIGHT NORTHERN HARD MAPLE. This grade is selected for light color. The color tones in individual strips will vary somewhat, but after laying, this grade provides a luxurious "light" appearing floor. It costs around $2.75 per sq. ft. for material.

SELECTED FIRST GRADE AMBER NORTHERN HARD MAPLE. This grade is selected for amber color. The color tones in individual strips will vary somewhat, but after laying, this grade provides a luxurious "amber" appearing floor. The cost is $2.50 per sq. ft. for material.

OAK FLOORING
Quarter Sawed

CLEAR.—The face shall be practically clear, admitting an average of ⅜ of an inch of bright sap. The question of color shall not be considered. Bundles to be 2 foot and up. Average length 3¾ feet.

SELECT.—The face may contain sap, small streaks, pin worm holes, burls, slight imperfections in working, and small tight knots which do not average more than one to every 3 feet. Bundles to be 2 foot and up. Average length 3¼ feet.

Plain Sawed

CLEAR.—The face shall be practically clear, admitting an average of ⅜ of an inch of bright sap. The question of color shall not be considered. Bundles to be 2 foot and up. Average length 3¾ feet.

SELECT.—The face may contain sap, small streaks, pin worm holes, burls, slight imperfections in working, and small tight knots which do not average more than one to every 3 feet. Bundles to be 2 foot and up. Average length 3¼ feet.

No. 1 COMMON.—Shall be of such nature that will lay a good residential floor and may contain varying wood characteristics, such as flags, heavy streaks and checks, worm holes, knots and minor imperfections in working. Bundles to be 2 foot and up. Average length 2¾ feet.

No. 2 COMMON.—May contain sound natural variations of the forest product and manufacturing imperfections. The purpose of this grade is to furnish an economical floor suitable for homes, general utility use, or where character marks and contrasting appearance is desired. Bundles to be 1¼ foot and up. Average length 2¼ feet.

1¼ SHORTS.—Pieces 9 to 18 inches long are to be bundled together and designated as 1¼' shorts. Pieces grading No. 1 Common, Select and Clear to be bundled together and designated No. 1 Common and Better with pieces grading No. 2 Common bundled separately and designated as such. Although pieces 6" under and only 3" over the nominal length of the bundle may be included, the pieces must average 1¼' which is achieved through the natural preponderance of longer lengths.

Standard Thicknesses and Widths

25/32" thickness; widths 1½" face, 2" face, 2¼" face, and 3¼" face.
⅜" thickness; width 1½" face and 2" face.
½" thickness; width 1½" face and 2" face.

Above tongue and groove and end matched.

Square Edge Strip Flooring

Grades as shown above but bundling and lineals as follows:
5/16x2", 5/16x1½", 5/16x1⅓", 5/16x1¼", 5/16x1⅛", 5/16x1", 5/16 x⅞".

Also made rough back in 11/32x2" and 11/32x1½".
CLEAR.—Bundled 2 feet and up. Average length 5½ feet.
SELECT.—Bundled 2 feet and up. Average length 4½ feet.
No. 1 COMMON.—Bundled 2 feet and up. Average length 3½ feet.
No. 2 COMMON.—May contain defects of all characters, but will lay a serviceable floor.
Bundles to be 1¼ feet and up. Average length 2½ feet.
All faces shown above in 5/16" square edge are finished 1/64" over face.

Approximate Prices of Oak Flooring

Oak flooring prices will vary widely throughout the country. A recent quote gives 25/32" x 2¼ oak at $1.70 per sq. ft., select, $1.50 common. Clear will run 30% more than select, if it is at all available.

Grading Rules on Prefinished Hardwood Flooring
OAK

White and Red Oak to be separated in each grade. Grades are established after the flooring has been sanded and finished.
PRIME GRADE: Face shall be selected for appearance after finishing, but sapwood and the natural variations of color are permitted. Minimum average length 4'. Bundles 2' and longer.
STANDARD GRADE: Will contain sound wood characteristics which are even and smooth after filling and finishing and will lay a sound floor without cutting. Minimum average length 3'. Bundles 1¼' and longer.
STANDARD AND BETTER GRADE: A combination of Prime and Standard to contain the full product of the board except that no pieces are to be lower than Standard Grade. Minimum average length 3½'. Bundles 1¼' and longer.
TAVERN GRADE: Shall be of such nature as will make and lay a serviceable floor without cutting, but purposely containing typical wood characteristics, which are to be properly filled and finished. Minimum average length 2½'. Bundles 1¼' and longer.

BEECH AND PECAN
(Will be furnished only in a combination grade of Tavern and Better)
TAVERN AND BETTER: A combination of Prime, Standard, and Tavern to contain the full product of the board, except that no pieces are to be lower than Tavern Grade. Minimum average length 3'. Bundles 1¼' and longer.

GENERAL RULES
(All Species)

Hardwood flooring is bundled by averaging the lengths. A bundle may include pieces from 6 inches under to 6 inches over the nominal length of the bundle. No piece shorter than 9 inches admitted.

The percentages under 4 feet referred to apply on total footage in any one shipment of the item.

¾ inch allowance shall be added to the face length when measuring the length of each piece of flooring.

Flooring shall not be considered of standard grade unless the lumber from which the flooring is manufactured has been properly kiln dried.

Laying and Finishing Wood Floors

The following pages contain detailed itemized costs of laying and finishing all kinds of soft and hardwood floors.

Labor Laying 100 Sq. Ft. of Hardwood Floors

Description of Work	Laying Floors		
	Sq. Ft. Laid 8-Hr. Day	Carp. Hours	Labor Hours
Ordinary Workmanship			
25/32"x31/4" face softwood floors for porches, kitchens, factories, stores, etc.	400-500	1.8	0.6
25/32"x21/4" face Third Grade Maple, for warehouse, factory and loft building floors..............	375-425	2.0	0.6
25/32"x21/4" face Oak or Birch in residences, apartments, stores, offices, etc	250-300	2.9	0.6
Same as above laid by experienced floor-layer....	300-350	2.5	0.6
25/32"x11/2" face Third Grade Maple for warehouse, factory and loft building floors..............	300-325	2.3	0.7
25/32"x11/2" face Oak or Birch in residences, apartments, stores, offices, etc	150-200	4.5	0.8
Same as above laid by experienced floor-layer....	225-275	3.2	0.8
First Grade Workmanship			
25/32"x21/4" face Oak or Birch in fine residences, apartments, hotels, stores and offices	200-225	3.8	0.6
Same as above laid by experienced floor-layer....	225-250	3.3	0.6
25/32"x11/2" face Oak or Birch, same class of work as described above	120-150	6.0	0.8
Same as above laid by experienced floor-layer....	175-200	4.3	0.8

In the larger cities there are carpenters who make a specialty of laying and finishing floors and they seldom do any other kind of work. Due to their long experience, they not only do a better job but accomplish more work per day than the average carpenter and they expect to be paid accordingly. For this reason production times are given for floors laid and finished by both carpenters and floorlayers.

Another reason some floorlayers will lay more floors than carpenters is because they do not nail the flooring 8" or 16" on centers but will nail at every other bearing unless watched continually.

The first or Ordinary Workmanship includes the grade of work found in moderate priced homes, apartments, stores, etc., where the floor laying permits some hammer marks. Considerable savings can be obtained by laying the strip flooring all the way through prior to the erection of partitions.

First Grade Workmanship includes all classes of high class buildings where the best in workmanship is required, with all flooring closely driven up, laid free from hammer marks and with all nails set.

Labor Laying ⅜" or ½" Hardwood Flooring.—Figure same costs as given for 25/32" flooring.

Setting Sleepers in Mastic on Concrete Floors.—A sound, fast, economical way to provide nailing for strip wood flooring installed over concrete subfloors is the sleeper-in-mastic method, used extensively in the construction of slab on grade houses.

Assuming concrete slab is smooth, level and dampproofed to prevent moisture seepage, 2"x4" wood sleepers are laid flat in hot or cold mastic and are immediately ready to receive strip flooring. Installation is as follows :

After priming concrete slab, hot or cold mastic is spread to a 3/32" depth over the entire slab or in ¼" deep rivers along parallel lines in positions where sleepers are to be placed.

Wood sleepers should be straight, flat, dry, 2"x4" lumber in random lengths from 18" to 48". For best results, sleepers should be treated with an approved non-creosote wood preservative. Lay sleepers in mastic, flat side down, in staggered rows on 12" centers at right angles to direction of flooring. Ends should overlap 4" and end joints should be staggered by alternating short and long lengths. Leave 1" clearance between sleepers and all vertical surfaces to allow for normal expansion.

Laying strip wood flooring over sleepers in mastic should follow the same general procedure as for over wood subfloors with the following additional precautions:

Use end-matched flooring only.

Individual strips of flooring should span at least two sleeper spacings, with no adjacent joints occurring in the same spacing.

Flooring strips passing over sleeper laps should be nailed to both sleepers.

Material coverages and approximate prices are as follows:

Asphalt primer, 200 to 400 sq. ft. per gal., depending upon porosity of concrete surface. Price $3.00 per gal.

Hot mastic, 25 sq. ft. per gal. Price $2.25 per gal.

Cold mastic, 40 to 50 sq. ft. per gal. Price $2.00 per gal.

For average size rooms, 100 sq. ft. of floor area will require 120 to 135 lin. ft. or 80 to 90 ft. b.m. of 2"x4" wood sleepers.

Under ordinary conditions, a floorlayer should prime concrete floor, spread mastic and place wood sleepers at the rate of 100 sq. ft. per hour.

Labor Laying 1,000 Ft. B.M. Soft and Hardwood Flooring

Description of Work	Ordinary Workmanship		
	Ft. B. M. Laid Per 8-Hr. Day	Carp. Hours	Labor Hours
25/32"x31/4" face softwood floors for porches, kitchens, factories, stores, etc	500-600	14.5	4
25/32"x21/4" face Third Grade Maple, used in warehouses, factory and loft buildings, etc......	500-575	15.0	4
25/32"x21/4" face Birch or Oak flooring in residences, apartments, stores etc	350-400	21.0	4
Same as above laid by experienced floor-layer....	400-450	19.0	4
25/32"x11/2" face Third Grade Maple, used in warehouses, factory and loft buildings, etc......	450-500	17.0	4
25/32"x11/2" face Birch or Oak flooring in residences, apartments, store and office buildings, etc..	250-280	30.0	5
Same as above laid by experienced floor-layer....	350-400	21.0	5
First Grade Workmanship 25/32"x11/4" face Oak or Birch flooring in fine residences, apartments, hotels, stores and offices............	275-300	28	4
Same as above laid by experienced floor-layer....	325-375	23	4
25/32"x11/2" face Oak or Birch, same class of work as described above for 21/4" face flooring	175-225	39	5
Same as above laid by experienced floor-layer....	275-300	28	5

Sanding Wood Floors by Machine

Practically all wood floors today are sanded and finished by machine. The cost varies, depending on the kind of flooring, whether old or new work, class of workmanship, size of rooms, etc.

The floors are first sanded with a floor sanding machine and then the edges are finished, using a disc type edging machine, which is capable of sanding up to the base shoe or quarter round.

The class of workmanship of number of operations governs production and cost but where just an ordinary grade of workmanship is required in average size rooms in houses, apartments, offices, etc., a good machine and operator should finish and edge 800 to 900 sq. ft. per 8-hr. day, at the following labor cost per 100 sq. ft. :

	Hours	Rate	Total	Rate	Total
Operator-finisher	1	$. . . .	$. . . .	$16.47	$16.47
Cost per sq. ft			. . . .		.17

When used in store rooms, auditoriums and other large floor areas, a machine and operator should finish and edge 1,000 to 1,250 sq. ft. of new floor per 8-hr. day, at the following labor cost per 100 sq. ft. :

	Hours	Rate	Total	Rate	Total
Operator-finisher	0.75	$. . . .	$. . . .	$16.47	$12.35
Cost per sq. ft			. . . .		.12

Surfacing Floors by Machine, First Grade Workmanship.—In residences, apartments, hotel, store and office buildings, etc., where first grade workmanship is required, eliminating all irregularities and waves in the floor, about four cuts are necessary; first with No. 1½ or 2 grit sandpaper, then with No. ½ or 1 grit paper and the last two cuts with No. 0 paper.

On work of this kind an experienced operator should finish and edge 400 to 500 sq. ft. of floor per 8-hr. day, at the following labor cost per 100 sq. ft. :

	Hours	Rate	Total	Rate	Total
Operator-finisher	1.75	$. . . .	$. . . .	$16.47	$28.82
Cost per sq. ft			. . . .		.29

Resurfacing Old Floors by Machine.—When resurfacing old wood floors, including the removal of old varnish, dirt, grease spots, etc., the cost will vary with the condition of the old floors. This variation will run from 300 to 1,000 sq. ft. per 8-hr. day and should be determined by the estimator after examining the old floors.

There is little or no difference in resurfacing an old floor or surfacing a new one except that under normal conditions one extra cut with a special open-faced abrasive is usually necessary to remove old varnish or paint.

Chart for Floor Surfacing Operations

The following chart gives the proper abrasives or grits for surfacing all classes of wood floors :

New Floors			
Kind of Floors	Operation	Floor Conditions	Proper Grit
Oak, Maple and Close-Grained Hardwood Floors	Roughing	Ordinary Floors	2-2½
		Well Laid Floors	1½
		Very Uneven Floors	2½
		Very Hard Floors	2½
	Finishing	Ordinary Finish	½
		Fine Finish	0
		Extra Fine Finish	00
		Rough Finish	1
Parquet Floors	Roughing		1½
	Finishing	Semi-Finish	½
		Finishing	0
Cork Tile Floors	Roughing		1½
	Finshing	Semi-Finish	½
		Finishing	0
Old Floors			
Resurfacing Old Floors— Soft and Hard Woods	Removing Varnish, Paints, etc.	Ordinary Conditions	3½
		Extra Heavy Coat Paint or Varnish	4
	Roughing	Ordinary Floors	2-2½
	Finishing	Ordinary Finish	½
		Fine Finish	0
		For Penetrating Sealer	00

Sanding Wood Floors

The following specifications list the operations necessary to produce a first class job. Cheaper grades of workmanship will require less sanding and consequently less labor—and lower costs.

Sanding Old Wood Floors Laid in Strips.—The first operation is to remove the surface finish by sanding in the direction of the grain of the wood until all surface finish has been removed, using a drum type sanding machine, with extra heavy open coat sandpaper grit No. 3½ or 4.

The second operation is to remove the effects of the first operation and level the floor. Sand as in first operation, using same type machine with sharp sandpaper, grit No. 2 or 2½, to remove the effects of the coarse open coat sandpaper and level the floor surface.

The third operation is to use a spinner along the edges of the floor and hand scrape all corners where the machines could not reach. Use a No. 3½ or 4 grit paper and repeat, using No. 2 or 2½ grit paper.

The fourth operation is to sweep the entire floor to remove the coarse grits and abrasives left by the previous operations.

The fifth operation is to sand as in the first operation using the same type machine with No. 0 or No. ½ fine grit.

The sixth operation is the final smoothing of the edges. Sand as in third operation using same type of machine with fine sandpaper, grit 0 or ½.

After sweeping the entire floor clean it is ready for any kind of floor finish.

Sanding New Wood Floors Laid in Strips.—The first operation is to sand to a level surface. Sand in the direction of the grain of the wood until all the high edges or over-wood on the floor has been sanded off to a level surface, using a drum type sanding machine with No. 1½ or 2 grit sandpaper.

The second operation is to remove the scratch pattern of the first sanding. Sand in the direction of the grain of the wood until the scratch pattern of the first operation is eliminated using the same type sanding machine with No. 1 or ½ sandpaper. Frequent changes of sandpaper must be made to insure cutting rather than sliding over the surface.

The third operation is to use a spinner around the edges and scrape by hand where it is impossible to use the machine, without crossing the grain of the wood. Use same grits as on drum sander.

The fourth operation is to sweep the entire floor free from any grit and abrasives and then sand in direction of the grain of the wood, using the drum type machine as in the first operation, with No. 0 sandpaper. Change the sandpaper frequently to insure cutting rather than sliding over the surface.

After sweeping the floor clean it is ready for any kind of floor finish.

Sanding Old Parquet and Herringbone Wood Floors.—The first operation is to sand lengthwise of the room or at 45 degrees, using a drum type sanding machine with No. 3½ or 4 open coat grit, using a very light pressure on the drum.

The second operation is to remove the coarse marks of the first operation by sanding crosswise of the room using same type machine with No. 1 or 1½ grit. Change paper frequently to assure cutting rather than sliding over the surface. Then sand lengthwise of the room using No. ½ or No. 1 grit.

The third operation is to use a portable edging machine or hand scrape all places where the drum machine and edger cannot reach. Sweep the entire floor to remove coarse grits and abrasives left by the previous operations.

The fourth operation is to use a large diameter rotary sanding machine across and then lengthwise of the room with No. ½ sandpaper to remove the cross marks of previous operations.

The fifth and final operation is to sweep the entire floor free from all grits, then sand using the large diameter rotary sander with No. 00 or No. 0 sandpaper, going over the floor once across and once lengthwise of the room.

After sweeping the floor clean it is ready for any kind of floor finish.

Finishing Hardwood Floors

The modern and approved method of finishing hardwood floors is to use wood sealers which are a special compound of ingredients blended so that they penetrate into the fiber and cells of the wood, sealing the pores with tough elastic materials, thereby building up a base resistance to water, dirt and stains. It wears only as the wood wears down.

Finishing Hardwood Floors in Natural Color.—Before finishing hardwood floors, they should be sanded as stated in previous paragraphs.

After the floor has been completely sanded and swept clean of all dust, a good grade of permanent penetrating seal should be applied. One such seal on the market is Pentra-Seal*. May be applied quickly with a brush, rag, mop, lamb's wool applicator or squeegee and does not leave brush or lap marks.

For an extra fast job, Fast-Drying Seal* is recommended. Apply a thin even coat and after it has become tack free remove the excess with a No. 2 steel wool pad. Sweep floor clean and apply a second coat in the same manner. Where a highly glossy finish is desired, apply a third coat and do not use steel wool. This will remain on the surface, since the preceding coats have thoroughly filled every cell and/or pore, and result in a glossy surface finish.

The floor should now be given a final protective coat of a good grade of paste wax or Lustre Finish*. Paste wax should be applied manually and Lustre Finish can be put on with a mop. Two coats are recommended, followed by buffing with a mixed fiber brush.

Fast-Drying Seal is available in natural as well as seven decorator type colors. Coverage ranges from 400 to 450 sq. ft. per gallon for the first coat and 650 to 750 sq. ft. per gallon for the second coat. Price, Fast-Drying Seal Natural, about $9.00 per gal.; colors, slightly higher.

Coverage of Lustre-Finish is about 1,250 sq. ft. per gallon. Price, about $7.00 per gal.

Labor Finishing Wood Floors.—Assuming the floor has been sanded and is ready for the seal, an experienced man can finish about 1,000 sq. ft. of natural finish floor per 8-hr. day. Colored seal takes more time as the excess pigment must be removed by hand wiping.

Labor Cost of 100 Sq. Ft. of Fast-Drying Seal* and Lustre Finish*
Floor Finish on Wood Floors, Natural Color

	Hours	Rate	Total	Rate	Total
Mechanic	0.8	$. . . .	$. . . .	$16.47	$13.18
Cost per sq. ft			. . . .		.13

PREFINISHED HARDWOOD PLANK FLOORING

Plank or strip flooring in ¾″ thickness is available in several widths and finishes, most furnished in random lengths and widths. It is generally of red oak and predrilled for nailing. Widths are in the 3″ to 8″ range. Special distressed finishes and pegging are also available.

Plain flooring in narrower widths will run around $1.75 per sq. ft. For pegging add 10¢ per sq. ft. Wider planks will run $2.50 per sq. ft. Premium finishes can raise costs to as much as $2.75 a sq. ft.

One carpenter should lay at least 150 sq. ft. per day.

---

*The American-Lincoln Corp., Toledo, Ohio.

WOOD PARQUET FLOORING

Wood parquet is available in stock designs of oak, walnut, teak, cherry and maple. Standard thickness is 5/16", although heavier parquet is available on custom orders in 11/16" and 13/16" thicknesses. Sizes will vary with patterns running from 9"x9", 12"x12", 16"x16", on up to 36"x36".

Parquet is set in mastic on any level floor above grade. Most parquet is ordered factory finished, but may be unfinished if desired. Many patterns are available in "prime grade" and "character marked," square or bevelled edges.

Costs of all wood flooring are changing rapidly, but square foot material costs will approximate $1.70 for a standard 5/16" oak, $2.00 for walnut, and $2.70 for teak. The prime grades in more elaborate patterns will run much higher. If prefinished, add 30 cts per sq. ft.

Laying Wood Parquet Floors in Mastic Over Concrete.—Concrete subfloor shall be primed with one coat of asphalt priming paint and allowed to dry before laying parquet. Average covering capacity of primer 200 to 400 sq. ft. per gal. depending upon porosity of cement floor finish. Price $2.50 per gal. in 5-gal. cans.

Where concrete floors are in direct contact with the ground, below grade, or where there is not sufficient ventilation to prevent condensation, a membrane waterproofing shall be applied to the slabs, extending same up the sides of all walls, 4" to 6", according to manufacturer's specifications. Where wood parquet is to be laid in hot mastic, a hot application, two-ply membrane shall be applied. Where parquet is to be laid in cold mastic, a cold application one-ply polyethylene film shall be applied.

Installation of polyethylene film is as follows: after concrete surface is primed, cold mastic is applied at the rate of 80 to 100 sq. ft. per gal., into which polyethylene film is rolled, lapping joints 4 inches. Laps do not have to be sealed.

Parquet shall be laid to a level surface bedded in a suitable mastic without the use of nails or screws. One gallon of mastic will cover 45–50 sq. ft. and will run around $12.00 per gallon.

Expansion joints must be left around all edges of rooms according to manufacturer's specifications.

Labor Laying Wood Parquet Flooring.—When laying wood parquet floors in mastic in ordinary size rooms such as residences, apartments, etc., an experienced floorlayer with a helper should lay 300 to 350 sq. ft. of floor per 8 hr. day, while in large spaces such as schools, auditoriums, gymnasiums, etc., an experienced floorlayer should lay 350 to 400 sq. ft. of floor per 8-hr. day. This applies to either finished or unfinished parquet, as the same care is required in laying either type of flooring. This does not include any time preparing or leveling old concrete floors, as this must be added extra if required.

Where unfinished parquet is used, add for sanding and finishing.

Various Designs in Wood Parquet Flooring.

Laying Wood Parquet Floors in Mastic over Other Surfaces.—Wood parquet flooring may also be laid in mastic over plywood or board subflooring, present wood floors, asphalt tile, etc., using the same general procedures as given above for over concrete, except subsurface will not require priming before applying mastic.

Subsurfaces must be sound, smooth, clean and dry—asphalt tile must be firmly bonded and in good condition—wood subflooring or present wood flooring may require re-nailing. Where installation is over square edged board subflooring, one layer of 15-lb. asphalt saturated felt should be laid to keep mastic from seeping through joints.

Prices of Wood Parquet Floors.—Owing to the variation in freight rates, dealer's mark-up, etc., it is practically impossible to give prices that will apply to all parts of the country. Dealer's mark-up will probably vary from 25 percent on large jobs to 40 percent on smaller installations, plus freight from factory to destination.

Material Cost of 100 Sq. Ft. of 9"x9" 5/16" Oak Parquet Floors Laid in Mastic

	Rate	Total	Rate	Total
0.33 gal. asphalt primer	$. ...	$. ...	$ 2.50	$ 0.83
100 sq. ft. oak parquet	....	....	1.70	170.00
2.5 gal. hot mastic	....	....	12.00	30.00
Cost per 100 sq. ft		$. ...		$200.83
Cost per sq. ft		....		2.01

Cost of 100 Sq. Ft. of 9"x9" 5/16" Oak Parquet Floors Laid in Mastic

	Hours	Rate	Total	Rate	Total
Floorlayer	3	$. ...	$. ...	$16.47	$49.41
Helper	3	....	....	12.54	37.62
Cost per 100 sq. ft			$. ...		$87.03
Cost per sq. ft			....		.87

If wood parquet is laid over wood or asphalt tile surfaces, omit asphalt primer.

If parquet floors are laid in large spaces, such as auditoriums, etc., deduct ¼ to ½ hr. floorlayer time per 100 sq. ft.

Add for membrane waterproofing, if required.

If unfinished parquet is used, change price of blocks and add for sanding and finish.

Above costs do not include any allowance for preparing subfloor, if necessary.

Laminated Block Flooring.—Laminated oak blocks are fabricated from 3-ply all oak plywood. Furnished in 9"x9"x½" size, factory finished.

A 27 sq. ft. carton (48 pcs.) will run around $30.00 in a light or dark finish.

Laminated block flooring is for mastic installation only and may be laid over any sound, smooth, clean and dry subfloor, using hot or cold mastic. General installation procedure is the same as for wood parquet floors given on previous pages, except that no allowance for expansion is necessary.

Labor costs for laying laminated block flooring should be about the same as for 9"x9"x5/16" wood parquet given above.

Wood block industrial flooring. Creosoted wood blocks 2" to 3" thick will run from $1.70 to $2.30 per sq. ft. One carpenter will install around 200 sq. ft. per day.

09650 RESILIENT FLOORING

Resilient flooring includes asphalt tile, rubber, cork, vinyl asbestos, pure vinyl tile and vinyl sheet flooring. Each type has its own qualities and should be selected on these considerations rather than cost alone. If material is to be installed on or below grade, cork tile and certain sheet vinyls are not suitable.

If grease is a problem, rubber, cork and asphalt tile perform poorly, while if resilience and quietness are important, cork and rubber tile are excellent, but sheet vinyl is poor.

A quick guide for approximate installed costs per sq. ft. is as follows :

Type	Thickness	Cost
Asphalt tile	3/32"-1/8"	$0.65-0.90
Vinyl asbestos	1/16"	0.80-0.90
	3/32"&1/8"	0.85-0.95
Cork tile-Prefinished	1/8", 3/16", 5/16"	1.20-1.90
Vinyl tile	.050"	1.40-2.00
Vinyl sheet	.070"-.075	1.60-2.50
Rubber tile	1/8", 3/16"	1.50-1.85

ASPHALT TILE FLOORS

Asphalt tile is one of several types of resilient flooring that can be used satisfactorily on concrete floors that are in direct contact with the ground, at or below grade. Hence it is suitable for basement rooms, as it is not affected by the moisture and alkali always present in concrete in contact with the ground. Asphalt tile is also used in residences, stores, offices and institutions where a low cost floor is desired. They are furnished in a variety of colors, and tile sizes are 9"x9"x⅛" thick.

When applied over approved wood subfloors, the wood floor shall be finished smooth and level and covered with a layer of lining felt bonded to the wood with mastic and then rolled thoroughly with a linoleum roller. The asphalt tile is then laid over the lining felt, using an approved asphalt emulsion for bonding the tile to the felt.

When applied over concrete floors above, at, or below grade, the concrete floor must be smooth, dry and even. The tile is laid directly on the concrete, using approved asphalt emulsion. At grade or subgrade, if dampness due to humidity is extreme, the concrete shall be given one coat of asphalt cut-back primer after which the tile is laid in asphalt—solvent type or cutback—cement.

How to Estimate Tile Floors

To Determine and Price the Quantity of Material Required. —Compute the actual square footage of surface to be covered, and add a percentage for waste. The percentage will depend on the size of the area and regularity of outline. A fair percentage of waste to add to the actual area is shown in the accompanying table.

After the gross area has been determined, multiply it by the square foot price of tile. The result will be the price of the material.

The square foot price of the material will depend on the colors and quantity.

Table of Approximate Percentages of Allowable Waste
for Various Areas

Up to	50 sq. ft.	14%
50 to	100 sq. ft.	10%
100 to	200 sq. ft.	7% to 9%
200 to	300 sq. ft.	6% to 7%
500 to	1,000 sq. ft.	4% to 5%
1,000 to	5,000 sq. ft.	4%
5,000 to	10,000 sq. ft.	3%
10,000 and up		1½% to 3%

Quantity of Primer and Asphalt Cement Required.—When priming concrete floors, one gallon of primer should cover 200 to 300 sq. ft. of surface, depending upon the porosity of the surface. Price $2.50 per gal.

One gallon of asphalt cement should lay 175 to 225 sq. ft. of asphalt tile over concrete floors. Costs about $8.00 per gal.

One gallon of asphalt emulsion for installing asphalt tile over lining felt on wood subfloors and direct to above grade concrete floors will cover approximately 140 to 175 sq. ft. Price $3.00 per gal.

Labor Laying Asphalt Tile.—The cost of laying asphalt tile varies with the size and shape of the room or space where the tile are to be laid, size of tile, kind of subfloor, etc., as a tile setter will lay more sq. ft. of tile in a large room than one cut up with pilasters, offsets, etc., also a man will lay more sq. ft. of large size tile than of the smaller sizes.

When placing lining felt over wood subfloors, an experienced man should spread paste, lay and roll 250 sq. ft. of felt an hr. depending upon the room sizes.

After the lining felt has been laid (over wood floors), an experienced tile setter should lay the following quantities of tile per 8-hr. day :

Size of Tile	No. Sq. Ft. Laid per 8-Hr. Day	Hours per 100 Sq. Ft.
9"x9"	400–425	2.0

The above quantities are based on fairly large rooms. If laid in small spaces, increase the labor costs accordingly.

Sq. Ft. Prices on Asphalt Floor Tile

The following are average prices on asphalt tile.

Group	1/8 "
A colors	$0.33
B colors	.35
C colors	.40
D colors	.44

Diagonal half tile of 9"x9", add 3 cts. per sq. ft. additional.

Feature strips in the standard colors are priced at their group price plus 10 cts. per sq. ft. additional.

Standard, die-cut design asphalt tile inserts, size 18"x18", cost about $3.00 each.

"A" colors—Dark red and black solid patterns.

"B" colors—Dark marbleized patterns.

"C" colors—Medium light marbleized patterns.

"D" colors—Light Marbleized patterns.

Greaseproof Asphalt Tile

Greaseproof colors are also acid resistant. Standard size 9"x9".

	Price per Square Foot	1/8" Thick
Greaseproof Asphalt Tile		$0.48

VINYL ASBESTOS TILE

Vinyl asbestos tiles are composed of vinyl resins and asbestos fillers. These tiles are relatively inexpensive, can be used below grade, come in a large variety of colors and designs far superior to asphalt tile and have good durability and low maintenance cost. They are the most widely used floors today.

They are generally available in 9" and 12" squares in 1/16", 3/32" and 1/8 " gauges, and range from $0.45–$0.70 per sq. ft. Installation costs will approximate those of asphalt tile.

VINYL PLASTIC TILE

Vinyl plastic tile is a more expensive resilient flooring material. It is very stable, will not shrink or expand and is highly resistant to oils, fats, greases, acids, alkalis, detergents, soaps and most solvents.

Furnished in 9"x9", 12"x12" and 18"x36" sizes and in 3/32" and ⅛" thicknesses.

Subfloor requirements are the same as for other resilient flooring. May also be laid over concrete floors, or on grade or below grade, using special adhesives.

Vinyl plastic tile is laid in the same manner as other resilient tile and labor costs are about the same.

Material costs vary according to pattern, ranging from 65 cts. per sq. ft. for marbleized patterns to $2.25 per sq. ft. for translucent patterns, for ⅛" thick tiles. 3/32" thick tiles cost about 15 to 20 percent less.

RUBBER TILE FLOORS AND BASE

Rubber tile is furnished in 9"x9", 12"x12", and 18"x36" tile, ⅛" thick and in a variety of colors.

Rubber floor tile may be laid over wood or above grade concrete floors. It may also be laid over clean, dry, concrete floors on ground or below grade providing a special adhesive, recommended by the manufacturer of the tile, is used.

When laying rubber tile over an old wood floor, all loose, defective or badly worn boards shall be replaced and face nailed. Any unevenness in the boards shall be planed or sanded smooth and even.

If the wood floors cannot be placed in proper condition by these methods, install Masonite or ¼" plywood over the original wood floor.

Over the wood base there shall be installed semi-saturated lining felt which shall be securely affixed to the base with mastic. Depending upon the condition of the subfloor, the spreading capacity should be about 140 sq. ft. per gallon. After the felt is laid, it should be thoroughly rolled with a three-section 100-lb. linoleum roller. The tile should be installed over the felt, using mastic.

The finished floor should be thoroughly rolled with a three-section 100-lb. linoleum roller and any high edges smoothed in place with a flat headed hammer.

Laying Rubber Tile.—The cost of laying rubber floor tile varies with the size and shape of the room or space where the tile are to be laid, size of tile, kind of subfloor, etc., as it is easier to lay tile in a large square room than one cut up with offsets, pilasters, bays, etc., and a man will lay more sq. ft. of large size tile than of the smaller sizes.

When laying lining felt over wood subfloors, an experienced man should spread paste, lay and roll 250 sq. ft. of felt an hr. depending upon the room sizes.

After the felt has been laid (over wood floors), an experienced tile setter should lay and finish the following sq. ft. of tile of various sizes per 8-hr. day :

Size of Tile	No. Sq. Ft. Laid per 8-Hr. Day	Hours per 100 Sq. Ft.
9"x9"	400–425	2.0
12"x12", 18"x36"	600–675	1.3

The above quantities are based on fair sized rooms. If laid in small spaces, increase labor costs accordingly.

Material Prices of Rubber Floor Tile, All Standard Colors

⅛" Thick..$0.90-$1.60 Per Sq. Ft.

Rubber tile has lost some of its market to the vinyls but is still used quite often as a stair tread covering.

Stair Treads

Size	Marbleized	Plain Black
12½"x36"x¼"	$ 9.75	$ 9.00
x42	11.50	10.50
x48	13.00	12.00
x54	14.50	13.50
x60	16.00	15.00
x72	19.50	18.00

Landing tile will run around $1.10 per sq. ft. and 3/32" riser material about .65 cts. per lin. ft.

Prices on Rubber Composition Top-Set Cove Base

Description	Price per Lin. Ft.
4" base	$0.35
6" base	.40

CORK TILE

Cork tile are made from high-grade pure cork shavings compressed and baked in molds. It is used for floors and wainscoting in offices, libraries, hospitals, schools and other places where footsteps must be muffled.

They are furnished in 6"x12", 9"x9" and 12"x12" size ⅛", 3/16" and 5/16" thick. Cork tile are supplied beveled or unbeveled. With unbeveled cork tile, sanding the finished floor is usually necessary to offset unevenness of the subfloor. With beveled cork tile, the beveling conceals such irregularities, with the result that sanding is unnecessary and the smooth surface given at the tile factory is retained.

Cork tile should never be placed over a single wood subfloor but should be first covered with Masonite or ⅜" plywood. Never install on grade or subgrade floors.

The wood or above grade concrete subfloor must be smooth and even before cork tile can be placed. Lining felt is used over wood subfloors but this is not necessary when applied over smooth, level concrete floors.

After the floor has been laid and sanded (if unbeveled tile are used), the floor should be given a coat of cork tile filler, followed by two coats of cork tile finish (glossy or flat) and subsequent waxings with an approved water emulsion wax, or the floor may be finished by merely applying 3 applications of water emulsion wax and vigorously buffed by machine.

Approximate Square Foot Prices of Cork Tile

Kind of Tile	1/8" Thick	5/16" Thick
Color according to manufacturer's specs., regular sizes	$0.90	$1.25

Border feature strips: In any width under 3" and any length between 18" and 36" at 3 cts. per sq. ft. additional.

Factory finished tile: Add 5 cts. per sq. ft. for beveled cork tile finished at factory with sealer, undercoat, and one application of wax. After installation, apply one application of wax and buff.

Material Cost of 100 Sq. Ft. 9"x9"x⅛" Cork Tile Floors Over Wood Flooring

	Rate	Total	Rate	Total
0.75-gal. adhesive	$....	$....	$8.00	$ 6.00
12 sq. yds. lining felt	....	....	.30	3.60
105 sq. ft. cork tile	....	....	.90	94.50
0.75-gal. adhesive	....	....	8.00	6.00
Finishing Floor				
0.33-gal. cork tile sealer	....	....	3.00	1.00
0.25-gal. cork tile undercoat	....	....	3.00	.75
0.13-gal. water emul. wax	....	....	3.00	.39
Cost 100 sq. ft	$....			$112.24
Cost per sq. ft	....			1.12

Labor Cost of 100 Sq. Ft. 9"x9"x⅛" Cork Tile Floors Over Wood Flooring

	Hours	Rate	Total	Rate	Total
Mechanic	3.5	$....	$....	$16.47	$ 57.65
Sanding 100 sq. ft	1.0	....	....	16.47	16.47
Finishing Floor					
Labor clean and wax floor	2.0	....	....	16.47	32.94
Cost 100 sq. ft		$....			$107.06
Cost per sq. ft		....			1.07

SHEET VINYL FLOORING

Sheet vinyl can be installed on concrete floors, below and above grade, and on wooden floors that have a smooth surface. If the wood floor is rough, then install a layer of hardboard or ¼ " plywood to provide a good surface for the sheet vinyl.

This material is available in 6 and 12 foot wide rolls up to 120 feet in length, depending on the manufacturer. Thicknesses will vary depending on the product design, type of backing material, and the manufacturer; however, the most recognized gauges will fall in the .065", .080" and .090" range for non-cushion backed sheet vinyl.

Most sheet vinyl flooring materials are designed to require a minimum of maintenance, which means that the flooring cleans easily, is resistant to normal abuse, and does not require waxing to maintain the luster appearance.

Installation is by full adhesive coating on the 6 foot wide material and adhesive coating on seams only for the 12 foot wide rolls.

Material prices will vary from $1.00 - $2.00 per sq. ft. depending on the gauge and backing, however, the estimator should be aware that this material is sold by the square yard and that proper floor layout should be determined to reduce the waste factor.

Labor for installation will vary depending on the size and shape of the area to be covered and the number of cuts to be made. One man will usually lay about 600 sq. ft. in large areas per 8 hour day, and about 350 sq. ft. in smaller areas per 8 hour day.

09680 CARPETING

With the development of man-made fibers special padding and backings, carpeting has become competitive with resilient tile in commercial construction. While almost always subcontracted to a firm specializing in the field, the general contractor should be acquainted with the basic choices available today.

Pure wool has set the high standards most people expect from carpet and no man-made fiber has yet equaled it in overall performance and appearance, although in some respects, such as abrasion resistance, texture retention and mildew resistance, man-made fibers excel. But wool carpeting is in the high cost range, running $15.00 to $25.00 per square yard in the 36 to 42 oz. commercial grades, without padding.

Nylon is the most often used of the man-made fibers. It has outstanding wear ability and resiliency, is mildew resistant and mothproof. But there are problems with static electricity unless it is specially treated or has grounding mesh woven into the backing. In the lighter weights, it is extremely economical, running around $6.50 per square yard for 15 oz.; $8.00 for 22 oz., both without padding.

The olefin (polypropylene) fibers are comparable in price to nylon, but they are found mostly in the indoor-outdoor type of carpet as they offer the lowest moisture absorbency and are exceptionally stain-resistant. Their appearance is dull, resiliency low; they have the lowest melting point, so are subject to cigarette burns. Their range is $5.00 to $9.00.

Acrylics surpass the other man-made fibers in looks and feel, most closely resembling wool, but they do not wear as well as nylon, and will pill some. They run $7.50 to $12.00 per square yard for 36 to 42 oz. weights, without padding.

Polyesters have poor resiliency and therefore are seldom used alone, except for shag-type carpets which are seldom encountered in contract work. They fall in a medium cost range. Padding may be laid separately or come with the carpet. It will add $1.00 to $2.50 to the material cost.

Carpets can be taped down or laid in mastic. Installation cost will run around $2.80 per square yard on open work. If not integral with carpet, add another $1.25 per square yard for laying padding.

Estimating Quantities.—Carpet is estimated by the square yard measure (9 sq. ft. per sq. yd.). Since most carpeting is manufactured in 12 foot wide material, the estimator must be careful to determine how the carpet will lay on the floor to minimize the waste factor. Bear in mind that carpet has a definite "lay" to the nap and the "lay" must be in the same direction when joining pieces together, or else the carpet pieces will appear to have a different color.

09900 PAINTING

Estimates and bids on painting and interior decorating vary to a greater extent than almost any other branch of building construction. There are several kinds of painters— good painters, fair painters, poor painters and others who are just plain "daubers." There are painters who figure on "skinning the job" by applying one or two fewer coats of paint or varnish than called for in the specifications; others who figure on substituting cheap and inferior materials for those called for; and others who do not know how to do good work, so naturally one should expect a wide variation in figures.

Of course, a cheap or inferior job costs less than a first class job, and the buyer usually gets what he pays for.

When preparing estimates on painting, consider the quality of materials and workmanship required and price the work accordingly.

METHODS OF MEASURING AND LISTING PAINTING QUANTITIES

When estimating quantities of painting or interior finishing, the actual surface to be painted should be measured as accurately as possible from the plans or from measurements taken at the building. From these measurements it should be possible to estimate the quantity of materials required, but the labor quantities present a much more difficult problem, owing to the different classes of work and the difficulties encountered in applying the paint. For instance, the covering capacities of paint will be the same for a plain surface or a cornice, but the labor cost painting the cornice will be considerably more owing to the height of the cornice above the ground and the amount of "cutting" or "trimming" necessary. Care must be used when pricing any piece of work, as conditions on each job are different.

Clapboard or Drop Siding Walls.—Obtain actual area of all walls and gables. Add 10 percent to actual surface measurement to allow for painting under edge of boards. Do not deduct for openings less than 10'-0"x10'-0".

Shingle Siding.—Obtain actual area of all walls and gables and multiply by 1½. Do not deduct for openings less than 10'-0"x10'-0"

Brick, Wood, Stucco, Cement and Stone Walls.—Obtain actual area of all walls and gables. Do not deduct for openings less than 10'-0"x10'-0".

Eaves.—Plain eaves painted same colors as side walls, obtain area and multiply by 1½. If eaves are painted a different color than side walls, obtain area and multiply by 2.

Eaves with rafters running through, obtain area and multiply by 3.

Eaves over brick, stucco or stone walls, obtain area and multiply by 3.

For eaves over 30'-0" above ground, add ½ for each additional 10'-0" in height.

Cornices, Exterior.—Plain cornices, obtain area and multiply by 2. Fancy cornices or cornices containing dentils, etc., obtain area and multiply by 3.

Downspouts and Gutters.—Plain downspouts and gutters measure the area and multiply by 2. For fancy downspouts and gutters, obtain area and multiply by 3.

Blinds and Shutters.—Plain blinds or shutters, measure one side and multiply by 2. Slatted blinds or shutters, measure one side and multiply by 4.

Columns and Pilasters.—Plain columns or pilasters, obtain area in sq. ft.; if fluted, obtain area and multiply by 1½ ; if paneled, obtain area and multiply by 2.

Lattice Work.—Measure one side and multiply by 2 if painted one side only ; if painted two sides, obtain area and multiply by 4.

Porch Rail, Balustrade, Balusters.—If solid balustrade, add one foot to height to cover top and bottom rail. Multiply the length by the height. If painted two sides, multiply by 2.

If balustrade consists of individual balusters and handrail, multiply length by the height, then multiply the result by 4.

Handrail only. If under 1'-0" girth, figure as 1'-0" by length and multiply by 2.

Moldings.—Cut in on both sides, figure 1 sq. ft. per lin. ft. if under 12" girth. If over 12" girth, take actual measurement.

Doors and Frames, Exterior.—Inasmuch as it costs almost as much to paint a small door as a large one, do not figure any door less than 3'-0"x7'-0". Allowing for frame, add 2'-0" to the width and 1'-0" to the height. For instance, a 3'-0"x7'-0" door would be figured as 5'-0"x8'-0", or 40 sq. ft.

If a sash door, containing small lights of glass, add 2 sq. ft. for each additional light. A 4-lt. door would contain 8 sq. ft. additional ; a 12-lt. door, 24 sq. ft. additional, etc.

If painted on both sides, obtain sq. ft. area of one side and multiply by 2.

Door frames only, where no door is hung, allow area of opening to take care of both sides.

Windows, Exterior.—Inasmuch as it costs almost as much to paint a small window as a large one, do not figure any window less than 3'-0"x6'0". Add 2'-0" to both the width and height of the opening, to take care of the sides and head of the frame and the outside casing or brick mold, and multiply to obtain the area. For instance, a window opening 3'-0"x6'-0", add 2'-0" to both the width and height, making 5'- 0"x8'-0", containing 40 sq. ft. of surface.

If sash contain more than one light each, such as casement sash, etc., add 2 sq. ft. for each additional light. A 6-lt. window would contain 12 sq. ft. additional; a 12-ft. window, 24 sq. ft. additional, etc.

Roofs.—For flat roofs or nearly flat, measure actual area. For roofs having a quarter pitch, measure actual area and add 25 percent; roofs having a one-third pitch, measure actual area and add 331/3, percent; roofs having a half pitch, measure actual area and add 50 percent.

Fences.—Plain fences, measure one side and multiply by 2 ; picket fences, measure one side and multiply by 4.

Picture Mold and Chair Rail.—On picture mold, chair rail, etc., less than 6" wide, obtain the number of lin. ft. to be painted or varnished and figure ¾ sq. ft. per lin. ft.

Wood and Metal Base.—Wood or metal base 6" to 1'-0" high should be figured as 1'-0" high; under 6", figure at .5 sq. ft. per lin. ft.

Wood Panel Strips, Cornices, Etc.—When estimating quantities of wood panel strips, wainscot rail and ceiling cornices, measure the girth of the member and if less than 1'-0" wide, figure as 1'-0".

If over 1'-0" wide, multiply the actual girth by the length, viz., 1'-3"x200'-0"=250 sq. ft.

Refer to rule for exterior cornices.

Interior Doors, Jambs and Casings.—When estimating quantities for interior doors, jambs and casings, add 2'-0" to the width and 1'-0" to the height of the opening. This allows for painting or varnishing the edges of the door, the door jambs which are usually 6" wide, and the casings on each side of the door which average from 4" to 6" wide. Example : on a 3'-0"x7'-0" door opening, add 2'-0" to the width and 1'-0" to the height, which gives an opening 5'-0"x8'-0", containing 40 sq. ft. on each side. Some painters figure all single doors at 40 sq. ft. per side, or 80 sq. ft. for both sides, while others figure them at 50 sq. ft. per side or 100 sq. ft. for both sides. Do not deduct for glass in doors.

If a sash door, containing small lights of glass, add 2 sq. ft. for each additional light. A 4-lt. door would contain 8 sq. ft. additional ; a 12-lt. door 24 sq. ft. additional, etc.

Interior Windows, Jamb Linings, Sash and Casings.—When estimating painting quantities for windows and window trim, add 2'-0" to the sides and length to allow for jamb linings, casing at the top and sides, and window stool and apron at the bottom. Example : If the window opening is 3'-0"x6'-0", adding 2'-0" to both the width and length gives a window 5'-0"x8'-0", containing 40 sq. ft. of surface.

If sash contain more than one light each, such as casement sash, etc., add 2 sq. ft. for each additional light. A 6-lt. window would contain 12 sq. ft. additional ; a 12 lt. window, 24 sq. ft. additional, etc.

Stairs.—When estimating quantities of paint or varnish for wood stairs, add 2'-0" to the length of treads and risers to allow for stair strings on each side of the stairs. A stair tread is 10" to 12" wide, riser is about 7" high, and the average cove and underside of a stair tread 2" in girth, so each tread should be figured 2'-0" wide. Multiply the width by the length to obtain the number of sq. ft. of paint or varnish required for each step. Multiply the area of each step by the number of treads in the stair, and the result will be the number of sq. ft. of finish. Example: A flight of stairs containing 20 treads, 4'-0" long. Adding 2'-0" to the length of the treads to allow for the stringers, gives 6 lin. ft. in each tread, which, multiplied by the girth of 2'-0" gives 12 sq. ft. of surface in each step. Since there are 20 treads in the stairs, 12x20=240 sq. ft.

For painting the soffits of stairs, use same rules of measurement as given above.

Balustrades and Handrails with Balusters.—When estimating balustrades around well holes or stair handrail and balusters, measure the distance from the top of the treads to the top of the handrail and add 6" for painting or varnishing the handrail. Multiply the height of the balustrade by the length and the result will be the number of sq. ft. of balustrade. An easy method for computing the length of the stairs is to allow 1 lin. ft. in length for each tread in the stairs. Example: If the handrail is 2'-6" above the treads and the stairs contain 20 treads, add 6" to the height of 2'-6" to take care of the extra work on the handrail proper, making a total height of 3'-0". 3x20=60 sq. ft. of surface for one side of the balustrade or 120 sq. ft. for two sides.

The balustrade around the stair well hole is easily estimated by multiplying the height by the length.

On fancy balustrades having square or turned spindles which require considerable additional labor, use actual measurements as given above and multiply by 4 to allow for extra labor.

Wood Ceilings.—When estimating quantities for wood ceilings, multiply the length by the width. Do not deduct for openings less than 10'-0"x10'-0".

Wainscoting.—Plain wainscoting, obtain actual area. Paneled wainscoting, obtain actual area and multiply by 2.

Floors.—To compute the quantities of floors to be finished, multiply the length of each room by the width, and the result will be the number of sq. ft. of floor to be finished.

Plastered Walls and Ceilings.—To obtain the area of any ceiling, multiply the length by the breadth and the result will be the number of sq. ft. of ceiling.

When estimating walls, measure the entire distance around the room and multiply by the room height. The result will be the number of sq. ft. of wall to be decorated. For instance, a 12'-0"x15'-0" room has two sides 12'-0" long and two sides 15'-0" long, giving a total of 54 lin. ft. If the ceilings are 9'-0" high, 54x9=486 sq. ft. the area of the walls.

Do not deduct for door and window openings.

Cases, Cupboards, Bookcases, Etc.—When estimating surfaces of cupboards, wardrobes, bookcases, closets, etc., to be painted or varnished, obtain the area of the front and multiply by 3 if the cases do not have doors.

If the cases have doors, obtain the area of the front and multiply by 5.

This takes care of painting or varnishing doors inside and out, shelves 2 sides, cabinet ends and backs.

Radiators.

—For each front foot, multiply the face area by 7.

Sanding and Puttying Interior Trim.—Sanding and puttying on high grade interior trim should be figured as one coat of paint or varnish; on medium grade trim, figure at 50 percent of one coat of paint or varnish. For industrial work figure about 25 percent of the cost of one coat of paint or varnish.

Emulsified Asphalt

Emulsified asphalt may be used as a floor surfacer, a bonding coat, or a quicksetting, hard-drying and durable patching material. It may be used with satisfactory results over either old or new floors. The following procedure is generally used. First, clean the old surface free from dirt, grease and foreign matter; second, cutting out the worn spots or cracks with perpendicular sides to avoid the formation of feather edges; third, a bonding coat of emulsified asphalt is applied, consisting of 1 part clean water to 4 parts of emulsified asphalt, dry overnight; fourth, the mixture of portland cement, sand, gravel and emulsified asphalt is applied and rolled.

When applied direct to a concrete surface, 1 gallon will cover about 75 sq. ft. one coat or 50 sq. ft. two coats.

When used instead of water as a material for binding cement, sand and gravel the quantity will vary with the thickness applied and will require 4 to 5 gallons per sack of cement.

PAINTING CONCRETE AND MASONRY SURFACES

Hydrocide Colorcoat (Sonneborn).—An alkyd-epoxy-silicone-oil base formula, decorative coating especially designed for exterior, highly porous surfaces. Properly applied, it effectively seals and protects such surfaces with a color-fast, decorative coating, which will not run or sag, in one application. It will fill pores and bridge hairline cracks on concrete or masonry surfaces which are completely cured.

One gallon of material, used as it comes in the container, will cover 60 to 70 sq. ft. depending upon porosity of the surface. Price, $9.50 per gal.

Hydrocide Super Colorcoat (Sonneborn).—Super Colorcoat is for use on cured, green, or slightly damp surfaces. It is waterdispersed acrylic emulsion formulation. It is also mildew and fungus resistant and non-flammable.

Super Colorcoat will cover approximately 80 to 100 square feet per gal. Cost per gal. is $10.50.

SikaGard Hi-bild (Sika).—Tough 100% solids epoxy resin coating for concrete, block, steel, wood and other structural materials. Can be applied to dry, damp, or wet surfaces. Coverage with short nap roller will be approximately 175 to 200 sq. ft. per gal.; when spraying, 250 to 400 sq. ft. per gal. Coverage varies with temperature, substrate and environment. Available in colors. Price, $22.50 per gal.

COVERING CAPACITY OF PAINTS

It is difficult to say how many sq. ft. of surface one gallon of paint will cover, as there are several items that influence the covering capacity. By brushing the paint out very thin it will cover more surface than when applied thick. Dark paint will hide the surface better than light paint, and for that reason it can be brushed out thinner than the lighter colors. Also, a rough surface will require considerable more paint than a smooth surface.

Soft and porous wood will absorb more oil and require more paint than close grained lumber.

Another thing that must be considered are the ingredients entering into the paint. Different brands of paint vary in covering capacity or hiding power.

Covering Capacity of Oil Base Paint on New Exterior Wood.—In figuring the number of square feet a gallon of oil base paint will cover, a great deal depends upon the surface to be painted ; that is, the kind of wood, the degree of roughness, etc. Some woods are more porous than others and absorb more paint. Much depends on the way the paint is brushed out as some painters brush the paint out more and thus cover more surface.

The priming coat, properly applied, should cover 450 to 500 sq. ft. per gallon. The second coat should cover 500 to 550 sq. ft. per gallon and the third coat should cover 575 to 625 sq. ft. per gallon.

When estimating exterior trim painting, with measurements taken in accordance with standard methods as given on previous pages, the priming coat should cover 750 to 850 sq. ft. per gallon ; the second coat should cover 800 to 900 sq. ft. per gallon and the third coat should cover 900 to 1,000 sq. ft. per gallon.

Bear in mind the above coverages are based on surfaced lumber, as the covering capacity will be greatly reduced when applied to rough boards.

Painting Old Exterior Wood Surfaces.—When estimating repaint work on old exterior wood surfaces, if the existing surface is sound and in good condition, two coats should be sufficient to give a good job and paint coverages should be about the same as for the second and third coats on similar new work.

If the old surface shows cracking, blistering, scaling and peeling, the old paint should be removed by using either a paste or liquid paint remover of the slow drying type, or a blow torch and scraper.

If the old paint is removed completely, three coats of paint should be applied the same as recommended for new work.

Painting Wood Shingle Siding.—Wood shingle siding should receive three coats of paint on new work. Paint coverages per gallon should be as follows: first coat, 300 to 325 sq. ft.; second coat, 400 to 450 sq. ft.; third coat, 500 to 550 sq. ft.

Wood shingles which have been previously painted with an oil paint and are in suitable condition for repainting, should receive two coats of paint. The first coat should cover 400 to 450 sq. ft. per gal. and the second coat 500 to 550 sq. ft. per gal.

Painting Exterior Wood Floors and Steps.—When painting exterior wood floors and steps with oil base paint, material coverages per gallon should be as follows : first coat, 325 to 375 sq. ft. ; second coat, 475 to 525 sq. ft. ; third coat, 525 to 575 sq. ft.

Painting Interior Wood.—New interior wood should receive three coats of paint, a priming coat, a second coat and a final or finish coat.

Paint coverage for interior wood surfaces will vary with the type of work, such as doors and windows, running trim, paneling, etc., in accordance with the standard method of measuring the quantities.

For doors and windows, coverage per gallon should be: first coat, 475 to 500 sq. ft.; second coat, 500 to 550 sq. ft.; third coat, 500 to 550 sq. ft.

For running trim up to 6" wide, coverage per gallon should be; first coat, 800 to 900 lin. ft.; second and third coats, 1,100 to 1,200 lin. ft.

For ordinary flat wood paneling, coverage per gallon should be: first coat, 450 to 500 sq. ft.; second and third coats, 500 to 550 sq. ft.

When painting interior woodwork that has been previously painted and is in good condition, one gallon of paint should cover approximately the same as given for the second and third coats on new wood.

Painting Brick, Stone, Stucco and Concrete.—When painting masonry surfaces that are reasonably smooth, one gallon of oil base paint should cover 170 to 200 sq. ft. for the priming coat ; 350 to 400 sq. ft. for the second coat and 375 to 425 sq. ft. for the third coat.

These quantities are subject to considerable variation due to the porosity of the surfaces to which the paint is applied, smoothness of walls, etc., as a rough porous surface will require considerably more paint than a smooth hard surface.

Painting Metal Work.—The area which any paint may be expected to cover on metal work will vary with the surface to be painted as badly pitted or rough metal will require more paint than a perfectly smooth surface. The covering capacity will also vary with the temperature, consistency of the paint and the effort behind the brush.

However, for ordinary smooth surfaces, one gallon of paint should cover 550 to 650 sq. ft., one coat.

Aluminum Paint.—Aluminum paint is used for both priming and finishing coats on metal surfaces.

The covering capacity varies with the condition of the surface but on smooth surfaces, one gallon should cover 600 to 700 sq. ft. with one coat.

Painting Interior Walls.—When painting interior walls, the covering capacity of the paint will depend upon the surface to be painted, whether smooth or sand finished plaster, porous or hard finished wallboard, etc.

Three coats are recommended for interior plaster which has never been painted—a priming coat, a second or body coat, and a third or finishing coat.

However, if a two coat job on unpainted plaster is desired, use a good wall primer or sealer for the first coat, followed by the second or finishing coat. To make two coats hide better, tint the first to nearly the same color as the second coat.

If the surface has been painted before and the old paint is still in good condition, two coats are sufficient, no priming coat being required.

On smooth plaster or hard finished wallboard, one gallon of good wall primer or sealer should cover 575 to 625 sq. ft. for the priming coat. For the second coat, figure 500 to 550 sq. ft. per gallon and for the third coat, 575 to 625 sq. ft. gallon.

On rough, porous sand finished plaster, one gallon of wall primer or sealer may cover only 275 to 300 sq. ft. while on very porous wallboard it may be only 150 to 200 sq. ft. per gallon.

For the second and third coats on rough surfaced plaster or wallboard, figure 400 to 475 sq. ft. per gallon.

One Coat Wall Finishes.—There are any number of "one-coat" wall finishes on the market, but when applied to new plastered surfaces or over surfaces that have been previously unpainted, it is advisable to use a primer or wall seal coat, which seals the pores in the plaster or other surface and provides a base suitable to receive the one coat finish. Fully described above.

Latex Base Paints.—There are any number of latex base paints on the market that are compatible to most surfaces.

When used over smooth plastered surfaces, one gallon should cover 450 to 500 sq. ft. first coat and 550 to 650 sq. ft. second coat.

When applied over rough sand finish plaster one gallon should cover 325 to 350 sq. ft. per coat.

Covering Capacity of Wall Size.—The covering capacity of wall size will vary greatly with the material used.

Water size consisting of ground or flake glue and water will cover 600 to 700 sq. ft. of surface per gallon.

Varnish size consisting of varnish, benzine or turpentine and a little paint, will cover 450 to 550 sq. ft. per gallon.

Hard oil or gloss oil size consisting of rosin and benzine will cover 450 to 500 sq. ft. per gallon.

The covering capacity of prepared wall primers and sealers is given on the previous pages under "Painting Interior Walls."

Shingle Stain.—When staining wood shingles after they have been laid, one gallon of good stain should cover 120 to 150 sq. ft. of surface, for the first coat.

The second and third coats will go farther, as the wood does not absorb as much stain as on the first coat and should cover 200 to 225 sq. ft. per gallon.

One gallon of good stain should cover about 70 sq. ft. of roof with 2-coats or 50 sq. ft. with 3 coats.

If the shingles are dipped in stain before laying, it will require 3 to 3½ gals, of stain per 1,000 shingles. When dipping shingles in stain, only two-thirds of the shingle need be dipped, as it is unnecessary to dip the end of the shingle that is not exposed.

Covering Capacity of Oil and Spirit Stains.—When staining finishing lumber, such as birch, mahogany, oak, gum, etc., one gallon of oil stain should cover 700 to 725 sq. ft.

One gallon of spirit stain should cover 500 to 600 sq. ft.

Covering Capacity of Wood Fillers.—Liquid filler is usually used with close grained woods, such as pine, birch, etc., and one gallon should cover 500 to 550 sq. ft.

It is necessary to use a paste filler to fill the pores of all open grained wood, such as oak, ash, walnut, mahogany, etc.

If paste filler is used on oak, one gallon of filler should cover about 450 sq. ft. of surface.

Covering Capacity of Shellac.—The covering capacity of shellac will vary with its purity but one gallon of good shellac should cover 550 to 750 sq. ft. of surface, depending upon whether first, second or third coat.

Covering Capacity of Varnish.—One gallon of good varnish should cover 400 to 450 sq. ft. of softwood floors with one coat ; 200 to 225 sq. ft. with 2 coats and 135 to 150 sq. ft. with 3 coats.

When applied to hardwood floors, one gallon of good varnish should cover 500 to 550 sq. ft. with one coat ; 250 to 275 sq. ft. with 2 coats and 170 to 185 sq. ft. with 3 coats.

When applied to softwood interior finish, one gallon of good spar finishing varnish should cover 400 to 425 sq. ft. with one coat ; 200 to 215 sq. ft. with 2 coats and 135 to 145 sq. ft. with 3 coats.

When applied to hardwood interior finish, one gallon of good spar varnish should cover 425 to 450 sq. ft. with one coat; 210 to 225 sq. ft. with 2 coats and 140 to 150 sq. ft. with 3 coats.

Covering Capacity of Wax.—One gallon of good liquid wax should cover 1,050 to 1,075 sq. ft. of surface.

Covering Capacity of Varnish Remover.—The amount of varnish remover required to remove old varnish from floors and interior finish will depend upon the condition of the old work, but one gallon of good varnish remover should be sufficient for 150 to 180 sq. ft. of surface.

Enamel Finish.—Where interior woodwork is to receive an enamel finish, at least three coats are required, as follows : first coat, oil base paint primer and sealer; second coat, prepared enamel undercoat ; third coat, enamel finish coat. If a four-coat job is specified, the third coat may be a mixture of ½ undercoat and ½ enamel, followed by the enamel finish coat.

For flat work, figure material coverages per gallon as follows : paint primer and sealer, 575 to 600 sq. ft. ; undercoat, 375 to 400 sq. ft. ; enamel finish, 475 to 500 sq. ft.

For running trim, one gallon of material should cover as follows: paint primer and sealer, 1,100 to 1,200 lin. ft.; undercoat, 775 to 800 lin. ft.; enamel finish, 775 to 800 lin. ft.

Glazed Finish.—One gallon of good glazing liquid, colored with oil colors, should cover 500 to 550 sq. ft. of surface.

For glazed finish on interior running trim, not over 6" wide, one gallon should cover 1,050 to 1,100 lin. ft.

ESTIMATING LABOR COSTS

Perhaps in no other trade will labor costs vary to a greater extent than in painting and decorating. On some classes of work the material costs are almost negligible, while the labor costs may make it one of the most expensive kinds of work.

Naturally, the cost of the work will depend upon the grade of workmanship specified or upon the class of work the painting contractor figures on doing.

In the large cities particularly, there are two classes of painters—those who make a specialty of ordinary or commercial work such as factory buildings, flat or apartment buildings, speculative work and work of this class, and those who specialize in high class work. Naturally, the painter who figures the lower grade of workmanship is bound to be lower than the man who figures a first class job, as a painter can cover more surface per day when he just hits the "high spots" than when he is concientious in his work and gives a good, honest job.

When preparing an estimate, the estimator should consider the class of work required and price his work accordingly.

Another thing that must be borne in mind by the estimator, is to be certain all measurements and quantities are listed and computed in strict accordance with the methods of measurement as given at the beginning of this section.

This applies particularly to the allowances for labor when estimating such items as sash doors, windows having divided lights, drop siding and shingle siding, fences, trellis and lattice work where the face and edges of all surfaces must be painted, stairs where it is necessary to paint all four sides or the entire circumference of newels, handrails, balusters, etc.

These are all exceedingly important because if the quantities are not computed according to the methods given, then the labor costs will be far too low in many instances.

The costs as given on the following pages are based on the methods of estimating quantities as given at the beginning of this section.

ITEMIZED COSTS OF PAINTING

The labor and material tables on the following pages give detailed information on material coverages and production times for various types of paint on different kinds of surfaces.

In preparing an estimate for painting work, the estimator, after obtaining quantities required for the project, need only to refer to the proper table to determine the required quantity of material and labor hours necessary to perform the work. Once the gallons of paint and man hours are computed, apply the appropriate unit costs applicable for your area.

Do not overlook the items of protection, scaffolding or other equipment costs in preparing the estimate, as these are not given in the following tables.

For example, let's assume that the estimator is performing an estimate for staining shingle siding on a residence and the quantity of work to be performed amounts to 3,300 sq. ft. of surface to be covered with two (2) coats of stain. Refer to the following chart for exterior residential work and find the proper work classification - "Stain - Shingle Siding". The material coverage is stated as 120 to 150 sq. ft. per gallon for the first coat and 200 to 225 sq. ft. per gallon for the second coat. The labor production is 1.20 hours per 100 sq. ft. for first coat and 0.82 hours per 100 sq. ft. for the second coat. The sample estimate for this 3,300 sq. ft. of surface would be as follows:

Material Costs

	Rate	Total	Rate	Total
1st Coat				
Stain 27 gal	$. . . .	$. . . .	$8.00	$216.00
2nd Coat				
Stain 16 gal	. . . .	. . . .	8.00	128.00
Cost of 3,000 sq. ft.		$. . . .		$344.00
Cost per sq ft.		. . . .		.11

Labor Costs

	Hours	Rate	Total	Rate	Total
1st Coat					
Painter	39.6	$. . . .	$. . . .	$17.43	$ 690.23
2nd Coat					
Painter	27.1	. . . .	. . . .	17.43	472.35
Cost of 3,300 sq. ft.			$. . . .		$1,162.58
Cost per sq. ft.			. . . .		.35

Labor and Materials Required for Various Painting Operations

Exterior Work—Residential Description of Work	Painter No. Sq. Ft. per Hr.	Painter Hours 100 Sq. Ft.	Material Coverage Sq. Ft. per Gal.
Sanding and Puttying Plain Siding and Trim	200-210	0.50	....
Sanding and Puttying Outside Trim Only	115-120	0.86	....
Burning Paint off Plain Surfaces	40-50	2.20	....
Burning Paint off Wood Siding	30-40	2.85	....
Exterior Brush Painting Plain Siding And Trim ... Priming Coat	100-110	0.95	400-450
Second Coat	115-125	0.85	500-550
Third Coat	125-135	0.75	575-625
Exterior House Painting Rubberized Woodbond ... One Coat	125-135	0.75	450-500
Exterior Trim Only ... Priming Coat	65-75	1.40	750-850
Second Coat	85-95	1.12	800-900
Third Coat	95-105	1.00	900-1,000
Oil Paint—Shingle Siding* ... First Coat	115-125	0.82	250-300
Second Coat	150-160	0.65	375-425
Oil Paint—Shingle Roofs* ... First Coat	100-120	0.90	120-150
Second Coat	155-175	0.60	220-250
Stain—Shingle Siding* ... First Coat	75-85	1.20	120-150
Second Coat	115-130	0.82	200-225
Stain—Shingle Roofs* ... First Coat	145-155	0.67	100-120
Second Coat	190-210	0.50	170-200
*If surfaces are exceptionally dry, increase labor hours and decrease covering capacity.			
Asbestos Wall Shingles ... First Coat	60-70	1.50	150-180
Second Coat	85-95	1.11	350-400
Brick Walls, Oil Paint ... First Coat	100-120	0.93	170-200
Second Coat	140-160	0.67	350-400
Third Coat	155-175	0.60	375-425
Porch Floors and Steps, Oil Paint ... First Coat	235-245	0.42	325-375
Second Coat	270-280	0.36	475-525
Third Coat	285-295	0.35	525-575

Labor and Materials Required for Various Painting Operations—Con't.

Exterior Work—Residential Description of Work		Painter No. Sq. Ft. per Hr.	Painter Hours 100 Sq. Ft.	Material Coverage Sq. Ft. per Gal.
Waterproof Cement Paint—Smooth Face Brick	First Coat	175-185	0.55	90-110
	Second Coat	260-270	0.38	140-160
Clear Waterproof Paint—Smooth Face Brick	First Coat	200-210	0.50	500-510
	Second Coat	200-210	0.50	590-610
Stucco, Medium Texture, Oil Paint	First Coat	90-100	1.10	140-160
	Second Coat	150-160	0.65	340-360
	Third Coat	150-160	0.65	340-360
Stucco, Medium Texture, Wpf. Cement Paint	First Coat	130-140	0.74	90-110
	Second Coat	190-210	0.50	125-145
Exterior Masonry, Stucco, Asbestos Cement Board and Shingle Siding— Rubberized Flat Finish	First Coat	100-120	0.91	325-375
	Second Coat	165-175	0.59	375-425
Concrete Walls, Smooth, Wpf. Cement Paint	First Coat	170-180	0.57	110-130
	Second Coat	275-285	0.36	150-170
Concrete Floors and Steps—Floor Enamel	First Coat	250-280	0.38	440-460
	Second Coat	190-210	0.50	575-625
	Third Coat	200-220	0.48	575-625
Cement Floors—Color Stain and Finish	First Coat	350-370	0.28	475-525
	Second Coat	280-300	0.35	450-500
Fences, Plain, Average	First Coat	125-135	0.77	450-470
	Second Coat	190-200	0.51	530-550
Fences, Picket, Average	First Coat	140-150	0.70	630-650
	Second Coat	160-170	0.60	650-675
Fences, Wire-Metal, Average	First Coat	100-110	0.95	900-1,000
	Second Coat	140-150	0.70	975-1,125
Shutters, Average, Each Coat	No. of Shutters	2-3	0.40	11-13
Downspouts and Gutters, Paint	First Coat	170-180	0.57	540-560
	Second Coat	185-195	0.53	575-600
Screens, Wood Only, Average, Each Coat	No. of Screens	6-8	0.14	45-55
Storm Sash, 2 Light, Average, Each Coat	No. of Sash	3-4	0.29	23-27

Labor and Material Required For Commercial and Industrial Painting

Description of Work	Painter No. Sq. Ft. per Hr.	Painter Hours 100 Sq. Ft.	Material Coverage Sq. Ft. per Gal.
Sheet Metal Work ....First Coat	230-240	0.43	500-550
Second Coat	260-270	0.38	550-600
Steel Factory Sash.......First Coat	80-90	1.18	950-1,000
For Old Work, add 10 to 20 percent to Prepare Surfaces ....Second Coat	100-110	0.95	950-1,000
Third Coat	125-135	0.77	1,250-1,350
Doors and Windows, Paint ....First Coat	175-185	0.55	475-500
Second Coat	130-140	0.75	500-550
Third Coat	165-175	0.60	500-550
Base, Chair Rail, Picture Mold, and Other Trim Less than 6" Wide. Lin. Ft.* ....First Coat	275-300*	0.35*	800-900*
Second Coat	175-200*	0.55*	1,100-1,200*
Third Coat	150-165*	0.65*	1,100-1,200*
Masonry Walls—Bonding and Penetrating Oil Paint.......First Coat	350-360	0.28	225-250
Second Coat	350-375	0.28	275-300
Waterproofing—Common Brick, Smooth Brick, Concrete, Concrete or Cinder Concrete Blocks, Cement Plaster.......First Coat	100-110	1.00	90-100
Second Coat	180-195	0.53	175-200
Waterproofing Wire Cut Tapestry Brick or Very Rough Cinder Conc. Block or Stucco.......First Coat	90-100	1.05	50-60
Second Coat	150-165	0.65	150-175

*Lineal Feet

Spray Painting—Commercial Work

Description of Work	Painter No. Sq. Ft. per Hr.	Painter Hours 100 Sq. Ft.	Material Coverage Sq. Ft. per Gal.
Flat Finish Paint—Smooth Plaster Walls............First Coat	450-475	0.22	500-550
Second Coat	475-500	0.21	500-550
Sand Finish Plaster Walls............First Coat	400-425	0.25	450-500
Second Coat	425-450	0.24	400-450
Industrial Enamel—Smooth Finish Plaster ....First Coat	450-475	0.22	500-550
Second Coat	500-550	0.20	400-450
Third Coat	500-550	0.20	450-500

Spray Painting—Residential Work

Description of Work		Painter No. Sq. Ft. per Hr.	Painter Hours 100 Sq. Ft.	Material Coverage Sq. Ft. per Gal.
Brick, Tile and Cement—Latex Paint	First Coat	500-525	.20	90-110
	Second Coat	525-550	.19	140-160
Brick, Tile and Cement—Oil Paint	First Coat	475-500	.21	250-275
	Second Coat	500-525	.20	400-450
Rough Brick, Tile, Cement and Stucco*—Silicone	One Coat	800-900	.13	80-90
Asbestos Wall Shingles*	First Coat	300-325	.34	150-175
	Second Coat	325-350	.31	225-250
Stucco*—Exterior Latex Paint	First Coat	400-425	.25	90-100
	Second Coat	450-500	.22	125-150
Stucco*—Oil Paint	First Coat	300-325	.34	200-225
	Second Coat	325-350	.30	350-400
Shingle Roofs—Oil Paint	First Coat	400-450	.25	125-150
	Second Coat	450-500	.22	200-225
Shingle Roofs—Stain	First Coat	450-475	.22	125-150
	Second Coat	475-500	.21	200-225
Shingle Siding*—Oil Paint	First Coat	350-375	.29	125-150
	Second Coat	400-425	.25	225-250
Shingle Siding*—Stain	First Coat	400-450	.25	125-150
	Second Coat	450-500	.22	200-225

*Trim Must be Figured Separately as Hand Work

Labor and Material Required for Interior Painting and Finishing

Finishing Interior Trim—Residential

Description of Work	Painter No. Sq. Ft. per Hr.	Painter Hours 100 Sq. Ft.	Material Coverage Sq. Ft. per Gal.
Preparatory Work for Painting.................Sanding	290-300	0.35	....
Sanding and Puttying	190-200	0.50	....
Light Sanding	340-350	0.30	....
.............Lin. Ft.*	475-500*	0.21*	1,100-1,200*
Back Priming Interior Trim Up to 6" Wide.......First Coat	140-150	0.70	475-500
Doors and Windows, Painting Interior...........Second Coat	115-125	0.85	500-550
Third Coat	115-125	0.85	500-550
Oiling or Priming Wood Sash.........No. of Sash	9-10	....	600-700
Preparatory Work for Enamel Finishes.........Sanding	290-300	0.35	....
Sanding and Puttying	120-130	0.80	....
Light Sanding	135-145	0.72	....
Doors and Windows, Enamel Finish.......Paint First Coat	140-150	0.70	575-600
...................Undercoat Second Coat	90-100	1.10	375-400
Enamel Third Coat	100-110	1.00	475-500
Four Coat Work. Add ½ Underc. ½ Enamel.......Add'l Coat	90-100	1.10	425-450
Base, Chair Rail, Picture Mold, and Other Trim Up to 6" Wide. All Quantities are in Lineal Feet.......First Coat	250-260*	0.40*	800-900
Second Coat	145-155*	0.67*	1,100-1,200*
Third Coat	125-135*	0.77*	1,100-1,200*
Four Coat Work, Add ½ Underc. ½ Enamel.......Add'l Coat	135-145*	0.72*	1,100-1,200*
Stain Interior Woodwork.......One Coat	215-225*	0.45	700-725
Stain and Fill Interior Woodwork, Wipe Off.......One Coat	65-70	1.50	400-450
Shellac Interior Woodwork.......One Coat	215-225	0.45	700-725
Varnish, Gloss, Interior Woodwork.......One Coat	165-175	0.60	425-450
Varnish, Flat, Interior Woodwork.......One Coat	170-180	0.57	600-625
Wax and Polish—Standing Trim.......One Coat	100-110	0.95	550-600
Penetrating Stainwax—Standing Trim.......First Coat	170-180	0.57	525-550
Second Coat	190-200	0.51	600-625
Polishing Second Coat	190-200	0.51	....

*Lineal Feet

Labor and Material Required For Interior Painting and Finishing—Con't.

Finishing Interior Trim—Residential

Description of Work	Painter No. Sq. Ft. per Hr.	Painter Hours 100 Sq. Ft.	Material Coverage Sq. Ft. per Gal.
Sanding for Extra Fine Varnish Finish............Sanding	40-45	2.25	550-600
Synthetic Resin Finish, Requires Wiping...........First Coat	190-200	0.50	625-675
Second Coat	210-220	0.47	140-150*
Spackling or Swedish Putty over Flat Trim............One Coat	60-65	1.60	1,050-1,100
Glazing and Wiping over Enamel Trim............One Coat	60-65	1.60	
Brush Stippling Interior Trim, Painted ....Varnish	85-90	1.15	600-625
Flat Varnishing and Brush Stippling over Glazed Trim ....Varnish	170-180	0.57	
Stipple	240-250	0.42	

Old Work

Description of Work	Painter No. Sq. Ft. per Hr.	Painter Hours 100 Sq. Ft.	Material Coverage Sq. Ft. per Gal.
Washing Average Enamel Finish............Washing	85-90	1.15	
Washing Better Grade Enamel Finish............Washing	60-65	1.60	1,700-1,800
Polishing Better Grade Enamel Finish............Polish	170-180	0.57	150-180
Removing Varnish with Liquid Remover ....Flat Surfaces	30-35	3.00	
Wash, Touch Up, One Coat Varnish............Wash-Touch Up	165-175	0.60	575-600
Varnish	100-110	0.95	
Wash, Touch Up, One Coat Enamel............Wash-Touch Up	140-150	0.70	375-400
Enamel	70-75	1.40	
Wash, Touch Up, One Coat Undercoat and............Wash-Touch Up	140-150	0.70	475-500
One Coat Enamel............Undercoat	75-85	1.25	475-500
Enamel	75-85	1.25	
Burning Off Interior Trim............	20-25	4.00	
Burning Off Plain Surfaces............	30-35	3.00	

*Per Lb.

Labor and Material Required For Interior Floors

Description of Work	Painter No. Sq. Ft. per Hr.	Painter Hours 100 Sq. Ft.	Material Coverage Sq. Ft. per Gal.
Painting Wood Floors...First Coat	290-300	0.35	525-550
Second Coat	260-270	0.38	425-450
Filling Wood Floors, Wiping...Fill-Wipe	170-180	0.57	425-450
Penetrating Stainwax—Hardwood Floors...First Coat	390-400	0.25	525-550
Second Coat	480-490	0.21	600-625
Floor Seal...First Coat	490-500	0.20	600-625
Second Coat	590-600	0.17	1,200-1,250
Shellac...First Coat	390-400	0.25	500-550
Second Coat	475-500	0.21	650-675
Stainfill, 1 Shellac, 1 Varnish...Stainfill	260-270	0.38	475-500
Shellac	390-400	0.25	650-675
Varnish	300-310	0.33	500-550
Stainfill, 1 Shellac, Wax and Polish...Stainfill	260-270	0.38	475-500
Shellac	390-400	0.25	650-675
Wax-Polish	200-210	0.50	1,050-1,075
Varnish, each Coat over Shellac...Varnish	300-310	0.33	500-550
Buffing Floors—by Machine...Buffing	390-400	0.25	....
Waxing over 2 Coats of Seal and Polish...Wax-Polish	200-210	0.50	1,050-1,075
Old Work			
Removing Varnish with Liquid Remover...Flat Surfaces	40-45	2.20	170-180
Clean, Touch Up and Varnish...Clean-Touch Up	125-135	0.77	....
Varnish	300-310	0.33	525-550
Clean, Touch Up, Wax and Polish...Clean-Touch Up	125-135	0.77	....
Wax-Polish	200-210	0.50	1,050-1,075

Labor and Material Required For Painting Interior Walls

Description of Work	Painter No. Sq. Ft. per Hr.	Painter Hours 100 Sq. Ft.	Material Coverage Sq. Ft. per Gal.
Taping, Beading, Spotting Nail Heads and Sanding Gypsum Wallboard ... One Coat	90-110	1.00	8-10*
Texture Over Gypsum Wallboard ... One Coat	250-270	0.39	325-350
Casein or Resin Emulsion over Textured Gypsum Wallboard ... One Coat	300-320	0.32	600-700
Sizing New Smooth Finish Walls ... Sizing	350-400	0.27	575-625
Wall Sealer or Primer on Smooth Walls ... Sealer	200-225	0.45	275-300
Wall Sealer or Primer on Sand Finish Walls ... Sealer	125-135	0.80	575-625
Smooth Finish Plaster—Flat Finish ... Sealer	200-225	0.45	500-550
Second Coat	165-175	0.60	575-625
Third Coat	175-185	0.55	
Stippling	190-200	0.52	
Sand Finish Plaster, Flat Finish ... Sealer	125-135	0.80	275-300
Second Coat	135-145	0.70	400-450
Third Coat	125-135	0.80	425-475
Smooth Finish Plaster, Gloss or Semi-Gloss ... Sealer	200-225	0.45	575-625
Second Coat	170-180	0.57	500-550
Third Coat	140-150	0.70	500-550
Stippling	85-95	1.10	
Sand Finish Plaster, Gloss or Semi-Gloss ... Sealer	125-135	0.80	275-300
Second Coat	115-125	0.83	375-400
Third Coat	125-135	0.80	400-450
Texture Plaster, Average, Semi-Gloss ... Sealer	100-110	0.95	250-275
Second Coat	115-125	0.83	325-350
Third Coat	125-135	0.80	375-400
Smooth Finish Plaster—Latex Rubber Paints ... First Coat	140-160	0.67	300-350
Second Coat	140-160	0.67	300-350
Sand Finish or Average Texture Plaster—Latex Rubber Paints ... First Coat	110-130	0.83	250-300
Second Coat	110-130	0.83	250-300
Glazing and Mottling over Smooth Finish Plaster	80-90	1.15	1,050-1,075
Glazing and Mottling over Sand Finish Plater	60-65	1.60	875-900
Glazing and Highlighting Textured Plaster	80-90	1.15	825-850
Starch and Brush Stipple over Painted-Glazed Surface	115-125	0.83	

*Per Lb.

Labor and Material Required For Painting Interior Walls—Con't.

Description of Work	Painter No. Sq. Ft. per Hr.	Painter Hours 100 Sq. Ft.	Material Coverage Sq. Ft. per Gal.
Flat Varnish and Brush Stipple over Painted-Glazed Surface	105-115	0.90	500-550
Texture Oil Paint over Smooth Finish Plaster...Size	215-225	0.45	675-725
Texture	40-45	2.30	125-150
Water Texture over Smooth Finish Plaster...Size	215-225	0.45	700-725
Texture	50-55	1.90	85-95
Latex Base Paint, New Smooth Plaster...First Coat	190-200	0.50	500-550
Second Coat	210-220	0.47	650-700
On Rough Sand Finish Plaster...One Coat	150-160	0.65	325-350
On Cement Blocks...One Coat	135-145	0.70	300-325
On Acoustical Surfaces...One Coat	135-145	0.70	200-225
Casein Paint, over Smooth Finish Plaster...First Coat	270-280	0.37	500-550
Second Coat	300-310	0.33	500-550
Over Rough Sand Finish Plaster...One Coat	250-260	0.40	325-350
Over Cement Blocks...One Coat	165-175	0.60	300-325
Over Acoustical Surfaces...One Coat	165-175	0.60	200-225
Over Cinder Concrete Blocks...One Coat	125-135	0.77	125-150

When applying paint to walls and ceilings using a roller applicator, increase above quantities 10 to 15 percent and reduce painter time by the same amount.

Labor and Material Required For Painting Interior Walls

Description of Work	Painter No. Sq. Ft. per Hr.	Painter Hours 100 Sq. Ft.	Material Coverage Sq. Ft. per Gal.
Old Work			
Washing off Calcimine—Average Surfaces	115-125	0.83	
Washing Smooth Finish Plaster Walls—Average	145-155	0.67	
Washing Sand Finish Plaster Walls—Average	100-110	1.00	
Washing Starched Surfaces and Restarching, Smooth Surfaces ... Washing	145-155	0.67	
Restarching	130-140	0.75	
Removing Old Wall Paper—Not Over 3 Layers	65-75	1.45	
Washing Off Glue After Removing Paper (including Fixing Average Cracks and Sizing)	125-135	0.77	
Cutting Hard Oil or Varnish Size Walls (including Fixing Average Cracks)	125-135	0.77	
Cutting Gloss Painted Walls (including Fixing Cracks)	125-135	0.77	
Washing, Touch Up, One Coat Gloss Paint to Smooth Plaster Surfaces ... Wash-Touch Up	125-135	0.77	
One Coat	125-135	0.77	450-500
Synthetic-Resin Emulsion Paint over Old Painted Walls ... One Coat	190-200	0.50	400-425
Wallpaper, Canvas, Coated Fabrics, Paper Hanging, Wood Veneer			
Canvas Sheeting	50-60	1.82	
Coated Fabrics ... Single Roll	2-2½		
Wallpaper—One Edge Work ... Single Roll	3½-4		
Wire Edge Work ... Single Roll	2½-3		
Butt Work ... Single Roll	2¼-2¾		
Scenic Paper 40"x60" ... Single Roll	1-1¼		
Wood Veneer Flexwood ... Sq. Ft.	18		
Lacquer Finish Over Wood Veneer ... First Coat	100	1.00	275-325
Second Coat	100	1.00	450-500
Penetrating Wax or Synthetic Resin Application over Wood Veneer ... First Coat	100	1.00	550-600
Second Coat	100	1.00	650-675

Colors in Oil

Colors ground fine in pure linseed oil. One-half pint cans.

Blacks	Price	Green	Price
Drop Black	$0.80	Palco Green	$1.75
Lamp Black	.70	Chrome Green, L.	1.05
		Chrome Green, M.	1.05
Blue		Chrome Green, D.	1.05
Palco Blue	1.60		
Prussian Blue	1.30	Red	
Cobalt Blue	1.30	Vermilion	1.35
Ultramarine Blue	1.30	Rose Lake	1.10
		Venetian Red	.90
Brown			
Raw Turkey Umber	.59	Yellow	
Burnt Turkey Umber	.59	Chrome Yellow, L.	1.60
Raw Italian Sienna	.66	Chrome Yellow, M.	1.60
Burnt Italian Sienna	.66	Chrome Yellow, O.	1.60
Vandyke Brown	.59	Ochre	1.60

Shellac and Varnish

Kind of Material	Per Gal.
Shellac, White, 4-lb. Cut	$ 9.00–$11.00
Shellac, Orange, 4-lb. Cut	8.00– 10.00
Cabinet Finish Varnish	10.00– 12.00
Floor Varnish	10.00– 12.00
Flat Varnish	10.00– 12.00
Spar Varnish	12.00– 14.00
Varnish Remover, Liquid	8.00– 12.00
Varnish Remover, Paste	10.00– 14.00
Wallpaper Lacquer	6.00– 7.00
Wallpaper Lacquer Thinner	4.00– 5.00
Lacquer	10.00– 12.00
Lacquer Thinner	5.00– 6.00

Enamels and Enamel Undercoating

	Per Gal.
Enamel, Non-Yellowing White	$10.00–$13.00
Architectural Enamel Paint	10.00– 15.00
Enamel Undercoater	8.00– 12.00
Enamel, White	8.00– 12.00
Enamel, Quick Drying	8.00– 12.00
Enamel, Industrial	10.00– 15.00

Wall Size, Sealer, Paste, Etc.

		Price
Sizing, prepared	per gal.	$3.50–$4.50
Glue, Dry for Sizing	per lb.	.35– .50
Wall Primer and Sealer	per gal.	4.75– 5.85
Wallpaper Cleaner	per lb.	.25
Starch, gloss	per lb.	.15
Patching Plaster	per lb.	.15
Spackle	per lb.	.20
Glazing Liquid	per gal.	4.00– 6.00

Wood Fillers and Stains

	Price Per Gal.
Oil Stains	$5.00–$6.50
Acid Stains	5.00– 6.50
Penetrating Stainwax	5.00– 6.00
Stain, Varnish, all colors	5.00– 7.40
Stain, Fill, and Seal	5.00– 6.50
Paste Wood Fillers	4.00– 5.50
Floor Seal	9.00–10.00

Floor Wax

Liquid Floor Wax	per gal.	$8.00–$9.00
Self-Polishing Wax	per gal.	7.00– 9.00
Paste Wax	per lb.	5.00– 6.00

Creosote Shingle Stains

	Price Per Gal.
Brown and Red	$6.00–$ 8.00
Greens and Grays	7.00– 10.00

Miscellaneous Paints

	Per Gal.
Ready Mixed House Paint	$10.00–$15.00
House Paint—One Coat	9.00– 12.00
Porch and Floor Paint	10.00– 12.00
Flat Wall Paint	8.00– 14.00
One Coat Wall Paint (oil base)	9.00– 12.00
Metal Paint for structural steel, bridges, stacks, etc.	9.00– 15.00
Cement Floor Paint	8.00– 11.00
Cement Floor Paint (Rubber Base)	8.00– 11.00
Aluminum Paint	8.00– 10.00
Vinyl-Chloride Co-Polymer	14.00

Lettering On Glass or Wood

The cost of lettering will vary with the style and size of letters and whether painted or gold leaf.

The costs given on the following pages include only Roman or Gothic style letters, such as are ordinarily used for signs on doors, transoms, show windows, etc., and are based on all work being performed by experienced sign writers.

Plain Black Letters 2" to 4" High.—For plain black Gothic style letters or numerals, such as used on windows, transoms, office doors, etc., an experienced sign writer should average about 25 letters an hr. at the following labor cost per 100 letters :

	Hours	Rate	Total	Rate	Total
Sign painter	4	$. . . .	$. . . .	$20.00	$80.00
Cost per letter			. . . .		.80

Plain Black Letters 6" to 12" High.—An experienced sign writer should average 12 to 14 plain Gothic letters 6" to 12" high, an hr. at the following labor cost per 100 letters :

PAINTING

	Hours	Rate	Total	Rate	Total
Sign painter	7.7	$....	$....	$20.00	$154.00
Cost per letter			....		1.54

Gold Leaf Letters 2" to 4" High.—An experienced sign writer should complete about 5 plain Roman or Gothic letters in gold leaf an hr. (this class of work proceeds much slower than painted letters) at the following labor cost per 100 letters :

	Hours	Rate	Total	Rate	Total
Sign painter	20	$....	$....	$20.00	$400.00
Cost per letter			....		4.00

Gold Leaf Letters 6" to 12" High.—An experienced sign writer should complete 2 to 3 plain Roman or Gothic letters 6" to 12" high, in gold leaf, an hr. at the following labor cost per 100 letters :

	Hours	Rate	Total	Rate	Total
Sign painter	40	$....	$....	$20.00	$800.00
Cost per letter			....		8.00

Painting Pipe

The cost of painting pipe will vary considerably, depending upon accessibility of work, layout of piping, scaffolding required, type of paint, whether one color or more, and in the case of old work, the amount of surface preparation required and whether or not pipes are in service.

When estimating painting pipe, quantities are expressed in square feet of surface. When pipe quantities are measured from drawings, determine whether or not pipes are covered and thickness of same, as this will greatly affect the diameter and consequently the surface area, especially in the smaller sizes. Most estimators convert lineal feet of pipe to square feet of surface according to the following table :

O.D. Pipe or Covering	Sq. Ft. Area per Lin. Ft.	O.D. Pipe or Covering	Sq. Ft. Area per Lin. Ft.
3" and under	1	16" to 19"	5
4" to 7"	2	20" to 22"	6
8" to 11"	3	23" to 26"	7
12" to 15"	4	27" to 30"	8

Various types of paint materials are specified for pipes, depending upon their function, exposure, whether bare or covered, etc. Average material coverages and approximate prices for a few of these materials are as follows :

Description	Sq. Ft. per Gal.	Price per Gal.
Bare Pipe		
Black asphalt paint	One Coat 130–150	$ 5.00–$ 6.00
Aluminum paint	Each Coat 550–600	8.00– 10.00
Radiator enamel	Each Coat 550–600	18.00– 20.00
Covered Pipe		
Sizing	One Coat 350–400	.70– .80
Paint (oil base)	Each Coat 450–500	9.00– 12.00
Industrial enamel	1st Coat 450–500	12.00– 15.00
Industrial enamel	2nd Coat 550–600	12.00– 15.00

Labor Painting Pipe.—Labor costs for painting pipes are extremely variable due to the many conditions which may be encountered.

For simple work, consisting of horizontal runs and risers within stepladder reach, all painted one color, a painter should paint 800 sq. ft. of pipe, one coat, per 8-hr. day.

For boiler room or steam power plant piping of average difficulty, where some high ladder work is required, a painter should paint 600 sq. ft. of pipe, one coat, per 8-hr. day.

For difficult industrial process piping or in very congested steam power plants, where most of the work requires high scaffolding, a painter will paint only 350 to 450 sq. ft. of pipe, one coat, per 8-hr. day.

If pipes are painted different colors, for identification, reduce the above production 10 to 15 percent.

If pipes are coded by color banding at 10 to 15 foot intervals, consider this work as another coat for labor, but practically no material will be required.

PAINTING STRUCTURAL STEEL

The cost of painting structural steel after erection will depend upon a number of factors, i.e., the weight of the steel, whether light or heavy sections; access to work and method of painting, whether brush painting by hand or using a spray.

Lightweight members cost more to paint than heavy members, and the less climbing required, the more paint a man will apply, either by brush or when using a spray.

Structural steel may be estimated, either by the ton of steel or by the number of square feet of surface to be painted.

The ton method is much easier to compute but the square feet of surface method is much more accurate.

The following tables giving the approximate number of square feet of area on various size beams, angles and channels will assist the estimator in computing accurate quantities where the square foot method is used.

Number of Sq. Ft. Area Per Lin. Ft. of Structural "S" Beams of Various Sizes

Depth Beam Inches	Weight per Lin. Ft. Lbs.	Sq. Ft. Area per Lin. Ft.	Depth of Beam Inches	Weight per Lin. Ft. Lbs.	Sq. Ft. Area per Lin. Ft.
3	5– 8	1.25	10	25– 40	3.40
4	7–10	1.40	12	31– 55	3.70
5	10–15	1.90	15	42– 75	4.50
6	12–17	2.00	18	54– 70	5.00
7	15–20	2.60	20	65–100	5.40
8	18–25	2.70	24	80–120	6.40
9	21–35	2.90	27	90–135	7.50

Number of Sq. Ft. Area Per Lin. Ft. of Structural Channels of Various Sizes

Depth of Channel Inches	Weight per Lin. Ft. Lbs.	Sq. Ft. Area per Lin. Ft.	Depth of Channel Inches	Weight per Lin. Ft. Lbs.	Sq. Ft. Area per Lin. Ft.
3	4– 6	1.00	9	14–25	2.35
4	5– 7	1.20	10	15–35	2.70
5	7–11	1.50	12	21–40	3.00
6	8–16	1.70	15	32–50	3.70
7	10–20	2.00	18	34–55	4.40
8	11–21	2.20	20	42–58	5.00

Number of Sq. Ft. Area Per Lin. Ft. of Structural Angles of Various Sizes

Size of Angle Inches	Weight per Lin. Ft. Lbs.	Sq. Ft. Area per Lin. Ft.	Size of Angle Inches	Weight per Lin. Ft. Lbs.	Sq. Ft. Area per Lin. Ft.
2x2	2– 5	0.70	4x3	4–17	1.20
3x3	2–11	1.00	5x3	6–19	1.40
4x4	5–20	1.35	6x4	10–30	1.70
5x5	8–31	1.70	7x4	13–34	1.90
6x6	13–40	2.00	8x4	17–37	2.00
8x8	26–57	2.70	8x6	20–50	2.40

Painting Lightweight Structural Steel After Erection.—When painting lightweight structural steel after erection, by hand, using a brush, a painter should paint 600 to 650 sq. ft. per 8-hr. day on the first coat; and 800 to 900 sq. ft. per 8-hr. day on the second and third coats.

On a tonnage basis, this would be at the rate of 2 to 3 tons per 8-hr. day on the first coat and 3 to 4 tons per 8-hr. day on the second and third coats.

When using a spray gun, a painter should paint 3,600 sq. ft. per 8-hr. day on the first coat and 4,800 sq. ft. on the second and third coats.

Painting Medium to Heavyweight Structural Steel After Erection.—When painting medium to heavyweight structural steel after erection, by hand, using a brush, a painter should paint 750 to 850 sq. ft. on the first coat and 900 to 1,000 sq. ft. per 8-hr. day on the second and third coats.

On a tonnage basis, this would be at the rate of 4 to 5 tons per 8-hr. day on the first coat and 5 to 6 tons per 8-hr. day on the second and third coats.

Using a spray gun, a painter should paint 4,400 sq. ft. per 8-hr. day on the first coat and 5,600 sq. ft. per 8-hr. day on the second and third coats.

Labor Cost of 100 Sq. Ft. (1 Sq.) 2-Coat Oil Paint Applied to Lightweight Structural Steel

	Hours	Rate	Total	Rate	Total
First Coat					
Painter	1.3	$....	$....	$18.20	$23.66
Second Coat					
Painter	1.0	....	....	18.20	18.20
Cost 100 sq. ft.			$....		$41.86

For third coat, add same as second coat.

Spray Painting

	Hours	Rate	Total	Rate	Total
First Coat					
Painter	.23	$....	$....	$18.20	$4.19
Second Coat					
Painter	.17	....	....	18.20	3.10
Cost 100 sq. ft.			$....		$7.29

For third coat, add same as second coat.

Add for air compressor and engineer, depending upon number of spray guns operated.

Labor Cost of 100 Sq. Ft. (1 Sq.) 2-Coat Oil Paint Applied to Medium to Heavyweight Structural Steel

	Hours	Rate	Total	Rate	Total
First Coat					
Painter	1.0	$....	$....	$18.20	$18.20

Second Coat	Hours	Rate	Total	Rate	Total
Painter..................................	0.9	....	....	18.20	16.38
Cost 100 sq. ft.............................			$....		$34.58

For third coat, add same as second coat.

Spray Painting

First Coat	Hours	Rate	Total	Rate	Total
Painter..................................	.18	$....	$....	$18.20	$3.28
Second Coat					
Painter..................................	.15	....	....	18.20	2.73
Cost 100 sq. ft.............................			$....		$6.01

For third coat, add same as second coat.

Add for air compressor and engineer, depending upon number of spray guns operated.

PREPARING STEEL SURFACES

Proper protection of steel depends upon proper adhesion of the protective paint. Thus, when rust, oils, grease, scale and other dirt are not completely removed, the paint coat or coats will fail no matter how skillfully they are applied.

The Steel Structures Painting Council (SSPC), 4400 5th Ave., Pittsburgh, Pa., 15213, publishes detailed specifications for cleaning methods.

The simplest method is hand cleaning using scrapers, sanders, wire brushes and chipping tools. These remove only loose rust and scale and effectiveness depends largely on the zeal of the worker.

Power tool cleaning, using power operated sanders and brushes is slightly more expensive and slightly more effective.

For a superior job, one of the blasting methods is suggested : commercial, near white, or white. The last specifies the metal be blasted down to a uniform "white" surface. It will cost from $1.00 - 1.30 per sq. ft. but should provide a base good for 7 to 8 years. Near white specifies that at least 95% of the metal will meet the "white" surface test. Its price range is 60¢ to 85¢ per sq. ft. and its durability almost equal to white blast.

Commercial blast is difficult to specify, but the results should leave at least two-thirds of the metal in the "white" stage. Costs range from 45¢ to 60¢ per sq. ft. with durability of from 6 to 7 years.

With all the above methods, it is essential to use a primer with a good "wetting" power.

In figuring all these methods, cleaning up and protection for adjacent surfaces can add materially to the cost.

Cleaning Rust and Scale Off Structural Steel by Compressed Air.—When cleaning rust and scale off structural steel members, bridges, gas holders, oil tanks, and other metal surfaces before repainting, the number of sq. ft. cleaned per day will vary with the surfaces and the accessibility of same.

On structural steel members, bridges, etc., a man should clean 1,200 to 1,500 sq. ft. per 8-hr. day, using air operated wire brushes, at the following cost per 100 sq. ft. :

	Hours	Rate	Total	Rate	Total
Painter.................................	0.6	$....	$....	$18.20	$10.92
Compressor operator........................	0.3	....	....	17.05	5.12
Compressor expense & tools............	0.3	....	....	15.00	4.50
Cost per 100 sq. ft........................			$....		$20.54
Cost per sq. ft.................................			....		.21

On gas and oil tanks and other large surfaces, a man should clean about 2,000 sq. ft. of surface per 8-hr. day, at the following cost per 100 sq. ft. :

	Hours	Rate	Total	Rate	Total
Painter	0.4	$....	$....	$18.20	$ 7.28
Compressor operator	0.2	....	....	17.05	3.41
Compressor expense & tools	0.2	....	....	15.00	3.00
Cost per 100 sq. ft.			$....		$13.69
Cost per sq. ft.				....	.14

Comparative Cost by Hand

When cleaning steel surfaces by hand with wire brushes, a man will clean about 300 sq. ft. of surface per 8-hr. day, at the following labor cost per 100 sq. ft. :

	Hours	Rate	Total	Rate	Total
Painter	2.67	$....	$....	$18.20	$48.60
Cost per sq. ft.				....	.49

Spray Painting.—Almost any painting job can be done quicker with spray equipment. This is particularly true where large surfaces are to be covered or where there are angles and corners that cannot be reached with brushes or rollers.

Practically all surfaces that are now being painted with a bristle brush can be coated by spray painting. On rough surfaces such as cement, stucco, rough plaster, brick, tile, shingles and rough boards, spray painting affects a considerable saving in time and labor.

It is now possible to obtain mechanical painting equipment that will apply any paint, varnish, graphite, mineral oxides, epoxy, etc.

Painting Oil Tanks and other Large Metal Surfaces by Compressed Air.— In painting oil tanks and other large metal surfaces, a man should cover 1,000 sq. ft. an hr. at the following cost per 100 sq. ft. :

	Hours	Rate	Total	Rate	Total
Painter	.10	$....	$....	$18.20	$1.82
Compressor operator	.05	....	....	17.05	.85
Compressor expense	.05	....	....	15.00	.75
Cost per 100 sq. ft.			$....		$3.42
Cost per sq. ft.				....	.04

Does not include paint, scaffold, or preparation.

Comparative Cost by Hand

On the same class of work, a man will brush about 200 sq. ft. of surface an hr. at the following cost per 100 sq. ft. :

	Hours	Rate	Total	Rate	Total
Painter	.5	$....	$....	$18.20	$9.10
Cost per sq. ft.				....	.09

Does not include paint, scaffold, or preparation.

09950 WALLCOVERING

Wallpaper is estimated by the roll containing 36 sq. ft. Most rolls are 18" or 20½" wide and contain 36 sq. ft. or 4 sq. yds. of paper. The double roll contains just twice as much as the single roll, viz.: 72 sq. ft. or 8 sq. yds. of paper.

Estimating the Quantity of Wallpaper Required for any Room.—When estimating the quantity of wallpaper required for any room, measure the entire distance around the room in lin. ft. and multiply by the height of the walls or the distance from the floor to the ceiling. The result will be the number of sq. ft. of surface in the four walls.

Make deductions in full for the area of all openings such as windows, doors, mantels, built-in bookcases, etc. The difference between the area of the four walls and the area of the openings will be the actual number of sq. ft. of surface to be papered.

When adding for waste in matching, cutting, fitting, etc., paper hangers use different methods and allow different percentages, contingent upon the height of the ceiling, the design of the paper, and the size of the pattern or figure, but an allowance of 15 to 20 per cent will prove sufficient in nearly all cases.

When estimating the quantity of paper required for walls, bear in mind it is not necessary to run the side wallpaper more than one or two inches above the border, when a border is used. The height of the wood base should also be deducted from the total height of the wall.

In rooms having drop ceilings, the depth of the drop should be deducted from the height of the wall, which will decrease the area of side wallpaper in the same amount as it increases the quantity of ceiling paper.

If a border is required, measure the distance around the room and the result will be the number of lin. ft. of border. Dividing by 3, gives the number of yards of border required.

When estimating the quantity of wallpaper required for ceilings, multiply the width of the room by the length, and the result will be the number of sq. ft. of ceiling to be papered.

In rooms having a drop ceiling, add the depth of the drop or the distance the ceiling paper extends down the side walls to the length and breadth of the room and multiply as described above.

After the areas of both side walls and ceiling have been computed in sq. ft. divide by 36, and the result will be the number of single rolls of paper required; or by dividing by 72 gives the number of double rolls of paper required.

Paste Required Hanging Wallpaper.—The quantity of paste required for hanging wallpaper will vary with the weight of the paper and the surface to which it is applied.

Where light or medium weight wallpaper is used, one gallon of paste should hang 12 single rolls of paper.

If heavy or rough texture paper is used, it will often be necessary to give it two or three applications of paste to obtain satisfactory results. On work of this kind one gallon of paste should hang 4 to 6 single rolls of paper.

There are prepared pastes on the market which require only the addition of cold water to make them ready for use.

One pound of prepared cold water dry paste should make 1½ to 2 gallons of ready to use paste.

Labor Hanging Wallpaper.—Different methods of hanging wallpaper are used in different sections of the country. In some localities a paper hanger and helper work together, one man trimming the paper and pasting while the other man hangs the paper, while in many cities the work is performed by one man, who cuts, trims, fits, pastes and hangs the paper.

The quantities and costs given on the following pages are based on the performance of one man but will prove a fair average for use in any locality.

Hanging Wallpaper on Walls and Ceilings, One Edge Work.—When hanging light or medium weight paper on ceilings and drops, a paper hanger should hang 28 to 32 single rolls of paper per 8-hr. day, at the following labor cost per roll :

	Hours	Rate	Total	Rate	Total
Paper hanger	0.27	$. . . .	$. . . .	$17.43	$4.71

Hanging Wallpaper on Walls, Butt Work.—Where light or medium weight paper is used for bed rooms, halls, etc., and where a good grade of workmanship is required with all paper trimmed on both edges and hung with butt joints, a paper hanger should trim, fit and hang 20 to 24 single rolls of paper per 8-hr. day, at the following labor cost per roll :

	Hours	Rate	Total	Rate	Total
Paper hanger	0.36	$....	$....	$17.43	$6.28

Another thing that will slow up the work is where new paper is applied over old rough textured wallpaper, such as oatmeal designs, etc.

Hanging Wallpaper, First Grade Workmanship.—Where a good grade of medium or heavy weight wallpaper is used, with all paper hung with butt joints, a paper hanger should trim, fit and hang 16 to 20 single rolls of paper per 8-hr. day, at the following labor cost per roll :

	Hours	Rate	Total	Rate	Total
Paper hanger	0.44	$....	$....	$17.43	$7.67

Labor Hanging Scenic Paper.—Where scenic paper or paper having mural designs are used over the wall or up stairways, a paper hanger will hang only 8 to 10 single rolls per 8-hr. day, at the following labor cost per roll :

	Hours	Rate	Total	Rate	Total
Paper hanger	0.9	$....	$....	$17.43	$15.69

Placing Wallpaper Borders.—Wallpaper borders are usually estimated by the lin. ft. or yd. and as the width varies from 3" to 18" there is never over one width required.

A paper hanger should place border at the rate of 100 to 125 lin. ft. an hr. The labor cost per 100 lin. ft. (33⅓ yds.) should run as follows :

	Hours	Rate	Total	Rate	Total
Paper hanger	0.9	$....	$....	$17.43	$15.69
per lin. ft.			....		.16
Cost per yd.			....		.47

Hanging Coated Fabrics.—Where coated fabrics, such as Walltex, Sanitas, etc., are used, a paper hanger should hang 16 to 20 single rolls per 8-hr. day, at the following labor cost per roll :

	Hours	Rate	Total	Rate	Total
Paper hanger	0.44	$....	$....	$17.43	$7.67

Canvassing Plastered Walls.—When applying canvas to plastered walls, an average mechanic should apply 50 to 60 sq. ft. an hour.

Labor Cost of 100 Sq. Ft. Canvassed Walls

	Hours	Rate	Total	Rate	Total
Paper hanger	1.82	$....	$....	$17.43	$31.72
Cost per sq. ft.			....		.32

Flexwood.*—Flexwood is a genuine wood veneer cut to 1/85 of an inch, glued under heat and hydraulic pressure to cotton sheeting with a waterproof adhesive. A patented flexing operation breaks the cellular unity of the wood to produce a limp, pliable sheet which may be applied to any smooth surface, flat or curved. Waterproof Flexwood 710 Adhesive, which makes a permanent bond, is used to apply

U.S. Plywood, Louisville, Ky.

Flexwood. Standard sizes of stock material are 18-in. and 24-in. widths and 8-ft. to 12-ft. lengths.

The mechanics of installing Flexwood are as follows: the background is sized with Flexwood cement. Another coating of cement is brushed on the Flexwood which is then hung in the manner of any sheet wall covering. A stiff, broad knife, used with considerable pressure, smooths out the Flexwood, removes air spaces and furnishes the necessary contact.

Some of the wood available in Flexwood are mahogany, walnut, oak, prima vera, knotty pine, orientalwood, satinwood, zebrawood, rosewood, English oak, maple, lacewood, etc. The cost of the material varies from $1.10 to $3.00 per sq. ft. depending upon the kind of wood. The majority of woods run from $1.25 to $1.50. Adhesive runs $7.50 per gallon.

It requires a skilled and thoroughly competent paper hanger to apply this material and the labor costs vary considerably with the design to be obtained.

Considerable time is usually required to lay out any room to obtain the desired effect, spacing of strips, etc. This is especially true where certain designs must be obtained on columns, walls and pilasters and where narrow strips of contrasting woods are used to obtain inlay effects.

On plain walls without inlays, a paper hanger experienced in this class of work should apply 150 sq. ft. of Flexwood per 8-hr. day, at the following labor cost per 100 sq. ft.:

	Hours	Rate	Total	Rate	Total
Paper hanger	5.4	$. . . .	$. . . .	$17.43	$94.12
Cost per sq. ft.			. . . .		.94

On pilasters, columns and walls inlaid with narrow strips of contrasting woods, the work is considerably slower, due to the additional cutting and fitting required. On work of this kind, a paper hanger, experienced in this class of work should apply 75 sq. ft. per 8-hr. day, at the following labor cost per 100 sq. ft.:

	Hours	Rate	Total	Rate	Total
Paper hanger	10.7	$. . . .	$. . . .	$17.43	$186.50
Cost per sq. ft.			. . . .		1.87

Additional time for paper hanger or foreman will be required laying out the work into patterns, marking off strips, etc. This will vary with the size of the job.

After the Flexwood is applied, the joints should be sanded lightly to remove any imperfections in the joints.

Finishing Flexwood.—Flexwood will take any wood finish but where the natural color of the wood is desired, the Flexwood is given one coat of lacquer sealer, sanded lightly between coats and then given one coat of lacquer for a finish coat.

One gallon of lacquer sealer will cover about 300 sq. ft. of surface, one coat.

After the sealer has been applied, one gallon of lacquer should cover 450 to 550 sq. ft.

An experienced painter should apply lacquer sealer, sand lightly between coats and apply one coat of lacquer to 90 to 110 sq. ft. of surface an hr. at the following labor cost per 100 sq. ft.:

	Hours	Rate	Total	Rate	Total
Painter	1	$. . . .	$. . . .	$17.43	$17.43
Cost per sq. ft.			. . . .		.18

Vinyl Wall Covering—Vinyl wall covering is composed of a woven cotton fabric to which a compound of vinyl resin, pigment and plasticizer is electronically fused to one side. It comes 24", 27", and 54" wide in three weights : heavy (36 oz. per lineal yard); medium (24-33 oz. per lineal yard); and light (22-24 oz. per lineal yard).

Some of the patterns available simulate linens, silks, moires, grasses, tweeds, Honduras mahogany, travertine, damasks and burlaps. Costs range from 35¢ to 90¢ per square foot.

Vinyl fabric is hung using regular wallpaper hanging procedures for hanging fabric baked wallcoverings. A broad knife is used to smooth out any wrinkles, air pockets and insures a good contact. Wash off any excess paste that remains on the surface of the material: No further finishing is necessary.

A competent paper hanger can apply approximately 600 sq. ft. per 8 hr. day at the following costs per 100 sq. ft.

	Hours	Rate	Total	Rate	Total
Paper hanger	1.34	$. . . .	$. . . .	$17.43	$23.36
Cost per sq. ft.					.23

If excessive cutting and fitting is required the above figures should be increased to allow for the above conditions.

SPECIALTIES

The Construction Specifications Institute, Inc., whose format we have followed in the general presentation of our text, follows the sections covering the usual building trades with four sections covering allied fields which may or may not be part of the usual general contract.

CSI Division 10 includes specialties which are prefabricated and added to the building such as chalkboards, toilet partitions, decorative grilles, freestanding fireplaces, flagpoles, directories and signs, lockers, mailboxes and chutes, movable partitions, sun control devices, telephone booths, and the like. Many of these items will be subcontracted and installed under the general contractor's supervision. Some may be covered by allowances.

CSI Division 11 includes equipment such as vacuum systems; powered window washers; and bank, checkroom, church, school, food service, vending, athletic, lab, laundry, library, medical, parking, waste handling, loading dock, detention and stage equipment. These are often purchased by the owner separately and installed by the general contractor. Even if the general contractor is not involved with the installation of such equipment, he should have a list of all equipment going into the building with cuts and shop drawings so he may point out any conflicts between the structure and such equipment. A floor that has to be broken up to conceal a conduit or duct becomes a permanent scar which will always reflect on the general contractor even though he might be an innocent partner.

CSI Division 12 covers furnishings. Separate contracts are usually let for this, but again some coordination is necessary and the general contractor should have a clear understanding regarding damages to the building by other parties working on the premises, especially if painting and decorating are part of the general work. Also, furnishings and draperies can play havoc with a carefully balanced air handling system. If the owner is uncomfortable, the general contractor will be the one he'll call even though it is the 10 foot long sofa that blocks the baseboard heater or the $2,500.00 chandelier that interrupts the ceiling outlet airflow and causes a draft on the boss's neck. The architect should be the coordinator of all this, but an alert contractor will try to forestall as many problems as he can.

CSI Division 13 is titled "Special Construction" and covers such items as air supported structures, audiometric rooms, hyperbaric rooms, etc., and is a bit too specialized for inclusion in this book.

Some of the items of the other divisions are covered below.

10100   CHALKBOARDS AND TACKBOARDS

Chalkboards are made of ¼" or ½" hardboard with special finishes applied. These include the following which are available in either black or green:

¼" Thick - Material Prices

Standard, weighing 1.75#/sq. ft.	$1.40/sq. ft.
Moisture sealed at 1.8#/sq. ft.	1.65/sq. ft.
Porcelain steel at 2.1#/sq. ft.	4.25/sq. ft.

These are available in 4' widths in sheets to 16' long and are applied to the wall with mastic at the rate of one gallon per 70 sq. ft. They can be set in custom trim or the following stock aluminum trim may be ordered which comes complete with all necessary attaching clips:

Perimeter trim 1½" wide	$1.80/lin. ft.
Maprail headtrim	2.00 " "
Joint strip	1.05 " "
Chalktrough 2¼" wide	4.15 " "

Bulletin board materials are ¼" cork and available as follows:

Natural cork	$1.90/sq. ft.
Vinyl cork	2.75 " "
Cork on ⅜" fiberboard	2.90 " "
⅛" " " " vinyl f'c'd	2.75 " "

The cork comes in rolls 4' wide by up to 90' long. The panels are 4' wide up to 16' long. Stock trim costs $1.80 for ¼" stock, $2.00 for ½" stock.

10150   COMPARMENTS AND CUBICLES

Toilet Partitions and Screens

Toilet partitions are most often metal units of the floor mounted type, with baked enamel finish and chrome hardware. One door and side partition for use in a corner or abutting other stall side partitions will cost around $165.00. A porcelain enamelled finish will add $275.00 to the cost. Ceiling hung units cost around 10% more, plus of course the cost of ceiling reinforcement if needed. One carpenter and laborer will set a unit in about two hours. Urinal screens hung from the wall cost $100.00 in baked enamel and will take a two man team half an hour to set. Wall hung toilet stalls will cost $320.00 per unit.

10301   PREFABRICATED FIREPLACES

Metal fireplaces are available in a wide range of styles for burning wood or gas. They are made for placing against walls, in corners and freestanding. However, there are requirements for clearances from walls, and for floor protection. They can be vented to an existing flue or have a prefabricated smoke pipe terminating above the roof. A simulated masonry chimney top is even available. Finishes are either matte black or porcelainized. A 24" wide unit will cost $350.00 in matte, $450.00 in porcelain. A 45" wide unit will cost $475.00 in matte and $575.00 in porcelain.

Details of Aluminum Moldings Used with Chalkboards and Corkboards

Freestanding units open on all sides will cost $425.00 for matte and $525.00 for porcelain in 36" diameter size; $500.00 and $600.00 in 42" diameter. If they are to rest on the floor, add $50.00 for the metal base. If hung from the ceiling, add $25.00 for chains.

8" metal flues cost about $20.00 per foot. Chimney top housings for single flues cost $75.00 to $100.00.

10350   FLAGPOLES

Typical price examples of several types of flagpoles are as follows : 40'-0" ground type; swaged sectional steel, including fittings, ground tube and flashing collar. Price $1,500.00 erected in place and painted.

40'-0" ground type; cone tapered steel, including fittings, ground tube and flashing collar. Price $1,700.00 erected in place.

To above prices, add cost of concrete foundation and ornamental base if required.

20'-0" roof type; swaged sectional steel, including fittings, socket plate, tension braces and flashing collars. Price $850.00, erected in place and painted. Add cost of setting anchor bolts.

12'-0" outrigger type; cone tapered aluminum, including fittings and base. Price $900.00 erected in place. Add cost of setting anchor bolts.

10400   IDENTIFYING DEVICES

Directories and Signs

Directory boards are available glass enclosed or open-faced. Glass door units cost $350.00 for 36"x30", $500.00 for 36"x48".

Open faced directory boards cost as follows, with 100 letters:

12"X18"	$ 75.00
24"X18"	85.00
24"X36"	115.00

Letters for directory boards are ordered in an assortment including numbers. An assortment of 200 will cost as follows: ½" - $30.00; ¾" - $35.00; 1" - $45.00; 1½" - $55.00.

Plaques

Building committees often will ask the contractor to order a plaque to commemorate a donor, an outstanding citizen, or even themselves. Cast bronze is ultimate in permanence and will cost about $130.00 per sq. ft., aluminum is 20% less, as are built-up plaques.

Signs

Cast letters are often applied to walls and copings to identify the building occupant. Letters are cast from aluminum or bronze and then given a variety of finishes. Prices will cost as follows:

Material & Finish	2"	4"	6"	12"	18"	24"
Cast Aluminum	$ 7.00	$12.00	$22.00	$34.00	$60.00	$ 85.00
Cast Bronze	11.00	22.00	30.00	60.00	90.00	120.00

A man should install 3 to 4 letters per hour.
Plastic letters will cost $10.00 each in 8" height.

10500 LOCKERS

Lockers are made in single, double and triple high units totaling 72" high (single tier also comes 60" high), in widths 9", 12", 15" and 18" and depths of 12", 15" and 18". There are also box lockers 6 units high.

A 12"x15"x72" unit in baked enamel will cost around $40.00 single tier; $55.00 double tier; $75.00 triple tier. For a sloping top add $2.50 per tier. For closed base, add $1.50 per foot. Open mesh or "athletic" types will cost $90 to $110 per tier 72" high. Basket racks 8 high, 5 wide will cost $125.00 plus $6.50 per basket. For a combination lock, add $10.00 per door, $3.00 for a key lock.

Benches for lockers are sold by the lineal foot. Benches made of 1¼" thick laminated maple cost $12.00 per foot plus $14.00 for each pedestal.

10550 POSTAL SPECIALTIES

Mail chutes with glass and aluminum enclosures 9"x 3½" will cost $35.00 per lin. ft., and one man can install two floor heights per day. A lobby collection box will cost from $800.00 on up depending on the size and design.

Keyed mail delivery boxes in gangs will cost around $45.00 per unit rear loading and $48.00 front loading. One man can set around 30 units per day.

10600 PARTITIONS

Movable partitioning is available in metal in both flush and panel construction, in laminated gypsum board and made of open wire mesh. Solid units can be specially constructed for sound and fire isolation, and can be prefinished with baked enamel, vinyl or wood veneers or painted in the shop or on the job.

Metal office type partitioning is usually a nominal 2¼" or 3½" thick with a height limitation of 12'. Free standing units have a standard cornice height of 7' 1½". "Banker" types are usually 1¾" thick and 30" high. A standard glass panel can be mounted on the top rail to bring the height to 42" or 54".

The base for movable partitioning is designed to form a raceway for both electrical and telephone lines. Switches can usually be built into the door frame panels, which with doors, glazing panels and louver panels can all be ordered as part of the partitioning.

Mesh partitioning is usually made of 10 ga. wire worked into a 1½" diamond mesh set in a channel frame. It will cost around $2.00 per square foot on straight runs using standard sections. A 3'x7' door will cost $170.00. A 5'x7' sliding panel, $300.00. A shop coat of paint is standard. A baked enamel coat will add 10¢ per sq. ft. to costs. 2 erectors can install some 100 lineal feet per day. Add 1.5 hrs. for each door. Ceiling panels will install at the rate of 75 sq. ft. per hour two men working together.

Office partitions come in so many grades and job conditions are so diverse, it is difficult to give any basic costs, but as the field is highly competitive, it is always a simple matter to get prices together. A standard 2⅜" 20 ga. flush steel floor to ceiling partition 8 ft. to 9 ft. high will cost around $4.00 per square foot, without installation. Two men should erect around 40 lineal feet per day. Each door with lock but no closer will cost $200.00. If partitions are glazed add for both labor and material.

2¼" thick gypsum partitions will run around $2.75 per sq. ft. installed, but unfinished. If glazed panels are desired add $10.00 per sq ft.

Freestanding "bank" type partitioning 3 ft. high will cost $25.00 to $35.00 per lin. ft. and two men will set 70 lin. ft. per day.

10650   SCALES

Built-in scales must be very carefully set, adjusted and maintained and should be set by a contractor licensed by the scale manufacturer. For general budget allowances the following may be a guide:

		Cost	
Platform Size	Capacity	Mechanical	Electronic
7'x 9'	5 ton	$ 7,500	....
10'x24'	10 ton	8,500	....
10'x50'	30 ton	10,000	....
10'x70'	50 ton	24,000	28,000
10'x70'	60 ton	25,000	30,000
10'x70'	80 ton	28,500	32,000
10'x70'	100 ton	....	36,000

10670   STORAGE SHELVING

Free standing, baked enamel industrial shelving comes knocked down in 3' sections, complete with all hardware for erection at the job site. Cost will approximate the following:

Depth	Number of Shelves	
	4	7
12"	$29.00 / lin. ft.	$48.00 / lin. ft.
18"	34.00 " "	56.00 " "
24"	40.00 " "	65.00 " "

One man can erect between nine and twelve lineal feet per day.

10700   SUN CONTROL DEVICES

Awnings and canopies of canvas, vinyl and aluminum are prefabricated in the shop and delivered to the job site ready for erection.

Continuous, wall hung, aluminum canopies will cost from $4.00 to $6.00 a square foot and two men can erect around 250 sq. ft. per day.

Awnings over individual openings 3' 6" wide will cost $250.00 in aluminum, $150.00 in cloth. Awnings are often used as decorative features and take forms unrelated to sun control and can easily cost up to $30.00 a square foot.

Two men will take two hours to hang an individual awning.

10750   TELEPHONE ENCLOSURES

The trend in telephone noise control has been to rely on sound absorbtion rather than physical separation. Where booths are still used a simple steel and wood unit will run $900.00 while an extruded metal unit will cost twice that. Two men will set three units in two days.

Outdoor post types will cost around $700.00 and take one man six hours to set. Indoor shelf types will average around $400.00 a station and take one man three hours to install. Directory shelves will cost around $300.00 for a size that holds up to four books, and will take one man one hour to set.

10800   TOILET AND BATH ACCESSORIES

The following list may serve as a guide in purchasing chrome plated bathroom accessories. Sometimes accessories are of ceramic tile, costs of which are discussed in Chapter 14. Also, some items may be furnished with toilet compartments or will be furnished by service companies who supply the owner. Check to avoid duplication.

Prices of Toilet and
Bath Accessories

Ash tray bowl, wall mtd.	$ 35.00
recessed	75.00
Clothes hook —single	3.00
—double	3.50
Grab bar —12"	10.00
—24"	16.00
Hand dryer, electric	225.00
Medicine cabinet, 18"x24", lighted	65.00
Paper towel dispenser, wall mtd.	25.00
recessed with waste receptacle	240.00
Sanitary napkin dispenser, coin operated & surf. mtd.	200.00
Shelf, stainless steel, 24" long	18.00
Shower rod —3'	9.00
—5'	12.00
Soap dish—recessed	7.00
Soap dish with grab bar	10.00
Soap dispenser —powder	15.00
—liquid	22.00
Toilet tissue dispenser	8.00
Towel bar —12"	12.00
—18"	12.50
—24"	14.00
Towel ring	9.00
Towel ladder, 16"x30"	42.00
Tumbler/toothbrush holder	8.00
Waste receptacle, 10 gal	85.00

10900   WARDROBE SPECIALTIES

Hat and coat checking storage units will cost approximately the following:

Material	Single Tier	Double Tier
Steel	$15.00 / lin. ft.	$18.00 per lin. ft.
Aluminum	24.00 " "	30.00 " " "
Stainless	30.00 " "	36.00 " " "

One man can set 45 lin. ft. of single tier and 32 lin. ft. of double tier rack per day.

11900   RESIDENTIAL EQUIPMENT:
KITCHEN CABINETS

The following cabinets are furnished completely assembled and finished. Frames are ¾" kiln dried hardwood; doors and drawer fronts are ⅞" thick solid lumber; shelves are ½" plywood or chipboard; backs of base cabinets are Masonite; backs of wall cabinets and drawer bottoms are plywood; ends of cabinets are flush with hardwood faces.

Cabinets are factory finished with a penetrating sealer coat and clear lacquer. Interiors are finished and drawer bottoms are clear lacquered.

Door and drawer hardware come in a variety of styles.

These cabinets are the type found in most residential construction. Economy grade cases for cheaper construction will range from 50 percent to 25 percent less for finished units ready for installation. For top quality kitchen cabinets, the prices will run up to 50 percent higher for stock sizes and 100 percent higher or more for custom made units, depending upon the design, materials, hardware and finish specified.

All prices are f.o.b. factory.

Wall Units

Wood Wall Units, 12" Deep

Width	No. Doors	12" High	15" High	24" High	30" High
12"	1	$....	$....	$....	$ 56.00
15"	1	....	....	....	62.00
18"	1	....	....	....	66.00
21"	1	....	....	....	70.00
24"	2	....	....	67.00	80.00
27"	2	....	....	70.00	90.00
30"	2	53.00	58.00	75.00	98.00
33"	2	58.00	60.00	81.00	103.00
36"	2	60.00	67.00	83.00	108.00
42"	2	....	....	89.00	113.00
48"	2	....	....	100.00	135.00

Corner Shelf and Wall Units, 30" High

Corner wall unit, 23" deep and wide, with 2
   shelves ....$ 95.00
Corner wall unit, 23" deep and wide, with revolving
   3 shelf "Lazy Susan"...135.00

Utility Cabinet and Built-In Oven Cabinets

Utility cabinet, 84" high, 24" deep, 18" wide....................................$210.00
Oven cabinet, 84" high, 24" deep, 24" wide 250.00
Oven cabinet, 84" high, 24" deep, 27" wide 265.00

Wood Cabinet Base Units

Base units are 34½" high and 24" deep without tops.

Width Inches	Description	Without Tops
15"	4 drawers	$120.00
18"	4 drawers	125.00
21"	4 drawers	132.00
24"	4 drawers	140.00
12"	1 top drawer, 1 door below	86.00
15"	1 top drawer, 1 door below	92.00
18"	1 top drawer, 1 door below	97.00
21"	1 top drawer, 1 door below	107.00
24"	1 top drawer, 1 door below	120.00
27"	1 top drawer, 2 doors below	130.00
36"	2 top drawers, 2 doors below	160.00
42"	2 top drawers, 2 doors below	170.00
48"	2 top drawers, 2 doors below	184.00
42"	Corner base cabinet	142.00
35"	Corner base cabinet with revolving "Lazy Susan"	150.00
	3" corner filler	5.00

Wood Cabinet Base Units

Wood Sink Fronts

Description		Width	Price
Sink front only	1 pair doors	.... 27"	$68.00
" " "	" "	.... 30"	70.00
" " "	" "	.... 33"	73.00
" " "	" "	.... 36"	78.00
" " "	" "	.... 42"	84.00
" " "	" "	.... 48"	91.00
Sink end		.... 24"	11.00

Wood Sink Base Units

Width Inches	Price Without Tops or Sinks
27"	$110.00
30"	115.00
33"	120.00
36"	125.00
42"	135.00
48"	145.00

Steel Kitchen Units

Prices on steel case units vary considerably with different manufacturers, due to weight of metal used, method of manufacturing, kind of finish, sprayed or baked and volume of production as many kinds of steel cases are on a mass production basis, while others are custom made to fit the job. Naturally, the custom cabinets cost more.

The cabinets priced below are completely finished and ready to be installed. They are fabricated from 18, 20 and 22 gauge cold rolled steel and are spot welded, bonderized and then finished with 2 coats of enamel baked on at high temperature. They are medium quality.

Base units are mounted on an integral black enameled subbase 4" high and recessed 3" in front for toe space.

Photographer—Bill Hedrich, Hedrich-Blessing

Doors are ¾" thick, double-walled, die formed of one piece of steel and spot welded. Drawers have ¾" fronts which make them flush with the doors. Drawers operate on roller bearings.

Hardware consists of semi-concealed hinges. Die-cast chromium plated handles are furnished on doors and drawers.

Steel Wall Cabinets

Wall units are 13" deep. 18" units have 1 removable shelf; 30" units have 2 shelves.

Width	No. Doors	18" High	30" High
12"	1	$. . . .	$37.00
15"	1	. . . .	39.00
18"	2	37.00	43.00
21"	2	40.00	48.00
24"	2	42.00	50.00
27"	2	. . . .	54.00
30"	2	45.00	56.00
36"	2	52.00	64.00

Corner Shelf and Wall Units, 30" High

Description	Price Each
Corner shelf unit, 12¼" deep, 6" wide, with 2 fixed shelves	$ 18.00
Corner shelf unit, 12¼" deep and wide, with 2 fixed shelves	22.00
Corner wall unit, 21" deep and wide, with revolving 3-shelf "Lazy Susan"	100.00

Steel Base Cabinets

Base units—34½" high and 24" deep

Width	Description	Without Top
12"	1 drawer above, 1 door below	$ 68.00
15"	1 drawer above, 1 door below	75.00
18"	1 drawer above, 1 door below	81.00
21"	1 drawer above, 2 doors below	85.00
24"	1 drawer above, 2 doors below	94.00
27"	1 drawer above, 2 doors below	101.00
30"	1 drawer above, 2 doors below	108.00
36"	1 drawer above, 2 doors below	121.00
18"	3 drawers	116.00
33"	Corner base cabinet with revolving "Lazy Susan"	125.00

Steel Sink Fronts

Description	Width	Price
Sink front only—one pair doors	24"	$41.00
Sink front only—one pair doors	30"	43.00
Sink front only—one pair doors	36"	48.00

Fillers and End Panels for Steel Cabinets

Description	Price Each
Corner wall filler, 30" high, 15" deep, 15" wide	$14.00
Straight wall filler, 18" high, 13" deep, 3" wide	6.00
Straight wall filler, 30" high, 13" deep, 3" wide	8.00
Corner base filler, 34½" high, 27" deep, 27" wide	20.00
Straight base filler, 34½" high, 3" wide	4.00
End panel, 34½" high, 25" deep	11.00

Steel Kitchen Sink Units

Sinks are 36" high, 25" deep. Constructed of 18, 20, 22 ga. furniture steel with a baked on finish. Sink tops are 14 ga. deep drawn enameling steel to which porcelain enamel is bonded.

Tops have grooved drainboards and 4" back-splash, all insulated to deaden noise. Sink tops are one piece construction, roll rimmed.

Length Unit	Size Bowl	Single or Dbl. Bowl	Price Each
42"	20"x16½"x7"	Single	$210.00
54"	20"x16½"x7"	Single	230.00
66"	2-14½"x17½"x7"	Double	270.00

KITCHEN COUNTERTOP

Most all countertops for kitchen cabinets are plastic laminate over a plywood, particleboard, or masonite sub-surface.

The plastic decorative laminate is offered in a wide variety of colors and patterns including simulated wood grain and butcher-block styles.

The tops are usually made in self-edge design (square edges and 90 angle for the backsplash), however they are available in post-form design (rolled edges and angles) at a slight decrease in cost.

To estimate the total linear feet of countertop, measure the length of the area to be covered along the longest side (usually the wall surface). Plastic laminate tops, made to size, ready to install will cost from $16.00 to $19.00 per linear foot.

13130   SPECIAL PURPOSE ROOMS

SAUNA ROOMS

The luxury residence or office is often fitted with a prefabricated sauna cabinet. These come in a wide variety of sizes and there are several manufacturers so that more than one source should be contacted.

The minimum sauna, for sitting position only will be around 4'x4', weigh 500 lbs. and will need a separate 120 volt 20 AMP service. The cost complete with door, floor, ceiling, bench and equipment will cost about $1,800.00. A 4'x7' unit for one lying down or 2 sitting will weigh 750 lbs., will require 220 volt 30 AMP service and cost about $2,400.00. Units for several people measuring up to 7'x10' will weigh 1200–1500 lbs., require 220–250 AMP service and cost $3,200.00 to $4,000.00.

As these units are prefabricated they are simple to install, but will require a carpenter, a helper and an electrician. Except for the largest units, or when preparatory work is required, the installation should be completed in a days' time.

CHAPTER 16

CONVEYING SYSTEMS

CSI DIVISION 14
14100 DUMBWAITERS
14200 ELEVATORS
14600 MOVING STAIRS

14100 DUMBWAITERS

Dumbwaiters must be carefully selected to provide the maximum in convenience and efficiency for the minimum cost. They range from simple hand operation to memory-bank controlled automated models.

Standard speeds are 50 fpm, but speeds to 150 fpm are possible on some models and, where speed is not a consideration, 25 fpm will save some initial cost. Car sizes are directly related to capacities. In general, a 100 lb. capacity cab will have a maximum size of 6¼ sq. ft. and a height of 36". 300 and 500 lb. capacity cars may have a maximum of 9 sq. ft. and a height of 4 ft.

Dumbwaiters may be mounted on channel guides which are attached to the building construction, but a packaged, self-supporting tower is also available.

An automatic electrically operated unit with 500 lb. capacity, running through one floor with two stops, will cost approximately $8000.00. Hand operated units will be about half that amount.

14200 ELEVATORS

Elevator construction is such a highly specialized business that it is practically out of the question to expect anyone except an experienced elevator contractor to prepare an estimate for the installation of such equipment.

There are many items entering into the construction of an elevator which affect the cost. All of the elevator manufacturers submit their quotations on an installed basis and the location of the job must be taken into consideration as well as the intended use, capacity, car speed, size of car, travel, number of stops and openings, design of the car and kind of current available.

In addition there are many types of controls, such as car switch, continuous pressure push button, single call push button, up down collective (for operation with an operator or automatic operation) signal control, duplex and others.

Almost all modern passenger elevator installations have some degree of automatic control if they are not completely automatic. In new office buildings computerized systems are available which recognize traffic demands and automatically program elevator service to these demands. With these programming systems, elevator service is made as efficient as possible during peak or down periods.

Elevator Safety Codes

Many states and cities have their own elevator codes, and all elevator installations must comply with the rules and regulations of that state or city. In general, all municipal or state codes follow closely the American Safety Code for Elevators, Dumbwaiters and Escalators, ASA Standard A 17.1, with its supplement ASA A17-

1a, and the American Standard Safety Code Rules for Moving Walks ASA A17.1.13. It would be well for anyone endeavoring to estimate the cost of an elevator to familiarize himself with these codes. For example, the capacity of a passenger elevator must be based upon the effective area of the car platform in accordance with the contract load graphs shown in the latest edition of the American Standard Safety Code for Elevators, Dumbwaiters and Escalators.

Essential Information for Estimating the Cost of Elevators.—The following information is essential to enable the elevator manufacturer to prepare an intelligent estimate on the elevators required for any building.

1. New or renovation project
2. Location of building
3. Number of elevators
4. Use of elevator (passenger, freight or general purpose)
5. Capacity of elevators
6. Car speed of elevators
7. Controls and operation of elevators
8. Type of motor (single speed, two speed, or variable voltage)
9. Total travel of elevator in feet
10. Number of stops and openings
11. Single or double entrance car
12. Location and size of elevator machinery room or penthouse
13. Kind of current available
14. Size of hatch or platform
15. Design of cab, doors and frames
16. Car and hatch door power operators
17. Signals
18. Access door location

Costs of Electric and Hydraulic Elevators. —The cost of an elevator is a function of its speed, capacity and type of control which in turn are determined by type of building to be served, the use of the elevator and the height of the building or feet of travel of the elevator.

There are so many factors that enter into the cost that an estimator should get definite prices from an elevator contractor whenever possible. The following budget figures can be used for rough estimating only.

Electric Elevators:

1. General Purpose and Hotels
   High rise passenger elevator
   Gearless machines
   Standard platform size
   Center-opening hoistway and car doors
   $4,000.00 car allowance
   Basic cost for 12-story building

CAPACITY	SPEED		
	500 fpm	800 fpm	1000 fpm
2500 lbs.	$120,000	$145,000	$160,000
3000 lbs.	127,000	152,000	167,000
3500 lbs.	134,000	159,000	174,000
4000 lbs.	146,000	170,000	186,000

For each floor above 12, add $4,000.00 per floor

Many tall building elevators now operate at speeds up to 1800 fpm but these require special pricing, depending on express and local runs.

2. General Purpose—Hospitals, Hotels and Apartments

Low rise elevators
Geared machines
Standard platform size
Center-opening hoistway and car doors
$4,000.00 car allowance
Basic cost for 8-story building

CAPACITY		SPEED	
	200 fpm	300 fpm	350 fpm
2000 lbs.	$72,000	$77,000	$82,000
2500 lbs.	75,000	80,000	85,000
3000 lbs.	80,000	85,000	90,000
4000 lbs.	88,000	93,000	98,000

For each floor more than 8—add $3,500 per floor.

Oil Hydraulic Elevators.—Rise is limited to about 50 ft. so that they can be used for 5 or 6-story buildings. They are more economical to install and require no overhead penthouse. They are available in speeds up to 200 fpm and nearly any capacity.

The cost of an oil hydraulic installation must include the drilling of the hole for the piston. This depends on local conditions such as soil, water, rock, etc. The hydraulic elevators range in price, depending on size, speed, control and purpose. A 2000 lb., 125 fpm hydraulic elevator, serving 4 stories with 4 openings and a vertical run of some 30 ft. complete with all motors, pumps, controls, piping, drilling, cab, doors and frames, etc., will cost about $45,000.

Garage Elevator.—Freight elevator, single call push button control, two-speed motor, automatic leveling, capacity 6,000 lbs., speed 100'-0" per minute. Total rise 36'-0" with 4 stops and 4 openings. Car platform 10'-0"x20'-0". Car built of steel wainscoting on 2 sides 6'-0" high, a steel car top and two wood or expanded metal lift-up gates with electric contacts. Price installed $55,000. No doors or hatchway gates included.

Residence Type Elevators.—During recent years residence type elevators have become quite popular, especially where there are older people or invalids in the house. These elevators are easily installed both in new homes and homes already built.

A typical residential elevator, 350 lbs. capacity, speed 20'-0" per min. Total travel 11'-0", 2 stops and 2 openings, 110 or 220 volts, single phase 60 cycles. Platform 34"x30". Price installed approximately $11,000.00. Add for each additional stop, $1500.00.

Same elevator as above with 40"x36" platform. Price installed approximately $12,000.00. Add for each additional stop, $1500.00.

14600 MOVING STAIRS

Escalators in single widths will cost about $70,000 per floor (12 ft. height); in double width (48"), about $85,000.00 per floor.

CHAPTER **17**

MECHANICAL

CSI DIVISION 15
15A PLUMBING AND
     SEWERAGE
15B HEATING AND
     AIR CONDITIONING
15C SPRINKLER SYSTEMS

15A PLUMBING AND SEWERAGE

To prepare an accurate detailed estimate on plumbing and sewerage requires a working knowledge of the trade as all work must be laid out on paper and the number of lineal feet of each kind of pipe estimated separately, such as tile or cast iron sewers, soil and vent pipe, water pipe, gas pipe, drains, valves, pipe fittings, etc. Inasmuch as very few, if any, general building contractors or estimators possess this knowledge, it makes it doubly difficult for them to prepare more than an approximate estimate on the various kinds of plumbing work.

Budget figures for average plumbing installations can be estimated in one of two ways; as a percentage of the total cost of a project; or, as the grand total of a predetermined cost allowance for the installation of each fixture.

Plumbing estimates as a percentage of the total job will vary from a minimum of 3% on up to 12% depending on the number and type of fixtures required and on their disposition within the building. Percentage estimates would be valid for general plumbing only; process and other special condition piping would of course have to be entered as a separate item.

Those buildings which would require a high percentage allowance would include motels and nursing homes which are generally spread out over a large area and require plumbing for each room, and apartment high rises with many studio units which require full baths and kitchens for relatively small living areas.

On the low side would be assembly, mercantile, warehouse and industrial buildings where plumbing can be concentrated in a central core. Recent OSHA requirements limiting the travel distance to a rest room have raised the costs on some of these types.

The private residence is difficult to peg. Where kitchens and baths can be back to back the percentage is on the low side. But in a large, rambling house with a central powder room area but with kitchens, family and master bedrooms all in separate wings and with hose bibs and lawn sprinkling systems on all four sides and sunken bathtubs, shell-shaped lavatories and gold plated fixtures even 12% may be too low.

However the average building types will fall in the following ranges;

apartments	- 9% to 12%	motels	- 9% to 12%
assembly	- 4% to 7%	office bldgs.	- 4% to 7%
banks	- 3% to 6%	shops	- 4% to 7%
dormatories	- 7% to 10%	schools	- 5% to 8%
factories	- 4% to 8%	stores	- 3% to 6%
hospitals	- 8% to 12%	warehouses	- 3% to 7%

The allowance per fixture estimate is generally more accurate. An experienced plumber can acquire very accurate prices on this although he will have to use some judgment on how much to add for bringing in the main waste and supply lines, depending on how spread out the project may be.

The per fixture price will be the cost of the fixture itself plus the cost of the immediate piping connections and installation within the room. Some per fixture costs have added to them a prorated cost of the central piping system. It would be more accurate to add this, together with the cost of bringing waste and supply lines from the street to the building and permit costs and overhead and profit, each as separate items, as some buildings as we have noted above require a much more extensive central system to supply fewer fixtures than others. Per fixture allowances are discussed later in this chapter.

15400   PLUMBING SYSTEMS

Estimating Quantities and Costs of Sewer Work.—Sewerage is estimated by the lineal foot, obtained by measuring the number of lineal feet of each size sewer pipe required, such as 4", 6", 8" and 12" pipe, and all fittings should also be listed in detail, giving the number of ells, Ys, Ts, etc.

Cast iron sewer pipe is estimated in the same manner as tile sewer pipe.

Most plumbers estimate their sewer work at a certain price per lineal foot, including excavating, sewer pipe and labor. This is all right on ordinary jobs that do not require more than the usual amount of deep excavation, but if there is an exceptionally large amount of deep excavation, (from 7 to 15 feet deep), then the excavating, shoring and bracing should be estimated separately.

Undoubtedly the best method is to estimate the number of cubic yards of excavating required for sewers, based on a trench 1'-6" wider than the pipe diameter by the required depth. Refer to chapter on Site Work for labor costs excavating for trenches.

Where 4 to 6-inch sewer pipe is used, an experienced sewer layer should lay 12 to 15 lineal feet of pipe an hour or 100 to 125 lineal feet per 8 hour day, at the following labor cost per 100 lin. ft.:

	Hours	Rate	Total	Rate	Total
Sewer layer	7	$. . . .	$. . . .	$18.70	$130.90
Average per lin. ft.			. . . .		1.31

Add for excavating and additional time for setting catch basins, gravel basins, triple basins, etc.

Estimating Quantities of Cast Iron Soil Pipe, Downspouts, Stacks and Vents.—When estimating soil pipe, stacks, vents, etc., note the number of stacks and length of each, listing the number of lineal feet of each size pipe; the number of pieces, size, and kind of fittings required, and price them at current market prices. The labor cost of placing cast iron pipe is given on the following pages.

Estimating Quantity of Pipe and Fittings. —When estimating the quantity of black or galvanized pipe, list the number of lineal feet of each size pipe required, such as ½", ¾", 1", 1¼", 1½", 2", 2½", 3", etc., and estimate at the current market price.

After the cost of the pipe has been computed, take 60 to 75 percent of the cost of the pipe to cover the cost of all fittings required.

Brass or copper pipe and fittings should be estimated in the same manner as galvanized pipe.

Estimating the Quantity and Prices of Valves.—All valves of the different sizes should be listed on the estimate separately, stating the number of each size required, and pricing them at the current market price.

Estimating the Quantity of Fixtures.— Each type of fixture such as sinks, lavatories, laundry trays, water closets, bathtubs, drinking fountains, shower baths, hot water tanks and heaters, house pumps, bilge pumps, etc., should be listed separately and priced at the current market prices.

Labor Roughing in for Plumbing.—The usual practice among plumbers is to allow a certain percentage of the cost of the roughing materials to cover the labor cost of installing them. This will vary with the type of building and grade of work.

The total cost of the roughing in materials, such as castiron soil pipe, downspouts, stacks, vents, and fittings; also all black and galvanized pipe and fittings, all valves, increasers, tees, Ys, ⅛ bends; nipples; in fact, all pipe required to rough in the job is computed. Then the labor cost is estimated at a certain percentage of the cost of the pipe.

In medium price one and two story residences, the labor cost of roughing in the job will vary from 75 to 85 percent of the cost of the roughing materials, with 80 percent a fair average.

In apartment buildings of non-fireproof construction, the labor cost of roughing in will average about 90 to 100 percent of the cost of the roughing in materials, while on high grade fireproof construction, the labor cost roughing in will run from 100 to 120 percent of the cost of the roughing in materials.

In public garages, the labor cost roughing in the job will average about 90 to 110 percent of the cost of the roughing in materials.

Sewer work is estimated separately from roughing in.

Labor Handling and Placing Plumbing Fixtures.—The labor cost handling and placing all kinds of fixtures, such as laundry tubs, kitchen sinks, lavatories, bathtubs, shower baths, water closets, drinking fountains, etc., is usually estimated at 25 to 30 percent of the cost of the fixtures.

Preparing Detailed Plumbing Estimates

When preparing detailed estimates on plumbing, it is necessary to list the quantity of each class of work separately, such as the number of lineal feet of 4", 6" and 8" sewers, and fittings; number of lineal feet of each size soil pipe, vents and fittings; number of lineal feet of water pipe and fittings of different materials and sizes; number of valves of the various kinds and sizes; number of each type of fixture required, such as hot water heaters, laundry trays, kitchen sinks, slop sinks, bathtubs, lavatories, water closets, shower baths, etc., together with a list of fittings and supplies required for each.

The following is a list of items the plumbing contractor should include in his estimate :

1. Sewerage, including double sewer system.
2. House tanks, foundations, etc.
3. Compression tanks, foundations, etc.
4. Sewer ejector and bilge pump and basins.
5. Iron, gravel, catch and other basins.
6. Water filters, foundations.
7. Water meter.
8. Soil and vent pipe and fittings.

9. Shower and urinal traps.
10. Closet bends.
11. Drum traps.
12. Lead, solder, sundries.
13. Inside downspouts.
14. Iron sewer and fittings.
15. Floor drains.
16. Back water gates.
17. Service pipe to building.
18. Mason hydrant.
19. Stop cock and box.
20. Galvanized pipe and fittings.
21. Brass or copper pipe and fittings.
22. Valves, check valves.
23. Pet cocks, sill cocks, hose bibbs.
24. Hot water tank and heater.
25. Pipe covering, tank covering, heater covering, filter covering, painting.
26. Gas fitting, mains, ranges, heaters, etc.
27. Manholes and covers.
28. Catch basins and covers.
29. Surface drain basins and covers.
30. Fire system, incl. pumps, hose, Siamese connection, meter, etc.
31. Roughing in material.
32. Fixtures, such as water closets, lavatories, bathtubs, shower baths, sinks, laundry tubs, drinking fountains, N. P. fittings.
33. Permits, insurance, trucking, freight, telephone, watchman, etc.
34. Overhead expense and profit.

15060   PIPE AND PIPE FITTINGS

Approximate Lineal Foot Prices of Cast Iron Soil Pipe

All prices on all types of pipe and pipe fittings and supplies should be checked from a supplier at the time of the job as they are subject to continual change.

	Single Hub		Double Hub	
Size	Service	Extra Heavy	Service	Extra Heavy
2"	$ 2.75	$ 3.85	$ 3.80	$ 5.20
3"	3.25	4.50	4.40	6.10
4"	4.25	6.00	5.90	8.00
5"	5.50	7.70	7.50	10.50
6"	6.50	9.00	8.70	12.30
8"	10.00	14.00	13.70	19.00
10"	16.00	22.50	22.00	30.00
12"	23.00	32.30	32.00	43.50

Prices of Cast Iron Soil Pipe Fittings

Description of Fitting	Size in Inches					
	2"	3"	4"	5"	6"	8"
Quarter bends, service	$ 3.00	$ 5.50	$ 8.00	$10.50	$ 14.00	$ 46.00
Extra heavy	3.50	6.50	9.20	13.00	15.70	52.00
Eighth bends, service	2.20	4.50	6.50	9.20	11.00	32.00
Eighth bends Extra heavy	2.50	5.10	7.50	10.50	12.50	37.00
Sanitary T & Y branches Extra heavy	6.20	12.00	15.00	22.00	34.00	70.00
Plain Traps S P or ¾ Service	8.50	11.50	16.00	31.00	40.00	91.00
T cleanout Extra heavy	27.00	36.00	51.00	92.00	132.00	165.00
Comb. Y & eighth bends Extra heavy	9.00	12.00	17.00	33.00	40.00	93.00

Closet bends slip collar type—4"x4"x12"—$20.00; 4"x4"x18"—$24.00

Lineal Foot Prices of Buttweld Steel Pipe

Size	Standard Weight		Extra Heavy	
	Black	Galvanized	Black	Galvanized
¼"	$0.30	$0.45	$0.45	$0.60
⅜"	.40	.50	.55	.65
½"	.42	.52	.57	.68
¾"	.60	.70	.75	.90
1"	.90	1.10	1.15	1.30
1¼"	1.05	1.40	1.45	1.70
1½"	1.35	1.65	1.75	2.05
2"	1.80	2.20	2.35	2.75
2½"	2.60	3.25	3.65	4.25
3"	3.40	4.25	4.80	5.60
3½"	4.20	5.20	5.95	7.00
4"	5.00	6.10	7.10	8.25

Prices on Standard Malleable Iron Fittings

Type of Fitting	Size								
	½"	¾"	1"	1¼"	1½"	2"	3"	4"	6"
Tee - Black	$.55	$ .80	$1.40	$2.30	$2.80	$4.15	$10.50	$21.50	$60.00
Galv.	.80	1.15	1.85	3.30	4.00	5.35	14.50	30.00	85.00
90° El Black	.45	.50	.95	1.50	2.00	2.85	9.50	16.00	48.00
Galv.	.65	.70	1.35	2.10	2.80	4.00	13.00	22.50	65.00
45° El Black	.65	.85	1.10	1.80	2.20	3.00	10.50	17.00	55.00
Galv.	.95	1.20	1.55	2.20	3.00	3.20	14.50	23.70	75.00

Lineal Foot Prices of Copper Water Tubing

Type	Size									
	¼"	⅜"	½"	⅝"	¾"	1"	1¼"	1½"	2"	3"
Light Type M	$.35	$.40	$.60	$ .70	$ .85	$1.15	$1.60	$2.10	$2.30	$6.00
Medium Type L	.37	.50	.75	.90	1.10	1.50	1.95	2.50	3.75	7.25
Heavy Type K	.40	.65	.85	1.00	1.40	1.80	2.25	2.90	4.35	8.50

Prices of Wrought Copper Soldered Joint Fittings

Type	Size							
	½"	¾"	1"	1¼"	1½"	2"	3"	4"
Tee	$.20	$.50	$1.50	$2.35	$3.00	$4.80	$15.70	$32.00
90° Elbow	.15	.30	.70	1.10	1.50	2.65	8.00	18.00
45° Elbow	.25	.45	.75	1.30	1.70	2.50	9.75	20.00

15080   PIPING SPECIALTIES

Cast Iron Drum Traps

Drum trap with cover and gasket, 4"x8", 2 inlets .... $ 9.20
Drum trap with cover and gasket, 4"x8", 3 inlets 10.75

Roof Drains

	3"	4"	5"	6"	8"
Cast Iron	$115.00	$120.00	$175.00	$190.00	$300.00
Galv.	195.00	200.00	250.00	270.00	380.00

Floor Drains

Type	Size		
	3" outlet	4" outlet	5" outlet
Flat round top cast iron	$ 35.00	$ 40.00	$ 75.00
Flat square top cast iron	50.00	50.00	85.00
Funnel type brass	70.00	70.00	
Drain with bucket cast iron	100.00	100.00	105.00
Trench drain 10"x24"	115.00	135.00	

Sizes and Weights of Steel Storage Tanks
Black, heavy weight, welded, 100 lbs. working pressure.

Capacity Gallons	Dimensions Diam. Length	Overall Length	Heavy Weight Welded
	Tanks with One Convex and One Concave Head		
66	20" x 4'	52"	225
82	20" x 5'	64"	265
120	24" x 5'	64"	340
140	24" x 6'	76"	395
180	30" x 5'	66"	480
220	30" x 6'	78"	550
295	30" x 8'	102"	680
	Tanks with Two Convex Heads		
69	20" x 4'	55"	220
85	20" x 5'	67"	260
125	24" x 5'	68"	320
147	24" x 6'	80"	375
200	30" x 5'	71"	450
235	30" x 6'	83"	520
305	30" x 8'	107"	650
350	36" x 6'	85"	850
450	36" x 8'	109"	1050
560	36" x 10'	133"	1250
550	42" x 7'	99"	1200
770	42" x 10'	135"	1550

900	42" x 12'	159"	1800
1050	42" x 14'	183"	2100
1000	48" x 10'	137"	2150
1200	48" x 12'	161"	2500
1500	48" x 15'	197"	3050
1800	48" x 18'	233"	3550

Septic Sewage Disposal Tanks

Septic tanks are used in rural or suburban districts where running water is available but no sewers, and are used for homes, hotels, summer resorts, etc.

The capacity of a septic tank should be at least equal to the maximum daily flow of sewage. This is normally estimated at 50 gallons per person per day. For part time service in factories, churches, schools, etc., 25 gallons per person per day is generally satisfactory for estimating tank capacities. In some localities, health authorities require at least a 500 gallon tank for residential installation.

They are commonly furnished in 12 and 14 gauge copper bearing steel, electrically welded and covered with a thick coating of asphalt to protect the tanks against corrosive action.

Size of Tank	Capacity Gallons	Size Tile Connection	Weight Pounds	Capacity Home Use No. Persons	Price Each
46" dia. x 48"	300	4"–6"	275	5– 6	$ 67.50
52" " x 60"	500	4"–6"	380	6– 7	102.00
46" " x 120"	750	6"	575	7–12	140.00
48" " x 144"	1000	6"	980	12–15	195.00

Prices of Brass Valves

Kind of Valve	½"	¾"	1"	1¼"	1½"	2"	2½"	3"
Gate valve, 100 lbs. pressure	$ 8.00	$10.00	$13.00	$16.50	$19.25	$30.00	....	....
Gate valve, 125 lbs. pressure	11.00	13.75	17.00	22.00	28.00	40.00	80.00	120.00
Globe valve, 100 lbs. pressure	9.50	12.50	17.00	22.00	29.00	43.00	....	....
Globe valve, 125 lbs. pressure	10.00	13.50	18.00	25.00	32.00	48.00	80.00	125.00
Check valve, horizontal, 125 lbs.	15.00	20.00	28.00	36.00	45.00	74.00	100.00	140.00
Check valve, 125 lb. swing	12.00	14.00	18.00	23.00	27.00	39.00	70.00	100.00

Angle valves, same price as Globe valves.

Gas Heaters

Capacity Gals.	Overall Dia.	Overall Ht.	Approx. Wt.	Prices 5 Year	10 Year
20	14½"	61"	144 Lbs.	$100.00	$120.00
30	16¼"	64"	174	120.00	170.00
40	18½"	65"	195	150.00	185.00
50	20¼"	65"	274	190.00	240.00

Electric Heaters

Capacity Gals.	Overall Dia.	Overall Ht.	Approx. Wt.	Prices 10 Yr.
40	20	50	160	$170.00
52	23	50	181	200.00
66	23	62	227	220.00
82	25	63	270	250.00
110	28	69	430	290.00

15450 PLUMBING FIXTURES AND TRIM

Residential kitchens and bathrooms have tended to go full circle. When plumbing first moved indoors it was generally installed in a full sized room and was arranged about the walls as so much "furniture". Later sanitary concerns became paramount and kitchens and baths were designed as machines — all chrome and tile and porcelain enamel and reduced to minimum dimensions to facilitate cleaning.

A few years ago, with the realization that maids were a thing of the past, the kitchen emerged as an entertainment center where family and friends could gather. Today it is often difficult to tell where the family room ends and the kitchen begins. All this has had its effect on the materials used with appliances and fixtures made as inconspicuous as possible.

The bathroom now seems to be following suit, especially the bathing area which is often in a separate room together with built in equipment for exercise and relaxation and sometimes even a sauna, greenhouse and fireplace. Often the contractor will be called in to convert a spare bedroom into a bath and dressing room suite.

The following discussion of fixtures and their prices give figures which include the fittings. Also are included averages for the piping hook-ups to the main risers and stacks that might be added to arrive at per fixture installed budget figures.

Bathtubs

The least expensive, standard, recessed tub is made of enamelled steel. This will cost around $115 for the 4' 6" length and $145 for the 5' length. Units made of cast iron will run $225 and $250 in those sizes in white and $260 and $290 in color. If these units are to be recessed into the floor add another $20. Square recessed tubs 42" x 48" will run $390 in cast iron.

Recessed Tub   Corner Tub

Square Recessed Tub

Illustrations courtesy Eljer Plumbingware Division Wallace-Murray Corp.

Units complete with three enclosing walls can be had in fiberglass for around $275.

A plumber with a helper should set a tub with shower in about two hours.

To arrive at a per-fixture installed budget figure add around $210.

Shower Stalls

These can range from simple baked enamel on steel units costing around $100 to custom marble enclosures costing six times that. Prefabricated fiberglass units are very popular and can be had with built-in seats and grab bars for the elderly. A 3' x 3' unit will run $235 complete.

Often the plumber will find his work limited to furnishing the receptor and mixing valve and outlet with other trades providing finished walls. A precast receptor pan in terrazzo together with chrome fittings will cost $90 in 3' x 3' size and $110 in 3' x 4'. One man can install this in half a day.

In industrial, recreational and educational buildings group showers with a center outlet column and room for some five persons around it are specified. Such a unit will cost around $500 without subdividing partitions, $900 with them.

Two plumbers will set the basic unit in four hours but will take a full two days if partitions are to be set.

Lavatories

Wall hung lavatories are available in porcelain enamelled cast iron or vitreous china. A 19" x 17" unit will run $95 in white, cast iron, $110 colored; $100 china white, $120 china colored. A 22" long unit will run $100 cast iron white, $115 cast iron colored and $130 china white and $150 china colored.

Allow two men one hour to install. Add $230 to arrive at a per fixture installed price.

Drop-in, counter top units are mainly self-rimming today. They come in cast iron, china, steel, stainless and molded plastic. An 18" round unit in cast iron or vitreous will run around $65 in white, $75 in color; in stainless, $120 and in enameled steel, $50. A 20" by 18" unit will cost some $10 more. Two men will take an hour and a half to set a unit

Washfountains

A three foot round, free standing unit will serve five to six persons at a cost of $500 per unit in precast terrazzo and $625 in fiberglass or stainless. A four foot six unit will serve up to 10 and costs $600 in terrazzo, $820 in fiberglass and $900 in stainless.

Half round units are also available. A four foot six unit in terrazzo will cost $575, in stainless $725. Since installation costs will remain almost the same as for circular units the cost per person served is much more with these units.

While terrazzo units are cheapest they are also the heaviest and one should allow another hour to set these units. Otherwise allow two men two and one half hours to set a 36" unit, three hours to set a 54" unit.

Also available for vanity units are precast, one piece, bowl and countertops in a variety of marbleized colors and patterns. These are generally locally manufactured and prices should be checked but a 2' long top and bowl should run around $75 and take two men an hour and a half to set and a six foot top with double bowl should cost around $200 and take three hours to set.

STEP 1

STEP 2

STEP 3

STEP 4

"J" BOLT ASSEMBLY

INSTALLED SECTIONAL VIEW

Setting A Self Rimming Lavatory

Courtesy Eljer Plumbing Ware Division Wallace Murray Corp.

Water Closets

Water closets can be either wall hung or floor mounted, have tank or flush valve water supply and the tank can be a separate piece or the unit can be cast in one piece with the tank dropped behind the seat.

A floor mounted one piece will cost $240 in white, $285 in color. A floor mounted two piece unit will cost $110 in white, $135 in color. For areas where water supply is a problem units designed to flush with as little as two quarts can be had but will cost around $400.

Floor mounted units will take two men one and one half hours to set. For a per fixture installed budget figure add $180.

In commercial work wall hung, flush valve units are the norm. These will cost $150 for the fixture, $125 for a standard carrier and take two men two hours to set.

Urinals

Urinals are available in cast iron or vitreous china, floor, pedestal or wall mounted, single or in gangs or as wall mounted troughs.

Vitreous china, floor mounted units 18" wide and 38" high with 4" recessed into the floor will cost around $200. If set in gangs add $60 for a spacer-cover 6" wide.

Wall hung china units 18" wide and projecting 12½" will cost $170.

A women's pedestal type will run $235, men's, $190.

Two men should install a urinal in two and one half hours.

For a per fixture installed budget figure add $200 per unit.

Floor Mounted Bidet

Courtesy Eljer Plumbingware Division Wallace Murray Corp.

Bidets

For the ultimate touch in the luxury bath install a bidet. It will cost about $250 in white, $290 in color and take two men two hours to set.

Sinks

Kitchen sinks are usually self-rimming and set into a cabinet top of plastic or tile. With electric dishwashers many are only single bowl. Even in luxury kitchens a single bowl may be set where the dishes are washed with a second bowl near the refrigerator for washing vegetables and a third near the ice making machine for making drinks.

A single bowl 24" x 21" will cost $60 in enamelled steel, $90 in white cast iron, $110 in colored cast iron and $125 in stainless. Figure two men men three hours to set. Add $375 for a per fixture installed price.

A double unit 42" x 21" will cost $125 in steel, $160 in cast iron white, $180 in cast iron color and $200 in stainless. Two men will take three and a half hours to set such a unit.

Cast iron sink tops with drain boards will run $225 for single bowl 42" unit, $260 for double bowl 60" unit.

Enamelled steel sink and drain board tops will run 150 for a 42" length, $200 for a double bowl 60" length.

Laundry sinks in enamelled cast iron will cost $150 for a single compartment, $225 for a double; in cast stone, $75 and $115; in fiberglass, $60 and $90; in stainless steel, $150 and $250. Two men will set a single unit in one and one half hours; a double in two hours.

Electric Dishwashers

Many modern homes and apartments are now being equipped with electric dishwashers. These dishwashers may be installed as separate units or they may form a part of a complete sink.

The various units may be purchased at the following prices:

Dishwasher unit only with floor cabinet but without top and backsplasher, where used under a continuous countertop. Size 24"x25'" and 34½" high. Price. $400.00

Dishwasher unit with porcelain enamel top and backsplasher. Cabinets made of electrically welded rust-resistant steel with baked on enamel finish. Size 23"x27" and 39" high. Price $430.00

Add installation charges to all of the above prices.

Garbage Disposal Units.—The disposal does away with the kitchen garbage can. It is a self-contained unit attached to the kitchen sink to form an enlarged drain into which all kitchen food wastes can be placed.

The unit consists of a housing in which is contained a propeller and shredding mechanism. A ¼ h.p. motor is directly connected to the propeller and shredding mechanism and supplies the power for operating the waste unit.

Waste material can be accumulated in the upper receptacle of the unit until a normal charge is collected or it may be operated to dispose of the material immediately. The capacity of the waste receptacle is 1 qt.

Waste material passes through a series of shredders where they are reduced to a fine pulp. This pulp then passes into a revolving strainer disc through which it is forced centrifugally into a chamber below and around the flywheel. Fins on the flywheel centrifugally force the pulp into the outlet passage connected to the drain line which carries the waste to the sewer. Cold water from the faucet flowing through the unit during the grinding operation thoroughly flushes the waste down the drain.

The price of this unit averages about $130.00

Add installation charge to above price.

15B HEATING AND AIR CONDITIONING

In recent years, much improvement has been made in the methods of heating and cooling buildings, especially in residential work, largely due to the research programs conducted by heating and air conditioning associations, engineering societies, equipment manufacturers and research departments of various universities.

Types of Conventional Heating Systems.—There are four basic types of conventional heating systems—warm air, hot water, steam and electric, including both resistance types and heat pumps, each of which may be used separately or in combination. Solar heating is still in the experimental stage although in certain areas of the country there are enough installations and research that soon reliable information should be available.

For small to medium size residences, stores, churches, etc., of similar size and construction, warm air systems are most widely used. For medium size to large residences and non-residential structures of comparable size, hot water systems are usually selected. For large scale installations, steam systems are generally required. Inasmuch as there are no definite rules stating just when to select one system in preference to another, there is much overlapping in their application. An owner's preference for a particular system may be governed by initial cost of installation; operation cost; convenience of operation; adaptability of system to other uses, such as summer cooling using ducts of a warm air system, snow melting using hot water piping connected to a hot water system; fuel availability, and aesthetics.

Estimating the Cost of Heating Systems.—There is no fast, easy, "rule of thumb" method for estimating the cost of a heating system. This requires expert knowledge, as most jobs must be designed and laid out before they can be estimated. In residential and small commercial work, architects seldom give more information than merely specifying the type of system desired, the performance expected, and the kinds of material to use. It is then up to the heating contractor to figure the heating loads and design a system to fit the conditions before estimating the job. On larger work, the heating system is usually designed and laid out diagrammatically, but the heating contractor must work out details for piping, equipment connections, controls, etc., which all influence the cost.

To acquire the necessary technical knowledge for this purpose requires everyday participation in this line of work plus constant study to keep up to date. Heating and air conditioning is too complex a subject to treat in a volume of this kind, however, much useful data and information can be gained from manuals published by the North American Heating & Air Conditioning Wholesalers Assn., 1200 W. 5th Ave., Columbus, Ohio 43212 and the "Handbook of Air Conditioning, Heating and Ventilating."

Anything appearing in this volume on this involved subject is only the most elementary and approximate—for "roughing" or approximate estimates only.

Boilers are now designed for burning special fuels, such as hard coal, soft coal, oil, gas, etc., the costs varying with boiler sizes and accessories, such as domestic hot water heaters, thermostatic controls, etc.

The ordinary cast iron radiators standing in the room are practically a thing of the past in modern heating systems and concealed convection type radiators or baseboard radiation are now used. These are all items that affect costs.

Copper piping is now used for both heating and plumbing. The cost of this work depends to a large extent upon the experience of the contractor installing same. Mechanics familiar with sweating joints and installing copper piping can install this work rapidly, while those who are not accustomed to its use, will spend considerable time on it.

For the above reasons, the estimator should be familiar with the requirements of the trade in order to prepare intelligent estimates.

Determining Heating Loads.—The first step in designing and estimating heating work is to determine the maximum heating load. For most residential work, the maximum heating load is the total heat loss of the structure figured at design temperatures. For commercial and industrial structures, special purpose heating loads and pick-up loads may be required which must be added to the total building heat loss to obtain maximum heating load.

Heating loads and heat losses are invariably expressed in Btu per hour. A Btu (British thermal unit) is the quantity of heat required to raise the temperature of 1 lb. of water 1 degree F.

Building heat losses vary considerably depending upon climatic location of job, exposure, building size, architectural design, purpose of structure, construction materials used, quality of construction workmanship and other factors.

Heat losses should be figured separately for each room so that the proper amount of radiation may be provided for comfort in all areas. Space limitations in this volume do not permit giving detailed information and data on figuring building heat losses and anything less would be misleading. Complete information and data on this subject are contained in the manuals and guide previously mentioned.

FORCED WARM AIR HEATING SYSTEMS

Forced warm air heating is the most widely used residential heating method in use at the present time. Generally speaking, the forced warm air heating method differs from the old gravity type by using blower equipment to circulate the air in the system.

It is practically impossible for the average general estimator to compute the cost of a modern forced warm air heating system, unless he has a working knowledge of this subject and is able to calculate building heat losses and design systems which will satisfy heating requirements under design conditions. In addition, the subject has become so complex it cannot be covered in a volume of this kind so as to provide adequate information for the preparation of a cost estimate, other than a rough approximation to be used for preliminary purposes only. Always obtain firm quotations from reputable warm air heating contractors before submitting bids on jobs containing this work.

Winter Air Conditioning Systems.—Forced warm air heating systems which are equipped with automatic humidifying devices and air filters are popularly known as Winter Air Conditioning Systems. Most forced air installations are of this type.

Each system consists of a direct fired heating unit, a blower, a system of warm air ducts and return air duct system. Air filters are located in the return air duct system just ahead of the blower—humidifying equipment is placed in the warm air plenum chamber over the furnace unit. Air cooling equipment may be added to make this a year 'round system.

The unit may be gas, oil or coal fired. Gas or oil fired units designed for that particular fuel operate at high efficiency, although satisfactory, if less efficient results may be obtained through the use of conversion burners installed in units primarily designed for solid fuel.

Thermostatic control of heating apparatus is highly desirable and invariably used. Automatic controls vary the supply of heat in accordance with demand set up by a thermostat located in the living portion of the home. Actuated by limit controls, the blower commences operation automatically when the heated air in the generating unit has reached a pre-determined temperature and cuts out after the source of heat has been shut down and the temperature in the bonnet dropped to a pre-determined point.

Design of air conveying systems of this kind presents an engineering problem which can only be solved with a reasonable degree of accuracy by a competent designer. The heat loss of the building, room by room, is established and the volume of heated and conditioned air required for indoor comfort establishes data from which a duct system can be designed with due regard for velocities, static pressures and delivery temperatures.

Items to be Included in the Estimate.—Costs vary considerably and methods of estimating also differ, but every estimate for a winter air conditioning system should include the following items:

1. Cartage of material and equipment from shop to job.
2. Winter air conditioning unit, complete with humidifier, air filters, and controls.
3. Labor assembling and setting up unit.

4. Oil storage tank installation, including fill and vent piping, oil booster pump if required, oil gauge, piping to unit, oil filter, etc., for oil fired units, or—piping for gas fired units.
5. Installation of smoke pipe, fittings and accessories.
6. Warm air and return air plenum chambers.
7. Warm air and return air duct systems.
8. Diffusers, registers, intakes, grilles, etc.
9. Special insulation.
10. Electrical work.
11. Labor starting plant in operation and balancing system.
12. Service allowance.
13. Miscellaneous costs.
14. Overhead and profit.

Cost of Winter Air Conditioning Units.—Sizes, capacities and prices of units vary with the different manufacturers, but the following listings are representative of models available for residential use.

Approximate Net Prices of Winter Air Conditioning Units
Complete with Automatic Controls, Humidifier and Air Filters

Bonnet Rating Btu per Hr.	Register Rating Btu per Hr.	Heating Only	Net Price	With Cooling
	Gas Fired Units			
60,000	48,000	$280.00	24 MBH	$1100.00
76,000	60,800	310.00	29 MBH	1350.00
90,000	72,000	330.00	36 MBH	1450.00
108,000	86,400	400.00	47 MBH	1750.00
120,000	96,000	480.00	47 MBH	1850.00
150,000	120,00	625.00	58 MBH	2300.00
180,000	144,000	700.00	....	....
	Oil Fired Units			
84,000	71,400	525.00	36 MBH	1450.00
112,000	95,200	600.00	42 MBH	1750.00
140,000	119,000	700.00	47 MBH	2000.00
175,000	148,750	800.00	....	....
210,000	178,500	900.00	....	....

Assembling and Setting up Unit.—Most gas fired units can be assembled and set up by a sheet metal worker and a helper in 3 to 4 hours. For an oil fired unit, figure 4 to 5 hours for this work.

Oil Storage Tanks.—Common practice for oil storage tanks in the use of one or two inside storage tanks with a capacity of 275 gals. each. Occasionally outside buried tanks of 550 or 1,000 gallon capacity are used. On a contract basis, these tanks installed, including all piping, cost as follows:

1–275 gallon inside tank $200.00
2–275 gallon inside tanks 390.00
1–550 gallon outside tank 300.00
1–1,000 gallon outside tank 500.00

In the use of an outside tank a basement wall type pump costing from $60.00 to $80.00, must be used with some types of burner equipment.

Gas Piping.—For average installations, an allowance of $100.00 to $150.00 should cover cost of material and labor for connecting gas to unit within the heating room.

Smoke Pipe.—Material and labor for smoke pipe connection from unit to flue should cost $40.00 to $60.00 for gas fired and oil fired units.

Plenum Chambers.—Plenum chambers must be fabricated to order in the shop and are usually made from light gauge galvanized steel. They are mounted on top of the furnace and serve to connect the warm air and return air duct systems to the unit. Two plenum chambers are required for each installation.

Plenums will run $4.50/sq. ft in 18 ga., $5.50 in 16 ga., $6.00 in 14 ga.

Warm Air and Return Air Duct Systems.—The cost of duct work will vary with the size of job and type of distribution.

Under ordinary conditions, the warm air heating contractor will measure and list each type and size of duct, fitting and accessory. These quantities are priced for material and then an estimate is made of labor required for installation. This is a lengthy operation and requires an experienced heating man.

When pressed for time, the duct system cost may be approximated on an outlet basis, using a unit price which covers both material and labor. For ordinary installations, the following costs per outlet are about average where sheet metal workers' wages are from $15.00 to $16.00 per hr. :

Type of System	Cost per Outlet
Conventional system—warm air outlets located on inside walls.	
1st floor outlets	$ 70.00 to $ 80.00
2nd floor outlets	95.00 to 105.00
Capped stacks	27.00 to 33.00
Radial perimeter system in small 1-sty. buildings with basement or crawl space.	
Low velocity—6" to 8" dia. ducts	$ 70.00 to $ 80.00
High velocity—4" and 5" dia. ducts	65.00 to 70.00
Large central return air duct	87.00 to 95.00
Radial perimeter system in small 1-sty. basementless buildings.	
Galv. steel sheet metal pipe and fittings	$ 88.00 to $ 93.00
Fiber duct with galv. steel fittings	88.00 to 93.00
Cement asbestos pipe and fittings	105.00 to 115.00
Trunk and branch perimeter system in medium-size, 1 sty. buildings with basement or crawl space.	
Std., galv. steel, rectangular ducts	$100.00 to $110.00

Add for diffusers, registers, intakes, special insulation, etc. Excavation, backfill and concrete encasement where required, not included.

Diffusers, Registers and Intakes.—Diffusers and registers generally used are of pressed steel construction, with prime coat finish, are fully adjustable and are available in floor, baseboard or wall types. Return air intakes are of same construction and finish, but are non-adjustable.

Floor type diffusers cost $11.00 to $14.00 for sizes commonly used. Baseboard type diffusers cost $10.00 for 2'-0" lengths and $14.00 for 4'-0" lengths. Wall type registers cost $5.00 to $10.00. Wall type intake grilles cost $6.00 to $12.00.

Special Insulation.—Ducts in unexcavated spaces or in unheated attic spaces, both supply and return, should be insulated with not less than 1" of adequate insulating material. This material is furnished in flexible form and averages $1.00 to $1.25 per sq. ft. applied.

Electrical Work.—Custom varies with respect to the supply and installation of electrical circuit from meter to fused safety switch adjacent to unit. Likewise for the connection of blower motor to 110 volt controls and also for the 22 volt wiring to thermostat when used. Most often the job electrician runs the 110 volt service to the safety switch adjacent to unit. From this point it is handled as a subcontract under the heating contractor and represents a cost from $60.00 to $150.00 for complete wiring installation of all controls and motors.

Labor Starting and Balancing System.—It is customary for the heating contractor to start the plant in operation, test all controls and balance the air distribution system. For jobs completed during summer months, this means a come-back call at the beginning of the heating system.

Labor costs for this work will run $75.00 to $125.00 depending upon size of job.

Service Allowance.—Most jobs carry a 1-year free service warranty and an allowance should be included in the estimate to cover this contingency. Average costs for service during the first year of operation are $65.00 for gas fired and $90.00 for oil fired systems.

Miscellaneous Costs.—Under miscellaneous costs are classified such items as federal tax, sales tax, permits for oil burner or for installation, if required.

Approximate Cost of a Complete System Installed.—Based on installations in residences of ordinary construction, located in the larger cities, the cost of average winter air conditioning systems ranges as follows for gas fired units. For oil fired units, add 10 to 15 percent.

Floor Area Sq. Ft.	Heat Loss Btu per Hr.	Price for Complete Gas Fired Installation
Conventional Systems, 1-Sty. Buildings with Basements or Crawl Space.		
1,000	65,000	$1500.00 to $2000.00
1,500	80,000	2000.00 to 2300.00
2,000	100,000	2300.00 to 2500.00
Conventional Systems, 2-Sty. Buildings with Basements		
1,500	75,000	$2000.00 to $2300.00
2,000	90,000	2500.00 to 3000.00
Perimeter Systems, 1-Sty. Buildings with Basements or Crawl Space.		
1,000	65,000	$1500.00 to $2000.00
1,500	80,000	2000.00 to 2300.00
2,000	100,000	2500.00 to 2800.00
Perimeter Systems, 1-Sty. Basementless Buildings.		
1,000	65,000	$1500.00 to $2000.00
1,500	80,000	2000.00 to 2300.00

The above prices are for individual jobs. For multiple housing projects, where large numbers of similar dwelling units are involved, much better prices may be obtained.

As previously mentioned, in this industry, equipment, design and practices, vary greatly even in localized areas. No safe guide may be set down for general use. So much depends upon correct design, proven equipment and experienced installation that it behooves the buyer to carefully investigate all of these items before awarding a contract for this important part of the building.

SUMMER AIR CONDITIONING

Residential summer air conditioning is available at a price the average home owner can afford. This applies to both existing homes and new homes under construction and is especially true where forced warm air heating is used, as the same air distribution duct system can be utilized for both heating and cooling. Various warm air heating equipment manufacturers make cooling units for this purpose and installation work is usually done by heating contractors.

The function of summer air conditioning equipment is the reverse of winter air conditioning—circulating air is cooled and de-humidified—but design problems are similar, although more complex. A basic requirement for designing a cooling system is calculating the total cooling load or total heat which must be removed from the structure to achieve a pre-determined inside temperature and humidity under design conditions. Heat gain calculations are much more involved than those for heat losses as many additional factors must be considered, such as sun effect on roof, wall and glass areas; internal heat gains from human occupancy, cooking, lighting, appliances, etc. and latent heat energy involved in de-humidification. This is a job for an expert.

Heat gain calculations are always expressed in Btu per hour. Cooling equipment capacities are usually stated per ton of refrigeration. A ton of refrigeration will remove heat at the rate of 12,000 Btu per hr. In a 24-hr. period, this is equivalent to the heat required to melt a ton of ice. With water-cooled compression refrigeration equipment, one horsepower usually equals about one ton of cooling capacity.

Types of Cooling Equipment for Residential Use.—There are several types of cooling units available and in popular use for residential summer air conditioning. In general, equipment may be classified as water-cooled or air-cooled, with each type available for several methods of application.

One method employs a remotely located condensing unit, with refrigerant lines connected to a cooling coil installed in the warm air plenum chamber of a forced warm air heating system. This method uses the heating system blower and ducts for air distribution.

Another method uses a self-contained unit, with the condenser, cooling coil and blower combined in one cabinet. This type may be cut in to the heating ducts, or may have its own duct system where hot water or steam heat is used.

Net Prices of Remote System Cooling Units.—Approximate prices are for units factory assembled with heating and cooling thermostat included. No pipe, duct materials or blower assembly included.

Capacity Tons	Rating Btu per Hr.	Net Price Air-cooled	Labor To Install
2	24,000	$ 600.00	8 hr
3	36,000	850.00	14 hr
5	60,000	1800.00	30 hr

Where a complete air circulation duct system is required, figure same as previously given for winter air conditioning systems.

Approximate Cost of Complete Cooling Unit Installation.—Based on conditions of average difficulty, complete residential cooling unit installations should cost approximately as follows :

Capacity Tons	Rating Btu per Hr.	Price for Complete Installation
	Remote, Water-cooled System	
2	24,000	$2200.00 to $2500.00
3	36,000	2500.00 to 2800.00
	Remote, Air-cooled System	
2	24,000	2000.00 to 2200.00
3	36,000	2500.00 to 2700.00
5	60,000	4500.00 to 5000.00
	Self-contained, Water-cooled System	
2	24,000	2300.00 to 2500.00
3	36,000	3500.00 to 3700.00
	Self-contained, Air-cooled System	
2	24,000	2600.00 to 2800.00
3	36,000	4500.00 to 4800.00
5	60,000	7000.00 to 7500.00

Room Size Air Conditioning Units.—The cost of portable room size air conditioning units vary according to size, ranging from ½-Hp., which cools room areas up to 400 sq. ft., to 2-Hp., for spaces up to 1,200 sq. ft.

All models are for window installation and include automatic thermostat controls.
Net prices run from $125.00 for a ½-Hp. unit to $600.00 for 2-Hp. units.
If electric wiring is necessary, add extra for installation.

Commercial Installations

A number of mechanical equipment manufacturers are offering self-contained, one-piece rooftop electric heat pump units for the smaller commercial projects.

These units which are made in numerous heating and cooling capacities are pre-wired and precharged at the factory so that when they are delivered to the project site, the installation is relatively simple. The unit is set on a roof curb, power and control wiring connected and duct connections hooked up—and the unit is in operation.

Standard construction of these units includes a manual damper that can be preset to admit up to 25% outdoor ventilation air for year-round fresh air. Some units come equipped with an "economizer cycle" that has a thermostatically-controlled, fully-modulating damper which admits up to 100% outdoor ventilation air to provide low-cost cooling on mild days.

For quick approximations of air cooling costs cooling units can be estimated from $1,000.00 to $2,000.00 per ton depending on the ductwork, job conditions, types of grilles, etc. 12,000 Btus requires one ton of air cooling. The following table will give an approximation of final requirements:

Type of Space	Btu per Sq. Ft. Floor Area
Large Stores	40
Offices	45
Specialty Shops	50–70
Bars	80
Restaurants	100–120

STEAM AND HOT WATER HEATING

Because of different conditions surrounding the installation of heating apparatus, it is impossible to give any set rule that can be accepted, without modification, for all kinds of buildings to be heated. It is necessary to take into consideration all of the conditions in and around any building, and additions or deductions made to suit the requirements, no matter what rule may be used for figuring.

Methods of Computing Heating.—The most advanced method of figuring heating and the one that is generally used by heating engineers is the Btu Method. This method is based on replacing the heat loss through exposed walls, doors and windows, floors, ceilings, etc., and uses so many factors, based on the type of construction that it is impossible to go into detail in a book of this kind, as intricate and elaborate formulas are used for both warm air, steam and hot water heating. In fact, if the entire subject of heating was thoroughly covered, it would require a book as large as this one. For this reason only approximates can be treated in this chapter.

Flow Control Hot Water Heating Systems.—This method of hot water heating has largely superseded the older gravity or "open" type of hot water heating system. It is a complete combined automatic water heating and domestic water supply system for year 'round use. It's major advantages are instantaneous heating response, high efficiency, fuel economy and completely automatic operation.

The flow control valve makes the boiler a reservoir of high temperature water, in that it prevents circulation through the system when heat is not required. A circulator or pump attached to the main near the boiler forces this hot water through the mains the instant the room thermostat calls for heat. When the thermostat is satisfied the circulator ceases to operate and the flow control valve stops all circulation of hot water, thus preventing overheating and needless heat losses.

The flow control hot water heating system requires approximately 20 percent more radiation than is figured for steam heat. The mains are smaller reducing the cost of pipe, fittings and labor installing same.

All in all, a flow control hot water heating system costs a trifle less than the older gravity systems requiring large radiators and piping.

A gravity hot water heating system can be converted into a flow control system at a nominal expense.

Cost of Special Fittings for Flow Control Hot Water Heating Systems.—The cost of fittings will vary with the amount of radiation in the job and whether or not a complete flow control system is used or merely a flow control valve and a circulating pump.

Where only the flow control valve and water circulating pump are used, the costs for the special equipment will run as follows based on the amount of radiation in the job.

Capacities and Prices of Flow-Control Valves

Valve Size	Radiation Capacity in Square Feet			Weight Pounds	Price Each
	150 Btu	200 Btu	240 Btu		
1 "	354	265	221	5	$ 13.00
1¼"	786	590	492	5	14.00
1½"	1170	875	730	10	17.00
2 "	2140	1610	1340	21	22.00
2½"	3680	2750	2300	33	48.00
3 "	6660	5000	4160	42	65.00
4 "	14000	10000	8750	57	155.00

Sizes and Prices of Water Circulators or Booster Pumps

Recommended for use on modern forced circulation systems or for improving gravity hot water systems.

Direct Radiation Capacity
in Sq. Ft.

Size Inches	Motor H.P.	Dely. Gals. per Min.	150 Btu	200 Btu	240 Btu	Price Each	
						Iron Body	All Bronze
1	1/12	31	750	562	469	$115.00	$110.00
1¼	1/12	33	1000	750	675	115.00	130.00
1½	1/12	33	1000	750	675	115.00	130.00
2	1/6	75	3500	2625	2187	200.00	325.00
2½	1/4	120	4500	3375	2820	290.00	525.00
3	1/4	120	4500	3375	2820	290.00	600.00

Sizes and Prices of Relief Valves

A.S.M.E. Capacity at 30 lbs.in Btu per Hr.	Size Inches	Net Price
175,000	¾	$ 8.50
250,000	¾	12.00
350,000	¾	14.00
480,000	¾	15.00
750,000	1	25.00
1,050,000	1¼	40.00

Sizes and Prices of Air Control Boiler and Tank Fittings

Air control fittings are used for the purpose of removing air from a hot water heating system and preventing its return.

Air control boiler fittings consist of a tube inserted into a T-shaped fitting. The tube, in effect, extends the supply main down into the boiler water, thus preventing air (which has accumulated at the top of the boiler) from rising into the piping and heating units. This air flows around the tube, up into the T-fitting and then to the pipe leading to the air control tank fitting.

The air control tank fittings provide two separate passages—one through which air can flow to the top of the tank, and the other for the water displaced by the air to flow back into the pipe connections to the system. These will cost from $15.00 to $30.00 in sizes 1" to 3" for boiler fittings, $15 for tank controls up to 24".

Coils for Heating Water.—Where coils are inserted in the boiler for heating water for domestic purposes, the size of the boiler should be increased, figuring each gallon of water tank capacity as equivalent to 1 sq. ft. of steam radiation or 2 sq. ft. of hot water radiation. For example, a 30-gal. tank is equivalent to 30 sq. ft. of steam or 60 sq. ft. of hot water radiation.

Rule for Computing Boiler Sizes.—After the total number of square feet of radiation of total heat loss in Btu per hour has been obtained, the size of the boiler should be computed.

Practically all boiler manufacturers rate their boilers, giving maximum heat output under full drive and also the recommended load representing the attached load, including piping, which may be placed on the boiler in accordance with accepted installation standards for economical operation.

Regarding Steam Boiler Installation.—It is recommended on all installations of steam boilers that drain valves be placed on the returns and that the condensation from such returns be discharged into the sewer for a period of 3 days to one week after starting fire, thereby clearing system of grease and dirt.

At the end of this period the boiler should be thoroughly washed and blown out.

Boiler Capacity.—Boiler output is defined as the number of sq. ft. of standard direct column radiation or its equivalent in Btu, which a boiler will supply at its outlet in one hour, under standard conditions.

240 Btu represents one sq. ft. of standard direct column steam radiation.

150 Btu represents one sq. ft. of standard direct column water radiation.

Boiler Horsepower.—One boiler horsepower is defined as the evaporation of 34½ lbs. of water per hour from a feed water temperature of 212 degrees into steam at 212 degrees. This is equal to 34.5x970.4=33478.8 Btu per hour.

One boiler horsepower is equal therefore to $\frac{33478.8}{240} = 139$ sq. ft of steam radiation.

Note that boiler horsepower is not derived from engine horsepower.

1 horsepower hour=2545 Btu.

To find the Btu output per hour of a steam boiler, multiply the output in sq. ft. by 240.

To find the Btu output per hour of a water boiler, multiply the output in sq. ft. by 150.

To find the output in sq. ft. of steam radiation, divide the output in Btu per hour by 240.

To find the output in sq. ft. of water radiation, divide the output in Btu per hour by 150.

Estimating Pipe and Fittings.—To prepare an intelligent estimate on the cost of any heating plant, the entire job should be laid out using the quantity of pipe of each size required; connections, such as elbows, tees, nipples, valves, etc., should all be figured separately. This requires considerable labor and a man must thoroughly understand his business to figure with any degree of accuracy.

When estimating the quantities of piping required for any heating plant, the risers, mains and returns must be figured, together with all connections to and from the boilers. These will vary according to the system used and the type of building.

A one-pipe steam heating system should have the highest point in the main directly over the boiler, and from this point it pitches down around the circuit until it drops to the floor or enters the boiler. The more pitch the better, but there should be at least one inch in 20 feet. The lowest point in the steam main, where it drops to enter the boiler or to the floor, should not be less than 14 inches above the water line of boiler and the more the better.

Branches for radiators should be taken from top of main, using a nipple and elbow or a nipple and a 45-degree ell, and should pitch up to the radiator.

Connections for steam should never be taken from side of main, as this will cause water hammer and syphoning of the water of condensation in the steam main into the radiator.

Each radiator should be supplied with a good disc radiator valve and a good automatic air valve, and each return main should also have an automatic air valve where it drops to enter boiler.

If branches are longer than 8 to 10 feet, they should be one size larger than the upright connecting pipes. Where mains drop to floor, a pipe two sizes smaller can be used to return to boiler. Risers for either steam or water should be plumb and straight. If exposed in a room, and it is necessary to use couplings on same, the couplings should be at a uniform height from floor.

The valve on a steam radiator should be either wide open or tightly closed to prevent radiator filling with water.

In a two-pipe gravity hot water heating system the highest return rises from boiler to radiator, the lowest point being at the boiler and the highest point at the end of the main. These pipes may be reduced as the branches are taken off, care being taken not to reduce too rapidly. Each radiator should be fitted with a hot water radiator valve, union elbow and a compression air valve. All pipes should be well supported, so that no pockets or depressions will occur to retard or trap the circulation. All mains, both for steam and hot water, should run about 4 feet from wall to allow for expansion.

Practically all hot water heating systems installed today are flow control systems where the hot water is circulated from the boiler through the mains and pipes, using a circulating or booster pump. When this method is used, the size of the mains can be greatly reduced, as will be noted from the following table. With the water being circulated through the system at 30 to 120 gallons a minute, it is very much faster than the old gravity systems. Also the radiators may be placed above or below the boiler level, as desired.

Sizes of Pipes.—When estimating steam heating, using a rule of thumb method, the size of the main may be determined by taking the total amount of direct radiation to which add 25 percent for piping and from this total extract the square root, dividing same by 10, which gives the size of the main to use. This is for one pipe work.

For two pipe work one size less is sufficient and the return can be one or two sizes less than the supply. A steam main should not decrease in size but very little according to the area of its branches.

An example of the above method is as follows: A job contains 400 feet of direct steam radiation. By adding 25 percent for piping, the result is 500 square feet. Extracting the square root of 500, gives 22.4. Divide 22.4 by 10 equals 2.24 or it will require a 2½-inch steam main.

Table Giving Number of Square Feet of Radiation in Exposed Heating
Pipes of Various Sizes and Lengths

Length of Pipe	Size of Pipe											
	¾	1	1¼	1½	2	2½	3	4	5	6	7	8
1	.275	.346	.434	.494	.622	.753	.916	1.175	1.455	1.739	1.996	2.257
2	.5	.7	.9	1.	1.2	1.5	1.8	2.4	2.9	3.5	4.	4.5
3	.8	1.	1.3	1.5	1.9	2.3	2.7	3.5	4.4	5.2	6.	6.8
4	1.1	1.4	1.7	2.	2.5	3.	3.6	4.7	5.8	7.	8.	9.
5	1.4	1.7	2.2	2.4	3.1	3.8	4.6	5.8	7.3	7.7	10.	11.3
6	1.6	2.1	2.6	2.9	3.7	4.5	5.5	7.	8.7	10.5	12.	13.5
7	1.9	2.4	3.	3.4	4.4	5.3	6.4	8.2	10.2	12.1	14.	15.8
8	2.2	2.8	3.5	3.9	5.	6.	7.3	9.4	11.6	13.9	16.	18.
9	2.5	3.1	3.9	4.4	5.6	6.8	8.2	10.6	13.1	15.7	18.	20.3
10	2.7	3.5	4.3	4.9	6.2	7.5	9.1	11.8	14.6	17.4	20.	22.6
11	3.	3.8	4.8	5.4	6.8	8.3	10.	12.9	16.	19.1	22.	24.9
12	3.3	4.1	5.2	5.9	7.5	9.	11.	14.1	17.4	20.9	24.	27.1
13	3.6	4.5	5.6	6.4	8.1	9.8	11.9	15.3	18.9	22.6	26.	29.4
14	3.8	4.8	6.1	6.9	8.7	10.5	12.8	16.5	20.3	24.3	28.	31.6
15	4.1	5.2	6.5	7.4	9.3	11.3	13.7	17.6	21.8	26.1	30.	33.9
16	4.4	5.5	6.9	7.9	10.	12.	14.6	18.8	23.2	27.8	32.	36.1
17	4.7	5.9	7.4	8.4	10.6	12.8	15.5	20.	24.7	29.5	34.	38.4
18	5.	6.2	7.8	8.9	11.2	13.5	16.5	21.2	26.2	31.3	36.	40.6
19	5.2	6.6	8.3	9.4	11.8	14.3	17.4	22.3	27.6	33.1	38.	42.9
20	5.5	6.9	8.7	9.9	12.5	15.	18.3	23.5	29.1	34.8	40.	45.2
21	5.8	7.3	9.1	10.4	13.	15.8	19.2	24.7	30.5	36.5	42.	47.4
22	6.	7.6	9.6	10.9	13.7	16.5	20.2	25.9	32.	38.3	44.	49.7
23	6.3	8.	10.	11.3	14.3	17.3	21.1	27.	33.5	40.	46.	52.
24	6.6	8.3	10.4	11.9	14.9	18.	22.	28.2	34.9	41.7	48.	54.2
25	6.9	8.6	10.9	12.3	15.6	18.8	22.9	29.3	36.3	43.5	50.	56.4
26	7.1	9.	11.3	12.8	16.2	19.5	23.8	30.5	37.8	45.2	52.	58.6
27	7.4	9.4	11.7	13.3	16.8	20.3	24.7	31.7	39.3	47.	54.	61.
28	7.7	9.7	12.2	13.8	17.4	21.	25.6	32.9	40.7	48.7	56.	63.2
29	8.	10.	12.6	14.3	18.	21.8	26.6	34.1	42.2	50.4	58.	65.5
30	8.3	10.4	13.	14.8	18.7	22.5	27.5	35.3	43.6	52.1	60.	67.7
31	8.5	10.7	13.5	15.3	19.3	23.3	28.4	36.4	45.1	53.9	62.	70.
32	8.8	11.1	13.9	15.8	19.9	24.1	29.3	37.6	46.5	55.6	64.	72.2
33	9.1	11.4	14.3	16.3	20.5	24.8	30.2	38.8	48.	57.4	66.	74.4
34	9.4	11.7	14.7	16.8	21.2	25.6	31.1	40.	49.5	59.1	68.	76.7
35	9.6	12.1	15.2	17.3	21.8	26.3	32.	41.1	50.9	60.8	70.	79.
36	9.9	12.5	15.6	17.8	22.4	27.	33.	42.3	52.4	62.6	72.	81.3
37	10.2	12.8	16.1	18.3	23.	27.8	33.9	43.5	53.8	64.3	74.	83.5
38	10.5	13.2	16.5	18.8	23.7	28.5	34.8	44.6	55.2	66.	76.	85.8
39	10.7	13.5	16.9	19.3	24.3	29.3	35.7	45.8	56.7	67.8	78.	88.
40	11.	13.8	17.4	19.8	24.9	30.1	36.6	47.	58.2	69.5	80.	90.2
41	11.3	14.2	17.8	20.3	25.5	30.8	37.6	48.2	59.6	71.3	82.	92.5
42	11.5	14.5	18.2	20.8	26.1	31.6	38.5	49.4	61.1	73.	84.	94.8
43	11.8	14.9	18.7	21.3	26.8	32.3	39.4	50.6	62.5	74.8	86.	97.
44	12.1	15.2	19.1	21.8	27.4	33.1	40.3	51.7	64.	76.5	88.	99.3
45	12.4	15.6	19.5	22.2	28.	33.8	41.2	52.9	65.5	78.2	90.	101.6
46	12.7	15.9	20.	22.7	28.6	34.6	42.2	54.	80.	80.	92.	103.8
47	12.9	16.3	20.4	23.2	29.2	35.3	43.	55.2	68.4	81.7	94.	106.
48	13.2	16.6	20.8	23.7	29.9	36.1	43.9	56.4	69.8	83.5	96.	108.4
49	13.5	17.	21.3	24.2	30.5	36.8	44.8	57.6	71.2	85.1	98.	110.5
50	13.8	17.3	21.7	24.7	31.1	37.6	45.8	58.7	72.7	87.	100.	112.8

List of Sizes of Steam Mains.—The following table gives the correct sizes of mains for both one and two pipe steam heating systems, as determined by the above method :

Radiation Square Feet	One-Pipe Work Inches	Two-Pipe Work Inches	Radiation Square Feet	One-Pipe Work Inches	Two-Pipe Work Inches
125	1½	1¼x1	2,050	5	4½x4
250	2	1½x1¼	2,500	6	5x4½
400	2½	2 x1½	3,600	7	6x5
650	3	2½x2	5,000	8	7x6
900	3½	3 x2½	6,500	9	8x6
1,250	4	3½x3	8,100	10	9x6
1,600	4½	4 x3½			

List of Sizes for Hot Water Mains.—The following table gives the correct sizes of mains for hot water heating systems:

Radiation Square Feet	Gravity Size Pipe Inches	Forced H.W. 2-Pipe System*	Radiation Square Feet	Gravity Size Pipe Inches	Forced H.W. 2-Pipe System*
75 to 125	1¼	1	950 to 1,200	4	1½
125 to 175	1½	1	1,200 to 1,575	4½	2
175 to 300	2	1	1,575 to 1,975	5	2
300 to 475	2½	1	1,975 to 2,375	5½	2½
475 to 700	3	1¼	2,375 to 2,850	6	2½
700 to 950	3½	1½			

*Based on 200 Btu, emission and circulators having 1,725 r.p.m. motors.

Hot water flow mains may be reduced in size in proportion to the branches taken off. They should, however, have as large areas as the sum of all branches beyond that point. Returns should be same as flows.

Tables of Mains and Branches.—The following table gives the size of mains and the number of branches of different sizes that each main will supply :

```
Mains                          Branches
1   in. will supply 2........................................................................  ¾ in.
1¼ in. will supply 2........................................................................ 1   in.
1½ in. will supply 2........................................................................ 1¼ in.
2   in. will supply 2........................................................................ 1½ in.
2   in. will supply 2 1½ in. and 1 1¼ in. or 1 2   in. and 1.1¼ in.
3   in. will supply 1 2½ in. and 1 2   in. or 2 2   in. and 1.1½ in.
3½ in. will supply 2 2½ in. or 1 3   in. or 1 2   in. or 3.2   in.
4   in. will supply 1 3½ in. and 1 2½ in. or 2 3   in. and 4.2   in.
4½ in. will supply 1 3½ in. and 1 3   in. or 1 4   in. and 1.2½ in.
5   in. will supply 1 4   in. and 1 3   in. or 1 4½ in. and 1.2½ in.
6   in. will supply 2 4   in. and 1 3   in. or 4 3   in. or 10.2   in.
7   in. will supply 1 6   in. and 1 4   in. or 3 4   in. and 1.2   in.
```

Radiator Tappings.—The following is a schedule of tappings or branches required for radiators of different sizes.

	Steam	One Pipe
Radiators containing up to 24 sq. ft. of radiation		1 in.
Radiators containing 24 to 60 ft. of radiation		1¼ in.
Radiators containing 60 to 100 ft of radiation		1½ in.
Radiators containing above 100 ft. of radiation		2 in.
	Gravity Hot Water	Two Pipe
Radiators containing up to 40 ft. of radiation		1 x1 in.

Radiators containing 40 to 72 ft. of radiation ....1¼ x1¼ in.
Radiators containing more than 72 ft. of radiation ....1½ x1½ in.

Forced Flow Hot Water

A ¾" branch will take care of most any ordinary size radiator where forced flow is used.

Sizes and Net Prices of Standard Wrought Steel Pipe
All Weights and Dimensions are Nominal

Size	Price per 100 Lin. Ft. Random T. & C. Black	Galv.	Diameter External	Internal	Weight per Foot Plain Ends	Threads and Couplings	No. of Threads Per In. of Screw
⅛	$ 50.00	$ 80.00	.405	.269	.24	.24	27
¼	70.00	90.00	.540	.364	.42	.42	18
⅜	80.00	100.00	.675	.493	.57	.57	18
½	80.00	90.00	.840	.622	.85	.85	14
¾	85.00	100.00	1.050	.824	1.13	1.13	14
1	115.00	135.00	1.315	1.049	1.68	1.68	11½
1¼	150.00	175.00	1.660	1.380	2.27	2.28	11½
1½	170.00	200.00	1.900	1.610	2.72	2.73	11½
2	235.00	280.00	2.375	2.067	3.65	3.68	11½
2½	325.00	375.00	2.875	2.469	5.79	5.82	8
3	425.00	500.00	3.500	3.068	7.58	7.62	8
3½	540.00	625.00	4.000	3.548	9.11	9.20	8
4	625.00	750.00	4.500	4.026	10.79	10.89	8
5	1150.00	1250.00	5.563	5.047	14.62	14.81	8
6	1350.00	1550.00	6.625	6.065	18.97	19.18	8
8	2000.00	2200.00	8.625	8.071	24.70	25.55	8
8	2200.00	2400.00	8.625	7.981	28.55	29.35	8
10	2800.00	3000.00	10.750	10.136	34.24	35.75	8
10	3000.00	3200.00	10.750	10.020	40.48	41.85	8
12	3600.00	3800.00	12.750	12.090	43.77	45.45	8
12	3800.00	4000.00	12.750	12.000	49.56	51.15	8

For lengths cut to order, add 15 percent to random length prices.

Sizes and Net Prices of Extra Strong Wrought Steel Pipe

Size	Price per 100 Lin. Ft. Random P.E. Black	Galv.	Diameter External	Internal	Thickness	Weight per Foot Plain Ends
1	$ 90.00	$ 110.00	1.315	.957	.179	$ 2.17
1¼	120.00	140.00	1.660	1.278	.191	3.00
1½	140.00	160.00	1.900	1.500	.200	3.63
2	185.00	220.00	2.375	1.939	.218	5.02
2½	270.00	325.00	2.875	2.323	.276	7.66
3	350.00	425.00	3.500	2.900	.300	10.25
3½	425.00	525.00	4.000	3.364	.318	12.51
4	500.00	625.00	4.500	3.826	.337	14.98
5	875.00	1100.00	5.563	4.813	.375	20.78
6	1050.00	1250.00	6.625	5.761	.432	28.57
8	1350.00	1650.00	8.625	7.625	.500	43.39
10	1850.00	2200.00	10.750	9.750	.500	54.74
12	2400.00	3000.00	12.750	11.750	.500	65.42

For lengths cut to order, add 33⅓ percent to random length prices.

Net Prices Per 100 Lin. Ft. of Youngstown "Yoloy" Alloy Steel

	Pipe Standard Random T. & C.		Extra Heavy Random P.E.
Size	Black	Galv.	Black
½	$ 97.00	$ 115.00	$ 130.00
¾	115.00	135.00	140.00
1	160.00	190.00	215.00
1¼	210.00	250.00	285.00
1½	245.00	285.00	330.00
2	315.00	375.00	455.00
2½	475.00	550.00	665.00
3	615.00	725.00	855.00
3½	760.00	910.00	. . . .
4	925.00	1100.00	1500.00
5	1350.00	1600.00	1940.00
6	1750.00	2100.00	2700.00
8–28.55 lb.	2800.00	3200.00	3100.00
10–35 lb.	4000.00	4700.00	5400.00
12–45 lb.	5000.00	6000.00	8000.00

For black pipe lengths cut to order, add to random length prices, 25 percent for standard pipe and 35 percent for extra heavy pipe.

Size of Expansion Tanks Required for Hot Water Heating Systems

The following table gives the sizes of expansion tanks required for hot water heating systems, complete with trimmings :

Capacity Gallons	Size Inches	Square Feet of Radiation	Price Each with Trimmings
15	12x30	350	$ 45.00
18	12x36	500	50.00
24	12x48	1000	55.00
30	12x60	2000	62.50
40	14x60	3000	80.00
80	20x60	4500	150.00

For two gauge glass tappings, add.. $10.00

Radiation capacities are based on forced circulation and small pipes. On old systems use next larger size tank.

Heights, Sizes and Ratings of Thin Tube Cast Iron Radiators

Standard cast iron radiation measures 1¾ " in length per section. Add ½ " to the length for each bushing.

		Sq. Ft. of Radiation per Section of Various Height Radiators			
Number Tubes	Width Inches	19"	22"	25"	32"
3	3½	. . .	. . .	1.6	. . .
4	4¾	1.6	1.8	2.0	. . .
5	6	. . .	2.1	2.4	. . .
6	7¼	2.3	. . .	3.0	3.7

Approximate Prices of Thin Tube Radiators

Sold only in even number of sections. Price per 100 sq. ft.

No. Tubes	Height			
	19"	22"	25"	32"
3	$270.00			
4	270.00	265.00	265.00	
5		260.00	240.00	
6	275.00		260.00	400.00

Radiant Baseboard Panels

Radiant baseboard panels are produced and marketed by a number of concerns. Some are made of cast iron, others having a finned pipe covered with a steel facing or cover.

They are made in various heights from 7" to 9" and usually extend out from the wall 1¾" to 2½", depending upon the manufacturer.

Rated Heat Emission of Radiant Baseboard Panels

Rating in sq. ft. is based on standard emission of 240 Btu per hour per sq. ft. at average temperature of 215°.

9⅞ Inch Radiant Panel Baseboard

Rating Sq. Ft.	Water Flow Lbs. per Hr.	Heat Output Btu per Hour per Lin. Ft. of Panel							
		220°	210°	200°	190°	180°	170°	160°	150°
1.95	500	460	420	380	350	310	270	240	200
1.95	2,500	490	450	410	370	330	290	250	210

9⅞ Inch Radiant Convector Panel

Rating based on standards given for baseboard panels.

Rating Sq. Ft.	Water Flow Lbs. per Hr.	Heat Output Btu per Hour per Lin. Ft. of Panel							
		220°	210°	200°	190°	180°	170°	160°	150°
3.25	500	780	720	660	600	530	470	420	350
3.25	2,500	830	760	700	640	570	510	440	380

Price of Baseboard Radiation and Accessories

Description	Weight Pounds	Approx. Price
Radiant Baseboard Panels 9⅞", Price per lin. ft.	12.25	$ 6.00
Radiant Convector Panels 9⅞", Price per lin. ft.	14.00	12.00
Plain ends, 9⅞", Price each	2.50	4.00
Corner Plate, 9⅞", Price each	.50	4.00
Angle Radiator Valve, ¾", Price each	1.00	12.00

Convector-Radiators

Radiant convectors combine convection heating with quick-acting and comfortable radiant heating. They are formed of cast iron with finned sections or of copper, with or without finned sections, depending upon the manufacturer.

There are so many types of convectors it is impossible to describe all of them in detail.

Convectors described below are 6¼" deep overall and may be installed as freestanding or semi-recessed units. Maximum depth of recess should not exceed 4¾".

Sizes, Capacity and Approximate Prices of Convectors

CAPACITIES

Height Inches	Length Inches	Steam Sq. Ft. E.D.R.	Forced Hot Water, 1,000 Btu per Hr.			Price Each
			180°	200°	220°	
20	16	13.0	1.74	2.17	2.60	$ 52.00
20	20	17.0	2.28	2.84	3.40	65.00
20	24	21.0	2.81	3.51	4.20	78.00
20	28	26.0	3.48	4.34	5.20	97.50
20	32	29.5	2.95	4.93	5.90	110.00
20	36	33.5	4.48	5.59	6.70	125.00
20	40	38.5	5.15	6.43	7.70	140.00
20	44	43.5	5.82	7.26	8.70	156.00
20	48	46.5	6.22	7.76	9.30	170.00
20	56	55.5	7.43	9.26	11.10	200.00
20	64	63.0	8.44	10.50	12.60	225.00
24	20	18.5	2.48	3.09	3.70	68.00
24	24	23.0	3.08	3.84	4.60	84.00
24	28	28.0	3.75	4.67	5.60	94.00
24	32	32.5	4.35	5.42	6.50	120.00
24	36	36.5	4.88	6.09	7.30	135.00
24	40	42.0	5.62	7.01	8.40	156.00
24	44	47.5	6.36	7.93	9.50	160.00
24	48	51.0	6.83	8.51	10.20	190.00
24	56	60.5	8.18	10.10	12.10	220.00
24	64	69.0	9.24	11.52	13.80	255.00

For dampers, add $5.00. For air chambers and accessories, less air valve, add $6.00.

Types of Standard Radiant Convectors and Cabinets—Cast Iron

Standard Convectors and Cabinets—Non-Ferrous

Convectors listed below are 4", 6", 8" and 10" deep plus ¼" overall, and may be installed as freestanding or semi-recessed units.

HEIGHT

		20"			24"			32"		
Dep.	Len.	EDR	Approx. Price	Shpg. wt. lbs.	EDR	Approx. Price	Shpg. wt. lbs.	EDR	Approx. Price	Shpg. wt. lbs.
4	24	13.1	$ 48.00	25	15.0	$ 50.00	25	—	$ 60.00	—
6	24	19.0	67.00	30	22.4	70.00	30	25.5	82.00	35
	32	27.0	70.00	40	31.6	75.00	35	36.0	90.00	50

HEIGHT

Dep.	Len.	20" EDR	20" Approx. Price	20" Shpg. wt. lbs.	24" EDR	24" Approx. Price	24" Shpg. wt. lbs.	32" EDR	32" Approx. Price	32" Shpg. wt. lbs.
6	36	30.8	$ 75.00	40	36.2	$ 80.00	45	41.2	$ 95.00	55
	40	34.9	82.00	45	40.8	90.00	45	46.5	105.00	60
	44	38.8	90.00	50	45.4	100.00	55	51.7	120.00	65
	48	42.7	96.00	55	50.0	105.00	60	56.9	125.00	75
	56	50.6	102.00	60	59.2	110.00	65	67.4	138.00	75
	64	58.3	110.00	65	68.4	120.00	70	77.9	140.00	80
8	32	34.0	82.00	45	37.5	90.00	45	41.6	105.00	55
	36	39.1	88.00	45	43.1	95.00	45	47.8	112.00	60
	40	44.2	102.00	50	48.7	110.00	50	54.0	138.00	65
	48	54.3	120.00	60	59.8	125.00	65	66.4	148.00	75
	56	64.4	130.00	65	70.9	140.00	70	78.8	168.00	115
	64	74.4	145.00	70	82.0	165.00	70	91.0	190.00	120
10	36	45.0	100.00	50	49.0	115.00	50	54.5	132.00	65
	40	50.8	105.00	55	55.3	125.00	55	61.5	148.00	70
	48	62.1	130.00	65	67.7	145.00	75	75.4	170.00	85
	56	73.6	145.00	70	80.2	165.00	80	89.3	200.00	120
	64	85.1	165.00	75	92.7	185.00	85	103.1	215.00	130

Stock Model Convector Accessories

Description		Price	Shpg. Wt. Lbs.
Damper Package (Up to and incl. 64")		$ 8.60	5
Snap-in Inlet Grille for Stock Models Only		10.20	5
CM Molding Trim—24" to 36" Long		12.50	6
CM Molding Trim—40" to 64" Long		16.00	8
Air Vent—Manual	Ea.	7.10	1
Air Vent—Auto	Ea.	7.10	1

Hot Water Rating in Btu per Sq. Ft. EDR

Temp. Drop	Average Water Temp. °F 180°	190°	200°
10°F	160	181	201
20°F	141	159	177
30°F	131	149	166

Sizes and Approximate Prices of Cast-Iron Sectional Boilers and Gas or Light Oil Burners With Jacket

Steam Installed Radiation Sq. Ft.*	Price Each	Installed Radiation Sq. Ft.* For Oil Firing	Hot Water Btu Rating	Price Each
845	$1220.00	1500	225,000	$1220.00
1090	1460.00	1910	287,000	1460.00
1340	1675.00	2320	348,000	1675.00
1600	1870.00	2735	410,000	1870.00
1860	1950.00	3145	472,000	1950.00
2120	2050.00	3560	534,000	2050.00
2380	2350.00	3960	594,000	2350.00
2650	2850.00	4375	656,000	2850.00

2920	2900.00	4785	718,000	2900.00
3200	3000.00	5205	781,000	3000.00
3480	3100.00	5620	843,000	3100.00

Above prices are net to the heating contractor, f.o.b. factory. Prices include boiler trimmings.
*Capacities are based on net installed radiation and on the presence of a quantity of net installed radiation sufficient for the requirements of the building. Nothing need be added for normal piping.
All ratings are based on radiation at 240 Btu per sq. ft. for steam and 150 Btu per sq. ft. for water.

Sizes and Approximate Prices of Oil Burning Boiler Units

Price includes boiler complete with oil burner, trombone tank type water heater or tankless type water heater, steel jacket with baked enamel finish, aircell type insulation furnished with jacket, plain room thermostat, limit control and combustion control.

In addition, steam boiler prices include: low water cut-off, pop safety valve, try cocks, water gauge, pressure and vacuum gauge and flue brush.

Water boiler prices include: combination altitude gauge and thermometer and flue brush.

	Steam Boiler			Water Boiler		
Installed	Jacketed Prices With		Installed	Net	Jacketed Prices With	
Radiation	Trombone	Tankless	Radiation	Btu	Trombone	Tankless
Sq. Ft.*	Heater	Heater	Sq. Ft.*	Rating	Heater	Heater
460	$ 770.00	$ 810.00	840	126,000	$ 770.00	$ 810.00
570	930.00	970.00	1030	155,000	930.00	970.00
680	1000.00	1050.00	1220	183,000	1000.00	1050.00
790	1100.00	1150.00	1405	211,000	1100.00	1150.00
900	1150.00	1200.00	1590	239,000	1150.00	1200.00

Above prices are net to the heating contractor, f.o.b. factory. Prices include boiler trimmings.
*Capacities are based on net installed radiation and on the presence of a quantity of net installed radiation sufficient for the requirements of the building. Nothing need be added for normal piping.
All ratings are based on radiation at 240 Btu per sq. ft. for steam and 150 Btu per sq. ft. for water.

If boilers are furnished without tankless heater, deduct $45.00; less trombone tank type heater, deduct $25.00.

Water Heater Capacities

Tankless type.. 4 gals. per minute
Trombone tank type ....66 gals. in 3 hours.

Above capacities are based on 40° to 140° rise with boiler water at 180°.
Above prices are f.o.b. factory. Prices include boiler trimmings as listed above.

Approximate Prices of Gas Fired Boilers for Hot Water Heating Systems

Furnished with finned cast iron sections to provide staggered heat travel, draft diverter, heavily insulated steel jacket with baked enamel finish.

Boiler price includes gas valve and hydraulic limit control, relay gas valve and recycling manual control switch, transformer, electric safety pilot, gas pressure regulator, manual gas shut-off valve, combination thermometer and altitude gauge, pilot pressure regulator, drain cock, pilot cock, draft diverter, gas manifold, tubing and fittings. Room thermostat not included. Prices are f.o.b. factory.

Net I-B-R Rating Radiation Sq. Ft.	Output Btu per hr.	A.G.A. Gross Output Btu per hr.	Price Boiler Complete
515	77,500	103,000	$ 660.00
860	129,000	172,000	870.00
1205	181,000	241,000	1140.00
1550	233,000	310,000	1300.00
1720	258,000	344,000	1450.00
1960	294,000	392,000	1600.00
2240	336,000	448,000	1900.00
2520	378,000	504,000	2100.00
2800	420,000	560,000	2250.00
3080	462,000	616,000	2600.00
3350	502,500	672,000	3000.00
3920	588,000	784,000	3400.00
4480	672,000	896,000	3600.00
5040	756,000	1,008,000	4050.00
5600	840,000	1,120,000	4300.00
6160	924,000	1,232,000	4600.00
6666	1,000,000	1,344,000	5000.00
7222	1,083,350	1,456,000	5400.00
7777	1,166,650	1,568,000	5600.00
8333	1,250,000	1,680,000	5800.00
8888	1,333,300	1,792,000	6200.00

Approximate Prices of Gas Fired Boilers for Steam and Vapor Heating Systems

Furnished with finned cast iron sections to provide staggered heat travel, draft diverter, heavily insulated steel jacket with baked enamel finish.

Boiler price includes gas valve, relay gas valve and recycling manual control switch, transformer, electric safety pilot, gas pressure regulator, pilot pressure regulator, manual gas shut-off valve, electric hot water cut-off, steam pressure regulator, retard steam gauge, pop safety valve, water gauge, try cocks, drain cock, pilot and burner cocks, draft diverter, gas manifold, tubing and fittings. Room thermostat not included. Prices are f.o.b. factory.

Net I-B-R Rating Radiation Sq. Ft.	Output Btu per hr.	A.G.A. Gross Output Btu per hr.	Price Boiler Complete
280	66,750	103,000	$ 540.00
475	113,500	172,000	780.00
670	161,000	241,000	1,150.00
875	210,000	310,000	1,450.00
980	235,000	344,000	1,750.00
1121	269,100	392,000	2,000.00
1293	310,250	448,000	2,100.00
1466	351,700	504,000	2,200.00
1641	393,800	560,000	2,400.00
1996	478,950	672,000	2,600.00
2353	565,250	784,000	3,000.00
2717	652,100	896,000	3,400.00
3081	739,550	1,008,000	3,600.00
3449	827,800	1,120,000	3,800.00
3814	915,300	1,232,000	4,200.00

4167	1,000,000	1,344,000	4,600.00
4514	1,083,350	1,456,000	4,900.00
4861	1,166,650	1,568,000	5,200.00
5208	1,250,000	1,680,000	5,400.00
5555	1,333,300	1,792,000	6,600.00
5903	1,416,650	1,904,000	8,000.00

Net boiler is the amount of actual installed radiation which may be attached to the boiler, based on a heat emission rate from the radiators of 240 Btu per sq. ft. for steam and 150 Btu for gravity hot water. For forced circulation for hot water systems, select the boiler from the net Btu column.

Standard Black and Galvanized Steel Pipe

Inside Diameter Inches	Black Pipe Price Per Lin. Ft.	Galv. Pipe Price Per Lin. Ft.	Threads Extra Each End
¼	$0.70	$0.90	$0.70
⅜	.80	1.00	.70
½	.80	.90	.70
¾	.85	1.00	.70
1	1.15	1.35	.70
1¼	1.50	1.75	.80
1½	1.70	2.00	1.00
2	2.35	2.75	1.20
2½	3.25	3.75	2.00
3	4.25	5.00	2.50

Above prices are for pipe cut to order. Add for threads as given.

Prices of Steam Radiator Valves

	Sizes in Inches					
	½	¾	1	1¼	1½	2
Angle steam radiator valve	$21.00	$26.00	$36.00	$49.00	$63.00	$100.00
Gate steam radiator valve	11.00	14.00	17.00	22.00	28.00	40.00

Prices of Water Radiator Valves

	½	¾	1	1¼	1½
Hot water radiator valve	$6.00	$8.00	$11.00	$18.00	$25.00
Hot water radiator union ell	2.50	3.00	4.50	6.00	8.00

Prices on Floor and Ceiling Plates

Floor and ceiling plates made of cold rolled steel, halves securely rivited by a concealed hinge. Can be opened or closed on pipe without effort.

For Pipe size in inches	½	¾	1	1¼	1½	2	2½	3
Chrome plated, ea.	$0.46	$0.46	$0.50	$0.54	$0.58	$0.64	$0.98	$1.22

Copper Pipe

The cost of copper pipe on heating work depends to a large extent upon the experience of the steam fitter installing same. The technique of installing copper pipe is entirely different from that of steel pipe. Inexperienced steam fitters figure more for labor than on a steel pipe job, while others figure the same for labor, whether steel or copper. Steam fitters who are thoroughly experienced in installing copper pipe and fittings and who are equipped to handle same efficiently state they can install copper pipe for 15 to 20 percent less than steel pipe.

The price of the pipe itself is subject to wide fluctuations, due to the price of copper so that prices on copper pipe and fittings should ALWAYS be obtained for each job.

Copper pipe is furnished in 3 weights, as follows :

Heavy Copper Pipe (Government Type "K") for underground service, such as water, gas, steam, oil lines, industrial uses, also interior plumbing.

Standard Copper Pipe (Government Type "L") for interior plumbing and heating.

Light Copper Pipe (Government Type "M") for low pressure interior plumbing.

Sizes and Weights of Copper Pipe

Nominal Size Inches	Type "K" Heavy Wall Thickness Inches	Weight per Lin, Ft. Lbs.	Type "L" Standard Wall Thickness Inches	Weight per Lin. Ft. Lbs.	Type "M" Light Wall Thickness Inches	Weight per Lin. Ft. Lbs.
⅜	.049	.269	.035	.198	.025	.145
½	.049	.344	.040	.285	.028	.204
⅝	.049	.418	.042	.362	....	....
¾	.065	.641	.045	.455	.032	.328
1	.065	.839	.050	.655	.035	.465
1¼	.065	1.04	.055	.884	.042	.681
1½	.072	1.36	.060	1.14	.049	.940
2	.083	2.06	.070	1.75	.058	1.46
2½	.095	2.93	.080	2.48	.065	2.03
3	.109	4.00	.090	3.33	.072	2.68
3½	.120	5.12	.100	4.29	.083	3.58
4	.134	6.51	.110	5.38	.095	4.66
5	.160	9.67	.125	7.61	.109	6.66
6	.192	13.87	.140	10.20	.122	8.91
8	.271	25.90	.200	19.30	.170	16.46
10	.338	40.30	.250	30.10	.212	25.60
12	.405	57.80	.280	40.40	.254	36.70

Because of the smooth inside surface of copper pipe, usually one size smaller pipe can be used than where steel pipe is used. Outside diameter of pipe is ⅛ " more than nominal size of tubing.

Approximate Prices on Copper Tubing

Prices on copper tube and fittings are subject to wide fluctuations due to the fluctuations in the price of copper. Always check prices before submitting bids.

Copper Tubing

Price per lin. ft.	¼ "	⅜ "	½ "	⅝ "	¾ "	1 "	1¼ "	1½ "
Nominal Size								
Type "K"	$0.40	$0.65	$0.85	$1.00	$1.46	$1.80	$2.25	$2.90
Type "L"	.37	.50	.75	.90	1.10	1.50	1.45	2.50

Price per lin. ft.	2"	2½ "	3"	3½ "	4"	5"	6"
Type "K"	$4.35	$6.20	$8.50	$12.00	$15.00	$35.00	$50.00
Type "L"	3.75	5.40	7.25	9.80	12.00	28.00	38.00

Prices of Brass Solder-Joint Valves

These valves are made for use with copper tubing and can be used for steam or water service.

Description of Fitting	½ "	¾ "	1"	1¼ "	1½ "	2"
	Size in Inches					
Gate value..............................	$11.15	$14.25	$17.25	$23.00	$29.75	$41.75

Approximate Quantity of Solder and Flux Required to
Make 100 Joints

Size, Inches	3/8"	1/2"	3/4"	1"	1 1/4"	1 1/2"	2"	2 1/2"	3"	3 1/2"	4"	5"	6"	8"	10"
Solder, Lbs	1/2	3/4	1	1 1/2	1 3/4	2	2 1/2	3 1/2	4 1/2	5	6 1/2	9	17	35	45
Flux, Oz.	1	1 1/2	2	3	3 1/2	4	4	7	9	10	13	18	34	70	90

Estimating the Cost of Pipe and Fittings.—To estimate the cost of pipe and fittings required for any heating plant, obtain the total number of square feet of radiation in the job (based on cast iron radiation) and figure at $1.20 to $1.35 per square foot. This is based on prices of pipe as given on the previous pages.

For one story residences having short pipe runs, figure the cost of the pipe and fittings at $1.25 per square foot of radiation (based on standard cast iron radiation).

An example of the method used in estimating the cost of pipe and fittings is as follows: For a job containing 1,400 square feet of standard radiation, take 120 per cent of 1,400 gives 1680, or the pipe and fittings would cost approximately $1700.00.

The above allowances for pipe and fittings do not include radiator valves, floor and ceiling plates, boiler fittings, expansion tanks, etc., but only the pipe and fittings required for roughing-in the job.

PIPE COVERING

Pipe covering is furnished in several kinds of materials, fiberglass sectional pipe insulation covering being the most common and cheapest for covering steam and hot water pipes; 85 percent magnesia pipe covering for high pressure steam work, and wool felt pipe covering for protecting pipes against freezing and pipe covering in general.

Pipe covering is usually furnished in 3 foot lengths, 3/4 to 1-inch thick and is covered with a canvas or fire safety jacket.

Labor Covering Pipes Up to 3 Inches in Diameter.—When placing covering on pipes up to 3 inches in diameter, a pipe coverer should apply 10 to 15 lineal feet of covering an hour, or 80 to 120 lineal feet per 8 hour day, and the labor cost per 100 lineal feet should average as follows :

	Hours	Rate	Total	Rate	Total
Pipe coverer	8	$. . . .	$. . . .	$16.94	$135.52
Cost per lin. ft			. . . .		1.36

Labor Covering Pipes 4 to 6 Inches in Diameter.—When placing covering on pipes 4 to 6 inches in diameter, a pipe coverer should apply 9 to 11 lineal feet of covering an hour or 72 to 88 lineal feet per 8 hour day, and the labor cost per 100 lineal feet should average as follows :

	Hours	Rate	Total	Rate	Total
Pipe coverer	10	$. . . .	$. . . .	$16.94	$169.40
Cost per lin. ft			. . . .		1.70

When estimating quantities of pipe covering, add one lineal foot of covering for each fitting up to 2-in. pipe and two lineal feet of covering for each fitting on pipe over 2 inches.

Temperature Regulators for Heating Plants

Temperature regulators and thermostats are furnished in a variety of different models for controlling the temperature.

There are so many different methods of controlling the inside temperature with straight thermostats, day and night thermostats, inside-outside controls, etc., that it is practically impossible to give even a fraction of them in a book of this kind.

Temperature Regulators With Clock.—The thermostat with a clock attachment automatically lowers the temperature at night and raises it in the morning. Electric clocks are now usually furnished with all thermostats of this type.

Models of this kind are furnished with thermometer, clock, electric motor, chain, pulleys, transformer and fittings, at the following approximate prices :
Thermostat complete with electric clock... $100.00

Temperature Regulators Without Clock.—Many thermostats or regulators are used without the clock attachment and are either manually operated or the same temperature maintained day and night.

There are a number of different types of regulators on the market and they vary in price from $30.00 to $80.00, depending upon the method of wiring and number of controls necessary.

Labor Installing Thermostats.—An experienced mechanic should install any of the above thermostat outfits in 1 to 2 hours. If they are installed by the manufacturers in cities where they maintain offices, a charge of $125.00 to $150.00 is usually made.

Making Up a Complete Estimate on a Steam Heating Plant.—When making up an estimate on a steam heating system, the following items should always be included in order that the estimate may be complete in every detail.

1. Boiler	11. Floor and ceiling plates.
2. Oil or gas burner.	12. Painting radiators.
3. Radiators or convectors.	13. Thermostat regulator.
4. Radiator Valves.	14. Freight and trucking.
5. Automatic air valves.	15. Water Connection.
6. Pipe and fittings.	16. Labor installing system.
7. Smoke pipe.	17. Bond and insurance.
8. Special foundation, if required.	18. Overhead and misc. expense.
9. Asbestos boiler covering.	19. Profit.
10. Pipe covering.	

Making Up a Complete Estimate on a Hot Water Heating Plant.—When making up an estimate on a hot water heating plant, the following items should always be included in order that the estimate may be complete in every detail.

1. Special boiler foundation, if required.	13. Painting radiators.
	14. Boiler covering.
2. Boiler.	15. Pipe covering.
3. Oil or gas burner.	16. Floor and ceiling plates.
4. Pipe and fittings.	17. Thermostat regulator.
5. Radiators or convectors.	18. Labor installing system.
6. Radiator valves-union ells.	19. Freight and trucking.
7. Circulating pumps.	20. Water connection.
8. Compression air valves.	21. Bond and insurance.
9. Expansion tank and gauge.	22. Overhead and miscellaneous expense.
10. Altitude gauge.	
11. Thermometer.	23. Profit.
12. Smoke pipe.	

Unit Heaters.—Unit heaters are often used to heat commercial and industrial areas, garages, and to supplement other heating units at entrances. They may be direct-fired by gas, oil or electricity, or have hot water or steam piped to them. Direct-fired are cheaper when only a few units are required and no central plant is available. Hot water and steam are the most common units in use where a central plant is available or its cost justified by the number of units it serves.

Units may have propeller or centrifugal fans blowing either horizontally, up or down. Air may be drawn through or blown through, the latter being used for direct-fired units.

The following costs do not include valves, fittings or piping which will vary widely with job conditions and are based on units served with 2 lbs. steam, 60° entering air, single phase 115 volts.

| Type | H.P. | Btu | Cost | | Installation Time |
			1 Speed	2 Speed	
Horizontal	1/40	25,000	$155.00	$ —	1 hour
Delivery	1/15	50,000	200.00	245.00	1 1/2 hours
	1/10	100,000	300.00	340.00	2 1/2 hours
	1/5	200,000	480.00	540.00	4 hours
Vertical	1/40	50,000	190.00	—	1 1/2 hours
Delivery	1/12	100,000	260.00	300.00	2 hours
	1/3	200,000	520.00	575.00	5 hours
Cabinet	1/20	25,000	160.00	—	1 1/2 hours
	1/12	60,000	230.00	—	2 hours
	1/8	120,000	475.00	—	3 hours

For 230 volts, add $16.00.
For ceiling suspension, add $25.00

15C SPRINKLER SYSTEMS

The construction industry and the general public has, for many years, been aware of the need and availability of fire protection systems, however, this protection and the associated costs were severely restricted to building areas where local building codes or extreme business risks dictated the demand of their use.

In general, there has existed a great insensitivity toward providing fire protection for the public even in areas of high concentration such as restaurants, theaters, hotels, sporting events, etc. This disregard for public safety has abruptly changed because of an increase in the number of fire disasters which have received national publicity and have become the catalyst to inspire total remedial action on the part of Federal and State authorities as well as insurance companies, building owners and the professional architect.

The fire protection industry responded promptly to provide the specialized technical design data and to produce trained personnel with the capability of educating the designers and specification writers in selecting the particular system, or combination of systems, most appropriate to provide the degree of protection desired.

As the need has developed to provide special protection for costly equipment such as computer rooms and tape storage libraries, or irreplaceable documents, art objects, sophisticated research projects, the aged and the handicapped, new alarm, control and protection systems have been developed, some of the alarm and control systems are mechanically operated while others are completely electrically controlled. In some cases alarm and control systems are designed as a combination of mechanical and electrical devices and, in these instances, the Estimator must determine, with great precision, the precise cost and installation responsibility which each trade will assume in the overall function of the system.

WET PIPE SPRINKLER SYSTEM

A system that meets most fire protection requirements for normal hazards where piping is not subject to freezing.
Water is constantly supplied under pressure to each sprinkler head.

DRY PIPE SPRINKLER SYSTEM

This system may be used in areas subject to freezing as the piping is filled with air under pressure which contains the water at the supply source until an activated sprinkler head allows the air pressure to disburse which then produces the water supply at the head.

Sprinkler systems are a specialized type of work requiring the services of a sub-contractor who has the access to equipment which, in many instances, is a sole source item, not available on the open market. It will usually be found that the contract documents require the service of such a specialist experienced in this type of work for a certain number of years and having the capabilities to design the sprinkler system.

For the wet and dry pipe systems, unit costs have been developed on a per head basis. It is assumed that an adequate water supply is available and that special alarm and control work is not included.

In the case of high rise buildings, booster pumps may be required which will add $10,000 - $12,000 cost to the installation.

Maximum area coverage per head is 130 sq. ft.

The average area covered per head is 80 sq. ft. to 130 sq. ft. under normal hazard.

Distance between sprinkler heads cannot exceed 15 ft.

WET SYSTEM	TOTAL COST PER HEAD
Warehouse	90.00
Hospital	150.00
Shopping Center	100.00
Municipal Buildings	125.00

DRY SYSTEM	TOTAL COST PER HEAD
	Add 15.00 - 20.00 per head

It may appear that a fire sprinkler system is costly to install, but that is not the total overview of the benefits which flow to the Owner as it tends to insulate the Owner from costly litigation resulting from fire loss and, in a more identifiable form, it produces an immediate and continuing savings in fire insurance premiums. Most sprinkler installations return their initial cost in reduced insurance premiums within 3 - 10 years.

ELECTRICAL

Electricity plays an increasingly more important role in construction each day. Some buildings towering 70 stories or more are completely dependent upon electricity, not only for light, power, and building services, but for completely electrified tenant equipment and most of their heat. Heating and lighting are dependent upon each other so that a maximum use of electric energy may be obtained at a minimum of expense.

Obviously such sophisticated designs are well beyond the scope of this book. But even a modest residence will involve electrical planning and power demands well beyond anything thought of 25 years ago.

For very preliminary estimates, electrical work can be assumed to run from 6 to 8 percent of the total job cost for residences, 8 to 11 percent for office buildings, and 8 to 12 percent for industrial buildings.

In order to obtain accurate costs and to check actual installation costs, an estimator should make a detailed electrical take off. The estimator must then have a thorough knowledge of the working drawings and specifications from which the job will be built as well as the building code under which it will be inspected, and also regulations set by the local utility company.

A rotometer which will scale distances directly off the drawings is handy if drawings include sufficient interior room elevations so that not only horizontal but vertical distances may be measured.

The electrical take off sheet should have a column for each of the following:
1. Ceiling outlets
2. Bracket outlets
3. Duplex convenience outlets
4. Triplex convenience outlets
5. Waterproof convenience outlets
6. Range outlets
7. Lighting outlet with lampholder
8. Special purpose outlets (dishwashers, clothes dryers, etc.)
9. Single pole switches
10. Three-way switches
11. Length of wiring

The estimator then should check off each item and enter the total on the sheet. This is best done floor by floor or, in large one-story residences, wing by wing, so that checking and changes ordered on the job will be simplified.

Enter next the circuit and feeder control centers, listing them by type and size, and then the branch circuits. As mentioned above, a rotometer will be helpful in measuring raceway runs, but be sure all drawings used are at the same scale. List separately two-wire and three-wire runs and empty conduit for telephones and such other special items as intercoms, buzzer systems, television outlets, and the like.

A final list of such items as fixture hangers, studs, locknuts, bushings, cable clamps and outlet covers will complete the take off; then all these items can be transferred to the pricing sheet item by item.

A pricing sheet should include a line for each type of item; the quantity involved; the material cost by unit and in total; the hours estimated for installation per unit and hourly rate; the total labor cost, and finally, the total labor and material cost.

Service Entrance Equipment

The service entrance equipment consists of the service drop, service entrance conductors, the meter, main control center, and the service ground. The local utility will govern the details of this installation and may or may not furnish the meter. Underground service is highly desirable and may be required by local ordinances. Underground service for residential use will run about $8.00 per lineal foot. Overhead lines will run around $400.00 per pole. The main control center for residential work will consist of a circuit breaker-fuse panel of at least 100 amp. capacity.

Circuit Breakers

Capacity	Cost	Hours to set
100 amp. 3 pole, gen. purpose	$160.00	4
200 amp. 3 pole, gen. purpose	290.00	7
400 amp. 3 pole, gen. purpose	575.00	9
600 amp. 3 pole, gen. purpose	950.00	12

16110　RACEWAYS

Rigid conduit is the safest system and, in some large cities, is required by code. Rigid systems may be galvanized steel, aluminum or electric metallic tubing, generally referred to as "thinwall" or "EMT".

Material Costs and Labor to Set 100 Lin. Ft. of Rigid Conduit
Including Supports and Fittings

Size	Galvanized		Aluminum		EMT	
	Cost	Hours to Set	Cost	Hours to Set	Cost	Hours to Set
½"	$ 60.00	8	$ 65.00	6	$ 25.00	6
1"	90.00	10	110.00	9	55.00	8
1½"	150.00	12	190.00	11	90.00	9
2"	200.00	14	250.00	12	120.00	10
3"	425.00	23	525.00	18	330.00	14
4"	650.00	36	800.00	26	525.00	18

Flexible conduit, or as it is often referred to, "Greenfield," is easier to install.

Size	Conduit Cost	Connector Cost	Hours to Set
½"	$ 30.00	$ 0.60	5
1"	75.00	1.50	8
1½"	120.00	2.75	14
2"	160.00	4.00	18
3"	250.00	11.00	30

Where codes allow the least expensive methods of wiring are prewired armored cable (BX) or non-metallic sheathed cable such as Romex.

Material Cost to Install 100 Lin. Ft. of Armored Cable

Gauge of Wire	No. of Wires	Cost	Hours to Set
#10	2	$60.00	5
#10	3	75.00	6
#12	2	40.00	4
#12	3	48.00	5
#14	2	30.00	4
#14	3	37.00	5

Material Cost and Labor to Install 100 Lin. Ft. of Non-Metallic Sheathed Cable with Ground

Gauge of Wire	No. of Wires	Cost	Hours to Set
#12	2	$15.00	3
#12	3	27.00	3.5
#14	2	11.00	2.5
#14	3	19.00	3

Duct System

Underfloor metal duct systems vary widely in their complexity. Some floor systems have raceways built in, others are added in the floor fill and may be a one or two level system. Standard duct is 1-⅜"x3-⅛". Super duct is 1-⅜"x7-¼". Lengths are 5', 6', 10' and 12'. Detailed take offs on these systems are beyond the scope of this book. The Square D Company catalog gives a suggested take off method with all the required fittings.

	Standard		Super	
	Material Cost	Hours Per 100 Lin. Ft.	Material Cost	Hours per 100 Lin. Ft.
Plain Duct Per L.F.	$ 2.60	10	$ 5.00	13.5
Duct with				
Inserts per L.F.	3.10	11.5	5.80	16
Single Box	45.00	2 ea.	70.00	3 ea.
Double Box	70.00	2.5 ea.	200.00	3.5 ea.
Support	4.00	.25 ea.	5.00	.5 ea.
Elbow	10.00	.75 ea.	19.00	1 ea.
Connector	4.00	.25 ea.	7.50	.3 ea.

Material Cost and Labor to Set 100' of Fiber Duct

Size	Duct Cost	Labor to Set	Coupling Cost	Hours to Set
2"	$ 65.00	4	$ 7.00	.4
3"	80.00	10	9.00	.6
4"	95.00	14	12.00	.75
6"	175.00	28	24.00	1.5

Material Cost and Labor to Pull 1000 Lin. Ft. of Wire in Conduit Already in Place (600 V Type THW; 3 Wire)

Gauge	Material Cost	Hours to Pull
14	$ 40.00	6
12	60.00	7.25
10	90.00	8
8	180.00	10
6	220.00	12.5
4	330.00	15
2	475.00	18
1/0	800.00	24
2/0	950.00	27.5
4/0	1,500.00	30

16130   OUTLET BOXES

Type	Cost per Unit	Hrs to Install
Switch Box	$ .65	.3
3 Gang Box	2.00	.6
4" Octagon	.45	.5
4" Square	.65	.5
4 Concrete	1.50	.4
1 Gang Cast WP	4.00	.66
3 " " "	10.00	1.0

16140   SWITCHES AND RECEPTACLES

Once the box has been set and connected to the conduit and the wiring pulled the proper switch or receptacle can be installed which will run as follows:

Type	Cost per Unit	Hours to Install
15 amp single pole toggle	$ 2.00	.2
20 " " " "	3.50	.3
15 amp 3-way toggle	3.00	.35
20 " " "	4.00	.5
15 amp 4-way toggle	8.00	.6
20 " " "	9.00	.75
600 watt single pole dimmer	9.00	.5
" " three " "	10.00	.66
1000 watt single pole dimmer	35.00	.5
" " three " "	45.00	.66
15 amp duplex receptacle	2.00	.2
20 " " "	3.00	.3
30 amp dryer	3.00	.5
50 amp range	3.00	.75

Cost of Toggle Switch and Duplex Receptacle Plates per 100

Description	Cost in Phenolic	Cost in Stainless
1 Gang Switch	$ 60.00	$110.00
2 Gang Switch	120.00	200.00
3 Gang Switch	180.00	280.00
4 Gang Switch	240.00	350.00
1 Gang Outlet	60.00	110.00
2 Gang Outlet	120.00	200.00
3 Gang Outlet	180.00	280.00
Blank-Single	60.00	110.00
Telephone	60.00	110.00

Weatherproof switches with grey Hypalon presswitch plates will run $400.00 a hundred.

16500   LIGHTING

The following is a sampling of some fixtures an electrician might encounter on average work.

Type	Wattage	Size	Shielding	Cost
Industrial Fluorescent	2–40	4'	Open	$35.00
Pendant Mtg.	4–40	8'	Open	65.00
	2–40	4'	Louvers	40.00
	4–40	8'	Louvers	80.00

Commercial Fluorescent	2–40	4'	35°x35°	40.00
Surface Mounted	4–40	4'	35°x35°	50.00
Average Quality	4–40	8'	35°x35°	85.00
Commercial Fluorescent	8–40	8'	35°x35°	100.00
Surface Mounted	2–40	4'	Acrylic	50.00
Premium Quality	4–40	8'	Acrylic	100.00
Commercial Fluorescent	2–40	4'	Holophane	60.00
Recessed Troffers	4–40	8"	Holophane	125.00
Commercial Fluorescent	4–20	2x2	Acrylic	70.00
Square Surface	6–30	3x3	Acrylic	130.00
	6–40	4x4	Acrylic	150.00
Commercial Fluorescent				
Corridor Unit	1–40	4'	Acrylic	32.00
Incandescent Industrial	1–100	12" diam.	Open	6.00
Dome Bowl	1–150	14" diam.	Open	7.00
	1–200	16" diam.	Open	8.00
	300–500	18" diam.	Open	9.00
	750–1000	20" diam.	Open	12.00
High Bay Mercury	400	24" diam.	Open	130.00
	1000	24" diam.	Open	195.00
Commercial Incandescent				
Recessed Fixed Spot	150	10" diam.	Open	12.50
Adj. Spot	150	10" diam.	Open	18.00
Punch Spot	150	8" diam.	Open	18.00
Recessed Square	100	8x8	Opal	28.00
Alum. Wall Bracket	75	5" diam.	Glass	19.00
Pendant Metal Shade	100	11" diam.	Glass	19.00
Pendant Sphere	40	16" diam.	Glass	44.00
Drum	100	18" diam.	Opal	26.00

For Stems add: 6" Drop $1.70
            18" Drop 1.90
            30" Drop 2.50
For plaster rings for spots add $1.50
Track Lighting

1 Circuit	. . .	8'	. . . .	$30.00
3 Circuit	. . .	12'	. . . .	45.00
150 W Spot	. . .	. . .	. . . .	20.00 ea.

16720 ALARM AND DETECTION EQUIPMENT

The largest single threat to property and life associated with buildings is fire. While the prime concern is always potential loss of life, damage to or loss of buildings, equipment, materials, files and records can put a company out of business. And in the case of a museum, part of history could be lost forever. Consequently, life safety and fire codes are being enacted across the country for all types of buildings, from institutional to office buildings, from hotels to single family residences.

The threat of property damage and loss of life due to fire can be substantially reduced through the proper use of smoke and fire detectors.

These detectors may either be wired into the building's electrical system, battery operated or both. The detector that is wired to the electrical system should be designed to receive back-up power from a battery source in the event of a power failure.

Battery operated detectors normally have a small light that flashes periodically to indicate that the battery is still functioning properly. They do not connect to the main electrical system, which makes this device ideal for use in existing buildings (primarily residences) where exposed wiring is objectionable and concealed wiring means costly "cut and patch" work. The one drawback to the battery operated detector is that while they will warn building occupants, they cannot be connected to an exterior alarm system.

Fire detectors and smoke detectors differ in that the former can be adjusted to alarm at a particular temperature and the latter is set to alarm at a certain percentage of smoke in the air surrounding the detector. Because most fire fatalities are due to smoke inhalation, in most cases the smoke detector gives the best protection and the quickest alarm.

Placement of detectors will vary depending upon the type of building and the interior partitioning. In institutional and commercial buildings, a detector should be located in every room and storage closet, one (1) for every 900 sq. ft. in large open areas, and at approximately 60 lin.ft. on center in corridors. In residential homes a detector should be placed near the kitchen, in hallways leading to bedrooms and at the top of all stairways at each floor level.

The cost of fire or smoke detectors varies widely depending upon the amount of sensitivity and sophistication of the alarm system. Residential detectors, either battery or hardwired type, range in price from $10.00 to $35.00 each. Detectors for institutional and commercial buildings will range from $65.00 to $130.00 each.

In addition to the cost of the detectors for non-residential use, there are other associated items of equipment that may be added to the alarm system, such as alarm station pull boxes, test panels and control console panels costing several thousand dollars each.

16850   ELECTRICAL HEATING

Heating electrically includes a broad range of methods—radiant units, convector units, baseboard units, electric furnaces with forced air, and heat pumps with forced air. Some of these methods are often included as simple supplemental heaters; others are full plants and involve quite sophisticated design in order to function most efficiently.

Wall heaters are the simplest, most commonly encountered means of heating electrically. Both natural and fan driven units are available and they are usually recessed and have built in thermostats. A 1500 watt ceiling fan unit will cost around $60; a 1500 watt ceiling radiant unit around $75; a 1000 watt ceiling radiant unit, $50. A 1200 watt fan forced unit will run $35 with manual control, $40 with thermostat.

Baseboard heaters are available in standard and high wattage models in surface mounted and semi-recessed designs. The wiring serving these units may be part of the common wiring system or fed from separate branch control centers.

In some areas electricity for heating is separately metered and must be on a separate system. For preliminary figures, baseboard units will run around $30 a lineal foot installed, but not including separate wiring system. Add $32.00 for each thermostat. Baseboards, if used as the sole heating source, are often supplemented by fan units at entrances.

Electric furnaces are similar to conventional warm air furnaces but substituting electric resistance-heating units for flame. They are very compact, require no flue or vent and can be placed anywhere so zone heating is simplified. Also, air cooling systems can be added to provide a year 'round system. A variation on this are through-the-wall units including both heating and cooling capacities. These are especially suitable for motels, dormitories and other similar buildings where individual control is important. Electric furnaces will run around $375 for a 1000 BTU/hr unit up to $700 for a 140 MBH model. Assuming wiring is already in the furnace room two men can install the unit in half a day.

Heat pumps however are more efficient than resistance type heaters because they are designed to utilize existing outside conditions to supplement inside heating and cooling. That is, in the winter it extracts heat from outside air, ground or water and conducts it to the inside and, in summer, removes heat from base to the outside. It is most generally placed outside to eliminate noise. Units are usually designed to handle air cooling loads. In areas of extreme cold where heating capacity of pump is inadequate, it may be supplemented by electric resistance heaters. The unit is connected to a duct system within the house.

REMODELING WORK

Remodeling work poses many problems for the estimator not found in new work, but it can be quite profitable, can lead to new clients and future work, and can provide the small carry-over jobs needed at times to keep crews busy during bad weather or slack periods.

The following suggestions for estimating concern themselves with items encountered in smaller remodeling work. But even here the estimator must be aware of all the job conditions and should investigate and make allowances for them, including:

a) Need for protecting adjacent work.
b) Removal of rubbish and/or salvage.
c) Restrictions during normal working hours.
d) Extent of daily cleaning up involved.
e) Is there enough work to keep the various trades busy for a full day's work or will there be call backs for only fractional periods?
f) Will owner be constantly supervising work?
g) Are plans of original structure available which will show mechanical and structural details to be encountered?
h) Is work governed by a strict building code which might require work beyond that immediately ordered, such as strengthening structural members, adding additional exits to a remodeled 3rd floor, etc.?
i) Do you have men available who can form a cooperative team as there are almost always changes from plans to be made to fit actual job conditions as they are uncovered?

A contractor is often asked whether he thinks it best to remodel, move, or build anew. On residential construction, a common rule is that if the cost of remodeling plus the original building cost equals or is less than the cost of a comparable new house, it is best to remodel. Not only are moving costs eliminated, but so are the percentages charged to sell the old house and to buy the new one. Also, investments in landscaping, drapes, carpets and the like, are salvaged.

On commercial work there are other considerations that may be favorable to remodeling rather than rebuilding. First, remodeling may eliminate zoning and building code restrictions that apply to new construction but not to remodeling. Two, the expense and time of wrecking is eliminated. Thirdly, if the basic structure is sound, this may be considered as equity for a construction loan. If this is so, financing may be simplified by eliminating preliminary loans before final project is under way.

The savings in time will mean less chance of cost escalation and earlier occupancy. These savings can be reflected in the rental scale making a remodeled structure more competitive with new. And finally, we are in an age of nostalgia where certain businesses, such as restaurants, may find a well remodeled older structure a definite plus in their public image.

Masonry Repairs

Removal of Brick by Hand Chipping.—A mason should hand chip around 200 pieces per 8 hour day at the following labor cost per 100 pieces:

	Hours	Rate	Total	Rate	Total
Mason	4.5	$....	$....	$18.91	$ 85.10
Helper	1.0	....	....	14.95	14.95
Cost per 100 pieces			$....		$100.05
Cost per brick					1.00

Removal of Concrete Block.—Block walls can usually be knocked down manually by laborers. For 12" or 8" block walls, figure per 100 sq. ft. as follows:

	Hours	Rate	Total	Rate	Total
Labor	8	$....	$....	$12.54	$100.32
Cost per sq. ft					1.00

To remove 4" partition block, figure per 100 sq. ft. as follows:

	Hours	Rate	Total	Rate	Total
Labor	5	$....	$....	$12.54	$62.70
Cost per sq. ft					.63

Where bearing walls are cut through, it will be necessary to provide needling and shoring, and to set in a new lintel. A 3' wide opening in an 8" wall will cost about $80.00 to shore and another $40.00 to build in lintels. A 6' wide opening will cost about $100.00 to shore and $50.00 to set a beam for a lintel.

To Patch Face Brick.—To patch 4" face brick, figure one mason and one helper can lay about 65 sq. ft. per day.

Labor Cost per 100 Sq. Ft.

	Labor	Rate	Total	Rate	Total
Mason	12	$....	$....	$18.91	$226.92
Helper	12	....	....	14.95	179.40
			$....		$406.32
Labor cost per sq. ft					4.06

Material cost per sq. ft. will add another $1.35 for brick in the $120.00 per 1000 range.

To Patch Building Brick.—Where openings are closed up or otherwise patched out in small areas, one mason and one helper should lay about 30 cu. ft. of building brick per day.

Labor Cost to Patch 100 Sq. Ft. of 12" Building Brick Wall

	Hours	Rate	Total	Rate	Total
Mason	27	$....	$....	$18.91	$510.57
Helper	27	....	....	14.95	403.65
			$....		$914.22
Cost per sq. ft					9.14

Material cost per cu. ft. will add another $1.45.

To Patch Concrete Block Partitions.—One mason and a helper should patch out 150 sq. ft. of 4" block partitioning per day.

Labor Cost to Patch 100 Sq. Ft. of 4" Block Partitioning

	Hours	Rate	Total	Rate	Total
Mason	5.5	$. . . .	$. . . .	$18.91	$104.01
Helper	5.5	. . . .	. . . .	14.95	82.23
			$. . . .		$186.24
Cost per sq. ft					1.86

Material costs will add about $0.55 per sq. ft.

To Rebuild Chimneys.—On small chimneys one mason with a helper will lay about 500 bricks per day.

To Add Ceramic Brick to Existing Walls.—1" thick ceramic units formed to the size of standard and Norman bricks may be applied over any nailable surface such as plywood, wood sheathing, furring strips and various types of concrete and masonry. One sq. ft. of wall will require 6 of the standard units which cost $90.00 per 1000 in red, $99.00 in golden blend and $112.50 in gray. On small jobs one mason (check local jurisdiction on this, as it varies) working alone should install around 700 standard units per 8 hour day. Mortar requirements are 1-½ cu. ft. per 100 sq. ft. of wall and a tuckpointer working alone should point about 450 standard units a day.

Special units are available for corners.

SIDING

To determine siding costs, it is best to sketch an outline of each elevation. For straight walls multiply the width by the height and deduct from the total the combined square foot area of the doors and windows. Gable area are found by multiplying the width by the height and dividing by two. Deduct window openings. Wall areas for dormers can be figured from the following table, the window areas being already deducted.

Side Wall Area for Dormers

Dormer Width	Lift Type	Type of Dormer Gable Type	Hip Type
5'-0" (One Window)	88 Sq. Ft.	60	52
6'-0"	94	70	58
7'-0"	100	80	64
8'-0" (Two Window)	88	70	52
9'-0"	94	75	58
10'-0"	100	85	64
11'-0"	106	90	70
12'-0" (Three Window)	93	80	57

Add the net square feet of each elevation, gable and dormer together, add 8% for waste and figure to the nearest larger half or full square or nearest bundle of shingles for material costs.

Old siding may be stripped off before new is applied, or may be covered over either by applying an underlayment board or by applying wood stripping and felt over present siding. Underlayment board would figure the same quantity as siding. Wood stripping can be figured per square, depending on the exposure of existing siding, as follows:

Exposure in inches	4"	4½"	5"	5½"	6"	6½"	8"
Lin. Ft. of Stripping req'd.	300	280	240	224	200	118	152

For corners, use full height for metal corners. For corner boards figure one 3" and one 4" board for each other. Add molding, flashing and caulking costs at doors and windows.

Labor Cost To Add ½" Plywood Sheathing Over Existing Boards, Per 100 Sq. Ft.

	Hours	Rate	Total	Rate	Total
Carpenter	1.0	$. . . .	$. . . .	$16.47	$16.47
Helper	.3	. . . .	. . . .	12.54	3.76
Cost Per 100 Sq. Ft			$. . . .		$20.23
Cost Per Sq. Ft					.20

Corner Boards and Beads

Allow one-half hour of carpenter's time to set one 1x3 and one 1x4 corner per story. Beads will take approximately the same time.

Often exterior moldings, corner boards and base molds must be replaced because of decay. Cedar, cypress or redwood, preferably the heartwood, will resist future decay.

Costs per lineal foot for #2 pine trim and labor would be 80¢ for 1 x 4; $1.25 for 1 x 6; $1.40 for 1 x 8; $2.20 for 1 x 12.

Remodeling For Energy Conservation

With the great rise in fuel costs, most buildings, if reappraised for their insulating value, would be found wanting. This has opened a whole new field to the remodeling contractor which is generally referred to as "retrofitting". As we have seen in Chapter 12, under the insulating division, deciding whether to invest in better methods of conserving energy is a simple matter of comparing dollars spent on new materials and equipment against the dollars saved in fuel consumption. Obviously when a gallon of oil increases from 14¢ to 90¢ or more in a six year period while a 3½" batt of insulation only goes from 8¢ a square foot to 12¢, the balance swings in favor of adding more insulation. It will not only pay for itself in a few years but will make a solid return for years to come, will help to preserve our natural resources and will create a more comfortable environment.

As most insulation is hidden within walls and under roofing one must become a "house detective" before one can become a "house doctor." One of the most reliable aids in determining just where the worst leaks are occuring is the infrared scanner on which escaping heat shows up as red, the worse the leak the brighter the color. Such studies can be made into photographs and submitted in reports, and can be retaken after the corrective measures have been taken to show their effectiveness. Even when the original drawings and specifications are available to show what was originally installed such scans can be useful as installation may have been poorly done, condensation within the walls may have rendered the insulation useless, or the insulation may have packed down in the case of loose types or shrunk in the case of sprayed or foamed types, leaving substantial gaps for free air passage.

While it is air passage through the walls and roofs that is usually attacked first, infrared scans will often show the greatest heat loss from infiltration around openings which is probably the first thing to correct.

Caulking

Most caulking materials are available in either knife grade for wide cracks and gun grade for narrow ones. They can be further classified into basic, intermediate and high performance groups lasting approximately up to five years, from five to twenty years and twenty years and over. On the low end are cord and rope types and those made of oil and resin. These are low in cost and fine for seasonal protection but will need constant replacement. The intermediate caulks are the rubbers, natural and synthetic, starting at the low end with latexes, with butyls in the middle and neoprene at the top. The high performance types are all synthetic including polyurethane, polysulfide and silicones. A building erected as late as 1970 and caulked with an intermediate grade could have major problems today with infiltration. All openings around windows, doors, trim, sills, cornice and verge boards and all through the wall piping, air conditioners, vents and intakes should be carefully checked.

When the space between the door and window frames and a masonry wall are caulked with oakum, an experienced man should crank 340-400 lin. ft. of opening per 8 hour day at the following labor cost per 100 lin. ft. :

	Hours	Rate	Total	Rate	Total
Caulker	2.1	$. . . .	$. . . .	$17.43	$36.60
Cost per lin. ft					.37

Where it is not necessary to pack in oakum, one man using a pressure gun should do about 600 lin. ft. per 8 hour day at the following labor cost per 100 lin. ft.

	Hours	Rate	Total	Rate	Total
Caulker	1.33	$. . . .	$. . . .	$17.43	$23.18
Cost per lin. ft					.23

Refer to Chapter 12 for the quantity of material that will be required for various sizes of joints.

Porch Repairs

To replace an existing porch floor with new 25/32"x3-¼" face soft wood, allow 2 carpenter hours and 1 laborer hour per 100 sq. ft. Painting not included. Cypress, Douglas fir, larch pine and redwood will resist decay and warping, and offer medium resistance to wear.

To replace a simple porch column, allow ½ hour to remove and shore existing column, 1 hour to replace column.

To replace a simple wood porch railing, figure a carpenter can remove and replace 15 lin. ft. per 8 hour day.

To erect an exterior utility stair of 2x10 stringers, allow 20 carpenter hours per average story in one run, 25 hours if stair has intermediate landing.

Roof Repairs

To Remove Built-Up Roofing.—To remove a 3 to 5 ply roof, figure ½ hour each of a roofer's and laborer's time per each square.

To Replace Built-Up Roofing.—Figure a crew of 5 per 8 hour day can install 18 squares of 5 ply roofing, 22 squares of 4 ply and 24 squares of 3 ply roofing over wood decks on low buildings.

To Replace Shingle Roofing.—To remove asphalt roof shingles and underlayment, figure 1 roofer and 1 laborer working together can strip around 20 squares per 8 hour day. A carpenter can install a square of 12"x36" strip asphalt shingles on plain double pitch or gable roofs without hips, valleys or dormers in 2 hours. New shingles are often applied over existing.

To replace sheet metal gutters, a tinner and a helper should replace 20 lin. ft. of gutter or 25 lin. ft. of downspout per hour.

To replace metal counterflashing, figure 1 tinner can replace 15 lin. ft. per hour providing a satisfactory reglet exists. If a reglet has to be chiseled into masonry, figure 1 laborer can cut 8 to 10 lin. ft. per hour. If an electric cutting saw can be used, figure 12 to 15 lin. ft. per hour.

Chimney Flashing Table

No. of Brick		Inches		To ⅓	Pitch of Roof To ½	To ¾
Wide	Long	Wide	Long		Lin. Ft. of Flashing	
2	2	18"	18"	11'	11'	12'
2	2½	18"	22"	12'	12'	13'
2	3	18"	26"	13'	13'	15'
2	3½	18"	30"	13'	14'	16'
2	4	18"	34"	14'	15'	17'
2½	4	22"	34"	15'	16'	18'
2½	4½	22"	38"	16'	17'	20'
2½	5	22"	43"	16'	18'	22'
2½	5½	22"	47"	17'	19'	23'
2½	6	22"	51"	18'	20'	24'

Material Requirements to Add a Gabled Dormer
(Roofing, Siding, Painting and Interior Finishes Not Included)

Materials & Size		5'	6'	Width of Dormer 8'	10'	12'
Rafter Headers	2"x6"	2-10'	2-12'	2-16'	4-10'	4'-12'
Rafters Plates	2"x6"	2-16'	2-16'	2-16'	2-16'	2-16'
Studs and Plates						
Side	2"x4"	2-10'	2-10'	2-10'	2-10'	2-10'
Side	2"x4"	2-12'	2-12'	2-12'	2-12'	2-12'
Front	2"x4"	7-10'	8-10'	11-10'	13-10'	11-12'
Gable	2"x4"	1-12'	1-12'	3-10'	2-10'	2-12'
Gable	2"x4"	1-8'	1-8'	1-10'	1-14'	1-14'
Rafters	2"x4"	6-8'	6-10'	6-12'	6-12'	6-14'
Ridge Boards	1"x6"	1-10'	1-10'	1-10'	1-10'	1-10'
Framing						
Total BM		179 BM	200 BM	237 BM	264 BM	283 BM
Sheathing	1"x6"	60 BM	70 BM	70 BM	85 BM	80 BM
Roof Boards	1"x6"	60 BM	72 BM	86 BM	100 BM	120 BM
Cornice Fascia	1"x6"	22 Lin. Ft.	24 Lin. Ft.	26 Lin. Ft.	28 Lin. Ft.	30 Lin. Ft.
Crown Mold	1"x3"	22 Lin. Ft.	24 Lin. Ft.	26 Lin. Ft.	28 Lin. Ft.	30 Lin. Ft.
Nails		10 lb.	11 lb.	12 lb.	13 lb.	14 lb.
Flashing	12" Wide	33 Lin. Ft.	35 Lin. Ft.	39 Lin. Ft.	42 Lin. Ft.	46 Lin. Ft.
Felt	30 lb.	1 Rl.	1 Rl.	1 Rl.	1 Rl.	2 Rl.
Window Frame	2/6x4/10	1	1	2	2	3
Window Sash	2/6x4/10	1	1	2	2	3
Labor To Install		26 Hours	27 Hrs.	32 Hrs.	36 Hrs.	40 Hrs.

Gabled Dormer Roof and Side Areas for ½ Pitch Roofs

	5' Width	6' Width	8' Width	10' Width	12' Width
Roof	60 Sq. Ft.	72 Sq. Ft.	86 Sq. Ft.	100 Sq. Ft.	120 Sq. Ft.
Wall	60 Sq. Ft.	70 Sq. Ft.	70 Sq. Ft.	85 Sq. Ft.	80 Sq. Ft.
Starter	10 Lin. Ft.	10 Lin. Ft.	10 Lin. Ft.	10 Lin. Ft.	10 Lin. Ft.
Ridge	12 Lin. Ft.	12 Lin. Ft.	12 Lin. Ft.	12 Lin. Ft.	12 Lin. Ft.

Weatherstripping

All doors and windows should be tightly sealed including those leading to un-heated attics, garages or basements. Check windows for sealing gaskets and tight closing latches. Interlocking devices at jambs and heads of doors are preferable to spring or pressure types and select a proper adjustable sill seal and threshold that will fit to floor finish.

Lineal foot cost of material weatherstrip for double hung windows will average as follows in zinc:

Descriptions	Thickness		
	1-⅜"	1-¾"	2-¼"
Side, head & sill strips	$0.40	$0.45	$0.50
Meeting rail strips	.60	.65	.75

Bronze will run about 100% more.

Single or double casement windows that swing in require a ¾" wide track strip on the hinged side, running in a groove in the sash. At the top, on the latch side, and where the sash meet, if in pairs, interlocking hook and flat strips are generally used. A trough strip and a hook strip is generally used at the bottom of the sash.

Weather strips for single wood doors will range from $4.50 for zinc to $7.00 for bronze per opening. If interlocking bronze rather than spring type is used, add $3.00 to the above figures. Extruded aluminum and neoprene spring stop mounted strips run $1.00 per foot for 1-⅛" x ¼" and $2.40 per foot for 1-¼"x25/32". Bottom rail gasket runs $1.35 per foot.

Labor Installing Weather Strips

	Hours
1⅜" D.H. sash to 3x5'6"	0.9
1¾" D.H. sash to 4x7	1.0
2¼" D.H. sash to 4x7	1.25
Single inswinging casement	1.3
Double " "	2.0
Single outswinging casement	1.0
Double outswinging casement	1.4
3x7x1¾" wood door, spring bronze	1.6
3x7x2¼" wood door " "	2.0
3x7x1¾" wood door interlock	2.5
1 pr. 1¾" French doors	2.3

It may pay to weatherstrip a door at the head of a stair leading to the attic. Chimneys should also be checked for tight fitting dampers.

Doors and windows in exposed locations should be fitted with storm sash if operable and double glazing if fixed. Openings on the north side and on the side toward prevailing winter winds especially need this extra protection.

A carpenter with helper should set an average preglazed storm window in 1½ hours. Storm windows, up to 12 sq. ft. in area, will average about $50.00 per unit installed. Larger units cost around $65.00 installed. Smaller units can be handled by a carpenter working alone.

Doors in exposed locations are best protected by adding a vestibule rather than just a storm door. In commercial work with heavy traffic a revolving door will offer the best protection although the initial cost will be $15,000 or more.

Windows that are not used for ventilation may be permanently sealed, but it should be remembered that if sealing it means having to rely completely on running air cooling in the summer months this could be self-defeating.

Fixed windows can often be fitted with an interior storm window. This is especially adaptable to curtain wall construction where tubular frames can be fitted with a spacer, new metal sash bars and an additional layer of glass, which can be of the solar type further cutting down on heat penetration and glare. Such additions can be added entirely from the inside but because the area between the old and new glass will be hermetically sealed only experienced installers should be considered. Pittsburg Plate Glass has a patented system which they will install on a custom basis.

Several manufacturers offer special replacement windows which can be pivoted, projected, double hung or casement types. To install these one first removes the existing window, but not the frame. A factory assembled subframe is then fitted over the existing frame and secured to it. The new window sash is then installed within the new subframe. These units are usually aluminum, can be had preglazed in either single or double glazing or with part of the window blocked off with insulated panels, and often can be set, window by window, from the inside. A two man crew should remove the existing sash, install the new subframe and add the new sash in approximately three hours per opening.

Unwanted windows had best be eliminated altogether by either bricking them up or covering them with insulated panels. A variation of this, which can retain partial light and ventilation, is to add an insulated panel over part of the window, usually the upper half. These are usually porcelainized metal laminated over a polystyrene core with an aluminum frame factory assembled to fit to the existing window frame. One such manufacturer is Mapes Industries, Lincoln, Nebraska who rate their add-on panels 1¼" thick at an "R" value of 6.56.

Head

Meeting Rail

Jamb

Insulated Panel Over Double Hung Window*

---

*Courtesy Mapes Industries

Solar Heat Controls

Heat gain by the rays of the sun penetrating into a building is a blessing in the winter but not in the summer. There are ways to take advantage of this by cooperating with the season. The sun is high in the summer, low in the winter and exterior screens and shades, such as canopies, awnings, operable shutters and even a few well placed deciduous trees, can be so placed that the summer rays are excluded while the low winter rays are welcomed in. Excluding the sun from the outside rather than blocking it once it has entered the inside should be the first consideration.

Continuous canopies will run around $5.00 per sq. ft. while individual window ones will cost $150 in canvas, $250 in aluminum per unit. With a three foot overhang two men should erect 80 lin. ft. of running canopy or four individual ones per day.

If windows are to be reglazed the sun effect can be counteracted by using tinted or reflective glass. Glass with a fused reflective coating will cost about $3.00 per sq. ft. more than plain plate.

Existing glass can be coated with a solar rejecting film for around $1.60 per sq. ft.

Insulation

Chapter 12 has a discussion on the types of heat loss and the insulation materials available to counteract them. In retrofitting ones choices are somewhat limited.

Ceilings as we have seen are prime areas for adding insulation. Few would argue that 6" is probably easily justifiable north of St. Louis although 3½" was standard just a few years ago. In most residences gabled roofs provide attic space where additional batts and blankets can be added. Where some insulation is already in place, blankets without any cover can be added so that no vapor barrier will be introduced which might cause condensation problems. 3" thick rolls will cost around 17¢ per sq. ft. and 6", 24¢. The cost of installing will be about the same with one man completing 1200 to 1400 sq. ft. per day.

Attics can serve as a buffer between outside conditions and interior spaces most effectively if they are ventilated. This is usually designed to allow gravity to carry off unwanted build-ups by providing low inlets in the eaves which exhaust through the roof at the high point. Eave vents 4" x 16" cost around $7.00 and will take one carpenter some 15 minutes to set. Some metal eaves are perforated to allow air to enter. Ridge vent strips will cost $1.50 per lin. ft. and will take one man one hour to install 20'.

Where no accessible attic space exists insulation must be added on top of the roofing. Assuming a relatively flat roof in good condition it is possible to add new non-absorbing rigid insulating board, such as Styrofoam, over the existing roofing, mop it in and hold it down with a new application of stone. This will help protect the existing roofing, but if there is any doubt as to the condition of what is there it should be properly repaired as the insulation is not to take the place of roofing membranes. The amount of insulation added can vary from 1½" to 3" in single sheets but can also be built up by adding additional plies. Should the old roofing eventually need replacement the stone can be removed from the insulation and new roofing applied over it. Roof drains and flashings of course would have to be reset to the new roof level.

Another possibility for covering existing roofs, and this is not limited to level ones, is to spray on a polystyrene or urethane insulating coat and then cover this with a butyl or elastomeric roofing coat. Costs of applying such coats and roofing will be similar to new work as discussed in Chapter 12 with the addition of costs to adapt adjacent drains and flashings to the new levels.

Walls present special problems. If the exterior and interior finishes are to remain there is little alternative to adding insulation to whatever cavity space exists in the center of the wall. In wood stud construction this is relatively simple. To get into the cavity between the studs it is necessary to cut holes between each stud at least every story height. These are usually cut from the outside by removing a strip of siding. Bevel and shiplap can be pried up easily, but tongue and grooved will have to be broken out. Most walls will have a sheathing course behind the siding. This will have to be drilled with holes large enough to accomodate the nozzle for blowing in the insulation, usually 2" to 3". Once the holes are cut the cavity must be plumbed to determine if any obstructions are present, and if any are found additional holes will have to be drilled. Material costs for fiberglass or mineral wool will run from 40¢ to 50¢ per cu. ft. One carpenter can install some 400 cu. ft. per day. To this must be added the cost of removing and replacing siding.

Where an owner contemplates new interior finish in a building or space the existing walls can be covered with rigid insulation board and then faced with whatever new finish is contemplated. In older buildings where the plaster is in poor condition this may be the better alternative. It will mean replacing all trim, but in removing the baseboards it will also allow batts to be packed between the floor joists to keep air currents from flowing down the outside walls and out over the floor area. Rigid board of fiberglass or extruded polystyrene will cost around 23¢ per sq. ft. and will have "R" values of 4.3 and 5.4 respectively. One carpenter should install 700 to 800 sq. ft. per day.

Wood framed structures can also be resided with vinyl or aluminum siding with insulation fitted to the back. Adding the insulation will cost around 10¢ per sq. ft. so that vinyl will run 75¢ and aluminum 85¢ per sq. ft. and one carpenter can install 250 sq. ft. per day.

Masonry veneer walls can have the cavities filled much the same as wood sided walls, removing brick to allow access. Patching a masonry wall will be more obvious than a wood one. Where the brick has concrete block as a back-up it will be easier to add rigid insulation to the inside face. If the wall is at all irregular it will have to be furred, but the air space will contribute to the insulation value.

Block walls can also be foamed by drilling block every other unit on four foot vertical runs. Foaming is tricky and should be subcontracted to someone experienced in its application. Cost will be around 45¢ per inch of thickness installed per sq. ft.

Basement and crawl space walls, especially those with portions above grade should be insulated. Rigid board or sprayed on types are applicable here although the sprayed on type will need protection if the space is subject to much traffic. The same treatments along with batts or blankets can be continued under the floor construction above these spaces. Ducts and piping should also be insulated.

INTERIOR WORK

In remodeling work it is especially important that new wood used with old be dry so that shrinkage will be minimized. Wood should be ordered that has been kiln dried. Edge grained wood has a shrinkage just half that of flat grained wood, and should be used.

Shrinkage is considerable in birch, gum, maple and oak; is average in cherry, hemlock, pine and walnut; and minimal in cedar and redwood.

In extending or closing up doorways and openings in existing walls, often stud dimensions will not match the existing. Special lumber or furring will be required.

New studs are ⅛" less in size than those in use a few years ago. In much older buildings studs will be found that are not even sanded. In demolishing old work, plan ahead to see if some of the old lumber should not be salvaged. It may be harder to work with, but the size may be irreplaceable.

Often contractors are asked to judge what caused cracks in walls and ceilings and whether to repair, replace or reface them. Many cracks are not attributable to the workmanship of the cracked surface, but to faults in the materials to which they are applied. Such cracks can be caused by lumber shrinkage, foundation settlement, inadequate bracing and expansion and contraction.

Plaster can crack, spall and disintegrate because of poor workmanship. A common condition occurs when the white coat is applied over a base coat that has been allowed to dry out too much. When the white coat is applied, the base coat soaks the moisture from the finish coat and a spidery pattern of cracking occurs. If the surface of the plaster seems soft, the plaster has probably been oversanded. If cracks occur at random on the white coat, it is probably too thin a coat.

Structural cracks usually show up in the first few years of occupancy. Unless major defects are involved, they reach an equilibrium and can be successfully patched if the plaster on either side is not loose. If the crack is less than ½" wide, dig out an inverted "V" joint, moisten the adjacent areas to reduce suction, and fill with prepared patching compound in two stages. For larger cracks, the crack should be made wider to accommodate a piece of metal lath. The resulting channel is then filled with a three coat application.

When the plaster is soft or cracking is prevalent over a large area, it is best to cover the plaster. This can be done with plasterboard, canvas, acoustical tile, or any of the decorative hardboards and plywoods. If the plaster is over sound wood lath, the applied board can be nailed through the plaster into the lath or stud. If the backing is unsound, a new furring system should be applied. Furring strips are nailed at right angles to studs or joists at intervals of 12", 18" or 24", depending on module of new finish.

Walls.—To strip a wall of plaster and leave ready to replaster if masonry, or relath and plaster, figure 1 laborer will strip from 60 to 70 sq. ft. per hour. Add for cost of clean up and debris removal if this is to be included.

Refer to Chapter 14 for costs on new plaster work.

To add ½" gypsum board over stripped wood studs, figure 1 carpenter will place around 50 sq. ft. per hour in the average room. If large unbroken expanses are involved, he may double this. If gypsum board is not a pre-finished type but is to be painted, it will require taping. Figure a carpenter will tape 100 lin. ft. of joint, which, in usual applications, would cover the needs of 300 sq. ft. of wall in about 4 hours.

When pre-finished plywood is specified for first class work, gypsum back-up board should be used as an underlayment.

To add wallboard over an existing wood stud and plaster wall, the wall should first be furred. Furring strips should be nailed at right angles to studs. Any convenient width may be used, but it must be wide enough to provide a nailing base to nail all edges of the decorative strips at least every 6". Furring may be of No. 2 soft wood. The success of the wall will depend largely on the success of leveling the furring strips.

With studs 16" on center, a carpenter should cut and place 450 to 500 lin. ft. of furring strips per 8 hour day at the following cost per 100 lin. ft. :

	Hours	Rate	Total	Rate	Total
Carpenter	1.8	$. . . .	$. . . .	$16.47	$29.65
Cost per lineal foot:					.30

When applying wallboard in smaller rooms requiring considerable cutting and fitting around doors, windows, etc., a carpenter should fit and place 400 sq. ft. per 8 hour day, at the following labor cost per 100 sq. ft.:

	Hours	Rate	Total	Rate	Total
Carpenter	2	$. . . .	$. . . .	$16.47	$32.94
Cost per sq. ft.					.33

In large rooms with unbroken walls, this cost would be reduced to about 20¢ per sq. ft.

To add thin board, such as 5/32" prefinished over an existing plaster base with adhesive, figure labor costs as follows per 100 sq. ft.:

	Hours	Rate	Total	Rate	Total
Carpenter	3.2	$. . . .	$. . . .	$16.47	$52.70
Cost per sq. ft.					.53

To remove ceramic tile from an existing wall, figure one laborer will remove around 30 to 40 sq. ft. per hour, not including removal of debris. As old existing tile is undoubtedly set in a mortar base, it will be necessary to set the new in the same way or to replaster the wall to provide a smooth base for applying tile with adhesive. 4"x4" tile applied with mortar will run about $4.75 per sq. ft. Adhesive set tile will run in the neighborhood of $4.00 per sq. ft. to which must be added to the cost of a new base.

Ceilings

To remove plaster from ceiling joists, figure 1 laborer will clear about 50 sq. ft. of area per hour. If the ceiling is suspended and the hangers are also to be removed, figure about 30 sq. ft. per hour.

To strip a plastered ceiling of acoustical tile, figure 1 laborer can clear 75 sq. ft. per hour. These figures do not include removal of debris.

Often a contractor is asked to patch plaster over concrete or old plastered ceilings. This can be accomplished with the use of a product such as Plaster-Weld* which is brushed, rolled or sprayed on, and in an hour or so, recoated with the desired plaster finish.

To add 12"x12" wood fiber tile to existing plaster ceiling, figure 2 mechanics working together on a scaffolding can set 800 tiles per day at the following labor cost per 100 sq. ft.:

	Hours	Rate	Total	Rate	Total
Mechanics	2	$. . . .	$. . . .	$16.47	$32.94
Cost per sq. ft					.33

Flooring

To Remove Wood Subflooring.—Where existing subfloors are badly worn and must be removed, figure 1 carpenter and 1 laborer can remove 100 sq. ft. per hour at the following cost:

	Hours	Rate	Total	Rate	Total
Carpenter	1	$. . . .	$. . . .	$16.47	$16.47
Laborer	1	. . . .	. . . .	12.54	12.54
Cost per 100 sq. ft.			$. . . .		$29.01
Cost per sq. ft.			. . . .		.29

*Larsen Products Corp.

To Replace Wood Subfloors.—In small cut-up areas, a carpenter should lay between 700 and 800 BF per 8 hour day of 1"x6"s or 1"x8"s at the following labor cost per 1000 BF.

	Hours	Rate	Total	Rate	Total
Carpenter	10.6	$. . . .	$. . . .	$16.47	$174.58
Laborer	6.0	. . . .	. . . .	12.54	75.24
			$. . . .		$249.82

To Add Plywood Over Existing Subfloors.—Labor for 100 sq. ft. of plywood added over existing subfloor would be as follows:

	Hours	Rate	Total	Rate	Total
Carpenter	1.0	$. . . .	$. . . .	$16.47	$16.47
Laborer	0.5	. . . .	. . . .	12.54	6.27
Cost per 100 sq. ft			$. . . .		$22.74
Cost per sq. ft			. . . .		.23

To Add New Finished Flooring.—Once an existing subfloor is put in good condition, replaced, or given a good underlayment, the following installed costs per sq. ft. may be used for various floor finishes:

1x1 Ceramic Tile, Adhesive Set	$4.00-5.00
Asphalt Tile	.65- .90
Vinyl Asbestos 1/16"	.80-1.00
Cork Tile	1.20-1.90
Vinyl Tile .050"	1.00-1.10
Vinyl Sheet .070"	1.30-1.75
Rubber Tile	1.50-1.85
9x9x5/16" Oak Parquet	2.75-2.90

Trim

To Remove Existing Wood Base.—A carpenter and helper working together should remove a one-piece wood base at the rate of 100 lin. ft. per hour.

	Hours	Rate	Total	Rate	Total
Carpenter	1	$. . . .	$. . . .	$16.47	$16.47
Laborer	1	. . . .	. . . .	12.54	12.54
Cost per 100 lin. ft			$. . . .		$29.01
Cost per lin. ft			. . . .		.29

To Replace Wood Base.—To place a single member base plus carpet strip, allow $1.20 per lin. ft. for 3½, $1.50 for 4½.

To add new interior trim, the following installed prices may be used.

Wood picture molding	$1.10 per lin. ft.
Wood chair rail	.95 " " "
Wood cornices 1"x8"	1.50 " " "
Panel strips	.60 " " "

To Replace Window Sash and Trim.—A carpenter should fit and hang, complete with sash cord, chain and weights or counterbalances, 10 to 12 single sash per 8 hour day, at the following labor cost per pair:

	Hours	Rate	Total	Rate	Total
Carpenter	1.5	$. . . .	$. . . .	$16.47	$24.71

A carpenter should fit and hang 10 to 13 casement sash per 8 hour day at the following labor cost per pair :

	Hours	Rate	Total	Rate	Total
Carpenter	1.4	$. . . .	$. . . .	$16.47	$23.06

A carpenter should set a one member window trim and cap at 1.5 hours per window at the following labor cost :

	Hours	Rate	Total	Rate	Total
Carpenter	1.5	$. . . .	$. . . .	$16.47	$24.71

To Replace Wood Doors and Trim.—A carpenter should set a plain wood jamb for a 3'x7' wood door in one hour, and trim same both sides in 1.5 hours. To hang the door will take another hour for a total cost as follows :

	Hours	Rate	Total	Rate	Total
Carpenter	3.5	$. . . .	$. . . .	$16.47	$57.65

A frame set for 3'x6'-8" opening will run about $15.00 in pine, $20.00 in birch. If door unit is prefit and knocked down, a carpenter should assemble and install it in one hour.

To Set Factory Assembled and Finished Kitchen Cabinets.—When setting factory assembled and finished kitchen cabinets, about 0.15 hours of carpenter time for each sq. ft. of cabinet face area should be allowed.

To Fit and Place Closet Shelving.—A carpenter should fit and place approximately 125 sq. ft. of closet shelving per 8 hour day at the following labor cost per 100 sq. ft:

	Hours	Rate	Total	Rate	Total
Carpenter	6.4	$. . . .	$. . . .	$16.47	$105.41
Cost per sq. ft					1.05

To Install Rod in Closet.—A carpenter should install about 3 hanging rods per hour at the following labor cost per rod :

	Hours	Rate	Total	Rate	Total
Carpenter	0.33	$. . . .	$. . . .	$16.47	$5.50

To Set Medicine Cabinets.—A carpenter should set 12 face mounted cabinets per 8 hour day at the following labor cost per cabinet :

	Hours	Rate	Total	Rate	Total
Carpenter	.66	$. . . .	$. . . .	$16.47	$10.98

To Set Bathroom Accessories.—A carpenter should set 30 to 34 bathroom accessories per 8 hour day at the following cost per accessory :

	Hours	Rate	Total	Rate	Total
Carpenter	.25	$. . . .	$. . . .	$16.47	$4.12

To Set Finish Hardware

Type	Carpenter Hrs. Each
Rim Lock	.5
Mortised Lock	1.0
Cylinder Lock	.5
Front Entrance Cylinder Lock	2.0
Surface Door Closer	1.0
Concealed Closer	3.0
Sash Lifts and Lock	.5
Kickplates	1.0

To Set Cabinet Hardware

Type	Carpenter Hrs. Each
Surface Bolts	.05
Offset Bolts	.1
Catches	.1
Drawer Pulls	.06
Knobs	.05
Rim Catch	.1
Mortise Lock	.4

PAINTING

To repaint exterior wood siding and trim:

	No. Sq. Ft./Hr.	Hrs./100 Sq. Ft.	Sq. Ft./Gal.
First Coat	115-125	.85	500-550
Second Coat	125-135	.75	575-625

If existing surface is in poor condition, add:

	No. Sq. Ft./Hr.	Hrs./100 Sq. Ft.	Sq. Ft./Gal.
Burning Paint Off	30-40	2.75	
Prime Coat	100-110	.95	400-450

To add 2 coats to wood shingles:

	No. Sq. Ft./Hr.	Hr./100 Sq. Ft.	Sq. Ft./Gal.
First Coat, Oil	115-125	.82	250-300
Second Coat, Oil	150-160	.65	375-425
First Coat, Stain	75-85	1.2	120-150
Second Coat, Stain	115-130	.82	200-225

To prepare interior work:

Type of Work	No. Sq. Ft. Per Hr.	Hrs. Per 100 Sq. Ft.	Sq. Ft. Per Gal.
Sanding & Puttying	190-200	0.5	
Washing Enamel	85-90	1.15	
Removing Varnish with			
Liquid Remover	30-35	3.00	150-180
Wash &			
Touch Up Varnish	165-175	.60	
Add 1 Coat Varnish	100-110	.95	575-600
Burn Off Interior Trim	20-25	4.00	
Burn Off Flat Surfaces	30-35	3.00	
Spackle Over			
Flat Surface	60-65	1.60	140-150
Filling Wood Floors,			
Wiping	170-180	.57	425-450
Taping, Spotting & Sanding			
Gypsum Wallboard	90-110	1.00	
Wash Off Calcimine	115-125	.83	
Wash Smooth Wall	145-155	.67	
Remove Old Wallpaper	65-75	1.45	
Wash Off Glue	125-135	.77	

To paint interior work:

Type of work		No Sq. Ft. Per Hr.	Hrs. Per 100 Sq. Ft.	Sq. Ft. Per Gal.
Wood Floors				
1st Coat Paint		290–300	0.35	525–550
2nd Coat Paint		260–270	0.38	425–450
1st Coat Stainwax		390–400	.25	525–550
2nd Coat Stainwax		480–490	.21	600–625
Varnish, Each Coat		300–310	.33	500–550
Buffing		390–400	.25	. . . .
Waxing		200–210	.50	1,050–1,075
Walls				
Sizing		350–400	.27	600–700
Sealer		200–225	.45	575–625
Flat Finish	1st Coat	165–175	.60	500–550
	2nd Coat	175–185	.55	575–625
Latex	1st Coat	140–160	.67	300–350
	2nd Coat	140–160	.67	300–350
Canvas		50–60	1.82	
Wallpaper				
	Per Single Roll	2½-3		
Trim				
Enamel				
Undercoat		90–100	1.10	375–400
Finish Coat		100–110	1.00	475–500
Stain Woodwork		215–225	.45	700–725
Shellac Woodwork		215–225	.45	700–725
Varnish Woodwork		165–175	.60	425–450
Wax and Polish		100–110	.95	550–600

PLUMBING

The following may be used as a guide for replacing existing fixtures. No roughing in or pipe replacement is figured. Fixture cost and roughing in is discussed in Chapter 17.

	Hours
To Replace Bathtub	8
To Add Shower-Surface Mtd.	2
To Replace Single Kitchen Sink	5
To Replace Double Kitchen Sink	7
To Replace Laundry Tub	5
To Replace Lavatory	5
To Replace Shower Stall	12
To Replace Water Closet	7
To Replace Water Heater	4

ELECTRICAL

To rewire, pulling in old conduit:
                    Per outlet—2.5 hours.
To add new, using BX cable in open wall:
                    Per outlet—1 hour.

Budget figures for remodeling residential work using armored cable and including labor and materials but not fixtures or patching, unless so noted, would run somewhat as follows:

To add a ceiling outlet $70.00
To add a wall switch 55.00
To add a wall outlet 65.00
Wire and connect a dishwasher... 80.00
Wire and connect a clothes washer.. 90.00
Wire and connect a disposal 100.00

To Furnish, wire and connect the following:

40 gal. electric water heater $140.00
52 gal. electric water heater 170.00
82 gal. electric water heater 220.00

Expanded programs to retain the shell of structures and to renovate them to meet the current needs of industry, as a method of reducing construction costs, has created a need for new products and methods for the preservation, adaptation and modification of existing components so they will integrate with the design concept and function of the new facility.

To achieve this goal, the epoxy industry has responded to the industry need by providing the research and guide specifications for the education of the Architect and Contractor in the proper selection and use of the various products which have been developed.

The use of expoxy products has produced excellent results in instances where efforts were closely coordinated with the manufacturers recommendations or the work was performed on a subcontract basis by a licensed representative of the manufacturer. It is of extreme importance to select the products which precisely meet the requirements of the specification and to then obtain costing information from the manufacturer as there exists wide variances in material costs and labor productivity.

In situations where it is found that the Architect or Owner is not well educated to the capabilities of the modern products and methods of application offered by the epoxy industry, the Contractor is in the position of offering a significant contribution by making this information available so that it may be considered as one alternative toward reducing construction costs, or as a remedy in developing solutions to problems encountered with concrete and masonry in restoration or renovation projects. Epoxy also serves as a valuable tool in restoring the structural integrity to structures.

SURFACE PREPARATION

Surface cleaning and/or preparation is frequently required to remove laitance or foreign matter so that concrete or masonry surfaces can be restored or prepared to receive some type finish requiring a positive bond with the existing surface. Sandblasting is also used extensively in producing an architectural finish for exposed concrete.

Sandblasting is considered as the most efficient and economical method of surface cleaning when using epoxies since the quality of the finished work is predicated upon proper preparation of the existing surface. While sandblasting is usually performed by a specialty subcontractor, the General Contractor/Owner should be aware of the need of providing protection for persons, adjacent properties, glass and metal surfaces, ornamental structures, trees, power lines, or other items subject to damage by the sandblasting operation. Clean up at the conclusion of each work day is another issue which should be included. By endorsement on an existing property damage insurance policy, or the issuance of a special policy covering this specific risk, protection can be obtained from the hazards of this work.

SAND BLASTING

Light penetration including protection and cleaning	$.50 sq. ft.
Heavy penetration including protection and cleaning	$1.50 to 1.60 sq. ft.
On Metal Surfaces - light penetration incl. protection & cleaning	$1.10 sq. ft.
On Metal Surfaces - heavy penetration incl. protection & cleaning	$1.75 sq. ft.

The following list of products and description is but a selection of products available to the industry:

A. Epoxy Penetrating Sealers. This system is a two component resin available in 15%, 30%, and 50% solids. Its use is to waterproof concrete, brick mortar, cement block and to present itself as a protective zone to increase the substrates resistance to damage caused by fuels, oils, alkali conditions, most known mild acids, de-icing chemicals, salts and aggressive waters. When properly applied on concrete bridge decks, skid resistance is not altered and some savings can be realized by reducing the amount of de-icing chemicals required per application.

This method is similar to the application of a coat of silicone on an exterior concrete or masonry wall surface with certain superior advantages, one of which is the epoxy provides a chemical adhesion in addition to the mechanical adhesion which is found in both the epoxy and silicone.

Application may be accomplished by rollers, brush, or the spray method.

General Guide Costs For Penetrating Sealer

	Cost Per Sq. Ft.
Clean surface by sand blasting method	$ .50
Epoxy sealer material	.10
Apply epoxy sealer	.40
TOTAL COST	$1.00

B. Epoxy Injection Systems. This system is a two component resin used to "weld" concrete together which has cracked or broken. Its other principal uses are: filling honeycombed concrete, "welding" cold joints, repairing delaminations, and stopping water seepage through foundation walls, floors, beams, decks etc. Some injection materials can be used underwater to structurally repair piling, retaining walls, water tanks, tunnel linings etc. Some products can cure on moist surfaces down to 33°F (10°C) and on dry surfaces down to temperatures of 0°F (-18°C) without any use of heaters or lamps to initiate curing.

1. Cartridge Method

For use on small hair line cracks with no honeycomb or delamination

	Cost Per Lin. Ft.
Material - 6 oz. tube @ 20.00 (4 L.F./Tube)	$ 5.00
Labor	20.00
TOTAL COST	$25.00

2. Injection Machine Method

For repair (welding) of structural cracks, delaminated concrete, or voids caused by honeycombs.

At best, this is an exercise in hazardous estimating as there is no way of making a determination, with any degree of precision, of interior quality of the concrete.

It is strongly suggested that several sample areas be repaired and, predicated upon the cost information developed from the tests, a unit price per gallon of epoxy material be used, or a square foot price be established, as the basis of determining the ultimate cost.

General Guide Costs For "Welding" Cracks In Concrete

Material cost is 45.00 per gallon which, under average conditions, will "weld" cracks not in excess of 1/16" at the rate of 40 L.F./Gal.

	Cost Per Lin. Ft.
Material	$ 1.13
Injection Machine Rental - 500.00 week/100 L.F. week	5.00
Injection Machine Operating Costs - 40 hrs. @ 4.00 = 160.00	1.60
Injection Machine Operator - 40 hrs. @ 17.05 = 682.00/wk/100L.F.	6.82
Labor Cost 20/L.F. per day = 8 hrs. 17.05 = 136.40/20	6.82
TOTAL COST	$21.37

C. Epoxy Mortar Systems. This system is a two component resin with an aggregate added to the resin. Some of the reasons the aggregate is used is to 1) reduce the cost of the application, 2) provide a textured surface, 3) increased abrasion resistance, 4) increase the mortars compatibility to the physical properties of concrete etc. Some epoxy products have the ability to be extended with aggregates to as much as 15 parts aggregate to 1 part resin for thicker applications and as much as 8 parts aggregate to 1 part resin for thin applications (parts are by weight). Some products are able to be used as a thin overlayment, as thin as ⅛" (.125") and feather to zero. Some products can be placed ¼" thick and be able to cure to 0°F (-18°C) on dry surfaces and 33°F (1°C) on moist surfaces. Some products can cure even below zero without the need of heaters or lamps to initiate curing.

This system is effective as an overlay over existing concrete slabs and requires the following work procedure with the applicable costs as follows:

General Guide Costs Applying ⅜" Mortar System

1. Sand blast or pressure clean surface	$ .50 sq. ft.
2. Repair spall areas - Material 2.00 + Labor 3.00	5.00 sq. ft.
3. Crack "Welding"	25.00 to 45.00 L. F.
4. Apply penetrating sealer	.98 sq. ft.
5. Apply ⅜" overlay	4.50 sq. ft.

Actual extent of work required can only be determined by a detailed site investigation, and because of the variance that may be encountered in the condition of an existing concrete slab, the tendering of a fixed price proposal for work of this nature is very risky and should not be attempted by an estimator who is not experienced in this type of work.

D. Epoxy Masonry Mortar. This system is a two component resin used instead of cement mortar for cement or concrete block construction. The use of this product saves construction labor and, when used in wall construction considerably increases the wall strength reducing the risk of toppling due to strong winds.

The use of the epoxy mortar system eliminates the cost of the cement mortar and this should be given consideration by the estimator in the preparation of the masonry cost estimate. The epoxy masonry mortar is applied with bulk type caulking guns in a ribbon form on each horizontal and vertical surface of the block with the masonry units then pressed into position forming a "welded" wall with virtually no visable joints. Block laying production will increase by about 100%.

The cost of epoxy mortar will be considerably more than cement mortar which is caused by the cost of the epoxy material and the use of special (round) sand which is required for this type of installation.

In general masonry walls constructed by this method are not frequently found as part of the usual construction project because of the cost and resistance encountered from the masons for such a radical departure from the normal building practice.

It would also be recommended to include an additional mortarman in the estimate to cover the costs for loading and cleaning of the bulk caulking guns.

E. Epoxy Acid Mortars. This system is a two component resin with aggregates similar to the epoxy mortars described above, except that it is used as a protective overlay of concrete where high resistance to acids and industrial chemicals is required.

F. Epoxy Coatings. Coatings are two component systems and are used in many applications. Some of the more popular uses are as follows:

    1) Floor Coatings
    2) Ramp Coatings
    3) Pool Coatings
    4) Tank Coatings
    5) Acid Coatings
    6) Food Processing Coatings
    7) Underwater Coatings
    8) Chemical Liner Coatings
    9) Abrasion Resistant Coatings
   10) All purpose Coatings

Many coatings are available in a variety of colors such as green, red, tile red, blue, white, light blue, dark blue, dark gray, and concrete gray. Some epoxy manufacturers will even special formulate for special colors. Some products will cure on moist surfaces down to 32°F (0°C) and on dry surfaces to 20°F (-7°C).

EPOXY COATINGS

Epoxy coatings require basically the same surface preparations as epoxy mortars. Knowing what preparations are required, sand blasting, steel shot blasting, etc. one can use the same costs per square foot that is used for the cleaning process.

The costs for cleaning substrates are either already known by the company doing the estimate as a result of historical costs, or the costs can be derived from other chapters in this book.

To estimate the quantity of 99 to 100% solids epoxy coatings required to cover a given area one can use the following formula:

$$\frac{\text{Square foot of area} \times 12 \times \text{Mil thickness}}{231}$$

Note that one mil is equal to .001", ten mils is .010", 100 mils is .100" etc. The following chart gives the coverage for 1,000 square feet.

Mils	Gallons Required per 1000 sq. ft.	Mils	Gallons Required per 1000 sq. ft.
1	.052 gal.	19	.987 gal.
2	.104 gal.	20	1.039 gal.
3	.156 gal.	21	1.091 gal.
4	.208 gal.	22	1.143 gal.

5	.260 gal.	23	1.195 gal.
6	.312 gal.	24	1.247 gal.
7	.364 gal.	25	1.299 gal.
8	.416 gal.	26	1.351 gal.
9	.468 gal.	27	1.403 gal.
10	.519 gal.	28	1.455 gal.
11	.571 gal.	29	1.506 gal.
12	.623 gal.	30	1.558 gal.
13	.675 gal.	35	1.818 gal.
14	.727 gal.	40	2.078 gal.
15	.779 gal.	45	2.338 gal.
16	.831 gal.	50	2.597 gal.
17	.883 gal.	55	2.857 gal.
18	.935 gal.	60	3.117 gal.

Most epoxy manufacturers have epoxy coatings available in a variety of colors. Some manufacturers will even special formulate colors with little or no added cost.

General Cost Estimate For 4 Mil./1,000 Sq. Ft.

Epoxy Material 45.00 Gal. x .208 =	$ 9.36
Labor - Painter (w/roller) 4 hrs. @ 17.43	69.72
	$79.08

G. Epoxy Bonding Systems. This system is a two component system whose principal use is to bond two products together, such as concrete to concrete, steel to concrete, wood to concrete, steel to steel, steel to wood, wood to wood etc. (their uses are not limited to the above). Some products can bond newly placed concrete to old concrete. Some bonders will cure down to 0°F (-18°C).

PRODUCT SELECTION

To estimate work proposed using epoxies it is very important to understand the limitations of the product selected for use.

The selection of a product whose curing capabilities are limited to 40°F to 50°F and rising could force the contractor into expensive aggregate, substrate and epoxy preheating, artificial environment heating and/or gas fired reflective or electric infra-red heating costs just to initiate curing of the epoxies. Likewise if the epoxies selected are moisture sensitive expensive drying will be required if drying is in fact able to be accomplished.

Also it is important to determine if there are any costs for the technical service from the manufacturer if required. Some companies charge a significant daily amount plus transportation for a "factory trained individual".

EQUIPMENT

The following is a general list of tools required for epoxy work (this list is a guide of most used tools and should not be considered to be a complete list).

BASIC TOOL AND EQUIPMENT GUIDE

Substrate Cleaning Equipment:

    a. Sand Blaster
    b. High Pressure Water Blaster
    c. Grinders

    d. Scabblers
    e. Broom, Mop and Pails
    f. Automatic Scrubbers

Measuring Tools:

    a. Scales
      1. Flat plate type up to 200 lbs.
      2. Gram or 10th pound balancing type
    b. Tape measure
    c. Empty 2, 5 and 6½ gallon pails

Mixing Equipment:

    a. Barrel Type Cement Mortar Mixer (entire
       drum turns with blades welded
       to inside of drum)
    b. Drill and Mixing Tool (200-600 rpm)
    c. Wooden Mixing Paddles

Application Equipment and Tools

    a. Airless Sprayer
    b. Air Sprayer
    c. Power Screed
    d. Manual Screed
    e. Squeegees
    f. Trowels, flat, various sizes
    g. Coving Trowels, ½, ¾ and 1 inch radius
    h. Roller Pans
    j. Fiber Bristle Brushes with Wood Handles
    k. Paint Brushes

Miscellaneous Items:

    a. Hand Tools, Pliers, Hammers, Screw drivers, etc.
    b. Kneeling Boards or Pads
    c. Plaster Spiked Shoes
    d. Rags and Paper towels
    e. Plastic or Masking Paper and Masking Tape
    f. Wet Film Gauge

Safety Equipment:

    a. Respirator for Organic Vapors
    b. Goggles or Face Shield
    c. Rubber or Plastic Gloves
    d. Skin Cream

TECHNIQUES

Injection.—There are three generally accepted methods of injecting epoxies into concrete cracks, honeycombing, delaminations, cold joints etc. and the selection of the method depends in great part on the quantity of injection required and/or the quality of injection required.

The basic difference between quantity and quality referred to herein is that quantity may be to fill voids such as honeycombing adequately to accomplish mass and quality might infer fine (.002") hair line cracks filled with a 95% to 100% integrity requirement.

Method A - Cartridges.—Some manufacturers provide an injection resin in plastic cartridges not unlike caulking tubes. They do differ from caulking tubes in that the "A" & "B" Components must be mixed immediately prior to their use within the tube itself.

Method B - Pressure Pot.—A pressure pot is a paint type pressure vessel used to inject epoxies. The major draw back to this type of procedure is that the epoxies must be mixed in bulk prior to injection and that the injection must be completed prior to the epoxies gelling within the container and lines.

The pressure pot method is practical where fairly large quantities of epoxies must be injected and it is known that no slowing of the injection will occur. This method is particularly adaptable to the filling of void areas such as honeycombing and large cracks such as delaminations.

Method C - Injection Machine.—An injection machine is necessary where small cracks (.125" and smaller) are to be injected, where there is a necessity to inject small (.002" and larger) cracks, and where the injection procedure is steady and slow. There are many commercially available injection machines on the market. The use of any type machine with any given epoxy product should be endorsed by the epoxy manufacturer. The injection machine must be able to properly proportion the "A" & "B" Components, completely mix the components and be able to inject the epoxies at the epoxy manufacturers recommended sustained operating pressures.

Any injection machine considered should be as self contained as possible to eliminate having to rent air compressors, power generators, etc.

Epoxy Mortars.—The selection of an epoxy for a mortar application should be selected with the job site conditions in mind. Some of the job conditions which are to be considered are as follows:

1. Temperature curing ranges of the epoxy.
2. Moisture insensitivity or sensitivity of the epoxies.
3. Can local aggregates be used or is it necessary to ship aggregates as a part of the epoxy system.
4. Can the aggregate/epoxy system actually perform as an overlay in the environment specified.

When selecting an epoxy for a mortar application be aware that epoxies are available that are capable of curing as low as -10°F to as high as 140°F. Some epoxies can be extended with an aggregate (aggregate to resin ratio, by weight) for thin overlays ⅛" to ¼" as high as 8 to 1, and most usually 6.5 to 1, using local sands and for thicker applications utilizing ¾" to 1½" stone as high as 15 to 1 using local aggregates.

Knowing these facts one can introduce economy into an estimate for materials. Some epoxy mortars have very poor workability and require power troweling while others can be places with a screed to line and grade. Some epoxy mortars can be self leveling extended at 3.5 to 1 while other mortars might require power trowels at 3.5 to 1 aggregate to resin ratio.

Knowing these facts one can introduce economy into an estimate by reducing labor and equipment costs.

Estimating Quantities:

The most common denominator is to reduce all the quantities to the cubic foot (the epoxy manufacturer will provide the contractor with a "mix design" ie., aggregate sizes and weights and weights of epoxy).

Example:

Gradation =	79.34#
Gradation =	33.9 #
Epoxy Resin	18.4 #
Yield	$= 132.05\#/Ft^3$

The epoxy weight = 9.2#/Gallon
thus
$1 \ ft^3$ = 2 Gallon epoxy resin
thus
Coverage for one cubic foot if the overlay is to be ¼" thick is 12" ÷ .25 = 48 sq. ft./cubic ft.

If a prime coat is required add the quantity of epoxy required. Usually the coating of epoxy as a primer is the mortar resin itself. If a separate resin is recommended as primer, it should not be less than 100% solids and should not contain solvents.

A typical crew size for a mortar overlayment job might be as follows:

1 Laborer	- Mixing
2 Laborers	- Screed
1 Finisher	- Finishing
2 Laborer	- Priming & Spreading
1 Foreman	
6 Man Crew	= 48 Man Hours/Day

Assuming all the substrate cleaning, sand blasting etc. has been completed, the above crew can apply a ⅛" or ¼" overlayment as follows:

6' 0" ribbons 400 lin. ft. (2400 ft²)
Each 8 hour day (48 man hours)
or
50 square feet/man hour

IMPORTANT CONSIDERATIONS IN APPLYING EPOXY MORTARS

1.0 Viscosity

Viscosity is a critical factor normally not considered curing applications, however, it controls part of the success of the project. Viscosity must be considered especially when applications are to be made when the weather is providing cold nights or during winter months. The following figures indicate typical variations in viscosity with temperatures:

Typical Initial Viscosities
"A" & "B" Components Mixed

Temperature		
°F	°C	Centipoises
90	32	300
77	25	500
68	20	1,000
60	15.5	3,000
50	10	4,000
33	1	7,500

As you can see, the warmer the temperature, the lower the viscosity. The chief requirement for adding the filler or silica sand in the proper quantities is that the viscosity of the epoxy be such that all surfaces of the filler are evenly wetted and bonded together. When laboratory mortar formulations are designed, the epoxy is maintained at an even temperature of 72°F (22°C). This is the balancing point for filler addition. When the viscosity raises or becomes thicker, it is not possible to maintain the same filler balance, thus the mortar may exhibit characteristics that are not desirable. During cold weather applications, it is recommended that the resin and curing agent be stored in areas above 40°F (4.4°C) and be heated to a minimum temperature of 90°F (32°C) before use. Epoxy heating jackets provide an easy and fast method of warming the materials. Upon pouring the warm components into the mortar mixer, experience has shown that the components will drop immediately to a temperature of 60° - 70°F (17° - 21°C). This range of temperature at the beginning of the mixing procedure will continue to drop with the addition of the cold sand and gravel. Heating both epoxy components will result in mortar mixtures providing proper aggregate proportions, thus giving the desired physical properties when cured. When temperatures are above 50° - 60°F (10° - 15.5°C), heating is optional.

2.0 Reactivity

Reactivity of the mixed system determines working characteristics, potlife, tack-free stage, cure speed, working schedule and method of application. The cure reaction starts as soon as components "A" & "B" are mixed. The higher the viscosity or colder the temperature of the resin system, the slower the reactivity. It shows as a longer potlife time and longer time to fully cure. In cold temperature applications, slower reactivity could possibly cause application defects or failure, because the substrate could go through a thermal expansion or contraction movement prior to the mortar developing its tack-free stage. Effects could be loss of adhesion, or cracks in the mortar. Thus by heating both components to a preselected temperature, desirable lower working viscosites are achieved and, ideally we are obeying the Arrhenius' law, which in effect states, "we reduce our potlife and shorten the span of time necessary to achieve tack-free stage." The shorter the time required to achieve this desirable stage of tack-freeness, the less chance of mortar damage prior to initial curing. Once the tack-free stage develops, common physical properties are obtained in the epoxy mortar which are similar to that of concrete.

3.0 Sand

The mortar mixture contains considerably more sand and larger aggregate than epoxy, so the proper choice of the sand is imperative. It may be necessary to blend different grades of sand on the job site to insure the proper aggregate blend. The sand should be bagged, clean, dry and of high quality silica or quartz, with a minimum hardness of 7.0 on the Mohs scale.

Particle shape, as well as gradation, is important to have a workable mix at the required sand to binder ratio. All these factors must be checked carefully before specifying a particular sand to assure a mix which will give the greatest ease of application and performance. All sands should be tested by the manufacturers laboratory prior to use. Remember, different applications require different graded blends to achieve the best results. The manufacturers representative should assist you in selecting the best and most economical blends.

PREPARATION OF CONCRETE AND OTHER SUBSTRATES
FOR EPOXY FLOOR SYSTEMS

Epoxy flooring systems require a certain amount of preparation of the substrate prior to actual application. These are general instructions on floor and job preparation; however, each job should be considered an individual and preparation be estimated on the actual existing conditions. Very few jobs are exactly alike even though you may use only a few of the following cleanup steps on many of your jobs.

Preparation

General: Careful attention should be paid to surface preparation of areas to be treated, mixing and storing of materials area and adjacent areas not being treated. Epoxy flooring products have excellent adhesion, so it is extremely important to prevent waste and spills. Materials allowed to cure where they are not wanted will be extremely difficult and time consuming to clean up later. It is a good practice to protect the flooring area where the mixing takes place if it is not going to receive a new overlay. Typical method of protecting the floor area would be to spread heavy cardboard or plastic over the floor.

Good working conditions and preparations are essential to have a smooth operation and high quality job.

Work Areas:—Plan your job well . . . provide proper material storage and mixing area. Allow sufficient space and ease of access to the area where the floor is to be treated. Locate the storage and mixing area as close to the areas to be treated as possible, and on large jobs, mix in many different locations so the hauling labor is reduced to the minimum. When mortar mixers are used, make sure the mixer is elevated high enough so that all the mix may be dumped into the wheelbarrow. Set out your tools so they may be cleaned as necessary in the solvent containers. Clean your mixer once every 30 to 40 minutes by adding a gallon of solvent mixture and rotating it until the drum cleans itself. It may be used many times over and over, but remember to keep a cover on the solvent pail. The mixing area is always a "no-smoking" area.

Moisture:—Moisture creates no problem with most epoxy flooring systems; however, some require less labor to apply than others when the floor surfaces are dry.

Epoxy flooring systems may be applied to moist or wet surfaces down to 33°F temperatures. Remember, do not apply epoxies on ice surfaces or frozen concrete.

Temperatures:—Air temperatures do not govern the concrete temperatures, as a general rule. Concrete temperatures are normally related to soil and soil moisture temperatures. A room with an average air temperature of 70°F (20°C) could still have substrate concrete temperatures of 40°-50°F (4.5° - 10°C). Cold temperatures do not affect the application of some epoxy flooring systems; however, some epoxies require surface temperatures higher than 40°F (4.5°C) to apply their materials

General Surface Preparation:—Epoxy flooring systems may be applied over any sound substrate with the exception of asphalt and petroleum base surfaces. Surfaces must be cleaned and properly prepared as outlined in this section. The ultimate success or failure of the job depends a great deal on proper bonding to the substrate.

Service life of mortars or coating systems are primarily dependent upon good surface preparation. The life will be extended considerably by following the suggested recommendations.

Steel Surfaces

Must be dry and clean in accordance with the following requirements:
a) Remove all debris from surfaces to be treated.
b) Removal of grease, oils and contaminants may be accomplished by wiping the surface with a recommended cleaner. The cleaner will dissolve and dilute most contaminants sufficiently to permit them to be wiped or washed off. Prior to wiping, scrape or wire brush all heavier areas of contaminants to remove thicker deposits. Rags or brushes may be required to adequately clean the surface. Wear protective clothing during cleaning operation.
c) Sandblast surface to white metal finish, removing all mill scale, rust, rust scale, paints or other foreign matter. Care must be taken to see that extreme peaks and valleys are not left on the surface because they require additional coating build-up. These peaks, if too high, also cause thin spots in the coating application and could lead to early coating failure. Vacuum all debris and sand from area to be treated. Immediately apply the epoxy system as recommended.
d) Remove or grind all sharp edges and welds to a rounded surface.
e) Follow manufacturer's recommendations for applying the epoxy system.

New Or Non-Contaminated Concrete Surfaces

a) Remove all debris from working surface.
b) Surface must be cleaned free of all curing compounds, from release oils, and dust. New concrete should be fully cured. Dry surfaces permit easier applications, however, most epoxy products will adhere to clean, moist surfaces.
c) Determine pH factor of concrete surface and record. Do not apply acid etching solution unless required.
d) Record substrate and ambient temperature.
e) Etch concrete surface with a 10% solution of commercial muriatic acid only when surface pH is 10 or over. Apply 1 gallon of 10% solution per 40 square feet. The acid solution should be mechanically worked into the surface for two or three minutes or until bubbling stops. Wash surface with clean water to remove all acid. Rinse second time if any bubbling is noticed. Check pH after rinsing.
f) Sandblast or waterblast surface to remove contaminants, heavy laitance, sharp edges or protrusions which will interfere with the uniform application.

Old Concrete Surfaces Containing Fatty Acids

a) Scrape or wire brush all contaminants from the surface. Note: Do not dissolve grease or oil with solvents.
b) Remove all discolored and deteriorated concrete. Remove the laitance if heavy loads and excessive traffic will be applied to the flooring system. High pressure waterblasting with water is recommended. A 2% solution of a recommended cleaner may be added to the water by a metering device. Do not sandblast, grind, or scarify the concrete surfaces.
c) Record substrate and ambient temperatures.
d) Determine pH of concrete surface and record. Do not acid or alkali etch unless necessary. No etching is required when pH readings are between 5.5 and 9.9. Acid etch when pH readings are 10 or over with muriatic acid (water should be tap water temperature). Alkali etch when pH readings are 5.4 or lower with caustic sodas (water should be 150°F to 200°F (66°C - 93°C).

e) Etch concrete surface with appropriate 10% solution. Apply solution at the rate of 1 gallon per 40 square feet of surface area to be treated. Thoroughly clean and rinse, then move on to the next area. Rinse floor a second time to insure cleanliness.

f) Recheck pH readings after etching process is completed. If desired readings were not achieved, re-etch area. Note: Do not increase chemical etching percentages.

g) Squeegee or air blow all water free from the surface of the concrete. Be sure air supply has proper oil filter to prevent oil contamination of the concrete.

h) Wash cleaned floor with 4% pbw solution of a recommended cleaner. One (1) gallon solution will cover 40 square feet. Mechanically brush into surface. Work area until penetration is evident. Check surface area with magnifying glass to see if emulsified fatty acids are present. If fatty acid particles are found, rinse with clean water, blow off all excess water and rewash same area. Rinse with clean water.

i) Immediately apply first step of protective system.

Tile Preparation:—Epoxy flooring systems may be applied over old ceramic tile floors. The floor should be prepared by removing all grease, oil, paint and other contaminants. Scarify the surface of the floor with a sander or grinder. Remove old loose material in the joints. Vacuum all dust and then acid etch as described in the concrete preparation section. Conditions should now be ready for application of the epoxy floor system.

Wooden Floor Preparation:—Wooden floors are the less desirable substrate of all surfaces; however, if careful preparations are followed, an excellent floor system can be produced. Make sure the floor is sound and subjected to a minimum of movement. Epoxy floor systems will provide an excellent surface and hold the wood together; however, if there is a significant movement, cracking could develop. It is very important to remove all greases, oils and any contaminants that are present including old paints. A floor sander is often the best method to clean the surface. Vacuum entire area so no dust is present. The floor must be free of all dust and dirt. Wood that has been subjected to weather elements must be sanded well prior to installation. All wooden floors that are subjected to heavier use must have a 1" x 1" wire mesh hardware cloth nailed to the floor after priming has been completed. The primer need not be dry to install the wire mesh or to start the application of epoxy flooring systems.

Other Coatings Or Epoxy Surfaces:—It is a must to remove all other coatings and epoxy surfaces by abrasive blasting or scarifying. Vacuum all dust and debris. If the previous system failed, you cannot depend on it to be between the substrate and your epoxy floor system, for the adhesion is only as good as the surface which previously failed.

EPOXY FLOOR COATING DESCRIPTION:

Epoxy floor coatings are a two component epoxy system designed to coat concrete, steel floor and deck areas for extra durability and wear resistance and provides an attractive, durable finish for light industrial and warehousing areas and parking deck surfaces where service is too severe for most paints.

The epoxy floor coating must be resistant to a variety of chemical spillage and exposures normally found in industrial plants and parking structures, such as alkalies, mild non-oxidizing acids, gasoline, de-icing salts and many solvents.

The epoxy floor coating must have an excellent resistance to abrasion and be available in smooth or non-skid formulations. Standard colors are gray, tile red, and green. The epoxy floor coating should be able to be applied on moist (no free standing water) surfaces down to freezing, 32°F/0°C, or on dry surfaces down to 20°F/-6.6°C.

Outstanding Features:

High film-build
May be applied at low ambient temperatures.
  Applicable on moist substrates to 33°F (0°C).
  Applicable on dry substrates down to 20°F(-6.6°C)
Rapid drying time
May be applied in neat form or with a sand filler.
Chemical and mechanical adhesion to concrete
Excellent resistance to chemical and petroleum spillages.
Tough-flexible, abrasive resistance to hard rubber wheels and slow moving vehicular traffic.
Excellent colors provide attractive finish, hiding patches and stains.

Typical Uses:

Chemical Plants
Equipment
Paper Mills
Parking Garage Decks
Pharmaceutical Plants
Machinery Areas
Marine Uses
Water Pollution Control Plants
Water Treatment Plants

Resistance Properties:

Excellent resistance to:
Abrasion
Fuel spillages
Industrial fumes
Ultraviolet rays
Machine tool cutting oil
Natural and synthetic oils
Alkalies
Gases (hydrogen sulfide, carbon
and sulphur dioxide)
Acid fumes
Fresh Water
Salt water
Sour crude oil

Limitations:

The epoxy floor coatings are usually not recommended where exposure will include direct contact with sulphuric, nitric, and acetic acids.

In addition to the typical preparation above, the following is offered for special requirements of the floor coating system.

1. Concrete:—Broomed finished surfaces are the easiest surfaces to prepare. Lightly abrasive blast or mechanically wire brush surface to remove curing compounds and other contaminants. Remove dust from surfaces by vacuum or wash with water and air blow or squeegee off puddles. Note: puddles indicate low spots and can create unsafe conditions during wet weather.

2. Aged Concrete:—All older concrete must be thoroughly cleaned by an abrasion method such as sandblasting or factory approved methods. All contaminants such as old sealers, grease, oil, paints, asphalt, tars and other coatings must be completely removed before applying the epoxy floor coating. Wash or vacuum all loose debris and dust after abrasive cleaning the surface (note puddles to locate low areas).

3. Acid Etching:—All concrete should be checked after the cleaning process to determine the correct pH of the substrate. Acid etching is not required unless the concrete indicates a pH of 10 or higher.

4. Cleaning solutions should not be used unless approved in writing by the epoxy manufacturer.

Structural Cracks:

The repair of structurally cracked and delaminated concrete is of paramount importance. The crack transmits contaminants and moisture from one level to another.

Hairline Cracks:

Small fissures, often called hairline cracks, are normally the result of the concrete laitance shrinkage during curing or the lack of proper strength or good bondage between the laitance and the main body of the concrete structure. Normal surface preparation will remove these hairline cracks and expose the cracks that are truly a structural problem.

Non-Moving Joint Sealing:

Saw cuts and construction joints should be filled with a flexible resin (do not cover the exposed surface area of the flexible resin with the epoxy floor coating).

Chuckholes Or Spalled Areas:

Repair all chuckholes and spalled areas with an epoxy mortar system.

Typical Application Tools:

1. 9" to 18" wide roller with extension handle. Use solvent resistant short covers.
2. Squeegee with flat edges.
3. Airless sprayer with extension spray gun.

Typical Rates Of Application For Neat Material:

1. 1st Coat - Is applied evenly over the surface at an average rate of 1 gallon per 160 to 250 square feet with squeegees, rollers or airless sprayer. Allow to become tack-free.
2. 2nd Coat - Within 24 hours after applying 1st coat, the second coat is applied at an average rate of 1 gallon per 160 to 250 square feet with squeegees, rollers or airless sprayer.

TYPICAL RATES OF APPLICATION FOR A SAND SEEDED SYSTEM FOR SKID RESISTANCE

1. 1st Coat - Apply epoxy floor coating at the rate of 100 to 125 square feet per gallon with squeegees, rollers or airless sprayer. Immediately seed dry clean silica sand, sand blasting grade, evenly over the tacky surface. The sand should be seeded until a thin layer of dry sand is on the surface. Use approximatley 30 lbs. of sand per 100 square feet. Allow to become tack-free.

2. After tack-free stage has developed, remove excess sand and vacuum surface free of all dust particles. Any areas where the sand is uneven or too high, grind the surface level to adjacent surfaces. Revacuum area.

3. Immediately apply the 2nd coat with squeegees, rollers and airless sprayer. The final anti-skid surface will be determined by the amount of epoxy applied to the surface. Typically, the second coat will cover 120 to 180 square feet per gallon. For actual quantities of epoxy to be used, a sample should be made because sand gradations and surface areas vary widely.

Technical data contributed by Thermal-Chem, Inc. is gratefully acknowledged.

CHAPTER 21

HANDICAPPED FACILITIES

The issue of providing facilities to enable the handicapped individual accessibility of the built environment was presented to the construction industry and Governmental agencies during the early 1950's in an unstructured and fragmented form by a number of civic groups. By 1961 a specification had been developed as a guide for making buildings and other facilities accessible to, and usable by, the physically handicapped. This standard was adopted by The American National Standards Institute (ANSI) and became the basis for future laws enacted to protect the interest of the handicapped. The first major Federal law was The Architectural Barrier Act of 1968 which provided that any building construction which was financed in any part with Federal funds, must make provisions for accessibility by the handicapped. Unfortunately, this law did not provide for enforcement, therefore, it had little impact on changing the design of buildings to incorporate new features.

In 1973 The Rehabilitation Act was passed and this law provided for the establishment of an Architectural and Transportation Compliance Board to oversee the application of the law. As a result of this act, numerous cities and States, most notably Massachusetts, enacted similar laws requiring the inclusion of design features in buildings to insure barrier free access and use by the physically handicapped.

The magnitude of the handicapped population is brought into focus when studies show that 12% of the entire population, or 27 million persons, suffer from some type of handicap. This figure continues to increase with the added longevity of 24 million aged persons who suffer some degree of disability associated with the normal process of aging which causes a reduction in perception, mobility and coordination.

Contractors are subjected to providing the specialized facilities and equipment necessary to allow the handicapped to function and perform in a manner compatible with their peers.

Studies conducted on new projects indicate that less than 1% will be added to the construction costs to make facilities acceptable to accomodate the needs of the handicapped. When modifying existing structures to provide the support services for the handicapped, considerable expense can be encountered especially if structural, electrical and vertical lift equipment must be modified.

A classic example is a project to provide free access and all support facilities for the use of a toilet area by the handicapped. In this case, coordination between trades is most difficult and considerable non-productive time will be encountered simply because of the large number of trades involved in a restricted area attempting to perform relatively small, but necessary, items of work. Under these conditions, the contractor should not attempt to rely on the usual unit cost method, but would be more accurate in identifying each item of work by trade and develop a daily work schedule for each crew. An allowance should be included for interferrance from other trades, and the added cost of demobilization and remobilization of equipment, and work crews to accommodate other trades in the performance of their work. Work production will be very poor under these conditions, therefore, the development and an adherence to a work schedule is the most positive means of cost control.

Even though clouded by the uncertainty of job conditions which could substantially alter the following unit prices, they have been utilized as a base cost figure in a large number of instances, thus, they will serve as a guide to the Contractor/Estimator.

SITE FACILITIES

1. Concrete ramps on grade, 1:12 pitch, 4" thick with mesh, broom finish, 4" base...$2.75 S.F.
2. Steel pipe handrail, 1½" galv., 2 rails 36" high ....28.00 L.F.
3. Non slip abrasives for walks and ramps.....................................36 S.F.
4. Parking signs for handicapped areas ....31.00 Ea.
5. Embedded abrasive warning strips ....6.65 L.F.
6. Exterior hazardous area warning grooves 1" deep....................1.68 L.F.
7. Depressed curb ramps...4.50 S.F.

INTERIOR STAIRS AND CORRIDORS

1. Hazardous area warning abrasive strips.................................$6.65 L.F.
2. 6" curbs with pipe rail at dangerous areas ....28.50 L.F.
3. Integrated stair nosings, non slip ....28.90 Ea.
4. Corridor wall rails, aluminum pipe ....16.00 L.F.
5. Balcony railings, aluminum 1½" post with pickets.................34.00 L.F.
6. Stair hand rails, steel, with newel and brackets ....28.00 L.F.
7. Floor grating, aluminum, ¼" maximum spacing, safety surface ....10.20 S.F.
8. Vinyl asbestos tile for sound reflection .... .75 to 1.00 S.F.
9. Ramps, wood not to exceed 12'-0" length and 4'-0" width....................6.25 S.F.
Note: For curved railings add 33%

Information Only

1. Corridors, minimum 36" width single with turn around area every 50 L.F.
2. 360° wheelchair turnaround is 64" dia. or 22.4 S.F.
3. Right angle grid pattern for corridors and public spaces are perceived more easily than curves or serpentines.
4. Diagonal reach from wheelchairs is 48" above floor level for telephones, etc., with forward reach of 18.5".
5. Overhead obstructions not less than 7'-0" above finish floor.

CASE WORK

1. Counter height should be 32" with minimum depth of 21"..............$36.00 L.F.
2. General work surfaces 25"-30" above finished floor with maximum depth of 25".. 36.00 to 55.00 L.F.
3. Built-in case work to have toe clearance of 7" recess and 10" above finished floor...38.00 L.F.
4. Closets to be partially accessible for wheelchairs should have adjustable clothes rod...22.00 Set
5. All case work to have recessed hardware ....N/C

DOORS AND OPENINGS

1. Electric operated automatic doors with delay mechanism,
   complete including hardware ....$5,500.00 Ea.
2. Access door adjacent to revolving door..1,200.00 Ea.
3. Vision light in solid doors..50.00 Ea.
4. Lever type hardware...30.00 Set
5. Knurled back handles for dangerous areas...30.00 Set
6. Corner protection, stainless steel, 16 ga..17.00 L.F.
7. Corner protection, vinyl, high impact with stainless frame................21.40 L.F.

IDENTIFYING DEVICES

1. International handicapped sign ....$31.00 Ea.
2. Plaques for rooms with raised letters..2.75/letter
3. Interior sign character 2½"-4"..5.15/letter
4. Exterior sign character 4"-8"..10.50/letter

ELEVATORS

1. Hall lanterns with visual and audible signals indicating
   elevator direction and floor location ....$250.00/Ea.
2. Elevator door jambs equipped with floor designations
   on each side having 2" characters...15.00/pair
3. Doors equipped with re-opening devices not requiring
   contact to be activated...500.00/car
4. Interior car position indicator with visual and audible
   signals for each floor level 250.00/car
5. Emergency telephone located 48" above floor...N/C
6. Call buttons with raised characters ....40.00 Ea.

KITCHEN

1. Front loading dishwashers..N/C
2. Front loading washers and dryers..N/C
3. Cooking surfaces with front controls ....N/C
4. Wall ovens mounted with top of door 31" above floor...............................N/C
5. Storage and utility areas to have peg boards, shelving, etc.
   at convenient level..N/C

MECHANICAL AND TOILET FACILITIES

1. Water closets, mounting height 20" above floor with 18" clearance each side
   (additional unit cost 75.00 Ea.)
   For complete installation, hook-up to existing waste & supply.............. $925.00
2. Stainless steel drinking fountain, dual level for handicapped
   Installed Complete..865.00
3. Drinking fountain, wheelchair handicapped unit
   Installed Complete..950.00
4. Shower, fiberglass with cast base 32" x 32", complete hook-up
   to existing waste and supply...610.00
5. Recessed drinking fountains - add for alcove...250.00
6. Water cooler, non-recessed stainless steel cabinet, wheelchair type–complete
   with hook-up to existing waste and supply...780.00

7. Grab bars, 24" stainless steel...28.00
8. Handicap shower head controls..75.00 Ea.
9. Shower transfer seat...70.00 Ea.
10. Toilet partitions, handicap units, porcelain enamel............................775.00 Ea.
11. Toilet partitions, handicap units, stainless steel.................................715.00 Ea.
12. Toilet screens, porcelain enamel...120.00 Ea.
13. Toilet screens, stainless steel...110.00 Ea.

Note: Type of anchoring system for toilet partitions will influence cost. Unit prices quoted above are for ceiling hung units

14. Tilting mirror - bracket only...70.00 Ea.

ELECTRICAL

1. Electric automatic door opening control located near door jamb...$400.00/Station
2. Audible signals at visual electric signals.............................350.00 Ea.
3. Electric door holder...125.00 Ea.
4. Public Address System...225.00/Speaker
5. Sound System, AM-FM, 12 outlets...8,000.00
6. Fire detection systems, including smoke and heat detectors For complete installation of a ten (10) detector system...........................2,500.00
7. Master clock system including clocks and bells - complete for a 20 unit system...14,250.00
8. Master TV antenna including VHF-UHF antenna and distribution to a 10 outlet system...4,800.00
9. Locating receptacles and switches at lower height for handicapped............N/C

CHAPTER 22

ENERGY MANAGEMENT

Energy management systems can no longer be considered a luxury. The impact of soaring energy costs has forced building owners to seek relief from these uncontrolled costs through management systems which control energy consumption, optimize comfort, interface with human performance and provide automatic data analysis.

This new market, which has only been lightly penetrated as of now, has attracted a great variety of manufacturers, consultants, vendors and suppliers, each attempting to capture a share of this lucrative business, and there has resulted many instances where owners have experienced excessive costs due to over design or systems did not provide results because of a lack of knowledge or expertise on the part of the installer.

Although energy management cannot be considered an organized industry as yet, competition and advanced technology are stablizing factors where it is now possible to make an analysis of the needs of a conservation package for a building and procure responsible bids from several specialists with a feeling of confidence that the system will be cost effective and produce the desired results.

The first major thrust into energy management was made by the national control and computer manufacturers. Through their established superior marketing and advertising capabilities, they were able to rapidly develop the large client market of hospitals, commercial and industrial buildings, educational institutions and major office buildings. With the advance of microprocessor technology, which sell for less than $1,000.00, the advantage of energy management is now available for use in smaller buildings now that the need for the $50,000 – $100,000 computer can be avoided.

In spite of the wide variety of electronic energy management systems available today, there also exists for relatively simple applications, the ability to design non-electronic control panels which can provide automatic setback of thermostats and restrictions on electrical or fuel consumption, both of which translate into operating economies of major significance. Other than the economy of initial installation costs, this simple system would be found attractive for use in the more remote areas where service could be supplied by the ordinary electrical or air conditioning service company. It is estimated that such panels could be installed for approximately $1,500 – $2,000 for an open space store area of about 25,000 square feet with limited glass area and insulated roof area.

The costs involved in the more complex computer controlled energy management systems cannot be pre-determined as each system is specifically designed to the needs of the Owner. It is not uncommon to include additional functions into the system such as fire and smoke alarms, security controls or other specialized features.

One of the hidden or ignored costs in building energy management systems is that they share a fundamental need with other complex contemporary technology; the continuous availability of maintenance service. Low cost microprocessors are reliable because of their relatively simple design, but these systems usually provide for service contracts which are available ranging in annual cost from 6% to 10% of the system purchase price. In the more complex computer controlled systems, service contracts are offered at approximately 10% of the purchase price with an annual escalation of 10–12%.

Consideration must be given to the costs associated with the training of personnel to operate the system as the maintenance staff should take an active part in the installation and commissioning procedure and actively participate with the operators in the initial training provided by the manufacturer of the system. While these costs are related to the operational cost factor that the Owner must provide for, they are many times not given consideration by the Owner as part of the installation cost of the energy management system. The Contractor/Estimator who is aware of these often times overlooked expenses and their relation to the efficient operation of the system, including minimizing the need for maintenance service by the manufacturer, will be in a position to extend an additional service to the Owner by offering certain guidelines for the type and duration of training for the operating personnel. In this way the Owner will become conscious of the need for operational training and alerted to the associated costs.

In the absence of a highly developed and organized energy management industry, there is no single source to obtain historical data relating to installation costs, performance reports, reliability studies or cost effectiveness of the various systems available. Much of the information available to the public is obtainable from the large control and computer manufacturers and is essentially sales oriented to their particular equipment and the accomplishments they have achieved. This situation places a burden on the contractor involved in an energy management system for a new building or modifying the energy systems in a existing building as he/she will be forced to evaluate technical proposals with extreme care before reaching a decision that the offer provides the maximum efficiency for the most reasonable price.

It can easily be concluded that, even when faced with some of the early financial disasters resulting from insufficient knowledge of the capabilities and limitations of energy management systems and the lack of expertise on the part of some companies, the industry is providing a significant contribution to the economic stability of the country.

It is not unusual to develop systems serving the needs of both small and large clients that will reduce energy consumption in a building by 25% and yield a payback of the initial installation cost in about two years. With such attractive figures as these, it must be contemplated that this industry will continue to mushroom and progressive contractors will recognize the potential of this market by initiating an in-house program to educate their personnel in the arts of this new industry.

From the position of establishing some basic method of determing a cost criteria to serve as an estimating guide, surveys indicate that it is inappropriate to attempt to develop such information at this time due to the absence of sufficient historical data cost bank. It has been found installations costs can be as low as 15 cents per square foot of floor area for industrial plants containing up to 2 million square feet, while in a 36,000 square feet store utilizing a non-electronic system, the cost was found to be 7 cents per square foot. In the instance of the institutional and office complex, installations which are frequently loaded with a series of other functions, in addition to energy management, it is virtually impossible to develop meaningful conceptual costs because of the individuality of the design.

The Frank R. Walker Company has developed this new division of The Building Estimator's Reference Book to serve the current needs of the Contractor/Estimator and will use it as a base to provide estimating techniques and technical information relating to energy management as it becomes available through future editions.

Energy management has rapidly become a major consideration of every business and industry as energy costs continue to rise. To lower energy usage, users are reducing the operation of lighting and HVAC equipment. However, additional help is often needed in the form of "load control". Load control is defined as automatically controlling electrical equipment loads to reduce the electrical consumption and usage rate.

Most electric utilities base the monthly charges to commercial users on two (2) separate items:

1. Consumption is the total kilowatt-hours of electricity used during any billing period.
2. Demand charges are based on the rate at which electricity is used. Most utility companies measure the electricity being used during a short period of time called a "demand interval". The monthly demand charge is then based on the maximum peak usage that occurred within the demand interval of each billing period.

Load control systems design requires the input of several important factors before the proper implementation can occur.

Every building has its own set of characteristics. Even for similar buildings, such as chain stores, each has unique situations that change the design and application of a load control system. Variables affecting the control system design include climate, building design and construction, type and capacity of HVAC and electrical equipment, business hours and traffic flow in and out of the building.

The building's electrical usage data (load profile) must be analyzed and all equipment items examined for their operating cycles in order to determine which items are in operation at peak demand times, lengths of continuous operation, etc. This information needs to be obtained for weekday, weekend and seasonal changes.

Local health codes are an important factor should the load control system be installed in food service facilities. It is possible for the load control system to reduce the operation of certain equipment to the point where a violation of the minimum health code requirements may occur. For example, if the load control system turns off the hot water heater too often or for too long a period of time, the temperature of the dishwasher water may fall below the minimum allowable by health codes.

Methods Of Load Control

Duty Cycling - Duty cycling accomplishes load control by turning on or off all loads on a programmed basis. The major function of this method is to reduce the kilowatt-hour consumption. Duty cycling equipment based on temperature (temperature selective duty cycling) increases the equipment's operational efficiency and longevity and lowers energy consumption. An example would be the shut-down of air conditioning equipment once the A-C unit has lowered the temperature to the prescribed level. Needless energy is wasted in allowing the unit to run nearly unloaded.

Each load (equipment item) that is connected to the duty cycler is programmed to turn off for a specific portion of time in a repeating time interval cycle. The amount of time a load is off is directly dependent upon the importance of that item to the overall operation of the facility.

Duty cycling can be accomplished by using two separate cycles, Long-Off-Time-Cycle (LOTC) and Short-Off-Time-Cycle (SOTC). When in the LOTC mode, the equipment load is turned off for longer periods of time in the repeating cycle. The duration of the LOTC, SOTC and the repeating time interval cycle is defined by the system user. Temperature set points are also used in conjunction with the LOTC and SOTC to give virtually any combination of temperature regulated on/off mode for the equipment load.

In applications where product temperatures are critical (food service), temperature override devices can be utilized to prevent the degree limit being reached.

The duty cycle function can be overriden by a demand limit control program should the actual demand level exceed the set points of the duty cycle load control center.

Demand Limiting - Demand limiting is a term that describes the process of turning off (shedding) loads temporarily to keep the kilowatt (KW) usage demand below a predetermined level. To accomplish this, the loads connected to the demand limit control are turned off, or shed, when the KW demand level reaches the set level, which is the lowest practical KW usage rate allowable to meet the facilities operating requirements.

The demand limit control is programmed to shed loads in a predetermined order when the demand set level is exceeded. This control will continue to shed loads until the KW demand level is less than the set level minus a "deadband" range. The deadband is intended to prevent unnecessary cycling of loads controlled by the demand limit control when the KW demand level is in an area around the set level. If the demand level is in the deadband range, additional loads are not shed or restored by the demand limit control.

Demand limit controls can monitor KW demand in several ways. One is the instantaneous rate method which compares the instantaneous rate of power usage to the predetermined set level. The other way is the average rate method which continuously samples and averages the KW demand level over a short period of time. The averaged usage range is then compared with the predetermined set level.

Time Of Day Programming - Another method of load control is time of day programming or load scheduling, which is the process of automatically turning on and off specific equipment loads based on the time of day.

Some loads are often left on because of failure to manually disengage the switch. These types of loads are the ones controlled by the load scheduling device.

Most all energy management load control systems incorporate all three methods of load control (duty cycling, demand limiting, and time of day programming) into one integrated package that gives the user a complete system for controlling all equipment loads in his facility.

CHAPTER 23

ASBESTOS ENCAPSULATION AND REMOVAL

For many years asbestos materials were used for acoustical insulation, fireproofing spray on structural steel framing members, fireproofing spray on ducts, pipe and duct insulation and boiler and breeching padding.

Early in the 1970's, scientists determined conclusively the extreme health hazard posed by asbestos fibers. The Government, at that point, immediately imposed strict regulations on the manufacture and use of asbestos products. The Occupational Safety and Health Administration (OSHA) has sets of regulations which control such things as permissable asbestos exposure limits and methods of complying with asbestos. Whenever work is performed involving the installation or removal of asbestos materials, it is required that instruments be used to measure and monitor the area's airborne fiber content. Special protective clothing shall be worn by all personnel exposed to the contaminated areas.

In 1978 asbestos was banned from all spray-on materials including those used for decorative purposes. The Environmental Protection Agency (EPA) contracted with Battelle Memorial Institute to conduct a test program to determine whether encapsulation or sealing of asbestos would be effective. As a result, several laws were proposed and in June, 1980 the Asbestos School Hazard Detection and Control Act was passed. This Act was to provide up to 50% of the funds for any abatement project. However, the program has not been funded. Any asbestos abatement projects now in progress are being funded at the local level.

Asbestos is considered to be hazardous in a friable state, which means that the material can be easily crushed or turned into powder by using hand pressure. The greatest hazard posed to humans by friable asbestos is airborne fibers. At present, OSHA has set levels of the allowable asbestos fibers in a certain quantity of air. The current Permissable Exposure Limit (PEL) is two (2) fibers, 5 micrometers in length, per cubic centimeters of air. A person inhales about one (1) cubic meter of air per hour which is 1 million cubic centimeters. This amounts to an intake of 2 million fibers per hour and that quantity does not include the fibers that are too small for detection by the current measuring equipment.

OSHA has proposed new standards on exposure to asbestos in the work place. Among other things, the new standard would set the Permissable Exposure Limit at either 0.5 fiber or 0.2 fibers per cubic centimeter of air. However, at this writing, the new standard has not yet been accepted as the prescribed standard. Certain states, however, have taken it upon themselves to issue stricter requirements for airborne asbestos fibers. For example, Virginia has recently passed a new guideline that reduces the acceptable asbestos exposure limits from 2 fibers per cubic centimeter of air to 0.5 fibers.

Asbestos is facing ultimate market extinction under a sweeping ban being contemplated by the EPA, and alternative product development is already underway. In late 1983, EPA advised asbestos producers, product manufacturers and environmental groups that it planned to issue a ruling in July, 1984 banning six (6) categories of asbestos containing products, and propose another rule in October 1984 setting production ceilings for remaining products. The limits will decline annually until all products containing asbestos, a known carcinogen, are eventually phased out. The products covered by the proposed ban are saturated asbestos roofing felt, unsaturated roofing felt, asbestos flooring felt, asbestos sheet flooring, vinyl asbestos floor tile, and cement asbestos pipe. A major factor in the schedule of planned phase-out of the remaining asbestos containing products will be the availability of a suitable substitute material. The proposed rulings are a continuation of the rule making that was begun in 1979 when the EPA announced its intent to ban such products under the Toxic Substance Control Act.

ABATEMENT METHODS

Once the problem of asbestos content had been discovered and analyzed, then the abatement program was initiated and continues to be expanded. There are presently three (3) methods of controlling asbestos.

The first is removal. This is by far the most expensive method, because it takes more time for removal and controlled disposal for final deposit in limited designated disposal sites around the country. Also consider that the material was originally installed for a purpose, whether it be thermal or sound, and therefore, usually needs to be replaced with a new non-asbestos material that performs the function of the removed asbestos.

The second method is the enclosure system which is simply installing another material, such as drywall, over the existing asbestos materials.

The third method is known as encapsulation which requires the spraying of a sealant material onto the asbestos either to form a bridge over the asbestos insulation, or to penetrate into the asbestos to bind the mass together. The effectiveness of encapsulation depends largely on the condition of the existing asbestos. For example, if the asbestos is very loose and becoming delaminated from the surface, encapsulation will not work because the material will simply fall off the structure once it is agitated by the application process of encapsulation sealants. These sealants are applied with an airless spray. The EPA has determined that brushes and rollers should not be used, because they create excessive disturbance on the surface of the asbestos materials. Using the airless spray also disturbs the surface of the asbestos, but it is the lesser of evils.

To be effective, the encapsulation sealant should virtually eliminate any asbestos fallout into the atmosphere (airborne materials), provide some resistance to impact and abrasion and be reasonably flexible to handle any normal movement of the building. It needs to have a flamespread resistance for building code compliance, and the material, in the dry state, has to be free of odor.

As stated, the encapsulation sealants basically fall into two (2) categories: bridging and penetrating.

A Bridging Sealant—is a coating that forms a barrier between the asbestos materials and its surrounding atmosphere. These sealants are applied at the rate of approximately 3 gallons per 100 sq. ft. for a dry film thickness of about 25 mils. There is very little surface penetration when using a bridging sealant. The bridging sealant is an application of two (2) coats, with half of the material applied in the first coat, and the remaining half on the second coat. Normally, the coats will be color coded for coat identification. Proper coverage is very important; if cracks develop in the finished surface, there is an escape point for the asbestos fibers and, therefore, the encapsulation will be rendered ineffective and the surface area must be retreated.

The Penetrating Sealant—is a very watery solution which penetrates into and hardens the asbestos materials. The penetrating sealant method is a little more difficult to control in the initial installation, since the purpose of the penetrating sealant is to absorb into the materials, cure and bind them together. If insufficient material is used on the first coat, and that coat is allowed to dry, a film is formed that interferes with the penetration of any subsequent coats. Therefore, the penetrating sealant is normally applied in a three (3) coat application with each coat being applied prior to the drying of the previous coat. It is more difficult to estimate the quantity of materials to be used in the penetrating method because of the porosity and thickness of the asbestos material. Normal coverage can be obtained with 2 to 3 gallons per 100 sq. ft., but in extreme conditions, 10 to 15 gallons per 100 sq. ft. may be required. This greatly increases the material cost and the amount of labor time required to spray on the additional material. The easiest method to determine the actual quantity of penetrating sealant to be used is to spray a test area of the asbestos, allowing it to cure, and then coring the portion of asbestos and examine it for proper penetration and bonding characteristics.

In any asbestos abatement program, whether removal, enclosure or encapsulation, the most expensive and time consuming job is the preparation of the area where the asbestos is to be removed or encapsulated. The walls and floors have to be completely covered with polyethylene. The entrance to the area has to be converted into an air lock (and preferably two (2) air locks at least 6' apart). Light fixtures have to be removed, other types of equipment and items have to be covered or removed, the return and supply air grills and diffusers have to be removed and the openings sealed in the duct. The work area itself should maintain a negative air pressure which helps prevent asbestos fibers from escaping into other areas.

Special change rooms are required where the workers can change into their work clothes and exit through an air lock into the work area. Upon return from the work area, the workmen will go into a room, strip, dispose of their work suits, enter the shower and wash, exit into the area containing their street clothes and exit the area. All workmen are prohibited from smoking, eating, or drinking in the contaminated work area, or the contaminated portions of the change rooms. Thus the average workman will normally exit the work area, dispose of the work suit, shower, and dress 3 to 4 times each work day.

One of the manufacturers of asbestos encapsulating sealant materials is the Foster Products Division of the H.B. Fuller Co. in Springhouse, PA. Its bridging encapsulant material is a latex elastomer, white in color, and has the consistency of heavy paint with a weight of 11.4 pounds per gallon. The material is applied by the airless spray method, and has a mild odor while wet but odorless when dry. The recommended coverages for this particular material is 3 gallons per 100 sq. ft. (or if applying in two (2) coats, 1½ gallons per 100 sq. ft., each coat) and this will depend on the texture and porosity of the asbestos.

The PROTEKTOR sealant material is an acrylic copolymer, comes in colors of yellow, blue, and untinted, and has a consistency of a thin fluid for proper penetration into the asbestos. Its weight per gallon is approximately 8.4 pounds. It is applied with the airless spray method the same as the bridging encapsulant. Its odor is mild while wet, odorless when dry. The recommended coverage of this material is 2 to 10 gallons per 100 sq. ft. depending on the absorbancy and thickness of the asbestos and is best applied in 2 to 3 coats.

During the process of asbestos removal or encapsulation there are two (2) main pieces of equipment which the contractor will need to have at his disposal during the entire process. The first is an exhaust filtration unit, available from a number of manufacturers, such as the Torit Division of the Donaldson Company, who has local representatives throughout the country. This unit is an exhaust fan made in varying sizes depending upon the amount of air, cubic feet per minute (CFM), that needs to be moved. The air that flows through this exhaust fan is filtered by means of a HEPA filter and deposits the asbestos fibers in a special drum below the exhaust fan. The typical drum size is a 55 gallon unit which contains a disposable-type plastic bag with a special marking, warning the handler of the contaminated hazardous waste material.

The size of the exhaust fan required for the project is determined by the number of air changes per hour that is specified (normally four (4)) and the size of the area that is to be exhausted. An example of this would be if you had a room 50' long x 50' wide x 15' high; the cubic content of this room would be 37,500 cubic feet. Because the exhaust unit is rated in cubic feet of air moved per minute (CFM), you would divide 37,500 cu.ft. by 60 minutes, then multiply by four (4) air changes, and the results would be 2,500 CFM capacity for the exhaust fan.

In recent quotations from the Torit Division, a 1000 CFM exhaust filtration unit, complete with HEPA filter, costs approximately $5,000.00. A 2000 CFM exhaust filtration unit, complete with HEPA filter, costs approximately $7,000.00. Units can be equipped with automatic alarms and other sensitive detection devices that can double the above prices.

The other item of equipment that will be required is the HEPA vacuum cleaner. One such unit is manufactured by the American Cleaning Equipment Corporation, in Addison, IL., which is handled by numerous sales representatives throughout the country. This particular vacuum cleaner contains a special five (5) stage filter unit similar to the exhaust filtration unit. The filter unit alone will cost in excess of $350. The drum that collects the asbestos fibers as they are vacuumed from a surface comes in 15, 30 and 55 gallon sizes. A specially marked plastic bag that the asbestos fibers are deposited into fits in the drum. When the bag is full, it is removed, sealed and placed ready for transport to an approved disposal site.

A recent price quote from a manufacturers representative for this vacuum unit placed the cost for a 15 gallon vacuum cleaner at $850, a 30 gallon vacuum cleaner at $1,100, and a 55 gallon vacuum cleaner at $1,150. In addition, they sell a special asbestos tool kit for about $50 which includes ten feet of hose and various shaped tools. Additional hose sections are approximately $35 per 10-ft. section.

To alert the Contractor/Estimator to the specific detailed procedures that all site personnel will be subjected to when involved with asbestos work the following standard "worker protection procedure" program has been included.

It is obvious that extremely low productivity will result from the work restrictions placed on employees. It is estimated that labor production could be reduced by 33% to 50% of normal production rates.

WORKER PROTECTION PROCEDURE

Arrival: Each worker or authorized visitor arriving at the job site shall remove street clothes in the Clean Change Room and put on a respirator with new filters and protective clothing and shoes before entering the Equipment Room or the Work Area.

Worker Decontamination: Each worker or authorized visitor shall, each time he leaves the Work Area, remove gross contamination from clothing, proceed to the Equipment Room and remove all clothing except respirators. Still wearing the respirator, proceed naked to the Shower Room, clean the outside of the respirator with soap and water, remove the respirator, shampoo and wash themselves, remove the respirator filter, wet them and dispose the filters in the container provided for that purpose, and wash and rinse the inside of the respirator.

4. Primary cloth filter

5. Paper filter protector

5 STAGE
FILTRATION SYSTEM

3. Final high efficiency filter certified to exceed OSHA requirements

2. impact filter

1. Coarse filter — disposable paper bag

Courtesy American Cleaning Equipment Corp.

Components of a Typical Asbestos Vacuum

Following showering and drying off, each worker and authorized visitor shall proceed directly to the Clean Change Room and dress in street clothes at the end of each day's work, or in clean coveralls for eating, smoking or drinking. Contaminated work footwear should be stored in the Equipment Room when not in use. Store contaminated work suits in the Equipment Room for re-use or place in container for disposal as asbestos contaminated material.

Upon completion of the asbestos abatement work, dispose of footwear as contaminated waste or clean thoroughly inside and out with soap and water before removing from the Equipment Room. Workers removing waste containers from the Equipment Decontamination Enclosure shall enter the Shower Room from outside wearing a respirator and dressed in clean coveralls.

All workers and visitors shall not eat, drink, smoke or chew gum or tobacco at the work site except in the established Clean Change Room. Workers shall be fully protected with respirators and protective clothing during preparation of system enclosure prior to commencing the actual asbestos abatement, throughout the entire abatement procedure, and until final clean-up is completed.

AIR MONITORING

The contractor shall be responsible for monitoring and testing to determine the fiber count content of the air and on surfaces in and adjoining the work area, during the abatement process and the clean-up phase. All air and surface sampling and testing is normally the expense of the Contractor, not the Owner. Air and surface samples shall be taken in accordance with OSHA and EPA requirements during each work day. Sampling shall begin by taking samples to establish a "background level" before the start of any construction, including the temporary facilities. Sampling may stop only when final acceptance of the area has been made in writing by the Owner.

The laboratory analyzing the samples must be accredited for asbestos analysis by the American Industrial Hygiene Association. Certified copies of the results must be delivered to the Owner normally within 24 hours of taking the samples. Continuous air monitoring equipment is used by the contractor to supplement the testing program, however, this air monitoring is not allowed to substitute for the air and surface sampling that is required.

AREA PROTECTION

Most contractors use polyethylene to completely cover the walls and floors (assuming the asbestos is on the ceiling) of all areas for decontamination. Normally a 6 mil poly is used for floors, the air locks, and sealing doors and windows. A 4 mil poly is used against wall surfaces. The poly should be as wide as possible to minimize joints, thereby minimizing leakage of airborne asbestos fibers to the outside area.

WORKER DECONTAMINATION FACILITIES

The contractor shall construct a worker decontamination facility consisting of four (4) chambers as follows:

1. EQUIPMENT ROOM - shall have two curtained doorways, one adjacent to the work area and one adjacent to the shower room. The Equipment Room shall be of sufficient size to accommodate at least one worker, allowing him enough room to remove his disposable uniform as well as house any equipment the contractor may need in the contaminated area during the work process.
2. SHOWER ROOM - shall have two (2) curtained doorways, one adjacent to the Equipment Room and one adjacent to the Clean Room. At least one shower with hot and cold water shall be installed in this room. The shower shall be constructed in such a manner that all water will be collected and pumped through a five micron filter. From the filter, water may be drained off in any conventional manner to a normal sewer system. Careful attention shall be paid to the construction of the shower to insure against leakage of any kind. The contractor maintains all soap and other drying supplies for the shower area.
3. CLEAN ROOM - shall have two curtained doorways, one adjacent to the Shower Room and one to the outside of the work area. The Clean Room shall be of sufficient size to accommodate at least one worker, towels for workers and a supply of clean overalls.
4. LOCKER ROOM - shall have two curtained doorways, one adjacent to the Clean Room and one to the outside. The Locker Room shall be of sufficient size to accommodate at least two (2) workers and one locker for each employee.

This Worker Decontamination Facility can be constructed of 2" x 4" wood framing with plywood for sheathing and completely lined with a 6 mil polyethylene. Doorways and air locks can be constructed using two (2) sheets of 6 mil poly hung between the head and the floor and jambs. Careful attention should be paid to the overlapping of poly joints and taping to insure an air tight structure.

On the occasion that you are bidding a project that enables access to the exterior directly from the contaminated area, there are firms such as the Williams Mobile Office Trailer Company that offer a unit specifically designed for asbestos abatement projects. It comes complete with all four (4) rooms including the shower facilities.

PLACEMENT OF EXHAUST AIR FILTERS

A negative air pressure shall be established in the work area and make-up air shall be obtained only through the decontamination chambers. The work area shall be under negative pressure whenever work is being performed, including preparation, removal, clean-up, or re-insulation.

Equipment shall be located so as to maximize air movement throughout the work area. Air movement shall be established in such a way that airborne fibers will be carried away from the immediate worker's area. At the end of each work day, the exhaust filtration unit shall be left on until the fiber count in the work area has reached a safe specified level, at which time the units can be turned off and the employee's decontamination disposal routes sealed. Note that an employee will need to be left at the project to accomplish this particular function.

REMOVAL OF ASBESTOS MATERIALS

The asbestos material shall be sprayed with water containing a wetting agent (amended water) to reduce airborne fibers preceding the removal of the asbestos material. The material shall be sufficiently saturated to prevent emission of airborne fibers in excess of the exposure limits prescribed in the OSHA standards.

The asbestos material shall be removed in small sections by two (2) man teams. Asbestos contaminated materials shall be handled carefully and deliberately. Unnecessary agitation of the material is strictly prohibited. The use of any high rpm power equipment, pressure washers or hydroblasters is not acceptable. The contractor shall not allow asbestos material to dry out or collect on the floors.

Removed materials shall be immediately placed in 6 mil plastic bags, sealed and placed into metal drums for transport to the disposal site. Bags and drums shall be marked with the OSHA label prescribed by the OSHA regulation referenced by their specification. The outside of all containers shall be cleaned by wet sponging prior to leaving the work area.

After the wet removal process has been completed, the existing substrate and any surface from which asbestos has been removed shall be thoroughly vacuumed with a HEPA filtered vacuum cleaner. A note of caution, a HEPA filter will fail if used on wet materials.

CLEAN-UP AFTER DECONTAMINATION

The following work is performed after decontamination has been accomplished. All plastic coverings shall be vacuumed using a HEPA vacuum cleaner to remove the loose asbestos fibers. The poly is then removed, folded carefully and disposed of in a proper container. The work area is then again vacuumed with a HEPA vacuum cleaner, wet wiped, and in some instances, a second wet wiping is specified. During this particular process, the air monitoring equipment and the exhaust filtration units are in full operation. Once all of the poly is picked up and the room has been vacuumed and wet wiped, the process of re-installing the registers and grilles, the light fixtures, smoke detectors, and any other equipment that was removed prior to the decontamination process can begin.

ESTIMATING CHECKLIST

Because each asbestos removal project is unique in the set of job performance parameters associated with it, cost data is impossible to supply. We offer the following checklist items that the Estimator needs to consider in preparing an estimate for asbestos removal work:

1. Work procedures submittal to required agencies.
2. Procedure approvals and permits.
3. Set-up of decontamination rooms.
4. Water, drain and electrical connections to the decontamination rooms.
5. Maintenance of decontamination rooms during the progress of work.
6. Dismantle and removal of decontamination rooms at completion of work.
7. Air monitoring equipment rental, set-up and dismantle.
8. Polyethylene covering for floors, walls and other large objects to remain in the work space; and the subsequent removal of this covering at completion of asbestos removal.
9. Removal and/or covering for lighting fixtures, smoke detectors, duct grills and registers, etc; and the subsequent reinstallation and/or uncovering at completion of asbestos removal.

10. Disposable clothing suits.
11. Respirators and filters.
12. Scaffolding, ladders and work platforms.
13. Surfactant for amended water application.
14. Removal of asbestos materials from substrate.
15. HEPA vacuum equipment rental and use.
16. Exhaust filtration units rental, set-up and dismantle.
17. Plastic disposal bags and transport drums for protection during haul to disposal site.
18. Transport costs to disposal site and dump fees.
19. Wet cleaning of all surfaces and fixtures.
20. Spray sealants to prevent release of residual asbestos fibers.
21. Replacement thermal or fireproofing materials where asbestos was removed.
22. Recordkeeping and final submittals.

GLOSSARY

Abatement: Procedures to eliminate fiber release from asbestos containing building materials including encapsulation, enclosure or removal methods.

Air Lock: A system for permitting ingress and egress from one room to another while permitting minimal air movement between rooms, typically constructed by placing two overlapping sheets of plastic over an existing or temporarily framed doorway. Two curtained doorways spaced a minimum of 6 ft. apart from an air lock.

Air Monitoring: The process of measuring fiber content of a specific volume of air in a stated period of time.

Amended Water: Water to which a surfactant is added.

Clean Room: An uncontaminated area or room which is part of the worker decontamination enclosure system, with provisions for storage of street clothes and protective equipment.

Curtained Doorway: A device to allow ingress or egress from one room to another while permitting minimal air movement between the rooms, typically constructed by placing two overlapping sheets of plastic over an existing or temporarily framed doorway. Two curtained doorways spaced a minimum of 6 ft. apart from an air lock.

Equipment Decontamination System: A decontamination enclosure system for materials and equipment, typically consisting of designated area of the work area, a wash room and an uncontaminated area.

Equipment Decontamination Enclosure System: A decontamination enclosure system for materials and equipment, typically consisting of designated area of the work area, a wash room and an uncontaminated area.

Equipment Room: A contaminated area or room which is part of the worker decontamination enclosure system with provisions for storage of contaminated clothing and equipment.

HEPA Filter: A High Efficiency Particulate Absolute (HEPA) filter capable of trapping and retaining 99.9% of asbestos fibers greater than 0.3 microns in length. These filters are used in vacuum cleaners or in air transfer units (exhaust fans).

HEPA Vacuum Equipment: High Efficiency Particulate Absolute filtered vacuuming equipment with a filter system capable of collecting and retaining asbestos fibers. Filters should be of a 99.9% efficiency for retaining fibers 0.3 microns (3/1000 of a millimeter) or larger.

Removal: The act of removing asbestos containing or contaminating materials from the building to an approved disposal site.

Sealant: A material spray applied to the substrate after the wet removal of the existing asbestos material. The purpose of this is to prevent the release of residual asbestos fibers left after removal and cleaning operations.

Shower Room: A room constituting an air lock between the clean room and the equipment room in the worker decontamination enclosure system, with hot and cold or warm running water suitably arranged for complete showering of workers during decontamination. Shower room always includes an air lock.

Specific Regulations: The following regulations govern work for asbestos removal:
(a) OSHA Title 29, Code of Federal Regulations, Section 1910.1001 Occupational Safety and Health Administration, U.S. Department of Labor.
(b) EPA Title 40, Code of Federal Regulations, Part 61, Subpart A and B, National Emmissions Standards for Hazardous Air Pollutants, U.S. Environmental Protection Agency.

Surfactant: A chemical wetting agent added to water to improve penetration into the asbestos thus reducing the amount of water required for a given operation or area. This surfactant is spray applied to the asbestos to decrease the amount of airborne fibers while removing the asbestos material from the substrate. (Wetting agent-50% polyglycol ester and 50% polyoxyethlene ether or approved equal shall be mixed with water to provide a concentration of 1 oz. surfactant to 5 gallons of water or the manufacturer's recommended concentration.

Wet Cleaning: The process of eliminating asbestos contamination from building surfaces and objects by using cloths, mops or other cleaning tools which have been dampened with water and by afterwards disposing of these cleaning tools as asbestos contaminated waste.

Worker Decontamination Enclosure System: A decontamination enclosure system for workers typically consisting of a clean room, a locker room, a shower room, and an equipment room.

CHAPTER 24

ROAD AND HIGHWAY CONSTRUCTION

This chapter will serve the needs of the General Contractor/Estimator and the specialty subcontractor engaged in preparing estimates and performing paving work associated with construction projects. It will serve as a guide and a reference in solving many of the problems encountered by providing practical information to expand the basic knowledge and understanding of the methods and techniques.

Much of the information contained in this chapter will be applicable to construction of major highway works and interstate transportation systems. However, this sophisticated and specialized form of highway construction is not specifically addressed in this chapter as it would produce information useful to relatively few construction organizations dedicated to this highly competitive and enormously complex segment of the construction industry.

Supplemental information is available through extensive publications devoted to research data, current practices, and standard specifications which have been produced through the efforts of the American Association of State Highway Officials, the Highway Research Board of the National Research Council and the Bureau of Public Roads. In addition, several material associations have provided factual publications for the benefit of the highway construction industry. The most notable and active associations are The American Concrete Institute, Association of Asphalt Paving Technologists, and The Asphalt Institute.

Road, highway and paving construction is a high risk business under the most favorable conditions, demanding a thorough knowledge of the technical skills involved, accompanied by managerial and administrative talent to support a totally efficient and coordinated operation. Like other types of contracting, the paving contractor is faced with the obligation of accurately evaluating the capabilities and resources of his/her own organization. This specifically includes the financial capacity of the company and the availability of external sources for interim financing; availability of construction equipment through ownership, lease or rental with emphasis on condition, suitability and sufficiency, facilities for maintenance and repair; existence of systems and personnel to efficiently provide cost control, fiscal control, scheduling, purchasing, expediting; and qualified estimators capable of producing accurate estimates. Once this evaluation is complete, it can then be determined the type, size and complexity of construction projects that would best match the resources and capabilities of the company. This matching process is the basis of selection of projects for bidding.

It should be noted that firms having sufficient financial capacity to qualify for bonding on projects of a certain magnitude may have a deficiency in providing sufficient experienced personnel to efficiently accomplish the work. Thus, contractors would be well advised to follow the formula:

FINANCIAL ASSETS + PEOPLE ASSETS = TOTAL CAPABILITY

Because of the impact of road and highway construction on the economic, social and political decisions of a community, there has to be a presumption that many laws and regulations affecting the conduct of a construction program involving roads and highways will exist. Each geographic location will have its own peculiarities requiring the Contractor/Estimator's investigation and familiarity with these requirements prior to becoming involved in the bidding process. Prequalification of contractors is often a prerequisite, as is compliance with local contractor license laws or registration as a foreign corporation if the company is not incorporated under that State's law.

The intent of this chapter is not to present extensive engineering and design theory, but rather, to provide sufficient examples of incidents where Contractors,

Estimators, Project Managers and other supervisory personnel have been confronted with problems relating to paving operations where without a source of some basic knowledge of design, they were placed at a disadvantage in the decision making process. This lack of knowledge could become an extremely important factor when attempting to establish a fair and reasonable cost factor in the initial estimating process and possibly have more impact when attempting to evaluate a remedial solution for defective paving which may be assigned as a Contractor's responsibility.

Contractors will generally be working within the requirements of the project specifications. Strict compliance normally will protect the Contractor from claims. However, not all specifications are complete and totally descriptive of the precise methods necessary in order to achieve the desired end results. In some instances the designing architect also lacks sufficient understanding of the capabilities and characteristics of the materials involved in the construction. Frequently, the problem is further clouded by including a general reference to the AASHO System (American Association of State Highway Officials) or an attempt to include elements extracted from the Unified Classification System for Soils and fortified with vague references to density, moisture content, compactness and gradation of materials.

While the road and paving may be a somewhat minor cost factor as it relates to an entire construction project, it would be unacceptable and unprofessional not to establish the most accurate cost possible for this work. Every estimator should be aware that failures are costly and properly developed estimates do not make provisions for re-performance of work. It is obvious that unacceptable work requiring remedial action escalates the cost of the project and lowers the profit margin.

Generally, it is found advantageous to utilize the talent, expertise and services of a specialty contractor in the paving operation. However, this subcontractor will demand certain supporting performances by the prime or general contractor. This may include the preparation of the subgrade, all line and grade engineering and any testing which may be required.

More extensive knowledge on the part of the Contractor/Estimator creates a position where methods, schedules, unit costs and other specific details can be discussed with authority and confidence prior to entering a contract for paving work, with reasonable assurance the ultimate objective will be achieved with minimal problems. This same theory should be applied with the same force and effect on each of the other divisions of work included in the total project.

CONSTRUCTION DOCUMENTS

1. PLANS

   Properly prepared plans provide sufficient information for an estimator to make a quantitative analysis of each discipline of work included in the project. This function should be performed even if the Owner has provided the *estimated quantities* in the Bid Form, because accuracy cannot be determined except by comparison with the Contractor's quantity survey.

2. SPECIFICATIONS

   The technical specification for highway work is usually a "guide specification" making many references to the *Standard Specification* requirements where the details appear for the various types of work to be performed. As an example, the guide specifications may identify curb and gutters as Types A, C, D and F at certain locations. To determine the specific design of each type, reference is made to the Standard Specification. Copies of Standard Specifications can be purchased from the agency issuing the plans and specifications.

3. BID FORMS

In the case of unit price bidding, which is the usual format in road and highway construction, the form will provide for each classification of work with estimated quantities included. The Contractor is required to insert a unit price, including allowance for overhead and profit on each item, and calculate the estimated quantity by the unit price to determine total cost on each line item. A typical example would be as follows:

Bid Form

Code	Item	Quantity	Unit	Unit price	Total
200	Clearing	6	Acres	$2,000.00	$12,000.00
303	Excavation, Class 2	100,000	Cu. Yds.	1.00	100,000.00
310	Seed	500	Lbs.	3.00	1,500.00
490	Asphalt Concrete Agg.	2,000	Ton	26.00	52,000.00
520	Cement Treated Base	2,000	Cu. Yds.	31.00	62,000.00
710	Modify Signs	11	Each	300.00	3,300.00
730	Pipe Culvert 24"	200	Lin. Ft.	40.00	8,000.00
732	RC Pipe 48"	200	Lin. Ft.	65.00	13,000.00
				Total Price	$251,800.00

Should an error appear in the extension to the total line amount, it has been ruled many times that the Owner has the right to use the unit price offered by the contractor and correct the mathematical error. Estimators should use extreme caution in developing the unit price bid item.

4. INSTRUCTION TO BIDDERS

These documents provide the precise information relating to bid guarantee, number of copies of the bid to be furnished, establishment of date and time bids are to be received, documentation to be included in the bid package, and other instructions appropriate to the project. These requirements are to be fulfilled *without deviation* or the bid may be considered non-responsive, making the contractor ineligible for award.

5. SPECIAL INSTRUCTIONS

1. Physically inspect every project site prior to bidding to determine working conditions.
2. Avoid accepting oral interpretation or instructions as they cannot be enforced.
3. If uncertainty or lack of confidence appears during the bidding process, the best decision would be to avoid bidding on the project rather than risk a serious mistake and possible loss.

SOIL STABILIZATION

The need for soil stabilization is widespread in the construction industry to support the designed contour configuration which provides beautification in the landscaping of buildings, parks, and other green areas having esthetic appeal. In addition there is an important functional aspect to soil stabilization. This applies to the prevention of soil erosion and providing a sound and stable base to support roads, walks, terraces, parking areas and other surfaces subjected to concentrated loading and heavy traffic wear.

Without sufficient stability of the subgrade, all finishes placed upon it are, to some degree, influenced by the credibility of the subgrade. It is not uncommon to encounter complete failure of paved areas which can be traced to poor or ineffective preparation of the subgrade prior to the installation of properly designed and installed finish surface.

Soils exist in a broad range of types with wide variation in the value of their properties and they react differently to the several methods available to improve their degree of stabilization. It is generally recognized that the ultimate resistance of a soil to the penetration of concentrated load depends on the following two (2) properties of the soil: (1) Friction between soil particles and (2) the Cohesion between the soil particles. Soil resistance due to Cohesion is dependent on the external pressure on the soil. However, Frictional Resistance increases in direct proportion to the external pressure.

EXAMPLE: Sand has a high Frictional Resistance without appreciable Cohesion and the bearing capacity is increased considerably when a concentrated load is surrounded by a distributed load.

Clay, having high Cohesive Resistance, but without appreciable Frictional Resistance, reacts quite differently as the bearing capacity is not increased to any great extent by applying a concentrated load surrounded by a distributed load.

The stability of a soil for subgrade purposes also depends on its Elasticity or the ability to rebound after a load has been applied and then removed. Excessive Elasticity makes a soil less stable because it cannot be permanently compacted.

Water content in the soil can be a significant factor in stabilization, particularly in the expansive soils which are subject to change in volume, because they possess cohesion and capillarity in appreciable amounts.

Often the quality of subgrades can be improved by installing some type of artificial drainage to prevent the accumulation of excessive moisture from the subgrade materials. Such a provision would also provide protection from "pumping" and "heaving" action during freezing and thaw periods.

EXCAVATION

Excavation, moving and depositing of existing materials to a new location are major cost items to the Contractor/Estimator involved in road and highway construction. All encountered materials are basically rock. However, because of the forces of nature, there are great variances in the size of particles, from materials classified as solid rock continuing down the scale to those identified as sand or clay.

This variation in material is so extensive it would be presumptuous and extremely dangerous for the excavator to attempt calculating the cost involved without engaging in preliminary studies. Each classification retains its own characteristics. In some instances blasting or ripping is required to separate the materials from the natural position, while other materials can be readily removed by loader or scraper. Dissimilar conditions will be encountered when attempting to compact the various types of materials.

On a given construction project, it would not be unusual to encounter several types or classifications of materials, requiring the contractor to modify methods and/or equipment to economically accommodate each condition. An understanding of geology and soils mechanics is necessary to effectively deal with the complexities of excavation. Experience in soils of a geographical location also contributes to the overall knowledge.

In excavating hard materials such as shales, sandstone and limestone, the strata of rock may be hard and strong, but if lamination thickness and vertical joint spacing are limited in size, production will be improved over instances where these planes of weakness produce greater cubic content of the material.

The most common practices used to determine the blasting and ripping zones for rock excavation is the track drill because of its availability, portability and ability to produce data relative to drilling costs and blasting techniques. The seismic timer is an instrumental means to determining the degree of consolidation of the rock which can be related to the cost of excavation.

When excavating the usual non-cementious materials, a variety of machines and methods are available to perform specific tasks at the most economical cost. Loaders, shovels, cranes, Gradalls, backhoes, draglines and pans are the most common machines in use, with each one having certain advantages and limitations. It is the duty of the Estimator to know the capabilities of these machines and to quantitatively prove their efficiency before making a judgment.

The following formula will provide hourly production for a front-end loader with a 3 cubic yard capacity bucket working average soil:

Loading time cycle through an angle of 120° is a 0.5 minutes, comprised of the following four (4) elements:
(1) Loading - 7 seconds (0.12 minutes)
(2) Swing-Position - 12 seconds (0.20 minutes)
(3) Dumping - 3 seconds (0.05 minutes)
(4) Return-Position - 8 seconds (0.13 minutes)
    or a total of 30 seconds (0.50 minutes)

Hourly production in cubic yards, bank measurement, per 50-minute working hour is determined by the following formula:

$$P = \frac{50}{CT} \times BF \times BC$$

P = Hourly Production, CY/BM
CT = Cycle Time in Minutes
BF = Bucket Factor
BC = Bucket Capacity

Bucket Factor Table

MATERIAL	Fill Factor	Conversion Factor	Bucket Factor
	Loose Measurement / Bucket Capacity	Bank Measurement / Loose Measurement	Bucket Measurement / Bucket Capacity
Sand - Gravel	1.05	0.88	0.92
Weathered Rock - Earth	0.90	0.78	0.72
Blasted Rock	0.80	0.67	0.54

The equation becomes:

$$P = \frac{50}{.50} \times BF \times BC$$

P = 100 x BF x BC
P = 100 x 0.92 x 3.0
P = 275 cubic yards - Bank measurements/50 min. work hour

Assume this 3 cu. yd. capacity front-end loader could be rented including operator, fuel, maintenance, and transportation charges for $80.00 per hour, then the cost for excavation would be :

$$Cost/CYBM = \frac{Rental\ Cost\ Per\ Hour}{CYBM\ Per\ Hour}$$

$$Cost/CYBM = \frac{\$80.00}{275}$$

$$Cost/CYBM = \$0.29$$

This cost is pure excavation and loading cost. To this must be added hauling, dumping charges, job supervision, establishment and maintenance of haul roads, compaction of fills and other miscellaneous costs. In making cost production analysis on each type of machine planned for use on a project, other valuable information will be produced. The hourly production rate on the excavating machine will determine the number of trucks required to haul the excavation to other locations within or external to the construction site. Truck costs can be established in similar analysis predicated upon loading time, capacity of the truck and the haul time to the disposal area.

To determine the number of trucks required to service and support the production rate of the previous front-end loader, it can be established as follows:

$$P = 275 \text{ CYBM}$$

$$PM = \text{CYBM Per Minute of 50 min. work hour}$$

$$WH = 50 \text{ minute/hour}$$

$$PM = \frac{P}{WH}$$

$$PM = \frac{275}{50}$$

$$PM = 5.5 \text{ CYBM}$$

A ten cubic yard truck could be loaded every two (2) minutes of work time, thus 25 trucks would be required to service the loader. Haul time for the trucks would have an influence on the number of trucks required.

HAULING

The hauling of excavated materials to a disposal area is the most important phase of excavation. It is usually the most costly and subject to many variables. In instances of off-site hauling by trucks, the traffic patterns will have a substantial influence on the haul time. More haul time will be consumed during morning and evening peak traffic periods than during the midday normal traffic flow.

On-site hauling by self propelled scrapers can be determined with more precision as the conditions can be controlled by the Contractor.

After selection of scrapers of a suitable capacity, there is a need to determine the travel time involved from the point of loading to the disposal destination. This can be accomplished by using the following equation:

$$\text{TRAVEL TIME (MIN.)} = \frac{\text{LENGTH OF HAUL (FEET)}}{88 \times \text{AVERAGE SPEED (MPH)}}$$

Eighty eight (88) represents a factor to convert miles per hour to feet per minute. Speed factor tables are available from equipment manufacturers for the conversion of maximum speed to average speed, but these are somewhat complicated to use. On smaller projects, past experience would serve the needs of the Contractor. In establishing an average speed factor, utilize the rated maximum speed of the machine and multiply by realistic percentages reflecting job conditions to determine average speed.

Short haul distances, 200 to 600 feet would be at a 50% level because of stop frequency and turn-around time loss. For haul distances of 2,000 to 3,000 feet, the percentage will increase to 75% to 80%. These percentages would be for average condition with level grades, no sharp turns or hazards, and without interference of cross traffic.

In using the equation to establish travel time after the loading operation, assume the length of haul to be 3,000 feet with an average speed of 12 mph.

$$\text{TRAVEL TIME (MIN.)} = \frac{3,000}{88 \times 12}$$

$$\text{TRAVEL TIME (MIN.)} = \frac{3,000}{1,056}$$

$$\text{TRAVEL TIME (MIN.)} = 2.84 \text{ Minutes}$$

To establish the number of haul cycles, divide the 50 minute work hour by the travel time:

$$\frac{50}{2.84}$$

Number of Haul Cycles/Hour = 17.6

Production could be calculated by allowing a 2 minute loading time plus the 2.84 minute haul time for a total 4.84 minute production cycle for a 24 cu. yd. scraper:

$$P = \frac{50}{4.84} \times 24$$

$$P = 10.33 \times 24$$

$$P = 247.92 \text{ cubic yards/hour}$$

Should materials or job conditions be encountered which would require a pusher machine, then it would be an additional cost to the cost of the basic scraper.

FILL COMPACTION

The distribution and compaction of fills to rigid specifications embraces the utilization of several machines designed for the densifying operation which is accomplished by spreading-mixing, wetting or drying, and the use of compactors.

Production and costs for spreading materials are complicated by the overlapping and sharing of work with other machines. The following general equations have been found acceptable for hourly production when using the horse power (HP) rating of the machine as a base:

DOZER $CYBM = 3 \times HP$

MOTOR GRADER $CYBM = 3 \times HP$

The use of water wagons is a function of the haul cycle time and can be estimated by the following equation:

$$\text{PRODUCTION (GAL/H)} = \frac{50}{\text{CYCLE TIME (MIN.)}} \times \text{CAPACITY (GAL.)}$$

Water wagons range in capacity from 1,500 to 14,000 gallons. The Contractor must decide on the most economical based on the anticipated needs at the project.

Laboratory test data provides the existing natural moisture content. The moisture-density curve indicates that the desired percent of relative density will be attained with a percentage of moisture by dry weight. With this information the Estimator can produce an estimated need for water and project this toward a water wagon cost.

In instances where excessive moisture is encountered, this can be reversed by aeration of existing materials or the importation and mixing of dry materials. In each instance, it becomes a costly operation and delays job progress.

CALCULATION OF EXCAVATION QUANTITIES

In establishing the quantities of excavation, which is subdivided into cuts and fills, the Estimator is actually determining the volume or cubic yards involved in each operation through calculations involving areas, distance and geometric figures.

    1.   VOLUME: 1 cubic yard = 27 cubic feet
    2.   DISTANCE: 1 linear foot = 12 inches
                   STATION = 100 linear feet
                   MILE = 5,280 feet
    3.   AREA: Square Yard = 9 square feet
               Acre = 43,560 square feet
               Square Mile = 640 ac.

The tables (pages 24.9-12) provide the formulas for calculating the quantities of the most frequently found geometric figures.

The usual methods of estimating are the Average End Area Method and the Cross Section Method.

Average End Area Method

FORMULA: $V = \dfrac{D(A_1 + A_2)}{2}$

                                 V = Volume in Cubic Feet
                                 D = Distance between stations or cross sections
                                 A_1 = Area of cross section at point 1, sq. ft.
                                 A_2 = Area of cross section at point 2, sq. ft.

This method produces sufficient precision for normal estimating purposes when the distances between cross sections are reduced at extreme changes in elevations. In most instances on level or rolling terrain, the stations or cross sections would be at 100'-0" intervals. However, in extreme changes in grades the distance would be decreased to 50'-0" or more to lessen the degree of error. The prismoidal formula found in the Earth Work Section (Page 7.14) will produce additional accuracy.

The following example will illustrate this method of estimating:
Assume the sketch to be a 300 foot portion of the topographic plan and profile plan of a proposed road. The topographic plan shows the undulations of the present ground elevations by the contours and the road location by the center and shoulder lines. The profile plan shows the relation between the present ground surface and

AREA (A : Area)

NAME FORMULA

SHAPE

Parallelogram

$A = B \times h$

Trapezoid

$A = \dfrac{B + C}{2} \times h$

Triangle

$A = \dfrac{B \times h}{2}$

Trapezium
(Divide into 2 triangles)
A = Sum of the 2 triangles
(See above)

Regular Polygon

$A = \dfrac{\text{Sum of sides (s)}}{2}$

x Inside Radius (R)

Circle

$A : \begin{array}{l} (1)\ \pi R^2 \\ (2)\ .7854 \times D^2 \\ (3)\ .0796 \times C^2 \end{array}$

AREA   (A : Area)

NAME FORMULA   SHAPE

Sector

$$A : \quad (1)\ \frac{a^2}{360\,o}\ \times \pi R^2$$

$$(2)\quad \text{Length of area} \ \times \ \frac{R}{2}$$

Segment

A : Area of sector minus

     triangle (See above)

Ellipse

$$A = M \times m \times .7854$$

Parabola

$$A = B \times \frac{2\ h}{3}$$

VOLUME (V : volume)

NAME FORMULA   SHAPE

Cube

$$V = a^3 \quad (\text{ in cubic units })$$

Rectangular Solids

$$V = L \times W \times h$$

Prisms

$$V\ (\ 1\ ) = \frac{B \times A}{2} \times h$$

$$V\ (\ 2\ ) = \frac{s \times R}{2} \times 6 \times h$$

$$V = \text{Area of end} \times h$$

Cylinder

$$V = x\pi R^2 \times h$$

VOLUME (V : volume)

NAME FORMULA	SHAPE
Cone $V = \dfrac{\pi R^2 \times h}{3}$	
Pyramids $V(1) = L \times W \times \dfrac{h}{3}$ $V(2) = \dfrac{B \times A}{2} \times \dfrac{h}{3}$ $V = \text{Area of Base} \times \dfrac{h}{3}$	
Sphere $V = \dfrac{1}{6}\pi D^3$	
Circular Ring (Torus) $V = 2\pi^2 \times R r^2$ $V = \text{Area of section}$ $\times 2\pi R$	

the proposed road bed grade at the centerline (¢) of the road. Also assume road bed of 20'-0" wide with shoulder slopes of 1½:1:

Topographic and Profile Plans.

Method of Plotting End Areas.

As each section is plotted at the stations, the area enclosed by the present and proposed grade lines are computed as cuts or fills. Final computations are made in the following form:

Station	Distance Between Stations	Actual Cut	Average Cut	Actual Fill	Average Fill	Volume in Cut	Cu. Ft. Fill
				Area of Sections in Sq. Ft.			
15+00		24.38		—			
	7.0 ft.		20.41		—	142.87	—
15+07		16.44		—			
	30.0 ft.		8.22		7.46	246.60	223.80
15+37		—		14.92			
	63.0 ft.		—		51.23	—	3,227.49
16+00		—		87.54			
	50.0 ft.		—		53.45	—	2,672.50
16+50		—		19.36			
	15.0 ft.		7.56		9.68	113.40	145.20
16+65		15.11		—			
	35.0 ft.		26.72		—	935.20	—
17+00		38.33		—			
	65.0 ft.		59.77		—	3,885.05	—
17+65		81.20		—			
	35.0 ft.		62.34		—	2,181.90	—
18+00		43.48		—			
						7,505.02	6,268.99

TOTAL CUT   7505.02/27   =   278 cu. yds.
TOTAL FILL   6268.99/27   =   232 cu. yds.

SURPLUS FOR DISPOSAL   =   46 cu. yds.

Cross Section Method

One of the easiest and most frequently used methods of computing grading cuts and fills when the plot plan shows both original and proposed contours is that of gridding or dividing up the area to be graded into squares, rectangles, triangles or combinations of these figures, of regular and convenient dimensions, thus forming a series of truncated prisms. The depths of cut or fill at the corners of these prisms are computed and the volume of each prism may be determined by averaging these depths and multiplying the average by the cross sectional area of the prism. The results of these computations are totaled, keeping the cuts separate from the fills, and the total excavation and/or fill required is obtained.

Example: Assume the contour sketch shown on page 24.16 to be a 400.0' x 200.0' portion of a site grading plan showing both present and proposed contour lines. Since both the present and proposed surfaces have fairly regular slopes, as indicated by the even spacing of contour lines, a grid spacing of 50 feet in both directions has been chosen and grid lines drawn as shown, labeled alphabetically for the vertical lines and numerically for the horizontal lines.

The next operation is to figure the present ground elevation for each grid line intersection by interpolating between the present contour lines. The results are marked on the drawing in the upper right quadrant of each respective intersection.

Following this, the proposed ground elevations are similarly figured for each intersection, interpolating between the proposed contour lines and entering the results in the upper left quadrant of each respective intersection.

When this is completed the depth of cut or fill at each intersection can be readily determined by subtracting one elevation from the other and noting whether a cut or fill is required. This determination should be made for each intersection and the results entered in the proper quadrant, using the lower left for cuts and the lower right for fills.

When all intersections have a depth value entered for either cut or fill the plan should be examined closely to see where the grading operation will change from cut to fill or vice versa. In the sketch it will be seen that advancing from left to right on horizontal grid line "1" a change from fill to cut occurs between vertical grid lines "D" and "E" and by interpolating between the fill and cut depths the zero point "a" can be established approximately 36 feet to the right of intersection "D-1."

Similarly, zero points "b, d, e and g" are established on the horizontal grid lines and zero points "c and f" on the vertical grid lines. These points are then connected by a line "a, d, c, d, e, f, g" and for all practical purposes it may be assumed that all points on this line are at the proposed elevation and require neither cut nor fill. In this case the line forms the boundary between cuts and fills. The change from cut to fill and vice versa could, of course, occur many times in actual problems depending on the irregularity of the present and/or proposed contour lines.

After the previously described operations have been completed, the actual volume of cut and/or fill required can be computed.

The volume of cut or fill in each prism in the grid system is obtained by averaging the corner depth values of each prism and multiplying by its cross sectional area. Thus, for the square prism "A-1, B-1, B-2, A-2" the volume of fill would be ¼ x (2.4'+1.6'+1.6'+2.5') x 50.0' x 50.0' or 5,063 cu. ft. All grid prisms are similarly treated, keeping cut and fill results separated, except those prisms through which the zero line passes. Therefore for square prism "D-1, E-1, E-2, D-2" it will be seen that this volume has been divided into two trapezoidal prisms "D-1, a, b, D-2" and "a, E-1, E-2, b." To compute the respective fill and cut volumes of these prisms they must be divided into triangular prisms first and then the corner depths may be averaged and multiplied by their respective cross sectional areas. Thus for the triangular prism "D-1, a, b" the volume of fill would be ⅓ x (0.5'+0.0'+0.0') x ½ x 50.0' x 36.0' or 150 cu. ft. Similarly, the volume of fill for the triangular prism "D-1, b, D-2" would be ⅓ x (0.5'+0.0'+0.3') x ½ x 50.0' x 21.0' or 140 cu. ft. The volume of cut required for the trapezoidal prism "a, E-1, E-2, b" is figured in the same manner. In the case of square cross section "C-2, D-2, D-3, C-3" it will be seen that the zero line crosses one corner of the square and divides it into a triangle "c, D-3, d" and a five-sided figure "C-2, D-2, c, d, C-3." The five-sided figure should be further divided into three triangles "C-2, D-2, c," "C-2, c, C-3" and "C-3, c, d" and the volumes of fill for each triangular prism would be as follows:

⅓ x (0.7'+0.3'+0.0') x ½ x 50.0' x 30.0' or 250 cu. ft.

⅓ x (0.7'+0.0'+0.5') x ½ x 50.0' x 50.0' or 500 cu. ft.

⅓ x (0.5'+0.0'+0.0') x ½ x 36.0' x 20.0' or   60 cu. ft.

When there are several prisms with the same cross section to be either cut or filled the volume of such can be computed as one solid by assembling them as follows: Multiply each corner depth by the number of typical prisms in which it occurs and then add these results and divide by 4. This is then multiplied by the cross sectional area of one prism. For example the volume bounded by A-1, D-1, D-2, C-2, C-4, B-4, B-5, A-5, A-1 can be obtained by one computation because it is composed of a series of prisms having the same cross section. In the summation of depths, those at A-1, D-1, D-2, C-4, B-5 and A-5 are taken but once, those at B-1, C-1, C-3, A-4, A-3 and A-2 are multiplied by 2, at C-2 and B-4 the depths are

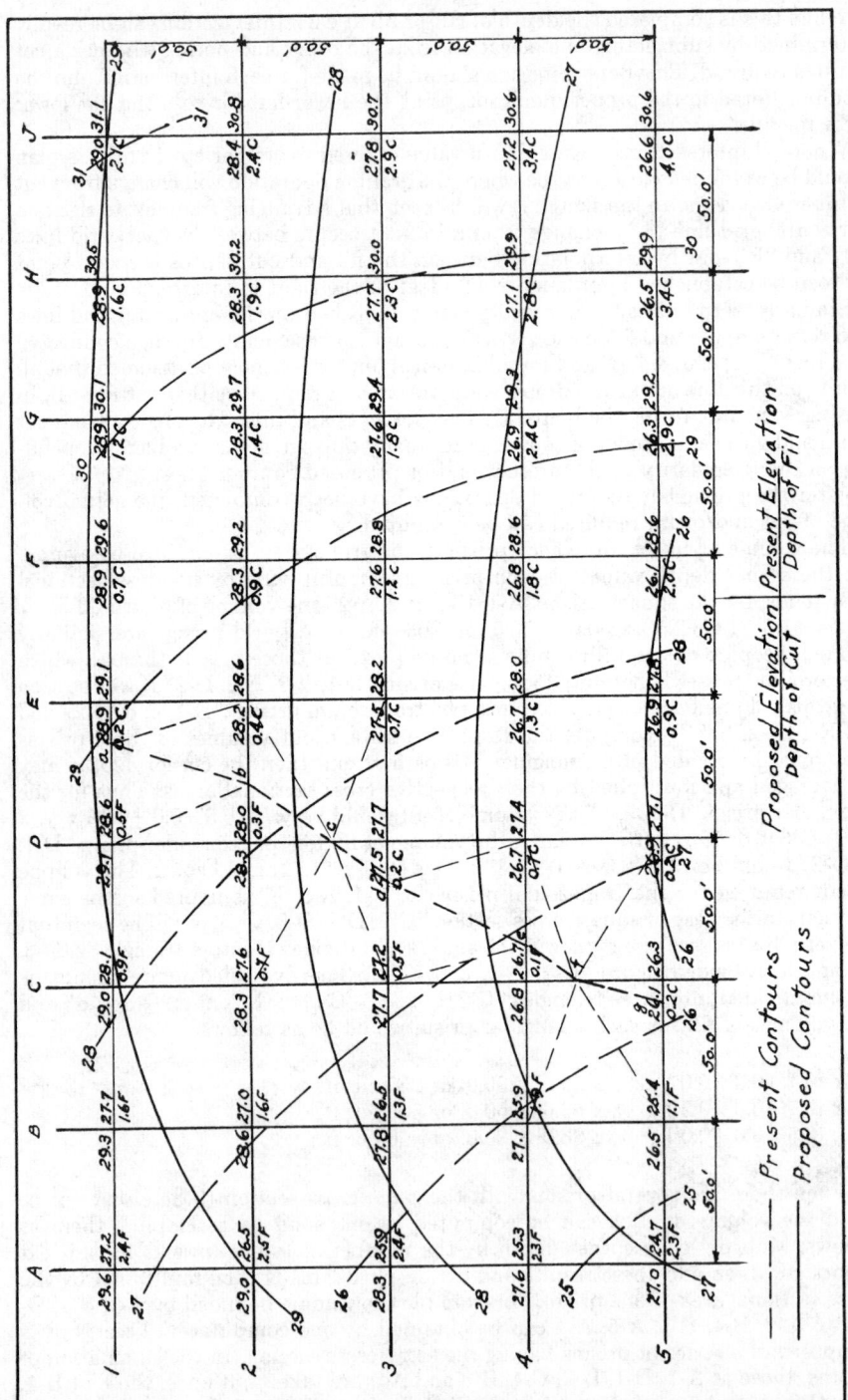

Method of Cross Sectioning a Contour Plan.

multiplied by 3 and at B-2 and B-3 they are multiplied by 4. All of these values are then totaled and divided by 4 and then multiplied by the cross sectional area of one prism giving the volume of fill required.

When a volume has been computed for each grid division of the site, either individually or collectively, it is only a matter of summing up the cuts and/or fills to obtain the total excavation and/or fill required to bring the site to the proposed grades.

Recordation of the detailed take off information can be assembled as shown on the following form for a permanent record showing the precise condition and calculation for each grid section of 50 feet x 50 feet or constant 2,500 square foot area of the project.

This information can be referenced at any time during construction for determination of the need of the type and quantity of excavation equipment.

Grid Section	Calculations	Grid Area 50 x 50	CUT cu. ft.	FILL cu. ft.
A1, B1, B-2, A-2	(2.4 + 1.6 + 1.6 + 2.5)= 8.1 x .25 = 2.025	2,500sq. ft.		

BITUMINOUS ROAD PAVING

This flexible paving method consists of several layers or courses of mineral materials placed upon a properly prepared subgrade of earthwork which performs as a foundation for the paving. Quality, quantity and methods of performance will be governed by the requirements of the contract specifications.

Subbase Course.—A layer of selected granular material of pre-determined depth applied to the subgrade for the support of the paving materials. It also prevents or reduces frost damage and provides under pavement drainage.

Base Course.—A high quality material such as crushed stone which is placed on the subgrade or the subbase. Gradation and depth controlled by specification. Also may be treated with portland cement or bituminous to reduce thickness and improve durability.

Leveling Course.—Used to eliminate irregularities in existing paved surfaces prior to resurfacing.

Binder Course.—In asphalt paving a binder course, which is often referred to as a leveling course, is installed at a prescribed thickness over the subbase. The strength of the binder course will be substantially equal to the designed surface or wearing course.

Surface Course.—The uppermost or exposed surface of the paving designed to withstand wear and polishing by traffic. Smooth surfaces reduce noise level but increase skid action.

Pumping.—Commonly referred to as mud pumping. Occurs when the subgrade contains excessive moisture and is a type containing a high percentage of fine particles. The surface becomes flexible and unstable to any load imposed upon it. This condition is unsuitable to support paving and must be removed and replaced with structural support for the paved surface.

Subgrade.—Foundation on which the subbase and paving materials are placed. The utilization of locally available materials for use in the subgrade has important economic benefit unless the material is of a poor quality requiring additives in the form of chemicals or modification in the soil content and gradation to achieve the desired quality of compaction. In such instances, consideration should be given to the economics of removal of a sufficient quantity of the existing soil and replacement by fill material requiring limited compaction effort to produce a suitable surface to receive the subbase material.

ASPHALT TERMS AND DEFINITIONS

1. **Natural Asphalt (or Lake Asphalt)** - Derived from petroleum by the natural process of evaporation leaving the asphalt material in open beds or lakes ready for extraction and processing.
2. **Petroleum Asphalt** - Asphalt refined from crude petroleum.
3. **Asphalt Cement** (A.C.) - Refined asphalt for paving purposes.
4. **Blown Asphalt** - Asphalt specially treated at high temperature levels with blown air to obtain specific characteristics necessary in products for roofing, pipe coating, membrane envelopes and hydraulic application.
5. **Flux Oil** - A thick petroleum product used to soften asphalt to a desired consistency.
6. **Cutback Asphalt** - Asphalt cement blended with petroleum solvents to produce a more liquid or fluid product thus improving handling and workability qualities. Exposure to atmosphere evaporates the solvents and the remaining asphalt performs the intended function. Variation in the types of solvent used produces the following products:
   A. Rapid Curing Asphalt (RC)
   B. Medium Curing Asphalt (MC)
   C. Slow Curing Asphalt (SC)
7. **Asphalt Primer** - Liquid asphalt applied to a non-bituminous surface to promote adhesion between existing surface and new surface application.
8. **Asphalt Joint Filler** - Liquid asphalt product used to seal cracks and joints in pavement to prevent moisture penetration.
9. **Preformed Asphalt Joint Filler** - Asphalt mixed with various fibrous materials and manufactured in solid form to a specific thickness. Widely used as an expansion joint material.
10. **Asphalt Planks** - Similar to preformed joint filler with variations to include reinforcing mesh, and surface treated with mineral grits, manufactured as planks or panels in sizes from 12 inches wide to 8 feet long. General usage provides planks or panels over built-up roofing to protect roof surfaces from hazards of pedestrian traffic.
11. **Asphalt Overlay** - Application of asphalt construction over an existing pavement. If the overlay is applied to a rigid concrete pavement, it is then referred to a combination or composite type.
12. **Asphalt Base Course** - Supporting or foundation course of granular materials bound together with asphalt material. Usually installed between the subbase and surface material.
13. **Asphalt Macadam** - A particular type of paving identified by a surface constructed of course screened aggregate consisting of stone, slag or gravel.
14. **Rigid Pavement** - Portland cement concrete pavement having capabilities of distribution of loads to the foundation over a large area.
15. **Flexible Pavement** - Consists of a relatively thin wearing or traffic surface applied over a designed base course and possibly a subbase course applied to prepared subgrade. The strength of the pavement is achieved in the distribution of the surface loads through the base and subbase courses to the subgrade. Failure in any of the elements produces unsatisfactory qualities in the paving suface which, if subject to excessive deflections, will fracture and delaminate

from the surface below allowing moisture penetration and deterioration of the entire paving system.

16. **Asphaltic Concrete** - Mixture of graded aggregate, a mineral filler and asphalt cement applied in layers.

17. **Sheet Asphalt Paving** - Fine aggregate mixed with mineral filler and asphalt to serve as a wearing surface. Provides a very smooth, noise free surface but will sag on steep grades and becomes slippery when wet.

18. **Mineral Filler** - Consists of finely ground particles of limestone, hydrated lime or portland cement.

19. **Undersealing** - The process of injecting heated asphalt through an aperture drilled in concrete paving to fill all voids under the slab.

20. **Mudjacking** - A method of filling, under pressure, voids under a concrete slab with a mixture of fine soil, cement and water through holes drilled in the slab. Sufficient pressure can be produced to elevate the slab to desired elevation.

21. **Concrete Curing** - A process of maintaining sufficient moisture and a favorable temperature in concrete during the period immediately following placement until attainment of a satisfactory percentage of ultimate strength. Duration of curing period is usually a specification requirement and varies with the exposure of the concrete surface to atmospheric conditions.

22. **Pugmill** - Performs same function as the asphalt batch mix plant, however, the mixing principle differs. In the batch plant mixing is restricted to the capacity of the mixing chamber and in the Pugmill there is a continuous flow of materials toward the point of discharge.

23. **Tack Coat** - Application of asphalt to an existing surface to improve bonding with new surface treatment such as an overlay.

24. **Seal Coat** - Seal and fill cracks in pavement to exclude penetration of moisture.

25. **Soil Cement** - A careful mixing of cement with existing soil to produce an improved quality in the soil. This treatment can be applied to granular soils and silty or clay soils to improve strength characteristics, reduce effect of frost on the subgrade and control shrinkage and expansion actions found in certain soils. It is frequently used as part of the pavement patching operation and for stabilization of shoulders.

The existing soil should be tested by a laboratory and a design established as a guide for preparing the subgrade to achieve the desired results.

GENERAL TERMS ASSOCIATED WITH PAVING

1. **Rip Rap** - Large or graded rocks either placed on earth slopes in loose form or grouted in place to prevent erosion from water currents or wave action.

2. **Cribbing** - Composed of interlocking members of wood, steel or pre-cast concrete in a design of "stretchers" and "headers" to stablize earth masses or embankments.

3. **Calcium Chloride Treatment** - Usually mixed with gravel as a surface treatment on unpaved roadways not subject to heavy traffic. Surface can be kept reasonably dust free by periodic surface treatment of additional calcium chloride.

4. **Shoulders** - The area paralleling a paved area which can be seeded, sodded, treated with a bituminous surface or soil cement. It is not intended to serve as a traffic area. The shoulder serves to slope or drain surface water from paved areas and provides protection of the subgrade from water infiltration and frost.

5. **Blow-Ups** - A raising of a concrete slab or paving usually occurring at a crack or joint in the surface of the slab. It is uneven in form and caused by expansion

resulting from high temperatures and moisture conditions. Requires immediate remedial action which can be extensive to avoid hazards to vehicular traffic.

6. **Mulching** - A 3 to 4" layer of peat, muck, straw, leaves or moss placed over a planting area to preserve the moisture content.

7. **Staking** - Guys attached to newly planted trees secured with ground stakes to prevent excessive movement in the plant until root establishment.

8. **Pruning** - To cut away or remove a portion of the limb and leaf structure of a tree type plant. When relocating plants loss of root structure will require a balance of the top structure by pruning process.

9. **Ground Cover** - A cover of plants with extensive rooting systems such as honeysuckle or vines generously planted on slopes to prevent erosion.

STANDARD DESIGN DETAILS

For general guidance and reference, the following standard design details are included, all of which have been approved by the Federal Highway Administration.

Certain states, counties or municipalities may require some slight modifications, but it is assumed that deviations will not significantly exceed the requirements of these standard details.

Index

	GRANULAR MATERIALS						SILT - CLAY MATERIALS						
GROUP CLASSIFICATION	A-2	A-3	A-2-4	A-4-4	A-2-7	A-7-2	A-4	A-4-7	A-7-4	A-7	A-6	A-5	A-8
GENERAL DESCRIPTION	SAND	SAND	SILTY-SAND	SANDY-SILT	CLAYEY-SAND	SANDY-CLAY	SILT	CLAYEY SILT	SILTY-CLAY	CLAY	COLLOIDAL CLAY	MICA, DIATOMS & SILT	SWAMP MUCK
STABILITY	WHEN N.P., HIGH; WHEN PLASTIC, GOOD WHEN DRY	IDEAL WHEN CONFINED	GOOD WHEN DRY	GOOD WHEN DRY	GOOD WHEN DRY & PROPERLY COMPACTED	GOOD WHEN DRY & PROPERLY COMPACTED	GOOD WHEN PROPERLY COMPACTED OR UNDISTURBED	GOOD WHEN PROPERLY COMPACTED OR UNDISTURBED	GOOD WHEN PROPERLY COMPACTED OR UNDISTURBED	GOOD WHEN PROPERLY COMPACTED OR UNDISTURBED	POOR	GOOD TO POOR	NONE
USE AS A BASE	FAIR	EXCELL.	EXCELL.	EXCELL.	FAIR	FAIR	POOR	UNSATISFACTORY	UNSATISFACTORY	UNSATISFACTORY	UNSATISFACTORY	UNSATISFACTORY	UNSATISFACTORY
USE AS A SUB-BASE	EXCELL.	EXCELL.	GOOD	GOOD	GOOD	GOOD	UNSATISFACTORY	UNSATISFACTORY	UNSATISFACTORY	UNSATISFACTORY	UNSATISFACTORY	UNSATISFACTORY	UNSATISFACTORY
USE AS A SUB-GRADE	EXCELL.	EXCELL.	GOOD	FAIR	FAIR	FAIR	POOR	POOR	POOR	VERY POOR	VERY POOR	POOR	UNSATISFACTORY
FILLS UNDER 50'	EXCELL.	GOOD	GOOD	POOR	FAIR	FAIR	GOOD TO POOR	POOR	POOR	VERY POOR	VERY POOR	POOR	WASTE
FILLS OVER 50'	GOOD	GOOD TO FAIR	GOOD	POOR	FAIR	POOR	POOR TO VERY GOOD	POOR	VERY POOR	VERY POOR	VERY POOR	GOOD TO POOR	WASTE
FROST ACTION	NONE TO LOW	NONE TO LOW	MEDIUM	HIGH	MEDIUM	MEDIUM	HIGH	MEDIUM TO HIGH	MEDIUM TO HIGH	MEDIUM	MEDIUM	HIGH	MEDIUM
RANGE OF MAX. DRY DENSITY (AASHO T-180) (PCF)	115-135	105-130	110-130	110-135	115-135	115-130	110-135	110-135	105-130	100-120	90-115	100-135	LESS THAN 100
RANGE OF OPTIMUM MOISTURE CONTENTS (AASHO T-180)(%)	9-12	8-15	8-15	9-15	6-12	9-15	8-15	10-15	10-17	12-25	14-30	11-18	
REQUIRED COMPACTION (AASHO T-180)(%)	92-95	92-95	92-95	92-95	92-95	92-95	92-95	92-95	92-95	≤ 95%	≤ 95%	≤ 95%	
COMPACTION METHODS	ROLLING WITH SMOOTH FACE TAMPING RUBBER TIRED ROLLER OR VIBRATORY COMPACTOR	TRACTOR DISKING VIBRATION	TAMPING OR RUBBER-TIRED ROLLER	TAMPING OR RUBBER-TIRED ROLLER	TAMPING OR RUBBER-TIRED ROLLER	TAMPING SHEEPFOOT ROLLER OR RUBBER-TIRED ROLLER	TAMPING SHEEPFOOT ROLLER OR RUBBER-TIRED ROLLER	TAMPING SHEEPFOOT ROLLER OR RUBBER-TIRED ROLLER	TAMPING SHEEPFOOT ROLLER OR RUBBER-TIRED ROLLER	TAMPING SHEEPFOOT ROLLER OR RUBBER-TIRED ROLLER	TAMPING OR RUBBER-TIRED ROLLER	TAMPING OR RUBBER-TIRED ROLLER	WASTE
COMPACTION ABILITIES	GOOD WITH CLOSE CONTROL	GOOD	GOOD TO POOR	GOOD TO POOR	GOOD TO POOR	GOOD TO POOR	GOOD TO POOR	FAIR TO POOR	FAIR TO POOR	FAIR TO POOR	POOR	POOR	WASTE
PUMPING ACTION	SLIGHT TO NONE	SLIGHT TO NONE	SLIGHT TO NONE	SLIGHT TO NONE	SLIGHT TO NONE	SLIGHT TO NONE	GOOD TO FAIR	GOOD TO FAIR	GOOD TO FAIR	GOOD TO FAIR	GOOD TO FAIR	GOOD TO FAIR	GOOD TO FAIR
BEARING VALUE	EXCELLENT TO FAIR	EXCELLENT TO FAIR	GOOD TO FAIR	GOOD TO FAIR	GOOD TO FAIR	GOOD TO FAIR	FAIR TO POOR	FAIR TO POOR	FAIR TO POOR	POOR	POOR	POOR	POOR
DRAINAGE	GOOD	DRAINS FREELY	GOOD	FAIR TO PRACTICALLY IMPERVIOUS	FAIR TO PRACTICALLY IMPERVIOUS	FAIR TO PRACTICALLY IMPERVIOUS	FAIR TO IMPERVIOUS	FAIR TO IMPERVIOUS	FAIR TO IMPERVIOUS	FAIR TO IMPERVIOUS	IMPERVIOUS	FAIR TO IMPERVIOUS	POOR

NOTES:

A-2 TO A-3 SOILS: WHEN USED AS A BASE, PLASTICITY INDEX AND LIQUID LIMIT SHOULD NOT EXCEED 6 AND 25 RESPECTIVELY. BEST FOR SOIL-CEMENT STABILIZATION, GENERALLY 8 TO 12 % CEMENT BY WEIGHT WILL BE SUFFICIENT.

NON-PLASTIC A-2 TO A-3 SOILS: MAY REQUIRE VIBRATION FOR COMPACTION.

A-4 TO A-7 SOILS: FILLS SHOULD BE PLACED IN DRY SEASON.

A-4 SILTS: SUSCEPTIBLE TO SETTLEMENT AND EROSION.

A-5 SOILS: WHEN MICA IS PRESENT, VERY DIFFICULT TO COMPACT BECAUSE OF EXPANSION AND REBOUND.

A-6 SOILS (CLAY): WILL PUMP IN POROUS BASES FORMING CRACKS FILLS WILL SETTLE OVER LONG PERIODS OF TIME. HIGH BANKS IN CUTS AND FILLS VERY LIABLE TO SLIDE.

Soils & Soil - Aggregate Mixtures Characteristics And Performance

TYPE	TREATMENT & USE
A-2 SOILS:	WELL GRADED TO POORLY GRADED SAND AND GRAVELS. GOOD BASE FOR MODERATE FLEXIBLE OR THIN RIGID PAVEMENT. GOOD FILL. FROST HEAVE, BREAK-UP IF PLASTIC. SOFTENS WHEN WET IF PLASTIC. USE BASE COURSE WHEN SUB-GRADE P.I. IS GREATER THAN 6. SUB-DRAINAGE EFFECTIVE. STABILIZE; WITH BITUMEN, CHLORIDES, CEMENT OR ADMIXTURE SOIL.
A-3 SOILS:	CLEAN SANDS AND GRAVELS. IDEAL BASE FOR MODERATE FLEXIBLE OR THIN RIGID PAVEMENT. GOOD FILL. NO FROST HEAVE OR BREAK-UP. SUB-DRAINAGE ONLY THROUGH IMPERVIOUS SHOULDERS. STABILIZE; WITH SOIL BINDER, BITUMINOUS, OR CHEMICAL ADMIXTURES.
A-4 SOILS:	SILTY SOILS. NOT GOOD FOR SURFACE. POOR BASE. ABSORBS WATER. UNSTABLE WHEN WET. BAD FROST HEAVE AND BREAK-UP. USE SUB-DRAINAGE AND/OR BASE AND SUB-BASE WITH FLEXIBLE PAVEMENT. USE BITUMINOUS SUB-GRADE PRIME. USE THICK CONCRETE PAVEMENT (7" TO 10") WITH STEEL REINFORCEMENT AND CRACK CONTROL.
A-5 SOILS:	ELASTIC SILTS. USE SUB-DRAINAGE AND/OR GRANULAR BASE AND SUB-BASE WITH BITUMINOUS SUB-GRADE PRIME. USE THICK CONCRETE PAVEMENT, REINFORCED WITH CRACK CONTROL.
A-6 SOILS:	CLAYS. IMPERMEABLE AND STABLE WHEN DRY AND UNDISTURBED (HARD CLAY). PLASTIC AND ABSORBENT IF DISTURBED. BAD PUMPING INTO POROUS BASE. MACADAM OR PAVEMENT JOINTS. SHRINKS OR CRACKS WHEN DRY. USE GRANULAR BASE AND SUBBASE. USE SUB-DRAINAGE ONLY WHEN MADE PERVIOUS BY CRACKS, ROOT HOLES AND LAMINATIONS. FROST HEAVE SLIGHT WHEN IMPERMEABLE; BAD WHEN PERVIOUS. USE SUB-GRADE PRIME. USE THICK, STRONG, DENSE FLEXIBLE PAVEMENT OR REINFORCED CRACK CONTROLLED CONCRETE.
A-7 SOILS:	EXPANSIVE, PLASTIC CLAYS. EXCESSIVE VOLUME CHANGE. BAD FROST HEAVE AND BREAK-UP. SUB-DRAINAGE NOT EFFECTIVE. USE THICK, DENSE, FLEXIBLE PAVEMENT WITH BASE AND SUB-BASE OVER SUB-GRADE PRIME OR REINFORCED CRACK CONTROLLED CONCRETE PLACED ON IMPERVIOUS PAPER.
A-8 SOILS:	MUCK AND PEAT. UNFIT FOR CONSTRUCTION PURPOSES. EXCAVATE TO SOLID STRATUM AND REPLACE WITH SELECTED FILL. DISPLACEMENT BY SUPERIMPOSED FILL IS DOUBTFUL. DISPLACEMENT BY EXPLOSIVE UNDER SUPERIMPOSED FILL IS SOMETIMES EFFECTIVE.

Soils & Soil - Aggregate Mixtures
Treatment & Use

	SYMBOLS	TYPICAL GRADING	TYPICAL PHYSICALS	REMARKS
A-3 SAND		C.S. =22% F.S. =48% SILT =20% CLAY = 8% COLL. = 2%	L.L. = N.P P.I. = N.P.	SAND-53% MIN. %-#200-20% MAX. P.I.- N.P. L.L.-MUST BE N.P
A-2 SAND & FINES		C.S. =20% F.S. =43% SILT =19% CLAY =10% COLL. = 8%	L.L. =22 P.I. = 2 S.L. =18	SAND-53% MIN. %-#200-20% MAX. P.I.-7 MAX. L.L.-34 MAX. (MUST HAVE L.L.)
A-2-4 SILTY SAND		C.S. =25% F.S. =30% SILT =32% CLAY = 7% COLL. = 6%	L.L. =24 P.I. = 2 S.L. =21	SAND-53% MIN. % #200-21% MIN.-30% MAX. P.I.-7 MAX. L.L.-34 MAX.(MAY BE N.P.)
A-4-2 SANDY SILT		C.S. =23% F.S. =28% SILT =33% CLAY =10% COLL. = 6%	L.L. =25 P.I. = 3 S.L. =21	SAND-48% MIN %-#200-31% MIN. P.I.-7 MAX. L.L.- 40 MAX.(MAY BE N.P)
A-2-7 CLAYEY SAND		C.S =38% F.S. =31% SILT =15% CLAY = 8% COLL. = 8%	L.L. =31 P.I. =10 S.L. =18	SAND-48% MIN. CLAY-29% MAX. P.I.- 8-14 L.L.- 40 MAX.
A-7-2 SANDY CLAY		C.S. =20% F.S. =29% SILT =17% CLAY =21% COLL. =13%	L.L. =39 P.I. =17 S.L. =16	SAND-48% MIN. CLAY-17%-35% P.I.-15 MIN. L.L.-30 MIN.
A-4 SILT		C.S. =20% F.S. =22% SILT =40% CLAY =10% COLL = 8%	L.L. =30 P.I. = 6 S.L. =19	SAND-47% MAX. CLAY-29% MAX. P.I-9 MAX. L.L.-40 MAX.
A-4-7 CLAYEY SILT		C.S. = 8% F.S. = 17% SILT = 40% CLAY = 23% COLL. = 12%	L.L. =33 P.I. =11 S.L. =18	SAND-47% MAX. CLAY-25% MIN. P.I.-14 MAX. L.L.-40 MAX.
A-7-4 SILTY CLAY		C.S. =18% F.S. =20% SILT =35% CLAY =12% COLL. =15%	L.L. =39 P.I. =15 S.L. =16	SAND-47% MAX. CLAY-29% MAX. P.I.-15 MIN. L.L.- 30 MIN.
A-7 CLAY		C.S. =18% F.S. =22% SILT =23% CLAY =22% COLL. =15%	L.L. =40 P.I. =17 S.L. =15	SAND-47% MAX. CLAY-30%-59% P.L-15 MIN. L.L.-35 MIN.
A-6 COLLOIDAL CLAY		C.S. = 6% F.S. = 7% SILT =18% CLAY =33% COLL. =36%	L.L. =50 P.I. =33 S.L. =14	CLAY-60% MIN. P.I.-25 MIN. L.L.- 45 MIN.
A-5 MICA, DIATOMS, DECOMPOSED ROCK		C.S. =15% F.S. =35% SILT =30% CLAY =15% COLL. = 5%	L.L. =35 P.I. = 4 S.L. =26	GRAD. NOT SIGNIFICANT P.I.- LOW L.L.-HIGH S.L.- 26 MIN. VISUAL INSPECTION NECESSARY TO DETERMINE TYPE
A-8 SWAMP MUCK		C.S. =18% F.S. =26% SILT =45% CLAY = 7% COLL. = 4%	L.L. =52 P.I. = 7 S.L. =38	ORGANIC CONTENT-4% MIN. P.I.- LOW L.L.-HIGH, WHEN OBTAINABLE S.L. - 26 MIN.
ROCK REFUSAL				

Soils & Soil - Aggregate Mixtures
Guide To Classification

RAILS TO BE UNTREATED TIMBER.

POSTS TO BE TREATED TIMBER. WHERE PAINT IS
CALLED FOR, TWO COATS TO BE APPLIED

ALL LUMBER TO BE SOUTHERN YELLOW PINE OF 1400# STRESS GRADE
OR BETTER.

TO BE PAINTED IN A COLOR SCHEME AND STRIPING DETAILS IN
ACCORDANCE WITH THE LATEST "MANUAL OF UNIFORM TRAFFIC CONTROL DEVICES."

TO BE USED WHERE NOTED ON PLANS

THE COST OF ALL EXCAVATION AND BACKFILL TO BE INCLUDED IN
LUMP SUM PRICE BID FOR TIMBER BARRICADE.

THE LUMP SUM PRICE BID SHALL INCLUDE THE FURNISHING OF
ALL MATERIALS, PRESERVATIVE TREATMENTS, TIMBER CONNECTORS
AND HARDWARE, PAINTING, GALVANIZING AND CONCRETE AS WELL AS ALL LABOR,
TOOLS AND EQUIPMENT AND ALL WORK INCIDENTAL THERETO.

Standard Timber Barricade

SEED & MULCH BERM DITCH 1:1
SOD BERM DITCH 1:1

CONCRETE BERM DITCH 1:1
"A"

SEED & MULCH BERM DITCH 2:1
SOD BERM DITCH 2:1

CONCRETE BERM DITCH 2:1
"B"

GENERAL NOTES

1. DITCH ELEVATIONS AS NOTED ON PLANS.
2. 2" TOP SOIL AND SEED & MULCH ALL CONSTRUCTION SLOPES NOT OTHERWISE TREATED.
3. WHERE DITCH LOCATION IS OTHER THAN SHOWN, DISTANCE FROM DITCH ℄ TO A REFERENCE POINT SHALL BE INDICATED ON THE PLANS.
4. EARTH BERM TO BE CONSTRUCTED FROM DITCH EXCAVATION AND COMPACTED AS DIRECTED BY THE ENGINEER. THIS WILL NOT BE A PAY ITEM.

Standard Berm Ditches

CONCRETE SURFACE DRAIN DITCH

SOD SURFACE DRAIN DITCH
"A"

CONCRETE SURFACE DRAIN DITCH

SOD SURFACE DRAIN DITCH
"B"

GENERAL NOTES

1. CONCRETE TO BE EXTENDED AS REQUIRED WHERE DEPTH OF FLOW IN DITCH EXCEEDS LIMIT OF CONCRETE AS INDICATED ABOVE.
2. OMIT SOD STRIP ON SHOULDER SIDE OF DITCH OF NON— TRENCHED SHOULDERS.

Standard Surface Drain Ditches
Concrete & Sod

CONCRETE SIDE DITCH

SOD SIDE DITCH
"A"

CONCRETE SIDE DITCH

SOD SIDE DITCH
"B"

GENERAL NOTES

WHEN DITCHES ARE CONSTRUCTED OTHER THAN WHAT IS
SHOWN, THE ELEVATIONS WILL BE NOTED ON THE PLANS.

Standard Side Ditches Cut Areas

SEED & MULCH MEDIAN DITCH
SOD MEDIAN DITCH

CONCRETE MEDIAN DITCH

"B"	"D"
1'-0"	0.24'
1'-6"	0.36'
2'-0"	0.49'
2'-6"	0.61'
3'-0"	0.73'
3'-6"	0.85'
4'-0"	0.97'

NOTES:
1. DITCH DEPTHS OTHER THAN 6" TO BE NOTED BY ELEVATIONS ON PLANS.
2. DIMENSION "B" MAY BE MODIFIED BY NOTE ON PLANS.
3. 4" TOP SOIL AND SEED & MULCH ALL CONSTRUCTION SLOPES NOT OTHERWISE TREATED.
4. WHERE DITCH IS OTHER THAN 0'-6" DEPTH, THE 4:1 SLOPE RATIO SHALL BE MAINTAINED, EXCEPT AT TRANSITIONS TO INLETS OF DRAINAGE STRUCTURES.
5. WHERE ℄ DITCH IS NOT AT ℄ MEDIAN, OFFSET DISTANCES SHALL BE INDICATED ON PLANS.

SEED & MULCH OR
SOD SIDE DITCH

CONCRETE SIDE DITCH

"B"	"D"
1'-6"	0.67'
1'-9"	0.78'
2'-0"	0.89'
2'-3"	1.01'
2'-6"	1.12'
2'-9"	1.23'
3'-0"	1.34'
3'-3"	1.45'
3'-6"	1.57'
3'-9"	1.68'
4'-0"	1.79'

NOTES:
1. ELEVATIONS AS NOTED ON PLANS.
2. DIMENSIONS "A" AND "B" MAY BE MODIFIED BY NOTE ON PLANS.
3. 2" TOP SOIL AND SEED & MULCH ALL CONSTRUCTION SLOPES NOT OTHERWISE TREATED
4. WHERE DITCH IS NOT ADJACENT TO TOE OF FILL, DISTANCES FROM ℄ DITCH TO A REFERENCE POINT SHALL BE INDICATED ON THE PLANS.

Standard Side Ditches
Standard Median Ditches - Flat

CONCRETE MEDIAN 'V' DITCH

SOD MEDIAN 'V' DITCH

Standard Median Ditches
V Slope

NOTE: IF POSSIBLE, DO NOT CONSTRUCT A CHIMNEY CONNECTION
FROM INLET TO X-DRAIN BUT MAKE A 45° CONNECTION TO
X-DRAIN. WHEN NECESSARY TO USE A CHIMNEY CONNECTION
POUR A 6" CONCRETE COLLAR AROUND THE VERTICAL
PIPE FOR ITS ENTIRE HEIGHT.

Standard Pipe Outlets
Surface Drainage On Fills

TRANSITION OF CURB FACE AND CROSS
SLOPE AT DEPRESSED INLETS

Method Of Depressing Pavement
At Inlets

ELEVATION

"S" DISTANCES FROM INSIDE SURFACE OF PIPE TO VERTICAL BARS IN FRONT AND REAR FACE
4" FOR 12" DIA. TO 18" DIA. PIPES INCLUSIVE
6" FOR 24" DIA. TO 36" DIA. PIPES INCLUSIVE
8" FOR 42" DIA. TO 60" DIA. PIPES INCLUSIVE

SECTION A-A

DISPOSITION OF BARS DETAIL

PLAN

GENERAL NOTES:
SPECIFICATIONS : LATEST S.H.A.
CONCRETE : SEE S.H.A. SPECIFICATIONS
REINFORCING : DEFORMED STEEL BARS #4 BARS ($\frac{1}{2}\phi$)
CHAMFER : ALL EXPOSED EDGES 1"X1"OR AS DIRECTED

OPENINGS		DIMENSIONS						VOLUME	STEEL
D	AREA							CONC.	
INCHES	SQ.FT.	A	B	C	E	H	L	C.Y.	LBS.
12	0.79	9"	6"	6"	1'-9"	0'-10"	4'-0"	0.27	19
15	1.23	9"	6"	6"	1'-9"	1'-0½"	4'-9"	0.34	23
18	1.78	9"	6"	6"	1'-9"	1'-3"	5'-6"	0.41	29
24	3.14	9"	14"	6"	2'-3"	1'-6"	7'-0"	0.69	38
30	4.91	9"	14"	6"	2'-5"	1'-9"	8'-6"	0.88	58
36	7.07	12"	16"	10"	3'-2"	2'-0"	10'-0"	1.74	90
42	9.62	12"	16"	10"	3'-2"	2'-3"	11'-6"	2.07	99
48	12.57	12"	16"	10"	3'-2"	2'-6"	13'-0"	2.43	119
54	15.90	12"	20"	12"	3'-8"	2'-9"	14'-6"	3.08	132
60	19.64	12"	20"	12"	3'-8"	3'-0"	16'-0"	3.50	158

QUANTITIES IN TABLE TO BE USED FOR ESTIMATING ONLY

Standard End Support Wall
Round Pipe

ELEVATION

SECTION A-A

PLAN

DISPOSITION OF BARS DETAIL

"S" DISTANCES
6" FOR 29"X18" TO 43"X27" INCLUSIVE
8" FOR 50"X31" TO 72"X44" INCLUSIVE

GENERAL NOTES:
SPECIFICATIONS: LATEST S.H.A.
CONCRETE : SEE S.H.A. SPECIFICATIONS
REINFORCING : DEFORMED STEEL BARS #4 BARS ($\frac{1}{2}$" ϕ)
CHAMFER : ALL EXPOSED EDGES 1"X1" OR AS DIRECTED

OPENINGS		DIMENSIONS						VOLUME	STEEL
SIZE	AREA	A	B	C	E	H	L	CONC. C.Y.	LBS.
INCHES	SQ. FT.								
29"x18"	2.2	9"	14"	6"	2'-5"	1'-0"	6'-0"	.51	34
36"x22"	4.4	9"	14"	6"	2'-5"	1'-3"	7'-6"	.67	41
43"x27"	4.8	12"	16"	10"	3'-2"	1'-6"	8'-11"	1.37	59
50"x31"	6.5	12"	16"	10"	3'-2"	1'-9"	10'-5"	1.66	103
58"x36"	8.7	12"	16"	10"	3'-2"	2'-0"	12'-2"	2.01	133
65"x40"	11.0	12"	20"	12"	3'-8"	2'-3"	13'-8"	2.59	156
72"x44"	13.5	12"	20"	12"	3'-8"	2'-6"	15'-2"	2.97	176

QUANTITIES IN TABLE TO BE USED FOR ESTIMATING ONLY

Standard End Support Wall
Arch Pipe

4 - #6 BENT BARS HORIZONTAL
2 EACH FACE

2'-6"   2"   2'-6"

1'-0"   1'-0"

2" 2"   2" 2"

ELEVATION

#4 STRAIGHT BARS VERTICAL
@ 1'-6" MIN. TO 2'-0" MAX.
FRONT FACE

1½ : 1, 2:1 OR 4:1 SLOPE

2"   3"

#4 BARS HORIZONTAL
@ 1'-0" C/C BELOW
OPENING

3"

SECTION A - A

45°   45°

D

PLAN

ISOMETRIC VIEW

J   D   I O   G   L   F   K   C   M   N   B   E

1'-0"   1'-0"

#4 BARS HORIZONTAL @ 1'-7" MAX. C/C BOTH
FACES BOTTOM BARS BENT ALONG ENDWALL
OTHERS STRAIGHT.

1 - #4 BENT BAR HORIZONTAL
1 - #4 BENT BAR HORIZONTAL

#4 BENT BARS @ 1'-0" C/C BOTH
WINGWALLS - ALL ENDWALLS

2 - #4 STRAIGHT BARS HORIZONTAL,
1 EACH WINGWALL

#4 STRAIGHT BARS HORIZONTAL
@ 1'-0" C/C @ BOTH SIDES OF
OPENING

SECTION A-A

2"   2"   2"   3"   3"
2"   0"   1'-3"

**DISPOSITION OF
BARS DETAIL**

GENERAL NOTES:
SPECIFICATIONS : LATEST S.H.A.
CONCRETE : SEE LATEST S.H.A. SPECIFICATIONS
REINFORCING: DEFORMED STEEL BARS #4 & #6
CHAMFER: ALL EXPOSED EDGES 1" X 1" OR AS DIRECTED

OPENING		DIMENSIONS												VOL. CONC. C.Y.	STEEL LBS.
D INCHES	AREA SQ. FT.	B	C	E	F	G	H	J	K	L	M	N	O		
48	12.57	1'-4"	10"	3'-2"	2'-9"	7'-0¾"	5'-0"	4'-10"	6'-3½"	6'-8½"	5'-9"	2'-10¾"	5'-6"	4.3	262
54	15.9	1'-8"	1'-0"	3'-8"	3'-0"	7'-8½"	5'-6"	5'-4"	6'-10½"	7'-3½"	6'-2¼"	3'-1½"	6'-2"	5.3	301
60	19.64	1'-8"	1'-0"	3'-8"	3'-3"	8'-5"	6'-0"	5'-10"	7'-7¼"	8'-0¼"	6'-11"	3'-7½"	6'-8"	6.0	361

QUANTITIES IN TABLE TO BE USED FOR ESTIMATING ONLY

Standard Headwalls

ELEVATION

SECTION A-A

PLAN

ISOMETRIC VIEW

DISPOSITION OF BARS DETAIL

SECTION B-B

GENERAL NOTES:
SPECIFICATIONS: LATEST S.H.A.
CONCRETE: SEE S.H.A. SPECIFICATIONS
REINFORCING: DEFORMED STEEL BARS #4 & #6
CHAMFER: ALL EXPOSED EDGES 1"X1" OR AS DIRECTED

OPENING		DIMENSIONS													VOL. CONC. C.Y.	STEEL LBS.
D INCHES	AREA SQ.FT.	B	C	E	F	G	H	J	K	L	M	N	O	P		
66	23.80	2'-6"	1'-3"	4'-9"	3'-0"	11'-2 1/2"	6'-8 1/2"	6'-4"	10'-3 1/4"	10'-8 1/2"	9'-3"	3'-5"	7'-4 1/2"	2'-0"	9.7	585
72	28.27	2'-6"	1'-3"	4'-9"	3'-3"	12'-1"	7'-3"	6'-10"	11'-1 3/4"	11'-6 3/4"	10'-1 1/4"	3'-11"	7'-10 1/2"	2'-0"	10.9	645
78	33.20	3'-0"	1'-6"	5'-6"	3'-6	13'-0 1/2"	7'-9 1/2"	7'-4"	12'-0"	12'-5"	10'-9"	4'-0"	8'-6 3/4"	2'-6"	13.3	865
84	38.48	3'-0"	1'-6"	5'-6"	3'-9"	13'-10"	8'-4"	7'-10"	12'-9 1/2"	13'-2 1/2"	11'-6 3/4"	4'-6"	9'-0 3/4"	2'-6"	14.7	984

QUANTITIES IN TABLE TO BE USED FOR ESTIMATING ONLY

Standard Headwalls

4 - #4 STRAIGHT BARS VERTICAL IN FRONT FACE
FOR 12" TO 18" DIA. PIPE ENDWALLS INCLUSIVE

2 - #4 STRAIGHT BARS HORIZONTAL I EACH FACE
FOR 36" DIAMETER TO 60" DIAMETER PIPE
ENDWALLS INCLUSIVE

#4 STRAIGHT BARS VERTICAL @ 1'-6" MIN. TO 2'-0" MAX.
FRONT FACE FOR 24" DIAMETER TO 60" DIAMETER.
PIPE ENDWALLS INCLUSIVE.

ELEVATION

#4 STRAIGHT BARS HORIZONTAL @ 1'-7" C/C BOTH FACES-
TOP AND BOTTOM BARS TO BE FULL LENGTH - ALL ENDWALLS.

#4 STRAIGHT BARS HORIZONTAL
@ 1'-0" C/C BOTH SIDES
OF OPENING FOR 36" TO
60" DIAMETER PIPE
ENDWALLS.

#4 BENT BARS @ 1'-0"
C/C - ALL ENDWALLS

1-#4 STRAIGHT BAR
HORIZONTAL - ALL
ENDWALLS

2 - #4 STRAIGHT BARS HORIZONTAL
FOR 36" TO 60" DIAMETER PIPE ENDWALLS.

SECTION A - A

"S" DISTANCES FROM INSIDE SURFACE OF PIPE
TO VERTICAL BARS IN FRONT AND REAR FACE.
4" FOR 12" DIAMETER TO 18" DIAMETER PIPES INCLUSIVE
6" FOR 24" DIAMETER TO 36" DIAMETER PIPES INCLUSIVE
8" FOR 42" DIAMETER TO 60" DIAMETER PIPES INCLUSIVE

GENERAL NOTES:
SPECIFICATIONS : LATEST S.H.A.
CONCRETE : SEE S.H.A. SPECIFICATIONS.
REINFORCING : DEFORMED STEEL BARS. #4 ϕ BARS
CHAMFER : ALL EXPOSED EDGES 1" X 1" OR
AS DIRECTED.

PLAN

DISPOSITION OF BARS - DETAIL

OPENINGS		DIMENSIONS							VOLUME	STEEL
D	AREA	A	B	C	E	F	H	L	CONC.	LBS.
INCHES	SQ. FT.								C.Y.	
12	0.79	9"	6"	6"	1'-9"	9"	1'-9"	6'-6"	0.61	38
15	1.23	9"	6"	6"	1'-9"	9"	2'-0"	7'-9"	0.78	55
18	1.78	9"	6"	6"	1'-9"	9"	2'-3"	9'-0"	0.95	62
24	3.14	9"	14"	6"	2'-5"	9"	2'-9"	11'-6"	1.56	91
30	4.91	9"	14"	6"	2'-5"	12"	3'-6"	14'-2"	2.19	136
36	7.07	12"	16"	10"	3'-2"	12"	4'-0"	16'-8"	4.18	222
42	9.62	12"	16"	10"	3'-2"	12"	4'-6"	19'-2"	5.13	265
48	12.57	12"	16"	10"	3'-2"	12"	5'-0"	21'-8"	6.12	326
54	15.90	12"	20"	12"	3'-8"	12"	5'-6"	24'-2"	7.68	384
60	19.64	12"	20"	12"	3'-8"	12"	6'-0"	26'-8"	8.86	428

QUANTITIES IN TABLE TO BE USED FOR ESTIMATING ONLY

Standard Type C Endwall
Round Pipe

4 - #4 STRAIGHT BARS VERTICAL IN FRONT FACE FOR
14" X 9" TO 29"X 18" PIPE ENDWALLS INCLUSIVE

2 - 4 STRAIGHT BARS HORIZONTAL, 1 EACH FACE FOR
50"X 31" TO 72"X 44" PIPE ENDWALLS INCLUSIVE

#4 STRAIGHT BARS VERTICAL @ 1'-6" MIN. TO
2'-0" MAX. FROM FACE FOR 36"X 22" TO
72"X 44" PIPE ENDWALLS INCLUSIVE

ELEVATION

#4 STRAIGHT BARS HORIZONTAL @ 1'-7" C/C BOTH FACES –
TOP 8 BOTTOM BARS TO BE FULL LENGTH - ALL ENDWALLS.

#4 BENT BARS @ 1'-0" C/C
ALL ENDWALLS.

#4 STRAIGHT BARS HORIZONTAL
@ 1'-0" C/C BOTH SIDES OF
OPENING FOR 50"X 31" TO L
72"X 44" PIPE ENDWALLS
INCLUSIVE

1- 4 STRAIGHT BAR
HORIZONTAL ALL
ENDWALLS

2 - 4 STRAIGHT BARS HORIZONTAL
FOR 50"X 31" TO 72"X 44" PIPE ENDWALLS INCLUSIVE

SECTION A - A

"S" DISTANCES
4" FOR 14"X 9" TO 22"X 13" INCLUSIVE
6" FOR 29"X 18" TO 43"X 27" INCLUSIVE
8" FOR 50"X 31" TO 72"X 44" INCLUSIVE

GENERAL NOTES:
SPECIFICATION : LATEST S.H.A.
CONCRETE   : SEE S.H.A. SPECIFICATIONS
REINFORCING : DEFORMED STEEL BARS : #4 φ BARS
CHAMFER   : ALL EXPOSED EDGES 1" X 1" OR
       AS DIRECTED.

PLAN

DISPOSITION OF BARS - DETAILS

OPENINGS		DIMENSIONS						VOLUME	STEEL
SIZE	AREA	A	B	C	E	F	H	L	CONC. C.Y.	LBS.
INCHES	SQ. FT.									
14"X 9"	0.8	9"	6"	6"	1'-9"	8"	1'-5"	6'-4"	.54	41
18"X11"	1.1	9"	6"	6"	1'-9"	6"	1'-5"	6'-4"	.53	41
22"X13"	1.6	9"	6"	6"	1'-9"	11"	2'-0"	9'-7"	.95	68
29"X18"	2.8	9"	6"	6"	1'-9"	6"	2'-0"	9'-7"	.92	67
36"X22"	4.4	9"	14"	6"	2'-5"	14"	3'-0"	13'-9"	1.95	120
43"X27"	6.4	9"	14"	6"	2'-5"	9"	3'-0"	13'-9"	1.89	126
50"X31"	8.7	12"	16"	10"	3'-2"	14"	3'-9"	18'-0"	4.29	226
58"X36"	11.4	12"	16"	10"	3'-2"	9"	3'-9"	18'-0"	4.19	226
65"X40"	14.3	12"	20"	12"	3'-8"	13"	4'-5"	21'-10"	6.01	310
72"X44"	17.6	12"	20"	12"	3'-8"	9"	4'-5"	21'-10"	5.89	310

QUANTITIES IN TABLE TO BE USED FOR ESTIMATING ONLY.

Standard Type C Endwall
Arch Pipe

2 #4 BENT BARS HORIZONTAL
I EACH FACE FOR 36" DIA. TO 60"
DIA. PIPE ENDWALLS INCLUSIVE

2 -#4 STRAIGHT BARS VERTICAL
IN FRONT FACE FOR 12" TO 18"
DIA. PIPE ENDWALLS INCLUSIVE

#4 STRAIGHT BARS HORIZONTAL @ I'-7" MAX.
C/C BOTH FACES - LAP I'-3" TOP & BOTTOM
BARS @ CORNER - BOTH WINGWALLS -
ALL ENDWALLS

#4 BENT BARS @ I'-0" C/C
BOTH WINGWALLS ALL ENDWALLS

2 -#4 STRAIGHT BARS VERTICAL IN
FRONT FACE FOR 12" TO 18" DIA.
PIPE ENDWALLS INCLUSIVE

I:I, I.2:I OR 4:I SLOPE

2:I, I.2:I OR 4:I SLOPE

#4 STRAIGHT BARS
HORIZONTAL @ I'-0" C/C
BOTH SIDES OF OPENING
BOTH WINGWALLS FOR 36" TO
60" DIA. PIPE ENDWALLS

4 -#4 STRAIGHT BARS HORIZONTAL,
2 EACH WINGWALL FOR 36" TO 60"
DIA. PIPE ENDWALL

#4 STRAIGHT BARS VERTICAL
@ I'-6" MIN. TO 2'-0" MAX.
FRONT FACE FOR 24" DIA. TO 60"
DIA. PIPE ENDWALLS INCLUSIVE

2 -#4 STRAIGHT BARS HORIZONTAL,
I EACH WINGWALL - ALL ENDWALLS

#4 STRAIGHT BARS VERTICAL @ I'-6" MIN.
TO 2'-0" MAX. FRONT FACE FOR 24" TO
60" DIA. PIPE ENDWALLS INCLUSIVE

SECTION A-A

SECTION B-B

DISPOSITION OF BARS DETAIL

PLAN

"S" DISTANCES FROM INSIDE SURFACE OF PIPE TO
VERTICAL BARS IN FRONT AND REAR FACE
4" FOR 12" DIA. TO 18" DIA. PIPES INCLUSIVE
6" FOR 24" DIA. TO 36" DIA. PIPES INCLUSIVE
8" FOR 42" DIA. TO 60" DIA. PIPES INCLUSIVE

GENERAL NOTES

SPECIFICATIONS : LATEST S.H.A.
CONCRETE : SEE S.H.A SPECIFICATIONS.
REINFORCING : DEFORMED STEEL BARS - 4 BARS.
CHAMFER : ALL EXPOSED EDGES I" X I"

OPENINGS		DIMENSIONS						VOL.	STEEL
D INCHES	AREA SQ.FT.	A	B	C	E	H	L	CONC. C.Y.	LBS.
12	0.79	9"	6"	6"	I'-9"	I'-9"	3'-6"	0.73	56
15	1.23	9"	6"	6"	I'-9"	2'-0"	4'-3"	0.93	77
18	1.78	9"	6"	6"	I'-9"	2'-3"	5'-0"	1.14	89
24	3.14	9"	14"	6"	2'-5"	2'-9"	6'-6"	1.8	124
30	4.91	9"	14"	6"	2'-5"	3'-6"	8'-0"	2.6	160
36	7.07	12"	16"	10"	3'-2"	4'-0"	9'-6"	5.0	287
42	9.62	12"	16"	10"	3'-2"	4'-6"	11'-0"	6.2	333
48	12.57	12"	16"	10"	3'-2"	5'-0"	12'-6"	7.4	387
54	15.90	12"	20"	12"	3'-8"	5'-6"	14'-0"	9.2	458
60	19.64	12"	20"	12"	3'-8"	6'-0"	15'-6"	10.7	519

QUANTITIES IN TABLE TO BE USED FOR ESTIMATING ONLY

Standard Type E Endwall
Round Pipe

SECTION A-A

SECTION B-B

DISPOSITION OF BARS DETAIL

PLAN

OPENINGS		DIMENSIONS						VOL.	STEEL
SIZE INCHES	AREA SQ.FT.	A	B	C	E	H	L	CONC. C.Y.	LBS.
14"X9"	0.8	9"	6"	6"	1'-9"	1'-5"	3'-10"	.72	56
18"X11"	1.1	9"	6"	6"	1'-9"	1'-5"	3'-10"	.71	56
22"X13"	1.6	9"	6"	6"	1'-9"	2'-0"	5'-11"	1.27	90
29"X18"	2.8	9"	6"	6"	1'-9"	2'-0"	5'-11"	1.23	87
36X22	4.4	9"	14"	6"	2'-5"	3'-0"	8'-7"	2.53	173
43X27	6.4	9"	14"	6"	2'-5"	3'-0"	8'-7"	2.47	167
50X31	8.7	12"	16"	10"	3'-2"	3'-9"	11'-4"	5.68	334
58X36	11.4	12"	16"	10"	3'-2"	3'-9"	11'-4"	5.58	331
65X40	14.3	12"	20"	12"	3'-8"	4'-5"	13'-10"	7.97	451
72X44	17.6	12"	20"	12"	3'-8"	4'-5"	13'-10"	7.85	442

QUANTITIES IN TABLE TO BE USED FOR ESTIMATING ONLY

"S" DISTANCES
4" FOR 14"X 9" TO 22"X13" INCLUSIVE
6" FOR 29"X 18" TO 43"X 27"INCLUSIVE
8" FOR 50"X 31" TO 72"X 44" INCLUSIVE

GENERAL NOTES
SPECIFICATIONS: LATEST S.H.A
CONCRETE : SEE S.H.A. SPECIFICATIONS
REINFORCING : DEFORMED STEEL BARS $\frac{3}{4}\phi$ BARS
CHAMFER : ALL EXPOSED EDGES 1" X 1" OR AS DIRECTED

Standard Type E Endwall
Arch Pipe

ISOMETRIC VIEW

ELEVATION

REINFORCEMENT, DEFORMED STEEL BARS
STEEL IN WINGS:
VERTICAL NO. 6 φ BARS 12" C/C
HORIZONTAL NO. 4 φ BARS 12" C/C
HOOKED ON ONE END
CONCRETE (SEE S.H.A. SPECIFICATIONS)
CHAMFER ALL EXPOSED EDGES 1" X 1" OR AS DIRECTED

TOE OF SLOPE

PLAN

1½ : 1 SLOPE = 9"
2 : 1 SLOPE = 12"
4 : 1 SLOPE = 24"

OPENING		DIMENSIONS FOR CONCRETE									QUANTITIES	
		ENDWALL			WINGS						1 ENDWALL 2 WINGS	
DIAM. PIPE D	AREA SQ.FT.	L	B	F	H	W	X	Z	K_1	K	P	CONC. CU.YDS.	STEEL LBS.

SLOPE 2:1

DIAM. PIPE D	AREA SQ.FT	L	B	F	H	W	X	Z	K_1	K	P	CONC. CU.YDS.	STEEL LBS.
12"	0.79	1'-9"	9"	9"	1'-7"	2'-3"	1'-10"	2'-0"	9"	7"	1'-3"	0.55	38
15"	1.23	2'-0"	9"	9"	1'-11"	2'-6"	2'-6"	2'-2"	9"	8"	1'-5"	0.73	46
16"	1.39	2'-1"	9"	9"	1'-11"	2'-7"	2'-6"	2'-3"	9"	8"	1'-5"	0.75	46
18"	1.78	2'-3"	9"	9"	2'-2"	3'-0"	3'-0"	2'-7"	9"	8"	1'-7"	0.89	57
24"	3.14	3'-0"	12"	9"	2'-9"	4'-0"	4'-0"	3'-5"	12"	11"	2'-1"	1.64	82
30"	4.91	3'-6"	12"	9"	3'-4"	4'-8"	5'-2"	4'-1"	12"	11"	2'-6"	2.14	113
36"	7.07	4'-0"	12"	9"	3'-10"	5'-4"	6'-2"	4'-7"	12"	11"	2'-10"	2.48	132
42"	9.62	4'-6"	12"	9"	4'-5"	6'-4"	7'-4"	5'-6"	12"	11"	3'-2"	3.30	163
48"	12.57	5'-0"	12"	9	4'-11"	7'-0"	8'-4"	6'-1"	12"	11"	3'-6"	3.85	193

SLOPE 1½ : 1

DIAM. PIPE D	AREA SQ.FT	L	B	F	H	W	X	Z	K_1	K	P	CONC. CU.YDS.	STEEL LBS.
12"	0.79	1'-9"	9"	9"	1'-8"	2'-0"	1'-3"	1'-9"	9"	7"	1'-4"	0.51	38
15"	1.23	2'-0"	9"	9"	2'-0"	2'-4"	1'-9"	2'-0"	9"	7"	1'-5"	0.63	42
16"	1.39	2'-1"	9"	9"	2'-0"	2'-4"	1'-9"	2'-0"	9"	7"	1'-5"	0.65	42
18"	1.78	2'-3"	9"	9"	2'-3"	2'-8"	2'-1"	2'-4"	9"	8"	1'-6"	0.77	48
24"	3.14	3'-0"	12"	9"	2'-11"	3'-6"	2'-11"	3'-0"	12"	11"	2'-1"	1.43	73
30"	4.91	3'-6"	12"	9"	3'-6"	4'-4"	3'-9"	3'-9"	12"	11"	2'-4"	1.88	96
36"	7.07	4'-0"	12"	9"	4'-0"	5'-0"	4'-6"	4'-4"	12"	11"	2'-11"	2.32	118
42"	9.62	4'-6"	12"	9"	4'-7"	5'-9"	5'-4"	5'-0"	12"	11"	2'-11"	2.85	144
48"	12.57	5'-0"	12"	9"	5'-1"	6'-4"	6'-2"	5'-6"	12"	11"	3'-3"	3.37	170

SLOPE 4:1

DIAM. PIPE D	AREA SQ.FT	L	B	F	H	W	X	Z	K_1	K	P	CONC. CU.YDS.	STEEL LBS.
12"	0.79	1'-9"	9"	9"	1'-5"	2'-6"	3'-8"	2'-2"	9"	8"	1'-3"	0.81	45
15"	1.23	2'-0"	9"	9"	1'-9"	3'-0"	5'-0"	2'-7"	9"	8"	1'-5"	1.04	68
16"	1.39	2'-1"	9"	9"	1'-9"	3'-0"	5'-0"	2'-7"	9"	8"	1'-5"	1.06	68
18"	1.78	2'-3"	9"	9"	2'-0"	3'-6"	6'-0"	3'-0"	9"	8"	1'-7"	1.26	74
24"	3.14	3'-0"	12"	9"	2'-6"	4'-6"	8'-0"	3'-11"	12"	11"	2'-2"	2.23	104
30"	4.91	3'-6"	12"	9"	3'-1"	5'-6"	10'-4"	4'-9"	12"	11"	2'-7"	3.04	147
36"	7.07	4'-0"	12"	9"	3'-7"	6'-6"	12'-4"	5'-8"	12"	11"	3'-0"	3.75	180
42"	9.62	4'-6"	12"	9"	4'-2"	7'-6"	14'-6"	6'-6"	12"	11"	3'-5"	4.67	218
48"	12.57	5'-0"	12"	9"	4'-8"	8'-3"	16'-8"	7'-2"	12"	11"	3'-0"	5.57	277

Standard Type F Endwall
Round Pipe

ISOMETRIC VIEW

ELEVATION

CONCRETE:(SEE S.H.A. SPECIFICATIONS)
STEEL: DEFORMED BARS
ENDWALLS - NO.6 φ U- BEND
WINGS - VERTICAL NO.6 φ 12" C/C - 18" BEND ONE END
        HORIZONTAL NO. 4 φ 12" C/C - 6" HOOK ONE END
FLOOR - NO. 4φ 12" C/C , 2 WAYS
CHAMFER ALL EXPOSED EDGES 1"X 1" OR AS DIRECTED

TOE OF SLOPE

PLAN

SLOPE 1½ : 1

OPENING		DIMENSIONS FOR CONCRETE									QUANTITIES 1-ENDWALL 2-WINGS	
		ENDWALL				WINGS						
DIAM. PIPE D	AREA SQ.FT	L	B	F	H	W	X	Z	K	P	CONC. CU.YDS	STEEL LBS.
12"	0.79	3-3	9"	9"	1-8	1-11	1-3	1-8	9"	1-0	0.47	33
15"	1.23	3-6	9"	9"	2-0	2-5	1-8	2-0	9"	1-2	0.61	50
16"	1.39	3-7	9"	9"	2-0	2-5	1-9	2-0	9"	1-2	0.63	50
18"	1.78	3-9	9"	9"	2-3	2-8	2-1	2-4	9"	1-4	0.75	53
24"	3.14	5-0	12"	9"	2-11	3-5	2-11	2-11	12"	1-9	1.44	89
30"	4.91	5-6	12"	9"	3-6	4-2	3-8	3-7	12"	2-1	1.93	112
36"	7.07	6-0	12"	9"	4-0	4-10	4-6	4-2	12"	2-4	2.51	154
42"	9.62	6-6	12"	9"	4-7	5-7	5-4	4-10	12"	2-8	3.20	189
48"	12.57	7-0	12"	9"	5-1	6-3	6-2	5-5	12"	2-11	3.92	216

SLOPE 2:1

OPENING		DIMENSIONS FOR CONCRETE									QUANTITIES 1-ENDWALL 2-WINGS	
		ENDWALL				WINGS						
DIAM. PIPE D	AREA SQ.FT	L	B	F	H	W	X	Z	K	P	CONC. CU.YDS	STEEL LBS.
12"	0.79	3-3	9"	9"	1-7	2-1	1-10	1-10	9"	1-0	0.53	37
15"	1.23	3-6	9"	9"	1-11	2-7	2-5	2-3	9"	1-3	0.70	55
16"	1.39	3-7	9"	9"	1-11	2-7	2-6	2-3	9"	1-3	0.72	55
18"	1.78	3-9	9"	9"	2-2	2-11	2-11	2-6	9"	1-5	0.82	68
24"	3.14	5-0	12"	9"	2-9	3-7	4-0	3-1	12"	1-10	1.63	92
30"	4.91	5-6	12"	9"	3-4	4-5	5-1	3-10	12"	2-2	2.24	125
36"	7.07	6-0	12"	9"	3-10	5-2	6-2	4-6	12"	2-6	2.88	166
42"	9.62	6-6	12"	9"	4-5	6-0	7-3	5-2	12"	2-10	3.74	221
48"	12.57	7-0	12"	9"	4-11	6-9	8-4	5-10	12"	3-2	4.61	262

SLOPE 4:1

OPENING		DIMENSIONS FOR CONCRETE									QUANTITIES 1-ENDWALL 2-WINGS	
		ENDWALL				WINGS						
DIAM. PIPE D	AREA SQ.FT	L	B	F	H	W	X	Z	K	P	CONC. CU.YDS	STEEL LBS.
12"	0.79	3-3	9"	9"	1-5	2-4	4-2	2-0	9"	1-2	0.76	61
15"	1.23	3-6	9"	9"	1-9	2-11	5-3	2-6	9"	1-5	1.03	81
16"	1.39	3-7	9"	9"	1-9	2-11	5-6	2-6	9"	1-5	1.10	81
18"	1.78	3-9	9"	9"	2-0	3-1	6-4	2-11	9"	1-7	1.33	98
24"	3.14	5-0	12"	9"	2-6	4-0	8-6	3-6	12"	2-0	2.45	161
30"	4.91	5-6	12"	9"	3-1	5-0	10-8	4-4	12"	2-5	3.44	225
36"	7.07	6-0	12"	9"	3-7	5-11	12-10	5-2	12"	2-10	4.59	246
42"	9.62	6-6	12"	9"	4-2	7-0	15-6	6-1	12"	3-4	5.98	367
48"	12.57	7-0	12"	9"	4-8	7-11	17-2	6-10	12"	3-9	7.46	499

Special Type F Endwall
Round Pipe

ELEVATION

STEEL:
VERTICAL #6 φ BARS 12" C/C
HORIZONTAL #4 φ BARS 12" C/C
HOOKED ON ONE END

#6 φ BARS BENT AS SHOWN

SECTION A-A

REINFORCEMENT, DEFORMED STEEL BARS
CONCRETE (SEE S.H.A SPECIFICATIONS)
CHAMFER ALL EXPOSED EDGES 1"x1"
OR AS DIRECTED.

PLAN

TOE OF SLOPE

$1\frac{1}{2}$: 1 SLOPE = 9"
2 : 1 SLOPE = 12"
4 : 1 SLOPE = 24"

ISOMETRIC VIEW

QUANTITIES APPROXIMATE

SLOPE 1½:1											SLOPE 2:1									
OPENING		DIMENSIONS FOR CONCRETE							QUANTITIES 1-ENDWALL 2-WINGS		OPENING		DIMENSIONS FOR CONCRETE							QUANTITIES 1-ENDWALL 2-WINGS	
									ENDWALL / WINGS				ENDWALL					WINGS			
DIAM OF PIPE D	AREA SQ.FT	L	B	F	H	W	K	K₁	CONC CU.YDS	STEEL LBS	DIAM OF PIPE D	AREA SQ.FT	L	B	F	H	W	K	K₁	CONC CU.YDS	STEEL LBS
12"	0.79	2'-0"	9"	9"	1'-8"	1'-0"	6"	6"	0.33	28	12"	0.79	2'-0"	9"	9"	1'-7	1'-4"	6"	6"	0.36	34
15"	1.23	2'-3"	9"	9"	2'-0"	1'-5"	6"	6"	0.42	31	15"	1.23	2'-3"	9"	9"	1'-11"	1'-11"	6"	6"	0.47	37
16"	1.39	2'-4"	9"	9"	2'-0"	1'-6"	6"	6"	0.44	35	16"	1.39	2'-4"	9"	9"	1'-11"	2'-0"	6"	6"	0.50	38
18"	1.78	2'-6"	9"	9"	2'-3"	1'-10"	6"	6"	0.52	41	18"	1.78	2'-6"	9"	9"	2'-2"	2'-6"	6"	6"	0.61	53
24"	3.14	3'-6"	12"	9"	2'-11"	2'-8"	9"	9"	1.02	53	24"	3.14	3'-6"	12"	9"	2'-9"	3'-6"	9"	9"	1.17	71
30"	4.91	4'-0"	12"	9"	3'-6"	3'-6"	9"	9"	1.39	73	30"	4.91	4'-0"	12"	9"	3'-4"	4'-7"	9"	9"	1.62	91
36"	7.07	4'-6"	12"	9"	4'-0"	4'-3"	9"	9"	1.77	104	36"	7.07	4'-6"	12"	9"	3'-10"	5'-8"	9"	9"	2.12	116
42"	9.62	5'-6"	12"	9"	4'-7"	5'-1"	9"	12"	2.61	126	42"	9.62	5'-6"	12"	9"	4'-5"	6'-9"	9"	12"	3.15	141
48"	12.57	6'-0"	12"	9"	5'-1"	5'-11"	9"	12"	3.21	155	48"	12.57	6'-0"	12"	9"	4'-11"	7'-10"	9"	12"	3.88	174

Standard Type G Endwall
Round Pipe

2-#4 STRAIGHT BARS VERTICAL IN FRONT FACE FOR 12" DIA. TO 18" DIA. PIPE END WALLS INCLUSIVE

#4 BARS HORIZONTAL @ 1'-7" MAX. C/C BOTH FACES - LAP 1'-3" TOP & BOTTOM BARS @ CORNER BOTH WINGWALLS ALL ENDWALLS

2 -#4 BENT BARS HORIZONTAL 1 EACH FACE

2-#4 STRAIGHT BARS VERTICAL IN FRONT FACE FOR 12"TO 18" DIA. PIPE ENDWALLS INCLUSIVE

#4 BENT BARS @ 1'-0"C/C BOTH WINGWALLS ALL ENDWALLS

ELEVATION OF INVERT AS DIRECTED

#4 STRAIGHT BARS HORIZONTAL @ 1'-0"C/C BOTH SIDE OF OPENING FOR 36" TO 60" DIA. PIPE ENDWALLS.

#4 STRAIGHT BARS VERTICAL @ 1'-6"MIN. TO 2'-0"MAX. FRONT FACE FOR 24" TO 60" DIA. PIPE ENDWALL INCLUSIVE

4 -#4 STRAIGHT BARS HORIZONTAL 2 EACH WING-WALL FOR 36"TO 60" DIA. PIPE ENDWALLS INCLUSIVE

#4 STRAIGHT BARS VERTICAL @ 1'-6" MIN. TO 2'-0" MAX. FRONT FACE FOR 24" TO 60" DIA. PIPE ENDWALLS INCLUSIVE.

SECTION A-A

2-#4 STRAIGHT BARS HORIZONTAL 1 EACH WINGWALL ALL ENDWALLS

SECTION B-B

'S' DISTANCES FROM INSIDE SURFACE OF PIPE TO VERTICAL BARS IN FRONT & REAR FACE.
4" FOR 12" DIA. TO 18" DIA. PIPES INCLUSIVE
6" FOR 24" DIA. TO 36" DIA. PIPES INCLUSIVE
8" FOR 42" DIA. TO 60" DIA. PIPES INCLUSIVE

1-#4 STRAIGHT BAR VERTICAL ALL ENDWALLS

2-#4 STRAIGHT BARS HORIZONTAL 1 TOP, 1 BOTTOM, ALL ENDWALLS

PLAN

GENERAL NOTES:
SPECIFICATIONS: LATEST S.H.A.
CONCRETE : SEE S.H.A. SPECIFICATIONS
REINFORCING : DEFORMED STEEL BARS #4 φ BARS
CHAMFER : ALL EXPOSED EDGES 1"X 1" OR AS DIRECTED
SOD : 3' OF SOD WILL BE CALLED FOR ON INDIVIDUAL JOBS.

DISPOSITION OF BARS DETAIL

QUANTITIES IN TABLE TO BE USED FOR ESTIMATING ONLY

D	D	A	B	C	E	H	L	G	VOL. CONC. C.Y.	STEEL LBS.
12"	12"	9"	6"	6"	1'-9"	1'-9"	3'-6"	1'-6"	0.7	62
15"	12"	9"	6"	6"	1'-9"	2'-0"	4'-3"	2'-0"	0.9	76
15"	15"	9"	6"	6"	1'-9"	2'-0"	4'-3"	2'-0"	0.9	76
18"	12"	9"	6"	6"	1'-9"	2'-3"	5'-0"	2'-3"	1.1	88
18"	15"	9"	6"	6"	1'-9"	2'-3"	5'-0"	2'-3"	1.1	85
18"	18"	9"	6"	6"	1'-9"	2'-3"	5'-0"	2'-3"	1.1	85
24"	12"	9"	14"	6"	2'-5"	2'-9"	6'-6"	3'-0"	1.8	123
24"	15"	9"	14"	6"	2'-5"	2'-9"	6'-6"	3'-0"	1.8	122
24"	18"	9"	14"	6"	2'-5"	2'-9"	6'-6"	3'-0"	1.8	117
24"	24"	9"	14"	6"	2'-5"	2'-9"	6'-6"	3'-0"	1.8	116
30"	12"	9"	14"	6"	2'-5"	3'-6"	8'-0"	3'-9"	2.6	172
30"	15"	9"	14"	6"	2'-5"	3'-6"	8'-0"	3'-9"	2.6	171
30"	18"	9"	14"	6"	2'-5"	3'-6"	8'-0"	3'-9"	2.5	171
30"	24"	9"	14"	6"	2'-5"	3'-6"	8'-0"	3'-9"	2.5	166
30"	30"	9"	14"	6"	2'-5"	3'-6"	8'-0"	3'-9"	2.5	164
36"	12"	12"	16"	10"	3'-2"	4'-0"	9'-6"	4'-6"	5.0	279
36"	15"	12"	16"	10"	3'-2"	4'-0"	9'-6"	4'-6"	5.0	279
36"	18"	12"	16"	10"	3'-2"	4'-0"	9'-6"	4'-6"	5.0	274
36"	24"	12"	16"	10"	3'-2"	4'-0"	9'-6"	4'-6"	4.9	272
36"	30"	12"	16"	10"	3'-2"	4'-0"	9'-6"	4'-6"	4.9	262
36"	36"	12"	16"	10"	3'-2"	4'-0"	9'-6"	4'-6"	4.8	260
42"	12"	12"	16"	10"	3'-2"	4'-6"	11'-0"	5'-3"	6.1	328
42"	15"	12"	16"	10"	3'-2"	4'-6"	11'-0"	5'-3"	6.1	328
42"	18"	12"	16"	10"	3'-2"	4'-6"	11'-0"	5'-3"	6.1	327
42"	24"	12"	16"	10"	3'-2"	4'-6"	11'-0"	5'-3"	6.1	317
42"	30"	12"	16"	10"	3'-2"	4'-6"	11'-0"	5'-3"	6.0	315
42"	36"	12"	16"	10"	3'-2	4'-6"	11'-0"	5'-3"	5.9	308
42"	42"	12"	16"	10"	3'-2"	4'-6"	11'-0"	5'-3"	5.8	307
48"	12"	12"	16"	10"	3'-2"	5'-0"	12'-6"	6'-0"	7.4	404
48"	15"	12"	16"	10"	3'-2"	5'-0"	12'-6"	6'-0"	7.4	404
48"	18"	12"	16"	10"	3'-2"	5'-0"	12'-6"	6'-0"	7.4	393
48"	24"	12"	16"	10"	3'-2"	5'-0"	12'-6"	6'-0"	7.3	391
48"	30"	12"	16"	10"	3'-2"	5'-0"	12'-6"	6'-0"	7.2	383
48"	36"	12"	16"	10"	3'-2"	5'-0"	12'-6"	6'-0"	7.2	381
48"	42"	12"	16"	10"	3'-2"	5'-0"	12'-6"	6'-0"	7.1	373
48"	48"	12"	16"	10"	3'-2"	5'-0"	12'-6"	6'-0"	7.0	371

Standard Type H Endwall
Round Pipe

GENERAL NOTES
1 CONCRETE TO BE CONCRETE MIX. NO. 2
2 REINFORCEMENT TO BE NO. 4 ($\frac{1}{2}\phi$) DEFORMED BARS AT 6"C/C, 2"COVER.
3 GRATE TO BE OF STEEL CONSTRUCTION & SHALL BE SQUARE, FLAT & TRUE.
4 STRUCTURAL STEEL SHALL BE A.S.T.M DESIGNATION A-36.
5 GRATE TO BE GALV. AFTER FABRICATION IN ACCORDANCE WITH A.S.T.M DESIGNATION A-123 EXCEPT FOR ADHERENCE WHICH SHALL BE IN ACCORDANCE WITH A.S.T.M. DESIGNATION A-153.
6 SEE LATEST S.H.A. SPECIFICATIONS.
7 INSTALL 4-$\frac{5}{8}\phi$ CONCRETE EXPANSION ANCHORS, OR APPROVED EQUIVALENT, WITH 4-$\frac{5}{8}\phi$ HEX. HEAD BOLTS (GALV.)

Standard Type K Inlet
Open-End Grate

Standard Underdrains

Sediment And Erosion Control
Fiberglass Erosion Stop

NOTE: AT INTERSECTIONS, THE THICKNESS IS TO BE 8", AND THE TYPICAL SECTION IS SUBJECT TO EXISTING CONDITIONS

3'-0" MINIMUM

3"

5"

VALLEY GUTTER

SHOULDER WIDTH VARIABLE

SLOPE $\frac{1}{4}$" / 1'-0"

18"

5"

CONCRETE SHOULDER & REBUT

SLOPE VARIABLE

DIA. OF PIPE

SLOPE VARIABLE

CONC. LUG 9" IN DEPTH AND 9" THICK TO BE CONSTRUCTED ON 6'-0" CENTERS

5"

9"

VARIABLE

CONCRETE FLUME

2'-0"

SLOPE 2:1

2'-0"

2:1 SLOPE

5"

FLUME

SLOPE 2:1

PIPE

CONC. FLUME - SEE DETAIL THIS SHEET

ORIGINAL GROUND

PIPE LOCATIONS UNDER DEEP FILL

Standard Concrete Valley Gutter, Flumes Concrete Shoulder & Rebut

SECTION A-A

PLACED RIP RAP (SEE PLANS)

PLACED RIP RAP DITCH REQUIRED TO
REDUCE WATER VELOCITY AND CONTROL EROSION.

5" CONCRETE GUTTER (WITH TOE WALL
AS SHOWN ON STANDARD MD - 389.02)

NOTE: FOR RIP RAP SEE LATEST
S.H.A. SPECIFICATIONS
OR SPECIAL PROVISIONS.

PLAN

LENGTH AS SHOWN ON PLANS
OR AS DIRECTED BY THE ENGINEER.

GENERAL NOTE:
ALL DIMENSIONS AND LOCATIONS
NOT INDICATED, FOR ITEMS APPEARING
ON THIS SHEET OR ON THE PLANS,
SHALL BE DIRECTED BY THE ENGINEER.

Sediment And Erosion Control
Placed Rip-Rap Ditch

Piled Rip-Rap Detail

NOTE: PLACE VERTICAL JOINT AT CENTER OF EVERY RAIL PANEL. JOINTS SHALL BE FORMED BY POURING ALTERNATE SECTIONS. THE POURS OF ADJACENT SECTIONS SHALL HAVE A TWO (2) DAY DELAY BETWEEN POURS. IN ORDER TO FORM A WATER TIGHT JOINT, A PARAFFIN COATING SHALL BE APPLIED AT EVERY JOINT. NO REINFORCING STEEL SHALL PASS THROUGH JOINT. JOINT SPACING, TREATMENT, ETC. APPLIES TO MEDIAN BARRIER ALSO.

STANDARD ONE STRAND BRIDGE RAILING.

NOTE: ALL LONGITUDINAL BARS ARE #5 BARS SPACED AS SHOWN (PARAPET AND MEDIAN BARRIERS)

NOTE: IF SCUPPERS ARE USED IN THIS BARRIER, EACH SCUPPER SHALL BE SET INTO BARRIER AS SHOWN, CHAMFER FROM BARRIER FACES OF CONCRETE TO FACE OF SCUPPER AS REQUIRED.

TYPICAL CROSS SECTION - BRIDGE MEDIAN BARRIER

TYPICAL CROSS SECTIONS - BRIDGE PARAPET

FOR BRIDGES OVER HIGHWAYS   FOR BRIDGES OVER STREAMS & RAILROADS

Concrete Bridge Parapet And Bridge
Median Barrier Details

Reinforced Concrete Pavement
Types Of Joints

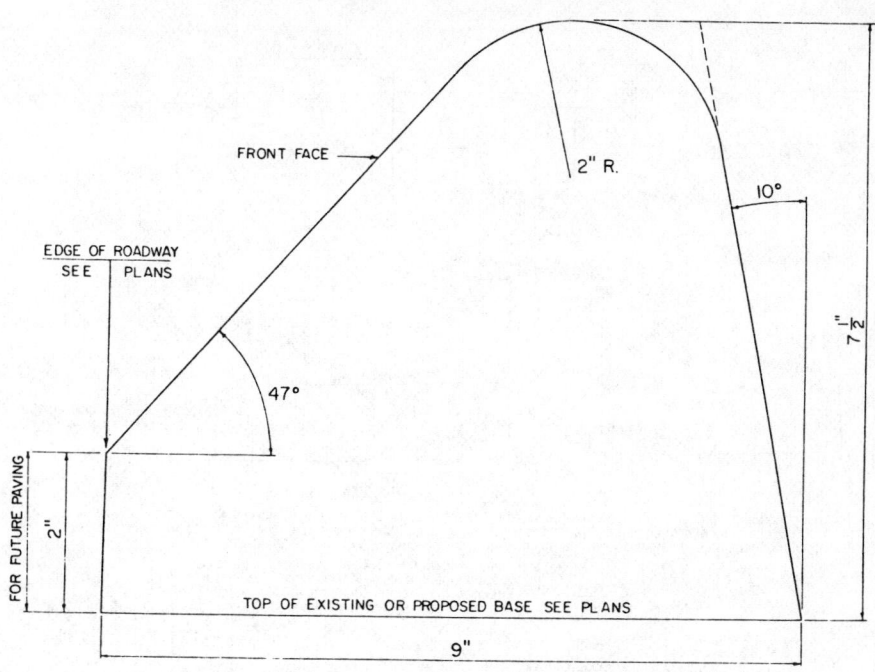

FRONT FACE

2" R.

10°

EDGE OF ROADWAY
SEE   PLANS

7 1/2"

47°

FOR FUTURE PAVING

2"

TOP OF EXISTING OR PROPOSED BASE SEE PLANS

9"

Standard Bituminous Concrete Curb

NOTE :- UNLESS OTHERWISE SPECIFIED, LONGITUDINAL TIE BAR DEVICE ; TYPE 'A' OR TYPE 'B'; PLACED AT MIDDLE OF KEYWAY (NOT SHOWN) & SPACED ACCORDING TO SPECIFICATIONS FOR CONCRETE PAVEMENT SHALL BE USED AT CONSTRUCTION JOINT BETWEEN COMBINATION CURB & GUTTER AND RIGID PAVEMENT. SEE STANDARD MD-572.61.

NOTE :- JOINT SPACING FOR CONCRETE CURB AND COMB. CURB & GUTTER = 10'. SEE SPECIFICATIONS FOR LOCATION AND DESCRIPTION OF TREATMENT FOR THE TYPES OF JOINTS USED.

Standard Types Of Concrete Curb
Combination Concrete Curb & Gutter

CHAPTER 25

CITY COST INDEXES

Tabulated in the following table are average construction labor cost factors for ninety seven (97) major U.S. cities.

The base city is Washington, DC using union wages as of June 1984 equaling 1.000. The weighted averages are predicated upon wage differentials and not on comparative productivity. The labor rates found in this book are those prevailing in the Washington, DC area. The labor rate for any of these 97 cities is computed by multiplying the labor rate shown by the appropriate city cost factor from the tables below.

CITY COST INDEX

CITY	INDEX	CITY	INDEX
Akron, OH	1.147	Cincinnati, OH	1.132
Albany, NY	.980	Cleveland, OH	1.254
Albequerque, NM	.945	Columbia, SC	.688
Anchorage, AK	1.606	Columbus, OH	1.051
Atlanta, GA	.851	Dallas, TX	1.026
Baltimore, MD	.970	Dayton, OH	1.086
Birmingham, AL	.826	Denver, CO	1.111
Boise, ID	1.011	Des Moines, IA	.907
Boston, MA	1.134	Detroit, MI	1.197
Buffalo, NY	1.200	Duluth, MN	.989
Burlington, VT	.846	El Paso, TX	.712
Butte, MT	.963	Erie, PA	1.010
Charleston, SC	.719	Fairbanks, AK	1.637
Charleston, WV	1.020	Fargo, ND	.786
Charlotte, NC	.685	Grand Rapids, MI	.920
Cheyenne, WY	.906	Hartford, CT	1.081
Chicago, IL	1.158	Honolulu, HI	1.168

CITY	INDEX	CITY	INDEX
Houston, TX	1.041	Pittsburg, PA	1.098
Indianapolis, IN	1.073	Portland, ME	.856
Jackson, MS	.730	Portland, OR	1.179
Jacksonville, FL	.876	Providence, RI	1.076
Kansas City, MO	1.062	Raleigh, NC	.611
Knoxville, TN	.788	Reading, PA	.972
Lansing, MI	1.026	Richmond, VA	.793
Las Vegas, NV	1.286	Rochester, NY	1.052
Little Rock, AR	.872	Sacramento, CA	1.306
Los Angeles, CA	1.360	St. Louis, MO	1.085
Louisville, KY	.952	St. Paul, MN	1.109
Madison, WI	.953	St. Petersburg, FL	.906
Manchester, NH	.947	Salt Lake City, UT	.932
Memphis, TN	.936	San Diego, CA	1.337
Miami, FL	.954	San Francisco, CA	1.473
Milwaukee, WI	1.076	Santa Fe, NM	1.001
Minneapolis, MN	1.113	Savannah, GA	.788
Montgomery, AL	.860	Scranton, PA	1.012
Nashville, TN	.762	Seattle, WA	1.220
Newark, NJ	1.155	Sioux Falls, SD	.756
New Haven, CT	1.062	South Bend, IN	1.042
New Orleans, LA	.978	Spokane, WA	1.156
New York, NY	1.346	Stamford, CT	1.066
Norfolk, VA	.783	Syracuse, NY	1.029
Oakland, CA	1.444	Tampa, FL	.937
Oklahoma City, OK	.902	Toledo, OH	1.199
Omaha, NB	.948	Topeka, KS	.904
Philadelphia, PA	1.109	Trenton, NJ	1.163
Phoenix, AZ	1.007	Tulsa, OK	.908

CITY	INDEX	CITY	INDEX
Washington, DC	1.000	York, PA	.966
Wichita, KS	.862	Youngstown, OH	1.110
Wilmington, DE	1.104		

CHAPTER 26

MENSURATION

The information on the following pages will enable contractors and estimators to estimate quantities and costs accurately and in a minimum of time.

In a number of tables, the quantities are stated in decimals, as by using decimals, it is possible to state the fractional parts of inches, feet and yards, in a smaller space than where regular fractions are used.

For the contractor or estimator who is not thoroughly familiar with the decimal system, complete explanations are given covering the use of all classes of decimal fractions, so they may be used rapidly and accurately.

Estimating is nearly all "figures" of one kind or another, so it is essential that the contractor and estimator possess a fair working knowledge of arithmetic, if his estimates are to be accurate and dependable.

Most of the estimator's computations involve measurements of surface and cubical contents and are stated in lineal feet (lin. ft.), square feet (sq. ft.), square yards (sq. yds.), squares (sqs.) containing 100 sq. ft., cubic feet (cu. ft.), and cubic yards (cu. yds.). These quantities are often further reduced to thousands of brick, board feet of lumber, etc.

The rules on the following pages will enable the user to compute areas in square feet or cubical contents easily, accurately and quickly.

The following abbreviations are used throughout the book to make all tables brief and concise and to insert them in the smallest possible space :

	Abbreviation	Symbols
Inches	= in.	= "
Lineal feet	= lin. ft. or l.f.	= '
Feet and inches	= ft. in.	= 2'-5"
Square feet	= sq. ft. or s.f.	
Square yards	= sq. yds. or s.y.	
Squares	= sqs.	
Cubic feet	= cu. ft. or c.f.	
Cubic yards	= cu. yds. or c.y.	
Board measure/board feet	= b.m. or b.f.	

Linear Measure

12	inches	= 12	in.=	12"	= 1 foot	= 1 ft.	= 1'-0"
3	feet	= 3	ft.=	3'- 0"	= 1 yard	= 1 yd.	
16½	feet	= 16½	ft.=	16'- 6"	= 1 rod	= 1 rd.	
40	rods	= 40 rds.			= 1 furlong	= 1 fur.	
8	furlongs	= 8 fur.			= 1 mile	= 1 mi.	
5,280	feet	= 1,760 yds.			= 1 mile	= 1 mi.	

Square Measure or Measure of Surfaces

144 square inches	= 144 sq. in.	= 1 square foot = 1 sq. ft.
9 square feet	= 9 sq. ft.	= 1 square yard = 1 sq. yd.
100 square feet	= 100 sq. ft.	= 1 square = 1 sq. (Architects' and Builders' Measure.)
30¼ square yards	= 30¼ sq. yds.	= 30.25 sq. yds. = 1 square rod = 1 sq. rd.
160 square rods	= 160 sq. rds.	= 1 Acre = 1 A.
43,560 sq. ft.	= 4,840 sq. yds.	= 1 Acre = 1 A.
640 acres	= 640 A.	= 1 square mile = 1 sq. mi.

Cubic Measure or Cubical Contents

1,728 cubic inches	= 1,728 cu. in.	= 1 cubic foot = 1 cu. Ft.
27 cubic feet	= 27 cu. ft.	= 1 cubic yard = 1 cu. yd.
128 cubic feet	= 128 cu. ft.	= 1 cord = 1 cd.
24¾ cubic feet	= 24¾ cu. ft.	= 24¾ cu. ft. = 24.75 cu. ft. = 1 perch* = 1P.

*A perch of stone is nominally 16½ ft. long, 1 ft. high and 1½ ft. thick, and contains 24¾ cu. ft. However, in some states, especially west of the Mississippi, rubble work is figured by the perch containing 16½ cu. ft. Before submitting prices on masonry by the perch, find out the common practice in your locality.

SQUARE   RECTANGLE   PARALLELOGRAM

To Compute the Area of a Square, Rectangle or Parallelogram.—Multiply the length by the breadth or height. Example : Obtain the area of a wall 22 ft. (22'-0") long and 9 ft. (9'-0") high. 22x9=198 sq. ft.

To Compute the Area of a Triangle.—Multiply the base by ½ the altitude or perpendicular height. Example : Find the area of the end gable of a house 24 ft. (24'-0") wide and 12 ft. (12'-0") high from the base to high point of roof. 24 ft. x 6 ft. (½ the height)=144 sq. ft.

To Compute the Circumference of a Circle.—Multiply the diameter by 3.1416. The diameter multiplied by 3 1/7 is close enough for all practical purposes. Example : Find the circumference or distance around a circle, the diameter of which is 12 ft. (12'-0"). 12x3.1416=37.6992 ft. the distance around the circle or 12x3 1/7=37.714 ft.

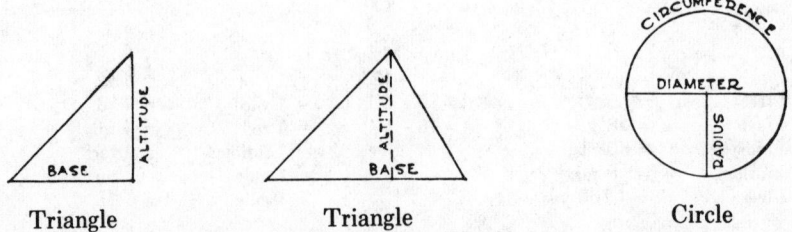

Triangle   Triangle   Circle

To Compute the Area of a Circle.—Multiply the square of the diameter by 0.7854 or multiply the square of the radius by 3.1416. Example : Find the area of a round concrete column 24 in. (24" or 2'-0") in diameter. The square of the diameter is 2x2=4. 4x0.7854=3.1416 sq. ft. the area of the circle.

The radius is ½ the diameter. If the diameter is 2 ft. (2'-0") the radius would be 1 ft. (1'-0"). To obtain the square of the radius, 1x1=1. Multiply the square of the radius, 1x3.1416=3.1416, the area of the circle.

To Compute the Cubical Contents of a Circular Column.—Multiply the area of the circle by the height. Example : Find the cubical contents of a round concrete column 2 ft. (2'-0") in diameter and 14 ft. (14'-0") long.

From the previous example, the area of a circle 2 ft. in diameter is 3.1416 sq. ft. 3.1416x14 ft. (the height)=43.9824 cu. ft. or for all practical purposes 44 cu. ft. of concrete in each column.

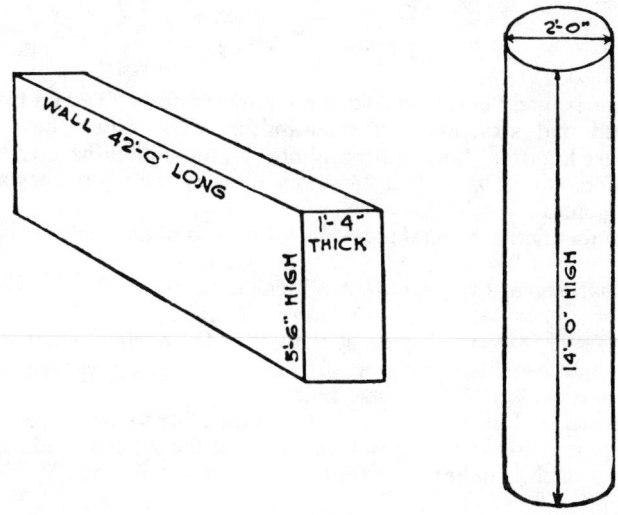

To Compute the Cubical Contents of any Solid.—Multiply the length by the breadth or height by the thickness. Computations of this kind are used extensively in estimating all classes of building work, such as excavating, concrete foundations, reinforced concrete, brick masonry, cut stone, granite, etc.

Example : Find the cubical contents of a wall 42 ft. (42'-0") long, 5 ft. 6 in. (5'-6") high, and 1 ft. 4 in. (1'-4") thick, 42'-0"x5'-6"x1'-4"=308 cu. ft. To reduce cu. ft. to cu. yds. divide 308 by 27, and the result is 11₁₁/₂₇ or 11½ or 11.41 cu. yds.

How to Use the Decimal System of Numerals in Estimating.—In construction work nearly all figures are either feet and inches or dollars and cents but inasmuch as it is convenient to use the decimal system in making many computations, a brief description of this system is given for the use of contractors and estimators who are not familiar with it.

A decimal is a fraction whose denominator is not written, but is some power of 10. They are often called decimal fractions but more often simply a decimal.

Example: We know 50 cents is 50/100 or 1/2 of a dollar. Writing the same thing in decimals would be $0.50.

Numerator	50
Denominator	100

When written as .50 or .5, it is a fraction whose denominator is not written, it being understood to be 10 from the fact that 5 occupies the first place to the right of the decimal point.
Therefore we have the following :

0.5	means	5/10, for the	5 extends to the	10th's place;
0.25	means	25/100 , for the	25 extends to the	100dth's place;
0.125	means	125/1000, for the	125 extends to the	1000dth's place.

The name of the places are, in part, as follows:

Thousands	Hundreds	Tens	Units	(Decimal Point)	Tenths	Hundredths	Thousandths	Ten-Thousandths
1	3	4	5	.	2	7	6	5

This number is read "one thousand three hundred forty-five and two thousand seven hundred and sixty-five ten thousandths." The orders beyond the ten thousandths are hundred thousandths, millionths, ten millionths, etc., but in figuring construction work it is seldom necessary to carry the figures beyond three or four decimal points.

A whole number and a decimal together, form a mixed decimal. Example : 2.25 is the same as 2$25/100$ or 2¼.

The period written at the left of tenths is called the decimal point. Example : 0.5 = 5/10; 0.25 = 25/100 = ¼; 0.375 = 375/1000 = ⅜, etc.

It is not necessary to write a zero at the left of the decimal point in the above examples, for 0.5 means the same as .5. The zero is often written there to call attention more quickly to the decimal point.

In the construction business, decimals are used chiefly to denote feet and inches, hours and minutes, the fractional working units of the various kinds of materials, and in money, which is dollars and cents or fractional parts of 100.

Table of Feet and Inches Reduced to Decimals

The following table illustrates how feet and inches may be expressed in four different ways, all meaning the same thing.

1	inch =	1	"=	1/12th	foot=0.083
1½	inches=	1½" =		1/8th	foot=0.125
2	inches=	2	"=	1/6th	foot=0.1667
2½	inches=	2½" =		5/24ths	foot=0.2087
3	inches=	3	"=	1/4th	foot=0.25
3½	inches=	3½" =		7/24ths	foot=0.2917
4	inches=	4	"=	1/3rd	foot=0.333
4½	inches=	4½" =		3/8ths	foot=0.375
5	inches=	5	"=	5/12ths	foot=0.417
5½	inches=	5½" =		11/24ths	foot=0.458
6	inches=	6	"=	1/2	foot=0.5
6½	inches=	6½" =		13/24ths	foot=0.5417
7	inches=	7	"=	7/12ths	foot=0.583
7½	inches=	7½" =		5/8ths	foot=0.625
8	inches=	8	"=	2/3rds	foot=0.667
8½	inches=	8½" =		17/24ths	foot=0.708
9	inches=	9	"=	3/4ths	foot=0.75

9 1/2 inches=	9 1/2"=	19/24ths	foot=	0.792
10	inches=10	"= 5/6ths	foot=	0.833
10 1/2 inches=	10 1/2"=	7/8ths	foot=	0.875
11	inches=11	"= 11/12ths	foot=	0.917
11 1/2 inches=	11 1/2"=	23/24ths	foot=	0.958
12	inches=12	"=1	foot=	1.0

Example : Write 5 feet, 7½ inches, in decimals. It would be written 5.625, which is equivalent to 5 5/8 feet.

Table of Common Fractions Stated in Decimals

The following table gives the decimal equivalents of common fractions frequently used in estimating :

$1/16 = \dfrac{6 1/4}{100} = 0.0625$ $7/16 = \dfrac{43 3/4}{100} = 0.4375$

$1/8 = \dfrac{12 1/2}{100} = 0.125$ $1/2 = \dfrac{50}{100} = 0.5$

$3/16 = \dfrac{18 3/4}{100}\ 0.1875$ $9/16 = \dfrac{56 1/4}{100} = 0.5625$

$1/4 = \dfrac{25}{100} = 0.25$ $5/8 = \dfrac{62 1/2}{100} = 0.625$

$5/16 = \dfrac{31 1/4}{100} = 0.3125$ $11/16 = \dfrac{68 3/4}{100} = 0.6875$

$3/8 = \dfrac{37 1/2}{100} = 0.375$ $3/4 = \dfrac{75}{100} = 0.75$

$13/16 = \dfrac{81 1/4}{100} = 0.8125$ $15/16 = \dfrac{93 3/4}{100} = 0.9375$

$7/8 = \dfrac{87 1/2}{100} = 0.875$ $8/8 = \dfrac{100}{100} = 1.0$

Annexing zeros to a number does not change its value; 0.5 is the same as 0.500.

Table of Hours and Minutes Reduced to Decimals

The following table illustrates how minutes may be reduced to fractional parts of hours and to decimal parts of hours.

Number of Minutes		Fractional Part of an Hour		Decimal Part of an Hour	Number of Minutes		Fractional Part of an Hour		Decimal Part of an Hour
1	=	1/60th	=	0.0167	31	=	31/60ths	=	0.5167
2	=	1/30th	=	0.0333	32	=	8/15ths	=	0.5333
3	=	1/20th	=	0.05	33	=	11/20ths	=	0.55
4	=	1/15th	=	0.0667	34	=	17/30ths	=	0.5667
5	=	1/12th	=	0.0833	35	=	7/12ths	=	0.5833
6	=	1/10th	=	0.10	36	=	3/5ths	=	0.60
7	=	7/60ths	=	0.1167	37	=	37/60ths	=	0.6167
8	=	2/15ths	=	0.1333	38	=	19/30ths	=	0.6333
9	=	3/20ths	=	0.15	39	=	13/20ths	=	0.65
10	=	1/6th	=	0.1667	40	=	2/3rds	=	0.6667
11	=	11/60ths	=	0.1833	41	=	41/60ths	=	0.6833
12	=	1/5th	=	0.20	42	=	7/10ths	=	0.70
13	=	13/60ths	=	0.2167	43	=	43/60ths	=	0.7167
14	=	7/30ths	=	0.2333	44	=	11/15ths	=	0.7333
15	=	1/4th	=	0.25	45	=	3/4ths	=	0.75
16	=	4/15ths	=	0.2667	46	=	23/30ths	=	0.7667
17	=	17/60ths	=	0.2833	47	=	47/60ths	=	0.7833
18	=	3/10ths	=	0.30	48	=	4/5ths	=	0.80
19	=	19/60ths	=	0.3167	49	=	49/60ths	=	0.8167

Number of Minutes		Fractional Part of an Hour		Decimal Part of an Hour	Number of Minutes		Fractional Part of an Hour		Decimal Part of an Hour
20	=	1/3rd	=	0.333	50	=	5/6ths	=	0.8333
21	=	7/20ths	=	0.35	51	=	51/60ths	=	0.85
22	=	11/30ths	=	0.3667	52	=	13/15ths	=	0.8667
23	=	23/60ths	=	0.3833	53	=	53/60ths	=	0.8833
24	=	2/5ths	=	0.40	54	=	9/10ths	=	0.90
25	=	5/12ths	=	0.4167	55	=	11/12ths	=	0.9167
26	=	13/30ths	=	0.4333	56	=	14/15ths	=	0.9333
27	=	9/20ths	=	0.45	57	=	19/20ths	=	0.95
28	=	7/15ths	=	0.4667	58	=	29/30ths	=	0.9667
29	=	29/60ths	=	0.4833	59	=	59/60ths	=	0.9833
30	=	1/2	=	0.5	60	=	1	=	1.0

Example: Write 4 hours and 37 minutes in decimals. It would be written 4.6167.

Now find the labor cost for 4.6167 hrs. at $7.50 an hr.

$$
\begin{array}{rl}
\text{Hours,} & 4.6167 \\
\text{X Hourly Rate,} & \underline{\$7.50} \\
& 2308350 \\
& \underline{323169} \\
\text{Total Cost,} & \$34.625250 \text{ or } \$34.63
\end{array}
$$

Similar Decimals.—Decimals that have the same number of decimal places are called similar decimals. Thus, 0.75 and 0.25 are similar decimals and so are 0.150 and 0.275; but 0.15 and 0.275 are dissimilar decimals.

To reduce dissimilar decimals to similar decimals, give them the same number of decimal places by annexing or cutting off zeros. Example, 0.125, 0.25, 0.375 and 0.5 may all be reduced to thousandths as follows : 0.125, 0.250, 0.375 and 0.500.

To Reduce a Decimal to a Common Fraction.—Omit the decimal point, write the denominator of the decimal, and then reduce the common fraction to its lowest terms.

Example : 0.375 equals 375/1000, which reduced to its lowest terms, equals 3/8.

$$25 \ \frac{375}{1000} \ = 5 \ \frac{15}{40} \ = \text{⅜}$$

How to Add Decimals.—To add numbers containing decimals, write like orders under one another, and then add as with whole numbers. Example, add 0.125, 0.25, 0.375, and 1.0.

The total of the addition is 1.750, which equals 1 750/1000, and which may be further reduced to 1¾

$$
\begin{array}{r}
0.125 \\
0.25 \\
0.375 \\
\underline{1.0} \\
1.750
\end{array}
$$

How to Subtract Decimals.—To subtract one number from another, when either or both contain decimals, reduce to similar decimals, write like orders under one another, and subtract as with whole numbers.

Subtract 20 hours and 37 minutes from 27 hours and 13 minutes, both being written in decimals, The difference is 6.6000 or 6.6 hours, which reduced to a common fraction is 63/5 hours or 6 hours and 36 minutes.

$$
\begin{array}{r}
27.2167 \\
\underline{20.6167} \\
6.6000
\end{array}
$$

How to Multiply Decimals.—The same method is used as in multiplying other numbers, and the result should contain as many decimal points as there are in both of the numbers multiplied.

Example: Multiply 3 feet 6 inches by 4 feet 9 inches

4 ft. 9 in.=4'-9"=4¾ ft.=4.75 Mutiplicand	4.75
3 ft. 6 in.=3'-6"=3½ ft.=3.5 Multiplier	3.5
To multiply, proceed as illustrated	2375
	1425
Product	16625 or 16.625

In the above example, there are 2 decimals in the multiplicand and 1 decimal in the multiplier. The result or product should contain the same number of decimals as the multiplicand and multiplier combined, which is three. Starting at the right and counting to the left three places, place the decimal point between the two sixes. The result would be 16.625, which equals 16625/1000 = 16⅝ sq. ft.

Practical Examples Using Decimals

The quantities of cement, sand and gravel required for one cu. yd. of concrete are ordinarily stated in decimals. For instance, a cu. yd. of concrete mixed in the proportions of 1 part cement; 2 parts sand; and 4 parts gravel, is usually stated 1: 2: 4 and requires the following materials:

1.50 bbls. cement=150/100 bbls.=11/2 bbls. =**6** sacks.
0.42 cu. yds. sand=42/100 cu. yds.=21/50 cu. yds.=**11**1/3 cu. ft.
0.84 cu. yds. gravel=84/100 cu. yds.=**22**2/3 cu. ft.

There are 4 sacks of cement to the bbl., and each sack weighs 94 lbs. and contains approximately 1 cu. ft. of cement.

There are 27 cu. ft. in a cu. yd.; to obtain the number of cu. ft. of sand required for a yard of concrete, 0.42 cu. yds=42/100 cu. yds.=21/50 cu. yds. and 21/50 of 27

cu. ft. = $\dfrac{21 \times 27}{50}$ = $\dfrac{567}{50}$ .567÷50=11.34 or 11⅓ cu. ft. sand.

To find the cost of a cu. yd. of concrete based on the above quantities, and assuming it requires 2¼ hrs. labor time to mix and place one cu. yd. of concrete, proceed as follows :

1½ bbls. cement	= 1.50 bbls. cement	@	**$9.00** per bbl.	=	**$13.5000**
11⅓ cu. ft. sand	= 0.42 cu. yd. sand	@	5.75 per cu. yd.	=	2.4150
22⅔ cu. ft. gravel	= 0.84 cu. yd. gravel	@	5.75 per cu. yd.	=	4.8300
2¼ hrs. labor	= 2.25 hrs. labor	@	9.80 per hr.	=	22.0500
Cost per cu. yd.					$42.7950

You will note the total is carried in four decimal places. This is to illustrate the actual figures obtained by multiplying. The total would be $42.80 per cu. yd. of concrete.

In making the above multiplications you will note there are 2 decimals in both the multiplicand and the multiplier of all the amounts multiplied, so the result should contain as many decimals as the sum of the multiplicand and multiplier, which is 4. You will note all the totals contain 4 decimal places. Always bear this in mind when making your multiplications because if the decimal place is wrong it makes a difference of 90 percent in your total. You know the correct result is $42.7950 but suppose by mistake you counted off 5 decimal places. The result would be $4.27950 or $4.28 per cu. yd. or just about 1/10 enough; or if you made a mistake the other way and counted off just 3 decimal places, the result would be $427.950 or $427.95 per cu. yd., or just about 10 times too much. Watch your decimal places.

The same method is used in estimating lumber. All kinds of framing lumber are ordinarily sold by the 1,000 ft., b.m., so 1,000 is the unit or decimal used when estimating lumber.

Suppose you buy 150 pcs. of 2"x8"-16'-0", which contains 3,200 ft. b.m. This is equivalent to $3200/1000=32/10=3$ 1/15 thousandths or stated in decimals it may be either 3.2 or 3.200. The cost of this lumber at $230.00 per 1,000 ft. b.m., would be obtained by multiplying 3.2 ft. at $230.00 per 1,000 ft. b.m., as follows:
3.2 M ft. of lumber @ $230.00 = $736.00.

Things to Remember When Using Decimals

at the left of a figure indicates a decimal.

.6	indicates tenths, thus .6=6/10=.60=60/100=.600=600/1000
.06	indicates hundredths, thus .06 =6/100=.060=60/1000
.006	indicates thousandths, thus .006 =6/1000
.0006	indicates ten thousandths, thus .0006 =6/10000

When multiplying decimals always remember that the result or product must contain as many decimal places as the sum of the decimal places in both the multiplicand and the multiplier; thus : Find the cost of 3 hrs. 20 min. laborer's time at $9.80 an hr.

3 hrs. 20 min. equal 3/⅓ hrs.   =3.3333 Multiplicand
9 dollars 80 cents an hr.   =   9.80 Multiplier
There are 3 decimals in the multiplicand   2666640
and 2 decimals in the multiplier, so the result   299997
or total should contain as many decimals as   32666340=32.666340=$32.67
the sum of the multiplicand and multiplier,
which is 5.

Be sure your decimal point is in the RIGHT place and the rest is easy.

Conversion Factors
S. I. Metric - English Systems

Multiply	by	to obtain
acres	0.404687	hectares
"	4.04687×10^{-3}	square kilometers
ares	1076.39	square feet
board feet	144 sq in. X 1 in.	cubic inches
" "	0.0833	cubic feet
bushels	0.3521	hectoliters
centimeters	3.28083×10^{-2}	feet
"	0.3937	inches
cubic centimeters	3.53145×10^{-5}	cubic feet
" "	6.102×10^{-2}	cubic inches
cubic feet	2.8317×10^{4}	cubic centimeters
" "	2.8317×10^{-2}	cubic meters
" "	6.22905	gallons, Imperial
" "	0.2832	hectoliters
" "	28.3170	liters
" "	2.38095×10^{-2}	tons, British shipping
" "	0.025	tons, U.S. shipping
cubic inches	16.38716	cubic centimeters

Conversion Factors (Cont'd)

Multiply	by	to obtain
cubic meters	35.3145	cubic feet
" "	1.30794	cubic yards
" "	264.2	gallons, U. S.
cubic yards	0.764559	cubic meters
" "	7.6336	hectoliters
degrees, angular	0.0174533	radians
degrees, F (less 32 F)	0.5556	degrees, C
" C	1.8	degrees, F (less 32 F)
foot pounds	0.13826	kilogram meters
feet	30.4801	centimeters
"	0.304801	meters
"	304.801	millimeters
"	1.64468×10^{-4}	miles, nautical
gallons, Imperial	0.160538	cubic feet
" "	1.20091	gallons, U. S.
" "	4.54596	liters
gallons, U.S	0.832702	gallons, Imperial
" "	0.13368	cubic feet
" "	231.	cubic inches
" "	0.0378	hectoliters
" "	3.78543	liters
grams, metric	2.20462×10^{-3}	pounds, avoirdupois
hectares	2.47104	acres
"	1.076387×10^{5}	square feet
hectares	3.86101×10^{-3}	square miles
hectoliters	3.531	cubic feet
"	2.84	bushels
"	0.131	cubic yards
hectoliters	26.42	gallons
horsepower, metric	0.98632	horsepower, U. S.
horsepower, U.S.	1.01387	horsepower, metric
inches	2.54001	centimeters
"	2.54001×10^{-2}	meters
"	25.4001	millimeters
kilograms	2.20462	pounds
"	9.84206×10^{-4}	long tons
"	1.10231×10^{-3}	short tons
kilogram meters	7.233	foot pounds
kilograms per m	0.671972	pounds per ft
kilograms per sq cm	14.2234	pounds per sq in.
kilograms per sq m	0.204817	pounds per sq ft
" " " "	9.14362×10^{-5}	long tons per sq ft
kilograms per sq mm	1422.34	pounds per sq in.
" " " "	0.634973	long tons per sq. in.
kilograms per cu m	6.24283×10^{-2}	pounds per cu ft
kilometers	0.62137	miles, statute
"	0.53959	miles, nautical
"	3280.7	feet
liters	0.219975	gallons, Imperial
"	0.26417	gallons, U.S.
"	3.53145×10^{-2}	cubic feet
"	61.022	cubic inches
meters	3.28083	feet
"	39.37	inches
"	1.09361	yards

Conversion Factors (Cont'd)

Multiply	by	to obtain
miles, statute	1.60935	kilometers
" "	0.8684	miles, nautical
miles, nautical	6080.204	feet
" "	1.85325	kilometers
" "	1.1516	miles, statute
millimeters	3.28083×10^{-3}	feet
"	3.937×10^{-2}	inches
pounds, avoirdupois	453.592	grams, metric
" "	0.453592	kilograms
" "	4.464×10^{-4}	tons, long
" "	4.53592×10^{-4}	tons, metric
pounds per ft	1.48816	kilograms per m
pounds per sq ft...............	4.88241	kilograms per sq m
pounds per sq in..............	7.031×10^{-2}	kilograms per sq cm
" " " "	7.031×10^{-4}	kilograms per sq mm
pounds per cu ft...............	16.0184	kilograms per cu m
radians.............................	57.29578	degrees, angular
square centimeters...........	0.1550	square inches
square feet........................	9.29034×10^{-4}	ares
square feet........................	9.29034×10^{-6}	hectares
" "	0.0929034	square meters
square inches....................	6.45163	square centimeters
" "	645.163	square millimeters
square kilometers.............	247.104	acres
" "	0.3861	square miles
square meters...................	10.7639	square feet
" "	1.19599	square yards
square miles.....................	259.0	hectares
" "	2.590	square kilometers
square millimeters...........	1.550×10^{-3}	square inches
square yards.....................	0.83613	square meters
tons, long...........................	1016.05	kilograms
" "	2240.	pounds
" "	1.01605	tons, metric
" ' "	1.120	tons, short
tons, long, per sq ft..........	1.09366×10^{-4}	kilograms per sq m
tons, long, per sq in.........	1.57494	kilograms per sq mm
tons, metric......................	2204.62	pounds
" "	0.98421	tons, long
" "	1.10231	tons, short
tons, short	907.185	kilograms
" "	0.892857	tons, long
" "	0.907185	tons, metric
tons, British shipping......	42.00	cubic feet
" " "	0.952381	tons, U. S. shipping
tons, U. S. shipping..........	40.00	cubic feet
" " "	1.050	tons, British shipping
yards................................	0.914402	meters

Functions of Numbers, 1 to 49

No.	Square	Cube	Square Root	Cubic Root	Logarithm
1	1	1	1.0000	1.0000	0.00000
2	4	8	1.4142	1.2599	0.30103
3	9	27	1.7321	1.4422	0.47712
4	16	64	2.0000	1.5874	0.60206
5	25	125	2.2361	1.7100	0.69897
6	36	216	2.4495	1.8171	0.77815
7	49	343	2.6458	1.9129	0.84510
8	64	512	2.8284	2.0000	0.90309
9	81	729	3.0000	2.0801	0.95424
10	100	1000	3.1623	2.1544	1.00000
11	121	1331	3.3166	2.2240	1.04139
12	144	1728	3.4641	2.2894	1.07918
13	169	2197	3.6056	2.3513	1.11394
14	196	2744	3.7417	2.4101	1.14613
15	225	3375	3.8730	2.4662	1.17609
16	256	4096	4.0000	2.5198	1.20412
17	289	4913	4.1231	2.5713	1.23045
18	324	5832	4.2426	2.6207	1.25527
19	361	6859	4.3589	2.6684	1.27875
20	400	8000	4.4721	2.7144	1.30103
21	441	9261	4.5826	2.7589	1.32222
22	484	10648	4.6904	2.8020	1.34242
23	529	12167	4.7958	2.8439	1.36173
24	576	13824	4.8990	2.8845	1.38021
25	625	15625	5.0000	2.9240	1.39794
26	676	17576	5.0990	2.9625	1.41497
27	729	19683	5.1962	3.0000	1.43136
28	784	21952	5.2915	3.0366	1.44716
29	841	24389	5.3852	3.0723	1.46240
30	900	27000	5.4772	3.1072	1.47712
31	961	29791	5.5678	3.1414	1.49136
32	1024	32768	5.6569	3.1748	1.50515
33	1089	35937	5.7446	3.2075	1.51851
34	1156	39304	5.8310	3.2396	1.53148
35	1225	42875	5.9161	3.2711	1.54407
36	1296	46656	6.0000	3.3019	1.55630
37	1369	50653	6.0828	3.3322	1.56820
38	1444	54872	6.1644	3.3620	1.57978
39	1521	59319	6.2450	3.3912	1.59106
40	1600	64000	6.3246	3.4200	1.60206
41	1681	68921	6.4031	3.4482	1.61278
42	1764	74088	6.4807	3.4760	1.62325
43	1849	79507	6.5574	3.5034	1.63347
44	1936	85184	6.6332	3.5303	1.64345
45	2025	91125	6.7082	3.5569	1.65321
46	2116	97336	6.7823	3.5830	1.66276
47	2209	103823	6.8557	3.6088	1.67210
48	2304	110592	6.9282	3.6342	1.68124
49	2401	117649	7.0000	3.6593	1.69020

Functions of Numbers, 50 to 99

No.	Square	Cube	Square Root	Cubic Root	Logarithm
50	2500	125000	7.0711	3.6840	1.69897
51	2601	132651	7.1414	3.7084	1.70757
52	2704	140608	7.2111	3.7325	1.71600
53	2809	148877	7.2801	3.7563	1.72428
54	2916	157464	7.3485	3.7798	1.73239
55	3025	166375	7.4162	3.8030	1.74036
56	3136	175616	7.4833	3.8259	1.74819
57	3249	185193	7.5498	3.8485	1.75587
58	3364	195112	7.6158	3.8709	1.76343
59	3481	205379	7.6811	3.8930	1.77085
60	3600	216000	7.7460	3.9149	1.77815
61	3721	226981	7.8102	3.9365	1.78533
62	3844	238328	7.8740	3.9579	1.79239
63	3969	250047	7.9373	3.9791	1.79934
64	4096	262144	8.0000	4.0000	1.80618
65	4225	274625	8.0623	4.0207	1.81291
66	4356	287496	8.1240	4.0412	1.81954
67	4489	300763	8.1854	4.0615	1.82607
68	4624	314432	8.2462	4.0817	1.83251
69	4761	328509	8.3066	4.1016	1.83885
70	4900	343000	8.3666	4.1213	1.84510
71	5041	357911	8.4261	4.1408	1.85126
72	5184	373248	8.4853	4.1602	1.85733
73	5329	389017	8.5440	4.1793	1.86332
74	5476	405224	8.6023	4.1983	1.86923
75	5625	421875	8.6603	4.2172	1.87506
76	5776	438976	8.7178	4.2358	1.88081
77	5929	456533	8.7750	4.2543	1.88649
78	6084	474552	8.8318	4.2727	1.89209
79	6241	493039	8.8882	4.2908	1.89763
80	6400	512000	8.9443	4.3089	1.90309
81	6561	531441	9.0000	4.3267	1.90849
82	6724	551368	9.0554	4.3445	1.91381
83	6889	571787	9.1104	4.3621	1.91908
84	7056	592704	9.1652	4.3795	1.92428
85	7225	614125	9.2195	4.3968	1.92942
86	7396	636056	9.2736	4.4140	1.93450
87	7569	658503	9.3274	4.4310	1.93952
88	7744	681472	9.3808	4.4480	1.94448
89	7921	704969	9.4340	4.4647	1.94939
90	8100	729000	9.4868	4.4814	1.95424
91	8281	753571	9.5394	4.4979	1.95904
92	8464	778688	9.5917	4.5144	1.96379
93	8649	804357	9.6437	4.5307	1.96848
94	8836	830584	9.6954	4.5468	1.97313
95	9025	857375	9.7468	4.5629	1.97772
96	9216	884736	9.7980	4.5789	1.98227
97	9409	912673	9.8489	4.5947	1.98677
98	9604	941192	9.8995	4.6104	1.99123
99	9801	970299	9.9499	4.6261	1.99564

Functions of Numbers, 100 to 149

No.	Square	Cube	Square Root	Cubic Root	Logarithm
100	10000	1000000	10.0000	4.6416	2.00000
101	10201	1030301	10.0499	4.6570	2.00432
102	10404	1061208	10.0995	4.6723	2.00860
103	10609	1092727	10.1489	4.6875	2.01284
104	10816	1124864	10.1980	4.7027	2.01703
105	11025	1157625	10.2470	4.7177	2.02119
106	11236	1191016	10.2956	4.7326	2.02531
107	11449	1225043	10.3441	4.7475	2.02938
108	11664	1259712	10.3923	4.7622	2.03342
109	11881	1295029	10.4403	4.7769	2.03743
110	12100	1331000	10.4881	4.7914	2.04139
111	12321	1367631	10.5357	4.8059	2.04532
112	12544	1404928	10.5830	4.8203	2.04922
113	12769	1442897	10.6301	4.8346	2.05308
114	12996	1481544	10.6771	4.8488	2.05690
115	13225	1520875	10.7238	4.8629	2.06070
116	13456	1560896	10.7703	4.8770	2.06446
117	13689	1601613	10.8167	4.8910	2.06819
118	13924	1643032	10.8628	4.9049	2.07188
119	14161	1685159	10.9087	4.9187	2.07555
120	14400	1728000	10.9545	4.9324	2.07918
121	14641	1771561	11.0000	4.9461	2.08279
122	14884	1815848	11.0454	4.9597	2.08636
123	15129	1860867	11.0905	4.9732	2.08991
124	15376	1906624	11.1355	4.9866	2.09342
125	15625	1953125	11.1803	5.0000	2.09691
126	15876	2000376	11.2250	5.0133	2.10037
127	16129	2048383	11.2694	5.0265	2.10380
128	16384	2097152	11.3137	5.0397	2.10721
129	16641	2146689	11.3578	5.0528	2.11059
130	16900	2197000	11.4018	5.0658	2.11394
131	17161	2248091	11.4455	5.0788	2.11727
132	17424	2299968	11.4891	5.0916	2.12057
133	17689	2352637	11.5326	5.1045	2.12385
134	17956	2406104	11.5758	5.1172	2.12710
135	18225	2460375	11.6190	5.1299	2.13033
136	18496	2515456	11.6619	5.1426	2.13354
137	18769	2571353	11.7047	5.1551	2.13672
138	19044	2628072	11.7473	5.1676	2.13988
139	19321	2685619	11.7898	5.1801	2.14301
140	19600	2744000	11.8322	5.1925	2.14613
141	19881	2803221	11.8743	5.2048	2.14922
142	20164	2863288	11.9164	5.2171	2.15229
143	20449	2924207	11.9583	5.2293	2.15534
144	20736	2985984	12.0000	5.2415	2.15836
145	21025	3048625	12.0416	5.2536	2.16137
146	21316	3112136	12.0830	5.2656	2.16435
147	21609	3176523	12.1244	5.2776	2.16732
148	21904	3241792	12.1655	5.2896	2.17026
149	22201	3307949	12.2066	5.3015	2.17319

Functions of Numbers, 150 to 199

No.	Square	Cube	Square Root	Cubic Root	Logarithm
150	22500	3375000	12.2474	5.3133	2.17609
151	22801	3442951	12.2882	5.3251	2.17898
152	23104	3511808	12.3288	5.3368	2.18184
153	23409	3581577	12.3693	5.3485	2.18469
154	23716	3652264	12.4097	5.3601	2.18752
155	24025	3723875	12.4499	5.3717	2.19033
156	24336	3796416	12.4900	5.3832	2.19312
157	24649	3869893	12.5300	5.3947	2.19590
158	24964	3944312	12.5698	5.4061	2.19866
159	25281	4019679	12.6095	5.4175	2.20140
160	25600	4096000	12.6491	5.4288	2.20412
161	25921	4173281	12.6886	5.4401	2.20683
162	26244	4251528	12.7279	5.4514	2.20952
163	26569	4330747	12.7671	5.4626	2.21219
164	26896	4410944	12.8062	5.4737	2.21484
165	27225	4492125	12.8452	5.4848	2.21748
166	27556	4574296	12.8841	5.4959	2.22011
167	27889	4657463	12.9228	5.5069	2.22272
168	28224	4741632	12.9615	5.5178	2.22531
169	28561	4826809	13.0000	5.5288	2.22789
170	28900	4913000	13.0384	5.5397	2.23045
171	29241	5000211	13.0767	5.5505	2.23300
172	29584	5088448	13.1149	5.5613	2.23553
173	29929	5177717	13.1529	5.5721	2.23805
174	30276	5268024	13.1909	5.5828	2.24055
175	30625	5359375	13.2288	5.5934	2.24304
176	30976	5451776	13.2665	5.6041	2.24551
177	31329	5545233	13.3041	5.6147	2.24797
178	31684	5639752	13.3417	5.6252	2.25042
179	32041	5735339	13.3791	5.6357	2.25285
180	32400	5832000	13.4164	5.6462	2.25527
181	32761	5929741	13.4536	5.6567	2.25768
182	33124	6028568	13.4907	5.6671	2.26007
183	33489	6128487	13.5277	5.6774	2.26245
184	33856	6229504	13.5647	5.6877	2.26482
185	34225	6331625	13.6015	5.6980	2.26717
186	34596	6434856	13.6382	5.7083	2.26951
187	34969	6539203	13.6748	5.7185	2.27184
188	35344	6644672	13.7113	5.7287	2.27416
189	35721	6751269	13.7477	5.7388	2.27646
190	36100	6859000	13.7840	5.7489	2.27875
191	36481	6967871	13.8203	5.7590	2.28103
192	36864	7077888	13.8564	5.7690	2.28330
193	37249	7189057	13.8924	5.7790	2.28556
194	37636	7301384	13.9284	5.7890	2.28780
195	38025	7414875	13.9642	5.7989	2.29003
196	38416	7529536	14.0000	5.8088	2.29226
197	38809	7645373	14.0357	5.8186	2.29447
198	39204	7762392	14.0712	5.8285	2.29667
199	39601	7880599	14.1067	5.8383	2.29885

Functions of Numbers, 200 to 249

No.	Square	Cube	Square Root	Cubic Root	Logarithm
200	40000	8000000	14.1421	5.8480	2.30103
201	40401	8120601	14.1774	5.8578	2.30320
202	40804	8242408	14.2127	5.8675	2.30535
203	41209	8365427	14.2478	5.8771	2.30750
204	41616	8489664	14.2829	5.8868	2.30963
205	42025	8615125	14.3178	5.8964	2.31175
206	42436	8741816	14.3527	5.9059	2.31387
207	42849	8869743	14.3875	5.9155	2.31597
208	43264	8998912	14.4222	5.9250	2.31806
209	43681	9129329	14.4568	5.9345	2.32015
210	44100	9261000	14.4914	5.9439	2.32222
211	44521	9393931	14.5258	5.9533	2.32428
212	44944	9528128	14.5602	5.9627	2.32634
213	45369	9663597	14.5945	5.9721	2.32838
214	45796	9800344	14.6287	5.9814	2.33041
215	46225	9938375	14.6629	5.9907	2.33244
216	46656	10077696	14.6969	6.0000	2.33445
217	47089	10218313	14.7309	6.0092	2.33646
218	47524	10360232	14.7648	6.0185	2.33846
219	47961	10503459	14.7986	6.0277	2.34044
220	48400	10648000	14.8324	6.0368	2.34242
221	48841	10793861	14.8661	6.0459	2.34439
222	49284	10941048	14.8997	6.0550	2.34635
223	49729	11089567	14.9332	6.0641	2.34830
224	50176	11239424	14.9666	6.0732	2.35025
225	50625	11390625	15.0000	6.0822	2.35218
226	51076	11543176	15.0333	6.0912	2.35411
227	51529	11697083	15.0665	6.1002	2.35603
228	51984	11852352	15.0997	6.1091	2.35793
229	52441	12008989	15.1327	6.1180	2.35984
230	52900	12167000	15.1658	6.1269	2.36173
231	53361	12326391	15.1987	6.1358	2.36361
232	53824	12487168	15.2315	6.1446	2.36549
233	54289	12649337	15.2643	6.1534	2.36736
234	54756	12812904	15.2971	6.1622	2.36922
235	55225	12977875	15.3297	6.1710	2.37107
236	55696	13144256	15.3623	6.1797	2.37291
237	56169	13312053	15.3948	6.1885	2.37475
238	56644	13481272	15.4272	6.1972	2.37658
239	57121	13651919	15.4596	6.2058	2.37840
240	57600	13824000	15.4919	6.2145	2.38021
241	58081	13997521	15.5242	6.2231	2.38202
242	58564	14172488	15.5563	6.2317	2.38382
243	59049	14348907	15.5885	6.2403	2.38561
244	59536	14526784	15.6205	6.2488	2.38739
245	60025	14706125	15.6525	6.2573	2.38917
246	60516	14886936	15.6844	6.2658	2.39094
247	61009	15069223	15.7162	6.2743	2.39270
248	61504	15252992	15.7480	6.2828	2.39445
249	62001	15438249	15.7797	6.2912	2.39620

Functions of Numbers, 250 to 299

No.	Square	Cube	Square Root	Cubic Root	Logarithm
250	62500	15625000	15.8114	6.2996	2.39794
251	63001	15813251	15.8430	6.3080	2.39967
252	63504	16003008	15.8745	6.3164	2.40140
253	64009	16194277	15.9060	6.3247	2.40312
254	64516	16387064	15.9374	6.3330	2.40483
255	65025	16581375	15.9687	6.3413	2.40654
256	65536	16777216	16.0000	6.3496	2.40824
257	66049	16974593	16.0312	6.3579	2.40993
258	66564	17173512	16.0624	6.3661	2.41162
259	67081	17373979	16.0935	6.3743	2.41330
260	67600	17576000	16.1245	6.3825	2.41497
261	68121	17779581	16.1555	6.3907	2.41664
262	68644	17984728	16.1864	6.3988	2.41830
263	69169	18191447	16.2173	6.4070	2.41996
264	69696	18399744	16.2481	6.4151	2.42160
265	70225	18609625	16.2788	6.4232	2.42325
266	70756	18821096	16.3095	6.4312	2.42488
267	71289	19034163	16.3401	6.4393	2.42651
268	71824	19248832	16.3707	6.4473	2.42813
269	72361	19465109	16.4012	6.4553	2.42975
270	72900	19683000	16.4317	6.4633	2.43136
271	73441	19902511	16.4621	6.4713	2.43297
272	73984	20123648	16.4924	6.4792	2.43457
273	74529	20346417	16.5227	6.4872	2.43616
274	75076	20570824	16.5529	6.4951	2.43775
275	75625	20796875	16.5831	6.5030	2.43933
276	76176	21024576	16.6132	6.5108	2.44091
277	76729	21253933	16.6433	6.5187	2.44248
278	77284	21484952	16.6733	6.5265	2.44404
279	77841	21717639	16.7033	6.5343	2.44560
280	78400	21952000	16.7332	6.5421	2.44716
281	78961	22188041	16.7631	6.5499	2.44871
282	79524	22425768	16.7929	6.5577	2.45025
283	80089	22665187	16.8226	6.5654	2.45179
284	80656	22906304	16.8523	6.5731	2.45332
285	81225	23149125	16.8819	6.5808	2.45484
286	81796	23393656	16.9115	6.5885	2.45637
287	82369	23639903	16.9411	6.5962	2.45788
288	82944	23887872	16.9706	6.6039	2.45939
289	83521	24137569	17.0000	6.6115	2.46090
290	84100	24389000	17.0294	6.6191	2.46240
291	84681	24642171	17.0587	6.6267	2.46389
292	85264	24897088	17.0880	6.6343	2.46538
293	85849	25153757	17.1172	6.6419	2.46687
294	86436	25412184	17.1464	6.6494	2.46835
295	87025	25672375	17.1756	6.6569	2.46982
296	87616	25934336	17.2047	6.6644	2.47129
297	88209	26198073	17.2337	6.6719	2.47276
298	88804	26463592	17.2627	6.6794	2.47422
299	89401	26730899	17.2916	6.6869	2.47567

Page

Page

Page

Page

Page

T

Page

WALKER PUBLICATIONS

WALKER'S
uilding Estimator's Reference Book

WALKER'S
Pocket Estimator

WALKER'S
Practical Accounting & Cost
Keeping For Contractors

WALKER'S
Insulator's Estimating Handbook

WALKER'S
Quantity Surveying & Basic
Construction Estimating

WALKER'S
Manual For Construction
Cost Estimating

WALKER'S
Remodeling Estimator's
Reference Book

Walker's Building Estimator's Reference Book
THE MOST USEFUL HANDBOOK
IN THE CONSTRUCTION INDUSTRY.

67 Years of Experience and Service to The Construction Industry. No matter what area of construction you are in. **Walker's Building Estimator's Reference Book** is an invaluable guide; because it covers the latest construction methods and estimating costs for all building trades... ranging from carpentry to plumbing, from excavating to sewerage. The 21st Edition revises and incorporates all of the valuable information from previous editions and adds new data to aid you in your construction and estimating techniques.

Covers key construction data such as:
- new construction methods
- types of materials available
- amount of materials required
- current cost of materials
- current labor costs
- application rates, and more

... in short, the kind of precise information which spells the difference between loss and profit in today's highly competitive construction market.

For additional copies of **Walker's Building Estimator's Reference Book,** including your copy of **Walker's Pocket Estimator,** contact:

Frank R. Walker Company
5030 North Harlem Avenue
Chicago, Illinois 60656

Revised and edited by
William S. Spradlin
Wash., D.C.